DATA COMMUNICATIONS AND NETWORKING

McGraw-Hill Forouzan Networking Series

Titles by Behrouz A. Forouzan:

Data Communications and Networking
TCP/IP Protocol Suite
Local Area Networks
Business Data Communications

DATA COMMUNICATIONS AND NETWORKING

Fourth Edition

Behrouz A. Forouzan

DeAnza College

with

Sophia Chung Fegan

Boston Burr Ridge, IL Dubuque, IA Madison, WI New York San Francisco St. Louis
Bangkok Bogotá Caracas Kuala Lumpur Lisbon London Madrid Mexico City
Milan Montreal New Delhi Santiago Seoul Singapore Sydney Taipei Toronto

The McGraw·Hill Companies

DATA COMMUNICATIONS AND NETWORKING, FOURTH EDITION

Published by McGraw-Hill, a business unit of The McGraw-Hill Companies, Inc., 1221 Avenue of the Americas, New York, NY 10020.

This book is printed on acid-free paper.

2 3 4 5 6 7 8 9 0 DOC / DOC 0 9 8 7

ISBN-13 978–0–07–296775–3
ISBN-10 0–07–296775–7

Publisher: *Alan R. Apt*
Developmental Editor: *Rebecca Olson*
Executive Marketing Manager: *Michael Weitz*
Senior Project Manager: *Sheila M. Frank*
Senior Production Supervisor: *Kara Kudronowicz*
Senior Media Project Manager: *Jodi K. Banowetz*
Associate Media Producer: *Christina Nelson*
Senior Designer: *David W. Hash*
Cover Designer: *Rokusek Design*
(USE) Cover Image: Women ascending Mount McKinley, Alaska. Mount McKinley (Denali) 12,000 feet, ©*Allan Kearney/Getty Images*
Compositor: *Interactive Composition Corporation*
Typeface: *10/12 Times Roman*
Printer: *R. R. Donnelley Crawfordsville, IN*

Library of Congress Cataloging-in-Publication Data

Forouzan, Behrouz A.
Data communications and networking / Behrouz A. Forouzan. — 4th ed.
p. cm. — (McGraw-Hill Forouzan networking series)
Includes index.
ISBN 978–0–07–296775–3 — ISBN 0–07–296775–7 (hard copy : alk. paper)
1. Data transmission systems. 2. Computer networks. I. Title. II. Series.

TK5105.F6617 2007
004.6—dc22 2006000013
CIP

www.mhhe.com

To my wife, Faezeh, with love

Behrouz Forouzan

BRIEF CONTENTS

CONTENTS

PART 4 Network Layer 547

Chapter 21 *Network Layer: Address Mapping, Error Reporting, and Multicasting 611*

Chapter 22 *Network Layer: Delivery, Forwarding, and Routing 647*

Chapter 24 *Congestion Control and Quality of Service 761*

PART 6 Application Layer 795

Chapter 25 *Domain Name System 797*

Preface

Data communications and networking may be the fastest growing technologies in our culture today. One of the ramifications of that growth is a dramatic increase in the number of professions where an understanding of these technologies is essential for success—and a proportionate increase in the number and types of students taking courses to learn about them.

Features of the Book

Several features of this text are designed to make it particularly easy for students to understand data communications and networking.

Structure

We have used the five-layer Internet model as the framework for the text not only because a thorough understanding of the model is essential to understanding most current networking theory but also because it is based on a structure of interdependencies: Each layer builds upon the layer beneath it and supports the layer above it. In the same way, each concept introduced in our text builds upon the concepts examined in the previous sections. The Internet model was chosen because it is a protocol that is fully implemented.

This text is designed for students with little or no background in telecommunications or data communications. For this reason, we use a bottom-up approach. With this approach, students learn first about data communications (lower layers) before learning about networking (upper layers).

Visual Approach

The book presents highly technical subject matter without complex formulas by using a balance of text and figures. More than 700 figures accompanying the text provide a visual and intuitive opportunity for understanding the material. Figures are particularly important in explaining networking concepts, which are based on connections and transmission. Both of these ideas are easy to grasp visually.

Highlighted Points

We emphasize important concepts in highlighted boxes for quick reference and immediate attention.

Examples and Applications

When appropriate, we have selected examples to reflect true-to-life situations. For example, in Chapter 6 we have shown several cases of telecommunications in current telephone networks.

Recommended Reading

Each chapter includes a list of books and sites that can be used for further reading.

Key Terms

Each chapter includes a list of key terms for the student.

Summary

Each chapter ends with a summary of the material covered in that chapter. The summary provides a brief overview of all the important points in the chapter.

Practice Set

Each chapter includes a practice set designed to reinforce and apply salient concepts. It consists of three parts: review questions, exercises, and research activities (only for appropriate chapters). Review questions are intended to test the student's first-level understanding of the material presented in the chapter. Exercises require deeper understanding of the material. Research activities are designed to create motivation for further study.

Appendixes

The appendixes are intended to provide quick reference material or a review of materials needed to understand the concepts discussed in the book.

Glossary and Acronyms

The book contains an extensive glossary and a list of acronyms.

Changes in the Fourth Edition

The Fourth Edition has major changes from the Third Edition, both in the organization and in the contents.

Organization

The following lists the changes in the organization of the book:

1. Chapter 6 now contains multiplexing as well as spreading.
2. Chapter 8 is now totally devoted to switching.
3. The contents of Chapter 12 are moved to Chapter 11.
4. Chapter 17 covers SONET technology.
5. Chapter 19 discusses IP addressing.
6. Chapter 20 is devoted to the Internet Protocol.
7. Chapter 21 discusses three protocols: ARP, ICMP, and IGMP.
8. Chapter 28 is new and devoted to network management in the Internet.
9. The previous Chapters 29 to 31 are now Chapters 30 to 32.

Contents

We have revised the contents of many chapters including the following:

1. The contents of Chapters 1 to 5 are revised and augmented. Examples are added to clarify the contents.
2. The contents of Chapter 10 are revised and augmented to include methods of error detection and correction.
3. Chapter 11 is revised to include a full discussion of several control link protocols.
4. Delivery, forwarding, and routing of datagrams are added to Chapter 22.
5. The new transport protocol, SCTP, is added to Chapter 23.
6. The contents of Chapters 30, 31, and 32 are revised and augmented to include additional discussion about security issues and the Internet.
7. New examples are added to clarify the understanding of concepts.

End Materials

1. A section is added to the end of each chapter listing additional sources for study.
2. The review questions are changed and updated.
3. The multiple-choice questions are moved to the book site to allow students to self-test their knowledge about the contents of the chapter and receive immediate feedback.
4. Exercises are revised and new ones are added to the appropriate chapters.
5. Some chapters contain research activities.

Instructional Materials

Instructional materials for both the student and the teacher are revised and augmented. The solutions to exercises contain both the explanation and answer including full colored figures or tables when needed. The Powerpoint presentations are more comprehensive and include text and figures.

Contents

The book is divided into seven parts. The first part is an overview; the last part concerns network security. The middle five parts are designed to represent the five layers of the Internet model. The following summarizes the contents of each part.

Part One: Overview

The first part gives a general overview of data communications and networking. Chapter 1 covers introductory concepts needed for the rest of the book. Chapter 2 introduces the Internet model.

Part Two: Physical Layer

The second part is a discussion of the physical layer of the Internet model. Chapters 3 to 6 discuss telecommunication aspects of the physical layer. Chapter 7 introduces the transmission media, which, although not part of the physical layer, is controlled by it. Chapter 8 is devoted to switching, which can be used in several layers. Chapter 9 shows how two public networks, telephone and cable TV, can be used for data transfer.

Part Three: Data Link Layer

The third part is devoted to the discussion of the data link layer of the Internet model. Chapter 10 covers error detection and correction. Chapters 11, 12 discuss issues related to data link control. Chapters 13 through 16 deal with LANs. Chapters 17 and 18 are about WANs. LANs and WANs are examples of networks operating in the first two layers of the Internet model.

Part Four: Network Layer

The fourth part is devoted to the discussion of the network layer of the Internet model. Chapter 19 covers IP addresses. Chapters 20 and 21 are devoted to the network layer protocols such as IP, ARP, ICMP, and IGMP. Chapter 22 discusses delivery, forwarding, and routing of packets in the Internet.

Part Five: Transport Layer

The fifth part is devoted to the discussion of the transport layer of the Internet model. Chapter 23 gives an overview of the transport layer and discusses the services and duties of this layer. It also introduces three transport-layer protocols: UDP, TCP, and SCTP. Chapter 24 discusses congestion control and quality of service, two issues related to the transport layer and the previous two layers.

Part Six: Application Layer

The sixth part is devoted to the discussion of the application layer of the Internet model. Chapter 25 is about DNS, the application program that is used by other application programs to map application layer addresses to network layer addresses. Chapter 26 to 29 discuss some common applications protocols in the Internet.

Part Seven: Security

The seventh part is a discussion of security. It serves as a prelude to further study in this subject. Chapter 30 briefly discusses cryptography. Chapter 31 introduces security aspects. Chapter 32 shows how different security aspects can be applied to three layers of the Internet model.

Online Learning Center

The McGraw-Hill Online Learning Center contains much additional material. Available at www.mhhe.com/forouzan. As students read through *Data Communications and Networking,* they can go online to take self-grading quizzes. They can also access lecture materials such as PowerPoint slides, and get additional review from animated figures from the book. Selected solutions are also available over the Web. The solutions to odd-numbered problems are provided to students, and instructors can use a password to access the complete set of solutions.

Additionally, McGraw-Hill makes it easy to create a website for your networking course with an exclusive McGraw-Hill product called PageOut. It requires no prior knowledge of HTML, no long hours, and no design skills on your part. Instead, PageOut offers a series of templates. Simply fill them with your course information and click on one of 16 designs. The process takes under an hour and leaves you with a professionally designed website.

Although PageOut offers "instant" development, the finished website provides powerful features. An interactive course syllabus allows you to post content to coincide with your lectures, so when students visit your PageOut website, your syllabus will direct them to components of Forouzan's Online Learning Center, or specific material of your own.

How to Use the Book

This book is written for both an academic and a professional audience. The book can be used as a self-study guide for interested professionals. As a textbook, it can be used for a one-semester or one-quarter course. The following are some guidelines.

- Parts one to three are strongly recommended.
- Parts four to six can be covered if there is no following course in TCP/IP protocol.
- Part seven is recommended if there is no following course in network security.

Acknowledgments

It is obvious that the development of a book of this scope needs the support of many people.

Peer Review

The most important contribution to the development of a book such as this comes from peer reviews. We cannot express our gratitude in words to the many reviewers who spent numerous hours reading the manuscript and providing us with helpful comments and ideas. We would especially like to acknowledge the contributions of the following reviewers for the third and fourth editions of this book.

Farid Ahmed, *Catholic University*
Kaveh Ashenayi, *University of Tulsa*
Yoris Au, *University of Texas, San Antonio*
Essie Bakhtiar, *Clayton College & State University*
Anthony Barnard, *University of Alabama, Brimingham*
A.T. Burrell, *Oklahoma State University*
Scott Campbell, *Miami University*
Teresa Carrigan, *Blackburn College*
Hwa Chang, *Tufts University*
Edward Chlebus, *Illinois Institute of Technology*
Peter Cooper, *Sam Houston State University*
Richard Coppins, *Virginia Commonwealth University*
Harpal Dhillon, *Southwestern Oklahoma State University*
Hans-Peter Dommel, *Santa Clara University*
M. Barry Dumas, *Baruch College, CUNY*
William Figg, *Dakota State University*
Dale Fox, *Quinnipiac University*
Terrence Fries, *Coastal Carolina University*
Errin Fulp, *Wake Forest University*

Sandeep Gupta, *Arizona State University*
George Hamer, *South Dakota State University*
James Henson, *California State University, Fresno*
Tom Hilton, *Utah State University*
Allen Holliday, *California State University, Fullerton*
Seyed Hossein Hosseini, *University of Wisconsin, Milwaukee*
Gerald Isaacs, *Carroll College, Waukesha*
Hrishikesh Joshi, *DeVry University*
E.S. Khosravi, *Southern University*
Bob Kinicki, *Worcester Polytechnic University*
Kevin Kwiat, *Hamilton College*
Ten-Hwang Lai, *Ohio State University*
Chung-Wei Lee, *Auburn University*
Ka-Cheong Leung, *Texas Tech University*
Gertrude Levine, *Fairleigh Dickinson University*
Alvin Sek See Lim, *Auburn University*
Charles Liu, *California State University, Los Angeles*
Wenhang Liu, *California State University, Los Angeles*
Mark Llewellyn, *University of Central Florida*
Sanchita Mal-Sarkar, *Cleveland State University*
Louis Marseille, *Harford Community College*
Kevin McNeill, *University of Arizona*
Arnold C. Meltzer, *George Washington University*
Rayman Meservy, *Brigham Young University*
Prasant Mohapatra, *University of California, Davis*
Hung Z Ngo, *SUNY, Buffalo*
Larry Owens, *California State University, Fresno*
Arnold Patton, *Bradley University*
Dolly Samson, *Hawaii Pacific University*
Joseph Sherif, *California State University, Fullerton*
Robert Simon, *George Mason University*
Ronald J. Srodawa, *Oakland University*
Daniel Tian, *California State University, Monterey Bay*
Richard Tibbs, *Radford University*
Christophe Veltsos, *Minnesota State University, Mankato*
Yang Wang, *University of Maryland, College Park*
Sherali Zeadally, *Wayne State University*

McGraw-Hill Staff

Special thanks go to the staff of McGraw-Hill. Alan Apt, our publisher, proved how a proficient publisher can make the impossible possible. Rebecca Olson, the developmental editor, gave us help whenever we needed it. Sheila Frank, our project manager, guided us through the production process with enormous enthusiasm. We also thank David Hash in design, Kara Kudronowicz in production, and Patti Scott, the copy editor.

PART 1

Overview

Objectives

Part 1 provides a general idea of what we will see in the rest of the book. Four major concepts are discussed: data communications, networking, protocols and standards, and networking models.

Networks exist so that data may be sent from one place to another—the basic concept of *data communications*. To fully grasp this subject, we must understand the data communication components, how different types of data can be represented, and how to create a data flow.

Data communications between remote parties can be achieved through a process called *networking,* involving the connection of computers, media, and networking devices. Networks are divided into two main categories: local area networks (LANs) and wide area networks (WANs). These two types of networks have different characteristics and different functionalities. The Internet, the main focus of the book, is a collection of LANs and WANs held together by internetworking devices.

Protocols and standards are vital to the implementation of data communications and networking. Protocols refer to the rules; a standard is a protocol that has been adopted by vendors and manufacturers.

Network models serve to organize, unify, and control the hardware and software components of data communications and networking. Although the term "network model" suggests a relationship to networking, the model also encompasses data communications.

Chapters

This part consists of two chapters: Chapter 1 and Chapter 2.

Chapter 1

In Chapter 1, we introduce the concepts of data communications and networking. We discuss data communications components, data representation, and data flow. We then move to the structure of networks that carry data. We discuss network topologies, categories of networks, and the general idea behind the Internet. The section on protocols and standards gives a quick overview of the organizations that set standards in data communications and networking.

Chapter 2

The two dominant networking models are the Open Systems Interconnection (OSI) and the Internet model (TCP/IP).The first is a theoretical framework; the second is the actual model used in today's data communications. In Chapter 2, we first discuss the OSI model to give a general background. We then concentrate on the Internet model, which is the foundation for the rest of the book.

CHAPTER 1

Introduction

Data communications and networking are changing the way we do business and the way we live. Business decisions have to be made ever more quickly, and the decision makers require immediate access to accurate information. Why wait a week for that report from Germany to arrive by mail when it could appear almost instantaneously through computer networks? Businesses today rely on computer networks and internetworks. But before we ask how quickly we can get hooked up, we need to know how networks operate, what types of technologies are available, and which design best fills which set of needs.

The development of the personal computer brought about tremendous changes for business, industry, science, and education. A similar revolution is occurring in data communications and networking. Technological advances are making it possible for communications links to carry more and faster signals. As a result, services are evolving to allow use of this expanded capacity. For example, established telephone services such as conference calling, call waiting, voice mail, and caller ID have been extended.

Research in data communications and networking has resulted in new technologies. One goal is to be able to exchange data such as text, audio, and video from all points in the world. We want to access the Internet to download and upload information quickly and accurately and at any time.

This chapter addresses four issues: data communications, networks, the Internet, and protocols and standards. First we give a broad definition of data communications. Then we define networks as a highway on which data can travel. The Internet is discussed as a good example of an internetwork (i.e., a network of networks). Finally, we discuss different types of protocols, the difference between protocols and standards, and the organizations that set those standards.

1.1 DATA COMMUNICATIONS

When we communicate, we are sharing information. This sharing can be local or remote. Between individuals, local communication usually occurs face to face, while remote communication takes place over distance. The term ***telecommunication,*** which includes telephony, telegraphy, and television, means communication at a distance (*tele* is Greek for "far").

The word ***data*** refers to information presented in whatever form is agreed upon by the parties creating and using the data.

Data communications are the exchange of data between two devices via some form of transmission medium such as a wire cable. For data communications to occur, the communicating devices must be part of a communication system made up of a combination of hardware (physical equipment) and software (programs). The effectiveness of a data communications system depends on four fundamental characteristics: delivery, accuracy, timeliness, and jitter.

1. **Delivery.** The system must deliver data to the correct destination. Data must be received by the intended device or user and only by that device or user.
2. **Accuracy.** The system must deliver the data accurately. Data that have been altered in transmission and left uncorrected are unusable.
3. **Timeliness.** The system must deliver data in a timely manner. Data delivered late are useless. In the case of video and audio, timely delivery means delivering data as they are produced, in the same order that they are produced, and without significant delay. This kind of delivery is called *real-time* transmission.
4. **Jitter.** Jitter refers to the variation in the packet arrival time. It is the uneven delay in the delivery of audio or video packets. For example, let us assume that video packets are sent every 30 ms. If some of the packets arrive with 30-ms delay and others with 40-ms delay, an uneven quality in the video is the result.

Components

A data communications system has five components (see Figure 1.1).

Figure 1.1 *Five components of data communication*

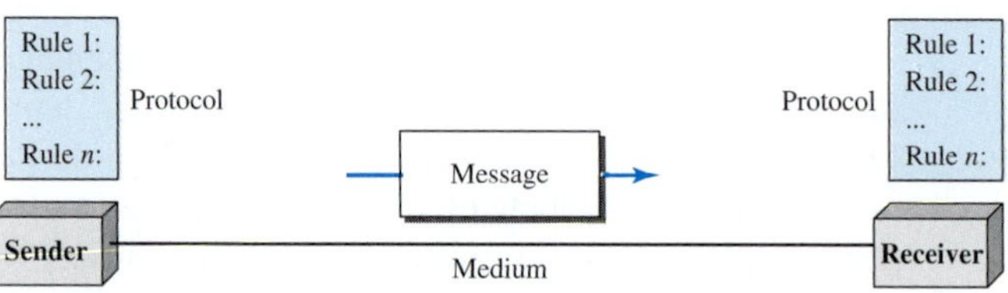

1. **Message.** The **message** is the information (data) to be communicated. Popular forms of information include text, numbers, pictures, audio, and video.
2. **Sender.** The **sender** is the device that sends the data message. It can be a computer, workstation, telephone handset, video camera, and so on.
3. **Receiver.** The **receiver** is the device that receives the message. It can be a computer, workstation, telephone handset, television, and so on.
4. **Transmission medium.** The **transmission medium** is the physical path by which a message travels from sender to receiver. Some examples of transmission media include twisted-pair wire, coaxial cable, fiber-optic cable, and radio waves.

5. **Protocol.** A **protocol** is a set of rules that govern data communications. It represents an agreement between the communicating devices. Without a protocol, two devices may be connected but not communicating, just as a person speaking French cannot be understood by a person who speaks only Japanese.

Data Representation

Information today comes in different forms such as text, numbers, images, audio, and video.

Text

In data communications, text is represented as a bit pattern, a sequence of bits (0s or 1s). Different sets of bit patterns have been designed to represent text symbols. Each set is called a **code,** and the process of representing symbols is called coding. Today, the prevalent coding system is called **Unicode**, which uses 32 bits to represent a symbol or character used in any language in the world. The **American Standard Code for Information Interchange (ASCII),** developed some decades ago in the United States, now constitutes the first 127 characters in Unicode and is also referred to as **Basic Latin.** Appendix A includes part of the Unicode.

Numbers

Numbers are also represented by bit patterns. However, a code such as ASCII is not used to represent numbers; the number is directly converted to a binary number to simplify mathematical operations. Appendix B discusses several different numbering systems.

Images

Images are also represented by bit patterns. In its simplest form, an image is composed of a matrix of pixels (picture elements), where each pixel is a small dot. The size of the pixel depends on the *resolution.* For example, an image can be divided into 1000 pixels or 10,000 pixels. In the second case, there is a better representation of the image (better resolution), but more memory is needed to store the image.

After an image is divided into pixels, each pixel is assigned a bit pattern. The size and the value of the pattern depend on the image. For an image made of only black-and-white dots (e.g., a chessboard), a 1-bit pattern is enough to represent a pixel.

If an image is not made of pure white and pure black pixels, you can increase the size of the bit pattern to include gray scale. For example, to show four levels of gray scale, you can use 2-bit patterns. A black pixel can be represented by 00, a dark gray pixel by 01, a light gray pixel by 10, and a white pixel by 11.

There are several methods to represent color images. One method is called **RGB,** so called because each color is made of a combination of three primary colors: *r*ed, *g*reen, and *b*lue. The intensity of each color is measured, and a bit pattern is assigned to it. Another method is called **YCM,** in which a color is made of a combination of three other primary colors: *y*ellow, *c*yan, and *m*agenta.

Audio

Audio refers to the recording or broadcasting of sound or music. Audio is by nature different from text, numbers, or images. It is continuous, not discrete. Even when we use a microphone to change voice or music to an electric signal, we create a continuous signal. In Chapters 4 and 5, we learn how to change sound or music to a digital or an analog signal.

Video

Video refers to the recording or broadcasting of a picture or movie. Video can either be produced as a continuous entity (e.g., by a TV camera), or it can be a combination of images, each a discrete entity, arranged to convey the idea of motion. Again we can change video to a digital or an analog signal, as we will see in Chapters 4 and 5.

Data Flow

Communication between two devices can be simplex, half-duplex, or full-duplex as shown in Figure 1.2.

Figure 1.2 *Data flow (simplex, half-duplex, and full-duplex)*

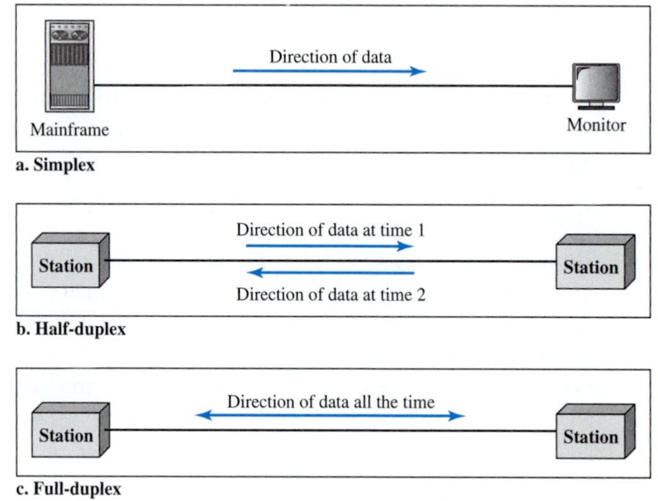

Simplex

In **simplex mode,** the communication is unidirectional, as on a one-way street. Only one of the two devices on a link can transmit; the other can only receive (see Figure 1.2a).

Keyboards and traditional monitors are examples of simplex devices. The keyboard can only introduce input; the monitor can only accept output. The simplex mode can use the entire capacity of the channel to send data in one direction.

Half-Duplex

In **half-duplex mode,** each station can both transmit and receive, but not at the same time. When one device is sending, the other can only receive, and vice versa (see Figure 1.2b).

The half-duplex mode is like a one-lane road with traffic allowed in both directions. When cars are traveling in one direction, cars going the other way must wait. In a half-duplex transmission, the entire capacity of a channel is taken over by whichever of the two devices is transmitting at the time. Walkie-talkies and CB (citizens band) radios are both half-duplex systems.

The half-duplex mode is used in cases where there is no need for communication in both directions at the same time; the entire capacity of the channel can be utilized for each direction.

Full-Duplex

In **full-duplex mode** (also called **duplex**), both stations can transmit and receive simultaneously (see Figure 1.2c).

The full-duplex mode is like a two-way street with traffic flowing in both directions at the same time. In full-duplex mode, signals going in one direction share the capacity of the link with signals going in the other direction. This sharing can occur in two ways: Either the link must contain two physically separate transmission paths, one for sending and the other for receiving; or the capacity of the channel is divided between signals traveling in both directions.

One common example of full-duplex communication is the telephone network. When two people are communicating by a telephone line, both can talk and listen at the same time.

The full-duplex mode is used when communication in both directions is required all the time. The capacity of the channel, however, must be divided between the two directions.

1.2 NETWORKS

A **network** is a set of devices (often referred to as *nodes*) connected by communication links. A node can be a computer, printer, or any other device capable of sending and/or receiving data generated by other nodes on the network.

Distributed Processing

Most networks use **distributed processing,** in which a task is divided among multiple computers. Instead of one single large machine being responsible for all aspects of a process, separate computers (usually a personal computer or workstation) handle a subset.

Network Criteria

A network must be able to meet a certain number of criteria. The most important of these are performance, reliability, and security.

Performance

Performance can be measured in many ways, including transit time and response time. Transit time is the amount of time required for a message to travel from one device to another. Response time is the elapsed time between an inquiry and a response. The performance of a network depends on a number of factors, including the number of users, the type of transmission medium, the capabilities of the connected hardware, and the efficiency of the software.

Performance is often evaluated by two networking metrics: **throughput** and **delay.** We often need more throughput and less delay. However, these two criteria are often contradictory. If we try to send more data to the network, we may increase throughput but we increase the delay because of traffic congestion in the network.

Reliability

In addition to accuracy of delivery, network **reliability** is measured by the frequency of failure, the time it takes a link to recover from a failure, and the network's robustness in a catastrophe.

Security

Network **security** issues include protecting data from unauthorized access, protecting data from damage and development, and implementing policies and procedures for recovery from breaches and data losses.

Physical Structures

Before discussing networks, we need to define some network attributes.

Type of Connection

A network is two or more devices connected through links. A link is a communications pathway that transfers data from one device to another. For visualization purposes, it is simplest to imagine any link as a line drawn between two points. For communication to occur, two devices must be connected in some way to the same link at the same time. There are two possible types of connections: point-to-point and multipoint.

Point-to-Point A **point-to-point connection** provides a dedicated link between two devices. The entire capacity of the link is reserved for transmission between those two devices. Most point-to-point connections use an actual length of wire or cable to connect the two ends, but other options, such as microwave or satellite links, are also possible (see Figure 1.3a). When you change television channels by infrared remote control, you are establishing a point-to-point connection between the remote control and the television's control system.

Multipoint A **multipoint** (also called **multidrop**) **connection** is one in which more than two specific devices share a single link (see Figure 1.3b).

In a multipoint environment, the capacity of the channel is shared, either spatially or temporally. If several devices can use the link simultaneously, it is a *spatially shared* connection. If users must take turns, it is a *timeshared* connection.

Physical Topology

The term ***physical topology*** refers to the way in which a network is laid out physically. Two or more devices connect to a link; two or more links form a topology. The topology

Figure 1.3 *Types of connections: point-to-point and multipoint*

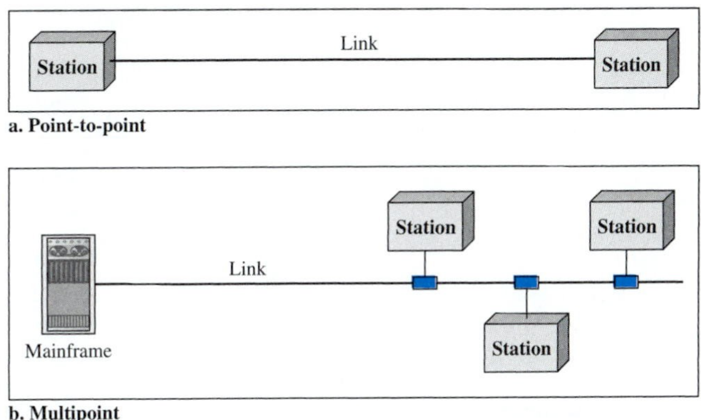

of a network is the geometric representation of the relationship of all the links and linking devices (usually called **nodes**) to one another. There are four basic topologies possible: mesh, star, bus, and ring (see Figure 1.4).

Figure 1.4 *Categories of topology*

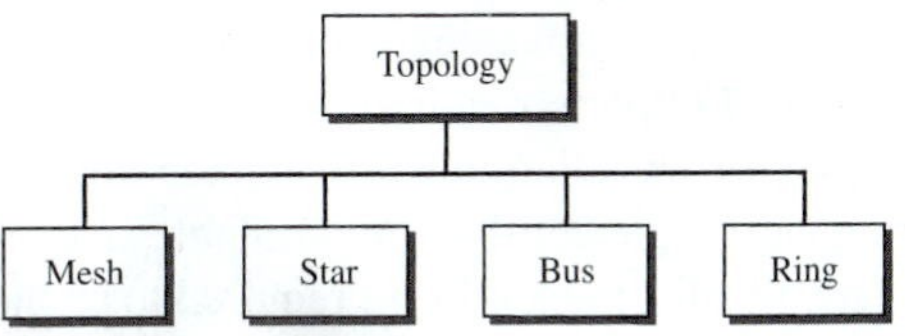

Mesh In a **mesh topology,** every device has a dedicated point-to-point link to every other device. The term *dedicated* means that the link carries traffic only between the two devices it connects. To find the number of physical links in a fully connected mesh network with n nodes, we first consider that each node must be connected to every other node. Node 1 must be connected to $n - 1$ nodes, node 2 must be connected to $n - 1$ nodes, and finally node n must be connected to $n - 1$ nodes. We need $n(n - 1)$ physical links. However, if each physical link allows communication in both directions (duplex mode), we can divide the number of links by 2. In other words, we can say that in a mesh topology, we need

$$n(n - 1) / 2$$

duplex-mode links.

To accommodate that many links, every device on the network must have $n - 1$ input/output (I/O) ports (see Figure 1.5) to be connected to the other $n - 1$ stations.

Figure 1.5 *A fully connected mesh topology (five devices)*

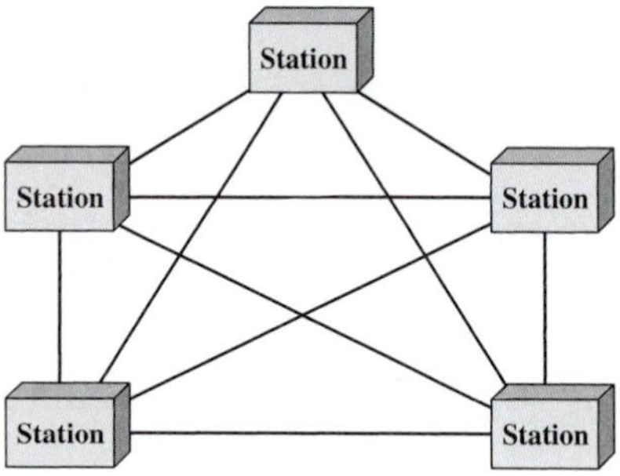

A mesh offers several advantages over other network topologies. First, the use of dedicated links guarantees that each connection can carry its own data load, thus eliminating the traffic problems that can occur when links must be shared by multiple devices. Second, a mesh topology is robust. If one link becomes unusable, it does not incapacitate the entire system. Third, there is the advantage of privacy or security. When every message travels along a dedicated line, only the intended recipient sees it. Physical boundaries prevent other users from gaining access to messages. Finally, point-to-point links make fault identification and fault isolation easy. Traffic can be routed to avoid links with suspected problems. This facility enables the network manager to discover the precise location of the fault and aids in finding its cause and solution.

The main disadvantages of a mesh are related to the amount of cabling and the number of I/O ports required. First, because every device must be connected to every other device, installation and reconnection are difficult. Second, the sheer bulk of the wiring can be greater than the available space (in walls, ceilings, or floors) can accommodate. Finally, the hardware required to connect each link (I/O ports and cable) can be prohibitively expensive. For these reasons a mesh topology is usually implemented in a limited fashion, for example, as a backbone connecting the main computers of a hybrid network that can include several other topologies.

One practical example of a mesh topology is the connection of telephone regional offices in which each regional office needs to be connected to every other regional office.

Star Topology In a **star topology,** each device has a dedicated point-to-point link only to a central controller, usually called a **hub.** The devices are not directly linked to one another. Unlike a mesh topology, a star topology does not allow direct traffic between devices. The controller acts as an exchange: If one device wants to send data to another, it sends the data to the controller, which then relays the data to the other connected device (see Figure 1.6) .

A star topology is less expensive than a mesh topology. In a star, each device needs only one link and one I/O port to connect it to any number of others. This factor also makes it easy to install and reconfigure. Far less cabling needs to be housed, and additions, moves, and deletions involve only one connection: between that device and the hub.

Other advantages include robustness. If one link fails, only that link is affected. All other links remain active. This factor also lends itself to easy fault identification and

Figure 1.6 *A star topology connecting four stations*

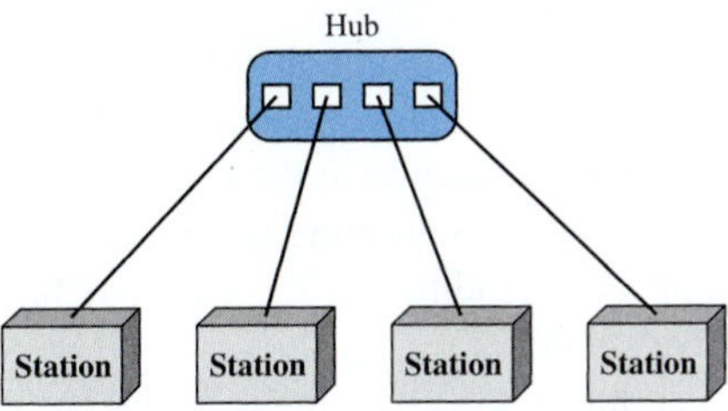

fault isolation. As long as the hub is working, it can be used to monitor link problems and bypass defective links.

One big disadvantage of a star topology is the dependency of the whole topology on one single point, the hub. If the hub goes down, the whole system is dead.

Although a star requires far less cable than a mesh, each node must be linked to a central hub. For this reason, often more cabling is required in a star than in some other topologies (such as ring or bus).

The star topology is used in local-area networks (LANs), as we will see in Chapter 13. High-speed LANs often use a star topology with a central hub.

Bus Topology The preceding examples all describe point-to-point connections. A **bus topology,** on the other hand, is multipoint. One long cable acts as a **backbone** to link all the devices in a network (see Figure 1.7).

Figure 1.7 *A bus topology connecting three stations*

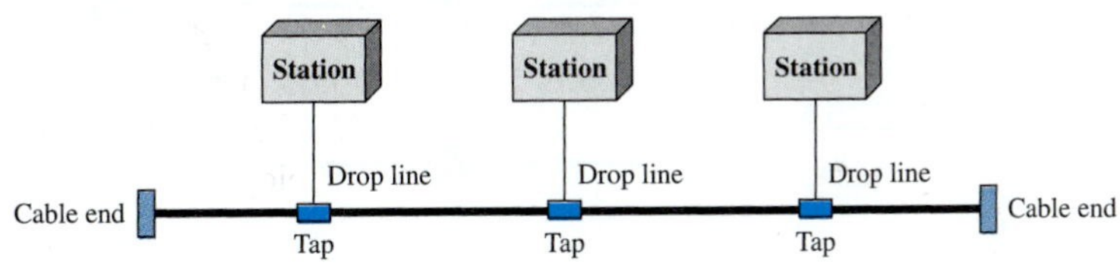

Nodes are connected to the bus cable by drop lines and taps. A drop line is a connection running between the device and the main cable. A tap is a connector that either splices into the main cable or punctures the sheathing of a cable to create a contact with the metallic core. As a signal travels along the backbone, some of its energy is transformed into heat. Therefore, it becomes weaker and weaker as it travels farther and farther. For this reason there is a limit on the number of taps a bus can support and on the distance between those taps.

Advantages of a bus topology include ease of installation. Backbone cable can be laid along the most efficient path, then connected to the nodes by drop lines of various lengths. In this way, a bus uses less cabling than mesh or star topologies. In a star, for example, four network devices in the same room require four lengths of cable reaching all the way to the hub. In a bus, this redundancy is eliminated. Only the backbone cable stretches through the entire facility. Each drop line has to reach only as far as the nearest point on the backbone.

Disadvantages include difficult reconnection and fault isolation. A bus is usually designed to be optimally efficient at installation. It can therefore be difficult to add new devices. Signal reflection at the taps can cause degradation in quality. This degradation can be controlled by limiting the number and spacing of devices connected to a given length of cable. Adding new devices may therefore require modification or replacement of the backbone.

In addition, a fault or break in the bus cable stops all transmission, even between devices on the same side of the problem. The damaged area reflects signals back in the direction of origin, creating noise in both directions.

Bus topology was the one of the first topologies used in the design of early local-area networks. Ethernet LANs can use a bus topology, but they are less popular now for reasons we will discuss in Chapter 13.

Ring Topology In a **ring topology,** each device has a dedicated point-to-point connection with only the two devices on either side of it. A signal is passed along the ring in one direction, from device to device, until it reaches its destination. Each device in the ring incorporates a repeater. When a device receives a signal intended for another device, its repeater regenerates the bits and passes them along (see Figure 1.8).

Figure 1.8 *A ring topology connecting six stations*

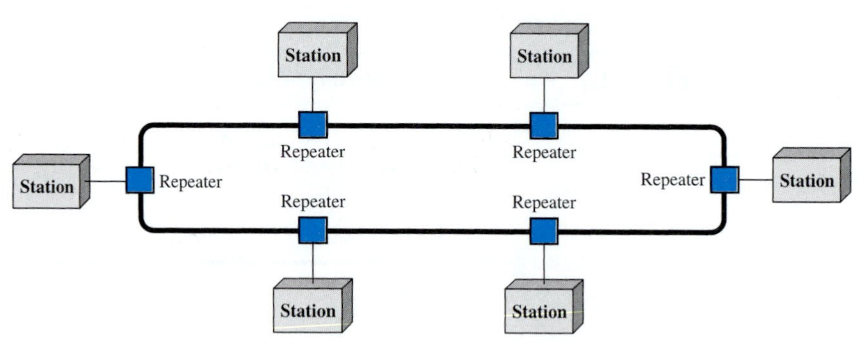

A ring is relatively easy to install and reconfigure. Each device is linked to only its immediate neighbors (either physically or logically). To add or delete a device requires changing only two connections. The only constraints are media and traffic considerations (maximum ring length and number of devices). In addition, fault isolation is simplified. Generally in a ring, a signal is circulating at all times. If one device does not receive a signal within a specified period, it can issue an alarm. The alarm alerts the network operator to the problem and its location.

However, unidirectional traffic can be a disadvantage. In a simple ring, a break in the ring (such as a disabled station) can disable the entire network. This weakness can be solved by using a dual ring or a switch capable of closing off the break.

Ring topology was prevalent when IBM introduced its local-area network Token Ring. Today, the need for higher-speed LANs has made this topology less popular.

Hybrid Topology A network can be hybrid. For example, we can have a main star topology with each branch connecting several stations in a bus topology as shown in Figure 1.9.

Figure 1.9 *A hybrid topology: a star backbone with three bus networks*

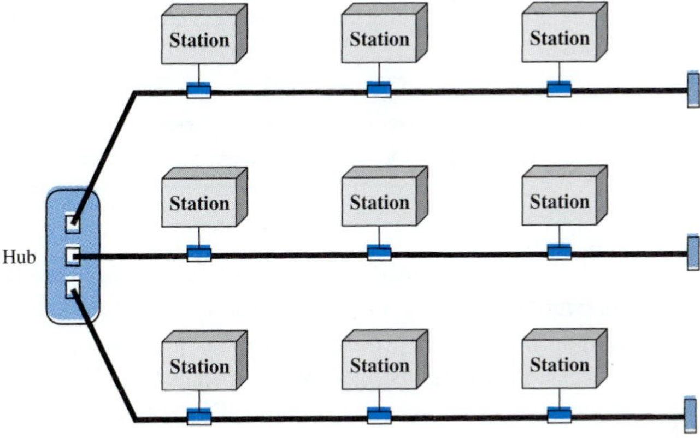

Network Models

Computer networks are created by different entities. Standards are needed so that these heterogeneous networks can communicate with one another. The two best-known standards are the OSI model and the Internet model. In Chapter 2 we discuss these two models. The OSI (Open Systems Interconnection) model defines a seven-layer network; the Internet model defines a five-layer network. This book is based on the Internet model with occasional references to the OSI model.

Categories of Networks

Today when we speak of networks, we are generally referring to two primary categories: local-area networks and wide-area networks. The category into which a network falls is determined by its size. A LAN normally covers an area less than 2 mi; a WAN can be worldwide. Networks of a size in between are normally referred to as metropolitan-area networks and span tens of miles.

Local Area Network

A **local area network (LAN)** is usually privately owned and links the devices in a single office, building, or campus (see Figure 1.10). Depending on the needs of an organization and the type of technology used, a LAN can be as simple as two PCs and a printer in someone's home office; or it can extend throughout a company and include audio and video peripherals. Currently, LAN size is limited to a few kilometers.

Figure 1.10 *An isolated LAN connecting 12 computers to a hub in a closet*

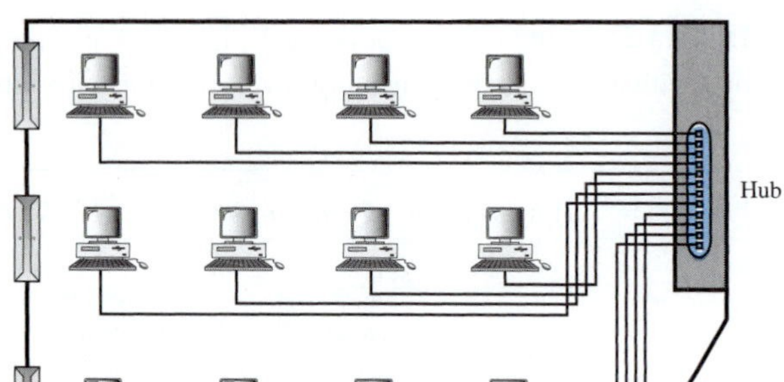

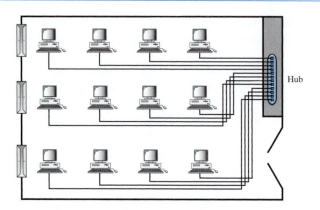

LANs are designed to allow resources to be shared between personal computers or workstations. The resources to be shared can include hardware (e.g., a printer), software (e.g., an application program), or data. A common example of a LAN, found in many business environments, links a workgroup of task-related computers, for example, engineering workstations or accounting PCs. One of the computers may be given a large-capacity disk drive and may become a server to clients. Software can be stored on this central server and used as needed by the whole group. In this example, the size of the LAN may be determined by licensing restrictions on the number of users per copy of software, or by restrictions on the number of users licensed to access the operating system.

In addition to size, LANs are distinguished from other types of networks by their transmission media and topology. In general, a given LAN will use only one type of transmission medium. The most common LAN topologies are bus, ring, and star.

Early LANs had data rates in the 4 to 16 megabits per second (Mbps) range. Today, however, speeds are normally 100 or 1000 Mbps. LANs are discussed at length in Chapters 13, 14, and 15.

Wireless LANs are the newest evolution in LAN technology. We discuss wireless LANs in detail in Chapter 14.

Wide Area Network

A **wide area network (WAN)** provides long-distance transmission of data, image, audio, and video information over large geographic areas that may comprise a country, a continent, or even the whole world. In Chapters 17 and 18 we discuss wide-area networks in greater detail. A WAN can be as complex as the backbones that connect the Internet or as simple as a dial-up line that connects a home computer to the Internet. We normally refer to the first as a switched WAN and to the second as a point-to-point WAN (Figure 1.11). The switched WAN connects the end systems, which usually comprise a router (internetworking connecting device) that connects to another LAN or WAN. The point-to-point WAN is normally a line leased from a telephone or cable TV provider that connects a home computer or a small LAN to an Internet service provider (ISP). This type of WAN is often used to provide Internet access.

Figure 1.11 *WANs: a switched WAN and a point-to-point WAN*

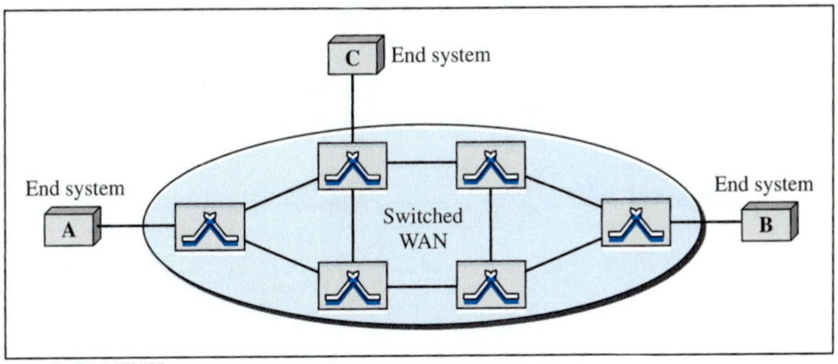

An early example of a switched WAN is X.25, a network designed to provide connectivity between end users. As we will see in Chapter 18, X.25 is being gradually replaced by a high-speed, more efficient network called Frame Relay. A good example of a switched WAN is the asynchronous transfer mode (ATM) network, which is a network with fixed-size data unit packets called cells. We will discuss ATM in Chapter 18. Another example of WANs is the wireless WAN that is becoming more and more popular. We discuss wireless WANs and their evolution in Chapter 16.

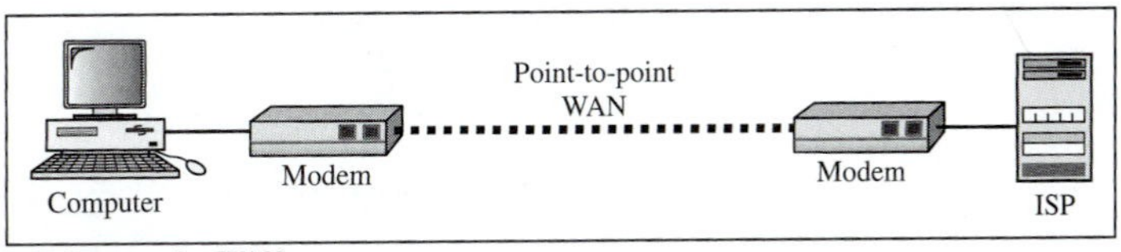

Metropolitan Area Networks

A **metropolitan area network (MAN)** is a network with a size between a LAN and a WAN. It normally covers the area inside a town or a city. It is designed for customers who need a high-speed connectivity, normally to the Internet, and have endpoints spread over a city or part of city. A good example of a MAN is the part of the telephone company network that can provide a high-speed DSL line to the customer. Another example is the cable TV network that originally was designed for cable TV, but today can also be used for high-speed data connection to the Internet. We discuss DSL lines and cable TV networks in Chapter 9.

Interconnection of Networks: Internetwork

Today, it is very rare to see a LAN, a MAN, or a LAN in isolation; they are connected to one another. When two or more networks are connected, they become an **internetwork,** or **internet.**

As an example, assume that an organization has two offices, one on the east coast and the other on the west coast. The established office on the west coast has a bus topology LAN; the newly opened office on the east coast has a star topology LAN. The president of the company lives somewhere in the middle and needs to have control over the company from her home. To create a backbone WAN for connecting these three entities (two LANs and the president's computer), a switched WAN (operated by a service provider such as a telecom company) has been leased. To connect the LANs to this switched WAN, however, three point-to-point WANs are required. These point-to-point WANs can be a high-speed DSL line offered by a telephone company or a cable modem line offered by a cable TV provider as shown in Figure 1.12.

Figure 1.12 *A heterogeneous network made of four WANs and two LANs*

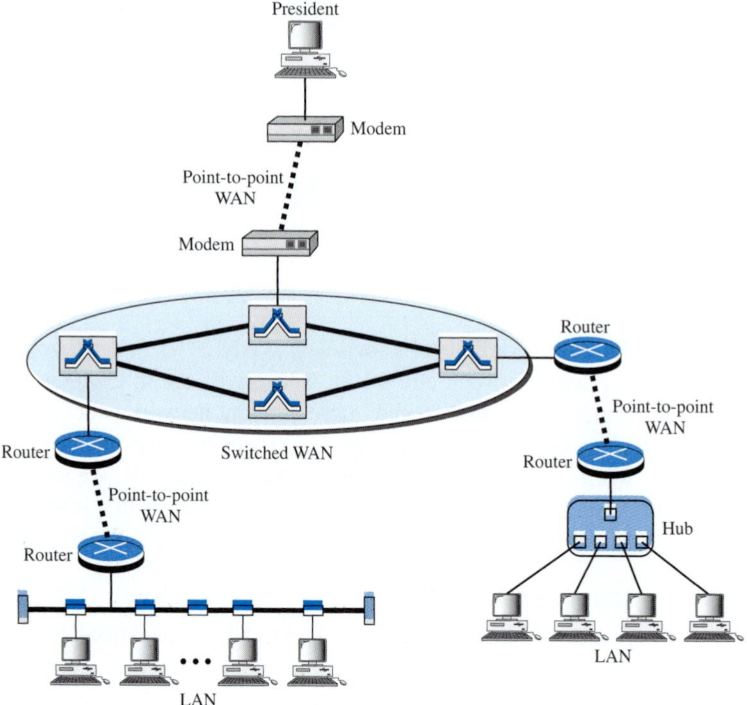

1.3 THE INTERNET

The Internet has revolutionized many aspects of our daily lives. It has affected the way we do business as well as the way we spend our leisure time. Count the ways you've used the Internet recently. Perhaps you've sent electronic mail (e-mail) to a business associate, paid a utility bill, read a newspaper from a distant city, or looked up a local movie schedule—all by using the Internet. Or maybe you researched a medical topic, booked a hotel reservation, chatted with a fellow Trekkie, or comparison-shopped for a car. The Internet is a communication system that has brought a wealth of information to our fingertips and organized it for our use.

The Internet is a structured, organized system. We begin with a brief history of the Internet. We follow with a description of the Internet today.

A Brief History

A network is a group of connected communicating devices such as computers and printers. An internet (note the lowercase letter i) is two or more networks that can communicate with each other. The most notable internet is called the **Internet** (uppercase letter I), a collaboration of more than hundreds of thousands of interconnected networks. Private individuals as well as various organizations such as government agencies, schools, research facilities, corporations, and libraries in more than 100 countries use the Internet. Millions of people are users. Yet this extraordinary communication system only came into being in 1969.

In the mid-1960s, mainframe computers in research organizations were standalone devices. Computers from different manufacturers were unable to communicate with one another. The **Advanced Research Projects Agency (ARPA)** in the Department of Defense (DoD) was interested in finding a way to connect computers so that the researchers they funded could share their findings, thereby reducing costs and eliminating duplication of effort.

In 1967, at an Association for Computing Machinery (ACM) meeting, ARPA presented its ideas for **ARPANET,** a small network of connected computers. The idea was that each host computer (not necessarily from the same manufacturer) would be attached to a specialized computer, called an *interface message processor* (IMP). The IMPs, in turn, would be connected to one another. Each IMP had to be able to communicate with other IMPs as well as with its own attached host.

By 1969, ARPANET was a reality. Four nodes, at the University of California at Los Angeles (UCLA), the University of California at Santa Barbara (UCSB), Stanford Research Institute (SRI), and the University of Utah, were connected via the IMPs to form a network. Software called the *Network Control Protocol* (NCP) provided communication between the hosts.

In 1972, Vint Cerf and Bob Kahn, both of whom were part of the core ARPANET group, collaborated on what they called the *Internetting Project*. Cerf and Kahn's landmark 1973 paper outlined the protocols to achieve end-to-end delivery of packets. This paper on Transmission Control Protocol (TCP) included concepts such as encapsulation, the datagram, and the functions of a gateway.

Shortly thereafter, authorities made a decision to split TCP into two protocols: **Transmission Control Protocol (TCP)** and **Internetworking Protocol (IP).** IP would handle datagram routing while TCP would be responsible for higher-level functions such as segmentation, reassembly, and error detection. The internetworking protocol became known as TCP/IP.

The Internet Today

The Internet has come a long way since the 1960s. The Internet today is not a simple hierarchical structure. It is made up of many wide- and local-area networks joined by connecting devices and switching stations. It is difficult to give an accurate representation of the Internet because it is continually changing—new networks are being added, existing networks are adding addresses, and networks of defunct companies are being removed. Today most end users who want Internet connection use the services of **Internet service providers (ISPs).** There are international service providers, national service providers, regional service providers, and local service providers. The Internet today is run by private companies, not the government. Figure 1.13 shows a conceptual (not geographic) view of the Internet.

Figure 1.13 *Hierarchical organization of the Internet*

a. Structure of a national ISP

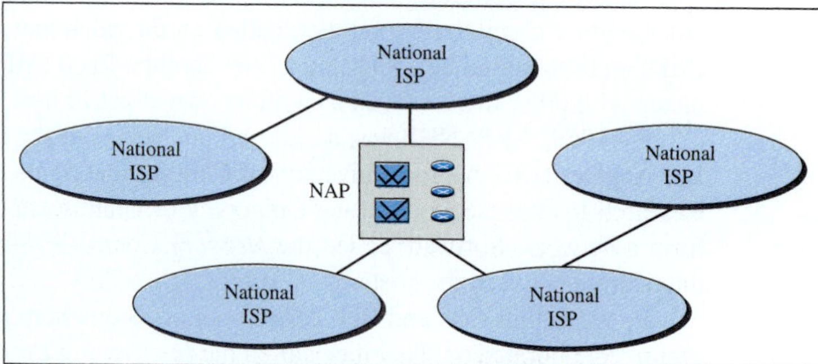

b. Interconnection of national ISPs

International Internet Service Providers

At the top of the hierarchy are the international service providers that connect nations together.

National Internet Service Providers

The **national Internet service providers** are backbone networks created and maintained by specialized companies. There are many national ISPs operating in North America; some of the most well known are SprintLink, PSINet, UUNet Technology, AGIS, and internet MCI. To provide connectivity between the end users, these backbone networks are connected by complex switching stations (normally run by a third party) called **network access points (NAPs).** Some national ISP networks are also connected to one another by private switching stations called *peering points*. These normally operate at a high data rate (up to 600 Mbps).

Regional Internet Service Providers

Regional internet service providers or **regional ISPs** are smaller ISPs that are connected to one or more national ISPs. They are at the third level of the hierarchy with a smaller data rate.

Local Internet Service Providers

Local Internet service providers provide direct service to the end users. The local ISPs can be connected to regional ISPs or directly to national ISPs. Most end users are connected to the local ISPs. Note that in this sense, a local ISP can be a company that just provides Internet services, a corporation with a network that supplies services to its own employees, or a nonprofit organization, such as a college or a university, that runs its own network. Each of these local ISPs can be connected to a regional or national service provider.

1.4 PROTOCOLS AND STANDARDS

In this section, we define two widely used terms: protocols and standards. First, we define *protocol,* which is synonymous with *rule.* Then we discuss *standards,* which are agreed-upon rules.

Protocols

In computer networks, communication occurs between entities in different systems. An **entity** is anything capable of sending or receiving information. However, two entities cannot simply send bit streams to each other and expect to be understood. For communication to occur, the entities must agree on a protocol. A protocol is a set of rules that govern data communications. A protocol defines what is communicated, how it is communicated, and when it is communicated. The key elements of a protocol are syntax, semantics, and timing.

- **Syntax.** The term ***syntax*** refers to the structure or format of the data, meaning the order in which they are presented. For example, a simple protocol might expect the first 8 bits of data to be the address of the sender, the second 8 bits to be the address of the receiver, and the rest of the stream to be the message itself.
- **Semantics.** The word ***semantics*** refers to the meaning of each section of bits. How is a particular pattern to be interpreted, and what action is to be taken based on that interpretation? For example, does an address identify the route to be taken or the final destination of the message?
- **Timing.** The term ***timing*** refers to two characteristics: when data should be sent and how fast they can be sent. For example, if a sender produces data at 100 Mbps but the receiver can process data at only 1 Mbps, the transmission will overload the receiver and some data will be lost.

Standards

Standards are essential in creating and maintaining an open and competitive market for equipment manufacturers and in guaranteeing national and international interoperability of data and telecommunications technology and processes. Standards provide guidelines to manufacturers, vendors, government agencies, and other service providers to ensure the kind of interconnectivity necessary in today's marketplace and in international communications. Data communication standards fall into two categories: *de facto* (meaning "by fact" or "by convention") and *de jure* (meaning "by law" or "by regulation").

- **De facto.** Standards that have not been approved by an organized body but have been adopted as standards through widespread use are **de facto standards.** De facto standards are often established originally by manufacturers who seek to define the functionality of a new product or technology.
- **De jure.** Those standards that have been legislated by an officially recognized body are **de jure standards.**

Standards Organizations

Standards are developed through the cooperation of standards creation committees, forums, and government regulatory agencies.

Standards Creation Committees

While many organizations are dedicated to the establishment of standards, data telecommunications in North America rely primarily on those published by the following:

- **International Organization for Standardization (ISO).** The ISO is a multinational body whose membership is drawn mainly from the standards creation committees of various governments throughout the world. The ISO is active in developing cooperation in the realms of scientific, technological, and economic activity.
- **International Telecommunication Union—Telecommunication Standards Sector (ITU-T).** By the early 1970s, a number of countries were defining national standards for telecommunications, but there was still little international compatibility. The United Nations responded by forming, as part of its International Telecommunication Union (ITU), a committee, the **Consultative Committee for International Telegraphy and Telephony (CCITT).** This committee was devoted to the research and establishment of standards for telecommunications in general and for phone and data systems in particular. On March 1, 1993, the name of this committee was changed to the International Telecommunication Union—Telecommunication Standards Sector (ITU-T).
- **American National Standards Institute (ANSI).** Despite its name, the American National Standards Institute is a completely private, nonprofit corporation not affiliated with the U.S. federal government. However, all ANSI activities are undertaken with the welfare of the United States and its citizens occupying primary importance.
- **Institute of Electrical and Electronics Engineers (IEEE).** The Institute of Electrical and Electronics Engineers is the largest professional engineering society in the world. International in scope, it aims to advance theory, creativity, and product quality in the fields of electrical engineering, electronics, and radio as well as in all related branches of engineering. As one of its goals, the IEEE oversees the development and adoption of international standards for computing and communications.
- **Electronic Industries Association (EIA).** Aligned with ANSI, the Electronic Industries Association is a nonprofit organization devoted to the promotion of

electronics manufacturing concerns. Its activities include public awareness education and lobbying efforts in addition to standards development. In the field of information technology, the EIA has made significant contributions by defining physical connection interfaces and electronic signaling specifications for data communication.

Forums

Telecommunications technology development is moving faster than the ability of standards committees to ratify standards. Standards committees are procedural bodies and by nature slow-moving. To accommodate the need for working models and agreements and to facilitate the standardization process, many special-interest groups have developed **forums** made up of representatives from interested corporations. The forums work with universities and users to test, evaluate, and standardize new technologies. By concentrating their efforts on a particular technology, the forums are able to speed acceptance and use of those technologies in the telecommunications community. The forums present their conclusions to the standards bodies.

Regulatory Agencies

All communications technology is subject to regulation by government agencies such as the **Federal Communications Commission (FCC)** in the United States. The purpose of these agencies is to protect the public interest by regulating radio, television, and wire/cable communications. The FCC has authority over interstate and international commerce as it relates to communications.

Internet Standards

An **Internet standard** is a thoroughly tested specification that is useful to and adhered to by those who work with the Internet. It is a formalized regulation that must be followed. There is a strict procedure by which a specification attains Internet standard status. A specification begins as an Internet draft. An **Internet draft** is a working document (a work in progress) with no official status and a 6-month lifetime. Upon recommendation from the Internet authorities, a draft may be published as a **Request for Comment (RFC).** Each RFC is edited, assigned a number, and made available to all interested parties. RFCs go through maturity levels and are categorized according to their requirement level.

1.5 RECOMMENDED READING

For more details about subjects discussed in this chapter, we recommend the following books and sites. The items enclosed in brackets [. . .] refer to the reference list at the end of the book.

Books

The introductory materials covered in this chapter can be found in [Sta04] and [PD03]. [Tan03] discusses standardization in Section 1.6.

Sites

The following sites are related to topics discussed in this chapter.

- www.acm.org/sigcomm/sos.html This site gives the status of various networking standards.
- www.ietf.org/ The Internet Engineering Task Force (IETF) home page.

RFCs

The following site lists all RFCs, including those related to IP and TCP. In future chapters we cite the RFCs pertinent to the chapter material.

- www.ietf.org/rfc.html

1.6 KEY TERMS

Advanced Research Projects Agency (ARPA)
American National Standards Institute (ANSI)
American Standard Code for Information Interchange (ASCII)
ARPANET
audio
backbone
Basic Latin
bus topology
code
Consultative Committee for International Telegraphy and Telephony (CCITT)
data
data communications
de facto standards
de jure standards
delay
distributed processing
Electronic Industries Association (EIA)
entity
Federal Communications Commission (FCC)
forum
full-duplex mode, or duplex
half-duplex mode
hub
image
Institute of Electrical and Electronics Engineers (IEEE)
International Organization for Standardization (ISO)
International Telecommunication Union—Telecommunication Standards Sector (ITU-T)
Internet
Internet draft
Internet service provider (ISP)
Internet standard
internetwork or internet
local area network (LAN)
local Internet service providers
mesh topology
message
metropolitan area network (MAN)
multipoint or multidrop connection
national Internet service provider
network

network access points (NAPs)
node
performance
physical topology
point-to-point connection
protocol
receiver
regional ISP
reliability
Request for Comment (RFC)
RGB
ring topology
security
semantics
sender
simplex mode
star topology
syntax
telecommunication
throughput
timing
Transmission Control Protocol/ Internetworking Protocol (TCP/IP)
transmission medium
Unicode
video
wide area network (WAN)
YCM

1.7 SUMMARY

- Data communications are the transfer of data from one device to another via some form of transmission medium.
- A data communications system must transmit data to the correct destination in an accurate and timely manner.
- The five components that make up a data communications system are the message, sender, receiver, medium, and protocol.
- Text, numbers, images, audio, and video are different forms of information.
- Data flow between two devices can occur in one of three ways: simplex, half-duplex, or full-duplex.
- A network is a set of communication devices connected by media links.
- In a point-to-point connection, two and only two devices are connected by a dedicated link. In a multipoint connection, three or more devices share a link.
- Topology refers to the physical or logical arrangement of a network. Devices may be arranged in a mesh, star, bus, or ring topology.
- A network can be categorized as a local area network or a wide area network.
- A LAN is a data communication system within a building, plant, or campus, or between nearby buildings.
- A WAN is a data communication system spanning states, countries, or the whole world.
- An internet is a network of networks.
- The Internet is a collection of many separate networks.
- There are local, regional, national, and international Internet service providers.
- A protocol is a set of rules that govern data communication; the key elements of a protocol are syntax, semantics, and timing.

- Standards are necessary to ensure that products from different manufacturers can work together as expected.
- The ISO, ITU-T, ANSI, IEEE, and EIA are some of the organizations involved in standards creation.
- Forums are special-interest groups that quickly evaluate and standardize new technologies.
- A Request for Comment is an idea or concept that is a precursor to an Internet standard.

1.8 PRACTICE SET

Review Questions

1. Identify the five components of a data communications system.
2. What are the advantages of distributed processing?
3. What are the three criteria necessary for an effective and efficient network?
4. What are the advantages of a multipoint connection over a point-to-point connection?
5. What are the two types of line configuration?
6. Categorize the four basic topologies in terms of line configuration.
7. What is the difference between half-duplex and full-duplex transmission modes?
8. Name the four basic network topologies, and cite an advantage of each type.
9. For n devices in a network, what is the number of cable links required for a mesh, ring, bus, and star topology?
10. What are some of the factors that determine whether a communication system is a LAN or WAN?
11. What is an internet? What is the Internet?
12. Why are protocols needed?
13. Why are standards needed?

Exercises

14. What is the maximum number of characters or symbols that can be represented by Unicode?
15. A color image uses 16 bits to represent a pixel. What is the maximum number of different colors that can be represented?
16. Assume six devices are arranged in a mesh topology. How many cables are needed? How many ports are needed for each device?
17. For each of the following four networks, discuss the consequences if a connection fails.
    a. Five devices arranged in a mesh topology
    b. Five devices arranged in a star topology (not counting the hub)
    c. Five devices arranged in a bus topology
    d. Five devices arranged in a ring topology

18. You have two computers connected by an Ethernet hub at home. Is this a LAN, a MAN, or a WAN? Explain your reason.
19. In the ring topology in Figure 1.8, what happens if one of the stations is unplugged?
20. In the bus topology in Figure 1.7, what happens if one of the stations is unplugged?
21. Draw a hybrid topology with a star backbone and three ring networks.
22. Draw a hybrid topology with a ring backbone and two bus networks.
23. Performance is inversely related to delay. When you use the Internet, which of the following applications are more sensitive to delay?
    a. Sending an e-mail
    b. Copying a file
    c. Surfing the Internet
24. When a party makes a local telephone call to another party, is this a point-to-point or multipoint connection? Explain your answer.
25. Compare the telephone network and the Internet. What are the similarities? What are the differences?

Research Activities

26. Using the site www.cne.gmu.edu/modules/network/osi.html, discuss the OSI model.
27. Using the site www.ansi.org, discuss ANSI's activities.
28. Using the site www.ieee.org, discuss IEEE's activities.
29. Using the site www.ietf.org/, discuss the different types of RFCs.

CHAPTER 2

Network Models

A network is a combination of hardware and software that sends data from one location to another. The hardware consists of the physical equipment that carries signals from one point of the network to another. The software consists of instruction sets that make possible the services that we expect from a network.

We can compare the task of networking to the task of solving a mathematics problem with a computer. The fundamental job of solving the problem with a computer is done by computer hardware. However, this is a very tedious task if only hardware is involved. We would need switches for every memory location to store and manipulate data. The task is much easier if software is available. At the highest level, a program can direct the problem-solving process; the details of how this is done by the actual hardware can be left to the layers of software that are called by the higher levels.

Compare this to a service provided by a computer network. For example, the task of sending an e-mail from one point in the world to another can be broken into several tasks, each performed by a separate software package. Each software package uses the services of another software package. At the lowest layer, a signal, or a set of signals, is sent from the source computer to the destination computer.

In this chapter, we give a general idea of the layers of a network and discuss the functions of each. Detailed descriptions of these layers follow in later chapters.

2.1 LAYERED TASKS

We use the concept of layers in our daily life. As an example, let us consider two friends who communicate through postal mail. The process of sending a letter to a friend would be complex if there were no services available from the post office. Figure 2.1 shows the steps in this task.

Figure 2.1 *Tasks involved in sending a letter*

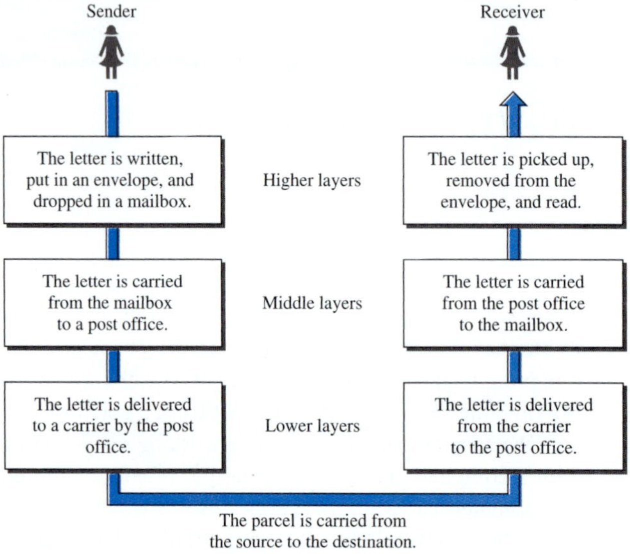

Sender, Receiver, and Carrier

In Figure 2.1 we have a sender, a receiver, and a carrier that transports the letter. There is a hierarchy of tasks.

At the Sender Site

Let us first describe, in order, the activities that take place at the sender site.

- **Higher layer.** The sender writes the letter, inserts the letter in an envelope, writes the sender and receiver addresses, and drops the letter in a mailbox.
- **Middle layer.** The letter is picked up by a letter carrier and delivered to the post office.
- **Lower layer.** The letter is sorted at the post office; a carrier transports the letter.

On the Way

The letter is then on its way to the recipient. On the way to the recipient's local post office, the letter may actually go through a central office. In addition, it may be transported by truck, train, airplane, boat, or a combination of these.

At the Receiver Site

- **Lower layer.** The carrier transports the letter to the post office.
- **Middle layer.** The letter is sorted and delivered to the recipient's mailbox.
- **Higher layer.** The receiver picks up the letter, opens the envelope, and reads it.

Hierarchy

According to our analysis, there are three different activities at the sender site and another three activities at the receiver site. The task of transporting the letter between the sender and the receiver is done by the carrier. Something that is not obvious immediately is that the tasks must be done in the order given in the hierarchy. At the sender site, the letter must be written and dropped in the mailbox before being picked up by the letter carrier and delivered to the post office. At the receiver site, the letter must be dropped in the recipient mailbox before being picked up and read by the recipient.

Services

Each layer at the sending site uses the services of the layer immediately below it. The sender at the higher layer uses the services of the middle layer. The middle layer uses the services of the lower layer. The lower layer uses the services of the carrier.

The layered model that dominated data communications and networking literature before 1990 was the **Open Systems Interconnection (OSI) model.** Everyone believed that the OSI model would become the ultimate standard for data communications, but this did not happen. The TCP/IP protocol suite became the dominant commercial architecture because it was used and tested extensively in the Internet; the OSI model was never fully implemented.

In this chapter, first we briefly discuss the OSI model, and then we concentrate on TCP/IP as a protocol suite.

2.2 THE OSI MODEL

Established in 1947, the International Standards Organization (ISO) is a multinational body dedicated to worldwide agreement on international standards. An ISO standard that covers all aspects of network communications is the Open Systems Interconnection model. It was first introduced in the late 1970s. An **open system** is a set of protocols that allows any two different systems to communicate regardless of their underlying architecture. The purpose of the OSI model is to show how to facilitate communication between different systems without requiring changes to the logic of the underlying hardware and software. The OSI model is not a protocol; it is a model for understanding and designing a network architecture that is flexible, robust, and interoperable.

ISO is the organization. OSI is the model.

The OSI model is a layered framework for the design of network systems that allows communication between all types of computer systems. It consists of seven separate but related layers, each of which defines a part of the process of moving information across a network (see Figure 2.2). An understanding of the fundamentals of the OSI model provides a solid basis for exploring data communications.

Figure 2.2 *Seven layers of the OSI model*

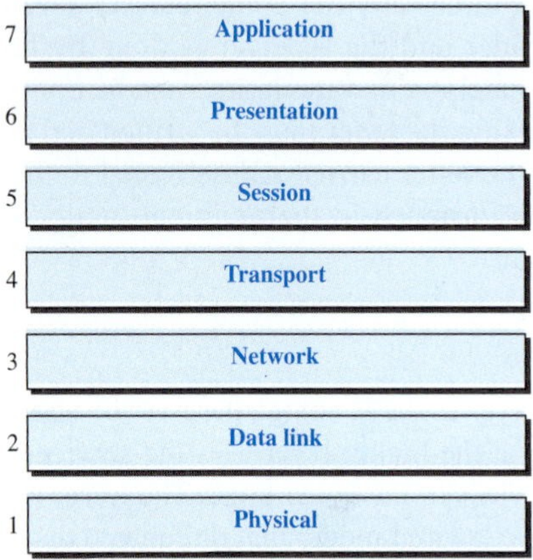

Layered Architecture

The OSI model is composed of seven ordered layers: physical (layer 1), data link (layer 2), network (layer 3), transport (layer 4), session (layer 5), presentation (layer 6), and application (layer 7). Figure 2.3 shows the layers involved when a message is sent from device A to device B. As the message travels from A to B, it may pass through many intermediate nodes. These intermediate nodes usually involve only the first three layers of the OSI model.

In developing the model, the designers distilled the process of transmitting data to its most fundamental elements. They identified which networking functions had related uses and collected those functions into discrete groups that became the layers. Each layer defines a family of functions distinct from those of the other layers. By defining and localizing functionality in this fashion, the designers created an architecture that is both comprehensive and flexible. Most importantly, the OSI model allows complete interoperability between otherwise incompatible systems.

Within a single machine, each layer calls upon the services of the layer just below it. Layer 3, for example, uses the services provided by layer 2 and provides services for layer 4. Between machines, layer x on one machine communicates with layer x on another machine. This communication is governed by an agreed-upon series of rules and conventions called protocols. The processes on each machine that communicate at a given layer are called **peer-to-peer processes.** Communication between machines is therefore a peer-to-peer process using the protocols appropriate to a given layer.

Peer-to-Peer Processes

At the physical layer, communication is direct: In Figure 2.3, device A sends a stream of bits to device B (through intermediate nodes). At the higher layers, however, communication must move down through the layers on device A, over to device B, and then

Figure 2.3 *The interaction between layers in the OSI model*

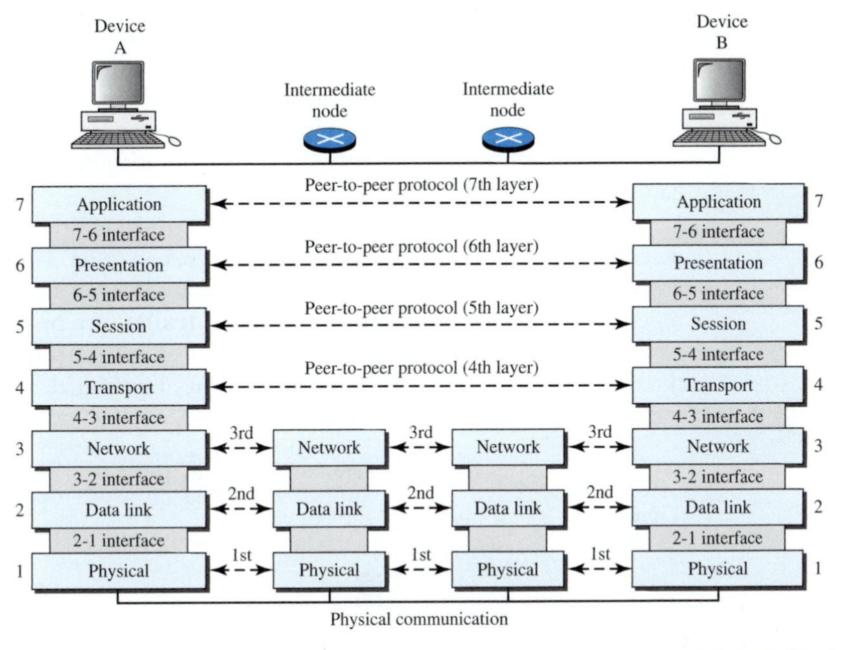

back up through the layers. Each layer in the sending device adds its own information to the message it receives from the layer just above it and passes the whole package to the layer just below it.

At layer 1 the entire package is converted to a form that can be transmitted to the receiving device. At the receiving machine, the message is unwrapped layer by layer, with each process receiving and removing the data meant for it. For example, layer 2 removes the data meant for it, then passes the rest to layer 3. Layer 3 then removes the data meant for it and passes the rest to layer 4, and so on.

Interfaces Between Layers

The passing of the data and network information down through the layers of the sending device and back up through the layers of the receiving device is made possible by an **interface** between each pair of adjacent layers. Each interface defines the information and services a layer must provide for the layer above it. Well-defined interfaces and layer functions provide modularity to a network. As long as a layer provides the expected services to the layer above it, the specific implementation of its functions can be modified or replaced without requiring changes to the surrounding layers.

Organization of the Layers

The seven layers can be thought of as belonging to three subgroups. Layers 1, 2, and 3—physical, data link, and network—are the network support layers; they deal with the physical aspects of moving data from one device to another (such as electrical specifications, physical connections, physical addressing, and transport timing and reliability). Layers 5, 6, and 7—session, presentation, and application—can be thought of as the user support layers; they allow interoperability among unrelated software systems. Layer 4, the transport layer, links the two subgroups and ensures that what the lower layers have transmitted is in a form that the upper layers can use. The upper OSI layers are almost always implemented in software; lower layers are a combination of hardware and software, except for the physical layer, which is mostly hardware.

In Figure 2.4, which gives an overall view of the OSI layers, D7 means the data unit at layer 7, D6 means the data unit at layer 6, and so on. The process starts at layer 7 (the application layer), then moves from layer to layer in descending, sequential order. At each layer, a **header,** or possibly a **trailer,** can be added to the data unit. Commonly, the trailer is added only at layer 2. When the formatted data unit passes through the physical layer (layer 1), it is changed into an electromagnetic signal and transported along a physical link.

Figure 2.4 *An exchange using the OSI model*

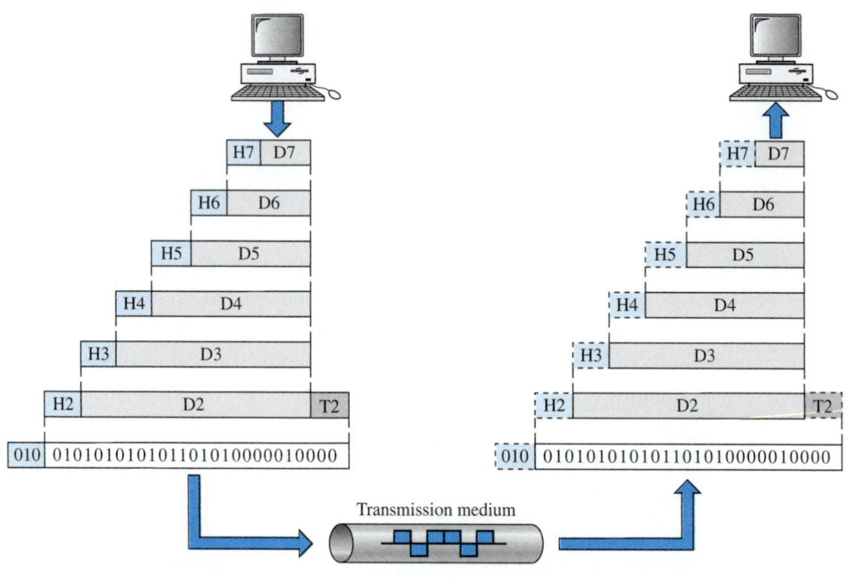

Upon reaching its destination, the signal passes into layer 1 and is transformed back into digital form. The data units then move back up through the OSI layers. As each block of data reaches the next higher layer, the headers and trailers attached to it at the corresponding sending layer are removed, and actions appropriate to that layer are taken. By the time it reaches layer 7, the message is again in a form appropriate to the application and is made available to the recipient.

Encapsulation

Figure 2.3 reveals another aspect of data communications in the OSI model: encapsulation. A packet (header and data) at level 7 is encapsulated in a packet at level 6. The whole packet at level 6 is encapsulated in a packet at level 5, and so on.

In other words, the data portion of a packet at level $N - 1$ carries the whole packet (data and header and maybe trailer) from level N. The concept is called *encapsulation;* level $N - 1$ is not aware of which part of the encapsulated packet is data and which part is the header or trailer. For level $N - 1$, the whole packet coming from level N is treated as one integral unit.

2.3 LAYERS IN THE OSI MODEL

In this section we briefly describe the functions of each layer in the OSI model.

Physical Layer

The **physical layer** coordinates the functions required to carry a bit stream over a physical medium. It deals with the mechanical and electrical specifications of the interface and transmission medium. It also defines the procedures and functions that physical devices and interfaces have to perform for transmission to occur. Figure 2.5 shows the position of the physical layer with respect to the transmission medium and the data link layer.

Figure 2.5 *Physical layer*

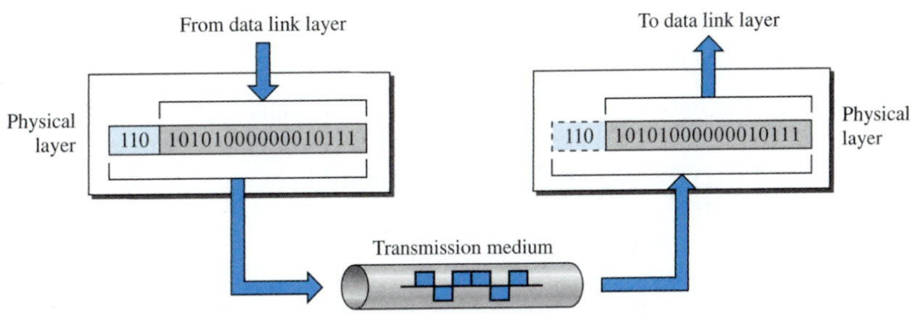

The physical layer is responsible for movements of individual bits from one hop (node) to the next.

The physical layer is also concerned with the following:

- **Physical characteristics of interfaces and medium.** The physical layer defines the characteristics of the interface between the devices and the transmission medium. It also defines the type of transmission medium.
- **Representation of bits.** The physical layer data consists of a stream of **bits** (sequence of 0s or 1s) with no interpretation. To be transmitted, bits must be

encoded into signals—electrical or optical. The physical layer defines the type of **encoding** (how 0s and 1s are changed to signals).

- **Data rate.** The **transmission rate**—the number of bits sent each second—is also defined by the physical layer. In other words, the physical layer defines the duration of a bit, which is how long it lasts.
- **Synchronization of bits.** The sender and receiver not only must use the same bit rate but also must be synchronized at the bit level. In other words, the sender and the receiver clocks must be synchronized.
- **Line configuration.** The physical layer is concerned with the connection of devices to the media. In a point-to-point configuration, two devices are connected through a dedicated link. In a multipoint configuration, a link is shared among several devices.
- **Physical topology.** The physical topology defines how devices are connected to make a network. Devices can be connected by using a mesh topology (every device is connected to every other device), a star topology (devices are connected through a central device), a ring topology (each device is connected to the next, forming a ring), a bus topology (every device is on a common link), or a hybrid topology (this is a combination of two or more topologies).
- **Transmission mode.** The physical layer also defines the direction of transmission between two devices: simplex, half-duplex, or full-duplex. In simplex mode, only one device can send; the other can only receive. The simplex mode is a one-way communication. In the half-duplex mode, two devices can send and receive, but not at the same time. In a full-duplex (or simply duplex) mode, two devices can send and receive at the same time.

Data Link Layer

The **data link layer** transforms the physical layer, a raw transmission facility, to a reliable link. It makes the physical layer appear error-free to the upper layer (network layer). Figure 2.6 shows the relationship of the data link layer to the network and physical layers.

Figure 2.6 *Data link layer*

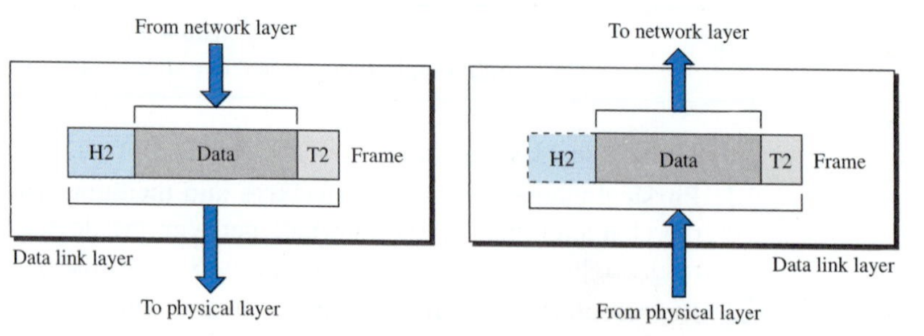

The data link layer is responsible for moving frames from one hop (node) to the next.

Other responsibilities of the data link layer include the following:

- **Framing.** The data link layer divides the stream of bits received from the network layer into manageable data units called **frames.**
- **Physical addressing.** If frames are to be distributed to different systems on the network, the data link layer adds a header to the frame to define the sender and/or receiver of the frame. If the frame is intended for a system outside the sender's network, the receiver address is the address of the device that connects the network to the next one.
- **Flow control.** If the rate at which the data are absorbed by the receiver is less than the rate at which data are produced in the sender, the data link layer imposes a flow control mechanism to avoid overwhelming the receiver.
- **Error control.** The data link layer adds reliability to the physical layer by adding mechanisms to detect and retransmit damaged or lost frames. It also uses a mechanism to recognize duplicate frames. Error control is normally achieved through a trailer added to the end of the frame.
- **Access control.** When two or more devices are connected to the same link, data link layer protocols are necessary to determine which device has control over the link at any given time.

Figure 2.7 illustrates **hop-to-hop (node-to-node) delivery** by the data link layer.

Figure 2.7 *Hop-to-hop delivery*

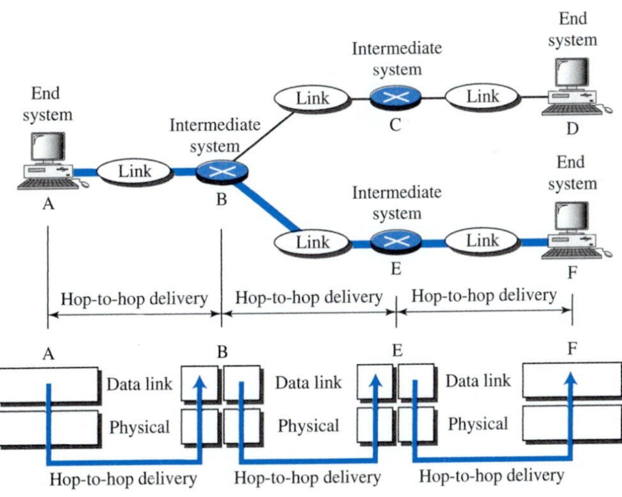

As the figure shows, communication at the data link layer occurs between two adjacent nodes. To send data from A to F, three partial deliveries are made. First, the data link layer at A sends a frame to the data link layer at B (a router). Second, the data link layer at B sends a new frame to the data link layer at E. Finally, the data link layer at E sends a new frame to the data link layer at F. Note that the frames that are exchanged between the three nodes have different values in the headers. The frame from A to B has B as the destination address and A as the source address. The frame from B to E has E as the destination address and B as the source address. The frame from E to F has F as the destination address and E as the source address. The values of the trailers can also be different if error checking includes the header of the frame.

Network Layer

The **network layer** is responsible for the source-to-destination delivery of a packet, possibly across multiple networks (links). Whereas the data link layer oversees the delivery of the packet between two systems on the same network (links), the network layer ensures that each packet gets from its point of origin to its final destination.

If two systems are connected to the same link, there is usually no need for a network layer. However, if the two systems are attached to different networks (links) with connecting devices between the networks (links), there is often a need for the network layer to accomplish source-to-destination delivery. Figure 2.8 shows the relationship of the network layer to the data link and transport layers.

Figure 2.8 *Network layer*

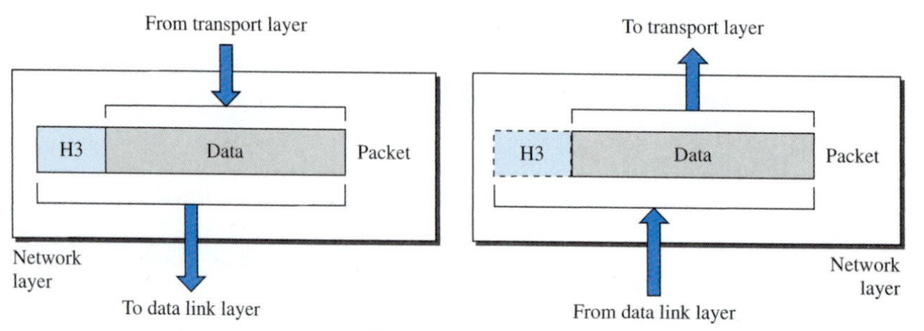

The network layer is responsible for the delivery of individual packets from the source host to the destination host.

Other responsibilities of the network layer include the following:

- **Logical addressing.** The physical addressing implemented by the data link layer handles the addressing problem locally. If a packet passes the network boundary, we need another addressing system to help distinguish the source and destination systems. The network layer adds a header to the packet coming from the upper layer that, among other things, includes the logical addresses of the sender and receiver. We discuss logical addresses later in this chapter.
- **Routing.** When independent networks or links are connected to create *internetworks* (network of networks) or a large network, the connecting devices (called *routers*

or *switches*) route or switch the packets to their final destination. One of the functions of the network layer is to provide this mechanism.

Figure 2.9 illustrates end-to-end delivery by the network layer.

Figure 2.9 *Source-to-destination delivery*

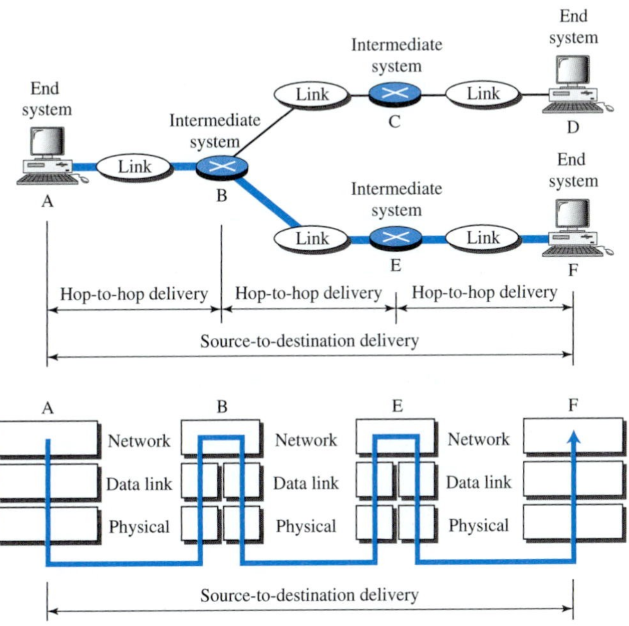

As the figure shows, now we need a source-to-destination delivery. The network layer at A sends the packet to the network layer at B. When the packet arrives at router B, the router makes a decision based on the final destination (F) of the packet. As we will see in later chapters, router B uses its routing table to find that the next hop is router E. The network layer at B, therefore, sends the packet to the network layer at E. The network layer at E, in turn, sends the packet to the network layer at F.

Transport Layer

The **transport layer** is responsible for **process-to-process delivery** of the entire message. A process is an application program running on a host. Whereas the network layer oversees **source-to-destination delivery** of individual packets, it does not recognize any relationship between those packets. It treats each one independently, as though each piece belonged to a separate message, whether or not it does. The transport layer, on the other hand, ensures that the whole message arrives intact and in order, overseeing both error control and flow control at the source-to-destination level. Figure 2.10 shows the relationship of the transport layer to the network and session layers.

Figure 2.10 *Transport layer*

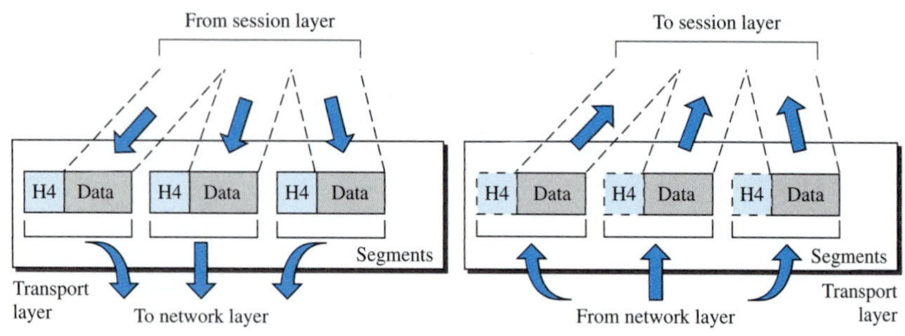

The transport layer is responsible for the delivery of a message from one process to another.

Other responsibilities of the transport layer include the following:

- **Service-point addressing.** Computers often run several programs at the same time. For this reason, source-to-destination delivery means delivery not only from one computer to the next but also from a specific process (running program) on one computer to a specific process (running program) on the other. The transport layer header must therefore include a type of address called a *service-point address* (or port address). The network layer gets each packet to the correct computer; the transport layer gets the entire message to the correct process on that computer.
- **Segmentation and reassembly.** A message is divided into transmittable segments, with each segment containing a sequence number. These numbers enable the transport layer to reassemble the message correctly upon arriving at the destination and to identify and replace packets that were lost in transmission.
- **Connection control.** The transport layer can be either connectionless or connection-oriented. A connectionless transport layer treats each segment as an independent packet and delivers it to the transport layer at the destination machine. A connection-oriented transport layer makes a connection with the transport layer at the destination machine first before delivering the packets. After all the data are transferred, the connection is terminated.
- **Flow control.** Like the data link layer, the transport layer is responsible for **flow control.** However, flow control at this layer is performed end to end rather than across a single link.
- **Error control.** Like the data link layer, the transport layer is responsible for **error control.** However, error control at this layer is performed process-to-process rather than across a single link. The sending transport layer makes sure that the entire message arrives at the receiving transport layer without **error** (damage, loss, or duplication). Error correction is usually achieved through retransmission.

Figure 2.11 illustrates process-to-process delivery by the transport layer.

Figure 2.11 *Reliable process-to-process delivery of a message*

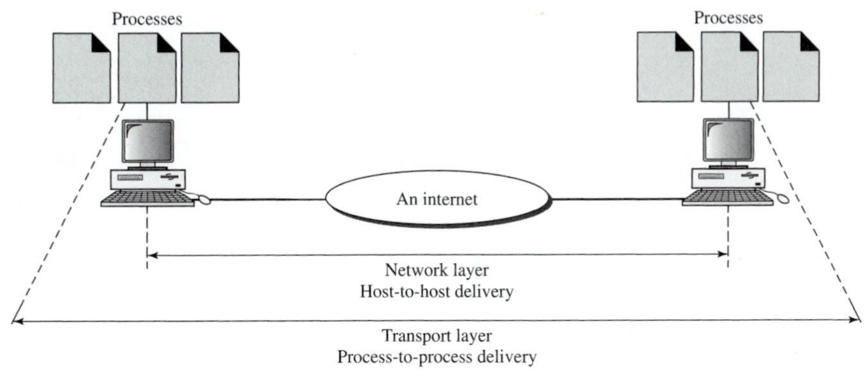

Session Layer

The services provided by the first three layers (physical, data link, and network) are not sufficient for some processes. The **session layer** is the network *dialog controller*. It establishes, maintains, and synchronizes the interaction among communicating systems.

The session layer is responsible for dialog control and synchronization.

Specific responsibilities of the session layer include the following:

- **Dialog control.** The session layer allows two systems to enter into a dialog. It allows the communication between two processes to take place in either half-duplex (one way at a time) or full-duplex (two ways at a time) mode.
- **Synchronization.** The session layer allows a process to add checkpoints, or **synchronization points,** to a stream of data. For example, if a system is sending a file of 2000 pages, it is advisable to insert checkpoints after every 100 pages to ensure that each 100-page unit is received and acknowledged independently. In this case, if a crash happens during the transmission of page 523, the only pages that need to be resent after system recovery are pages 501 to 523. Pages previous to 501 need not be resent. Figure 2.12 illustrates the relationship of the session layer to the transport and presentation layers.

Presentation Layer

The **presentation layer** is concerned with the syntax and semantics of the information exchanged between two systems. Figure 2.13 shows the relationship between the presentation layer and the application and session layers.

Figure 2.12 *Session layer*

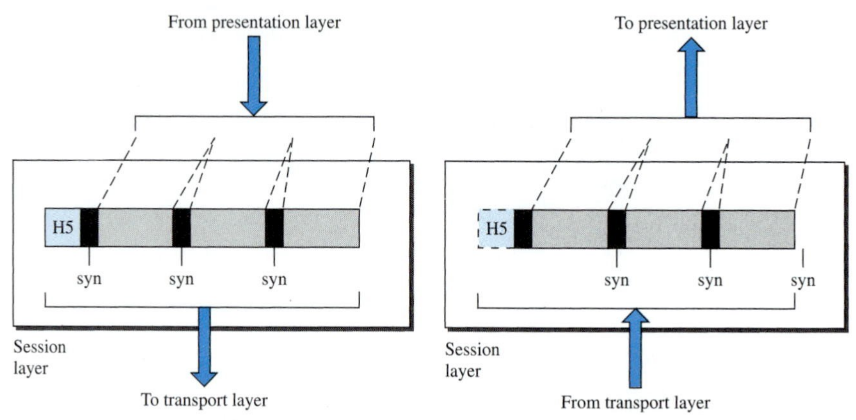

Figure 2.13 *Presentation layer*

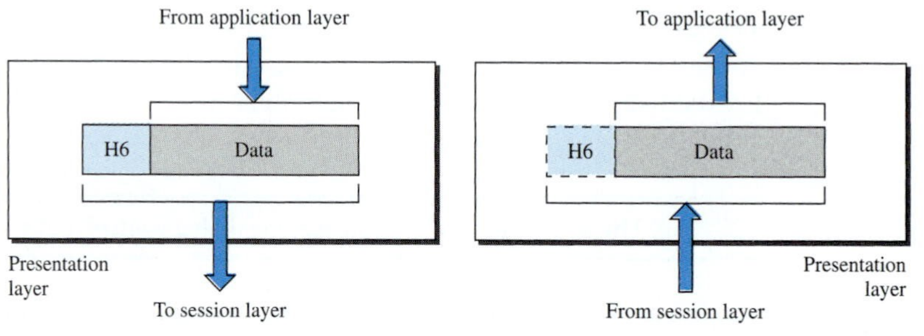

The presentation layer is responsible for translation, compression, and encryption.

Specific responsibilities of the presentation layer include the following:

- **Translation.** The processes (running programs) in two systems are usually exchanging information in the form of character strings, numbers, and so on. The information must be changed to bit streams before being transmitted. Because different computers use different encoding systems, the presentation layer is responsible for interoperability between these different encoding methods. The presentation layer at the sender changes the information from its sender-dependent format into a common format. The presentation layer at the receiving machine changes the common format into its receiver-dependent format.
- **Encryption.** To carry sensitive information, a system must be able to ensure privacy. Encryption means that the sender transforms the original information to

another form and sends the resulting message out over the network. Decryption reverses the original process to transform the message back to its original form.

- **Compression.** Data compression reduces the number of bits contained in the information. Data compression becomes particularly important in the transmission of multimedia such as text, audio, and video.

Application Layer

The **application layer** enables the user, whether human or software, to access the network. It provides user interfaces and support for services such as electronic mail, remote file access and transfer, shared database management, and other types of distributed information services.

Figure 2.14 shows the relationship of the application layer to the user and the presentation layer. Of the many application services available, the figure shows only three: X.400 (message-handling services), X.500 (directory services), and file transfer, access, and management (FTAM). The user in this example employs X.400 to send an e-mail message.

Figure 2.14 *Application layer*

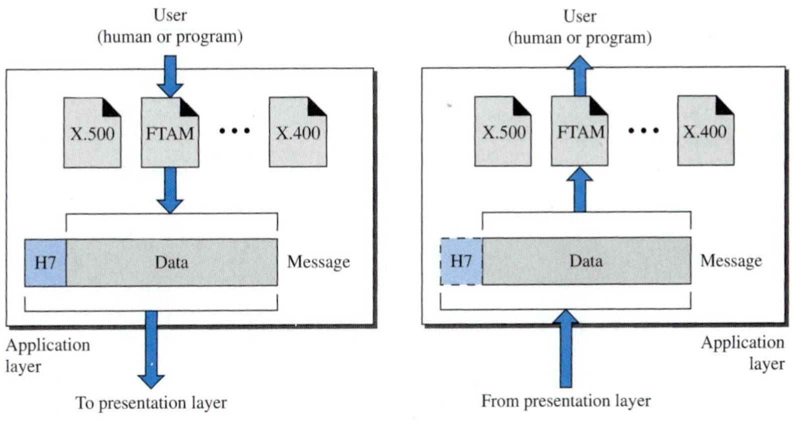

The application layer is responsible for providing services to the user.

Specific services provided by the application layer include the following:

- **Network virtual terminal.** A network virtual terminal is a software version of a physical terminal, and it allows a user to log on to a remote host. To do so, the application creates a software emulation of a terminal at the remote host. The user's computer talks to the software terminal which, in turn, talks to the host, and vice versa. The remote host believes it is communicating with one of its own terminals and allows the user to log on.

- **File transfer, access, and management.** This application allows a user to access files in a remote host (to make changes or read data), to retrieve files from a remote computer for use in the local computer, and to manage or control files in a remote computer locally.
- **Mail services.** This application provides the basis for e-mail forwarding and storage.
- **Directory services.** This application provides distributed database sources and access for global information about various objects and services.

Summary of Layers

Figure 2.15 shows a summary of duties for each layer.

Figure 2.15 *Summary of layers*

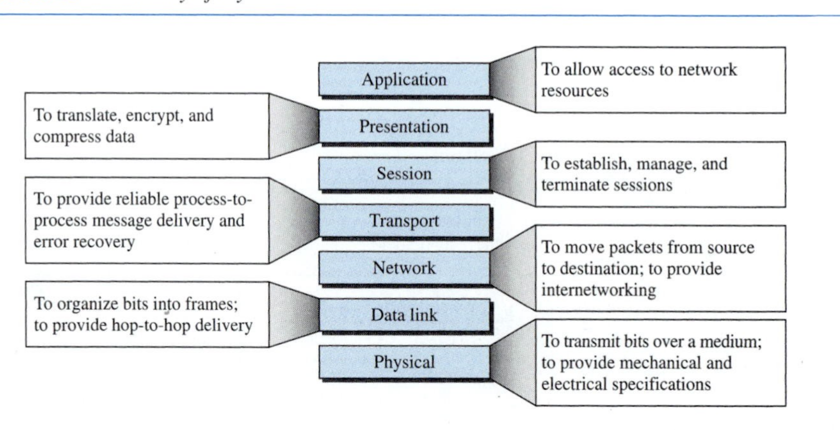

2.4 TCP/IP PROTOCOL SUITE

The **TCP/IP protocol suite** was developed prior to the OSI model. Therefore, the layers in the TCP/IP protocol suite do not exactly match those in the OSI model. The original TCP/IP protocol suite was defined as having four layers: host-to-network, internet, transport, and application. However, when TCP/IP is compared to OSI, we can say that the host-to-network layer is equivalent to the combination of the physical and data link layers. The internet layer is equivalent to the network layer, and the application layer is roughly doing the job of the session, presentation, and application layers with the transport layer in TCP/IP taking care of part of the duties of the session layer. So in this book, we assume that the TCP/IP protocol suite is made of five layers: physical, data link, network, transport, and application. The first four layers provide physical standards, network interfaces, internetworking, and transport functions that correspond to the first four layers of the OSI model. The three topmost layers in the OSI model, however, are represented in TCP/IP by a single layer called the *application layer* (see Figure 2.16).

Figure 2.16 *TCP/IP and OSI model*

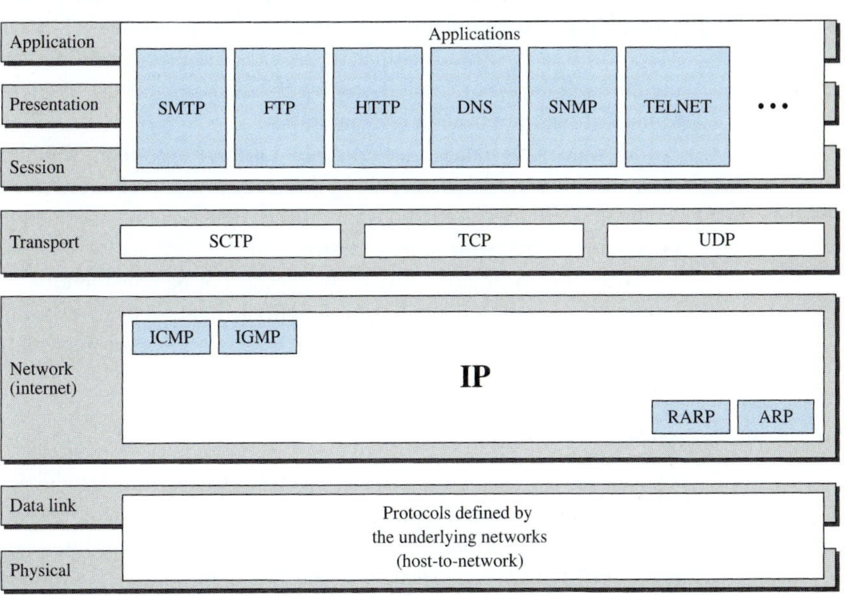

TCP/IP is a hierarchical protocol made up of interactive modules, each of which provides a specific functionality; however, the modules are not necessarily interdependent. Whereas the OSI model specifies which functions belong to each of its layers, the layers of the TCP/IP protocol suite contain relatively independent protocols that can be mixed and matched depending on the needs of the system. The term *hierarchical* means that each upper-level protocol is supported by one or more lower-level protocols.

At the transport layer, TCP/IP defines three protocols: Transmission Control Protocol (TCP), User Datagram Protocol (UDP), and Stream Control Transmission Protocol (SCTP). At the network layer, the main protocol defined by TCP/IP is the Internetworking Protocol (IP); there are also some other protocols that support data movement in this layer.

Physical and Data Link Layers

At the physical and data link layers, TCP/IP does not define any specific protocol. It supports all the standard and proprietary protocols. A network in a TCP/IP internetwork can be a local-area network or a wide-area network.

Network Layer

At the network layer (or, more accurately, the internetwork layer), TCP/IP supports the Internetworking Protocol. IP, in turn, uses four supporting protocols: ARP, RARP, ICMP, and IGMP. Each of these protocols is described in greater detail in later chapters.

Internetworking Protocol (IP)

The Internetworking Protocol (IP) is the transmission mechanism used by the TCP/IP protocols. It is an unreliable and connectionless protocol—a **best-effort delivery** service. The term *best effort* means that IP provides no error checking or tracking. IP assumes the unreliability of the underlying layers and does its best to get a transmission through to its destination, but with no guarantees.

IP transports data in packets called *datagrams,* each of which is transported separately. Datagrams can travel along different routes and can arrive out of sequence or be duplicated. IP does not keep track of the routes and has no facility for reordering datagrams once they arrive at their destination.

The limited functionality of IP should not be considered a weakness, however. IP provides bare-bones transmission functions that free the user to add only those facilities necessary for a given application and thereby allows for maximum efficiency. IP is discussed in Chapter 20.

Address Resolution Protocol

The **Address Resolution Protocol (ARP)** is used to associate a logical address with a physical address. On a typical physical network, such as a LAN, each device on a link is identified by a physical or station address, usually imprinted on the network interface card (NIC). ARP is used to find the physical address of the node when its Internet address is known. ARP is discussed in Chapter 21.

Reverse Address Resolution Protocol

The **Reverse Address Resolution Protocol (RARP)** allows a host to discover its Internet address when it knows only its physical address. It is used when a computer is connected to a network for the first time or when a diskless computer is booted. We discuss RARP in Chapter 21.

Internet Control Message Protocol

The **Internet Control Message Protocol (ICMP)** is a mechanism used by hosts and gateways to send notification of datagram problems back to the sender. ICMP sends query and error reporting messages. We discuss ICMP in Chapter 21.

Internet Group Message Protocol

The **Internet Group Message Protocol (IGMP)** is used to facilitate the simultaneous transmission of a message to a group of recipients. We discuss IGMP in Chapter 22.

Transport Layer

Traditionally the transport layer was represented in TCP/IP by two protocols: TCP and UDP. IP is a **host-to-host protocol,** meaning that it can deliver a packet from one physical device to another. UDP and TCP are **transport level protocols** responsible for delivery of a message from a process (running program) to another process. A new transport layer protocol, SCTP, has been devised to meet the needs of some newer applications.

User Datagram Protocol

The **User Datagram Protocol (UDP)** is the simpler of the two standard TCP/IP transport protocols. It is a process-to-process protocol that adds only port addresses, checksum error control, and length information to the data from the upper layer. UDP is discussed in Chapter 23.

Transmission Control Protocol

The **Transmission Control Protocol (TCP)** provides full transport-layer services to applications. TCP is a reliable stream transport protocol. The term *stream,* in this context, means connection-oriented: A connection must be established between both ends of a transmission before either can transmit data.

At the sending end of each transmission, TCP divides a stream of data into smaller units called *segments.* Each segment includes a sequence number for reordering after receipt, together with an acknowledgment number for the segments received. Segments are carried across the internet inside of IP datagrams. At the receiving end, TCP collects each datagram as it comes in and reorders the transmission based on sequence numbers. TCP is discussed in Chapter 23.

Stream Control Transmission Protocol

The **Stream Control Transmission Protocol (SCTP)** provides support for newer applications such as voice over the Internet. It is a transport layer protocol that combines the best features of UDP and TCP. We discuss SCTP in Chapter 23.

Application Layer

The *application layer* in TCP/IP is equivalent to the combined session, presentation, and application layers in the OSI model. Many protocols are defined at this layer. We cover many of the standard protocols in later chapters.

2.5 ADDRESSING

Four levels of addresses are used in an internet employing the TCP/IP protocols: **physical** (link) **addresses, logical** (IP) **addresses, port addresses,** and specific addresses (see Figure 2.17).

Figure 2.17 *Addresses in TCP/IP*

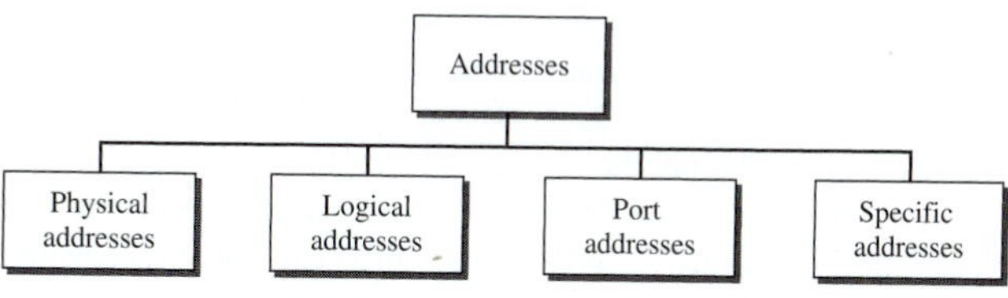

Each address is related to a specific layer in the TCP/IP architecture, as shown in Figure 2.18.

Figure 2.18 *Relationship of layers and addresses in TCP/IP*

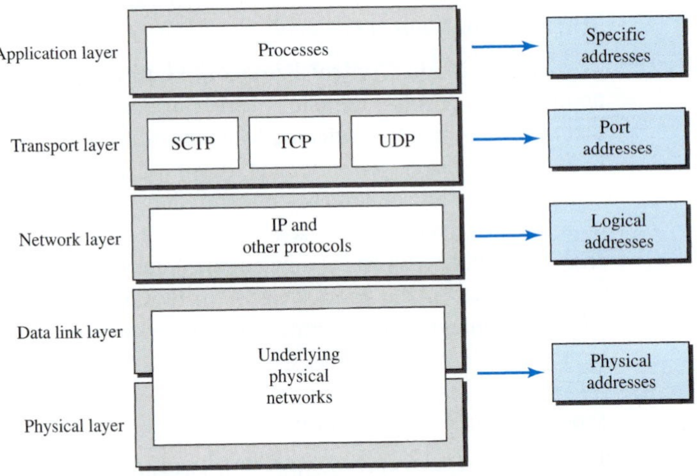

Physical Addresses

The physical address, also known as the link address, is the address of a node as defined by its LAN or WAN. It is included in the frame used by the data link layer. It is the lowest-level address.

The physical addresses have authority over the network (LAN or WAN). The size and format of these addresses vary depending on the network. For example, Ethernet uses a 6-byte (48-bit) physical address that is imprinted on the network interface card (NIC). LocalTalk (Apple), however, has a 1-byte dynamic address that changes each time the station comes up.

Example 2.1

In Figure 2.19 a node with physical address 10 sends a frame to a node with physical address 87. The two nodes are connected by a link (bus topology LAN). At the data link layer, this frame contains physical (link) addresses in the header. These are the only addresses needed. The rest of the header contains other information needed at this level. The trailer usually contains extra bits needed for error detection. As the figure shows, the computer with physical address 10 is the sender, and the computer with physical address 87 is the receiver. The data link layer at the sender receives data from an upper layer. It encapsulates the data in a frame, adding a header and a trailer. The header, among other pieces of information, carries the receiver and the sender physical (link) addresses. Note that in most data link protocols, the destination address, 87 in this case, comes before the source address (10 in this case).

We have shown a bus topology for an isolated LAN. In a bus topology, the frame is propagated in both directions (left and right). The frame propagated to the left dies when it reaches the end of the cable if the cable end is terminated appropriately. The frame propagated to the right is

Figure 2.19 *Physical addresses*

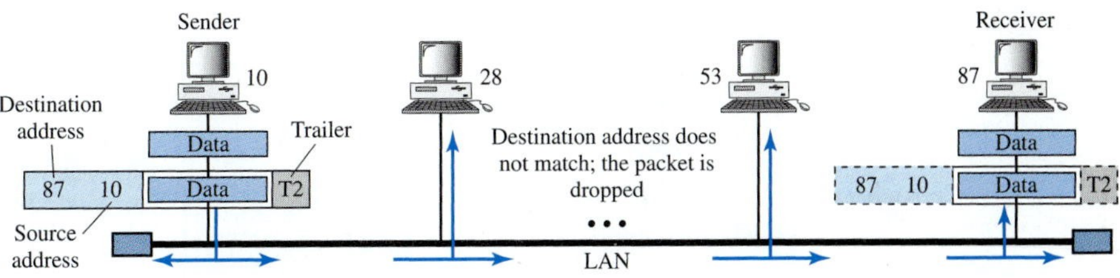

sent to every station on the network. Each station with a physical addresses other than 87 drops the frame because the destination address in the frame does not match its own physical address. The intended destination computer, however, finds a match between the destination address in the frame and its own physical address. The frame is checked, the header and trailer are dropped, and the data part is decapsulated and delivered to the upper layer.

Example 2.2

As we will see in Chapter 13, most local-area networks use a 48-bit (6-byte) physical address written as 12 hexadecimal digits; every byte (2 hexadecimal digits) is separated by a colon, as shown below:

07:01:02:01:2C:4B
A 6-byte (12 hexadecimal digits) physical address

Logical Addresses

Logical addresses are necessary for universal communications that are independent of underlying physical networks. Physical addresses are not adequate in an internetwork environment where different networks can have different address formats. A universal addressing system is needed in which each host can be identified uniquely, regardless of the underlying physical network.

The logical addresses are designed for this purpose. A logical address in the Internet is currently a 32-bit address that can uniquely define a host connected to the Internet. No two publicly addressed and visible hosts on the Internet can have the same IP address.

Example 2.3

Figure 2.20 shows a part of an internet with two routers connecting three LANs. Each device (computer or router) has a pair of addresses (logical and physical) for each connection. In this case, each computer is connected to only one link and therefore has only one pair of addresses. Each router, however, is connected to three networks (only two are shown in the figure). So each router has three pairs of addresses, one for each connection. Although it may obvious that each router must have a separate physical address for each connection, it may not be obvious why it needs a logical address for each connection. We discuss these issues in Chapter 22 when we discuss routing.

Figure 2.20 *IP addresses*

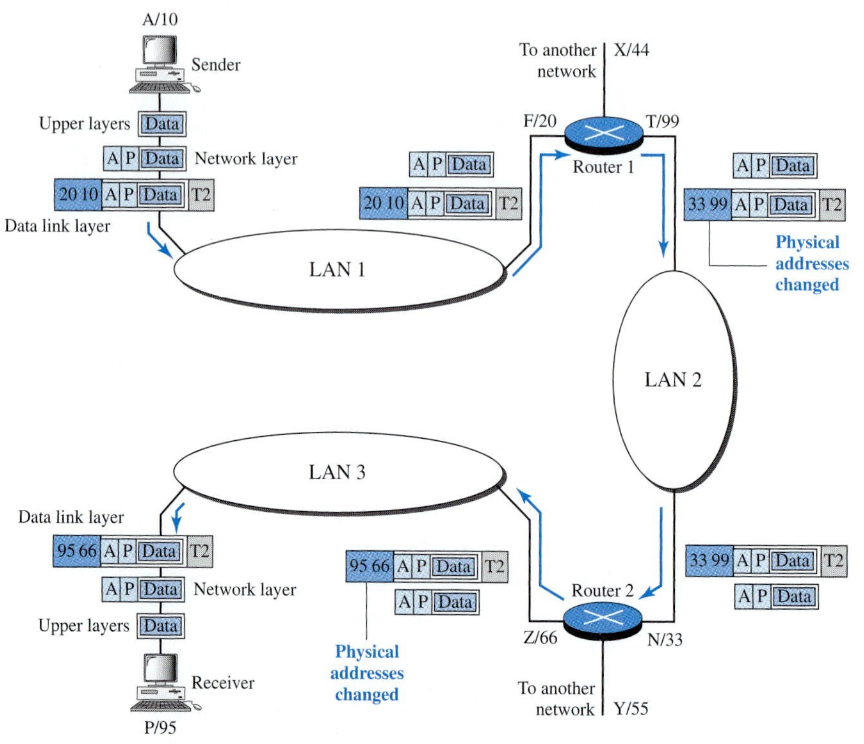

The computer with logical address A and physical address 10 needs to send a packet to the computer with logical address P and physical address 95. We use letters to show the logical addresses and numbers for physical addresses, but note that both are actually numbers, as we will see later in the chapter.

The sender encapsulates its data in a packet at the network layer and adds two logical addresses (A and P). Note that in most protocols, the logical source address comes before the logical destination address (contrary to the order of physical addresses). The network layer, however, needs to find the physical address of the next hop before the packet can be delivered. The network layer consults its routing table (see Chapter 22) and finds the logical address of the next hop (router 1) to be F. The ARP discussed previously finds the physical address of router 1 that corresponds to the logical address of 20. Now the network layer passes this address to the data link layer, which in turn, encapsulates the packet with physical destination address 20 and physical source address 10.

The frame is received by every device on LAN 1, but is discarded by all except router 1, which finds that the destination physical address in the frame matches with its own physical address. The router decapsulates the packet from the frame to read the logical destination address P. Since the logical destination address does not match the router's logical address, the router knows that the packet needs to be forwarded. The router consults its routing table and ARP to find the physical destination address of the next hop (router 2), creates a new frame, encapsulates the packet, and sends it to router 2.

Note the physical addresses in the frame. The source physical address changes from 10 to 99. The destination physical address changes from 20 (router 1 physical address) to 33 (router 2 physical address). The logical source and destination addresses must remain the same; otherwise the packet will be lost.

At router 2 we have a similar scenario. The physical addresses are changed, and a new frame is sent to the destination computer. When the frame reaches the destination, the packet is decapsulated. The destination logical address P matches the logical address of the computer. The data are decapsulated from the packet and delivered to the upper layer. Note that although physical addresses will change from hop to hop, logical addresses remain the same from the source to destination. There are some exceptions to this rule that we discover later in the book.

The physical addresses will change from hop to hop, but the logical addresses usually remain the same.

Port Addresses

The IP address and the physical address are necessary for a quantity of data to travel from a source to the destination host. However, arrival at the destination host is not the final objective of data communications on the Internet. A system that sends nothing but data from one computer to another is not complete. Today, computers are devices that can run multiple processes at the same time. The end objective of Internet communication is a process communicating with another process. For example, computer A can communicate with computer C by using TELNET. At the same time, computer A communicates with computer B by using the File Transfer Protocol (FTP). For these processes to receive data simultaneously, we need a method to label the different processes. In other words, they need addresses. In the TCP/IP architecture, the label assigned to a process is called a port address. A port address in TCP/IP is 16 bits in length.

Example 2.4

Figure 2.21 shows two computers communicating via the Internet. The sending computer is running three processes at this time with port addresses a, b, and c. The receiving computer is running two processes at this time with port addresses j and k. Process a in the sending computer needs to communicate with process j in the receiving computer. Note that although both computers are using the same application, FTP, for example, the port addresses are different because one is a client program and the other is a server program, as we will see in Chapter 23. To show that data from process a need to be delivered to process j, and not k, the transport layer encapsulates data from the application layer in a packet and adds two port addresses (a and j), source and destination. The packet from the transport layer is then encapsulated in another packet at the network layer with logical source and destination addresses (A and P). Finally, this packet is encapsulated in a frame with the physical source and destination addresses of the next hop. We have not shown the physical addresses because they change from hop to hop inside the cloud designated as the Internet. Note that although physical addresses change from hop to hop, logical and port addresses remain the same from the source to destination. There are some exceptions to this rule that we discuss later in the book.

Figure 2.21 *Port addresses*

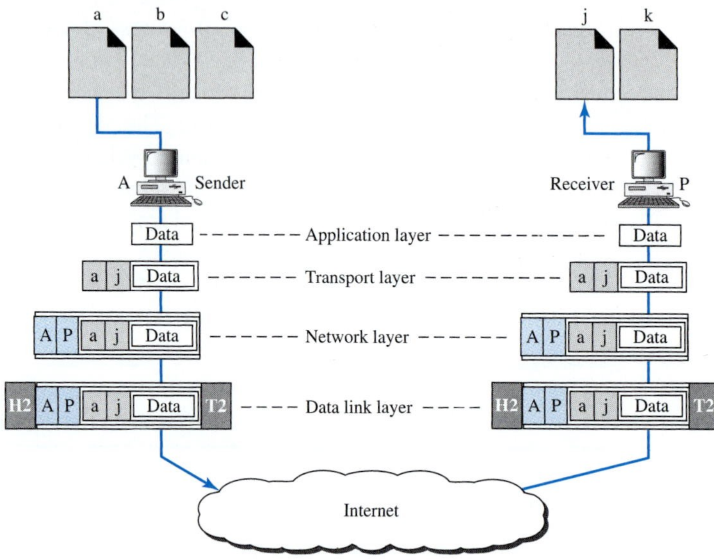

The physical addresses change from hop to hop, but the logical and port addresses usually remain the same.

Example 2.5

As we will see in Chapter 23, a port address is a 16-bit address represented by one decimal number as shown.

753
A 16-bit port address represented as one single number

Specific Addresses

Some applications have user-friendly addresses that are designed for that specific address. Examples include the e-mail address (for example, forouzan@fhda.edu) and the Universal Resource Locator (URL) (for example, www.mhhe.com). The first defines the recipient of an e-mail (see Chapter 26); the second is used to find a document on the World Wide Web (see Chapter 27). These addresses, however, get changed to the corresponding port and logical addresses by the sending computer, as we will see in Chapter 25.

2.6 RECOMMENDED READING

For more details about subjects discussed in this chapter, we recommend the following books and sites. The items enclosed in brackets, [. . .] refer to the reference list at the end of the text.

Books

Network models are discussed in Section 1.3 of [Tan03], Chapter 2 of [For06], Chapter 2 of [Sta04], Sections 2.2 and 2.3 of [GW04], Section 1.3 of [PD03], and Section 1.7 of [KR05]. A good discussion about addresses can be found in Section 1.7 of [Ste94].

Sites

The following site is related to topics discussed in this chapter.

- www.osi.org/ Information about OSI.

RFCs

The following site lists all RFCs, including those related to IP and port addresses.

- www.ietf.org/rfc.html

2.7 KEY TERMS

access control
Address Resolution Protocol (ARP)
application layer
best-effort delivery
bits
connection control
data link layer
encoding
error
error control
flow control
frame
header
hop-to-hop delivery
host-to-host protocol
interface
Internet Control Message Protocol (ICMP)
Internet Group Message Protocol (IGMP)
logical addressing
mail service
network layer
node-to-node delivery
open system
Open Systems Interconnection (OSI) model
peer-to-peer process
physical addressing
physical layer
port address
presentation layer
process-to-process delivery
Reverse Address Resolution Protocol (RARP)
routing
segmentation
session layer
source-to-destination delivery
Stream Control Transmission Protocol (SCTP)
synchronization point
TCP/IP protocol suite
trailer
Transmission Control Protocol (TCP)
transmission rate
transport layer
transport level protocols
User Datagram Protocol (UDP)

2.8 SUMMARY

- The International Standards Organization created a model called the Open Systems Interconnection, which allows diverse systems to communicate.
- The seven-layer OSI model provides guidelines for the development of universally compatible networking protocols.
- The physical, data link, and network layers are the network support layers.
- The session, presentation, and application layers are the user support layers.
- The transport layer links the network support layers and the user support layers.
- The physical layer coordinates the functions required to transmit a bit stream over a physical medium.
- The data link layer is responsible for delivering data units from one station to the next without errors.
- The network layer is responsible for the source-to-destination delivery of a packet across multiple network links.
- The transport layer is responsible for the process-to-process delivery of the entire message.
- The session layer establishes, maintains, and synchronizes the interactions between communicating devices.
- The presentation layer ensures interoperability between communicating devices through transformation of data into a mutually agreed upon format.
- The application layer enables the users to access the network.
- TCP/IP is a five-layer hierarchical protocol suite developed before the OSI model.
- The TCP/IP application layer is equivalent to the combined session, presentation, and application layers of the OSI model.
- Four levels of addresses are used in an internet following the TCP/IP protocols: physical (link) addresses, logical (IP) addresses, port addresses, and specific addresses.
- The physical address, also known as the link address, is the address of a node as defined by its LAN or WAN.
- The IP address uniquely defines a host on the Internet.
- The port address identifies a process on a host.
- A specific address is a user-friendly address.

2.9 PRACTICE SET

Review Questions

1. List the layers of the Internet model.
2. Which layers in the Internet model are the network support layers?
3. Which layer in the Internet model is the user support layer?
4. What is the difference between network layer delivery and transport layer delivery?

5. What is a peer-to-peer process?
6. How does information get passed from one layer to the next in the Internet model?
7. What are headers and trailers, and how do they get added and removed?
8. What are the concerns of the physical layer in the Internet model?
9. What are the responsibilities of the data link layer in the Internet model?
10. What are the responsibilities of the network layer in the Internet model?
11. What are the responsibilities of the transport layer in the Internet model?
12. What is the difference between a port address, a logical address, and a physical address?
13. Name some services provided by the application layer in the Internet model.
14. How do the layers of the Internet model correlate to the layers of the OSI model?

Exercises

15. How are OSI and ISO related to each other?
16. Match the following to one or more layers of the OSI model:
    a. Route determination
    b. Flow control
    c. Interface to transmission media
    d. Provides access for the end user
17. Match the following to one or more layers of the OSI model:
    a. Reliable process-to-process message delivery
    b. Route selection
    c. Defines frames
    d. Provides user services such as e-mail and file transfer
    e. Transmission of bit stream across physical medium
18. Match the following to one or more layers of the OSI model:
    a. Communicates directly with user's application program
    b. Error correction and retransmission
    c. Mechanical, electrical, and functional interface
    d. Responsibility for carrying frames between adjacent nodes
19. Match the following to one or more layers of the OSI model:
    a. Format and code conversion services
    b. Establishes, manages, and terminates sessions
    c. Ensures reliable transmission of data
    d. Log-in and log-out procedures
    e. Provides independence from differences in data representation
20. In Figure 2.22, computer A sends a message to computer D via LAN1, router R1, and LAN2. Show the contents of the packets and frames at the network and data link layer for each hop interface.

Figure 2.22 *Exercise 20*

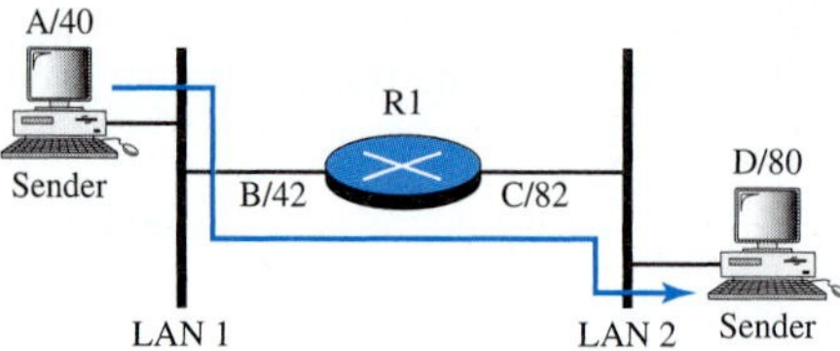

21. In Figure 2.22, assume that the communication is between a process running at computer A with port address *i* and a process running at computer D with port address *j*. Show the contents of packets and frames at the network, data link, and transport layer for each hop.
22. Suppose a computer sends a frame to another computer on a bus topology LAN. The physical destination address of the frame is corrupted during the transmission. What happens to the frame? How can the sender be informed about the situation?
23. Suppose a computer sends a packet at the network layer to another computer somewhere in the Internet. The logical destination address of the packet is corrupted. What happens to the packet? How can the source computer be informed of the situation?
24. Suppose a computer sends a packet at the transport layer to another computer somewhere in the Internet. There is no process with the destination port address running at the destination computer. What will happen?
25. If the data link layer can detect errors between hops, why do you think we need another checking mechanism at the transport layer?

Research Activities

26. Give some advantages and disadvantages of combining the session, presentation, and application layer in the OSI model into one single application layer in the Internet model.
27. Dialog control and synchronization are two responsibilities of the session layer in the OSI model. Which layer do you think is responsible for these duties in the Internet model? Explain your answer.
28. Translation, encryption, and compression are some of the duties of the presentation layer in the OSI model. Which layer do you think is responsible for these duties in the Internet model? Explain your answer.
29. There are several transport layer models proposed in the OSI model. Find all of them. Explain the differences between them.
30. There are several network layer models proposed in the OSI model. Find all of them. Explain the differences between them.

PART

2

Physical Layer and Media

Objectives

We start the discussion of the Internet model with the bottom-most layer, the physical layer. It is the layer that actually interacts with the transmission media, the physical part of the network that connects network components together. This layer is involved in physically carrying information from one node in the network to the next.

The physical layer has complex tasks to perform. One major task is to provide services for the data link layer. The data in the data link layer consists of 0s and 1s organized into frames that are ready to be sent across the transmission medium. This stream of 0s and 1s must first be converted into another entity: signals. One of the services provided by the physical layer is to create a signal that represents this stream of bits.

The physical layer must also take care of the physical network, the transmission medium. The transmission medium is a passive entity; it has no internal program or logic for control like other layers. The transmission medium must be controlled by the physical layer. The physical layer decides on the directions of data flow. The physical layer decides on the number of logical channels for transporting data coming from different sources.

In Part 2 of the book, we discuss issues related to the physical layer and the transmission medium that is controlled by the physical layer. In the last chapter of Part 2, we discuss the structure and the physical layers of the telephone network and the cable network.

Part 2 of the book is devoted to the physical layer and the transmission media.

Chapters

This part consists of seven chapters: Chapters 3 to 9.

Chapter 3

Chapter 3 discusses the relationship between data, which are created by a device, and electromagnetic signals, which are transmitted over a medium.

Chapter 4

Chapter 4 deals with digital transmission. We discuss how we can covert digital or analog data to digital signals.

Chapter 5

Chapter 5 deals with analog transmission. We discuss how we can covert digital or analog data to analog signals.

Chapter 6

Chapter 6 shows how we can use the available bandwidth efficiently. We discuss two separate, but related topics, multiplexing and spreading.

Chapter 7

After explaining some ideas about data and signals and how we can use them efficiently, we discuss the characteristics of transmission media, both guided and unguided, in this chapter. Although transmission media operates under the physical layer, they are controlled by the physical layer.

Chapter 8

Although the previous chapters in this part are issues related to the physical layer or transmission media, Chapter 8 discusses switching, a topic that can be related to several layers. We have included this topic in this part of the book to avoid repeating the discussion for each layer.

Chapter 9

Chapter 9 shows how the issues discussed in the previous chapters can be used in actual networks. In this chapter, we first discuss the telephone network as designed to carry voice. We then show how it can be used to carry data. Second, we discuss the cable network as a television network. We then show how it can also be used to carry data.

CHAPTER 3

Data and Signals

One of the major functions of the physical layer is to move data in the form of electromagnetic signals across a transmission medium. Whether you are collecting numerical statistics from another computer, sending animated pictures from a design workstation, or causing a bell to ring at a distant control center, you are working with the transmission of **data** across network connections.

Generally, the data usable to a person or application are not in a form that can be transmitted over a network. For example, a photograph must first be changed to a form that transmission media can accept. Transmission media work by conducting energy along a physical path.

To be transmitted, data must be transformed to electromagnetic signals.

3.1 ANALOG AND DIGITAL

Both data and the signals that represent them can be either **analog** or **digital** in form.

Analog and Digital Data

Data can be analog or digital. The term **analog data** refers to information that is continuous; **digital data** refers to information that has discrete states. For example, an analog clock that has hour, minute, and second hands gives information in a continuous form; the movements of the hands are continuous. On the other hand, a digital clock that reports the hours and the minutes will change suddenly from 8:05 to 8:06.

Analog data, such as the sounds made by a human voice, take on continuous values. When someone speaks, an analog wave is created in the air. This can be captured by a microphone and converted to an analog signal or sampled and converted to a digital signal.

Digital data take on discrete values. For example, data are stored in computer memory in the form of 0s and 1s. They can be converted to a digital signal or modulated into an analog signal for transmission across a medium.

Data can be analog or digital. Analog data are continuous and take continuous values. Digital data have discrete states and take discrete values.

Analog and Digital Signals

Like the data they represent, **signals** can be either analog or digital. An **analog signal** has infinitely many levels of intensity over a period of time. As the wave moves from value *A* to value *B*, it passes through and includes an infinite number of values along its path. A **digital signal,** on the other hand, can have only a limited number of defined values. Although each value can be any number, it is often as simple as 1 and 0.

The simplest way to show signals is by plotting them on a pair of perpendicular axes. The vertical axis represents the value or strength of a signal. The horizontal axis represents time. Figure 3.1 illustrates an analog signal and a digital signal. The curve representing the analog signal passes through an infinite number of points. The vertical lines of the digital signal, however, demonstrate the sudden jump that the signal makes from value to value.

Signals can be analog or digital. Analog signals can have an infinite number of values in a range; digital signals can have only a limited number of values.

Figure 3.1 *Comparison of analog and digital signals*

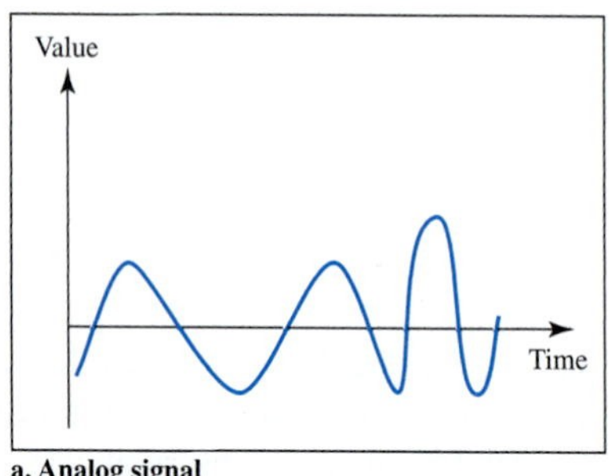

a. Analog signal

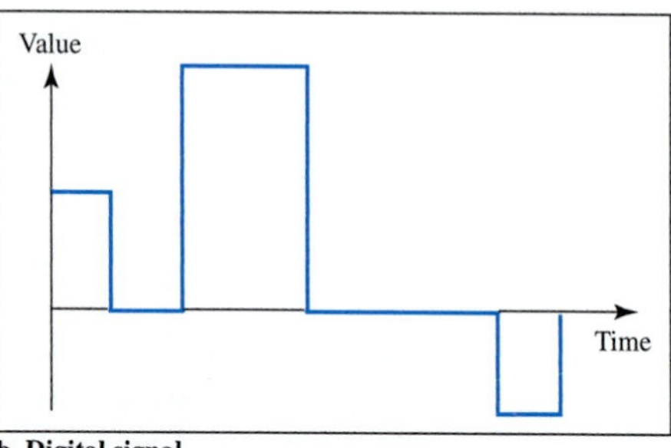

b. Digital signal

Periodic and Nonperiodic Signals

Both analog and digital signals can take one of two forms: *periodic* or *nonperiodic* (sometimes refer to as *aperiodic,* because the prefix *a* in Greek means "non").

A **periodic signal** completes a pattern within a measurable time frame, called a **period,** and repeats that pattern over subsequent identical periods. The completion of one full pattern is called a **cycle.** A **nonperiodic signal** changes without exhibiting a pattern or cycle that repeats over time.

Both analog and digital signals can be periodic or nonperiodic. In data communications, we commonly use periodic analog signals (because they need less bandwidth, as we will see in Chapter 5) and nonperiodic digital signals (because they can represent variation in data, as we will see in Chapter 6).

In data communications, we commonly use periodic analog signals and nonperiodic digital signals.

3.2 PERIODIC ANALOG SIGNALS

Periodic analog signals can be classified as simple or composite. A simple periodic analog signal, a **sine wave,** cannot be decomposed into simpler signals. A composite periodic analog signal is composed of multiple sine waves.

Sine Wave

The sine wave is the most fundamental form of a periodic analog signal. When we visualize it as a simple oscillating curve, its change over the course of a cycle is smooth and consistent, a continuous, rolling flow. Figure 3.2 shows a sine wave. Each cycle consists of a single arc above the time axis followed by a single arc below it.

Figure 3.2 *A sine wave*

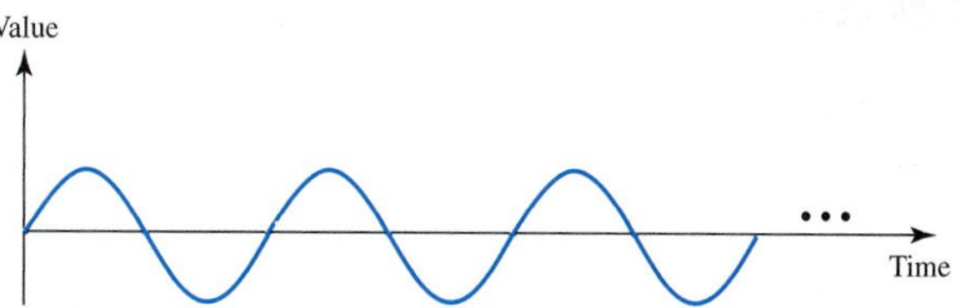

We discuss a mathematical approach to sine waves in Appendix C.

A sine wave can be represented by three parameters: the *peak amplitude,* the *frequency,* and the *phase*. These three parameters fully describe a sine wave.

Peak Amplitude

The **peak amplitude** of a signal is the absolute value of its highest intensity, proportional to the energy it carries. For electric signals, peak amplitude is normally measured in *volts*. Figure 3.3 shows two signals and their peak amplitudes.

Example 3.1

The power in your house can be represented by a sine wave with a peak amplitude of 155 to 170 V. However, it is common knowledge that the voltage of the power in U.S. homes is 110 to 120 V.

Figure 3.3 *Two signals with the same phase and frequency, but different amplitudes*

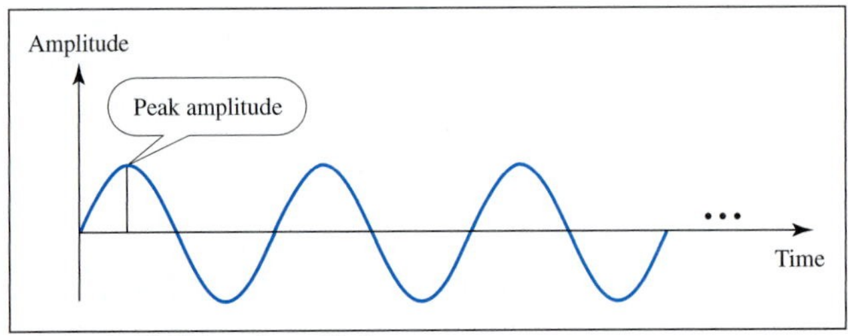

a. A signal with high peak amplitude

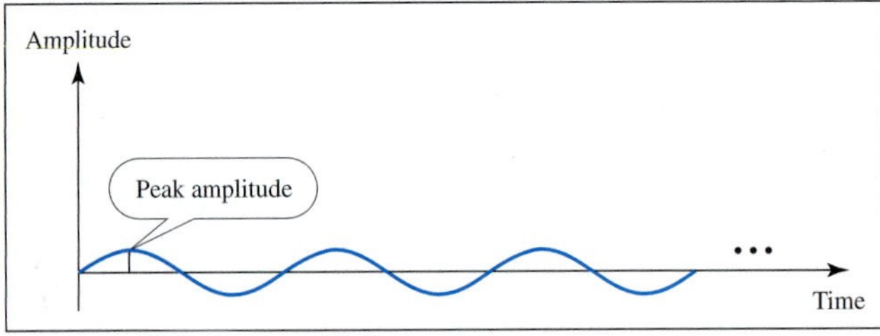

b. A signal with low peak amplitude

This discrepancy is due to the fact that these are root mean square (rms) values. The signal is squared and then the average amplitude is calculated. The peak value is equal to $2^{1/2} \times$ rms value.

Example 3.2

The voltage of battery is a constant; this constant value can be considered a sine wave, as we will see later. For example, the peak value of an AA battery is normally 1.5 V.

Period and Frequency

Period refers to the amount of time, in seconds, a signal needs to complete 1 cycle. **Frequency** refers to the number of periods in 1 s. Note that period and frequency are just one characteristic defined in two ways. Period is the inverse of frequency, and frequency is the inverse of period, as the following formulas show.

$$f = \frac{1}{T} \quad \text{and} \quad T = \frac{1}{f}$$

Frequency and period are the inverse of each other.

Figure 3.4 shows two signals and their frequencies.

Figure 3.4 *Two signals with the same amplitude and phase, but different frequencies*

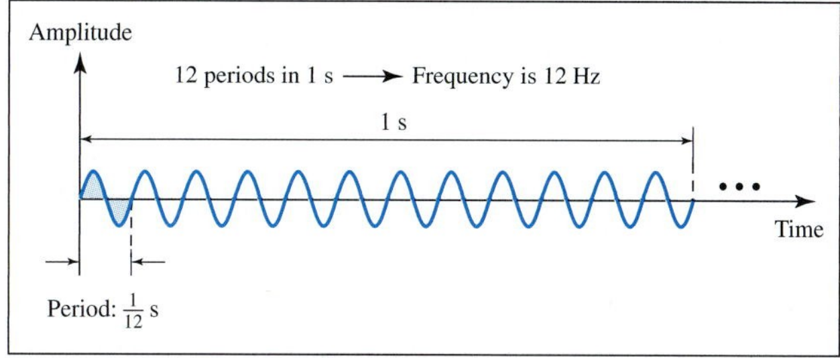

a. A signal with a frequency of 12 Hz

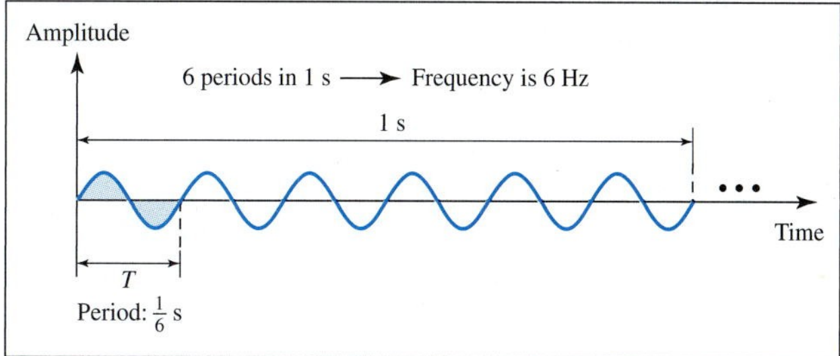

b. A signal with a frequency of 6 Hz

Period is formally expressed in seconds. Frequency is formally expressed in **Hertz (Hz),** which is cycle per second. Units of period and frequency are shown in Table 3.1.

Table 3.1 *Units of period and frequency*

Unit	*Equivalent*	*Unit*	*Equivalent*
Seconds (s)	1 s	Hertz (Hz)	1 Hz
Milliseconds (ms)	10^{-3} s	Kilohertz (kHz)	10^{3} Hz
Microseconds (μs)	10^{-6} s	Megahertz (MHz)	10^{6} Hz
Nanoseconds (ns)	10^{-9} s	Gigahertz (GHz)	10^{9} Hz
Picoseconds (ps)	10^{-12} s	Terahertz (THz)	10^{12} Hz

Example 3.3

The power we use at home has a frequency of 60 Hz (50 Hz in Europe). The period of this sine wave can be determined as follows:

$$T = \frac{1}{f} = \frac{1}{60} = 0.0166 \text{ s} = 0.0166 \times 10^3 \text{ ms} = 16.6 \text{ ms}$$

This means that the period of the power for our lights at home is 0.0116 s, or 16.6 ms. Our eyes are not sensitive enough to distinguish these rapid changes in amplitude.

Example 3.4

Express a period of 100 ms in microseconds.

Solution

From Table 3.1 we find the equivalents of 1 ms (1 ms is 10^{-3} s) and 1 s (1 s is 10^6 μs). We make the following substitutions:

$$100 \text{ ms} = 100 \times 10^{-3} \text{ s} = 100 \times 10^{-3} \times 10^{6} \text{ μs} = 10^{2} \times 10^{-3} \times 10^{6} \text{ μs} = 10^{5} \text{ μs}$$

Example 3.5

The period of a signal is 100 ms. What is its frequency in kilohertz?

Solution

First we change 100 ms to seconds, and then we calculate the frequency from the period (1 Hz = 10^{-3} kHz).

$$100 \text{ ms} = 100 \times 10^{-3} \text{ s} = 10^{-1} \text{ s}$$

$$f = \frac{1}{T} = \frac{1}{10^{-1}} \text{ Hz} = 10 \text{ Hz} = 10 \times 10^{-3} \text{ kHz} = 10^{-2} \text{ kHz}$$

More About Frequency

We already know that frequency is the relationship of a signal to time and that the frequency of a wave is the number of cycles it completes in 1 s. But another way to look at frequency is as a measurement of the rate of change. Electromagnetic signals are oscillating waveforms; that is, they fluctuate continuously and predictably above and below a mean energy level. A 40-Hz signal has one-half the frequency of an 80-Hz signal; it completes 1 cycle in twice the time of the 80-Hz signal, so each cycle also takes twice as long to change from its lowest to its highest voltage levels. Frequency, therefore, though described in cycles per second (hertz), is a general measurement of the rate of change of a signal with respect to time.

Frequency is the rate of change with respect to time. Change in a short span of time means high frequency. Change over a long span of time means low frequency.

If the value of a signal changes over a very short span of time, its frequency is high. If it changes over a long span of time, its frequency is low.

Two Extremes

What if a signal does not change at all? What if it maintains a constant voltage level for the entire time it is active? In such a case, its frequency is zero. Conceptually, this idea is a simple one. If a signal does not change at all, it never completes a cycle, so its frequency is 0 Hz.

But what if a signal changes instantaneously? What if it jumps from one level to another in no time? Then its frequency is infinite. In other words, when a signal changes instantaneously, its period is zero; since frequency is the inverse of period, in this case, the frequency is 1/0, or infinite (unbounded).

If a signal does not change at all, its frequency is zero.
If a signal changes instantaneously, its frequency is infinite.

Phase

The term **phase** describes the position of the waveform relative to time 0. If we think of the wave as something that can be shifted backward or forward along the time axis, phase describes the amount of that shift. It indicates the status of the first cycle.

Phase describes the position of the waveform relative to time 0.

Phase is measured in degrees or radians [360° is 2π rad; 1° is 2π/360 rad, and 1 rad is 360/(2π)]. A phase shift of 360° corresponds to a shift of a complete period; a phase shift of 180° corresponds to a shift of one-half of a period; and a phase shift of 90° corresponds to a shift of one-quarter of a period (see Figure 3.5).

Figure 3.5 *Three sine waves with the same amplitude and frequency, but different phases*

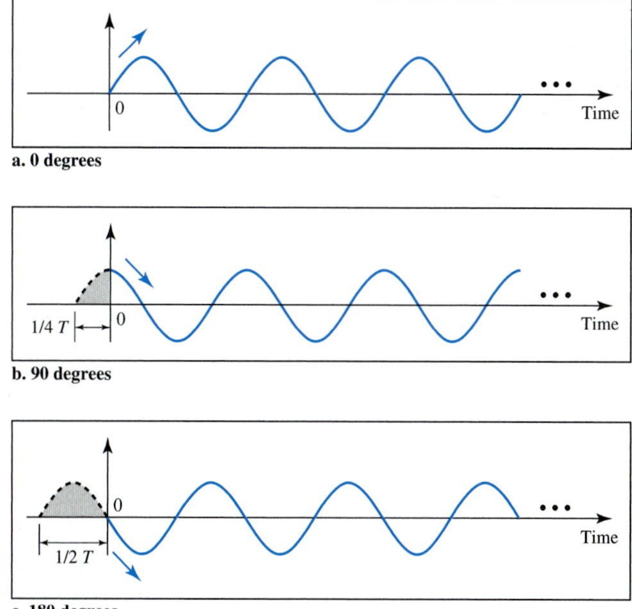

Looking at Figure 3.5, we can say that

1. A sine wave with a phase of 0° starts at time 0 with a zero amplitude. The amplitude is increasing.
2. A sine wave with a phase of 90° starts at time 0 with a peak amplitude. The amplitude is decreasing.

3. A sine wave with a phase of 180° starts at time 0 with a zero amplitude. The amplitude is decreasing.

Another way to look at the phase is in terms of shift or offset. We can say that

1. A sine wave with a phase of 0° is not shifted.
2. A sine wave with a phase of 90° is shifted to the left by $\frac{1}{4}$ cycle. However, note that the signal does not really exist before time 0.
3. A sine wave with a phase of 180° is shifted to the left by $\frac{1}{2}$ cycle. However, note that the signal does not really exist before time 0.

Example 3.6

A sine wave is offset $\frac{1}{6}$ cycle with respect to time 0. What is its phase in degrees and radians?

Solution

We know that 1 complete cycle is 360°. Therefore, $\frac{1}{6}$ cycle is

$$\frac{1}{6} \times 360 = 60° = 60 \times \frac{2\pi}{360} \text{ rad} = \frac{\pi}{3} \text{ rad} = 1.046 \text{ rad}$$

Wavelength

Wavelength is another characteristic of a signal traveling through a transmission medium. Wavelength binds the period or the frequency of a simple sine wave to the **propagation speed** of the medium (see Figure 3.6).

Figure 3.6 *Wavelength and period*

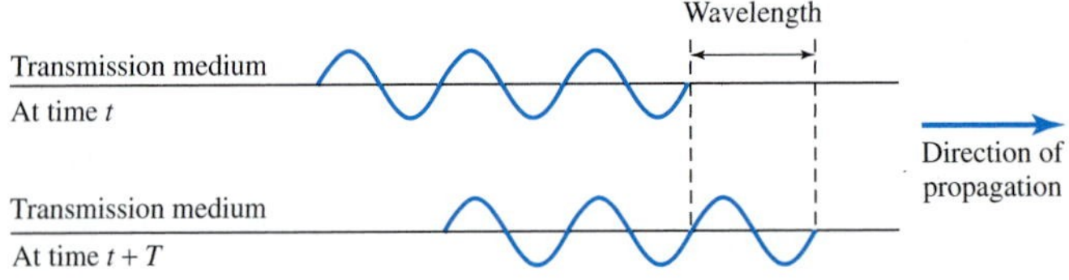

While the frequency of a signal is independent of the medium, the wavelength depends on both the frequency and the medium. Wavelength is a property of any type of signal. In data communications, we often use wavelength to describe the transmission of light in an optical fiber. The wavelength is the distance a simple signal can travel in one period.

Wavelength can be calculated if one is given the propagation speed (the speed of light) and the period of the signal. However, since period and frequency are related to each other, if we represent wavelength by λ, propagation speed by c (speed of light), and frequency by f, we get

$$\textbf{Wavelength} = \textbf{propagation speed} \times \textbf{period} = \frac{\textbf{propagation speed}}{\textbf{frequency}}$$

$$\lambda = \frac{c}{f}$$

The propagation speed of electromagnetic signals depends on the medium and on the frequency of the signal. For example, in a vacuum, light is propagated with a speed of 3×10^8 m/s. That speed is lower in air and even lower in cable.

The wavelength is normally measured in micrometers (microns) instead of meters. For example, the wavelength of red light (frequency $= 4 \times 10^{14}$) in air is

$$\lambda = \frac{c}{f} = \frac{3 \times 10^8}{4 \times 10^{14}} = 0.75 \times 10^{-6}\ \text{m} = 0.75\ \mu\text{m}$$

In a coaxial or fiber-optic cable, however, the wavelength is shorter (0.5 μm) because the propagation speed in the cable is decreased.

Time and Frequency Domains

A sine wave is comprehensively defined by its amplitude, frequency, and phase. We have been showing a sine wave by using what is called a **time-domain** plot. The time-domain plot shows changes in signal amplitude with respect to time (it is an amplitude-versus-time plot). Phase is not explicitly shown on a time-domain plot.

To show the relationship between amplitude and frequency, we can use what is called a **frequency-domain** plot. A frequency-domain plot is concerned with only the peak value and the frequency. Changes of amplitude during one period are not shown. Figure 3.7 shows a signal in both the time and frequency domains.

Figure 3.7 *The time-domain and frequency-domain plots of a sine wave*

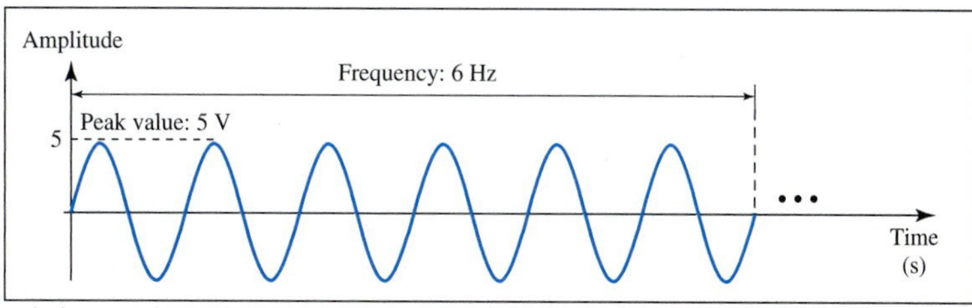

a. A sine wave in the time domain (peak value: 5 V, frequency: 6 Hz)

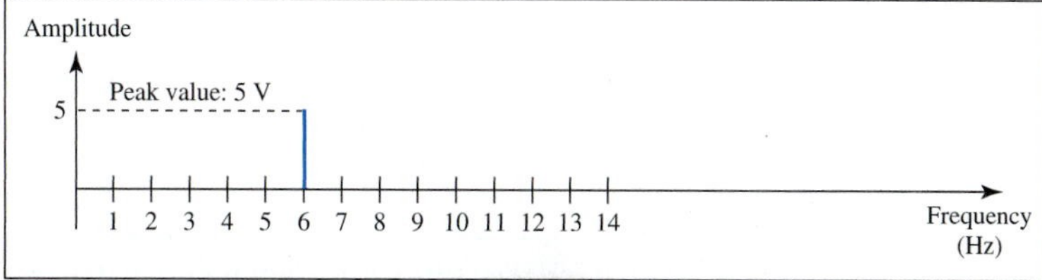

b. The same sine wave in the frequency domain (peak value: 5 V, frequency: 6 Hz)

It is obvious that the frequency domain is easy to plot and conveys the information that one can find in a time domain plot. The advantage of the frequency domain is that we can immediately see the values of the frequency and peak amplitude. A complete sine wave is represented by one spike. The position of the spike shows the frequency; its height shows the peak amplitude.

A complete sine wave in the time domain can be represented by one single spike in the frequency domain.

Example 3.7

The frequency domain is more compact and useful when we are dealing with more than one sine wave. For example, Figure 3.8 shows three sine waves, each with different amplitude and frequency. All can be represented by three spikes in the frequency domain.

Figure 3.8 *The time domain and frequency domain of three sine waves*

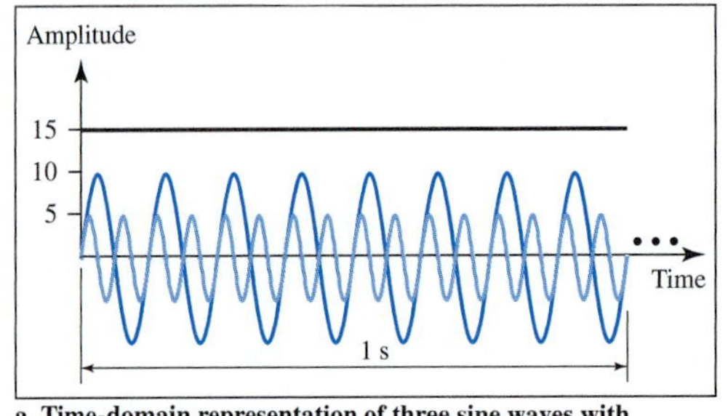

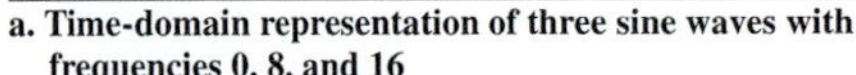

a. Time-domain representation of three sine waves with frequencies 0, 8, and 16

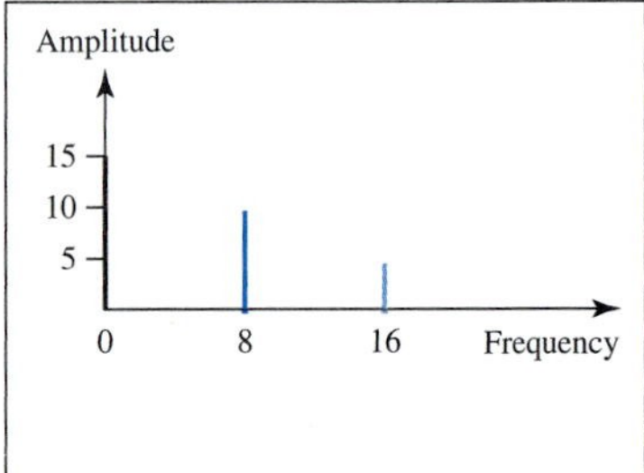

b. Frequency-domain representation of the same three signals

Composite Signals

So far, we have focused on simple sine waves. Simple sine waves have many applications in daily life. We can send a single sine wave to carry electric energy from one place to another. For example, the power company sends a single sine wave with a frequency of 60 Hz to distribute electric energy to houses and businesses. As another example, we can use a single sine wave to send an alarm to a security center when a burglar opens a door or window in the house. In the first case, the sine wave is carrying energy; in the second, the sine wave is a signal of danger.

If we had only one single sine wave to convey a conversation over the phone, it would make no sense and carry no information. We would just hear a buzz. As we will see in Chapters 4 and 5, we need to send a composite signal to communicate data. A **composite signal** is made of many simple sine waves.

A single-frequency sine wave is not useful in data communications; we need to send a composite signal, a signal made of many simple sine waves.

In the early 1900s, the French mathematician Jean-Baptiste Fourier showed that any composite signal is actually a combination of simple sine waves with different frequencies, amplitudes, and phases. **Fourier analysis** is discussed in Appendix C; for our purposes, we just present the concept.

According to Fourier analysis, any composite signal is a combination of simple sine waves with different frequencies, amplitudes, and phases. Fourier analysis is discussed in Appendix C.

A composite signal can be periodic or nonperiodic. A periodic composite signal can be decomposed into a series of simple sine waves with discrete frequencies—frequencies that have integer values (1, 2, 3, and so on). A nonperiodic composite signal can be decomposed into a combination of an infinite number of simple sine waves with continuous frequencies, frequencies that have real values.

If the composite signal is periodic, the decomposition gives a series of signals with discrete frequencies; if the composite signal is nonperiodic, the decomposition gives a combination of sine waves with continuous frequencies.

Example 3.8

Figure 3.9 shows a periodic composite signal with frequency *f*. This type of signal is not typical of those found in data communications. We can consider it to be three alarm systems, each with a different frequency. The analysis of this signal can give us a good understanding of how to decompose signals.

Figure 3.9 *A composite periodic signal*

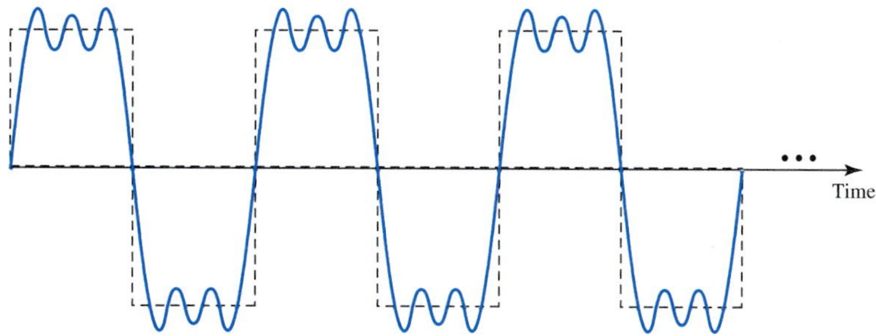

It is very difficult to manually decompose this signal into a series of simple sine waves. However, there are tools, both hardware and software, that can help us do the job. We are not concerned about how it is done; we are only interested in the result. Figure 3.10 shows the result of decomposing the above signal in both the time and frequency domains.

The amplitude of the sine wave with frequency *f* is almost the same as the peak amplitude of the composite signal. The amplitude of the sine wave with frequency 3*f* is one-third of that of the first, and the amplitude of the sine wave with frequency 9*f* is one-ninth of the first. The frequency

Figure 3.10 *Decomposition of a composite periodic signal in the time and frequency domains*

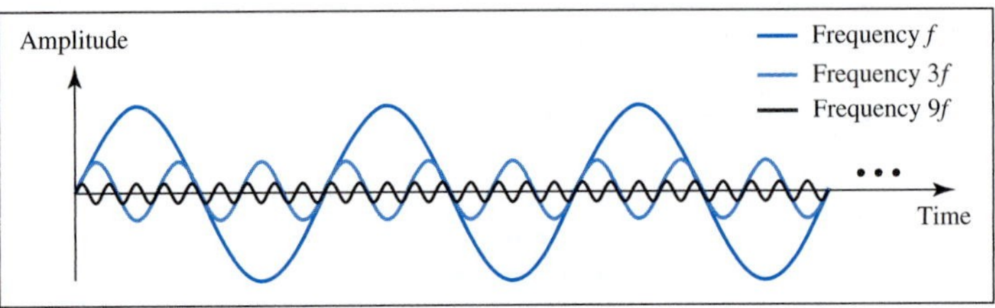

a. Time-domain decomposition of a composite signal

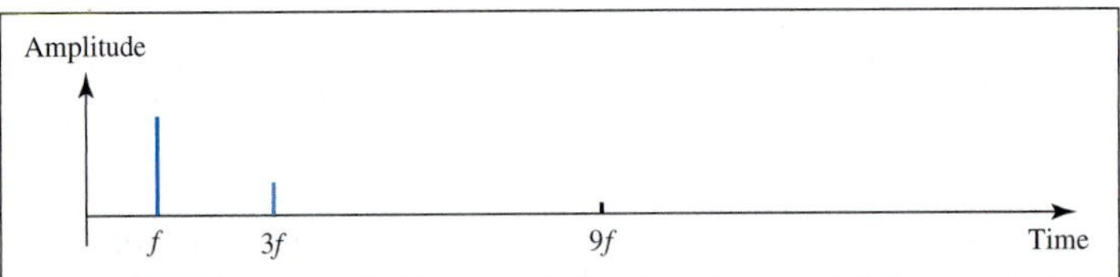

b. Frequency-domain decomposition of the composite signal

of the sine wave with frequency f is the same as the frequency of the composite signal; it is called the **fundamental frequency,** or first **harmonic.** The sine wave with frequency $3f$ has a frequency of 3 times the fundamental frequency; it is called the third harmonic. The third sine wave with frequency $9f$ has a frequency of 9 times the fundamental frequency; it is called the ninth harmonic.

Note that the frequency decomposition of the signal is discrete; it has frequencies f, $3f$, and $9f$. Because f is an integral number, $3f$ and $9f$ are also integral numbers. There are no frequencies such as $1.2f$ or $2.6f$. The frequency domain of a periodic composite signal is always made of discrete spikes.

Example 3.9

Figure 3.11 shows a nonperiodic composite signal. It can be the signal created by a microphone or a telephone set when a word or two is pronounced. In this case, the composite signal cannot be periodic, because that implies that we are repeating the same word or words with exactly the same tone.

Figure 3.11 *The time and frequency domains of a nonperiodic signal*

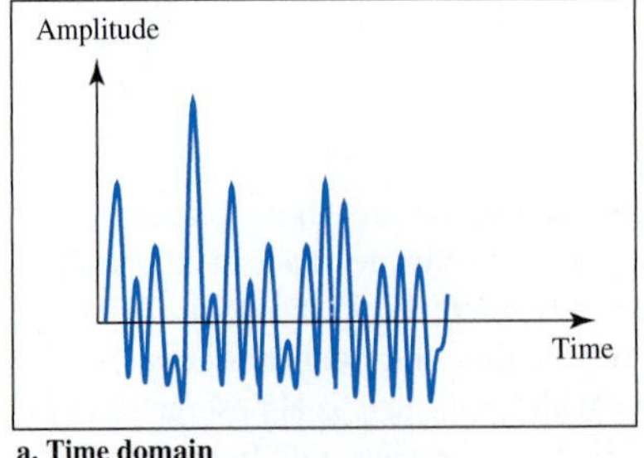

a. Time domain

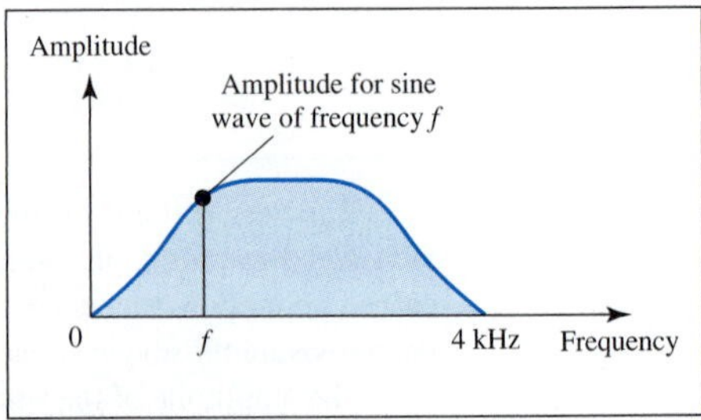

b. Frequency domain

In a time-domain representation of this composite signal, there are an infinite number of simple sine frequencies. Although the number of frequencies in a human voice is infinite, the range is limited. A normal human being can create a continuous range of frequencies between 0 and 4 kHz.

Note that the frequency decomposition of the signal yields a continuous curve. There are an infinite number of frequencies between 0.0 and 4000.0 (real values). To find the amplitude related to frequency *f*, we draw a vertical line at *f* to intersect the envelope curve. The height of the vertical line is the amplitude of the corresponding frequency.

Bandwidth

The range of frequencies contained in a composite signal is its **bandwidth.** The bandwidth is normally a difference between two numbers. For example, if a composite signal contains frequencies between 1000 and 5000, its bandwidth is 5000 – 1000, or 4000.

The bandwidth of a composite signal is the difference between the highest and the lowest frequencies contained in that signal.

Figure 3.12 shows the concept of bandwidth. The figure depicts two composite signals, one periodic and the other nonperiodic. The bandwidth of the periodic signal contains all integer frequencies between 1000 and 5000 (1000, 1001, 1002, . . .). The bandwidth of the nonperiodic signals has the same range, but the frequencies are continuous.

Figure 3.12 *The bandwidth of periodic and nonperiodic composite signals*

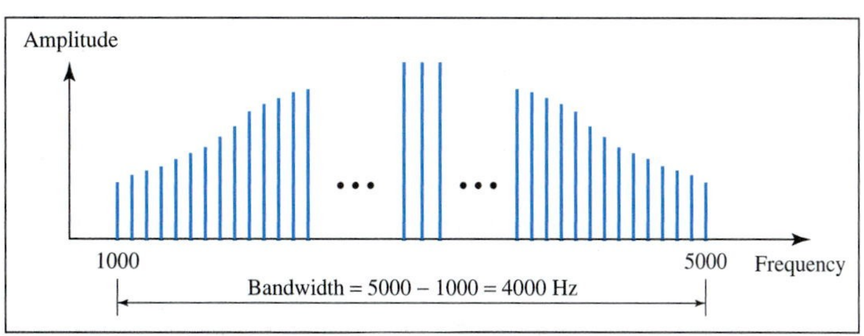

a. Bandwidth of a periodic signal

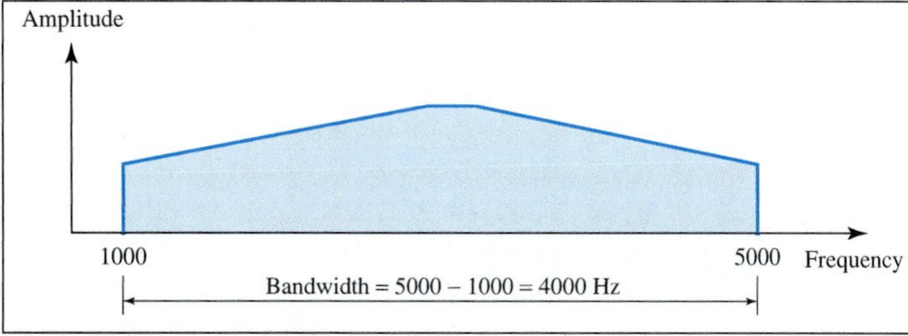

b. Bandwidth of a nonperiodic signal

Example 3.10

If a periodic signal is decomposed into five sine waves with frequencies of 100, 300, 500, 700, and 900 Hz, what is its bandwidth? Draw the spectrum, assuming all components have a maximum amplitude of 10 V.

Solution

Let f_h be the highest frequency, f_l the lowest frequency, and B the bandwidth. Then

$$B = f_h - f_l = 900 - 100 = 800 \text{ Hz}$$

The spectrum has only five spikes, at 100, 300, 500, 700, and 900 Hz (see Figure 3.13).

Figure 3.13 *The bandwidth for Example 3.10*

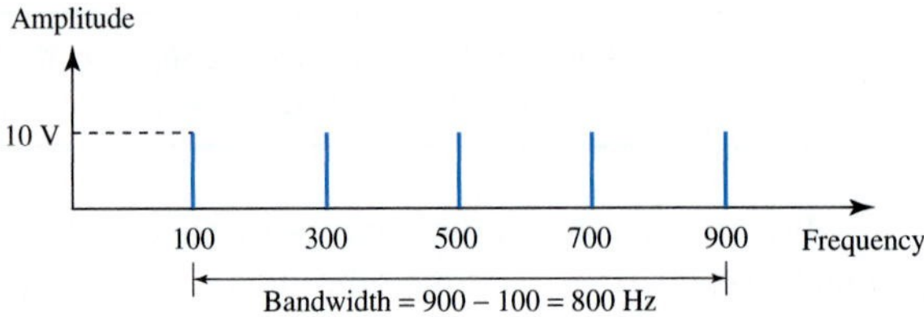

Example 3.11

A periodic signal has a bandwidth of 20 Hz. The highest frequency is 60 Hz. What is the lowest frequency? Draw the spectrum if the signal contains all frequencies of the same amplitude.

Solution

Let f_h be the highest frequency, f_l the lowest frequency, and B the bandwidth. Then

$$B = f_h - f_l \;\Rightarrow\; 20 = 60 - f_l \;\Rightarrow\; f_l = 60 - 20 = 40 \text{ Hz}$$

The spectrum contains all integer frequencies. We show this by a series of spikes (see Figure 3.14).

Figure 3.14 *The bandwidth for Example 3.11*

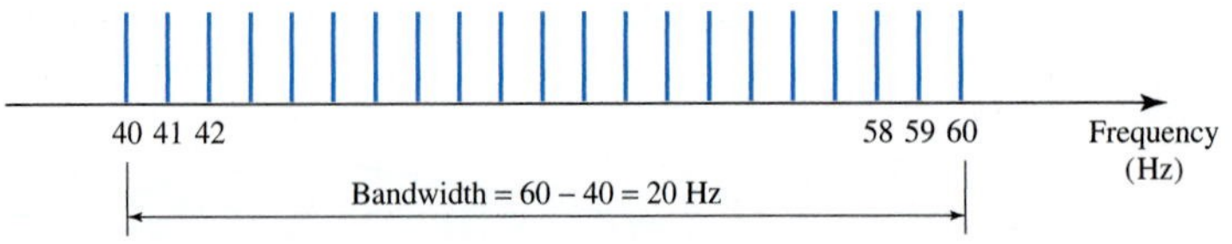

Example 3.12

A nonperiodic composite signal has a bandwidth of 200 kHz, with a middle frequency of 140 kHz and peak amplitude of 20 V. The two extreme frequencies have an amplitude of 0. Draw the frequency domain of the signal.

Solution

The lowest frequency must be at 40 kHz and the highest at 240 kHz. Figure 3.15 shows the frequency domain and the bandwidth.

Figure 3.15 *The bandwidth for Example 3.12*

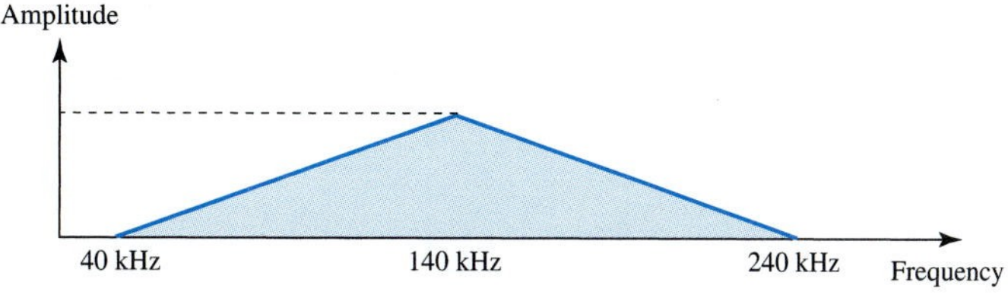

Example 3.13

An example of a nonperiodic composite signal is the signal propagated by an AM radio station. In the United States, each AM radio station is assigned a 10-kHz bandwidth. The total bandwidth dedicated to AM radio ranges from 530 to 1700 kHz. We will show the rationale behind this 10-kHz bandwidth in Chapter 5.

Example 3.14

Another example of a nonperiodic composite signal is the signal propagated by an FM radio station. In the United States, each FM radio station is assigned a 200-kHz bandwidth. The total bandwidth dedicated to FM radio ranges from 88 to 108 MHz. We will show the rationale behind this 200-kHz bandwidth in Chapter 5.

Example 3.15

Another example of a nonperiodic composite signal is the signal received by an old-fashioned analog black-and-white TV. A TV screen is made up of pixels (picture elements) with each pixel being either white or black. The screen is scanned 30 times per second. (Scanning is actually 60 times per second, but odd lines are scanned in one round and even lines in the next and then interleaved.) If we assume a resolution of 525 × 700 (525 vertical lines and 700 horizontal lines), which is a ratio of 3:4, we have 367,500 pixels per screen. If we scan the screen 30 times per second, this is 367,500 × 30 = 11,025,000 pixels per second. The worst-case scenario is alternating black and white pixels. In this case, we need to represent one color by the minimum amplitude and the other color by the maximum amplitude. We can send 2 pixels per cycle. Therefore, we need 11,025,000 / 2 = 5,512,500 cycles per second, or Hz. The bandwidth needed is 5.5124 MHz. This worst-case scenario has such a low probability of occurrence that the assumption is that we need only 70 percent of this bandwidth, which is 3.85 MHz. Since audio and synchronization signals are also needed, a 4-MHz bandwidth has been set aside for each black and white TV channel. An analog color TV channel has a 6-MHz bandwidth.

3.3 DIGITAL SIGNALS

In addition to being represented by an analog signal, information can also be represented by a digital signal. For example, a 1 can be encoded as a positive voltage and a 0 as zero voltage. A digital signal can have more than two levels. In this case, we can send more than 1 bit for each level. Figure 3.16 shows two signals, one with two levels and the other with four.

Figure 3.16 *Two digital signals: one with two signal levels and the other with four signal levels*

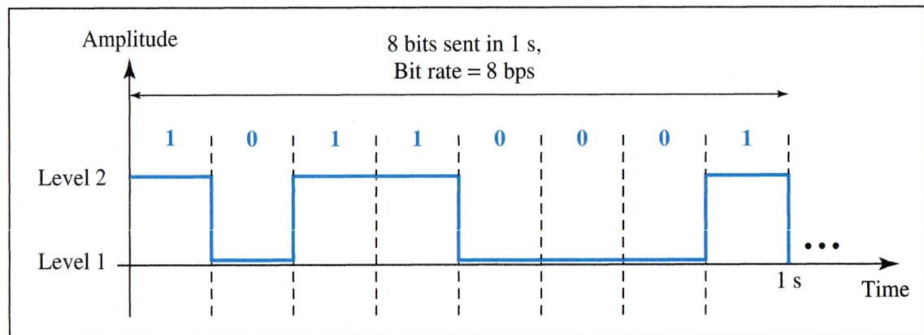

a. A digital signal with two levels

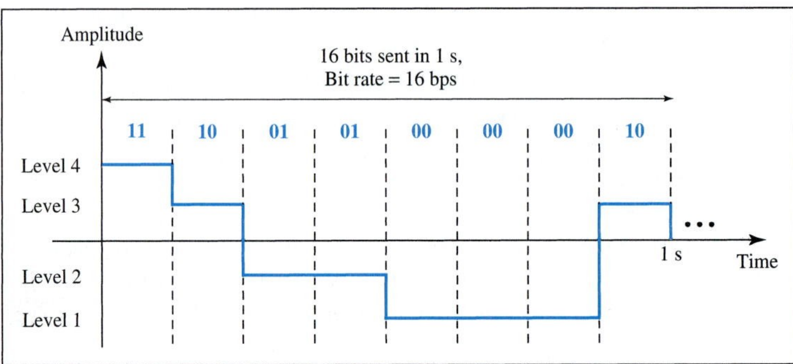

b. A digital signal with four levels

We send 1 bit per level in part a of the figure and 2 bits per level in part b of the figure. In general, if a signal has L levels, each level needs $\log_2 L$ bits.

Appendix C reviews information about exponential and logarithmic functions.

Example 3.16

A digital signal has eight levels. How many bits are needed per level? We calculate the number of bits from the formula

$$\text{Number of bits per level} = \log_2 8 = 3$$

Each signal level is represented by 3 bits.

Example 3.17

A digital signal has nine levels. How many bits are needed per level? We calculate the number of bits by using the formula. Each signal level is represented by 3.17 bits. However, this answer is not realistic. The number of bits sent per level needs to be an integer as well as a power of 2. For this example, 4 bits can represent one level.

Bit Rate

Most digital signals are nonperiodic, and thus period and frequency are not appropriate characteristics. Another term—*bit rate* (instead of *frequency*)—is used to describe digital signals. The **bit rate** is the number of bits sent in 1s, expressed in **bits per second (bps).** Figure 3.16 shows the bit rate for two signals.

Example 3.18

Assume we need to download text documents at the rate of 100 pages per minute. What is the required bit rate of the channel?

Solution

A page is an average of 24 lines with 80 characters in each line. If we assume that one character requires 8 bits, the bit rate is

$$100 \times 24 \times 80 \times 8 = 1{,}636{,}000 \text{ bps} = 1.636 \text{ Mbps}$$

Example 3.19

A digitized voice channel, as we will see in Chapter 4, is made by digitizing a 4-kHz bandwidth analog voice signal. We need to sample the signal at twice the highest frequency (two samples per hertz). We assume that each sample requires 8 bits. What is the required bit rate?

Solution

The bit rate can be calculated as

$$2 \times 4000 \times 8 = 64{,}000 \text{ bps} = 64 \text{ kbps}$$

Example 3.20

What is the bit rate for high-definition TV (HDTV)?

Solution

HDTV uses digital signals to broadcast high quality video signals. The HDTV screen is normally a ratio of 16 : 9 (in contrast to 4 : 3 for regular TV), which means the screen is wider. There are 1920 by 1080 pixels per screen, and the screen is renewed 30 times per second. Twenty-four bits represents one color pixel. We can calculate the bit rate as

$$1920 \times 1080 \times 30 \times 24 = 1{,}492{,}992{,}000 \text{ or } 1.5 \text{ Gbps}$$

The TV stations reduce this rate to 20 to 40 Mbps through compression.

Bit Length

We discussed the concept of the wavelength for an analog signal: the distance one cycle occupies on the transmission medium. We can define something similar for a digital signal: the bit length. The **bit length** is the distance one bit occupies on the transmission medium.

Bit length = propagation speed × bit duration

Digital Signal as a Composite Analog Signal

Based on Fourier analysis, a digital signal is a composite analog signal. The bandwidth is infinite, as you may have guessed. We can intuitively come up with this concept when we consider a digital signal. A digital signal, in the time domain, comprises connected vertical and horizontal line segments. A vertical line in the time domain means a frequency of infinity (sudden change in time); a horizontal line in the time domain means a frequency of zero (no change in time). Going from a frequency of zero to a frequency of infinity (and vice versa) implies all frequencies in between are part of the domain.

Fourier analysis can be used to decompose a digital signal. If the digital signal is periodic, which is rare in data communications, the decomposed signal has a frequency-domain representation with an infinite bandwidth and discrete frequencies. If the digital signal is nonperiodic, the decomposed signal still has an infinite bandwidth, but the frequencies are continuous. Figure 3.17 shows a periodic and a nonperiodic digital signal and their bandwidths.

Figure 3.17 *The time and frequency domains of periodic and nonperiodic digital signals*

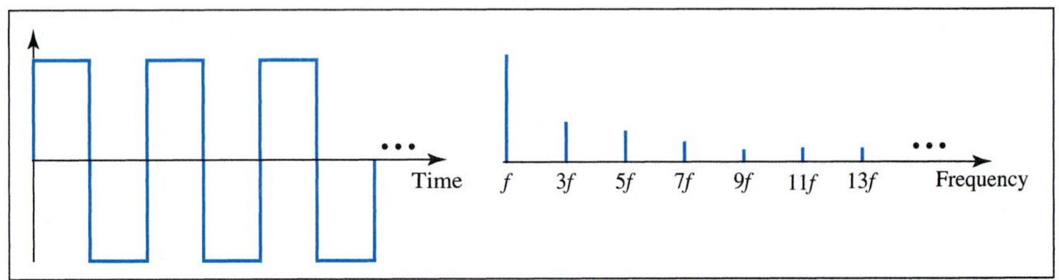

a. Time and frequency domains of periodic digital signal

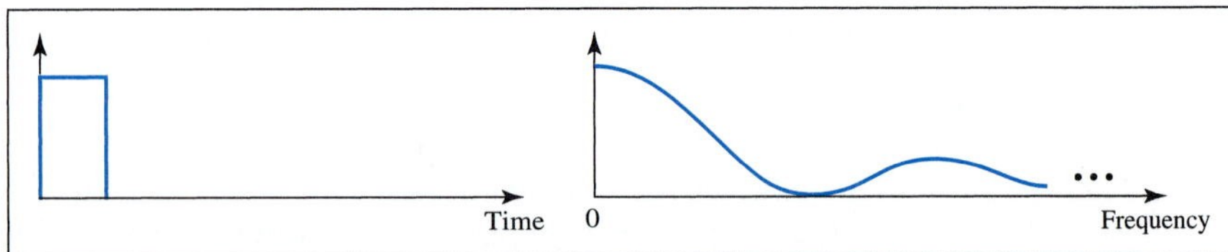

b. Time and frequency domains of nonperiodic digital signal

Note that both bandwidths are infinite, but the periodic signal has discrete frequencies while the nonperiodic signal has continuous frequencies.

Transmission of Digital Signals

The previous discussion asserts that a digital signal, periodic or nonperiodic, is a composite analog signal with frequencies between zero and infinity. For the remainder of the discussion, let us consider the case of a nonperiodic digital signal, similar to the ones we encounter in data communications. The fundamental question is, How can we send a digital signal from point *A* to point *B*? We can transmit a digital signal by using one of two different approaches: baseband transmission or broadband transmission (using modulation).

Baseband Transmission

Baseband transmission means sending a digital signal over a channel without changing the digital signal to an analog signal. Figure 3.18 shows **baseband** transmission.

Figure 3.18 *Baseband transmission*

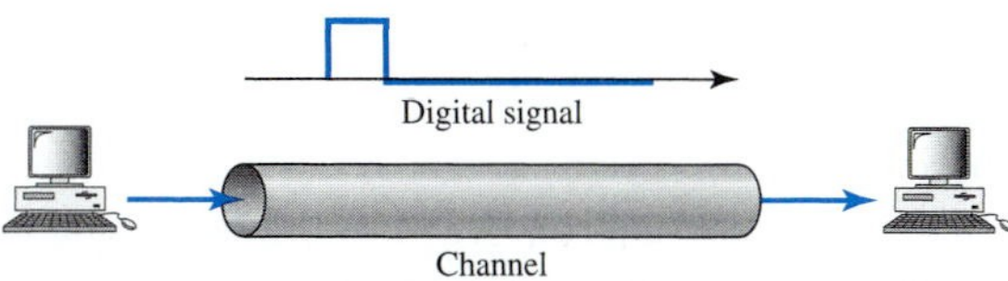

A digital signal is a composite analog signal with an infinite bandwidth.

Baseband transmission requires that we have a **low-pass channel,** a channel with a bandwidth that starts from zero. This is the case if we have a dedicated medium with a bandwidth constituting only one channel. For example, the entire bandwidth of a cable connecting two computers is one single channel. As another example, we may connect several computers to a bus, but not allow more than two stations to communicate at a time. Again we have a low-pass channel, and we can use it for baseband communication. Figure 3.19 shows two low-pass channels: one with a narrow bandwidth and the other with a wide bandwidth. We need to remember that a low-pass channel with infinite bandwidth is ideal, but we cannot have such a channel in real life. However, we can get close.

Figure 3.19 *Bandwidths of two low-pass channels*

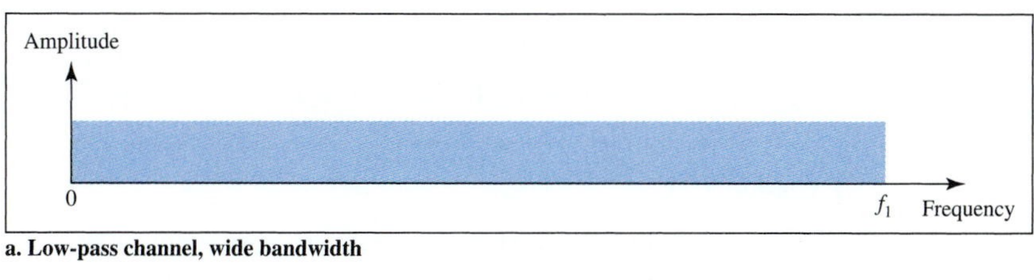

a. Low-pass channel, wide bandwidth

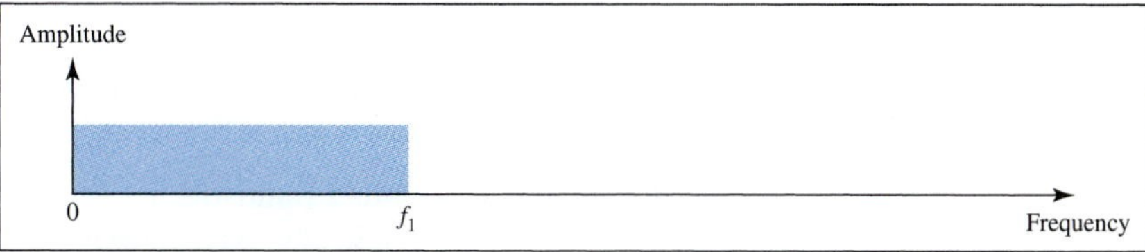

b. Low-pass channel, narrow bandwidth

Let us study two cases of a baseband communication: a low-pass channel with a wide bandwidth and one with a limited bandwidth.

Case 1: Low-Pass Channel with Wide Bandwidth

If we want to preserve the exact form of a nonperiodic digital signal with vertical segments vertical and horizontal segments horizontal, we need to send the entire spectrum, the continuous range of frequencies between zero and infinity. This is possible if we have a dedicated medium with an infinite bandwidth between the sender and receiver that preserves the exact amplitude of each component of the composite signal. Although this may be possible inside a computer (e.g., between CPU and memory), it is not possible between two devices. Fortunately, the amplitudes of the frequencies at the border of the bandwidth are so small that they can be ignored. This means that if we have a medium, such as a coaxial cable or fiber optic, with a very wide bandwidth, two stations can communicate by using digital signals with very good accuracy, as shown in Figure 3.20. Note that f_1 is close to zero, and f_2 is very high.

Figure 3.20 *Baseband transmission using a dedicated medium*

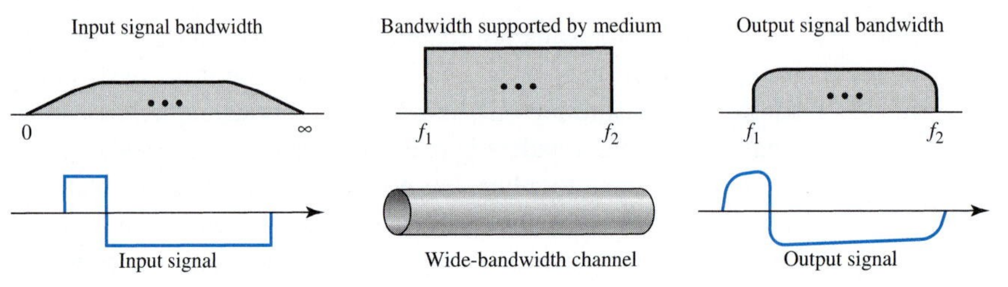

Although the output signal is not an exact replica of the original signal, the data can still be deduced from the received signal. Note that although some of the frequencies are blocked by the medium, they are not critical.

Baseband transmission of a digital signal that preserves the shape of the digital signal is possible only if we have a low-pass channel with an infinite or very wide bandwidth.

Example 3.21

An example of a dedicated channel where the entire bandwidth of the medium is used as one single channel is a LAN. Almost every wired LAN today uses a dedicated channel for two stations communicating with each other. In a bus topology LAN with multipoint connections, only two stations can communicate with each other at each moment in time (timesharing); the other stations need to refrain from sending data. In a star topology LAN, the entire channel between each station and the hub is used for communication between these two entities. We study LANs in Chapter 14.

Case 2: Low-Pass Channel with Limited Bandwidth

In a low-pass channel with limited bandwidth, we approximate the digital signal with an analog signal. The level of approximation depends on the bandwidth available.

Rough Approximation Let us assume that we have a digital signal of bit rate *N*. If we want to send analog signals to roughly simulate this signal, we need to consider the worst case, a maximum number of changes in the digital signal. This happens when the signal carries the sequence 01010101 · · · or the sequence 10101010· · · · To simulate these two cases, we need an analog signal of frequency $f = N/2$. Let 1 be the positive peak value and 0 be the negative peak value. We send 2 bits in each cycle; the frequency of the analog signal is one-half of the bit rate, or $N/2$. However, just this one frequency cannot make all patterns; we need more components. The maximum frequency is $N/2$. As an example of this concept, let us see how a digital signal with a 3-bit pattern can be simulated by using analog signals. Figure 3.21 shows the idea. The two similar cases (000 and 111) are simulated with a signal with frequency $f = 0$ and a phase of 180° for 000 and a phase of 0° for 111. The two worst cases (010 and 101) are simulated with an analog signal with frequency $f = N/2$ and phases of 180° and 0°. The other four cases can only be simulated with an analog signal with $f = N/4$ and phases of 180°, 270°, 90°, and 0°. In other words, we need a channel that can handle frequencies 0, $N/4$, and $N/2$. This rough approximation is referred to as using the first harmonic ($N/2$) frequency. The required bandwidth is

$$\text{Bandwidth} = \frac{N}{2} - 0 = \frac{N}{2}$$

Figure 3.21 *Rough approximation of a digital signal using the first harmonic for worst case*

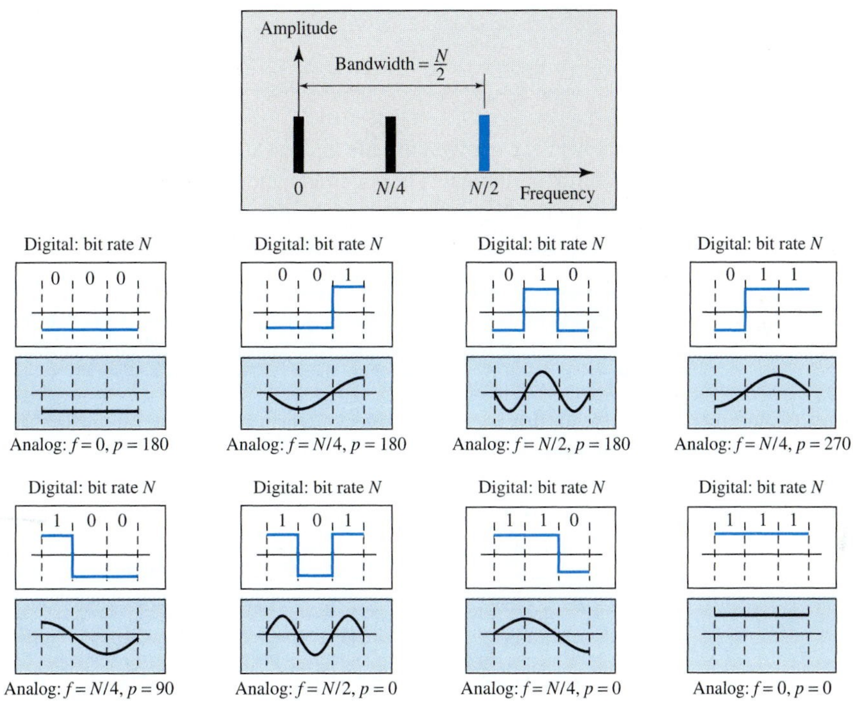

Better Approximation To make the shape of the analog signal look more like that of a digital signal, we need to add more harmonics of the frequencies. We need to increase the bandwidth. We can increase the bandwidth to $3N/2$, $5N/2$, $7N/2$, and so on. Figure 3.22 shows the effect of this increase for one of the worst cases, the pattern 010.

Figure 3.22 *Simulating a digital signal with three first harmonics*

Amplitude

Bandwidth = $\frac{5N}{2}$

0 $N/4$ $N/2$ $3N/4$ $3N/2$ $5N/4$ $5N/2$ Frequency

Digital: bit rate N

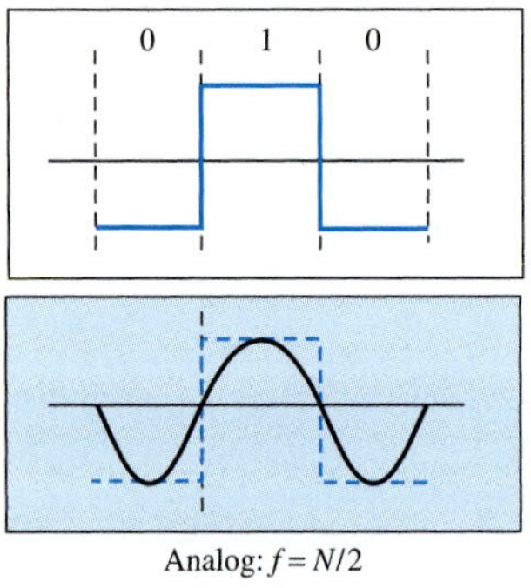

Analog: $f = N/2$ and $3N/2$

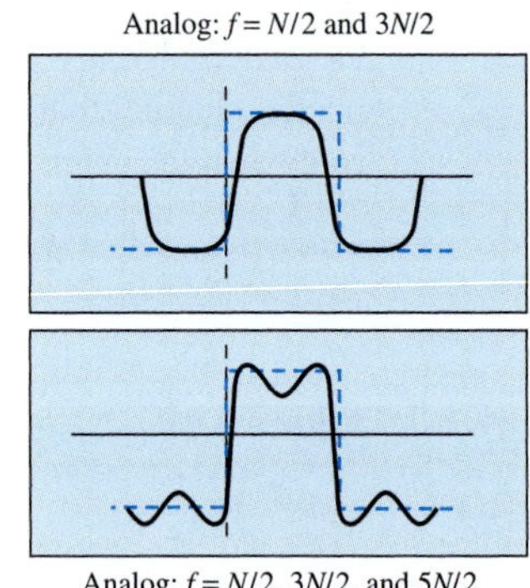

Analog: $f = N/2$

Analog: $f = N/2$, $3N/2$, and $5N/2$

Note that we have shown only the highest frequency for each harmonic. We use the first, third, and fifth harmonics. The required bandwidth is now $5N/2$, the difference between the lowest frequency 0 and the highest frequency $5N/2$. As we emphasized before, we need to remember that the required bandwidth is proportional to the bit rate.

In baseband transmission, the required bandwidth is proportional to the bit rate; if we need to send bits faster, we need more bandwidth.

By using this method, Table 3.2 shows how much bandwidth we need to send data at different rates.

Table 3.2 *Bandwidth requirements*

Bit Rate	*Harmonic 1*	*Harmonics 1, 3*	*Harmonics 1, 3, 5*
n = 1 kbps	B = 500 Hz	B = 1.5 kHz	B = 2.5 kHz
n = 10 kbps	B = 5 kHz	B = 15 kHz	B = 25 kHz
n = 100 kbps	B = 50 kHz	B = 150 kHz	B = 250 kHz

Example 3.22

What is the required bandwidth of a low-pass channel if we need to send 1 Mbps by using baseband transmission?

Solution

The answer depends on the accuracy desired.

a. The minimum bandwidth, a rough approximation, is B = bit rate /2, or 500 kHz. We need a low-pass channel with frequencies between 0 and 500 kHz.

b. A better result can be achieved by using the first and the third harmonics with the required bandwidth $B = 3 \times 500$ kHz = 1.5 MHz.

c. Still a better result can be achieved by using the first, third, and fifth harmonics with $B = 5 \times 500$ kHz = 2.5 MHz.

Example 3.23

We have a low-pass channel with bandwidth 100 kHz. What is the maximum bit rate of this channel?

Solution

The maximum bit rate can be achieved if we use the first harmonic. The bit rate is 2 times the available bandwidth, or 200 kbps.

Broadband Transmission (Using Modulation)

Broadband transmission or modulation means changing the digital signal to an analog signal for transmission. Modulation allows us to use a **bandpass channel**—a channel with a bandwidth that does not start from zero. This type of channel is more available than a low-pass channel. Figure 3.23 shows a bandpass channel.

Figure 3.23 *Bandwidth of a bandpass channel*

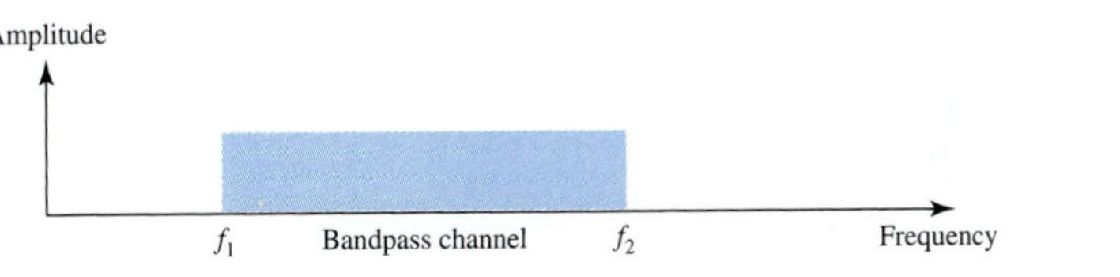

Note that a low-pass channel can be considered a bandpass channel with the lower frequency starting at zero.

Figure 3.24 shows the modulation of a digital signal. In the figure, a digital signal is converted to a composite analog signal. We have used a single-frequency analog signal (called a carrier); the amplitude of the carrier has been changed to look like the digital signal. The result, however, is not a single-frequency signal; it is a composite signal, as we will see in Chapter 5. At the receiver, the received analog signal is converted to digital, and the result is a replica of what has been sent.

If the available channel is a bandpass channel, we cannot send the digital signal directly to the channel; we need to convert the digital signal to an analog signal before transmission.

Figure 3.24 *Modulation of a digital signal for transmission on a bandpass channel*

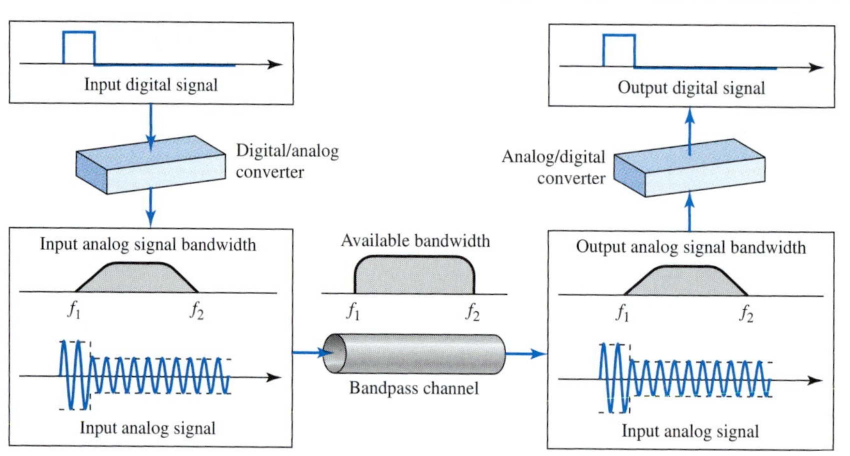

Example 3.24

An example of broadband transmission using modulation is the sending of computer data through a telephone subscriber line, the line connecting a resident to the central telephone office. These lines, installed many years ago, are designed to carry voice (analog signal) with a limited bandwidth (frequencies between 0 and 4 kHz). Although this channel can be used as a low-pass channel, it is normally considered a bandpass channel. One reason is that the bandwidth is so narrow (4 kHz) that if we treat the channel as low-pass and use it for baseband transmission, the maximum bit rate can be only 8 kbps. The solution is to consider the channel a bandpass channel, convert the digital signal from the computer to an analog signal, and send the analog signal. We can install two converters to change the digital signal to analog and vice versa at the receiving end. The converter, in this case, is called a *modem* (*mo*dulator/*dem*odulator), which we discuss in detail in Chapter 5.

Example 3.25

A second example is the digital cellular telephone. For better reception, digital cellular phones convert the analog voice signal to a digital signal (see Chapter 16). Although the bandwidth allocated to a company providing digital cellular phone service is very wide, we still cannot send the digital signal without conversion. The reason is that we only have a bandpass channel available between caller and callee. For example, if the available bandwidth is W and we allow 1000 couples to talk simultaneously, this means the available channel is W/1000, just part of the entire bandwidth. We need to convert the digitized voice to a composite analog signal before sending. The digital cellular phones convert the analog audio signal to digital and then convert it again to analog for transmission over a bandpass channel.

3.4 TRANSMISSION IMPAIRMENT

Signals travel through transmission media, which are not perfect. The imperfection causes signal impairment. This means that the signal at the beginning of the medium is not the same as the signal at the end of the medium. What is sent is not what is received. Three causes of impairment are attenuation, distortion, and noise (see Figure 3.25).

Figure 3.25 *Causes of impairment*

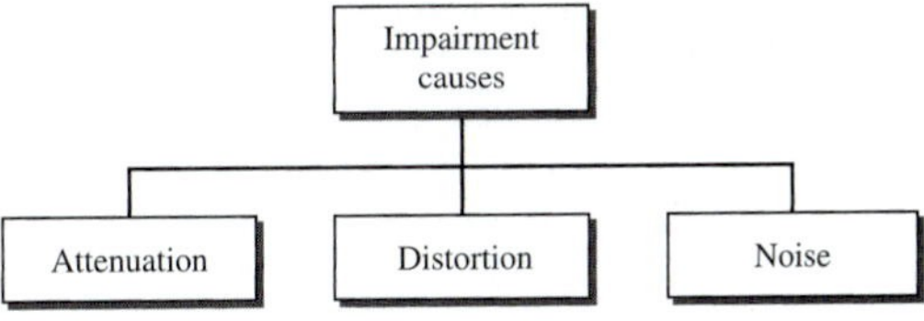

Attenuation

Attenuation means a loss of energy. When a signal, simple or composite, travels through a medium, it loses some of its energy in overcoming the resistance of the medium. That is why a wire carrying electric signals gets warm, if not hot, after a while. Some of the electrical energy in the signal is converted to heat. To compensate for this loss, amplifiers are used to amplify the signal. Figure 3.26 shows the effect of attenuation and amplification.

Figure 3.26 *Attenuation*

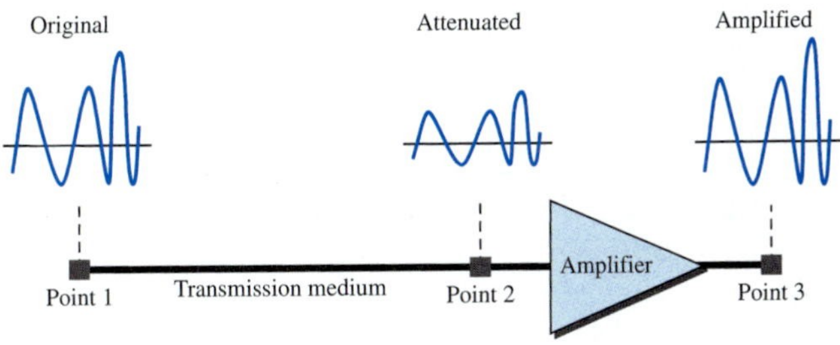

Decibel

To show that a signal has lost or gained strength, engineers use the unit of the decibel. The **decibel (dB)** measures the relative strengths of two signals or one signal at two different points. Note that the decibel is negative if a signal is attenuated and positive if a signal is amplified.

$$\mathbf{dB = 10 \log_{10} \frac{P_2}{P_1}}$$

Variables P_1 and P_2 are the powers of a signal at points 1 and 2, respectively. Note that some engineering books define the decibel in terms of voltage instead of power. In this case, because power is proportional to the square of the voltage, the formula is dB = $20 \log_{10} (V_2/V_1)$. In this text, we express dB in terms of power.

Example 3.26

Suppose a signal travels through a transmission medium and its power is reduced to one-half. This means that $P_2 = \frac{1}{2} P_1$. In this case, the attenuation (loss of power) can be calculated as

$$10 \log_{10} \frac{P_2}{P_1} = 10 \log_{10} \frac{0.5P_1}{P_1} = 10 \log_{10} 0.5 = 10(-0.3) = -3 \text{ dB}$$

A loss of 3 dB (–3 dB) is equivalent to losing one-half the power.

Example 3.27

A signal travels through an amplifier, and its power is increased 10 times. This means that $P_2 = 10P_1$. In this case, the amplification (gain of power) can be calculated as

$$10 \log_{10} \frac{P_2}{P_1} = 10 \log_{10} \frac{10P_1}{P_1} = 10 \log_{10} 10 = 10(1) = 10 \text{ dB}$$

Example 3.28

One reason that engineers use the decibel to measure the changes in the strength of a signal is that decibel numbers can be added (or subtracted) when we are measuring several points (cascading) instead of just two. In Figure 3.27 a signal travels from point 1 to point 4. The signal is attenuated by the time it reaches point 2. Between points 2 and 3, the signal is amplified. Again, between points 3 and 4, the signal is attenuated. We can find the resultant decibel value for the signal just by adding the decibel measurements between each set of points.

Figure 3.27 *Decibels for Example 3.28*

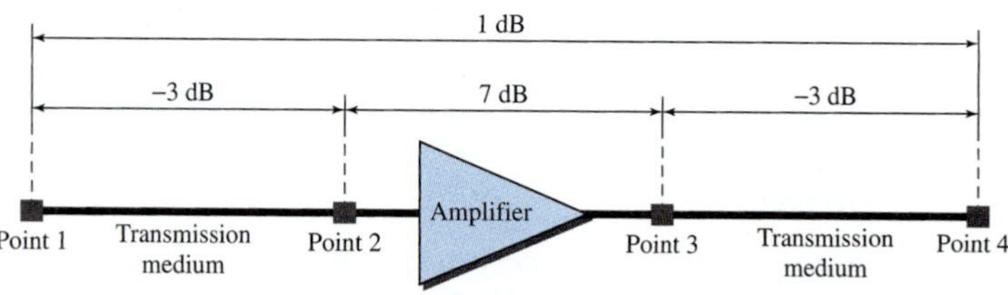

In this case, the decibel value can be calculated as

$$\text{dB} = -3 + 7 - 3 = +1$$

The signal has gained in power.

Example 3.29

Sometimes the decibel is used to measure signal power in milliwatts. In this case, it is referred to as dB_m and is calculated as $\text{dB}_\text{m} = 10 \log_{10} P_m$, where P_m is the power in milliwatts. Calculate the power of a signal if its $\text{dB}_\text{m} = -30$.

Solution

We can calculate the power in the signal as

$$dB_m = 10 \log_{10} P_m = -30$$
$$\log_{10} P_m = -3 \qquad P_m = 10^{-3}\ \text{mW}$$

Example 3.30

The loss in a cable is usually defined in decibels per kilometer (dB/km). If the signal at the beginning of a cable with −0.3 dB/km has a power of 2 mW, what is the power of the signal at 5 km?

Solution

The loss in the cable in decibels is $5 \times (-0.3) = -1.5$ dB. We can calculate the power as

$$dB = 10 \log_{10} \frac{P_2}{P_1} = -1.5$$
$$\frac{P_2}{P_1} = 10^{-0.15} = 0.71$$
$$P_2 = 0.71 P_1 = 0.7 \times 2 = 1.4\ \text{mW}$$

Distortion

Distortion means that the signal changes its form or shape. Distortion can occur in a composite signal made of different frequencies. Each signal component has its own propagation speed (see the next section) through a medium and, therefore, its own delay in arriving at the final destination. Differences in delay may create a difference in phase if the delay is not exactly the same as the period duration. In other words, signal components at the receiver have phases different from what they had at the sender. The shape of the composite signal is therefore not the same. Figure 3.28 shows the effect of distortion on a composite signal.

Figure 3.28 *Distortion*

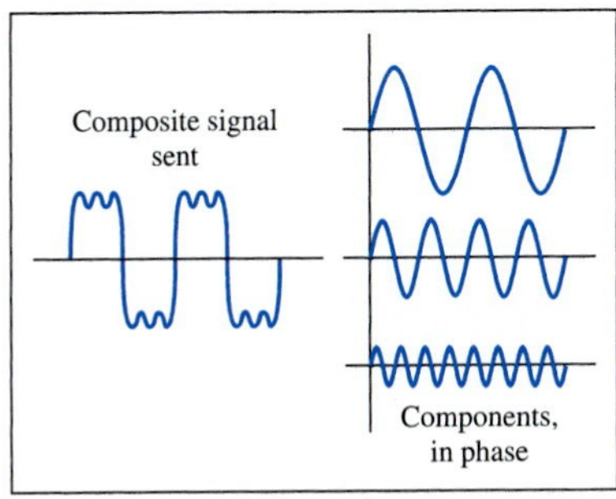

At the sender

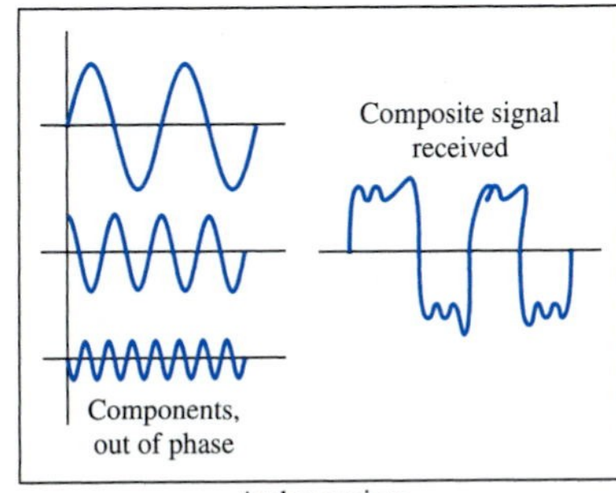

At the receiver

Noise

Noise is another cause of impairment. Several types of noise, such as thermal noise, induced noise, crosstalk, and impulse noise, may corrupt the signal. Thermal noise is the random motion of electrons in a wire which creates an extra signal not originally sent by the transmitter. Induced noise comes from sources such as motors and appliances. These devices act as a sending antenna, and the transmission medium acts as the receiving antenna. Crosstalk is the effect of one wire on the other. One wire acts as a sending antenna and the other as the receiving antenna. Impulse noise is a spike (a signal with high energy in a very short time) that comes from power lines, lightning, and so on. Figure 3.29 shows the effect of noise on a signal. We discuss error in Chapter 10.

Figure 3.29 *Noise*

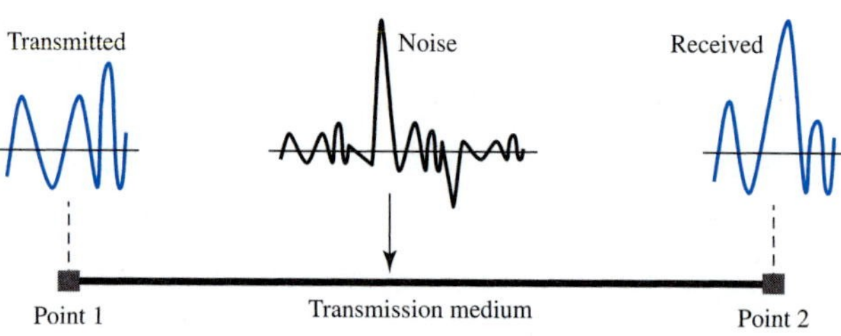

Signal-to-Noise Ratio (SNR)

As we will see later, to find the theoretical bit rate limit, we need to know the ratio of the signal power to the noise power. The **signal-to-noise ratio** is defined as

$$\textbf{SNR} = \frac{\textbf{average signal power}}{\textbf{average noise power}}$$

We need to consider the average signal power and the average noise power because these may change with time. Figure 3.30 shows the idea of SNR.

SNR is actually the ratio of what is wanted (signal) to what is not wanted (noise). A high SNR means the signal is less corrupted by noise; a low SNR means the signal is more corrupted by noise.

Because SNR is the ratio of two powers, it is often described in decibel units, SNR_{dB}, defined as

$$\textbf{SNR}_{\textbf{dB}} = \textbf{10} \log_{10} \textbf{SNR}$$

Example 3.31

The power of a signal is 10 mW and the power of the noise is 1 μW; what are the values of SNR and SNR_{dB}?

Figure 3.30 *Two cases of SNR: a high SNR and a low SNR*

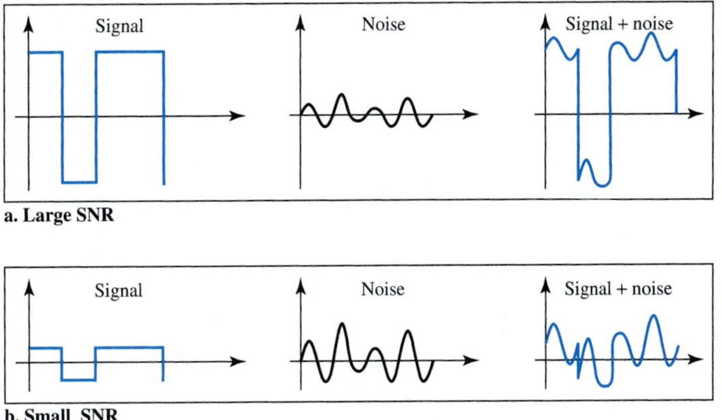

Solution

The values of SNR and SNR_{dB} can be calculated as follows:

$$SNR = \frac{10{,}000\ \mu W}{1\ mW} = 10{,}000$$
$$SNR_{dB} = 10 \log_{10} 10{,}000 = 10 \log_{10} 10^4 = 40$$

Example 3.32

The values of SNR and SNR_{dB} for a noiseless channel are

$$SNR = \frac{\text{signal power}}{0} = \infty$$
$$SNR_{dB} = 10 \log_{10} \infty = \infty$$

We can never achieve this ratio in real life; it is an ideal.

3.5 DATA RATE LIMITS

A very important consideration in data communications is how fast we can send data, in bits per second, over a channel. Data rate depends on three factors:

1. The bandwidth available
2. The level of the signals we use
3. The quality of the channel (the level of noise)

Two theoretical formulas were developed to calculate the data rate: one by Nyquist for a noiseless channel, another by Shannon for a noisy channel.

Noiseless Channel: Nyquist Bit Rate

For a noiseless channel, the **Nyquist bit rate** formula defines the theoretical maximum bit rate

$$\text{BitRate} = 2 \times \text{bandwidth} \times \log_2 L$$

In this formula, bandwidth is the bandwidth of the channel, L is the number of signal levels used to represent data, and BitRate is the bit rate in bits per second.

According to the formula, we might think that, given a specific bandwidth, we can have any bit rate we want by increasing the number of signal levels. Although the idea is theoretically correct, practically there is a limit. When we increase the number of signal levels, we impose a burden on the receiver. If the number of levels in a signal is just 2, the receiver can easily distinguish between a 0 and a 1. If the level of a signal is 64, the receiver must be very sophisticated to distinguish between 64 different levels. In other words, increasing the levels of a signal reduces the reliability of the system.

Increasing the levels of a signal may reduce the reliability of the system.

Example 3.33

Does the Nyquist theorem bit rate agree with the intuitive bit rate described in baseband transmission?

Solution

They match when we have only two levels. We said, in baseband transmission, the bit rate is 2 times the bandwidth if we use only the first harmonic in the worst case. However, the Nyquist formula is more general than what we derived intuitively; it can be applied to baseband transmission and modulation. Also, it can be applied when we have two or more levels of signals.

Example 3.34

Consider a noiseless channel with a bandwidth of 3000 Hz transmitting a signal with two signal levels. The maximum bit rate can be calculated as

$$\text{BitRate} = 2 \times 3000 \times \log_2 2 = 6000 \text{ bps}$$

Example 3.35

Consider the same noiseless channel transmitting a signal with four signal levels (for each level, we send 2 bits). The maximum bit rate can be calculated as

$$\text{BitRate} = 2 \times 3000 \times \log_2 4 = 12{,}000 \text{ bps}$$

Example 3.36

We need to send 265 kbps over a noiseless channel with a bandwidth of 20 kHz. How many signal levels do we need?

Solution

We can use the Nyquist formula as shown:

$$265{,}000 = 2 \times 20{,}000 \times \log_2 L$$
$$\log_2 L = 6.625 \qquad L = 2^{6.625} = 98.7 \text{ levels}$$

Since this result is not a power of 2, we need to either increase the number of levels or reduce the bit rate. If we have 128 levels, the bit rate is 280 kbps. If we have 64 levels, the bit rate is 240 kbps.

Noisy Channel: Shannon Capacity

In reality, we cannot have a noiseless channel; the channel is always noisy. In 1944, Claude Shannon introduced a formula, called the **Shannon capacity,** to determine the theoretical highest data rate for a noisy channel:

$$\textbf{Capacity} = \textbf{bandwidth} \times \log_2 (1 + \textbf{SNR})$$

In this formula, bandwidth is the bandwidth of the channel, SNR is the signal-to-noise ratio, and capacity is the capacity of the channel in bits per second. Note that in the Shannon formula there is no indication of the signal level, which means that no matter how many levels we have, we cannot achieve a data rate higher than the capacity of the channel. In other words, the formula defines a characteristic of the channel, not the method of transmission.

Example 3.37

Consider an extremely noisy channel in which the value of the signal-to-noise ratio is almost zero. In other words, the noise is so strong that the signal is faint. For this channel the capacity C is calculated as

$$C = B \log_2 (1 + \text{SNR}) = B \log_2 (1 + 0) = B \log_2 1 = B \times 0 = 0$$

This means that the capacity of this channel is zero regardless of the bandwidth. In other words, we cannot receive any data through this channel.

Example 3.38

We can calculate the theoretical highest bit rate of a regular telephone line. A telephone line normally has a bandwidth of 3000 Hz (300 to 3300 Hz) assigned for data communications. The signal-to-noise ratio is usually 3162. For this channel the capacity is calculated as

$$C = B \log_2 (1 + \text{SNR}) = 3000 \log_2 (1 + 3162) = 3000 \log_2 3163$$
$$= 3000 \times 11.62 = 34{,}860 \text{ bps}$$

This means that the highest bit rate for a telephone line is 34.860 kbps. If we want to send data faster than this, we can either increase the bandwidth of the line or improve the signal-to-noise ratio.

Example 3.39

The signal-to-noise ratio is often given in decibels. Assume that $SNR_{dB} = 36$ and the channel bandwidth is 2 MHz. The theoretical channel capacity can be calculated as

$$SNR_{dB} = 10 \log_{10} SNR \rightarrow SNR = 10^{SNR_{dB}/10} \rightarrow SNR = 10^{3.6} = 3981$$
$$C = B \log_2 (1 + SNR) = 2 \times 10^6 \times \log_2 3982 = 24 \text{ Mbps}$$

Example 3.40

For practical purposes, when the SNR is very high, we can assume that SNR + 1 is almost the same as SNR. In these cases, the theoretical channel capacity can be simplified to

$$C = B \times \frac{SNR_{dB}}{3}$$

For example, we can calculate the theoretical capacity of the previous example as

$$C = 2 \text{ MHz} \times \frac{36}{3} = 24 \text{ Mbps}$$

Using Both Limits

In practice, we need to use both methods to find the limits and signal levels. Let us show this with an example.

Example 3.41

We have a channel with a 1-MHz bandwidth. The SNR for this channel is 63. What are the appropriate bit rate and signal level?

Solution

First, we use the Shannon formula to find the upper limit.

$$C = B \log_2 (1 + SNR) = 10^6 \log_2 (1 + 63) = 10^6 \log_2 64 = 6 \text{ Mbps}$$

The Shannon formula gives us 6 Mbps, the upper limit. For better performance we choose something lower, 4 Mbps, for example. Then we use the Nyquist formula to find the number of signal levels.

$$4 \text{ Mbps} = 2 \times 1 \text{ MHz} \times \log_2 L \rightarrow L = 4$$

**The Shannon capacity gives us the upper limit;
the Nyquist formula tells us how many signal levels we need.**

3.6 PERFORMANCE

Up to now, we have discussed the tools of transmitting data (signals) over a network and how the data behave. One important issue in networking is the performance of the network—how good is it? We discuss quality of service, an overall measurement of network performance, in greater detail in Chapter 24. In this section, we introduce terms that we need for future chapters.

Bandwidth

One characteristic that measures network performance is bandwidth. However, the term can be used in two different contexts with two different measuring values: bandwidth in hertz and bandwidth in bits per second.

Bandwidth in Hertz

We have discussed this concept. Bandwidth in hertz is the range of frequencies contained in a composite signal or the range of frequencies a channel can pass. For example, we can say the bandwidth of a subscriber telephone line is 4 kHz.

Bandwidth in Bits per Seconds

The term *bandwidth* can also refer to the number of bits per second that a channel, a link, or even a network can transmit. For example, one can say the bandwidth of a Fast Ethernet network (or the links in this network) is a maximum of 100 Mbps. This means that this network can send 100 Mbps.

Relationship

There is an explicit relationship between the bandwidth in hertz and bandwidth in bits per seconds. Basically, an increase in bandwidth in hertz means an increase in bandwidth in bits per second. The relationship depends on whether we have baseband transmission or transmission with modulation. We discuss this relationship in Chapters 4 and 5.

In networking, we use the term *bandwidth* in two contexts.

- The first, *bandwidth in hertz,* refers to the range of frequencies in a composite signal or the range of frequencies that a channel can pass.
- The second, *bandwidth in bits per second,* refers to the speed of bit transmission in a channel or link.

Example 3.42

The bandwidth of a subscriber line is 4 kHz for voice or data. The bandwidth of this line for data transmission can be up to 56,000 bps using a sophisticated modem to change the digital signal to analog.

Example 3.43

If the telephone company improves the quality of the line and increases the bandwidth to 8 kHz, we can send 112,000 bps by using the same technology as mentioned in Example 3.42.

Throughput

The **throughput** is a measure of how fast we can actually send data through a network. Although, at first glance, bandwidth in bits per second and throughput seem the same, they are different. A link may have a bandwidth of B bps, but we can only send T bps through this link with T always less than B. In other words, the bandwidth is a potential measurement of a link; the throughput is an actual measurement of how fast we can send data. For example, we may have a link with a bandwidth of 1 Mbps, but the devices connected to the end of the link may handle only 200 kbps. This means that we cannot send more than 200 kbps through this link.

Imagine a highway designed to transmit 1000 cars per minute from one point to another. However, if there is congestion on the road, this figure may be reduced to 100 cars per minute. The bandwidth is 1000 cars per minute; the throughput is 100 cars per minute.

Example 3.44

A network with bandwidth of 10 Mbps can pass only an average of 12,000 frames per minute with each frame carrying an average of 10,000 bits. What is the throughput of this network?

Solution

We can calculate the throughput as

$$\text{Throughput} = \frac{12{,}000 \times 10{,}000}{60} = 2 \text{ Mbps}$$

The throughput is almost one-fifth of the bandwidth in this case.

Latency (Delay)

The latency or delay defines how long it takes for an entire message to completely arrive at the destination from the time the first bit is sent out from the source. We can say that latency is made of four components: **propagation time, transmission time, queuing time** and **processing delay.**

Latency = propagation time + transmission time + queuing time + processing delay

Propagation Time

Propagation time measures the time required for a bit to travel from the source to the destination. The propagation time is calculated by dividing the distance by the propagation speed.

$$\textbf{Propagation time} = \frac{\textbf{Distance}}{\textbf{Propagation speed}}$$

The propagation speed of electromagnetic signals depends on the medium and on the frequency of the signal. For example, in a vacuum, light is propagated with a speed of 3×10^8 m/s. It is lower in air; it is much lower in cable.

Example 3.45

What is the propagation time if the distance between the two points is 12,000 km? Assume the propagation speed to be 2.4×10^8 m/s in cable.

Solution

We can calculate the propagation time as

$$\text{Propagation time} = \frac{12{,}000 \times 1000}{2.4 \times 10^8} = 50 \text{ ms}$$

The example shows that a bit can go over the Atlantic Ocean in only 50 ms if there is a direct cable between the source and the destination.

Transmission Time

In data communications we don't send just 1 bit, we send a message. The first bit may take a time equal to the propagation time to reach its destination; the last bit also may take the same amount of time. However, there is a time between the first bit leaving the sender and the last bit arriving at the receiver. The first bit leaves earlier and arrives earlier; the last bit leaves later and arrives later. The time required for transmission of a message depends on the size of the message and the bandwidth of the channel.

$$\textbf{Transmission time} = \frac{\textbf{Message size}}{\textbf{Bandwidth}}$$

Example 3.46

What are the propagation time and the transmission time for a 2.5-kbyte message (an e-mail) if the bandwidth of the network is 1 Gbps? Assume that the distance between the sender and the receiver is 12,000 km and that light travels at 2.4×10^8 m/s.

Solution

We can calculate the propagation and transmission time as

$$\text{Propagation time} = \frac{12{,}000 \times 1000}{2.4 \times 10^8} = 50 \text{ ms}$$

$$\text{Transmission time} = \frac{2500 \times 8}{10^9} = 0.020 \text{ ms}$$

Note that in this case, because the message is short and the bandwidth is high, the dominant factor is the propagation time, not the transmission time. The transmission time can be ignored.

Example 3.47

What are the propagation time and the transmission time for a 5-Mbyte message (an image) if the bandwidth of the network is 1 Mbps? Assume that the distance between the sender and the receiver is 12,000 km and that light travels at 2.4×10^8 m/s.

Solution

We can calculate the propagation and transmission times as

$$\text{Propagation time} = \frac{12{,}000 \times 1000}{2.4 \times 10^8} = 50 \text{ ms}$$

$$\text{Transmission time} = \frac{5{,}000{,}000 \times 8}{10^6} = 40 \text{ s}$$

Note that in this case, because the message is very long and the bandwidth is not very high, the dominant factor is the transmission time, not the propagation time. The propagation time can be ignored.

Queuing Time

The third component in latency is the queuing time, the time needed for each intermediate or end device to hold the message before it can be processed. The queuing time is not a fixed factor; it changes with the load imposed on the network. When there is heavy traffic on the network, the queuing time increases. An intermediate device, such as a router, queues the arrived messages and processes them one by one. If there are many messages, each message will have to wait.

Bandwidth-Delay Product

Bandwidth and delay are two performance metrics of a link. However, as we will see in this chapter and future chapters, what is very important in data communications is the product of the two, the bandwidth-delay product. Let us elaborate on this issue, using two hypothetical cases as examples.

- **Case 1.** Figure 3.31 shows case 1.

Figure 3.31 *Filling the link with bits for case 1*

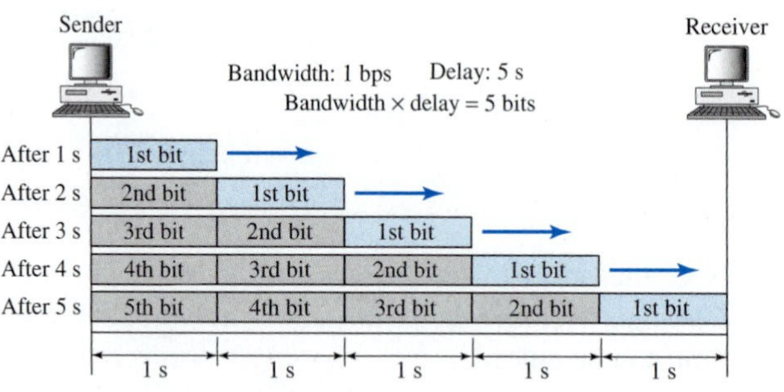

Let us assume that we have a link with a bandwidth of 1 bps (unrealistic, but good for demonstration purposes). We also assume that the delay of the link is 5 s (also unrealistic). We want to see what the bandwidth-delay product means in this case.

Looking at figure, we can say that this product 1×5 is the maximum number of bits that can fill the link. There can be no more than 5 bits at any time on the link.

- **Case 2.** Now assume we have a bandwidth of 4 bps. Figure 3.32 shows that there can be maximum $4 \times 5 = 20$ bits on the line. The reason is that, at each second, there are 4 bits on the line; the duration of each bit is 0.25 s.

Figure 3.32 *Filling the link with bits in case 2*

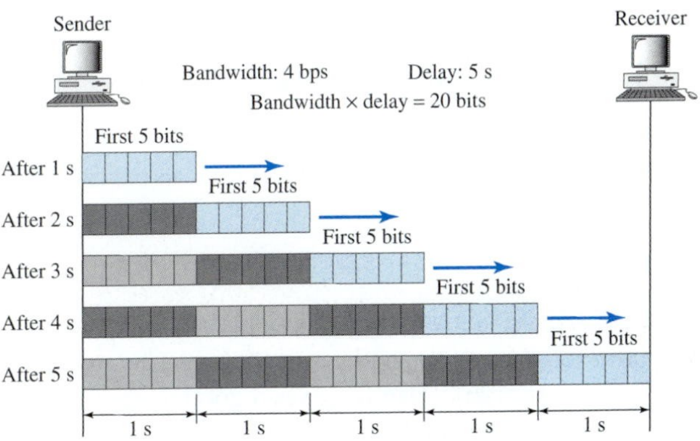

The above two cases show that the product of bandwidth and delay is the number of bits that can fill the link. This measurement is important if we need to send data in bursts and wait for the acknowledgment of each burst before sending the next one. To use the maximum capability of the link, we need to make the size of our burst 2 times the product of bandwidth and delay; we need to fill up the full-duplex channel (two directions). The sender should send a burst of data of ($2 \times$ bandwidth $\times$ delay) bits. The sender then waits for receiver acknowledgment for part of the burst before sending another burst. The amount $2 \times$ bandwidth $\times$ delay is the number of bits that can be in transition at any time.

The bandwidth-delay product defines the number of bits that can fill the link.

Example 3.48

We can think about the link between two points as a pipe. The cross section of the pipe represents the bandwidth, and the length of the pipe represents the delay. We can say the volume of the pipe defines the bandwidth-delay product, as shown in Figure 3.33.

Figure 3.33 *Concept of bandwidth-delay product*

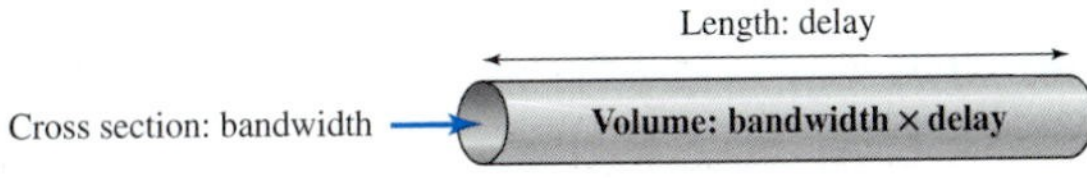

Jitter

Another performance issue that is related to delay is **jitter.** We can roughly say that jitter is a problem if different packets of data encounter different delays and the application using the data at the receiver site is time-sensitive (audio and video data, for example). If the delay for the first packet is 20 ms, for the second is 45 ms, and for the third is 40 ms, then the real-time application that uses the packets endures jitter. We discuss jitter in greater detail in Chapter 29.

3.7 RECOMMENDED READING

For more details about subjects discussed in this chapter, we recommend the following books. The items in brackets [. . .] refer to the reference list at the end of the text.

Books

Data and signals are elegantly discussed in Chapters 1 to 6 of [Pea92]. [Cou01] gives an excellent coverage about signals in Chapter 2. More advanced materials can be found in [Ber96]. [Hsu03] gives a good mathematical approach to signaling. Complete coverage of Fourier Analysis can be found in [Spi74]. Data and signals are discussed in Chapter 3 of [Sta04] and Section 2.1 of [Tan03].

3.8 KEY TERMS

analog
analog data
analog signal
attenuation
bandpass channel
bandwidth
baseband transmission
bit rate
bits per second (bps)
broadband transmission
composite signal
cycle
decibel (dB)
digital
digital data
digital signal
distortion
Fourier analysis
frequency
frequency-domain
fundamental frequency
harmonic
Hertz (Hz)
jitter
low-pass channel
noise
nonperiodic signal
Nyquist bit rate
peak amplitude
period
periodic signal
phase
processing delay
propagation speed

propagation time
queuing time
Shannon capacity
signal
signal-to-noise ratio (SNR)
sine wave
throughput
time-domain
transmission time
wavelength

3.9 SUMMARY

- Data must be transformed to electromagnetic signals to be transmitted.
- Data can be analog or digital. Analog data are continuous and take continuous values. Digital data have discrete states and take discrete values.
- Signals can be analog or digital. Analog signals can have an infinite number of values in a range; digital signals can have only a limited number of values.
- In data communications, we commonly use periodic analog signals and nonperiodic digital signals.
- Frequency and period are the inverse of each other.
- Frequency is the rate of change with respect to time.
- Phase describes the position of the waveform relative to time 0.
- A complete sine wave in the time domain can be represented by one single spike in the frequency domain.
- A single-frequency sine wave is not useful in data communications; we need to send a composite signal, a signal made of many simple sine waves.
- According to Fourier analysis, any composite signal is a combination of simple sine waves with different frequencies, amplitudes, and phases.
- The bandwidth of a composite signal is the difference between the highest and the lowest frequencies contained in that signal.
- A digital signal is a composite analog signal with an infinite bandwidth.
- Baseband transmission of a digital signal that preserves the shape of the digital signal is possible only if we have a low-pass channel with an infinite or very wide bandwidth.
- If the available channel is a bandpass channel, we cannot send a digital signal directly to the channel; we need to convert the digital signal to an analog signal before transmission.
- For a noiseless channel, the Nyquist bit rate formula defines the theoretical maximum bit rate. For a noisy channel, we need to use the Shannon capacity to find the maximum bit rate.
- Attenuation, distortion, and noise can impair a signal.
- Attenuation is the loss of a signal's energy due to the resistance of the medium.
- Distortion is the alteration of a signal due to the differing propagation speeds of each of the frequencies that make up a signal.
- Noise is the external energy that corrupts a signal.
- The bandwidth-delay product defines the number of bits that can fill the link.

3.10 PRACTICE SET

Review Questions

1. What is the relationship between period and frequency?
2. What does the amplitude of a signal measure? What does the frequency of a signal measure? What does the phase of a signal measure?
3. How can a composite signal be decomposed into its individual frequencies?
4. Name three types of transmission impairment.
5. Distinguish between baseband transmission and broadband transmission.
6. Distinguish between a low-pass channel and a band-pass channel.
7. What does the Nyquist theorem have to do with communications?
8. What does the Shannon capacity have to do with communications?
9. Why do optical signals used in fiber optic cables have a very short wave length?
10. Can we say if a signal is periodic or nonperiodic by just looking at its frequency domain plot? How?
11. Is the frequency domain plot of a voice signal discrete or continuous?
12. Is the frequency domain plot of an alarm system discrete or continuous?
13. We send a voice signal from a microphone to a recorder. Is this baseband or broadband transmission?
14. We send a digital signal from one station on a LAN to another station. Is this baseband or broadband transmission?
15. We modulate several voice signals and send them through the air. Is this baseband or broadband transmission?

Exercises

16. Given the frequencies listed below, calculate the corresponding periods.
    a. 24 Hz
    b. 8 MHz
    c. 140 KHz
17. Given the following periods, calculate the corresponding frequencies.
    a. 5 s
    b. 12 μs
    c. 220 ns
18. What is the phase shift for the following?
    a. A sine wave with the maximum amplitude at time zero
    b. A sine wave with maximum amplitude after 1/4 cycle
    c. A sine wave with zero amplitude after 3/4 cycle and increasing
19. What is the bandwidth of a signal that can be decomposed into five sine waves with frequencies at 0, 20, 50, 100, and 200 Hz? All peak amplitudes are the same. Draw the bandwidth.

20. A periodic composite signal with a bandwidth of 2000 Hz is composed of two sine waves. The first one has a frequency of 100 Hz with a maximum amplitude of 20 V; the second one has a maximum amplitude of 5 V. Draw the bandwidth.
21. Which signal has a wider bandwidth, a sine wave with a frequency of 100 Hz or a sine wave with a frequency of 200 Hz?
22. What is the bit rate for each of the following signals?
    a. A signal in which 1 bit lasts 0.001 s
    b. A signal in which 1 bit lasts 2 ms
    c. A signal in which 10 bits last 20 μs
23. A device is sending out data at the rate of 1000 bps.
    a. How long does it take to send out 10 bits?
    b. How long does it take to send out a single character (8 bits)?
    c. How long does it take to send a file of 100,000 characters?
24. What is the bit rate for the signal in Figure 3.34?

Figure 3.34 *Exercise 24*

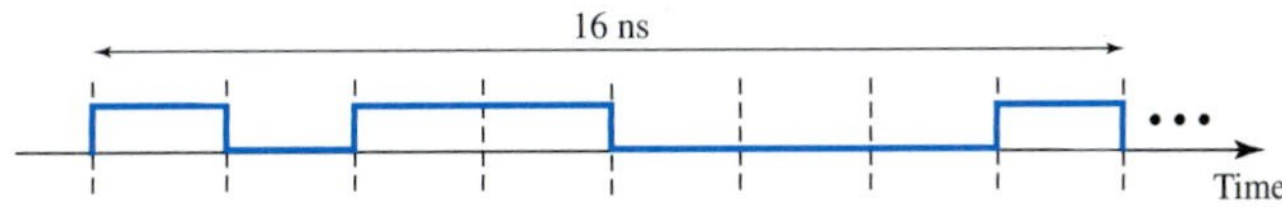

25. What is the frequency of the signal in Figure 3.35?

Figure 3.35 *Exercise 25*

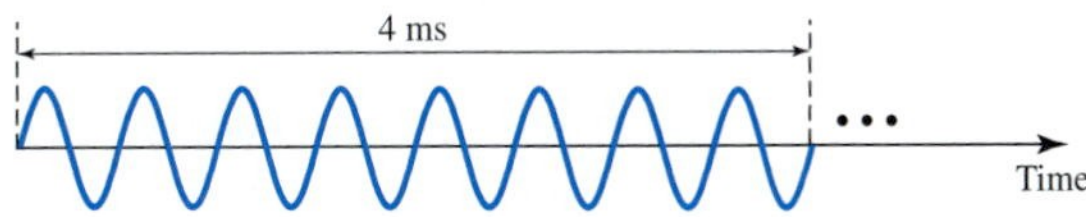

26. What is the bandwidth of the composite signal shown in Figure 3.36.

Figure 3.36 *Exercise 26*

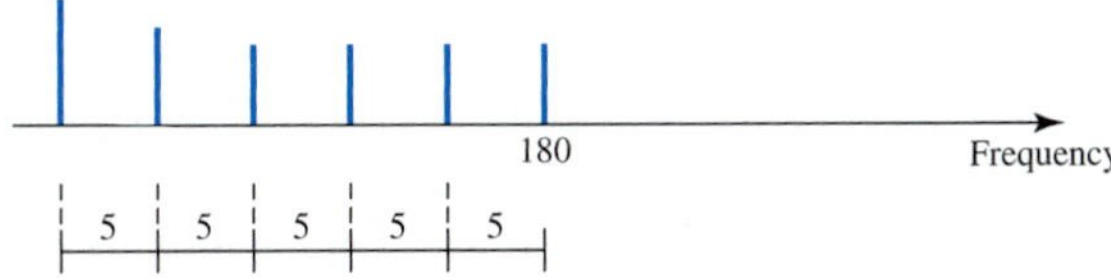

27. A periodic composite signal contains frequencies from 10 to 30 KHz, each with an amplitude of 10 V. Draw the frequency spectrum.
28. A non-periodic composite signal contains frequencies from 10 to 30 KHz. The peak amplitude is 10 V for the lowest and the highest signals and is 30 V for the 20-KHz signal. Assuming that the amplitudes change gradually from the minimum to the maximum, draw the frequency spectrum.
29. A TV channel has a bandwidth of 6 MHz. If we send a digital signal using one channel, what are the data rates if we use one harmonic, three harmonics, and five harmonics?
30. A signal travels from point A to point B. At point A, the signal power is 100 W. At point B, the power is 90 W. What is the attenuation in decibels?
31. The attenuation of a signal is –10 dB. What is the final signal power if it was originally 5 W?
32. A signal has passed through three cascaded amplifiers, each with a 4 dB gain. What is the total gain? How much is the signal amplified?
33. If the bandwidth of the channel is 5 Kbps, how long does it take to send a frame of 100,000 bits out of this device?
34. The light of the sun takes approximately eight minutes to reach the earth. What is the distance between the sun and the earth?
35. A signal has a wavelength of 1 μm in air. How far can the front of the wave travel during 1000 periods?
36. A line has a signal-to-noise ratio of 1000 and a bandwidth of 4000 KHz. What is the maximum data rate supported by this line?
37. We measure the performance of a telephone line (4 KHz of bandwidth). When the signal is 10 V, the noise is 5 mV. What is the maximum data rate supported by this telephone line?
38. A file contains 2 million bytes. How long does it take to download this file using a 56-Kbps channel? 1-Mbps channel?
39. A computer monitor has a resolution of 1200 by 1000 pixels. If each pixel uses 1024 colors, how many bits are needed to send the complete contents of a screen?
40. A signal with 200 milliwatts power passes through 10 devices, each with an average noise of 2 microwatts. What is the SNR? What is the SNR_{dB}?
41. If the peak voltage value of a signal is 20 times the peak voltage value of the noise, what is the SNR? What is the SNR_{dB}?
42. What is the theoretical capacity of a channel in each of the following cases:
    a. Bandwidth: 20 KHz $SNR_{dB} = 40$
    b. Bandwidth: 200 KHz $SNR_{dB} = 4$
    c. Bandwidth: 1 MHz $SNR_{dB} = 20$
43. We need to upgrade a channel to a higher bandwidth. Answer the following questions:
    a. How is the rate improved if we double the bandwidth?
    b. How is the rate improved if we double the SNR?

44. We have a channel with 4 KHz bandwidth. If we want to send data at 100 Kbps, what is the minimum SNR_{dB}? What is SNR?
45. What is the transmission time of a packet sent by a station if the length of the packet is 1 million bytes and the bandwidth of the channel is 200 Kbps?
46. What is the length of a bit in a channel with a propagation speed of 2×10^8 m/s if the channel bandwidth is
    a. 1 Mbps?
    b. 10 Mbps?
    c. 100 Mbps?
47. How many bits can fit on a link with a 2 ms delay if the bandwidth of the link is
    a. 1 Mbps?
    b. 10 Mbps?
    c. 100 Mbps?
48. What is the total delay (latency) for a frame of size 5 million bits that is being sent on a link with 10 routers each having a queuing time of 2 μs and a processing time of 1 μs. The length of the link is 2000 Km. The speed of light inside the link is 2×10^8 m/s. The link has a bandwidth of 5 Mbps. Which component of the total delay is dominant? Which one is negligible?

CHAPTER 4

Digital Transmission

A computer network is designed to send information from one point to another. This information needs to be converted to either a digital signal or an analog signal for transmission. In this chapter, we discuss the first choice, conversion to digital signals; in Chapter 5, we discuss the second choice, conversion to analog signals.

We discussed the advantages and disadvantages of digital transmission over analog transmission in Chapter 3. In this chapter, we show the schemes and techniques that we use to transmit data digitally. First, we discuss **digital-to-digital conversion** techniques, methods which convert digital data to digital signals. Second, we discuss **analog-to-digital conversion** techniques, methods which change an analog signal to a digital signal. Finally, we discuss **transmission modes.**

4.1 DIGITAL-TO-DIGITAL CONVERSION

In Chapter 3, we discussed data and signals. We said that data can be either digital or analog. We also said that signals that represent data can also be digital or analog. In this section, we see how we can represent digital data by using digital signals. The conversion involves three techniques: line coding, block coding, and scrambling. Line coding is always needed; block coding and scrambling may or may not be needed.

Line Coding

Line coding is the process of converting digital data to digital signals. We assume that data, in the form of text, numbers, graphical images, audio, or video, are stored in computer memory as sequences of bits (see Chapter 1). Line coding converts a sequence of bits to a digital signal. At the sender, digital data are encoded into a digital signal; at the receiver, the digital data are recreated by decoding the digital signal. Figure 4.1 shows the process.

Characteristics

Before discussing different line coding schemes, we address their common characteristics.

Figure 4.1 *Line coding and decoding*

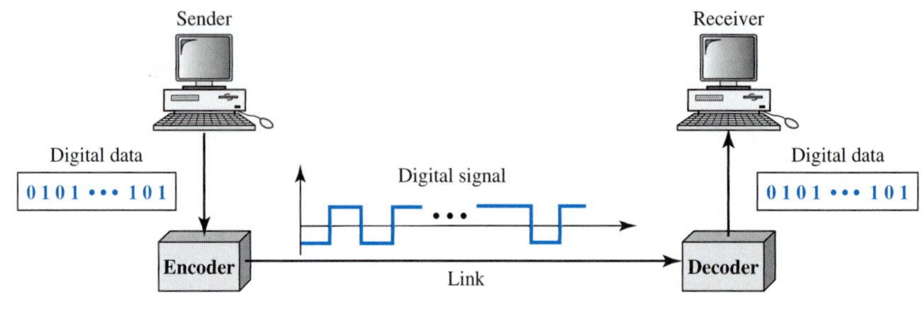

Signal Element Versus Data Element Let us distinguish between a **data element** and a **signal element.** In data communications, our goal is to send data elements. A data element is the smallest entity that can represent a piece of information: this is the bit. In digital data communications, a signal element carries data elements. A signal element is the shortest unit (timewise) of a digital signal. In other words, data elements are what we need to send; signal elements are what we can send. Data elements are being carried; signal elements are the carriers.

We define a ratio r which is the number of data elements carried by each signal element. Figure 4.2 shows several situations with different values of r.

Figure 4.2 *Signal element versus data element*

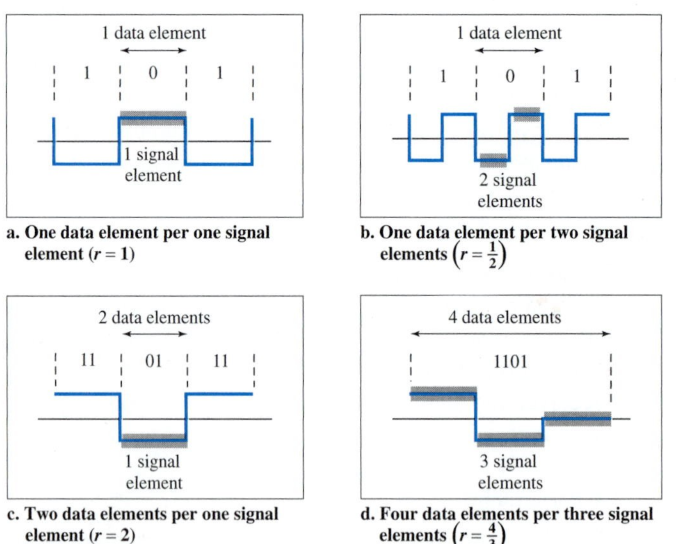

In part a of the figure, one data element is carried by one signal element ($r = 1$). In part b of the figure, we need two signal elements (two transitions) to carry each data element ($r = \frac{1}{2}$). We will see later that the extra signal element is needed to guarantee synchronization. In part c of the figure, a signal element carries two data elements ($r = 2$). Finally, in part d, a group of 4 bits is being carried by a group of three signal elements ($r = \frac{4}{3}$). For every line coding scheme we discuss, we will give the value of r.

An analogy may help here. Suppose each data element is a person who needs to be carried from one place to another. We can think of a signal element as a vehicle that can carry people. When $r = 1$, it means each person is driving a vehicle. When $r > 1$, it means more than one person is travelling in a vehicle (a carpool, for example). We can also have the case where one person is driving a car and a trailer ($r = \frac{1}{2}$).

Data Rate Versus Signal Rate The **data rate** defines the number of data elements (bits) sent in 1s. The unit is bits per second (bps). The **signal rate** is the number of signal elements sent in 1s. The unit is the baud. There are several common terminologies used in the literature. The data rate is sometimes called the **bit rate;** the signal rate is sometimes called the **pulse rate,** the **modulation rate,** or the **baud rate.**

One goal in data communications is to increase the data rate while decreasing the signal rate. Increasing the data rate increases the speed of transmission; decreasing the signal rate decreases the bandwidth requirement. In our vehicle-people analogy, we need to carry more people in fewer vehicles to prevent traffic jams. We have a limited *bandwidth* in our transportation system.

We now need to consider the relationship between data rate and signal rate (bit rate and baud rate). This relationship, of course, depends on the value of r. It also depends on the data pattern. If we have a data pattern of all 1s or all 0s, the signal rate may be different from a data pattern of alternating 0s and 1s. To derive a formula for the relationship, we need to define three cases: the worst, best, and average. The worst case is when we need the maximum signal rate; the best case is when we need the minimum. In data communications, we are usually interested in the average case. We can formulate the relationship between data rate and signal rate as

$$S = c \times N \times \frac{1}{r} \quad \text{baud}$$

where N is the data rate (bps); c is the case factor, which varies for each case; S is the number of signal elements; and r is the previously defined factor.

Example 4.1

A signal is carrying data in which one data element is encoded as one signal element ($r = 1$). If the bit rate is 100 kbps, what is the average value of the baud rate if c is between 0 and 1?

Solution

We assume that the average value of c is $\frac{1}{2}$. The baud rate is then

$$S = c \times N \times \frac{1}{r} = \frac{1}{2} \times 100{,}000 \times \frac{1}{1} = 50{,}000 = 50 \text{ kbaud}$$

Bandwidth We discussed in Chapter 3 that a digital signal that carries information is nonperiodic. We also showed that the bandwidth of a nonperiodic signal is continuous with an infinite range. However, most digital signals we encounter in real life have a bandwidth with finite values. In other words, the bandwidth is theoretically infinite, but many of the components have such a small amplitude that they can be ignored. The effective bandwidth is finite. From now on, when we talk about the bandwidth of a digital signal, we need to remember that we are talking about this effective bandwidth.

Although the actual bandwidth of a digital signal is infinite, the effective bandwidth is finite.

We can say that the baud rate, not the bit rate, determines the required bandwidth for a digital signal. If we use the transportation analogy, the number of vehicles affects the traffic, not the number of people being carried. More changes in the signal mean injecting more frequencies into the signal. (Recall that frequency means change and change means frequency.) The bandwidth reflects the range of frequencies we need. There is a relationship between the baud rate (signal rate) and the bandwidth. Bandwidth is a complex idea. When we talk about the bandwidth, we normally define a range of frequencies. We need to know where this range is located as well as the values of the lowest and the highest frequencies. In addition, the amplitude (if not the phase) of each component is an important issue. In other words, we need more information about the bandwidth than just its value; we need a diagram of the bandwidth. We will show the bandwidth for most schemes we discuss in the chapter. For the moment, we can say that the bandwidth (range of frequencies) is proportional to the signal rate (baud rate). The minimum bandwidth can be given as

$$B_{min} = c \times N \times \frac{1}{r}$$

We can solve for the maximum data rate if the bandwidth of the channel is given.

$$N_{max} = \frac{1}{c} \times B \times r$$

Example 4.2

The maximum data rate of a channel (see Chapter 3) is $N_{max} = 2 \times B \times \log_2 L$ (defined by the Nyquist formula). Does this agree with the previous formula for N_{max}?

Solution

A signal with L levels actually can carry $\log_2 L$ bits per level. If each level corresponds to one signal element and we assume the average case ($c = \frac{1}{2}$), then we have

$$N_{max} = \frac{1}{c} \times B \times r = 2 \times B \times \log_2 L$$

Baseline Wandering In decoding a digital signal, the receiver calculates a running average of the received signal power. This average is called the ***baseline.*** The incoming signal power is evaluated against this baseline to determine the value of the data element. A long string of 0s or 1s can cause a drift in the baseline (**baseline wandering**) and make it difficult for the receiver to decode correctly. A good line coding scheme needs to prevent baseline wandering.

DC Components When the voltage level in a digital signal is constant for a while, the spectrum creates very low frequencies (results of Fourier analysis). These frequencies around zero, called DC (direct-current) *components,* present problems for a system that cannot pass low frequencies or a system that uses electrical coupling (via a transformer). For example, a telephone line cannot pass frequencies below 200 Hz. Also a long-distance link may use one or more transformers to isolate different parts of the line electrically. For these systems, we need a scheme with no **DC component.**

Self-synchronization To correctly interpret the signals received from the sender, the receiver's bit intervals must correspond exactly to the sender's bit intervals. If the receiver clock is faster or slower, the bit intervals are not matched and the receiver might misinterpret the signals. Figure 4.3 shows a situation in which the receiver has a shorter bit duration. The sender sends 10110001, while the receiver receives 110111000011.

Figure 4.3 *Effect of lack of synchronization*

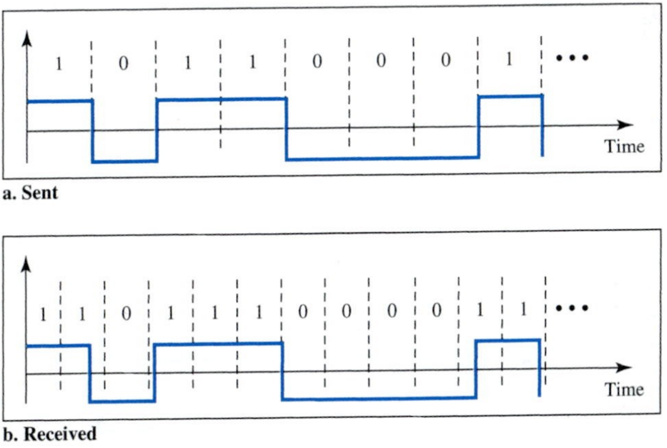

A **self-synchronizing** digital signal includes timing information in the data being transmitted. This can be achieved if there are transitions in the signal that alert the receiver to the beginning, middle, or end of the pulse. If the receiver's clock is out of synchronization, these points can reset the clock.

Example 4.3

In a digital transmission, the receiver clock is 0.1 percent faster than the sender clock. How many extra bits per second does the receiver receive if the data rate is 1 kbps? How many if the data rate is 1 Mbps?

Solution

At 1 kbps, the receiver receives 1001 bps instead of 1000 bps.

1000 bits sent	1001 bits received	1 extra bps

At 1 Mbps, the receiver receives 1,001,000 bps instead of 1,000,000 bps.

1,000,000 bits sent	1,001,000 bits received	1000 extra bps

Built-in Error Detection It is desirable to have a built-in error-detecting capability in the generated code to detect some of or all the errors that occurred during transmission. Some encoding schemes that we will discuss have this capability to some extent.

Immunity to Noise and Interference Another desirable code characteristic is a code that is immune to noise and other interferences. Some encoding schemes that we will discuss have this capability.

Complexity A complex scheme is more costly to implement than a simple one. For example, a scheme that uses four signal levels is more difficult to interpret than one that uses only two levels.

Line Coding Schemes

We can roughly divide line coding schemes into five broad categories, as shown in Figure 4.4.

Figure 4.4 *Line coding schemes*

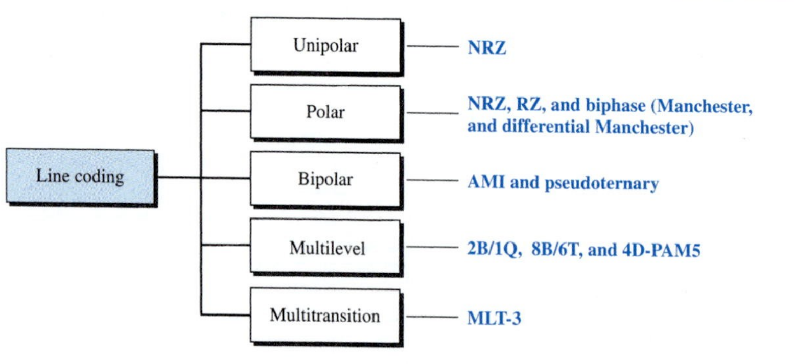

There are several schemes in each category. We need to be familiar with all schemes discussed in this section to understand the rest of the book. This section can be used as a reference for schemes encountered later.

Unipolar Scheme

In a unipolar scheme, all the signal levels are on one side of the time axis, either above or below.

NRZ (Non-Return-to-Zero) Traditionally, a unipolar scheme was designed as a **non-return-to-zero (NRZ)** scheme in which the positive voltage defines bit 1 and the zero voltage defines bit 0. It is called NRZ because the signal does not return to zero at the middle of the bit. Figure 4.5 show a unipolar NRZ scheme.

Figure 4.5 *Unipolar NRZ scheme*

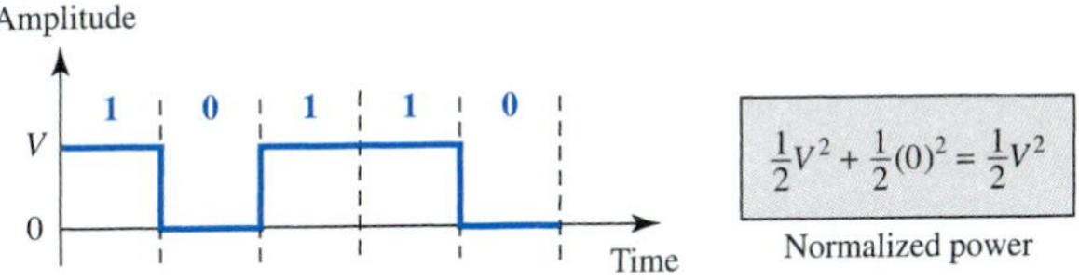

Compared with its polar counterpart (see the next section), this scheme is very costly. As we will see shortly, the normalized power (power needed to send 1 bit per unit line resistance) is double that for polar NRZ. For this reason, this scheme is normally not used in data communications today.

Polar Schemes

In polar schemes, the voltages are on the both sides of the time axis. For example, the voltage level for 0 can be positive and the voltage level for 1 can be negative.

Non-Return-to-Zero (NRZ) In **polar NRZ** encoding, we use two levels of voltage amplitude. We can have two versions of polar NRZ: NRZ-L and NRZ-I, as shown in Figure 4.6. The figure also shows the value of *r*, the average baud rate, and the bandwidth. In the first variation, NRZ-L (**NRZ-Level**), the level of the voltage determines the value of the bit. In the second variation, NRZ-I (**NRZ-Invert**), the change or lack of change in the level of the voltage determines the value of the bit. If there is no change, the bit is 0; if there is a change, the bit is 1.

Figure 4.6 *Polar NRZ-L and NRZ-I schemes*

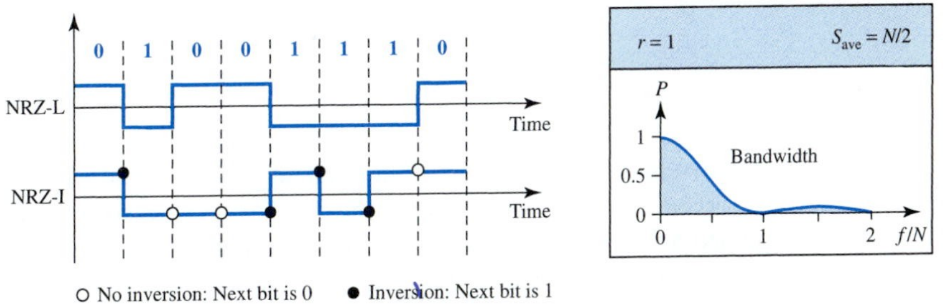

In NRZ-L the level of the voltage determines the value of the bit. In NRZ-I the inversion or the lack of inversion determines the value of the bit.

Let us compare these two schemes based on the criteria we previously defined. Although baseline wandering is a problem for both variations, it is twice as severe in NRZ-L. If there is a long sequence of 0s or 1s in NRZ-L, the average signal power becomes skewed. The receiver might have difficulty discerning the bit value. In NRZ-I this problem occurs only for a long sequence of 0s. If somehow we can eliminate the long sequence of 0s, we can avoid baseline wandering. We will see shortly how this can be done.

The synchronization problem (sender and receiver clocks are not synchronized) also exists in both schemes. Again, this problem is more serious in NRZ-L than in NRZ-I. While a long sequence of 0s can cause a problem in both schemes, a long sequence of 1s affects only NRZ-L.

Another problem with NRZ-L occurs when there is a sudden change of polarity in the system. For example, if twisted-pair cable is the medium, a change in the polarity of the wire results in all 0s interpreted as 1s and all 1s interpreted as 0s. NRZ-I does not have this problem. Both schemes have an average signal rate of $N/2$ Bd.

NRZ-L and NRZ-I both have an average signal rate of $N/2$ Bd.

Let us discuss the bandwidth. Figure 4.6 also shows the normalized bandwidth for both variations. The vertical axis shows the power density (the power for each 1 Hz of bandwidth); the horizontal axis shows the frequency. The bandwidth reveals a very serious problem for this type of encoding. The value of the power density is very high around frequencies close to zero. This means that there are DC components that carry a high level of energy. As a matter of fact, most of the energy is concentrated in frequencies between 0 and $N/2$. This means that although the average of the signal rate is $N/2$, the energy is not distributed evenly between the two halves.

NRZ-L and NRZ-I both have a DC component problem.

Example 4.4

A system is using NRZ-I to transfer 10-Mbps data. What are the average signal rate and minimum bandwidth?

Solution

The average signal rate is $S = N/2 = 500$ kbaud. The minimum bandwidth for this average baud rate is $B_{min} = S = 500$ kHz.

Return to Zero (RZ) The main problem with NRZ encoding occurs when the sender and receiver clocks are not synchronized. The receiver does not know when one bit has ended and the next bit is starting. One solution is the **return-to-zero (RZ)** scheme, which uses three values: positive, negative, and zero. In RZ, the signal changes not between bits but during the bit. In Figure 4.7 we see that the signal goes to 0 in the middle of each bit. It remains there until the beginning of the next bit. The main disadvantage of RZ encoding is that it requires two signal changes to encode a bit and therefore occupies greater bandwidth. The same problem we mentioned, a sudden change of polarity resulting in all 0s interpreted as 1s and all 1s interpreted as 0s, still exist here, but there is no DC component problem. Another problem is the complexity: RZ uses three levels of voltage, which is more complex to create and discern. As a result of all these deficiencies, the scheme is not used today. Instead, it has been replaced by the better-performing Manchester and differential Manchester schemes (discussed next).

Figure 4.7 *Polar RZ scheme*

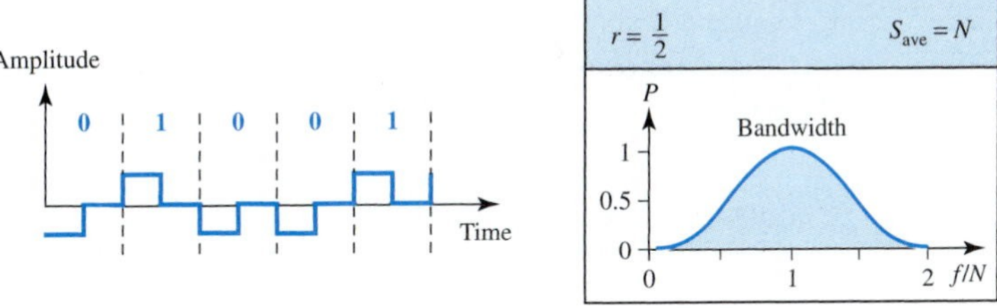

Biphase: Manchester and Differential Manchester The idea of RZ (transition at the middle of the bit) and the idea of NRZ-L are combined into the **Manchester** scheme. In Manchester encoding, the duration of the bit is divided into two halves. The voltage remains at one level during the first half and moves to the other level in the second half. The transition at the middle of the bit provides synchronization. **Differential Manchester,** on the other hand, combines the ideas of RZ and NRZ-I. There is always a transition at the middle of the bit, but the bit values are determined at the beginning of the bit. If the next bit is 0, there is a transition; if the next bit is 1, there is none. Figure 4.8 shows both Manchester and differential Manchester encoding.

Figure 4.8 *Polar biphase: Manchester and differential Manchester schemes*

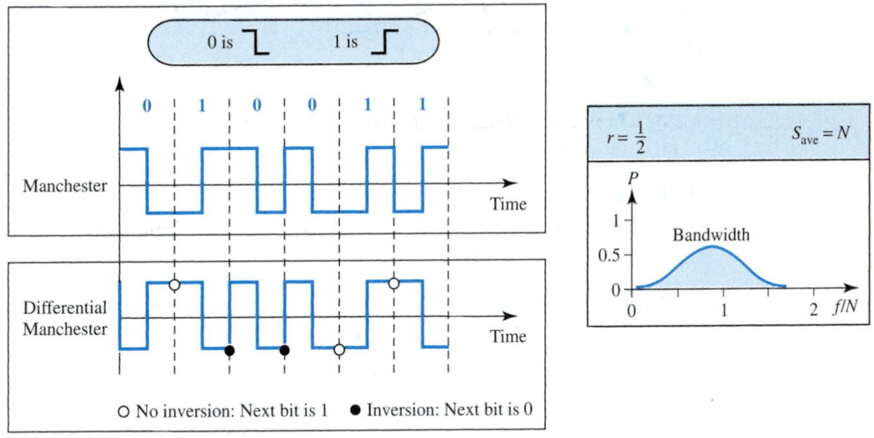

In Manchester and differential Manchester encoding, the transition at the middle of the bit is used for synchronization.

The Manchester scheme overcomes several problems associated with NRZ-L, and differential Manchester overcomes several problems associated with NRZ-I. First, there is no baseline wandering. There is no DC component because each bit has a positive and negative voltage contribution. The only drawback is the signal rate. The signal rate for Manchester and differential Manchester is double that for NRZ. The reason is that there is always one transition at the middle of the bit and maybe one transition at the end of each bit. Figure 4.8 shows both Manchester and differential Manchester encoding schemes. Note that Manchester and differential Manchester schemes are also called **biphase** schemes.

The minimum bandwidth of Manchester and differential Manchester is 2 times that of NRZ.

Bipolar Schemes

In **bipolar** encoding (sometimes called ***multilevel binary***), there are three voltage levels: positive, negative, and zero. The voltage level for one data element is at zero, while the voltage level for the other element alternates between positive and negative.

In bipolar encoding, we use three levels: positive, zero, and negative.

AMI and Pseudoternary Figure 4.9 shows two variations of bipolar encoding: AMI and pseudoternary. A common bipolar encoding scheme is called bipolar **alternate mark inversion (AMI).** In the term *alternate mark inversion,* the word *mark* comes from telegraphy and means 1. So AMI means alternate 1 inversion. A neutral zero voltage represents binary 0. Binary 1s are represented by alternating positive and negative voltages. A variation of AMI encoding is called **pseudoternary** in which the 1 bit is encoded as a zero voltage and the 0 bit is encoded as alternating positive and negative voltages.

Figure 4.9 *Bipolar schemes: AMI and pseudoternary*

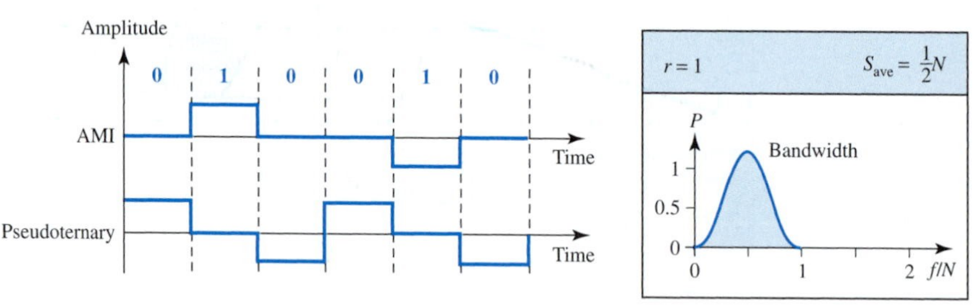

The bipolar scheme was developed as an alternative to NRZ. The bipolar scheme has the same signal rate as NRZ, but there is no DC component. The NRZ scheme has most of its energy concentrated near zero frequency, which makes it unsuitable for transmission over channels with poor performance around this frequency. The concentration of the energy in bipolar encoding is around frequency $N/2$. Figure 4.9 shows the typical energy concentration for a bipolar scheme.

One may ask why we do not have DC component in bipolar encoding. We can answer this question by using the Fourier transform, but we can also think about it intuitively. If we have a long sequence of 1s, the voltage level alternates between positive and negative; it is not constant. Therefore, there is no DC component. For a long sequence of 0s, the voltage remains constant, but its amplitude is zero, which is the same as having no DC component. In other words, a sequence that creates a constant zero voltage does not have a DC component.

AMI is commonly used for long-distance communication, but it has a synchronization problem when a long sequence of 0s is present in the data. Later in the chapter, we will see how a scrambling technique can solve this problem.

Multilevel Schemes

The desire to increase the data speed or decrease the required bandwidth has resulted in the creation of many schemes. The goal is to increase the number of bits per baud by encoding a pattern of m data elements into a pattern of n signal elements. We only have two types of data elements (0s and 1s), which means that a group of m data elements can produce a combination of 2^m data patterns. We can have different types of signal elements by allowing different signal levels. If we have L different levels, then we can produce L^n combinations of signal patterns. If $2^m = L^n$, then each data pattern is encoded into one signal pattern. If $2^m < L^n$, data patterns occupy only a subset of signal patterns. The subset can be carefully designed to prevent baseline wandering, to provide synchronization, and to detect errors that occurred during data transmission. Data encoding is not possible if $2^m > L^n$ because some of the data patterns cannot be encoded.

The code designers have classified these types of coding as *mBnL*, where m is the length of the binary pattern, B means binary data, n is the length of the signal pattern, and L is the number of levels in the signaling. A letter is often used in place of L: B (binary) for $L = 2$, T (ternary) for $L = 3$, and Q (quaternary) for $L = 4$. Note that the first two letters define the data pattern, and the second two define the signal pattern.

In *mBnL* schemes, a pattern of m data elements is encoded as a pattern of n signal elements in which $2^m \leq L^n$.

2B1Q The first *mBnL* scheme we discuss, **two binary, one quaternary (2B1Q),** uses data patterns of size 2 and encodes the 2-bit patterns as one signal element belonging to a four-level signal. In this type of encoding $m = 2$, $n = 1$, and $L = 4$ (quaternary). Figure 4.10 shows an example of a 2B1Q signal.

The average signal rate of 2B1Q is $S = N/4$. This means that using 2B1Q, we can send data 2 times faster than by using NRZ-L. However, 2B1Q uses four different signal levels, which means the receiver has to discern four different thresholds. The reduced bandwidth comes with a price. There are no redundant signal patterns in this scheme because $2^2 = 4^1$.

As we will see in Chapter 9, 2B1Q is used in DSL (Digital Subscriber Line) technology to provide a high-speed connection to the Internet by using subscriber telephone lines.

Figure 4.10 *Multilevel: 2B1Q scheme*

	Previous level: positive	Previous level: negative
Next bits	Next level	Next level
00	+1	−1
01	+3	−3
10	−1	+1
11	−3	+3

Transition table

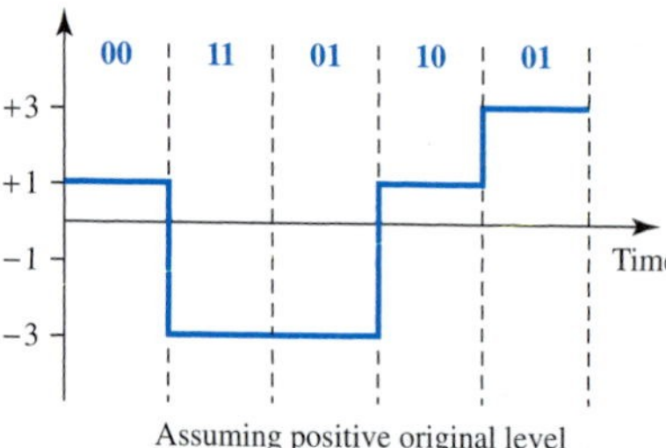

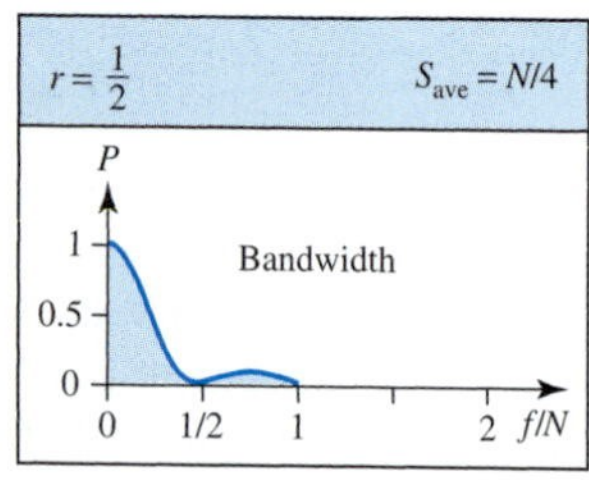

8B6T A very interesting scheme is **eight binary, six ternary (8B6T).** This code is used with 100BASE-4T cable, as we will see in Chapter 13. The idea is to encode a pattern of 8 bits as a pattern of 6 signal elements, where the signal has three levels (ternary). In this type of scheme, we can have $2^8 = 256$ different data patterns and $3^6 = 478$ different signal patterns. The mapping table is shown in Appendix D. There are $478 - 256 = 222$ redundant signal elements that provide synchronization and error detection. Part of the redundancy is also used to provide DC balance. Each signal pattern has a weight of 0 or +1 DC values. This means that there is no pattern with the weight −1. To make the whole stream DC-balanced, the sender keeps track of the weight. If two groups of weight 1 are encountered one after another, the first one is sent as is, while the next one is totally inverted to give a weight of −1.

Figure 4.11 shows an example of three data patterns encoded as three signal patterns. The three possible signal levels are represented as –, 0, and +. The first 8-bit pattern 00010001 is encoded as the signal pattern –0–0++ with weight 0; the second 8-bit pattern 01010011 is encoded as – + – + + 0 with weight +1. The third bit pattern should be encoded as + – – + 0 + with weight +1. To create DC balance, the sender inverts the actual signal. The receiver can easily recognize that this is an inverted pattern because the weight is –1. The pattern is inverted before decoding.

Figure 4.11 *Multilevel: 8B6T scheme*

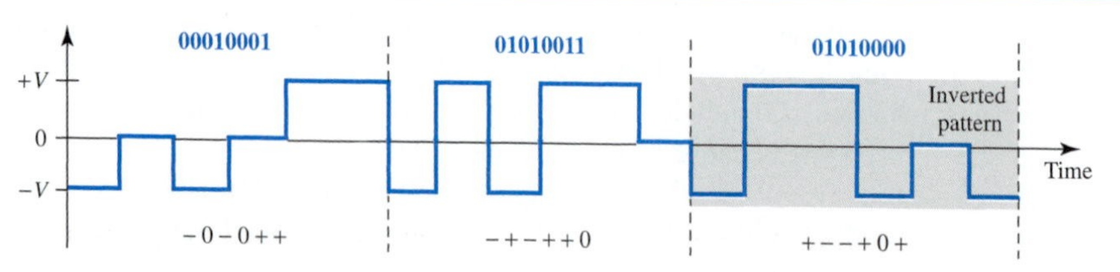

The average signal rate of the scheme is theoretically $S_{ave} = \frac{1}{2} \times N \times \frac{6}{8}$; in practice the minimum bandwidth is very close to $6N/8$.

4D-PAM5 The last signaling scheme we discuss in this category is called **four-dimensional five-level pulse amplitude modulation (4D-PAM5).** The 4D means that data is sent over four wires at the same time. It uses five voltage levels, such as −2, −1, 0, 1, and 2. However, one level, level 0, is used only for forward error detection (discussed in Chapter 10). If we assume that the code is just one-dimensional, the four levels create something similar to 8B4Q. In other words, an 8-bit word is translated to a signal element of four different levels. The worst signal rate for this imaginary one-dimensional version is $N \times 4/8$, or $N/2$.

The technique is designed to send data over four channels (four wires). This means the signal rate can be reduced to $N/8$, a significant achievement. All 8 bits can be fed into a wire simultaneously and sent by using one signal element. The point here is that the four signal elements comprising one signal group are sent simultaneously in a four-dimensional setting. Figure 4.12 shows the imaginary one-dimensional and the actual four-dimensional implementation. Gigabit LANs (see Chapter 13) use this technique to send 1-Gbps data over four copper cables that can handle 125 Mbaud. This scheme has a lot of redundancy in the signal pattern because 2^8 data patterns are matched to $4^4 = 256$ signal patterns. The extra signal patterns can be used for other purposes such as error detection.

Figure 4.12 *Multilevel: 4D-PAM5 scheme*

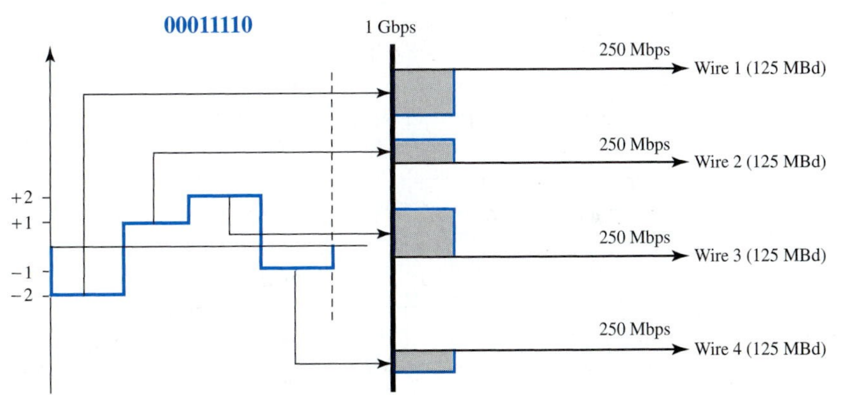

Multiline Transmission: MLT-3

NRZ-I and differential Manchester are classified as differential encoding but use two transition rules to encode binary data (no inversion, inversion). If we have a signal with more than two levels, we can design a differential encoding scheme with more than two transition rules. MLT-3 is one of them. The **multiline transmission, three level (MLT-3)** scheme uses three levels ($+V$, 0, and $-V$) and three transition rules to move between the levels.

1. If the next bit is 0, there is no transition.
2. If the next bit is 1 and the current level is not 0, the next level is 0.
3. If the next bit is 1 and the current level is 0, the next level is the opposite of the last nonzero level.

The behavior of MLT-3 can best be described by the state diagram shown in Figure 4.13. The three voltage levels (−*V*, 0, and +*V*) are shown by three states (ovals). The transition from one state (level) to another is shown by the connecting lines. Figure 4.13 also shows two examples of an MLT-3 signal.

Figure 4.13 *Multitransition: MLT-3 scheme*

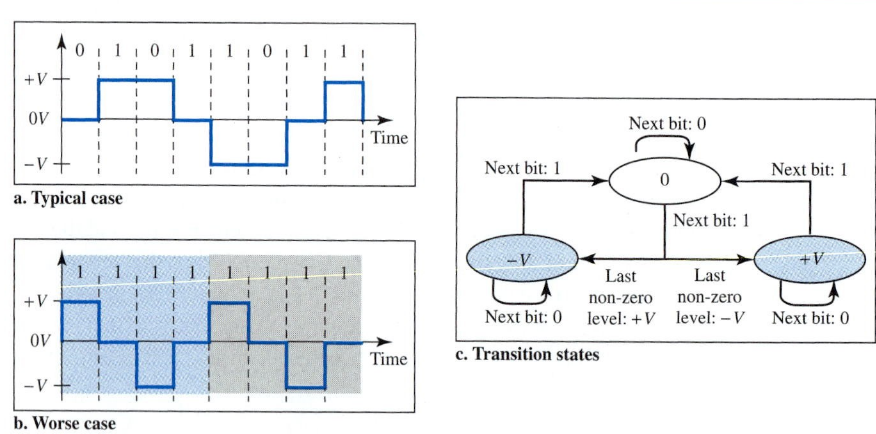

One might wonder why we need to use MLT-3, a scheme that maps one bit to one signal element. The signal rate is the same as that for NRZ-I, but with greater complexity (three levels and complex transition rules). It turns out that the shape of the signal in this scheme helps to reduce the required bandwidth. Let us look at the worst-case scenario, a sequence of 1s. In this case, the signal element pattern $+V0-V0$ is repeated every 4 bits. A nonperiodic signal has changed to a periodic signal with the period equal to 4 times the bit duration. This worst-case situation can be simulated as an analog signal with a frequency one-fourth of the bit rate. In other words, the signal rate for MLT-3 is one-fourth the bit rate. This makes MLT-3 a suitable choice when we need to send 100 Mbps on a copper wire that cannot support more than 32 MHz (frequencies above this level create electromagnetic emissions). MLT-3 and LANs are discussed in Chapter 13.

Summary of Line Coding Schemes

We summarize in Table 4.1 the characteristics of the different schemes discussed.

Table 4.1 *Summary of line coding schemes*

Category	*Scheme*	*Bandwidth (average)*	*Characteristics*
Unipolar	NRZ	$B = N/2$	Costly, no self-synchronization if long 0s or 1s, DC
Unipolar	NRZ-L	$B = N/2$	No self-synchronization if long 0s or 1s, DC
	NRZ-I	$B = N/2$	No self-synchronization for long 0s, DC
	Biphase	$B = N$	Self-synchronization, no DC, high bandwidth

Table 4.1 *Summary of line coding schemes (continued)*

Category	*Scheme*	*Bandwidth (average)*	*Characteristics*
Bipolar	AMI	$B = N/2$	No self-synchronization for long 0s, DC
Multilevel	2B1Q	$B = N/4$	No self-synchronization for long same double bits
	8B6T	$B = 3N/4$	Self-synchronization, no DC
	4D-PAM5	$B = N/8$	Self-synchronization, no DC
Multiline	MLT-3	$B = N/3$	No self-synchronization for long 0s

Block Coding

We need redundancy to ensure synchronization and to provide some kind of inherent error detecting. Block coding can give us this redundancy and improve the performance of line coding. In general, **block coding** changes a block of m bits into a block of n bits, where n is larger than m. Block coding is referred to as an *mB/nB* encoding technique.

Block coding is normally referred to as *mB/nB* coding; it replaces each *m*-bit group with an *n*-bit group.

The slash in block encoding (for example, 4B/5B) distinguishes block encoding from multilevel encoding (for example, 8B6T), which is written without a slash. Block coding normally involves three steps: division, substitution, and combination. In the division step, a sequence of bits is divided into groups of m bits. For example, in 4B/5B encoding, the original bit sequence is divided into 4-bit groups. The heart of block coding is the substitution step. In this step, we substitute an m-bit group for an n-bit group. For example, in 4B/5B encoding we substitute a 4-bit code for a 5-bit group. Finally, the n-bit groups are combined together to form a stream. The new stream has more bits than the original bits. Figure 4.14 shows the procedure.

Figure 4.14 *Block coding concept*

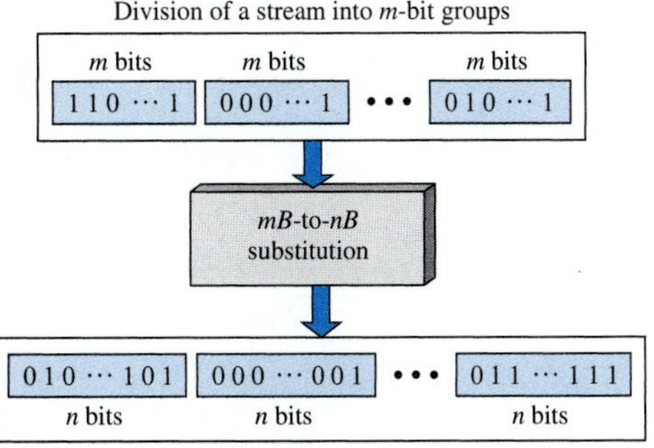

4B/5B

The **four binary/five binary (4B/5B)** coding scheme was designed to be used in combination with NRZ-I. Recall that NRZ-I has a good signal rate, one-half that of the biphase, but it has a synchronization problem. A long sequence of 0s can make the receiver clock lose synchronization. One solution is to change the bit stream, prior to encoding with NRZ-I, so that it does not have a long stream of 0s. The 4B/5B scheme achieves this goal. The block-coded stream does not have more that three consecutive 0s, as we will see later. At the receiver, the NRZ-I encoded digital signal is first decoded into a stream of bits and then decoded to remove the redundancy. Figure 4.15 shows the idea.

Figure 4.15 *Using block coding 4B/5B with NRZ-I line coding scheme*

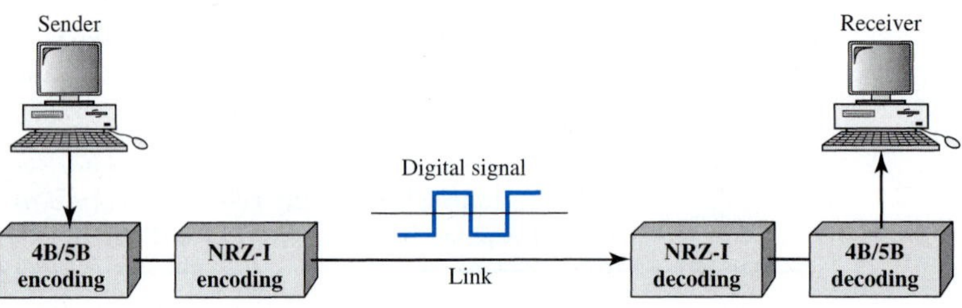

In 4B/5B, the 5-bit output that replaces the 4-bit input has no more than one leading zero (left bit) and no more than two trailing zeros (right bits). So when different groups are combined to make a new sequence, there are never more than three consecutive 0s. (Note that NRZ-I has no problem with sequences of 1s.) Table 4.2 shows the corresponding pairs used in 4B/5B encoding. Note that the first two columns pair a 4-bit group with a 5-bit group. A group of 4 bits can have only 16 different combinations while a group of 5 bits can have 32 different combinations. This means that there are 16 groups that are not used for 4B/5B encoding. Some of these unused groups are used for control purposes; the others are not used at all. The latter provide a kind of error detection. If a 5-bit group arrives that belongs to the unused portion of the table, the receiver knows that there is an error in the transmission.

Table 4.2 *4B/5B mapping codes*

Data Sequence	*Encoded Sequence*	*Control Sequence*	*Encoded Sequence*
0000	11110	Q (Quiet)	00000
0001	01001	I (Idle)	11111
0010	10100	H (Halt)	00100
0011	10101	J (Start delimiter)	11000
0100	01010	K (Start delimiter)	10001
0101	01011	T (End delimiter)	01101

Table 4.2 *4B/5B mapping codes (continued)*

Data Sequence	*Encoded Sequence*	*Control Sequence*	*Encoded Sequence*
0110	01110	S (Set)	11001
0111	01111	R (Reset)	00111
1000	10010		
1001	10011		
1010	10110		
1011	10111		
1100	11010		
1101	11011		
1110	11100		
1111	11101		

Figure 4.16 shows an example of substitution in 4B/5B coding. 4B/5B encoding solves the problem of synchronization and overcomes one of the deficiencies of NRZ-I. However, we need to remember that it increases the signal rate of NRZ-I. The redundant bits add 20 percent more baud. Still, the result is less than the biphase scheme which has a signal rate of 2 times that of NRZ-I. However, 4B/5B block encoding does not solve the DC component problem of NRZ-I. If a DC component is unacceptable, we need to use biphase or bipolar encoding.

Figure 4.16 *Substitution in 4B/5B block coding*

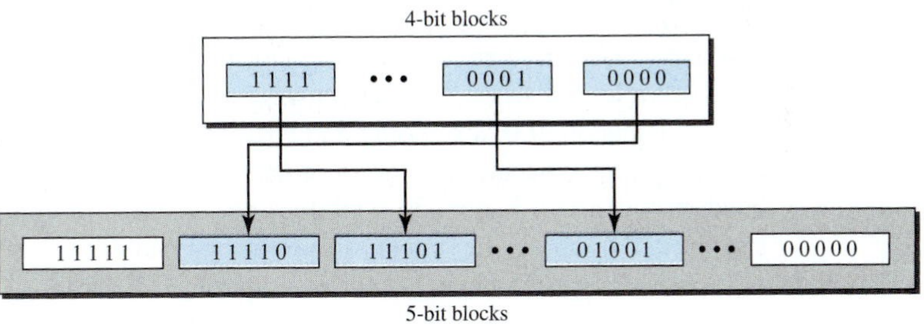

Example 4.5

We need to send data at a 1-Mbps rate. What is the minimum required bandwidth, using a combination of 4B/5B and NRZ-I or Manchester coding?

Solution

First 4B/5B block coding increases the bit rate to 1.25 Mbps. The minimum bandwidth using NRZ-I is $N/2$ or 625 kHz. The Manchester scheme needs a minimum bandwidth of 1 MHz. The first choice needs a lower bandwidth, but has a DC component problem; the second choice needs a higher bandwidth, but does not have a DC component problem.

8B/10B

The **eight binary/ten binary (8B/10B)** encoding is similar to 4B/5B encoding except that a group of 8 bits of data is now substituted by a 10-bit code. It provides greater error detection capability than 4B/5B. The 8B/10B block coding is actually a combination of 5B/6B and 3B/4B encoding, as shown in Figure 4.17.

Figure 4.17 *8B/10B block encoding*

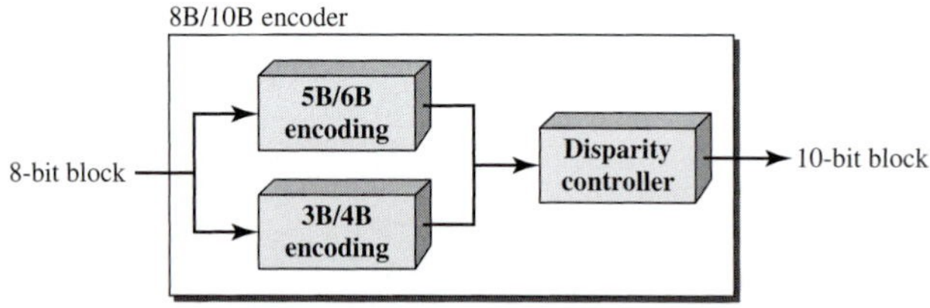

The most five significant bits of a 10-bit block is fed into the 5B/6B encoder; the least 3 significant bits is fed into a 3B/4B encoder. The split is done to simplify the mapping table. To prevent a long run of consecutive 0s or 1s, the code uses a disparity controller which keeps track of excess 0s over 1s (or 1s over 0s). If the bits in the current block create a disparity that contributes to the previous disparity (either direction), then each bit in the code is complemented (a 0 is changed to a 1 and a 1 is changed to a 0). The coding has $2^{10} - 2^8 = 768$ redundant groups that can be used for disparity checking and error detection. In general, the technique is superior to 4B/5B because of better built-in error-checking capability and better synchronization.

Scrambling

Biphase schemes that are suitable for dedicated links between stations in a LAN are not suitable for long-distance communication because of their wide bandwidth requirement. The combination of block coding and NRZ line coding is not suitable for long-distance encoding either, because of the DC component. Bipolar AMI encoding, on the other hand, has a narrow bandwidth and does not create a DC component. However, a long sequence of 0s upsets the synchronization. If we can find a way to avoid a long sequence of 0s in the original stream, we can use bipolar AMI for long distances. We are looking for a technique that does not increase the number of bits and does provide synchronization. We are looking for a solution that substitutes long zero-level pulses with a combination of other levels to provide synchronization. One solution is called **scrambling.** We modify part of the AMI rule to include scrambling, as shown in Figure 4.18. Note that scrambling, as opposed to block coding, is done at the same time as encoding. The system needs to insert the required pulses based on the defined scrambling rules. Two common scrambling techniques are B8ZS and HDB3.

B8ZS

Bipolar with 8-zero substitution (B8ZS) is commonly used in North America. In this technique, eight consecutive zero-level voltages are replaced by the sequence

Figure 4.18 *AMI used with scrambling*

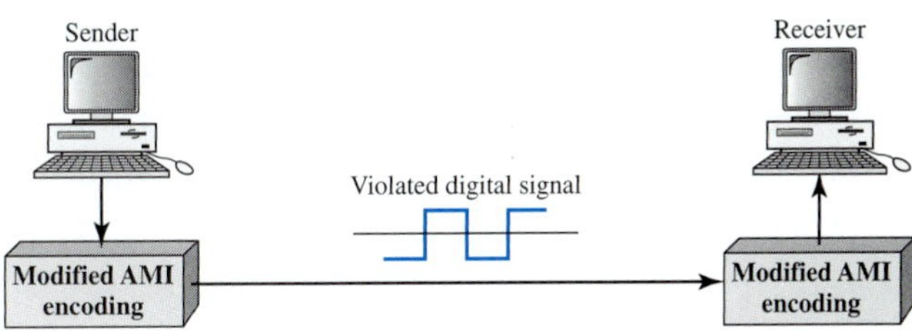

000VB0VB. The V in the sequence denotes *violation;* this is a nonzero voltage that breaks an AMI rule of encoding (opposite polarity from the previous). The B in the sequence denotes *bipolar,* which means a nonzero level voltage in accordance with the AMI rule. There are two cases, as shown in Figure 4.19.

Figure 4.19 *Two cases of B8ZS scrambling technique*

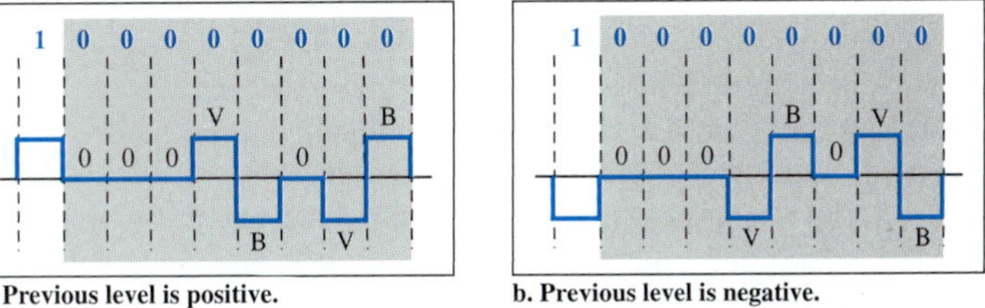

Note that the scrambling in this case does not change the bit rate. Also, the technique balances the positive and negative voltage levels (two positives and two negatives), which means that the DC balance is maintained. Note that the substitution may change the polarity of a 1 because, after the substitution, AMI needs to follow its rules.

B8ZS substitutes eight consecutive zeros with 000VB0VB.

One more point is worth mentioning. The letter V (violation) or B (bipolar) here is relative. The V means the same polarity as the polarity of the previous nonzero pulse; B means the polarity opposite to the polarity of the previous nonzero pulse.

HDB3

High-density bipolar 3-zero (HDB3) is commonly used outside of North America. In this technique, which is more conservative than B8ZS, four consecutive zero-level voltages are replaced with a sequence of **000V** or **B00V.** The reason for two different substitutions is to maintain the even number of nonzero pulses after each substitution. The two rules can be stated as follows:

1. If the number of nonzero pulses after the last substitution is odd, the substitution pattern will be **000V**, which makes the total number of nonzero pulses even.
2. If the number of nonzero pulses after the last substitution is even, the substitution pattern will be **B00V**, which makes the total number of nonzero pulses even.

Figure 4.20 shows an example.

Figure 4.20 *Different situations in HDB3 scrambling technique*

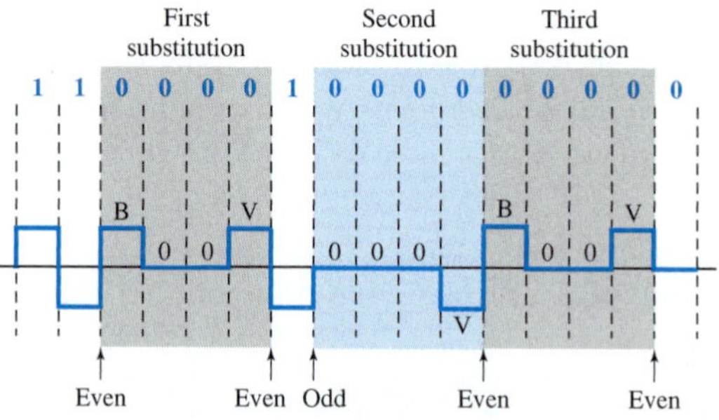

There are several points we need to mention here. First, before the first substitution, the number of nonzero pulses is even, so the first substitution is B00V. After this substitution, the polarity of the 1 bit is changed because the AMI scheme, after each substitution, must follow its own rule. After this bit, we need another substitution, which is 000V because we have only one nonzero pulse (odd) after the last substitution. The third substitution is B00V because there are no nonzero pulses after the second substitution (even).

HDB3 substitutes four consecutive zeros with 000V or B00V depending on the number of nonzero pulses after the last substitution.

4.2 ANALOG-TO-DIGITAL CONVERSION

The techniques described in Section 4.1 convert digital data to digital signals. Sometimes, however, we have an analog signal such as one created by a microphone or camera. We have seen in Chapter 3 that a digital signal is superior to an analog signal. The tendency today is to change an analog signal to digital data. In this section we describe two techniques, pulse code modulation and delta modulation. After the digital data are created (digitization), we can use one of the techniques described in Section 4.1 to convert the digital data to a digital signal.

Pulse Code Modulation (PCM)

The most common technique to change an analog signal to digital data (**digitization**) is called **pulse code modulation (PCM).** A PCM encoder has three processes, as shown in Figure 4.21.

Figure 4.21 *Components of PCM encoder*

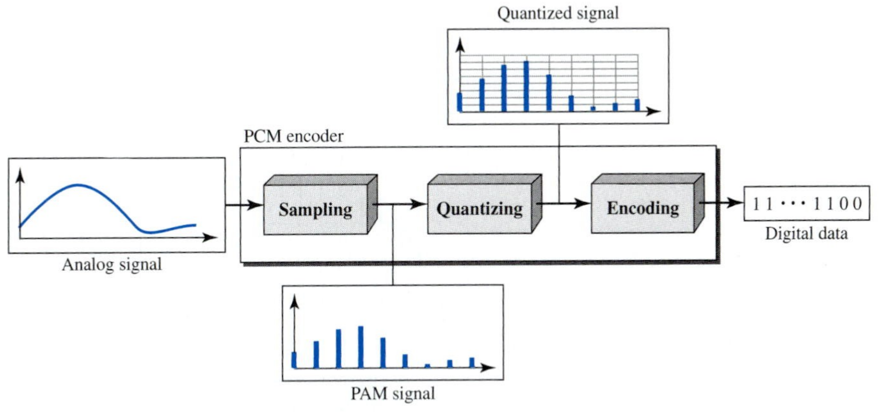

1. The analog signal is sampled.
2. The sampled signal is quantized.
3. The quantized values are encoded as streams of bits.

Sampling

The first step in PCM is **sampling.** The analog signal is sampled every T_s s, where T_s is the sample interval or period. The inverse of the sampling interval is called the **sampling rate** or **sampling frequency** and denoted by f_s, where $f_s = 1/T_s$. There are three sampling methods—ideal, natural, and flat-top—as shown in Figure 4.22.

In ideal sampling, pulses from the analog signal are sampled. This is an ideal sampling method and cannot be easily implemented. In natural sampling, a high-speed switch is turned on for only the small period of time when the sampling occurs. The result is a sequence of samples that retains the shape of the analog signal. The most common sampling method, called **sample and hold,** however, creates flat-top samples by using a circuit.

The sampling process is sometimes referred to as **pulse amplitude modulation (PAM).** We need to remember, however, that the result is still an analog signal with nonintegral values.

Sampling Rate One important consideration is the sampling rate or frequency. What are the restrictions on T_s? This question was elegantly answered by Nyquist. According to the **Nyquist theorem,** to reproduce the original analog signal, one necessary condition is that the **sampling rate** be at least twice the highest frequency in the original signal.

Figure 4.22 *Three different sampling methods for PCM*

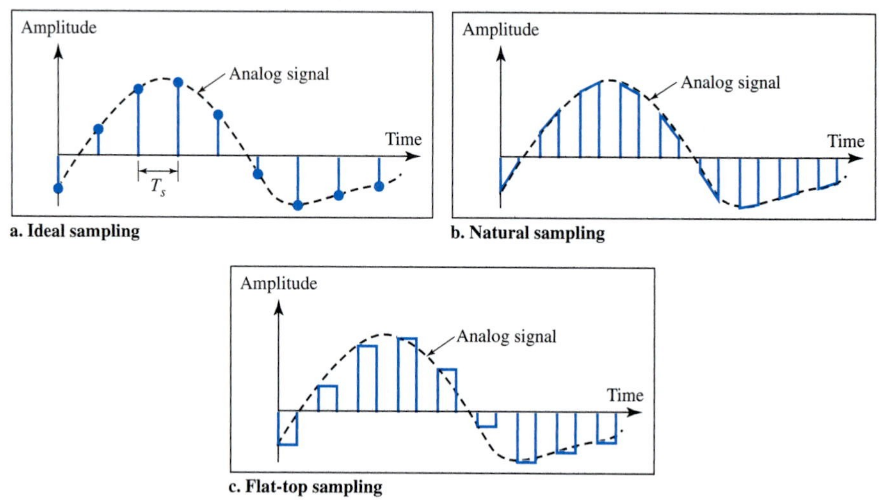

According to the Nyquist theorem, the sampling rate must be at least 2 times the highest frequency contained in the signal.

We need to elaborate on the theorem at this point. First, we can sample a signal only if the signal is band-limited. In other words, a signal with an infinite bandwidth cannot be sampled. Second, the sampling rate must be at least 2 times the highest frequency, not the bandwidth. If the analog signal is low-pass, the bandwidth and the highest frequency are the same value. If the analog signal is bandpass, the bandwidth value is lower than the value of the maximum frequency. Figure 4.23 shows the value of the sampling rate for two types of signals.

Figure 4.23 *Nyquist sampling rate for low-pass and bandpass signals*

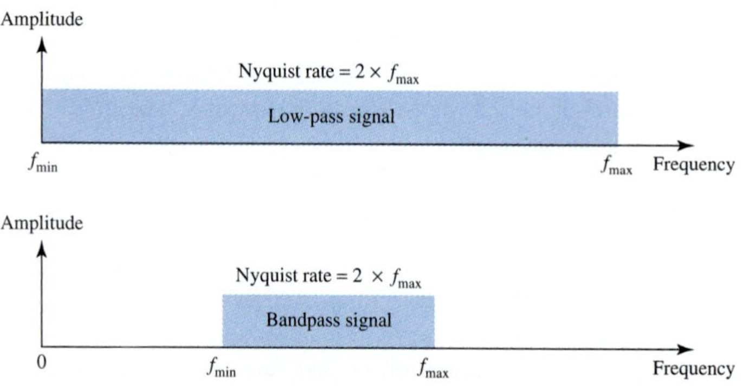

Example 4.6

For an intuitive example of the Nyquist theorem, let us sample a simple sine wave at three sampling rates: $f_s = 4f$ (2 times the Nyquist rate), $f_s = 2f$ (Nyquist rate), and $f_s = f$ (one-half the Nyquist rate). Figure 4.24 shows the sampling and the subsequent recovery of the signal.

Figure 4.24 *Recovery of a sampled sine wave for different sampling rates*

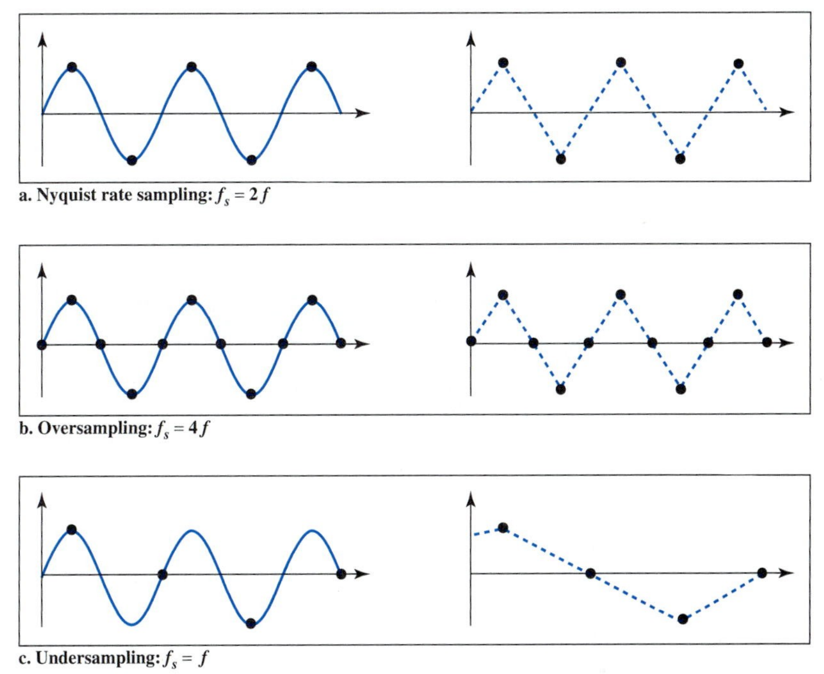

It can be seen that sampling at the Nyquist rate can create a good approximation of the original sine wave (part a). Oversampling in part b can also create the same approximation, but it is redundant and unnecessary. Sampling below the Nyquist rate (part c) does not produce a signal that looks like the original sine wave.

Example 4.7

As an interesting example, let us see what happens if we sample a periodic event such as the revolution of a hand of a clock. The second hand of a clock has a period of 60 s. According to the Nyquist theorem, we need to sample the hand (take and send a picture) every 30 s ($T_s = \frac{1}{2}T$ or $f_s = 2f$). In Figure 4.25a, the sample points, in order, are 12, 6, 12, 6, 12, and 6. The receiver of the samples cannot tell if the clock is moving forward or backward. In part b, we sample at double the Nyquist rate (every 15 s). The sample points, in order, are 12, 3, 6, 9, and 12. The clock is moving forward. In part c, we sample below the Nyquist rate ($T_s = \frac{3}{4}T$ or $f_s = \frac{4}{3}f$). The sample points, in order, are 12, 9, 6, 3, and 12. Although the clock is moving forward, the receiver thinks that the clock is moving backward.

Figure 4.25 *Sampling of a clock with only one hand*

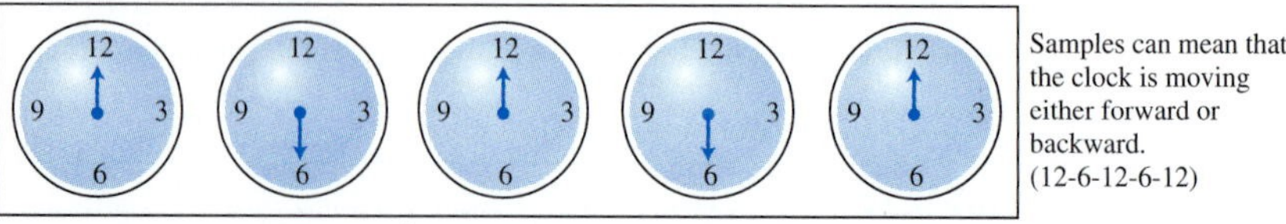

Samples can mean that the clock is moving either forward or backward.
(12-6-12-6-12)

a. Sampling at Nyquist rate: $T_s = \frac{1}{2}T$

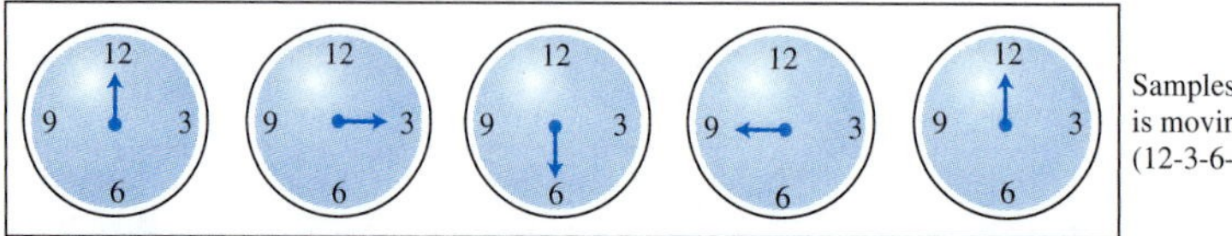

Samples show clock is moving forward.
(12-3-6-9-12)

b. Oversampling (above Nyquist rate): $T_s = \frac{1}{4}T$

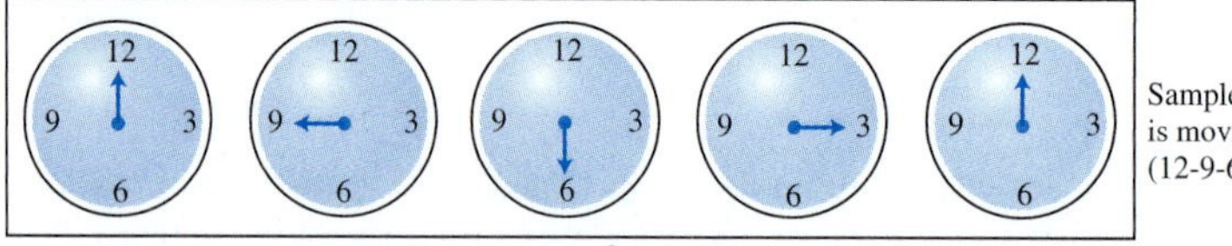

Samples show clock is moving backward.
(12-9-6-3-12)

c. Undersampling (below Nyquist rate): $T_s = \frac{3}{4}T$

Example 4.8

An example related to Example 4.7 is the seemingly backward rotation of the wheels of a forward-moving car in a movie. This can be explained by undersampling. A movie is filmed at 24 frames per second. If a wheel is rotating more than 12 times per second, the undersampling creates the impression of a backward rotation.

Example 4.9

Telephone companies digitize voice by assuming a maximum frequency of 4000 Hz. The sampling rate therefore is 8000 samples per second.

Example 4.10

A complex low-pass signal has a bandwidth of 200 kHz. What is the minimum sampling rate for this signal?

Solution

The bandwidth of a low-pass signal is between 0 and f, where f is the maximum frequency in the signal. Therefore, we can sample this signal at 2 times the highest frequency (200 kHz). The sampling rate is therefore 400,000 samples per second.

Example 4.11

A complex bandpass signal has a bandwidth of 200 kHz. What is the minimum sampling rate for this signal?

Solution

We cannot find the minimum sampling rate in this case because we do not know where the bandwidth starts or ends. We do not know the maximum frequency in the signal.

Quantization

The result of sampling is a series of pulses with amplitude values between the maximum and minimum amplitudes of the signal. The set of amplitudes can be infinite with nonintegral values between the two limits. These values cannot be used in the encoding process. The following are the steps in quantization:

1. We assume that the original analog signal has instantaneous amplitudes between V_{min} and V_{max}.
2. We divide the range into L zones, each of height Δ (delta).

$$\Delta = \frac{V_{max} - V_{min}}{L}$$

3. We assign quantized values of 0 to $L - 1$ to the midpoint of each zone.
4. We approximate the value of the sample amplitude to the quantized values.

As a simple example, assume that we have a sampled signal and the sample amplitudes are between −20 and +20 V. We decide to have eight levels ($L = 8$). This means that $\Delta = 5$ V. Figure 4.26 shows this example.

Figure 4.26 *Quantization and encoding of a sampled signal*

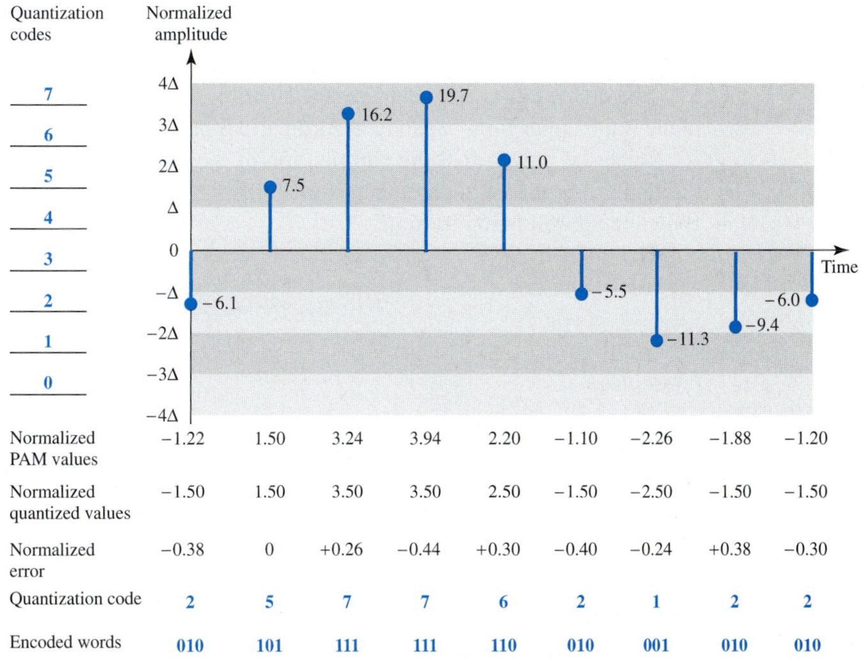

Normalized PAM values	−1.22	1.50	3.24	3.94	2.20	−1.10	−2.26	−1.88	−1.20
Normalized quantized values	−1.50	1.50	3.50	3.50	2.50	−1.50	−2.50	−1.50	−1.50
Normalized error	−0.38	0	+0.26	−0.44	+0.30	−0.40	−0.24	+0.38	−0.30
Quantization code	2	5	7	7	6	2	1	2	2
Encoded words	010	101	111	111	110	010	001	010	010

We have shown only nine samples using ideal sampling (for simplicity). The value at the top of each sample in the graph shows the actual amplitude. In the chart, the first row is the normalized value for each sample (actual amplitude/Δ). The quantization process selects the quantization value from the middle of each zone. This means that the normalized quantized values (second row) are different from the normalized amplitudes. The difference is called the *normalized error* (third row). The fourth row is the quantization code for each sample based on the quantization levels at the left of the graph. The encoded words (fifth row) are the final products of the conversion.

Quantization Levels In the previous example, we showed eight quantization levels. The choice of L, the number of levels, depends on the range of the amplitudes of the analog signal and how accurately we need to recover the signal. If the amplitude of a signal fluctuates between two values only, we need only two levels; if the signal, like voice, has many amplitude values, we need more quantization levels. In audio digitizing, L is normally chosen to be 256; in video it is normally thousands. Choosing lower values of L increases the quantization error if there is a lot of fluctuation in the signal.

Quantization Error One important issue is the error created in the quantization process. (Later, we will see how this affects high-speed modems.) Quantization is an approximation process. The input values to the quantizer are the real values; the output values are the approximated values. The output values are chosen to be the middle value in the zone. If the input value is also at the middle of the zone, there is no quantization error; otherwise, there is an error. In the previous example, the normalized amplitude of the third sample is 3.24, but the normalized quantized value is 3.50. This means that there is an error of +0.26. The value of the error for any sample is less than Δ/2. In other words, we have $-\Delta/2 \leq \text{error} \leq \Delta/2$.

The quantization error changes the signal-to-noise ratio of the signal, which in turn reduces the upper limit capacity according to Shannon.

It can be proven that the contribution of the **quantization error** to the SNR_{dB} of the signal depends on the number of quantization levels L, or the bits per sample n_b, as shown in the following formula:

$$\mathbf{SNR_{dB} = 6.02}n_b \mathbf{+ 1.76 \quad dB}$$

Example 4.12

What is the SNR_{dB} in the example of Figure 4.26?

Solution

We can use the formula to find the quantization. We have eight levels and 3 bits per sample, so $SNR_{dB} = 6.02(3) + 1.76 = 19.82$ dB. Increasing the number of levels increases the SNR.

Example 4.13

A telephone subscriber line must have an SNR_{dB} above 40. What is the minimum number of bits per sample?

Solution

We can calculate the number of bits as

$$SNR_{dB} = 6.02n_b + 1.76 = 40 \rightarrow n = 6.35$$

Telephone companies usually assign 7 or 8 bits per sample.

Uniform Versus Nonuniform Quantization For many applications, the distribution of the instantaneous amplitudes in the analog signal is not uniform. Changes in amplitude often occur more frequently in the lower amplitudes than in the higher ones. For these types of applications it is better to use nonuniform zones. In other words, the height of Δ is not fixed; it is greater near the lower amplitudes and less near the higher amplitudes. Nonuniform quantization can also be achieved by using a process called **companding** and **expanding.** The signal is companded at the sender before conversion; it is expanded at the receiver after conversion. Companding means reducing the instantaneous voltage amplitude for large values; expanding is the opposite process. Companding gives greater weight to strong signals and less weight to weak ones. It has been proved that nonuniform quantization effectively reduces the SNR_{dB} of quantization.

Encoding

The last step in PCM is encoding. After each sample is quantized and the number of bits per sample is decided, each sample can be changed to an n_b-bit code word. In Figure 4.26 the encoded words are shown in the last row. A quantization code of 2 is encoded as 010; 5 is encoded as 101; and so on. Note that the number of bits for each sample is determined from the number of quantization levels. If the number of quantization levels is L, the number of bits is $n_b = \log_2 L$. In our example L is 8 and n_b is therefore 3. The bit rate can be found from the formula

$$\textbf{Bit rate} = \textbf{sampling rate} \times \textbf{number of bits per sample} = f_s \times n_b$$

Example 4.14

We want to digitize the human voice. What is the bit rate, assuming 8 bits per sample?

Solution

The human voice normally contains frequencies from 0 to 4000 Hz. So the sampling rate and bit rate are calculated as follows:

$$\text{Sampling rate} = 4000 \times 2 = 8000 \text{ samples/s}$$
$$\text{Bit rate} = 8000 \times 8 = 64{,}000 \text{ bps} = 64 \text{ kbps}$$

Original Signal Recovery

The recovery of the original signal requires the PCM decoder. The decoder first uses circuitry to convert the code words into a pulse that holds the amplitude until the next pulse. After the staircase signal is completed, it is passed through a low-pass filter to smooth the staircase signal into an analog signal. The filter has the same cutoff frequency as the original signal at the sender. If the signal has been sampled at (or greater than) the Nyquist sampling rate and if there are enough quantization levels, the original signal will be recreated. Note that the maximum and minimum values of the original signal can be achieved by using amplification. Figure 4.27 shows the simplified process.

Figure 4.27 *Components of a PCM decoder*

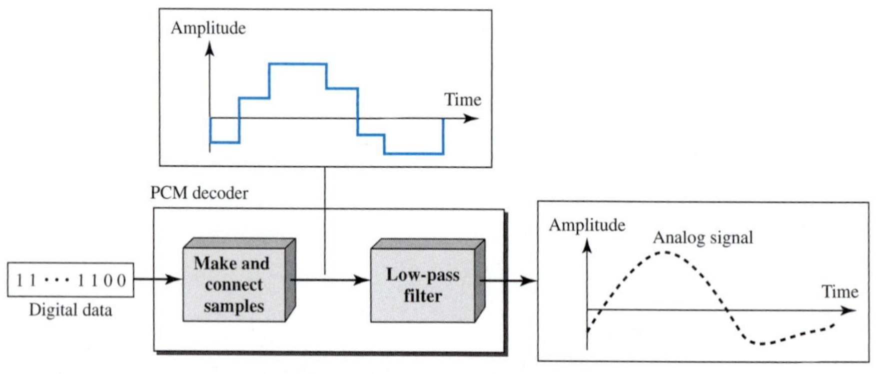

PCM Bandwidth

Suppose we are given the bandwidth of a low-pass analog signal. If we then digitize the signal, what is the new minimum bandwidth of the channel that can pass this digitized signal? We have said that the minimum bandwidth of a line-encoded signal is $B_{min} = c \times N \times (1/r)$. We substitute the value of N in this formula:

$$B_{min} = c \times N \times \frac{1}{r} = c \times n_b \times f_s \times \frac{1}{r} = c \times n_b \times 2 \times B_{analog} \times \frac{1}{r}$$

When $1/r = 1$ (for a NRZ or bipolar signal) and $c = (1/2)$ (the average situation), the minimum bandwidth is

$$\mathbf{B_{min} = n_b \times B_{analog}}$$

This means the minimum bandwidth of the digital signal is n_b times greater than the bandwidth of the analog signal. This is the price we pay for digitization.

Example 4.15

We have a low-pass analog signal of 4 kHz. If we send the analog signal, we need a channel with a minimum bandwidth of 4 kHz. If we digitize the signal and send 8 bits per sample, we need a channel with a minimum bandwidth of 8×4 kHz = 32 kHz.

Maximum Data Rate of a Channel

In Chapter 3, we discussed the Nyquist theorem which gives the data rate of a channel as $N_{max} = 2 \times B \times \log_2 L$. We can deduce this rate from the Nyquist sampling theorem by using the following arguments.

1. We assume that the available channel is low-pass with bandwidth B.
2. We assume that the digital signal we want to send has L levels, where each level is a signal element. This means $r = 1/\log_2 L$.
3. We first pass the digital signal through a low-pass filter to cut off the frequencies above B Hz.
4. We treat the resulting signal as an analog signal and sample it at $2 \times B$ samples per second and quantize it using L levels. Additional quantization levels are useless because the signal originally had L levels.
5. The resulting bit rate is $N = f_s \times n_b = 2 \times B \times \log_2 L$. This is the maximum bandwidth; if the case factor c increases, the data rate is reduced.

$$\mathbf{N_{max} = 2 \times B \times \log_2 L \quad bps}$$

Minimum Required Bandwidth

The previous argument can give us the minimum bandwidth if the data rate and the number of signal levels are fixed. We can say

$$\mathbf{B_{min} = \frac{N}{2 \times \log_2 L} \quad Hz}$$

Delta Modulation (DM)

PCM is a very complex technique. Other techniques have been developed to reduce the complexity of PCM. The simplest is *delta modulation*. PCM finds the value of the signal amplitude for each sample; DM finds the change from the previous sample. Figure 4.28 shows the process. Note that there are no code words here; bits are sent one after another.

Figure 4.28 *The process of delta modulation*

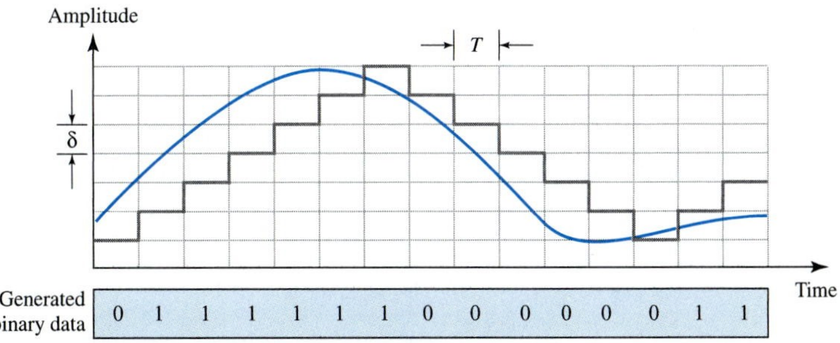

Modulator

The modulator is used at the sender site to create a stream of bits from an analog signal. The process records the small positive or negative changes, called delta δ. If the delta is positive, the process records a 1; if it is negative, the process records a 0. However, the process needs a base against which the analog signal is compared. The modulator builds a second signal that resembles a staircase. Finding the change is then reduced to comparing the input signal with the gradually made staircase signal. Figure 4.29 shows a diagram of the process.

Figure 4.29 *Delta modulation components*

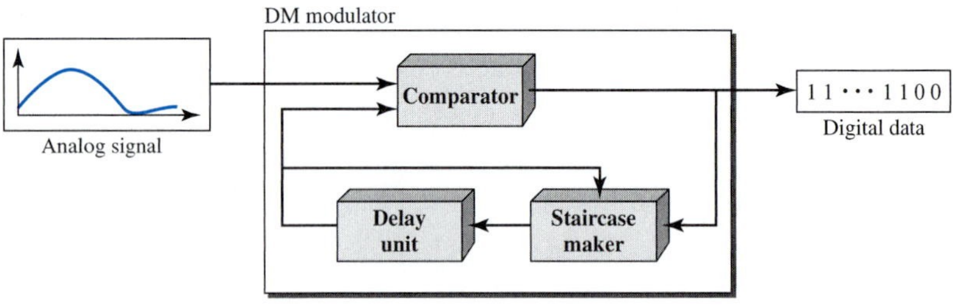

The modulator, at each sampling interval, compares the value of the analog signal with the last value of the staircase signal. If the amplitude of the analog signal is larger, the next bit in the digital data is 1; otherwise, it is 0. The output of the comparator, however, also makes the staircase itself. If the next bit is 1, the staircase maker moves the last point of the staircase signal δ up; it the next bit is 0, it moves it δ down. Note that we need a delay unit to hold the staircase function for a period between two comparisons.

Demodulator

The demodulator takes the digital data and, using the staircase maker and the delay unit, creates the analog signal. The created analog signal, however, needs to pass through a low-pass filter for smoothing. Figure 4.30 shows the schematic diagram.

Figure 4.30 *Delta demodulation components*

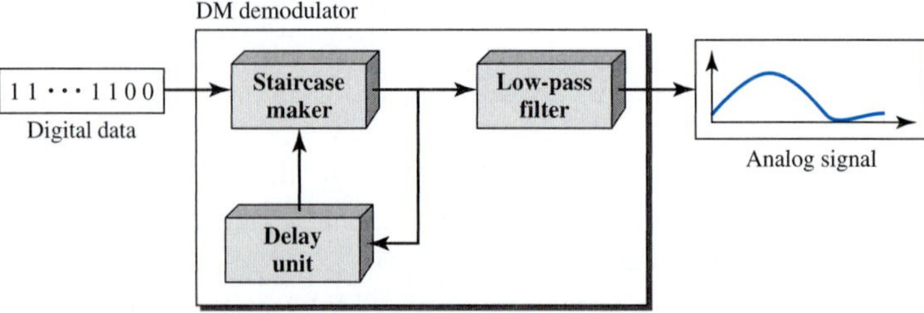

Adaptive DM

A better performance can be achieved if the value of δ is not fixed. In **adaptive delta modulation,** the value of δ changes according to the amplitude of the analog signal.

Quantization Error

It is obvious that DM is not perfect. Quantization error is always introduced in the process. The quantization error of DM, however, is much less than that for PCM.

4.3 TRANSMISSION MODES

Of primary concern when we are considering the transmission of data from one device to another is the wiring, and of primary concern when we are considering the wiring is the data stream. Do we send 1 bit at a time; or do we group bits into larger groups and, if so, how? The transmission of binary data across a link can be accomplished in either parallel or serial mode. In parallel mode, multiple bits are sent with each clock tick. In serial mode, 1 bit is sent with each clock tick. While there is only one way to send parallel data, there are three subclasses of serial transmission: asynchronous, synchronous, and isochronous (see Figure 4.31).

Figure 4.31 *Data transmission and modes*

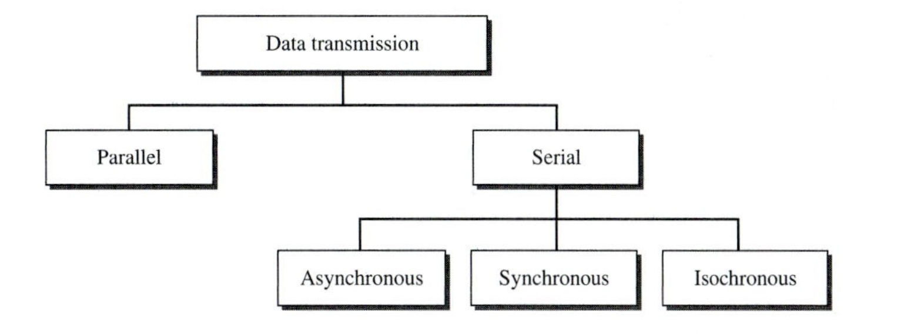

Parallel Transmission

Binary data, consisting of 1s and 0s, may be organized into groups of n bits each. Computers produce and consume data in groups of bits much as we conceive of and use spoken language in the form of words rather than letters. By grouping, we can send data n bits at a time instead of 1. This is called **parallel transmission.**

The mechanism for parallel transmission is a conceptually simple one: Use n wires to send n bits at one time. That way each bit has its own wire, and all n bits of one group can be transmitted with each clock tick from one device to another. Figure 4.32 shows how parallel transmission works for $n = 8$. Typically, the eight wires are bundled in a cable with a connector at each end.

The advantage of parallel transmission is speed. All else being equal, parallel transmission can increase the transfer speed by a factor of n over serial transmission.

Figure 4.32 *Parallel transmission*

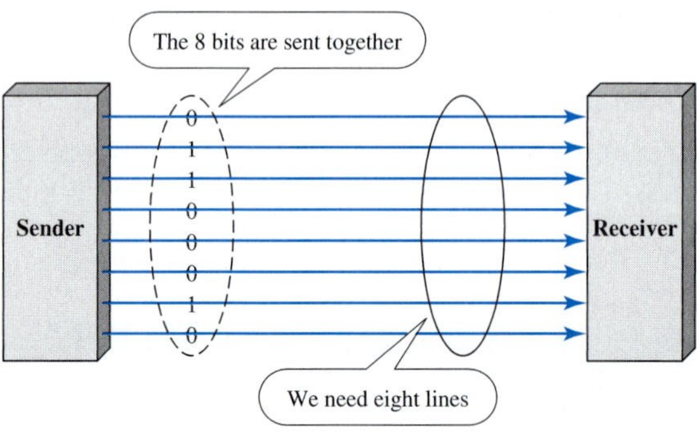

But there is a significant disadvantage: cost. Parallel transmission requires *n* communication lines (wires in the example) just to transmit the data stream. Because this is expensive, parallel transmission is usually limited to short distances.

Serial Transmission

In **serial transmission** one bit follows another, so we need only one communication channel rather than *n* to transmit data between two communicating devices (see Figure 4.33).

Figure 4.33 *Serial transmission*

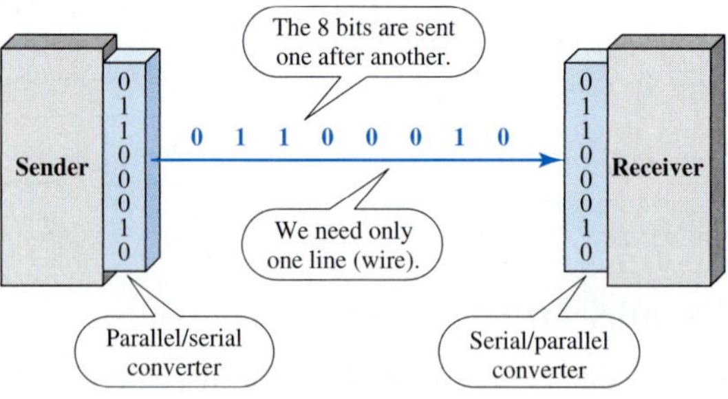

The advantage of serial over parallel transmission is that with only one communication channel, serial transmission reduces the cost of transmission over parallel by roughly a factor of *n*.

Since communication within devices is parallel, conversion devices are required at the interface between the sender and the line (parallel-to-serial) and between the line and the receiver (serial-to-parallel).

Serial transmission occurs in one of three ways: asynchronous, synchronous, and isochronous.

Asynchronous Transmission

Asynchronous transmission is so named because the timing of a signal is unimportant. Instead, information is received and translated by agreed upon patterns. As long as those patterns are followed, the receiving device can retrieve the information without regard to the rhythm in which it is sent. Patterns are based on grouping the bit stream into bytes. Each group, usually 8 bits, is sent along the link as a unit. The sending system handles each group independently, relaying it to the link whenever ready, without regard to a timer.

Without synchronization, the receiver cannot use timing to predict when the next group will arrive. To alert the receiver to the arrival of a new group, therefore, an extra bit is added to the beginning of each byte. This bit, usually a 0, is called the **start bit.** To let the receiver know that the byte is finished, 1 or more additional bits are appended to the end of the byte. These bits, usually 1s, are called **stop bits.** By this method, each byte is increased in size to at least 10 bits, of which 8 bits is information and 2 bits or more are signals to the receiver. In addition, the transmission of each byte may then be followed by a gap of varying duration. This gap can be represented either by an idle channel or by a stream of additional stop bits.

In asynchronous transmission, we send 1 start bit (0) at the beginning and 1 or more stop bits (1s) at the end of each byte. There may be a gap between each byte.

The start and stop bits and the gap alert the receiver to the beginning and end of each byte and allow it to synchronize with the data stream. This mechanism is called *asynchronous* because, at the byte level, the sender and receiver do not have to be synchronized. But within each byte, the receiver must still be synchronized with the incoming bit stream. That is, some synchronization is required, but only for the duration of a single byte. The receiving device resynchronizes at the onset of each new byte. When the receiver detects a start bit, it sets a timer and begins counting bits as they come in. After n bits, the receiver looks for a stop bit. As soon as it detects the stop bit, it waits until it detects the next start bit.

Asynchronous here means "asynchronous at the byte level," but the bits are still synchronized; their durations are the same.

Figure 4.34 is a schematic illustration of asynchronous transmission. In this example, the start bits are 0s, the stop bits are 1s, and the gap is represented by an idle line rather than by additional stop bits.

The addition of stop and start bits and the insertion of gaps into the bit stream make asynchronous transmission slower than forms of transmission that can operate without the addition of control information. But it is cheap and effective, two advantages that make it an attractive choice for situations such as low-speed communication. For example, the connection of a keyboard to a computer is a natural application for asynchronous transmission. A user types only one character at a time, types extremely slowly in data processing terms, and leaves unpredictable gaps of time between each character.

Figure 4.34 *Asynchronous transmission*

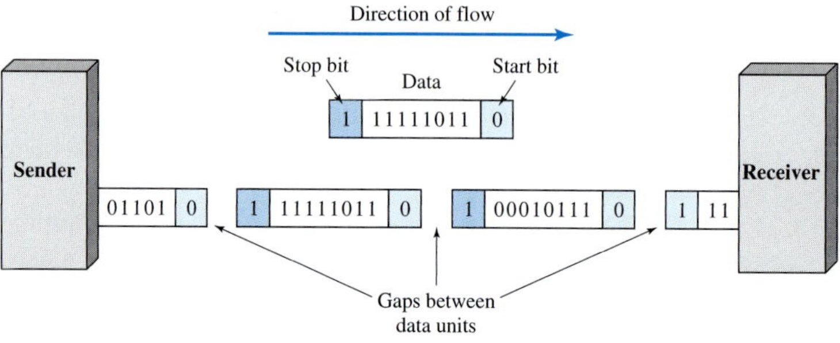

Synchronous Transmission

In **synchronous transmission,** the bit stream is combined into longer "frames," which may contain multiple bytes. Each byte, however, is introduced onto the transmission link without a gap between it and the next one. It is left to the receiver to separate the bit stream into bytes for decoding purposes. In other words, data are transmitted as an unbroken string of 1s and 0s, and the receiver separates that string into the bytes, or characters, it needs to reconstruct the information.

In synchronous transmission, we send bits one after another without start or stop bits or gaps. It is the responsibility of the receiver to group the bits.

Figure 4.35 gives a schematic illustration of synchronous transmission. We have drawn in the divisions between bytes. In reality, those divisions do not exist; the sender puts its data onto the line as one long string. If the sender wishes to send data in separate bursts, the gaps between bursts must be filled with a special sequence of 0s and 1s that means *idle*. The receiver counts the bits as they arrive and groups them in 8-bit units.

Figure 4.35 *Synchronous transmission*

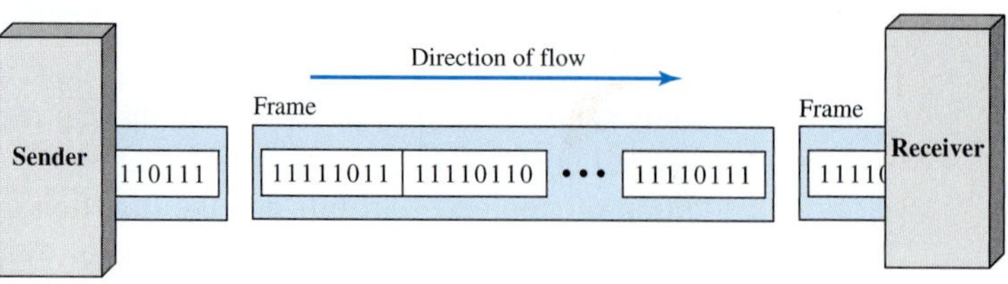

Without gaps and start and stop bits, there is no built-in mechanism to help the receiving device adjust its bit synchronization midstream. Timing becomes very important, therefore, because the accuracy of the received information is completely dependent on the ability of the receiving device to keep an accurate count of the bits as they come in.

The advantage of synchronous transmission is speed. With no extra bits or gaps to introduce at the sending end and remove at the receiving end, and, by extension, with fewer bits to move across the link, synchronous transmission is faster than asynchronous transmission. For this reason, it is more useful for high-speed applications such as the transmission of data from one computer to another. Byte synchronization is accomplished in the data link layer.

We need to emphasize one point here. Although there is no gap between characters in synchronous serial transmission, there may be uneven gaps between frames.

Isochronous

In real-time audio and video, in which uneven delays between frames are not acceptable, synchronous transmission fails. For example, TV images are broadcast at the rate of 30 images per second; they must be viewed at the same rate. If each image is sent by using one or more frames, there should be no delays between frames. For this type of application, synchronization between characters is not enough; the entire stream of bits must be synchronized. The **isochronous transmission** guarantees that the data arrive at a fixed rate.

4.4 RECOMMENDED READING

For more details about subjects discussed in this chapter, we recommend the following books. The items in brackets [. . .] refer to the reference list at the end of the text.

Books

Digital to digital conversion is discussed in Chapter 7 of [Pea92], Chapter 3 of [Cou01], and Section 5.1 of [Sta04]. Sampling is discussed in Chapters 15, 16, 17, and 18 of [Pea92], Chapter 3 of [Cou01], and Section 5.3 of [Sta04]. [Hsu03] gives a good mathematical approach to modulation and sampling. More advanced materials can be found in [Ber96].

4.5 KEY TERMS

adaptive delta modulation
alternate mark inversion (AMI)
analog-to-digital conversion
asynchronous transmission
baseline
baseline wandering
baud rate
biphase
bipolar
bipolar with *8*-zero substitution (B*8*ZS)
bit rate
block coding
companding and expanding
data element
data rate
DC component
delta modulation (DM)
differential Manchester
digital-to-digital conversion
digitization

eight binary/ten binary (8B/10B)
eight-binary, six-ternary (8B6T)
four binary/five binary (4B/5B)
four dimensional, five-level pulse amplitude modulation (4D-PAM5)
high-density bipolar 3-zero (HDB3)
isochronous transmission
line coding
Manchester
modulation rate
multilevel binary
multiline transmission, 3 level (MLT-3)
nonreturn to zero (NRZ)
nonreturn to zero, invert (NRZ-I)
nonreturn to zero, level (NRZ-L)
Nyquist theorem
parallel transmission
polar
pseudoternary
pulse amplitude modulation (PAM)
pulse code modulation (PCM)
pulse rate
quantization
quantization error
return to zero (RZ)
sampling
sampling rate
scrambling
self-synchronizing
serial transmission
signal element
signal rate
start bit
stop bit
synchronous transmission
transmission mode
two-binary, one quaternary (2B1Q)
unipolar

4.6 SUMMARY

- Digital-to-digital conversion involves three techniques: line coding, block coding, and scrambling.
- Line coding is the process of converting digital data to a digital signal.
- We can roughly divide line coding schemes into five broad categories: unipolar, polar, bipolar, multilevel, and multitransition.
- Block coding provides redundancy to ensure synchronization and inherent error detection. Block coding is normally referred to as *mB/nB* coding; it replaces each *m*-bit group with an *n*-bit group.
- Scrambling provides synchronization without increasing the number of bits. Two common scrambling techniques are B8ZS and HDB3.
- The most common technique to change an analog signal to digital data (digitization) is called pulse code modulation (PCM).
- The first step in PCM is sampling. The analog signal is sampled every T_s s, where T_s is the sample interval or period. The inverse of the sampling interval is called the *sampling rate* or *sampling frequency* and denoted by f_s, where $f_s = 1/T_s$. There are three sampling methods—ideal, natural, and flat-top.
- According to the *Nyquist theorem,* to reproduce the original analog signal, one necessary condition is that the *sampling rate* be at least twice the highest frequency in the original signal.

- Other sampling techniques have been developed to reduce the complexity of PCM. The simplest is *delta modulation*. PCM finds the value of the signal amplitude for each sample; DM finds the change from the previous sample.
- While there is only one way to send parallel data, there are three subclasses of serial transmission: asynchronous, synchronous, and isochronous.
- In asynchronous transmission, we send 1 start bit (0) at the beginning and 1 or more stop bits (1s) at the end of each byte.
- In synchronous transmission, we send bits one after another without start or stop bits or gaps. It is the responsibility of the receiver to group the bits.
- The isochronous mode provides synchronized for the entire stream of bits must. In other words, it guarantees that the data arrive at a fixed rate.

4.7 PRACTICE SET

Review Questions

1. List three techniques of digital-to-digital conversion.
2. Distinguish between a signal element and a data element.
3. Distinguish between data rate and signal rate.
4. Define baseline wandering and its effect on digital transmission.
5. Define a DC component and its effect on digital transmission.
6. Define the characteristics of a self-synchronizing signal.
7. List five line coding schemes discussed in this book.
8. Define block coding and give its purpose.
9. Define scrambling and give its purpose.
10. Compare and contrast PCM and DM.
11. What are the differences between parallel and serial transmission?
12. List three different techniques in serial transmission and explain the differences.

Exercises

13. Calculate the value of the signal rate for each case in Figure 4.2 if the data rate is 1 Mbps and $c = 1/2$.
14. In a digital transmission, the sender clock is 0.2 percent faster than the receiver clock. How many extra bits per second does the sender send if the data rate is 1 Mbps?
15. Draw the graph of the NRZ-L scheme using each of the following data streams, assuming that the last signal level has been positive. From the graphs, guess the bandwidth for this scheme using the average number of changes in the signal level. Compare your guess with the corresponding entry in Table 4.1.
    a. 00000000
    b. 11111111
    c. 01010101
    d. 00110011

16. Repeat Exercise 15 for the NRZ-I scheme.
17. Repeat Exercise 15 for the Manchester scheme.
18. Repeat Exercise 15 for the differential Manchester scheme.
19. Repeat Exercise 15 for the 2B1Q scheme, but use the following data streams.
    a. 0000000000000000
    b. 1111111111111111
    c. 0101010101010101
    d. 0011001100110011
20. Repeat Exercise 15 for the MLT-3 scheme, but use the following data streams.
    a. 00000000
    b. 11111111
    c. 01010101
    d. 00011000
21. Find the 8-bit data stream for each case depicted in Figure 4.36.

Figure 4.36 *Exercise 21*

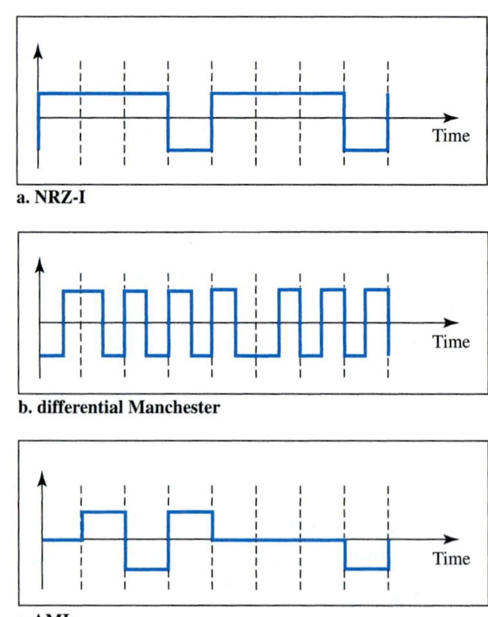

22. An NRZ-I signal has a data rate of 100 Kbps. Using Figure 4.6, calculate the value of the normalized energy (P) for frequencies at 0 Hz, 50 KHz, and 100 KHz.
23. A Manchester signal has a data rate of 100 Kbps. Using Figure 4.8, calculate the value of the normalized energy (P) for frequencies at 0 Hz, 50 KHz, 100 KHz.

24. The input stream to a 4B/5B block encoder is 0100 0000 0000 0000 0000 0001. Answer the following questions:
    a. What is the output stream?
    b. What is the length of the longest consecutive sequence of 0s in the input?
    c. What is the length of the longest consecutive sequence of 0s in the output?
25. How many invalid (unused) code sequences can we have in 5B/6B encoding? How many in 3B/4B encoding?
26. What is the result of scrambling the sequence 11100000000000 using one of the following scrambling techniques? Assume that the last non-zero signal level has been positive.
    a. B8ZS
    b. HDB3 (The number of nonzero pules is odd after the last substitution)
27. What is the Nyquist sampling rate for each of the following signals?
    a. A low-pass signal with bandwidth of 200 KHz?
    b. A band-pass signal with bandwidth of 200 KHz if the lowest frequency is 100 KHz?
28. We have sampled a low-pass signal with a bandwidth of 200 KHz using 1024 levels of quantization.
    a. Calculate the bit rate of the digitized signal.
    b. Calculate the SNR_{dB} for this signal.
    c. Calculate the PCM bandwidth of this signal.
29. What is the maximum data rate of a channel with a bandwidth of 200 KHz if we use four levels of digital signaling.
30. An analog signal has a bandwidth of 20 KHz. If we sample this signal and send it through a 30 Kbps channel what is the SNR_{dB}?
31. We have a baseband channel with a 1-MHz bandwidth. What is the data rate for this channel if we use one of the following line coding schemes?
    a. NRZ-L
    b. Manchester
    c. MLT-3
    d. 2B1Q
32. We want to transmit 1000 characters with each character encoded as 8 bits.
    a. Find the number of transmitted bits for synchronous transmission.
    b. Find the number of transmitted bits for asynchronous transmission.
    c. Find the redundancy percent in each case.

CHAPTER 5

Analog Transmission

In Chapter 3, we discussed the advantages and disadvantages of digital and analog transmission. We saw that while digital transmission is very desirable, a low-pass channel is needed. We also saw that analog transmission is the only choice if we have a bandpass channel. Digital transmission was discussed in Chapter 4; we discuss analog transmission in this chapter.

Converting digital data to a bandpass analog signal is traditionally called digital-to-analog conversion. Converting a low-pass analog signal to a bandpass analog signal is traditionally called analog-to-analog conversion. In this chapter, we discuss these two types of conversions.

5.1 DIGITAL-TO-ANALOG CONVERSION

Digital-to-analog conversion is the process of changing one of the characteristics of an analog signal based on the information in digital data. Figure 5.1 shows the relationship between the digital information, the digital-to-analog modulating process, and the resultant analog signal.

Figure 5.1 *Digital-to-analog conversion*

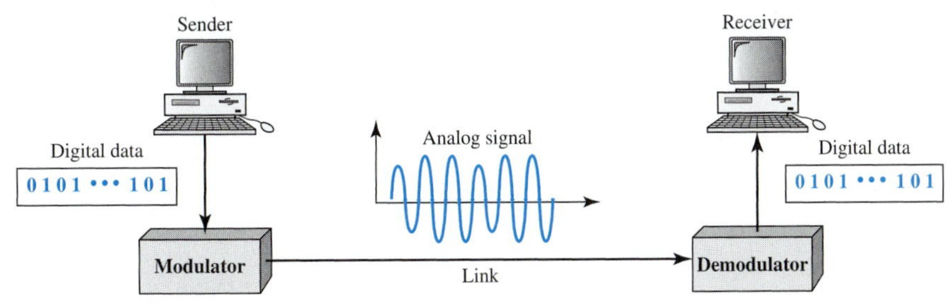

As discussed in Chapter 3, a sine wave is defined by three characteristics: amplitude, frequency, and phase. When we vary any one of these characteristics, we create a different version of that wave. So, by changing one characteristic of a simple electric signal, we can use it to represent digital data. Any of the three characteristics can be altered in this way, giving us at least three mechanisms for modulating digital data into an analog signal: **amplitude shift keying (ASK), frequency shift keying (FSK),** and **phase shift keying (PSK).** In addition, there is a fourth (and better) mechanism that combines changing both the amplitude and phase, called **quadrature amplitude modulation (QAM).** QAM is the most efficient of these options and is the mechanism commonly used today (see Figure 5.2).

Figure 5.2 *Types of digital-to-analog conversion*

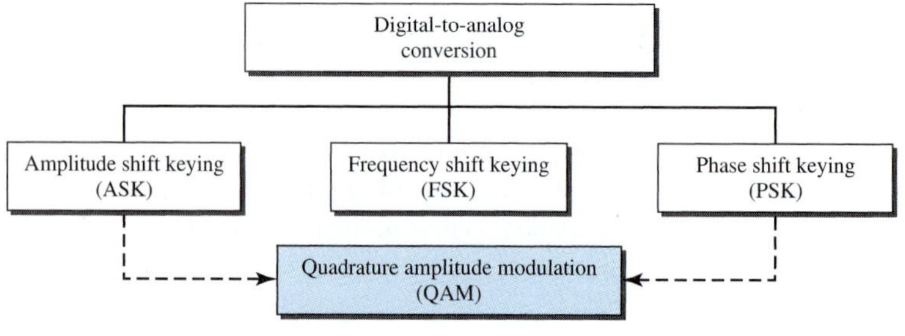

Aspects of Digital-to-Analog Conversion

Before we discuss specific methods of digital-to-analog modulation, two basic issues must be reviewed: bit and baud rates and the carrier signal.

Data Element Versus Signal Element

In Chapter 4, we discussed the concept of the data element versus the signal element. We defined a data element as the smallest piece of information to be exchanged, the bit. We also defined a signal element as the smallest unit of a signal that is constant. Although we continue to use the same terms in this chapter, we will see that the nature of the signal element is a little bit different in analog transmission.

Data Rate Versus Signal Rate

We can define the data rate (bit rate) and the signal rate (baud rate) as we did for digital transmission. The relationship between them is

$$S = N \times \frac{1}{r} \quad \textbf{baud}$$

where N is the data rate (bps) and r is the number of data elements carried in one signal element. The value of r in analog transmission is $r = \log_2 L$, where L is the type of signal element, not the level. The same nomenclature is used to simplify the comparisons.

Bit rate is the number of bits per second. Baud rate is the number of signal elements per second. In the analog transmission of digital data, the baud rate is less than or equal to the bit rate.

The same analogy we used in Chapter 4 for bit rate and baud rate applies here. In transportation, a baud is analogous to a vehicle, and a bit is analogous to a passenger. We need to maximize the number of people per car to reduce the traffic.

Example 5.1

An analog signal carries 4 bits per signal element. If 1000 signal elements are sent per second, find the bit rate.

Solution

In this case, $r = 4$, $S = 1000$, and N is unknown. We can find the value of N from

$$S = N \times \frac{1}{r} \quad \text{or} \quad N = S \times r = 1000 \times 4 = 4000 \text{ bps}$$

Example 5.2

An analog signal has a bit rate of 8000 bps and a baud rate of 1000 baud. How many data elements are carried by each signal element? How many signal elements do we need?

Solution

In this example, $S = 1000$, $N = 8000$, and r and L are unknown. We find first the value of r and then the value of L.

$$S = N \times \frac{1}{r} \rightarrow r = \frac{N}{S} = \frac{8000}{1000} = 8 \text{ bits/baud}$$

$$r = \log_2 L \rightarrow L = 2^r = 2^8 = 256$$

Bandwidth

The required bandwidth for analog transmission of digital data is proportional to the signal rate except for FSK, in which the difference between the carrier signals needs to be added. We discuss the bandwidth for each technique.

Carrier Signal

In analog transmission, the sending device produces a high-frequency signal that acts as a base for the information signal. This base signal is called the **carrier signal** or carrier frequency. The receiving device is tuned to the frequency of the carrier signal that it expects from the sender. Digital information then changes the carrier signal by modifying one or more of its characteristics (amplitude, frequency, or phase). This kind of modification is called modulation (shift keying).

Amplitude Shift Keying

In amplitude shift keying, the amplitude of the carrier signal is varied to create signal elements. Both frequency and phase remain constant while the amplitude changes.

Binary ASK (BASK)

Although we can have several levels (kinds) of signal elements, each with a different amplitude, ASK is normally implemented using only two levels. This is referred to as binary amplitude shift keying or *on-off keying* (OOK). The peak amplitude of one signal level is 0; the other is the same as the amplitude of the carrier frequency. Figure 5.3 gives a conceptual view of binary ASK.

Figure 5.3 *Binary amplitude shift keying*

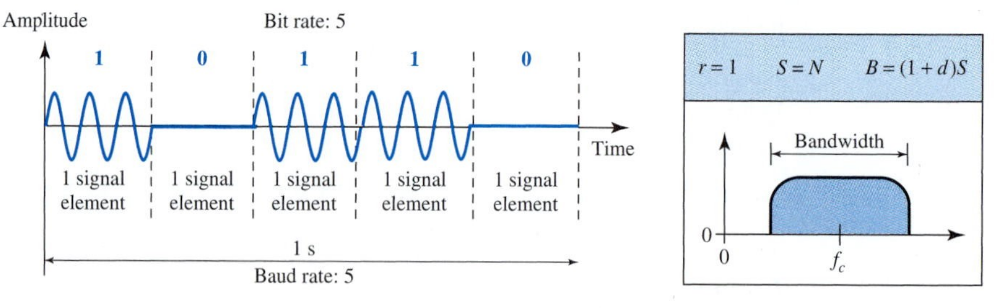

Bandwidth for ASK Figure 5.3 also shows the bandwidth for ASK. Although the carrier signal is only one simple sine wave, the process of modulation produces a nonperiodic composite signal. This signal, as was discussed in Chapter 3, has a continuous set of frequencies. As we expect, the bandwidth is proportional to the signal rate (baud rate). However, there is normally another factor involved, called *d*, which depends on the modulation and filtering process. The value of *d* is between 0 and 1. This means that the bandwidth can be expressed as shown, where *S* is the signal rate and the *B* is the bandwidth.

$$B = (1 + d) \times S$$

The formula shows that the required bandwidth has a minimum value of *S* and a maximum value of 2*S*. The most important point here is the location of the bandwidth. The middle of the bandwidth is where f_c, the carrier frequency, is located. This means if we have a bandpass channel available, we can choose our f_c so that the modulated signal occupies that bandwidth. This is in fact the most important advantage of digital-to-analog conversion. We can shift the resulting bandwidth to match what is available.

Implementation The complete discussion of ASK implementation is beyond the scope of this book. However, the simple ideas behind the implementation may help us to better understand the concept itself. Figure 5.4 shows how we can simply implement binary ASK.

If digital data are presented as a unipolar NRZ (see Chapter 4) digital signal with a high voltage of 1 V and a low voltage of 0 V, the implementation can achieved by multiplying the NRZ digital signal by the carrier signal coming from an oscillator. When the amplitude of the NRZ signal is 1, the amplitude of the carrier frequency is

Figure 5.4 *Implementation of binary ASK*

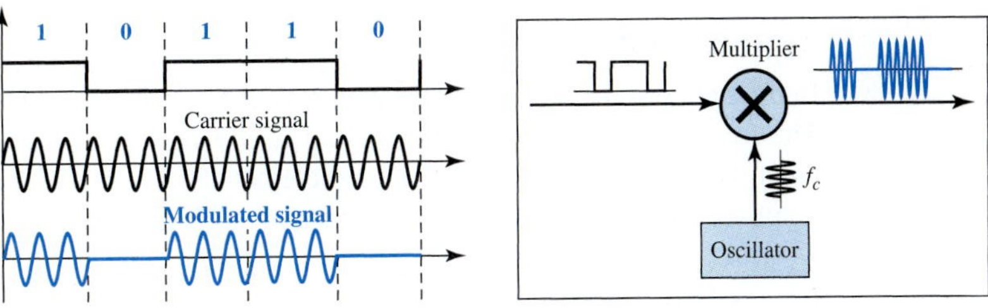

held; when the amplitude of the NRZ signal is 0, the amplitude of the carrier frequency is zero.

Example 5.3

We have an available bandwidth of 100 kHz which spans from 200 to 300 kHz. What are the carrier frequency and the bit rate if we modulated our data by using ASK with $d = 1$?

Solution

The middle of the bandwidth is located at 250 kHz. This means that our carrier frequency can be at $f_c = 250$ kHz. We can use the formula for bandwidth to find the bit rate (with $d = 1$ and $r = 1$).

$$B = (1 + d) \times S = 2 \times N \times \frac{1}{r} = 2 \times N = 100 \text{ kHz} \quad \Rightarrow \quad N = 50 \text{ kbps}$$

Example 5.4

In data communications, we normally use full-duplex links with communication in both directions. We need to divide the bandwidth into two with two carrier frequencies, as shown in Figure 5.5. The figure shows the positions of two carrier frequencies and the bandwidths.The available bandwidth for each direction is now 50 kHz, which leaves us with a data rate of 25 kbps in each direction.

Figure 5.5 *Bandwidth of full-duplex ASK used in Example 5.4*

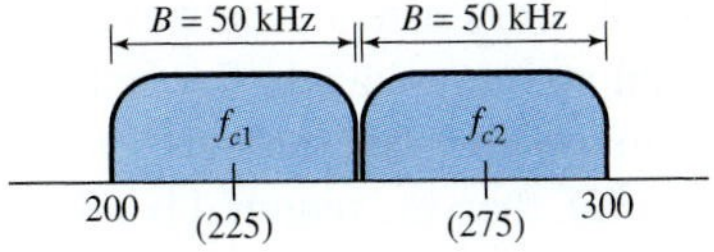

Multilevel ASK

The above discussion uses only two amplitude levels. We can have multilevel ASK in which there are more than two levels. We can use 4, 8, 16, or more different amplitudes for the signal and modulate the data using 2, 3, 4, or more bits at a time. In these cases,

$r = 2$, $r = 3$, $r = 4$, and so on. Although this is not implemented with pure ASK, it is implemented with QAM (as we will see later).

Frequency Shift Keying

In frequency shift keying, the frequency of the carrier signal is varied to represent data. The frequency of the modulated signal is constant for the duration of one signal element, but changes for the next signal element if the data element changes. Both peak amplitude and phase remain constant for all signal elements.

Binary FSK (BFSK)

One way to think about binary FSK (or BFSK) is to consider two carrier frequencies. In Figure 5.6, we have selected two carrier frequencies, f_1 and f_2. We use the first carrier if the data element is 0; we use the second if the data element is 1. However, note that this is an unrealistic example used only for demonstration purposes. Normally the carrier frequencies are very high, and the difference between them is very small.

Figure 5.6 *Binary frequency shift keying*

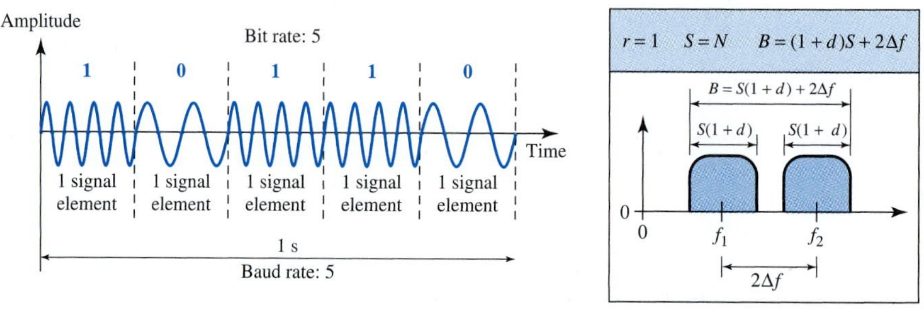

As Figure 5.6 shows, the middle of one bandwidth is f_1 and the middle of the other is f_2. Both f_1 and f_2 are Δf apart from the midpoint between the two bands. The difference between the two frequencies is $2\Delta f$.

Bandwidth for BFSK Figure 5.6 also shows the bandwidth of FSK. Again the carrier signals are only simple sine waves, but the modulation creates a nonperiodic composite signal with continuous frequencies. We can think of FSK as two ASK signals, each with its own carrier frequency (f_1 or f_2). If the difference between the two frequencies is $2\Delta f$, then the required bandwidth is

$$B = (1 + d) \times S + 2\Delta f$$

What should be the minimum value of $2\Delta f$? In Figure 5.6, we have chosen a value greater than $(1 + d)S$. It can be shown that the minimum value should be at least S for the proper operation of modulation and demodulation.

Example 5.5

We have an available bandwidth of 100 kHz which spans from 200 to 300 kHz. What should be the carrier frequency and the bit rate if we modulated our data by using FSK with $d = 1$?

Solution

This problem is similar to Example 5.3, but we are modulating by using FSK. The midpoint of the band is at 250 kHz. We choose $2\Delta f$ to be 50 kHz; this means

$$B = (1 + d) \times S + 2\Delta f = 100 \quad \rightarrow \quad 2S = 50 \text{ kHz} \quad S = 25 \text{ kbaud} \quad N = 25 \text{ kbps}$$

Compared to Example 5.3, we can see the bit rate for ASK is 50 kbps while the bit rate for FSK is 25 kbps.

Implementation There are two implementations of BFSK: noncoherent and coherent. In noncoherent BFSK, there may be discontinuity in the phase when one signal element ends and the next begins. In coherent BFSK, the phase continues through the boundary of two signal elements. Noncoherent BFSK can be implemented by treating BFSK as two ASK modulations and using two carrier frequencies. Coherent BFSK can be implemented by using one *voltage-controlled oscillator* (VCO) that changes its frequency according to the input voltage. Figure 5.7 shows the simplified idea behind the second implementation. The input to the oscillator is the unipolar NRZ signal. When the amplitude of NRZ is zero, the oscillator keeps its regular frequency; when the amplitude is positive, the frequency is increased.

Figure 5.7 *Implementation of BFSK*

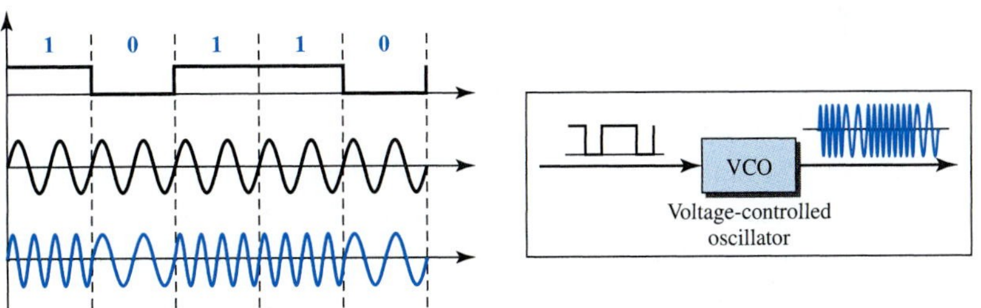

Multilevel FSK

Multilevel modulation (MFSK) is not uncommon with the FSK method. We can use more than two frequencies. For example, we can use four different frequencies f_1, f_2, f_3, and f_4 to send 2 bits at a time. To send 3 bits at a time, we can use eight frequencies. And so on. However, we need to remember that the frequencies need to be $2\Delta f$ apart. For the proper operation of the modulator and demodulator, it can be shown that the minimum value of $2\Delta f$ needs to be S. We can show that the bandwidth with $d = 0$ is

$$B = (1 + d) \times S + (L - 1)2\Delta f \quad \rightarrow \quad B = L \times S$$

Example 5.6

We need to send data 3 bits at a time at a bit rate of 3 Mbps. The carrier frequency is 10 MHz. Calculate the number of levels (different frequencies), the baud rate, and the bandwidth.

Solution

We can have $L = 2^3 = 8$. The baud rate is $S = 3$ MHz/3 = 1000 Mbaud. This means that the carrier frequencies must be 1 MHz apart ($2\Delta f = 1$ MHz). The bandwidth is $B = 8 \times 1000 = 8000$. Figure 5.8 shows the allocation of frequencies and bandwidth.

Figure 5.8 *Bandwidth of MFSK used in Example 5.6*

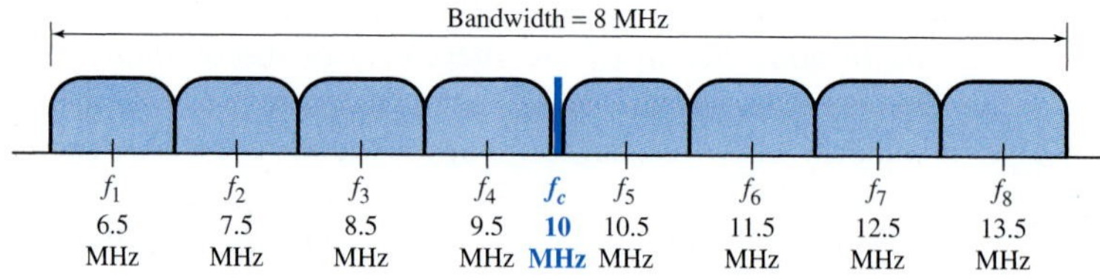

Phase Shift Keying

In phase shift keying, the phase of the carrier is varied to represent two or more different signal elements. Both peak amplitude and frequency remain constant as the phase changes. Today, PSK is more common than ASK or FSK. However, we will see shortly that QAM, which combines ASK and PSK, is the dominant method of digital-to-analog modulation.

Binary PSK (BPSK)

The simplest PSK is binary PSK, in which we have only two signal elements, one with a phase of 0°, and the other with a phase of 180°. Figure 5.9 gives a conceptual view of PSK. Binary PSK is as simple as binary ASK with one big advantage—it is less

Figure 5.9 *Binary phase shift keying*

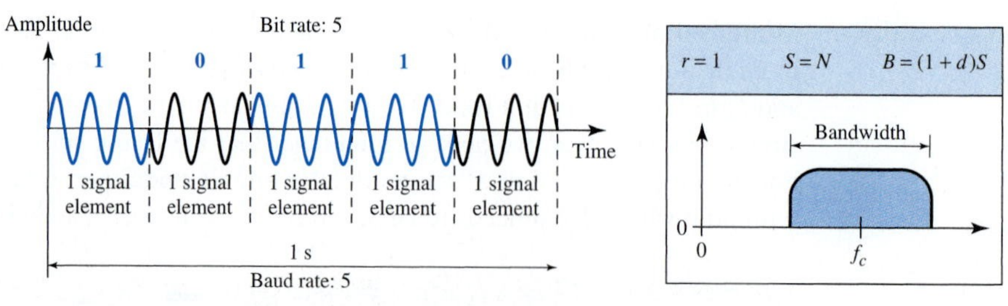

susceptible to noise. In ASK, the criterion for bit detection is the amplitude of the signal; in PSK, it is the phase. Noise can change the amplitude easier than it can change the phase. In other words, PSK is less susceptible to noise than ASK. PSK is superior to FSK because we do not need two carrier signals.

Bandwidth Figure 5.9 also shows the bandwidth for BPSK. The bandwidth is the same as that for binary ASK, but less than that for BFSK. No bandwidth is wasted for separating two carrier signals.

Implementation The implementation of BPSK is as simple as that for ASK. The reason is that the signal element with phase 180° can be seen as the complement of the signal element with phase 0°. This gives us a clue on how to implement BPSK. We use the same idea we used for ASK but with a polar NRZ signal instead of a unipolar NRZ signal, as shown in Figure 5.10. The polar NRZ signal is multiplied by the carrier frequency; the 1 bit (positive voltage) is represented by a phase starting at 0°; the 0 bit (negative voltage) is represented by a phase starting at 180°.

Figure 5.10 *Implementation of BASK*

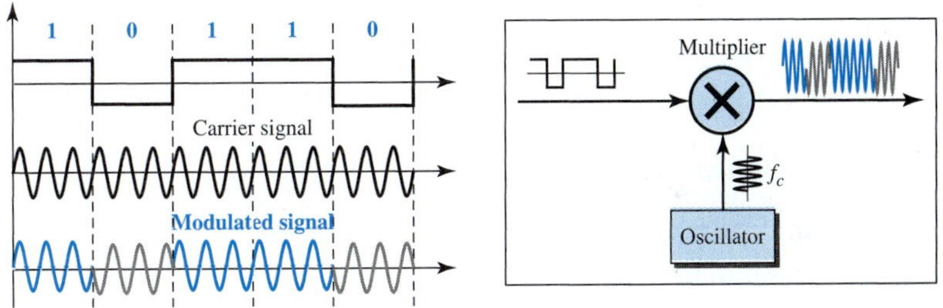

Quadrature PSK (QPSK)

The simplicity of BPSK enticed designers to use 2 bits at a time in each signal element, thereby decreasing the baud rate and eventually the required bandwidth. The scheme is called quadrature PSK or QPSK because it uses two separate BPSK modulations; one is in-phase, the other quadrature (out-of-phase). The incoming bits are first passed through a serial-to-parallel conversion that sends one bit to one modulator and the next bit to the other modulator. If the duration of each bit in the incoming signal is T, the duration of each bit sent to the corresponding BPSK signal is $2T$. This means that the bit to each BPSK signal has one-half the frequency of the original signal. Figure 5.11 shows the idea.

The two composite signals created by each multiplier are sine waves with the same frequency, but different phases. When they are added, the result is another sine wave, with one of four possible phases: 45°, −45°, 135°, and −135°. There are four kinds of signal elements in the output signal ($L = 4$), so we can send 2 bits per signal element ($r = 2$).

Figure 5.11 *QPSK and its implementation*

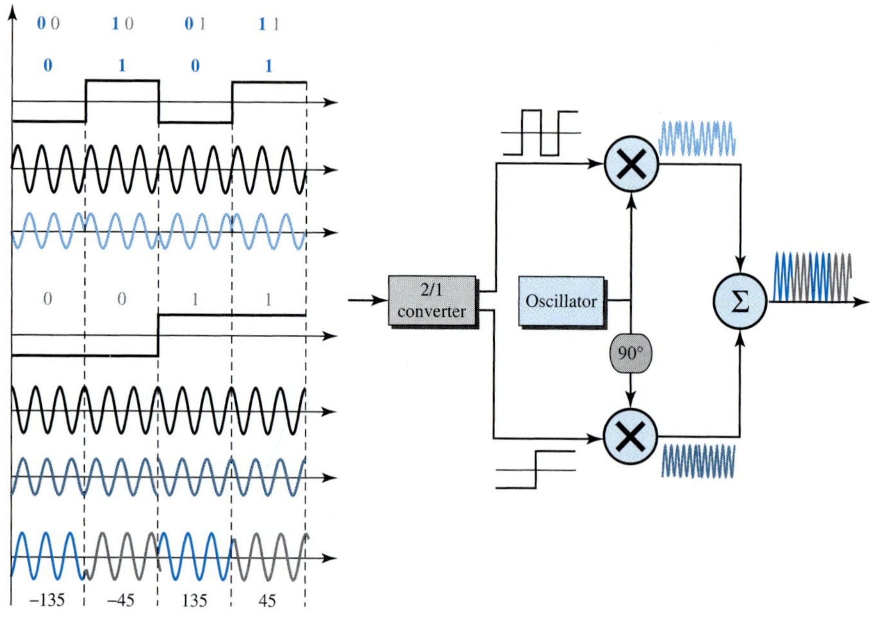

Example 5.7

Find the bandwidth for a signal transmitting at 12 Mbps for QPSK. The value of $d = 0$.

Solution

For QPSK, 2 bits is carried by one signal element. This means that $r = 2$. So the signal rate (baud rate) is $S = N \times (1/r) = 6$ Mbaud. With a value of $d = 0$, we have $B = S = 6$ MHz.

Constellation Diagram

A **constellation diagram** can help us define the amplitude and phase of a signal element, particularly when we are using two carriers (one in-phase and one quadrature). The diagram is useful when we are dealing with multilevel ASK, PSK, or QAM (see next section). In a constellation diagram, a signal element type is represented as a dot. The bit or combination of bits it can carry is often written next to it.

The diagram has two axes. The horizontal X axis is related to the in-phase carrier; the vertical Y axis is related to the quadrature carrier. For each point on the diagram, four pieces of information can be deduced. The projection of the point on the X axis defines the peak amplitude of the in-phase component; the projection of the point on the Y axis defines the peak amplitude of the quadrature component. The length of the line (vector) that connects the point to the origin is the peak amplitude of the signal element (combination of the X and Y components); the angle the line makes with the X axis is the phase of the signal element. All the information we need, can easily be found on a constellation diagram. Figure 5.12 shows a constellation diagram.

Figure 5.12 *Concept of a constellation diagram*

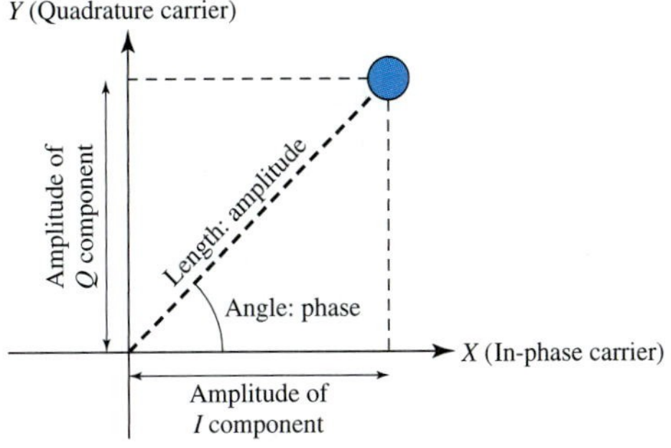

Example 5.8

Show the constellation diagrams for an ASK (OOK), BPSK, and QPSK signals.

Solution

Figure 5.13 shows the three constellation diagrams.

Figure 5.13 *Three constellation diagrams*

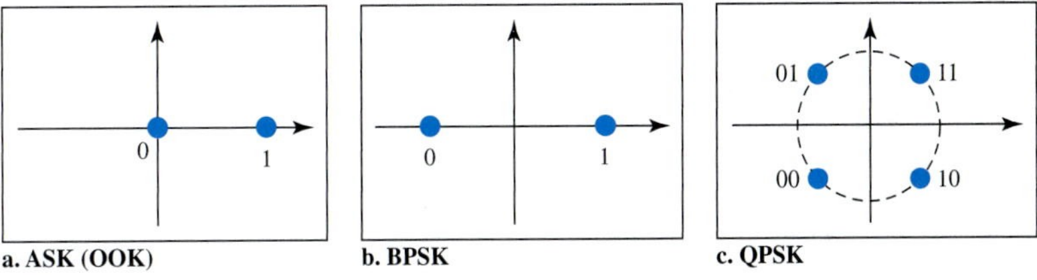

Let us analyze each case separately:

a. For ASK, we are using only an in-phase carrier. Therefore, the two points should be on the X axis. Binary 0 has an amplitude of 0 V; binary 1 has an amplitude of 1 V (for example). The points are located at the origin and at 1 unit.

b. BPSK also uses only an in-phase carrier. However, we use a polar NRZ signal for modulation. It creates two types of signal elements, one with amplitude 1 and the other with amplitude −1. This can be stated in other words: BPSK creates two different signal elements, one with amplitude 1 V and in phase and the other with amplitude 1 V and 180° out of phase.

c. QPSK uses two carriers, one in-phase and the other quadrature. The point representing 11 is made of two combined signal elements, both with an amplitude of 1 V. One element is represented by an in-phase carrier, the other element by a quadrature carrier. The amplitude of the final signal element sent for this 2-bit data element is $2^{1/2}$, and the phase is 45°. The argument is similar for the other three points. All signal elements have an amplitude of $2^{1/2}$, but their phases are different (45°, 135°, −135°, and −45°). Of course, we could have chosen the amplitude of the carrier to be $1/(2^{1/2})$ to make the final amplitudes 1 V.

Quadrature Amplitude Modulation

PSK is limited by the ability of the equipment to distinguish small differences in phase. This factor limits its potential bit rate. So far, we have been altering only one of the three characteristics of a sine wave at a time; but what if we alter two? Why not combine ASK and PSK? The idea of using two carriers, one in-phase and the other quadrature, with different amplitude levels for each carrier is the concept behind **quadrature amplitude modulation (QAM).**

Quadrature amplitude modulation is a combination of ASK and PSK.

The possible variations of QAM are numerous. Figure 5.14 shows some of these schemes. Figure 5.14a shows the simplest 4-QAM scheme (four different signal element types) using a unipolar NRZ signal to modulate each carrier. This is the same mechanism we used for ASK (OOK). Part b shows another 4-QAM using polar NRZ, but this is exactly the same as QPSK. Part c shows another QAM-4 in which we used a signal with two positive levels to modulate each of the two carriers. Finally, Figure 5.14d shows a 16-QAM constellation of a signal with eight levels, four positive and four negative.

Figure 5.14 *Constellation diagrams for some QAMs*

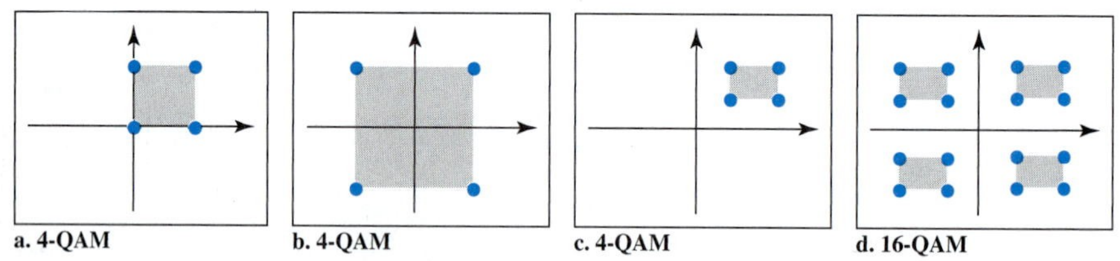

Bandwidth for QAM

The minimum bandwidth required for QAM transmission is the same as that required for ASK and PSK transmission. QAM has the same advantages as PSK over ASK.

5.2 ANALOG-TO-ANALOG CONVERSION

Analog-to-analog conversion, or analog modulation, is the representation of analog information by an analog signal. One may ask why we need to modulate an analog signal; it is already analog. Modulation is needed if the medium is bandpass in nature or if only a bandpass channel is available to us. An example is radio. The government assigns a narrow bandwidth to each radio station. The analog signal produced by each station is a low-pass signal, all in the same range. To be able to listen to different stations, the low-pass signals need to be shifted, each to a different range.

Analog-to-analog conversion can be accomplished in three ways: **amplitude modulation (AM), frequency modulation (FM),** and **phase modulation (PM).** FM and PM are usually categorized together. See Figure 5.15.

Figure 5.15 *Types of analog-to-analog modulation*

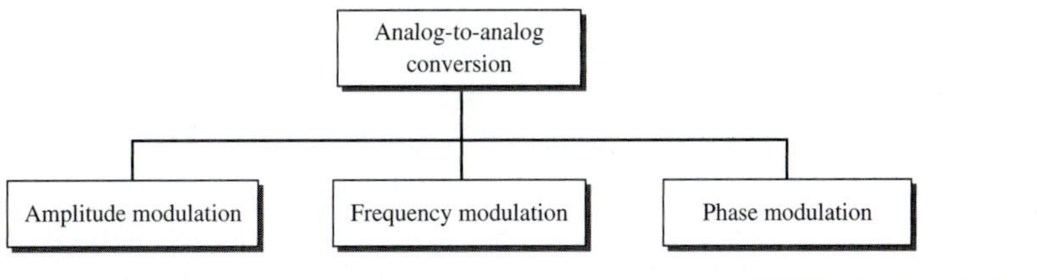

Amplitude Modulation

In AM transmission, the carrier signal is modulated so that its amplitude varies with the changing amplitudes of the modulating signal. The frequency and phase of the carrier remain the same; only the amplitude changes to follow variations in the information. Figure 5.16 shows how this concept works. The modulating signal is the envelope of the carrier.

Figure 5.16 *Amplitude modulation*

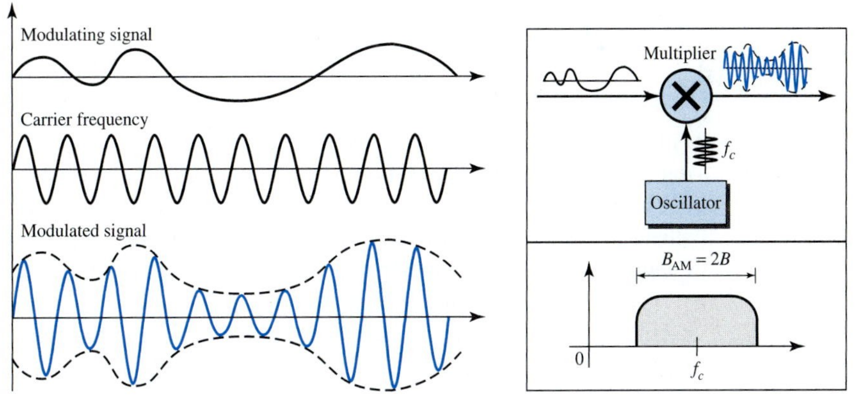

As Figure 5.16 shows, AM is normally implemented by using a simple multiplier because the amplitude of the carrier signal needs to be changed according to the amplitude of the modulating signal.

AM Bandwidth

Figure 5.16 also shows the bandwidth of an AM signal. The modulation creates a bandwidth that is twice the bandwidth of the modulating signal and covers a range centered on the carrier frequency. However, the signal components above and below the carrier frequency carry exactly the same information. For this reason, some implementations discard one-half of the signals and cut the bandwidth in half.

The total bandwidth required for AM can be determined from the bandwidth of the audio signal: $B_{AM} = 2B$.

Standard Bandwidth Allocation for AM Radio

The bandwidth of an audio signal (speech and music) is usually 5 kHz. Therefore, an AM radio station needs a bandwidth of 10 kHz. In fact, the Federal Communications Commission (FCC) allows 10 kHz for each AM station.

AM stations are allowed carrier frequencies anywhere between 530 and 1700 kHz (1.7 MHz). However, each station's carrier frequency must be separated from those on either side of it by at least 10 kHz (one AM bandwidth) to avoid interference. If one station uses a carrier frequency of 1100 kHz, the next station's carrier frequency cannot be lower than 1110 kHz (see Figure 5.17).

Figure 5.17 *AM band allocation*

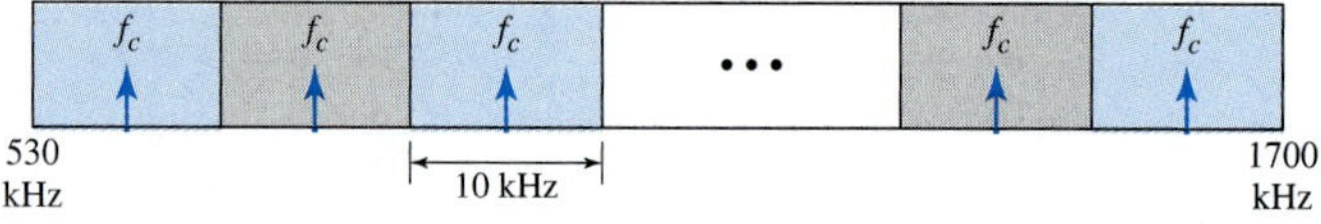

Frequency Modulation

In FM transmission, the frequency of the carrier signal is modulated to follow the changing voltage level (amplitude) of the modulating signal. The peak amplitude and phase of the carrier signal remain constant, but as the amplitude of the information signal changes, the frequency of the carrier changes correspondingly. Figure 5.18 shows the relationships of the modulating signal, the carrier signal, and the resultant FM signal.

As Figure 5.18 shows, FM is normally implemented by using a voltage-controlled oscillator as with FSK. The frequency of the oscillator changes according to the input voltage which is the amplitude of the modulating signal.

FM Bandwidth

Figure 5.18 also shows the bandwidth of an FM signal. The actual bandwidth is difficult to determine exactly, but it can be shown empirically that it is several times that of the analog signal or $2(1 + \beta)B$ where β is a factor depends on modulation technique with a common value of 4.

The total bandwidth required for FM can be determined from the bandwidth of the audio signal: $B_{FM} = 2(1 + \beta)B$.

Figure 5.18 *Frequency modulation*

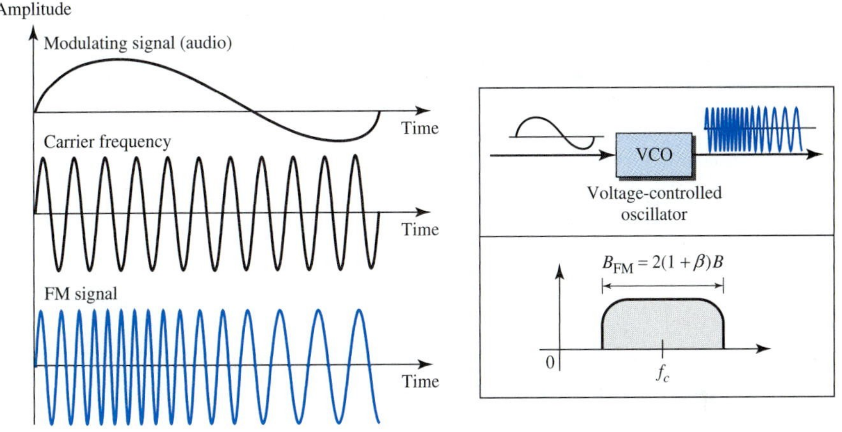

Standard Bandwidth Allocation for FM Radio

The bandwidth of an audio signal (speech and music) broadcast in stereo is almost 15 kHz. The FCC allows 200 kHz (0.2 MHz) for each station. This mean $\beta = 4$ with some extra guard band. FM stations are allowed carrier frequencies anywhere between 88 and 108 MHz. Stations must be separated by at least 200 kHz to keep their bandwidths from overlapping. To create even more privacy, the FCC requires that in a given area, only alternate bandwidth allocations may be used. The others remain unused to prevent any possibility of two stations interfering with each other. Given 88 to 108 MHz as a range, there are 100 potential FM bandwidths in an area, of which 50 can operate at any one time. Figure 5.19 illustrates this concept.

Figure 5.19 *FM band allocation*

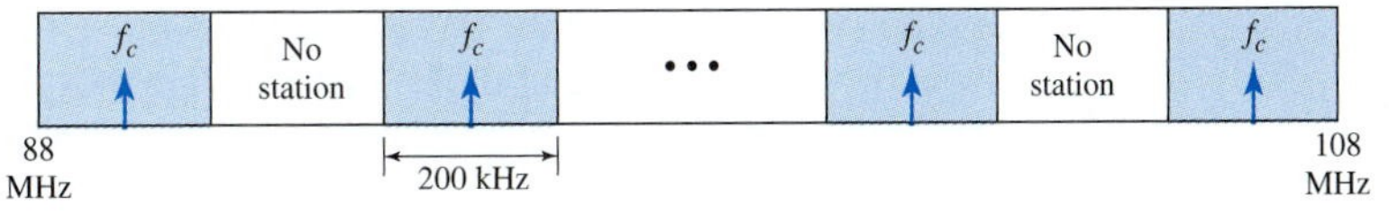

Phase Modulation

In PM transmission, the phase of the carrier signal is modulated to follow the changing voltage level (amplitude) of the modulating signal. The peak amplitude and frequency of the carrier signal remain constant, but as the amplitude of the information signal changes, the phase of the carrier changes correspondingly. It can proved mathematically (see Appendix C) that PM is the same as FM with one difference. In FM, the instantaneous change in the carrier frequency is proportional to the amplitude of the modulating signal; in PM the instantaneous change in the carrier frequency is proportional to the derivative of the amplitude of the modulating signal. Figure 5.20 shows the relationships of the modulating signal, the carrier signal, and the resultant PM signal.

Figure 5.20 *Phase modulation*

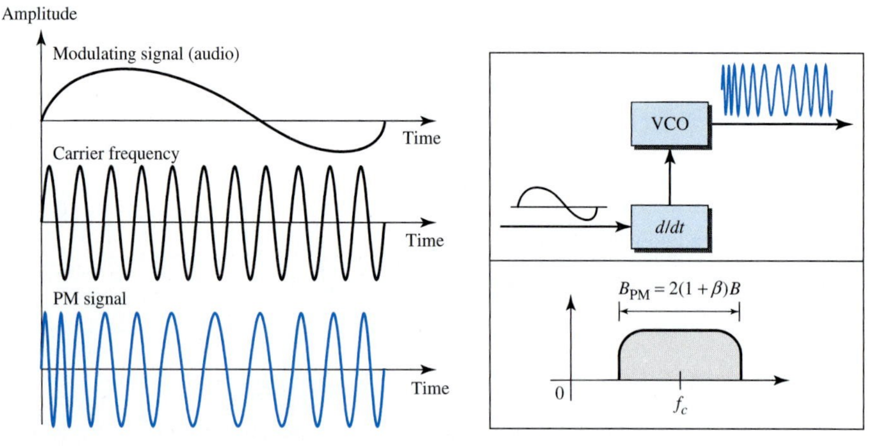

As Figure 5.20 shows, PM is normally implemented by using a voltage-controlled oscillator along with a derivative. The frequency of the oscillator changes according to the derivative of the input voltage which is the amplitude of the modulating signal.

PM Bandwidth

Figure 5.20 also shows the bandwidth of a PM signal. The actual bandwidth is difficult to determine exactly, but it can be shown empirically that it is several times that of the analog signal. Although, the formula shows the same bandwidth for FM and PM, the value of β is lower in the case of PM (around 1 for narrowband and 3 for wideband).

The total bandwidth required for PM can be determined from the bandwidth and maximum amplitude of the modulating signal: $B_{PM} = 2(1 + \beta)B$.

5.3 RECOMMENDED READING

For more details about subjects discussed in this chapter, we recommend the following books. The items in brackets [. . .] refer to the reference list at the end of the text.

Books

Digital-to-analog conversion is discussed in Chapter 14 of [Pea92], Chapter 5 of [Cou01], and Section 5.2 of [Sta04]. Analog-to-analog conversion is discussed in Chapters 8 to 13 of [Pea92], Chapter 5 of [Cou01], and Section 5.4 of [Sta04]. [Hsu03] gives a good mathematical approach to all materials discussed in this chapter. More advanced materials can be found in [Ber96].

5.4 KEY TERMS

amplitude modulation (AM)
amplitude shift keying (ASK)
analog-to-analog conversion
carrier signal
constellation diagram
digital-to-analog conversion
frequency modulation (FM)
frequency shift keying (FSK)
phase modulation (PM)
phase shift keying (PSK)
quadrature amplitude modulation (QAM)

5.5 SUMMARY

- Digital-to-analog conversion is the process of changing one of the characteristics of an analog signal based on the information in the digital data.
- Digital-to-analog conversion can be accomplished in several ways: amplitude shift keying (ASK), frequency shift keying (FSK), and phase shift keying (PSK). Quadrature amplitude modulation (QAM) combines ASK and PSK.
- In amplitude shift keying, the amplitude of the carrier signal is varied to create signal elements. Both frequency and phase remain constant while the amplitude changes.
- In frequency shift keying, the frequency of the carrier signal is varied to represent data. The frequency of the modulated signal is constant for the duration of one signal element, but changes for the next signal element if the data element changes. Both peak amplitude and phase remain constant for all signal elements.
- In phase shift keying, the phase of the carrier is varied to represent two or more different signal elements. Both peak amplitude and frequency remain constant as the phase changes.
- A constellation diagram shows us the amplitude and phase of a signal element, particularly when we are using two carriers (one in-phase and one quadrature).
- Quadrature amplitude modulation (QAM) is a combination of ASK and PSK. QAM uses two carriers, one in-phase and the other quadrature, with different amplitude levels for each carrier.
- Analog-to-analog conversion is the representation of analog information by an analog signal. Conversion is needed if the medium is bandpass in nature or if only a bandpass bandwidth is available to us.
- Analog-to-analog conversion can be accomplished in three ways: amplitude modulation (AM), frequency modulation (FM), and phase modulation (PM).
- In AM transmission, the carrier signal is modulated so that its amplitude varies with the changing amplitudes of the modulating signal. The frequency and phase of the carrier remain the same; only the amplitude changes to follow variations in the information.
- In FM transmission, the frequency of the carrier signal is modulated to follow the changing voltage level (amplitude) of the modulating signal. The peak amplitude

and phase of the carrier signal remain constant, but as the amplitude of the information signal changes, the frequency of the carrier changes correspondingly.

- In PM transmission, the phase of the carrier signal is modulated to follow the changing voltage level (amplitude) of the modulating signal. The peak amplitude and frequency of the carrier signal remain constant, but as the amplitude of the information signal changes, the phase of the carrier changes correspondingly.

5.6 PRACTICE SET

Review Questions

1. Define analog transmission.
2. Define carrier signal and its role in analog transmission.
3. Define digital-to-analog conversion.
4. Which characteristics of an analog signal are changed to represent the digital signal in each of the following digital-to-analog conversion?
   a. ASK
   b. FSK
   c. PSK
   d. QAM
5. Which of the four digital-to-analog conversion techniques (ASK, FSK, PSK or QAM) is the most susceptible to noise? Defend your answer.
6. Define constellation diagram and its role in analog transmission.
7. What are the two components of a signal when the signal is represented on a constellation diagram? Which component is shown on the horizontal axis? Which is shown on the vertical axis?
8. Define analog-to-analog conversion?
9. Which characteristics of an analog signal are changed to represent the lowpass analog signal in each of the following analog-to-analog conversions?
   a. AM
   b. FM
   c. PM
10. Which of the three analog-to-analog conversion techniques (AM, FM, or PM) is the most susceptible to noise? Defend your answer.

Exercises

11. Calculate the baud rate for the given bit rate and type of modulation.
    a. 2000 bps, FSK
    b. 4000 bps, ASK
    c. 6000 bps, QPSK
    d. 36,000 bps, 64-QAM

12. Calculate the bit rate for the given baud rate and type of modulation.
    a. 1000 baud, FSK
    b. 1000 baud, ASK
    c. 1000 baud, BPSK
    d. 1000 baud, 16-QAM
13. What is the number of bits per baud for the following techniques?
    a. ASK with four different amplitudes
    b. FSK with 8 different frequencies
    c. PSK with four different phases
    d. QAM with a constellation of 128 points.
14. Draw the constellation diagram for the following:
    a. ASK, with peak amplitude values of 1 and 3
    b. BPSK, with a peak amplitude value of 2
    c. QPSK, with a peak amplitude value of 3
    d. 8-QAM with two different peak amplitude values, 1 and 3, and four different phases.
15. Draw the constellation diagram for the following cases. Find the peak amplitude value for each case and define the type of modulation (ASK, FSK, PSK, or QAM). The numbers in parentheses define the values of I and Q respectively.
    a. Two points at (2, 0) and (3, 0).
    b. Two points at (3, 0) and (−3, 0).
    c. Four points at (2, 2), (−2, 2), (−2, −2), and (2, −2).
    d. Two points at (0 , 2) and (0, −2).
16. How many bits per baud can we send in each of the following cases if the signal constellation has one of the following number of points?
    a. 2
    b. 4
    c. 16
    d. 1024
17. What is the required bandwidth for the following cases if we need to send 4000 bps? Let d = 1.
    a. ASK
    b. FSK with $2\Delta f$ = 4 KHz
    c. QPSK
    d. 16-QAM
18. The telephone line has 4 KHz bandwidth. What is the maximum number of bits we can send using each of the following techniques? Let d = 0.
    a. ASK
    b. QPSK
    c. 16-QAM
    d. 64-QAM

19. A corporation has a medium with a 1-MHz bandwidth (lowpass). The corporation needs to create 10 separate independent channels each capable of sending at least 10 Mbps. The company has decided to use QAM technology. What is the minimum number of bits per baud for each channel? What is the number of points in the constellation diagram for each channel? Let d = 0.

20. A cable company uses one of the cable TV channels (with a bandwidth of 6 MHz) to provide digital communication for each resident. What is the available data rate for each resident if the company uses a 64-QAM technique?

21. Find the bandwidth for the following situations if we need to modulate a 5-KHz voice.

    a. AM

    b. FM (set $\beta = 5$)

    c. PM (set $\beta = 1$)

22. Find the total number of channels in the corresponding band allocated by FCC.

    a. AM

    b. FM

CHAPTER 6

Bandwidth Utilization: Multiplexing and Spreading

In real life, we have links with limited bandwidths. The wise use of these bandwidths has been, and will be, one of the main challenges of electronic communications. However, the meaning of *wise* may depend on the application. Sometimes we need to combine several low-bandwidth channels to make use of one channel with a larger bandwidth. Sometimes we need to expand the bandwidth of a channel to achieve goals such as privacy and antijamming. In this chapter, we explore these two broad categories of bandwidth utilization: multiplexing and spreading. In multiplexing, our goal is efficiency; we combine several channels into one. In spreading, our goals are privacy and antijamming; we expand the bandwidth of a channel to insert redundancy, which is necessary to achieve these goals.

Bandwidth utilization is the wise use of available bandwidth to achieve specific goals.

Efficiency can be achieved by multiplexing;
privacy and antijamming can be achieved by spreading.

6.1 MULTIPLEXING

Whenever the bandwidth of a medium linking two devices is greater than the bandwidth needs of the devices, the link can be shared. **Multiplexing** is the set of techniques that allows the simultaneous transmission of multiple signals across a single data link. As data and telecommunications use increases, so does traffic. We can accommodate this increase by continuing to add individual links each time a new channel is needed; or we can install higher-bandwidth links and use each to carry multiple signals. As described in Chapter 7, today's technology includes high-bandwidth media such as optical fiber and terrestrial and satellite microwaves. Each has a bandwidth far in excess of that needed for the average transmission signal. If the bandwidth of a link is greater than the bandwidth needs of the devices connected to it, the bandwidth is wasted. An efficient system maximizes the utilization of all resources; bandwidth is one of the most precious resources we have in data communications.

In a multiplexed system, n lines share the bandwidth of one link. Figure 6.1 shows the basic format of a multiplexed system. The lines on the left direct their transmission streams to a **multiplexer (MUX),** which combines them into a single stream (many-to-one). At the receiving end, that stream is fed into a **demultiplexer (DEMUX),** which separates the stream back into its component transmissions (one-to-many) and directs them to their corresponding lines. In the figure, the word **link** refers to the physical path. The word **channel** refers to the portion of a link that carries a transmission between a given pair of lines. One link can have many (n) channels.

Figure 6.1 *Dividing a link into channels*

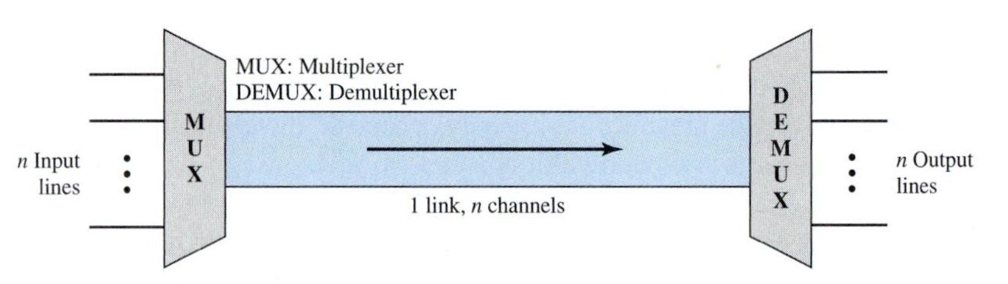

There are three basic multiplexing techniques: frequency-division multiplexing, wavelength-division multiplexing, and time-division multiplexing. The first two are techniques designed for analog signals, the third, for digital signals (see Figure 6.2).

Figure 6.2 *Categories of multiplexing*

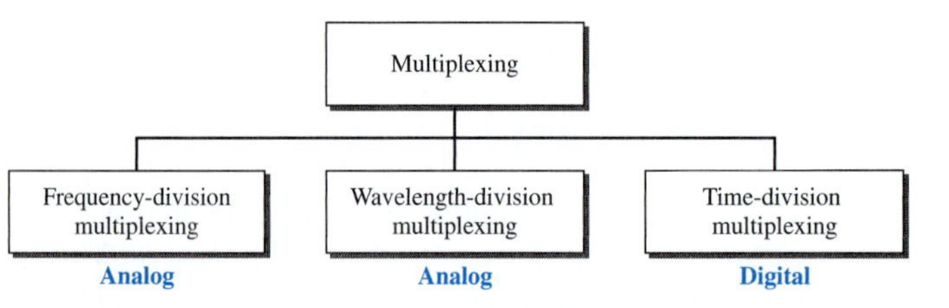

Although some textbooks consider *carrier division multiple access* (CDMA) as a fourth multiplexing category, we discuss CDMA as an access method (see Chapter 12).

Frequency-Division Multiplexing

Frequency-division multiplexing (FDM) is an analog technique that can be applied when the bandwidth of a link (in hertz) is greater than the combined bandwidths of the signals to be transmitted. In FDM, signals generated by each sending device modulate different carrier frequencies. These modulated signals are then combined into a single composite signal that can be transported by the link. Carrier frequencies are separated by sufficient bandwidth to accommodate the modulated signal. These bandwidth ranges are the channels through which the various signals travel. Channels can be separated by strips of unused bandwidth—**guard bands**—to prevent signals from overlapping. In addition, carrier frequencies must not interfere with the original data frequencies.

Figure 6.3 gives a conceptual view of FDM. In this illustration, the transmission path is divided into three parts, each representing a channel that carries one transmission.

Figure 6.3 *Frequency-division multiplexing*

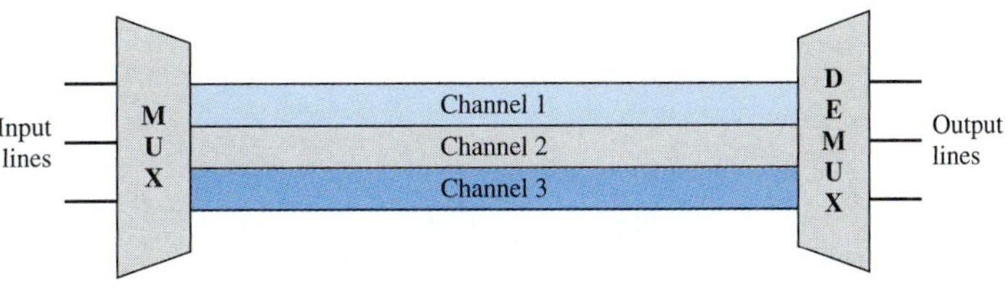

We consider FDM to be an analog multiplexing technique; however, this does not mean that FDM cannot be used to combine sources sending digital signals. A digital signal can be converted to an analog signal (with the techniques discussed in Chapter 5) before FDM is used to multiplex them.

FDM is an analog multiplexing technique that combines analog signals.

Multiplexing Process

Figure 6.4 is a conceptual illustration of the multiplexing process. Each source generates a signal of a similar frequency range. Inside the multiplexer, these similar signals modulates different carrier frequencies (f_1, f_2, and f_3). The resulting modulated signals are then combined into a single composite signal that is sent out over a media link that has enough bandwidth to accommodate it.

Figure 6.4 *FDM process*

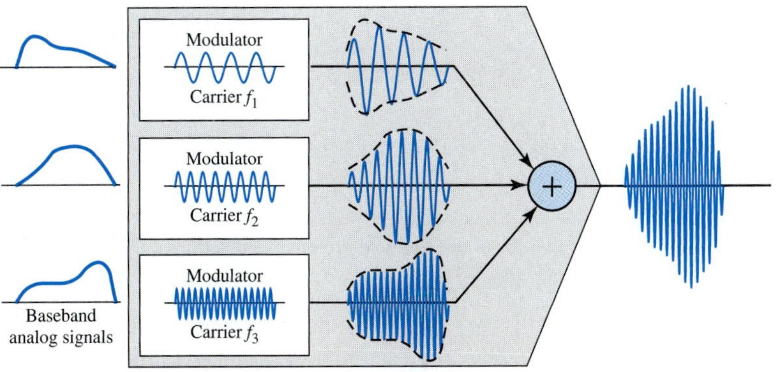

Demultiplexing Process

The demultiplexer uses a series of filters to decompose the multiplexed signal into its constituent component signals. The individual signals are then passed to a demodulator that separates them from their carriers and passes them to the output lines. Figure 6.5 is a conceptual illustration of demultiplexing process.

Figure 6.5 *FDM demultiplexing example*

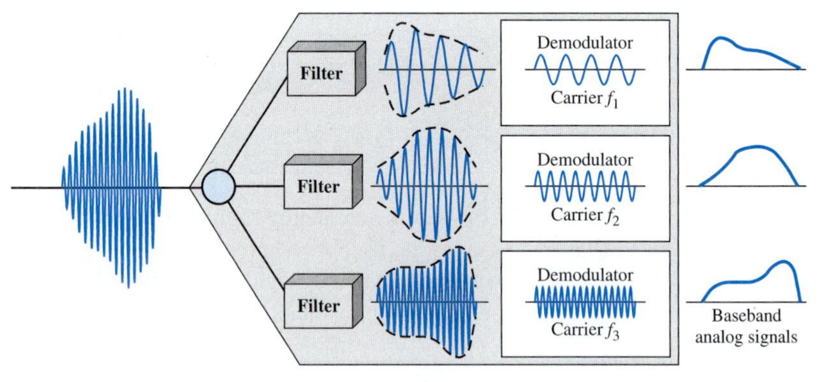

Example 6.1

Assume that a voice channel occupies a bandwidth of 4 kHz. We need to combine three voice channels into a link with a bandwidth of 12 kHz, from 20 to 32 kHz. Show the configuration, using the frequency domain. Assume there are no guard bands.

Solution

We shift (modulate) each of the three voice channels to a different bandwidth, as shown in Figure 6.6. We use the 20- to 24-kHz bandwidth for the first channel, the 24- to 28-kHz bandwidth for the second channel, and the 28- to 32-kHz bandwidth for the third one. Then we combine them as shown in Figure 6.6. At the receiver, each channel receives the entire signal, using a filter to separate out its own signal. The first channel uses a filter that passes frequencies between 20 and 24 kHz and filters out (discards) any other frequencies. The second channel uses a filter that passes frequencies between 24 and 28 kHz, and the third channel uses a filter that passes frequencies between 28 and 32 kHz. Each channel then shifts the frequency to start from zero.

Example 6.2

Five channels, each with a 100-kHz bandwidth, are to be multiplexed together. What is the minimum bandwidth of the link if there is a need for a guard band of 10 kHz between the channels to prevent interference?

Solution

For five channels, we need at least four guard bands. This means that the required bandwidth is at least $5 \times 100 + 4 \times 10 = 540$ kHz, as shown in Figure 6.7.

Figure 6.6 *Example 6.1*

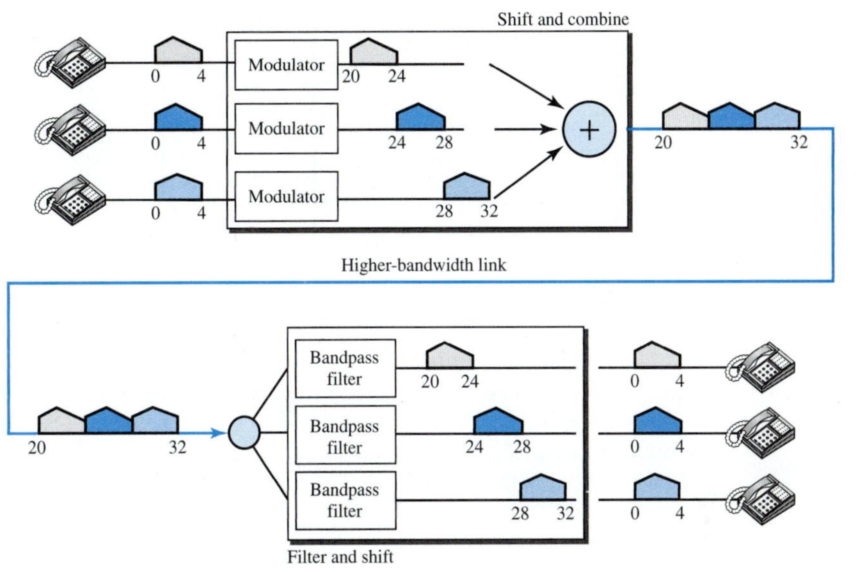

Figure 6.7 *Example 6.2*

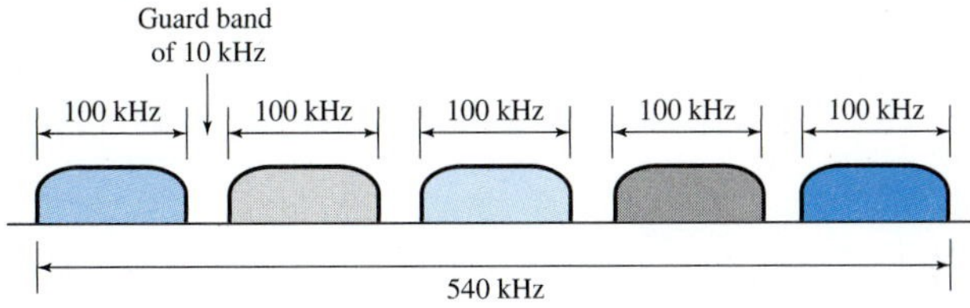

Example 6.3

Four data channels (digital), each transmitting at 1 Mbps, use a satellite channel of 1 MHz. Design an appropriate configuration, using FDM.

Solution

The satellite channel is analog. We divide it into four channels, each channel having a 250-kHz bandwidth. Each digital channel of 1 Mbps is modulated such that each 4 bits is modulated to 1 Hz. One solution is 16-QAM modulation. Figure 6.8 shows one possible configuration.

The Analog Carrier System

To maximize the efficiency of their infrastructure, telephone companies have traditionally multiplexed signals from lower-bandwidth lines onto higher-bandwidth lines. In this way, many switched or leased lines can be combined into fewer but bigger channels. For analog lines, FDM is used.

Figure 6.8 *Example 6.3*

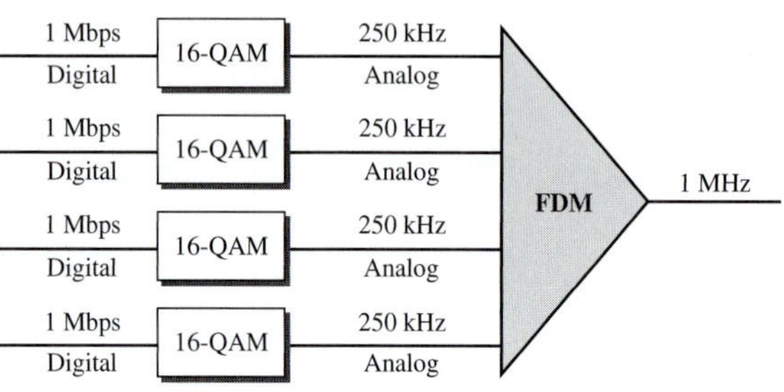

One of these hierarchical systems used by AT&T is made up of groups, supergroups, master groups, and jumbo groups (see Figure 6.9).

Figure 6.9 *Analog hierarchy*

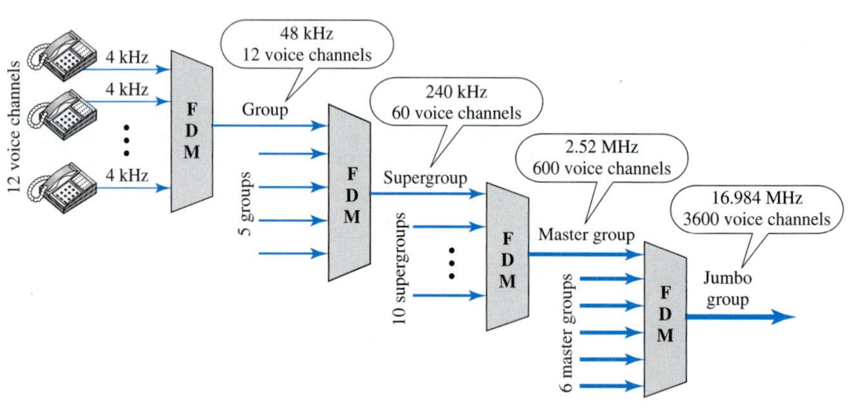

In this **analog hierarchy,** 12 voice channels are multiplexed onto a higher-bandwidth line to create a **group.** A group has 48 kHz of bandwidth and supports 12 voice channels.

At the next level, up to five groups can be multiplexed to create a composite signal called a **supergroup.** A supergroup has a bandwidth of 240 kHz and supports up to 60 voice channels. Supergroups can be made up of either five groups or 60 independent voice channels.

At the next level, 10 supergroups are multiplexed to create a **master group.** A master group must have 2.40 MHz of bandwidth, but the need for guard bands between the supergroups increases the necessary bandwidth to 2.52 MHz. Master groups support up to 600 voice channels.

Finally, six master groups can be combined into a **jumbo group.** A jumbo group must have 15.12 MHz (6 × 2.52 MHz) but is augmented to 16.984 MHz to allow for guard bands between the master groups.

Other Applications of FDM

A very common application of FDM is AM and FM radio broadcasting. Radio uses the air as the transmission medium. A special band from 530 to 1700 kHz is assigned to AM radio. All radio stations need to share this band. As discussed in Chapter 5, each AM station needs 10 kHz of bandwidth. Each station uses a different carrier frequency, which means it is shifting its signal and multiplexing. The signal that goes to the air is a combination of signals. A receiver receives all these signals, but filters (by tuning) only the one which is desired. Without multiplexing, only one AM station could broadcast to the common link, the air. However, we need to know that there is physical multiplexer or demultiplexer here. As we will see in Chapter 12 multiplexing is done at the data link layer.

The situation is similar in FM broadcasting. However, FM has a wider band of 88 to 108 MHz because each station needs a bandwidth of 200 kHz.

Another common use of FDM is in television broadcasting. Each TV channel has its own bandwidth of 6 MHz.

The first generation of cellular telephones (still in operation) also uses FDM. Each user is assigned two 30-kHz channels, one for sending voice and the other for receiving. The voice signal, which has a bandwidth of 3 kHz (from 300 to 3300 Hz), is modulated by using FM. Remember that an FM signal has a bandwidth 10 times that of the modulating signal, which means each channel has 30 kHz (10×3) of bandwidth. Therefore, each user is given, by the base station, a 60-kHz bandwidth in a range available at the time of the call.

Example 6.4

The *Advanced Mobile Phone System* (AMPS) uses two bands. The first band of 824 to 849 MHz is used for sending, and 869 to 894 MHz is used for receiving. Each user has a bandwidth of 30 kHz in each direction. The 3-kHz voice is modulated using FM, creating 30 kHz of modulated signal. How many people can use their cellular phones simultaneously?

Solution

Each band is 25 MHz. If we divide 25 MHz by 30 kHz, we get 833.33. In reality, the band is divided into 832 channels. Of these, 42 channels are used for control, which means only 790 channels are available for cellular phone users. We discuss AMPS in greater detail in Chapter 16.

Implementation

FDM can be implemented very easily. In many cases, such as radio and television broadcasting, there is no need for a physical multiplexer or demultiplexer. As long as the stations agree to send their broadcasts to the air using different carrier frequencies, multiplexing is achieved. In other cases, such as the cellular telephone system, a base station needs to assign a carrier frequency to the telephone user. There is not enough bandwidth in a cell to permanently assign a bandwidth range to every telephone user. When a user hangs up, her or his bandwidth is assigned to another caller.

Wavelength-Division Multiplexing

Wavelength-division multiplexing (WDM) is designed to use the high-data-rate capability of fiber-optic cable. The optical fiber data rate is higher than the data rate of metallic transmission cable. Using a fiber-optic cable for one single line wastes the available bandwidth. Multiplexing allows us to combine several lines into one.

WDM is conceptually the same as FDM, except that the multiplexing and demultiplexing involve optical signals transmitted through fiber-optic channels. The idea is the same: We are combining different signals of different frequencies. The difference is that the frequencies are very high.

Figure 6.10 gives a conceptual view of a WDM multiplexer and demultiplexer. Very narrow bands of light from different sources are combined to make a wider band of light. At the receiver, the signals are separated by the demultiplexer.

Figure 6.10 *Wavelength-division multiplexing*

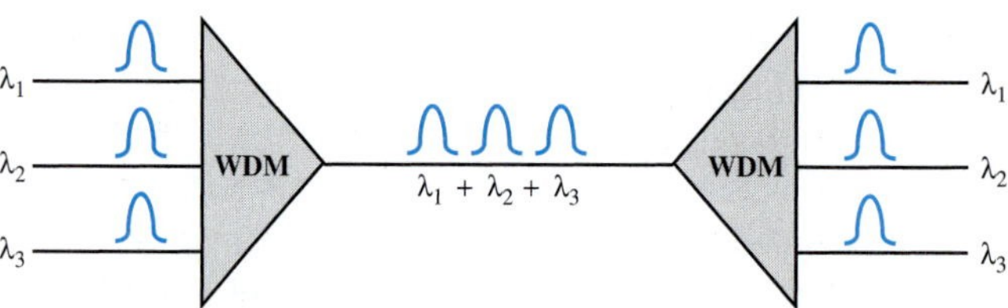

WDM is an analog multiplexing technique to combine optical signals.

Although WDM technology is very complex, the basic idea is very simple. We want to combine multiple light sources into one single light at the multiplexer and do the reverse at the demultiplexer. The combining and splitting of light sources are easily handled by a prism. Recall from basic physics that a prism bends a beam of light based on the angle of incidence and the frequency. Using this technique, a multiplexer can be made to combine several input beams of light, each containing a narrow band of frequencies, into one output beam of a wider band of frequencies. A demultiplexer can also be made to reverse the process. Figure 6.11 shows the concept.

Figure 6.11 *Prisms in wavelength-division multiplexing and demultiplexing*

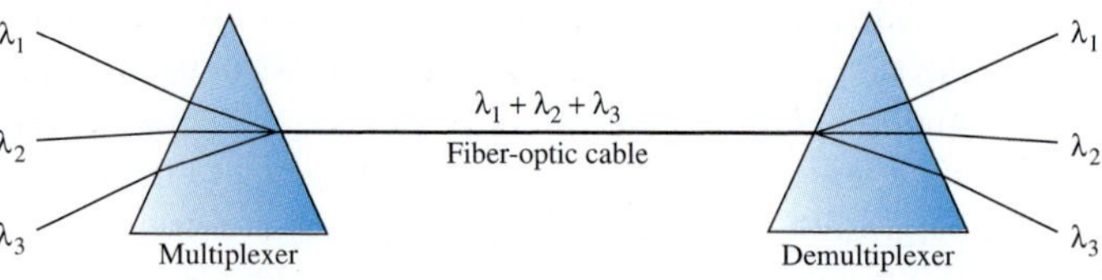

One application of WDM is the SONET network in which multiple optical fiber lines are multiplexed and demultiplexed. We discuss SONET in Chapter 17.

A new method, called **dense WDM (DWDM),** can multiplex a very large number of channels by spacing channels very close to one another. It achieves even greater efficiency.

Synchronous Time-Division Multiplexing

Time-division multiplexing (TDM) is a digital process that allows several connections to share the high bandwidth of a link. Instead of sharing a portion of the bandwidth as in FDM, time is shared. Each connection occupies a portion of time in the link. Figure 6.12 gives a conceptual view of TDM. Note that the same link is used as in FDM; here, however, the link is shown sectioned by time rather than by frequency. In the figure, portions of signals 1, 2, 3, and 4 occupy the link sequentially.

Figure 6.12 *TDM*

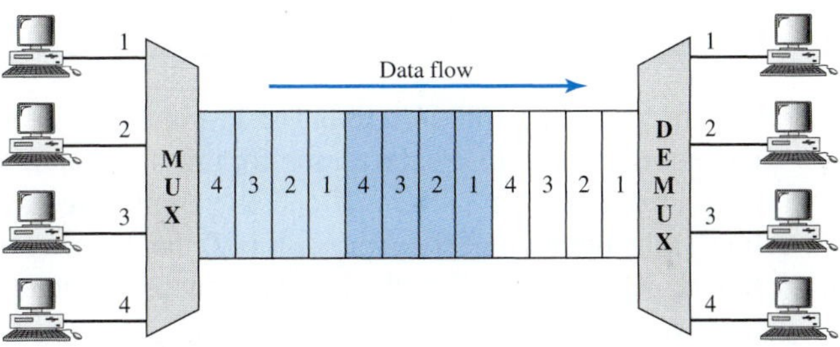

Note that in Figure 6.12 we are concerned with only multiplexing, not switching. This means that all the data in a message from source 1 always go to one specific destination, be it 1, 2, 3, or 4. The delivery is fixed and unvarying, unlike switching.

We also need to remember that TDM is, in principle, a digital multiplexing technique. Digital data from different sources are combined into one timeshared link. However, this does not mean that the sources cannot produce analog data; analog data can be sampled, changed to digital data, and then multiplexed by using TDM.

TDM is a digital multiplexing technique for combining several low-rate channels into one high-rate one.

We can divide TDM into two different schemes: synchronous and statistical. We first discuss **synchronous TDM** and then show how **statistical TDM** differs. In synchronous TDM, each input connection has an allotment in the output even if it is not sending data.

Time Slots and Frames

In synchronous TDM, the data flow of each input connection is divided into units, where each input occupies one input time slot. A unit can be 1 bit, one character, or one block of data. Each input unit becomes one output unit and occupies one output time slot. However, the duration of an output time slot is n times shorter than the duration of an input time slot. If an input time slot is T s, the output time slot is T/n s, where n is the number of connections. In other words, a unit in the output connection has a shorter duration; it travels faster. Figure 6.13 shows an example of synchronous TDM where n is 3.

Figure 6.13 *Synchronous time-division multiplexing*

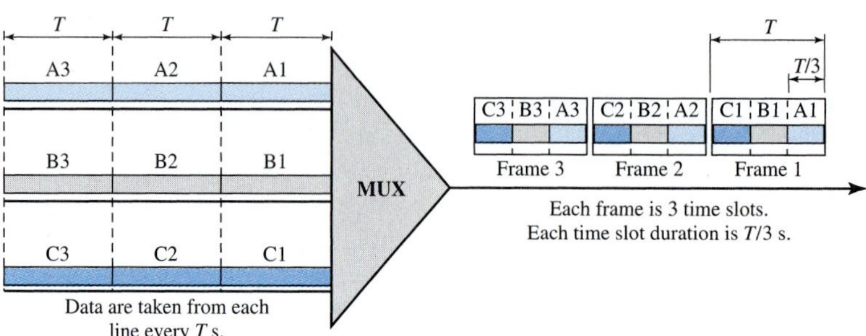

In synchronous TDM, a round of data units from each input connection is collected into a frame (we will see the reason for this shortly). If we have n connections, a frame is divided into n time slots and one slot is allocated for each unit, one for each input line. If the duration of the input unit is T, the duration of each slot is T/n and the duration of each frame is T (unless a frame carries some other information, as we will see shortly).

The data rate of the output link must be n times the data rate of a connection to guarantee the flow of data. In Figure 6.13, the data rate of the link is 3 times the data rate of a connection; likewise, the duration of a unit on a connection is 3 times that of the time slot (duration of a unit on the link). In the figure we represent the data prior to multiplexing as 3 times the size of the data after multiplexing. This is just to convey the idea that each unit is 3 times longer in duration before multiplexing than after.

In synchronous TDM, the data rate of the link is *n* times faster, and the unit duration is *n* times shorter.

Time slots are grouped into frames. A frame consists of one complete cycle of time slots, with one slot dedicated to each sending device. In a system with n input lines, each frame has n slots, with each slot allocated to carrying data from a specific input line.

Example 6.5

In Figure 6.13, the data rate for each input connection is 3 kbps. If 1 bit at a time is multiplexed (a unit is 1 bit), what is the duration of (a) each input slot, (b) each output slot, and (c) each frame?

Solution

We can answer the questions as follows:

a. The data rate of each input connection is 1 kbps. This means that the bit duration is 1/1000 s or 1 ms. The duration of the input time slot is 1 ms (same as bit duration).

b. The duration of each output time slot is one-third of the input time slot. This means that the duration of the output time slot is 1/3 ms.

c. Each frame carries three output time slots. So the duration of a frame is $3 \times 1/3$ ms, or 1 ms. The duration of a frame is the same as the duration of an input unit.

Example 6.6

Figure 6.14 shows synchronous TDM with a data stream for each input and one data stream for the output. The unit of data is 1 bit. Find (a) the input bit duration, (b) the output bit duration, (c) the output bit rate, and (d) the output frame rate.

Figure 6.14 *Example 6.6*

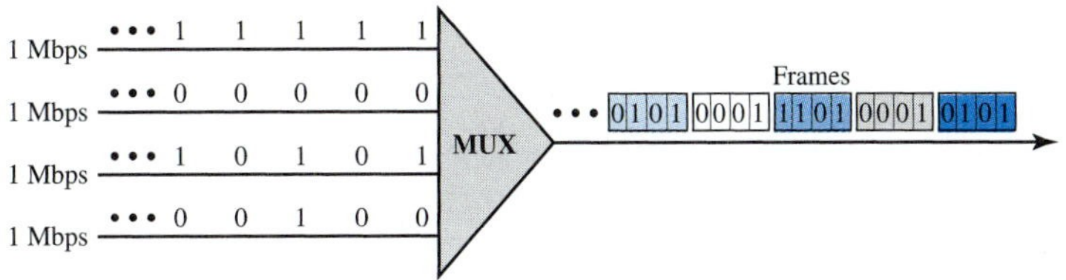

Solution

We can answer the questions as follows:

a. The input bit duration is the inverse of the bit rate: 1/1 Mbps = 1 μs.
b. The output bit duration is one-fourth of the input bit duration, or 1/4 μs.
c. The output bit rate is the inverse of the output bit duration or 1/4 μs, or 4 Mbps. This can also be deduced from the fact that the output rate is 4 times as fast as any input rate; so the output rate = 4 × 1 Mbps = 4 Mbps.
d. The frame rate is always the same as any input rate. So the frame rate is 1,000,000 frames per second. Because we are sending 4 bits in each frame, we can verify the result of the previous question by multiplying the frame rate by the number of bits per frame.

Example 6.7

Four 1-kbps connections are multiplexed together. A unit is 1 bit. Find (a) the duration of 1 bit before multiplexing, (b) the transmission rate of the link, (c) the duration of a time slot, and (d) the duration of a frame.

Solution

We can answer the questions as follows:

a. The duration of 1 bit before multiplexing is 1/1 kbps, or 0.001 s (1 ms).
b. The rate of the link is 4 times the rate of a connection, or 4 kbps.
c. The duration of each time slot is one-fourth of the duration of each bit before multiplexing, or 1/4 ms or 250 μs. Note that we can also calculate this from the data rate of the link, 4 kbps. The bit duration is the inverse of the data rate, or 1/4 kbps or 250 μs.
d. The duration of a frame is always the same as the duration of a unit before multiplexing, or 1 ms. We can also calculate this in another way. Each frame in this case has four time slots. So the duration of a frame is 4 times 250 μs, or 1 ms.

Interleaving

TDM can be visualized as two fast-rotating switches, one on the multiplexing side and the other on the demultiplexing side. The switches are synchronized and rotate at the same speed, but in opposite directions. On the multiplexing side, as the switch opens in front of a connection, that connection has the opportunity to send a unit onto the path. This process is called **interleaving.** On the demultiplexing side, as the switch opens in front of a connection, that connection has the opportunity to receive a unit from the path.

Figure 6.15 shows the interleaving process for the connection shown in Figure 6.13. In this figure, we assume that no switching is involved and that the data from the first connection at the multiplexer site go to the first connection at the demultiplexer. We discuss switching in Chapter 8.

Figure 6.15 *Interleaving*

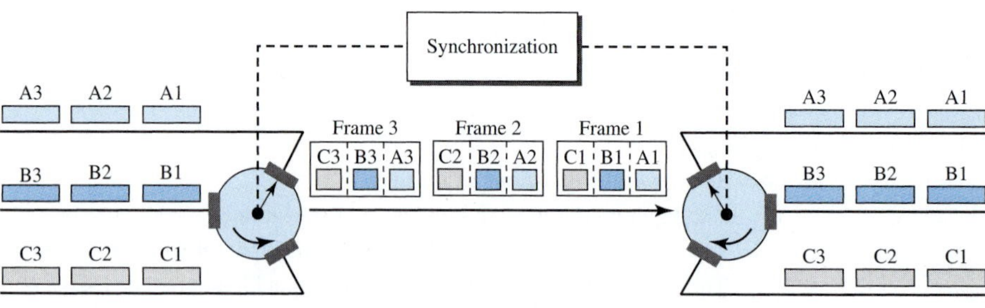

Example 6.8

Four channels are multiplexed using TDM. If each channel sends 100 bytes/s and we multiplex 1 byte per channel, show the frame traveling on the link, the size of the frame, the duration of a frame, the frame rate, and the bit rate for the link.

Solution

The multiplexer is shown in Figure 6.16. Each frame carries 1 byte from each channel; the size of each frame, therefore, is 4 bytes, or 32 bits. Because each channel is sending 100 bytes/s and a frame carries 1 byte from each channel, the frame rate must be 100 frames per second. The duration of a frame is therefore 1/100 s. The link is carrying 100 frames per second, and since each frame contains 32 bits, the bit rate is 100×32, or 3200 bps. This is actually 4 times the bit rate of each channel, which is $100 \times 8 = 800$ bps.

Figure 6.16 *Example 6.8*

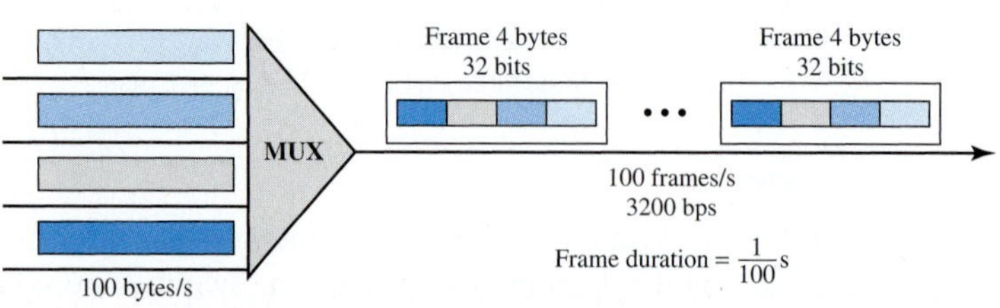

Example 6.9

A multiplexer combines four 100-kbps channels using a time slot of 2 bits. Show the output with four arbitrary inputs. What is the frame rate? What is the frame duration? What is the bit rate? What is the bit duration?

Solution

Figure 6.17 shows the output for four arbitrary inputs. The link carries 50,000 frames per second since each frame contains 2 bits per channel. The frame duration is therefore 1/50,000 s or 20 μs. The frame rate is 50,000 frames per second, and each frame carries 8 bits; the bit rate is 50,000 × 8 = 400,000 bits or 400 kbps. The bit duration is 1/400,000 s, or 2.5 μs. Note that the frame duration is 8 times the bit duration because each frame is carrying 8 bits.

Figure 6.17 *Example 6.9*

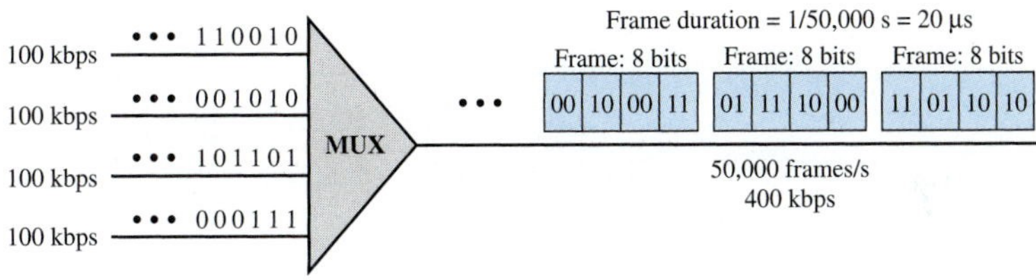

Empty Slots

Synchronous TDM is not as efficient as it could be. If a source does not have data to send, the corresponding slot in the output frame is empty. Figure 6.18 shows a case in which one of the input lines has no data to send and one slot in another input line has discontinuous data.

Figure 6.18 *Empty slots*

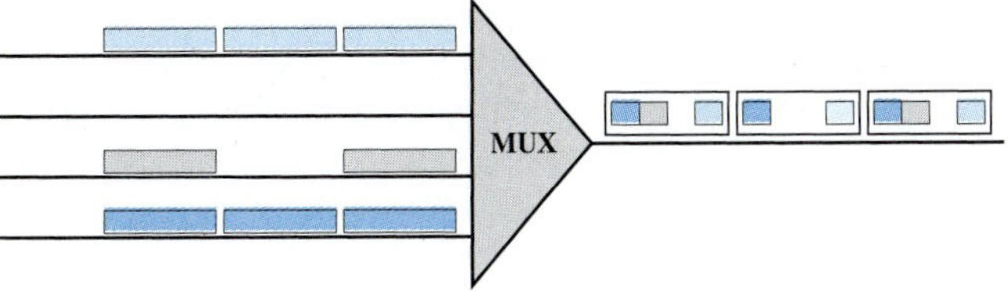

The first output frame has three slots filled, the second frame has two slots filled, and the third frame has three slots filled. No frame is full. We learn in the next section that statistical TDM can improve the efficiency by removing the empty slots from the frame.

Data Rate Management

One problem with TDM is how to handle a disparity in the input data rates. In all our discussion so far, we assumed that the data rates of all input lines were the same. However, if data rates are not the same, three strategies, or a combination of them, can be used. We call these three strategies **multilevel multiplexing, multiple-slot allocation,** and **pulse stuffing.**

Multilevel Multiplexing Multilevel multiplexing is a technique used when the data rate of an input line is a multiple of others. For example, in Figure 6.19, we have two inputs of 20 kbps and three inputs of 40 kbps. The first two input lines can be multiplexed together to provide a data rate equal to the last three. A second level of multiplexing can create an output of 160 kbps.

Figure 6.19 *Multilevel multiplexing*

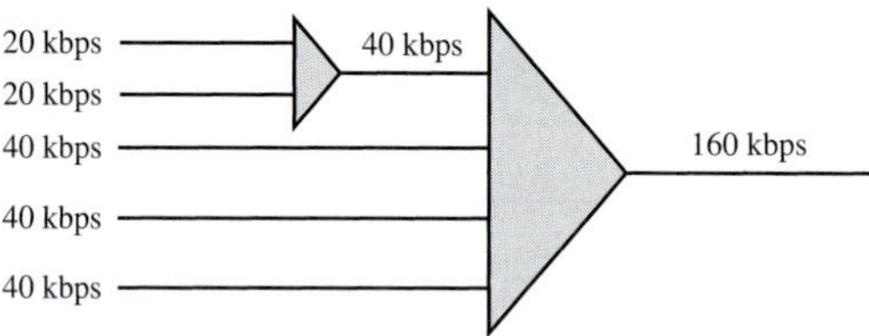

Multiple-Slot Allocation Sometimes it is more efficient to allot more than one slot in a frame to a single input line. For example, we might have an input line that has a data rate that is a multiple of another input. In Figure 6.20, the input line with a 50-kbps data rate can be given two slots in the output. We insert a serial-to-parallel converter in the line to make two inputs out of one.

Figure 6.20 *Multiple-slot multiplexing*

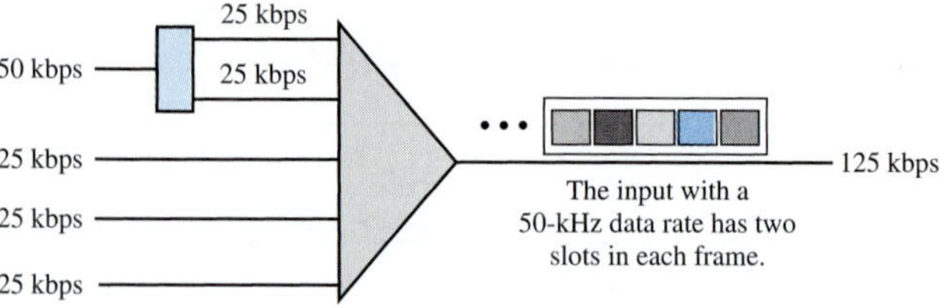

Pulse Stuffing Sometimes the bit rates of sources are not multiple integers of each other. Therefore, neither of the above two techniques can be applied. One solution is to make the highest input data rate the dominant data rate and then add dummy bits to the input lines with lower rates. This will increase their rates. This technique is called pulse stuffing, bit padding, or bit stuffing. The idea is shown in Figure 6.21. The input with a data rate of 46 is pulse-stuffed to increase the rate to 50 kbps. Now multiplexing can take place.

Figure 6.21 *Pulse stuffing*

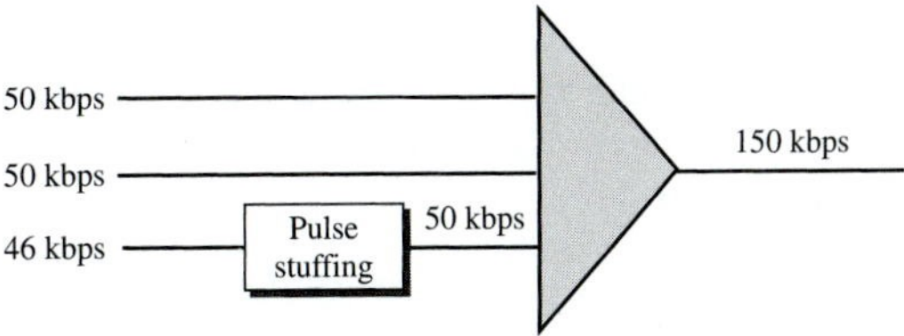

Frame Synchronizing

The implementation of TDM is not as simple as that of FDM. Synchronization between the multiplexer and demultiplexer is a major issue. If the multiplexer and the demultiplexer are not synchronized, a bit belonging to one channel may be received by the wrong channel. For this reason, one or more synchronization bits are usually added to the beginning of each frame. These bits, called **framing bits,** follow a pattern, frame to frame, that allows the demultiplexer to synchronize with the incoming stream so that it can separate the time slots accurately. In most cases, this synchronization information consists of 1 bit per frame, alternating between 0 and 1, as shown in Figure 6.22.

Figure 6.22 *Framing bits*

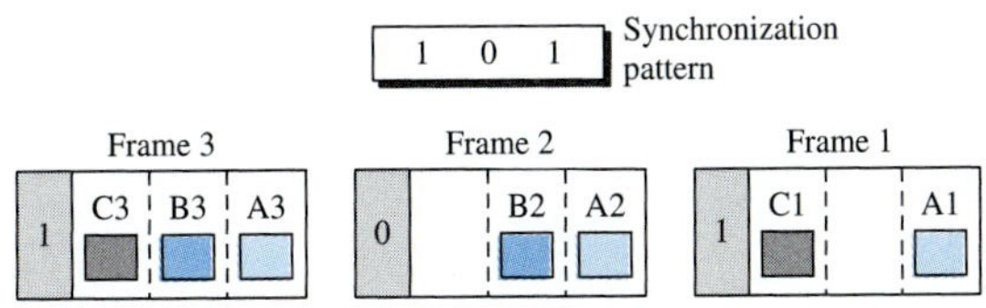

Example 6.10

We have four sources, each creating 250 characters per second. If the interleaved unit is a character and 1 synchronizing bit is added to each frame, find (a) the data rate of each source, (b) the duration of each character in each source, (c) the frame rate, (d) the duration of each frame, (e) the number of bits in each frame, and (f) the data rate of the link.

Solution

We can answer the questions as follows:

a. The data rate of each source is $250 \times 8 = 2000$ bps = 2 kbps.
b. Each source sends 250 characters per second; therefore, the duration of a character is 1/250 s, or 4 ms.
c. Each frame has one character from each source, which means the link needs to send 250 frames per second to keep the transmission rate of each source.
d. The duration of each frame is 1/250 s, or 4 ms. Note that the duration of each frame is the same as the duration of each character coming from each source.
e. Each frame carries 4 characters and 1 extra synchronizing bit. This means that each frame is $4 \times 8 + 1 = 33$ bits.

f. The link sends 250 frames per second, and each frame contains 33 bits. This means that the data rate of the link is 250 × 33, or 8250 bps. Note that the bit rate of the link is greater than the combined bit rates of the four channels. If we add the bit rates of four channels, we get 8000 bps. Because 250 frames are traveling per second and each contains 1 extra bit for synchronizing, we need to add 250 to the sum to get 8250 bps.

Example 6.11

Two channels, one with a bit rate of 100 kbps and another with a bit rate of 200 kbps, are to be multiplexed. How this can be achieved? What is the frame rate? What is the frame duration? What is the bit rate of the link?

Solution

We can allocate one slot to the first channel and two slots to the second channel. Each frame carries 3 bits. The frame rate is 100,000 frames per second because it carries 1 bit from the first channel. The frame duration is 1/100,000 s, or 10 ms. The bit rate is 100,000 frames/s × 3 bits per frame, or 300 kbps. Note that because each frame carries 1 bit from the first channel, the bit rate for the first channel is preserved. The bit rate for the second channel is also preserved because each frame carries 2 bits from the second channel.

Digital Signal Service

Telephone companies implement TDM through a hierarchy of digital signals, called **digital signal (DS) service** or **digital hierarchy.** Figure 6.23 shows the data rates supported by each level.

Figure 6.23 *Digital hierarchy*

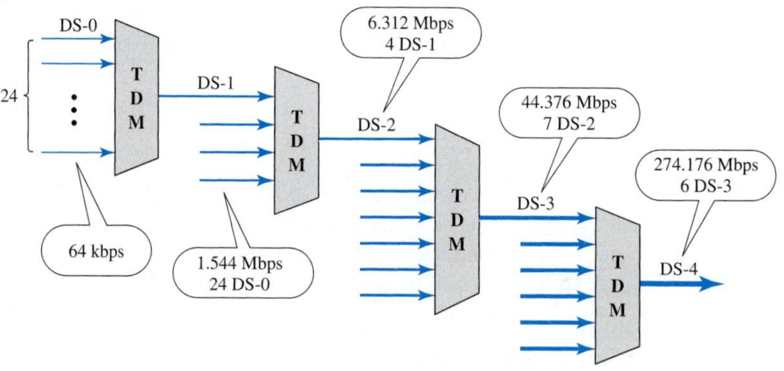

- A **DS-0** service is a single digital channel of 64 kbps.
- **DS-1** is a 1.544-Mbps service; 1.544 Mbps is 24 times 64 kbps plus 8 kbps of overhead. It can be used as a single service for 1.544-Mbps transmissions, or it can be used to multiplex 24 DS-0 channels or to carry any other combination desired by the user that can fit within its 1.544-Mbps capacity.
- **DS-2** is a 6.312-Mbps service; 6.312 Mbps is 96 times 64 kbps plus 168 kbps of overhead. It can be used as a single service for 6.312-Mbps transmissions; or it can

be used to multiplex 4 DS-1 channels, 96 DS-0 channels, or a combination of these service types.

- **DS-3** is a 44.376-Mbps service; 44.376 Mbps is 672 times 64 kbps plus 1.368 Mbps of overhead. It can be used as a single service for 44.376-Mbps transmissions; or it can be used to multiplex 7 DS-2 channels, 28 DS-1 channels, 672 DS-0 channels, or a combination of these service types.
- **DS-4** is a 274.176-Mbps service; 274.176 is 4032 times 64 kbps plus 16.128 Mbps of overhead. It can be used to multiplex 6 DS-3 channels, 42 DS-2 channels, 168 DS-1 channels, 4032 DS-0 channels, or a combination of these service types.

T Lines

DS-0, DS-1, and so on are the names of services. To implement those services, the telephone companies use **T lines** (T-1 to T-4). These are lines with capacities precisely matched to the data rates of the DS-1 to DS-4 services (see Table 6.1). So far only T-1 and T-3 lines are commercially available.

Table 6.1 *DS and T line rates*

Service	*Line*	*Rate (Mbps)*	*Voice Channels*
DS-1	T-1	1.544	24
DS-2	T-2	6.312	96
DS-3	T-3	44.736	672
DS-4	T-4	274.176	4032

The T-1 line is used to implement DS-1; T-2 is used to implement DS-2; and so on. As you can see from Table 6.1, DS-0 is not actually offered as a service, but it has been defined as a basis for reference purposes.

T Lines for Analog Transmission

T lines are digital lines designed for the transmission of digital data, audio, or video. However, they also can be used for analog transmission (regular telephone connections), provided the analog signals are first sampled, then time-division multiplexed.

The possibility of using T lines as analog carriers opened up a new generation of services for the telephone companies. Earlier, when an organization wanted 24 separate telephone lines, it needed to run 24 twisted-pair cables from the company to the central exchange. (Remember those old movies showing a busy executive with 10 telephones lined up on his desk? Or the old office telephones with a big fat cable running from them? Those cables contained a bundle of separate lines.) Today, that same organization can combine the 24 lines into one T-1 line and run only the T-1 line to the exchange. Figure 6.24 shows how 24 voice channels can be multiplexed onto one T-1 line. (Refer to Chapter 5 for PCM encoding.)

The T-1 Frame As noted above, DS-1 requires 8 kbps of overhead. To understand how this overhead is calculated, we must examine the format of a 24-voice-channel frame.

The frame used on a T-1 line is usually 193 bits divided into 24 slots of 8 bits each plus 1 extra bit for synchronization ($24 \times 8 + 1 = 193$); see Figure 6.25. In other words,

Figure 6.24 *T-1 line for multiplexing telephone lines*

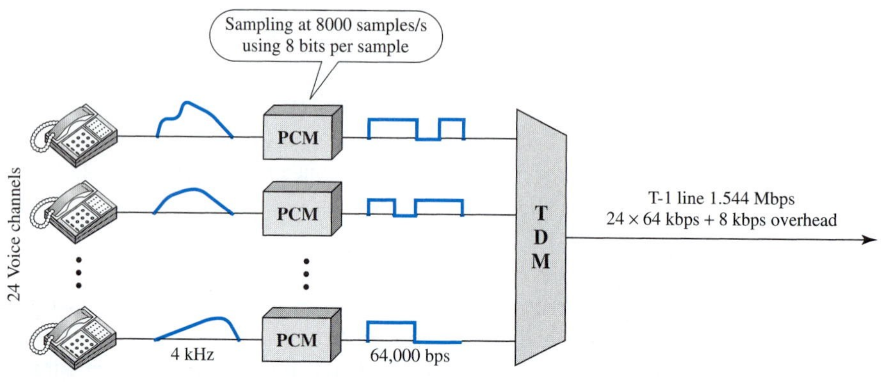

Figure 6.25 *T-1 frame structure*

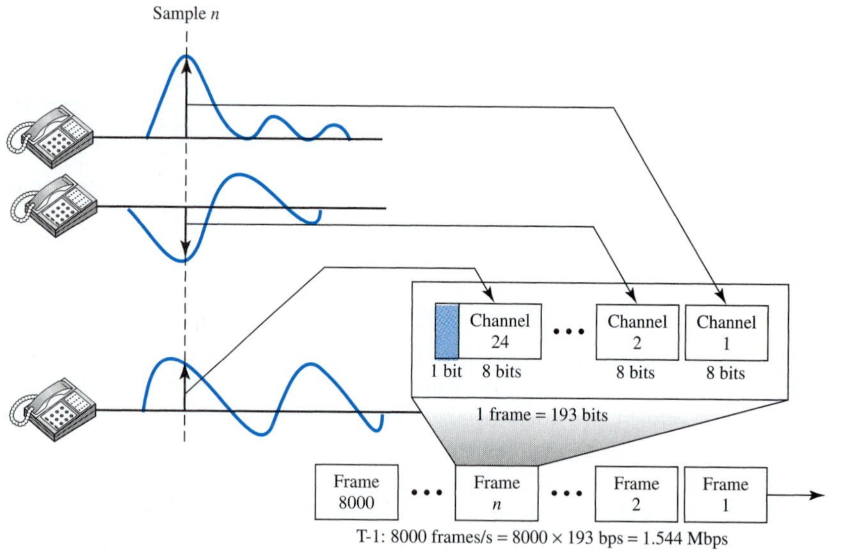

each slot contains one signal segment from each channel; 24 segments are interleaved in one frame. If a T-1 line carries 8000 frames, the data rate is 1.544 Mbps (193 × 8000 = 1.544 Mbps)—the capacity of the line.

E Lines

Europeans use a version of T lines called **E lines.** The two systems are conceptually identical, but their capacities differ. Table 6.2 shows the E lines and their capacities.

Table 6.2 *E line rates*

Line	*Rate (Mbps)*	*Voice Channels*
E-1	2.048	30
E-2	8.448	120
E-3	34.368	480
E-4	139.264	1920

More Synchronous TDM Applications

Some second-generation cellular telephone companies use synchronous TDM. For example, the digital version of cellular telephony divides the available bandwidth into 30-kHz bands. For each band, TDM is applied so that six users can share the band. This means that each 30-kHz band is now made of six time slots, and the digitized voice signals of the users are inserted in the slots. Using TDM, the number of telephone users in each area is now 6 times greater. We discuss second-generation cellular telephony in Chapter 16.

Statistical Time-Division Multiplexing

As we saw in the previous section, in synchronous TDM, each input has a reserved slot in the output frame. This can be inefficient if some input lines have no data to send. In statistical time-division multiplexing, slots are dynamically allocated to improve bandwidth efficiency. Only when an input line has a slot's worth of data to send is it given a slot in the output frame. In statistical multiplexing, the number of slots in each frame is less than the number of input lines. The multiplexer checks each input line in round-robin fashion; it allocates a slot for an input line if the line has data to send; otherwise, it skips the line and checks the next line.

Figure 6.26 shows a synchronous and a statistical TDM example. In the former, some slots are empty because the corresponding line does not have data to send. In the latter, however, no slot is left empty as long as there are data to be sent by any input line.

Addressing

Figure 6.26 also shows a major difference between slots in synchronous TDM and statistical TDM. An output slot in synchronous TDM is totally occupied by data; in statistical TDM, a slot needs to carry data as well as the address of the destination. In synchronous TDM, there is no need for addressing; synchronization and preassigned relationships between the inputs and outputs serve as an address. We know, for example, that input 1 always goes to input 2. If the multiplexer and the demultiplexer are synchronized, this is guaranteed. In statistical multiplexing, there is no fixed relationship between the inputs and outputs because there are no preassigned or reserved slots. We need to include the address of the receiver inside each slot to show where it is to be delivered. The addressing in its simplest form can be n bits to define N different output lines with $n = \log_2 N$. For example, for eight different output lines, we need a 3-bit address.

Figure 6.26 *TDM slot comparison*

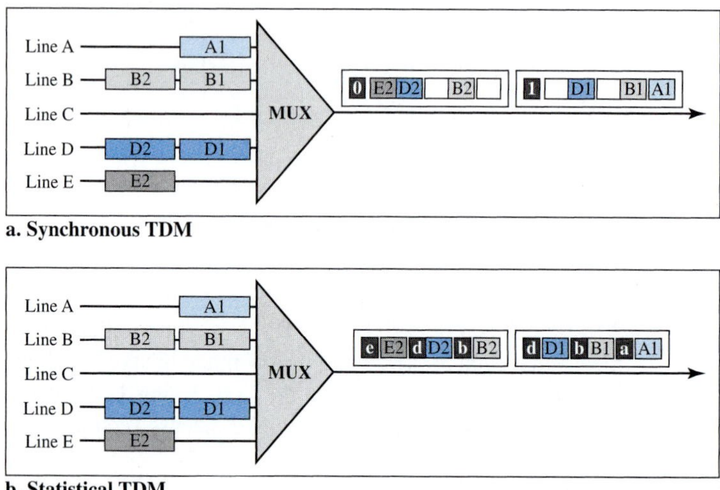

Slot Size

Since a slot carries both data and an address in statistical TDM, the ratio of the data size to address size must be reasonable to make transmission efficient. For example, it would be inefficient to send 1 bit per slot as data when the address is 3 bits. This would mean an overhead of 300 percent. In statistical TDM, a block of data is usually many bytes while the address is just a few bytes.

No Synchronization Bit

There is another difference between synchronous and statistical TDM, but this time it is at the frame level. The frames in statistical TDM need not be synchronized, so we do not need synchronization bits.

Bandwidth

In statistical TDM, the capacity of the link is normally less than the sum of the capacities of each channel. The designers of statistical TDM define the capacity of the link based on the statistics of the load for each channel. If on average only x percent of the input slots are filled, the capacity of the link reflects this. Of course, during peak times, some slots need to wait.

6.2 SPREAD SPECTRUM

Multiplexing combines signals from several sources to achieve bandwidth efficiency; the available bandwidth of a link is divided between the sources. In **spread spectrum (SS),** we also combine signals from different sources to fit into a larger bandwidth, but our goals are somewhat different. Spread spectrum is designed to be used in wireless applications (LANs and WANs). In these types of applications, we have some concerns that outweigh bandwidth efficiency. In wireless applications, all stations use air (or a vacuum) as the medium for communication. Stations must be able to share this medium without interception by an eavesdropper and without being subject to jamming from a malicious intruder (in military operations, for example).

To achieve these goals, spread spectrum techniques add redundancy; they spread the original spectrum needed for each station. If the required bandwidth for each station is B, spread spectrum expands it to B_{SS}, such that $B_{SS} >> B$. The expanded bandwidth allows the source to wrap its message in a protective envelope for a more secure transmission. An analogy is the sending of a delicate, expensive gift. We can insert the gift in a special box to prevent it from being damaged during transportation, and we can use a superior delivery service to guarantee the safety of the package.

Figure 6.27 shows the idea of spread spectrum. Spread spectrum achieves its goals through two principles:

1. The bandwidth allocated to each station needs to be, by far, larger than what is needed. This allows redundancy.
2. The expanding of the original bandwidth B to the bandwidth B_{SS} must be done by a process that is independent of the original signal. In other words, the spreading process occurs after the signal is created by the source.

Figure 6.27 *Spread spectrum*

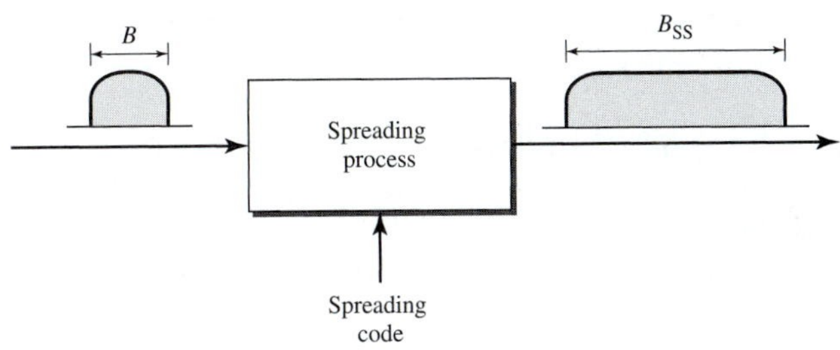

After the signal is created by the source, the spreading process uses a spreading code and spreads the bandwidth. The figure shows the original bandwidth B and the spreaded bandwidth B_{SS}. The spreading code is a series of numbers that look random, but are actually a pattern.

There are two techniques to spread the bandwidth: frequency hopping spread spectrum (FHSS) and direct sequence spread spectrum (DSSS).

Frequency Hopping Spread Spectrum (FHSS)

The **frequency hopping spread spectrum (FHSS)** technique uses M different carrier frequencies that are modulated by the source signal. At one moment, the signal modulates one carrier frequency; at the next moment, the signal modulates another carrier frequency. Although the modulation is done using one carrier frequency at a time, M frequencies are used in the long run. The bandwidth occupied by a source after spreading is $B_{\text{FHSS}} >> B$.

Figure 6.28 shows the general layout for FHSS. A **pseudorandom code generator,** called **pseudorandom noise (PN),** creates a k-bit pattern for every **hopping period T_h.** The frequency table uses the pattern to find the frequency to be used for this hopping period and passes it to the frequency synthesizer. The frequency synthesizer creates a carrier signal of that frequency, and the source signal modulates the carrier signal.

Figure 6.28 *Frequency hopping spread spectrum (FHSS)*

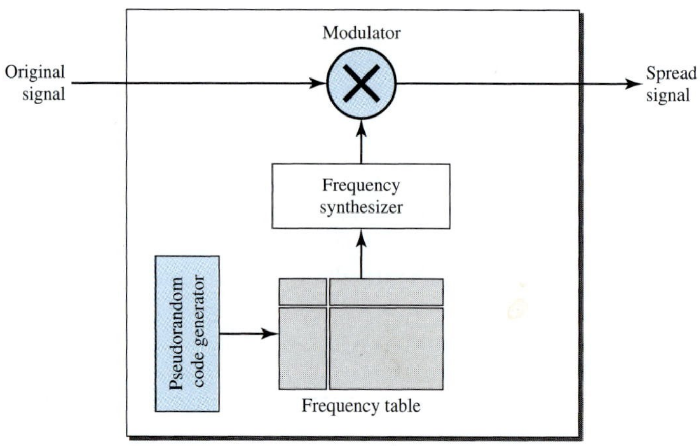

Suppose we have decided to have eight hopping frequencies. This is extremely low for real applications and is just for illustration. In this case, M is 8 and k is 3. The pseudorandom code generator will create eight different 3-bit patterns. These are mapped to eight different frequencies in the frequency table (see Figure 6.29).

Figure 6.29 *Frequency selection in FHSS*

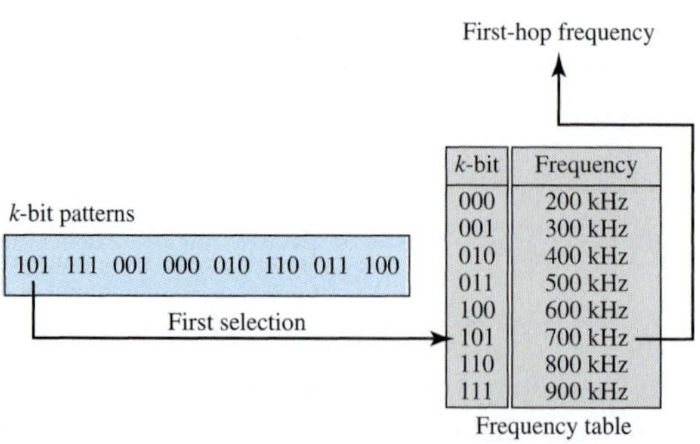

The pattern for this station is 101, 111, 001, 000, 010, 011, 100. Note that the pattern is pseudorandom it is repeated after eight hoppings. This means that at hopping period 1, the pattern is 101. The frequency selected is 700 kHz; the source signal modulates this carrier frequency. The second k-bit pattern selected is 111, which selects the 900-kHz carrier; the eighth pattern is 100, the frequency is 600 kHz. After eight hoppings, the pattern repeats, starting from 101 again. Figure 6.30 shows how the signal hops around from carrier to carrier. We assume the required bandwidth of the original signal is 100 kHz.

Figure 6.30 *FHSS cycles*

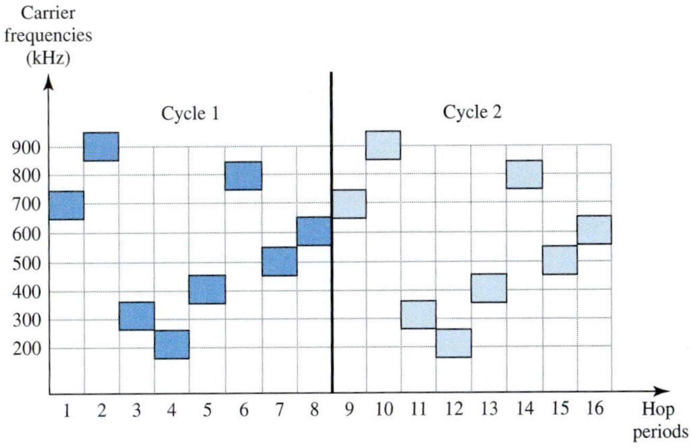

It can be shown that this scheme can accomplish the previously mentioned goals. If there are many k-bit patterns and the hopping period is short, a sender and receiver can have privacy. If an intruder tries to intercept the transmitted signal, she can only access a small piece of data because she does not know the spreading sequence to quickly adapt herself to the next hop. The scheme has also an antijamming effect. A malicious sender may be able to send noise to jam the signal for one hopping period (randomly), but not for the whole period.

Bandwidth Sharing

If the number of hopping frequencies is M, we can multiplex M channels into one by using the same B_{SS} bandwidth. This is possible because a station uses just one frequency in each hopping period; $M - 1$ other frequencies can be used by other $M - 1$ stations. In other words, M different stations can use the same B_{SS} if an appropriate modulation technique such as multiple FSK (MFSK) is used. FHSS is similar to FDM, as shown in Figure 6.31.

Figure 6.31 shows an example of four channels using FDM and four channels using FHSS. In FDM, each station uses $1/M$ of the bandwidth, but the allocation is fixed; in FHSS, each station uses $1/M$ of the bandwidth, but the allocation changes hop to hop.

Figure 6.31 *Bandwidth sharing*

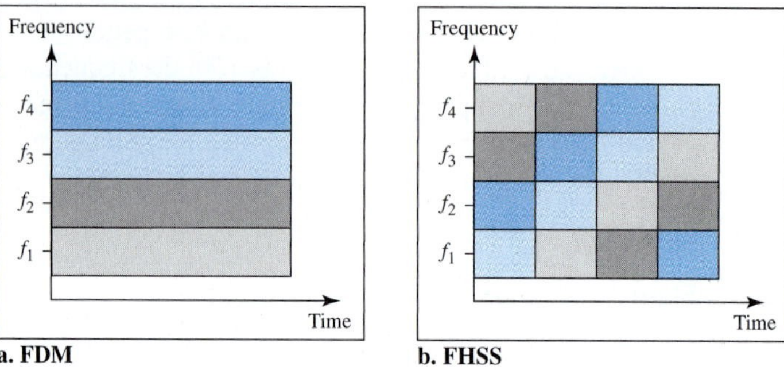

Direct Sequence Spread Spectrum

The **direct sequence spread spectrum (DSSS)** technique also expands the bandwidth of the original signal, but the process is different. In DSSS, we replace each data bit with n bits using a spreading code. In other words, each bit is assigned a code of n bits, called chips, where the chip rate is n times that of the data bit. Figure 6.32 shows the concept of DSSS.

Figure 6.32 *DSSS*

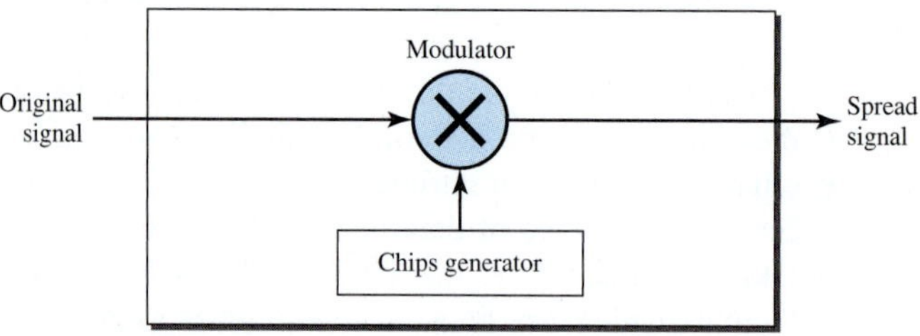

As an example, let us consider the sequence used in a wireless LAN, the famous **Barker sequence** where n is 11. We assume that the original signal and the chips in the chip generator use polar NRZ encoding. Figure 6.33 shows the chips and the result of multiplying the original data by the chips to get the spread signal.

In Figure 6.33, the spreading code is 11 chips having the pattern 10110111000 (in this case). If the original signal rate is N, the rate of the spread signal is $11N$. This means that the required bandwidth for the spread signal is 11 times larger than the bandwidth of the original signal. The spread signal can provide privacy if the intruder does not know the code. It can also provide immunity against interference if each station uses a different code.

Figure 6.33 *DSSS example*

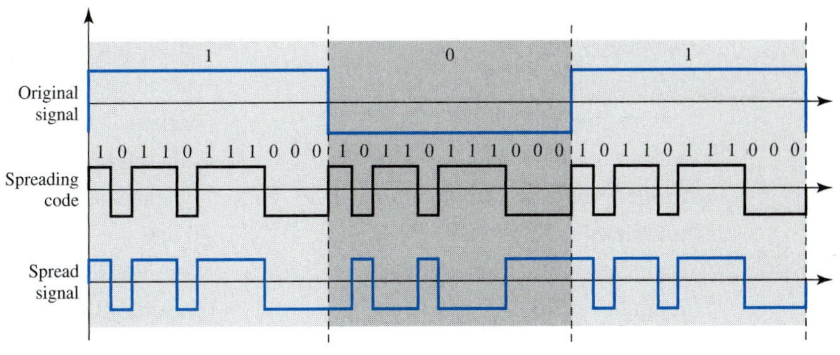

Bandwidth Sharing

Can we share a bandwidth in DSSS as we did in FHSS? The answer is no and yes. If we use a spreading code that spreads signals (from different stations) that cannot be combined and separated, we cannot share a bandwidth. For example, as we will see in Chapter 14, some wireless LANs use DSSS and the spread bandwidth cannot be shared. However, if we use a special type of sequence code that allows the combining and separating of spread signals, we can share the bandwidth. As we will see in Chapter 16, a special spreading code allows us to use DSSS in cellular telephony and share a bandwidth between several users.

6.3 RECOMMENDED READING

For more details about subjects discussed in this chapter, we recommend the following books. The items in brackets [. . .] refer to the reference list at the end of the text.

Books

Multiplexing is elegantly discussed in Chapters 19 of [Pea92]. [Cou01] gives excellent coverage of TDM and FDM in Sections 3.9 to 3.11. More advanced materials can be found in [Ber96]. Multiplexing is discussed in Chapter 8 of [Sta04]. A good coverage of spread spectrum can be found in Section 5.13 of [Cou01] and Chapter 9 of [Sta04].

6.4 KEY TERMS

analog hierarchy
Barker sequence
channel
chip
demultiplexer (DEMUX)
dense WDM (DWDM)
digital signal (DS) service
direct sequence spread spectrum (DSSS)
E line
framing bit
frequency hopping spread spectrum (FSSS)

frequency-division multiplexing (FDM)
group
guard band
hopping period
interleaving
jumbo group
link
master group
multilevel multiplexing
multiple-slot multiplexing
multiplexer (MUX)
multiplexing
pseudorandom code generator
pseudorandom noise (PN)
pulse stuffing
spread spectrum (SS)
statistical TDM
supergroup
synchronous TDM
T line
time-division multiplexing (TDM)
wavelength-division multiplexing (WDM)

6.5 SUMMARY

- Bandwidth utilization is the use of available bandwidth to achieve specific goals. Efficiency can be achieved by using multiplexing; privacy and antijamming can be achieved by using spreading.
- Multiplexing is the set of techniques that allows the simultaneous transmission of multiple signals across a single data link. In a multiplexed system, n lines share the bandwidth of one link. The word link refers to the physical path. The word channel refers to the portion of a link that carries a transmission.
- There are three basic multiplexing techniques: frequency-division multiplexing, wavelength-division multiplexing, and time-division multiplexing. The first two are techniques designed for analog signals, the third, for digital signals
- Frequency-division multiplexing (FDM) is an analog technique that can be applied when the bandwidth of a link (in hertz) is greater than the combined bandwidths of the signals to be transmitted.
- Wavelength-division multiplexing (WDM) is designed to use the high bandwidth capability of fiber-optic cable. WDM is an analog multiplexing technique to combine optical signals.
- Time-division multiplexing (TDM) is a digital process that allows several connections to share the high bandwidth of a link. TDM is a digital multiplexing technique for combining several low-rate channels into one high-rate one.
- We can divide TDM into two different schemes: synchronous or statistical. In synchronous TDM, each input connection has an allotment in the output even if it is not sending data. In statistical TDM, slots are dynamically allocated to improve bandwidth efficiency.
- In spread spectrum (SS), we combine signals from different sources to fit into a larger bandwidth. Spread spectrum is designed to be used in wireless applications in which stations must be able to share the medium without interception by an eavesdropper and without being subject to jamming from a malicious intruder.
- The frequency hopping spread spectrum (FHSS) technique uses M different carrier frequencies that are modulated by the source signal. At one moment, the signal

modulates one carrier frequency; at the next moment, the signal modulates another carrier frequency.

- The direct sequence spread spectrum (DSSS) technique expands the bandwidth of a signal by replacing each data bit with n bits using a spreading code. In other words, each bit is assigned a code of n bits, called chips.

6.6 PRACTICE SET

Review Questions

1. Describe the goals of multiplexing.
2. List three main multiplexing techniques mentioned in this chapter.
3. Distinguish between a link and a channel in multiplexing.
4. Which of the three multiplexing techniques is (are) used to combine analog signals? Which of the three multiplexing techniques is (are) used to combine digital signals?
5. Define the analog hierarchy used by telephone companies and list different levels of the hierarchy.
6. Define the digital hierarchy used by telephone companies and list different levels of the hierarchy.
7. Which of the three multiplexing techniques is common for fiber optic links? Explain the reason.
8. Distinguish between multilevel TDM, multiple slot TDM, and pulse-stuffed TDM.
9. Distinguish between synchronous and statistical TDM.
10. Define spread spectrum and its goal. List the two spread spectrum techniques discussed in this chapter.
11. Define FHSS and explain how it achieves bandwidth spreading.
12. Define DSSS and explain how it achieves bandwidth spreading.

Exercises

13. Assume that a voice channel occupies a bandwidth of 4 kHz. We need to multiplex 10 voice channels with guard bands of 500 Hz using FDM. Calculate the required bandwidth.
14. We need to transmit 100 digitized voice channels using a pass-band channel of 20 KHz. What should be the ratio of bits/Hz if we use no guard band?
15. In the analog hierarchy of Figure 6.9, find the overhead (extra bandwidth for guard band or control) in each hierarchy level (group, supergroup, master group, and jumbo group).
16. We need to use synchronous TDM and combine 20 digital sources, each of 100 Kbps. Each output slot carries 1 bit from each digital source, but one extra bit is added to each frame for synchronization. Answer the following questions:
    a. What is the size of an output frame in bits?
    b. What is the output frame rate?

c. What is the duration of an output frame?

d. What is the output data rate?

e. What is the efficiency of the system (ratio of useful bits to the total bits).

17. Repeat Exercise 16 if each output slot carries 2 bits from each source.

18. We have 14 sources, each creating 500 8-bit characters per second. Since only some of these sources are active at any moment, we use statistical TDM to combine these sources using character interleaving. Each frame carries 6 slots at a time, but we need to add four-bit addresses to each slot. Answer the following questions:

    a. What is the size of an output frame in bits?

    b. What is the output frame rate?

    c. What is the duration of an output frame?

    d. What is the output data rate?

19. Ten sources, six with a bit rate of 200 kbps and four with a bit rate of 400 kbps are to be combined using multilevel TDM with no synchronizing bits. Answer the following questions about the final stage of the multiplexing:

    a. What is the size of a frame in bits?

    b. What is the frame rate?

    c. What is the duration of a frame?

    d. What is the data rate?

20. Four channels, two with a bit rate of 200 kbps and two with a bit rate of 150 kbps, are to be multiplexed using multiple slot TDM with no synchronization bits. Answer the following questions:

    a. What is the size of a frame in bits?

    b. What is the frame rate?

    c. What is the duration of a frame?

    d. What is the data rate?

21. Two channels, one with a bit rate of 190 kbps and another with a bit rate of 180 kbps, are to be multiplexed using pulse stuffing TDM with no synchronization bits. Answer the following questions:

    a. What is the size of a frame in bits?

    b. What is the frame rate?

    c. What is the duration of a frame?

    d. What is the data rate?

22. Answer the following questions about a T-1 line:

    a. What is the duration of a frame?

    b. What is the overhead (number of extra bits per second)?

23. Show the contents of the five output frames for a synchronous TDM multiplexer that combines four sources sending the following characters. Note that the characters are sent in the same order that they are typed. The third source is silent.

    a. Source 1 message: HELLO

    b. Source 2 message: HI

c. Source 3 message:
d. Source 4 message: BYE

24. Figure 6.34 shows a multiplexer in a synchronous TDM system. Each output slot is only 10 bits long (3 bits taken from each input plus 1 framing bit). What is the output stream? The bits arrive at the multiplexer as shown by the arrows.

Figure 6.34 *Exercise 24*

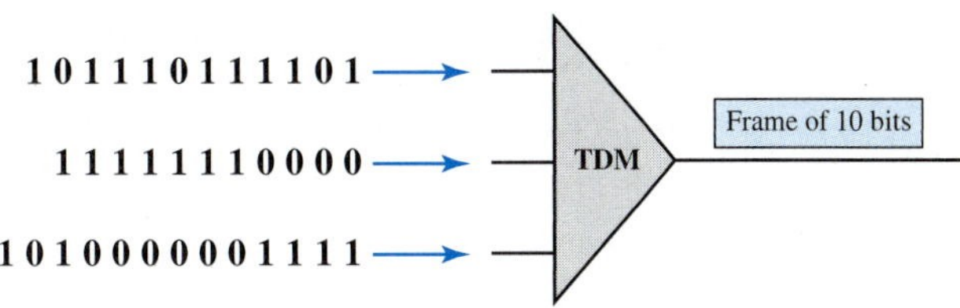

25. Figure 6.35 shows a demultiplexer in a synchronous TDM. If the input slot is 16 bits long (no framing bits), what is the bit stream in each output? The bits arrive at the demultiplexer as shown by the arrows.

Figure 6.35 *Exercise 25*

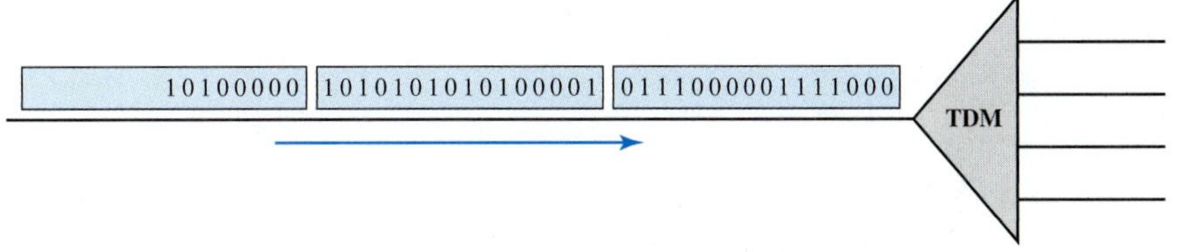

26. Answer the following questions about the digital hierarchy in Figure 6.23:
    a. What is the overhead (number of extra bits) in the DS-1 service?
    b. What is the overhead (number of extra bits) in the DS-2 service?
    c. What is the overhead (number of extra bits) in the DS-3 service?
    d. What is the overhead (number of extra bits) in the DS-4 service?

27. What is the minimum number of bits in a PN sequence if we use FHSS with a channel bandwidth of B = 4 KHz and B_{SS} = 100 KHz?

28. An FHSS system uses a 4-bit PN sequence. If the bit rate of the PN is 64 bits per second, answer the following questions:
    a. What is the total number of possible hops?
    b. What is the time needed to finish a complete cycle of PN?

29. A pseudorandom number generator uses the following formula to create a random series:

$$\mathbf{N_{i+1} = (5 + 7N_i) \bmod 17 - 1}$$

In which N_i defines the current random number and N_{i+1} defines the next random number. The term *mod* means the value of the remainder when dividing $(5 + 7N_i)$ by 17.

30. We have a digital medium with a data rate of 10 Mbps. How many 64-kbps voice channels can be carried by this medium if we use DSSS with the Barker sequence?

CHAPTER 7

Transmission Media

We discussed many issues related to the physical layer in Chapters 3 through 6. In this chapter, we discuss transmission media. Transmission media are actually located below the physical layer and are directly controlled by the physical layer. You could say that transmission media belong to layer zero. Figure 7.1 shows the position of transmission media in relation to the physical layer.

Figure 7.1 *Transmission medium and physical layer*

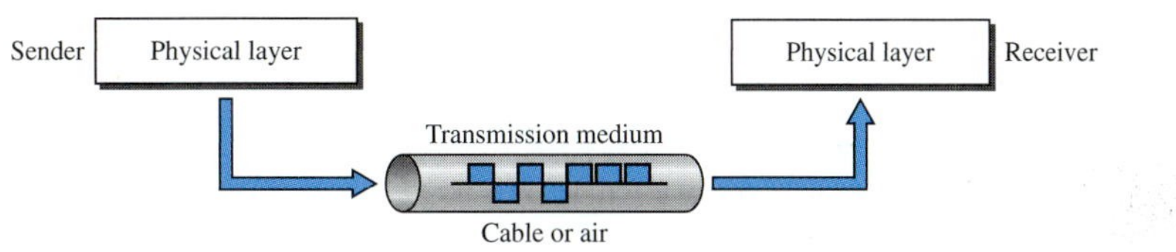

A **transmission medium** can be broadly defined as anything that can carry information from a source to a destination. For example, the transmission medium for two people having a dinner conversation is the air. The air can also be used to convey the message in a smoke signal or semaphore. For a written message, the transmission medium might be a mail carrier, a truck, or an airplane.

In data communications the definition of the information and the transmission medium is more specific. The transmission medium is usually free space, metallic cable, or fiber-optic cable. The information is usually a signal that is the result of a conversion of data from another form.

The use of long-distance communication using electric signals started with the invention of the telegraph by Morse in the 19th century. Communication by telegraph was slow and dependent on a metallic medium.

Extending the range of the human voice became possible when the telephone was invented in 1869. Telephone communication at that time also needed a metallic medium to carry the electric signals that were the result of a conversion from the human voice.

The communication was, however, unreliable due to the poor quality of the wires. The lines were often noisy and the technology was unsophisticated.

Wireless communication started in 1895 when Hertz was able to send high-frequency signals. Later, Marconi devised a method to send telegraph-type messages over the Atlantic Ocean.

We have come a long way. Better metallic media have been invented (twisted-pair and coaxial cables, for example). The use of optical fibers has increased the data rate incredibly. Free space (air, vacuum, and water) is used more efficiently, in part due to the technologies (such as modulation and multiplexing) discussed in the previous chapters.

As discussed in Chapter 3, computers and other telecommunication devices use signals to represent data. These signals are transmitted from one device to another in the form of electromagnetic energy, which is propagated through transmission media.

Electromagnetic energy, a combination of electric and magnetic fields vibrating in relation to each other, includes power, radio waves, infrared light, visible light, ultraviolet light, and X, gamma, and cosmic rays. Each of these constitutes a portion of the **electromagnetic spectrum.** Not all portions of the spectrum are currently usable for telecommunications, however. The media to harness those that are usable are also limited to a few types.

In telecommunications, transmission media can be divided into two broad categories: guided and unguided. Guided media include twisted-pair cable, coaxial cable, and fiber-optic cable. Unguided medium is free space. Figure 7.2 shows this taxonomy.

Figure 7.2 *Classes of transmission media*

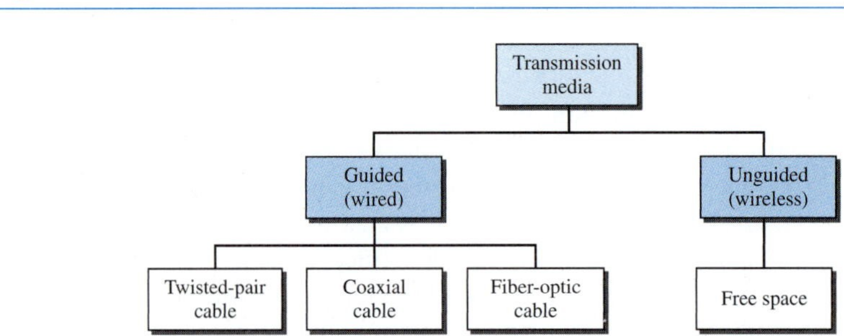

7.1 GUIDED MEDIA

Guided media, which are those that provide a conduit from one device to another, include **twisted-pair cable, coaxial cable,** and **fiber-optic cable.** A signal traveling along any of these media is directed and contained by the physical limits of the medium. Twisted-pair and coaxial cable use metallic (copper) conductors that accept and transport signals in the form of electric current. **Optical fiber** is a cable that accepts and transports signals in the form of light.

Twisted-Pair Cable

A twisted pair consists of two conductors (normally copper), each with its own plastic insulation, twisted together, as shown in Figure 7.3.

Figure 7.3 *Twisted-pair cable*

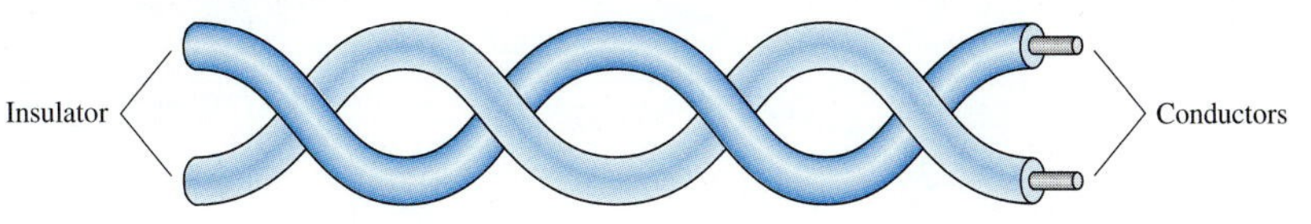

One of the wires is used to carry signals to the receiver, and the other is used only as a ground reference. The receiver uses the difference between the two.

In addition to the signal sent by the sender on one of the wires, interference (noise) and crosstalk may affect both wires and create unwanted signals.

If the two wires are parallel, the effect of these unwanted signals is not the same in both wires because they are at different locations relative to the noise or crosstalk sources (e.g., one is closer and the other is farther). This results in a difference at the receiver. By twisting the pairs, a balance is maintained. For example, suppose in one twist, one wire is closer to the noise source and the other is farther; in the next twist, the reverse is true. Twisting makes it probable that both wires are equally affected by external influences (noise or crosstalk). This means that the receiver, which calculates the difference between the two, receives no unwanted signals. The unwanted signals are mostly canceled out. From the above discussion, it is clear that the number of twists per unit of length (e.g., inch) has some effect on the quality of the cable.

Unshielded Versus Shielded Twisted-Pair Cable

The most common twisted-pair cable used in communications is referred to as **unshielded twisted-pair (UTP).** IBM has also produced a version of twisted-pair cable for its use called **shielded twisted-pair (STP).** STP cable has a metal foil or braided-mesh covering that encases each pair of insulated conductors. Although metal casing improves the quality of cable by preventing the penetration of noise or crosstalk, it is bulkier and more expensive. Figure 7.4 shows the difference between UTP and STP. Our discussion focuses primarily on UTP because STP is seldom used outside of IBM.

Categories

The Electronic Industries Association (EIA) has developed standards to classify unshielded twisted-pair cable into seven categories. Categories are determined by cable quality, with 1 as the lowest and 7 as the highest. Each EIA category is suitable for specific uses. Table 7.1 shows these categories.

Connectors

The most common UTP connector is **RJ45** (RJ stands for registered jack), as shown in Figure 7.5. The RJ45 is a keyed connector, meaning the connector can be inserted in only one way.

Figure 7.4 *UTP and STP cables*

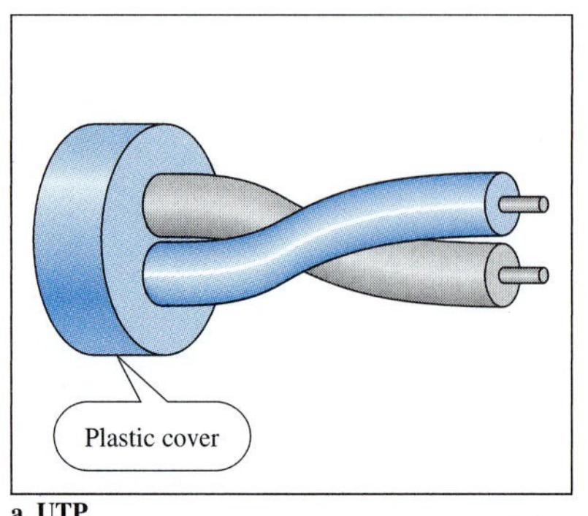

a. UTP

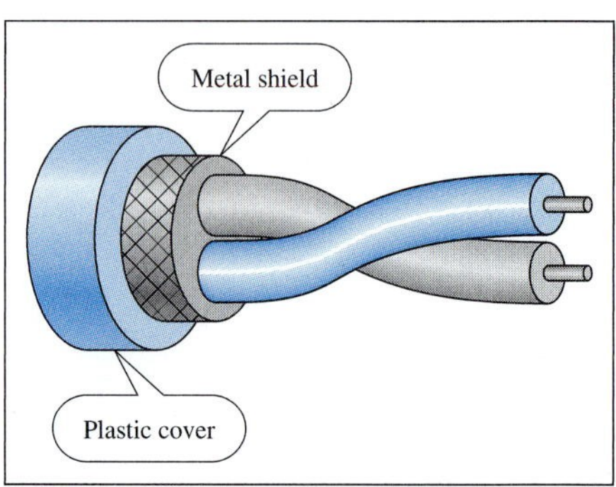

b. STP

Table 7.1 *Categories of unshielded twisted-pair cables*

Category	*Specification*	*Data Rate (Mbps)*	*Use*
1	Unshielded twisted-pair used in telephone	< 0.1	Telephone
2	Unshielded twisted-pair originally used in T-lines	2	T-1 lines
3	Improved CAT 2 used in LANs	10	LANs
4	Improved CAT 3 used in Token Ring networks	20	LANs
5	Cable wire is normally 24 AWG with a jacket and outside sheath	100	LANs
5E	An extension to category 5 that includes extra features to minimize the crosstalk and electromagnetic interference	125	LANs
6	A new category with matched components coming from the same manufacturer. The cable must be tested at a 200-Mbps data rate.	200	LANs
7	Sometimes called SSTP (shielded screen twisted-pair). Each pair is individually wrapped in a helical metallic foil followed by a metallic foil shield in addition to the outside sheath. The shield decreases the effect of crosstalk and increases the data rate.	600	LANs

Performance

One way to measure the performance of twisted-pair cable is to compare attenuation versus frequency and distance. A twisted-pair cable can pass a wide range of frequencies. However, Figure 7.6 shows that with increasing frequency, the attenuation, measured in decibels per kilometer (dB/km), sharply increases with frequencies above 100 kHz. Note that ***gauge*** is a measure of the thickness of the wire.

Figure 7.5 *UTP connector*

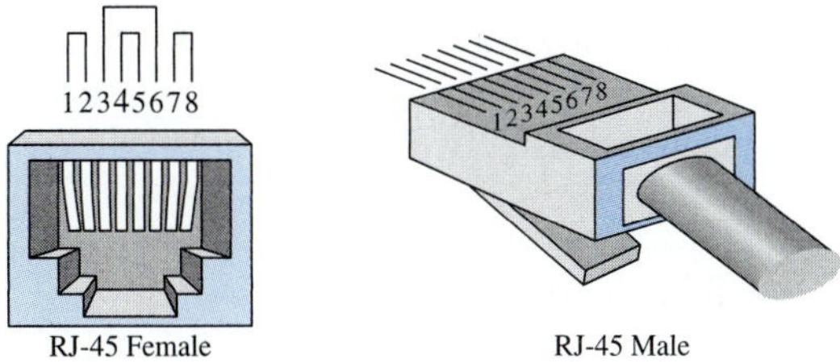

Figure 7.6 *UTP performance*

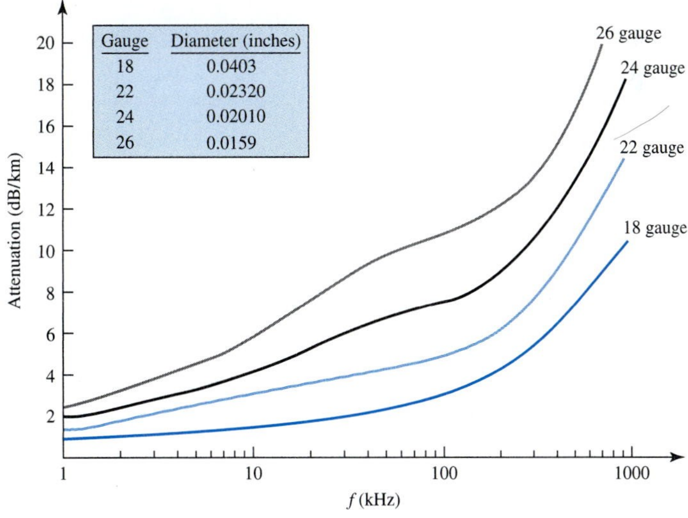

Applications

Twisted-pair cables are used in telephone lines to provide voice and data channels. The local loop—the line that connects subscribers to the central telephone office—commonly consists of unshielded twisted-pair cables. We discuss telephone networks in Chapter 9.

The DSL lines that are used by the telephone companies to provide high-data-rate connections also use the high-bandwidth capability of unshielded twisted-pair cables. We discuss DSL technology in Chapter 9.

Local-area networks, such as 10Base-T and 100Base-T, also use twisted-pair cables. We discuss these networks in Chapter 13.

Coaxial Cable

Coaxial cable (or *coax*) carries signals of higher frequency ranges than those in twisted-pair cable, in part because the two media are constructed quite differently. Instead of having two wires, coax has a central core conductor of solid or stranded wire (usually copper) enclosed in an insulating sheath, which is, in turn, encased in an outer conductor of metal foil, braid, or a combination of the two. The outer metallic wrapping serves both as a shield against noise and as the second conductor, which completes the circuit. This outer conductor is also enclosed in an insulating sheath, and the whole cable is protected by a plastic cover (see Figure 7.7).

Figure 7.7 *Coaxial cable*

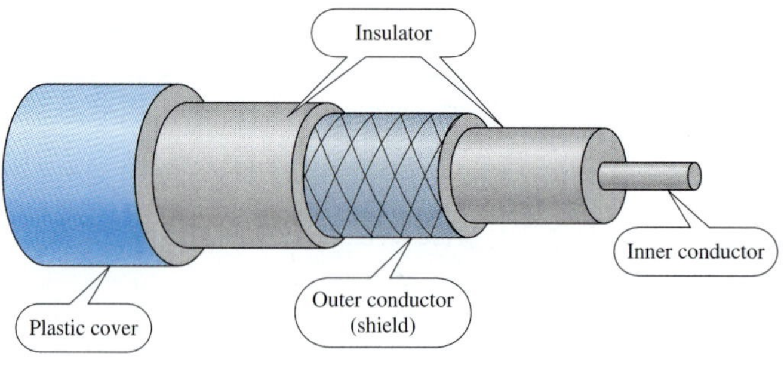

Coaxial Cable Standards

Coaxial cables are categorized by their **radio government (RG)** ratings. Each RG number denotes a unique set of physical specifications, including the wire gauge of the inner conductor, the thickness and type of the inner insulator, the construction of the shield, and the size and type of the outer casing. Each cable defined by an RG rating is adapted for a specialized function, as shown in Table 7.2.

Table 7.2 *Categories of coaxial cables*

Category	*Impedance*	*Use*
RG-59	75 Ω	Cable TV
RG-58	50 Ω	Thin Ethernet
RG-11	50 Ω	Thick Ethernet

Coaxial Cable Connectors

To connect coaxial cable to devices, we need coaxial connectors. The most common type of connector used today is the **Bayone-Neill-Concelman (BNC),** connector. Figure 7.8 shows three popular types of these connectors: the BNC connector, the BNC T connector, and the BNC terminator.

The BNC connector is used to connect the end of the cable to a device, such as a TV set. The BNC T connector is used in Ethernet networks (see Chapter 13) to branch out to a connection to a computer or other device. The BNC terminator is used at the end of the cable to prevent the reflection of the signal.

Figure 7.8 *BNC connectors*

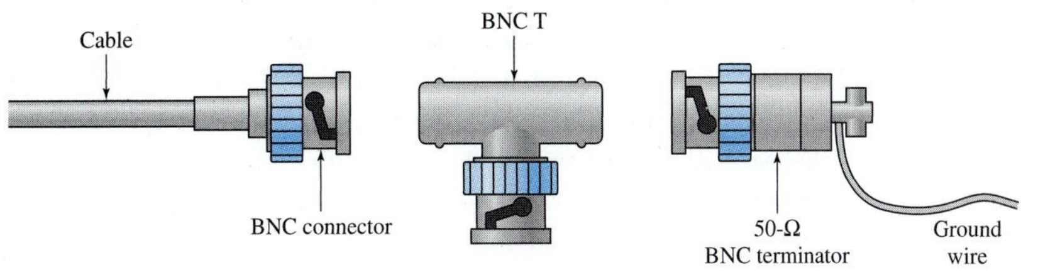

Performance

As we did with twisted-pair cables, we can measure the performance of a coaxial cable. We notice in Figure 7.9 that the attenuation is much higher in coaxial cables than in twisted-pair cable. In other words, although coaxial cable has a much higher bandwidth, the signal weakens rapidly and requires the frequent use of repeaters.

Figure 7.9 *Coaxial cable performance*

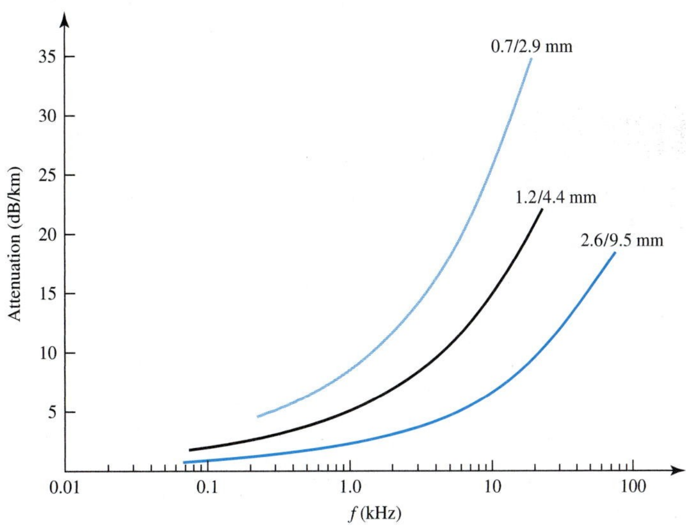

Applications

Coaxial cable was widely used in analog telephone networks where a single coaxial network could carry 10,000 voice signals. Later it was used in digital telephone networks where a single coaxial cable could carry digital data up to 600 Mbps. However, coaxial cable in telephone networks has largely been replaced today with fiber-optic cable.

Cable TV networks (see Chapter 9) also use coaxial cables. In the traditional cable TV network, the entire network used coaxial cable. Later, however, cable TV providers replaced most of the media with fiber-optic cable; hybrid networks use coaxial cable only at the network boundaries, near the consumer premises. Cable TV uses RG-59 coaxial cable.

Another common application of coaxial cable is in traditional Ethernet LANs (see Chapter 13). Because of its high bandwidth, and consequently high data rate, coaxial cable was chosen for digital transmission in early Ethernet LANs. The 10Base-2, or Thin Ethernet, uses RG-58 coaxial cable with BNC connectors to transmit data at 10 Mbps with a range of 185 m. The 10Base5, or Thick Ethernet, uses RG-11 (thick coaxial cable) to transmit 10 Mbps with a range of 5000 m. Thick Ethernet has specialized connectors.

Fiber-Optic Cable

A fiber-optic cable is made of glass or plastic and transmits signals in the form of light. To understand optical fiber, we first need to explore several aspects of the nature of light.

Light travels in a straight line as long as it is moving through a single uniform substance. If a ray of light traveling through one substance suddenly enters another substance (of a different density), the ray changes direction. Figure 7.10 shows how a ray of light changes direction when going from a more dense to a less dense substance.

Figure 7.10 *Bending of light ray*

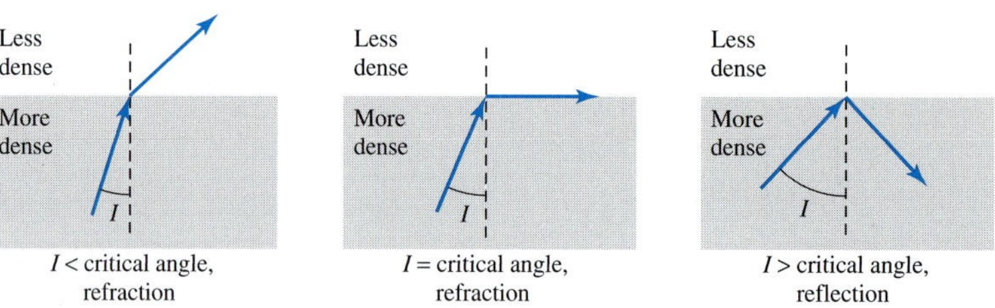

As the figure shows, if the **angle of incidence** *I* (the angle the ray makes with the line perpendicular to the interface between the two substances) is less than the **critical angle,** the ray **refracts** and moves closer to the surface. If the angle of incidence is equal to the critical angle, the light bends along the interface. If the angle is greater than the critical angle, the ray **reflects** (makes a turn) and travels again in the denser substance. Note that the critical angle is a property of the substance, and its value differs from one substance to another.

Optical fibers use reflection to guide light through a channel. A glass or plastic **core** is surrounded by a **cladding** of less dense glass or plastic. The difference in density of the two materials must be such that a beam of light moving through the core is reflected off the cladding instead of being refracted into it. See Figure 7.11.

Propagation Modes

Current technology supports two modes (multimode and single mode) for propagating light along optical channels, each requiring fiber with different physical characteristics. Multimode can be implemented in two forms: step-index or graded-index (see Figure 7.12).

Figure 7.11 *Optical fiber*

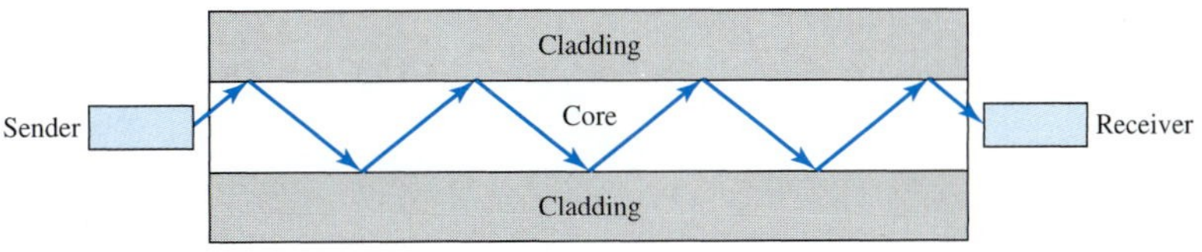

Figure 7.12 *Propagation modes*

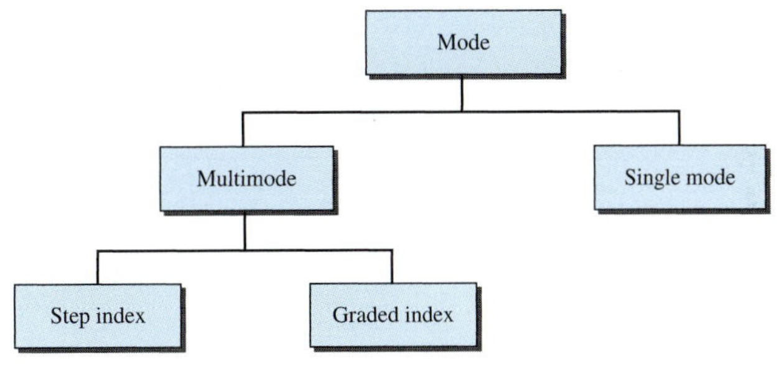

Multimode Multimode is so named because multiple beams from a light source move through the core in different paths. How these beams move within the cable depends on the structure of the core, as shown in Figure 7.13.

In **multimode step-index fiber,** the density of the core remains constant from the center to the edges. A beam of light moves through this constant density in a straight line until it reaches the interface of the core and the cladding. At the interface, there is an abrupt change due to a lower density; this alters the angle of the beam's motion. The term *step index* refers to the suddenness of this change, which contributes to the distortion of the signal as it passes through the fiber.

A second type of fiber, called **multimode graded-index fiber,** decreases this distortion of the signal through the cable. The word *index* here refers to the index of refraction. As we saw above, the index of refraction is related to density. A graded-index fiber, therefore, is one with varying densities. Density is highest at the center of the core and decreases gradually to its lowest at the edge. Figure 7.13 shows the impact of this variable density on the propagation of light beams.

Single-Mode Single-mode uses step-index fiber and a highly focused source of light that limits beams to a small range of angles, all close to the horizontal. The **single-mode fiber** itself is manufactured with a much smaller diameter than that of multimode fiber, and with substantially lower density (index of refraction). The decrease in density results in a critical angle that is close enough to 90° to make the propagation of beams almost horizontal. In this case, propagation of different beams is almost identical, and delays are negligible. All the beams arrive at the destination "together" and can be recombined with little distortion to the signal (see Figure 7.13).

Figure 7.13 *Modes*

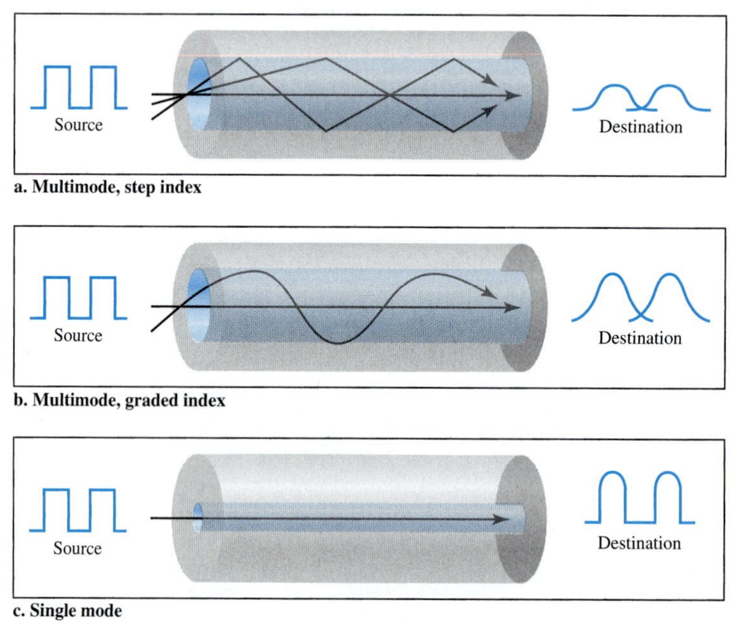

Fiber Sizes

Optical fibers are defined by the ratio of the diameter of their core to the diameter of their cladding, both expressed in micrometers. The common sizes are shown in Table 7.3. Note that the last size listed is for single-mode only.

Table 7.3 *Fiber types*

Type	*Core* (μm)	*Cladding* (μm)	*Mode*
50/125	50.0	125	Multimode, graded index
62.5/125	62.5	125	Multimode, graded index
100/125	100.0	125	Multimode, graded index
7/125	7.0	125	Single mode

Cable Composition

Figure 7.14 shows the composition of a typical fiber-optic cable. The outer jacket is made of either PVC or Teflon. Inside the jacket are Kevlar strands to strengthen the cable. Kevlar is a strong material used in the fabrication of bulletproof vests. Below the Kevlar is another plastic coating to cushion the fiber. The fiber is at the center of the cable, and it consists of cladding and core.

Fiber-Optic Cable Connectors

There are three types of connectors for fiber-optic cables, as shown in Figure 7.15.

Figure 7.14 *Fiber construction*

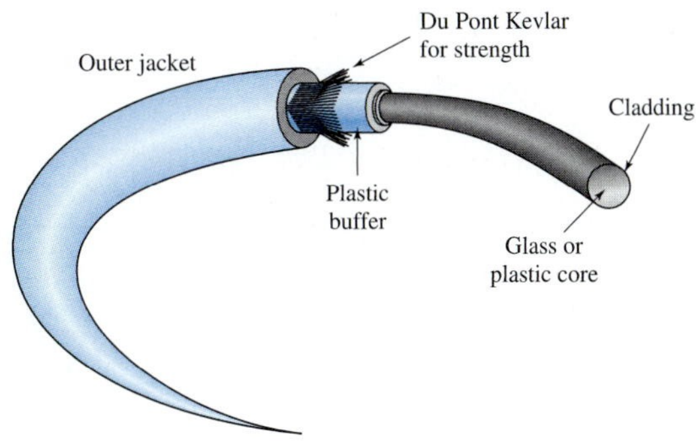

Figure 7.15 *Fiber-optic cable connectors*

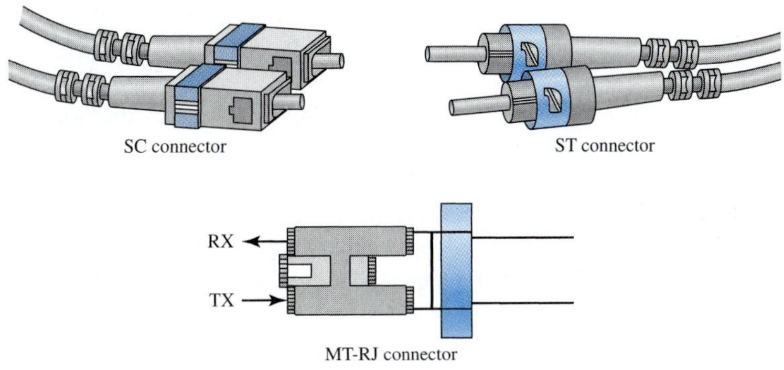

The **subscriber channel (SC) connector** is used for cable TV. It uses a push/pull locking system. The **straight-tip (ST) connector** is used for connecting cable to networking devices. It uses a bayonet locking system and is more reliable than SC. **MT-RJ** is a connector that is the same size as RJ45.

Performance

The plot of attenuation versus wavelength in Figure 7.16 shows a very interesting phenomenon in fiber-optic cable. Attenuation is flatter than in the case of twisted-pair cable and coaxial cable. The performance is such that we need fewer (actually 10 times less) repeaters when we use fiber-optic cable.

Applications

Fiber-optic cable is often found in backbone networks because its wide bandwidth is cost-effective. Today, with wavelength-division multiplexing (WDM), we can transfer

Figure 7.16 *Optical fiber performance*

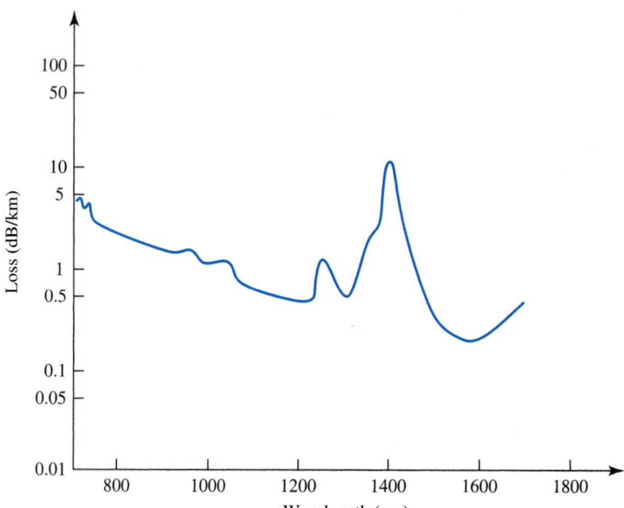

data at a rate of 1600 Gbps. The SONET network that we discuss in Chapter 17 provides such a backbone.

Some cable TV companies use a combination of optical fiber and coaxial cable, thus creating a hybrid network. Optical fiber provides the backbone structure while coaxial cable provides the connection to the user premises. This is a cost-effective configuration since the narrow bandwidth requirement at the user end does not justify the use of optical fiber.

Local-area networks such as 100Base-FX network (Fast Ethernet) and 1000Base-X also use fiber-optic cable.

Advantages and Disadvantages of Optical Fiber

Advantages Fiber-optic cable has several advantages over metallic cable (twisted-pair or coaxial).

- **Higher bandwidth.** Fiber-optic cable can support dramatically higher bandwidths (and hence data rates) than either twisted-pair or coaxial cable. Currently, data rates and bandwidth utilization over fiber-optic cable are limited not by the medium but by the signal generation and reception technology available.
- **Less signal attenuation.** Fiber-optic transmission distance is significantly greater than that of other guided media. A signal can run for 50 km without requiring regeneration. We need repeaters every 5 km for coaxial or twisted-pair cable.
- **Immunity to electromagnetic interference.** Electromagnetic noise cannot affect fiber-optic cables.
- **Resistance to corrosive materials.** Glass is more resistant to corrosive materials than copper.

- **Light weight.** Fiber-optic cables are much lighter than copper cables.
- **Greater immunity to tapping.** Fiber-optic cables are more immune to tapping than copper cables. Copper cables create antenna effects that can easily be tapped.

Disadvantages There are some disadvantages in the use of optical fiber.

- **Installation and maintenance.** Fiber-optic cable is a relatively new technology. Its installation and maintenance require expertise that is not yet available everywhere.
- **Unidirectional light propagation.** Propagation of light is unidirectional. If we need bidirectional communication, two fibers are needed.
- **Cost.** The cable and the interfaces are relatively more expensive than those of other guided media. If the demand for bandwidth is not high, often the use of optical fiber cannot be justified.

7.2 UNGUIDED MEDIA: WIRELESS

Unguided media transport electromagnetic waves without using a physical conductor. This type of communication is often referred to as **wireless communication.** Signals are normally broadcast through free space and thus are available to anyone who has a device capable of receiving them.

Figure 7.17 shows the part of the electromagnetic spectrum, ranging from 3 kHz to 900 THz, used for wireless communication.

Figure 7.17 *Electromagnetic spectrum for wireless communication*

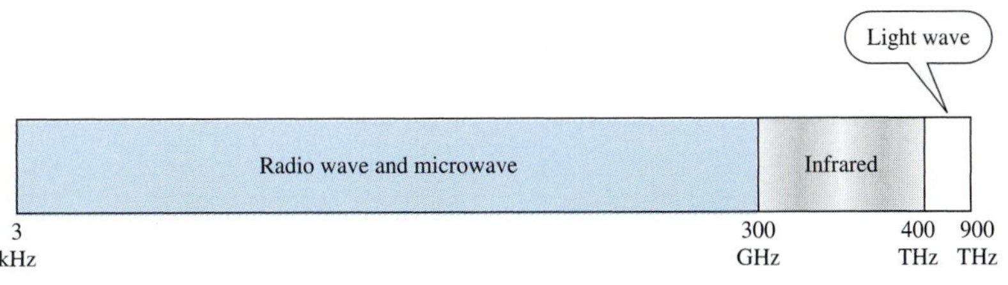

Unguided signals can travel from the source to destination in several ways: ground propagation, sky propagation, and line-of-sight propagation, as shown in Figure 7.18.

In **ground propagation,** radio waves travel through the lowest portion of the atmosphere, hugging the earth. These low-frequency signals emanate in all directions from the transmitting antenna and follow the curvature of the planet. Distance depends on the amount of power in the signal: The greater the power, the greater the distance. In **sky propagation,** higher-frequency radio waves radiate upward into the ionosphere (the layer of atmosphere where particles exist as ions) where they are reflected back to earth. This type of transmission allows for greater distances with lower output power. In **line-of-sight propagation,** very high-frequency signals are transmitted in straight lines directly from antenna to antenna. Antennas must be directional, facing each other,

Figure 7.18 *Propagation methods*

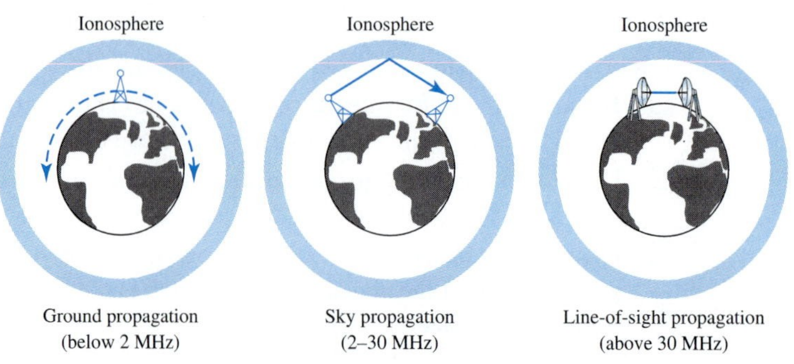

and either tall enough or close enough together not to be affected by the curvature of the earth. Line-of-sight propagation is tricky because radio transmissions cannot be completely focused.

The section of the electromagnetic spectrum defined as radio waves and microwaves is divided into eight ranges, called *bands,* each regulated by government authorities. These bands are rated from *very low frequency* (VLF) to *extremely high frequency* (EHF). Table 7.4 lists these bands, their ranges, propagation methods, and some applications.

Table 7.4 *Bands*

Band	*Range*	*Propagation*	*Application*
VLF (very low frequency)	3–30 kHz	Ground	Long-range radio navigation
LF (low frequency)	30–300 kHz	Ground	Radio beacons and navigational locators
MF (middle frequency)	300 kHz–3 MHz	Sky	AM radio
HF (high frequency)	3–30 MHz	Sky	Citizens band (CB), ship/aircraft communication
VHF (very high frequency)	30–300 MHz	Sky and line-of-sight	VHF TV, FM radio
UHF (ultrahigh frequency)	300 MHz–3 GHz	Line-of-sight	UHF TV, cellular phones, paging, satellite
SHF (superhigh frequency)	3–30 GHz	Line-of-sight	Satellite communication
EHF (extremely high frequency)	30–300 GHz	Line-of-sight	Radar, satellite

We can divide wireless transmission into three broad groups: radio waves, microwaves, and infrared waves. See Figure 7.19.

Figure 7.19 *Wireless transmission waves*

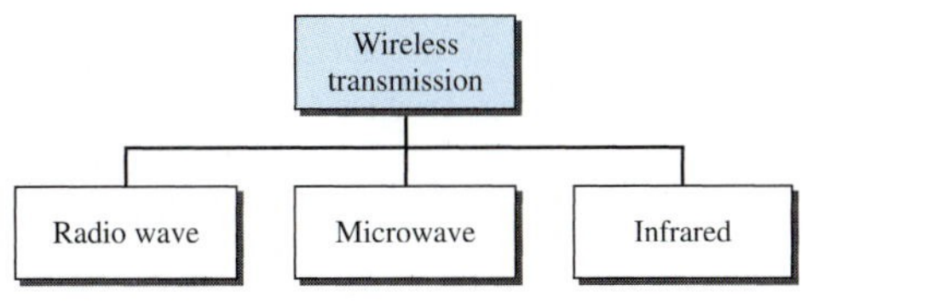

Radio Waves

Although there is no clear-cut demarcation between radio waves and microwaves, electromagnetic waves ranging in frequencies between 3 kHz and 1 GHz are normally called **radio waves;** waves ranging in frequencies between 1 and 300 GHz are called **microwaves.** However, the behavior of the waves, rather than the frequencies, is a better criterion for classification.

Radio waves, for the most part, are omnidirectional. When an antenna transmits radio waves, they are propagated in all directions. This means that the sending and receiving antennas do not have to be aligned. A sending antenna sends waves that can be received by any receiving antenna. The omnidirectional property has a disadvantage, too. The radio waves transmitted by one antenna are susceptible to interference by another antenna that may send signals using the same frequency or band.

Radio waves, particularly those waves that propagate in the sky mode, can travel long distances. This makes radio waves a good candidate for long-distance broadcasting such as AM radio.

Radio waves, particularly those of low and medium frequencies, can penetrate walls. This characteristic can be both an advantage and a disadvantage. It is an advantage because, for example, an AM radio can receive signals inside a building. It is a disadvantage because we cannot isolate a communication to just inside or outside a building. The radio wave band is relatively narrow, just under 1 GHz, compared to the microwave band. When this band is divided into subbands, the subbands are also narrow, leading to a low data rate for digital communications.

Almost the entire band is regulated by authorities (e.g., the FCC in the United States). Using any part of the band requires permission from the authorities.

Omnidirectional Antenna

Radio waves use **omnidirectional antennas** that send out signals in all directions. Based on the wavelength, strength, and the purpose of transmission, we can have several types of antennas. Figure 7.20 shows an omnidirectional antenna.

Applications

The omnidirectional characteristics of radio waves make them useful for multicasting, in which there is one sender but many receivers. AM and FM radio, television, maritime radio, cordless phones, and paging are examples of multicasting.

Figure 7.20 *Omnidirectional antenna*

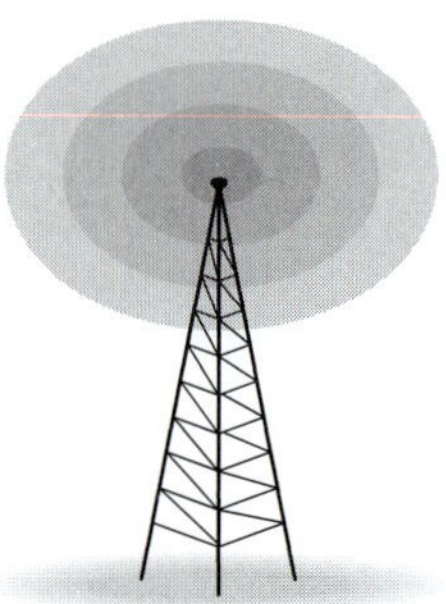

Radio waves are used for multicast communications, such as radio and television, and paging systems.

Microwaves

Electromagnetic waves having frequencies between 1 and 300 GHz are called microwaves.

Microwaves are unidirectional. When an antenna transmits microwave waves, they can be narrowly focused. This means that the sending and receiving antennas need to be aligned. The unidirectional property has an obvious advantage. A pair of antennas can be aligned without interfering with another pair of aligned antennas. The following describes some characteristics of microwave propagation:

- Microwave propagation is line-of-sight. Since the towers with the mounted antennas need to be in direct sight of each other, towers that are far apart need to be very tall. The curvature of the earth as well as other blocking obstacles do not allow two short towers to communicate by using microwaves. Repeaters are often needed for long-distance communication.
- Very high-frequency microwaves cannot penetrate walls. This characteristic can be a disadvantage if receivers are inside buildings.
- The microwave band is relatively wide, almost 299 GHz. Therefore wider subbands can be assigned, and a high data rate is possible
- Use of certain portions of the band requires permission from authorities.

Unidirectional Antenna

Microwaves need **unidirectional antennas** that send out signals in one direction. Two types of antennas are used for microwave communications: the parabolic dish and the horn (see Figure 7.21).

A **parabolic dish antenna** is based on the geometry of a parabola: Every line parallel to the line of symmetry (line of sight) reflects off the curve at angles such that all the lines intersect in a common point called the focus. The parabolic dish works as a

Figure 7.21 *Unidirectional antennas*

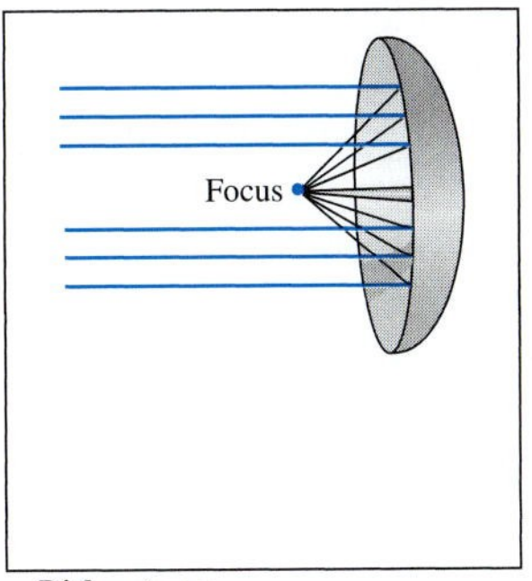

a. Dish antenna

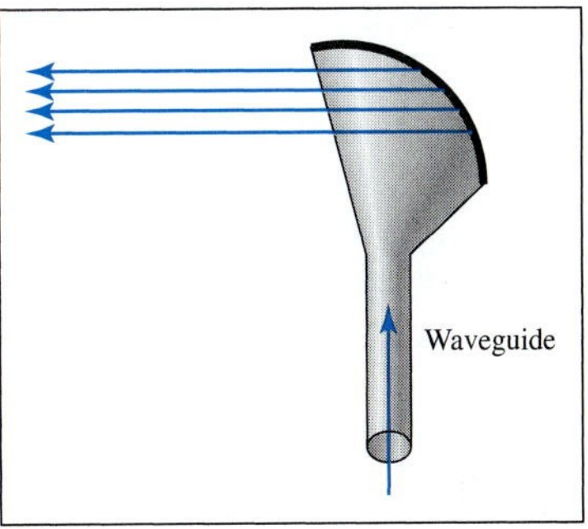

b. Horn antenna

funnel, catching a wide range of waves and directing them to a common point. In this way, more of the signal is recovered than would be possible with a single-point receiver.

Outgoing transmissions are broadcast through a horn aimed at the dish. The microwaves hit the dish and are deflected outward in a reversal of the receipt path.

A **horn antenna** looks like a gigantic scoop. Outgoing transmissions are broadcast up a stem (resembling a handle) and deflected outward in a series of narrow parallel beams by the curved head. Received transmissions are collected by the scooped shape of the horn, in a manner similar to the parabolic dish, and are deflected down into the stem.

Applications

Microwaves, due to their unidirectional properties, are very useful when unicast (one-to-one) communication is needed between the sender and the receiver. They are used in cellular phones (Chapter 16), satellite networks (Chapter 16), and wireless LANs (Chapter 14).

Microwaves are used for unicast communication such as cellular telephones, satellite networks, and wireless LANs.

Infrared

Infrared waves, with frequencies from 300 GHz to 400 THz (wavelengths from 1 mm to 770 nm), can be used for short-range communication. Infrared waves, having high frequencies, cannot penetrate walls. This advantageous characteristic prevents interference between one system and another; a short-range communication system in one room cannot be affected by another system in the next room. When we use our infrared remote control, we do not interfere with the use of the remote by our neighbors. However, this same characteristic makes infrared signals useless for long-range communication. In addition, we cannot use infrared waves outside a building because the sun's rays contain infrared waves that can interfere with the communication.

Applications

The infrared band, almost 400 THz, has an excellent potential for data transmission. Such a wide bandwidth can be used to transmit digital data with a very high data rate. The *Infrared Data Association* (IrDA), an association for sponsoring the use of infrared waves, has established standards for using these signals for communication between devices such as keyboards, mice, PCs, and printers. For example, some manufacturers provide a special port called the **IrDA port** that allows a wireless keyboard to communicate with a PC. The standard originally defined a data rate of 75 kbps for a distance up to 8 m. The recent standard defines a data rate of 4 Mbps.

Infrared signals defined by IrDA transmit through line of sight; the IrDA port on the keyboard needs to point to the PC for transmission to occur.

Infrared signals can be used for short-range communication in a closed area using line-of-sight propagation.

7.3 RECOMMENDED READING

For more details about subjects discussed in this chapter, we recommend the following books. The items in brackets [. . .] refer to the reference list at the end of the text.

Books

Transmission media is discussed in Section 3.8 of [GW04], Chapter 4 of [Sta04], Section 2.2 and 2.3 of [Tan03]. [SSS05] gives a full coverage of transmission media.

7.4 KEY TERMS

angle of incidence
Bayone-Neil-Concelman (BNC) connector
cladding
coaxial cable
core
critical angle
electromagnetic spectrum
fiber-optic cable
gauge
ground propagation
guided media
horn antenna
infrared wave
IrDA port
line-of-sight propagation
microwave
MT-RJ
multimode graded-index fiber
multimode step-index fiber
omnidirectional antenna
optical fiber
parabolic dish antenna
Radio Government (RG) number
radio wave
reflection
refraction
RJ45

shielded twisted-pair (STP)
single-mode fiber
sky propagation
straight-tip (ST) connector
subscriber channel (SC) connector
transmission medium
twisted-pair cable
unguided medium
unidirectional antenna
unshielded twisted-pair (UTP)
wireless communication

7.5 SUMMARY

- Transmission media lie below the physical layer.
- A guided medium provides a physical conduit from one device to another. Twisted-pair cable, coaxial cable, and optical fiber are the most popular types of guided media.
- Twisted-pair cable consists of two insulated copper wires twisted together. Twisted-pair cable is used for voice and data communications.
- Coaxial cable consists of a central conductor and a shield. Coaxial cable can carry signals of higher frequency ranges than twisted-pair cable. Coaxial cable is used in cable TV networks and traditional Ethernet LANs.
- Fiber-optic cables are composed of a glass or plastic inner core surrounded by cladding, all encased in an outside jacket. Fiber-optic cables carry data signals in the form of light. The signal is propagated along the inner core by reflection. Fiber-optic transmission is becoming increasingly popular due to its noise resistance, low attenuation, and high-bandwidth capabilities. Fiber-optic cable is used in backbone networks, cable TV networks, and Fast Ethernet networks.
- Unguided media (free space) transport electromagnetic waves without the use of a physical conductor.
- Wireless data are transmitted through ground propagation, sky propagation, and line-of-sight propagation.Wireless waves can be classified as radio waves, microwaves, or infrared waves. Radio waves are omnidirectional; microwaves are unidirectional. Microwaves are used for cellular phone, satellite, and wireless LAN communications.
- Infrared waves are used for short-range communications such as those between a PC and a peripheral device. It can also be used for indoor LANs.

7.6 PRACTICE SET

Review Questions

1. What is the position of the transmission media in the OSI or the Internet model?
2. Name the two major categories of transmission media.
3. How do guided media differ from unguided media?
4. What are the three major classes of guided media?
5. What is the significance of the twisting in twisted-pair cable?

6. What is refraction? What is reflection?
7. What is the purpose of cladding in an optical fiber?
8. Name the advantages of optical fiber over twisted-pair and coaxial cable.
9. How does sky propagation differ from line-of-sight propagation?
10. What is the difference between omnidirectional waves and unidirectional waves?

Exercises

11. Using Figure 7.6, tabulate the attenuation (in dB) of a 18-gauge UTP for the indicated frequencies and distances.

Table 7.5 *Attenuation for 18-gauge UTP*

Distance	*dB at 1 KHz*	*dB at 10 KHz*	*dB at 100 KHz*
1 Km			
10 Km			
15 Km			
20 Km			

12. Use the result of Exercise 11 to infer that the bandwidth of a UTP cable decreases with an increase in distance.
13. If the power at the beginning of a 1 Km 18-gauge UTP is 200 mw, what is the power at the end for frequencies 1 KHz, 10 KHz, and 100 KHz? Use the result of Exercise 11.
14. Using Figure 7.9, tabulate the attenuation (in dB) of a 2.6/9.5 mm coaxial cable for the indicated frequencies and distances.

Table 7.6 *Attenuation for 2.6/9.5 mm coaxial cable*

Distance	*dB at 1 KHz*	*dB at 10 KHz*	*dB at 100 KHz*
1 Km			
10 Km			
15 Km			
20 Km			

15. Use the result of Exercise 14 to infer that the bandwidth of a coaxial cable decreases with the increase in distance.
16. If the power at the beginning of a 1 Km 2.6/9.5 mm coaxial cable is 200 mw, what is the power at the end for frequencies 1 KHz, 10 KHz, and 100 KHz? Use the result of Exercise 14.
17. Calculate the bandwidth of the light for the following wavelength ranges (assume a propagation speed of 2×10^8 m):
    a. 1000 to 1200 nm
    b. 1000 to 1400 nm

18. The horizontal axes in Figure 7.6 and 7.9 represent frequencies. The horizontal axis in Figure 7.16 represents wavelength. Can you explain the reason? If the propagation speed in an optical fiber is 2×10^8 m, can you change the units in the horizontal axis to frequency? Should the vertical-axis units be changed too? Should the curve be changed too?
19. Using Figure 7.16, tabulate the attenuation (in dB) of an optical fiber for the indicated wavelength and distances.

Table 7.7 *Attenuation for optical fiber*

Distance	*dB at 800 nm*	*dB at 1000 nm*	*dB at 1200 nm*
1 Km			
10 Km			
15 Km			
20 Km			

20. A light signal is travelling through a fiber. What is the delay in the signal if the length of the fiber-optic cable is 10 m, 100 m, and 1 Km (assume a propagation speed of 2×10^8 m)?
21. A beam of light moves from one medium to another medium with less density. The critical angle is 60°. Do we have refraction or reflection for each of the following incident angles? Show the bending of the light ray in each case.
    a. 40°
    b. 60°
    c. 80°

CHAPTER 8

Switching

A network is a set of connected devices. Whenever we have multiple devices, we have the problem of how to connect them to make one-to-one communication possible. One solution is to make a point-to-point connection between each pair of devices (a mesh topology) or between a central device and every other device (a star topology). These methods, however, are impractical and wasteful when applied to very large networks. The number and length of the links require too much infrastructure to be cost-efficient, and the majority of those links would be idle most of the time. Other topologies employing multipoint connections, such as a bus, are ruled out because the distances between devices and the total number of devices increase beyond the capacities of the media and equipment.

A better solution is **switching.** A switched network consists of a series of interlinked nodes, called **switches.** Switches are devices capable of creating temporary connections between two or more devices linked to the switch. In a switched network, some of these nodes are connected to the end systems (computers or telephones, for example). Others are used only for routing. Figure 8.1 shows a switched network.

Figure 8.1 *Switched network*

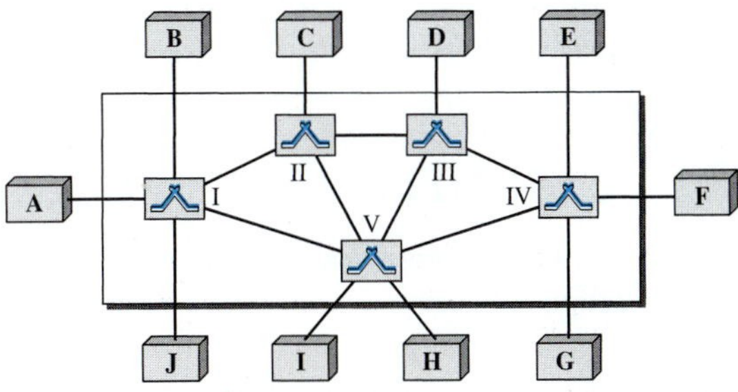

The **end systems** (communicating devices) are labeled A, B, C, D, and so on, and the switches are labeled I, II, III, IV, and V. Each switch is connected to multiple links.

Traditionally, three methods of switching have been important: circuit switching, packet switching, and message switching. The first two are commonly used today. The third has been phased out in general communications but still has networking applications. We can then divide today's networks into three broad categories: circuit-switched networks, packet-switched networks, and message-switched. Packet-switched networks can further be divided into two subcategories—virtual-circuit networks and datagram networks—as shown in Figure 8.2.

Figure 8.2 *Taxonomy of switched networks*

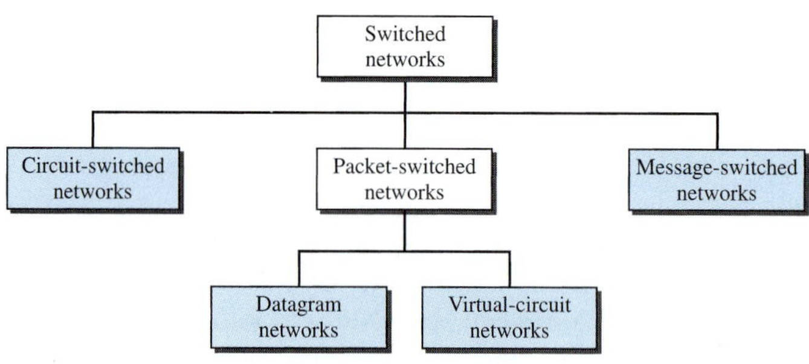

We can say that the virtual-circuit networks have some common characteristics with circuit-switched and datagram networks. Thus, we first discuss circuit-switched networks, then datagram networks, and finally virtual-circuit networks.

Today the tendency in packet switching is to combine datagram networks and virtual-circuit networks. Networks route the first packet based on the datagram addressing idea, but then create a virtual-circuit network for the rest of the packets coming from the same source and going to the same destination. We will see some of these networks in future chapters.

In message switching, each switch stores the whole message and forwards it to the next switch. Although, we don't see message switching at lower layers, it is still used in some applications like electronic mail (e-mail). We will not discuss this topic in this book.

8.1 CIRCUIT-SWITCHED NETWORKS

A **circuit-switched network** consists of a set of switches connected by physical links. A connection between two stations is a dedicated path made of one or more links. However, each connection uses only one dedicated channel on each link. Each link is normally divided into n channels by using FDM or TDM as discussed in Chapter 6.

A circuit-switched network is made of a set of switches connected by physical links, in which each link is divided into n channels.

Figure 8.3 shows a trivial circuit-switched network with four switches and four links. Each link is divided into n (n is 3 in the figure) channels by using FDM or TDM.

Figure 8.3 *A trivial circuit-switched network*

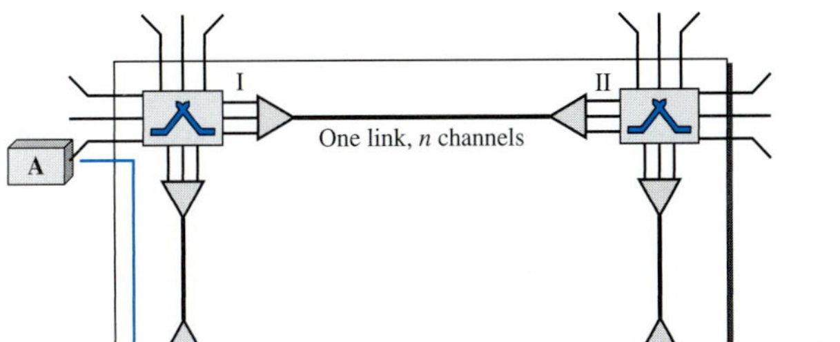

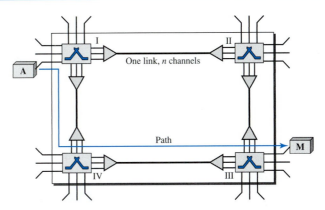

We have explicitly shown the multiplexing symbols to emphasize the division of the link into channels even though multiplexing can be implicitly included in the switch fabric.

The end systems, such as computers or telephones, are directly connected to a switch. We have shown only two end systems for simplicity. When end system A needs to communicate with end system M, system A needs to request a connection to M that must be accepted by all switches as well as by M itself. This is called the **setup phase;** a circuit (channel) is reserved on each link, and the combination of circuits or channels defines the dedicated path. After the dedicated path made of connected circuits (channels) is established, **data transfer** can take place. After all data have been transferred, the circuits are torn down.

We need to emphasize several points here:

- Circuit switching takes place at the physical layer.
- Before starting communication, the stations must make a reservation for the resources to be used during the communication. These resources, such as channels (bandwidth in FDM and time slots in TDM), switch buffers, switch processing time, and switch input/output ports, must remain dedicated during the entire duration of data transfer until the **teardown phase.**
- Data transferred between the two stations are not packetized (physical layer transfer of the signal). The data are a continuous flow sent by the source station and received by the destination station, although there may be periods of silence.
- There is no addressing involved during data transfer. The switches route the data based on their occupied band (FDM) or time slot (TDM). Of course, there is end-to-end addressing used during the setup phase, as we will see shortly.

In circuit switching, the resources need to be reserved during the setup phase; the resources remain dedicated for the entire duration of data transfer until the teardown phase.

Example 8.1

As a trivial example, let us use a circuit-switched network to connect eight telephones in a small area. Communication is through 4-kHz voice channels. We assume that each link uses FDM to connect a maximum of two voice channels. The bandwidth of each link is then 8 kHz. Figure 8.4 shows the situation. Telephone 1 is connected to telephone 7; 2 to 5; 3 to 8; and 4 to 6. Of course the situation may change when new connections are made. The switch controls the connections.

Figure 8.4 *Circuit-switched network used in Example 8.1*

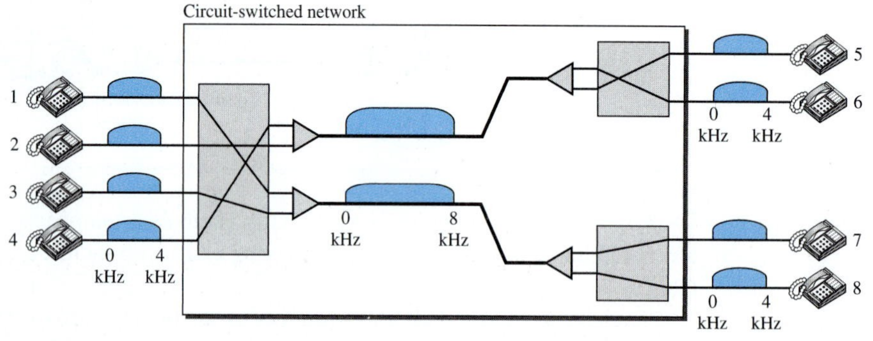

Example 8.2

As another example, consider a circuit-switched network that connects computers in two remote offices of a private company. The offices are connected using a T-1 line leased from a communication service provider. There are two 4 × 8 (4 inputs and 8 outputs) switches in this network. For each switch, four output ports are folded into the input ports to allow communication between computers in the same office. Four other output ports allow communication between the two offices. Figure 8.5 shows the situation.

Figure 8.5 *Circuit-switched network used in Example 8.2*

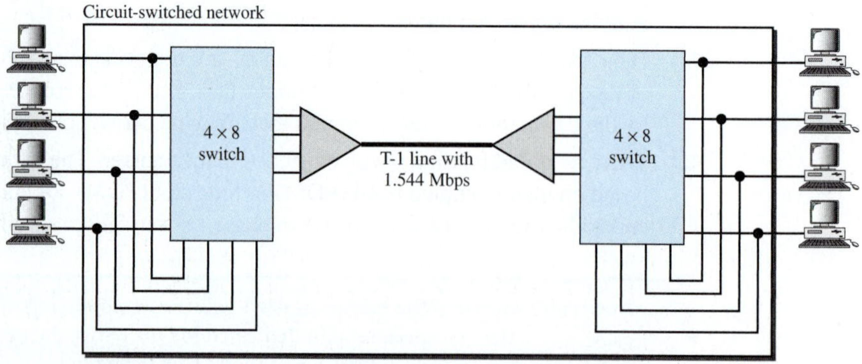

Three Phases

The actual communication in a circuit-switched network requires three phases: connection setup, data transfer, and connection teardown.

Setup Phase

Before the two parties (or multiple parties in a conference call) can communicate, a dedicated circuit (combination of channels in links) needs to be established. The end systems are normally connected through dedicated lines to the switches, so connection setup means creating dedicated channels between the switches. For example, in Figure 8.3, when system A needs to connect to system M, it sends a setup request that includes the address of system M, to switch I. Switch I finds a channel between itself and switch IV that can be dedicated for this purpose. Switch I then sends the request to switch IV, which finds a dedicated channel between itself and switch III. Switch III informs system M of system A's intention at this time.

In the next step to making a connection, an acknowledgment from system M needs to be sent in the opposite direction to system A. Only after system A receives this acknowledgment is the connection established.

Note that end-to-end addressing is required for creating a connection between the two end systems. These can be, for example, the addresses of the computers assigned by the administrator in a TDM network, or telephone numbers in an FDM network.

Data Transfer Phase

After the establishment of the dedicated circuit (channels), the two parties can transfer data.

Teardown Phase

When one of the parties needs to disconnect, a signal is sent to each switch to release the resources.

Efficiency

It can be argued that circuit-switched networks are not as efficient as the other two types of networks because resources are allocated during the entire duration of the connection. These resources are unavailable to other connections. In a telephone network, people normally terminate the communication when they have finished their conversation. However, in computer networks, a computer can be connected to another computer even if there is no activity for a long time. In this case, allowing resources to be dedicated means that other connections are deprived.

Delay

Although a circuit-switched network normally has low efficiency, the delay in this type of network is minimal. During data transfer the data are not delayed at each switch; the resources are allocated for the duration of the connection. Figure 8.6 shows the idea of delay in a circuit-switched network when only two switches are involved.

As Figure 8.6 shows, there is no waiting time at each switch. The total delay is due to the time needed to create the connection, transfer data, and disconnect the circuit. The

Figure 8.6 *Delay in a circuit-switched network*

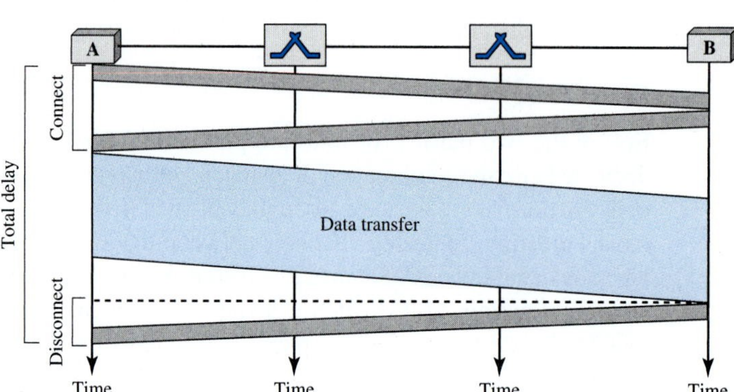

delay caused by the setup is the sum of four parts: the propagation time of the source computer request (slope of the first gray box), the request signal transfer time (height of the first gray box), the propagation time of the acknowledgment from the destination computer (slope of the second gray box), and the signal transfer time of the acknowledgment (height of the second gray box). The delay due to data transfer is the sum of two parts: the propagation time (slope of the colored box) and data transfer time (height of the colored box), which can be very long. The third box shows the time needed to tear down the circuit. We have shown the case in which the receiver requests disconnection, which creates the maximum delay.

Circuit-Switched Technology in Telephone Networks

As we will see in Chapter 9, the telephone companies have previously chosen the circuit-switched approach to switching in the physical layer; today the tendency is moving toward other switching techniques. For example, the telephone number is used as the global address, and a signaling system (called SS7) is used for the setup and teardown phases.

Switching at the physical layer in the traditional telephone network uses the circuit-switching approach.

8.2 DATAGRAM NETWORKS

In data communications, we need to send messages from one end system to another. If the message is going to pass through a packet-switched network, it needs to be divided into packets of fixed or variable size. The size of the packet is determined by the network and the governing protocol.

In packet switching, there is no resource allocation for a packet. This means that there is no reserved bandwidth on the links, and there is no scheduled processing time for each packet. Resources are allocated on demand. The allocation is done on a first-come, first-served basis. When a switch receives a packet, no matter what is the source or destination, the packet must wait if there are other packets being processed. As with other systems in our daily life, this lack of reservation may create delay. For example, if we do not have a reservation at a restaurant, we might have to wait.

In a packet-switched network, there is no resource reservation; resources are allocated on demand.

In a **datagram network,** each packet is treated independently of all others. Even if a packet is part of a multipacket transmission, the network treats it as though it existed alone. Packets in this approach are referred to as **datagrams.**

Datagram switching is normally done at the network layer. We briefly discuss datagram networks here as a comparison with circuit-switched and virtual-circuit-switched networks. In Part 4 of this text, we go into greater detail.

Figure 8.7 shows how the datagram approach is used to deliver four packets from station A to station X. The switches in a datagram network are traditionally referred to as routers. That is why we use a different symbol for the switches in the figure.

Figure 8.7 *A datagram network with four switches (routers)*

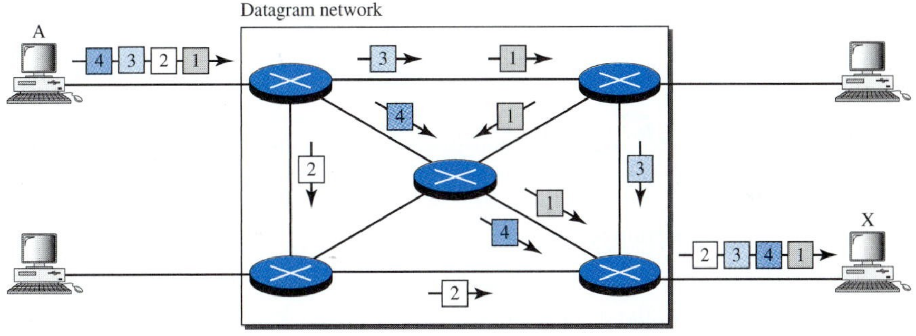

In this example, all four packets (or datagrams) belong to the same message, but may travel different paths to reach their destination. This is so because the links may be involved in carrying packets from other sources and do not have the necessary bandwidth available to carry all the packets from A to X. This approach can cause the datagrams of a transmission to arrive at their destination out of order with different delays between the packets. Packets may also be lost or dropped because of a lack of resources. In most protocols, it is the responsibility of an upper-layer protocol to reorder the datagrams or ask for lost datagrams before passing them on to the application.

The datagram networks are sometimes referred to as **connectionless networks.** The term *connectionless* here means that the switch (packet switch) does not keep information about the connection state. There are no setup or teardown phases. Each packet is treated the same by a switch regardless of its source or destination.

Routing Table

If there are no setup or teardown phases, how are the packets routed to their destinations in a datagram network? In this type of network, each switch (or packet switch) has a routing table which is based on the destination address. The routing tables are dynamic and are updated periodically. The destination addresses and the corresponding forwarding output ports are recorded in the tables. This is different from the table of a circuit-switched network in which each entry is created when the setup phase is completed and deleted when the teardown phase is over. Figure 8.8 shows the routing table for a switch.

Figure 8.8 *Routing table in a datagram network*

Destination address	Output port
1232	1
4150	2
⋮	⋮
9130	3

A switch in a datagram network uses a routing table that is based on the destination address.

Destination Address

Every packet in a datagram network carries a header that contains, among other information, the destination address of the packet. When the switch receives the packet, this destination address is examined; the routing table is consulted to find the corresponding port through which the packet should be forwarded. This address, unlike the address in a virtual-circuit-switched network, remains the same during the entire journey of the packet.

The destination address in the header of a packet in a datagram network remains the same during the entire journey of the packet.

Efficiency

The efficiency of a datagram network is better than that of a circuit-switched network; resources are allocated only when there are packets to be transferred. If a source sends a packet and there is a delay of a few minutes before another packet can be sent, the resources can be reallocated during these minutes for other packets from other sources.

Delay

There may be greater delay in a datagram network than in a virtual-circuit network. Although there are no setup and teardown phases, each packet may experience a wait at a switch before it is forwarded. In addition, since not all packets in a message necessarily travel through the same switches, the delay is not uniform for the packets of a message. Figure 8.9 gives an example of delay in a datagram network for one single packet.

Figure 8.9 *Delay in a datagram network*

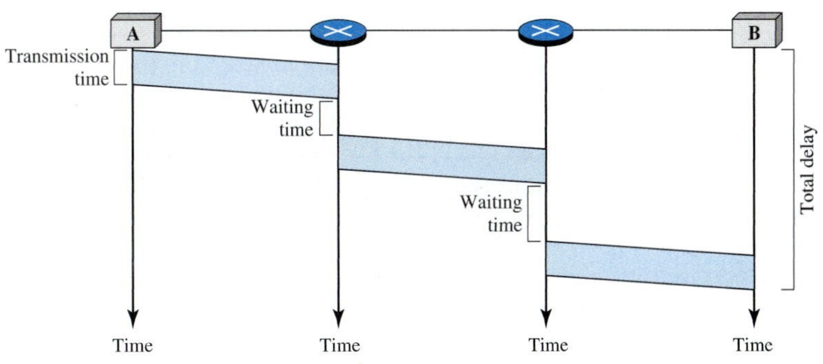

The packet travels through two switches. There are three transmission times ($3T$), three propagation delays (slopes 3τ of the lines), and two waiting times ($w_1 + w_2$). We ignore the processing time in each switch. The total delay is

$$\text{Total delay} = 3T + 3\tau + w_1 + w_2$$

Datagram Networks in the Internet

As we will see in future chapters, the Internet has chosen the datagram approach to switching at the network layer. It uses the universal addresses defined in the network layer to route packets from the source to the destination.

Switching in the Internet is done by using the datagram approach to packet switching at the network layer.

8.3 VIRTUAL-CIRCUIT NETWORKS

A **virtual-circuit network** is a cross between a circuit-switched network and a datagram network. It has some characteristics of both.

1. As in a circuit-switched network, there are setup and teardown phases in addition to the data transfer phase.

2. Resources can be allocated during the setup phase, as in a circuit-switched network, or on demand, as in a datagram network.
3. As in a datagram network, data are packetized and each packet carries an address in the header. However, the address in the header has local jurisdiction (it defines what should be the next switch and the channel on which the packet is being carried), not end-to-end jurisdiction. The reader may ask how the intermediate switches know where to send the packet if there is no final destination address carried by a packet. The answer will be clear when we discuss virtual-circuit identifiers in the next section.
4. As in a circuit-switched network, all packets follow the same path established during the connection.
5. A virtual-circuit network is normally implemented in the data link layer, while a circuit-switched network is implemented in the physical layer and a datagram network in the network layer. But this may change in the future.

Figure 8.10 is an example of a virtual-circuit network. The network has switches that allow traffic from sources to destinations. A source or destination can be a computer, packet switch, bridge, or any other device that connects other networks.

Figure 8.10 *Virtual-circuit network*

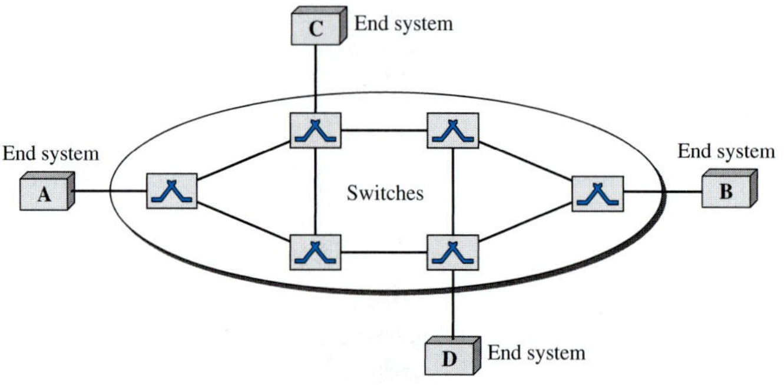

Addressing

In a virtual-circuit network, two types of addressing are involved: global and local (virtual-circuit identifier).

Global Addressing

A source or a destination needs to have a global address—an address that can be unique in the scope of the network or internationally if the network is part of an international network. However, we will see that a global address in virtual-circuit networks is used only to create a virtual-circuit identifier, as discussed next.

Virtual-Circuit Identifier

The identifier that is actually used for data transfer is called the **virtual-circuit identifier (VCI).** A VCI, unlike a global address, is a small number that has only switch scope; it is used by a frame between two switches. When a frame arrives at a switch, it has a VCI; when it leaves, it has a different VCI. Figure 8.11 shows how the VCI in a data frame changes from one switch to another. Note that a VCI does not need to be a large number since each switch can use its own unique set of VCIs.

Figure 8.11 *Virtual-circuit identifier*

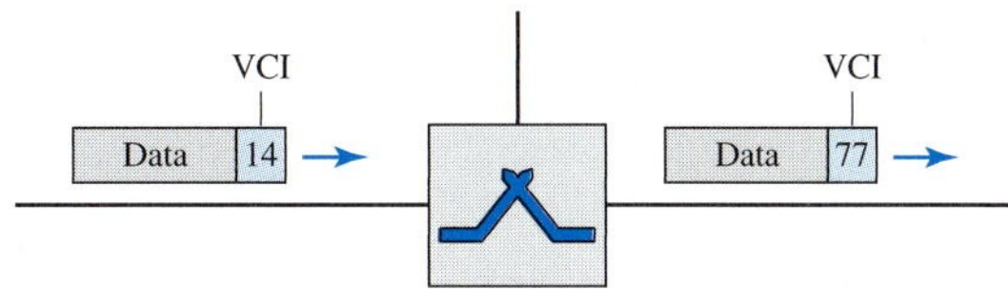

Three Phases

As in a circuit-switched network, a source and destination need to go through three phases in a virtual-circuit network: setup, data transfer, and teardown. In the setup phase, the source and destination use their global addresses to help switches make table entries for the connection. In the teardown phase, the source and destination inform the switches to delete the corresponding entry. Data transfer occurs between these two phases. We first discuss the data transfer phase, which is more straightforward; we then talk about the setup and teardown phases.

Data Transfer Phase

To transfer a frame from a source to its destination, all switches need to have a table entry for this virtual circuit. The table, in its simplest form, has four columns. This means that the switch holds four pieces of information for each virtual circuit that is already set up. We show later how the switches make their table entries, but for the moment we assume that each switch has a table with entries for all active virtual circuits. Figure 8.12 shows such a switch and its corresponding table.

Figure 8.12 shows a frame arriving at port 1 with a VCI of 14. When the frame arrives, the switch looks in its table to find port 1 and a VCI of 14. When it is found, the switch knows to change the VCI to 22 and send out the frame from port 3.

Figure 8.13 shows how a frame from source A reaches destination B and how its VCI changes during the trip. Each switch changes the VCI and routes the frame.

The data transfer phase is active until the source sends all its frames to the destination. The procedure at the switch is the same for each frame of a message. The process creates a virtual circuit, not a real circuit, between the source and destination.

Setup Phase

In the setup phase, a switch creates an entry for a virtual circuit. For example, suppose source A needs to create a virtual circuit to B. Two steps are required: the setup request and the acknowledgment.

Figure 8.12 *Switch and tables in a virtual-circuit network*

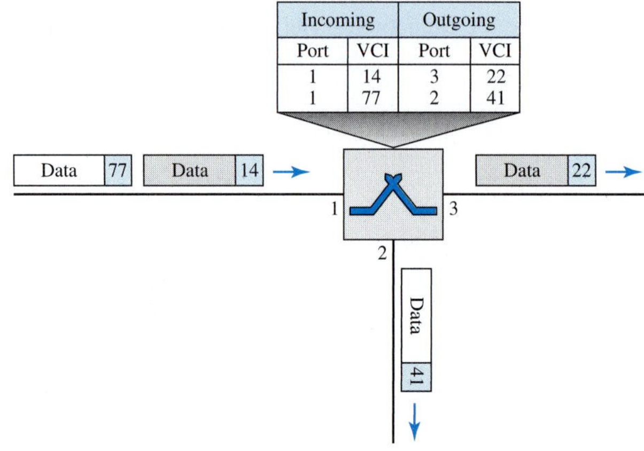

Figure 8.13 *Source-to-destination data transfer in a virtual-circuit network*

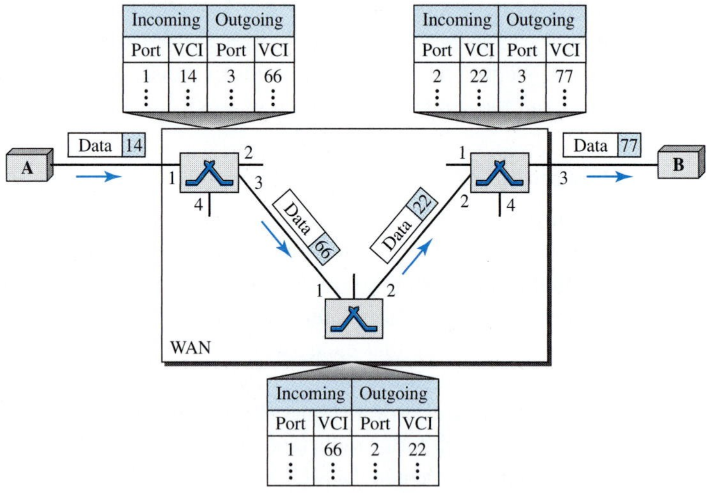

Setup Request A setup request frame is sent from the source to the destination. Figure 8.14 shows the process.

a. Source A sends a setup frame to switch 1.

b. Switch 1 receives the setup request frame. It knows that a frame going from A to B goes out through port 3. How the switch has obtained this information is a point covered in future chapters. The switch, in the setup phase, acts as a packet switch; it has a routing table which is different from the switching table. For the moment, assume that it knows the output port. The switch creates an entry in its table for

Figure 8.14 *Setup request in a virtual-circuit network*

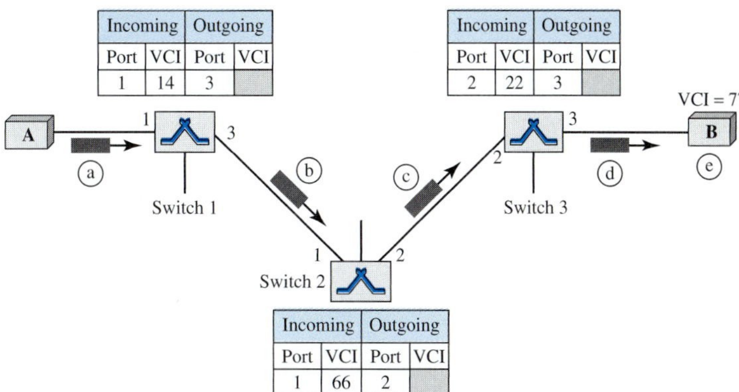

this virtual circuit, but it is only able to fill three of the four columns. The switch assigns the incoming port (1) and chooses an available incoming VCI (14) and the outgoing port (3). It does not yet know the outgoing VCI, which will be found during the acknowledgment step. The switch then forwards the frame through port 3 to switch 2.

c. Switch 2 receives the setup request frame. The same events happen here as at switch 1; three columns of the table are completed: in this case, incoming port (1), incoming VCI (66), and outgoing port (2).

d. Switch 3 receives the setup request frame. Again, three columns are completed: incoming port (2), incoming VCI (22), and outgoing port (3).

e. Destination B receives the setup frame, and if it is ready to receive frames from A, it assigns a VCI to the incoming frames that come from A, in this case 77. This VCI lets the destination know that the frames come from A, and not other sources.

Acknowledgment A special frame, called the acknowledgment frame, completes the entries in the switching tables. Figure 8.15 shows the process.

a. The destination sends an acknowledgment to switch 3. The acknowledgment carries the global source and destination addresses so the switch knows which entry in the table is to be completed. The frame also carries VCI 77, chosen by the destination as the incoming VCI for frames from A. Switch 3 uses this VCI to complete the outgoing VCI column for this entry. Note that 77 is the incoming VCI for destination B, but the outgoing VCI for switch 3.

b. Switch 3 sends an acknowledgment to switch 2 that contains its incoming VCI in the table, chosen in the previous step. Switch 2 uses this as the outgoing VCI in the table.

c. Switch 2 sends an acknowledgment to switch 1 that contains its incoming VCI in the table, chosen in the previous step. Switch 1 uses this as the outgoing VCI in the table.

d. Finally switch 1 sends an acknowledgment to source A that contains its incoming VCI in the table, chosen in the previous step.

e. The source uses this as the outgoing VCI for the data frames to be sent to destination B.

Figure 8.15 *Setup acknowledgment in a virtual-circuit network*

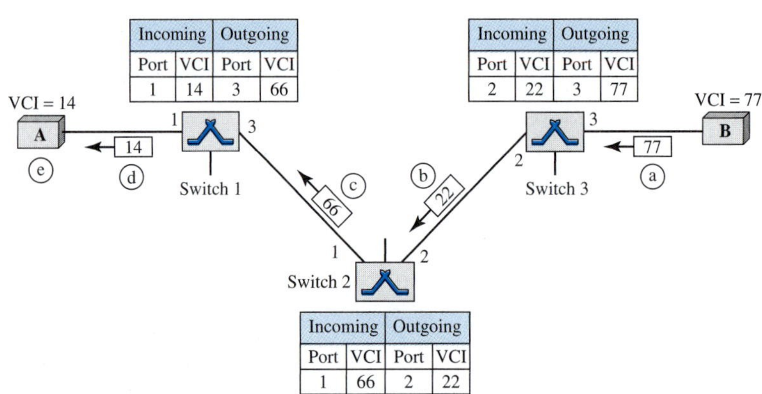

Teardown Phase

In this phase, source A, after sending all frames to B, sends a special frame called a *teardown request*. Destination B responds with a teardown confirmation frame. All switches delete the corresponding entry from their tables.

Efficiency

As we said before, resource reservation in a virtual-circuit network can be made during the setup or can be on demand during the data transfer phase. In the first case, the delay for each packet is the same; in the second case, each packet may encounter different delays. There is one big advantage in a virtual-circuit network even if resource allocation is on demand. The source can check the availability of the resources, without actually reserving it. Consider a family that wants to dine at a restaurant. Although the restaurant may not accept reservations (allocation of the tables is on demand), the family can call and find out the waiting time. This can save the family time and effort.

In virtual-circuit switching, all packets belonging to the same source and destination travel the same path; but the packets may arrive at the destination with different delays if resource allocation is on demand.

Delay in Virtual-Circuit Networks

In a virtual-circuit network, there is a one-time delay for setup and a one-time delay for teardown. If resources are allocated during the setup phase, there is no wait time for individual packets. Figure 8.16 shows the delay for a packet traveling through two switches in a virtual-circuit network.

The packet is traveling through two switches (routers). There are three transmission times ($3T$), three propagation times (3τ), data transfer depicted by the sloping lines, a setup delay (which includes transmission and propagation in two directions),

Figure 8.16 *Delay in a virtual-circuit network*

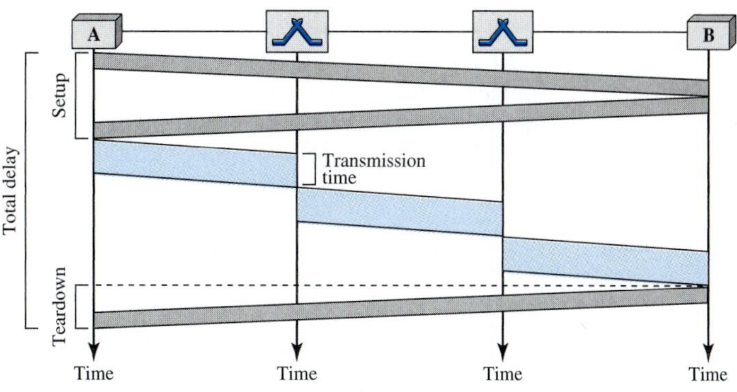

and a teardown delay (which includes transmission and propagation in one direction). We ignore the processing time in each switch. The total delay time is

$$\text{Total delay} = 3T + 3\tau + \text{setup delay} + \text{teardown delay}$$

Circuit-Switched Technology in WANs

As we will see in Chapter 18, virtual-circuit networks are used in switched WANs such as Frame Relay and ATM networks. The data link layer of these technologies is well suited to the virtual-circuit technology.

Switching at the data link layer in a switched WAN is normally implemented by using virtual-circuit techniques.

8.4 STRUCTURE OF A SWITCH

We use switches in circuit-switched and packet-switched networks. In this section, we discuss the structures of the switches used in each type of network.

Structure of Circuit Switches

Circuit switching today can use either of two technologies: the space-division switch or the time-division switch.

Space-Division Switch

In **space-division switching,** the paths in the circuit are separated from one another spatially. This technology was originally designed for use in analog networks but is used currently in both analog and digital networks. It has evolved through a long history of many designs.

Crossbar Switch A **crossbar switch** connects n inputs to m outputs in a grid, using electronic microswitches (transistors) at each **crosspoint** (see Figure 8.17). The major limitation of this design is the number of crosspoints required. To connect n inputs to m outputs using a crossbar switch requires $n \times m$ crosspoints. For example, to connect 1000 inputs to 1000 outputs requires a switch with 1,000,000 crosspoints. A crossbar with this number of crosspoints is impractical. Such a switch is also inefficient because statistics show that, in practice, fewer than 25 percent of the crosspoints are in use at any given time. The rest are idle.

Figure 8.17 *Crossbar switch with three inputs and four outputs*

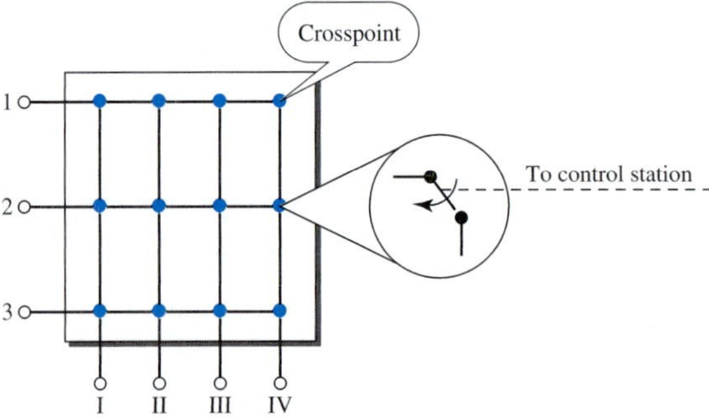

Multistage Switch The solution to the limitations of the crossbar switch is the **multistage switch,** which combines crossbar switches in several (normally three) stages, as shown in Figure 8.18. In a single crossbar switch, only one row or column (one path) is active for any connection. So we need $N \times N$ crosspoints. If we can allow multiple paths inside the switch, we can decrease the number of crosspoints. Each crosspoint in the middle stage can be accessed by multiple crosspoints in the first or third stage.

Figure 8.18 *Multistage switch*

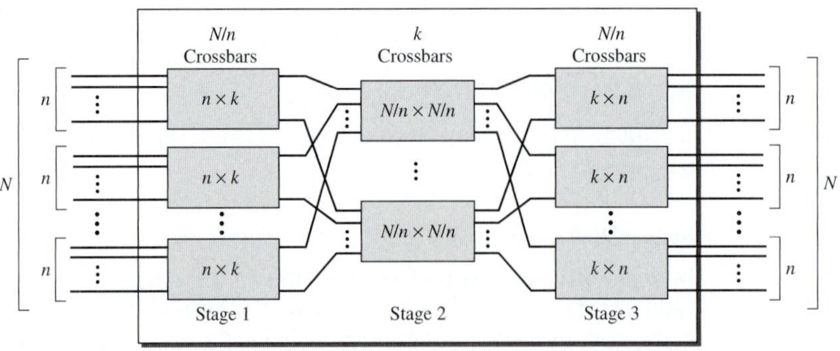

To design a three-stage switch, we follow these steps:

1. We divide the N input lines into groups, each of n lines. For each group, we use one crossbar of size $n \times k$, where k is the number of crossbars in the middle stage. In other words, the first stage has N/n crossbars of $n \times k$ crosspoints.
2. We use k crossbars, each of size $(N/n) \times (N/n)$ in the middle stage.
3. We use N/n crossbars, each of size $k \times n$ at the third stage.

We can calculate the total number of crosspoints as follows:

$$\frac{N}{n}(n \times k) + k\left(\frac{N}{n} \times \frac{N}{n}\right) + \frac{N}{n}(k \times n) = 2kN + k\left(\frac{N}{n}\right)^2$$

In a three-stage switch, the total number of crosspoints is

$$2kN + k\left(\frac{N}{n}\right)^2$$

which is much smaller than the number of crosspoints in a single-stage switch (N^2).

Example 8.3

Design a three-stage, 200×200 switch ($N = 200$) with $k = 4$ and $n = 20$.

Solution

In the first stage we have N/n or 10 crossbars, each of size 20×4. In the second stage, we have 4 crossbars, each of size 10×10. In the third stage, we have 10 crossbars, each of size 4×20. The total number of crosspoints is $2kN + k(N/n)^2$, or 2000 crosspoints. This is 5 percent of the number of crosspoints in a single-stage switch ($200 \times 200 = 40{,}000$).

The multistage switch in Example 8.3 has one drawback—**blocking** during periods of heavy traffic. The whole idea of multistage switching is to share the crosspoints in the middle-stage crossbars. Sharing can cause a lack of availability if the resources are limited and all users want a connection at the same time. Blocking refers to times when one input cannot be connected to an output because there is no path available between them—all the possible intermediate switches are occupied.

In a single-stage switch, blocking does not occur because every combination of input and output has its own crosspoint; there is always a path. (Cases in which two inputs are trying to contact the same output do not count. That path is not blocked; the output is merely busy.) In the multistage switch described in Example 8.3, however, only 4 of the first 20 inputs can use the switch at a time, only 4 of the second 20 inputs can use the switch at a time, and so on. The small number of crossbars at the middle stage creates blocking.

In large systems, such as those having 10,000 inputs and outputs, the number of stages can be increased to cut down on the number of crosspoints required. As the number of stages increases, however, possible blocking increases as well. Many people have experienced blocking on public telephone systems in the wake of a natural disaster when the calls being made to check on or reassure relatives far outnumber the regular load of the system.

Clos investigated the condition of nonblocking in multistage switches and came up with the following formula. In a nonblocking switch, the number of middle-stage switches must be at least $2n - 1$. In other words, we need to have $k \geq 2n - 1$.

Note that the number of crosspoints is still smaller than that in a single-stage switch. Now we need to minimize the number of crosspoints with a fixed N by using the Clos criteria. We can take the derivative of the equation with respect to n (the only variable) and find the value of n that makes the result zero. This n must be equal to or greater than $(N/2)^{1/2}$. In this case, the total number of crosspoints is greater than or equal to $4N\,[(2N)^{1/2} - 1]$. In other words, the minimum number of crosspoints according to the Clos criteria is proportional to $N^{3/2}$.

According to Clos criterion:
$n = (N/2)^{1/2}$
$k > 2n - 1$
Total number of crosspoints $\geq 4N\,[(2N)^{1/2} - 1]$

Example 8.4

Redesign the previous three-stage, 200 × 200 switch, using the Clos criteria with a minimum number of crosspoints.

Solution

We let $n = (200/2)^{1/2}$, or $n = 10$. We calculate $k = 2n - 1 = 19$. In the first stage, we have 200/10, or 20, crossbars, each with 10 × 19 crosspoints. In the second stage, we have 19 crossbars, each with 10 × 10 crosspoints. In the third stage, we have 20 crossbars each with 19 × 10 crosspoints. The total number of crosspoints is 20(10 × 19) + 19(10 × 10) + 20(19 ×10) = 9500. If we use a single-stage switch, we need 200 × 200 = 40,000 crosspoints. The number of crosspoints in this three-stage switch is 24 percent that of a single-stage switch. More points are needed than in Example 8.3 (5 percent). The extra crosspoints are needed to prevent blocking.

A multistage switch that uses the Clos criteria and a minimum number of crosspoints still requires a huge number of crosspoints. For example, to have a 100,000 input/output switch, we need something close to 200 million crosspoints (instead of 10 billion). This means that if a telephone company needs to provide a switch to connect 100,000 telephones in a city, it needs 200 million crosspoints. The number can be reduced if we accept blocking. Today, telephone companies use time-division switching or a combination of space- and time-division switches, as we will see shortly.

Time-Division Switch

Time-division switching uses time-division multiplexing (TDM) inside a switch. The most popular technology is called the **time-slot interchange (TSI).**

Time-Slot Interchange Figure 8.19 shows a system connecting four input lines to four output lines. Imagine that each input line wants to send data to an output line according to the following pattern:

1 → 3 2 → 4 3 → 1 4 → 2

Figure 8.22 *Input port*

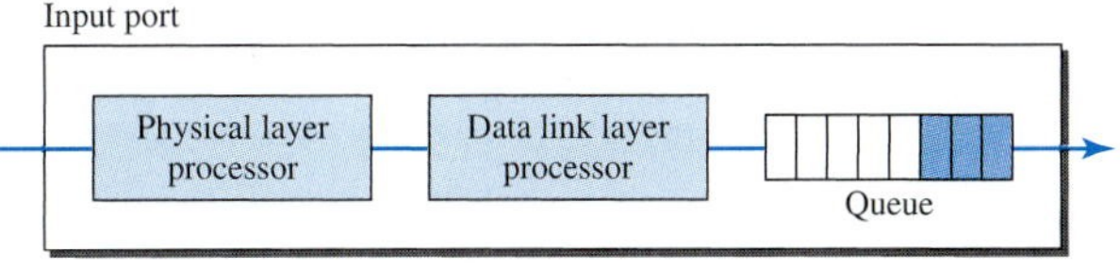

Output Port

The output port performs the same functions as the input port, but in the reverse order. First the outgoing packets are queued, then the packet is encapsulated in a frame, and finally the physical layer functions are applied to the frame to create the signal to be sent on the line. Figure 8.23 shows a schematic diagram of an output port.

Figure 8.23 *Output port*

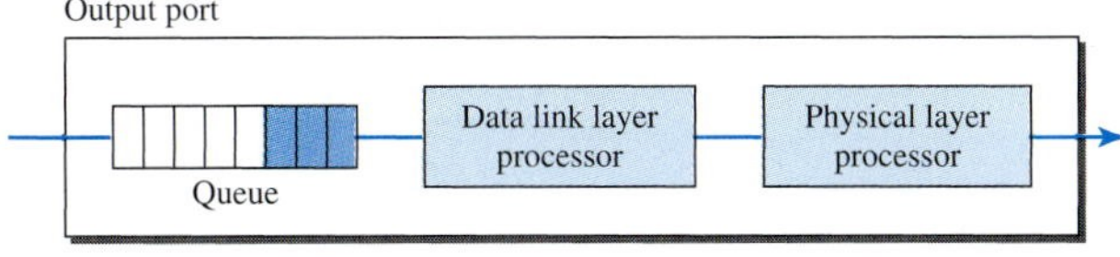

Routing Processor

The routing processor performs the functions of the network layer. The destination address is used to find the address of the next hop and, at the same time, the output port number from which the packet is sent out. This activity is sometimes referred to as **table lookup** because the routing processor searches the routing table. In the newer packet switches, this function of the routing processor is being moved to the input ports to facilitate and expedite the process.

Switching Fabrics

The most difficult task in a packet switch is to move the packet from the input queue to the output queue. The speed with which this is done affects the size of the input/output queue and the overall delay in packet delivery. In the past, when a packet switch was actually a dedicated computer, the memory of the computer or a bus was used as the switching fabric. The input port stored the packet in memory; the output port retrieved the packet from memory. Today, packet switches are specialized mechanisms that use a variety of switching fabrics. We briefly discuss some of these fabrics here.

Crossbar Switch The simplest type of switching fabric is the crossbar switch, discussed in the previous section.

Banyan Switch A more realistic approach than the crossbar switch is the **banyan switch** (named after the banyan tree). A banyan switch is a multistage switch with microswitches at each stage that route the packets based on the output port represented as a binary string. For n inputs and n outputs, we have $\log_2 n$ stages with $n/2$ microswitches at each stage. The first stage routes the packet based on the high-order bit of the binary string. The second stage routes the packet based on the second high-order bit, and so on. Figure 8.24 shows a banyan switch with eight inputs and eight outputs. The number of stages is $\log_2(8) = 3$.

Figure 8.24 *A banyan switch*

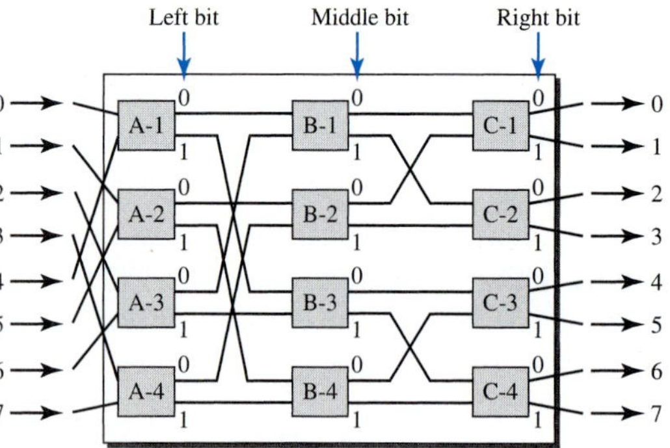

Figure 8.25 shows the operation. In part a, a packet has arrived at input port 1 and must go to output port 6 (110 in binary). The first microswitch (A-2) routes the packet based on the first bit (1), the second microswitch (B-4) routes the packet based on the second bit (1), and the third microswitch (C-4) routes the packet based on the third bit (0). In part b, a packet has arrived at input port 5 and must go to output port 2 (010 in binary). The first microswitch (A-2) routes the packet based on the first bit (0), the second microswitch (B-2) routes the packet based on the second bit (1), and the third microswitch (C-2) routes the packet based on the third bit (0).

Figure 8.25 *Examples of routing in a banyan switch*

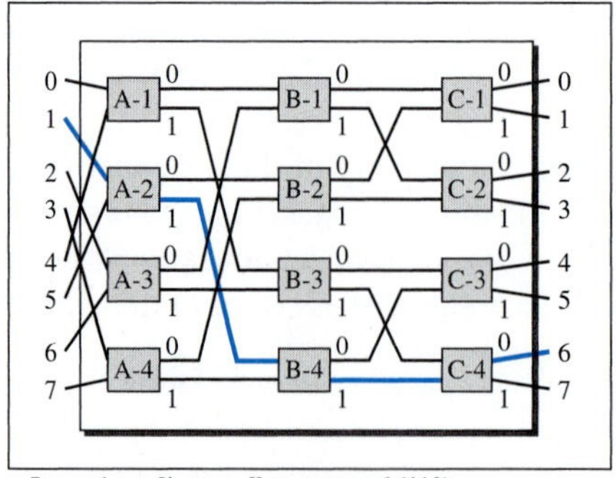

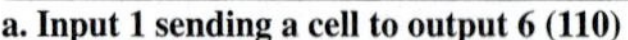
a. Input 1 sending a cell to output 6 (110)

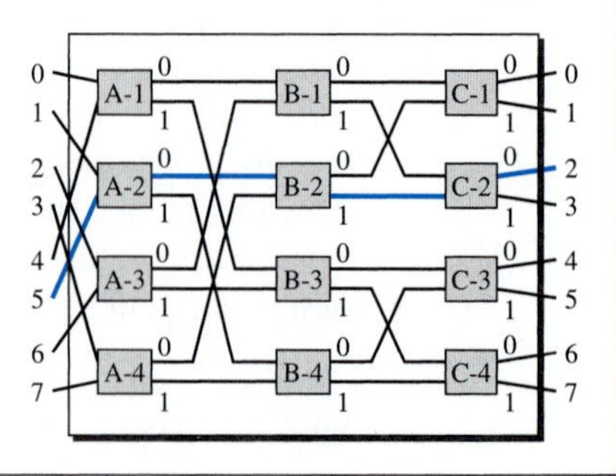

b. Input 5 sending a cell to output 2 (010)

Figure 8.26 *Batcher-banyan switch*

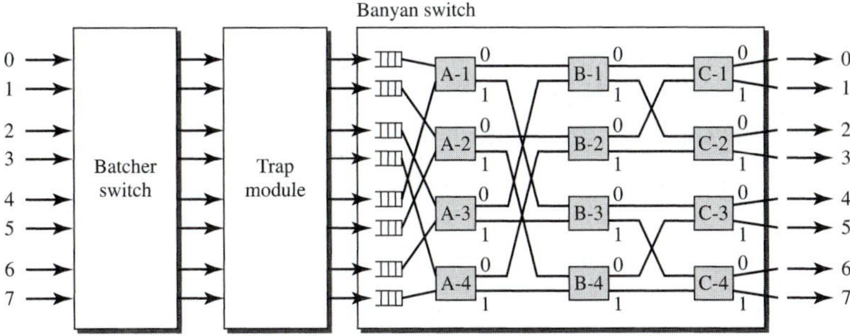

Batcher-Banyan Switch The problem with the banyan switch is the possibility of internal collision even when two packets are not heading for the same output port. We can solve this problem by sorting the arriving packets based on their destination port.

K. E. Batcher designed a switch that comes before the banyan switch and sorts the incoming packets according to their final destinations. The combination is called the **Batcher-banyan switch.** The sorting switch uses hardware merging techniques, but we do not discuss the details here. Normally, another hardware module called a **trap** is added between the Batcher switch and the banyan switch (see Figure 8.26) The trap module prevents duplicate packets (packets with the same output destination) from passing to the banyan switch simultaneously. Only one packet for each destination is allowed at each tick; if there is more than one, they wait for the next tick.

8.5 RECOMMENDED READING

For more details about subjects discussed in this chapter, we recommend the following books. The items in brackets [. . .] refer to the reference list at the end of the text.

Books

Switching is discussed in Chapter 10 of [Sta04] and Chapters 4 and 7 of [GW04]. Circuit-switching is fully discussed in [BEL00].

8.6 KEY TERMS

banyan switch
Batcher-banyan switch
blocking
circuit switching
circuit-switched network
crossbar switch
crosspoint
data transfer phase
datagram
datagram network
end system
input port

multistage switch
output port
packet-switched network
routing processor
setup phase
space-division switching
switch
switching
switching fabric
table lookup
teardown phase
time-division switching
time-slot interchange (TSI)
time-space-time (TST) switch
trap
virtual-circuit identifier (VCI)
virtual-circuit network

8.7 SUMMARY

- A switched network consists of a series of interlinked nodes, called switches. Traditionally, three methods of switching have been important: circuit switching, packet switching, and message switching.
- We can divide today's networks into three broad categories: circuit-switched networks, packet-switched networks, and message-switched. Packet-switched networks can also be divided into two subcategories: virtual-circuit networks and datagram networks
- A circuit-switched network is made of a set of switches connected by physical links, in which each link is divided into n channels. Circuit switching takes place at the physical layer. In circuit switching, the resources need to be reserved during the setup phase; the resources remain dedicated for the entire duration of data transfer phase until the teardown phase.
- In packet switching, there is no resource allocation for a packet. This means that there is no reserved bandwidth on the links, and there is no scheduled processing time for each packet. Resources are allocated on demand.
- In a datagram network, each packet is treated independently of all others. Packets in this approach are referred to as datagrams. There are no setup or teardown phases.
- A virtual-circuit network is a cross between a circuit-switched network and a datagram network. It has some characteristics of both.
- Circuit switching uses either of two technologies: the space-division switch or the time-division switch.
- A switch in a packet-switched network has a different structure from a switch used in a circuit-switched network.We can say that a packet switch has four types of components: input ports, output ports, a routing processor, and switching fabric.

8.8 PRACTICE SET

Review Questions

1. Describe the need for switching and define a switch.
2. List the three traditional switching methods. What are the most common today?

3. What are the two approaches to packet-switching?
4. Compare and contrast a circuit-switched network and a packet-switched network.
5. What is the role of the address field in a packet traveling through a datagram network?
6. What is the role of the address field in a packet traveling through a virtual-circuit network?
7. Compare space-division and time-division switches.
8. What is TSI and its role in a time-division switching?
9. Define blocking in a switched network.
10. List four major components of a packet switch and their functions.

Exercises

11. A path in a digital circuit-switched network has a data rate of 1 Mbps. The exchange of 1000 bits is required for the setup and teardown phases. The distance between two parties is 5000 km. Answer the following questions if the propagataion speed is 2×10^8 m:
    a. What is the total delay if 1000 bits of data are exchanged during the data transfer phase?
    b. What is the total delay if 100,000 bits of data are exchanged during the data transfer phase?
    c. What is the total delay if 1,000,000 bits of data are exchanged during the data transfer phase?
    d. Find the delay per 1000 bits of data for each of the above cases and compare them. What can you infer?
12. Five equal-size datagrams belonging to the same message leave for the destination one after another. However, they travel through different paths as shown in Table 8.1.

Table 8.1 *Exercise 12*

Datagram	*Path Length*	*Visited Switches*
1	3200 Km	1, 3, 5
2	11,700 Km	1, 2, 5
3	12,200 Km	1, 2, 3, 5
4	10,200 Km	1, 4, 5
5	10,700 Km	1, 4, 3, 5

We assume that the delay for each switch (including waiting and processing) is 3, 10, 20, 7, and 20 ms respectively. Assuming that the propagation speed is 2×10^8 m, find the order the datagrams arrive at the destination and the delay for each. Ignore any other delays in transmission.

13. Transmission of information in any network involves end-to-end addressing and sometimes local addressing (such as VCI). Table 8.2 shows the types of networks and the addressing mechanism used in each of them.

Table 8.2 *Exercise 13*

Network	*Setup*	*Data Transfer*	*Teardown*
Circuit-switched	End-to-end		End-to-end
Datagram		End-to-end	
Virtual-circuit	End-to-end	Local	End-to-end

Answer the following questions:

a. Why does a circuit-switched network need end-to-end addressing during the setup and teardown phases? Why are no addresses needed during the data transfer phase for this type of network?

b. Why does a datagram network need only end-to-end addressing during the data transfer phase, but no addressing during the setup and teardown phases?

c. Why does a virtual-circuit network need addresses during all three phases?

14. We mentioned that two types of networks, datagram and virtual-circuit, need a routing or switching table to find the output port from which the information belonging to a destination should be sent out, but a circuit-switched network has no need for such a table. Give the reason for this difference.

15. An entry in the switching table of a virtual-circuit network is normally created during the setup phase and deleted during the teardown phase. In other words, the entries in this type of network reflect the current connections, the activity in the network. In contrast, the entries in a routing table of a datagram network do not depend on the current connections; they show the configuration of the network and how any packet should be routed to a final destination. The entries may remain the same even if there is no activity in the network. The routing tables, however, are updated if there are changes in the network. Can you explain the reason for these two different characteristics? Can we say that a virtual-circuit is a *connection-oriented* network and a datagram network is a *connectionless* network because of the above characteristics?

16. The minimum number of columns in a datagram network is two; the minimum number of columns in a virtual-circuit network is four. Can you explain the reason? Is the difference related to the type of addresses carried in the packets of each network?

17. Figure 8.27 shows a switch (router) in a datagram network.

    Find the output port for packets with the following destination addresses:
    Packet 1: 7176
    Packet 2: 1233
    Packet 3: 8766
    Packet 4: 9144

18. Figure 8.28 shows a switch in a virtual circuit network.

Figure 8.27 *Exercise 17*

Destination address	Output port
1233	3
1456	2
3255	1
4470	4
7176	2
8766	3
9144	2

Figure 8.28 *Exercise 18*

Incoming		Outgoing	
Port	VCI	Port	VCI
1	14	3	22
2	71	4	41
2	92	1	45
3	58	2	43
3	78	2	70
4	56	3	11

Find the output port and the output VCI for packets with the following input port and input VCI addresses:
Packet 1: 3, 78
Packet 2: 2, 92
Packet 3: 4, 56
Packet 4: 2, 71

19. Answer the following questions:
    a. Can a routing table in a datagram network have two entries with the same destination address? Explain.
    b. Can a switching table in a virtual-circuit network have two entries with the same input port number? With the same output port number? With the same incoming VCIs? With the same outgoing VCIs? With the same incoming values (port, VCI)? With the same outgoing values (port, VCI)?
20. It is obvious that a router or a switch needs to do searching to find information in the corresponding table. The searching in a routing table for a datagram network is based on the destination address; the searching in a switching table in a virtual-circuit network is based on the combination of incoming port and incoming VCI. Explain the reason and define how these tables must be ordered (sorted) based on these values.
21. Consider an $n \times k$ crossbar switch with n inputs and k outputs.
    a. Can we say that switch acts as a multiplexer if $n > k$?
    b. Can we say that switch acts as a demultiplexer if $n < k$?

22. We need a three-stage space-division switch with N = 100. We use 10 crossbars at the first and third stages and 4 crossbars at the middle stage.
    a. Draw the configuration diagram.
    b. Calculate the total number of crosspoints.
    c. Find the possible number of simultaneous connections.
    d. Find the possible number of simultaneous connections if we use one single crossbar (100 × 100).
    e. Find the blocking factor, the ratio of the number of connections in c and in d.
23. Repeat Exercise 22 if we use 6 crossbars at the middle stage.
24. Redesign the configuration of Exercise 22 using the Clos criteria.
25. We need to have a space-division switch with 1000 inputs and outputs. What is the total number of crosspoints in each of the following cases?
    a. Using one single crossbar.
    b. Using a multi-stage switch based on the Clos criteria
26. We need a three-stage time-space-time switch with N = 100. We use 10 TSIs at the first and third stages and 4 crossbars at the middle stage.
    a. Draw the configuration diagram.
    b. Calculate the total number of crosspoints.
    c. Calculate the total number of memory locations we need for the TSIs.

CHAPTER 9

Using Telephone and Cable Networks for Data Transmission

Telephone networks were originally created to provide voice communication. The need to communicate digital data resulted in the invention of the dial-up modem. With the advent of the Internet came the need for high-speed downloading and uploading; the modem was just too slow. The telephone companies added a new technology, the *digital subscriber line* (DSL). Although dial-up modems still exist in many places all over the world, DSL provides much faster access to the Internet through the telephone network. In this chapter, we first discuss the basic structure of the telephone network. We then see how dial-up modems and DSL technology use these networks to access the Internet.

Cable networks were originally created to provide access to TV programs for those subscribers who had no reception because of natural obstructions such as mountains. Later the cable network became popular with people who just wanted a better signal. In addition, cable networks enabled access to remote broadcasting stations via microwave connections. Cable TV also found a good market in Internet access provision using some of the channels originally designed for video. After discussing the basic structure of cable networks, we discuss how cable modems can provide a high-speed connection to the Internet.

9.1 TELEPHONE NETWORK

Telephone networks use circuit switching. The telephone network had its beginnings in the late 1800s. The entire network, which is referred to as the **plain old telephone system (POTS),** was originally an analog system using analog signals to transmit voice. With the advent of the computer era, the network, in the 1980s, began to carry data in addition to voice. During the last decade, the telephone network has undergone many technical changes. The network is now digital as well as analog.

Major Components

The telephone network, as shown in Figure 9.1, is made of three major components: local loops, trunks, and switching offices. The telephone network has several levels of switching offices such as **end offices, tandem offices,** and **regional offices.**

Figure 9.1 *A telephone system*

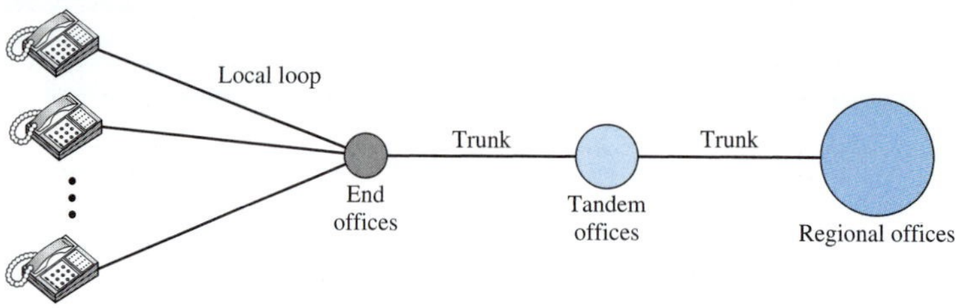

Local Loops

One component of the telephone network is the **local loop,** a twisted-pair cable that connects the subscriber telephone to the nearest end office or local central office. The local loop, when used for voice, has a bandwidth of 4000 Hz (4 kHz). It is interesting to examine the telephone number associated with each local loop. The first three digits of a local telephone number define the office, and the next four digits define the local loop number.

Trunks

Trunks are transmission media that handle the communication between offices. A trunk normally handles hundreds or thousands of connections through multiplexing. Transmission is usually through optical fibers or satellite links.

Switching Offices

To avoid having a permanent physical link between any two subscribers, the telephone company has switches located in a **switching office.** A switch connects several local loops or trunks and allows a connection between different subscribers.

LATAs

After the divestiture of 1984 (see Appendix E), the United States was divided into more than 200 **local-access transport areas (LATAs).** The number of LATAs has increased since then. A LATA can be a small or large metropolitan area. A small state may have one single LATA; a large state may have several LATAs. A LATA boundary may overlap the boundary of a state; part of a LATA can be in one state, part in another state.

Intra-LATA Services

The services offered by the **common carriers** (telephone companies) inside a LATA are called *intra-LATA* services. The carrier that handles these services is called a **local exchange carrier (LEC).** Before the Telecommunications Act of 1996 (see Appendix E), intra-LATA services were granted to one single carrier. This was a monopoly. After 1996, more than one carrier could provide services inside a LATA. The carrier that provided services before 1996 owns the cabling system (local loops) and is called the **incumbent local exchange carrier (ILEC).** The new carriers that can provide services are called **competitive local exchange carriers (CLECs).** To avoid the costs of new cabling, it was agreed that the ILECs would continue to provide the main services, and the CLECs would provide other services such as mobile telephone service, toll calls inside a LATA, and so on. Figure 9.2 shows a LATA and switching offices.

Intra-LATA services are provided by local exchange carriers. Since 1996, there are two types of LECs: incumbent local exchange carriers and competitive local exchange carriers.

Figure 9.2 *Switching offices in a LATA*

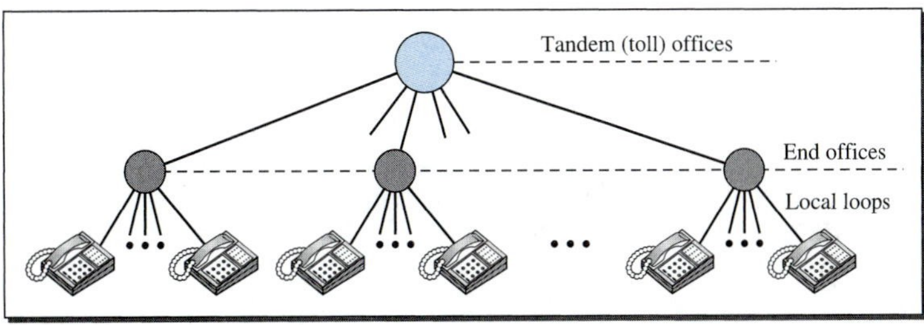

Communication inside a LATA is handled by end switches and tandem switches. A call that can be completed by using only end offices is considered toll-free. A call that has to go through a tandem office (intra-LATA toll office) is charged.

Inter-LATA Services

The services between LATAs are handled by **interexchange carriers (IXCs).** These carriers, sometimes called **long-distance companies,** provide communication services between two customers in different LATAs. After the act of 1996 (see Appendix E), these services can be provided by any carrier, including those involved in intra-LATA services. The field is wide open. Carriers providing inter-LATA services include AT&T, MCI, WorldCom, Sprint, and Verizon.

The IXCs are long-distance carriers that provide general data communications services including telephone service. A telephone call going through an IXC is normally digitized, with the carriers using several types of networks to provide service.

Points of Presence

As we discussed, intra-LATA services can be provided by several LECs (one ILEC and possibly more than one CLEC). We also said that inter-LATA services can be provided by several IXCs. How do these carriers interact with one another? The answer is, via a switching office called a **point of presence (POP).** Each IXC that wants to provide inter-LATA services in a LATA must have a POP in that LATA. The LECs that provide services inside the LATA must provide connections so that every subscriber can have access to all POPs. Figure 9.3 illustrates the concept.

A subscriber who needs to make a connection with another subscriber is connected first to an end switch and then, either directly or through a tandem switch, to a POP. The

Figure 9.3 *Point of presences (POPs)*

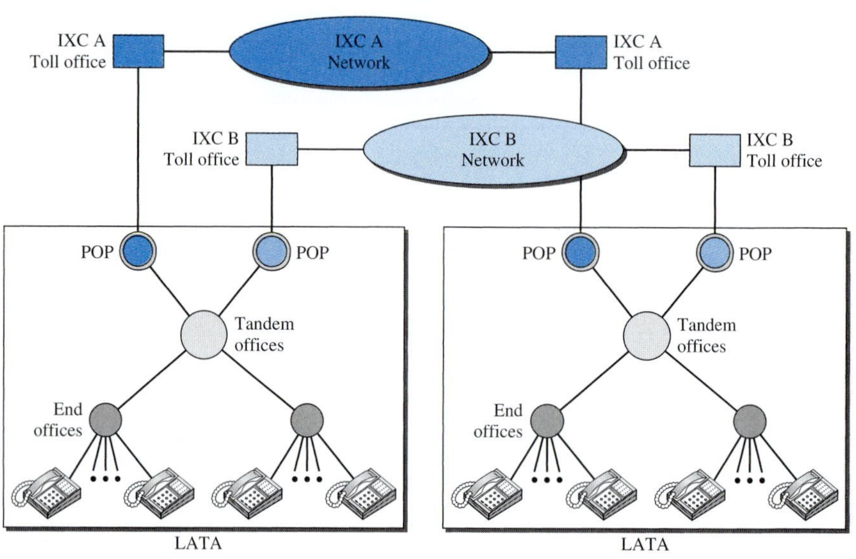

call now goes from the POP of an IXC (the one the subscriber has chosen) in the source LATA to the POP of the same IXC in the destination LATA. The call is passed through the toll office of the IXC and is carried through the network provided by the IXC.

Signaling

The telephone network, at its beginning, used a circuit-switched network with dedicated links (multiplexing had not yet been invented) to transfer voice communication. As we saw in Chapter 8, a circuit-switched network needs the setup and teardown phases to establish and terminate paths between the two communicating parties. In the beginning, this task was performed by human operators. The operator room was a center to which all subscribers were connected. A subscriber who wished to talk to another subscriber picked up the receiver (off-hook) and rang the operator. The operator, after listening to the caller and getting the identifier of the called party, connected the two by using a wire with two plugs inserted into the corresponding two jacks. A dedicated circuit was created in this way. One of the parties, after the conversation ended, informed the operator to disconnect the circuit. This type of signaling is called **in-band signaling** because the same circuit can be used for both signaling and voice communication.

Later, the signaling system became automatic. Rotary telephones were invented that sent a digital signal defining each digit in a multidigit telephone number. The switches in the telephone companies used the digital signals to create a connection between the caller and the called parties. Both in-band and **out-of-band signaling** were used. In in-band signaling, the 4-kHz voice channel was also used to provide signaling. In out-of-band signaling, a portion of the voice channel bandwidth was used for signaling; the voice bandwidth and the signaling bandwidth were separate.

As telephone networks evolved into a complex network, the functionality of the signaling system increased. The signaling system was required to perform other tasks such as

1. Providing dial tone, ring tone, and busy tone
2. Transferring telephone numbers between offices
3. Maintaining and monitoring the call
4. Keeping billing information
5. Maintaining and monitoring the status of the telephone network equipment
6. Providing other functions such as caller ID, voice mail, and so on

These complex tasks resulted in the provision of a separate network for signaling. This means that a telephone network today can be thought of as two networks: a signaling network and a data transfer network.

The tasks of data transfer and signaling are separated in modern telephone networks: data transfer is done by one network, signaling by another.

However, we need to emphasize a point here. Although the two networks are separate, this does not mean that there are separate physical links everywhere; the two networks may use separate channels of the same link in parts of the system.

Data Transfer Network

The data transfer network that can carry multimedia information today is, for the most part, a circuit-switched network, although it can also be a packet-switched network. This network follows the same type of protocols and model as other networks discussed in this book.

Signaling Network

The signaling network, which is our main concern in this section, is a packet-switched network involving the layers similar to those in the OSI model or Internet model, discussed in Chapter 2. The nature of signaling makes it more suited to a packet-switching network with different layers. For example, the information needed to convey a telephone address can easily be encapsulated in a packet with all the error control and addressing information. Figure 9.4 shows a simplified situation of a telephone network in which the two networks are separated.

The user telephone or computer is connected to the **signal points (SPs).** The link between the telephone set and SP is common for the two networks. The signaling network uses nodes called **signal transport ports (STPs)** that receive and forward signaling messages. The signaling network also includes a **service control point (SCP)** that controls the whole operation of the network. Other systems such as a database center may be included to provide stored information about the entire signaling network.

Signaling System Seven (SS7)

The protocol that is used in the signaling network is called **Signaling System Seven (SS7).** It is very similar to the five-layer Internet model we saw in Chapter 2, but the layers have different names, as shown in Figure 9.5.

Figure 9.4 *Data transfer and signaling networks*

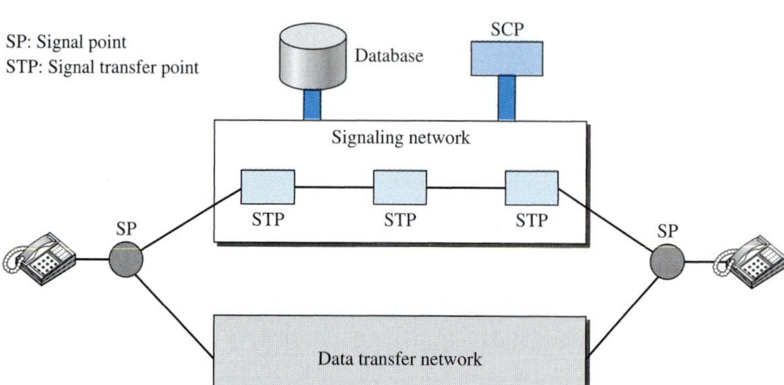

Figure 9.5 *Layers in SS7*

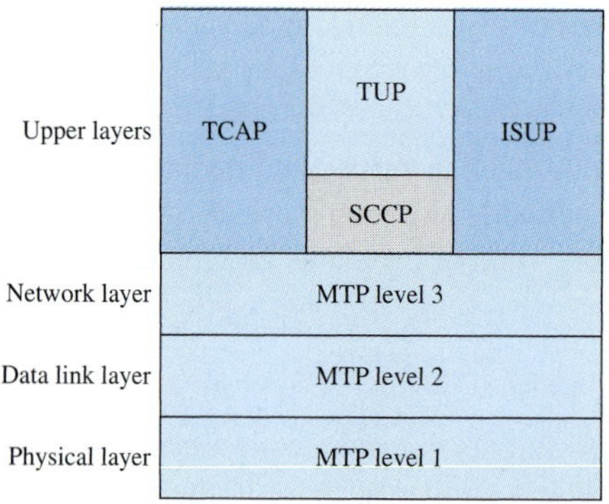

MTP: Message transfer part
SCCP: Signaling connection control point
TCAP: Transaction capabilities application port
TUP: Telephone user port
ISUP: ISDN user port

Physical Layer: MTP Level 1 The physical layer in SS7 called **message transport part (MTP) level** 1 uses several physical layer specifications such as T-1 (1.544 Mbps) and DC0 (64 kbps).

Data Link Layer: MTP Level 2 The MTP level 2 layer provides typical data link layer services such as packetizing, using source and destination address in the packet header, and CRC for error checking.

Network Layer: MTP Level 3 The MTP level 3 layer provides end-to-end connectivity by using the datagram approach to switching. Routers and switches route the signal packets from the source to the destination.

Transport Layer: SCCP The **signaling connection control point (SCCP)** is used for special services such as 800-call processing.

Upper Layers: TUP, TCAP, and ISUP There are three protocols at the upper layers. **Telephone user port (TUP)** is responsible for setting up voice calls. It receives the dialed digits and routes the calls. **Transaction capabilities application port (TCAP)** provides remote calls that let an application program on a computer invoke a procedure on another computer. **ISDN user port (ISUP)** can replace TUP to provide services similar to those of an ISDN network.

Services Provided by Telephone Networks

Telephone companies provide two types of services: analog and digital.

Analog Services

In the beginning, telephone companies provided their subscribers with analog services. These services still continue today. We can categorize these services as either **analog switched services** or **analog leased services.**

Analog Switched Services This is the familiar dial-up service most often encountered when a home telephone is used. The signal on a local loop is analog, and the bandwidth is usually between 0 and 4000 Hz. A local call service is normally provided for a flat monthly rate, although in some LATAs, the carrier charges for each call or a set of calls. The rationale for a non flat-rate charge is to provide cheaper service for those customers who do not make many calls. A toll call can be intra-LATA or inter-LATA. If the LATA is geographically large, a call may go through a tandem office (toll office) and the subscriber will pay a fee for the call. The inter-LATA calls are long-distance calls and are charged as such.

Another service is called 800 service. If a subscriber (normally an organization) needs to provide free connections for other subscribers (normally customers), it can request the **800 service.** In this case, the call is free for the caller, but it is paid by the callee. An organization uses this service to encourage customers to call. The rate is less expensive than that for a normal long-distance call.

The **wide-area telephone service (WATS)** is the opposite of the 800 service. The latter are inbound calls paid by the organization; the former are outbound calls paid by the organization. This service is a less expensive alternative to regular toll calls; charges are based on the number of calls. The service can be specified as outbound calls to the same state, to several states, or to the whole country, with rates charged accordingly.

The **900 services** are like the 800 service, in that they are inbound calls to a subscriber. However, unlike the 800 service, the call is paid by the caller and is normally much more expensive than a normal long-distance call. The reason is that the carrier charges *two* fees: the first is the long-distance toll, and the second is the fee paid to the callee for each call.

Analog Leased Service An **analog leased service** offers customers the opportunity to lease a line, sometimes called a *dedicated line,* that is permanently connected to another customer. Although the connection still passes through the switches in the telephone network, subscribers experience it as a single line because the switch is always closed; no dialing is needed.

Digital Services

Recently telephone companies began offering **digital services** to their subscribers. Digital services are less sensitive than analog services to noise and other forms of interference.

The two most common digital services are switched/56 service and **digital data service (DDS).** We already discussed high-speed digital services—the T lines—in Chapter 6. We discuss the other services in this chapter.

Switched/56 Service **Switched/56 service** is the digital version of an analog switched line. It is a switched digital service that allows data rates of up to 56 kbps. To communicate through this service, both parties must subscribe. A caller with normal telephone service cannot connect to a telephone or computer with switched/56 service even if the caller is using a modem. On the whole, digital and analog services represent two completely different domains for the telephone companies. Because the line in a switched/56 service is already digital, subscribers do not need modems to transmit digital data. However, they do need another device called a **digital service unit (DSU).**

Digital Data Service **Digital data service (DDS)** is the digital version of an analog leased line; it is a digital leased line with a maximum data rate of 64 kbps.

9.2 DIAL-UP MODEMS

Traditional telephone lines can carry frequencies between 300 and 3300 Hz, giving them a bandwidth of 3000 Hz. All this range is used for transmitting voice, where a great deal of interference and distortion can be accepted without loss of intelligibility. As we have seen, however, data signals require a higher degree of accuracy to ensure integrity. For safety's sake, therefore, the edges of this range are not used for data communications. In general, we can say that the signal bandwidth must be smaller than the cable bandwidth. The effective bandwidth of a telephone line being used for data transmission is 2400 Hz, covering the range from 600 to 3000 Hz. Note that today some telephone lines are capable of handling greater bandwidth than traditional lines. However, modem design is still based on traditional capability (see Figure 9.6).

Figure 9.6 *Telephone line bandwidth*

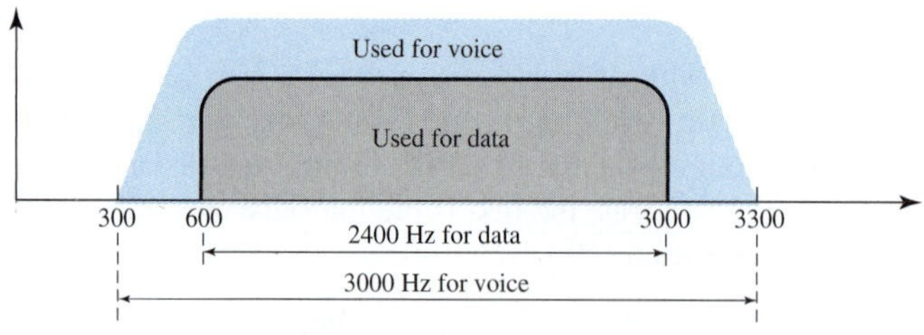

The term **modem** is a composite word that refers to the two functional entities that make up the device: a signal *mo*dulator and a signal *dem*odulator. A **modulator** creates a bandpass analog signal from binary data. A **demodulator** recovers the binary data from the modulated signal.

***Modem* stands for modulator/demodulator.**

Figure 9.7 shows the relationship of modems to a communications link. The computer on the left sends a digital signal to the modulator portion of the modem; the data are sent as an analog signal on the telephone lines. The modem on the right receives the analog signal, demodulates it through its demodulator, and delivers data to the computer on the right. The communication can be bidirectional, which means the computer on the right can simultaneously send data to the computer on the left, using the same modulation/demodulation processes.

Figure 9.7 *Modulation/demodulation*

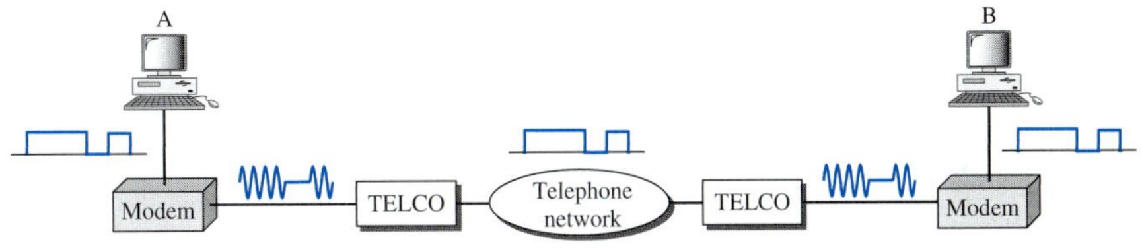

Modem Standards

Today, many of the most popular modems available are based on the **V-series** standards published by the ITU-T. We discuss just the most recent series.

V.32 and V.32bis

The **V.32** modem uses a combined modulation and encoding technique called **trellis-coded modulation.** Trellis is essentially QAM plus a redundant bit. The data stream is divided into 4-bit sections. Instead of a quadbit (4-bit pattern), however, a *pentabit* (5-bit pattern) is transmitted. The value of the extra bit is calculated from the values of the data bits. The extra bit is used for error detection.

The V.32 calls for 32-QAM with a baud rate of 2400. Because only 4 bits of each pentabit represent data, the resulting data rate is $4 \times 2400 = 9600$ bps. The constellation diagram and bandwidth are shown in Figure 9.8.

The **V.32bis** modem was the first of the ITU-T standards to support 14,400-bps transmission. The V.32bis uses 128-QAM transmission (7 bits/baud with 1 bit for error control) at a rate of 2400 baud ($2400 \times 6 = 14{,}400$ bps).

An additional enhancement provided by V.32bis is the inclusion of an automatic fall-back and fall-forward feature that enables the modem to adjust its speed upward or downward depending on the quality of the line or signal. The constellation diagram and bandwidth are also shown in Figure 9.8.

V.34bis

The **V.34bis** modem provides a bit rate of 28,800 with a 960-point constellation and a bit rate of 33,600 bps with a 1664-point constellation.

Figure 9.8 *The V.32 and V.32bis constellation and bandwidth*

90
180
0
270
Full-duplex 2400-baud
9600-bps 2-wire
600
1800
3000

a. Constellation and bandwidth for V.32

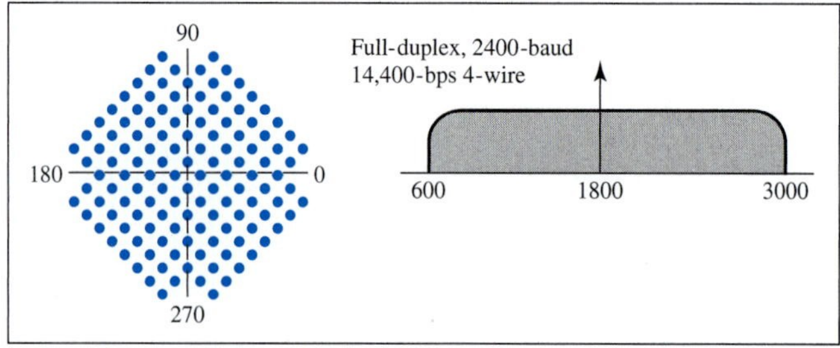

b. Constellation and bandwidth for V.32bis

V.90

Traditional modems have a data rate limitation of 33.6 kbps, as determined by the Shannon capacity (see Chapter 3). However, **V.90** modems with a bit rate of 56,000 bps are available; these are called **56K modems.** These modems may be used only if one party is using digital signaling (such as through an Internet provider). They are asymmetric in that the downloading rate (flow of data from the Internet service provider to the PC) is a maximum of 56 kbps, while the uploading rate (flow of data from the PC to the Internet provider) can be a maximum of 33.6 kbps. Do these modems violate the Shannon capacity principle? No, in the downstream direction, the SNR ratio is higher because there is no quantization error (see Figure 9.9).

In **uploading,** the analog signal must still be sampled at the switching station. In this direction, quantization noise (as we saw in Chapter 4) is introduced into the signal, which reduces the SNR ratio and limits the rate to 33.6 kbps.

However, there is no sampling in the **downloading.** The signal is not affected by quantization noise and not subject to the Shannon capacity limitation. The maximum data rate in the uploading direction is still 33.6 kbps, but the data rate in the downloading direction is now 56 kbps.

One may wonder how we arrive at the 56-kbps figure. The telephone companies sample 8000 times per second with 8 bits per sample. One of the bits in each sample is used for control purposes, which means each sample is 7 bits. The rate is therefore 8000×7, or 56,000 bps or 56 kbps.

Figure 9.9 *Uploading and downloading in 56K modems*

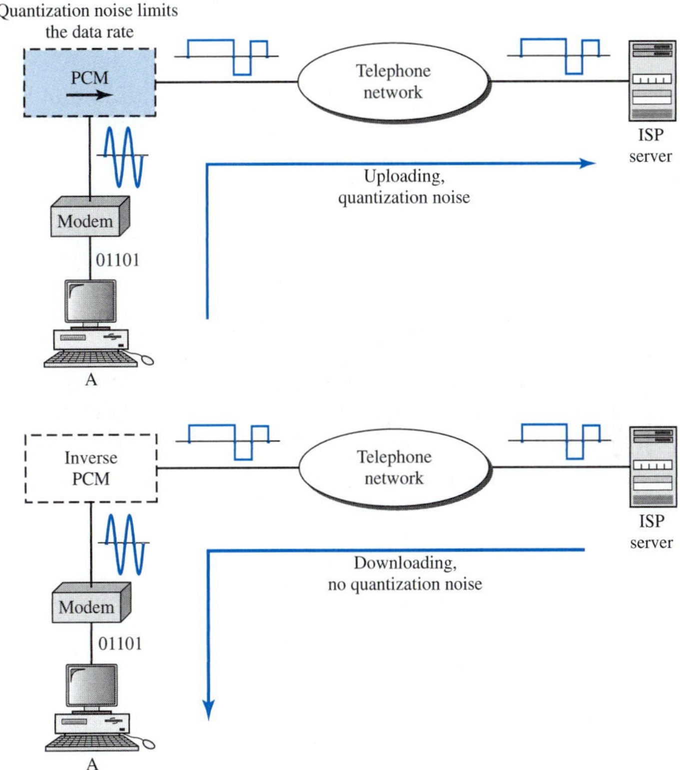

V.92

The standard above V.90 is called **V.92.** These modems can adjust their speed, and if the noise allows, they can upload data at the rate of 48 kbps. The downloading rate is still 56 kbps. The modem has additional features. For example, the modem can interrupt the Internet connection when there is an incoming call if the line has call-waiting service.

9.3 DIGITAL SUBSCRIBER LINE

After traditional modems reached their peak data rate, telephone companies developed another technology, DSL, to provide higher-speed access to the Internet. **Digital subscriber line (DSL)** technology is one of the most promising for supporting high-speed digital communication over the existing local loops. DSL technology is a set of technologies, each differing in the first letter (ADSL, VDSL, HDSL, and SDSL). The set is often referred to as *x*DSL, where *x* can be replaced by A, V, H, or S.

ADSL

The first technology in the set is **asymmetric DSL (ADSL).** ADSL, like a 56K modem, provides higher speed (bit rate) in the downstream direction (from the Internet to the resident) than in the upstream direction (from the resident to the Internet). That is the reason it is called asymmetric. Unlike the asymmetry in 56K modems, the designers of ADSL specifically divided the available bandwidth of the local loop unevenly for the residential customer. The service is not suitable for business customers who need a large bandwidth in both directions.

ADSL is an asymmetric communication technology designed for residential users; it is not suitable for businesses.

Using Existing Local Loops

One interesting point is that ADSL uses the existing local loops. But how does ADSL reach a data rate that was never achieved with traditional modems? The answer is that the twisted-pair local loop is actually capable of handling bandwidths up to 1.1 MHz, but the filter installed at the end office of the telephone company where each local loop terminates limits the bandwidth to 4 kHz (sufficient for voice communication). If the filter is removed, however, the entire 1.1 MHz is available for data and voice communications.

The existing local loops can handle bandwidths up to 1.1 MHz.

Adaptive Technology

Unfortunately, 1.1 MHz is just the theoretical bandwidth of the local loop. Factors such as the distance between the residence and the switching office, the size of the cable, the signaling used, and so on affect the bandwidth. The designers of ADSL technology were aware of this problem and used an adaptive technology that tests the condition and bandwidth availability of the line before settling on a data rate. The data rate of ADSL is not fixed; it changes based on the condition and type of the local loop cable.

ADSL is an adaptive technology. The system uses a data rate based on the condition of the local loop line.

Discrete Multitone Technique

The modulation technique that has become standard for ADSL is called the **discrete multitone technique (DMT)** which combines QAM and FDM. There is no set way that the bandwidth of a system is divided. Each system can decide on its bandwidth division. Typically, an available bandwidth of 1.104 MHz is divided into 256 channels. Each channel uses a bandwidth of 4.312 kHz, as shown in Figure 9.10. Figure 9.11 shows how the bandwidth can be divided into the following:

- **Voice.** Channel 0 is reserved for voice communication.
- **Idle.** Channels 1 to 5 are not used and provide a gap between voice and data communication.

Figure 9.10 *Discrete multitone technique*

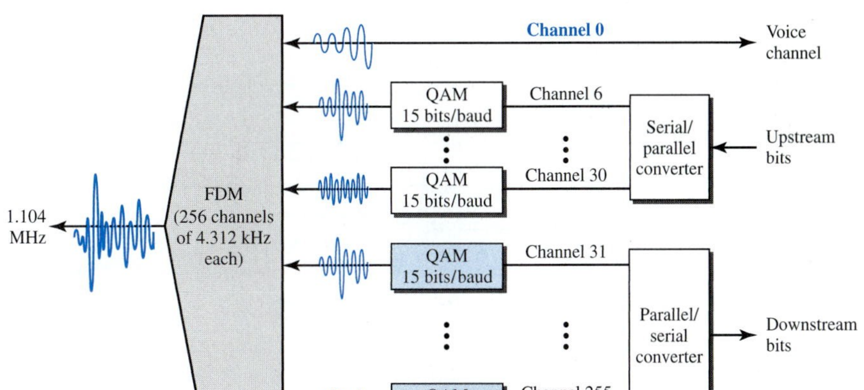

Figure 9.11 *Bandwidth division in ADSL*

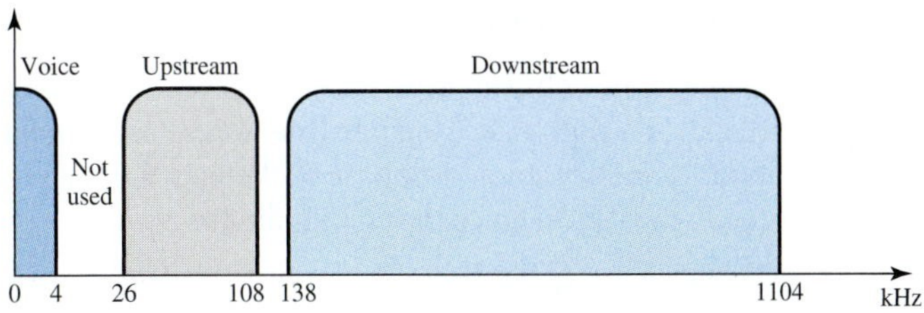

- **Upstream data and control.** Channels 6 to 30 (25 channels) are used for upstream data transfer and control. One channel is for control, and 24 channels are for data transfer. If there are 24 channels, each using 4 kHz (out of 4.312 kHz available) with QAM modulation, we have 24 × 4000 × 15, or a 1.44-Mbps bandwidth, in the upstream direction. However, the data rate is normally below 500 kbps because some of the carriers are deleted at frequencies where the noise level is large. In other words, some of channels may be unused.
- **Downstream data and control.** Channels 31 to 255 (225 channels) are used for downstream data transfer and control. One channel is for control, and 224 channels are for data. If there are 224 channels, we can achieve up to 224 × 4000 × 15, or 13.4 Mbps. However, the data rate is normally below 8 Mbps because some of the carriers are deleted at frequencies where the noise level is large. In other words, some of channels may be unused.

Customer Site: ADSL Modem

Figure 9.12 shows an **ADSL modem** installed at a customer's site. The local loop connects to a splitter which separates voice and data communications. The ADSL modem modulates and demodulates the data, using DMT, and creates downstream and upstream channels.

Figure 9.12 *ADSL modem*

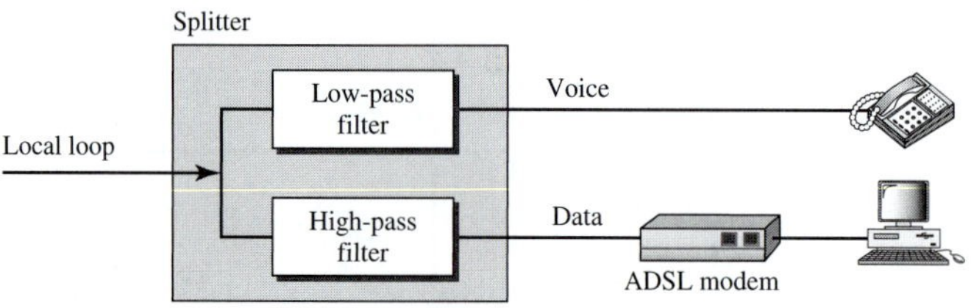

Note that the splitter needs to be installed at the customer's premises, normally by a technician from the telephone company. The voice line can use the existing telephone wiring in the house, but the data line needs to be installed by a professional. All this makes the ADSL line expensive. We will see that there is an alternative technology, Universal ADSL (or ADSL Lite).

Telephone Company Site: DSLAM

At the telephone company site, the situation is different. Instead of an ADSL modem, a device called a **digital subscriber line access multiplexer (DSLAM)** is installed that functions similarly. In addition, it packetizes the data to be sent to the Internet (ISP server). Figure 9.13 shows the configuration.

Figure 9.13 *DSLAM*

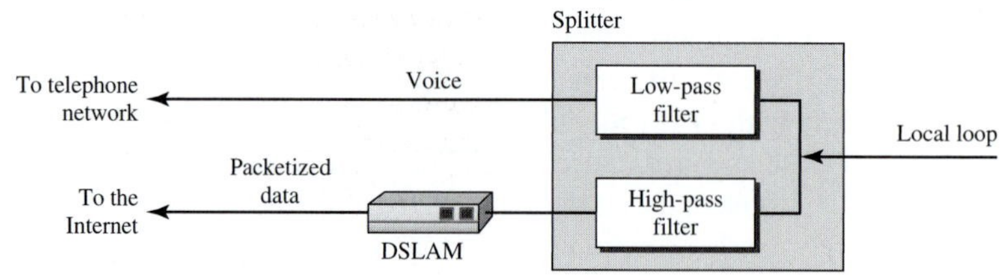

ADSL Lite

The installation of splitters at the border of the premises and the new wiring for the data line can be expensive and impractical enough to dissuade most subscribers. A new version of ADSL technology called **ADSL Lite** (or Universal ADSL or splitterless ADSL) is available for these subscribers. This technology allows an ASDL Lite modem to be plugged directly into a telephone jack and connected to the computer. The splitting is done at the telephone company. ADSL Lite uses 256 DMT carriers with 8-bit modulation (instead of 15-bit). However, some of the carriers may not be available because errors created by the voice signal might mingle with them. It can provide a maximum downstream data rate of 1.5 Mbps and an upstream data rate of 512 kbps.

HDSL

The **high-bit-rate digital subscriber line (HDSL)** was designed as an alternative to the T-1 line (1.544 Mbps). The T-1 line uses alternate mark inversion (AMI) encoding, which is very susceptible to attenuation at high frequencies. This limits the length of a T-1 line to 3200 ft (1 km). For longer distances, a repeater is necessary, which means increased costs.

HDSL uses 2B1Q encoding (see Chapter 4), which is less susceptible to attenuation. A data rate of 1.544 Mbps (sometimes up to 2 Mbps) can be achieved without repeaters up to a distance of 12,000 ft (3.86 km). HDSL uses two twisted pairs (one pair for each direction) to achieve full-duplex transmission.

SDSL

The **symmetric digital subscriber line (SDSL)** is a one twisted-pair version of HDSL. It provides full-duplex symmetric communication supporting up to 768 kbps in each direction. SDSL, which provides symmetric communication, can be considered an alternative to ADSL. ADSL provides asymmetric communication, with a downstream bit rate that is much higher than the upstream bit rate. Although this feature meets the needs of most residential subscribers, it is not suitable for businesses that send and receive data in large volumes in both directions.

VDSL

The **very high-bit-rate digital subscriber line (VDSL),** an alternative approach that is similar to ADSL, uses coaxial, fiber-optic, or twisted-pair cable for short distances. The modulating technique is DMT. It provides a range of bit rates (25 to 55 Mbps) for upstream communication at distances of 3000 to 10,000 ft. The downstream rate is normally 3.2 Mbps.

Summary

Table 9.1 shows a summary of DSL technologies. Note that the data rate and distances are approximations and can vary from one implementation to another.

Table 9.1 *Summary of DSL technologies*

Technology	*Downstream Rate*	*Upstream Rate*	*Distance (ft)*	*Twisted Pairs*	*Line Code*
ADSL	1.5–6.1 Mbps	16–640 kbps	12,000	1	DMT
ADSL Lite	1.5 Mbps	500 kbps	18,000	1	DMT
HDSL	1.5–2.0 Mbps	1.5–2.0 Mbps	12,000	2	2B1Q
SDSL	768 kbps	768 kbps	12,000	1	2B1Q
VDSL	25–55 Mbps	3.2 Mbps	3000–10,000	1	DMT

9.4 CABLE TV NETWORKS

The **cable TV network** started as a video service provider, but it has moved to the business of Internet access. In this section, we discuss cable TV networks per se; in Section 9.5 we discuss how this network can be used to provide high-speed access to the Internet.

Traditional Cable Networks

Cable TV started to distribute broadcast video signals to locations with poor or no reception in the late 1940s. It was called **community antenna TV (CATV)** because an antenna at the top of a tall hill or building received the signals from the TV stations and distributed them, via coaxial cables, to the community. Figure 9.14 shows a schematic diagram of a traditional cable TV network.

Figure 9.14 *Traditional cable TV network*

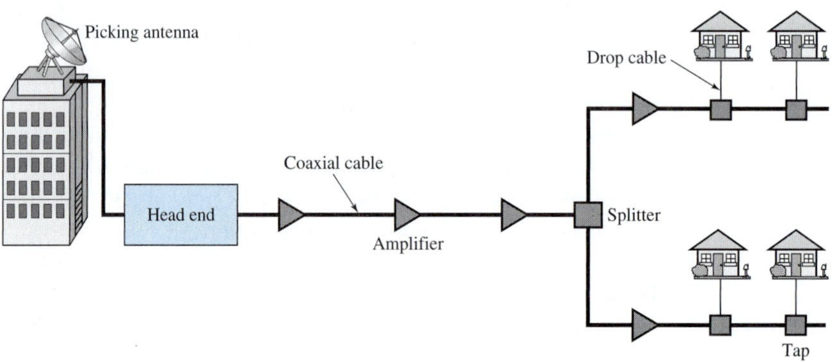

The cable TV office, called the **head end,** receives video signals from broadcasting stations and feeds the signals into coaxial cables. The signals became weaker and weaker with distance, so amplifiers were installed through the network to renew the signals. There could be up to 35 amplifiers between the head end and the subscriber premises. At the other end, splitters split the cable, and taps and drop cables make the connections to the subscriber premises.

The traditional cable TV system used coaxial cable end to end. Due to attenuation of the signals and the use of a large number of amplifiers, communication in the traditional network was unidirectional (one-way). Video signals were transmitted downstream, from the head end to the subscriber premises.

Communication in the traditional cable TV network is unidirectional.

Hybrid Fiber-Coaxial (HFC) Network

The second generation of cable networks is called a **hybrid fiber-coaxial (HFC) network.** The network uses a combination of fiber-optic and coaxial cable. The transmission medium from the cable TV office to a box, called the **fiber node,** is optical fiber; from the fiber node through the neighborhood and into the house is still coaxial cable. Figure 9.15 shows a schematic diagram of an HFC network.

Figure 9.15 *Hybrid fiber-coaxial (HFC) network*

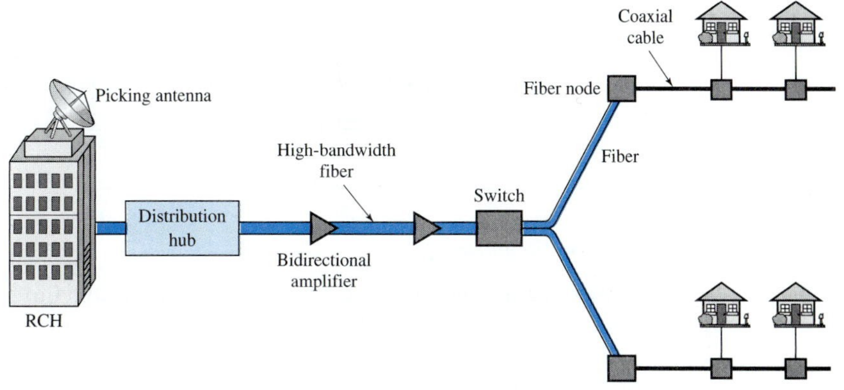

The **regional cable head (RCH)** normally serves up to 400,000 subscribers. The RCHs feed the **distribution hubs,** each of which serves up to 40,000 subscribers. The distribution hub plays an important role in the new infrastructure. Modulation and distribution of signals are done here; the signals are then fed to the fiber nodes through fiber-optic cables. The fiber node splits the analog signals so that the same signal is sent to each coaxial cable. Each coaxial cable serves up to 1000 subscribers. The use of fiber-optic cable reduces the need for amplifiers down to eight or less.

One reason for moving from traditional to hybrid infrastructure is to make the cable network bidirectional (two-way).

Communication in an HFC cable TV network can be bidirectional.

9.5 CABLE TV FOR DATA TRANSFER

Cable companies are now competing with telephone companies for the residential customer who wants high-speed data transfer. DSL technology provides high-data-rate connections for residential subscribers over the local loop. However, DSL uses the existing unshielded twisted-pair cable, which is very susceptible to interference. This imposes an upper limit on the data rate. Another solution is the use of the cable TV network. In this section, we briefly discuss this technology.

Bandwidth

Even in an HFC system, the last part of the network, from the fiber node to the subscriber premises, is still a coaxial cable. This coaxial cable has a bandwidth that ranges from 5 to 750 MHz (approximately). To provide Internet access, the cable company has divided this bandwidth into three bands: video, downstream data, and upstream data, as shown in Figure 9.16.

Figure 9.16 *Division of coaxial cable band by CATV*

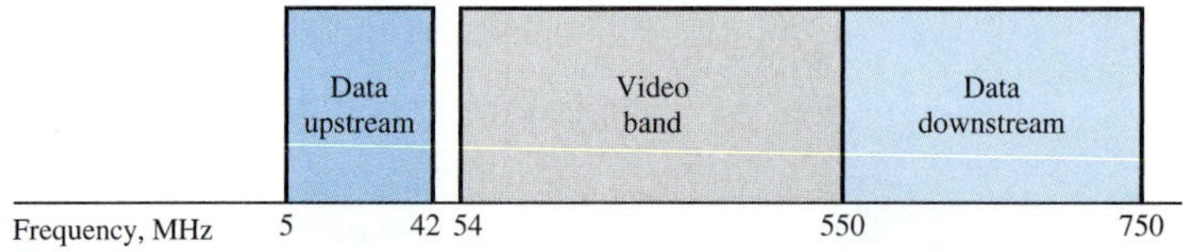

Downstream Video Band

The **downstream video band** occupies frequencies from 54 to 550 MHz. Since each TV channel occupies 6 MHz, this can accommodate more than 80 channels.

Downstream Data Band

The downstream data (from the Internet to the subscriber premises) occupies the upper band, from 550 to 750 MHz. This band is also divided into 6-MHz channels.

Modulation **Downstream data band** uses the 64-QAM (or possibly 256-QAM) modulation technique.

Downstream data are modulated using the 64-QAM modulation technique.

Data Rate There is 6 bits/baud in 64-QAM. One bit is used for forward error correction; this leaves 5 bits of data per baud. The standard specifies 1 Hz for each baud; this means that, theoretically, downstream data can be received at 30 Mbps (5 bits/Hz $\times$ 6 MHz). The standard specifies only 27 Mbps. However, since the cable modem is normally connected to the computer through a 10Base-T cable (see Chapter 13), this limits the data rate to 10 Mbps.

The theoretical downstream data rate is 30 Mbps.

Upstream Data Band

The upstream data (from the subscriber premises to the Internet) occupies the lower band, from 5 to 42 MHz. This band is also divided into 6-MHz channels.

Modulation The **upstream data band** uses lower frequencies that are more susceptible to noise and interference. For this reason, the QAM technique is not suitable for this band. A better solution is QPSK.

Upstream data are modulated using the QPSK modulation technique.

Data Rate There are 2 bits/baud in QPSK. The standard specifies 1 Hz for each baud; this means that, theoretically, upstream data can be sent at 12 Mbps (2 bits/Hz × 6 MHz). However, the data rate is usually less than 12 Mbps.

The theoretical upstream data rate is 12 Mbps.

Sharing

Both upstream and downstream bands are shared by the subscribers.

Upstream Sharing

The upstream data bandwidth is 37 MHz. This means that there are only six 6-MHz channels available in the upstream direction. A subscriber needs to use one channel to send data in the upstream direction. The question is, "How can six channels be shared in an area with 1000, 2000, or even 100,000 subscribers?" The solution is timesharing. The band is divided into channels using FDM; these channels must be shared between subscribers in the same neighborhood. The cable provider allocates one channel, statically or dynamically, for a group of subscribers. If one subscriber wants to send data, she or he contends for the channel with others who want access; the subscriber must wait until the channel is available.

Downstream Sharing

We have a similar situation in the downstream direction. The downstream band has 33 channels of 6 MHz. A cable provider probably has more than 33 subscribers; therefore, each channel must be shared between a group of subscribers. However, the situation is different for the downstream direction; here we have a multicasting situation. If there are data for any of the subscribers in the group, the data are sent to that channel. Each subscriber is sent the data. But since each subscriber also has an address registered with the provider; the cable modem for the group matches the address carried with the data to the address assigned by the provider. If the address matches, the data are kept; otherwise, they are discarded.

CM and CMTS

To use a cable network for data transmission, we need two key devices: a **cable modem (CM)** and a **cable modem transmission system (CMTS).**

CM

The **cable modem (CM)** is installed on the subscriber premises. It is similar to an ADSL modem. Figure 9.17 shows its location.

CMTS

The **cable modem transmission system (CMTS)** is installed inside the distribution hub by the cable company. It receives data from the Internet and passes them to the combiner, which sends them to the subscriber. The CMTS also receives data from the subscriber and passes them to the Internet. Figure 9.18 shows the location of the CMTS.

Figure 9.17 *Cable modem (CM)*

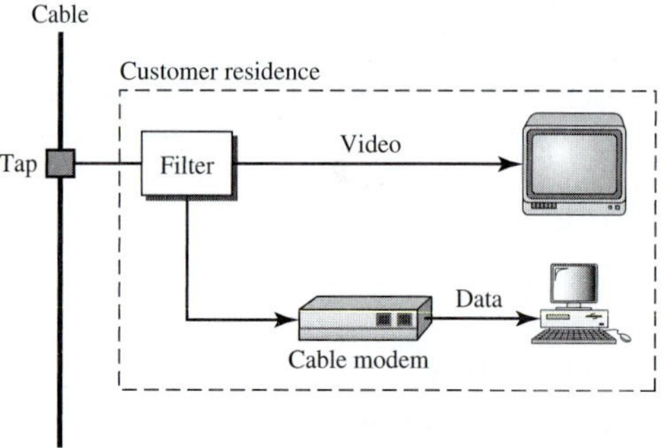

Figure 9.18 *Cable modem transmission system (CMTS)*

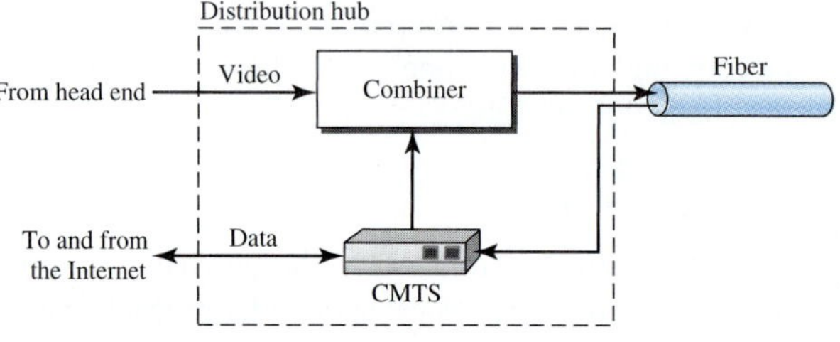

Data Transmission Schemes: DOCSIS

During the last few decades, several schemes have been designed to create a standard for data transmission over an HFC network. Prevalent is the one devised by Multimedia Cable Network Systems (MCNS), called **Data Over Cable System Interface Specification (DOCSIS).** DOCSIS defines all the protocols necessary to transport data from a CMTS to a CM.

Upstream Communication

The following is a very simplified version of the protocol defined by DOCSIS for upstream communication. It describes the steps that must be followed by a CM:

1. The CM checks the downstream channels for a specific packet periodically sent by the CMTS. The packet asks any new CM to announce itself on a specific upstream channel.
2. The CMTS sends a packet to the CM, defining its allocated downstream and upstream channels.
3. The CM then starts a process, called **ranging,** which determines the distance between the CM and CMTS. This process is required for synchronization between all

CMs and CMTSs for the minislots used for timesharing of the upstream channels. We will learn about this timesharing when we discuss contention protocols in Chapter 12.

4. The CM sends a packet to the ISP, asking for the Internet address.
5. The CM and CMTS then exchange some packets to establish security parameters, which are needed for a public network such as cable TV.
6. The CM sends its unique identifier to the CMTS.
7. Upstream communication can start in the allocated upstream channel; the CM can contend for the minislots to send data.

Downstream Communication

In the downstream direction, the communication is much simpler. There is no contention because there is only one sender. The CMTS sends the packet with the address of the receiving CM, using the allocated downstream channel.

9.6 RECOMMENDED READING

For more details about subjects discussed in this chapter, we recommend the following books. The items in brackets [. . .] refer to the reference list at the end of the text.

Books

[Cou01] gives an interesting discussion about telephone systems, DSL technology, and CATV in Chapter 8. [Tan03] discusses telephone systems and DSL technology in Section 2.5 and CATV in Section 2.7. [GW04] discusses telephone systems in Section 1.1.1 and standard modems in Section 3.7.3. A complete coverage of residential broadband (DSL and CATV) can be found in [Max99].

9.7 KEY TERMS

56K modem
800 service
900 service
ADSL Lite
ADSL modem
analog leased service
analog switched service
asymmetric DSL (ADSL)
cable modem (CM)
cable modem transmission system (CMTS)
cable TV network
common carrier
community antenna TV (CATV)
competitive local exchange carrier (CLEC)
Data Over Cable System Interface Specification (DOCSIS)
demodulator
digital data service (DDS)
digital service
digital subscriber line (DSL)
digital subscriber line access multiplexer (DSLAM)
discrete multitone technique (DMT)
distribution hub

downloading
downstream data band
end office
fiber node
head end
high-bit-rate DSL (HDSL)
hybrid fiber-coaxial (HFC) network
in-band signaling
incumbent local exchange carrier (ILEC)
interexchange carrier (IXC)
ISDN user port (ISUP)
local access transport area (LATA)
local exchange carrier (LEC)
local loop
long distance company
message transport port (MTP) level
modem
modulator
out-of-band signaling
plain old telephone system (POTS)
point of presence (POP)
ranging
regional cable head (RCH)
regional office
server control point (SCP)
signal point (SP)
signal transport port (STP)
signaling connection control point (SCCP)
Signaling System Seven (SS7)
switched/56 service
switching office
symmetric DSL (SDSL)
tandem office
telephone user port (TUP)
transaction capabilities application port (TCAP)
trunk
uploading
upstream data band
V.32
V.32bis
V.34bis
V.90
V.92
very-high-bit-rate DSL (VDSL)
video band
V-series
wide-area telephone service (WATS)

9.8 SUMMARY

- The telephone, which is referred to as the plain old telephone system (POTS), was originally an analog system. During the last decade, the telephone network has undergone many technical changes. The network is now digital as well as analog.
- The telephone network is made of three major components: local loops, trunks, and switching offices. It has several levels of switching offices such as end offices, tandem offices, and regional offices.
- The United States is divided into many local access transport areas (LATAs). The services offered inside a LATA are called intra-LATA services. The carrier that handles these services is called a local exchange carrier (LEC). The services between LATAs are handled by interexchange carriers (IXCs).
- In in-band signaling, the same circuit is used for both signaling and data. In out-of-band signaling, a portion of the bandwidth is used for signaling and another portion

for data. The protocol that is used for signaling in the telephone network is called Signaling System Seven (SS7).

- Telephone companies provide two types of services: analog and digital. We can categorize analog services as either analog switched services or analog leased services. The two most common digital services are switched/56 service and digital data service (DDS).
- Data transfer using the telephone local loop was traditionally done using a dial-up modem. The term *modem* is a composite word that refers to the two functional entities that make up the device: a signal modulator and a signal demodulator.
- Most popular modems available are based on the V-series standards. The V.32 modem has a data rate of 9600 bps. The V.32bis modem supports 14,400-bps transmission. V.90 modems, called 56K modems, with a downloading rate of 56 kbps and uploading rate of 33.6 kbps are very common. The standard above V.90 is called V.92. These modems can adjust their speed, and if the noise allows, they can upload data at the rate of 48 kbps.
- Telephone companies developed another technology, digital subscriber line (DSL), to provide higher-speed access to the Internet. DSL technology is a set of technologies, each differing in the first letter (ADSL, VDSL, HDSL, and SDSL. ADSL provides higher speed in the downstream direction than in the upstream direction. The high-bit-rate digital subscriber line (HDSL) was designed as an alternative to the T-1 line (1.544 Mbps). The symmetric digital subscriber line (SDSL) is a one twisted-pair version of HDSL. The very high-bit-rate digital subscriber line (VDSL) is an alternative approach that is similar to ADSL.
- Community antenna TV (CATV) was originally designed to provide video services for the community. The traditional cable TV system used coaxial cable end to end. The second generation of cable networks is called a hybrid fiber-coaxial (HFC) network. The network uses a combination of fiber-optic and coaxial cable.
- Cable companies are now competing with telephone companies for the residential customer who wants high-speed access to the Internet. To use a cable network for data transmission, we need two key devices: a cable modem (CM) and a cable modem transmission system (CMTS).

9.9 PRACTICE SET

Review Questions

1. What are the three major components of a telephone network?
2. Give some hierarchical switching levels of a telephone network.
3. What is LATA? What are intra-LATA and inter-LATA services?
4. Describe the SS7 service and its relation to the telephone network.
5. What are the two major services provided by telephone companies in the United States?
6. What is dial-up modem technology? List some of the common modem standards discussed in this chapter and give their data rates.

7. What is DSL technology? What are the services provided by the telephone companies using DSL? Distinguish between a DSL modem and a DSLAM.
8. Compare and contrast a traditional cable network with a hybrid fiber-coaxial network.
9. How is data transfer achieved using CATV channels?
10. Distinguish between CM and CMTS.

Exercises

11. Using the discussion of circuit-switching in Chapter 8, explain why this type of switching was chosen for telephone networks.
12. In Chapter 8, we discussed the three communication phases involved in a circuit-switched network. Match these phases with the phases in a telephone call between two parties.
13. In Chapter 8, we learned that a circuit-switched network needs end-to-end addressing during the setup and teardown phases. Define end-to-end addressing in a telephone network when two parties communicate.
14. When we have an overseas telephone conversation, we sometimes experience a delay. Can you explain the reason?
15. Draw a barchart to compare the different downloading data rates of common modems.
16. Draw a barchart to compare the different downloading data rates of common DSL technology implementations (use minimum data rates).
17. Calculate the minimum time required to download one million bytes of information using each of the following technologies:
    a. V.32 modem
    b. V.32bis modem
    c. V.90 modem
18. Repeat Exercise 17 using different DSL implementations (consider the minimum rates).
19. Repeat Exercise 17 using a cable modem (consider the minimum rates).
20. What type of topology is used when customers in an area use DSL modems for data transfer purposes? Explain.
21. What type of topology is used when customers in an area use cable modems for data transfer purposes? Explain.

PART 3

Data Link Layer

Objectives

The data link layer transforms the physical layer, a raw transmission facility, to a link responsible for node-to-node (hop-to-hop) communication. Specific responsibilities of the data link layer include *framing, addressing, flow control, error control,* and *media access control.* The data link layer divides the stream of bits received from the network layer into manageable data units called frames. The data link layer adds a header to the frame to define the addresses of the sender and receiver of the frame. If the rate at which the data are absorbed by the receiver is less than the rate at which data are produced in the sender, the data link layer imposes a flow control mechanism to avoid overwhelming the receiver. The data link layer also adds reliability to the physical layer by adding mechanisms to detect and retransmit damaged, duplicate, or lost frames. When two or more devices are connected to the same link, data link layer protocols are necessary to determine which device has control over the link at any given time.

In Part 3 of the book, we first discuss services provided by the data link layer. We then discuss the implementation of these services in local area networks (LANs). Finally we discuss how wide area networks (WANs) use these services.

Part 3 of the book is devoted to the data link layer and the services provided by this layer.

Chapters

This part consists of nine chapters: Chapters 10 to 18.

Chapter 10

Chapter 10 discusses error detection and correction. Although the quality of devices and media have been improved during the last decade, we still need to check for errors and correct them in most applications.

Chapter 11

Chapter 11 is named data link control, which involves flow and error control. It discusses some protocols that are designed to handle the services required from the data link layer in relation to the network layer.

Chapter 12

Chapter 12 is devoted to access control, the duties of the data link layer that are related to the use of the physical layer.

Chapter 13

This chapter introduces wired local area networks. A wired LAN, viewed as a link, is mostly involved in the physical and data link layers. We have devoted the chapter to the discussion of Ethernet and its evolution, a dominant technology today.

Chapter 14

This chapter introduces wireless local area networks. The wireless LAN is a growing technology in the Internet. We devote one chapter to this topic.

Chapter 15

After discussing wired and wireless LANs, we show how they can be connected together using connecting devices.

Chapter 16

This is the first chapter on wide area networks (WANs). We start with wireless WANs and then move on to satellite networks and mobile telephone networks.

Chapter 17

To demonstrate the operation of a high-speed wide area network that can be used as a backbone for other WANs or for the Internet, we have chosen to devote all of Chapter 17 to SONET, a wide area network that uses fiber-optic technology.

Chapter 18

This chapter concludes our discussion on wide area networks. Two switched WANs, Frame Relay and ATM, are discussed here.

CHAPTER 10

Error Detection and Correction

Networks must be able to transfer data from one device to another with acceptable accuracy. For most applications, a system must guarantee that the data received are identical to the data transmitted. Any time data are transmitted from one node to the next, they can become corrupted in passage. Many factors can alter one or more bits of a message. Some applications require a mechanism for detecting and correcting **errors.**

Data can be corrupted during transmission.
Some applications require that errors be detected and corrected.

Some applications can tolerate a small level of error. For example, random errors in audio or video transmissions may be tolerable, but when we transfer text, we expect a very high level of accuracy.

10.1 INTRODUCTION

Let us first discuss some issues related, directly or indirectly, to error detection and correcion.

Types of Errors

Whenever bits flow from one point to another, they are subject to unpredictable changes because of **interference.** This interference can change the shape of the signal. In a single-bit error, a 0 is changed to a 1 or a 1 to a 0. In a burst error, multiple bits are changed. For example, a 1/100 s burst of impulse noise on a transmission with a data rate of 1200 bps might change all or some of the12 bits of information.

Single-Bit Error

The term ***single-bit error*** means that only 1 bit of a given data unit (such as a byte, character, or packet) is changed from 1 to 0 or from 0 to 1.

In a single-bit error, only 1 bit in the data unit has changed.

Figure 10.1 shows the effect of a single-bit error on a data unit. To understand the impact of the change, imagine that each group of 8 bits is an ASCII character with a 0 bit added to the left. In Figure 10.1, 00000010 (ASCII *STX*) was sent, meaning *start of text,* but 00001010 (ASCII *LF*) was received, meaning *line feed.* (For more information about ASCII code, see Appendix A.)

Figure 10.1 *Single-bit error*

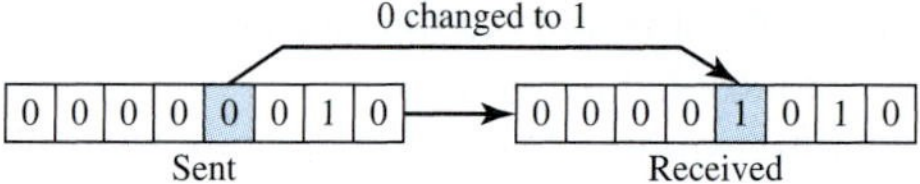

Single-bit errors are the least likely type of error in serial data transmission. To understand why, imagine data sent at 1 Mbps. This means that each bit lasts only 1/1,000,000 s, or 1 μs. For a single-bit error to occur, the noise must have a duration of only 1 μs, which is very rare; noise normally lasts much longer than this.

Burst Error

The term ***burst error*** means that 2 or more bits in the data unit have changed from 1 to 0 or from 0 to 1.

A burst error means that 2 or more bits in the data unit have changed.

Figure 10.2 shows the effect of a burst error on a data unit. In this case, 0100010001000011 was sent, but 0101110101100011 was received. Note that a burst error does not necessarily mean that the errors occur in consecutive bits. The length of the burst is measured from the first corrupted bit to the last corrupted bit. Some bits in between may not have been corrupted.

Figure 10.2 *Burst error of length 8*

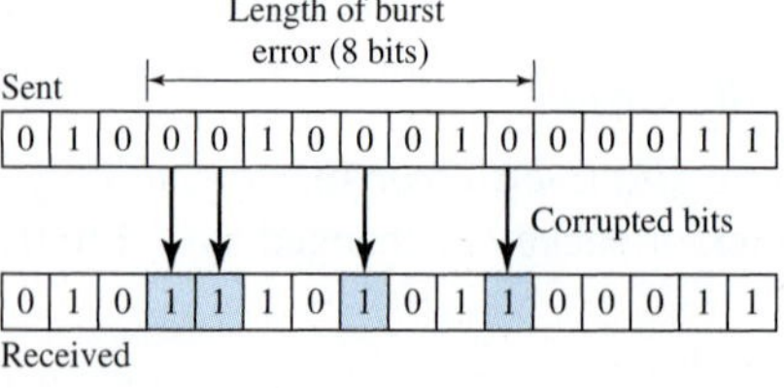

A burst error is more likely to occur than a single-bit error. The duration of noise is normally longer than the duration of 1 bit, which means that when noise affects data, it affects a set of bits. The number of bits affected depends on the data rate and duration of noise. For example, if we are sending data at 1 kbps, a noise of 1/100 s can affect 10 bits; if we are sending data at 1 Mbps, the same noise can affect 10,000 bits.

Redundancy

The central concept in detecting or correcting errors is **redundancy.** To be able to detect or correct errors, we need to send some extra bits with our data. These redundant bits are added by the sender and removed by the receiver. Their presence allows the receiver to detect or correct corrupted bits.

To detect or correct errors, we need to send extra (redundant) bits with data.

Detection Versus Correction

The correction of errors is more difficult than the detection. In **error detection,** we are looking only to see if any error has occurred. The answer is a simple yes or no. We are not even interested in the number of errors. A single-bit error is the same for us as a burst error.

In **error correction,** we need to know the exact number of bits that are corrupted and more importantly, their location in the message. The number of the errors and the size of the message are important factors. If we need to correct one single error in an 8-bit data unit, we need to consider eight possible error locations; if we need to correct two errors in a data unit of the same size, we need to consider 28 possibilities. You can imagine the receiver's difficulty in finding 10 errors in a data unit of 1000 bits.

Forward Error Correction Versus Retransmission

There are two main methods of error correction. **Forward error correction** is the process in which the receiver tries to guess the message by using redundant bits. This is possible, as we see later, if the number of errors is small. Correction by **retransmission** is a technique in which the receiver detects the occurrence of an error and asks the sender to resend the message. Resending is repeated until a message arrives that the receiver believes is error-free (usually, not all errors can be detected).

Coding

Redundancy is achieved through various coding schemes. The sender adds redundant bits through a process that creates a relationship between the redundant bits and the actual data bits. The receiver checks the relationships between the two sets of bits to detect or correct the errors. The ratio of redundant bits to the data bits and the robustness of the process are important factors in any coding scheme. Figure 10.3 shows the general idea of coding.

We can divide coding schemes into two broad categories: **block coding** and **convolution coding.** In this book, we concentrate on block coding; convolution coding is more complex and beyond the scope of this book.

Figure 10.3 *The structure of encoder and decoder*

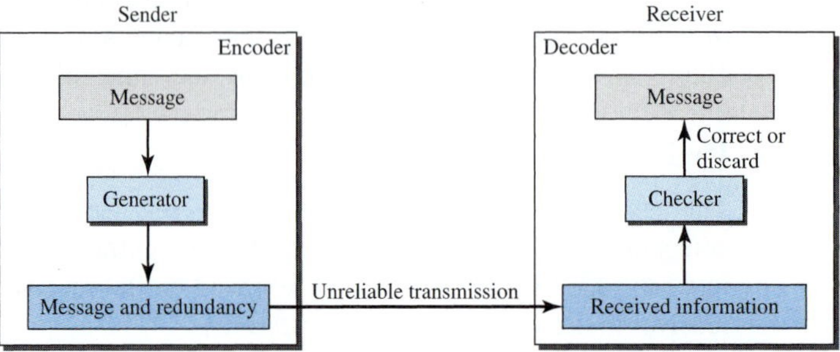

In this book, we concentrate on block codes; we leave convolution codes to advanced texts.

Modular Arithmetic

Before we finish this section, let us briefly discuss a concept basic to computer science in general and to error detection and correction in particular: modular arithmetic. Our intent here is not to delve deeply into the mathematics of this topic; we present just enough information to provide a background to materials discussed in this chapter.

In **modular arithmetic,** we use only a limited range of integers. We define an upper limit, called a **modulus** N. We then use only the integers 0 to $N - 1$, inclusive. This is modulo-N arithmetic. For example, if the modulus is 12, we use only the integers 0 to 11, inclusive. An example of modulo arithmetic is our clock system. It is based on modulo-12 arithmetic, substituting the number 12 for 0. In a modulo-N system, if a number is greater than N, it is divided by N and the remainder is the result. If it is negative, as many Ns as needed are added to make it positive. Consider our clock system again. If we start a job at 11 A.M. and the job takes 5 h, we can say that the job is to be finished at 16:00 if we are in the military, or we can say that it will be finished at 4 P.M. (the remainder of 16/12 is 4).

In modulo-N arithmetic, we use only the integers in the range 0 to $N - 1$, inclusive.

Addition and subtraction in modulo arithmetic are simple. There is no carry when you add two digits in a column. There is no carry when you subtract one digit from another in a column.

Modulo-2 Arithmetic

Of particular interest is modulo-2 arithmetic. In this arithmetic, the modulus N is 2. We can use only 0 and 1. Operations in this arithmetic are very simple. The following shows how we can add or subtract 2 bits.

Adding:	$0 + 0 = 0$	$0 + 1 = 1$	$1 + 0 = 1$	$1 + 1 = 0$
Subtracting:	$0 - 0 = 0$	$0 - 1 = 1$	$1 - 0 = 1$	$1 - 1 = 0$

Notice particularly that addition and subtraction give the same results. In this arithmetic we use the XOR (exclusive OR) operation for both addition and subtraction. The result of an XOR operation is 0 if two bits are the same; the result is 1 if two bits are different. Figure 10.4 shows this operation.

Figure 10.4 *XORing of two single bits or two words*

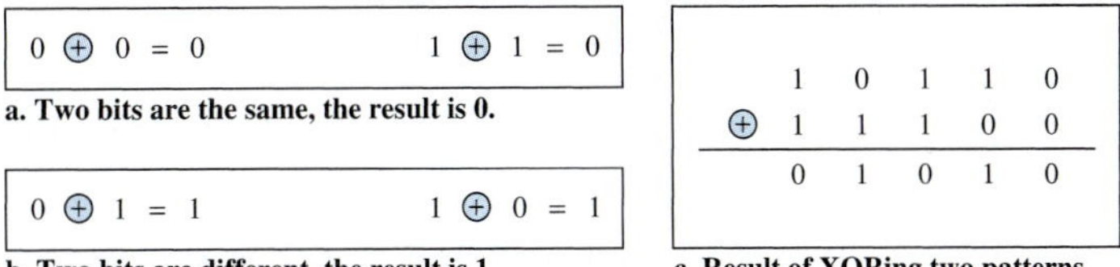

a. Two bits are the same, the result is 0.

b. Two bits are different, the result is 1.

c. Result of XORing two patterns

Other Modulo Arithmetic

We also use, modulo-N arithmetic through the book. The principle is the same; we use numbers between 0 and $N - 1$. If the modulus is not 2, addition and subtraction are distinct. If we get a negative result, we add enough multiples of N to make it positive.

10.2 BLOCK CODING

In block coding, we divide our message into blocks, each of k bits, called **datawords.** We add r redundant bits to each block to make the length $n = k + r$. The resulting n-bit blocks are called **codewords.** How the extra r bits is chosen or calculated is something we will discuss later. For the moment, it is important to know that we have a set of datawords, each of size k, and a set of codewords, each of size of n. With k bits, we can create a combination of 2^k datawords; with n bits, we can create a combination of 2^n codewords. Since $n > k$, the number of possible codewords is larger than the number of possible datawords. The block coding process is one-to-one; the same dataword is always encoded as the same codeword. This means that we have $2^n - 2^k$ codewords that are not used. We call these codewords invalid or illegal. Figure 10.5 shows the situation.

Figure 10.5 *Datawords and codewords in block coding*

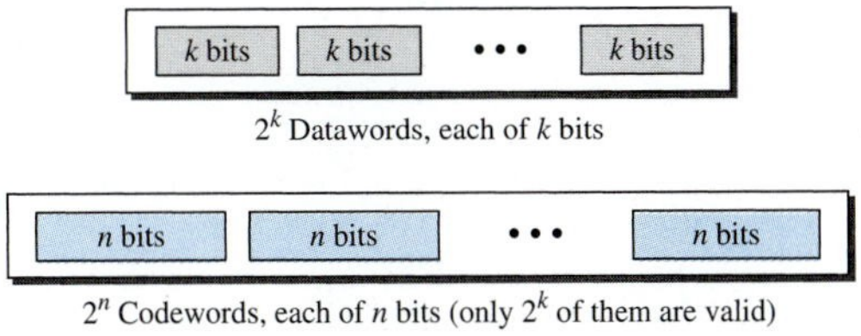

k bits | k bits | • • • | k bits

2^k Datawords, each of k bits

n bits | n bits | • • • | n bits

2^n Codewords, each of n bits (only 2^k of them are valid)

Example 10.1

The 4B/5B block coding discussed in Chapter 4 is a good example of this type of coding. In this coding scheme, $k = 4$ and $n = 5$. As we saw, we have $2^k = 16$ datawords and $2^n = 32$ codewords. We saw that 16 out of 32 codewords are used for message transfer and the rest are either used for other purposes or unused.

Error Detection

How can errors be detected by using block coding? If the following two conditions are met, the receiver can detect a change in the original codeword.

1. The receiver has (or can find) a list of valid codewords.
2. The original codeword has changed to an invalid one.

Figure 10.6 shows the role of block coding in error detection.

Figure 10.6 *Process of error detection in block coding*

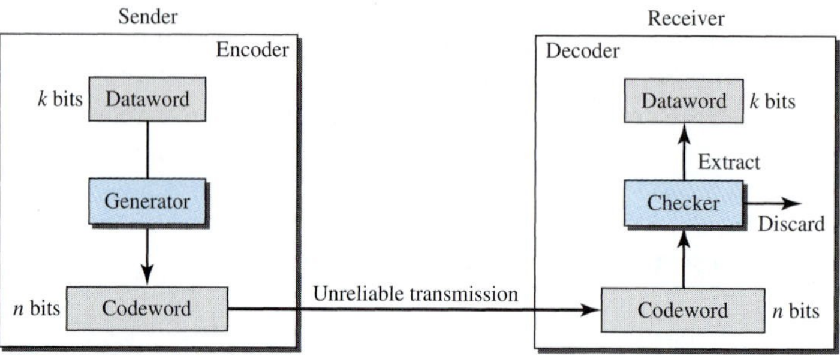

The sender creates codewords out of datawords by using a generator that applies the rules and procedures of encoding (discussed later). Each codeword sent to the receiver may change during transmission. If the received codeword is the same as one of the valid codewords, the word is accepted; the corresponding dataword is extracted for use. If the received codeword is not valid, it is discarded. However, if the codeword is corrupted during transmission but the received word still matches a valid codeword, the error remains undetected. This type of coding can detect only single errors. Two or more errors may remain undetected.

Example 10.2

Let us assume that $k = 2$ and $n = 3$. Table 10.1 shows the list of datawords and codewords. Later, we will see how to derive a codeword from a dataword.

Table 10.1 *A code for error detection (Example 10.2)*

Datawords	*Codewords*
00	000
01	011
10	101
11	110

Assume the sender encodes the dataword 01 as 011 and sends it to the receiver. Consider the following cases:

1. The receiver receives 011. It is a valid codeword. The receiver extracts the dataword 01 from it.
2. The codeword is corrupted during transmission, and 111 is received (the leftmost bit is corrupted). This is not a valid codeword and is discarded.
3. The codeword is corrupted during transmission, and 000 is received (the right two bits are corrupted). This is a valid codeword. The receiver incorrectly extracts the dataword 00. Two corrupted bits have made the error undetectable.

An error-detecting code can detect only the types of errors for which it is designed; other types of errors may remain undetected.

Error Correction

As we said before, error correction is much more difficult than error detection. In error detection, the receiver needs to know only that the received codeword is invalid; in error correction the receiver needs to find (or guess) the original codeword sent. We can say that we need more redundant bits for error correction than for error detection. Figure 10.7 shows the role of block coding in error correction. We can see that the idea is the same as error detection but the checker functions are much more complex.

Figure 10.7 *Structure of encoder and decoder in error correction*

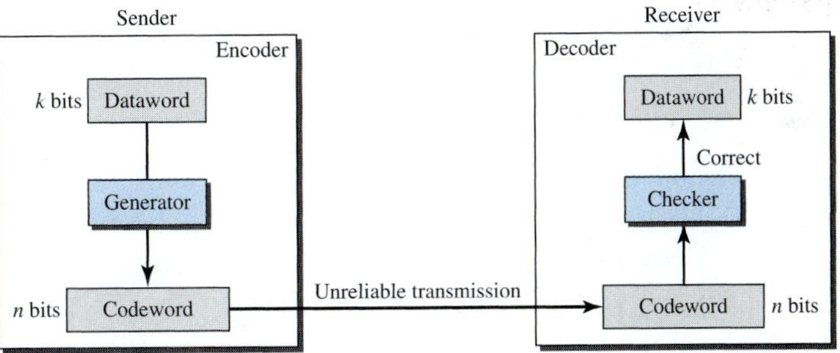

Example 10.3

Let us add more redundant bits to Example 10.2 to see if the receiver can correct an error without knowing what was actually sent. We add 3 redundant bits to the 2-bit dataword to make 5-bit codewords. Again, later we will show how we chose the redundant bits. For the moment let us concentrate on the error correction concept. Table 10.2 shows the datawords and codewords.

Assume the dataword is 01. The sender consults the table (or uses an algorithm) to create the codeword 01011. The codeword is corrupted during transmission, and 01001 is received (error in the second bit from the right). First, the receiver finds that the received codeword is not in the table. This means an error has occurred. (Detection must come before correction.) The receiver, assuming that there is only 1 bit corrupted, uses the following strategy to guess the correct dataword.

Table 10.2 *A code for error correction (Example 10.3)*

Dataword	*Codeword*
00	00000
01	01011
10	10101
11	11110

1. Comparing the received codeword with the first codeword in the table (01001 versus 00000), the receiver decides that the first codeword is not the one that was sent because there are two different bits.
2. By the same reasoning, the original codeword cannot be the third or fourth one in the table.
3. The original codeword must be the second one in the table because this is the only one that differs from the received codeword by 1 bit. The receiver replaces 01001 with 01011 and consults the table to find the dataword 01.

Hamming Distance

One of the central concepts in coding for error control is the idea of the Hamming distance. The **Hamming distance** between two words (of the same size) is the number of differences between the corresponding bits. We show the Hamming distance between two words x and y as $d(x, y)$.

The Hamming distance can easily be found if we apply the XOR operation (⊕) on the two words and count the number of 1s in the result. Note that the Hamming distance is a value greater than zero.

The Hamming distance between two words is the number of differences between corresponding bits.

Example 10.4

Let us find the Hamming distance between two pairs of words.

1. The Hamming distance $d(000, 011)$ is 2 because $000 \oplus 011$ is 011 (two 1s).
2. The Hamming distance $d(10101, 11110)$ is 3 because $10101 \oplus 11110$ is 01011 (three 1s).

Minimum Hamming Distance

Although the concept of the Hamming distance is the central point in dealing with error detection and correction codes, the measurement that is used for designing a code is the minimum Hamming distance. In a set of words, the **minimum Hamming distance** is the smallest Hamming distance between all possible pairs. We use d_{min} to define the minimum Hamming distance in a coding scheme. To find this value, we find the Hamming distances between all words and select the smallest one.

The minimum Hamming distance is the smallest Hamming distance between all possible pairs in a set of words.

Example 10.5

Find the minimum Hamming distance of the coding scheme in Table 10.1.

Solution

We first find all Hamming distances.

$d(000, 011) = 2$ $d(000, 101) = 2$ $d(000, 110) = 2$ $d(011, 101) = 2$
$d(011, 110) = 2$ $d(101, 110) = 2$

The d_{min} in this case is 2.

Example 10.6

Find the minimum Hamming distance of the coding scheme in Table 10.2.

Solution

We first find all the Hamming distances.

$d(00000, 01011) = 3$ $d(00000, 10101) = 3$ $d(00000, 11110) = 4$
$d(01011, 10101) = 4$ $d(01011, 11110) = 3$ $d(10101, 11110) = 3$

The d_{min} in this case is 3.

Three Parameters

Before we continue with our discussion, we need to mention that any coding scheme needs to have at least three parameters: the codeword size n, the dataword size k, and the minimum Hamming distance d_{min}. A coding scheme C is written as $C(n, k)$ with a separate expression for d_{min}. For example, we can call our first coding scheme $C(3, 2)$ with $d_{min} = 2$ and our second coding scheme $C(5, 2)$ with $d_{min} = 3$.

Hamming Distance and Error

Before we explore the criteria for error detection or correction, let us discuss the relationship between the Hamming distance and errors occurring during transmission. When a codeword is corrupted during transmission, the Hamming distance between the sent and received codewords is the number of bits affected by the error. In other words, the Hamming distance between the received codeword and the sent codeword is the number of bits that are corrupted during transmission. For example, if the codeword 00000 is sent and 01101 is received, 3 bits are in error and the Hamming distance between the two is $d(00000, 01101) = 3$.

Minimum Distance for Error Detection

Now let us find the minimum Hamming distance in a code if we want to be able to detect up to s errors. If s errors occur during transmission, the Hamming distance between the sent codeword and received codeword is s. If our code is to detect up to s errors, the minimum distance between the valid codes must be $s + 1$, so that the received codeword does not match a valid codeword. In other words, if the minimum distance between all valid codewords is $s + 1$, the received codeword cannot be erroneously mistaken for another codeword. The distances are not enough $(s + 1)$ for the receiver to accept it as valid. The error will be detected. We need to clarify a point here: Although a code with $d_{min} = s + 1$

may be able to detect more than s errors in some special cases, only s or fewer errors are guaranteed to be detected.

To guarantee the detection of up to s errors in all cases, the minimum Hamming distance in a block code must be $d_{min} = s + 1$.

Example 10.7

The minimum Hamming distance for our first code scheme (Table 10.1) is 2. This code guarantees detection of only a single error. For example, if the third codeword (101) is sent and one error occurs, the received codeword does not match any valid codeword. If two errors occur, however, the received codeword may match a valid codeword and the errors are not detected.

Example 10.8

Our second block code scheme (Table 10.2) has $d_{min} = 3$. This code can detect up to two errors. Again, we see that when any of the valid codewords is sent, two errors create a codeword which is not in the table of valid codewords. The receiver cannot be fooled. However, some combinations of three errors change a valid codeword to another valid codeword. The receiver accepts the received codeword and the errors are undetected.

We can look at this geometrically. Let us assume that the sent codeword x is at the center of a circle with radius s. All other received codewords that are created by 1 to s errors are points inside the circle or on the perimeter of the circle. All other valid codewords must be outside the circle, as shown in Figure 10.8.

Figure 10.8 *Geometric concept for finding d_{min} in error detection*

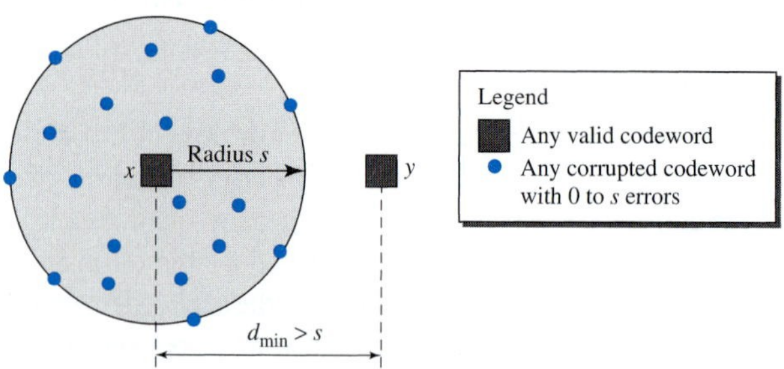

In Figure 10.8, d_{min} must be an integer greater than s; that is, $d_{min} = s + 1$.

Minimum Distance for Error Correction

Error correction is more complex than error detection; a decision is involved. When a received codeword is not a valid codeword, the receiver needs to decide which valid codeword was actually sent. The decision is based on the concept of territory, an exclusive area surrounding the codeword. Each valid codeword has its own territory.

We use a geometric approach to define each territory. We assume that each valid codeword has a circular territory with a radius of t and that the valid codeword is at the center. For example, suppose a codeword x is corrupted by t bits or less. Then this corrupted codeword is located either inside or on the perimeter of this circle. If the receiver receives a codeword that belongs to this territory, it decides that the original codeword is the one at the center. Note that we assume that only up to t errors have occurred; otherwise, the decision is wrong. Figure 10.9 shows this geometric interpretation. Some texts use a sphere to show the distance between all valid block codes.

Figure 10.9 *Geometric concept for finding d_{min} in error correction*

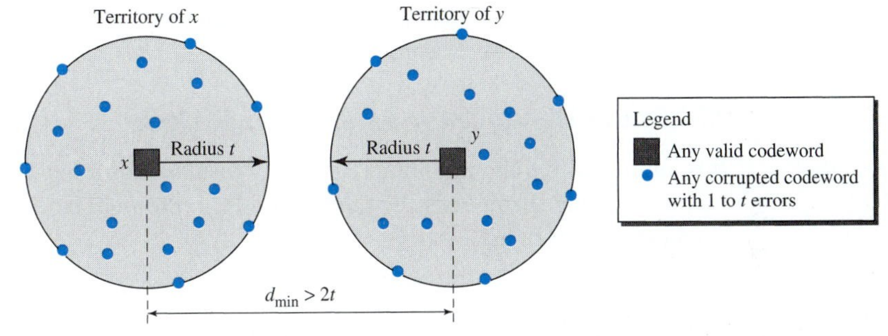

In Figure 10.9, $d_{\text{min}} > 2t$; since the next integer increment is 1, we can say that $d_{\text{min}} = 2t + 1$.

To guarantee correction of up to t errors in all cases, the minimum Hamming distance in a block code must be $d_{\text{min}} = 2t + 1$.

Example 10.9

A code scheme has a Hamming distance $d_{\text{min}} = 4$. What is the error detection and correction capability of this scheme?

Solution

This code guarantees the detection of up to three errors ($s = 3$), but it can correct up to one error. In other words, if this code is used for error correction, part of its capability is wasted. Error correction codes need to have an odd minimum distance (3, 5, 7, . . .).

10.3 LINEAR BLOCK CODES

Almost all block codes used today belong to a subset called **linear block codes.** The use of nonlinear block codes for error detection and correction is not as widespread because their structure makes theoretical analysis and implementation difficult. We therefore concentrate on linear block codes.

The formal definition of linear block codes requires the knowledge of abstract algebra (particularly Galois fields), which is beyond the scope of this book. We therefore give an informal definition. For our purposes, a linear block code is a code in which the exclusive OR (addition modulo-2) of two valid codewords creates another valid codeword.

In a linear block code, the exclusive OR (XOR) of any two valid codewords creates another valid codeword.

Example 10.10

Let us see if the two codes we defined in Table 10.1 and Table 10.2 belong to the class of linear block codes.

1. The scheme in Table 10.1 is a linear block code because the result of XORing any codeword with any other codeword is a valid codeword. For example, the XORing of the second and third codewords creates the fourth one.
2. The scheme in Table 10.2 is also a linear block code. We can create all four codewords by XORing two other codewords.

Minimum Distance for Linear Block Codes

It is simple to find the minimum Hamming distance for a linear block code. The minimum Hamming distance is the number of 1s in the nonzero valid codeword with the smallest number of 1s.

Example 10.11

In our first code (Table 10.1), the numbers of 1s in the nonzero codewords are 2, 2, and 2. So the minimum Hamming distance is $d_{min} = 2$. In our second code (Table 10.2), the numbers of 1s in the nonzero codewords are 3, 3, and 4. So in this code we have $d_{min} = 3$.

Some Linear Block Codes

Let us now show some linear block codes. These codes are trivial because we can easily find the encoding and decoding algorithms and check their performances.

Simple Parity-Check Code

Perhaps the most familiar error-detecting code is the **simple parity-check code.** In this code, a k-bit dataword is changed to an n-bit codeword where $n = k + 1$. The extra bit, called the parity bit, is selected to make the total number of 1s in the codeword even. Although some implementations specify an odd number of 1s, we discuss the even case. The minimum Hamming distance for this category is $d_{min} = 2$, which means that the code is a single-bit error-detecting code; it cannot correct any error.

A simple parity-check code is a single-bit error-detecting code in which $n = k + 1$ with $d_{min} = 2$.

Our first code (Table 10.1) is a parity-check code with $k = 2$ and $n = 3$. The code in Table 10.3 is also a parity-check code with $k = 4$ and $n = 5$.

Figure 10.10 shows a possible structure of an encoder (at the sender) and a decoder (at the receiver).

The encoder uses a generator that takes a copy of a 4-bit dataword (a_0, a_1, a_2, and a_3) and generates a parity bit r_0. The dataword bits and the **parity bit** create the 5-bit codeword. The parity bit that is added makes the number of 1s in the codeword even.

Table 10.3 *Simple parity-check code C(5, 4)*

Datawords	*Codewords*	*Datawords*	*Codewords*
0000	00000	1000	10001
0001	00011	1001	10010
0010	00101	1010	10100
0011	00110	1011	10111
0100	01001	1100	11000
0101	01010	1101	11011
0110	01100	1110	11101
0111	01111	1111	11110

Figure 10.10 *Encoder and decoder for simple parity-check code*

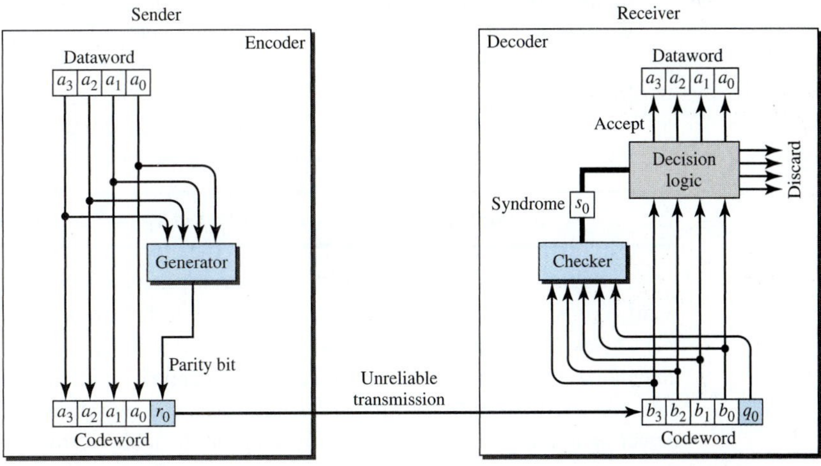

This is normally done by adding the 4 bits of the dataword (modulo-2); the result is the parity bit. In other words,

$$r_0 = a_3 + a_2 + a_1 + a_0 \qquad \text{(modulo-2)}$$

If the number of 1s is even, the result is 0; if the number of 1s is odd, the result is 1. In both cases, the total number of 1s in the codeword is even.

The sender sends the codeword which may be corrupted during transmission. The receiver receives a 5-bit word. The checker at the receiver does the same thing as the generator in the sender with one exception: The addition is done over all 5 bits. The result, which is called the **syndrome,** is just 1 bit. The syndrome is 0 when the number of 1s in the received codeword is even; otherwise, it is 1.

$$s_0 = b_3 + b_2 + b_1 + b_0 + q_0 \qquad \text{(modulo-2)}$$

The syndrome is passed to the decision logic analyzer. If the syndrome is 0, there is no error in the received codeword; the data portion of the received codeword is accepted as the dataword; if the syndrome is 1, the data portion of the received codeword is discarded. The dataword is not created.

Example 10.12

Let us look at some transmission scenarios. Assume the sender sends the dataword 1011. The codeword created from this dataword is 10111, which is sent to the receiver. We examine five cases:

1. No error occurs; the received codeword is 10111. The syndrome is 0. The dataword 1011 is created.
2. One single-bit error changes a_1. The received codeword is 10011. The syndrome is 1. No dataword is created.
3. One single-bit error changes r_0. The received codeword is 10110. The syndrome is 1. No dataword is created. Note that although none of the dataword bits are corrupted, no dataword is created because the code is not sophisticated enough to show the position of the corrupted bit.
4. An error changes r_0 and a second error changes a_3. The received codeword is 00110. The syndrome is 0. The dataword 0011 is created at the receiver. Note that here the dataword is wrongly created due to the syndrome value. The simple parity-check decoder cannot detect an even number of errors. The errors cancel each other out and give the syndrome a value of 0.
5. Three bits—a_3, a_2, and a_1—are changed by errors. The received codeword is 01011. The syndrome is 1. The dataword is not created. This shows that the simple parity check, guaranteed to detect one single error, can also find any odd number of errors.

A simple parity-check code can detect an odd number of errors.

A better approach is the **two-dimensional parity check.** In this method, the dataword is organized in a table (rows and columns). In Figure 10.11, the data to be sent, five 7-bit bytes, are put in separate rows. For each row and each column, 1 parity-check bit is calculated. The whole table is then sent to the receiver, which finds the syndrome for each row and each column. As Figure 10.11 shows, the two-dimensional parity check can detect up to three errors that occur anywhere in the table (arrows point to the locations of the created nonzero syndromes). However, errors affecting 4 bits may not be detected.

Hamming Codes

Now let us discuss a category of error-correcting codes called **Hamming codes.** These codes were originally designed with $d_{min} = 3$, which means that they can detect up to two errors or correct one single error. Although there are some Hamming codes that can correct more than one error, our discussion focuses on the single-bit error-correcting code.

First let us find the relationship between n and k in a Hamming code. We need to choose an integer $m >= 3$. The values of n and k are then calculated from m as $n = 2^m - 1$ and $k = n - m$. The number of check bits $r = m$.

All Hamming codes discussed in this book have $d_{min} = 3$.
The relationship between m and n in these codes is $n = 2^m - 1$.

For example, if $m = 3$, then $n = 7$ and $k = 4$. This is a Hamming code $C(7, 4)$ with $d_{min} = 3$. Table 10.4 shows the datawords and codewords for this code.

Figure 10.11 *Two-dimensional parity-check code*

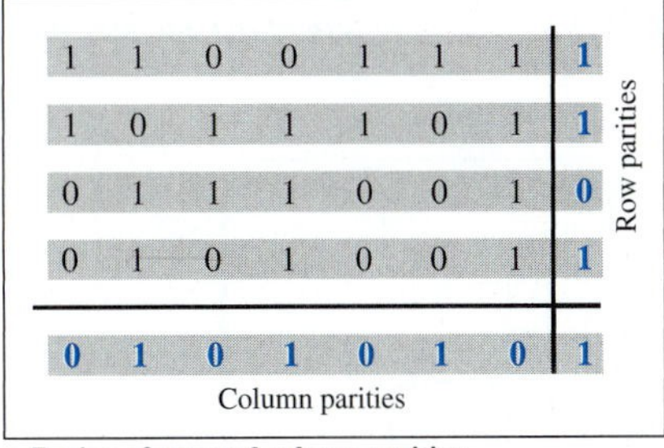

a. Design of row and column parities

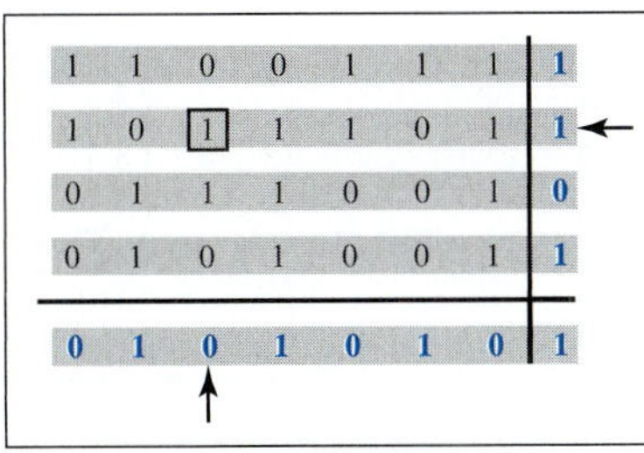

b. One error affects two parities

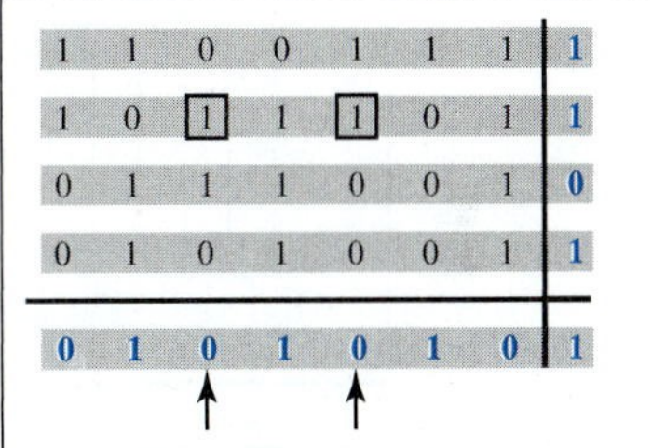

c. Two errors affect two parities

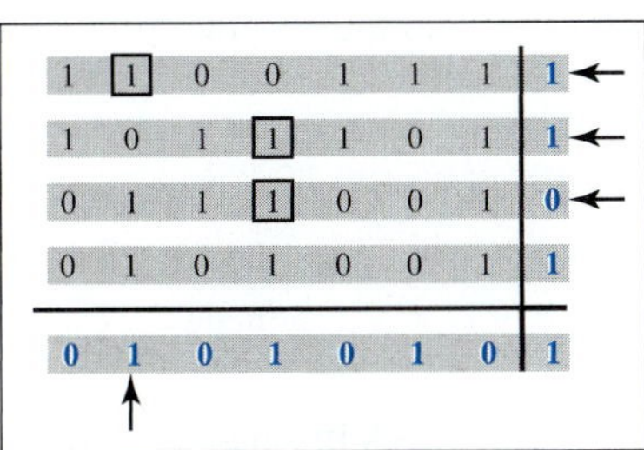

d. Three errors affect four parities

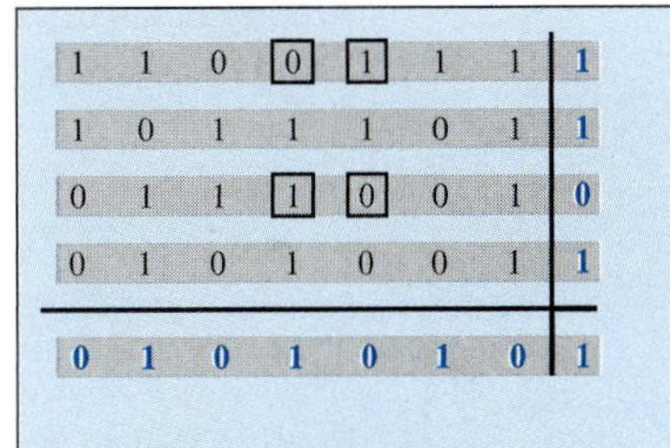

e. Four errors cannot be detected

Table 10.4 *Hamming code C(7, 4)*

Datawords	*Codewords*	*Datawords*	*Codewords*
0000	0000000	1000	1000110
0001	0001101	1001	1001011
0010	0010111	1010	1010001
0011	0011010	1011	1011100
0100	0100011	1100	1100101
0101	0101110	1101	1101000
0110	0110100	1110	1110010
0111	0111001	1111	1111111

Figure 10.12 shows the structure of the encoder and decoder for this example.

Figure 10.12 *The structure of the encoder and decoder for a Hamming code*

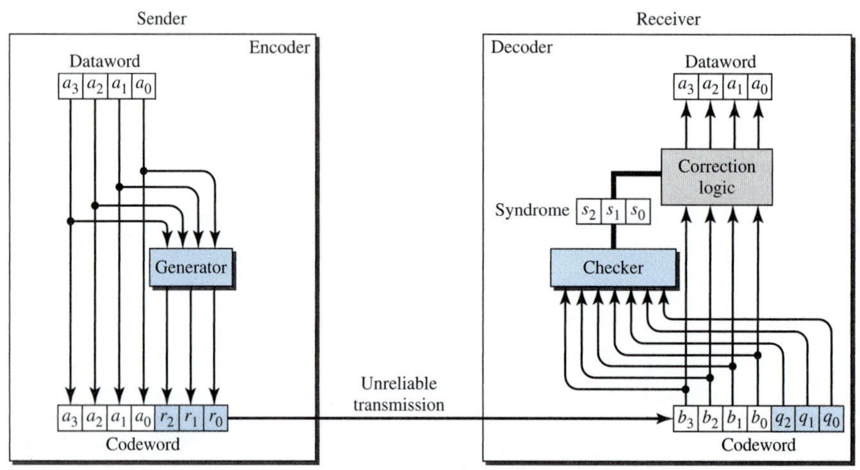

A copy of a 4-bit dataword is fed into the generator that creates three parity checks r_0, r_1, and r_2, as shown below:

$$r_0 = a_2 + a_1 + a_0 \quad \text{modulo-2}$$
$$r_1 = a_3 + a_2 + a_1 \quad \text{modulo-2}$$
$$r_2 = a_1 + a_0 + a_3 \quad \text{modulo-2}$$

In other words, each of the parity-check bits handles 3 out of the 4 bits of the dataword. The total number of 1s in each 4-bit combination (3 dataword bits and 1 parity bit) must be even. We are not saying that these three equations are unique; any three equations that involve 3 of the 4 bits in the dataword and create independent equations (a combination of two cannot create the third) are valid.

The checker in the decoder creates a 3-bit syndrome ($s_2s_1s_0$) in which each bit is the parity check for 4 out of the 7 bits in the received codeword:

$$s_0 = b_2 + b_1 + b_0 + q_0 \quad \text{modulo-2}$$
$$s_1 = b_3 + b_2 + b_1 + q_1 \quad \text{modulo-2}$$
$$s_2 = b_1 + b_0 + b_3 + q_2 \quad \text{modulo-2}$$

The equations used by the checker are the same as those used by the generator with the parity-check bits added to the right-hand side of the equation. The 3-bit syndrome creates eight different bit patterns (000 to 111) that can represent eight different conditions. These conditions define a lack of error or an error in 1 of the 7 bits of the received codeword, as shown in Table 10.5.

Table 10.5 *Logical decision made by the correction logic analyzer of the decoder*

Syndrome	000	001	010	011	100	101	110	111
Error	None	q_0	q_1	b_2	q_2	b_0	b_3	b_1

Note that the generator is not concerned with the four cases shaded in Table 10.5 because there is either no error or an error in the parity bit. In the other four cases, 1 of the bits must be flipped (changed from 0 to 1 or 1 to 0) to find the correct dataword.

The syndrome values in Table 10.5 are based on the syndrome bit calculations. For example, if q_0 is in error, s_0 is the only bit affected; the syndrome, therefore, is 001. If b_2 is in error, s_0 and s_1 are the bits affected; the syndrome, therefore is 011. Similarly, if b_1 is in error, all 3 syndrome bits are affected and the syndrome is 111.

There are two points we need to emphasize here. First, if two errors occur during transmission, the created dataword might not be the right one. Second, if we want to use the above code for error detection, we need a different design.

Example 10.13

Let us trace the path of three datawords from the sender to the destination:

1. The dataword 0100 becomes the codeword 0100011. The codeword 0100011 is received. The syndrome is 000 (no error), the final dataword is 0100.
2. The dataword 0111 becomes the codeword 0111001. The codeword 0011001 is received. The syndrome is 011. According to Table 10.5, b_2 is in error. After flipping b_2 (changing the 1 to 0), the final dataword is 0111.
3. The dataword 1101 becomes the codeword 1101000. The codeword 0001000 is received (two errors). The syndrome is 101, which means that b_0 is in error. After flipping b_0, we get 0000, the wrong dataword. This shows that our code cannot correct two errors.

Example 10.14

We need a dataword of at least 7 bits. Calculate values of k and n that satisfy this requirement.

Solution

We need to make $k = n - m$ greater than or equal to 7, or $2^m - 1 - m \geq 7$.

1. If we set $m = 3$, the result is $n = 2^3 - 1$ and $k = 7 - 3$, or 4, which is not acceptable.
2. If we set $m = 4$, then $n = 2^4 - 1 = 15$ and $k = 15 - 4 = 11$, which satisfies the condition. So the code is $C(15, 11)$. There are methods to make the dataword a specific size, but the discussion and implementation are beyond the scope of this book.

Performance

A Hamming code can only correct a single error or detect a double error. However, there is a way to make it detect a burst error, as shown in Figure 10.13.

The key is to split a burst error between several codewords, one error for each codeword. In data communications, we normally send a packet or a frame of data. To make the Hamming code respond to a burst error of size N, we need to make N codewords out of our frame. Then, instead of sending one codeword at a time, we arrange the codewords in a table and send the bits in the table a column at a time. In Figure 10.13, the bits are sent column by column (from the left). In each column, the bits are sent from the bottom to the top. In this way, a frame is made out of the four codewords and sent to the receiver. Figure 10.13 shows

Figure 10.13 *Burst error correction using Hamming code*

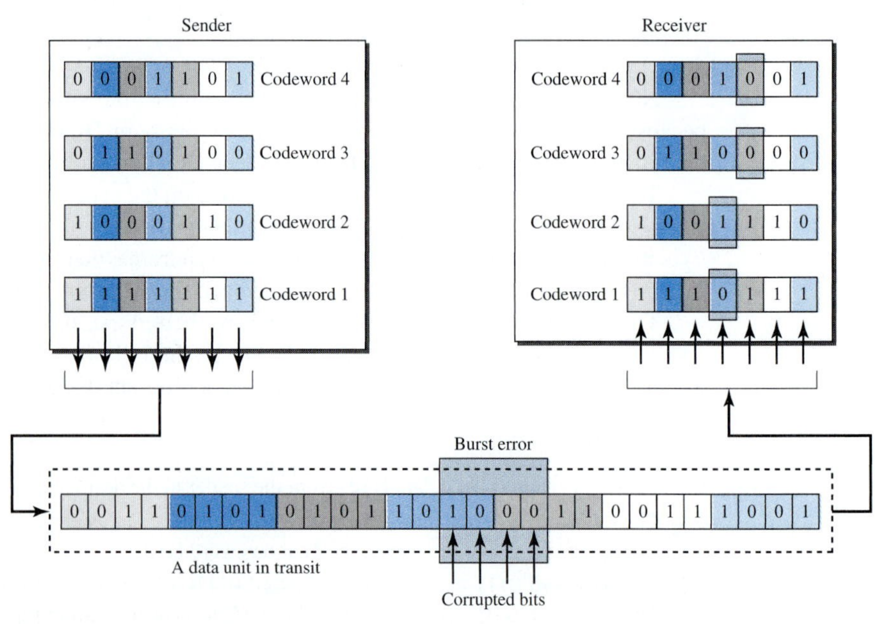

that when a burst error of size 4 corrupts the frame, only 1 bit from each codeword is corrupted. The corrupted bit in each codeword can then easily be corrected at the receiver.

10.4 CYCLIC CODES

Cyclic codes are special linear block codes with one extra property. In a **cyclic code,** if a codeword is cyclically shifted (rotated), the result is another codeword. For example, if 1011000 is a codeword and we cyclically left-shift, then 0110001 is also a codeword. In this case, if we call the bits in the first word a_0 to a_6, and the bits in the second word b_0 to b_6, we can shift the bits by using the following:

$$b_1 = a_0 \quad b_2 = a_1 \quad b_3 = a_2 \quad b_4 = a_3 \quad b_5 = a_4 \quad b_6 = a_5 \quad b_0 = a_6$$

In the rightmost equation, the last bit of the first word is wrapped around and becomes the first bit of the second word.

Cyclic Redundancy Check

We can create cyclic codes to correct errors. However, the theoretical background required is beyond the scope of this book. In this section, we simply discuss a category of cyclic codes called the **cyclic redundancy check (CRC)** that is used in networks such as LANs and WANs.

Table 10.6 shows an example of a CRC code. We can see both the linear and cyclic properties of this code.

Table 10.6 *A CRC code with C(7, 4)*

Dataword	*Codeword*	*Dataword*	*Codeword*
0000	0000000	1000	1000101
0001	0001011	1001	1001110
0010	0010110	1010	1010011
0011	0011101	1011	1011000
0100	0100111	1100	1100010
0101	0101100	1101	1101001
0110	0110001	1110	1110100
0111	0111010	1111	1111111

Figure 10.14 shows one possible design for the encoder and decoder.

Figure 10.14 *CRC encoder and decoder*

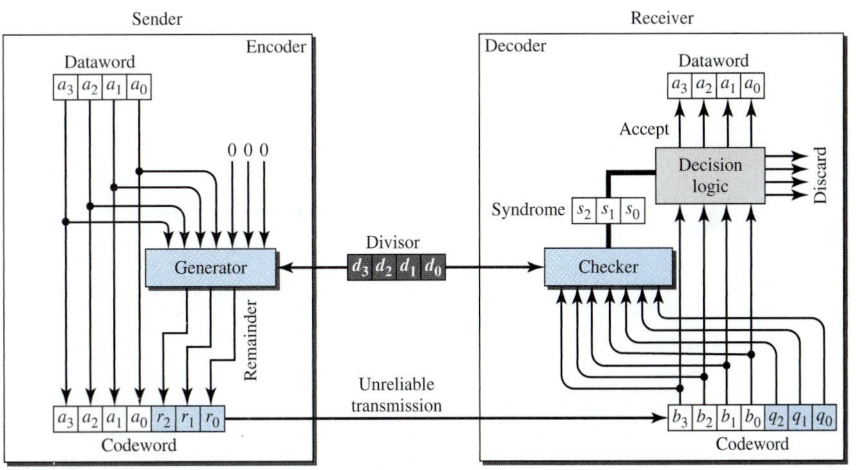

In the encoder, the dataword has k bits (4 here); the codeword has n bits (7 here). The size of the dataword is augmented by adding $n - k$ (3 here) 0s to the right-hand side of the word. The n-bit result is fed into the generator. The generator uses a divisor of size $n - k + 1$ (4 here), predefined and agreed upon. The generator divides the augmented dataword by the divisor (modulo-2 division). The quotient of the division is discarded; the remainder ($r_2r_1r_0$) is appended to the dataword to create the codeword.

The decoder receives the possibly corrupted codeword. A copy of all n bits is fed to the checker which is a replica of the generator. The remainder produced by the checker is a syndrome of $n - k$ (3 here) bits, which is fed to the decision logic analyzer. The analyzer has a simple function. If the syndrome bits are all 0s, the 4 leftmost bits of the codeword are accepted as the dataword (interpreted as no error); otherwise, the 4 bits are discarded (error).

Encoder

Let us take a closer look at the encoder. The encoder takes the dataword and augments it with $n - k$ number of 0s. It then divides the augmented dataword by the divisor, as shown in Figure 10.15.

Figure 10.15 *Division in CRC encoder*

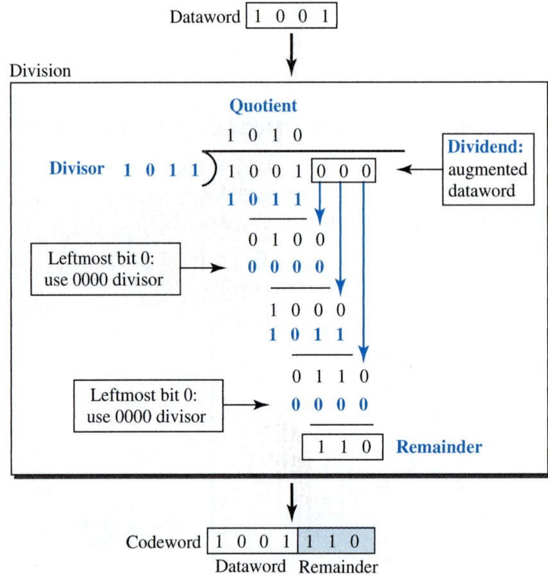

The process of modulo-2 binary division is the same as the familiar division process we use for decimal numbers. However, as mentioned at the beginning of the chapter, in this case addition and subtraction are the same. We use the XOR operation to do both.

As in decimal division, the process is done step by step. In each step, a copy of the divisor is XORed with the 4 bits of the dividend. The result of the XOR operation (remainder) is 3 bits (in this case), which is used for the next step after 1 extra bit is pulled down to make it 4 bits long. There is one important point we need to remember in this type of division. If the leftmost bit of the dividend (or the part used in each step) is 0, the step cannot use the regular divisor; we need to use an all-0s divisor.

When there are no bits left to pull down, we have a result. The 3-bit remainder forms the check bits (r_2, r_1, and r_0). They are appended to the dataword to create the codeword.

Decoder

The codeword can change during transmission. The decoder does the same division process as the encoder. The remainder of the division is the syndrome. If the syndrome is all 0s, there is no error; the dataword is separated from the received codeword and accepted. Otherwise, everything is discarded. Figure 10.16 shows two cases: The left-hand figure shows the value of syndrome when no error has occurred; the syndrome is 000. The right-hand part of the figure shows the case in which there is one single error. The syndrome is not all 0s (it is 011).

Figure 10.16 *Division in the CRC decoder for two cases*

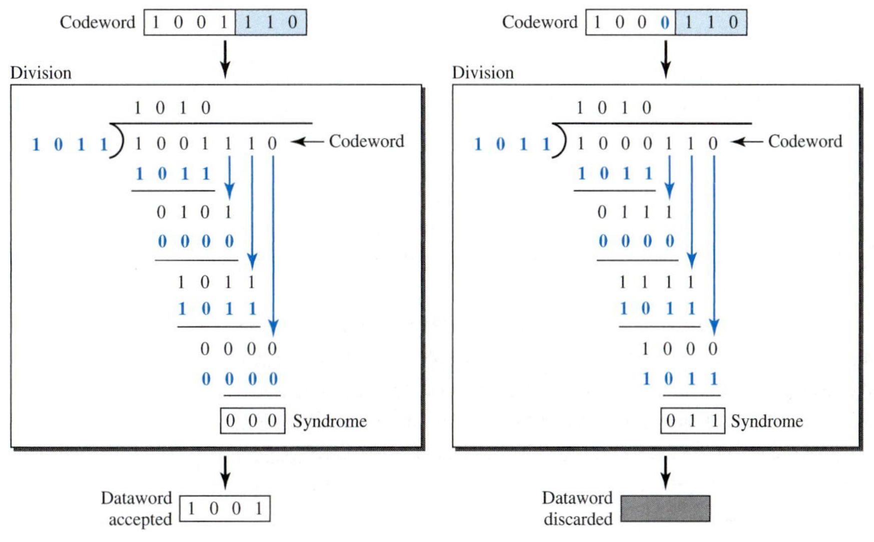

Divisor

You may be wondering how the divisor 1011 is chosen. Later in the chapter we present some criteria, but in general it involves abstract algebra.

Hardware Implementation

One of the advantages of a cyclic code is that the encoder and decoder can easily and cheaply be implemented in hardware by using a handful of electronic devices. Also, a hardware implementation increases the rate of check bit and syndrome bit calculation. In this section, we try to show, step by step, the process. The section, however, is optional and does not affect the understanding of the rest of the chapter.

Divisor

Let us first consider the divisor. We need to note the following points:

1. The divisor is repeatedly XORed with part of the dividend.

2. The divisor has $n - k + 1$ bits which either are predefined or are all 0s. In other words, the bits do not change from one dataword to another. In our previous example, the divisor bits were either 1011 or 0000. The choice was based on the leftmost bit of the part of the augmented data bits that are active in the XOR operation.
3. A close look shows that only $n - k$ bits of the divisor is needed in the XOR operation. The leftmost bit is not needed because the result of the operation is always 0, no matter what the value of this bit. The reason is that the inputs to this XOR operation are either both 0s or both 1s. In our previous example, only 3 bits, not 4, is actually used in the XOR operation.

Using these points, we can make a fixed (hardwired) divisor that can be used for a cyclic code if we know the divisor pattern. Figure 10.17 shows such a design for our previous example. We have also shown the XOR devices used for the operation.

Figure 10.17 *Hardwired design of the divisor in CRC*

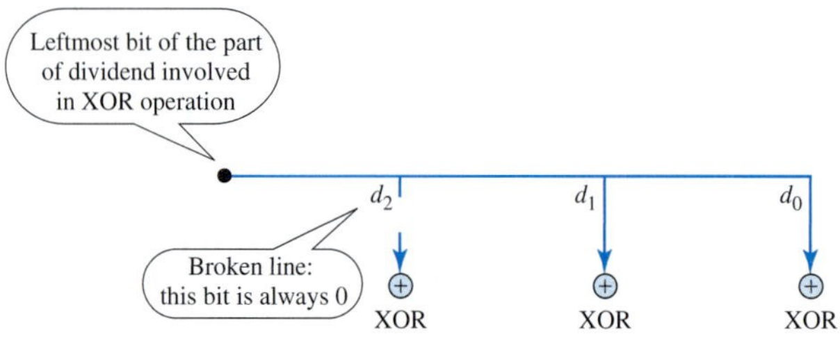

Note that if the leftmost bit of the part of dividend to be used in this step is 1, the divisor bits ($d_2d_1d_0$) are 011; if the leftmost bit is 0, the divisor bits are 000. The design provides the right choice based on the leftmost bit.

Augmented Dataword

In our paper-and-pencil division process in Figure 10.15, we show the augmented dataword as fixed in position with the divisor bits shifting to the right, 1 bit in each step. The divisor bits are aligned with the appropriate part of the augmented dataword. Now that our divisor is fixed, we need instead to shift the bits of the augmented dataword to the left (opposite direction) to align the divisor bits with the appropriate part. There is no need to store the augmented dataword bits.

Remainder

In our previous example, the remainder is 3 bits ($n - k$ bits in general) in length. We can use three **registers** (single-bit storage devices) to hold these bits. To find the final remainder of the division, we need to modify our division process. The following is the step-by-step process that can be used to simulate the division process in hardware (or even in software).

1. We assume that the remainder is originally all 0s (000 in our example).

2. At each time click (arrival of 1 bit from an augmented dataword), we repeat the following two actions:
   a. We use the leftmost bit to make a decision about the divisor (011 or 000).
   b. The other 2 bits of the remainder and the next bit from the augmented dataword (total of 3 bits) are XORed with the 3-bit divisor to create the next remainder.

Figure 10.18 shows this simulator, but note that this is not the final design; there will be more improvements.

Figure 10.18 *Simulation of division in CRC encoder*

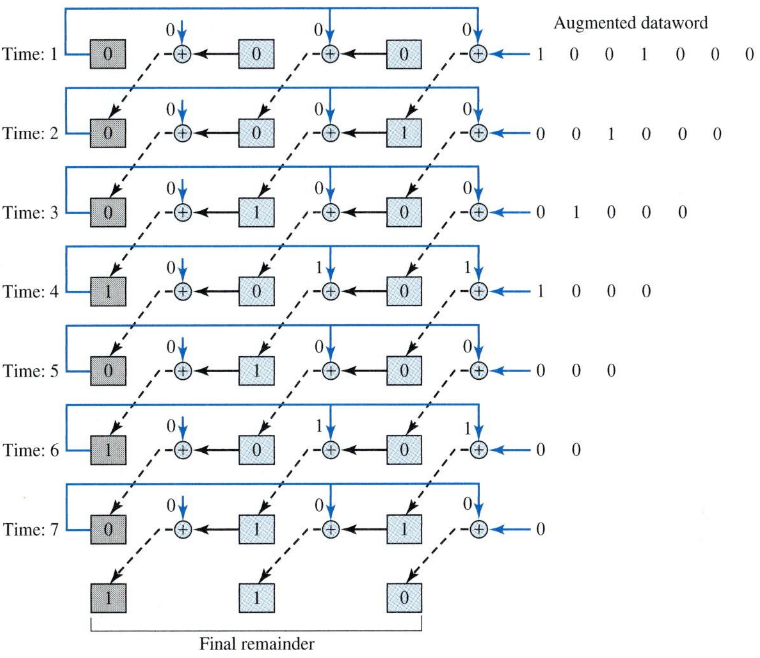

At each clock tick, shown as different times, one of the bits from the augmented dataword is used in the XOR process. If we look carefully at the design, we have seven steps here, while in the paper-and-pencil method we had only four steps. The first three steps have been added here to make each step equal and to make the design for each step the same. Steps 1, 2, and 3 push the first 3 bits to the remainder registers; steps 4, 5, 6, and 7 match the paper-and-pencil design. Note that the values in the remainder register in steps 4 to 7 exactly match the values in the paper-and-pencil design. The final remainder is also the same.

The above design is for demonstration purposes only. It needs simplification to be practical. First, we do not need to keep the intermediate values of the remainder bits; we need only the final bits. We therefore need only 3 registers instead of 24. After the XOR operations, we do not need the bit values of the previous remainder. Also, we do not need 21 XOR devices; two are enough because the output of an XOR operation in which one of the bits is 0 is simply the value of the other bit. This other bit can be used as the output. With these two modifications, the design becomes tremendously simpler and less expensive, as shown in Figure 10.19.

Figure 10.19 *The CRC encoder design using shift registers*

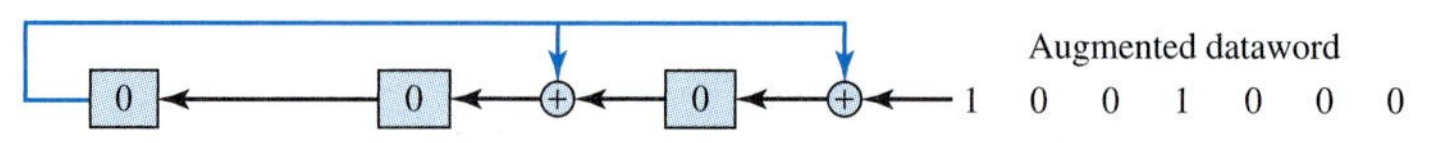

We need, however, to make the registers shift registers. A 1-bit shift register holds a bit for a duration of one clock time. At a time click, the shift register accepts the bit at its input port, stores the new bit, and displays it on the output port. The content and the output remain the same until the next input arrives. When we connect several 1-bit shift registers together, it looks as if the contents of the register are shifting.

General Design

A general design for the encoder and decoder is shown in Figure 10.20.

Figure 10.20 *General design of encoder and decoder of a CRC code*

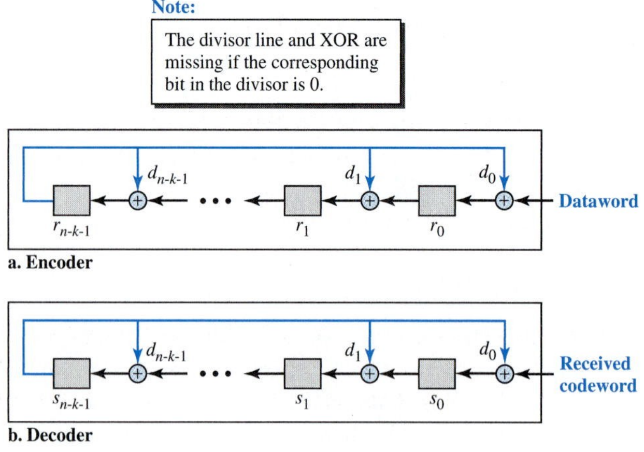

Note that we have $n - k$ 1-bit shift registers in both the encoder and decoder. We have up to $n - k$ XOR devices, but the divisors normally have several 0s in their pattern, which reduces the number of devices. Also note that, instead of augmented datawords, we show the dataword itself as the input because after the bits in the dataword are all fed into the encoder, the extra bits, which all are 0s, do not have any effect on the right-most XOR. Of course, the process needs to be continued for another $n - k$ steps before the check bits are ready. This fact is one of the criticisms of this design. Better schemes have been designed to eliminate this waiting time (the check bits are ready after k steps), but we leave this as a research topic for the reader. In the decoder, however, the entire codeword must be fed to the decoder before the syndrome is ready.

Polynomials

A better way to understand cyclic codes and how they can be analyzed is to represent them as polynomials. Again, this section is optional.

A pattern of 0s and 1s can be represented as a **polynomial** with coefficients of 0 and 1. The power of each term shows the position of the bit; the coefficient shows the value of the bit. Figure 10.21 shows a binary pattern and its polynomial representation. In Figure 10.21a we show how to translate a binary pattern to a polynomial; in Figure 10.21b we show how the polynomial can be shortened by removing all terms with zero coefficients and replacing x^1 by x and x^0 by 1.

Figure 10.21 *A polynomial to represent a binary word*

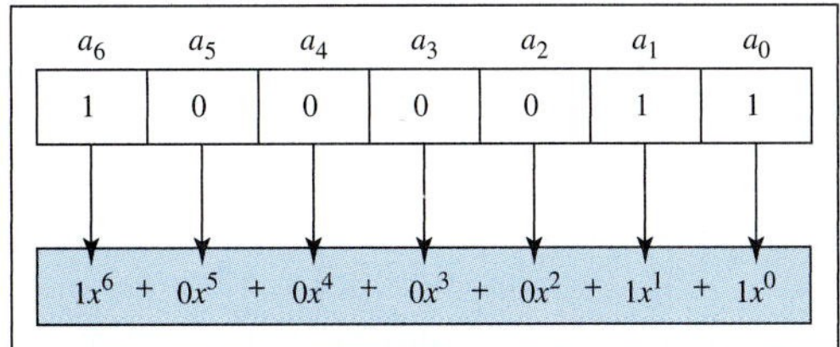

a. Binary pattern and polynomial

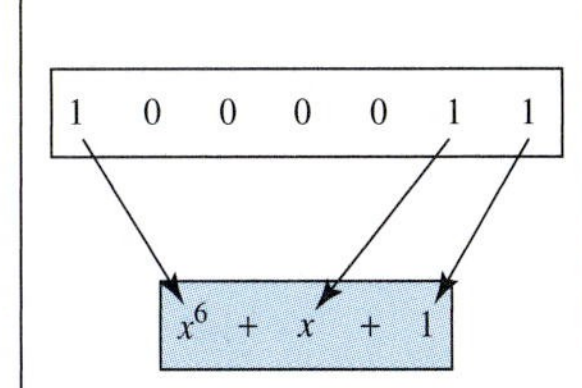

b. Short form

Figure 10.21 shows one immediate benefit; a 7-bit pattern can be replaced by three terms. The benefit is even more conspicuous when we have a polynomial such as $x^{23} + x^3 + 1$. Here the bit pattern is 24 bits in length (three 1s and twenty-one 0s) while the polynomial is just three terms.

Degree of a Polynomial

The degree of a polynomial is the highest power in the polynomial. For example, the degree of the polynomial $x^6 + x + 1$ is 6. Note that the degree of a polynomial is 1 less that the number of bits in the pattern. The bit pattern in this case has 7 bits.

Adding and Subtracting Polynomials

Adding and subtracting polynomials in mathematics are done by adding or subtracting the coefficients of terms with the same power. In our case, the coefficients are only 0 and 1, and adding is in modulo-2. This has two consequences. First, addition and subtraction are the same. Second, adding or subtracting is done by combining terms and deleting pairs of identical terms. For example, adding $x^5 + x^4 + x^2$ and $x^6 + x^4 + x^2$ gives just $x^6 + x^5$. The terms x^4 and x^2 are deleted. However, note that if we add, for example, three polynomials and we get x^2 three times, we delete a pair of them and keep the third.

Multiplying or Dividing Terms

In this arithmetic, multiplying a term by another term is very simple; we just add the powers. For example, $x^3 \times x^4$ is x^7. For dividing, we just subtract the power of the second term from the power of the first. For example, x^5/x^2 is x^3.

Multiplying Two Polynomials

Multiplying a polynomial by another is done term by term. Each term of the first polynomial must be multiplied by all terms of the second. The result, of course, is then simplified, and pairs of equal terms are deleted. The following is an example:

$$\begin{aligned}
&(x^5 + x^3 + x^2 + x)(x^2 + x + 1)\\
&\quad = x^7 + x^6 + x^5 + x^5 + x^4 + x^3 + x^4 + x^3 + x^2 + x^3 + x^2 + x\\
&\quad = x^7 + x^6 + x^3 + x
\end{aligned}$$

Dividing One Polynomial by Another

Division of polynomials is conceptually the same as the binary division we discussed for an encoder. We divide the first term of the dividend by the first term of the divisor to get the first term of the quotient. We multiply the term in the quotient by the divisor and subtract the result from the dividend. We repeat the process until the dividend degree is less than the divisor degree. We will show an example of division later in this chapter.

Shifting

A binary pattern is often shifted a number of bits to the right or left. Shifting to the left means adding extra 0s as rightmost bits; shifting to the right means deleting some rightmost bits. Shifting to the left is accomplished by multiplying each term of the polynomial by x^m, where m is the number of shifted bits; shifting to the right is accomplished by dividing each term of the polynomial by x^m. The following shows shifting to the left and to the right. Note that we do not have negative powers in the polynomial representation.

Shifting left 3 bits:	10011 becomes 10011000	$x^4 + x + 1$ becomes $x^7 + x^4 + x^3$
Shifting right 3 bits:	10011 becomes 10	$x^4 + x + 1$ becomes x

When we augmented the dataword in the encoder of Figure 10.15, we actually shifted the bits to the left. Also note that when we concatenate two bit patterns, we shift the first polynomial to the left and then add the second polynomial.

Cyclic Code Encoder Using Polynomials

Now that we have discussed operations on polynomials, we show the creation of a codeword from a dataword. Figure 10.22 is the polynomial version of Figure 10.15. We can see that the process is shorter. The dataword 1001 is represented as $x^3 + 1$. The divisor 1011 is represented as $x^3 + x + 1$. To find the augmented dataword, we have left-shifted the dataword 3 bits (multiplying by x^3). The result is $x^6 + x^3$. Division is straightforward. We divide the first term of the dividend, x^6, by the first term of the divisor, x^3. The first term of the quotient is then x^6/x^3, or x^3. Then we multiply x^3 by the divisor and subtract (according to our previous definition of subtraction) the result from the dividend. The result is x^4, with a degree greater than the divisor's degree; we continue to divide until the degree of the remainder is less than the degree of the divisor.

Figure 10.22 *CRC division using polynomials*

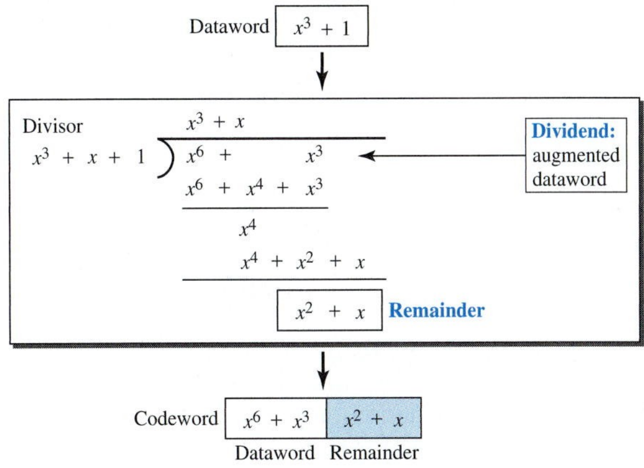

It can be seen that the polynomial representation can easily simplify the operation of division in this case, because the two steps involving all-0s divisors are not needed here. (Of course, one could argue that the all-0s divisor step can also be eliminated in binary division.) In a polynomial representation, the divisor is normally referred to as the **generator polynomial** $t(x)$.

The divisor in a cyclic code is normally called the generator polynomial or simply the generator.

Cyclic Code Analysis

We can analyze a cyclic code to find its capabilities by using polynomials. We define the following, where $f(x)$ is a polynomial with binary coefficients.

Dataword: $d(x)$	Codeword: $c(x)$	Generator: $g(x)$
Syndrome: $s(x)$	Error: $e(x)$	

If $s(x)$ is not zero, then one or more bits is corrupted. However, if $s(x)$ is zero, either no bit is corrupted or the decoder failed to detect any errors.

In a cyclic code,

1. If $s(x) \neq 0$, one or more bits is corrupted.
2. If $s(x) = 0$, either
   a. No bit is corrupted. or
   b. Some bits are corrupted, but the decoder failed to detect them.

In our analysis we want to find the criteria that must be imposed on the generator, $g(x)$ to detect the type of error we especially want to be detected. Let us first find the relationship among the sent codeword, error, received codeword, and the generator. We can say

$$\text{Received codeword} = c(x) + e(x)$$

In other words, the received codeword is the sum of the sent codeword and the error. The receiver divides the received codeword by $g(x)$ to get the syndrome. We can write this as

$$\frac{\text{Received codeword}}{g(x)} = \frac{c(x)}{g(x)} + \frac{e(x)}{g(x)}$$

The first term at the right-hand side of the equality does not have a remainder (according to the definition of codeword). So the syndrome is actually the remainder of the second term on the right-hand side. If this term does not have a remainder (syndrome = 0), either $e(x)$ is 0 or $e(x)$ is divisible by $g(x)$. We do not have to worry about the first case (there is no error); the second case is very important. Those errors that are divisible by $g(x)$ are not caught.

In a cyclic code, those $e(x)$ errors that are divisible by $g(x)$ are not caught.

Let us show some specific errors and see how they can be caught by a well-designed $g(x)$.

Single-Bit Error

What should be the structure of $g(x)$ to guarantee the detection of a single-bit error? A single-bit error is $e(x) = x^i$, where i is the position of the bit. If a single-bit error is caught, then x^i is not divisible by $g(x)$. (Note that when we say *not divisible,* we mean that there is a remainder.) If $g(x)$ has at least two terms (which is normally the case) and the coefficient of x^0 is not zero (the rightmost bit is 1), then $e(x)$ cannot be divided by $g(x)$.

If the generator has more than one term and the coefficient of x^0 is 1, all single errors can be caught.

Example 10.15

Which of the following $g(x)$ values guarantees that a single-bit error is caught? For each case, what is the error that cannot be caught?

a. $x + 1$

b. x^3

c. 1

Solution

a. No x^i can be divisible by $x + 1$. In other words, $x^i/(x + 1)$ always has a remainder. So the syndrome is nonzero. Any single-bit error can be caught.

b. If i is equal to or greater than 3, x^i is divisible by $g(x)$. The remainder of x^i/x^3 is zero, and the receiver is fooled into believing that there is no error, although there might be one. Note that in this case, the corrupted bit must be in position 4 or above. All single-bit errors in positions 1 to 3 are caught.

c. All values of i make x^i divisible by $g(x)$. No single-bit error can be caught. In addition, this $g(x)$ is useless because it means the codeword is just the dataword augmented with $n - k$ zeros.

Two Isolated Single-Bit Errors

Now imagine there are two single-bit isolated errors. Under what conditions can this type of error be caught? We can show this type of error as $e(x) = x^j + x^i$. The values of i and j define the positions of the errors, and the difference $j - i$ defines the distance between the two errors, as shown in Figure 10.23.

Figure 10.23 *Representation of two isolated single-bit errors using polynomials*

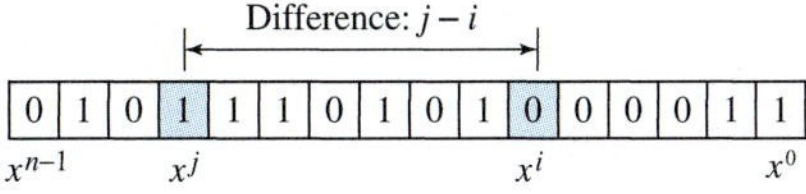

We can write $e(x) = x^i(x^{j-i} + 1)$. If $g(x)$ has more than one term and one term is x^0, it cannot divide x^i, as we saw in the previous section. So if $g(x)$ is to divide $e(x)$, it must divide $x^{j-i} + 1$. In other words, $g(x)$ must not divide $x^t + 1$, where t is between 0 and $n - 1$. However, $t = 0$ is meaningless and $t = 1$ is needed as we will see later. This means t should be between 2 and $n - 1$.

If a generator cannot divide $x^t + 1$ (t between 0 and $n - 1$), then all isolated double errors can be detected.

Example 10.16

Find the status of the following generators related to two isolated, single-bit errors.

a. $x + 1$

b. $x^4 + 1$

c. $x^7 + x^6 + 1$

d. $x^{15} + x^{14} + 1$

Solution

a. This is a very poor choice for a generator. Any two errors next to each other cannot be detected.

b. This generator cannot detect two errors that are four positions apart. The two errors can be anywhere, but if their distance is 4, they remain undetected.

c. This is a good choice for this purpose.

d. This polynomial cannot divide any error of type $x^t + 1$ if t is less than 32,768. This means that a codeword with two isolated errors that are next to each other or up to 32,768 bits apart can be detected by this generator.

Odd Numbers of Errors

A generator with a factor of $x + 1$ can catch all odd numbers of errors. This means that we need to make $x + 1$ a factor of any generator. Note that we are not saying that the generator itself should be $x + 1$; we are saying that it should have a factor of $x + 1$. If it is only $x + 1$, it cannot catch the two adjacent isolated errors (see the previous section). For example, $x^4 + x^2 + x + 1$ can catch all odd-numbered errors since it can be written as a product of the two polynomials $x + 1$ and $x^3 + x^2 + 1$.

A generator that contains a factor of $x + 1$ can detect all odd-numbered errors.

Burst Errors

Now let us extend our analysis to the burst error, which is the most important of all. A burst error is of the form $e(x) = (x^j + \cdots + x^i)$. Note the difference between a burst error and two isolated single-bit errors. The first can have two terms or more; the second can only have two terms. We can factor out x^i and write the error as $x^i(x^{j-i} + \cdots + 1)$. If our generator can detect a single error (minimum condition for a generator), then it cannot divide x^i. What we should worry about are those generators that divide $x^{j-i} + \cdots + 1$. In other words, the remainder of $(x^{j-i} + \cdots + 1)/(x^r + \cdots + 1)$ must not be zero. Note that the denominator is the generator polynomial. We can have three cases:

1. If $j - i < r$, the remainder can never be zero. We can write $j - i = L - 1$, where L is the length of the error. So $L - 1 < r$ or $L < r + 1$ or $L \leq r$. This means all burst errors with length smaller than or equal to the number of check bits r will be detected.
2. In some rare cases, if $j - i = r$, or $L = r + 1$, the syndrome is 0 and the error is undetected. It can be proved that in these cases, the probability of undetected burst error of length $r + 1$ is $(1/2)^{r-1}$. For example, if our generator is $x^{14} + x^3 + 1$, in which $r = 14$, a burst error of length $L = 15$ can slip by undetected with the probability of $(1/2)^{14-1}$ or almost 1 in 10,000.
3. In some rare cases, if $j - i > r$, or $L > r + 1$, the syndrome is 0 and the error is undetected. It can be proved that in these cases, the probability of undetected burst error of length greater than $r + 1$ is $(1/2)^r$. For example, if our generator is $x^{14} + x^3 + 1$, in which $r = 14$, a burst error of length greater than 15 can slip by undetected with the probability of $(1/2)^{14}$ or almost 1 in 16,000 cases.

- **All burst errors with $L \leq r$ will be detected.**
- **All burst errors with $L = r + 1$ will be detected with probability $1 - (1/2)^{r-1}$.**
- **All burst errors with $L > r + 1$ will be detected with probability $1 - (1/2)^r$.**

Example 10.17

Find the suitability of the following generators in relation to burst errors of different lengths.

a. $x^6 + 1$

b. $x^{18} + x^7 + x + 1$

c. $x^{32} + x^{23} + x^7 + 1$

Solution

a. This generator can detect all burst errors with a length less than or equal to 6 bits; 3 out of 100 burst errors with length 7 will slip by; 16 out of 1000 burst errors of length 8 or more will slip by.

b. This generator can detect all burst errors with a length less than or equal to 18 bits; 8 out of 1 million burst errors with length 19 will slip by; 4 out of 1 million burst errors of length 20 or more will slip by.

c. This generator can detect all burst errors with a length less than or equal to 32 bits; 5 out of 10 billion burst errors with length 33 will slip by; 3 out of 10 billion burst errors of length 34 or more will slip by.

Summary

We can summarize the criteria for a good polynomial generator:

A good polynomial generator needs to have the following characteristics:

1. **It should have at least two terms.**
2. **The coefficient of the term x^0 should be 1.**
3. **It should not divide $x^t + 1$, for t between 2 and $n - 1$.**
4. **It should have the factor $x + 1$.**

Standard Polynomials

Some standard polynomials used by popular protocols for CRC generation are shown in Table 10.7.

Table 10.7 *Standard polynomials*

Name	*Polynomial*	*Application*
CRC-8	$x^8 + x^2 + x + 1$	ATM header
CRC-10	$x^{10} + x^9 + x^5 + x^4 + x^2 + 1$	ATM AAL
CRC-16	$x^{16} + x^{12} + x^5 + 1$	HDLC
CRC-32	$x^{32} + x^{26} + x^{23} + x^{22} + x^{16} + x^{12} + x^{11} + x^{10} + x^8 + x^7 + x^5 + x^4 + x^2 + x + 1$	LANs

Advantages of Cyclic Codes

We have seen that cyclic codes have a very good performance in detecting single-bit errors, double errors, an odd number of errors, and burst errors. They can easily be implemented in hardware and software. They are especially fast when implemented in hardware. This has made cyclic codes a good candidate for many networks.

Other Cyclic Codes

The cyclic codes we have discussed in this section are very simple. The check bits and syndromes can be calculated by simple algebra. There are, however, more powerful polynomials that are based on abstract algebra involving Galois fields. These are beyond the scope of this book. One of the most interesting of these codes is the **Reed-Solomon code** used today for both detection and correction.

10.5 CHECKSUM

The last error detection method we discuss here is called the **checksum.** The checksum is used in the Internet by several protocols although not at the data link layer. However, we briefly discuss it here to complete our discussion on error checking.

Like linear and cyclic codes, the checksum is based on the concept of redundancy. Several protocols still use the checksum for error detection as we will see in future chapters, although the tendency is to replace it with a CRC. This means that the CRC is also used in layers other than the data link layer.

Idea

The concept of the checksum is not difficult. Let us illustrate it with a few examples.

Example 10.18

Suppose our data is a list of five 4-bit numbers that we want to send to a destination. In addition to sending these numbers, we send the sum of the numbers. For example, if the set of numbers is (7, 11, 12, 0, 6), we send (7, 11, 12, 0, 6, **36**), where 36 is the sum of the original numbers. The receiver adds the five numbers and compares the result with the sum. If the two are the same, the receiver assumes no error, accepts the five numbers, and discards the sum. Otherwise, there is an error somewhere and the data are not accepted.

Example 10.19

We can make the job of the receiver easier if we send the negative (complement) of the sum, called the *checksum*. In this case, we send (7, 11, 12, 0, 6, **−36**). The receiver can add all the numbers received (including the checksum). If the result is 0, it assumes no error; otherwise, there is an error.

One's Complement

The previous example has one major drawback. All of our data can be written as a 4-bit word (they are less than 15) except for the checksum. One solution is to use **one's complement** arithmetic. In this arithmetic, we can represent unsigned numbers between 0 and $2^n - 1$ using only n bits.[†] If the number has more than n bits, the extra leftmost bits need to be added to the n rightmost bits (wrapping). In one's complement arithmetic, a negative number can be represented by inverting all bits (changing a 0 to a 1 and a 1 to a 0). This is the same as subtracting the number from $2^n - 1$.

Example 10.20

How can we represent the number 21 in one's complement arithmetic using only four bits?

[†]Although one's complement can represent both positive and negative numbers, we are concerned only with unsigned representation here.

Solution

The number 21 in binary is 10101 (it needs five bits). We can wrap the leftmost bit and add it to the four rightmost bits. We have (0101 + 1) = 0110 or 6.

Example 10.21

How can we represent the number −6 in one's complement arithmetic using only four bits?

Solution

In one's complement arithmetic, the negative or complement of a number is found by inverting all bits. Positive 6 is 0110; negative 6 is 1001. If we consider only unsigned numbers, this is 9. In other words, the complement of 6 is 9. Another way to find the complement of a number in one's complement arithmetic is to subtract the number from $2^n - 1$ (16 − 1 in this case).

Example 10.22

Let us redo Exercise 10.19 using one's complement arithmetic. Figure 10.24 shows the process at the sender and at the receiver. The sender initializes the checksum to 0 and adds all data items and the checksum (the checksum is considered as one data item and is shown in color). The result is 36. However, 36 cannot be expressed in 4 bits. The extra two bits are wrapped and added with the sum to create the wrapped sum value 6. In the figure, we have shown the details in binary. The sum is then complemented, resulting in the checksum value 9 (15 − 6 = 9). The sender now sends six data items to the receiver including the checksum 9. The receiver follows the same procedure as the sender. It adds all data items (including the checksum); the result is 45. The sum is wrapped and becomes 15. The wrapped sum is complemented and becomes 0. Since the value of the checksum is 0, this means that the data is not corrupted. The receiver drops the checksum and keeps the other data items. If the checksum is not zero, the entire packet is dropped.

Figure 10.24

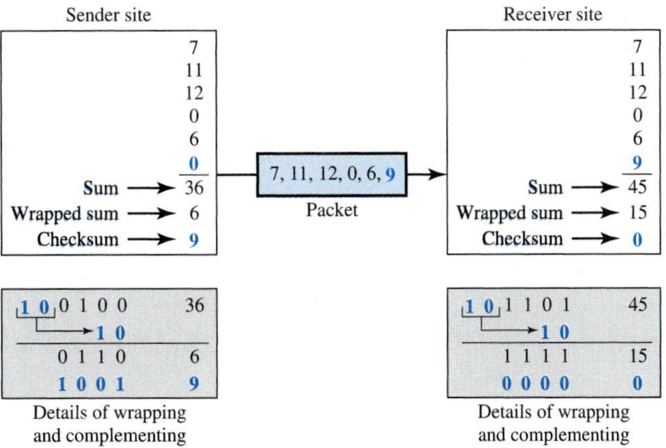

Internet Checksum

Traditionally, the Internet has been using a 16-bit checksum. The sender calculates the checksum by following these steps.

Sender site:

1. **The message is divided into 16-bit words.**
2. **The value of the checksum word is set to 0.**
3. **All words including the checksum are added using one's complement addition.**
4. **The sum is complemented and becomes the checksum.**
5. **The checksum is sent with the data.**

The receiver uses the following steps for error detection.

Receiver site:

1. **The message (including checksum) is divided into 16-bit words.**
2. **All words are added using one's complement addition.**
3. **The sum is complemented and becomes the new checksum.**
4. **If the value of checksum is 0, the message is accepted; otherwise, it is rejected.**

The nature of the checksum (treating words as numbers and adding and complementing them) is well-suited for software implementation. Short programs can be written to calculate the checksum at the receiver site or to check the validity of the message at the receiver site.

Example 10.23

Let us calculate the checksum for a text of 8 characters ("Forouzan"). The text needs to be divided into 2-byte (16-bit) words. We use ASCII (see Appendix A) to change each byte to a 2-digit hexadecimal number. For example, F is represented as 0x46 and o is represented as 0x6F. Figure 10.25 shows how the checksum is calculated at the sender and receiver sites. In part a of the figure, the value of partial sum for the first column is 0x36. We keep the rightmost digit (6) and insert the

Figure 10.25

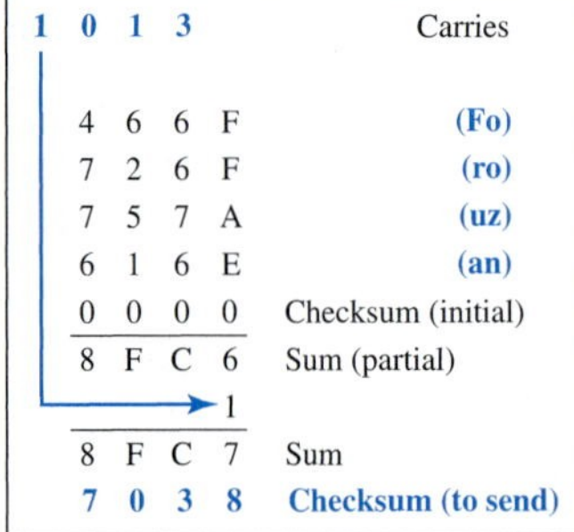

a. Checksum at the sender site

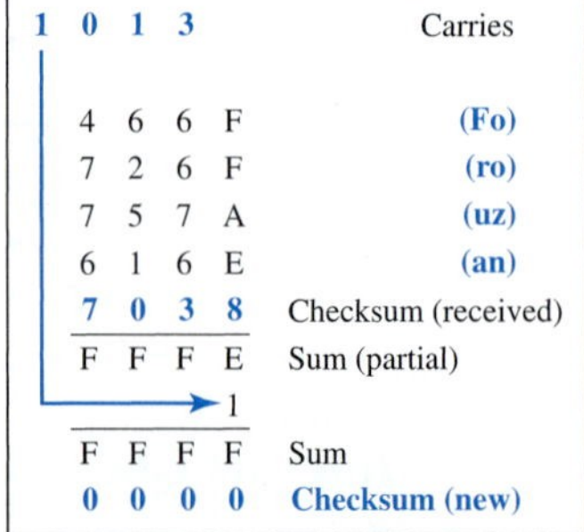

b. Checksum at the receiver site

leftmost dight (3) as the carry in the second column. The process is repeated for each column. Hexadecimal numbers are reviewed in Appendix B.

Note that if there is any corruption, the checksum recalculated by the receiver is not all 0s. We leave this an exercise.

Performance

The traditional checksum uses a small number of bits (16) to detect errors in a message of any size (sometimes thousands of bits). However, it is not as strong as the CRC in error-checking capability. For example, if the value of one word is incremented and the value of another word is decremented by the same amount, the two errors cannot be detected because the sum and checksum remain the same. Also if the values of several words are incremented but the total change is a multiple of 65535, the sum and the checksum does not change, which means the errors are not detected. Fletcher and Adler have proposed some weighted checksums, in which each word is multiplied by a number (its weight) that is related to its position in the text. This will eliminate the first problem we mentioned. However, the tendency in the Internet, particularly in designing new protocols, is to replace the checksum with a CRC.

10.6 RECOMMENDED READING

For more details about subjects discussed in this chapter, we recommend the following books. The items in brackets [. . .] refer to the reference list at the end of the text.

Books

Several excellent book are devoted to error coding. Among them we recommend [Ham80], [Zar02], [Ror96], and [SWE04].

RFCs

A discussion of the use of the checksum in the Internet can be found in RFC 1141.

10.7 KEY TERMS

block code
burst error
check bit
checksum
codeword
convolution code
cyclic code
cyclic redundancy check (CRC)
dataword
error
error correction
error detection
forward error correction
generator polynomial
Hamming code
Hamming distance
interference
linear block code
minimum Hamming distance
modular arithmetic

modulus
one's complement
parity bit
parity-check code
polynomial
redundancy
Reed-Solomon
register
retransmission
shift register
single-bit error
syndrome
two-dimensional parity check

10.8 SUMMARY

- Data can be corrupted during transmission. Some applications require that errors be detected and corrected.
- In a single-bit error, only one bit in the data unit has changed. A burst error means that two or more bits in the data unit have changed.
- To detect or correct errors, we need to send extra (redundant) bits with data.
- There are two main methods of error correction: forward error correction and correction by retransmission.
- We can divide coding schemes into two broad categories: *block coding* and *convolution* coding.
- In coding, we need to use modulo-2 arithmetic. Operations in this arithmetic are very simple; addition and subtraction give the same results. we use the XOR (exclusive OR) operation for both addition and subtraction.
- In block coding, we divide our message into blocks, each of k bits, called datawords. We add r redundant bits to each block to make the length $n = k + r$. The resulting n-bit blocks are called codewords.
- In block coding, errors be detected by using the following two conditions:
  a. The receiver has (or can find) a list of valid codewords.
  b. The original codeword has changed to an invalid one.
- The Hamming distance between two words is the number of differences between corresponding bits. The minimum Hamming distance is the smallest Hamming distance between all possible pairs in a set of words.
- To guarantee the detection of up to s errors in all cases, the minimum Hamming distance in a block code must be $d_{min} = s + 1$. To guarantee correction of up to t errors in all cases, the minimum Hamming distance in a block code must be $d_{min} = 2t + 1$.
- In a linear block code, the exclusive OR (XOR) of any two valid codewords creates another valid codeword.
- A simple parity-check code is a single-bit error-detecting code in which $n = k + 1$ with $d_{min} = 2$. A simple parity-check code can detect an odd number of errors.
- All Hamming codes discussed in this book have $d_{min} = 3$. The relationship between m and n in these codes is $n = 2m - 1$.
- Cyclic codes are special linear block codes with one extra property. In a cyclic code, if a codeword is cyclically shifted (rotated), the result is another codeword.

- A category of cyclic codes called the cyclic redundancy check (CRC) is used in networks such as LANs and WANs.
- A pattern of 0s and 1s can be represented as a polynomial with coefficients of 0 and 1.
- Traditionally, the Internet has been using a 16-bit checksum, which uses *one's complement* arithmetic. In this arithmetic, we can represent unsigned numbers between 0 and $2^n - 1$ using only n bits.

10.9 PRACTICE SET

Review Questions

1. How does a single-bit error differ from a burst error?
2. Discuss the concept of redundancy in error detection and correction.
3. Distinguish between forward error correction versus error correction by retransmission.
4. What is the definition of a linear block code? What is the definition of a cyclic code?
5. What is the Hamming distance? What is the minimum Hamming distance?
6. How is the simple parity check related to the two-dimensional parity check?
7. In CRC, show the relationship between the following entities (size means the number of bits):
   a. The size of the dataword and the size of the codeword
   b. The size of the divisor and the remainder
   c. The degree of the polynomial generator and the size of the divisor
   d. The degree of the polynomial generator and the size of the remainder
8. What kind of arithmetic is used to add data items in checksum calculation?
9. What kind of error is undetectable by the checksum?
10. Can the value of a checksum be all 0s (in binary)? Defend your answer. Can the value be all 1s (in binary)? Defend your answer.

Exercises

11. What is the maximum effect of a 2-ms burst of noise on data transmitted at the following rates?
    a. 1500 bps
    b. 12 kbps
    c. 100 kbps
    d. 100 Mbps
12. Apply the exclusive-or operation on the following pair of patterns (the symbol ⊕ means XOR):
    a. (10001) ⊕ (10000)
    b. (10001) ⊕ (10001) (What do you infer from the result?)
    c. (11100) ⊕ (00000) (What do you infer from the result?)
    d. (10011) ⊕ (11111) (What do you infer from the result?)

13. In Table 10.1, the sender sends dataword 10. A 3-bit burst error corrupts the codeword. Can the receiver detect the error? Defend your answer.
14. In Table 10.2, the sender sends dataword 10. If a 3-bit burst error corrupts the first three bits of the codeword, can the receiver detect the error? Defend your answer.
15. What is the Hamming distance for each of the following codewords:
    a. d (10000, 00000)
    b. d (10101, 10000)
    c. d (11111, 11111)
    d. d (000, 000)
16. Find the minimum Hamming distance for the following cases:
    a. Detection of two errors.
    b. Correction of two errors.
    c. Detection of 3 errors or correction of 2 errors.
    d. Detection of 6 errors or correction of 2 errors.
17. Using the code in Table 10.2, what is the dataword if one of the following codewords is received?
    a. 01011
    b. 11111
    c. 00000
    d. 11011
18. Prove that the code represented by Table 10.8 is not a linear code. You need to find only one case that violates the linearity.

Table 10.8 *Table for Exercise 18*

Dataword	*Codeword*
00	00000
01	01011
10	10111
11	11111

19. Although it can mathematically be proved that a simple parity check code is a linear code, use manual testing of linearity for five pairs of the codewords in Table 10.3 to partially prove this fact.
20. Show that the Hamming code C(7,4) of Table 10.4 can detect two-bit errors but not necessarily three-bit error by testing the code in the following cases. The character "∇" in the burst error means no error; the character "**E**" means an error.
    a. Dataword: 0100 Burst error: ∇**EE**∇∇∇∇
    b. Dataword: 0111 Burst error: **E**∇∇∇∇∇**E**
    c. Dataword: 1111 Burst error: **E**∇**E**∇∇∇**E**
    d. Dataword: 0000 Burst error: **EE**∇**E**∇∇∇

21. Show that the Hamming code C(7,4) of Table 10.4 can correct one-bit errors but not more by testing the code in the following cases. The character "∇" in the burst error means no error; the character "**E**" means an error.
    a. Dataword: 0100 Burst error: **E**∇∇∇∇∇∇
    b. Dataword: 0111 Burst error: ∇**E**∇∇∇∇∇
    c. Dataword: 1111 Burst error: **E**∇∇∇∇∇**E**
    d. Dataword: 0000 Burst error: **EE**∇∇∇∇**E**
22. Although it can be proved that code in Table 10.6 is both linear and cyclic, use only two tests to partially prove the fact:
    a. Test the cyclic property on codeword 0101100.
    b. Test the linear property on codewords 0010110 and 1111111.
23. We need a dataword of at least 11 bits. Find the values of k and n in the Hamming code C(n, k) with $d_{min} = 3$.
24. Apply the following operations on the corresponding polynomials:
    a. $(x^3 + x^2 + x + 1) + (x^4 + x^2 + x + 1)$
    b. $(x^3 + x^2 + x + 1) - (x^4 + x^2 + x + 1)$
    c. $(x^3 + x^2) \times (x^4 + x^2 + x + 1)$
    d. $(x^3 + x^2 + x + 1) / (x^2 + 1)$
25. Answer the following questions:
    a. What is the polynomial representation of 101110?
    b. What is the result of shifting 101110 three bits to the left?
    c. Repeat part b using polynomials.
    d. What is the result of shifting 101110 four bits to the right?
    e. Repeat part d using polynomials.
26. Which of the following CRC generators guarantee the detection of a single bit error?
    a. $x^3 + x + 1$
    b. $x^4 + x^2$
    c. 1
    d. $x^2 + 1$
27. Referring to the CRC-8 polynomial in Table 10.7, answer the following questions:
    a. Does it detect a single error? Defend your answer.
    b. Does it detect a burst error of size 6? Defend your answer.
    c. What is the probability of detecting a burst error of size 9?
    d. What is the probability of detecting a burst error of size 15?
28. Referring to the CRC-32 polynomial in Table 10.7, answer the following questions:
    a. Does it detect a single error? Defend your answer.
    b. Does it detect a burst error of size 16? Defend your answer.
    c. What is the probability of detecting a burst error of size 33?
    d. What is the probability of detecting a burst error of size 55?

29. Assuming even parity, find the parity bit for each of the following data units.
    a. 1001011
    b. 0001100
    c. 1000000
    d. 1110111
30. Given the dataword 1010011110 and the divisor 10111,
    a. Show the generation of the codeword at the sender site (using binary division).
    b. Show the checking of the codeword at the receiver site (assume no error).
31. Repeat Exercise 30 using polynomials.
32. A sender needs to send the four data items 0x3456, 0xABCC, 0x02BC, and 0xEEEE. Answer the following:
    a. Find the checksum at the sender site.
    b. Find the checksum at the receiver site if there is no error.
    c. Find the checksum at the receiver site if the second data item is changed to 0xABCE.
    d. Find the checksum at the receiver site if the second data item is changed to 0xABCE and the third data item is changed to 0x02BA.
33. This problem shows a special case in checksum handling. A sender has two data items to send: 0x4567 and 0xBA98. What is the value of the checksum?

CHAPTER 11

Data Link Control

The two main functions of the data link layer are data link control and media access control. The first, data link control, deals with the design and procedures for communication between two adjacent nodes: node-to-node communication. We discuss this functionality in this chapter. The second function of the data link layer is media access control, or how to share the link. We discuss this functionality in Chapter 12.

Data link control functions include framing, flow and error control, and software-implemented protocols that provide smooth and reliable transmission of frames between nodes. In this chapter, we first discuss framing, or how to organize the bits that are carried by the physical layer. We then discuss flow and error control. A subset of this topic, techniques for error detection and correction, was discussed in Chapter 10.

To implement data link control, we need protocols. Each protocol is a set of rules that need to be implemented in software and run by the two nodes involved in data exchange at the data link layer. We discuss five protocols: two for noiseless (ideal) channels and three for noisy (real) channels. Those in the first category are not actually implemented, but provide a foundation for understanding the protocols in the second category.

After discussing the five protocol designs, we show how a bit-oriented protocol is actually implemented by using the High-level Data Link Control (HDLC) Protocol as an example. We also discuss a popular byte-oriented protocol, Point-to-Point Protocol (PPP).

11.1 FRAMING

Data transmission in the physical layer means moving bits in the form of a signal from the source to the destination. The physical layer provides bit synchronization to ensure that the sender and receiver use the same bit durations and timing.

The data link layer, on the other hand, needs to pack bits into frames, so that each frame is distinguishable from another. Our postal system practices a type of framing. The simple act of inserting a letter into an envelope separates one piece of information from another; the envelope serves as the delimiter. In addition, each envelope defines the sender and receiver addresses since the postal system is a many-to-many carrier facility.

Framing in the data link layer separates a message from one source to a destination, or from other messages to other destinations, by adding a sender address and a destination address. The destination address defines where the packet is to go; the sender address helps the recipient acknowledge the receipt.

Although the whole message could be packed in one frame, that is not normally done. One reason is that a frame can be very large, making flow and error control very inefficient. When a message is carried in one very large frame, even a single-bit error would require the retransmission of the whole message. When a message is divided into smaller frames, a single-bit error affects only that small frame.

Fixed-Size Framing

Frames can be of fixed or variable size. In **fixed-size framing,** there is no need for defining the boundaries of the frames; the size itself can be used as a delimiter. An example of this type of framing is the ATM wide-area network, which uses frames of fixed size called cells. We discuss ATM in Chapter 18.

Variable-Size Framing

Our main discussion in this chapter concerns **variable-size framing,** prevalent in local-area networks. In variable-size framing, we need a way to define the end of the frame and the beginning of the next. Historically, two approaches were used for this purpose: a character-oriented approach and a bit-oriented approach.

Character-Oriented Protocols

In a **character-oriented protocol,** data to be carried are 8-bit characters from a coding system such as ASCII (see Appendix A). The header, which normally carries the source and destination addresses and other control information, and the trailer, which carries error detection or error correction redundant bits, are also multiples of 8 bits. To separate one frame from the next, an 8-bit (1-byte) **flag** is added at the beginning and the end of a frame. The flag, composed of protocol-dependent special characters, signals the start or end of a frame. Figure 11.1 shows the format of a frame in a character-oriented protocol.

Figure 11.1 *A frame in a character-oriented protocol*

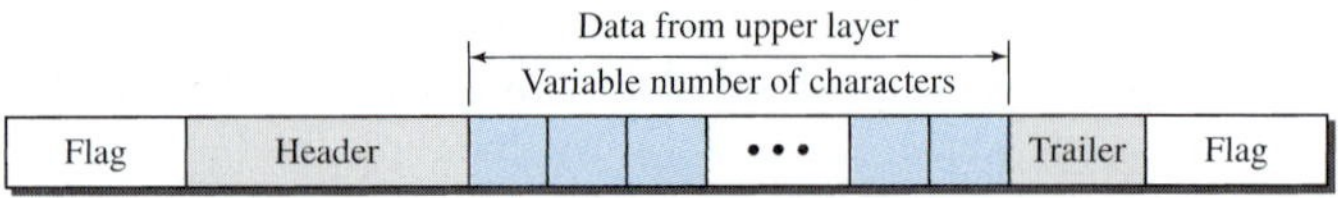

Character-oriented framing was popular when only text was exchanged by the data link layers. The flag could be selected to be any character not used for text communication. Now, however, we send other types of information such as graphs, audio, and video. Any pattern used for the flag could also be part of the information. If this happens, the receiver, when it encounters this pattern in the middle of the data, thinks it has reached the end of the frame. To fix this problem, a **byte-stuffing** strategy was added to character-oriented framing. In byte stuffing (or character stuffing), a special byte is added to the data section of the frame when there is a character with the same pattern as the flag. The data section is stuffed with an extra byte. This byte is usually called the **escape character (ESC),** which has a predefined bit pattern. Whenever the receiver encounters the ESC character, it removes it from the data section and treats the next character as data, not a delimiting flag.

Byte stuffing by the escape character allows the presence of the flag in the data section of the frame, but it creates another problem. What happens if the text contains one or more escape characters followed by a flag? The receiver removes the escape character, but keeps the flag, which is incorrectly interpreted as the end of the frame. To solve this problem, the escape characters that are part of the text must also be marked by another escape character. In other words, if the escape character is part of the text, an extra one is added to show that the second one is part of the text. Figure 11.2 shows the situation.

Figure 11.2 *Byte stuffing and unstuffing*

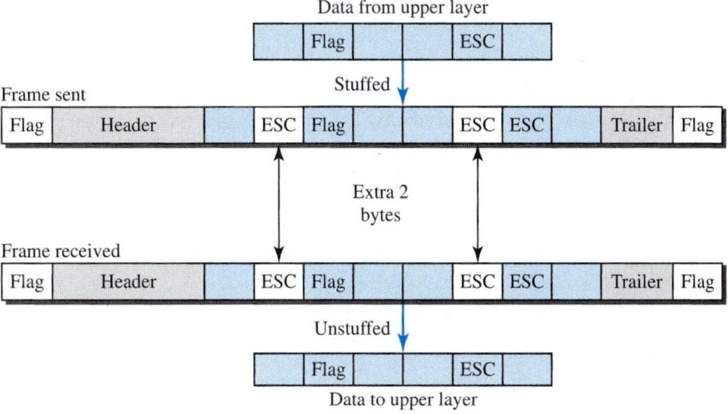

Byte stuffing is the process of adding 1 extra byte whenever there is a flag or escape character in the text.

Character-oriented protocols present another problem in data communications. The universal coding systems in use today, such as Unicode, have 16-bit and 32-bit characters that conflict with 8-bit characters. We can say that in general, the tendency is moving toward the bit-oriented protocols that we discuss next.

Bit-Oriented Protocols

In a **bit-oriented protocol,** the data section of a frame is a sequence of bits to be interpreted by the upper layer as text, graphic, audio, video, and so on. However, in addition to headers (and possible trailers), we still need a delimiter to separate one frame from the other. Most protocols use a special 8-bit pattern flag 01111110 as the delimiter to define the beginning and the end of the frame, as shown in Figure 11.3.

Figure 11.3 *A frame in a bit-oriented protocol*

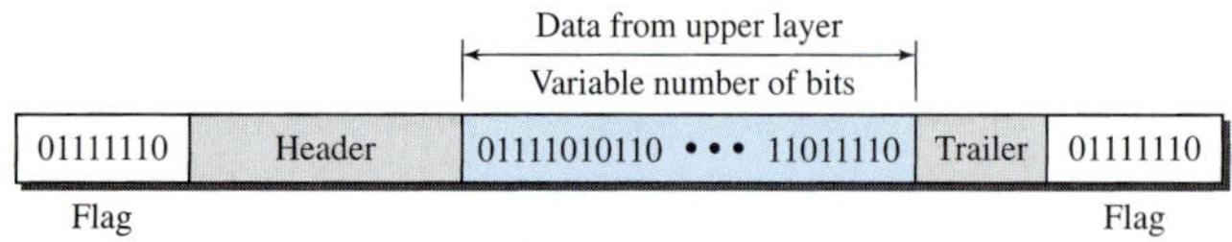

This flag can create the same type of problem we saw in the byte-oriented protocols. That is, if the flag pattern appears in the data, we need to somehow inform the receiver that this is not the end of the frame. We do this by stuffing 1 single bit (instead of 1 byte) to prevent the pattern from looking like a flag. The strategy is called **bit stuffing.** In bit stuffing, if a 0 and five consecutive 1 bits are encountered, an extra 0 is added. This extra stuffed bit is eventually removed from the data by the receiver. Note that the extra bit is added after one 0 followed by five 1s regardless of the value of the next bit. This guarantees that the flag field sequence does not inadvertently appear in the frame.

Bit stuffing is the process of adding one extra 0 whenever five consecutive 1s follow a 0 in the data, so that the receiver does not mistake the pattern 0111110 for a flag.

Figure 11.4 shows bit stuffing at the sender and bit removal at the receiver. Note that even if we have a 0 after five 1s, we still stuff a 0. The 0 will be removed by the receiver.

Figure 11.4 *Bit stuffing and unstuffing*

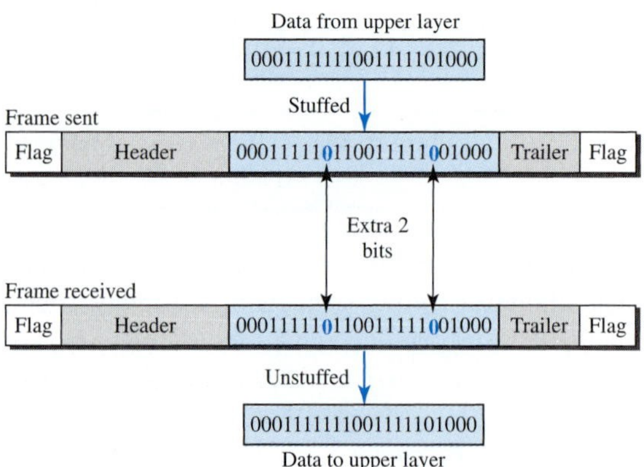

This means that if the flaglike pattern 01111110 appears in the data, it will change to 011111010 (stuffed) and is not mistaken as a flag by the receiver. The real flag 01111110 is not stuffed by the sender and is recognized by the receiver.

11.2 FLOW AND ERROR CONTROL

Data communication requires at least two devices working together, one to send and the other to receive. Even such a basic arrangement requires a great deal of coordination for an intelligible exchange to occur. The most important responsibilities of the data link layer are **flow control** and **error control.** Collectively, these functions are known as **data link control.**

Flow Control

Flow control coordinates the amount of data that can be sent before receiving an acknowledgment and is one of the most important duties of the data link layer. In most protocols, flow control is a set of procedures that tells the sender how much data it can transmit before it must wait for an acknowledgment from the receiver. The flow of data must not be allowed to overwhelm the receiver. Any receiving device has a limited speed at which it can process incoming data and a limited amount of memory in which to store incoming data. The receiving device must be able to inform the sending device before those limits are reached and to request that the transmitting device send fewer frames or stop temporarily. Incoming data must be checked and processed before they can be used. The rate of such processing is often slower than the rate of transmission. For this reason, each receiving device has a block of memory, called a *buffer,* reserved for storing incoming data until they are processed. If the buffer begins to fill up, the receiver must be able to tell the sender to halt transmission until it is once again able to receive.

Flow control refers to a set of procedures used to restrict the amount of data that the sender can send before waiting for acknowledgment.

Error Control

Error control is both error detection and error correction. It allows the receiver to inform the sender of any frames lost or damaged in transmission and coordinates the retransmission of those frames by the sender. In the data link layer, the term *error control* refers primarily to methods of error detection and retransmission. Error control in the data link layer is often implemented simply: Any time an error is detected in an exchange, specified frames are retransmitted. This process is called **automatic repeat request (ARQ).**

Error control in the data link layer is based on automatic repeat request, which is the retransmission of data.

11.3 PROTOCOLS

Now let us see how the data link layer can combine framing, flow control, and error control to achieve the delivery of data from one node to another. The protocols are normally implemented in software by using one of the common programming languages. To make our discussions language-free, we have written in pseudocode a version of each protocol that concentrates mostly on the procedure instead of delving into the details of language rules.

We divide the discussion of protocols into those that can be used for noiseless (error-free) channels and those that can be used for noisy (error-creating) channels. The protocols in the first category cannot be used in real life, but they serve as a basis for understanding the protocols of noisy channels. Figure 11.5 shows the classifications.

Figure 11.5 *Taxonomy of protocols discussed in this chapter*

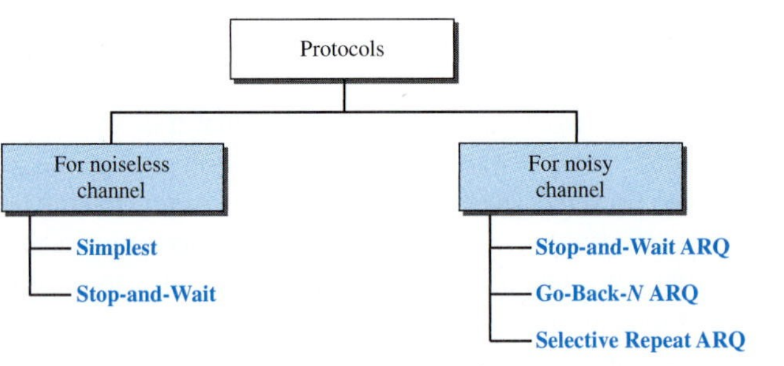

There is a difference between the protocols we discuss here and those used in real networks. All the protocols we discuss are unidirectional in the sense that the data frames travel from one node, called the sender, to another node, called the receiver. Although special frames, called **acknowledgment (ACK)** and **negative acknowledgment (NAK)** can flow in the opposite direction for flow and error control purposes, data flow in only one direction.

In a real-life network, the data link protocols are implemented as bidirectional; data flow in both directions. In these protocols the flow and error control information such as ACKs and NAKs is included in the data frames in a technique called **piggybacking.** Because bidirectional protocols are more complex than unidirectional ones, we chose the latter for our discussion. If they are understood, they can be extended to bidirectional protocols. We leave this extension as an exercise.

11.4 NOISELESS CHANNELS

Let us first assume we have an ideal channel in which no frames are lost, duplicated, or corrupted. We introduce two protocols for this type of channel. The first is a protocol that does not use flow control; the second is the one that does. Of course, neither has error control because we have assumed that the channel is a perfect **noiseless channel.**

Simplest Protocol

Our first protocol, which we call the **Simplest Protocol** for lack of any other name, is one that has no flow or error control. Like other protocols we will discuss in this chapter, it is a unidirectional protocol in which data frames are traveling in only one direction—from the sender to receiver. We assume that the receiver can immediately handle any frame it receives with a processing time that is small enough to be negligible. The data link layer of the receiver immediately removes the header from the frame and hands the data packet to its network layer, which can also accept the packet immediately. In other words, the receiver can never be overwhelmed with incoming frames.

Design

There is no need for flow control in this scheme. The data link layer at the sender site gets data from its network layer, makes a frame out of the data, and sends it. The data link layer at the receiver site receives a frame from its physical layer, extracts data from the frame, and delivers the data to its network layer. The data link layers of the sender and receiver provide transmission services for their network layers. The data link layers use the services provided by their physical layers (such as signaling, multiplexing, and so on) for the physical transmission of bits. Figure 11.6 shows a design.

Figure 11.6 *The design of the simplest protocol with no flow or error control*

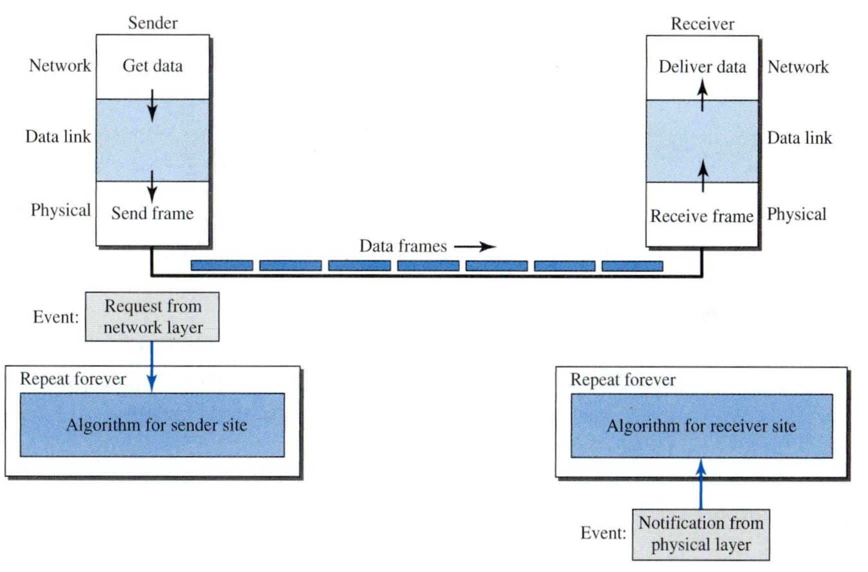

We need to elaborate on the procedure used by both data link layers. The sender site cannot send a frame until its network layer has a data packet to send. The receiver site cannot deliver a data packet to its network layer until a frame arrives. If the protocol is implemented as a procedure, we need to introduce the idea of **events** in the protocol. The procedure at the sender site is constantly running; there is no action until there is a request from the network layer. The procedure at the receiver site is also constantly running, but there is no action until notification from the physical layer arrives. Both procedures are constantly running because they do not know when the corresponding events will occur.

Algorithms

Algorithm 11.1 shows the procedure at the sender site.

Algorithm 11.1 ***Sender-site algorithm for the simplest protocol***

```
1  while(true)                          // Repeat forever
2  {
3    WaitForEvent();                    // Sleep until an event occurs
4    if(Event(RequestToSend))           //There is a packet to send
5    {
6       GetData();
7       MakeFrame();
8       SendFrame();                    //Send the frame
9    }
10 }
```

Analysis The algorithm has an infinite loop, which means lines 3 to 9 are repeated forever once the program starts. The algorithm is an event-driven one, which means that it *sleeps* (line 3) until an event *wakes* it *up* (line 4). This means that there may be an undefined span of time between the execution of line 3 and line 4; there is a gap between these actions. When the event, a request from the network layer, occurs, lines 6 though 8 are executed. The program then repeats the loop and again sleeps at line 3 until the next occurrence of the event. We have written pseudocode for the main process. We do not show any details for the modules GetData, Make-Frame, and SendFrame. GetData() takes a data packet from the network layer, MakeFrame() adds a header and delimiter flags to the data packet to make a frame, and SendFrame() delivers the frame to the physical layer for transmission.

Algorithm 11.2 shows the procedure at the receiver site.

Algorithm 11.2 ***Receiver-site algorithm for the simplest protocol***

```
while(true)                          // Repeat forever
{
  WaitForEvent();                    // Sleep until an event occurs
  if(Event(ArrivalNotification))     //Data frame arrived
  {
     ReceiveFrame();
     ExtractData();
     DeliverData();                  //Deliver data to network layer
  }
}
```

Analysis This algorithm has the same format as Algorithm 11.1, except that the direction of the frames and data is upward. The event here is the arrival of a data frame. After the event occurs, the data link layer receives the frame from the physical layer using the ReceiveFrame() process, extracts the data from the frame using the ExtractData() process, and delivers the data to the network layer using the DeliverData() process. Here, we also have an event-driven algorithm because the algorithm never knows when the data frame will arrive.

Example 11.1

Figure 11.7 shows an example of communication using this protocol. It is very simple. The sender sends a sequence of frames without even thinking about the receiver. To send three frames, three events occur at the sender site and three events at the receiver site. Note that the data frames are shown by tilted boxes; the height of the box defines the transmission time difference between the first bit and the last bit in the frame.

Figure 11.7 *Flow diagram for Example 11.1*

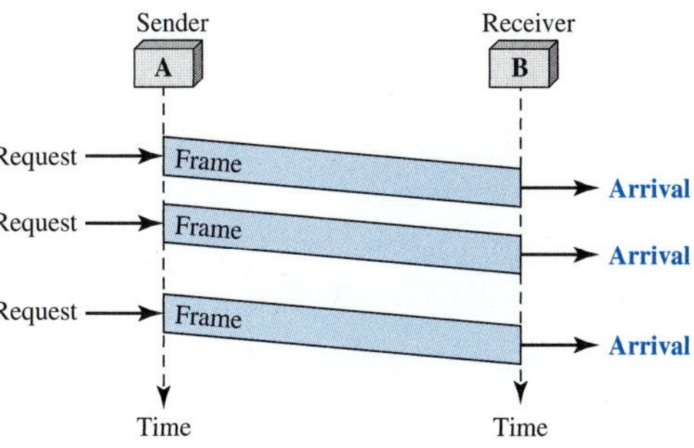

Stop-and-Wait Protocol

If data frames arrive at the receiver site faster than they can be processed, the frames must be stored until their use. Normally, the receiver does not have enough storage space, especially if it is receiving data from many sources. This may result in either the discarding of frames or denial of service. To prevent the receiver from becoming overwhelmed with frames,we somehow need to tell the sender to slow down. There must be feedback from the receiver to the sender.

The protocol we discuss now is called the **Stop-and-Wait Protocol** because the sender sends one frame, stops until it receives confirmation from the receiver (okay to go ahead), and then sends the next frame. We still have unidirectional communication for data frames, but auxiliary ACK frames (simple tokens of acknowledgment) travel from the other direction. We add flow control to our previous protocol.

Design

Figure 11.8 illustrates the mechanism. Comparing this figure with Figure 11.6, we can see the traffic on the forward channel (from sender to receiver) and the reverse channel. At any time, there is either one data frame on the forward channel or one ACK frame on the reverse channel. We therefore need a half-duplex link.

Algorithms

Algorithm 11.3 is for the sender site.

Figure 11.8 *Design of Stop-and-Wait Protocol*

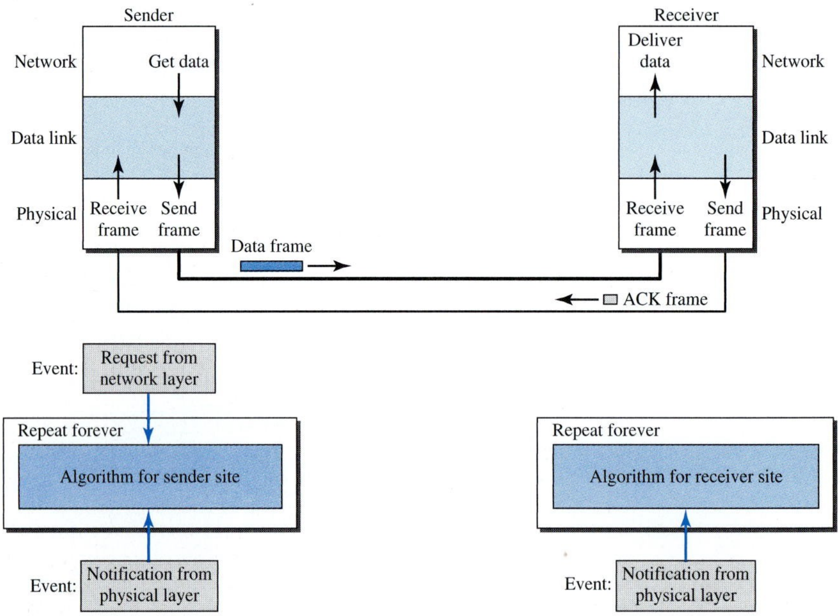

Algorithm 11.3 ***Sender-site algorithm for Stop-and-Wait Protocol***

```
while(true)                          //Repeat forever
canSend = true                       //Allow the first frame to go
{
  WaitForEvent();                    // Sleep until an event occurs
  if(Event(RequestToSend) AND canSend)
  {
     GetData();
     MakeFrame();
     SendFrame();                    //Send the data frame
     canSend = false;                //Cannot send until ACK arrives
  }
  WaitForEvent();                    // Sleep until an event occurs
  if(Event(ArrivalNotification)  // An ACK has arrived
   {
     ReceiveFrame();                 //Receive the ACK frame
     canSend = true;
   }
}
```

Analysis Here two events can occur: a request from the network layer or an arrival notification from the physical layer. The responses to these events must alternate. In other words, after a frame is sent, the algorithm must ignore another network layer request until that frame is acknowledged. We know that two arrival events cannot happen one after another because the channel is error-free and does not duplicate the frames. The requests from the network layer, however, may happen one after another without an arrival event in between. We need somehow to prevent the immediate sending of the data frame. Although there are several methods, we have used a simple *canSend* variable that can either be true or false. When a frame is sent, the variable is set to false to indicate that a new network request cannot be sent until *canSend* is true. When an ACK is received, canSend is set to true to allow the sending of the next frame.

Algorithm 11.4 shows the procedure at the receiver site.

Algorithm 11.4 ***Receiver-site algorithm for Stop-and-Wait Protocol***

```
1  while(true)                          //Repeat forever
2  {
3    WaitForEvent();                    // Sleep until an event occurs
4    if(Event(ArrivalNotification))     //Data frame arrives
5    {
6       ReceiveFrame();
7       ExtractData();
8       Deliver(data);                  //Deliver data to network layer
9       SendFrame();                    //Send an ACK frame
10   }
11 }
```

Analysis This is very similar to Algorithm 11.2 with one exception. After the data frame arrives, the receiver sends an ACK frame (line 9) to acknowledge the receipt and allow the sender to send the next frame.

Example 11.2

Figure 11.9 shows an example of communication using this protocol. It is still very simple. The sender sends one frame and waits for feedback from the receiver. When the ACK arrives, the sender sends the next frame. Note that sending two frames in the protocol involves the sender in four events and the receiver in two events.

Figure 11.9 *Flow diagram for Example 11.2*

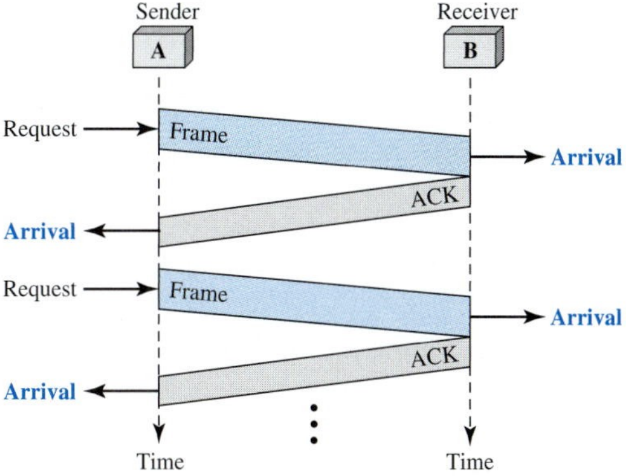

11.5 NOISY CHANNELS

Although the Stop-and-Wait Protocol gives us an idea of how to add flow control to its predecessor, noiseless channels are nonexistent. We can ignore the error (as we sometimes do), or we need to add error control to our protocols. We discuss three protocols in this section that use error control.

Stop-and-Wait Automatic Repeat Request

Our first protocol, called the **Stop-and-Wait Automatic Repeat Request (Stop-and-Wait ARQ),** adds a simple error control mechanism to the Stop-and-Wait Protocol. Let us see how this protocol detects and corrects errors.

To detect and correct corrupted frames, we need to add redundancy bits to our data frame (see Chapter 10). When the frame arrives at the receiver site, it is checked and if it is corrupted, it is silently discarded. The detection of errors in this protocol is manifested by the silence of the receiver.

Lost frames are more difficult to handle than corrupted ones. In our previous protocols, there was no way to identify a frame. The received frame could be the correct one, or a duplicate, or a frame out of order. The solution is to number the frames. When the receiver receives a data frame that is out of order, this means that frames were either lost or duplicated.

The corrupted and lost frames need to be resent in this protocol. If the receiver does not respond when there is an error, how can the sender know which frame to resend? To remedy this problem, the sender keeps a copy of the sent frame. At the same time, it starts a timer. If the timer expires and there is no ACK for the sent frame, the frame is resent, the copy is held, and the timer is restarted. Since the protocol uses the stop-and-wait mechanism, there is only one specific frame that needs an ACK even though several copies of the same frame can be in the network.

Error correction in Stop-and-Wait ARQ is done by keeping a copy of the sent frame and retransmitting of the frame when the timer expires.

Since an ACK frame can also be corrupted and lost, it too needs redundancy bits and a sequence number. The ACK frame for this protocol has a sequence number field. In this protocol, the sender simply discards a corrupted ACK frame or ignores an out-of-order one.

Sequence Numbers

As we discussed, the protocol specifies that frames need to be numbered. This is done by using **sequence numbers.** A field is added to the data frame to hold the sequence number of that frame.

One important consideration is the range of the sequence numbers. Since we want to minimize the frame size, we look for the smallest range that provides unambiguous communication. The sequence numbers of course can wrap around. For example, if we decide that the field is m bits long, the sequence numbers start from 0, go to $2^m - 1$, and then are repeated.

Let us reason out the range of sequence numbers we need. Assume we have used x as a sequence number; we only need to use $x + 1$ after that. There is no need for $x + 2$. To show this, assume that the sender has sent the frame numbered x. Three things can happen.

1. The frame arrives safe and sound at the receiver site; the receiver sends an acknowledgment. The acknowledgment arrives at the sender site, causing the sender to send the next frame numbered $x + 1$.
2. The frame arrives safe and sound at the receiver site; the receiver sends an acknowledgment, but the acknowledgment is corrupted or lost. The sender resends the frame (numbered x) after the time-out. Note that the frame here is a duplicate. The receiver can recognize this fact because it expects frame $x + 1$ but frame x was received.
3. The frame is corrupted or never arrives at the receiver site; the sender resends the frame (numbered x) after the time-out.

We can see that there is a need for sequence numbers x and $x + 1$ because the receiver needs to distinguish between case 1 and case 2. But there is no need for a frame to be numbered $x + 2$. In case 1, the frame can be numbered x again because frames x and $x + 1$ are acknowledged and there is no ambiguity at either site. In cases 2 and 3, the new frame is $x + 1$, not $x + 2$. If only x and $x + 1$ are needed, we can let $x = 0$ and $x + 1 = 1$. This means that the sequence is 0, 1, 0, 1, 0, and so on. Is this pattern familiar? This is modulo-2 arithmetic as we saw in Chapter 10.

In Stop-and-Wait ARQ, we use sequence numbers to number the frames. The sequence numbers are based on modulo-2 arithmetic.

Acknowledgment Numbers

Since the sequence numbers must be suitable for both data frames and ACK frames, we use this convention: The acknowledgment numbers always announce the sequence number of the next frame expected by the receiver. For example, if frame 0 has arrived safe and sound, the receiver sends an ACK frame with acknowledgment 1 (meaning frame 1 is expected next). If frame 1 has arrived safe and sound, the receiver sends an ACK frame with acknowledgment 0 (meaning frame 0 is expected).

In Stop-and-Wait ARQ, the acknowledgment number always announces in modulo-2 arithmetic the sequence number of the next frame expected.

Design

Figure 11.10 shows the design of the Stop-and-Wait ARQ Protocol. The sending device keeps a copy of the last frame transmitted until it receives an acknowledgment for that frame. A data frames uses a seqNo (sequence number); an ACK frame uses an ackNo (acknowledgment number). The sender has a control variable, which we call S_n (sender, next frame to send), that holds the sequence number for the next frame to be sent (0 or 1).

Figure 11.10 *Design of the Stop-and-Wait ARQ Protocol*

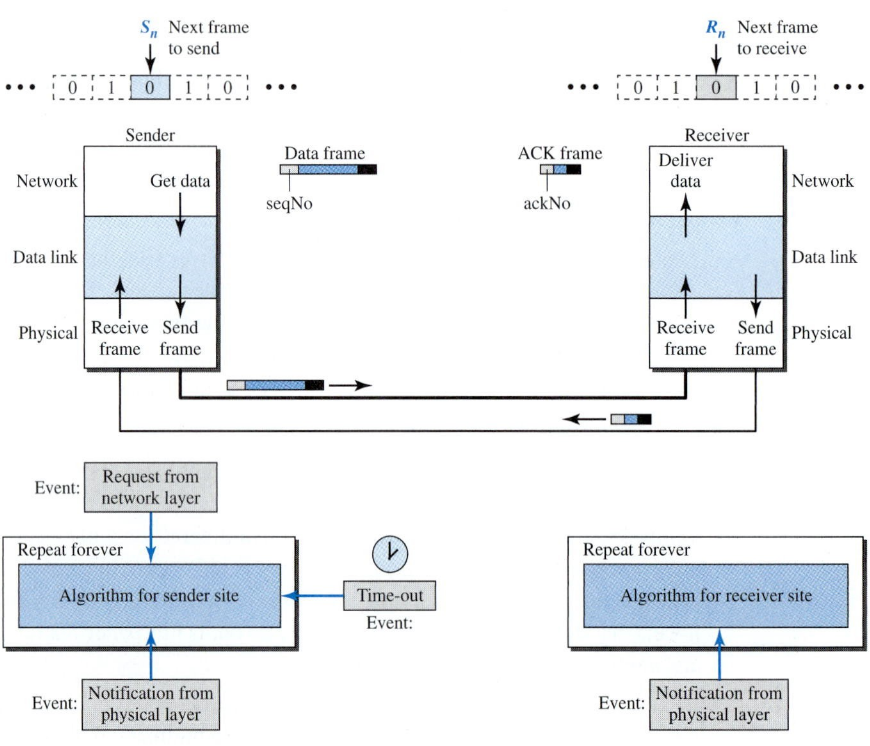

The receiver has a control variable, which we call R_n (receiver, next frame expected), that holds the number of the next frame expected. When a frame is sent, the value of S_n is incremented (modulo-2), which means if it is 0, it becomes 1 and vice versa. When a frame is received, the value of R_n is incremented (modulo-2), which means if it is 0, it becomes 1 and vice versa. Three events can happen at the sender site; one event can happen at the receiver site. Variable S_n points to the slot that matches the sequence number of the frame that has been sent, but not acknowledged; R_n points to the slot that matches the sequence number of the expected frame.

Algorithms

Algorithm 11.5 is for the sender site.

Algorithm 11.5 ***Sender-site algorithm for Stop-and-Wait ARQ***

```
Sn = 0;                        // Frame 0 should be sent first
canSend = true;                // Allow the first request to go
while(true)                    // Repeat forever
{
  WaitForEvent();              // Sleep until an event occurs
```

Algorithm 11.5 ***Sender-site algorithm for Stop-and-Wait ARQ (continued)***

```
 6    if(Event(RequestToSend) AND canSend)
 7    {
 8       GetData();
 9       MakeFrame(Sn);                         //The seqNo is Sn
10       StoreFrame(Sn);                        //Keep copy
11       SendFrame(Sn);
12       StartTimer();
13       Sn = Sn + 1;
14       canSend = false;
15    }
16    WaitForEvent();                           // Sleep
17      if(Event(ArrivalNotification)           // An ACK has arrived
18      {
19        ReceiveFrame(ackNo);                  //Receive the ACK frame
20        if(not corrupted AND ackNo == Sn)     //Valid ACK
21          {
22            Stoptimer();
23            PurgeFrame(Sn-1);                 //Copy is not needed
24            canSend = true;
25          }
26       }
27
28      if(Event(TimeOut)                       // The timer expired
29      {
30       StartTimer();
31       ResendFrame(Sn-1);                     //Resend a copy check
32      }
33 }
```

Analysis We first notice the presence of S_n, the sequence number of the next frame to be sent. This variable is initialized once (line 1), but it is incremented every time a frame is sent (line 13) in preparation for the next frame. However, since this is modulo-2 arithmetic, the sequence numbers are 0, 1, 0, 1, and so on. Note that the processes in the first event (SendFrame, StoreFrame, and PurgeFrame) use an S_n defining the frame sent out. We need at least one buffer to hold this frame until we are sure that it is received safe and sound. Line 10 shows that before the frame is sent, it is stored. The copy is used for resending a corrupt or lost frame. We are still using the canSend variable to prevent the network layer from making a request before the previous frame is received safe and sound. If the frame is not corrupted and the ackNo of the ACK frame matches the sequence number of the next frame to send, we stop the timer and purge the copy of the data frame we saved. Otherwise, we just ignore this event and wait for the next event to happen. After each frame is sent, a timer is started. When the timer expires (line 28), the frame is resent and the timer is restarted.

Algorithm 11.6 shows the procedure at the receiver site.

Algorithm 11.6 ***Receiver-site algorithm for Stop-and-Wait ARQ Protocol***

```
Rn = 0;                       // Frame 0 expected to arrive first
while(true)
{
  WaitForEvent();             // Sleep until an event occurs
```

Algorithm 11.6 ***Receiver-site algorithm for Stop-and-Wait ARQ Protocol (continued)***

```
 if(Event(ArrivalNotification))  //Data frame arrives
 {
    ReceiveFrame();
    if(corrupted(frame));
       sleep();
    if(seqNo == Rn)              //Valid data frame
    {
     ExtractData();
      DeliverData();             //Deliver data
      Rn = Rn + 1;
    }
     SendFrame(Rn);              //Send an ACK
 }
}
```

Analysis This is noticeably different from Algorithm 11.4. First, all arrived data frames that are corrupted are ignored. If the seqNo of the frame is the one that is expected (R_n), the frame is accepted, the data are delivered to the network layer, and the value of R_n is incremented. However, there is one subtle point here. Even if the sequence number of the data frame does not match the next frame expected, an ACK is sent to the sender. This ACK, however, just reconfirms the previous ACK instead of confirming the frame received. This is done because the receiver assumes that the previous ACK might have been lost; the receiver is sending a duplicate frame. The resent ACK may solve the problem before the time-out does it.

Example 11.3

Figure 11.11 shows an example of Stop-and-Wait ARQ. Frame 0 is sent and acknowledged. Frame 1 is lost and resent after the time-out. The resent frame 1 is acknowledged and the timer stops. Frame 0 is sent and acknowledged, but the acknowledgment is lost. The sender has no idea if the frame or the acknowledgment is lost, so after the time-out, it resends frame 0, which is acknowledged.

Efficiency

The Stop-and-Wait ARQ discussed in the previous section is very inefficient if our channel is *thick* and *long*. By *thick,* we mean that our channel has a large bandwidth; by *long,* we mean the round-trip delay is long. The product of these two is called the **bandwidth-delay product,** as we discussed in Chapter 3. We can think of the channel as a pipe. The bandwidth-delay product then is the volume of the pipe in bits. The pipe is always there. If we do not use it, we are inefficient. The bandwidth-delay product is a measure of the number of bits we can send out of our system while waiting for news from the receiver.

Example 11.4

Assume that, in a Stop-and-Wait ARQ system, the bandwidth of the line is 1 Mbps, and 1 bit takes 20 ms to make a round trip. What is the bandwidth-delay product? If the system data frames are 1000 bits in length, what is the utilization percentage of the link?

Solution

The bandwidth-delay product is

$$(1 \times 10^6) \times (20 \times 10^{-3}) = 20{,}000 \text{ bits}$$

Figure 11.11 *Flow diagram for Example 11.3*

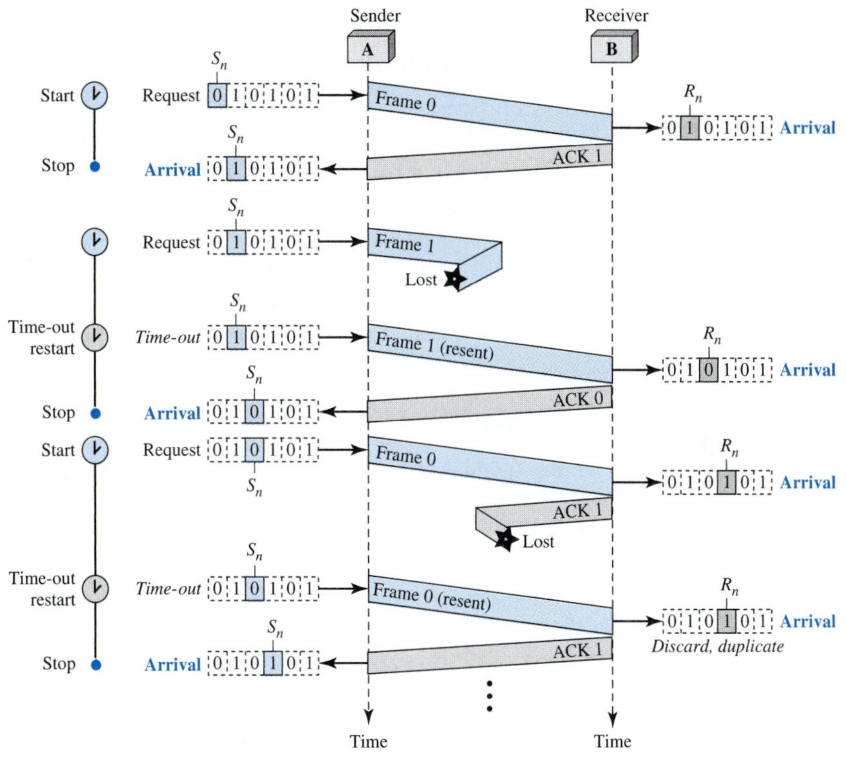

The system can send 20,000 bits during the time it takes for the data to go from the sender to the receiver and then back again. However, the system sends only 1000 bits. We can say that the link utilization is only 1000/20,000, or 5 percent. For this reason, for a link with a high bandwidth or long delay, the use of Stop-and-Wait ARQ wastes the capacity of the link.

Example 11.5

What is the utilization percentage of the link in Example 11.4 if we have a protocol that can send up to 15 frames before stopping and worrying about the acknowledgments?

Solution

The bandwidth-delay product is still 20,000 bits. The system can send up to 15 frames or 15,000 bits during a round trip. This means the utilization is 15,000/20,000, or 75 percent. Of course, if there are damaged frames, the utilization percentage is much less because frames have to be resent.

Pipelining

In networking and in other areas, a task is often begun before the previous task has ended. This is known as **pipelining.** There is no pipelining in Stop-and-Wait ARQ because we need to wait for a frame to reach the destination and be acknowledged before the next frame can be sent. However, pipelining does apply to our next two protocols because several frames can be sent before we receive news about the previous frames. Pipelining improves the efficiency of the transmission if the number of bits in transition is large with respect to the bandwidth-delay product.

Go-Back-*N* Automatic Repeat Request

To improve the efficiency of transmission (filling the pipe), multiple frames must be in transition while waiting for acknowledgment. In other words, we need to let more than one frame be outstanding to keep the channel busy while the sender is waiting for acknowledgment. In this section, we discuss one protocol that can achieve this goal; in the next section, we discuss a second.

The first is called **Go-Back-*N* Automatic Repeat Request** (the rationale for the name will become clear later). In this protocol we can send several frames before receiving acknowledgments; we keep a copy of these frames until the acknowledgments arrive.

Sequence Numbers

Frames from a sending station are numbered sequentially. However, because we need to include the sequence number of each frame in the header, we need to set a limit. If the header of the frame allows m bits for the sequence number, the sequence numbers range from 0 to $2^m - 1$. For example, if m is 4, the only sequence numbers are 0 through 15 inclusive. However, we can repeat the sequence. So the sequence numbers are

0, 1, 2, 3, 4, 5, 6, 7, 8, 9, 10, 11, 12, 13, 14, 15, **0, 1, 2, 3, 4, 5, 6, 7, 8, 9, 10, 11, . . .**

In other words, the sequence numbers are modulo-2^m.

In the Go-Back-*N* Protocol, the sequence numbers are modulo 2^m, where m is the size of the sequence number field in bits.

Sliding Window

In this protocol (and the next), the **sliding window** is an abstract concept that defines the range of sequence numbers that is the concern of the sender and receiver. In other words, the sender and receiver need to deal with only part of the possible sequence numbers. The range which is the concern of the sender is called the **send sliding window;** the range that is the concern of the receiver is called the **receive sliding window.** We discuss both here.

The send window is an imaginary box covering the sequence numbers of the data frames which can be in transit. In each window position, some of these sequence numbers define the frames that have been sent; others define those that can be sent. The maximum size of the window is $2^m - 1$ for reasons that we discuss later. In this chapter, we let the size be fixed and set to the maximum value, but we will see in future chapters that some protocols may have a variable window size. Figure 11.12 shows a sliding window of size 15 ($m = 4$).

The window at any time divides the possible sequence numbers into four regions. The first region, from the far left to the left wall of the window, defines the sequence

Figure 11.12 *Send window for Go-Back-N ARQ*

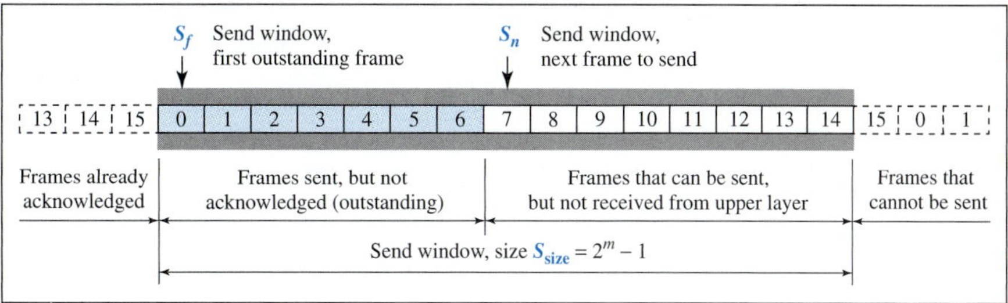

a. Send window before sliding

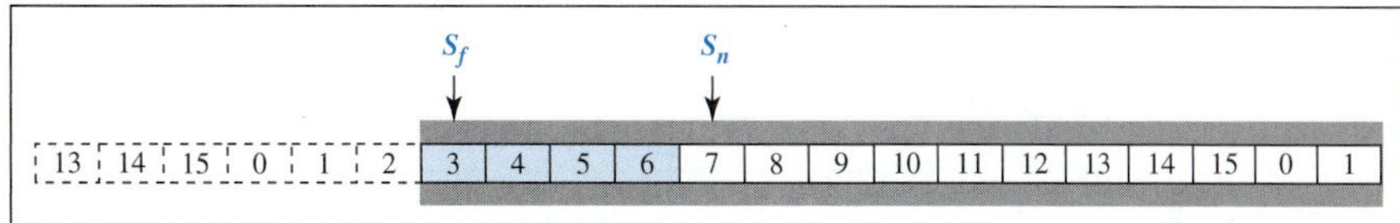

b. Send window after sliding

numbers belonging to frames that are already acknowledged. The sender does not worry about these frames and keeps no copies of them. The second region, colored in Figure 11.12a, defines the range of sequence numbers belonging to the frames that are sent and have an unknown status. The sender needs to wait to find out if these frames have been received or were lost. We call these outstanding frames. The third range, white in the figure, defines the range of sequence numbers for frames that can be sent; however, the corresponding data packets have not yet been received from the network layer. Finally, the fourth region defines sequence numbers that cannot be used until the window slides, as we see next.

The window itself is an abstraction; three variables define its size and location at any time. We call these variables S_f (send window, the first outstanding frame), S_n (send window, the next frame to be sent), and S_{size} (send window, size). The variable S_f defines the sequence number of the first (oldest) outstanding frame. The variable S_n holds the sequence number that will be assigned to the next frame to be sent. Finally, the variable S_{size} defines the size of the window, which is fixed in our protocol.

The send window is an abstract concept defining an imaginary box of size $2^m - 1$ with three variables: S_f, S_n, and S_{size}.

Figure 11.12b shows how a send window can slide one or more slots to the right when an acknowledgment arrives from the other end. As we will see shortly, the acknowledgments in this protocol are cumulative, meaning that more than one frame can be acknowledged by an ACK frame. In Figure 11.12b, frames 0, 1, and 2 are acknowledged, so the window has slid to the right three slots. Note that the value of S_f is 3 because frame 3 is now the first outstanding frame.

The send window can slide one or more slots when a valid acknowledgment arrives.

The receive window makes sure that the correct data frames are received and that the correct acknowledgments are sent. The size of the receive window is always 1. The receiver is always looking for the arrival of a specific frame. Any frame arriving out of order is discarded and needs to be resent. Figure 11.13 shows the receive window.

Figure 11.13 *Receive window for Go-Back-N ARQ*

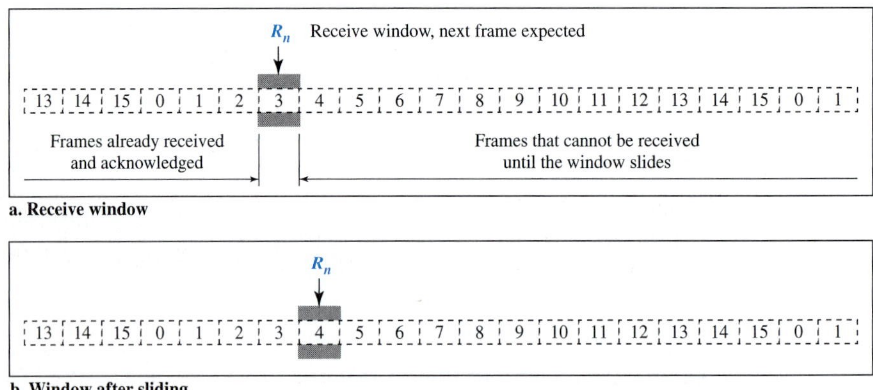

The receive window is an abstract concept defining an imaginary box of size 1 with one single variable R_n. The window slides when a correct frame has arrived; sliding occurs one slot at a time.

Note that we need only one variable R_n (receive window, next frame expected) to define this abstraction. The sequence numbers to the left of the window belong to the frames already received and acknowledged; the sequence numbers to the right of this window define the frames that cannot be received. Any received frame with a sequence number in these two regions is discarded. Only a frame with a sequence number matching the value of R_n is accepted and acknowledged.

The receive window also slides, but only one slot at a time. When a correct frame is received (and a frame is received only one at a time), the window slides.

Timers

Although there can be a timer for each frame that is sent, in our protocol we use only one. The reason is that the timer for the first outstanding frame always expires first; we send all outstanding frames when this timer expires.

Acknowledgment

The receiver sends a positive acknowledgment if a frame has arrived safe and sound and in order. If a frame is damaged or is received out of order, the receiver is silent and will discard all subsequent frames until it receives the one it is expecting. The silence of the receiver causes the timer of the unacknowledged frame at the sender site to expire. This, in turn, causes the sender to go back and resend all frames, beginning with the one with the expired timer. The receiver does not have to acknowledge each frame received. It can send one cumulative acknowledgment for several frames.

Resending a Frame

When the timer expires, the sender resends all outstanding frames. For example, suppose the sender has already sent frame 6, but the timer for frame 3 expires. This means that frame 3 has not been acknowledged; the sender goes back and sends frames 3, 4, 5, and 6 again. That is why the protocol is called Go-Back-*N* ARQ.

Design

Figure 11.14 shows the design for this protocol. As we can see, multiple frames can be in transit in the forward direction, and multiple acknowledgments in the reverse direction. The idea is similar to Stop-and-Wait ARQ; the difference is that the send

Figure 11.14 *Design of Go-Back-N ARQ*

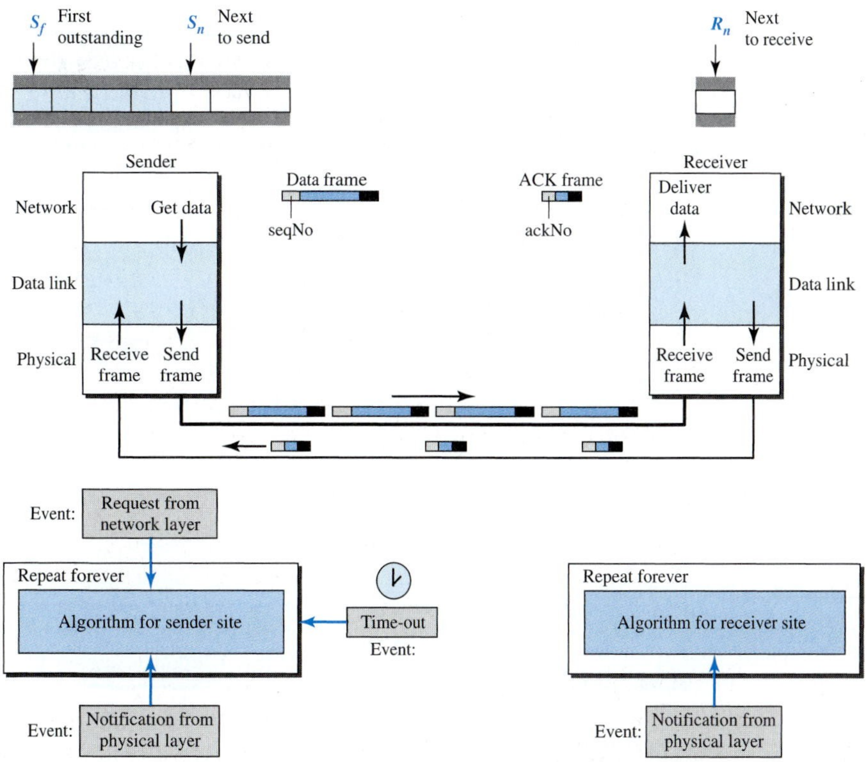

window allows us to have as many frames in transition as there are slots in the send window.

Send Window Size

We can now show why the size of the send window must be less than 2^m. As an example, we choose $m = 2$, which means the size of the window can be $2^m - 1$, or 3. Figure 11.15 compares a window size of 3 against a window size of 4. If the size of the window is 3 (less than 2^2) and all three acknowledgments are lost, the frame 0 timer expires and all three frames are resent. The receiver is now expecting frame 3, not frame 0, so the duplicate frame is correctly discarded. On the other hand, if the size of the window is 4 (equal to 2^2) and all acknowledgments are lost, the sender will send a duplicate of frame 0. However, this time the window of the receiver expects to receive frame 0, so it accepts frame 0, not as a duplicate, but as the first frame in the next cycle. This is an error.

Figure 11.15 *Window size for Go-Back-N ARQ*

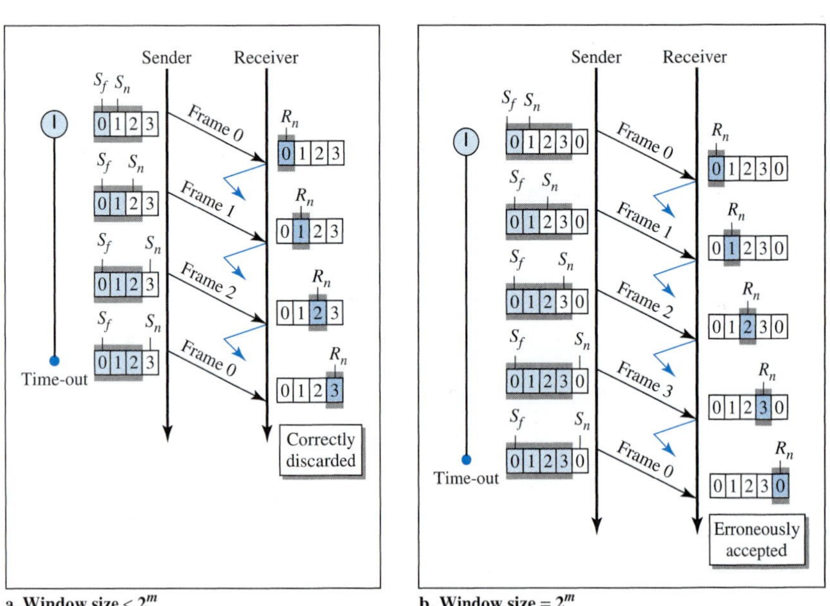

In Go-Back-*N* ARQ, the size of the send window must be less than 2^m; the size of the receiver window is always 1.

Algorithms

Algorithm 11.7 shows the procedure for the sender in this protocol.

Algorithm 11.7 *Go-Back-N sender algorithm*

```
1  Sw = 2^m - 1;
2  Sf = 0;
3  Sn = 0;
4
5  while (true)                              //Repeat forever
6  {
7   WaitForEvent();
8     if(Event(RequestToSend))               //A packet to send
9     {
10       if(Sn-Sf >= Sw)                     //If window is full
11             Sleep();
12       GetData();
13       MakeFrame(Sn);
14       StoreFrame(Sn);
15       SendFrame(Sn);
16       Sn = Sn + 1;
17       if(timer not running)
18            StartTimer();
19    }
20
21    if(Event(ArrivalNotification)) //ACK arrives
22    {
23       Receive(ACK);
24       if(corrupted(ACK))
25            Sleep();
26       if((ackNo>Sf)&&(ackNo<=Sn))  //If a valid ACK
27       While(Sf <= ackNo)
28        {
29         PurgeFrame(Sf);
30         Sf = Sf + 1;
31        }
32        StopTimer();
33    }
34
35    if(Event(TimeOut))                     //The timer expires
36    {
37     StartTimer();
38     Temp = Sf;
39     while(Temp < Sn);
40       {
41        SendFrame(Sf);
42        Sf = Sf + 1;
43       }
44    }
45 }
```

Analysis This algorithm first initializes three variables. Unlike Stop-and-Wait ARQ, this protocol allows several requests from the network layer without the need for other events to occur; we just need to be sure that the window is not full (line 12). In our approach, if the window is full, the request is just ignored and the network layer needs to try again. Some implementations use other methods such as enabling or disabling the network layer. The handling of the arrival event is more complex than in the previous protocol. If we receive a corrupted ACK, we ignore it. If the ackNo belongs to one of the outstanding frames, we use a loop to purge the buffers and move the left wall to the right. The time-out event is also more complex. We first start a new timer. We then resend all outstanding frames.

Algorithm 11.8 is the procedure at the receiver site.

Algorithm 11.8 ***Go-Back-N receiver algorithm***

```
Rn = 0;

while (true)                              //Repeat forever
{
  WaitForEvent();

  if(Event(ArrivalNotification)) /Data frame arrives
  {
     Receive(Frame);
     if(corrupted(Frame))
          Sleep();
     if(seqNo == Rn)                      //If expected frame
     {
       DeliverData();                     //Deliver data
       Rn = Rn + 1;                       //Slide window
       SendACK(Rn);
     }
  }
}
```

Analysis This algorithm is simple. We ignore a corrupt or out-of-order frame. If a frame arrives with an expected sequence number, we deliver the data, update the value of R_n, and send an ACK with the ackNo showing the next frame expected.

Example 11.6

Figure 11.16 shows an example of Go-Back-*N*. This is an example of a case where the forward channel is reliable, but the reverse is not. No data frames are lost, but some ACKs are delayed and one is lost. The example also shows how cumulative acknowledgments can help if acknowledgments are delayed or lost.

After initialization, there are seven sender events. Request events are triggered by data from the network layer; arrival events are triggered by acknowledgments from the physical layer. There is no time-out event here because all outstanding frames are acknowledged before the timer expires. Note that although ACK 2 is lost, ACK 3 serves as both ACK 2 and ACK 3.

There are four receiver events, all triggered by the arrival of frames from the physical layer.

Figure 11.16 *Flow diagram for Example 11.6*

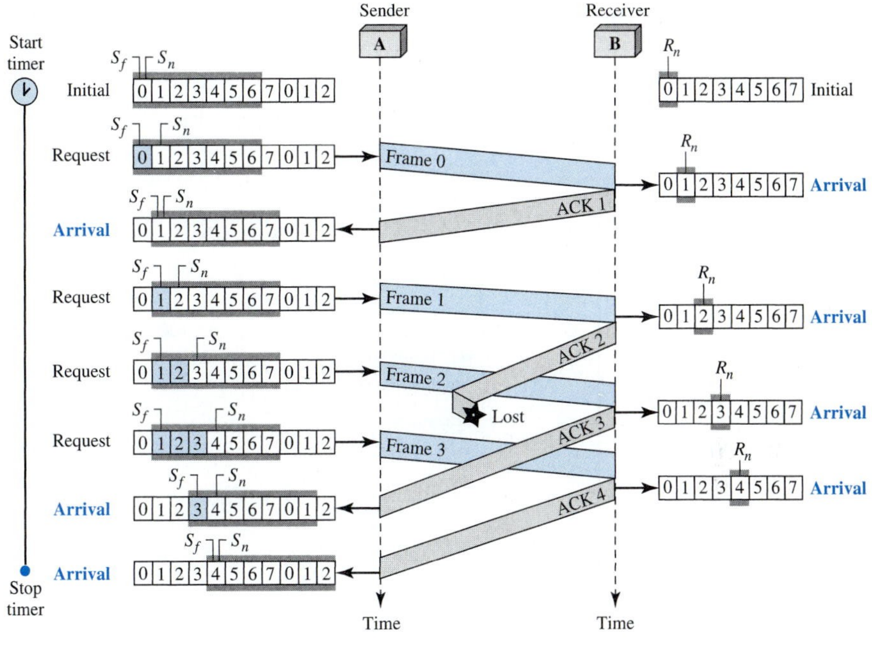

Example 11.7

Figure 11.17 shows what happens when a frame is lost. Frames 0, 1, 2, and 3 are sent. However, frame 1 is lost. The receiver receives frames 2 and 3, but they are discarded because they are received out of order (frame 1 is expected). The sender receives no acknowledgment about frames 1, 2, or 3. Its timer finally expires. The sender sends all outstanding frames (1, 2, and 3) because it does not know what is wrong. Note that the resending of frames 1, 2, and 3 is the response to one single event. When the sender is responding to this event, it cannot accept the triggering of other events. This means that when ACK 2 arrives, the sender is still busy with sending frame 3. The physical layer must wait until this event is completed and the data link layer goes back to its sleeping state. We have shown a vertical line to indicate the delay. It is the same story with ACK 3; but when ACK 3 arrives, the sender is busy responding to ACK 2. It happens again when ACK 4 arrives. Note that before the second timer expires, all outstanding frames have been sent and the timer is stopped.

Go-Back-N ARQ Versus Stop-and-Wait ARQ

The reader may find that there is a similarity between Go-Back-*N* ARQ and Stop-and-Wait ARQ. We can say that the Stop-and-Wait ARQ Protocol is actually a Go-Back-*N* ARQ in which there are only two sequence numbers and the send window size is 1. In other words, $m = 1$, $2^m - 1 = 1$. In Go-Back-*N* ARQ, we said that the addition is modulo-2^m; in Stop-and-Wait ARQ it is 2, which is the same as 2^m when $m = 1$.

Figure 11.17 *Flow diagram for Example 11.7*

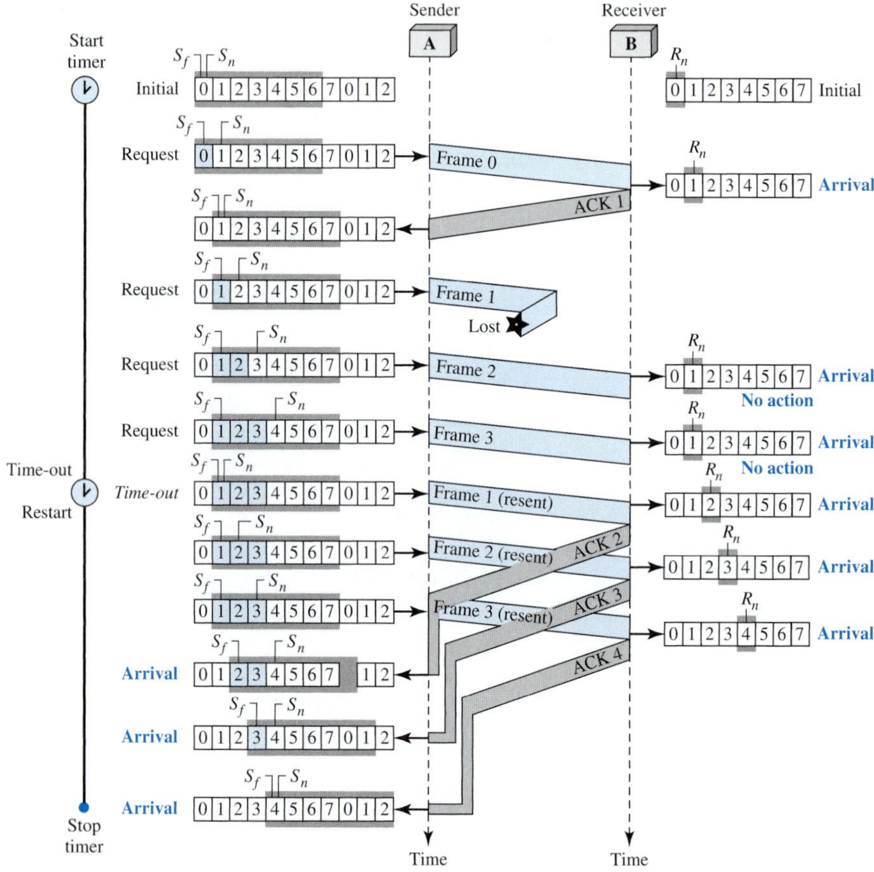

Stop-and-Wait ARQ is a special case of Go-Back-*N* ARQ in which the size of the send window is 1.

Selective Repeat Automatic Repeat Request

Go-Back-*N* ARQ simplifies the process at the receiver site. The receiver keeps track of only one variable, and there is no need to buffer out-of-order frames; they are simply discarded. However, this protocol is very inefficient for a noisy link. In a noisy link a frame has a higher probability of damage, which means the resending of multiple frames. This resending uses up the bandwidth and slows down the transmission. For noisy links, there is another mechanism that does not resend *N* frames when just one frame is damaged; only the damaged frame is resent. This mechanism is called **Selective Repeat ARQ.** It is more efficient for noisy links, but the processing at the receiver is more complex.

Windows

The Selective Repeat Protocol also uses two windows: a send window and a receive window. However, there are differences between the windows in this protocol and the ones in Go-Back-*N*. First, the size of the send window is much smaller; it is 2^{m-1}. The reason for this will be discussed later. Second, the receive window is the same size as the send window.

The send window maximum size can be 2^{m-1}. For example, if $m = 4$, the sequence numbers go from 0 to 15, but the size of the window is just 8 (it is 15 in the Go-Back-*N* Protocol). The smaller window size means less efficiency in filling the pipe, but the fact that there are fewer duplicate frames can compensate for this. The protocol uses the same variables as we discussed for Go-Back-*N*. We show the Selective Repeat send window in Figure 11.18 to emphasize the size. Compare it with Figure 11.12.

Figure 11.18 *Send window for Selective Repeat ARQ*

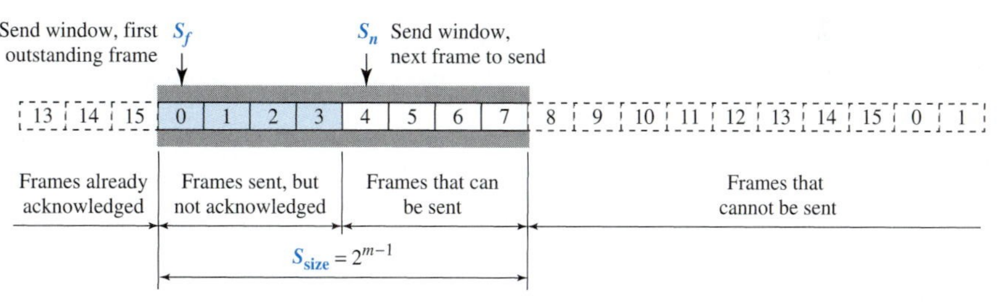

The receive window in Selective Repeat is totally different from the one in Go-Back-*N*. First, the size of the receive window is the same as the size of the send window (2^{m-1}). The Selective Repeat Protocol allows as many frames as the size of the receive window to arrive out of order and be kept until there is a set of in-order frames to be delivered to the network layer. Because the sizes of the send window and receive window are the same, all the frames in the send frame can arrive out of order and be stored until they can be delivered. We need, however, to mention that the receiver never delivers packets out of order to the network layer. Figure 11.19 shows the receive window in this

Figure 11.19 *Receive window for Selective Repeat ARQ*

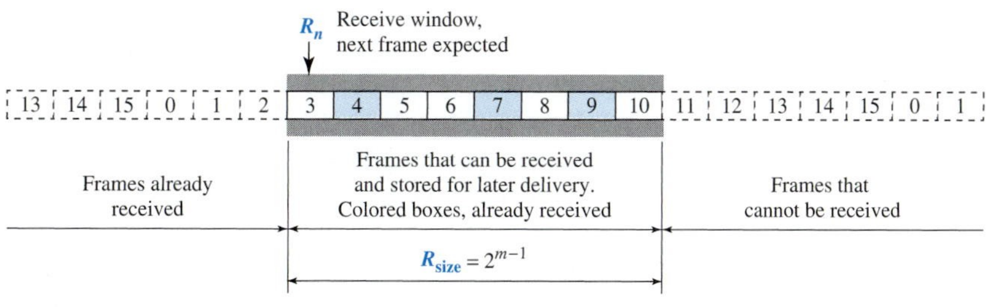

protocol. Those slots inside the window that are colored define frames that have arrived out of order and are waiting for their neighbors to arrive before delivery to the network layer.

Design

The design in this case is to some extent similar to the one we described for the Go-Back-*N*, but more complicated, as shown in Figure 11.20.

Figure 11.20 *Design of Selective Repeat ARQ*

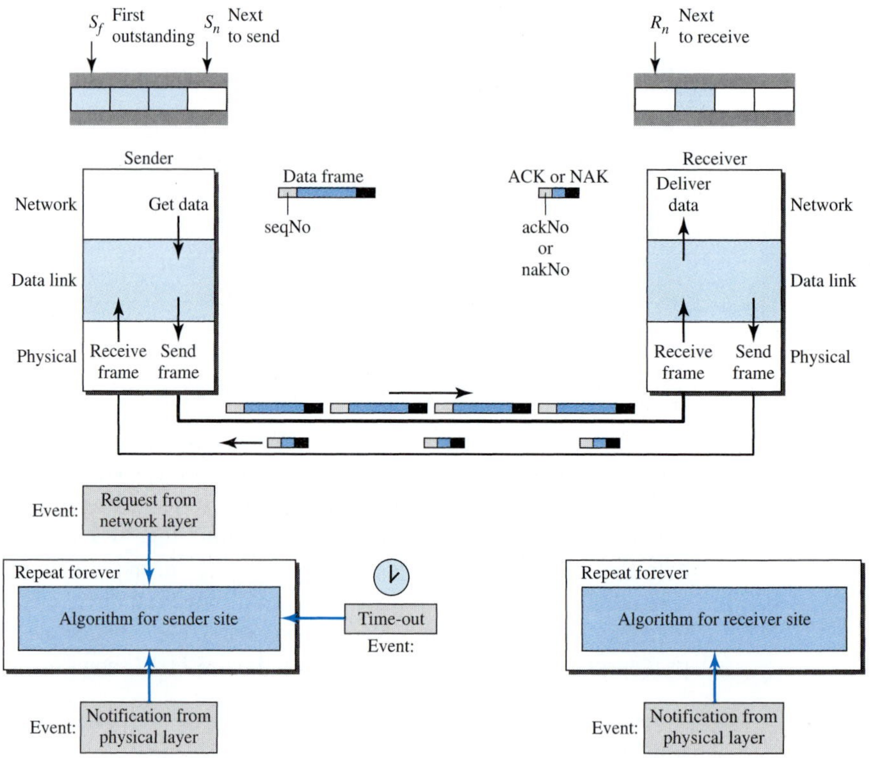

Window Sizes

We can now show why the size of the sender and receiver windows must be at most one-half of 2^m. For an example, we choose $m = 2$, which means the size of the window is $2^m/2$, or 2. Figure 11.21 compares a window size of 2 with a window size of 3.

If the size of the window is 2 and all acknowledgments are lost, the timer for frame 0 expires and frame 0 is resent. However, the window of the receiver is now expecting

Figure 11.21 *Selective Repeat ARQ, window size*

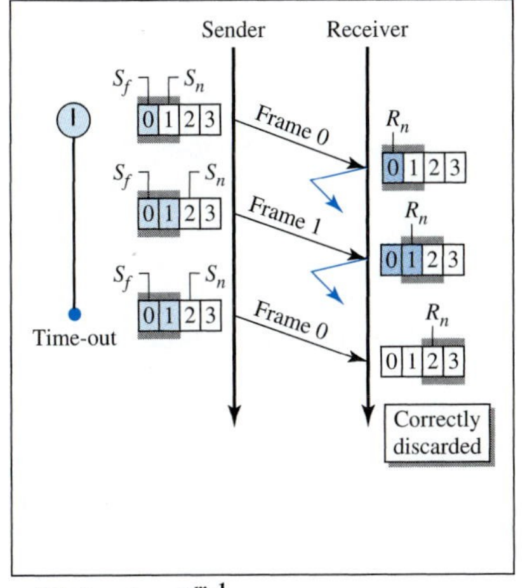

a. Window size = 2^{m-1}

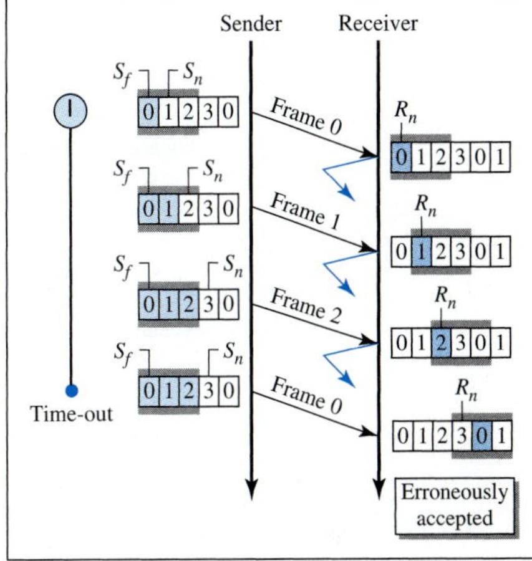

b. Window size > 2^{m-1}

frame 2, not frame 0, so this duplicate frame is correctly discarded. When the size of the window is 3 and all acknowledgments are lost, the sender sends a duplicate of frame 0. However, this time, the window of the receiver expects to receive frame 0 (0 is part of the window), so it accepts frame 0, not as a duplicate, but as the first frame in the next cycle. This is clearly an error.

In Selective Repeat ARQ, the size of the sender and receiver window must be at most one-half of 2^m.

Algorithms

Algorithm 11.9 shows the procedure for the sender.

Algorithm 11.9 ***Sender-site Selective Repeat algorithm***

```
Sw = 2^(m-1) ;
Sf = 0;
Sn = 0;

while (true)                          //Repeat forever
{
  WaitForEvent();
  if(Event(RequestToSend))            //There is a packet to send
  {
```

Algorithm 11.9 ***Sender-site Selective Repeat algorithm (continued)***

```
      if(Sn-Sf >= Sw)                        //If window is full
            Sleep();
      GetData();
      MakeFrame(Sn);
      StoreFrame(Sn);
      SendFrame(Sn);
      Sn = Sn + 1;
      StartTimer(Sn);
   }

   if(Event(ArrivalNotification)) //ACK arrives
   {
      Receive(frame);                        //Receive ACK or NAK
      if(corrupted(frame))
           Sleep();
      if (FrameType == NAK)
         if (nakNo between Sf and Sn)
         {
          resend(nakNo);
          StartTimer(nakNo);
         }
      if (FrameType == ACK)
         if (ackNo between Sf and Sn)
         {
           while(sf < ackNo)
           {
            Purge(sf);
            StopTimer(sf);
            Sf = Sf + 1;
           }
         }
   }

   if(Event(TimeOut(t)))                     //The timer expires
   {
    StartTimer(t);
    SendFrame(t);
   }
}
```

Analysis The handling of the request event is similar to that of the previous protocol except that one timer is started for each frame sent. The arrival event is more complicated here. An ACK or a NAK frame may arrive. If a valid NAK frame arrives, we just resend the corresponding frame. If a valid ACK arrives, we use a loop to purge the buffers, stop the corresponding timer, and move the left wall of the window. The time-out event is simpler here; only the frame which times out is resent.

Algorithm 11.10 shows the procedure for the receiver.

Algorithm 11.10 ***Receiver-site Selective Repeat algorithm***

```
Rn = 0;
NakSent = false;
AckNeeded = false;
Repeat(for all slots)
    Marked(slot) = false;

while (true)                                    //Repeat forever
{
  WaitForEvent();

  if(Event(ArrivalNotification))               /Data frame arrives
  {
     Receive(Frame);
     if(corrupted(Frame))&& (NOT NakSent)
     {
      SendNAK(Rn);
      NakSent = true;
      Sleep();
     }
     if(seqNo <> Rn)&& (NOT NakSent)
     {
      SendNAK(Rn);
      NakSent = true;
      if ((seqNo in window)&&(!Marked(seqNo))
      {
       StoreFrame(seqNo)
       Marked(seqNo)= true;
       while(Marked(Rn))
       {
        DeliverData(Rn);
        Purge(Rn);
        Rn = Rn + 1;
        AckNeeded = true;
       }
        if(AckNeeded);
        {
        SendAck(Rn);
        AckNeeded = false;
        NakSent = false;
        }
      }
     }
  }
}
```

Analysis Here we need more initialization. In order not to overwhelm the other side with NAKs, we use a variable called NakSent. To know when we need to send an ACK, we use a variable called AckNeeded. Both of these are initialized to false. We also use a set of variables to mark the slots in the receive window once the corresponding frame has arrived and is stored. If we receive a corrupted frame and a NAK has not yet been sent, we send a NAK to tell the other site that we have not received the frame we expected. If the frame is not corrupted and the sequence number is in the window, we store the frame and mark the slot. If contiguous frames, starting from R_n have been marked, we deliver their data to the network layer and slide the window. Figure 11.22 shows this situation.

Figure 11.22 *Delivery of data in Selective Repeat ARQ*

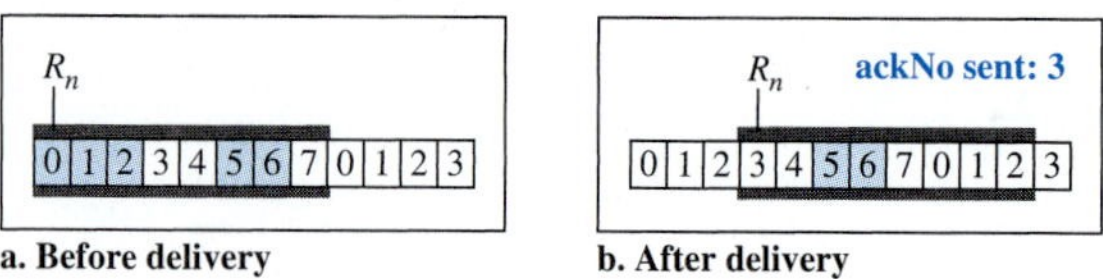

Example 11.8

This example is similar to Example 11.3 in which frame 1 is lost. We show how Selective Repeat behaves in this case. Figure 11.23 shows the situation.

Figure 11.23 *Flow diagram for Example 11.8*

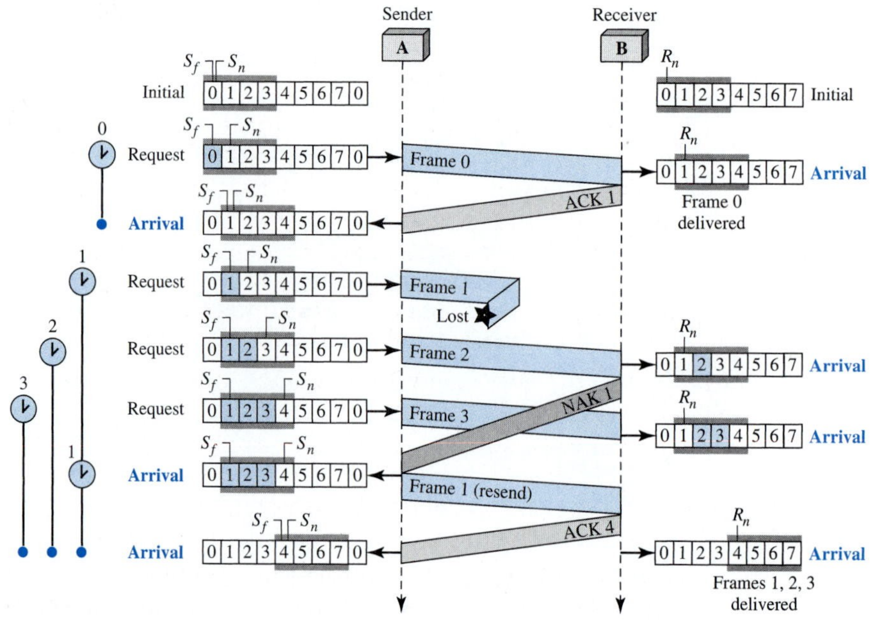

One main difference is the number of timers. Here, each frame sent or resent needs a timer, which means that the timers need to be numbered (0, 1, 2, and 3). The timer for frame 0 starts at the first request, but stops when the ACK for this frame arrives. The timer for frame 1 starts at the second request, restarts when a NAK arrives, and finally stops when the last ACK arrives. The other two timers start when the corresponding frames are sent and stop at the last arrival event.

At the receiver site we need to distinguish between the acceptance of a frame and its delivery to the network layer. At the second arrival, frame 2 arrives and is stored and marked (colored slot), but it cannot be delivered because frame 1 is missing. At the next arrival, frame 3 arrives and is marked and stored, but still none of the frames can be delivered. Only at the last arrival, when finally a copy of frame 1 arrives, can frames 1, 2, and 3 be delivered to the network layer. There are two conditions for the delivery of frames to the network layer: First, a set of consecutive frames must have arrived. Second, the set starts from the beginning of the window. After the first arrival, there was only one frame and it started from the beginning of the window. After the last arrival, there are three frames and the first one starts from the beginning of the window.

Another important point is that a NAK is sent after the second arrival, but not after the third, although both situations look the same. The reason is that the protocol does not want to crowd the network with unnecessary NAKs and unnecessary resent frames. The second NAK would still be NAK1 to inform the sender to resend frame 1 again; this has already been done. The first NAK sent is remembered (using the nakSent variable) and is not sent again until the frame slides. A NAK is sent once for each window position and defines the first slot in the window.

The next point is about the ACKs. Notice that only two ACKs are sent here. The first one acknowledges only the first frame; the second one acknowledges three frames. In Selective Repeat, ACKs are sent when data are delivered to the network layer. If the data belonging to n frames are delivered in one shot, only one ACK is sent for all of them.

Piggybacking

The three protocols we discussed in this section are all unidirectional: data frames flow in only one direction although control information such as ACK and NAK frames can travel in the other direction. In real life, data frames are normally flowing in both directions: from node A to node B and from node B to node A. This means that the control information also needs to flow in both directions. A technique called **piggybacking** is used to improve the efficiency of the bidirectional protocols. When a frame is carrying data from A to B, it can also carry control information about arrived (or lost) frames from B; when a frame is carrying data from B to A, it can also carry control information about the arrived (or lost) frames from A.

We show the design for a Go-Back-*N* ARQ using piggybacking in Figure 11.24. Note that each node now has two windows: one send window and one receive window. Both also need to use a timer. Both are involved in three types of events: request, arrival, and time-out. However, the arrival event here is complicated; when a frame arrives, the site needs to handle control information as well as the frame itself. Both of these concerns must be taken care of in one event, the arrival event. The request event uses only the send window at each site; the arrival event needs to use both windows.

An important point about piggybacking is that both sites must use the same algorithm. This algorithm is complicated because it needs to combine two arrival events into one. We leave this task as an exercise.

Figure 11.24 *Design of piggybacking in Go-Back-N ARQ*

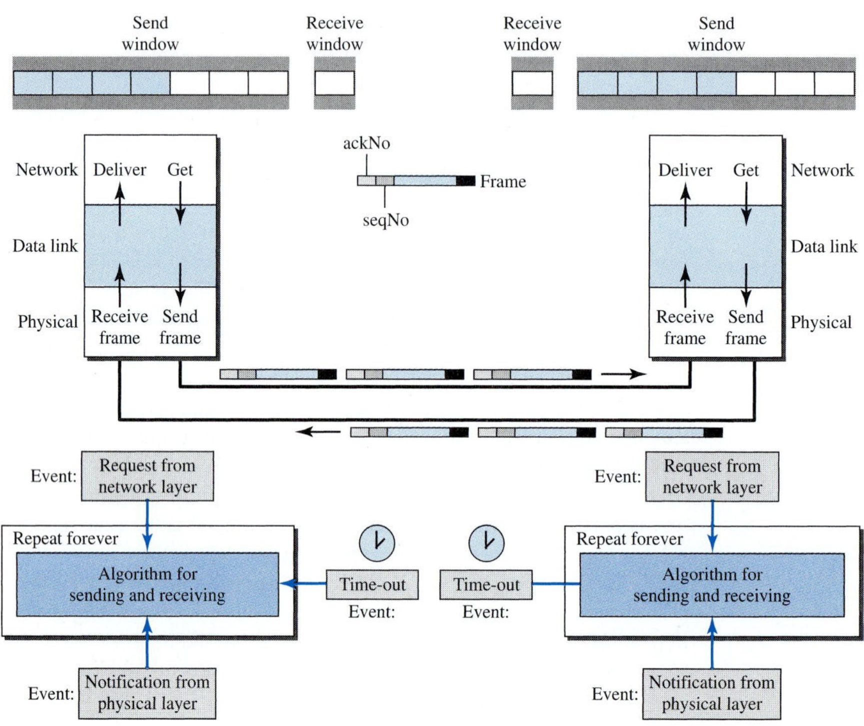

11.6 HDLC

High-level Data Link Control (HDLC) is a bit-oriented protocol for communication over point-to-point and multipoint links. It implements the ARQ mechanisms we discussed in this chapter.

Configurations and Transfer Modes

HDLC provides two common transfer modes that can be used in different configurations: **normal response mode (NRM)** and **asynchronous balanced mode (ABM).**

Normal Response Mode

In normal response mode (NRM), the station configuration is unbalanced. We have one primary station and multiple secondary stations. A **primary station** can send commands; a **secondary station** can only respond. The NRM is used for both point-to-point and multiple-point links, as shown in Figure 11.25.

Figure 11.25 *Normal response mode*

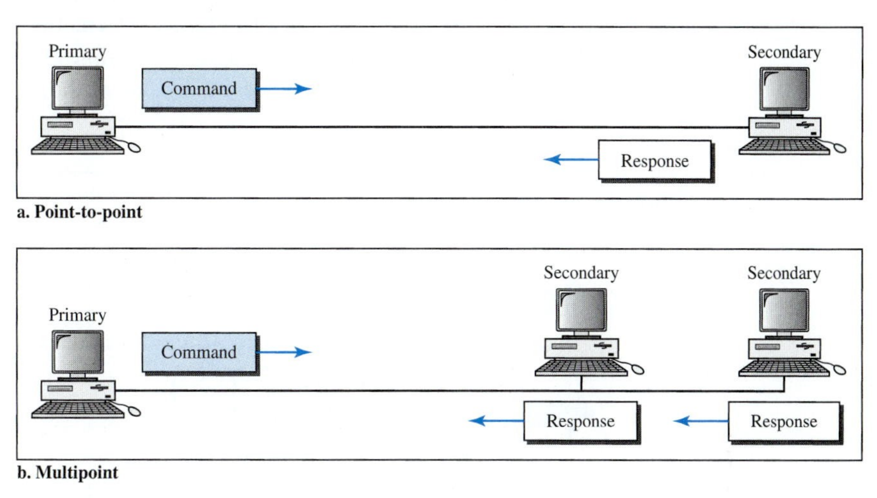

Asynchronous Balanced Mode

In asynchronous balanced mode (ABM), the configuration is balanced. The link is point-to-point, and each station can function as a primary and a secondary (acting as peers), as shown in Figure 11.26. This is the common mode today.

Figure 11.26 *Asynchronous balanced mode*

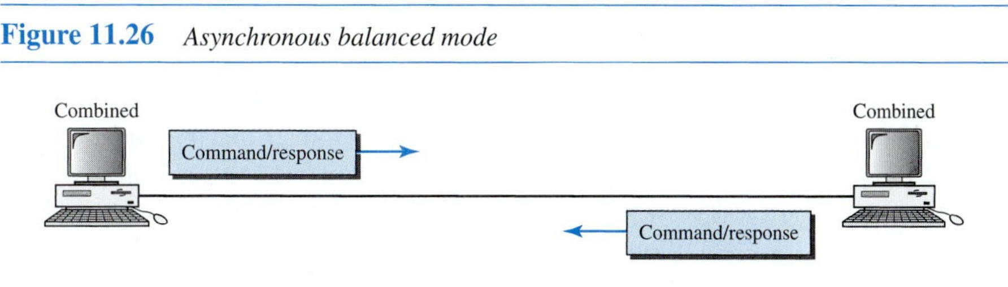

Frames

To provide the flexibility necessary to support all the options possible in the modes and configurations just described, HDLC defines three types of frames: **information frames (I-frames), supervisory frames (S-frames),** and **unnumbered frames (U-frames).** Each type of frame serves as an envelope for the transmission of a different type of message. I-frames are used to transport user data and control information relating to user data (piggybacking). S-frames are used only to transport control information. U-frames are reserved for system management. Information carried by U-frames is intended for managing the link itself.

Frame Format

Each frame in HDLC may contain up to six fields, as shown in Figure 11.27: a beginning flag field, an address field, a control field, an information field, a frame check sequence (FCS) field, and an ending flag field. In multiple-frame transmissions, the ending flag of one frame can serve as the beginning flag of the next frame.

Figure 11.27 *HDLC frames*

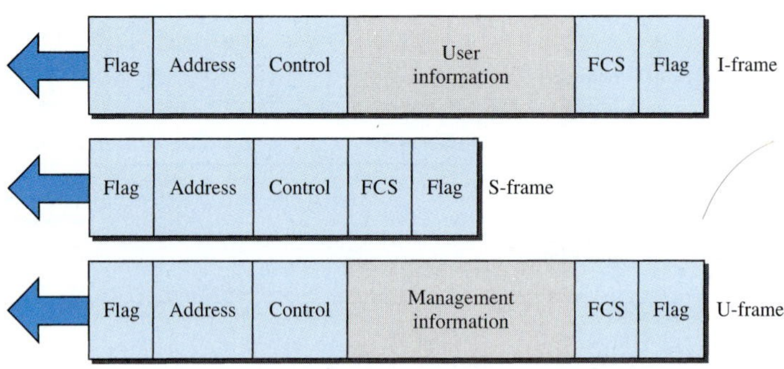

Fields

Let us now discuss the fields and their use in different frame types.

- **Flag field.** The flag field of an HDLC frame is an 8-bit sequence with the bit pattern 01111110 that identifies both the beginning and the end of a frame and serves as a synchronization pattern for the receiver.
- **Address field.** The second field of an HDLC frame contains the address of the secondary station. If a primary station created the frame, it contains a *to* address. If a secondary creates the frame, it contains a *from* address. An **address field** can be 1 byte or several bytes long, depending on the needs of the network. One byte can identify up to 128 stations (1 bit is used for another purpose). Larger networks require multiple-byte address fields. If the address field is only 1 byte, the last bit is always a 1. If the address is more than 1 byte, all bytes but the last one will end with 0; only the last will end with 1. Ending each intermediate byte with 0 indicates to the receiver that there are more address bytes to come.
- **Control field.** The control field is a 1- or 2-byte segment of the frame used for flow and error control. The interpretation of bits in this field depends on the frame type. We discuss this field later and describe its format for each frame type.
- **Information field.** The information field contains the user's data from the network layer or management information. Its length can vary from one network to another.
- **FCS field.** The frame check sequence (FCS) is the HDLC error detection field. It can contain either a 2- or 4-byte ITU-T CRC.

Control Field

The control field determines the type of frame and defines its functionality. So let us discuss the format of this field in greater detail. The format is specific for the type of frame, as shown in Figure 11.28.

Figure 11.28 *Control field format for the different frame types*

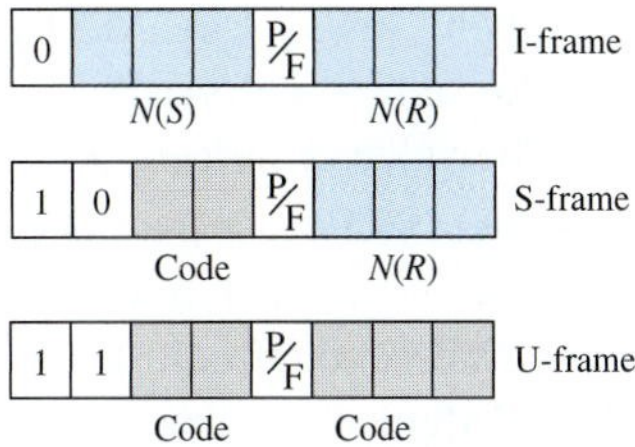

Control Field for I-Frames

I-frames are designed to carry user data from the network layer. In addition, they can include flow and error control information (piggybacking). The subfields in the control field are used to define these functions. The first bit defines the type. If the first bit of the control field is 0, this means the frame is an I-frame. The next 3 bits, called $N(S)$, define the sequence number of the frame. Note that with 3 bits, we can define a sequence number between 0 and 7; but in the extension format, in which the control field is 2 bytes, this field is larger. The last 3 bits, called $N(R)$, correspond to the acknowledgment number when piggybacking is used. The single bit between $N(S)$ and $N(R)$ is called the P/F bit. The P/F field is a single bit with a dual purpose. It has meaning only when it is set (bit = 1) and can mean poll or final. It means *poll* when the frame is sent by a primary station to a secondary (when the address field contains the address of the receiver). It means *final* when the frame is sent by a secondary to a primary (when the address field contains the address of the sender).

Control Field for S-Frames

Supervisory frames are used for flow and error control whenever piggybacking is either impossible or inappropriate (e.g., when the station either has no data of its own to send or needs to send a command or response other than an acknowledgment). S-frames do not have information fields. If the first 2 bits of the control field is 10, this means the frame is an S-frame. The last 3 bits, called $N(R)$, corresponds to the acknowledgment number (ACK) or negative acknowledgment number (NAK) depending on the type of S-frame. The 2 bits called code is used to define the type of S-frame itself. With 2 bits, we can have four types of S-frames, as described below:

- **Receive ready (RR).** If the value of the code subfield is 00, it is an RR S-frame. This kind of frame acknowledges the receipt of a safe and sound frame or group of frames. In this case, the value $N(R)$ field defines the acknowledgment number.

- **Receive not ready (RNR).** If the value of the code subfield is 10, it is an RNR S-frame. This kind of frame is an RR frame with additional functions. It acknowledges the receipt of a frame or group of frames, and it announces that the receiver is busy and cannot receive more frames. It acts as a kind of congestion control mechanism by asking the sender to slow down. The value of $N(R)$ is the acknowledgment number.
- **Reject (REJ).** If the value of the code subfield is 01, it is a REJ S-frame. This is a NAK frame, but not like the one used for Selective Repeat ARQ. It is a NAK that can be used in Go-Back-N ARQ to improve the efficiency of the process by informing the sender, before the sender time expires, that the last frame is lost or damaged. The value of $N(R)$ is the negative acknowledgment number.
- **Selective reject (SREJ).** If the value of the code subfield is 11, it is an SREJ S-frame. This is a NAK frame used in Selective Repeat ARQ. Note that the HDLC Protocol uses the term *selective reject* instead of *selective repeat*. The value of $N(R)$ is the negative acknowledgment number.

Control Field for U-Frames

Unnumbered frames are used to exchange session management and control information between connected devices. Unlike S-frames, U-frames contain an information field, but one used for system management information, not user data. As with S-frames, however, much of the information carried by U-frames is contained in codes included in the control field. U-frame codes are divided into two sections: a 2-bit prefix before the P/F bit and a 3-bit suffix after the P/F bit. Together, these two segments (5 bits) can be used to create up to 32 different types of U-frames. Some of the more common types are shown in Table 11.1.

Table 11.1 *U-frame control command and response*

Code	*Command*	*Response*	*Meaning*
00 001	SNRM		Set normal response mode
11 011	SNRME		Set normal response mode, extended
11 100	SABM	**DM**	Set asynchronous balanced mode or **disconnect mode**
11 110	SABME		Set asynchronous balanced mode, extended
00 000	UI	**UI**	Unnumbered information
00 110		**UA**	**Unnumbered acknowledgment**
00 010	DISC	**RD**	Disconnect or **request disconnect**
10 000	SIM	**RIM**	Set initialization mode or **request information mode**
00 100	UP		Unnumbered poll
11 001	RSET		Reset
11 101	XID	**XID**	Exchange ID
10 001	FRMR	**FRMR**	Frame reject

Example 11.9: Connection/Disconnection

Figure 11.29 shows how U-frames can be used for connection establishment and connection release. Node A asks for a connection with a set asynchronous balanced mode (SABM) frame; node B gives a positive response with an unnumbered acknowledgment (UA) frame. After these two exchanges, data can be transferred between the two nodes (not shown in the figure). After data transfer, node A sends a DISC (disconnect) frame to release the connection; it is confirmed by node B responding with a UA (unnumbered acknowledgment).

Figure 11.29 *Example of connection and disconnection*

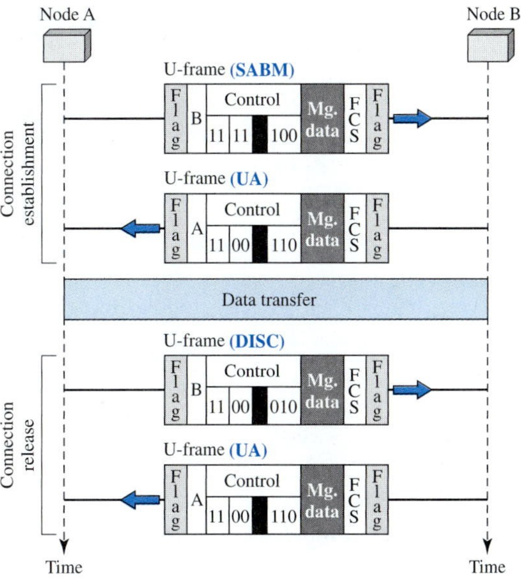

Example 11.10: Piggybacking without Error

Figure 11.30 shows an exchange using piggybacking. Node A begins the exchange of information with an I-frame numbered 0 followed by another I-frame numbered 1. Node B piggybacks its acknowledgment of both frames onto an I-frame of its own. Node B's first I-frame is also numbered 0 [$N(S)$ field] and contains a 2 in its $N(R)$ field, acknowledging the receipt of A's frames 1 and 0 and indicating that it expects frame 2 to arrive next. Node B transmits its second and third I-frames (numbered 1 and 2) before accepting further frames from node A. Its $N(R)$ information, therefore, has not changed: B frames 1 and 2 indicate that node B is still expecting A's frame 2 to arrive next. Node A has sent all its data. Therefore, it cannot piggyback an acknowledgment onto an I-frame and sends an S-frame instead. The RR code indicates that A is still ready to receive. The number 3 in the $N(R)$ field tells B that frames 0, 1, and 2 have all been accepted and that A is now expecting frame number 3.

Figure 11.30 *Example of piggybacking without error*

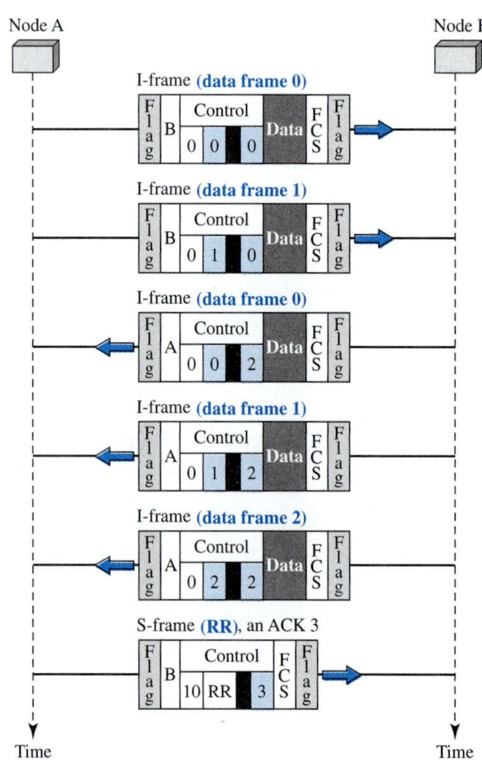

Example 11.11: Piggybacking with Error

Figure 11.31 shows an exchange in which a frame is lost. Node B sends three data frames (0, 1, and 2), but frame 1 is lost. When node A receives frame 2, it discards it and sends a REJ frame for frame 1. Note that the protocol being used is Go-Back-*N* with the special use of an REJ frame as a NAK frame. The NAK frame does two things here: It confirms the receipt of frame 0 and declares that frame 1 and any following frames must be resent. Node B, after receiving the REJ frame, resends frames 1 and 2. Node A acknowledges the receipt by sending an RR frame (ACK) with acknowledgment number 3.

11.7 POINT-TO-POINT PROTOCOL

Although HDLC is a general protocol that can be used for both point-to-point and multipoint configurations, one of the most common protocols for point-to-point access is the **Point-to-Point Protocol (PPP).** Today, millions of Internet users who need to connect their home computers to the server of an Internet service provider use PPP. The majority of these users have a traditional modem; they are connected to the Internet through a telephone line, which provides the services of the physical layer. But to control and

Figure 11.31 *Example of piggybacking with error*

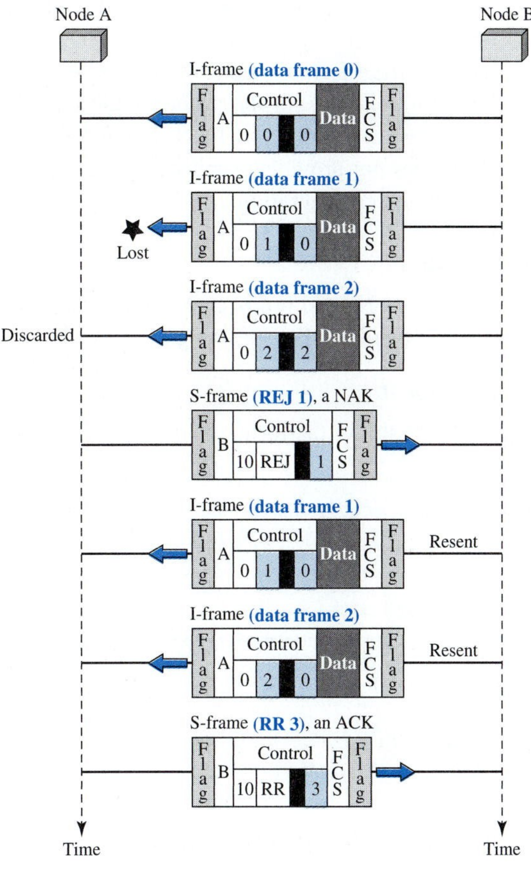

manage the transfer of data, there is a need for a point-to-point protocol at the data link layer. PPP is by far the most common.

PPP provides several services:

1. PPP defines the format of the frame to be exchanged between devices.
2. PPP defines how two devices can negotiate the establishment of the link and the exchange of data.
3. PPP defines how network layer data are encapsulated in the data link frame.
4. PPP defines how two devices can authenticate each other.
5. PPP provides multiple network layer services supporting a variety of network layer protocols.
6. PPP provides connections over multiple links.
7. PPP provides network address configuration. This is particularly useful when a home user needs a temporary network address to connect to the Internet.

On the other hand, to keep PPP simple, several services are missing:

1. PPP does not provide flow control. A sender can send several frames one after another with no concern about overwhelming the receiver.
2. PPP has a very simple mechanism for error control. A CRC field is used to detect errors. If the frame is corrupted, it is silently discarded; the upper-layer protocol needs to take care of the problem. Lack of error control and sequence numbering may cause a packet to be received out of order.
3. PPP does not provide a sophisticated addressing mechanism to handle frames in a multipoint configuration.

Framing

PPP is a byte-oriented protocol. Framing is done according to the discussion of byte-oriented protocols at the beginning of this chapter.

Frame Format

Figure 11.32 shows the format of a PPP frame. The description of each field follows:

Figure 11.32 *PPP frame format*

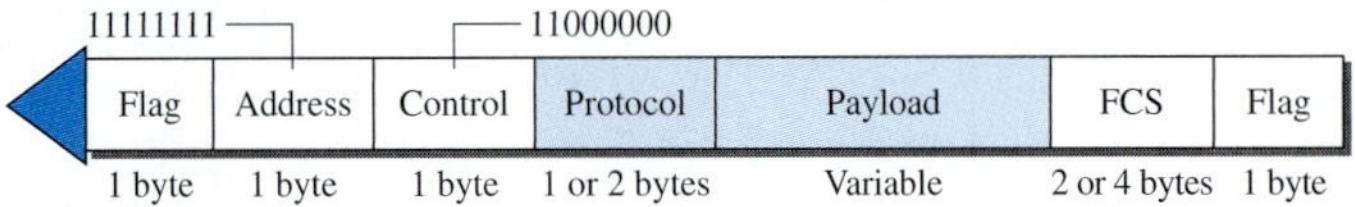

- **Flag.** A PPP frame starts and ends with a 1-byte flag with the bit pattern 01111110. Although this pattern is the same as that used in HDLC, there is a big difference. PPP is a byte-oriented protocol; HDLC is a bit-oriented protocol. The flag is treated as a byte, as we will explain later.
- **Address.** The address field in this protocol is a constant value and set to 11111111 (broadcast address). During negotiation (discussed later), the two parties may agree to omit this byte.
- **Control.** This field is set to the constant value 11000000 (imitating unnumbered frames in HDLC). As we will discuss later, PPP does not provide any flow control. Error control is also limited to error detection. This means that this field is not needed at all, and again, the two parties can agree, during negotiation, to omit this byte.
- **Protocol.** The protocol field defines what is being carried in the data field: either user data or other information. We discuss this field in detail shortly. This field is by default 2 bytes long, but the two parties can agree to use only 1 byte.
- **Payload field.** This field carries either the user data or other information that we will discuss shortly. The data field is a sequence of bytes with the default of a maximum of 1500 bytes; but this can be changed during negotiation. The data field is byte-stuffed if the flag byte pattern appears in this field. Because there is no field defining the size of the data field, padding is needed if the size is less than the maximum default value or the maximum negotiated value.
- **FCS.** The frame check sequence (FCS) is simply a 2-byte or 4-byte standard CRC.

Byte Stuffing

The similarity between PPP and HDLC ends at the frame format. PPP, as we discussed before, is a byte-oriented protocol totally different from HDLC. As a byte-oriented protocol, the flag in PPP is a byte and needs to be escaped whenever it appears in the data section of the frame. The escape byte is 01111101, which means that every time the flaglike pattern appears in the data, this extra byte is stuffed to tell the receiver that the next byte is not a flag.

PPP is a byte-oriented protocol using byte stuffing with the escape byte 01111101.

Transition Phases

A PPP connection goes through phases which can be shown in a **transition phase** diagram (see Figure 11.33).

Figure 11.33 *Transition phases*

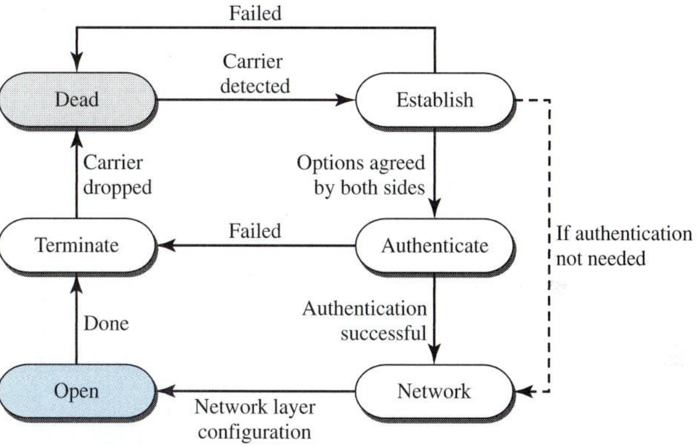

- **Dead.** In the dead phase the link is not being used. There is no active carrier (at the physical layer) and the line is quiet.
- **Establish.** When one of the nodes starts the communication, the connection goes into this phase. In this phase, options are negotiated between the two parties. If the negotiation is successful, the system goes to the authentication phase (if authentication is required) or directly to the networking phase. The link control protocol packets, discussed shortly, are used for this purpose. Several packets may be exchanged here.
- **Authenticate.** The authentication phase is optional; the two nodes may decide, during the establishment phase, not to skip this phase. However, if they decide to proceed with authentication, they send several authentication packets, discussed later. If the result is successful, the connection goes to the networking phase; otherwise, it goes to the termination phase.
- **Network.** In the network phase, negotiation for the network layer protocols takes place. PPP specifies that two nodes establish a network layer agreement before data at

the network layer can be exchanged. The reason is that PPP supports multiple protocols at the network layer. If a node is running multiple protocols simultaneously at the network layer, the receiving node needs to know which protocol will receive the data.

- **Open.** In the open phase, data transfer takes place. When a connection reaches this phase, the exchange of data packets can be started. The connection remains in this phase until one of the endpoints wants to terminate the connection.
- **Terminate.** In the termination phase the connection is terminated. Several packets are exchanged between the two ends for house cleaning and closing the link.

Multiplexing

Although PPP is a data link layer protocol, PPP uses another set of other protocols to establish the link, authenticate the parties involved, and carry the network layer data. Three sets of protocols are defined to make PPP powerful: the Link Control Protocol (LCP), two Authentication Protocols (APs), and several Network Control Protocols (NCPs). At any moment, a PPP packet can carry data from one of these protocols in its data field, as shown in Figure 11.34. Note that there is one LCP, two APs, and several NCPs. Data may also come from several different network layers.

Figure 11.34 *Multiplexing in PPP*

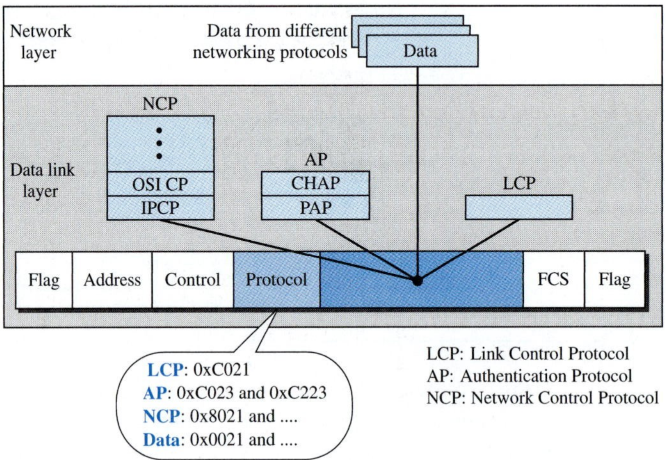

Link Control Protocol

The **Link Control Protocol (LCP)** is responsible for establishing, maintaining, configuring, and terminating links. It also provides negotiation mechanisms to set options between the two endpoints. Both endpoints of the link must reach an agreement about the options before the link can be established. See Figure 11.35.

All LCP packets are carried in the payload field of the PPP frame with the protocol field set to C021 in hexadecimal.

The code field defines the type of LCP packet. There are 11 types of packets as shown in Table 11.2.

Figure 11.35 *LCP packet encapsulated in a frame*

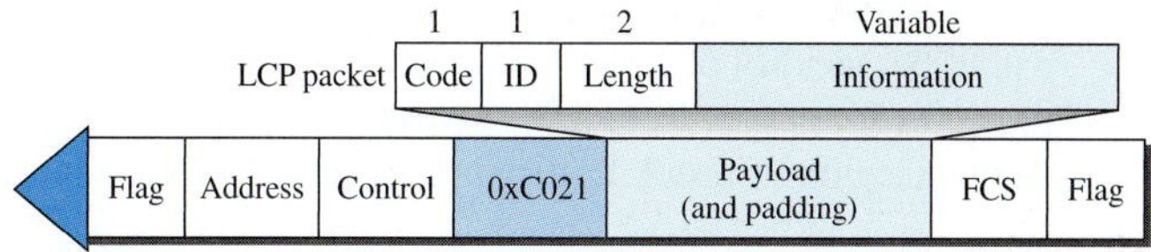

Table 11.2 *LCP packets*

Code	*Packet Type*	*Description*
0x01	Configure-request	Contains the list of proposed options and their values
0x02	Configure-ack	Accepts all options proposed
0x03	Configure-nak	Announces that some options are not acceptable
0x04	Configure-reject	Announces that some options are not recognized
0x05	Terminate-request	Request to shut down the line
0x06	Terminate-ack	Accept the shutdown request
0x07	Code-reject	Announces an unknown code
0x08	Protocol-reject	Announces an unknown protocol
0x09	Echo-request	A type of hello message to check if the other end is alive
0x0A	Echo-reply	The response to the echo-request message
0x0B	Discard-request	A request to discard the packet

There are three categories of packets. The first category, comprising the first four packet types, is used for link configuration during the establish phase. The second category, comprising packet types 5 and 6, is used for link termination during the termination phase. The last five packets are used for link monitoring and debugging.

The ID field holds a value that matches a request with a reply. One endpoint inserts a value in this field, which will be copied into the reply packet. The length field defines the length of the entire LCP packet. The information field contains information, such as options, needed for some LCP packets.

There are many options that can be negotiated between the two endpoints. Options are inserted in the information field of the configuration packets. In this case, the information field is divided into three fields: option type, option length, and option data. We list some of the most common options in Table 11.3.

Table 11.3 *Common options*

Option	*Default*
Maximum receive unit (payload field size)	1500
Authentication protocol	None
Protocol field compression	Off
Address and control field compression	Off

Authentication Protocols

Authentication plays a very important role in PPP because PPP is designed for use over dial-up links where verification of user identity is necessary. **Authentication** means validating the identity of a user who needs to access a set of resources. PPP has created two protocols for authentication: Password Authentication Protocol and Challenge Handshake Authentication Protocol. Note that these protocols are used during the authentication phase.

PAP The **Password Authentication Protocol (PAP)** is a simple authentication procedure with a two-step process:

1. The user who wants to access a system sends an authentication identification (usually the user name) and a password.
2. The system checks the validity of the identification and password and either accepts or denies connection.

Figure 11.36 shows the three types of packets used by PAP and how they are actually exchanged. When a PPP frame is carrying any PAP packets, the value of the protocol field is 0xC023. The three PAP packets are authenticate-request, authenticate-ack, and authenticate-nak. The first packet is used by the user to send the user name and password. The second is used by the system to allow access. The third is used by the system to deny access.

Figure 11.36 *PAP packets encapsulated in a PPP frame*

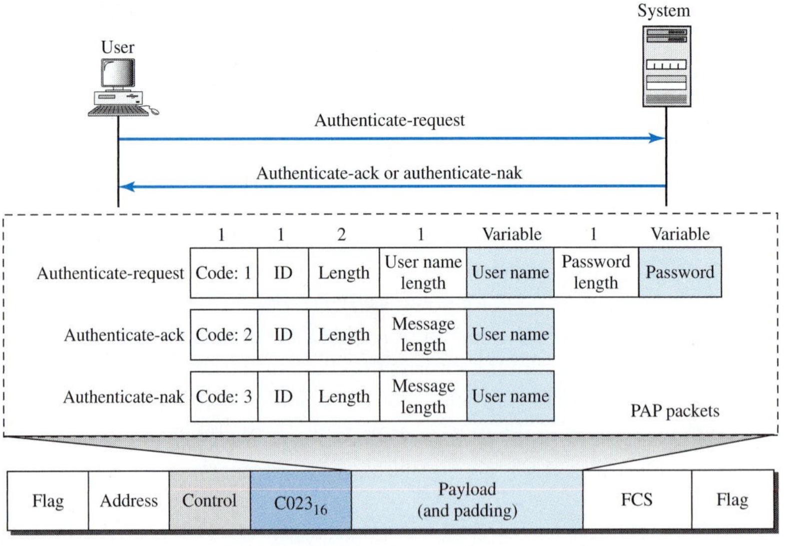

CHAP The **Challenge Handshake Authentication Protocol (CHAP)** is a three-way hand-shaking authentication protocol that provides greater security than PAP. In this method, the password is kept secret; it is never sent online.

1. The system sends the user a challenge packet containing a challenge value, usually a few bytes.
2. The user applies a predefined function that takes the challenge value and the user's own password and creates a result. The user sends the result in the response packet to the system.
3. The system does the same. It applies the same function to the password of the user (known to the system) and the challenge value to create a result. If the result created is the same as the result sent in the response packet, access is granted; otherwise, it is denied. CHAP is more secure than PAP, especially if the system continuously changes the challenge value. Even if the intruder learns the challenge value and the result, the password is still secret. Figure 11.37 shows the packets and how they are used.

Figure 11.37 *CHAP packets encapsulated in a PPP frame*

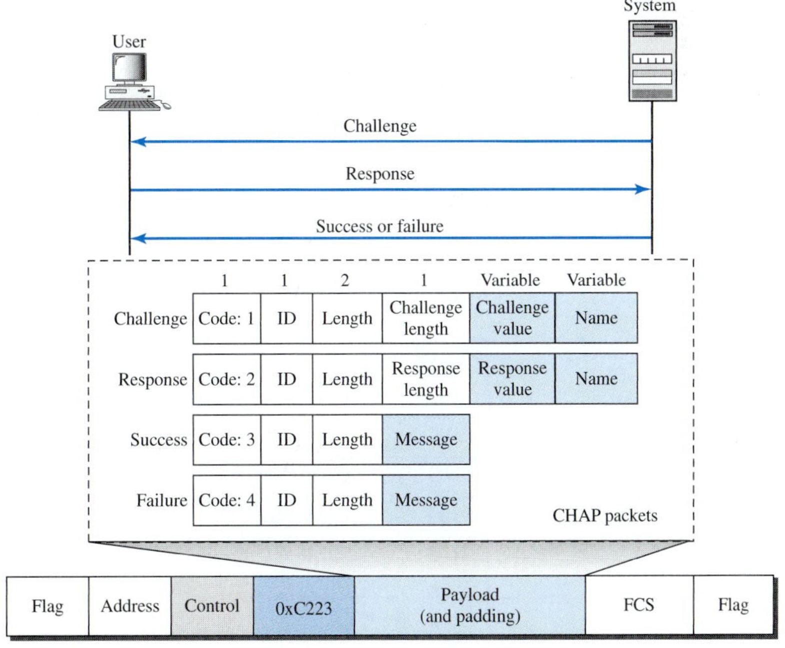

CHAP packets are encapsulated in the PPP frame with the protocol value C223 in hexadecimal. There are four CHAP packets: challenge, response, success, and failure. The first packet is used by the system to send the challenge value. The second is used by the user to return the result of the calculation. The third is used by the system to allow access to the system. The fourth is used by the system to deny access to the system.

Network Control Protocols

PPP is a multiple-network layer protocol. It can carry a network layer data packet from protocols defined by the Internet, OSI, Xerox, DECnet, AppleTalk, Novel, and so on.

To do this, PPP has defined a specific Network Control Protocol for each network protocol. For example, IPCP (Internet Protocol Control Protocol) configures the link for carrying IP data packets. Xerox CP does the same for the Xerox protocol data packets, and so on. Note that none of the NCP packets carry network layer data; they just configure the link at the network layer for the incoming data.

IPCP One NCP protocol is the **Internet Protocol Control Protocol (IPCP).** This protocol configures the link used to carry IP packets in the Internet. IPCP is especially of interest to us. The format of an IPCP packet is shown in Figure 11.38. Note that the value of the protocol field in hexadecimal is 8021.

Figure 11.38 *IPCP packet encapsulated in PPP frame*

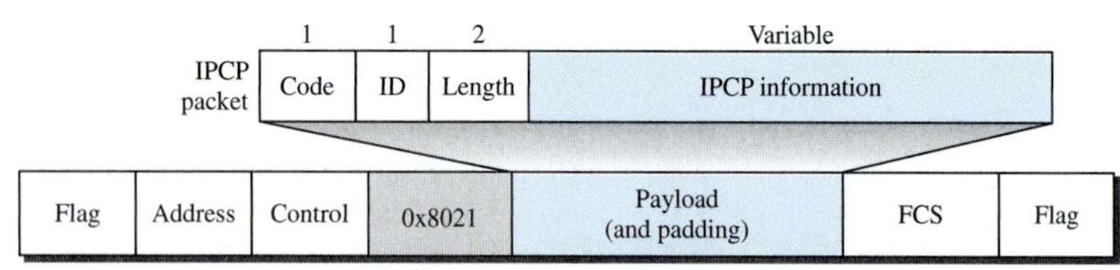

IPCP defines seven packets, distinguished by their code values, as shown in Table 11.4.

Table 11.4 *Code value for IPCP packets*

Code	*IPCP Packet*
0x01	Configure-request
0x02	Configure-ack
0x03	Configure-nak
0x04	Configure-reject
0x05	Terminate-request
0x06	Terminate-ack
0x07	Code-reject

Other Protocols There are other NCP protocols for other network layer protocols. The OSI Network Layer Control Protocol has a protocol field value of 8023; the Xerox NS IDP Control Protocol has a protocol field value of 8025; and so on. The value of the code and the format of the packets for these other protocols are the same as shown in Table 11.4.

Data from the Network Layer

After the network layer configuration is completed by one of the NCP protocols, the users can exchange data packets from the network layer. Here again, there are different protocol fields for different network layers. For example, if PPP is carrying data from the IP network layer, the field value is 0021 (note that the three rightmost digits are the same as for IPCP). If PPP is carrying data from the OSI network layer, the value of the protocol field is 0023, and so on. Figure 11.39 shows the frame for IP.

Figure 11.39 *IP datagram encapsulated in a PPP frame*

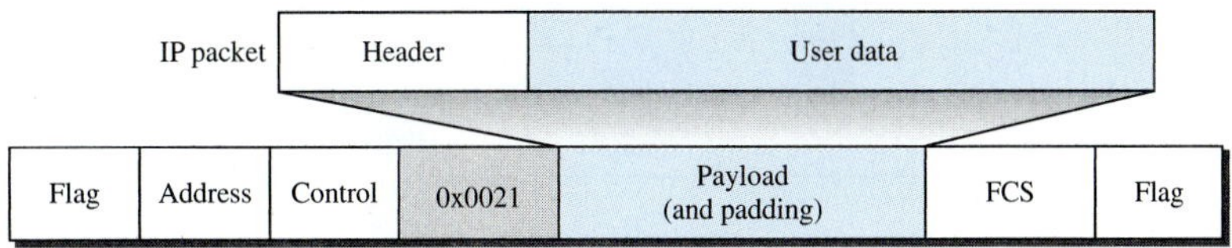

Multilink PPP

PPP was originally designed for a single-channel point-to-point physical link. The availability of multiple channels in a single point-to-point link motivated the development of Multilink PPP. In this case, a logical PPP frame is divided into several actual PPP frames. A segment of the logical frame is carried in the payload of an actual PPP frame, as shown in Figure 11.40. To show that the actual PPP frame is carrying a fragment of a

Figure 11.40 *Multilink PPP*

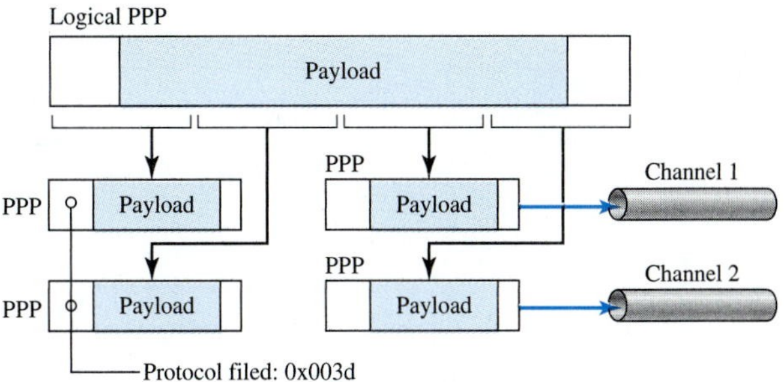

logical PPP frame, the protocol field is set to 0x003d. This new development adds complexity. For example, a sequence number needs to be added to the actual PPP frame to show a fragment's position in the logical frame.

Example 11.12

Let us go through the phases followed by a network layer packet as it is transmitted through a PPP connection. Figure 11.41 shows the steps. For simplicity, we assume unidirectional movement of data from the user site to the system site (such as sending an e-mail through an ISP).

Figure 11.41 *An example*

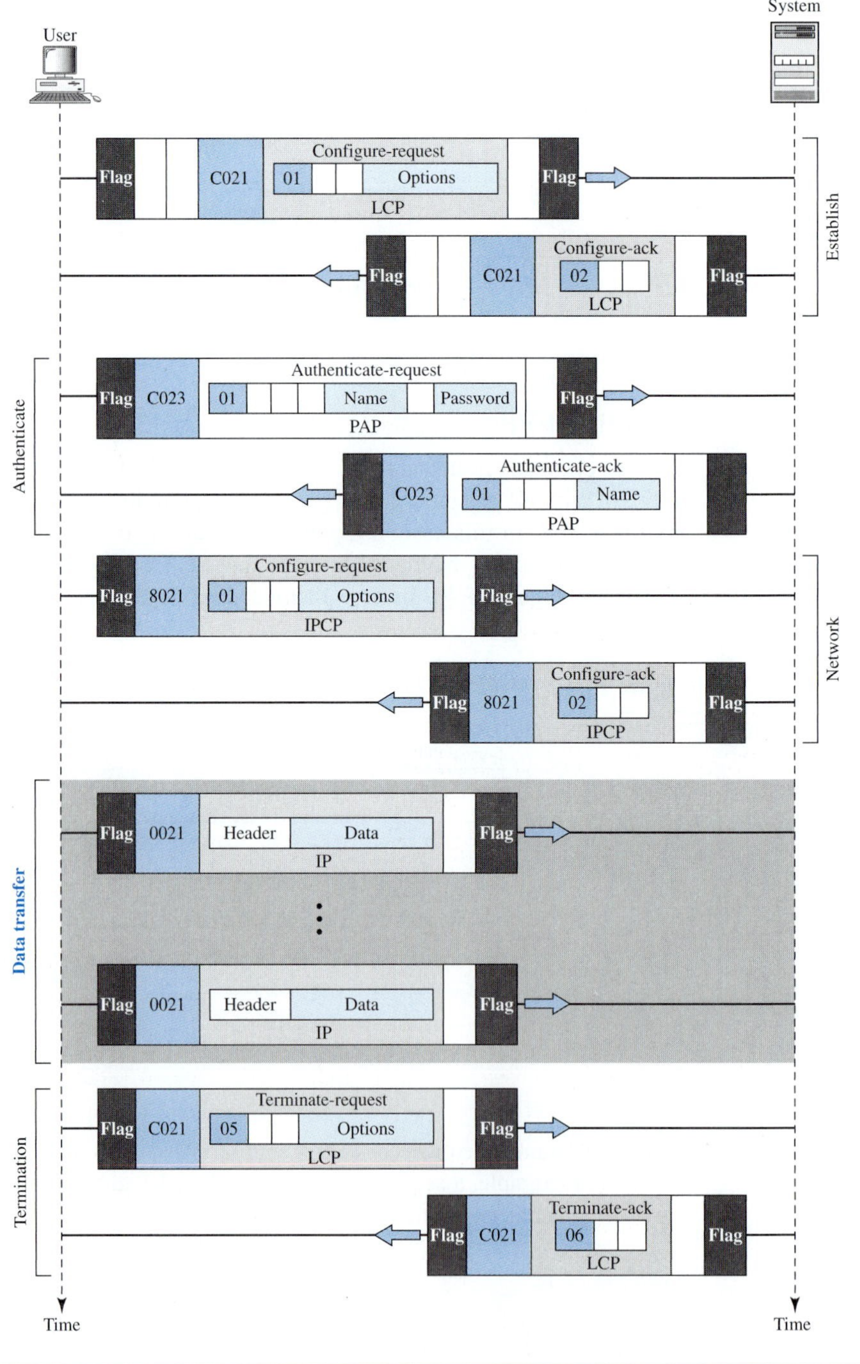

The first two frames show link establishment. We have chosen two options (not shown in the figure): using PAP for authentication and suppressing the address control fields. Frames 3 and 4 are for authentication. Frames 5 and 6 establish the network layer connection using IPCP.

The next several frames show that some IP packets are encapsulated in the PPP frame. The system (receiver) may have been running several network layer protocols, but it knows that the incoming data must be delivered to the IP protocol because the NCP protocol used before the data transfer was IPCP.

After data transfer, the user then terminates the data link connection, which is acknowledged by the system. Of course the user or the system could have chosen to terminate the network layer IPCP and keep the data link layer running if it wanted to run another NCP protocol.

The example is trivial, but it points out the similarities of the packets in LCP, AP, and NCP. It also shows the protocol field values and code numbers for particular protocols.

11.8 RECOMMENDED READING

For more details about subjects discussed in this chapter, we recommend the following books. The items in brackets [. . .] refer to the reference list at the end of the text.

Books

A discussion of data link control can be found in [GW04], Chapter 3 of [Tan03], Chapter 7 of [Sta04], Chapter 12 of [Kes97], and Chapter 2 of [PD03]. More advanced materials can be found in [KMK04].

11.9 KEY TERMS

acknowledgment (ACK)
asynchronous balanced mode (ABM)
automatic repeat request (ARQ)
bandwidth-delay product
bit-oriented protocol
bit stuffing
byte stuffing
Challenge Handshake Authentication Protocol (CHAP)
character-oriented protocol
data link control
error control
escape character (ESC)
event
fixed-size framing
flag
flow control
framing
Go-Back-*N* ARQ Protocol
High-level Data Link Control (HDLC)
information frame (I-frame)
Internet Protocol Control Protocol (IPCP)
Link Control Protocol (LCP)
negative acknowledgment (NAK)
noiseless channel
noisy channel
normal response mode (NRM)
Password Authentication Protocol (PAP)
piggybacking
pipelining
Point-to-Point Protocol (PPP)
primary station

receive sliding window
secondary station
Selective Repeat ARQ Protocol
send sliding window
sequence number
Simplest Protocol
sliding window
Stop-and-Wait ARQ Protocol
Stop-and-Wait Protocol
supervisory frame (S-frame)
transition phase
unnumbered frame (U-frame)
variable-size framing

11.10 SUMMARY

- Data link control deals with the design and procedures for communication between two adjacent nodes: node-to-node communication.
- Framing in the data link layer separates a message from one source to a destination, or from other messages going from other sources to other destinations,
- Frames can be of fixed or variable size. In fixed-size framing, there is no need for defining the boundaries of frames; in variable-size framing, we need a delimiter (flag) to define the boundary of two frames.
- Variable-size framing uses two categories of protocols: byte-oriented (or character-oriented) and bit-oriented. In a byte-oriented protocol, the data section of a frame is a sequence of bytes; in a bit-oriented protocol, the data section of a frame is a sequence of bits.
- In byte-oriented (or character-oriented) protocols, we use byte stuffing; a special byte added to the data section of the frame when there is a character with the same pattern as the flag.
- In bit-oriented protocols, we use bit stuffing; an extra 0 is added to the data section of the frame when there is a sequence of bits with the same pattern as the flag.
- Flow control refers to a set of procedures used to restrict the amount of data that the sender can send before waiting for acknowledgment. Error control refers to methods of error detection and correction.
- For the noiseless channel, we discussed two protocols: the Simplest Protocol and the Stop-and-Wait Protocol. The first protocol has neither flow nor error control; the second has no error control. In the Simplest Protocol, the sender sends its frames one after another with no regards to the receiver. In the Stop-and-Wait Protocol, the sender sends one frame, stops until it receives confirmation from the receiver, and then sends the next frame.
- For the noisy channel, we discussed three protocols: Stop-and-Wait ARQ, Go-Back-*N*, and Selective Repeat ARQ. The Stop-and-Wait ARQ Protocol, adds a simple error control mechanism to the Stop-and-Wait Protocol. In the Go-Back-*N* ARQ Protocol, we can send several frames before receiving acknowledgments, improving the efficiency of transmission. In the Selective Repeat ARQ protocol we avoid unnecessary transmission by sending only frames that are corrupted.
- Both Go-Back-*N* and Selective-Repeat Protocols use a sliding window. In Go-Back-*N* ARQ, if m is the number of bits for the sequence number, then the size of

the send window must be less than 2^m; the size of the receiver window is always 1. In Selective Repeat ARQ, the size of the sender and receiver window must be at most one-half of 2^m.

- A technique called piggybacking is used to improve the efficiency of the bidirectional protocols. When a frame is carrying data from A to B, it can also carry control information about frames from B; when a frame is carrying data from B to A, it can also carry control information about frames from A.
- High-level Data Link Control (HDLC) is a bit-oriented protocol for communication over point-to-point and multipoint links. However, the most common protocols for point-to-point access is the Point-to-Point Protocol (PPP), which is a byte-oriented protocol.

11.11 PRACTICE SET

Review Questions

1. Briefly describe the services provided by the data link layer.
2. Define framing and the reason for its need.
3. Compare and contrast byte-oriented and bit-oriented protocols. Which category has been popular in the past (explain the reason)? Which category is popular now (explain the reason)?
4. Compare and contrast byte-stuffing and bit-stuffing. Which technique is used in byte-oriented protocols? Which technique is used in bit-oriented protocols?
5. Compare and contrast flow control and error control.
6. What are the two protocols we discussed for noiseless channels in this chapter?
7. What are the three protocols we discussed for noisy channels in this chapter?
8. Explain the reason for moving from the Stop-and-Wait ARQ Protocol to the Go-Back-*N* ARQ Protocol.
9. Compare and contrast the Go-Back-*N* ARQ Protocol with Selective-Repeat ARQ.
10. Compare and contrast HDLC with PPP. Which one is byte-oriented; which one is bit-oriented?
11. Define piggybacking and its usefulness.
12. Which of the protocols described in this chapter utilize pipelining?

Exercises

13. Byte-stuff the data in Figure 11.42.

Figure 11.42 *Exercise 13*

ESC			Flag			ESC	ESC	ESC		Flag	

14. Bit-stuff the data in Figure 11.43.

Figure 11.43 *Exercise 14*

000111111100111110100011111111111000011111

15. Design two simple algorithms for byte-stuffing. The first adds bytes at the sender; the second removes bytes at the receiver.
16. Design two simple algorithms for bit-stuffing. The first adds bits at the sender; the second removes bits at the receiver.
17. A sender sends a series of packets to the same destination using 5-bit sequence numbers. If the sequence number starts with 0, what is the sequence number after sending 100 packets?
18. Using 5-bit sequence numbers, what is the maximum size of the send and receive windows for each of the following protocols?
    a. Stop-and-Wait ARQ
    b. Go-Back-*N* ARQ
    c. Selective-Repeat ARQ
19. Design a bidirectional algorithm for the Simplest Protocol using piggybacking. Note that the both parties need to use the same algorithm.
20. Design a bidirectional algorithm for the Stop-and-Wait Protocol using piggybacking. Note that both parties need to use the same algorithm.
21. Design a bidirectional algorithm for the Stop-and-Wait ARQ Protocol using piggybacking. Note that both parties need to use the same algorithm.
22. Design a bidirectional algorithm for the Go-Back-*N* ARQ Protocol using piggybacking. Note that both parties need to use the same algorithm.
23. Design a bidirectional algorithm for the Selective-Repeat ARQ Protocol using piggybacking. Note that both parties need to use the same algorithm.
24. Figure 11.44 shows a state diagram to simulate the behavior of Stop-and-Wait ARQ at the sender site.

Figure 11.44 *Exercise 24*

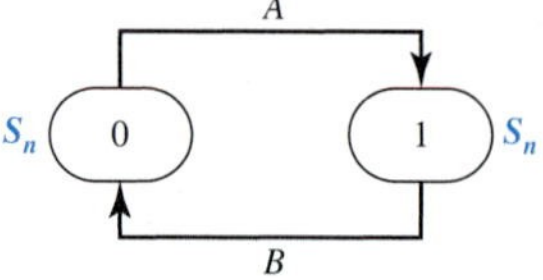

The states have a value of S_n (0 or 1). The arrows shows the transitions. Explain the events that cause the two transitions labeled A and B.

25. Figure 11.45 shows a state diagram to simulate the behavior of Stop-and-Wait ARQ at the receiver site.

Figure 11.45 *Exercise 25*

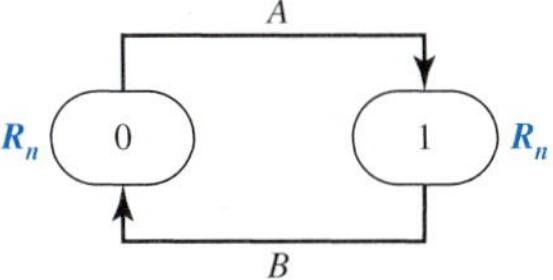

The states have a value of R_n (0 or 1). The arrows shows the transitions. Explain the events that cause the two transitions labeled A and B.

26. In Stop-and-Wait ARQ, we can combine the state diagrams of the sender and receiver in Exercises 24 and 25. One state defines the combined values of R_n and S_n. This means that we can have four states, each defined by (x, y), where x defines the value of S_n and y defines the value of R_n. In other words, we can have the four states shown in Figure 11.46. Explain the events that cause the four transitions labeled A, B, C, and D.

Figure 11.46 *Exercise 26*

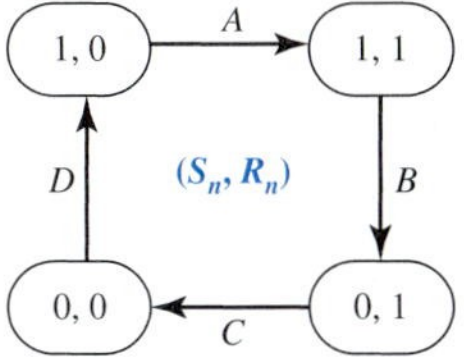

27. The timer of a system using the Stop-and-Wait ARQ Protocol has a time-out of 6 ms. Draw the flow diagram similar to Figure 11.11 for four frames if the round trip delay is 4 ms. Assume no data frame or control frame is lost or damaged.
28. Repeat Exercise 27 if the time-out is 4 ms and the round trip delay is 6.
29. Repeat Exercise 27 if the first frame (frame 0) is lost.
30. A system uses the Stop-and-Wait ARQ Protocol. If each packet carries 1000 bits of data, how long does it take to send 1 million bits of data if the distance between the sender and receiver is 5000 Km and the propagation speed is 2×10^8 m? Ignore transmission, waiting, and processing delays. We assume no data or control frame is lost or damaged.
31. Repeat Exercise 30 using the Go-back-*N* ARQ Protocol with a window size of 7. Ignore the overhead due to the header and trailer.
32. Repeat Exercise 30 using the Selective-Repeat ARQ Protocol with a window size of 4. Ignore the overhead due to the header and the trailer.

CHAPTER 12

Multiple Access

In Chapter 11 we discussed data link control, a mechanism which provides a link with reliable communication. In the protocols we described, we assumed that there is an available dedicated link (or channel) between the sender and the receiver. This assumption may or may not be true. If, indeed, we have a dedicated link, as when we connect to the Internet using PPP as the data link control protocol, then the assumption is true and we do not need anything else.

On the other hand, if we use our cellular phone to connect to another cellular phone, the channel (the band allocated to the vendor company) is not dedicated. A person a few feet away from us may be using the same channel to talk to her friend.

We can consider the data link layer as two sublayers. The upper sublayer is responsible for data link control, and the lower sublayer is responsible for resolving access to the shared media. If the channel is dedicated, we do not need the lower sublayer. Figure 12.1 shows these two sublayers in the data link layer.

Figure 12.1 *Data link layer divided into two functionality-oriented sublayers*

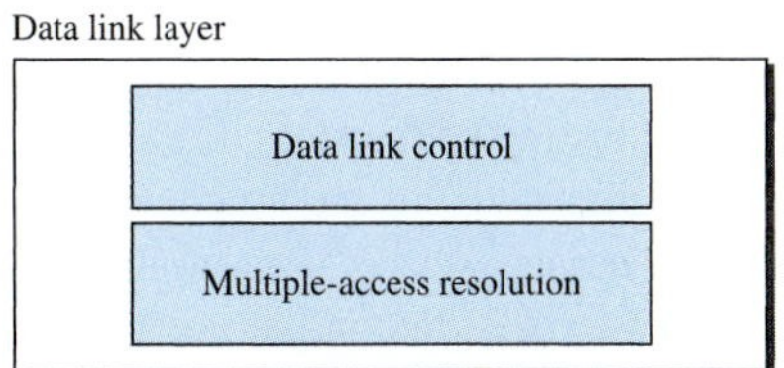

We will see in Chapter 13 that the IEEE has actually made this division for LANs. The upper sublayer that is responsible for flow and error control is called the logical link control (LLC) layer; the lower sublayer that is mostly responsible for multiple-access resolution is called the media access control (MAC) layer.

When nodes or stations are connected and use a common link, called a multipoint or broadcast link, we need a multiple-access protocol to coordinate access to the link. The problem of controlling the access to the medium is similar to the rules of speaking in an assembly. The procedures guarantee that the right to speak is upheld and ensure that two people do not speak at the same time, do not interrupt each other, do not monopolize the discussion, and so on.

The situation is similar for multipoint networks. Many formal protocols have been devised to handle access to a shared link. We categorize them into three groups. Protocols belonging to each group are shown in Figure 12.2.

Figure 12.2 *Taxonomy of multiple-access protocols discussed in this chapter*

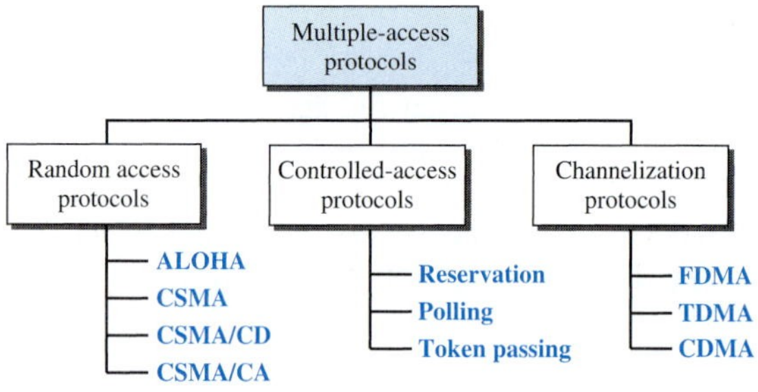

12.1 RANDOM ACCESS

In **random access** or **contention** methods, no station is superior to another station and none is assigned the control over another. No station permits, or does not permit, another station to send. At each instance, a station that has data to send uses a procedure defined by the protocol to make a decision on whether or not to send. This decision depends on the state of the medium (idle or busy). In other words, each station can transmit when it desires on the condition that it follows the predefined procedure, including the testing of the state of the medium.

Two features give this method its name. First, there is no scheduled time for a station to transmit. Transmission is random among the stations. That is why these methods are called *random access*. Second, no rules specify which station should send next. Stations compete with one another to access the medium. That is why these methods are also called *contention* methods.

In a random access method, each station has the right to the medium without being controlled by any other station. However, if more than one station tries to send, there is an access conflict—**collision**—and the frames will be either destroyed or modified. To avoid access conflict or to resolve it when it happens, each station follows a procedure that answers the following questions:

- When can the station access the medium?
- What can the station do if the medium is busy?
- How can the station determine the success or failure of the transmission?
- What can the station do if there is an access conflict?

The random access methods we study in this chapter have evolved from a very interesting protocol known as ALOHA, which used a very simple procedure called **multiple access (MA).** The method was improved with the addition of a procedure that forces the station to sense the medium before transmitting. This was called carrier sense multiple access. This method later evolved into two parallel methods: **carrier sense multiple access with collision detection (CSMA/CD)** and **carrier sense multiple access with collision avoidance (CSMA/CA).** CSMA/CD tells the station what to do when a collision is detected. CSMA/CA tries to avoid the collision.

ALOHA

ALOHA, the earliest random access method, was developed at the University of Hawaii in early 1970. It was designed for a radio (wireless) LAN, but it can be used on any shared medium.

It is obvious that there are potential collisions in this arrangement. The medium is shared between the stations. When a station sends data, another station may attempt to do so at the same time. The data from the two stations collide and become garbled.

Pure ALOHA

The original ALOHA protocol is called **pure ALOHA.** This is a simple, but elegant protocol. The idea is that each station sends a frame whenever it has a frame to send. However, since there is only one channel to share, there is the possibility of collision between frames from different stations. Figure 12.3 shows an example of frame collisions in pure ALOHA.

Figure 12.3 *Frames in a pure ALOHA network*

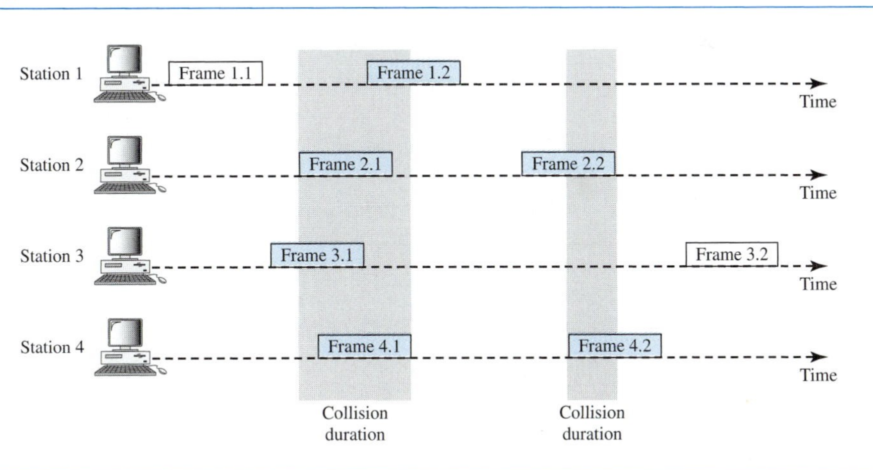

There are four stations (unrealistic assumption) that contend with one another for access to the shared channel. The figure shows that each station sends two frames; there are a total of eight frames on the shared medium. Some of these frames collide because multiple frames are in contention for the shared channel. Figure 12.3 shows that only two frames survive: frame 1.1 from station 1 and frame 3.2 from station 3. We need to mention that even if one bit of a frame coexists on the channel with one bit from another frame, there is a collision and both will be destroyed.

It is obvious that we need to resend the frames that have been destroyed during transmission. The pure ALOHA protocol relies on acknowledgments from the receiver. When a station sends a frame, it expects the receiver to send an acknowledgment. If the acknowledgment does not arrive after a time-out period, the station assumes that the frame (or the acknowledgment) has been destroyed and resends the frame.

A collision involves two or more stations. If all these stations try to resend their frames after the time-out, the frames will collide again. Pure ALOHA dictates that when the time-out period passes, each station waits a random amount of time before resending its frame. The randomness will help avoid more collisions. We call this time the back-off time T_B.

Pure ALOHA has a second method to prevent congesting the channel with retransmitted frames. After a maximum number of retransmission attempts K_{max}, a station must give up and try later. Figure 12.4 shows the procedure for pure ALOHA based on the above strategy.

Figure 12.4 *Procedure for pure ALOHA protocol*

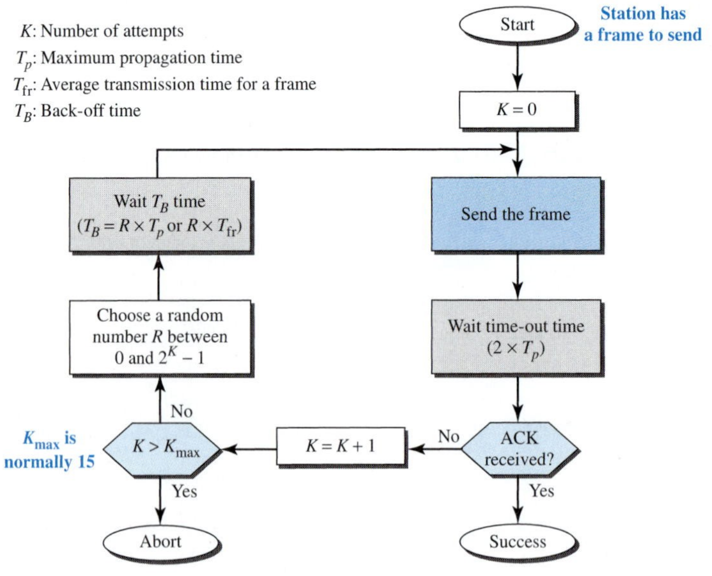

The time-out period is equal to the maximum possible round-trip propagation delay, which is twice the amount of time required to send a frame between the two most widely separated stations ($2 \times T_p$). The back-off time T_B is a random value that normally depends on K (the number of attempted unsuccessful transmissions). The formula for T_B depends on the implementation. One common formula is the **binary exponential back-off.** In this method, for each retransmission, a multiplier in the range 0 to $2^K - 1$ is randomly chosen and multiplied by T_p (maximum propagation time) or T_{fr} (the average time required to send out a frame) to find T_B. Note that in this procedure, the range of the random numbers increases after each collision. The value of K_{max} is usually chosen as 15.

Example 12.1

The stations on a wireless ALOHA network are a maximum of 600 km apart. If we assume that signals propagate at 3×10^8 m/s, we find $T_p = (600 \times 10^5) / (3 \times 10^8) = 2$ ms. Now we can find the value of T_B for different values of K.

a. For $K = 1$, the range is {0, 1}. The station needs to generate a random number with a value of 0 or 1. This means that T_B is either 0 ms (0 × 2) or 2 ms (1 × 2), based on the outcome of the random variable.

b. For $K = 2$, the range is {0, 1, 2, 3}. This means that T_B can be 0, 2, 4, or 6 ms, based on the outcome of the random variable.

c. For $K = 3$, the range is {0, 1, 2, 3, 4, 5, 6, 7}. This means that T_B can be 0, 2, 4, . . . , 14 ms, based on the outcome of the random variable.

d. We need to mention that if $K > 10$, it is normally set to 10.

Vulnerable time Let us find the length of time, the **vulnerable time,** in which there is a possibility of collision. We assume that the stations send fixed-length frames with each frame taking T_{fr} s to send. Figure 12.5 shows the vulnerable time for station A.

Figure 12.5 *Vulnerable time for pure ALOHA protocol*

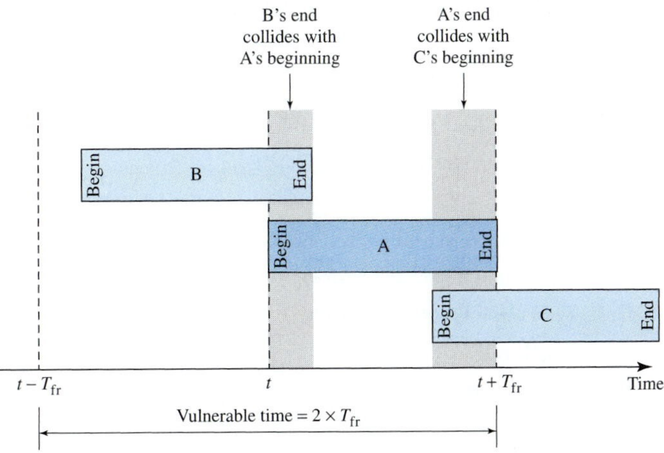

Station A sends a frame at time t. Now imagine station B has already sent a frame between $t - T_{fr}$ and t. This leads to a collision between the frames from station A and station B. The end of B's frame collides with the beginning of A's frame. On the other hand, suppose that station C sends a frame between t and $t + T_{fr}$. Here, there is a collision between frames from station A and station C. The beginning of C's frame collides with the end of A's frame.

Looking at Figure 12.5, we see that the vulnerable time, during which a collision may occur in pure ALOHA, is 2 times the frame transmission time.

Pure ALOHA vulnerable time = $2 \times T_{fr}$

Example 12.2

A pure ALOHA network transmits 200-bit frames on a shared channel of 200 kbps. What is the requirement to make this frame collision-free?

Solution

Average frame transmission time T_{fr} is 200 bits/200 kbps or 1 ms. The vulnerable time is 2×1 ms = 2 ms. This means no station should send later than 1 ms before this station starts transmission and no station should start sending during the one 1-ms period that this station is sending.

Throughput Let us call G the average number of frames generated by the system during one frame transmission time. Then it can be proved that the average number of successful transmissions for pure ALOHA is $S = G \times e^{-2G}$. The maximum throughput S_{max} is 0.184, for $G = \frac{1}{2}$. In other words, if one-half a frame is generated during one frame transmission time (in other words, one frame during two frame transmission times), then 18.4 percent of these frames reach their destination successfully. This is an expected result because the vulnerable time is 2 times the frame transmission time. Therefore, if a station generates only one frame in this vulnerable time (and no other stations generate a frame during this time), the frame will reach its destination successfully.

The throughput for pure ALOHA is $S = G \times e^{-2G}$.
The maximum throughput S_{max} = 0.184 when G = (1/2).

Example 12.3

A pure ALOHA network transmits 200-bit frames on a shared channel of 200 kbps. What is the throughput if the system (all stations together) produces

a. 1000 frames per second
b. 500 frames per second
c. 250 frames per second

Solution

The frame transmission time is 200/200 kbps or 1 ms.

a. If the system creates 1000 frames per second, this is 1 frame per millisecond. The load is 1. In this case $S = G \times e^{-2G}$ or $S = 0.135$ (13.5 percent). This means that the throughput is $1000 \times 0.135 = 135$ frames. Only 135 frames out of 1000 will probably survive.
b. If the system creates 500 frames per second, this is (1/2) frame per millisecond. The load is (1/2). In this case $S = G \times e^{-2G}$ or $S = 0.184$ (18.4 percent). This means that the throughput is $500 \times 0.184 = 92$ and that only 92 frames out of 500 will probably survive. Note that this is the maximum throughput case, percentagewise.
c. If the system creates 250 frames per second, this is (1/4) frame per millisecond. The load is (1/4). In this case $S = G \times e^{-2G}$ or $S = 0.152$ (15.2 percent). This means that the throughput is $250 \times 0.152 = 38$. Only 38 frames out of 250 will probably survive.

Slotted ALOHA

Pure ALOHA has a vulnerable time of $2 \times T_{fr}$. This is so because there is no rule that defines when the station can send. A station may send soon after another station has started or soon before another station has finished. Slotted ALOHA was invented to improve the efficiency of pure ALOHA.

In **slotted ALOHA** we divide the time into slots of T_{fr} s and force the station to send only at the beginning of the time slot. Figure 12.6 shows an example of frame collisions in slotted ALOHA.

Figure 12.6 *Frames in a slotted ALOHA network*

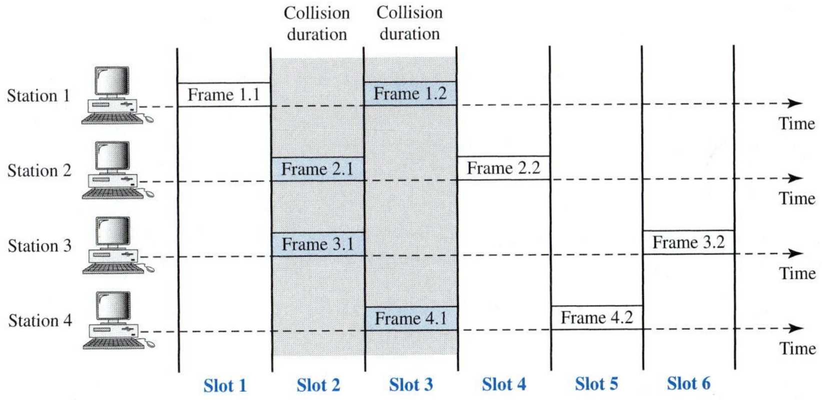

Because a station is allowed to send only at the beginning of the synchronized time slot, if a station misses this moment, it must wait until the beginning of the next time slot. This means that the station which started at the beginning of this slot has already finished sending its frame. Of course, there is still the possibility of collision if two stations try to send at the beginning of the same time slot. However, the vulnerable time is now reduced to one-half, equal to T_{fr}. Figure 12.7 shows the situation.

Figure 12.7 shows that the vulnerable time for slotted ALOHA is one-half that of pure ALOHA.

Slotted ALOHA vulnerable time = T_{fr}

Throughput It can be proved that the average number of successful transmissions for slotted ALOHA is $S = G \times e^{-G}$. The maximum throughput S_{max} is 0.368, when $G = 1$. In other words, if a frame is generated during one frame transmission time, then 36.8 percent of these frames reach their destination successfully. This result can be expected because the vulnerable time is equal to the frame transmission time. Therefore, if a station generates only one frame in this vulnerable time (and no other station generates a frame during this time), the frame will reach its destination successfully.

The throughput for slotted ALOHA is $S = G \times e^{-G}$.
The maximum throughput $S_{max} = 0.368$ when $G = 1$.

Figure 12.7 *Vulnerable time for slotted ALOHA protocol*

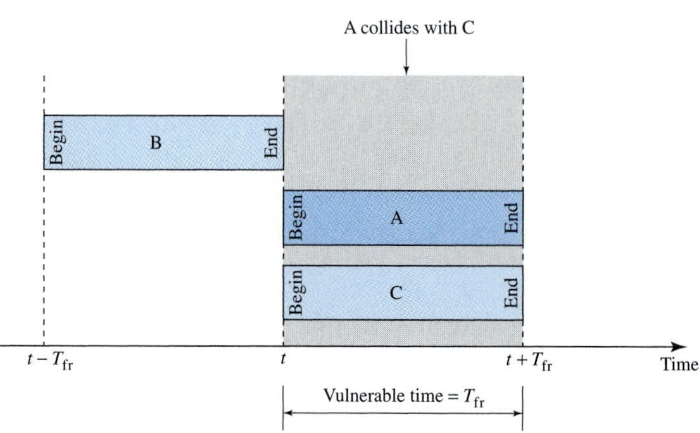

Example 12.4

A slotted ALOHA network transmits 200-bit frames using a shared channel with a 200-kbps bandwidth. Find the throughput if the system (all stations together) produces

a. 1000 frames per second

b. 500 frames per second

c. 250 frames per second

Solution

This situation is similar to the previous exercise except that the network is using slotted ALOHA instead of pure ALOHA. The frame transmission time is 200/200 kbps or 1 ms.

a. In this case G is 1. So $S = G \times e^{-G}$ or $S = 0.368$ (36.8 percent). This means that the throughput is $1000 \times 0.0368 = 368$ frames. Only 368 out of 1000 frames will probably survive. Note that this is the maximum throughput case, percentagewise.

b. Here G is $\frac{1}{2}$. In this case $S = G \times e^{-G}$ or $S = 0.303$ (30.3 percent). This means that the throughput is $500 \times 0.0303 = 151$. Only 151 frames out of 500 will probably survive.

c. Now G is $\frac{1}{4}$. In this case $S = G \times e^{-G}$ or $S = 0.195$ (19.5 percent). This means that the throughput is $250 \times 0.195 = 49$. Only 49 frames out of 250 will probably survive.

Carrier Sense Multiple Access (CSMA)

To minimize the chance of collision and, therefore, increase the performance, the CSMA method was developed. The chance of collision can be reduced if a station senses the medium before trying to use it. **Carrier sense multiple access (CSMA)** requires that each station first listen to the medium (or check the state of the medium) before sending. In other words, CSMA is based on the principle "sense before transmit" or "listen before talk."

CSMA can reduce the possibility of collision, but it cannot eliminate it. The reason for this is shown in Figure 12.8, a space and time model of a CSMA network. Stations are connected to a shared channel (usually a dedicated medium).

The possibility of collision still exists because of propagation delay; when a station sends a frame, it still takes time (although very short) for the first bit to reach every station

Figure 12.8 *Space/time model of the collision in CSMA*

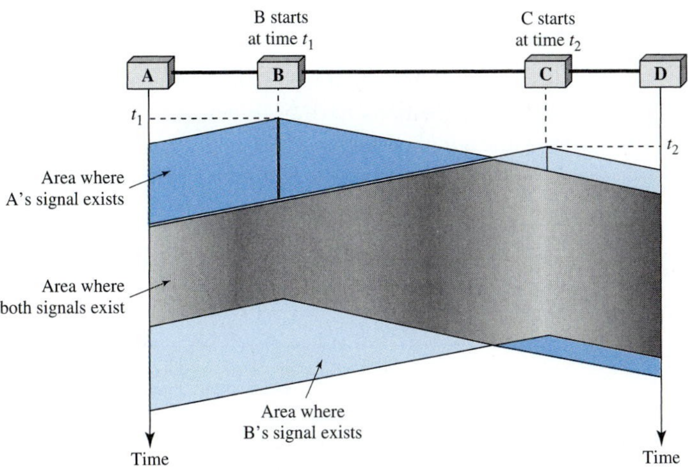

and for every station to sense it. In other words, a station may sense the medium and find it idle, only because the first bit sent by another station has not yet been received.

At time t_1, station B senses the medium and finds it idle, so it sends a frame. At time t_2 ($t_2 > t_1$), station C senses the medium and finds it idle because, at this time, the first bits from station B have not reached station C. Station C also sends a frame. The two signals collide and both frames are destroyed.

Vulnerable Time

The vulnerable time for CSMA is the **propagation time** T_p. This is the time needed for a signal to propagate from one end of the medium to the other. When a station sends a frame, and any other station tries to send a frame during this time, a collision will result. But if the first bit of the frame reaches the end of the medium, every station will already have heard the bit and will refrain from sending. Figure 12.9 shows the worst

Figure 12.9 *Vulnerable time in CSMA*

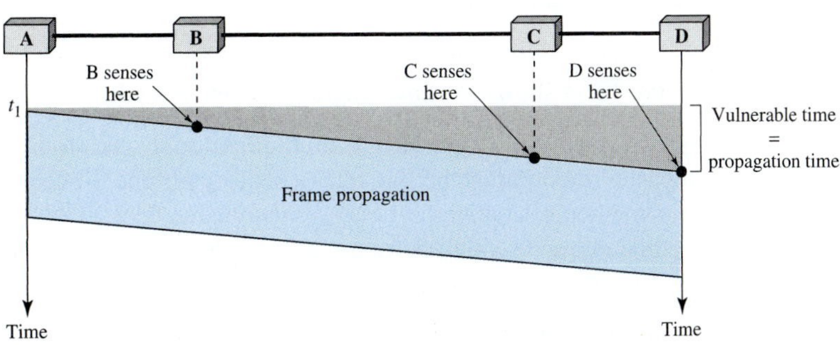

case. The leftmost station A sends a frame at time t_1, which reaches the rightmost station D at time $t_1 + T_p$. The gray area shows the vulnerable area in time and space.

Persistence Methods

What should a station do if the channel is busy? What should a station do if the channel is idle? Three methods have been devised to answer these questions: the 1-persistent method, the nonpersistent method, and the *p*-persistent method. Figure 12.10 shows the behavior of three persistence methods when a station finds a channel busy.

Figure 12.10 *Behavior of three persistence methods*

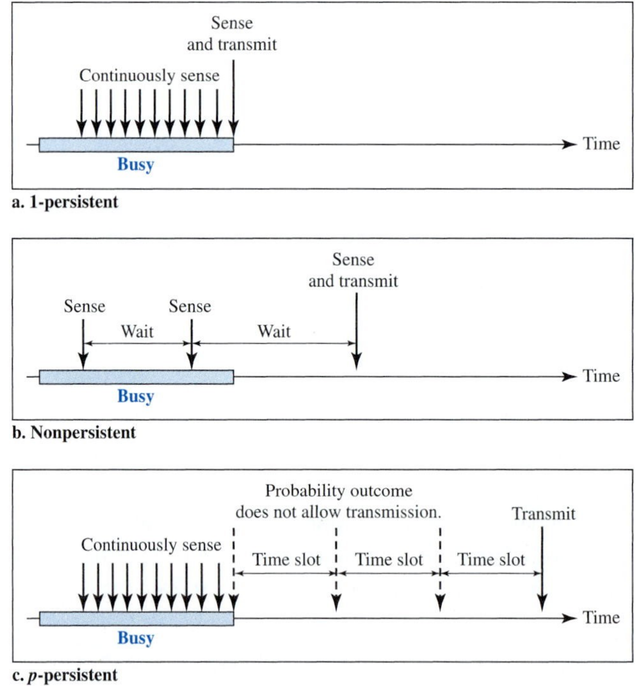

Figure 12.11 shows the flow diagrams for these methods.

1-Persistent The **1-persistent method** is simple and straightforward. In this method, after the station finds the line idle, it sends its frame immediately (with probability 1). This method has the highest chance of collision because two or more stations may find the line idle and send their frames immediately. We will see in Chapter 13 that Ethernet uses this method.

Nonpersistent In the **nonpersistent method,** a station that has a frame to send senses the line. If the line is idle, it sends immediately. If the line is not idle, it waits a

Figure 12.11 *Flow diagram for three persistence methods*

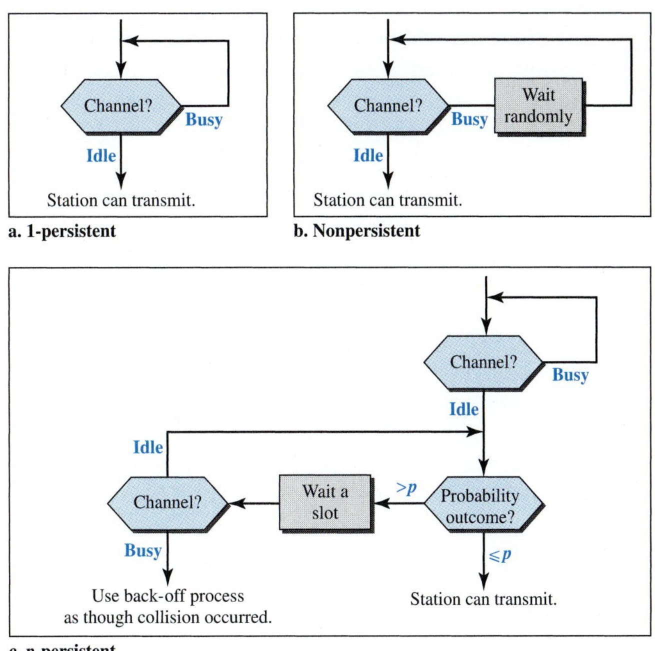

random amount of time and then senses the line again. The nonpersistent approach reduces the chance of collision because it is unlikely that two or more stations will wait the same amount of time and retry to send simultaneously. However, this method reduces the efficiency of the network because the medium remains idle when there may be stations with frames to send.

p-Persistent The ***p*-persistent method** is used if the channel has time slots with a slot duration equal to or greater than the maximum propagation time. The p-persistent approach combines the advantages of the other two strategies. It reduces the chance of collision and improves efficiency. In this method, after the station finds the line idle it follows these steps:

1. With probability p, the station sends its frame.
2. With probability $q = 1 - p$, the station waits for the beginning of the next time slot and checks the line again.
   a. If the line is idle, it goes to step 1.
   b. If the line is busy, it acts as though a collision has occurred and uses the back-off procedure.

Carrier Sense Multiple Access with Collision Detection (CSMA/CD)

The CSMA method does not specify the procedure following a collision. Carrier sense multiple access with collision detection (CSMA/CD) augments the algorithm to handle the collision.

In this method, a station monitors the medium after it sends a frame to see if the transmission was successful. If so, the station is finished. If, however, there is a collision, the frame is sent again.

To better understand CSMA/CD, let us look at the first bits transmitted by the two stations involved in the collision. Although each station continues to send bits in the frame until it detects the collision, we show what happens as the first bits collide. In Figure 12.12, stations A and C are involved in the collision.

Figure 12.12 *Collision of the first bit in CSMA/CD*

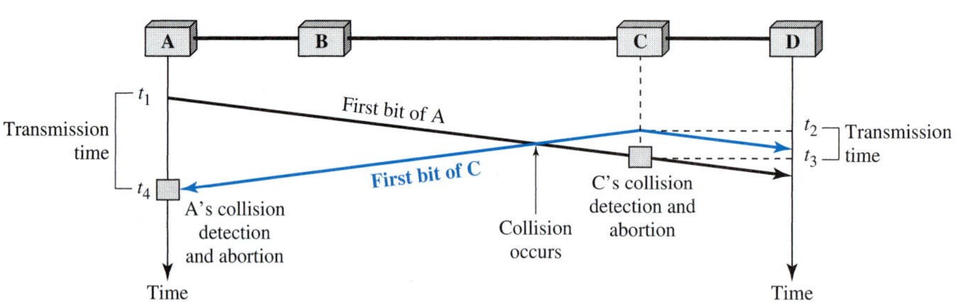

At time t_1, station A has executed its persistence procedure and starts sending the bits of its frame. At time t_2, station C has not yet sensed the first bit sent by A. Station C executes its persistence procedure and starts sending the bits in its frame, which propagate both to the left and to the right. The collision occurs sometime after time t_2. Station C detects a collision at time t_3 when it receives the first bit of A's frame. Station C immediately (or after a short time, but we assume immediately) aborts transmission. Station A detects collision at time t_4 when it receives the first bit of C's frame; it also immediately aborts transmission. Looking at the figure, we see that A transmits for the duration $t_4 - t_1$; C transmits for the duration $t_3 - t_2$. Later we show that, for the protocol to work, the length of any frame divided by the bit rate in this protocol must be more than either of these durations. At time t_4, the transmission of A's frame, though incomplete, is aborted; at time t_3, the transmission of B's frame, though incomplete, is aborted.

Now that we know the time durations for the two transmissions, we can show a more complete graph in Figure 12.13.

Minimum Frame Size

For CSMA/CD to work, we need a restriction on the frame size. Before sending the last bit of the frame, the sending station must detect a collision, if any, and abort the transmission. This is so because the station, once the entire frame is sent, does not keep a copy of the frame and does not monitor the line for collision detection. Therefore, the frame transmission time T_{fr} must be at least two times the maximum propagation time T_p. To understand the reason, let us think about the worst-case scenario. If the two stations involved in a collision are the maximum distance apart, the signal from the first takes time T_p to reach the second, and the effect of the collision takes another time T_P to reach the first. So the requirement is that the first station must still be transmitting after $2T_p$.

Figure 12.13 *Collision and abortion in CSMA/CD*

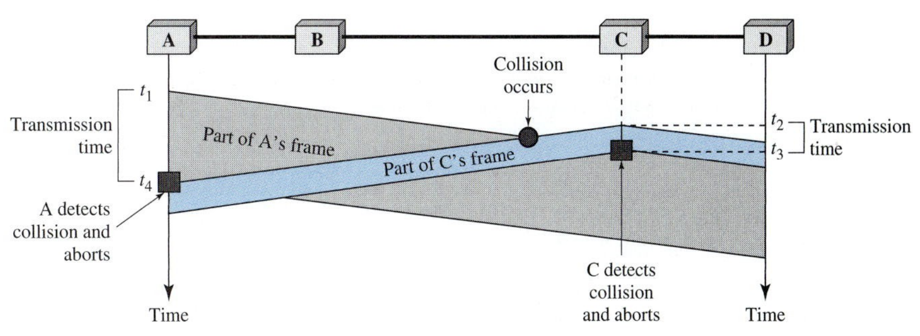

Example 12.5

A network using CSMA/CD has a bandwidth of 10 Mbps. If the maximum propagation time (including the delays in the devices and ignoring the time needed to send a jamming signal, as we see later) is 25.6 μs, what is the minimum size of the frame?

Solution

The frame transmission time is $T_{fr} = 2 \times T_p = 51.2$ μs. This means, in the worst case, a station needs to transmit for a period of 51.2 μs to detect the collision. The minimum size of the frame is 10 Mbps × 51.2 μs = 512 bits or 64 bytes. This is actually the minimum size of the frame for Standard Ethernet, as we will see in Chapter 13.

Procedure

Now let us look at the flow diagram for CSMA/CD in Figure 12.14. It is similar to the one for the ALOHA protocol, but there are differences.

The first difference is the addition of the persistence process. We need to sense the channel before we start sending the frame by using one of the persistence processes we discussed previously (nonpersistent, 1-persistent, or *p*-persistent). The corresponding box can be replaced by one of the persistence processes shown in Figure 12.11.

The second difference is the frame transmission. In ALOHA, we first transmit the entire frame and then wait for an acknowledgment. In CSMA/CD, transmission and collision detection is a continuous process. We do not send the entire frame and then look for a collision. The station transmits and receives continuously and simultaneously (using two different ports). We use a loop to show that transmission is a continuous process. We constantly monitor in order to detect one of two conditions: either transmission is finished or a collision is detected. Either event stops transmission. When we come out of the loop, if a collision has not been detected, it means that transmission is complete; the entire frame is transmitted. Otherwise, a collision has occurred.

The third difference is the sending of a short **jamming signal** that enforces the collision in case other stations have not yet sensed the collision.

Energy Level

We can say that the level of energy in a channel can have three values: zero, normal, and abnormal. At the zero level, the channel is idle. At the normal level, a station has

Figure 12.14 *Flow diagram for the CSMA/CD*

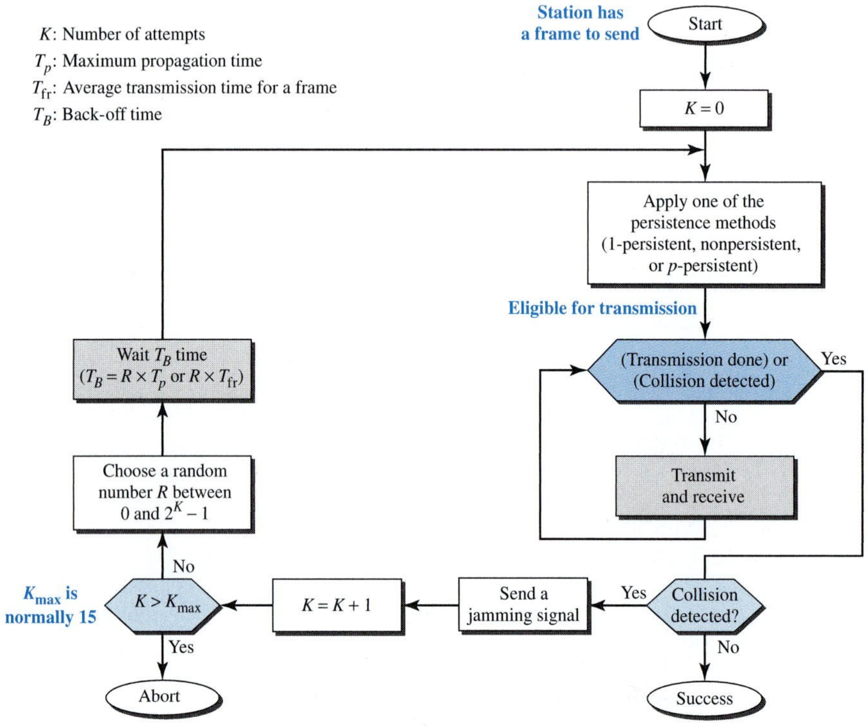

K: Number of attempts
T_p: Maximum propagation time
T_{fr}: Average transmission time for a frame
T_B: Back-off time

Station has a frame to send

Start

$K = 0$

Apply one of the persistence methods (1-persistent, nonpersistent, or p-persistent)

Eligible for transmission

(Transmission done) or (Collision detected)

Yes

No

Transmit and receive

Collision detected?

Yes

No

Success

Send a jamming signal

$K = K + 1$

$K > K_{max}$

K_{max} is normally 15

Yes

Abort

No

Choose a random number R between 0 and $2^K - 1$

Wait T_B time ($T_B = R \times T_p$ or $R \times T_{fr}$)

successfully captured the channel and is sending its frame. At the abnormal level, there is a collision and the level of the energy is twice the normal level. A station that has a frame to send or is sending a frame needs to monitor the energy level to determine if the channel is idle, busy, or in collision mode. Figure 12.15 shows the situation.

Figure 12.15 *Energy level during transmission, idleness, or collision*

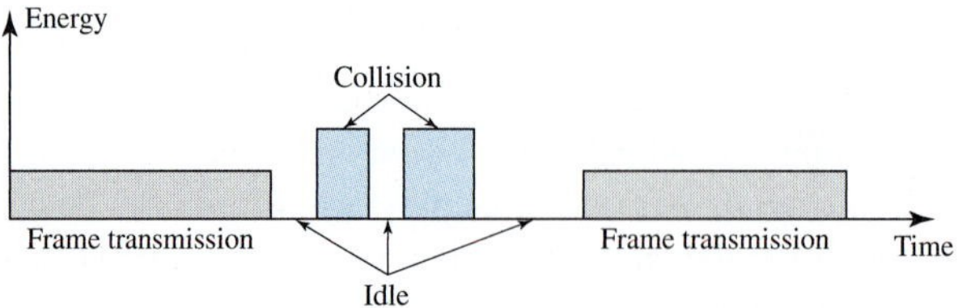

Throughput

The throughput of CSMA/CD is greater than that of pure or slotted ALOHA. The maximum throughput occurs at a different value of G and is based on the persistence method and the value of p in the p-persistent approach. For 1-persistent method the maximum throughput is around 50 percent when $G = 1$. For nonpersistent method, the maximum throughput can go up to 90 percent when G is between 3 and 8.

Carrier Sense Multiple Access with Collision Avoidance (CSMA/CA)

The basic idea behind CSMA/CD is that a station needs to be able to receive while transmitting to detect a collision. When there is no collision, the station receives one signal: its own signal. When there is a collision, the station receives two signals: its own signal and the signal transmitted by a second station. To distinguish between these two cases, the received signals in these two cases must be significantly different. In other words, the signal from the second station needs to add a significant amount of energy to the one created by the first station.

In a wired network, the received signal has almost the same energy as the sent signal because either the length of the cable is short or there are repeaters that amplify the energy between the sender and the receiver. This means that in a collision, the detected energy almost doubles.

However, in a wireless network, much of the sent energy is lost in transmission. The received signal has very little energy. Therefore, a collision may add only 5 to 10 percent additional energy. This is not useful for effective collision detection.

We need to avoid collisions on wireless networks because they cannot be detected. Carrier sense multiple access with collision avoidance (CSMA/CA) was invented for this network. Collisions are avoided through the use of CSMA/CA's three strategies: the interframe space, the contention window, and acknowledgments, as shown in Figure 12.16.

Figure 12.16 *Timing in CSMA/CA*

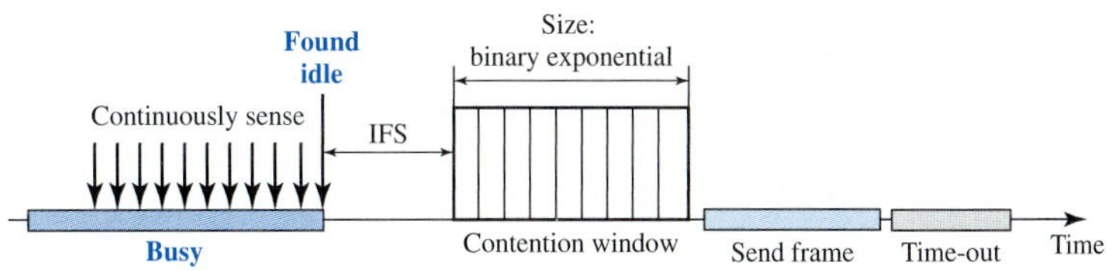

Interframe Space (IFS)

First, collisions are avoided by deferring transmission even if the channel is found idle. When an idle channel is found, the station does not send immediately. It waits for a period of time called the **interframe space** or **IFS.** Even though the channel may appear idle when it is sensed, a distant station may have already started transmitting. The distant station's signal has not yet reached this station. The IFS time allows the front of the transmitted signal by the distant station to reach this station. If after the IFS time the channel is still idle, the station can send, but it still needs to wait a time equal to the contention time (described next). The IFS variable can also be used to prioritize stations or frame types. For example, a station that is assigned a shorter IFS has a higher priority.

In CSMA/CA, the IFS can also be used to define the priority of a station or a frame.

Contention Window

The contention window is an amount of time divided into slots. A station that is ready to send chooses a random number of slots as its wait time. The number of slots in the window changes according to the binary exponential back-off strategy. This means that it is set to one slot the first time and then doubles each time the station cannot detect an idle channel after the IFS time. This is very similar to the *p*-persistent method except that a random outcome defines the number of slots taken by the waiting station. One interesting point about the contention window is that the station needs to sense the channel after each time slot. However, if the station finds the channel busy, it does not restart the process; it just stops the timer and restarts it when the channel is sensed as idle. This gives priority to the station with the longest waiting time.

In CSMA/CA, if the station finds the channel busy, it does not restart the timer of the contention window; it stops the timer and restarts it when the channel becomes idle.

Acknowledgment

With all these precautions, there still may be a collision resulting in destroyed data. In addition, the data may be corrupted during the transmission. The positive acknowledgment and the time-out timer can help guarantee that the receiver has received the frame.

Procedure

Figure 12.17 shows the procedure. Note that the channel needs to be sensed before and after the IFS. The channel also needs to be sensed during the contention time. For each time slot of the contention window, the channel is sensed. If it is found idle, the timer continues; if the channel is found busy, the timer is stopped and continues after the timer becomes idle again.

CSMA/CA and Wireless Networks

CSMA/CA was mostly intended for use in wireless networks. The procedure described above, however, is not sophisticated enough to handle some particular issues related to wireless networks, such as hidden terminals or exposed terminals. We will see how these issues are solved by augmenting the above protocol with hand-shaking features. The use of CSMA/CA in wireless networks will be discussed in Chapter 14.

Figure 12.17 *Flow diagram for CSMA/CA*

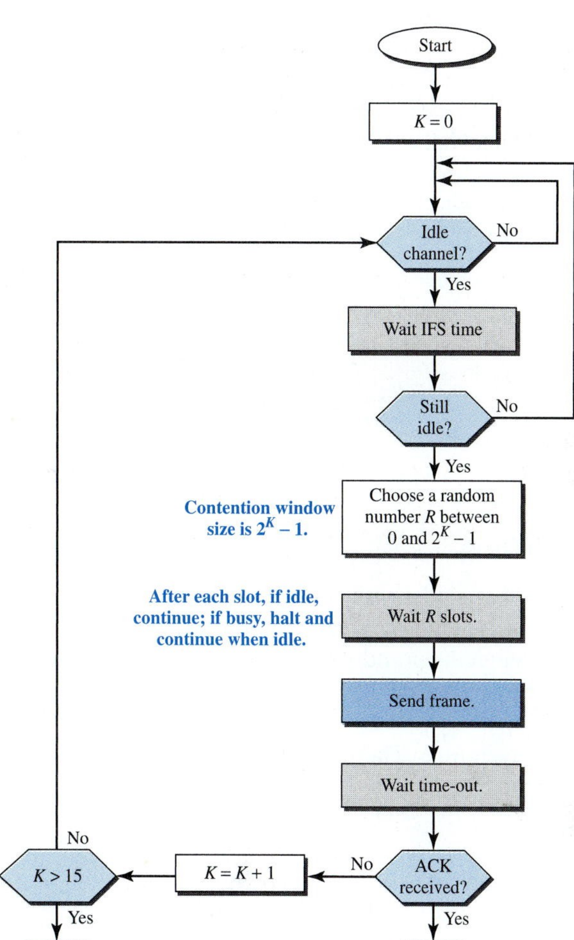

12.2 CONTROLLED ACCESS

In **controlled access,** the stations consult one another to find which station has the right to send. A station cannot send unless it has been authorized by other stations. We discuss three popular controlled-access methods.

Reservation

In the **reservation** method, a station needs to make a reservation before sending data. Time is divided into intervals. In each interval, a reservation frame precedes the data frames sent in that interval.

If there are *N* stations in the system, there are exactly *N* reservation minislots in the reservation frame. Each minislot belongs to a station. When a station needs to send a data frame, it makes a reservation in its own minislot. The stations that have made reservations can send their data frames after the reservation frame.

Figure 12.18 shows a situation with five stations and a five-minislot reservation frame. In the first interval, only stations 1, 3, and 4 have made reservations. In the second interval, only station 1 has made a reservation.

Figure 12.18 *Reservation access method*

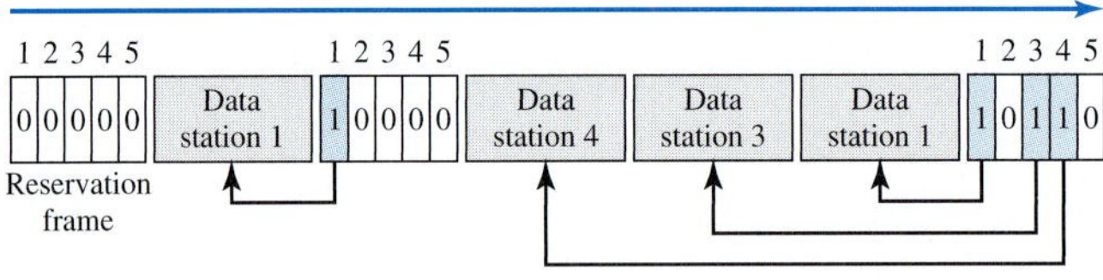

Polling

Polling works with topologies in which one device is designated as a **primary station** and the other devices are **secondary stations.** All data exchanges must be made through the primary device even when the ultimate destination is a secondary device. The primary device controls the link; the secondary devices follow its instructions. It is up to the primary device to determine which device is allowed to use the channel at a given time. The primary device, therefore, is always the initiator of a session (see Figure 12.19).

Figure 12.19 *Select and poll functions in polling access method*

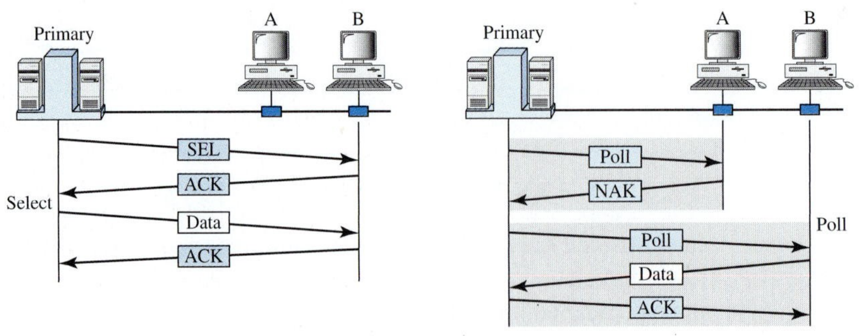

If the primary wants to receive data, it asks the secondaries if they have anything to send; this is called poll function. If the primary wants to send data, it tells the secondary to get ready to receive; this is called select function.

Select

The *select* function is used whenever the primary device has something to send. Remember that the primary controls the link. If the primary is neither sending nor receiving data, it knows the link is available.

If it has something to send, the primary device sends it. What it does not know, however, is whether the target device is prepared to receive. So the primary must alert the secondary to the upcoming transmission and wait for an acknowledgment of the secondary's ready status. Before sending data, the primary creates and transmits a select (SEL) frame, one field of which includes the address of the intended secondary.

Poll

The *poll* function is used by the primary device to solicit transmissions from the secondary devices. When the primary is ready to receive data, it must ask (poll) each device in turn if it has anything to send. When the first secondary is approached, it responds either with a NAK frame if it has nothing to send or with data (in the form of a data frame) if it does. If the response is negative (a NAK frame), then the primary polls the next secondary in the same manner until it finds one with data to send. When the response is positive (a data frame), the primary reads the frame and returns an acknowledgment (ACK frame), verifying its receipt.

Token Passing

In the **token-passing** method, the stations in a network are organized in a logical ring. In other words, for each station, there is a *predecessor* and a *successor*. The predecessor is the station which is logically before the station in the ring; the successor is the station which is after the station in the ring. The current station is the one that is accessing the channel now. The right to this access has been passed from the predecessor to the current station. The right will be passed to the successor when the current station has no more data to send.

But how is the right to access the channel passed from one station to another? In this method, a special packet called a **token** circulates through the ring. The possession of the token gives the station the right to access the channel and send its data. When a station has some data to send, it waits until it receives the token from its predecessor. It then holds the token and sends its data. When the station has no more data to send, it releases the token, passing it to the next logical station in the ring. The station cannot send data until it receives the token again in the next round. In this process, when a station receives the token and has no data to send, it just passes the data to the next station.

Token management is needed for this access method. Stations must be limited in the time they can have possession of the token. The token must be monitored to ensure it has not been lost or destroyed. For example, if a station that is holding the token fails, the token will disappear from the network. Another function of token management is to assign priorities to the stations and to the types of data being transmitted. And finally, token management is needed to make low-priority stations release the token to high-priority stations.

Logical Ring

In a token-passing network, stations do not have to be physically connected in a ring; the ring can be a logical one. Figure 12.20 show four different physical topologies that can create a logical ring.

Figure 12.20 *Logical ring and physical topology in token-passing access method*

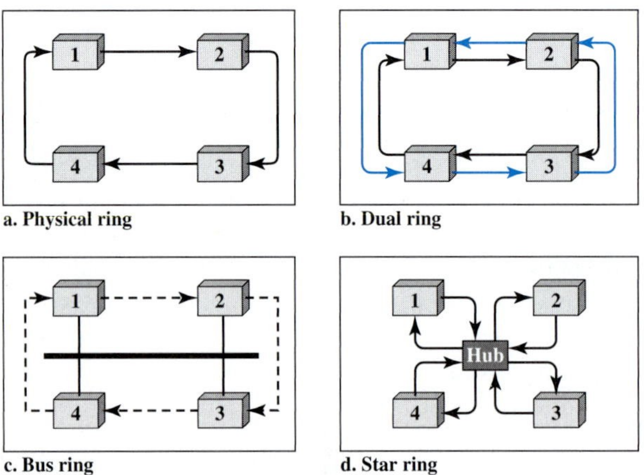

In the physical ring topology, when a station sends the token to its successor, the token cannot be seen by other stations; the successor is the next one in line. This means that the token does not have to have the address of the next successor. The problem with this topology is that if one of the links—the medium between two adjacent stations—fails, the whole system fails.

The dual ring topology uses a second (auxiliary) ring which operates in the reverse direction compared with the main ring. The second ring is for emergencies only (such as a spare tire for a car). If one of the links in the main ring fails, the system automatically combines the two rings to form a temporary ring. After the failed link is restored, the auxiliary ring becomes idle again. Note that for this topology to work, each station needs to have two transmitter ports and two receiver ports. The high-speed Token Ring networks called FDDI (Fiber Distributed Data Interface) and CDDI (Copper Distributed Data Interface) use this topology.

In the bus ring topology, also called a token bus, the stations are connected to a single cable called a bus. They, however, make a logical ring, because each station knows the address of its successor (and also predecessor for token management purposes). When a station has finished sending its data, it releases the token and inserts the address of its successor in the token. Only the station with the address matching the destination address of the token gets the token to access the shared media. The Token Bus LAN, standardized by IEEE, uses this topology.

In a star ring topology, the physical topology is a star. There is a hub, however, that acts as the connector. The wiring inside the hub makes the ring; the stations are connected to this ring through the two wire connections. This topology makes the network less prone to failure because if a link goes down, it will be bypassed by the hub and the rest of the stations can operate. Also adding and removing stations from the ring is easier. This topology is still used in the Token Ring LAN designed by IBM.

12.3 CHANNELIZATION

Channelization is a multiple-access method in which the available bandwidth of a link is shared in time, frequency, or through code, between different stations. In this section, we discuss three channelization protocols: FDMA, TDMA, and CDMA.

We see the application of all these methods in Chapter 16 when we discuss cellular phone systems.

Frequency-Division Multiple Access (FDMA)

In **frequency-division multiple access (FDMA),** the available bandwidth is divided into frequency bands. Each station is allocated a band to send its data. In other words, each band is reserved for a specific station, and it belongs to the station all the time. Each station also uses a bandpass filter to confine the transmitter frequencies. To prevent station interferences, the allocated bands are separated from one another by small *guard bands.* Figure 12.21 shows the idea of FDMA.

Figure 12.21 *Frequency-division multiple access (FDMA)*

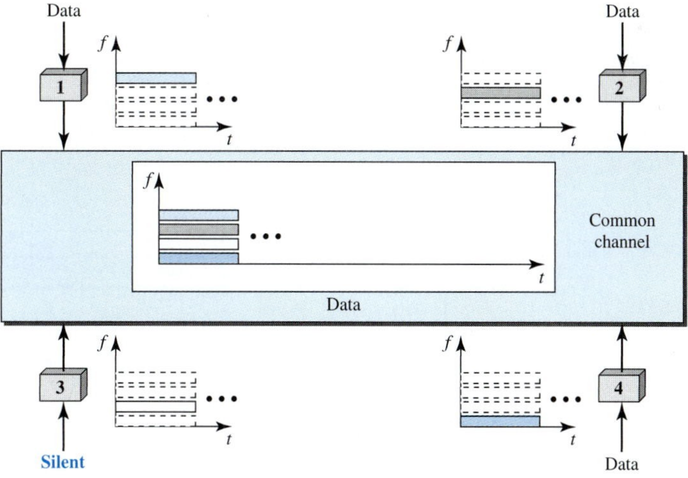

In FDMA, the available bandwidth of the common channel is divided into bands that are separated by guard bands.

FDMA specifies a predetermined frequency band for the entire period of communication. This means that stream data (a continuous flow of data that may not be packetized) can easily be used with FDMA. We will see in Chapter 16 how this feature can be used in cellular telephone systems.

We need to emphasize that although FDMA and FDM conceptually seem similar, there are differences between them. FDM, as we saw in Chapter 6, is a physical layer technique that combines the loads from low-bandwidth channels and transmits them by using a high-bandwidth channel. The channels that are combined are low-pass. The multiplexer modulates the signals, combines them, and creates a bandpass signal. The bandwidth of each channel is shifted by the multiplexer.

FDMA, on the other hand, is an access method in the data link layer. The data link layer in each station tells its physical layer to make a bandpass signal from the data passed to it. The signal must be created in the allocated band. There is no physical multiplexer at the physical layer. The signals created at each station are automatically bandpass-filtered. They are mixed when they are sent to the common channel.

Time-Division Multiple Access (TDMA)

In **time-division multiple access (TDMA),** the stations share the bandwidth of the channel in time. Each station is allocated a time slot during which it can send data. Each station transmits its data in is assigned time slot. Figure 12.22 shows the idea behind TDMA.

Figure 12.22 *Time-division multiple access (TDMA)*

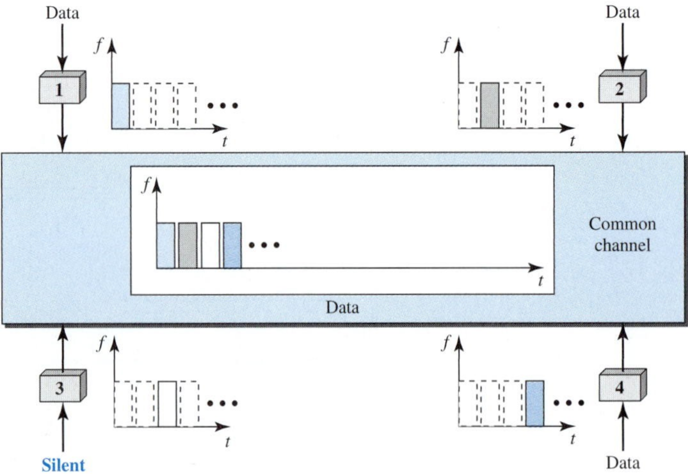

The main problem with TDMA lies in achieving synchronization between the different stations. Each station needs to know the beginning of its slot and the location of its slot. This may be difficult because of propagation delays introduced in the system if the stations are spread over a large area. To compensate for the delays, we can insert *guard*

times. Synchronization is normally accomplished by having some synchronization bits (normally referred to as preamble bits) at the beginning of each slot.

In TDMA, the bandwidth is just one channel that is timeshared between different stations.

We also need to emphasize that although TDMA and TDM conceptually seem the same, there are differences between them. TDM, as we saw in Chapter 6, is a physical layer technique that combines the data from slower channels and transmits them by using a faster channel. The process uses a physical multiplexer that interleaves data units from each channel.

TDMA, on the other hand, is an access method in the data link layer. The data link layer in each station tells its physical layer to use the allocated time slot. There is no physical multiplexer at the physical layer.

Code-Division Multiple Access (CDMA)

Code-division multiple access (CDMA) was conceived several decades ago. Recent advances in electronic technology have finally made its implementation possible. CDMA differs from FDMA because only one channel occupies the entire bandwidth of the link. It differs from TDMA because all stations can send data simultaneously; there is no timesharing.

In CDMA, one channel carries all transmissions simultaneously.

Analogy

Let us first give an analogy. CDMA simply means communication with different codes. For example, in a large room with many people, two people can talk in English if nobody else understands English. Another two people can talk in Chinese if they are the only ones who understand Chinese, and so on. In other words, the common channel, the space of the room in this case, can easily allow communication between several couples, but in different languages (codes).

Idea

Let us assume we have four stations 1, 2, 3, and 4 connected to the same channel. The data from station 1 are d_1, from station 2 are d_2, and so on. The code assigned to the first station is c_1, to the second is c_2, and so on. We assume that the assigned codes have two properties.

1. If we multiply each code by another, we get 0.
2. If we multiply each code by itself, we get 4 (the number of stations).

With these two properties in mind, let us see how the above four stations can send data using the same common channel, as shown in Figure 12.23.

Station 1 multiplies (a special kind of multiplication, as we will see) its data by its code to get $d_1 \cdot c_1$. Station 2 multiplies its data by its code to get $d_2 \cdot c_2$. And so on. The

Figure 12.23 *Simple idea of communication with code*

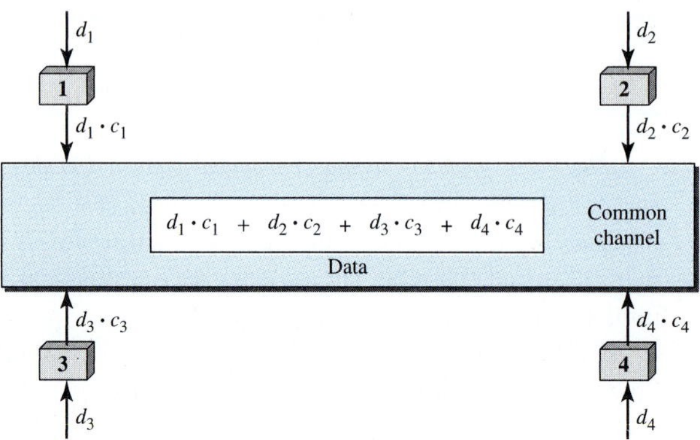

data that go on the channel are the sum of all these terms, as shown in the box. Any station that wants to receive data from one of the other three multiplies the data on the channel by the code of the sender. For example, suppose stations 1 and 2 are talking to each other. Station 2 wants to hear what station 1 is saying. It multiplies the data on the channel by c_1, the code of station 1.

Because $(c_1 \cdot c_1)$ is 4, but $(c_2 \cdot c_1)$, $(c_3 \cdot c_1)$, and $(c_4 \cdot c_1)$ are all 0s, station 2 divides the result by 4 to get the data from station 1.

$$\begin{aligned} \text{data} &= (d_1 \cdot c_1 + d_2 \cdot c_2 + d_3 \cdot c_3 + d_4 \cdot c_4) \cdot c_1 \\ &= d_1 \cdot c_1 \cdot c_1 + d_2 \cdot c_2 \cdot c_1 + d_3 \cdot c_3 \cdot c_1 + d_4 \cdot c_4 \cdot c_1 = 4 \times d_1 \end{aligned}$$

Chips

CDMA is based on coding theory. Each station is assigned a code, which is a sequence of numbers called chips, as shown in Figure 12.24. The codes are for the previous example.

Figure 12.24 *Chip sequences*

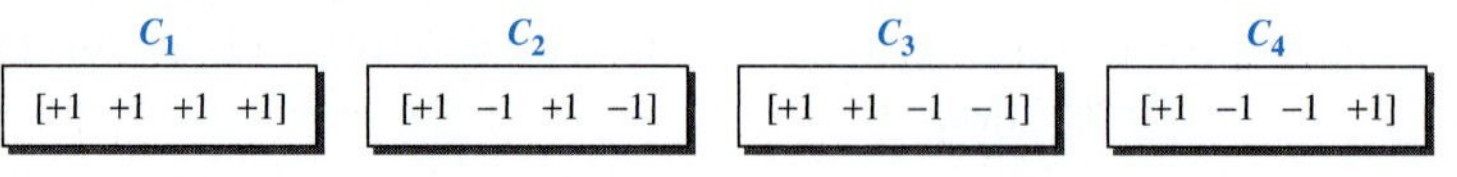

Later in this chapter we show how we chose these sequences. For now, we need to know that we did not choose the sequences randomly; they were carefully selected. They are called **orthogonal sequences** and have the following properties:

1. Each sequence is made of N elements, where N is the number of stations.

2. If we multiply a sequence by a number, every element in the sequence is multiplied by that element. This is called multiplication of a sequence by a scalar. For example,

$$2 \cdot [+1\ +1\ -1\ -1] = [+2\ +2\ -2\ -2]$$

3. If we multiply two equal sequences, element by element, and add the results, we get N, where N is the number of elements in the each sequence. This is called the **inner product** of two equal sequences. For example,

$$[+1\ +1\ -1\ -1] \cdot [+1\ +1\ -1\ -1] = 1 + 1 + 1 + 1 = 4$$

4. If we multiply two different sequences, element by element, and add the results, we get 0. This is called inner product of two different sequences. For example,

$$[+1\ +1\ -1\ -1] \cdot [+1\ +1\ +1\ +1] = 1 + 1 - 1 - 1 = 0$$

5. Adding two sequences means adding the corresponding elements. The result is another sequence. For example,

$$[+1\ +1\ -1\ -1] + [+1\ +1\ +1\ +1] = [+2\ +2\ 0\ 0]$$

Data Representation

We follow these rules for encoding: If a station needs to send a 0 bit, it encodes it as −1; if it needs to send a 1 bit, it encodes it as +1. When a station is idle, it sends no signal, which is interpreted as a 0. These are shown in Figure 12.25.

Figure 12.25 *Data representation in CDMA*

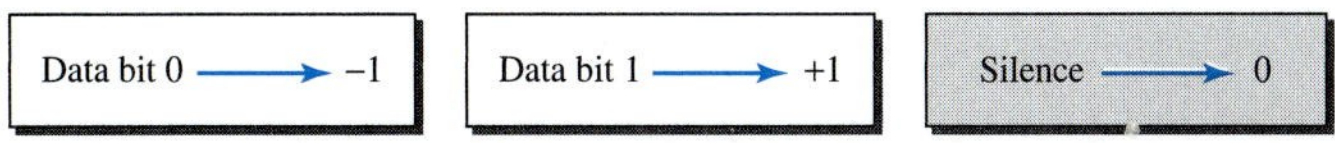

Encoding and Decoding

As a simple example, we show how four stations share the link during a 1-bit interval. The procedure can easily be repeated for additional intervals. We assume that stations 1 and 2 are sending a 0 bit and channel 4 is sending a 1 bit. Station 3 is silent. The data at the sender site are translated to −1, −1, 0, and +1. Each station multiplies the corresponding number by its chip (its orthogonal sequence), which is unique for each station. The result is a new sequence which is sent to the channel. For simplicity, we assume that all stations send the resulting sequences at the same time. The sequence on the channel is the sum of all four sequences as defined before. Figure 12.26 shows the situation.

Now imagine station 3, which we said is silent, is listening to station 2. Station 3 multiplies the total data on the channel by the code for station 2, which is [+1 −1 +1 −1], to get

$$[-1\ -1\ -3\ +1] \cdot [+1\ -1\ +1\ -1] = -4/4 = -1 \longrightarrow \text{bit 1}$$

Figure 12.26 *Sharing channel in CDMA*

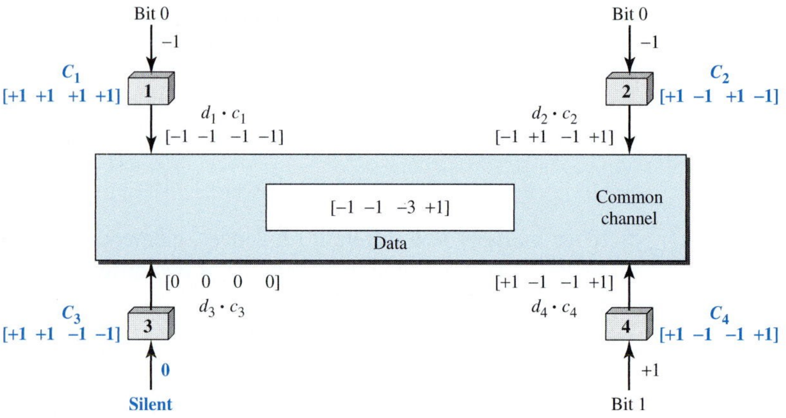

Signal Level

The process can be better understood if we show the digital signal produced by each station and the data recovered at the destination (see Figure 12.27). The figure shows the corresponding signals for each station (using NRZ-L for simplicity) and the signal that is on the common channel.

Figure 12.27 *Digital signal created by four stations in CDMA*

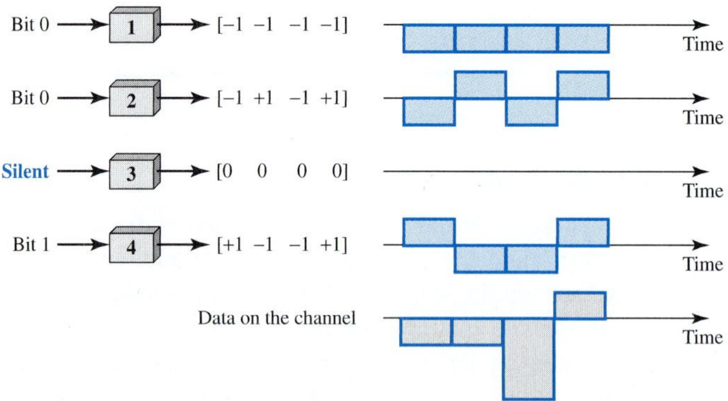

Figure 12.28 shows how station 3 can detect the data sent by station 2 by using the code for station 2. The total data on the channel are multiplied (inner product operation) by the signal representing station 2 chip code to get a new signal. The station then integrates and adds the area under the signal, to get the value −4, which is divided by 4 and interpreted as bit 0.

Figure 12.28 *Decoding of the composite signal for one in CDMA*

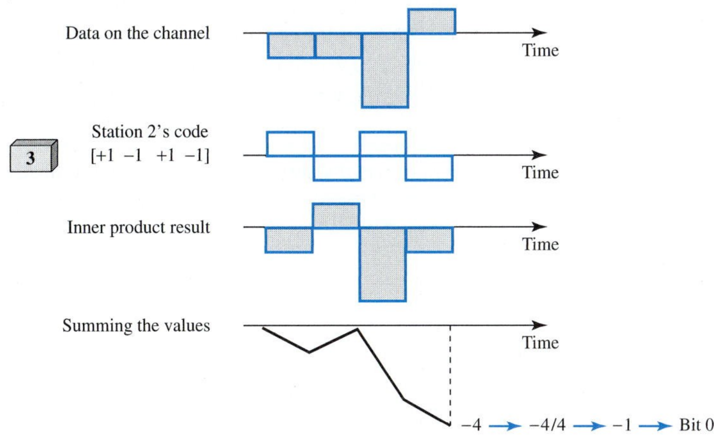

Sequence Generation

To generate chip sequences, we use a **Walsh table,** which is a two-dimensional table with an equal number of rows and columns, as shown in Figure 12.29.

Figure 12.29 *General rule and examples of creating Walsh tables*

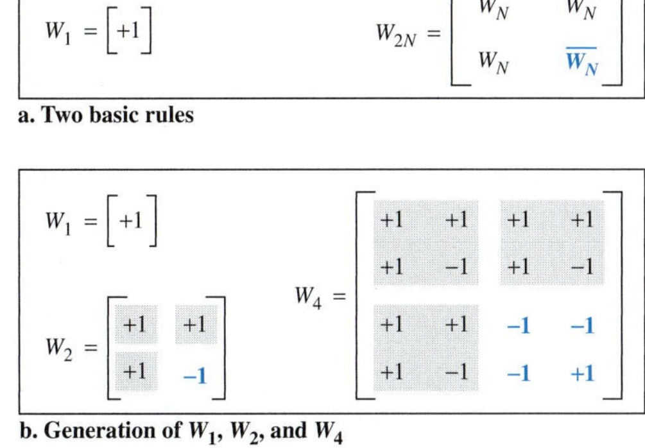

$W_1 = [+1]$
$W_{2N} = \begin{bmatrix} W_N & W_N \\ W_N & \overline{W_N} \end{bmatrix}$
a. Two basic rules
$W_1 = [+1]$
$W_2 = \begin{bmatrix} +1 & +1 \\ +1 & -1 \end{bmatrix}$
$W_4 = \begin{bmatrix} +1 & +1 & +1 & +1 \\ +1 & -1 & +1 & -1 \\ +1 & +1 & -1 & -1 \\ +1 & -1 & -1 & +1 \end{bmatrix}$
b. Generation of W_1, W_2, and W_4

In the Walsh table, each row is a sequence of chips. W_1 for a one-chip sequence has one row and one column. We can choose −1 or +1 for the chip for this trivial table (we chose +1). According to Walsh, if we know the table for N sequences W_N, we can create the table for $2N$ sequences W_{2N}, as shown in Figure 12.29. The W_N with the overbar $\overline{W_N}$ stands for the complement of W_N, where each +1 is changed to −1 and vice versa. Figure 12.29 also shows how we can create W_2 and W_4 from W_1. After we select W_1, W_2

can be made from four W_1's, with the last one the complement of W_1. After W_2 is generated, W_4 can be made of four W_2's, with the last one the complement of W_2. Of course, W_8 is composed of four W_4's, and so on. Note that after W_N is made, each station is assigned a chip corresponding to a row.

Something we need to emphasize is that the number of sequences N needs to be a power of 2. In other words, we need to have $N = 2^m$.

The number of sequences in a Walsh table needs to be $N = 2^m$.

Example 12.6

Find the chips for a network with

a. Two stations
b. Four stations

Solution

We can use the rows of W_2 and W_4 in Figure 12.29:

a. For a two-station network, we have [+1 +1] and [+1 −1].
b. For a four-station network we have [+1 +1 +1 +1], [+1 −1 +1 −1], [+1 +1 −1 −1], and [+1 −1 −1 +1].

Example 12.7

What is the number of sequences if we have 90 stations in our network?

Solution

The number of sequences needs to be 2^m. We need to choose $m = 7$ and $N = 2^7$ or 128. We can then use 90 of the sequences as the chips.

Example 12.8

Prove that a receiving station can get the data sent by a specific sender if it multiplies the entire data on the channel by the sender's chip code and then divides it by the number of stations.

Solution

Let us prove this for the first station, using our previous four-station example. We can say that the data on the channel $D = (d_1 \cdot c_1 + d_2 \cdot c_2 + d_3 \cdot c_3 + d_4 \cdot c_4)$. The receiver which wants to get the data sent by station 1 multiplies these data by c_1.

$$\begin{aligned} D \cdot c_1 &= (d_1 \cdot c_1 + d_2 \cdot c_2 + d_3 \cdot c_3 + d_4 \cdot c_4) \cdot c_1 \\ &= d_1 \cdot c_1 \cdot c_1 + d_2 \cdot c_2 \cdot c_1 + d_3 \cdot c_3 \cdot c_1 + d_4 \cdot c_4 \cdot c_1 \\ &= d_1 \times N + d_2 \times 0 + d_3 \times 0 + d_4 \times 0 \\ &= d_1 \times N \end{aligned}$$

When we divide the result by N, we get d_1.

12.4 RECOMMENDED READING

For more details about subjects discussed in this chapter, we recommend the following books. The items in brackets [. . .] refer to the reference list at the end of the text.

Books

Multiple access is discussed in Chapter 4 of [Tan03], Chapter 16 of [Sta04], Chapter 6 of [GW04], and Chapter 8 of [For03]. More advanced materials can be found in [KMK04].

12.5 KEY TERMS

1-persistent method
ALOHA
binary exponential backup
carrier sense multiple access (CSMA)
carrier sense multiple access with collision avoidance (CSMA/CA)
carrier sense multiple access with collision detection (CSMA/CD)
channelization
code-division multiple access (CDMA)
collision
contention
controlled access
frequency-division multiple access (FDMA)
inner product
interframe space (IFS)
jamming signal
multiple access (MA)
nonpersistent method
orthogonal sequence
polling
p-persistent method
primary station
propagation time
pure ALOHA
random access
reservation
secondary station
slotted ALOHA
time-division multiple access (TDMA)
token
token passing
vulnerable time
Walsh table

12.6 SUMMARY

- We can consider the data link layer as two sublayers. The upper sublayer is responsible for data link control, and the lower sublayer is responsible for resolving access to the shared media.
- Many formal protocols have been devised to handle access to a shared link. We categorize them into three groups: random access protocols, controlled access protocols, and channelization protocols.
- In random access or contention methods, no station is superior to another station and none is assigned the control over another.
- ALOHA allows multiple access (MA) to the shared medium.There are potential collisions in this arrangement. When a station sends data, another station may attempt to do so at the same time. The data from the two stations collide and become garbled.
- To minimize the chance of collision and, therefore, increase the performance, the CSMA method was developed. The chance of collision can be reduced if a station

senses the medium before trying to use it. Carrier sense multiple access (CSMA) requires that each station first listen to the medium before sending. Three methods have been devised for carrier sensing: 1-persistent, nonpersistent, and *p*-persistent.

- Carrier sense multiple access with collision detection (CSMA/CD) augments the CSMA algorithm to handle collision. In this method, a station monitors the medium after it sends a frame to see if the transmission was successful. If so, the station is finished. If, however, there is a collision, the frame is sent again.
- To avoid collisions on wireless networks, carrier sense multiple access with collision avoidance (CSMA/CA) was invented. Collisions are avoided through the use three strategies: the interframe space, the contention window, and acknowledgments.
- In controlled access, the stations consult one another to find which station has the right to send. A station cannot send unless it has been authorized by other stations. We discussed three popular controlled-access methods: reservation, polling, and token passing.
- In the reservation access method, a station needs to make a reservation before sending data. Time is divided into intervals. In each interval, a reservation frame precedes the data frames sent in that interval.
- In the polling method, all data exchanges must be made through the primary device even when the ultimate destination is a secondary device. The primary device controls the link; the secondary devices follow its instructions.
- In the token-passing method, the stations in a network are organized in a logical ring. Each station has a predecessor and a successor. A special packet called a token circulates through the ring.
- Channelization is a multiple-access method in which the available bandwidth of a link is shared in time, frequency, or through code, between different stations. We discussed three channelization protocols: FDMA, TDMA, and CDMA.
- In frequency-division multiple access (FDMA), the available bandwidth is divided into frequency bands. Each station is allocated a band to send its data. In other words, each band is reserved for a specific station, and it belongs to the station all the time.
- In time-division multiple access (TDMA), the stations share the bandwidth of the channel in time. Each station is allocated a time slot during which it can send data. Each station transmits its data in its assigned time slot.
- In code-division multiple access (CDMA), the stations use different codes to achieve multiple access. CDMA is based on coding theory and uses sequences of numbers called chips. The sequences are generated using orthogonal codes such the Walsh tables.

12.7 PRACTICE SET

Review Questions

1. List three categories of multiple access protocols discussed in this chapter.
2. Define random access and list three protocols in this category.

3. Define controlled access and list three protocols in this category.
4. Define channelization and list three protocols in this category.
5. Explain why collision is an issue in a random access protocol but not in controlled access or channelizing protocols.
6. Compare and contrast a random access protocol with a controlled access protocol.
7. Compare and contrast a random access protocol with a channelizing protocol.
8. Compare and contrast a controlled access protocol with a channelizing protocol.
9. Do we need a multiple access protocol when we use the local loop of the telephone company to access the Internet? Why?
10. Do we need a multiple access protocol when we use one CATV channel to access the Internet? Why?

Exercises

11. We have a pure ALOHA network with 100 stations. If T_{fr} = 1 μs, what is the number of frames/s each station can send to achieve the maximum efficiency.
12. Repeat Exercise 11 for slotted ALOHA.
13. One hundred stations on a pure ALOHA network share a 1-Mbps channel. If frames are 1000 bits long, find the throughput if each station is sending 10 frames per second.
14. Repeat Exercise 13 for slotted ALOHA.
15. In a CDMA/CD network with a data rate of 10 Mbps, the minimum frame size is found to be 512 bits for the correct operation of the collision detection process. What should be the minimum frame size if we increase the data rate to 100 Mbps? To 1 Gbps? To 10 Gbps?
16. In a CDMA/CD network with a data rate of 10 Mbps, the maximum distance between any station pair is found to be 2500 m for the correct operation of the collision detection process. What should be the maximum distance if we increase the data rate to 100 Mbps? To 1 Gbps? To 10 Gbps?
17. In Figure 12.12, the data rate is 10 Mbps, the distance between station A and C is 2000 m, and the propagation speed is 2×10^8 m/s. Station A starts sending a long frame at time $t_1 = 0$; station C starts sending a long frame at time t_2 = 3 μs. The size of the frame is long enough to guarantee the detection of collision by both stations. Find:
    a. The time when station C hears the collision (t_3).
    b. The time when station A hears the collision (t_4).
    c. The number of bits station A has sent before detecting the collision.
    d. The number of bits station C has sent before detecting the collision.
18. Repeat Exercise 17 if the data rate is 100 Mbps.
19. Calculate the Walsh table W_8 from W_4 in Figure 12.29.
20. Recreate the W_2 and W_4 tables in Figure 12.29 using $W_1 = [-1]$. Compare the recreated tables with the ones in Figure 12.29.
21. Prove the third and fourth orthogonal properties of Walsh chips for W_4 in Figure 12.29.

22. Prove the third and fourth orthogonal properties of Walsh chips for W_4 recreated in Exercise 20.
23. Repeat the scenario depicted in Figures 12.27 to 12.28 if both stations 1 and 3 are silent.
24. A network with one primary and four secondary stations uses polling. The size of a data frame is 1000 bytes. The size of the poll, ACK, and NAK frames are 32 bytes each. Each station has 5 frames to send. How many total bytes are exchanged if there is no limitation on the number of frames a station can send in response to a poll?
25. Repeat Exercise 24 if each station can send only one frame in response to a poll.

Research Activities

26. Can you explain why the vulnerable time in ALOHA depends on T_{Fr}, but in CSMA depends on T_p?
27. In analyzing ALOHA, we use only one parameter, time; in analyzing CSMA, we use two parameters, space and time. Can you explain the reason?

CHAPTER 13

Wired LANs: Ethernet

In Chapter 1, we learned that a local area network (LAN) is a computer network that is designed for a limited geographic area such as a building or a campus. Although a LAN can be used as an isolated network to connect computers in an organization for the sole purpose of sharing resources, most LANs today are also linked to a wide area network (WAN) or the Internet.

The LAN market has seen several technologies such as Ethernet, Token Ring, Token Bus, FDDI, and ATM LAN. Some of these technologies survived for a while, but Ethernet is by far the dominant technology.

In this chapter, we first briefly discuss the IEEE Standard Project 802, designed to regulate the manufacturing and interconnectivity between different LANs. We then concentrate on the Ethernet LANs.

Although Ethernet has gone through a four-generation evolution during the last few decades, the main concept has remained. Ethernet has changed to meet the market needs and to make use of the new technologies.

13.1 IEEE STANDARDS

In 1985, the Computer Society of the IEEE started a project, called **Project 802,** to set standards to enable intercommunication among equipment from a variety of manufacturers. Project 802 does not seek to replace any part of the OSI or the Internet model. Instead, it is a way of specifying functions of the physical layer and the data link layer of major LAN protocols.

The standard was adopted by the American National Standards Institute (ANSI). In 1987, the International Organization for Standardization (ISO) also approved it as an international standard under the designation ISO 8802.

The relationship of the 802 Standard to the traditional OSI model is shown in Figure 13.1. The IEEE has subdivided the data link layer into two sublayers: **logical link control (LLC)** and **media access control (MAC).** IEEE has also created several physical layer standards for different LAN protocols.

Figure 13.1 *IEEE standard for LANs*

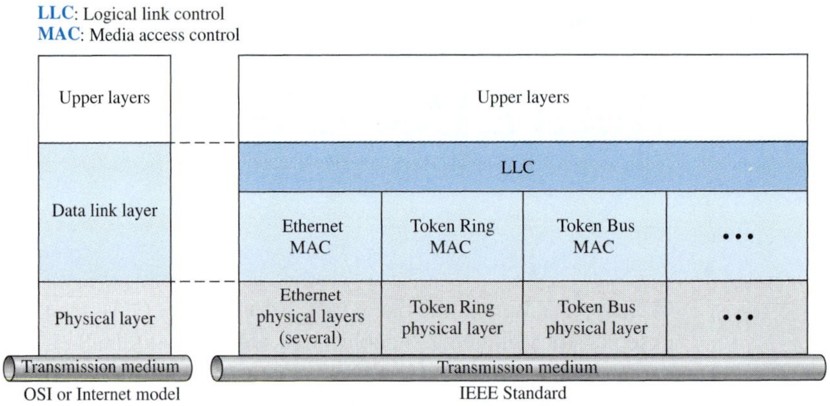

Data Link Layer

As we mentioned before, the data link layer in the IEEE standard is divided into two sublayers: LLC and MAC.

Logical Link Control (LLC)

In Chapter 11, we discussed data link control. We said that data link control handles framing, flow control, and error control. In IEEE Project 802, flow control, error control, and part of the framing duties are collected into one sublayer called the logical link control. Framing is handled in both the LLC sublayer and the MAC sublayer.

The LLC provides one single data link control protocol for all IEEE LANs. In this way, the LLC is different from the media access control sublayer, which provides different protocols for different LANs. A single LLC protocol can provide interconnectivity between different LANs because it makes the MAC sublayer transparent. Figure 13.1 shows one single LLC protocol serving several MAC protocols.

Framing LLC defines a protocol data unit (PDU) that is somewhat similar to that of HDLC. The header contains a control field like the one in HDLC; this field is used for flow and error control. The two other header fields define the upper-layer protocol at the source and destination that uses LLC. These fields are called the **destination service access point (DSAP)** and the **source service access point (SSAP).** The other fields defined in a typical data link control protocol such as HDLC are moved to the MAC sublayer. In other words, a frame defined in HDLC is divided into a PDU at the LLC sublayer and a frame at the MAC sublayer, as shown in Figure 13.2.

Need for LLC The purpose of the LLC is to provide flow and error control for the upper-layer protocols that actually demand these services. For example, if a LAN or several LANs are used in an isolated system, LLC may be needed to provide flow and error control for the application layer protocols. However, most upper-layer protocols

Figure 13.2 *HDLC frame compared with LLC and MAC frames*

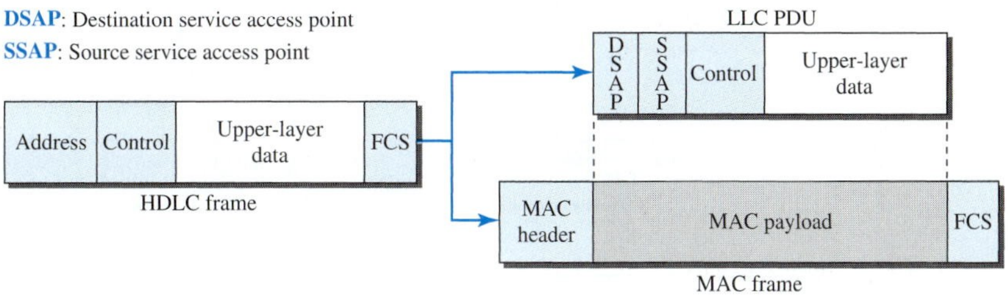

such as IP (discussed in Chapter 20), do not use the services of LLC. For this reason, we end our discussion of LLC.

Media Access Control (MAC)

In Chapter 12, we discussed multiple access methods including random access, controlled access, and channelization. IEEE Project 802 has created a sublayer called media access control that defines the specific access method for each LAN. For example, it defines CSMA/CD as the media access method for Ethernet LANs and the token-passing method for Token Ring and Token Bus LANs. As we discussed in the previous section, part of the framing function is also handled by the MAC layer.

In contrast to the LLC sublayer, the MAC sublayer contains a number of distinct modules; each defines the access method and the framing format specific to the corresponding LAN protocol.

Physical Layer

The physical layer is dependent on the implementation and type of physical media used. IEEE defines detailed specifications for each LAN implementation. For example, although there is only one MAC sublayer for Standard Ethernet, there is a different physical layer specifications for each Ethernet implementations as we will see later.

13.2 STANDARD ETHERNET

The original Ethernet was created in 1976 at Xerox's Palo Alto Research Center (PARC). Since then, it has gone through four generations: **Standard Ethernet** (10[†] Mbps), **Fast Ethernet** (100 Mbps), **Gigabit Ethernet** (1 Gbps), and **Ten-Gigabit Ethernet** (10 Gbps), as shown in Figure 13.3. We briefly discuss all these generations starting with the first, Standard (or traditional) Ethernet.

[†] Ethernet defined some 1-Mbps protocols, but they did not survive.

Figure 13.3 *Ethernet evolution through four generations*

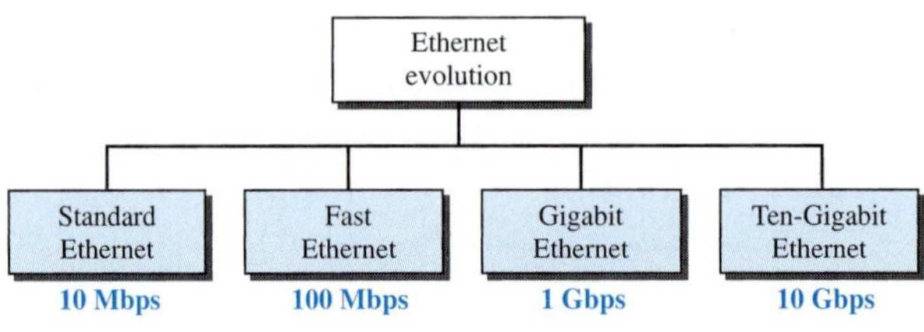

MAC Sublayer

In Standard Ethernet, the MAC sublayer governs the operation of the access method. It also frames data received from the upper layer and passes them to the physical layer.

Frame Format

The Ethernet frame contains seven fields: preamble, SFD, DA, SA, length or type of protocol data unit (PDU), upper-layer data, and the CRC. Ethernet does not provide any mechanism for acknowledging received frames, making it what is known as an unreliable medium. Acknowledgments must be implemented at the higher layers. The format of the MAC frame is shown in Figure 13.4.

Figure 13.4 *802.3 MAC frame*

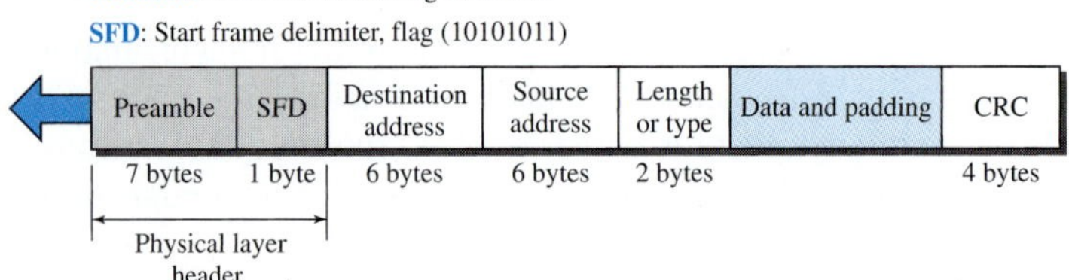

- **Preamble.** The first field of the 802.3 frame contains 7 bytes (56 bits) of alternating 0s and 1s that alerts the receiving system to the coming frame and enables it to synchronize its input timing. The pattern provides only an alert and a timing pulse. The 56-bit pattern allows the stations to miss some bits at the beginning of the frame. The **preamble** is actually added at the physical layer and is not (formally) part of the frame.
- **Start frame delimiter (SFD).** The second field (1 byte: 10101011) signals the beginning of the frame. The SFD warns the station or stations that this is the last chance for synchronization. The last 2 bits is 11 and alerts the receiver that the next field is the destination address.

- **Destination address (DA).** The DA field is 6 bytes and contains the physical address of the destination station or stations to receive the packet. We will discuss addressing shortly.
- **Source address (SA).** The SA field is also 6 bytes and contains the physical address of the sender of the packet. We will discuss addressing shortly.
- **Length or type.** This field is defined as a type field or length field. The original Ethernet used this field as the type field to define the upper-layer protocol using the MAC frame. The IEEE standard used it as the length field to define the number of bytes in the data field. Both uses are common today.
- **Data.** This field carries data encapsulated from the upper-layer protocols. It is a minimum of 46 and a maximum of 1500 bytes, as we will see later.
- **CRC.** The last field contains error detection information, in this case a CRC-32 (see Chapter 10).

Frame Length

Ethernet has imposed restrictions on both the minimum and maximum lengths of a frame, as shown in Figure 13.5.

Figure 13.5 *Minimum and maximum lengths*

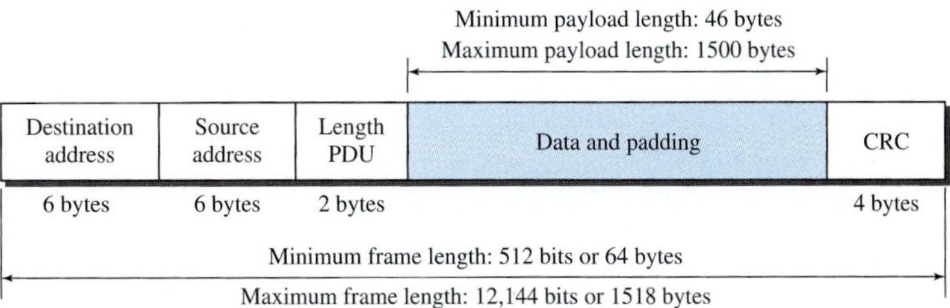

The minimum length restriction is required for the correct operation of CSMA/CD as we will see shortly. An Ethernet frame needs to have a minimum length of 512 bits or 64 bytes. Part of this length is the header and the trailer. If we count 18 bytes of header and trailer (6 bytes of source address, 6 bytes of destination address, 2 bytes of length or type, and 4 bytes of CRC), then the minimum length of data from the upper layer is $64 - 18 = 46$ bytes. If the upper-layer packet is less than 46 bytes, padding is added to make up the difference.

The standard defines the maximum length of a frame (without preamble and SFD field) as 1518 bytes. If we subtract the 18 bytes of header and trailer, the maximum length of the payload is 1500 bytes. The maximum length restriction has two historical reasons. First, memory was very expensive when Ethernet was designed: a maximum length restriction helped to reduce the size of the buffer. Second, the maximum length restriction prevents one station from monopolizing the shared medium, blocking other stations that have data to send.

Frame length:
Minimum: 64 bytes (512 bits) Maximum: 1518 bytes (12,144 bits)

Addressing

Each station on an Ethernet network (such as a PC, workstation, or printer) has its own **network interface card (NIC).** The NIC fits inside the station and provides the station with a 6-byte physical address. As shown in Figure 13.6, the Ethernet address is 6 bytes (48 bits), normally written in **hexadecimal notation,** with a colon between the bytes.

Figure 13.6 *Example of an Ethernet address in hexadecimal notation*

06 : 01 : 02 : 01 : 2C : 4B

6 bytes = 12 hex digits = 48 bits

Unicast, Multicast, and Broadcast Addresses A source address is always a unicast address—the frame comes from only one station. The destination address, however, can be unicast, multicast, or broadcast. Figure 13.7 shows how to distinguish a unicast address from a multicast address. If the least significant bit of the first byte in a destination address is 0, the address is unicast; otherwise, it is multicast.

Figure 13.7 *Unicast and multicast addresses*

The least significant bit of the first byte defines the type of address.
If the bit is 0, the address is unicast; otherwise, it is multicast.

A unicast destination address defines only one recipient; the relationship between the sender and the receiver is one-to-one. A multicast destination address defines a group of addresses; the relationship between the sender and the receivers is one-to-many.

The broadcast address is a special case of the multicast address; the recipients are all the stations on the LAN. A broadcast destination address is forty-eight 1s.

The broadcast destination address is a special case of
the multicast address in which all bits are 1s.

Example 13.1

Define the type of the following destination addresses:

a. 4A:30:10:21:10:1A
b. 47:20:1B:2E:08:EE
c. FF:FF:FF:FF:FF:FF

Solution

To find the type of the address, we need to look at the second hexadecimal digit from the left. If it is even, the address is unicast. If it is odd, the address is multicast. If all digits are F's, the address is broadcast. Therefore, we have the following:

a. This is a unicast address because A in binary is 1010 (even).
b. This is a multicast address because 7 in binary is 0111 (odd).
c. This is a broadcast address because all digits are F's.

The way the addresses are sent out on line is different from the way they are written in hexadecimal notation. The transmission is left-to-right, byte by byte; however, for each byte, the least significant bit is sent first and the most significant bit is sent last. This means that the bit that defines an address as unicast or multicast arrives first at the receiver.

Example 13.2

Show how the address 47:20:1B:2E:08:EE is sent out on line.

Solution

The address is sent left-to-right, byte by byte; for each byte, it is sent right-to-left, bit by bit, as shown below:

← 11100010 00000100 11011000 01110100 00010000 01110111

Access Method: CSMA/CD

Standard Ethernet uses 1-persistent CSMA/CD (see Chapter 12).

Slot Time In an Ethernet network, the round-trip time required for a frame to travel from one end of a maximum-length network to the other plus the time needed to send the jam sequence is called the slot time.

Slot time = round-trip time + time required to send the jam sequence

The slot time in Ethernet is defined in bits. It is the time required for a station to send 512 bits. This means that the actual slot time depends on the data rate; for traditional 10-Mbps Ethernet it is 51.2 μs.

Slot Time and Collision The choice of a 512-bit slot time was not accidental. It was chosen to allow the proper functioning of CSMA/CD. To understand the situation, let us consider two cases.

In the first case, we assume that the sender sends a minimum-size packet of 512 bits. Before the sender can send the entire packet out, the signal travels through the network and reaches the end of the network. If there is another signal at the end of the network (worst case), a collision occurs. The sender has the opportunity to abort the sending of the frame and to send a jam sequence to inform other stations of the collision. The round-trip time plus the time required to send the jam sequence should be less than the time needed for the sender to send the minimum frame, 512 bits. The sender needs to be aware of the collision before it is too late, that is, before it has sent the entire frame.

In the second case, the sender sends a frame larger than the minimum size (between 512 and 1518 bits). In this case, if the station has sent out the first 512 bits and has not heard a collision, it is guaranteed that collision will never occur during the transmission of this frame. The reason is that the signal will reach the end of the network in less than one-half the slot time. If all stations follow the CSMA/CD protocol, they have already sensed the existence of the signal (carrier) on the line and have refrained from sending. If they sent a signal on the line before one-half of the slot time expired, a collision has occurred and the sender has sensed the collision. In other words, collision can only occur during the first half of the slot time, and if it does, it can be sensed by the sender during the slot time. This means that after the sender sends the first 512 bits, it is guaranteed that collision will not occur during the transmission of this frame. The medium belongs to the sender, and no other station will use it. In other words, the sender needs to listen for a collision only during the time the first 512 bits are sent.

Of course, all these assumptions are invalid if a station does not follow the CSMA/CD protocol. In this case, we do not have a collision, we have a corrupted station.

Slot Time and Maximum Network Length There is a relationship between the slot time and the maximum length of the network (collision domain). It is dependent on the propagation speed of the signal in the particular medium. In most transmission media, the signal propagates at 2×10^8 m/s (two-thirds of the rate for propagation in air). For traditional Ethernet, we calculate

$$\text{MaxLength} = \text{PropagationSpeed} \times \frac{\text{SlotTime}}{2}$$

$$\text{MaxLength} = (2 \times 10^8) \times (51.2 \times 10^{-6} / 2) = 5120 \text{ m}$$

Of course, we need to consider the delay times in repeaters and interfaces, and the time required to send the jam sequence. These reduce the maximum-length of a traditional Ethernet network to 2500 m, just 48 percent of the theoretical calculation.

$$\text{MaxLength} = 2500 \text{ m}$$

Physical Layer

The Standard Ethernet defines several physical layer implementations; four of the most common, are shown in Figure 13.8.

Encoding and Decoding

All standard implementations use digital signaling (baseband) at 10 Mbps. At the sender, data are converted to a digital signal using the Manchester scheme; at the receiver, the

Figure 13.8 *Categories of Standard Ethernet*

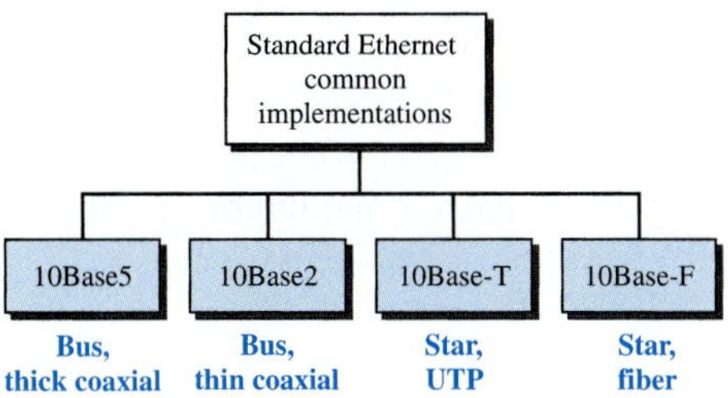

received signal is interpreted as Manchester and decoded into data. As we saw in Chapter 4, Manchester encoding is self-synchronous, providing a transition at each bit interval. Figure 13.9 shows the encoding scheme for Standard Ethernet.

Figure 13.9 *Encoding in a Standard Ethernet implementation*

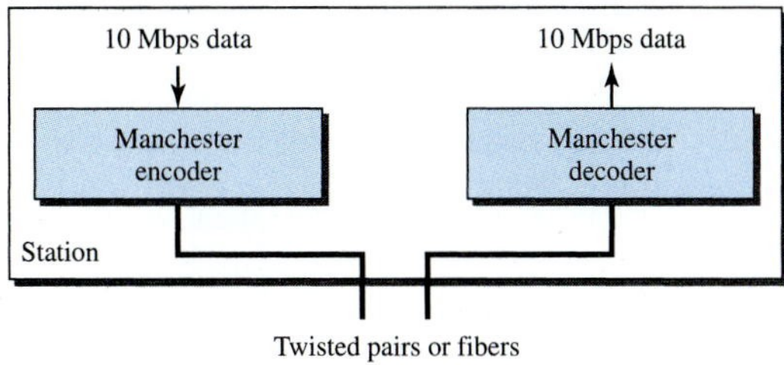

10Base5: Thick Ethernet

The first implementation is called **10Base5, thick Ethernet,** or **Thicknet.** The nickname derives from the size of the cable, which is roughly the size of a garden hose and too stiff to bend with your hands. 10Base5 was the first Ethernet specification to use a bus topology with an external **transceiver** (transmitter/receiver) connected via a tap to a thick coaxial cable. Figure 13.10 shows a schematic diagram of a 10Base5 implementation.

Figure 13.10 *10Base5 implementation*

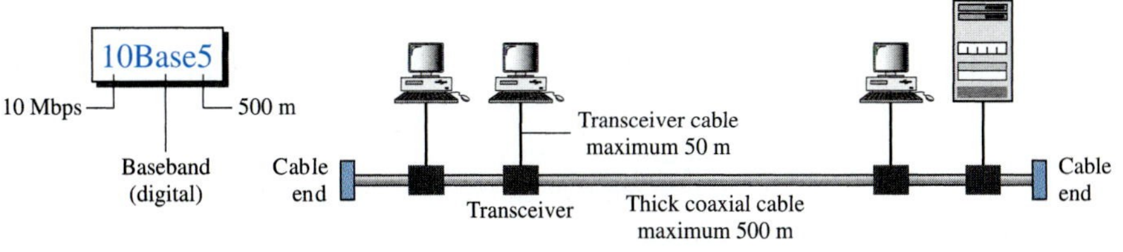

The transceiver is responsible for transmitting, receiving, and detecting collisions. The **transceiver** is connected to the station via a transceiver cable that provides separate paths for sending and receiving. This means that collision can only happen in the coaxial cable.

The maximum length of the coaxial cable must not exceed 500 m, otherwise, there is excessive degradation of the signal. If a length of more than 500 m is needed, up to five segments, each a maximum of 500-meter, can be connected using repeaters. Repeaters will be discussed in Chapter 15.

10Base2: Thin Ethernet

The second implementation is called **10Base2, thin Ethernet,** or **Cheapernet.** 10Base2 also uses a bus topology, but the cable is much thinner and more flexible. The cable can be bent to pass very close to the stations. In this case, the transceiver is normally part of the network interface card (NIC), which is installed inside the station. Figure 13.11 shows the schematic diagram of a 10Base2 implementation.

Figure 13.11 *10Base2 implementation*

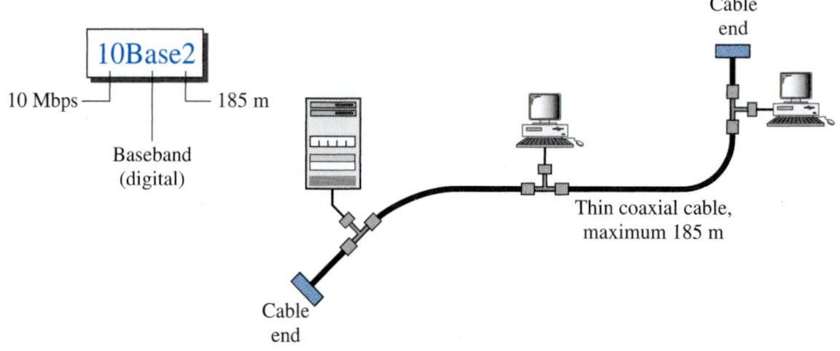

Note that the collision here occurs in the thin coaxial cable. This implementation is more cost effective than 10Base5 because thin coaxial cable is less expensive than thick coaxial and the tee connections are much cheaper than taps. Installation is simpler because the thin coaxial cable is very flexible. However, the length of each segment cannot exceed 185 m (close to 200 m) due to the high level of attenuation in thin coaxial cable.

10Base-T: Twisted-Pair Ethernet

The third implementation is called **10Base-T** or **twisted-pair Ethernet.** 10Base-T uses a physical star topology. The stations are connected to a hub via two pairs of twisted cable, as shown in Figure 13.12.

Note that two pairs of twisted cable create two paths (one for sending and one for receiving) between the station and the hub. Any collision here happens in the hub. Compared to 10Base5 or 10Base2, we can see that the hub actually replaces the coaxial

Figure 13.12 *10Base-T implementation*

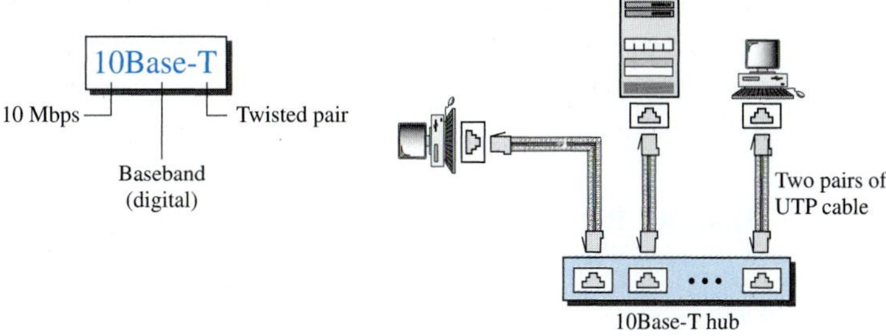

cable as far as a collision is concerned. The maximum length of the twisted cable here is defined as 100 m, to minimize the effect of attenuation in the twisted cable.

10Base-F: Fiber Ethernet

Although there are several types of optical fiber 10-Mbps Ethernet, the most common is called **10Base-F.** 10Base-F uses a star topology to connect stations to a hub. The stations are connected to the hub using two fiber-optic cables, as shown in Figure 13.13.

Figure 13.13 *10Base-F implementation*

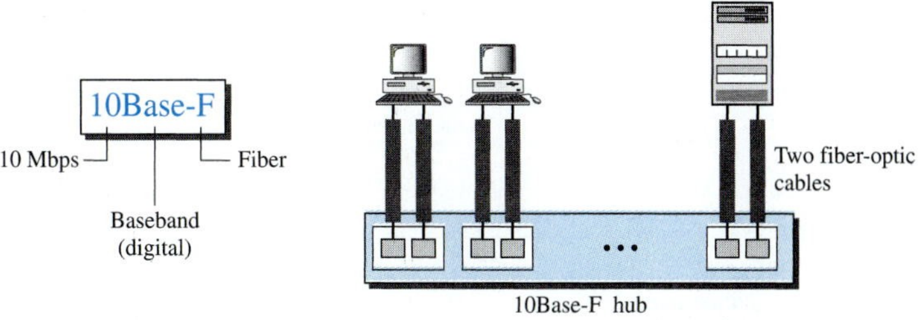

Summary

Table 13.1 shows a summary of Standard Ethernet implementations.

Table 13.1 *Summary of Standard Ethernet implementations*

Characteristics	*10Base5*	*10Base2*	*10Base-T*	*10Base-F*
Media	Thick coaxial cable	Thin coaxial cable	2 UTP	2 Fiber
Maximum length	500 m	185 m	100 m	2000 m
Line encoding	Manchester	Manchester	Manchester	Manchester

13.3 CHANGES IN THE STANDARD

The 10-Mbps Standard Ethernet has gone through several changes before moving to the higher data rates. These changes actually opened the road to the evolution of the Ethernet to become compatible with other high-data-rate LANs. We discuss some of these changes in this section.

Bridged Ethernet

The first step in the Ethernet evolution was the division of a LAN by **bridges.** Bridges have two effects on an Ethernet LAN: They raise the bandwidth and they separate collision domains. We discuss bridges in Chapter 15.

Raising the Bandwidth

In an unbridged Ethernet network, the total capacity (10 Mbps) is shared among all stations with a frame to send; the stations share the bandwidth of the network. If only one station has frames to send, it benefits from the total capacity (10 Mbps). But if more than one station needs to use the network, the capacity is shared. For example, if two stations have a lot of frames to send, they probably alternate in usage. When one station is sending, the other one refrains from sending. We can say that, in this case, each station on average, sends at a rate of 5 Mbps. Figure 13.14 shows the situation.

Figure 13.14 *Sharing bandwidth*

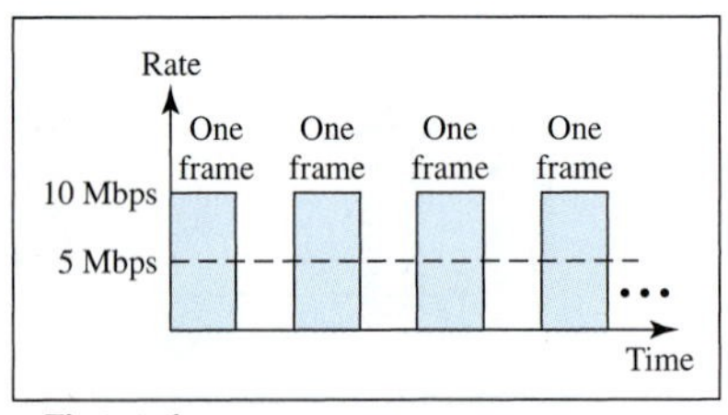

a. First station

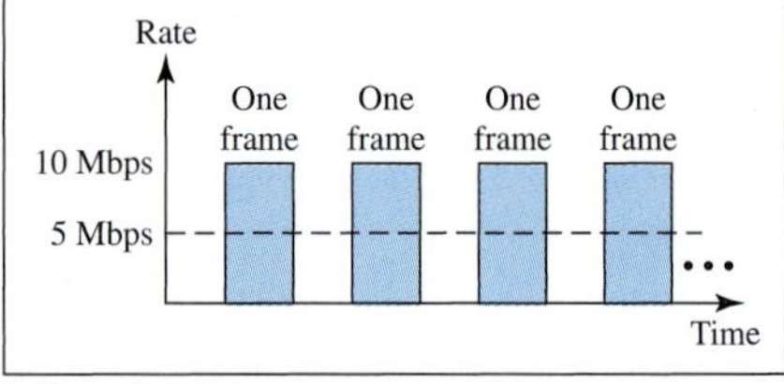

b. Second station

The bridge, as we will learn in Chapter 15, can help here. A bridge divides the network into two or more networks. Bandwidth-wise, each network is independent. For example, in Figure 13.15, a network with 12 stations is divided into two networks, each with 6 stations. Now each network has a capacity of 10 Mbps. The 10-Mbps capacity in each segment is now shared between 6 stations (actually 7 because the bridge acts as a station in each segment), not 12 stations. In a network with a heavy load, each station theoretically is offered 10/6 Mbps instead of 10/12 Mbps, assuming that the traffic is not going through the bridge.

It is obvious that if we further divide the network, we can gain more bandwidth for each segment. For example, if we use a four-port bridge, each station is now offered 10/3 Mbps, which is 4 times more than an unbridged network.

Figure 13.15 *A network with and without a bridge*

a. Without bridging

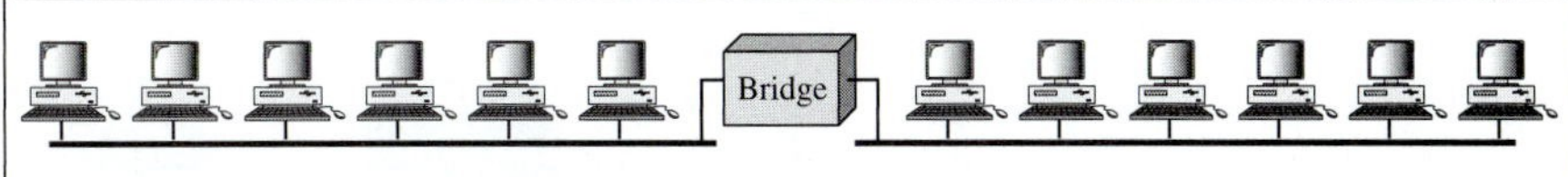

b. With bridging

Separating Collision Domains

Another advantage of a bridge is the separation of the **collision domain.** Figure 13.16 shows the collision domains for an unbridged and a bridged network. You can see that the collision domain becomes much smaller and the probability of collision is reduced tremendously. Without bridging, 12 stations contend for access to the medium; with bridging only 3 stations contend for access to the medium.

Figure 13.16 *Collision domains in an unbridged network and a bridged network*

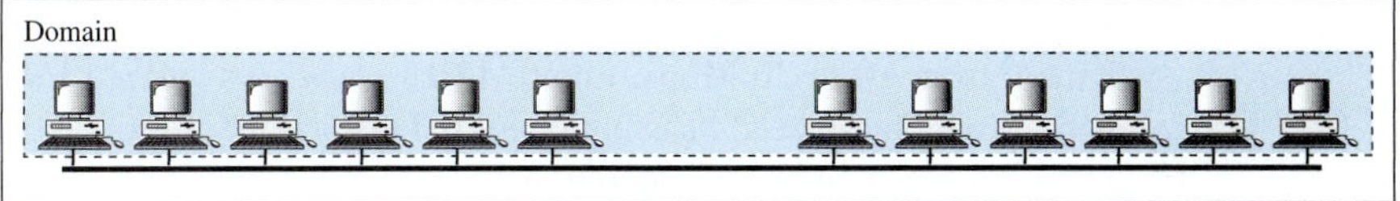

a. Without bridging

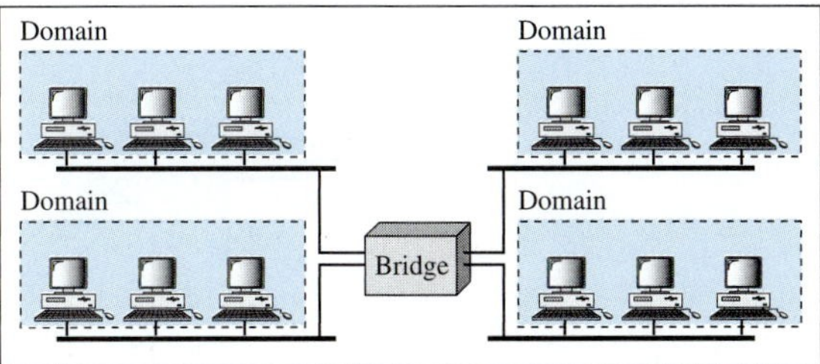

b. With bridging

Switched Ethernet

The idea of a bridged LAN can be extended to a switched LAN. Instead of having two to four networks, why not have N networks, where N is the number of stations on the LAN? In other words, if we can have a multiple-port bridge, why not have an N-port switch? In this way, the bandwidth is shared only between the station and the switch (5 Mbps each). In addition, the collision domain is divided into *N* domains.

A layer 2 **switch** is an *N*-port bridge with additional sophistication that allows faster handling of the packets. Evolution from a bridged Ethernet to a **switched Ethernet** was a big step that opened the way to an even faster Ethernet, as we will see. Figure 13.17 shows a switched LAN.

Figure 13.17 *Switched Ethernet*

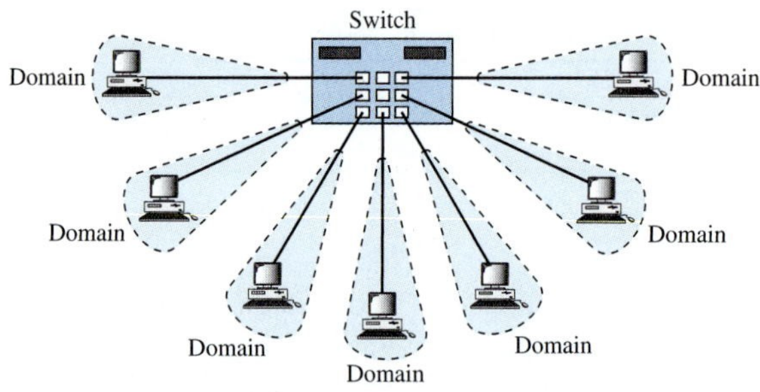

Full-Duplex Ethernet

One of the limitations of 10Base5 and 10Base2 is that communication is half-duplex (10Base-T is always full-duplex); a station can either send or receive, but may not do both at the same time. The next step in the evolution was to move from switched Ethernet to **full-duplex switched Ethernet.** The full-duplex mode increases the capacity of each domain from 10 to 20 Mbps. Figure 13.18 shows a switched Ethernet in full-duplex mode. Note that instead of using one link between the station and the switch, the configuration uses two links: one to transmit and one to receive.

Figure 13.18 *Full-duplex switched Ethernet*

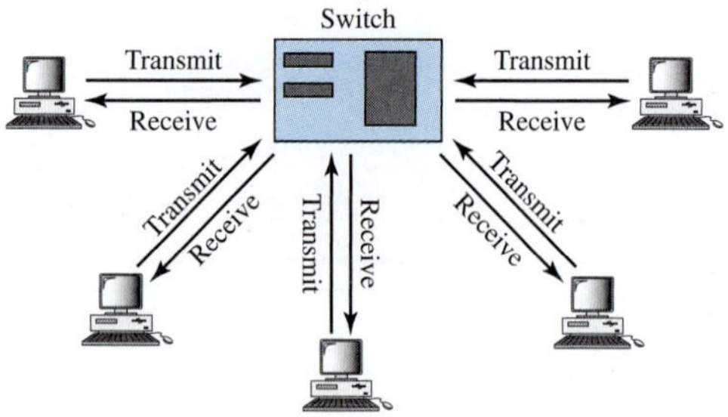

No Need for CSMA/CD

In full-duplex switched Ethernet, there is no need for the CSMA/CD method. In a full-duplex switched Ethernet, each station is connected to the switch via two separate links.

Each station or switch can send and receive independently without worrying about collision. Each link is a point-to-point dedicated path between the station and the switch. There is no longer a need for carrier sensing; there is no longer a need for collision detection. The job of the MAC layer becomes much easier. The carrier sensing and collision detection functionalities of the MAC sublayer can be turned off.

MAC Control Layer

Standard Ethernet was designed as a connectionless protocol at the MAC sublayer. There is no explicit flow control or error control to inform the sender that the frame has arrived at the destination without error. When the receiver receives the frame, it does not send any positive or negative acknowledgment.

To provide for flow and error control in full-duplex switched Ethernet, a new sublayer, called the MAC control, is added between the LLC sublayer and the MAC sublayer.

13.4 FAST ETHERNET

Fast Ethernet was designed to compete with LAN protocols such as FDDI or Fiber Channel (or Fibre Channel, as it is sometimes spelled). IEEE created Fast Ethernet under the name 802.3u. Fast Ethernet is backward-compatible with Standard Ethernet, but it can transmit data 10 times faster at a rate of 100 Mbps. The goals of Fast Ethernet can be summarized as follows:

1. Upgrade the data rate to 100 Mbps.
2. Make it compatible with Standard Ethernet.
3. Keep the same 48-bit address.
4. Keep the same frame format.
5. Keep the same minimum and maximum frame lengths.

MAC Sublayer

A main consideration in the evolution of Ethernet from 10 to 100 Mbps was to keep the MAC sublayer untouched. However, a decision was made to drop the bus topologies and keep only the star topology. For the star topology, there are two choices, as we saw before: half duplex and full duplex. In the half-duplex approach, the stations are connected via a hub; in the full-duplex approach, the connection is made via a switch with buffers at each port.

The access method is the same (CSMA/CD) for the half-duplex approach; for full-duplex Fast Ethernet, there is no need for CSMA/CD. However, the implementations keep CSMA/CD for backward compatibility with Standard Ethernet.

Autonegotiation

A new feature added to Fast Ethernet is called **autonegotiation.** It allows a station or a hub a range of capabilities. Autonegotiation allows two devices to negotiate the mode or data rate of operation. It was designed particularly for the following purposes:

- To allow incompatible devices to connect to one another. For example, a device with a maximum capacity of 10 Mbps can communicate with a device with a 100 Mbps capacity (but can work at a lower rate).
- To allow one device to have multiple capabilities.
- To allow a station to check a hub's capabilities.

Physical Layer

The physical layer in Fast Ethernet is more complicated than the one in Standard Ethernet. We briefly discuss some features of this layer.

Topology

Fast Ethernet is designed to connect two or more stations together. If there are only two stations, they can be connected point-to-point. Three or more stations need to be connected in a star topology with a hub or a switch at the center, as shown in Figure 13.19.

Figure 13.19 *Fast Ethernet topology*

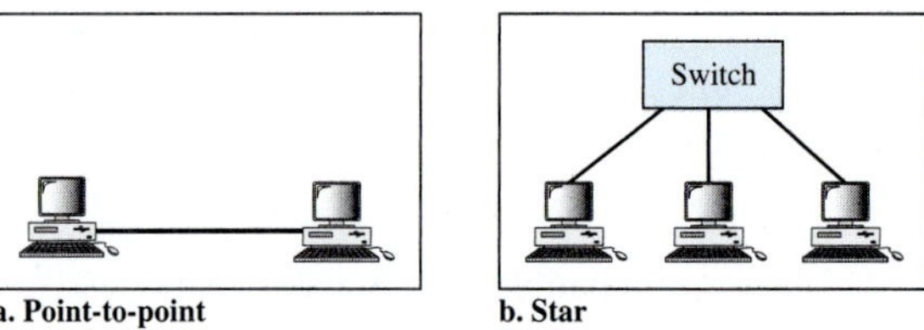

Implementation

Fast Ethernet implementation at the physical layer can be categorized as either two-wire or four-wire. The two-wire implementation can be either category 5 UTP (100Base-TX) or fiber-optic cable (100Base-FX). The four-wire implementation is designed only for category 3 UTP (100Base-T4). See Figure 13.20.

Figure 13.20 *Fast Ethernet implementations*

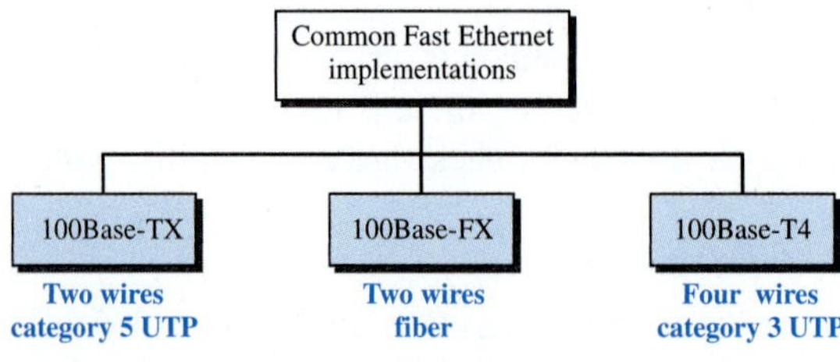

Encoding

Manchester encoding needs a 200-Mbaud bandwidth for a data rate of 100 Mbps, which makes it unsuitable for a medium such as twisted-pair cable. For this reason, the Fast Ethernet designers sought some alternative encoding/decoding scheme. However, it was found that one scheme would not perform equally well for all three implementations. Therefore, three different encoding schemes were chosen (see Figure 13.21).

Figure 13.21 *Encoding for Fast Ethernet implementation*

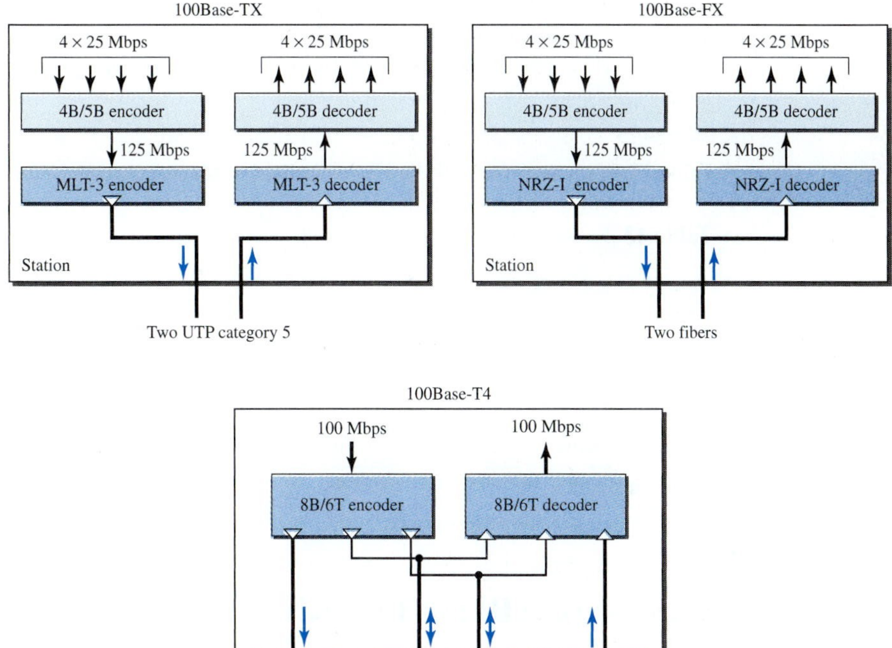

100Base-TX uses two pairs of twisted-pair cable (either category 5 UTP or STP). For this implementation, the MLT-3 scheme was selected since it has good bandwidth performance (see Chapter 4). However, since MLT-3 is not a self-synchronous line coding scheme, 4B/5B block coding is used to provide bit synchronization by preventing the occurrence of a long sequence of 0s and 1s (see Chapter 4). This creates a data rate of 125 Mbps, which is fed into MLT-3 for encoding.

100Base-FX uses two pairs of fiber-optic cables. Optical fiber can easily handle high bandwidth requirements by using simple encoding schemes. The designers of 100Base-FX selected the NRZ-I encoding scheme (see Chapter 4) for this implementation. However, NRZ-I has a bit synchronization problem for long sequences of 0s (or 1s, based on the encoding), as we saw in Chapter 4. To overcome this problem, the designers used 4B/5B block encoding as we described for 100Base-TX. The block encoding increases the bit rate from 100 to 125 Mbps, which can easily be handled by fiber-optic cable.

A 100Base-TX network can provide a data rate of 100 Mbps, but it requires the use of category 5 UTP or STP cable. This is not cost-efficient for buildings that have already been wired for voice-grade twisted-pair (category 3). A new standard, called **100Base-T4,** was designed to use category 3 or higher UTP. The implementation uses four pairs of UTP for transmitting 100 Mbps. Encoding/decoding in 100Base-T4 is more complicated. As this implementation uses category 3 UTP, each twisted-pair cannot easily handle more than 25 Mbaud. In this design, one pair switches between sending and receiving. Three pairs of UTP category 3, however, can handle only 75 Mbaud (25 Mbaud) each. We need to use an encoding scheme that converts 100 Mbps to a 75 Mbaud signal. As we saw in Chapter 4, 8B/6T satisfies this requirement. In 8B/6T, eight data elements are encoded as six signal elements. This means that 100 Mbps uses only (6/8) × 100 Mbps, or 75 Mbaud.

Summary

Table 13.2 is a summary of the Fast Ethernet implementations.

Table 13.2 *Summary of Fast Ethernet implementations*

Characteristics	*100Base-TX*	*100Base-FX*	*100Base-T4*
Media	Cat 5 UTP or STP	Fiber	Cat 4 UTP
Number of wires	2	2	4
Maximum length	100 m	100 m	100 m
Block encoding	4B/5B	4B/5B	
Line encoding	MLT-3	NRZ-I	8B/6T

13.5 GIGABIT ETHERNET

The need for an even higher data rate resulted in the design of the Gigabit Ethernet protocol (1000 Mbps). The IEEE committee calls the Standard 802.3z. The goals of the Gigabit Ethernet design can be summarized as follows:

1. Upgrade the data rate to 1 Gbps.
2. Make it compatible with Standard or Fast Ethernet.
3. Use the same 48-bit address.
4. Use the same frame format.
5. Keep the same minimum and maximum frame lengths.
6. To support autonegotiation as defined in Fast Ethernet.

MAC Sublayer

A main consideration in the evolution of Ethernet was to keep the MAC sublayer untouched. However, to achieve a data rate 1 Gbps, this was no longer possible. Gigabit Ethernet has two distinctive approaches for medium access: half-duplex and full-duplex.

Almost all implementations of Gigabit Ethernet follow the full-duplex approach. However, we briefly discuss the half-duplex approach to show that Gigabit Ethernet can be compatible with the previous generations.

Full-Duplex Mode

In full-duplex mode, there is a central switch connected to all computers or other switches. In this mode, each switch has buffers for each input port in which data are stored until they are transmitted. There is no collision in this mode, as we discussed before. This means that CSMA/CD is not used. Lack of collision implies that the maximum length of the cable is determined by the signal attenuation in the cable, not by the collision detection process.

In the full-duplex mode of Gigabit Ethernet, there is no collision; the maximum length of the cable is determined by the signal attenuation in the cable.

Half-Duplex Mode

Gigabit Ethernet can also be used in half-duplex mode, although it is rare. In this case, a switch can be replaced by a hub, which acts as the common cable in which a collision might occur. The half-duplex approach uses CSMA/CD. However, as we saw before, the maximum length of the network in this approach is totally dependent on the minimum frame size. Three methods have been defined: traditional, carrier extension, and frame bursting.

Traditional In the traditional approach, we keep the minimum length of the frame as in traditional Ethernet (512 bits). However, because the length of a bit is 1/100 shorter in Gigabit Ethernet than in 10-Mbps Ethernet, the slot time for Gigabit Ethernet is 512 bits × 1/1000 μs, which is equal to 0.512 μs. The reduced slot time means that collision is detected 100 times earlier. This means that the maximum length of the network is 25 m. This length may be suitable if all the stations are in one room, but it may not even be long enough to connect the computers in one single office.

Carrier Extension To allow for a longer network, we increase the minimum frame length. The **carrier extension** approach defines the minimum length of a frame as 512 bytes (4096 bits). This means that the minimum length is 8 times longer. This method forces a station to add extension bits (padding) to any frame that is less than 4096 bits. In this way, the maximum length of the network can be increased 8 times to a length of 200 m. This allows a length of 100 m from the hub to the station.

Frame Bursting Carrier extension is very inefficient if we have a series of short frames to send; each frame carries redundant data. To improve efficiency, **frame bursting** was proposed. Instead of adding an extension to each frame, multiple frames are sent. However, to make these multiple frames look like one frame, padding is added between the frames (the same as that used for the carrier extension method) so that the channel is not idle. In other words, the method deceives other stations into thinking that a very large frame has been transmitted.

Physical Layer

The physical layer in Gigabit Ethernet is more complicated than that in Standard or Fast Ethernet. We briefly discuss some features of this layer.

Topology

Gigabit Ethernet is designed to connect two or more stations. If there are only two stations, they can be connected point-to-point. Three or more stations need to be connected in a star topology with a hub or a switch at the center. Another possible configuration is to connect several star topologies or let a star topology be part of another as shown in Figure 13.22.

Figure 13.22 *Topologies of Gigabit Ethernet*

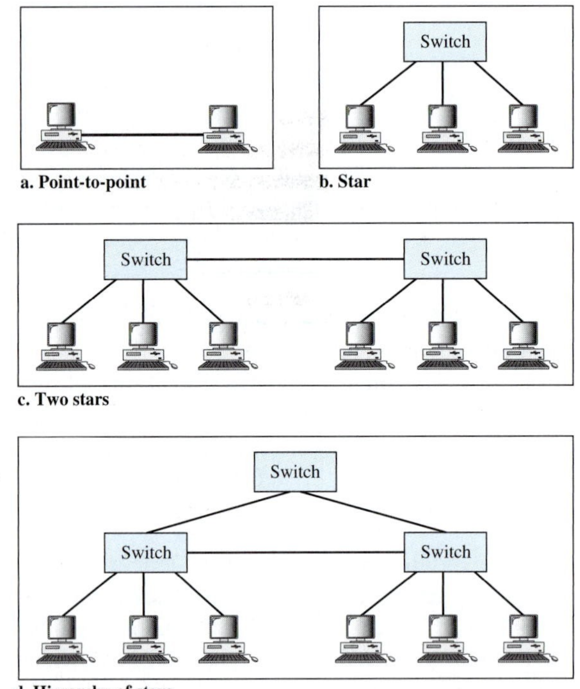

Implementation

Gigabit Ethernet can be categorized as either a two-wire or a four-wire implementation. The two-wire implementations use fiber-optic cable (**1000Base-SX, short-wave,** or **1000Base-LX, long-wave**), or STP (**1000Base-CX**). The four-wire version uses category 5 twisted-pair cable (**1000Base-T**). In other words, we have four implementations, as shown in Figure 13.23. 1000Base-T was designed in response to those users who had already installed this wiring for other purposes such as Fast Ethernet or telephone services.

Figure 13.23 *Gigabit Ethernet implementations*

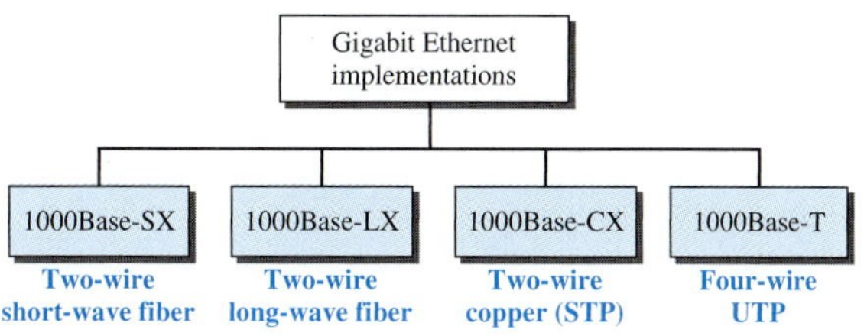

Encoding

Figure 13.24 shows the encoding/decoding schemes for the four implementations.

Figure 13.24 *Encoding in Gigabit Ethernet implementations*

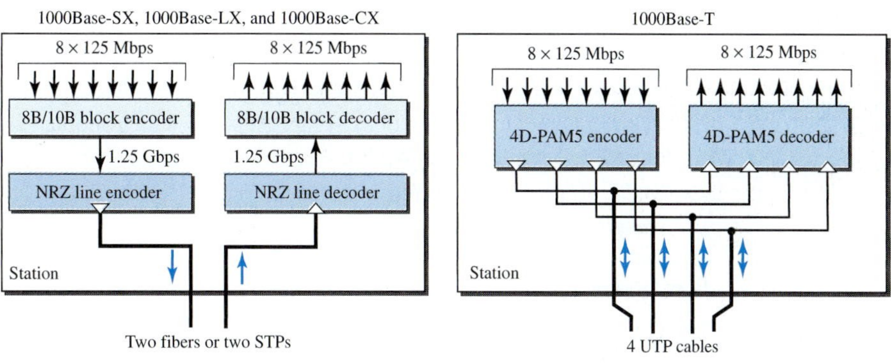

Gigabit Ethernet cannot use the Manchester encoding scheme because it involves a very high bandwidth (2 GBaud). The two-wire implementations use an NRZ scheme, but NRZ does not self-synchronize properly. To synchronize bits, particularly at this high data rate, 8B/10B block encoding, discussed in Chapter 4, is used.

This block encoding prevents long sequences of 0s or 1s in the stream, but the resulting stream is 1.25 Gbps. Note that in this implementation, one wire (fiber or STP) is used for sending and one for receiving.

In the four-wire implementation it is not possible to have 2 wires for input and 2 for output, because each wire would need to carry 500 Mbps, which exceeds the capacity for category 5 UTP. As a solution, 4D-PAM5 encoding, as discussed in Chapter 4, is used to reduce the bandwidth. Thus, all four wires are involved in both input and output; each wire carries 250 Mbps, which is in the range for category 5 UTP cable.

Summary

Table 13.3 is a summary of the Gigabit Ethernet implementations.

Table 13.3 *Summary of Gigabit Ethernet implementations*

Characteristics	*1000Base-SX*	*1000Base-LX*	*1000Base-CX*	*1000Base-T*
Media	Fiber short-wave	Fiber long-wave	STP	Cat 5 UTP
Number of wires	2	2	2	4
Maximum length	550 m	5000 m	25 m	100 m
Block encoding	8B/10B	8B/10B	8B/10B	
Line encoding	NRZ	NRZ	NRZ	4D-PAM5

Ten-Gigabit Ethernet

The IEEE committee created Ten-Gigabit Ethernet and called it Standard 802.3ae. The goals of the Ten-Gigabit Ethernet design can be summarized as follows:

1. Upgrade the data rate to 10 Gbps.
2. Make it compatible with Standard, Fast, and Gigabit Ethernet.
3. Use the same 48-bit address.
4. Use the same frame format.
5. Keep the same minimum and maximum frame lengths.
6. Allow the interconnection of existing LANs into a metropolitan area network (MAN) or a wide area network (WAN).
7. Make Ethernet compatible with technologies such as Frame Relay and ATM (see Chapter 18).

MAC Sublayer

Ten-Gigabit Ethernet operates only in full duplex mode which means there is no need for contention; CSMA/CD is not used in Ten-Gigabit Ethernet.

Physical Layer

The physical layer in Ten-Gigabit Ethernet is designed for using fiber-optic cable over long distances. Three implementations are the most common: **10GBase-S, 10GBase-L,** and **10GBase-E.** Table 13.4 shows a summary of the Ten-Gigabit Ethernet implementaions.

Table 13.4 *Summary of Ten-Gigabit Ethernet implementations*

Characteristics	*10GBase-S*	*10GBase-L*	*10GBase-E*
Media	Short-wave 850-nm multimode	Long-wave 1310-nm single mode	Extended 1550-mm single mode
Maximum length	300 m	10 km	40 km

13.6 RECOMMENDED READING

For more details about subjects discussed in this chapter, we recommend the following books. The items in brackets [. . .] refer to the reference list at the end of the text.

Books

Ethernet is discussed in Chapters 10, 11, and 12 of [For03], Chapter 5 of [Kei02], Section 4.3 of [Tan03], and Chapters 15 and 16 of [Sta04]. [Spu00] is a book about Ethernet. A complete discussion of Gigabit Ethernet can be found in [KCK98] and [Sau98]. Chapter 2 of [Izz00] has a good comparison between different generations of Ethernet.

13.7 KEY TERMS

1000Base-CX
1000Base-LX
1000Base-SX
1000Base-T
100Base-FX
100Base-T4
100Base-TX
10Base2
10Base5
10Base-F
10Base-T
10GBase-E
10GBase-L
10GBase-S
autonegotiation
bridge
carrier extension
Cheapernet
collision domain
destination service access point (DSAP)
Fast Ethernet
frame bursting
full-duplex switched Ethernet
Gigabit Ethernet
hexadecimal notation
logical link control (LLC)
media access control (MAC)
network interface card (NIC)
preamble
Project 802
source service access point (SSAP)
Standard Ethernet
switch
switched Ethernet
Ten-Gigabit Ethernet
thick Ethernet
Thicknet
thin Ethernet
transceiver
twisted-pair Ethernet

13.8 SUMMARY

- Ethernet is the most widely used local area network protocol.
- The IEEE 802.3 Standard defines 1-persistent CSMA/CD as the access method for first-generation 10-Mbps Ethernet.
- The data link layer of Ethernet consists of the LLC sublayer and the MAC sublayer.

- The MAC sublayer is responsible for the operation of the CSMA/CD access method and framing.
- Each station on an Ethernet network has a unique 48-bit address imprinted on its network interface card (NIC).
- The minimum frame length for 10-Mbps Ethernet is 64 bytes; the maximum is 1518 bytes.
- The common implementations of 10-Mbps Ethernet are 10Base5 (thick Ethernet), 10Base2 (thin Ethernet), 10Base-T (twisted-pair Ethernet), and 10Base-F (fiber Ethernet).
- The 10Base5 implementation of Ethernet uses thick coaxial cable. 10Base2 uses thin coaxial cable. 10Base-T uses four twisted-pair cables that connect each station to a common hub. 10Base-F uses fiber-optic cable.
- A bridge can increase the bandwidth and separate the collision domains on an Ethernet LAN.
- A switch allows each station on an Ethernet LAN to have the entire capacity of the network to itself.
- Full-duplex mode doubles the capacity of each domain and removes the need for the CSMA/CD method.
- Fast Ethernet has a data rate of 100 Mbps.
- In Fast Ethernet, autonegotiation allows two devices to negotiate the mode or data rate of operation.
- The common Fast Ethernet implementations are 100Base-TX (two pairs of twisted-pair cable), 100Base-FX (two fiber-optic cables), and 100Base-T4 (four pairs of voice-grade, or higher, twisted-pair cable).
- Gigabit Ethernet has a data rate of 1000 Mbps.
- Gigabit Ethernet access methods include half-duplex mode using traditional CSMA/CD (not common) and full-duplex mode (most popular method).
- The common Gigabit Ethernet implementations are 1000Base-SX (two optical fibers and a short-wave laser source), 1000Base-LX (two optical fibers and a long-wave laser source), and 1000Base-T (four twisted pairs).
- The latest Ethernet standard is Ten-Gigabit Ethernet that operates at 10 Gbps. The three common implementations are 10GBase-S, 10GBase-L, and 10GBase-E. These implementations use fiber-optic cables in full-duplex mode.

13.9 PRACTICE SET

Review Questions

1. How is the preamble field different from the SFD field?
2. What is the purpose of an NIC?
3. What is the difference between a unicast, multicast, and broadcast address?
4. What are the advantages of dividing an Ethernet LAN with a bridge?
5. What is the relationship between a switch and a bridge?

6. Why is there no need for CSMA/CD on a full-duplex Ethernet LAN?
7. Compare the data rates for Standard Ethernet, Fast Ethernet, Gigabit Ethernet, and Ten-Gigabit Ethernet.
8. What are the common Standard Ethernet implementations?
9. What are the common Fast Ethernet implementations?
10. What are the common Gigabit Ethernet implementations?
11. What are the common Ten-Gigabit Ethernet implementations?

Exercises

12. What is the hexadecimal equivalent of the following Ethernet address?

    01011010 00010001 01010101 00011000 10101010 00001111

13. How does the Ethernet address 1A:2B:3C:4D:5E:6F appear on the line in binary?
14. If an Ethernet destination address is 07:01:02:03:04:05, what is the type of the address (unicast, multicast, or broadcast)?
15. The address 43:7B:6C:DE:10:00 has been shown as the source address in an Ethernet frame. The receiver has discarded the frame. Why?
16. An Ethernet MAC sublayer receives 42 bytes of data from the upper layer. How many bytes of padding must be added to the data?
17. An Ethernet MAC sublayer receives 1510 bytes of data from the upper layer. Can the data be encapsulated in one frame? If not, how many frames need to be sent? What is the size of the data in each frame?
18. What is the ratio of useful data to the entire packet for the smallest Ethernet frame? What is the ratio for the largest frame?
19. Suppose the length of a 10Base5 cable is 2500 m. If the speed of propagation in a thick coaxial cable is 200,000,000 m/s, how long does it take for a bit to travel from the beginning to the end of the network? Assume there are 10 μs delay in the equipment.
20. The data rate of 10Base5 is 10 Mbps. How long does it take to create the smallest frame? Show your calculation.

CHAPTER 14

Wireless LANs

Wireless communication is one of the fastest-growing technologies. The demand for connecting devices without the use of cables is increasing everywhere. **Wireless LANs** can be found on college campuses, in office buildings, and in many public areas.

In this chapter, we concentrate on two promising wireless technologies for LANs: IEEE 802.11 wireless LANs, sometimes called wireless Ethernet, and Bluetooth, a technology for small wireless LANs. Although both protocols need several layers to operate, we concentrate mostly on the physical and data link layers.

14.1 IEEE 802.11

IEEE has defined the specifications for a wireless LAN, called **IEEE 802.11,** which covers the physical and data link layers.

Architecture

The standard defines two kinds of services: the basic service set (BSS) and the extended service set (ESS).

Basic Service Set

IEEE 802.11 defines the **basic service set (BSS)** as the building block of a wireless LAN. A basic service set is made of stationary or mobile wireless stations and an optional central base station, known as the **access point (AP).** Figure 14.1 shows two sets in this standard.

The BSS without an AP is a stand-alone network and cannot send data to other BSSs. It is called an *ad hoc architecture*. In this architecture, stations can form a network without the need of an AP; they can locate one another and agree to be part of a BSS. A BSS with an AP is sometimes referred to as an *infrastructure* network.

A BSS without an AP is called an ad hoc network;
a BSS with an AP is called an infrastructure network.

Figure 14.1 *Basic service sets (BSSs)*

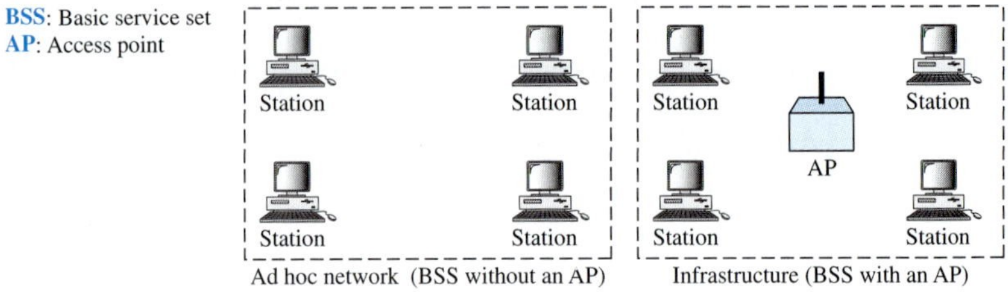

BSS: Basic service set
AP: Access point
Station Station Station Station
Ad hoc network (BSS without an AP)
Station Station AP Station Station
Infrastructure (BSS with an AP)

Extended Service Set

An **extended service set (ESS)** is made up of two or more BSSs with APs. In this case, the BSSs are connected through a *distribution system,* which is usually a wired LAN. The distribution system connects the APs in the BSSs. IEEE 802.11 does not restrict the distribution system; it can be any IEEE LAN such as an Ethernet. Note that the extended service set uses two types of stations: mobile and stationary. The mobile stations are normal stations inside a BSS. The stationary stations are AP stations that are part of a wired LAN. Figure 14.2 shows an ESS.

Figure 14.2 *Extended service sets (ESSs)*

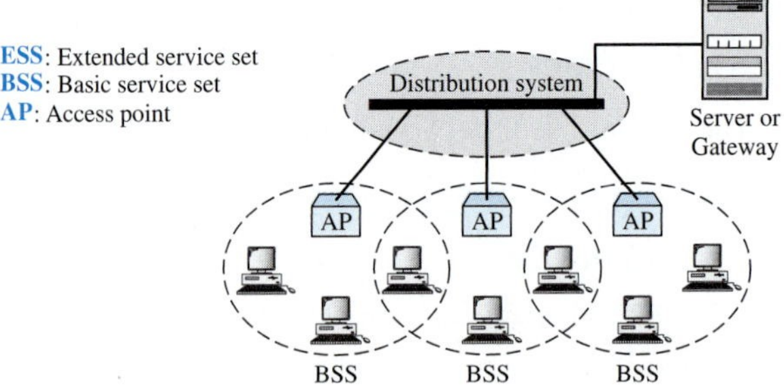

When BSSs are connected, the stations within reach of one another can communicate without the use of an AP. However, communication between two stations in two different BSSs usually occurs via two APs. The idea is similar to communication in a cellular network if we consider each BSS to be a cell and each AP to be a base station. Note that a mobile station can belong to more than one BSS at the same time.

Station Types

IEEE 802.11 defines three types of stations based on their mobility in a wireless LAN: **no-transition, BSS-transition,** and **ESS-transition mobility.** A station with no-transition mobility is either stationary (not moving) or moving only inside a BSS. A station with BSS-transition mobility can move from one BSS to another, but the movement is confined inside one ESS. A station with ESS-transition mobility can move from one ESS to another. However, IEEE 802.11 does not guarantee that communication is continuous during the move.

MAC Sublayer

IEEE 802.11 defines two MAC sublayers: the distributed coordination function (DCF) and point coordination function (PCF). Figure 14.3 shows the relationship between the two MAC sublayers, the LLC sublayer, and the physical layer. We discuss the physical layer implementations later in the chapter and will now concentrate on the MAC sublayer.

Figure 14.3 *MAC layers in IEEE 802.11 standard*

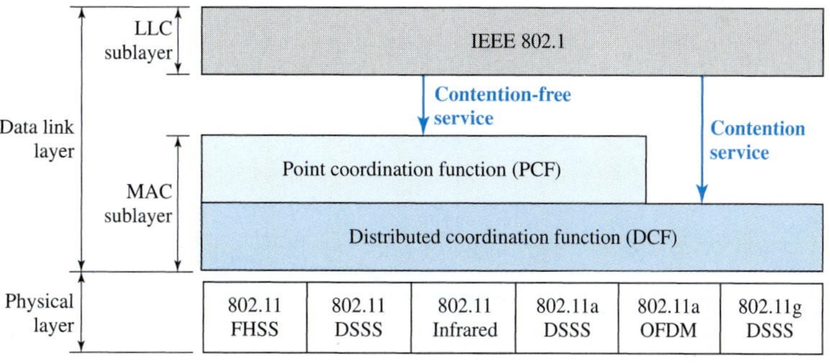

Distributed Coordination Function

One of the two protocols defined by IEEE at the MAC sublayer is called the **distributed coordination function (DCF).** DCF uses CSMA/CA (as defined in Chapter 12) as the access method. Wireless LANs cannot implement CSMA/CD for three reasons:

1. For collision detection a station must be able to send data and receive collision signals at the same time. This can mean costly stations and increased bandwidth requirements.
2. Collision may not be detected because of the hidden station problem. We will discuss this problem later in the chapter.
3. The distance between stations can be great. Signal fading could prevent a station at one end from hearing a collision at the other end.

Process Flowchart Figure 14.4 shows the process flowchart for CSMA/CA as used in wireless LANs. We will explain the steps shortly.

Frame Exchange Time Line Figure 14.5 shows the exchange of data and control frames in time.

Figure 14.4 *CSMA/CA flowchart*

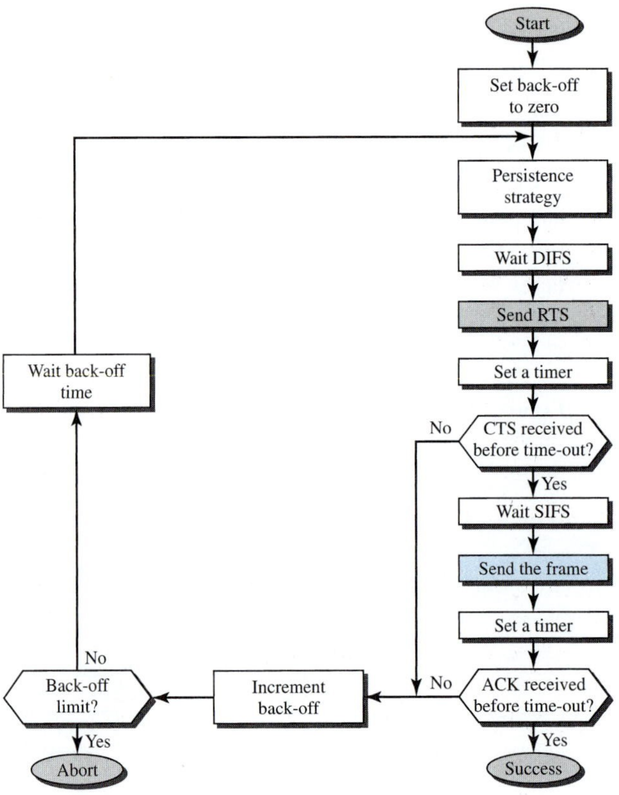

Figure 14.5 *CSMA/CA and NAV*

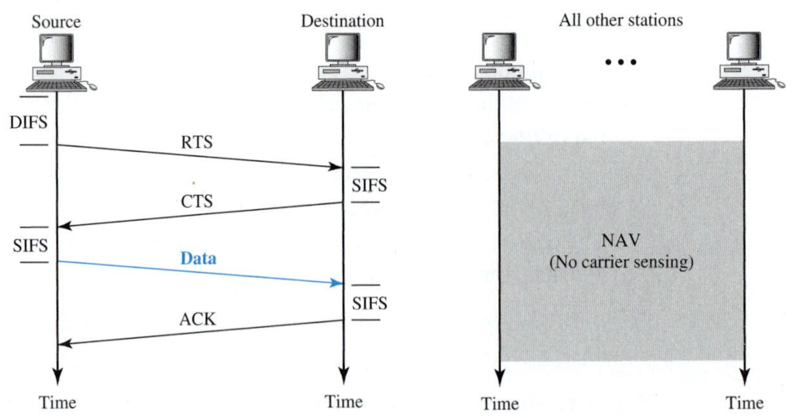

1. Before sending a frame, the source station senses the medium by checking the energy level at the carrier frequency.
   a. The channel uses a persistence strategy with back-off until the channel is idle.
   b. After the station is found to be idle, the station waits for a period of time called the **distributed interframe space (DIFS);** then the station sends a control frame called the request to send (RTS).
2. After receiving the RTS and waiting a period of time called the **short interframe space (SIFS),** the destination station sends a control frame, called the clear to send (CTS), to the source station. This control frame indicates that the destination station is ready to receive data.
3. The source station sends data after waiting an amount of time equal to SIFS.
4. The destination station, after waiting an amount of time equal to SIFS, sends an acknowledgment to show that the frame has been received. Acknowledgment is needed in this protocol because the station does not have any means to check for the successful arrival of its data at the destination. On the other hand, the lack of collision in CSMA/CD is a kind of indication to the source that data have arrived.

Network Allocation Vector How do other stations defer sending their data if one station acquires access? In other words, how is the *collision avoidance* aspect of this protocol accomplished? The key is a feature called NAV.

When a station sends an RTS frame, it includes the duration of time that it needs to occupy the channel. The stations that are affected by this transmission create a timer called a **network allocation vector (NAV)** that shows how much time must pass before these stations are allowed to check the channel for idleness. Each time a station accesses the system and sends an RTS frame, other stations start their NAV. In other words, each station, before sensing the physical medium to see if it is idle, first checks its NAV to see if it has expired. Figure 14.5 shows the idea of NAV.

Collision During Handshaking What happens if there is collision during the time when RTS or CTS control frames are in transition, often called the **handshaking period**? Two or more stations may try to send RTS frames at the same time. These control frames may collide. However, because there is no mechanism for collision detection, the sender assumes there has been a collision if it has not received a CTS frame from the receiver. The back-off strategy is employed, and the sender tries again.

Point Coordination Function (PCF)

The **point coordination function (PCF)** is an optional access method that can be implemented in an infrastructure network (not in an ad hoc network). It is implemented on top of the DCF and is used mostly for time-sensitive transmission.

PCF has a centralized, contention-free polling access method. The AP performs polling for stations that are capable of being polled. The stations are polled one after another, sending any data they have to the AP.

To give priority to PCF over DCF, another set of interframe spaces has been defined: PIFS and SIFS. The SIFS is the same as that in DCF, but the PIFS (PCF IFS) is shorter than the DIFS. This means that if, at the same time, a station wants to use only DCF and an AP wants to use PCF, the AP has priority.

Due to the priority of PCF over DCF, stations that only use DCF may not gain access to the medium. To prevent this, a repetition interval has been designed to cover both contention-free (PCF) and contention-based (DCF) traffic. The **repetition interval,** which is repeated continuously, starts with a special control frame, called a **beacon frame.** When the stations hear the beacon frame, they start their NAV for the duration of the contention-free period of the repetition interval. Figure 14.6 shows an example of a repetition interval.

Figure 14.6 *Example of repetition interval*

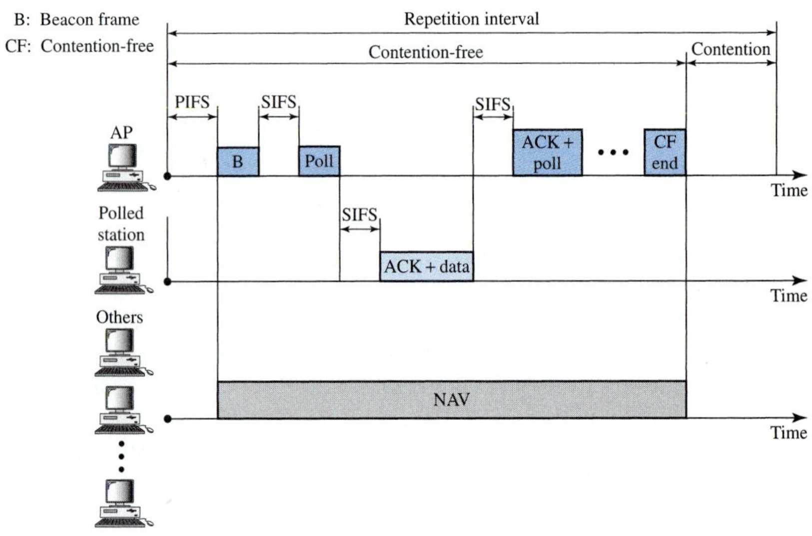

During the repetition interval, the PC (point controller) can send a poll frame, receive data, send an ACK, receive an ACK, or do any combination of these (802.11 uses piggybacking). At the end of the contention-free period, the PC sends a CF end (contention-free end) frame to allow the contention-based stations to use the medium.

Fragmentation

The wireless environment is very noisy; a corrupt frame has to be retransmitted. The protocol, therefore, recommends fragmentation—the division of a large frame into smaller ones. It is more efficient to resend a small frame than a large one.

Frame Format

The MAC layer frame consists of nine fields, as shown in Figure 14.7.

- **Frame control (FC).** The FC field is 2 bytes long and defines the type of frame and some control information. Table 14.1 describes the subfields. We will discuss each frame type later in this chapter.

Figure 14.7 *Frame format*

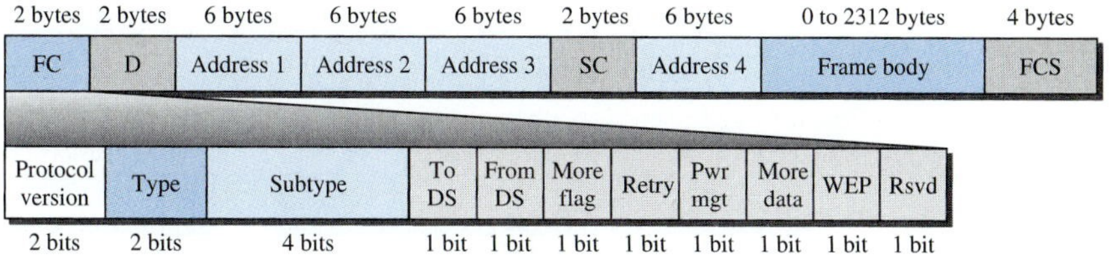

Table 14.1 *Subfields in FC field*

Field	*Explanation*
Version	Current version is 0
Type	Type of information: management (00), control (01), or data (10)
Subtype	Subtype of each type (see Table 14.2)
To DS	Defined later
From DS	Defined later
More flag	When set to 1, means more fragments
Retry	When set to 1, means retransmitted frame
Pwr mgt	When set to 1, means station is in power management mode
More data	When set to 1, means station has more data to send
WEP	Wired equivalent privacy (encryption implemented)
Rsvd	Reserved

- **D.** In all frame types except one, this field defines the duration of the transmission that is used to set the value of NAV. In one control frame, this field defines the ID of the frame.
- **Addresses.** There are four address fields, each 6 bytes long. The meaning of each address field depends on the value of the *To DS* and *From DS* subfields and will be discussed later.
- **Sequence control.** This field defines the sequence number of the frame to be used in flow control.
- **Frame body.** This field, which can be between 0 and 2312 bytes, contains information based on the type and the subtype defined in the FC field.
- **FCS.** The FCS field is 4 bytes long and contains a CRC-32 error detection sequence.

Frame Types

A wireless LAN defined by IEEE 802.11 has three categories of frames: management frames, control frames, and data frames.

Management Frames Management frames are used for the initial communication between stations and access points.

Control Frames Control frames are used for accessing the channel and acknowledging frames. Figure 14.8 shows the format.

Figure 14.8 *Control frames*

2 bytes	2 bytes	6 bytes	6 bytes	4 bytes
FC	D	Address 1	Address 2	FCS

RTS

2 bytes	2 bytes	6 bytes	4 bytes
FC	D	Address 1	FCS

CTS or ACK

For control frames the value of the type field is 01; the values of the subtype fields for frames we have discussed are shown in Table 14.2.

Table 14.2 *Values of subfields in control frames*

Subtype	*Meaning*
1011	Request to send (RTS)
1100	Clear to send (CTS)
1101	Acknowledgment (ACK)

Data Frames Data frames are used for carrying data and control information.

Addressing Mechanism

The IEEE 802.11 addressing mechanism specifies four cases, defined by the value of the two flags in the FC field, *To DS* and *From DS*. Each flag can be either 0 or 1, resulting in four different situations. The interpretation of the four addresses (address 1 to address 4) in the MAC frame depends on the value of these flags, as shown in Table 14.3.

Table 14.3 *Addresses*

To DS	*From DS*	*Address 1*	*Address 2*	*Address 3*	*Address 4*
0	0	Destination	Source	BSS ID	N/A
0	1	Destination	Sending AP	Source	N/A
1	0	Receiving AP	Source	Destination	N/A
1	1	Receiving AP	Sending AP	Destination	Source

Note that address 1 is always the address of the next device. Address 2 is always the address of the previous device. Address 3 is the address of the final destination station if it is not defined by address 1. Address 4 is the address of the original source station if it is not the same as address 2.

- **Case 1: 00** In this case, *To DS* = 0 and *From DS* = 0. This means that the frame is not going to a distribution system (*To DS* = 0) and is not coming from a distribution system (*From DS* = 0). The frame is going from one station in a BSS to another without passing through the distribution system. The ACK frame should be sent to the original sender. The addresses are shown in Figure 14.9.

Figure 14.9 *Addressing mechanisms*

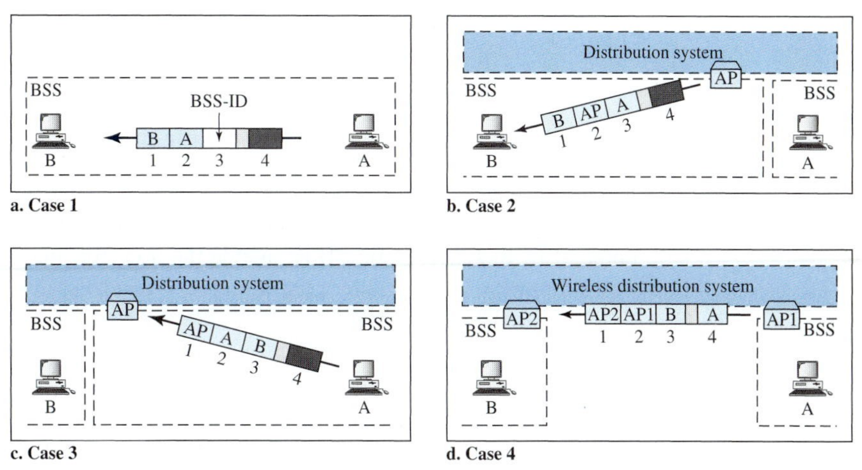

- **Case 2: 01** In this case, *To DS* = 0 and *From DS* = 1. This means that the frame is coming from a distribution system (*From DS* = 1). The frame is coming from an AP and going to a station. The ACK should be sent to the AP. The addresses are as shown in Figure 14.9. Note that address 3 contains the original sender of the frame (in another BSS).
- **Case 3: 10** In this case, *To DS* = 1 and *From DS* = 0. This means that the frame is going to a distribution system (*To DS* = 1). The frame is going from a station to an AP. The ACK is sent to the original station. The addresses are as shown in Figure 14.9. Note that address 3 contains the final destination of the frame (in another BSS).
- **Case 4:11** In this case, *To DS* = 1 and *From DS* = 1. This is the case in which the distribution system is also wireless. The frame is going from one AP to another AP in a wireless distribution system. We do not need to define addresses if the distribution system is a wired LAN because the frame in these cases has the format of a wired LAN frame (Ethernet, for example). Here, we need four addresses to define the original sender, the final destination, and two intermediate APs. Figure 14.9 shows the situation.

Hidden and Exposed Station Problems

We referred to hidden and exposed station problems in the previous section. It is time now to dicuss these problems and their effects.

Hidden Station Problem Figure 14.10 shows an example of the hidden station problem. Station B has a transmission range shown by the left oval (sphere in space); every station in this range can hear any signal transmitted by station B. Station C has

Figure 14.10 *Hidden station problem*

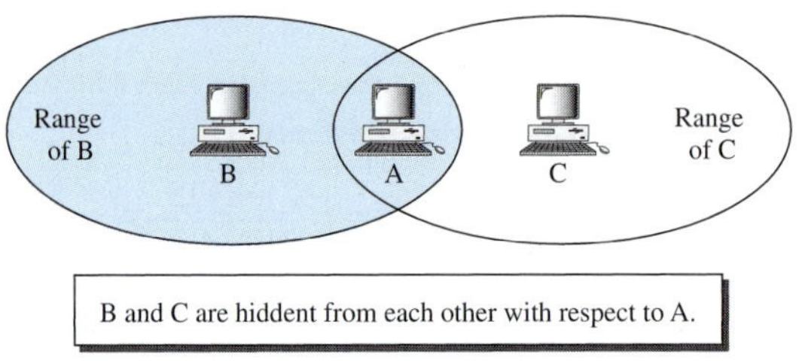

a transmission range shown by the right oval (sphere in space); every station located in this range can hear any signal transmitted by C. Station C is outside the transmission range of B; likewise, station B is outside the transmission range of C. Station A, however, is in the area covered by both B and C; it can hear any signal transmitted by B or C.

Assume that station B is sending data to station A. In the middle of this transmission, station C also has data to send to station A. However, station C is out of B's range and transmissions from B cannot reach C. Therefore C thinks the medium is free. Station C sends its data to A, which results in a collision at A because this station is receiving data from both B and C. In this case, we say that stations B and C are hidden from each other with respect to A. Hidden stations can reduce the capacity of the network because of the possibility of collision.

The solution to the hidden station problem is the use of the handshake frames (RTS and CTS) that we discussed earlier. Figure 14.11 shows that the RTS message from B reaches A, but not C. However, because both B and C are within the range of A, the CTS message, which contains the duration of data transmission from B to A reaches C. Station C knows that some hidden station is using the channel and refrains from transmitting until that duration is over.

The CTS frame in CSMA/CA handshake can prevent collision from a hidden station.

Figure 14.11 *Use of handshaking to prevent hidden station problem*

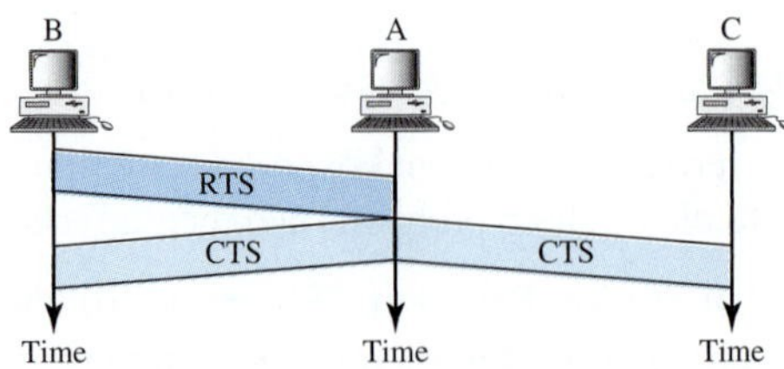

Exposed Station Problem Now consider a situation that is the inverse of the previous one: the exposed station problem. In this problem a station refrains from using a channel when it is, in fact, available. In Figure 14.12, station A is transmitting to station B. Station C has some data to send to station D, which can be sent without interfering with the transmission from A to B. However, station C is exposed to transmission from A; it hears what A is sending and thus refrains from sending. In other words, C is too conservative and wastes the capacity of the channel.

Figure 14.12 *Exposed station problem*

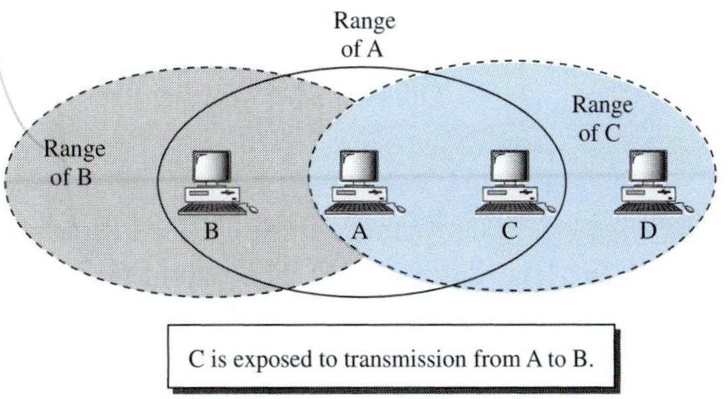

The handshaking messages RTS and CTS cannot help in this case, despite what you might think. Station C hears the RTS from A, but does not hear the CTS from B. Station C, after hearing the RTS from A, can wait for a time so that the CTS from B reaches A; it then sends an RTS to D to show that it needs to communicate with D. Both stations B and A may hear this RTS, but station A is in the sending state, not the receiving state. Station B, however, responds with a CTS. The problem is here. If station A has started sending its data, station C cannot hear the CTS from station D because of the collision; it cannot send its data to D. It remains exposed until A finishes sending its data as Figure 14.13 shows.

Figure 14.13 *Use of handshaking in exposed station problem*

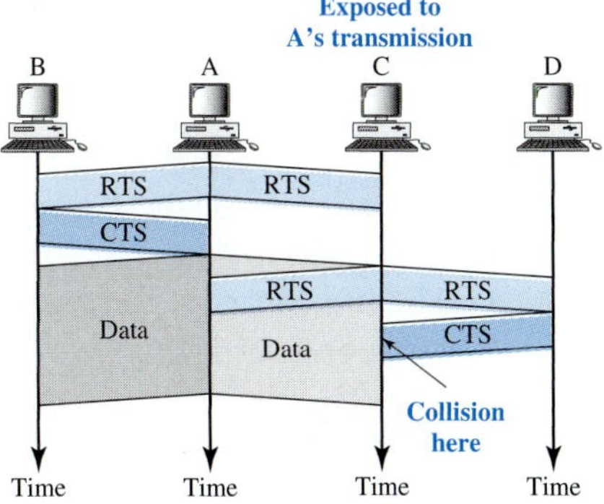

Physical Layer

We discuss six specifications, as shown in Table 14.4.

Table 14.4 *Physical layers*

IEEE	*Technique*	*Band*	*Modulation*	*Rate (Mbps)*
802.11	FHSS	2.4 GHz	FSK	1 and 2
	DSSS	2.4 GHz	PSK	1 and 2
		Infrared	PPM	1 and 2
802.11a	OFDM	5.725 GHz	PSK or QAM	6 to 54
802.11b	DSSS	2.4 GHz	PSK	5.5 and 11
802.11g	OFDM	2.4 GHz	Different	22 and 54

All implementations, except the infrared, operate in the *industrial, scientific,* and *medical (ISM)* band, which defines three unlicensed bands in the three ranges 902–928 MHz, 2.400–4.835 GHz, and 5.725–5.850 GHz, as shown in Figure 14.14.

Figure 14.14 *Industrial, scientific, and medical (ISM) band*

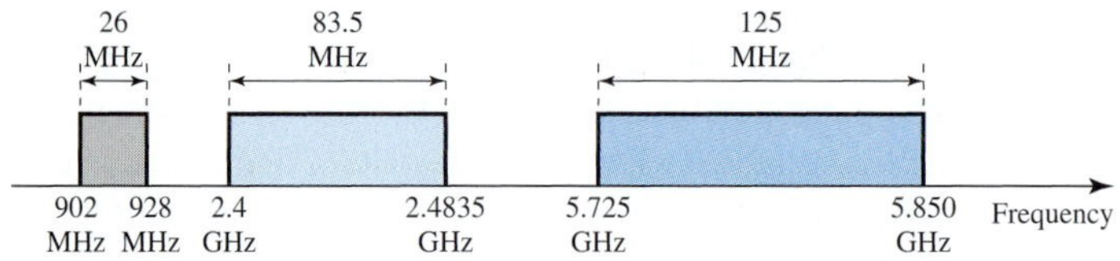

IEEE 802.11 FHSS

IEEE 802.11 FHSS uses the frequency-hopping spread spectrum (FHSS) method as discussed in Chapter 6. FHSS uses the 2.4-GHz ISM band. The band is divided into 79 subbands of 1 MHz (and some guard bands). A pseudorandom number generator selects the hopping sequence. The modulation technique in this specification is either two-level FSK or four-level FSK with 1 or 2 bits/baud, which results in a data rate of 1 or 2 Mbps, as shown in Figure 14.15.

IEEE 802.11 DSSS

IEEE 802.11 DSSS uses the direct sequence spread spectrum (DSSS) method as discussed in Chapter 6. DSSS uses the 2.4-GHz ISM band. The modulation technique in this specification is PSK at 1 Mbaud/s. The system allows 1 or 2 bits/baud (BPSK or QPSK), which results in a data rate of 1 or 2 Mbps, as shown in Figure 14.16.

IEEE 802.11 Infrared

IEEE 802.11 infrared uses infrared light in the range of 800 to 950 nm. The modulation technique is called **pulse position modulation (PPM)**. For a 1-Mbps data rate, a 4-bit

Figure 14.15 *Physical layer of IEEE 802.11 FHSS*

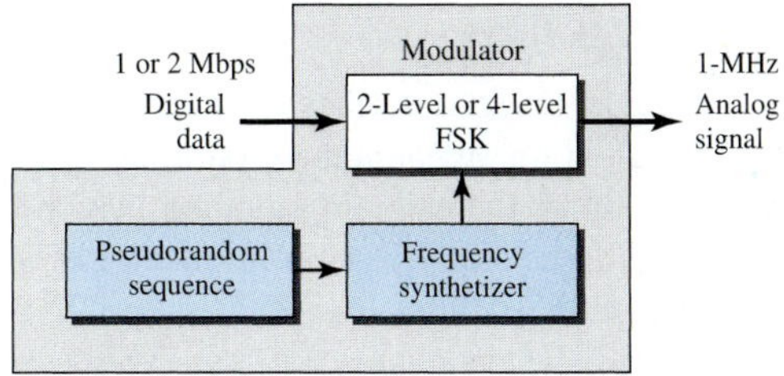

Figure 14.16 *Physical layer of IEEE 802.11 DSSS*

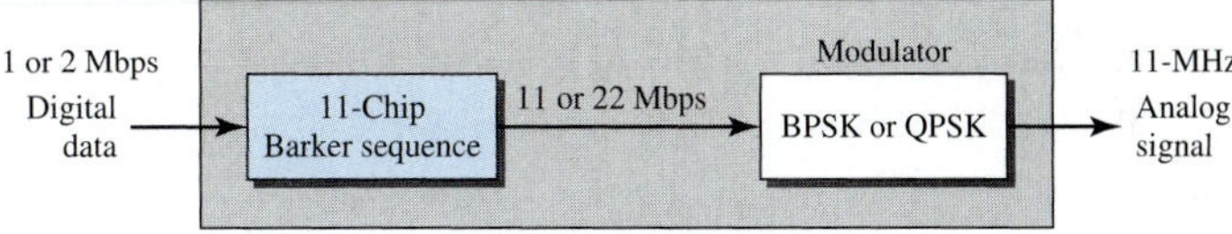

sequence is first mapped into a 16-bit sequence in which only one bit is set to 1 and the rest are set to 0. For a 2-Mbps data rate, a 2-bit sequence is first mapped into a 4-bit sequence in which only one bit is set to 1 and the rest are set to 0. The mapped sequences are then converted to optical signals; the presence of light specifies 1, the absence of light specifies 0. See Figure 14.17.

Figure 14.17 *Physical layer of IEEE 802.11 infrared*

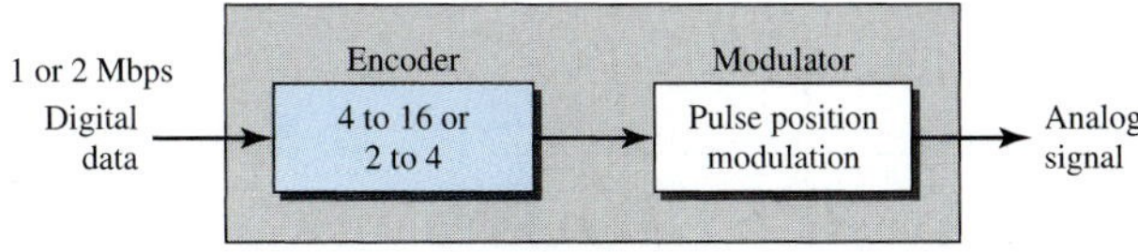

IEEE 802.11a OFDM

IEEE 802.11a OFDM describes the **orthogonal frequency-division multiplexing (OFDM)** method for signal generation in a 5-GHz ISM band. OFDM is similar to FDM as discussed in Chapter 6, with one major difference: All the subbands are used by one source at a given time. Sources contend with one another at the data link layer for access. The band is divided into 52 subbands, with 48 subbands for sending 48 groups of bits at a time and 4 subbands for control information. The scheme is similar to ADSL, as discussed in Chapter 9. Dividing the band into subbands diminishes the effects of interference. If the subbands are used randomly, security can also be increased.

OFDM uses PSK and QAM for modulation. The common data rates are 18 Mbps (PSK) and 54 Mbps (QAM).

IEEE 802.11b DSSS

IEEE 802.11b DSSS describes the **high-rate direct sequence spread spectrum (HR-DSSS)** method for signal generation in the 2.4-GHz ISM band. HR-DSSS is similar to DSSS except for the encoding method, which is called **complementary code keying (CCK).** CCK encodes 4 or 8 bits to one CCK symbol. To be backward compatible with DSSS, HR-DSSS defines four data rates: 1, 2, 5.5, and 11 Mbps. The first two use the same modulation techniques as DSSS. The 5.5-Mbps version uses BPSK and transmits at 1.375 Mbaud/s with 4-bit CCK encoding. The 11-Mbps version uses QPSK and transmits at 1.375 Mbps with 8-bit CCK encoding. Figure 14.18 shows the modulation technique for this standard.

Figure 14.18 *Physical layer of IEEE 802.11b*

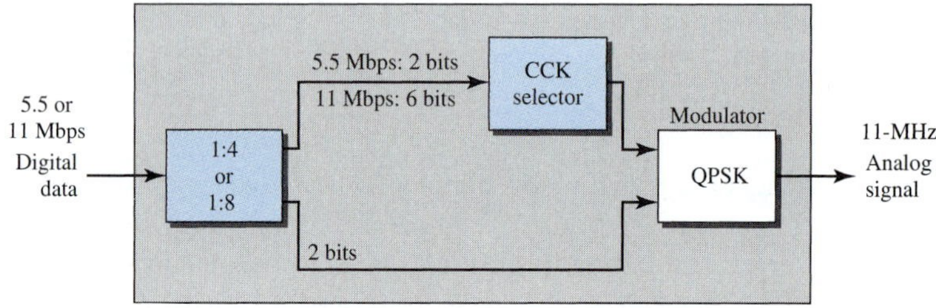

IEEE 802.11g

This new specification defines forward error correction and OFDM using the 2.4-GHz ISM band. The modulation technique achieves a 22- or 54-Mbps data rate. It is backward-compatible with 802.11b, but the modulation technique is OFDM.

14.2 BLUETOOTH

Bluetooth is a wireless LAN technology designed to connect devices of different functions such as telephones, notebooks, computers (desktop and laptop), cameras, printers, coffee makers, and so on. A Bluetooth LAN is an ad hoc network, which means that the network is formed spontaneously; the devices, sometimes called gadgets, find each other and make a network called a piconet. A Bluetooth LAN can even be connected to the Internet if one of the gadgets has this capability. A Bluetooth LAN, by nature, cannot be large. If there are many gadgets that try to connect, there is chaos.

Bluetooth technology has several applications. Peripheral devices such as a wireless mouse or keyboard can communicate with the computer through this technology. Monitoring devices can communicate with sensor devices in a small health care center. Home security devices can use this technology to connect different sensors to the main security controller. Conference attendees can synchronize their laptop computers at a conference.

Bluetooth was originally started as a project by the Ericsson Company. It is named for Harald Blaatand, the king of Denmark (940–981) who united Denmark and Norway. *Blaatand* translates to *Bluetooth* in English.

Today, Bluetooth technology is the implementation of a protocol defined by the IEEE 802.15 standard. The standard defines a wireless personal-area network (PAN) operable in an area the size of a room or a hall.

Architecture

Bluetooth defines two types of networks: piconet and scatternet.

Piconets

A Bluetooth network is called a **piconet,** or a small net. A piconet can have up to eight stations, one of which is called the **primary;**[†] the rest are called **secondaries.** All the secondary stations synchronize their clocks and hopping sequence with the primary. Note that a piconet can have only one primary station. The communication between the primary and the secondary can be one-to-one or one-to-many. Figure 14.19 shows a piconet.

Figure 14.19 *Piconet*

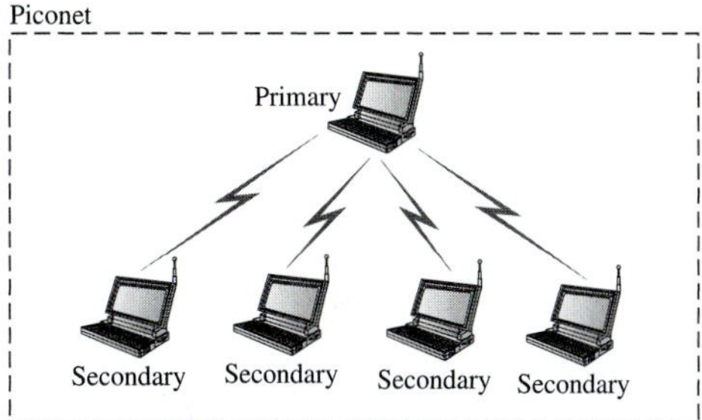

Although a piconet can have a maximum of seven secondaries, an additional eight secondaries can be in the *parked state*. A secondary in a parked state is synchronized with the primary, but cannot take part in communication until it is moved from the parked state. Because only eight stations can be active in a piconet, activating a station from the parked state means that an active station must go to the parked state.

Scatternet

Piconets can be combined to form what is called a **scatternet.** A secondary station in one piconet can be the primary in another piconet. This station can receive messages

[†]The literature sometimes uses the terms *master* and *slave* instead of *primary* and *secondary.* We prefer the latter.

from the primary in the first piconet (as a secondary) and, acting as a primary, deliver them to secondaries in the second piconet. A station can be a member of two piconets. Figure 14.20 illustrates a scatternet.

Figure 14.20 *Scatternet*

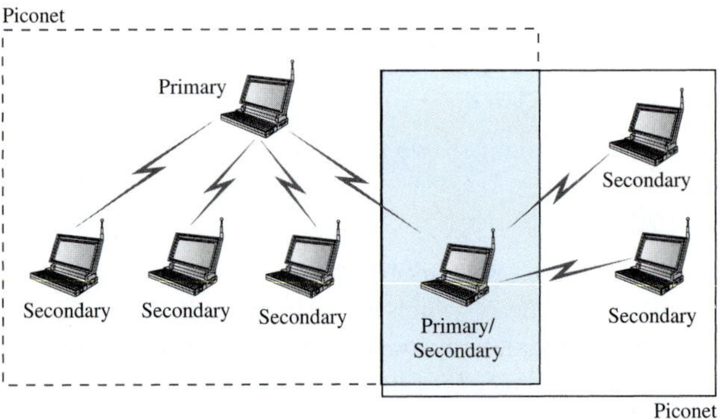

Bluetooth Devices

A Bluetooth device has a built-in short-range radio transmitter. The current data rate is 1 Mbps with a 2.4-GHz bandwidth. This means that there is a possibility of interference between the IEEE 802.11b wireless LANs and Bluetooth LANs.

Bluetooth Layers

Bluetooth uses several layers that do not exactly match those of the Internet model we have defined in this book. Figure 14.21 shows these layers.

Figure 14.21 *Bluetooth layers*

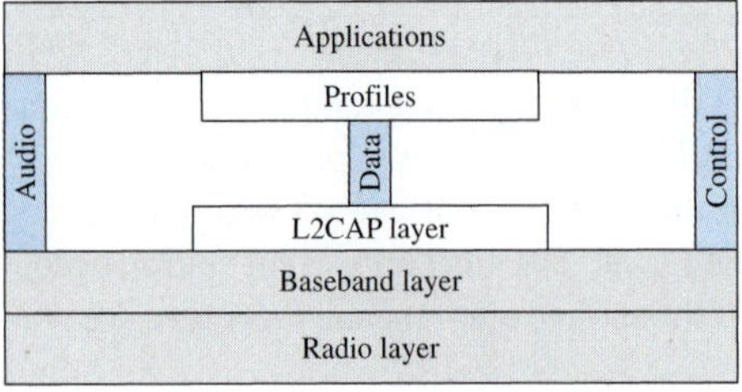

Radio Layer

The radio layer is roughly equivalent to the physical layer of the Internet model. Bluetooth devices are low-power and have a range of 10 m.

Band

Bluetooth uses a 2.4-GHz ISM band divided into 79 channels of 1 MHz each.

FHSS

Bluetooth uses the **frequency-hopping spread spectrum (FHSS)** method in the physical layer to avoid interference from other devices or other networks. Bluetooth hops 1600 times per second, which means that each device changes its modulation frequency 1600 times per second. A device uses a frequency for only 625 μs (1/1600 s) before it hops to another frequency; the dwell time is 625 μs.

Modulation

To transform bits to a signal, Bluetooth uses a sophisticated version of FSK, called GFSK (FSK with Gaussian bandwidth filtering; a discussion of this topic is beyond the scope of this book). GFSK has a carrier frequency. Bit 1 is represented by a frequency deviation above the carrier; bit 0 is represented by a frequency deviation below the carrier. The frequencies, in megahertz, are defined according to the following formula for each channel:

$$f_c = 2402 + n \qquad n = 0, 1, 2, 3, \ldots, 78$$

For example, the first channel uses carrier frequency 2402 MHz (2.402 GHz), and the second channel uses carrier frequency 2403 MHz (2.403 GHz).

Baseband Layer

The baseband layer is roughly equivalent to the MAC sublayer in LANs. The access method is TDMA (see Chapter 12). The primary and secondary communicate with each other using time slots. The length of a time slot is exactly the same as the dwell time, 625 μs. This means that during the time that one frequency is used, a sender sends a frame to a secondary, or a secondary sends a frame to the primary. Note that the communication is only between the primary and a secondary; secondaries cannot communicate directly with one another.

TDMA

Bluetooth uses a form of TDMA (see Chapter 12) that is called **TDD-TDMA (time-division duplex TDMA).** TDD-TDMA is a kind of half-duplex communication in which the secondary and receiver send and receive data, but not at the same time (half-duplex); however, the communication for each direction uses different hops. This is similar to walkie-talkies using different carrier frequencies.

Single-Secondary Communication If the piconet has only one secondary, the TDMA operation is very simple. The time is divided into slots of 625 μs. The primary uses even-numbered slots (0, 2, 4, . . .); the secondary uses odd-numbered slots (1, 3, 5, . . .). TDD-TDMA allows the primary and the secondary to communicate in half-duplex mode.

In slot 0, the primary sends, and the secondary receives; in slot 1, the secondary sends, and the primary receives. The cycle is repeated. Figure 14.22 shows the concept.

Figure 14.22 *Single-secondary communication*

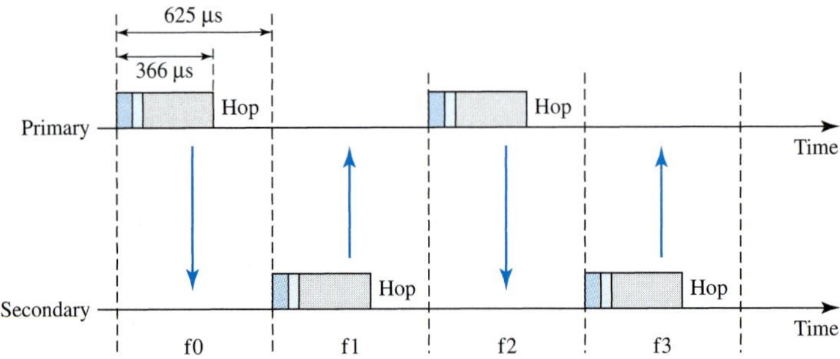

Multiple-Secondary Communication The process is a little more involved if there is more than one secondary in the piconet. Again, the primary uses the even-numbered slots, but a secondary sends in the next odd-numbered slot if the packet in the previous slot was addressed to it. All secondaries listen on even-numbered slots, but only one secondary sends in any odd-numbered slot. Figure 14.23 shows a scenario.

Figure 14.23 *Multiple-secondary communication*

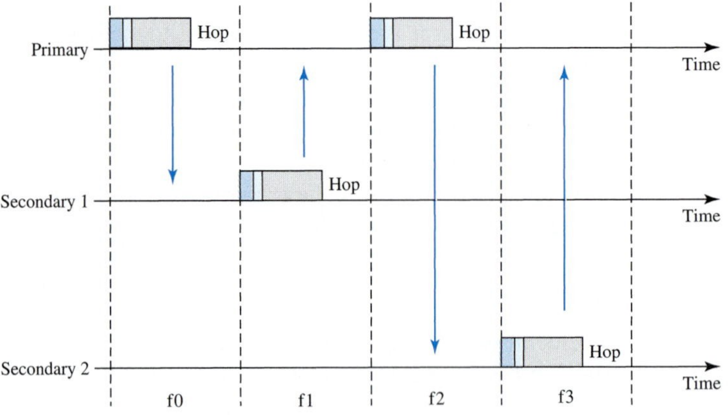

Let us elaborate on the figure.

1. In slot 0, the primary sends a frame to secondary 1.
2. In slot 1, only secondary 1 sends a frame to the primary because the previous frame was addressed to secondary 1; other secondaries are silent.

3. In slot 2, the primary sends a frame to secondary 2.
4. In slot 3, only secondary 2 sends a frame to the primary because the previous frame was addressed to secondary 2; other secondaries are silent.
5. The cycle continues.

We can say that this access method is similar to a poll/select operation with reservations. When the primary selects a secondary, it also polls it. The next time slot is reserved for the polled station to send its frame. If the polled secondary has no frame to send, the channel is silent.

Physical Links

Two types of links can be created between a primary and a secondary: SCO links and ACL links.

SCO A **synchronous connection-oriented (SCO)** link is used when avoiding latency (delay in data delivery) is more important than integrity (error-free delivery). In an SCO link, a physical link is created between the primary and a secondary by reserving specific slots at regular intervals. The basic unit of connection is two slots, one for each direction. If a packet is damaged, it is never retransmitted. SCO is used for real-time audio where avoiding delay is all-important. A secondary can create up to three SCO links with the primary, sending digitized audio (PCM) at 64 kbps in each link.

ACL An **asynchronous connectionless link (ACL)** is used when data integrity is more important than avoiding latency. In this type of link, if a payload encapsulated in the frame is corrupted, it is retransmitted. A secondary returns an ACL frame in the available odd-numbered slot if and only if the previous slot has been addressed to it. ACL can use one, three, or more slots and can achieve a maximum data rate of 721 kbps.

Frame Format

A frame in the baseband layer can be one of three types: one-slot, three-slot, or five-slot. A slot, as we said before, is 625 μs. However, in a one-slot frame exchange, 259 μs is needed for hopping and control mechanisms. This means that a one-slot frame can last only 625 − 259, or 366 μs. With a 1-MHz bandwidth and 1 bit/Hz, the size of a one-slot frame is 366 bits.

A three-slot frame occupies three slots. However, since 259 μs is used for hopping, the length of the frame is $3 \times 625 - 259 = 1616$ μs or 1616 bits. A device that uses a three-slot frame remains at the same hop (at the same carrier frequency) for three slots. Even though only one hop number is used, three hop numbers are consumed. That means the hop number for each frame is equal to the first slot of the frame.

A five-slot frame also uses 259 bits for hopping, which means that the length of the frame is $5 \times 625 - 259 = 2866$ bits.

Figure 14.24 shows the format of the three frame types.

The following describes each field:

- **Access code.** This 72-bit field normally contains synchronization bits and the identifier of the primary to distinguish the frame of one piconet from another.

Figure 14.24 *Frame format types*

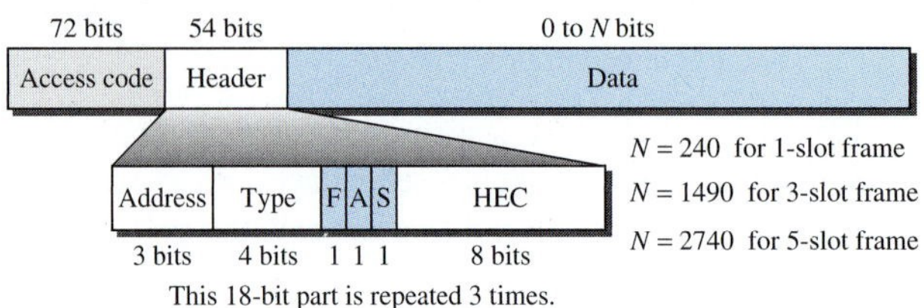

- **Header.** This 54-bit field is a repeated 18-bit pattern. Each pattern has the following subfields:
  1. **Address.** The 3-bit address subfield can define up to seven secondaries (1 to 7). If the address is zero, it is used for broadcast communication from the primary to all secondaries.
  2. **Type.** The 4-bit type subfield defines the type of data coming from the upper layers. We discuss these types later.
  3. **F.** This 1-bit subfield is for flow control. When set (1), it indicates that the device is unable to receive more frames (buffer is full).
  4. **A.** This 1-bit subfield is for acknowledgment. Bluetooth uses Stop-and-Wait ARQ; 1 bit is sufficient for acknowledgment.
  5. **S.** This 1-bit subfield holds a sequence number. Bluetooth uses Stop-and-Wait ARQ; 1 bit is sufficient for sequence numbering.
  6. **HEC.** The 8-bit header error correction subfield is a checksum to detect errors in each 18-bit header section.

  The header has three identical 18-bit sections. The receiver compares these three sections, bit by bit. If each of the corresponding bits is the same, the bit is accepted; if not, the majority opinion rules. This is a form of forward error correction (for the header only). This double error control is needed because the nature of the communication, via air, is very noisy. Note that there is no retransmission in this sublayer.
- **Payload.** This subfield can be 0 to 2740 bits long. It contains data or control information coming from the upper layers.

L2CAP

The **Logical Link Control and Adaptation Protocol,** or **L2CAP** (L2 here means LL), is roughly equivalent to the LLC sublayer in LANs. It is used for data exchange on an ACL link; SCO channels do not use L2CAP. Figure 14.25 shows the format of the data packet at this level.

The 16-bit length field defines the size of the data, in bytes, coming from the upper layers. Data can be up to 65,535 bytes. The channel ID (CID) defines a unique identifier for the virtual channel created at this level (see below).

The L2CAP has specific duties: multiplexing, segmentation and reassembly, quality of service (QoS), and group management.

Figure 14.25 *L2CAP data packet format*

2 bytes	2 bytes	0 to 65,535 bytes
Length	Channel ID	Data and control

Multiplexing

The L2CAP can do multiplexing. At the sender site, it accepts data from one of the upper-layer protocols, frames them, and delivers them to the baseband layer. At the receiver site, it accepts a frame from the baseband layer, extracts the data, and delivers them to the appropriate protocol layer. It creates a kind of virtual channel that we will discuss in later chapters on higher-level protocols.

Segmentation and Reassembly

The maximum size of the payload field in the baseband layer is 2774 bits, or 343 bytes. This includes 4 bytes to define the packet and packet length. Therefore, the size of the packet that can arrive from an upper layer can only be 339 bytes. However, application layers sometimes need to send a data packet that can be up to 65,535 bytes (an Internet packet, for example). The L2CAP divides these large packets into segments and adds extra information to define the location of the segments in the original packet. The L2CAP segments the packet at the source and reassembles them at the destination.

QoS

Bluetooth allows the stations to define a quality-of-service level. We discuss quality of service in Chapter 24. For the moment, it is sufficient to know that if no quality-of-service level is defined, Bluetooth defaults to what is called *best-effort* service; it will do its best under the circumstances.

Group Management

Another functionality of L2CAP is to allow devices to create a type of logical addressing between themselves. This is similar to multicasting. For example, two or three secondary devices can be part of a multicast group to receive data from the primary.

Other Upper Layers

Bluetooth defines several protocols for the upper layers that use the services of L2CAP; these protocols are specific for each purpose.

14.3 RECOMMENDED READING

For more details about subjects discussed in this chapter, we recommend the following books and sites. The items in brackets [. . .] refer to the reference list at the end of the text.

Books

Wireless LANs and Bluetooth are discussed in several books including [Sch03] and [Gas02]. Wireless LANs are discussed in Chapter 15 of [For03], Chapter 17 of [Sta04], Chapters 13 and 14 of [Sta02], and Chapter 8 of [Kei02]. Bluetooth is discussed in Chapter 15 of [Sta02] and Chapter 15 of [For03].

14.4 KEY TERMS

access point (AP)
asynchronous connectionless link (ACL)
basic service set (BSS)
beacon frame
Bluetooth
BSS-transition mobility
complementary code keying (CCK)
direct sequence spread spectrum (DSSS)
distributed coordination function (DCF)
distributed interframe space (DIFS)
ESS-transition mobility
extended service set (ESS)
frequency-hopping spread spectrum (FHSS)
handshaking period
high-rate direct sequence spread spectrum (HR-DSSS)
IEEE 802.11
Logical Link Control and Adaptation Protocol (L2CAP)
network allocation vector (NAV)
no-transition mobility
orthogonal frequency-division multiplexing (OFDM)
piconet
point coordination function (PCF)
primary
pulse position modulation (PPM)
repetition interval
scatternet
secondary
short interframe space (SIFS)
synchronous connection-oriented (SCO)
TDD-TDMA (time-division duplex TDMA)
wireless LAN

14.5 SUMMARY

- The IEEE 802.11 standard for wireless LANs defines two services: basic service set (BSS) and extended service set (ESS).
- The access method used in the distributed coordination function (DCF) MAC sublayer is CSMA/CA.
- The access method used in the point coordination function (PCF) MAC sublayer is polling.
- The network allocation vector (NAV) is a timer used for collision avoidance.
- The MAC layer frame has nine fields. The addressing mechanism can include up to four addresses.
- Wireless LANs use management frames, control frames, and data frames.

- IEEE 802.11 defines several physical layers, with different data rates and modulating techniques.
- Bluetooth is a wireless LAN technology that connects devices (called gadgets) in a small area.
- A Bluetooth network is called a piconet. Multiple piconets form a network called a scatternet.
- A Bluetooth network consists of one primary device and up to seven secondary devices.

14.6 PRACTICE SET

Review Questions

1. What is the difference between a BSS and an ESS?
2. Discuss the three types of mobility in a wireless LAN.
3. How is OFDM different from FDM?
4. What is the access method used by wireless LANs?
5. What is the purpose of the NAV?
6. Compare a piconet and a scatternet.
7. Match the layers in Bluetooth and the Internet model.
8. What are the two types of links between a Bluetooth primary and a Bluetooth secondary?
9. In multiple-secondary communication, who uses the even-numbered slots and who uses the odd-numbered slots?
10. How much time in a Bluetooth one-slot frame is used for the hopping mechanism? What about a three-slot frame and a five-slot frame?

Exercises

11. Compare and contrast CSMA/CD with CSMA/CA.
12. Use Table 14.5 to compare and contrast the fields in IEEE 802.3 and 802.11.

Table 14.5 *Exercise 12*

Fields	*IEEE 802.3 Field Size*	*IEEE 802.11 Field Size*
Destination address		
Source address		
Address 1		
Address 2		
Address 3		
Address 4		
FC		

Table 14.5 *Exercise 12 (continued)*

Fields	*IEEE 802.3 Field Size*	*IEEE 802.11 Field Size*
D/ID		
SC		
PDU length		
Data and padding		
Frame body		
FCS (CRC)		

CHAPTER 15

Connecting LANs, Backbone Networks, and Virtual LANs

LANs do not normally operate in isolation. They are connected to one another or to the Internet. To connect LANs, or segments of LANs, we use connecting devices. Connecting devices can operate in different layers of the Internet model. In this chapter, we discuss only those that operate in the physical and data link layers; we discuss those that operate in the first three layers in Chapter 19.

After discussing some connecting devices, we show how they are used to create backbone networks. Finally, we discuss virtual local area networks (VLANs).

15.1 CONNECTING DEVICES

In this section, we divide **connecting devices** into five different categories based on the layer in which they operate in a network, as shown in Figure 15.1.

Figure 15.1 *Five categories of connecting devices*

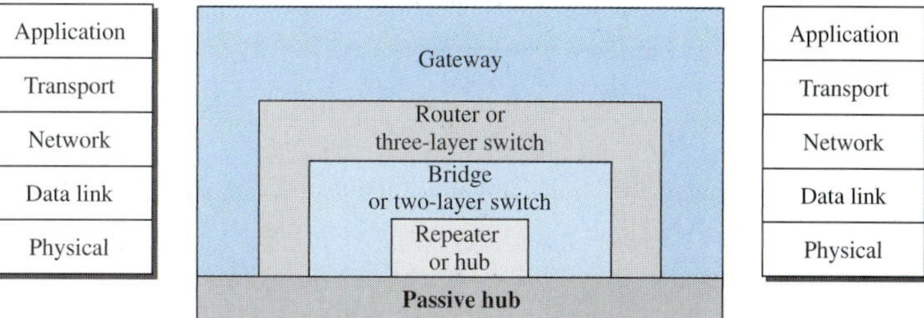

The five categories contain devices which can be defined as

1. Those which operate below the physical layer such as a passive hub.
2. Those which operate at the physical layer (a repeater or an active hub).
3. Those which operate at the physical and data link layers (a bridge or a two-layer switch).

4. Those which operate at the physical, data link, and network layers (a router or a three-layer switch).
5. Those which can operate at all five layers (a gateway).

Passive Hubs

A passive hub is just a connector. It connects the wires coming from different branches. In a star-topology Ethernet LAN, a passive hub is just a point where the signals coming from different stations collide; the hub is the collision point. This type of a hub is part of the media; its location in the Internet model is below the physical layer.

Repeaters

A **repeater** is a device that operates only in the physical layer. Signals that carry information within a network can travel a fixed distance before attenuation endangers the integrity of the data. A repeater receives a signal and, before it becomes too weak or corrupted, regenerates the original bit pattern. The repeater then sends the refreshed signal. A repeater can extend the physical length of a LAN, as shown in Figure 15.2.

Figure 15.2 *A repeater connecting two segments of a LAN*

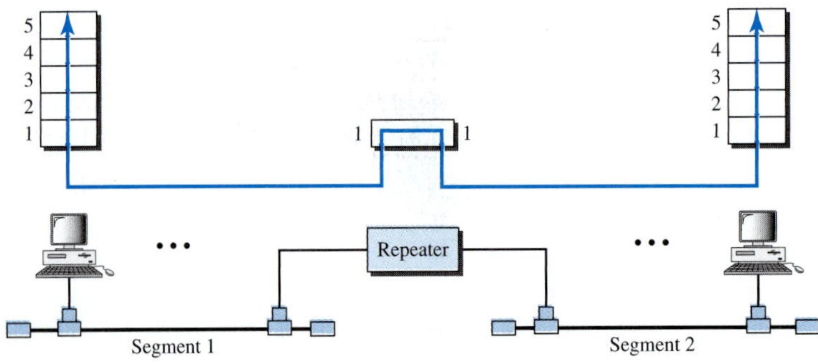

A repeater does not actually connect two LANs; it connects two segments of the same LAN. The segments connected are still part of one single LAN. A repeater is not a device that can connect two LANs of different protocols.

A repeater connects segments of a LAN.

A repeater can overcome the 10Base5 Ethernet length restriction. In this standard, the length of the cable is limited to 500 m. To extend this length, we divide the cable into segments and install repeaters between segments. Note that the whole network is still considered one LAN, but the portions of the network separated by repeaters are called **segments.** The repeater acts as a two-port node, but operates only in the physical layer. When it receives a frame from any of the ports, it regenerates and forwards it to the other port.

A repeater forwards every frame; it has no filtering capability.

It is tempting to compare a repeater to an amplifier, but the comparison is inaccurate. An **amplifier** cannot discriminate between the intended signal and noise; it amplifies equally everything fed into it. A repeater does not amplify the signal; it regenerates the signal. When it receives a weakened or corrupted signal, it creates a copy, bit for bit, at the original strength.

A repeater is a regenerator, not an amplifier.

The location of a repeater on a link is vital. A repeater must be placed so that a signal reaches it before any noise changes the meaning of any of its bits. A little noise can alter the precision of a bit's voltage without destroying its identity (see Figure 15.3). If the corrupted bit travels much farther, however, accumulated noise can change its meaning completely. At that point, the original voltage is not recoverable, and the error needs to be corrected. A repeater placed on the line before the legibility of the signal becomes lost can still read the signal well enough to determine the intended voltages and replicate them in their original form.

Figure 15.3 *Function of a repeater*

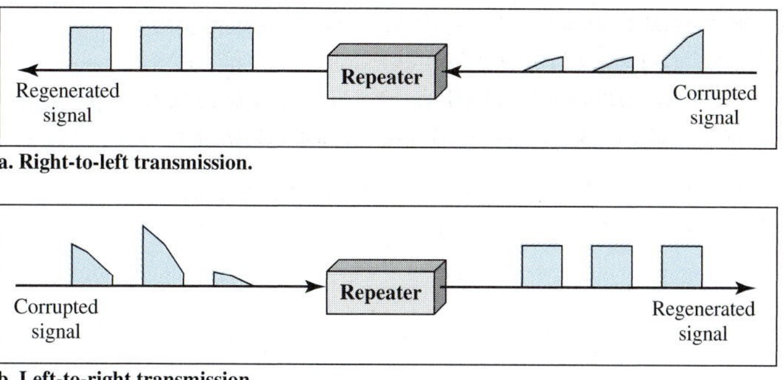

Active Hubs

An active **hub** is actually a multiport repeater. It is normally used to create connections between stations in a physical star topology. We have seen examples of hubs in some Ethernet implementations (10Base-T, for example). However, hubs can also be used to create multiple levels of hierarchy, as shown in Figure 15.4. The hierarchical use of hubs removes the length limitation of 10Base-T (100 m).

Bridges

A **bridge** operates in both the physical and the data link layer. As a physical layer device, it regenerates the signal it receives. As a data link layer device, the bridge can check the physical (MAC) addresses (source and destination) contained in the frame.

Figure 15.4 *A hierarchy of hubs*

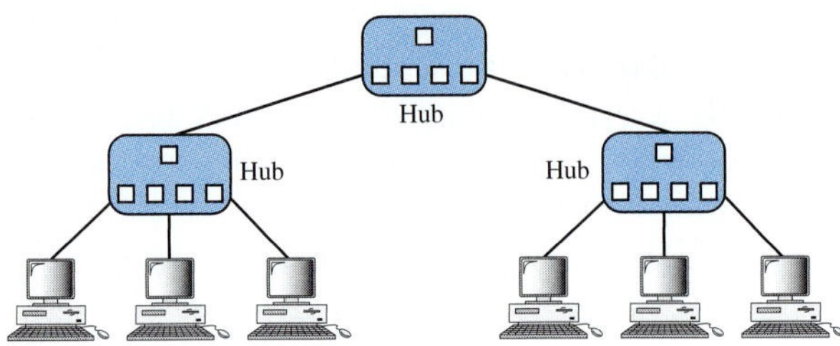

Filtering

One may ask, What is the difference in functionality between a bridge and a repeater? A bridge has **filtering** capability. It can check the destination address of a frame and decide if the frame should be forwarded or dropped. If the frame is to be forwarded, the decision must specify the port. A bridge has a table that maps addresses to ports.

A bridge has a table used in filtering decisions.

Let us give an example. In Figure 15.5, two LANs are connected by a bridge. If a frame destined for station 712B13456142 arrives at port 1, the bridge consults its table to find the departing port. According to its table, frames for 712B13456142 leave through port 1; therefore, there is no need for forwarding, and the frame is dropped. On the other hand, if a frame for 712B13456141 arrives at port 2, the departing port is port 1

Figure 15.5 *A bridge connecting two LANs*

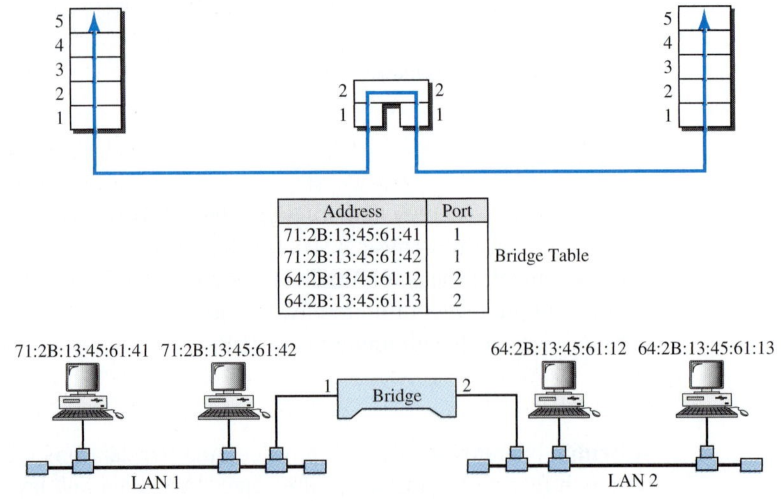

Address	Port
71:2B:13:45:61:41	1
71:2B:13:45:61:42	1
64:2B:13:45:61:12	2
64:2B:13:45:61:13	2

and the frame is forwarded. In the first case, LAN 2 remains free of traffic; in the second case, both LANs have traffic. In our example, we show a two-port bridge; in reality a bridge usually has more ports.

Note also that a bridge does not change the physical addresses contained in the frame.

A bridge does not change the physical (MAC) addresses in a frame.

Transparent Bridges

A **transparent bridge** is a bridge in which the stations are completely unaware of the bridge's existence. If a bridge is added or deleted from the system, reconfiguration of the stations is unnecessary. According to the IEEE 802.1d specification, a system equipped with transparent bridges must meet three criteria:

1. Frames must be forwarded from one station to another.
2. The forwarding table is automatically made by learning frame movements in the network.
3. Loops in the system must be prevented.

Forwarding A transparent bridge must correctly forward the frames, as discussed in the previous section.

Learning The earliest bridges had forwarding tables that were static. The systems administrator would manually enter each table entry during bridge setup. Although the process was simple, it was not practical. If a station was added or deleted, the table had to be modified manually. The same was true if a station's MAC address changed, which is not a rare event. For example, putting in a new network card means a new MAC address.

A better solution to the static table is a dynamic table that maps addresses to ports automatically. To make a table dynamic, we need a bridge that gradually learns from the frame movements. To do this, the bridge inspects both the destination and the source addresses. The destination address is used for the forwarding decision (table lookup); the source address is used for adding entries to the table and for updating purposes. Let us elaborate on this process by using Figure 15.6.

1. When station A sends a frame to station D, the bridge does not have an entry for either D or A. The frame goes out from all three ports; the frame floods the network. However, by looking at the source address, the bridge learns that station A must be located on the LAN connected to port 1. This means that frames destined for A, in the future, must be sent out through port 1. The bridge adds this entry to its table. The table has its first entry now.
2. When station E sends a frame to station A, the bridge has an entry for A, so it forwards the frame only to port 1. There is no flooding. In addition, it uses the source address of the frame, E, to add a second entry to the table.
3. When station B sends a frame to C, the bridge has no entry for C, so once again it floods the network and adds one more entry to the table.
4. The process of learning continues as the bridge forwards frames.

Figure 15.6 *A learning bridge and the process of learning*

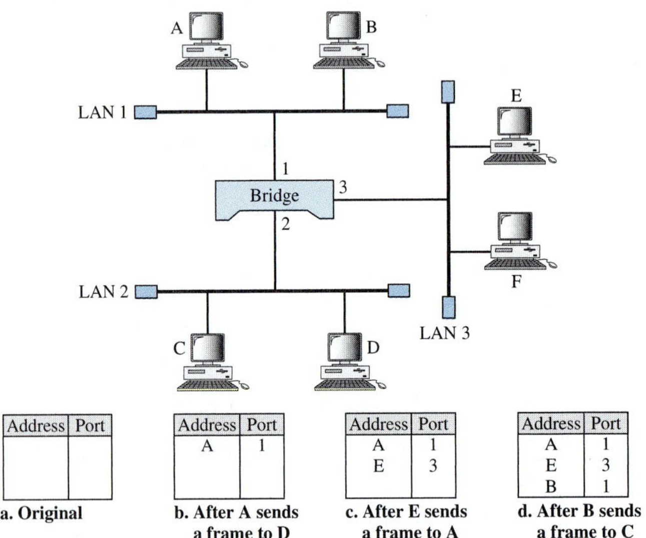

Loop Problem Transparent bridges work fine as long as there are no redundant bridges in the system. Systems administrators, however, like to have redundant bridges (more than one bridge between a pair of LANs) to make the system more reliable. If a bridge fails, another bridge takes over until the failed one is repaired or replaced. Redundancy can create loops in the system, which is very undesirable. Figure 15.7 shows a very simple example of a loop created in a system with two LANs connected by two bridges.

1. Station A sends a frame to station D. The tables of both bridges are empty. Both forward the frame and update their tables based on the source address A.
2. Now there are two copies of the frame on LAN 2. The copy sent out by bridge 1 is received by bridge 2, which does not have any information about the destination address D; it floods the bridge. The copy sent out by bridge 2 is received by bridge 1 and is sent out for lack of information about D. Note that each frame is handled separately because bridges, as two nodes on a network sharing the medium, use an access method such as CSMA/CD. The tables of both bridges are updated, but still there is no information for destination D.
3. Now there are two copies of the frame on LAN 1. Step 2 is repeated, and both copies flood the network.
4. The process continues on and on. Note that bridges are also repeaters and regenerate frames. So in each iteration, there are newly generated fresh copies of the frames.

To solve the looping problem, the IEEE specification requires that bridges use the spanning tree algorithm to create a loopless topology.

Figure 15.7 *Loop problem in a learning bridge*

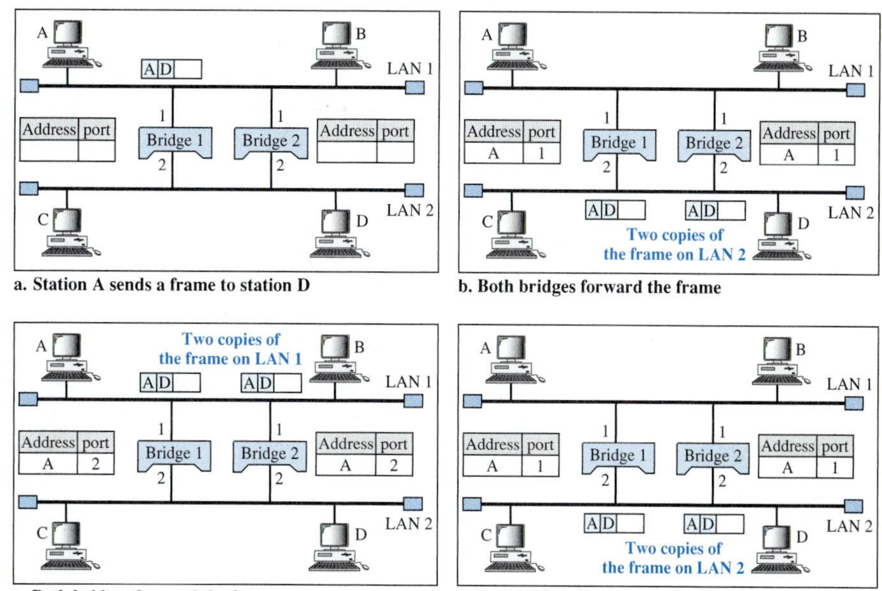

Spanning Tree

In graph theory, a **spanning tree** is a graph in which there is no loop. In a bridged LAN, this means creating a topology in which each LAN can be reached from any other LAN through one path only (no loop). We cannot change the physical topology of the system because of physical connections between cables and bridges, but we can create a logical topology that overlays the physical one. Figure 15.8 shows a system with four LANs and five bridges. We have shown the physical system and its representation in graph theory. Although some textbooks represent the LANs as nodes and the bridges as the connecting arcs, we have shown both LANs and bridges as nodes. The connecting arcs show the connection of a LAN to a bridge and vice versa. To find the spanning tree, we need to assign a cost (metric) to each arc. The interpretation of the cost is left up to the systems administrator. It may be the path with minimum hops (nodes), the path with minimum delay, or the path with maximum bandwidth. If two ports have the same shortest value, the systems administrator just chooses one. We have chosen the minimum hops. However, as we will see in Chapter 22, the hop count is normally 1 from a bridge to the LAN and 0 in the reverse direction.

The process to find the spanning tree involves three steps:

1. Every bridge has a built-in ID (normally the serial number, which is unique). Each bridge broadcasts this ID so that all bridges know which one has the smallest ID. The bridge with the smallest ID is selected as the *root* bridge (root of the tree). We assume that bridge B1 has the smallest ID. It is, therefore, selected as the root bridge.

Figure 15.8 *A system of connected LANs and its graph representation*

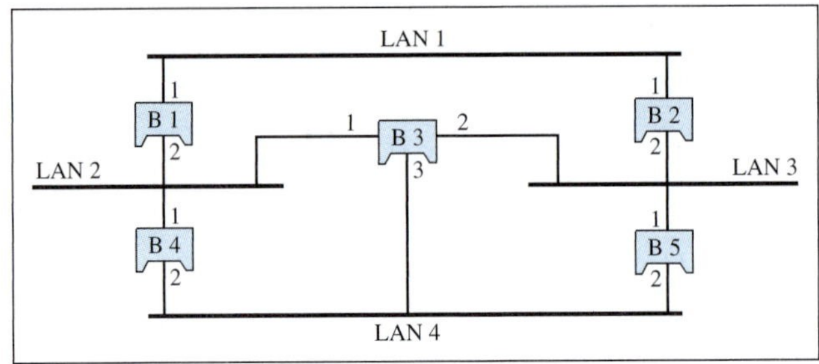

a. Actual system

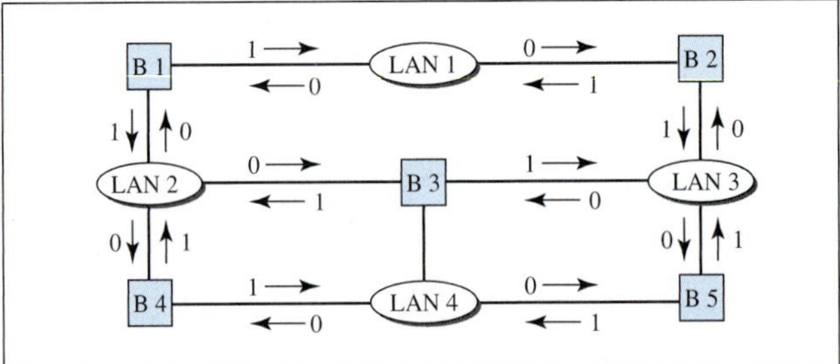

b. Graph representation with cost assigned to each arc

2. The algorithm tries to find the shortest path (a path with the shortest cost) from the root bridge to every other bridge or LAN. The shortest path can be found by examining the total cost from the root bridge to the destination. Figure 15.9 shows the shortest paths.
3. The combination of the shortest paths creates the shortest tree, which is also shown in Figure 15.9.
4. Based on the spanning tree, we mark the ports that are part of the spanning tree, the **forwarding ports,** which forward a frame that the bridge receives. We also mark those ports that are not part of the spanning tree, the **blocking ports,** which block the frames received by the bridge. Figure 15.10 shows the physical systems of LANs with forwarding points (solid lines) and blocking ports (broken lines).

Note that there is only one single path from any LAN to any other LAN in the spanning tree system. This means there is only one single path from one LAN to any other LAN. No loops are created. You can prove to yourself that there is only one path from LAN 1 to LAN 2, LAN 3, or LAN 4. Similarly, there is only one path from LAN 2 to LAN 1, LAN 3, and LAN 4. The same is true for LAN 3 and LAN 4.

Dynamic Algorithm We have described the spanning tree algorithm as though it required manual entries. This is not true. Each bridge is equipped with a software package that carries out this process dynamically. The bridges send special messages to one another, called bridge protocol data units (BPDUs), to update the spanning tree. The spanning tree is updated when there is a change in the system such as a failure of a bridge or an addition or deletion of bridges.

Figure 15.9 *Finding the shortest paths and the spanning tree in a system of bridges*

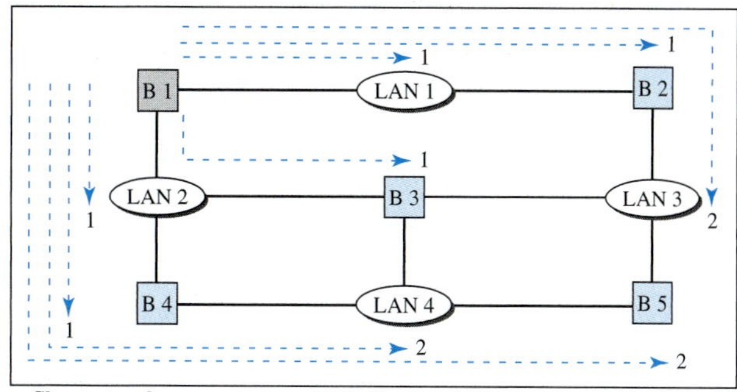

a. Shortest paths

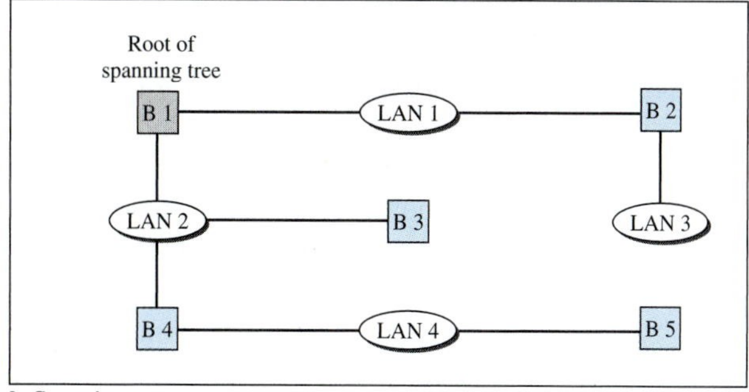

b. Spanning tree

Figure 15.10 *Forwarding and blocking ports after using spanning tree algorithm*

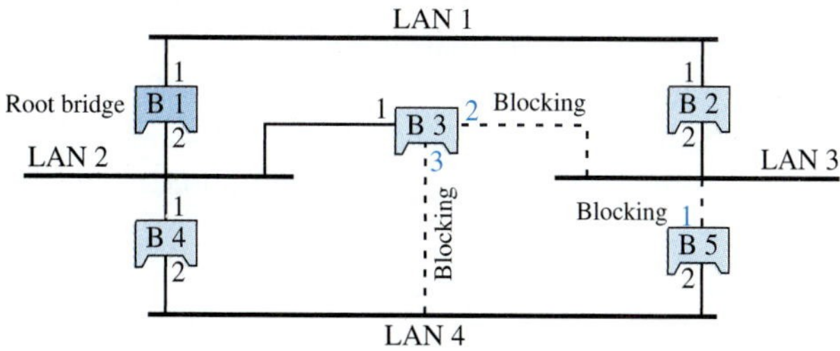

Source Routing Bridges

Another way to prevent loops in a system with redundant bridges is to use **source routing bridges.** A transparent bridge's duties include filtering frames, forwarding, and blocking. In a system that has source routing bridges, these duties are performed by the source station and, to some extent, the destination station.

In source routing, a sending station defines the bridges that the frame must visit. The addresses of these bridges are included in the frame. In other words, the frame contains not only the source and destination addresses, but also the addresses of all bridges to be visited.

The source gets these bridge addresses through the exchange of special frames with the destination prior to sending the data frame.

Source routing bridges were designed by IEEE to be used with Token Ring LANs. These LANs are not very common today.

Bridges Connecting Different LANs

Theoretically a bridge should be able to connect LANs using different protocols at the data link layer, such as an Ethernet LAN to a wireless LAN. However, there are many issues to be considered:

- **Frame format.** Each LAN type has its own frame format (compare an Ethernet frame with a wireless LAN frame).
- **Maximum data size.** If an incoming frame's size is too large for the destination LAN, the data must be fragmented into several frames. The data then need to be reassembled at the destination. However, no protocol at the data link layer allows the fragmentation and reassembly of frames. We will see in Chapter 19 that this is allowed in the network layer. The bridge must therefore discard any frames too large for its system.
- **Data rate.** Each LAN type has its own data rate. (Compare the 10-Mbps data rate of an Ethernet with the 1-Mbps data rate of a wireless LAN.) The bridge must buffer the frame to compensate for this difference.
- **Bit order.** Each LAN type has its own strategy in the sending of bits. Some send the most significant bit in a byte first; others send the least significant bit first.
- **Security.** Some LANs, such as wireless LANs, implement security measures in the data link layer. Other LANs, such as Ethernet, do not. Security often involves encryption (see Chapter 30). When a bridge receives a frame from a wireless LAN, it needs to decrypt the message before forwarding it to an Ethernet LAN.
- **Multimedia support.** Some LANs support multimedia and the quality of services needed for this type of communication; others do not.

Two-Layer Switches

When we use the term *switch,* we must be careful because a switch can mean two different things. We must clarify the term by adding the level at which the device operates. We can have a two-layer switch or a three-layer switch. A **three-layer switch** is used at the network layer; it is a kind of router. The **two-layer switch** performs at the physical and data link layers.

A two-layer switch is a bridge, a bridge with many ports and a design that allows better (faster) performance. A bridge with a few ports can connect a few LANs together. A bridge with many ports may be able to allocate a unique port to each station, with each station on its own independent entity. This means no competing traffic (no collision, as we saw in Ethernet).

A two-layer switch, as a bridge does, makes a filtering decision based on the MAC address of the frame it received. However, a two-layer switch can be more sophisticated. It can have a buffer to hold the frames for processing. It can have a switching factor that forwards the frames faster. Some new two-layer switches, called *cut-through* switches, have been designed to forward the frame as soon as they check the MAC addresses in the header of the frame.

Routers

A **router** is a three-layer device that routes packets based on their logical addresses (host-to-host addressing). A router normally connects LANs and WANs in the Internet and has a routing table that is used for making decisions about the route. The routing tables are normally dynamic and are updated using routing protocols. We discuss routers and routing in greater detail in Chapters 19 and 21. Figure 15.11 shows a part of the Internet that uses routers to connect LANs and WANs.

Figure 15.11 *Routers connecting independent LANs and WANs*

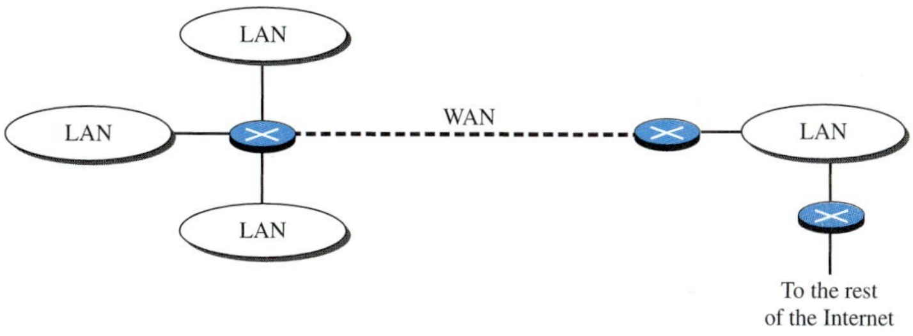

Three-Layer Switches

A three-layer switch is a router, but a faster and more sophisticated. The switching fabric in a three-layer switch allows faster table lookup and forwarding. In this book, we use the terms *router* and *three-layer switch* interchangeably.

Gateway

Although some textbooks use the terms *gateway* and *router* interchangeably, most of the literature distinguishes between the two. A gateway is normally a computer that operates in all five layers of the Internet or seven layers of OSI model. A gateway takes an application message, reads it, and interprets it. This means that it can be used as a connecting device between two internetworks that use different models. For example, a network designed to use the OSI model can be connected to another network using the Internet model. The gateway connecting the two systems can take a frame as it arrives from the first system, move it up to the OSI application layer, and remove the message.

Gateways can provide security. In Chapter 32, we learn that the gateway is used to filter unwanted application-layer messages.

15.2 BACKBONE NETWORKS

Some connecting devices discussed in this chapter can be used to connect LANs in a backbone network. A backbone network allows several LANs to be connected. In a backbone network, no station is directly connected to the backbone; the stations are part of a LAN, and the backbone connects the LANs. The backbone is itself a LAN that uses a LAN protocol such as Ethernet; each connection to the backbone is itself another LAN.

Although many different architectures can be used for a backbone, we discuss only the two most common: the bus and the star.

Bus Backbone

In a **bus backbone,** the topology of the backbone is a bus. The backbone itself can use one of the protocols that support a bus topology such as 10Base5 or 10Base2.

In a bus backbone, the topology of the backbone is a bus.

Bus backbones are normally used as a distribution backbone to connect different buildings in an organization. Each building can comprise either a single LAN or another backbone (normally a star backbone). A good example of a bus backbone is one that connects single- or multiple-floor buildings on a campus. Each single-floor building usually has a single LAN. Each multiple-floor building has a backbone (usually a star) that connects each LAN on a floor. A bus backbone can interconnect these LANs and backbones. Figure 15.12 shows an example of a bridge-based backbone with four LANs.

Figure 15.12 *Bus backbone*

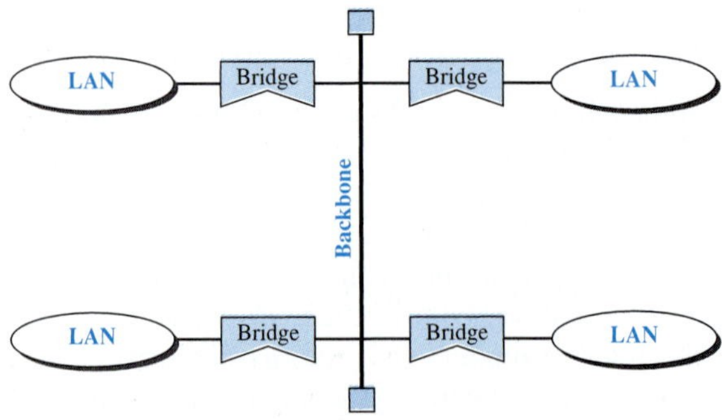

In Figure 15.12, if a station in a LAN needs to send a frame to another station in the same LAN, the corresponding bridge blocks the frame; the frame never reaches the backbone. However, if a station needs to send a frame to a station in another LAN, the bridge passes the frame to the backbone, which is received by the appropriate bridge and is delivered to the destination LAN. Each bridge connected to the backbone has a table that shows the stations on the LAN side of the bridge. The blocking or delivery of a frame is based on the contents of this table.

Star Backbone

In a **star backbone,** sometimes called a collapsed or switched backbone, the topology of the backbone is a star. In this configuration, the backbone is just one switch (that is why it is called, erroneously, a collapsed backbone) that connects the LANs.

> **In a star backbone, the topology of the backbone is a star; the backbone is just one switch.**

Figure 15.13 shows a star backbone. Note that, in this configuration, the switch does the job of the backbone and at the same time connects the LANs.

Figure 15.13 *Star backbone*

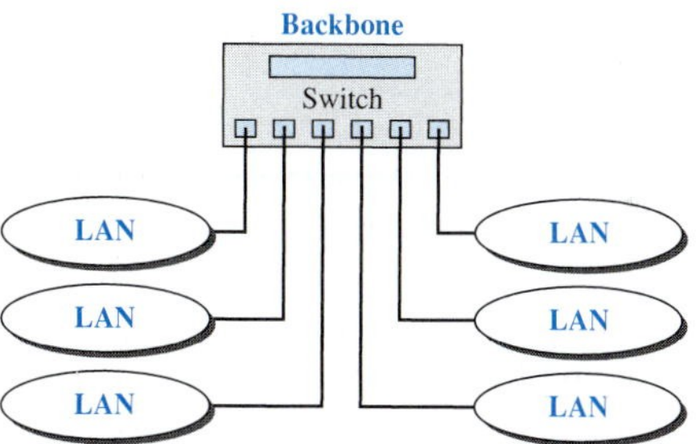

Star backbones are mostly used as a distribution backbone inside a building. In a multifloor building, we usually find one LAN that serves each particular floor. A star backbone connects these LANs. The backbone network, which is just a switch, can be installed in the basement or the first floor, and separate cables can run from the switch to each LAN. If the individual LANs have a physical star topology, either the hubs (or switches) can be installed in a closet on the corresponding floor, or all can be installed close to the switch. We often find a rack or chassis in the basement where the backbone switch and all hubs or switches are installed.

Connecting Remote LANs

Another common application for a backbone network is to connect remote LANs. This type of backbone network is useful when a company has several offices with LANs and needs to connect them. The connection can be done through bridges, sometimes called **remote bridges.** The bridges act as connecting devices connecting LANs and point-to-point networks, such as leased telephone lines or ADSL lines. The point-to-point network in this case is considered a LAN without stations. The point-to-point link can use a protocol such as PPP. Figure 15.14 shows a backbone connecting remote LANs.

Figure 15.14 *Connecting remote LANs with bridges*

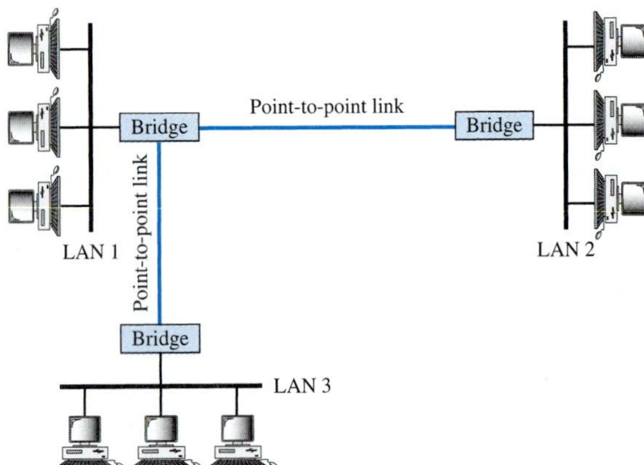

A point-to-point link acts as a LAN in a remote backbone connected by remote bridges.

15.3 VIRTUAL LANs

A station is considered part of a LAN if it physically belongs to that LAN. The criterion of membership is geographic. What happens if we need a virtual connection between two stations belonging to two different physical LANs? We can roughly define a **virtual local area network (VLAN)** as a local area network configured by software, not by physical wiring.

Let us use an example to elaborate on this definition. Figure 15.15 shows a switched LAN in an engineering firm in which 10 stations are grouped into three LANs that are connected by a switch. The first four engineers work together as the first group, the next three engineers work together as the second group, and the last three engineers work together as the third group. The LAN is configured to allow this arrangement.

But what would happen if the administrators needed to move two engineers from the first group to the third group, to speed up the project being done by the third group? The LAN configuration would need to be changed. The network technician must rewire. The problem is repeated if, in another week, the two engineers move back to their previous group. In a switched LAN, changes in the work group mean physical changes in the network configuration.

Figure 15.15 *A switch connecting three LANs*

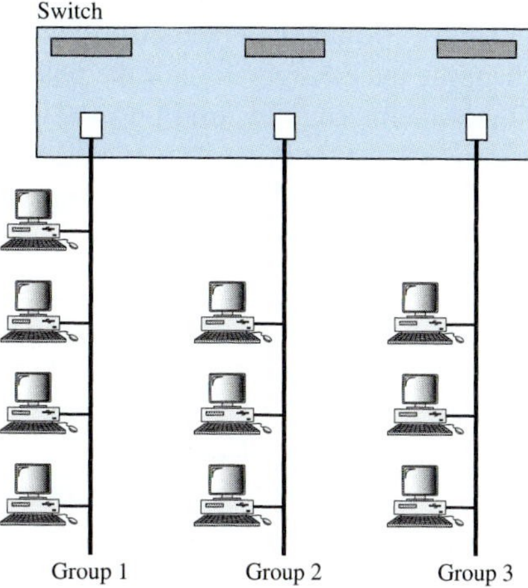

Figure 15.16 shows the same switched LAN divided into VLANs. The whole idea of VLAN technology is to divide a LAN into logical, instead of physical, segments. A LAN can be divided into several logical LANs called VLANs. Each VLAN is a work group in the organization. If a person moves from one group to another, there is no need to change the physical configuration. The group membership in VLANs is defined by software, not hardware. Any station can be logically moved to another VLAN. All members belonging to a VLAN can receive broadcast messages sent to that particular VLAN.

Figure 15.16 *A switch using VLAN software*

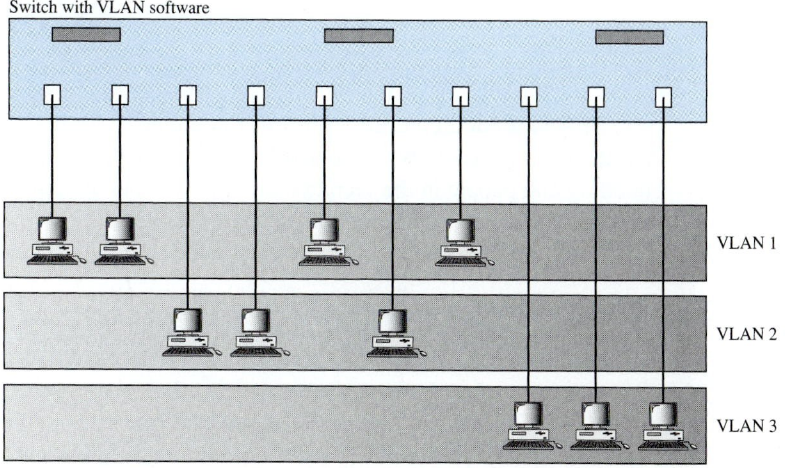

This means if a station moves from VLAN 1 to VLAN 2, it receives broadcast messages sent to VLAN 2, but no longer receives broadcast messages sent to VLAN 1.

It is obvious that the problem in our previous example can easily be solved by using VLANs. Moving engineers from one group to another through software is easier than changing the configuration of the physical network.

VLAN technology even allows the grouping of stations connected to different switches in a VLAN. Figure 15.17 shows a backbone local area network with two switches and three VLANs. Stations from switches A and B belong to each VLAN.

Figure 15.17 *Two switches in a backbone using VLAN software*

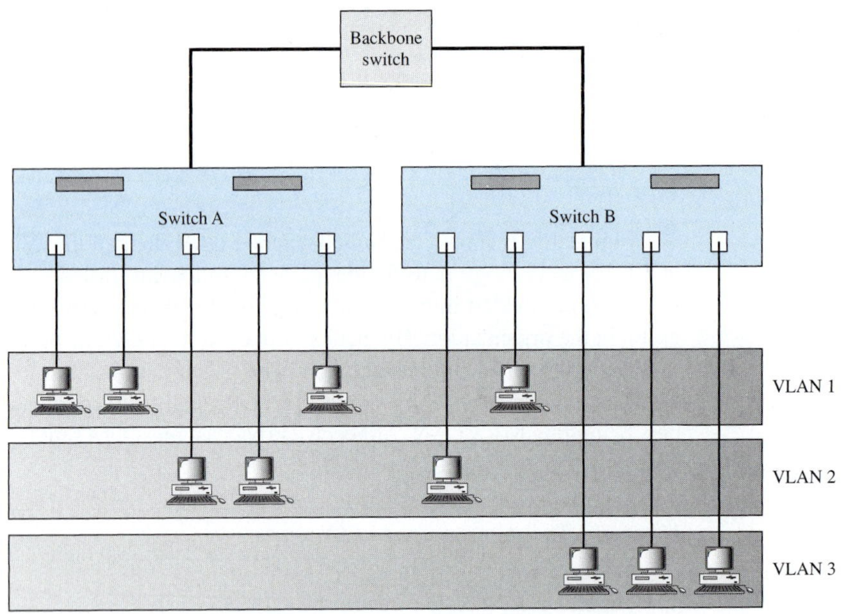

This is a good configuration for a company with two separate buildings. Each building can have its own switched LAN connected by a backbone. People in the first building and people in the second building can be in the same work group even though they are connected to different physical LANs.

From these three examples, we can define a VLAN characteristic:

VLANs create broadcast domains.

VLANs group stations belonging to one or more physical LANs into broadcast domains. The stations in a VLAN communicate with one another as though they belonged to a physical segment.

Membership

What characteristic can be used to group stations in a VLAN? Vendors use different characteristics such as port numbers, MAC addresses, IP addresses, IP multicast addresses, or a combination of two or more of these.

Port Numbers

Some VLAN vendors use switch port numbers as a membership characteristic. For example, the administrator can define that stations connecting to ports 1, 2, 3, and 7 belong to VLAN 1; stations connecting to ports 4, 10, and 12 belong to VLAN 2; and so on.

MAC Addresses

Some VLAN vendors use the 48-bit MAC address as a membership characteristic. For example, the administrator can stipulate that stations having MAC addresses E21342A12334 and F2A123BCD341 belong to VLAN 1.

IP Addresses

Some VLAN vendors use the 32-bit IP address (see Chapter 19) as a membership characteristic. For example, the administrator can stipulate that stations having IP addresses 181.34.23.67, 181.34.23.72, 181.34.23.98, and 181.34.23.112 belong to VLAN 1.

Multicast IP Addresses

Some VLAN vendors use the multicast IP address (see Chapter 19) as a membership characteristic. Multicasting at the IP layer is now translated to multicasting at the data link layer.

Combination

Recently, the software available from some vendors allows all these characteristics to be combined. The administrator can choose one or more characteristics when installing the software. In addition, the software can be reconfigured to change the settings.

Configuration

How are the stations grouped into different VLANs? Stations are configured in one of three ways: manual, semiautomatic, and automatic.

Manual Configuration

In a manual configuration, the network administrator uses the VLAN software to manually assign the stations into different VLANs at setup. Later migration from one VLAN to another is also done manually. Note that this is not a physical configuration; it is a logical configuration. The term *manually* here means that the administrator types the port numbers, the IP addresses, or other characteristics, using the VLAN software.

Automatic Configuration

In an automatic configuration, the stations are automatically connected or disconnected from a VLAN using criteria defined by the administrator. For example, the administrator can define the project number as the criterion for being a member of a group. When a user changes the project, he or she automatically migrates to a new VLAN.

Semiautomatic Configuration

A semiautomatic configuration is somewhere between a manual configuration and an automatic configuration. Usually, the initializing is done manually, with migrations done automatically.

Communication Between Switches

In a multiswitched backbone, each switch must know not only which station belongs to which VLAN, but also the membership of stations connected to other switches. For example, in Figure 15.17, switch A must know the membership status of stations connected to switch B, and switch B must know the same about switch A. Three methods have been devised for this purpose: table maintenance, frame tagging, and time-division multiplexing.

Table Maintenance

In this method, when a station sends a broadcast frame to its group members, the switch creates an entry in a table and records station membership. The switches send their tables to one another periodically for updating.

Frame Tagging

In this method, when a frame is traveling between switches, an extra header is added to the MAC frame to define the destination VLAN. The frame tag is used by the receiving switches to determine the VLANs to be receiving the broadcast message.

Time-Division Multiplexing (TDM)

In this method, the connection (trunk) between switches is divided into timeshared channels (see TDM in Chapter 6). For example, if the total number of VLANs in a backbone is five, each trunk is divided into five channels. The traffic destined for VLAN 1 travels in channel 1, the traffic destined for VLAN 2 travels in channel 2, and so on. The receiving switch determines the destination VLAN by checking the channel from which the frame arrived.

IEEE Standard

In 1996, the IEEE 802.1 subcommittee passed a standard called 802.1Q that defines the format for frame tagging. The standard also defines the format to be used in multiswitched backbones and enables the use of multivendor equipment in VLANs. IEEE 802.1Q has opened the way for further standardization in other issues related to VLANs. Most vendors have already accepted the standard.

Advantages

There are several advantages to using VLANs.

Cost and Time Reduction

VLANs can reduce the migration cost of stations going from one group to another. Physical reconfiguration takes time and is costly. Instead of physically moving one station to another segment or even to another switch, it is much easier and quicker to move it by using software.

Creating Virtual Work Groups

VLANs can be used to create virtual work groups. For example, in a campus environment, professors working on the same project can send broadcast messages to one another without the necessity of belonging to the same department. This can reduce traffic if the multicasting capability of IP was previously used.

Security

VLANs provide an extra measure of security. People belonging to the same group can send broadcast messages with the guaranteed assurance that users in other groups will not receive these messages.

15.4 RECOMMENDED READING

For more details about subjects discussed in this chapter, we recommend the following books and sites. The items in brackets [. . .] refer to the reference list at the end of the text.

Books

A book devoted to connecting devices is [Per00]. Connecting devices and VLANs are discussed in Section 4.7 of [Tan03]. Switches, bridges, and hubs are discussed in [Sta03] and [Sta04].

Site

 IEEE 802 LAN/MAN Standards Committee

15.5 KEY TERMS

amplifier
blocking port
bridge
bus backbone
connecting device
filtering
forwarding port
hub
remote bridge
repeater
router
segment

source routing bridge
spanning tree
star backbone
three-layer switch
transparent bridge
two-layer switch
virtual local area network (VLAN)

15.6 SUMMARY

- A repeater is a connecting device that operates in the physical layer of the Internet model. A repeater regenerates a signal, connects segments of a LAN, and has no filtering capability.
- A bridge is a connecting device that operates in the physical and data link layers of the Internet model.
- A transparent bridge can forward and filter frames and automatically build its forwarding table.
- A bridge can use the spanning tree algorithm to create a loopless topology.
- A backbone LAN allows several LANs to be connected.
- A backbone is usually a bus or a star.
- A virtual local area network (VLAN) is configured by software, not by physical wiring.
- Membership in a VLAN can be based on port numbers, MAC addresses, IP addresses, IP multicast addresses, or a combination of these features.
- VLANs are cost- and time-efficient, can reduce network traffic, and provide an extra measure of security.

15.7 PRACTICE SET

Review Questions

1. How is a repeater different from an amplifier?
2. What do we mean when we say that a bridge can filter traffic? Why is filtering important?
3. What is a transparent bridge?
4. How does a repeater extend the length of a LAN?
5. How is a hub related to a repeater?
6. What is the difference between a forwarding port and a blocking port?
7. What is the difference between a bus backbone and a star backbone?
8. How does a VLAN save a company time and money?
9. How does a VLAN provide extra security for a network?
10. How does a VLAN reduce network traffic?
11. What is the basis for membership in a VLAN?

Exercises

12. Complete the table in Figure 15.6 after each station has sent a packet to another station.
13. Find the spanning tree for the system in Figure 15.7.
14. Find the spanning tree for the system in Figure 15.8 if bridge B5 is removed.
15. Find the spanning tree for the system in Figure 15.8 if bridge B2 is removed.
16. Find the spanning tree for the system in Figure 15.8 if B5 is selected as the root bridge.
17. In Figure 15.6, we are using a bridge. Can we replace the bridge with a router? Explain the consequences.
18. A bridge uses a filtering table; a router uses a routing table. Can you explain the difference?
19. Create a system of three LANs with four bridges. The bridges (B1 to B4) connect the LANs as follows:
    a. B1 connects LAN 1 and LAN 2.
    b. B2 connects LAN 1 and LAN 3.
    c. B3 connects LAN 2 and LAN 3.
    d. B4 connects LAN 1, LAN 2, and LAN 3.

    Choose B1 as the root bridge. Show the forwarding and blocking ports, after applying the spanning tree procedure.
20. Which one has more overhead, a bridge or a router? Explain your answer.
21. Which one has more overhead, a repeater or a bridge? Explain your answer.
22. Which one has more overhead, a router or a gateway? Explain your answer.

CHAPTER 16

Wireless WANs: Cellular Telephone and Satellite Networks

We discussed wireless LANs in Chapter 14. Wireless technology is also used in cellular telephony and satellite networks. We discuss the former in this chapter as well as examples of channelization access methods (see Chapter 12). We also briefly discuss satellite networks, a technology that eventually will be linked to cellular telephony to access the Internet directly.

16.1 CELLULAR TELEPHONY

Cellular telephony is designed to provide communications between two moving units, called mobile stations (MSs), or between one mobile unit and one stationary unit, often called a land unit. A service provider must be able to locate and track a caller, assign a channel to the call, and transfer the channel from base station to base station as the caller moves out of range.

To make this tracking possible, each cellular service area is divided into small regions called cells. Each cell contains an antenna and is controlled by a solar or AC powered network station, called the base station (BS). Each base station, in turn, is controlled by a switching office, called a **mobile switching center (MSC).** The MSC coordinates communication between all the base stations and the telephone central office. It is a computerized center that is responsible for connecting calls, recording call information, and billing (see Figure 16.1).

Cell size is not fixed and can be increased or decreased depending on the population of the area. The typical radius of a cell is 1 to 12 mi. High-density areas require more, geographically smaller cells to meet traffic demands than do low-density areas. Once determined, cell size is optimized to prevent the interference of adjacent cell signals. The transmission power of each cell is kept low to prevent its signal from interfering with those of other cells.

Frequency-Reuse Principle

In general, neighboring cells cannot use the same set of frequencies for communication because it may create interference for the users located near the cell boundaries. However, the set of frequencies available is limited, and frequencies need to be reused. A

Figure 16.1 *Cellular system*

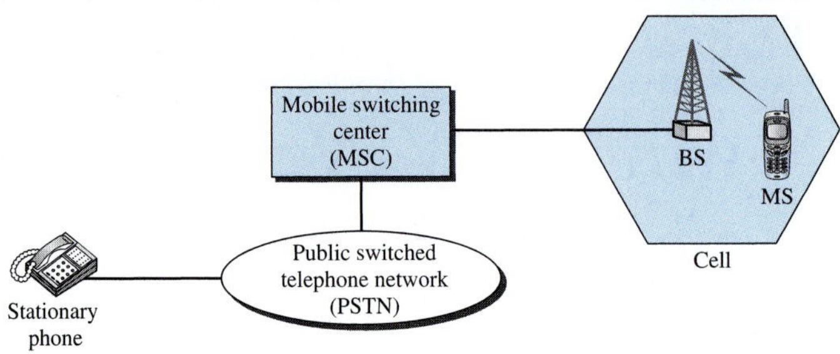

frequency reuse pattern is a configuration of N cells, N being the **reuse factor,** in which each cell uses a unique set of frequencies. When the pattern is repeated, the frequencies can be reused. There are several different patterns. Figure 16.2 shows two of them.

Figure 16.2 *Frequency reuse patterns*

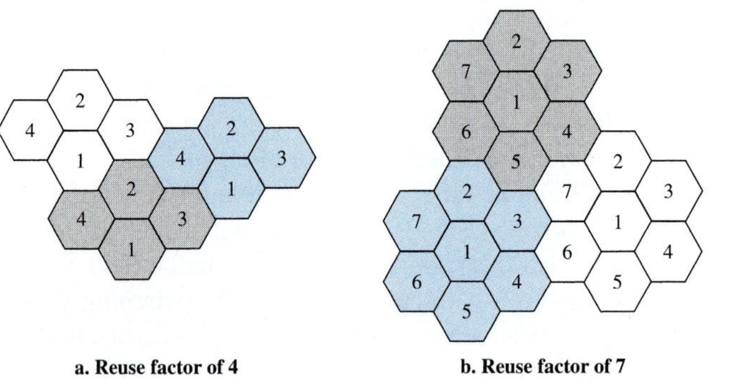

a. Reuse factor of 4
b. Reuse factor of 7

The numbers in the cells define the pattern. The cells with the same number in a pattern can use the same set of frequencies. We call these cells the *reusing cells.* As Figure 16.2 shows, in a pattern with reuse factor 4, only one cell separates the cells using the same set of frequencies. In the pattern with reuse factor 7, two cells separate the reusing cells.

Transmitting

To place a call from a mobile station, the caller enters a code of 7 or 10 digits (a phone number) and presses the send button. The mobile station then scans the band, seeking a setup channel with a strong signal, and sends the data (phone number) to the closest base station using that channel. The base station relays the data to the MSC. The MSC sends the data on to the telephone central office. If the called party is available, a connection is made and the result is relayed back to the MSC. At this point, the MSC assigns an unused voice channel to the call, and a connection is established. The mobile station automatically adjusts its tuning to the new channel, and communication can begin.

Receiving

When a mobile phone is called, the telephone central office sends the number to the MSC. The MSC searches for the location of the mobile station by sending query signals to each cell in a process called *paging*. Once the mobile station is found, the MSC transmits a ringing signal and, when the mobile station answers, assigns a voice channel to the call, allowing voice communication to begin.

Handoff

It may happen that, during a conversation, the mobile station moves from one cell to another. When it does, the signal may become weak. To solve this problem, the MSC monitors the level of the signal every few seconds. If the strength of the signal diminishes, the MSC seeks a new cell that can better accommodate the communication. The MSC then changes the channel carrying the call (hands the signal off from the old channel to a new one).

Hard Handoff Early systems used a hard **handoff.** In a hard handoff, a mobile station only communicates with one base station. When the MS moves from one cell to another, communication must first be broken with the previous base station before communication can be established with the new one. This may create a rough transition.

Soft Handoff New systems use a soft handoff. In this case, a mobile station can communicate with two base stations at the same time. This means that, during handoff, a mobile station may continue with the new base station before breaking off from the old one.

Roaming

One feature of cellular telephony is called **roaming.** Roaming means, in principle, that a user can have access to communication or can be reached where there is coverage. A service provider usually has limited coverage. Neighboring service providers can provide extended coverage through a roaming contract. The situation is similar to snail mail between countries. The charge for delivery of a letter between two countries can be divided upon agreement by the two countries.

First Generation

Cellular telephony is now in its second generation with the third on the horizon. The first generation was designed for voice communication using analog signals. We discuss one first-generation mobile system used in North America, AMPS.

AMPS

Advanced Mobile Phone System (AMPS) is one of the leading analog cellular systems in North America. It uses FDMA (see Chapter 12) to separate channels in a link.

AMPS is an analog cellular phone system using FDMA.

Bands AMPS operates in the ISM 800-MHz band. The system uses two separate analog channels, one for forward (base station to mobile station) communication and one for reverse (mobile station to base station) communication. The band between 824 and 849 MHz carries reverse communication; the band between 869 and 894 MHz carries forward communication (see Figure 16.3).

Figure 16.3 *Cellular bands for AMPS*

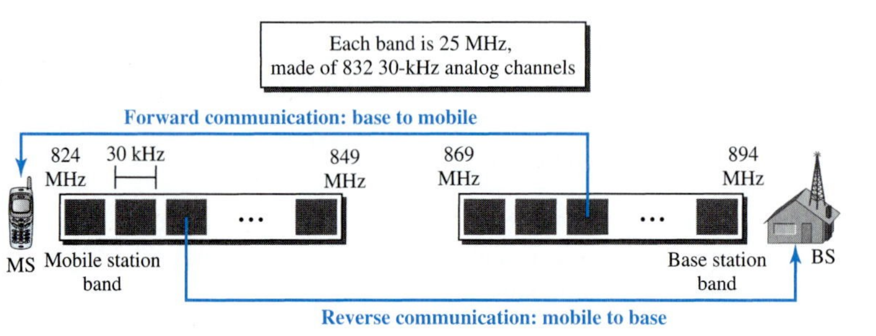

Each band is divided into 832 channels. However, two providers can share an area, which means 416 channels in each cell for each provider. Out of these 416, 21 channels are used for control, which leaves 395 channels. AMPS has a frequency reuse factor of 7; this means only one-seventh of these 395 traffic channels are actually available in a cell.

Transmission AMPS uses FM and FSK for modulation. Figure 16.4 shows the transmission in the reverse direction. Voice channels are modulated using FM, and control channels use FSK to create 30-kHz analog signals. AMPS uses FDMA to divide each 25-MHz band into 30-kHz channels.

Second Generation

To provide higher-quality (less noise-prone) mobile voice communications, the second generation of the cellular phone network was developed. While the first generation was designed for analog voice communication, the second generation was mainly designed for digitized voice. Three major systems evolved in the second generation, as shown in Figure 16.5. We will discuss each system separately.

Figure 16.4 *AMPS reverse communication band*

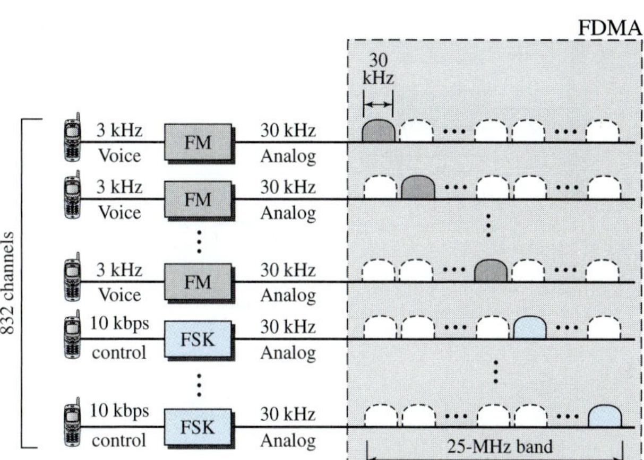

Figure 16.5 *Second-generation cellular phone systems*

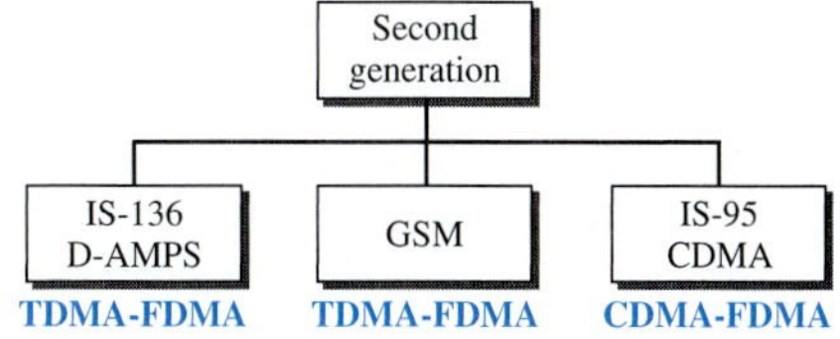

D-AMPS

The product of the evolution of the analog AMPS into a digital system is **digital AMPS (D-AMPS).** D-AMPS was designed to be backward-compatible with AMPS. This means that in a cell, one telephone can use AMPS and another D-AMPS. D-AMPS was first defined by IS-54 (Interim Standard 54) and later revised by IS-136.

Band D-AMPS uses the same bands and channels as AMPS.

Transmission Each voice channel is digitized using a very complex PCM and compression technique. A voice channel is digitized to 7.95 kbps. Three 7.95-kbps digital voice channels are combined using TDMA. The result is 48.6 kbps of digital data; much of this is overhead. As Figure 16.6 shows, the system sends 25 frames per second, with 1944 bits per frame. Each frame lasts 40 ms (1/25) and is divided into six slots shared by three digital channels; each channel is allotted two slots.

Each slot holds 324 bits. However, only 159 bits comes from the digitized voice; 64 bits are for control and 101 bits are for error correction. In other words, each channel drops 159 bits of data into each of the two channels assigned to it. The system adds 64 control bits and 101 error-correcting bits.

Figure 16.6 *D-AMPS*

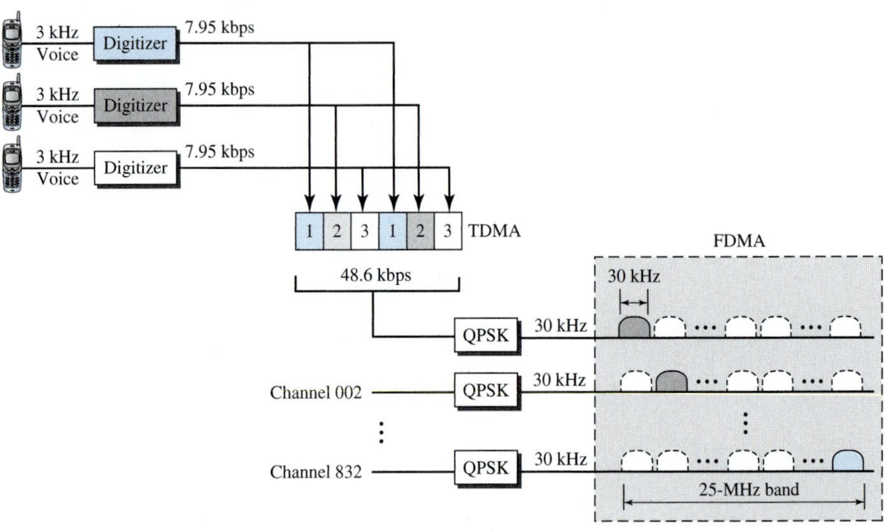

The resulting 48.6 kbps of digital data modulates a carrier using QPSK; the result is a 30-kHz analog signal. Finally, the 30-kHz analog signals share a 25-MHz band (FDMA). D-AMPS has a frequency reuse factor of 7.

D-AMPS, or IS-136, is a digital cellular phone system using TDMA and FDMA.

GSM

The **Global System for Mobile Communication (GSM)** is a European standard that was developed to provide a common second-generation technology for all Europe. The aim was to replace a number of incompatible first-generation technologies.

Bands GSM uses two bands for duplex communication. Each band is 25 MHz in width, shifted toward 900 MHz, as shown in Figure 16.7. Each band is divided into 124 channels of 200 kHz separated by guard bands.

Figure 16.7 *GSM bands*

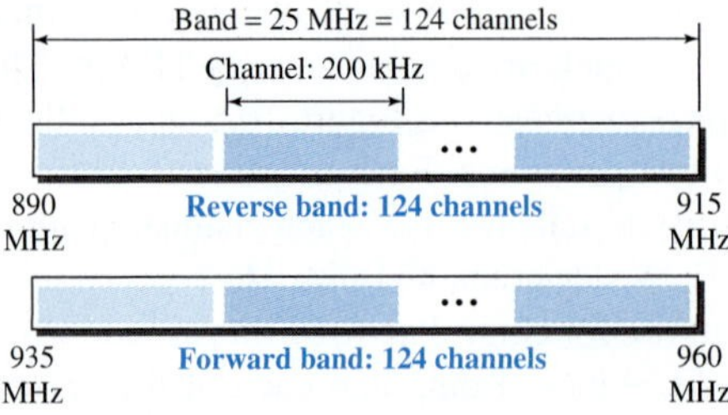

Transmission Figure 16.8 shows a GSM system. Each voice channel is digitized and compressed to a 13-kbps digital signal. Each slot carries 156.25 bits (see Figure 16.9). Eight slots share a frame (TDMA). Twenty-six frames also share a multiframe (TDMA). We can calculate the bit rate of each channel as follows:

$$\text{Channel data rate} = (1/120 \text{ ms}) \times 26 \times 8 \times 156.25 = 270.8 \text{ kbps}$$

Figure 16.8 *GSM*

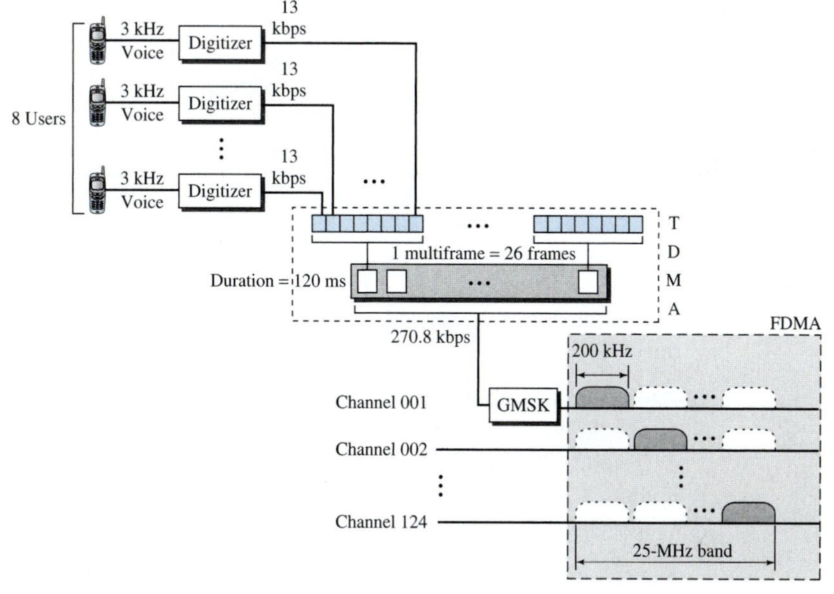

Each 270.8-kbps digital channel modulates a carrier using GMSK (a form of FSK used mainly in European systems); the result is a 200-kHz analog signal. Finally 124 analog channels of 200 kHz are combined using FDMA. The result is a 25-MHz band. Figure 16.9 shows the user data and overhead in a multiframe.

The reader may have noticed the large amount of overhead in TDMA. The user data are only 65 bits per slot. The system adds extra bits for error correction to make it 114 bits per slot. To this, control bits are added to bring it up to 156.25 bits per slot. Eight slots are encapsulated in a frame. Twenty-four traffic frames and two additional control frames make a multiframe. A multiframe has a duration of 120 ms. However, the architecture does define superframes and hyperframes that do not add any overhead; we will not discuss them here.

Reuse Factor Because of the complex error correction mechanism, GSM allows a reuse factor as low as 3.

Figure 16.9 *Multiframe components*

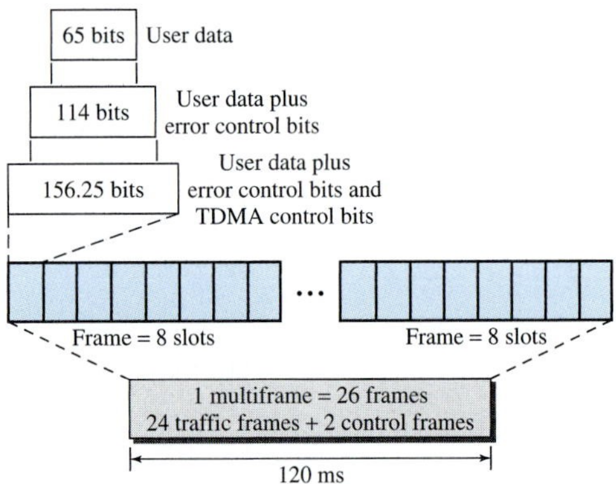

GSM is a digital cellular phone system using TDMA and FDMA.

IS-95

One of the dominant second-generation standards in North America is **Interim Standard 95 (IS-95).** It is based on CDMA and DSSS.

Bands and Channels IS-95 uses two bands for duplex communication. The bands can be the traditional ISM 800-MHz band or the ISM 1900-MHz band. Each band is divided into 20 channels of 1.228 MHz separated by guard bands. Each service provider is allotted 10 channels. IS-95 can be used in parallel with AMPS. Each IS-95 channel is equivalent to 41 AMPS channels (41×30 kHz = 1.23 MHz).

Synchronization All base channels need to be synchronized to use CDMA. To provide synchronization, bases use the services of GPS (Global Positioning System), a satellite system that we discuss in the next section.

Forward Transmission IS-95 has two different transmission techniques: one for use in the forward (base to mobile) direction and another for use in the reverse (mobile to base) direction. In the forward direction, communications between the base and all mobiles are synchronized; the base sends synchronized data to all mobiles. Figure 16.10 shows a simplified diagram for the forward direction.

Each voice channel is digitized, producing data at a basic rate of 9.6 kbps. After adding error-correcting and repeating bits, and interleaving, the result is a signal of 19.2 ksps (kilosignals per second). This output is now scrambled using a 19.2-ksps signal. The scrambling signal is produced from a long code generator that uses the electronic serial number (ESN) of the mobile station and generates 2^{42} pseudorandom chips, each chip having 42 bits. Note that the chips are generated pseudorandomly, not randomly, because the pattern repeats itself. The output of the long code generator is fed to a decimator, which chooses 1 bit out of 64 bits. The output of the decimator is used for scrambling. The scrambling is used to create privacy; the ESN is unique for each station.

Figure 16.10 *IS-95 forward transmission*

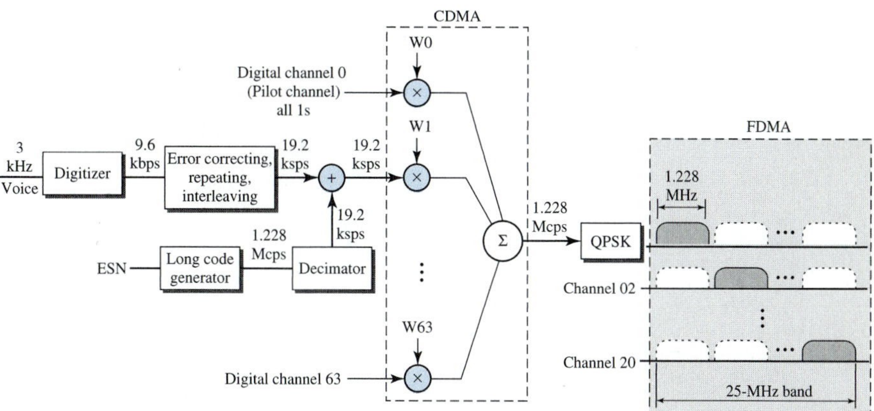

The result of the scrambler is combined using CDMA. For each traffic channel, one Walsh 64 × 64 row chip is selected. The result is a signal of 1.228 Mcps (megachips per second).

$$19.2 \text{ ksps} \times 64 \text{ cps} = 1.228 \text{ Mcps}$$

The signal is fed into a QPSK modulator to produce a signal of 1.228 MHz. The resulting bandwidth is shifted appropriately, using FDMA. An analog channel creates 64 digital channels, of which 55 channels are traffic channels (carrying digitized voice). Nine channels are used for control and synchronization:

- Channel 0 is a pilot channel. This channel sends a continuous stream of 1s to mobile stations. The stream provides bit synchronization, serves as a phase reference for demodulation, and allows the mobile station to compare the signal strength of neighboring bases for handoff decisions.
- Channel 32 gives information about the system to the mobile station.
- Channels 1 to 7 are used for paging, to send messages to one or more mobile stations.
- Channels 8 to 31 and 33 to 63 are traffic channels carrying digitized voice from the base station to the corresponding mobile station.

Reverse Transmission The use of CDMA in the forward direction is possible because the pilot channel sends a continuous sequence of 1s to synchronize transmission. The synchronization is not used in the reverse direction because we need an entity to do that, which is not feasible. Instead of CDMA, the reverse channels use DSSS (direct sequence spread spectrum), which we discussed in Chapter 8. Figure 16.11 shows a simplified diagram for reverse transmission.

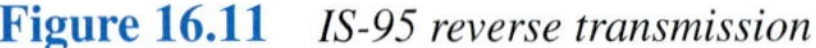

Figure 16.11 *IS-95 reverse transmission*

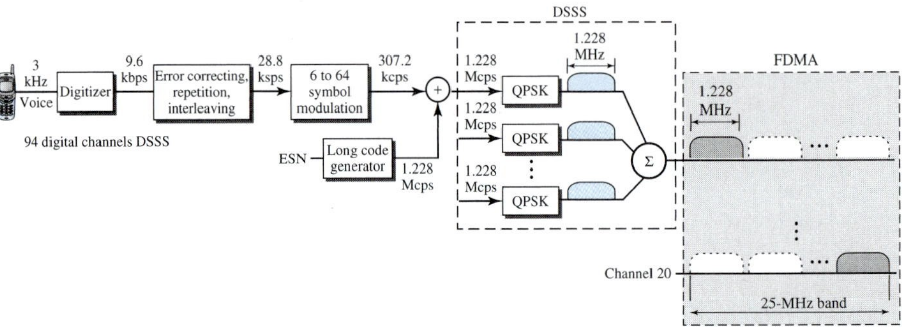

Each voice channel is digitized, producing data at a rate of 9.6 kbps. However, after adding error-correcting and repeating bits, plus interleaving, the result is a signal of 28.8 ksps. The output is now passed through a 6/64 symbol modulator. The symbols are divided into six-symbol chunks, and each chunk is interpreted as a binary number (from 0 to 63). The binary number is used as the index to a 64 × 64 Walsh matrix for selection of a row of chips. Note that this procedure is not CDMA; each bit is not multiplied by the chips in a row. Each six-symbol chunk is replaced by a 64-chip code. This is done to provide a kind of orthogonality; it differentiates the streams of chips from the different mobile stations. The result creates a signal of 307.2 kcps or (28.8/6) × 64.

Spreading is the next step; each chip is spread into 4. Again the ESN of the mobile station creates a long code of 42 bits at a rate of 1.228 Mcps, which is 4 times 307.2. After spreading, each signal is modulated using QPSK, which is slightly different from the one used in the forward direction; we do not go into details here. Note that there is no multiple-access mechanism here; all reverse channels send their analog signal into the air, but the correct chips will be received by the base station due to spreading.

Although we can create $2^{42} - 1$ digital channels in the reverse direction (because of the long code generator), normally 94 channels are used; 62 are traffic channels, and 32 are channels used to gain access to the base station.

IS-95 is a digital cellular phone system using CDMA/DSSS and FDMA.

Two Data Rate Sets IS-95 defines two data rate sets, with four different rates in each set. The first set defines 9600, 4800, 2400, and 1200 bps. If, for example, the selected rate is 1200 bps, each bit is repeated 8 times to provide a rate of 9600 bps. The second set defines 14,400, 7200, 3600, and 1800 bps. This is possible by reducing the number of bits used for error correction. The bit rates in a set are related to the activity of the channel. If the channel is silent, only 1200 bits can be transferred, which improves the spreading by repeating each bit 8 times.

Frequency-Reuse Factor In an IS-95 system, the frequency-reuse factor is normally 1 because the interference from neighboring cells cannot affect CDMA or DSSS transmission.

Soft Handoff Every base station continuously broadcasts signals using its pilot channel. This means a mobile station can detect the pilot signal from its cell and neighboring cells. This enables a mobile station to do a soft handoff in contrast to a hard handoff.

PCS

Before we leave the discussion of second-generation cellular telephones, let us explain a term generally heard in relation to this generation: PCS. **Personal communications system (PCS)** does not refer to a single technology such as GSM, IS-136, or IS-95. It is a generic name for a commercial system that offers several kinds of communication services. Common features of these systems can be summarized:

1. They may use any second-generation technology (GSM, IS-136, or IS-95).
2. They use the 1900-MHz band, which means that a mobile station needs more power because higher frequencies have a shorter range than lower ones. However, since a station's power is limited by the FCC, the base station and the mobile station need to be close to each other (smaller cells).
3. They offer communication services such as short message service (SMS) and limited Internet access.

Third Generation

The third generation of cellular telephony refers to a combination of technologies that provide a variety of services. Ideally, when it matures, the third generation can provide both digital data and voice communication. Using a small portable device, a person should be able to talk to anyone else in the world with a voice quality similar to that of the existing fixed telephone network. A person can download and watch a movie, can download and listen to music, can surf the Internet or play games, can have a video conference, and can do much more. One of the interesting characteristics of a third-generation system is that the portable device is always connected; you do not need to dial a number to connect to the Internet.

The third-generation concept started in 1992, when ITU issued a blueprint called the **Internet Mobile Communication 2000 (IMT-2000).** The blueprint defines some criteria for third-generation technology as outlined below:

- Voice quality comparable to that of the existing public telephone network.
- Data rate of 144 kbps for access in a moving vehicle (car), 384 kbps for access as the user walks (pedestrians), and 2 Mbps for the stationary user (office or home).
- Support for packet-switched and circuit-switched data services.
- A band of 2 GHz.
- Bandwidths of 2 MHz.
- Interface to the Internet.

The main goal of third-generation cellular telephony is to provide universal personal communication.

IMT-2000 Radio Interface

Figure 16.12 shows the radio interfaces (wireless standards) adopted by IMT-2000. All five are developed from second-generation technologies. The first two evolve from CDMA technology. The third evolves from a combination of CDMA and TDMA. The fourth evolves from TDMA, and the last evolves from both FDMA and TDMA.

Figure 16.12 *IMT-2000 radio interfaces*

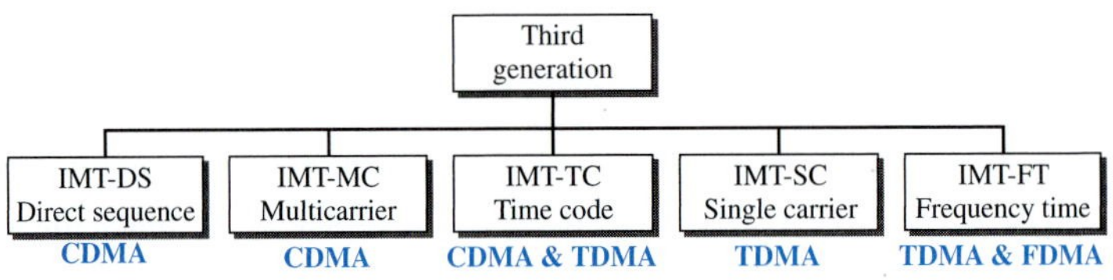

IMT-DS This approach uses a version of CDMA called wideband CDMA or W-CDMA. W-CDMA uses a 5-MHz bandwidth. It was developed in Europe, and it is compatible with the CDMA used in IS-95.

IMT-MC This approach was developed in North America and is known as CDMA 2000. It is an evolution of CDMA technology used in IS-95 channels. It combines the new wideband (15-MHz) spread spectrum with the narrowband (1.25-MHz) CDMA of IS-95. It is backward-compatible with IS-95. It allows communication on multiple 1.25-MHz channels (1, 3, 6, 9, 12 times), up to 15 MHz. The use of the wider channels allows it to reach the 2-Mbps data rate defined for the third generation.

IMT-TC This standard uses a combination of W-CDMA and TDMA. The standard tries to reach the IMT-2000 goals by adding TDMA multiplexing to W-CDMA.

IMT-SC This standard only uses TDMA.

IMT-FT This standard uses a combination of FDMA and TDMA.

16.2 SATELLITE NETWORKS

A **satellite network** is a combination of nodes, some of which are satellites, that provides communication from one point on the Earth to another. A node in the network can be a satellite, an Earth station, or an end-user terminal or telephone. Although a natural satellite, such as the Moon, can be used as a relaying node in the network, the use of artificial satellites is preferred because we can install electronic equipment on the satellite to regenerate the signal that has lost its energy during travel. Another restriction on using natural satellites is their distances from the Earth, which create a long delay in communication.

Satellite networks are like cellular networks in that they divide the planet into cells. Satellites can provide transmission capability to and from any location on Earth, no matter how remote. This advantage makes high-quality communication available to undeveloped parts of the world without requiring a huge investment in ground-based infrastructure.

Orbits

An artificial satellite needs to have an **orbit,** the path in which it travels around the Earth. The orbit can be equatorial, inclined, or polar, as shown in Figure 16.13.

Figure 16.13 *Satellite orbits*

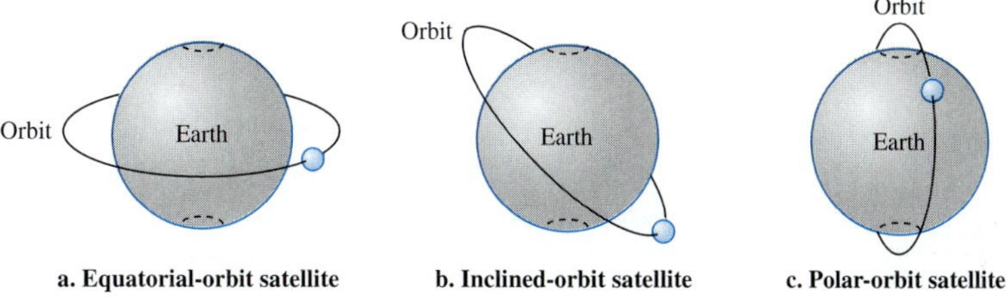

The period of a satellite, the time required for a satellite to make a complete trip around the Earth, is determined by Kepler's law, which defines the period as a function of the distance of the satellite from the center of the Earth.

Example 16.1

What is the period of the Moon, according to Kepler's law?

$$\text{Period} = C \times \text{distance}^{1.5}$$

Here C is a constant approximately equal to 1/100. The period is in seconds and the distance in kilometers.

Solution

The Moon is located approximately 384,000 km above the Earth. The radius of the Earth is 6378 km. Applying the formula, we get

$$\text{Period} = \frac{1}{100}(384{,}000 + 6378)^{1.5} = 2{,}439{,}090 \text{ s} = 1 \text{ month}$$

Example 16.2

According to Kepler's law, what is the period of a satellite that is located at an orbit approximately 35,786 km above the Earth?

Solution

Applying the formula, we get

$$\text{Period} = \frac{1}{100}(35{,}786 + 6378)^{1.5} = 86{,}579 \text{ s} = 24 \text{ h}$$

This means that a satellite located at 35,786 km has a period of 24 h, which is the same as the rotation period of the Earth. A satellite like this is said to be *stationary* to the Earth. The orbit, as we will see, is called a geosynchronous orbit.

Footprint

Satellites process microwaves with bidirectional antennas (line-of-sight). Therefore, the signal from a satellite is normally aimed at a specific area called the **footprint.** The signal power at the center of the footprint is maximum. The power decreases as we move out from the footprint center. The boundary of the footprint is the location where the power level is at a predefined threshold.

Three Categories of Satellites

Based on the location of the orbit, satellites can be divided into three categories: **geostationary Earth orbit (GEO), low-Earth-orbit (LEO),** and **middle-Earth-orbit (MEO).** Figure 16.14 shows the taxonomy.

Figure 16.14 *Satellite categories*

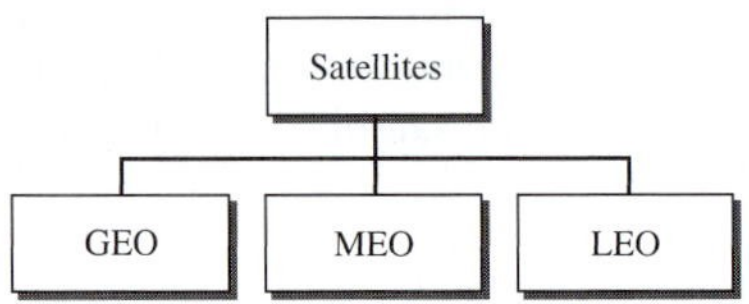

Figure 16.15 shows the satellite altitudes with respect to the surface of the Earth. There is only one orbit, at an altitude of 35,786 km for the GEO satellite. MEO satellites are located at altitudes between 5000 and 15,000 km. LEO satellites are normally below an altitude of 2000 km.

Figure 16.15 *Satellite orbit altitudes*

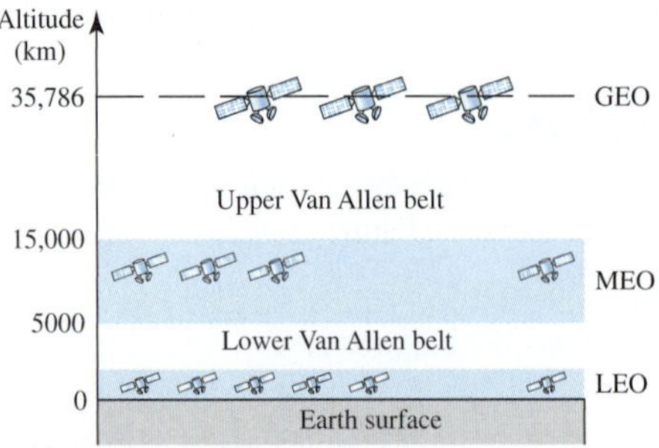

One reason for having different orbits is due to the existence of two Van Allen belts. A Van Allen belt is a layer that contains charged particles. A satellite orbiting in one of these two belts would be totally destroyed by the energetic charged particles. The MEO orbits are located between these two belts.

Frequency Bands for Satellite Communication

The frequencies reserved for satellite microwave communication are in the gigahertz (GHz) range. Each satellite sends and receives over two different bands. Transmission from the Earth to the satellite is called the **uplink.** Transmission from the satellite to the Earth is called the **downlink.** Table 16.1 gives the band names and frequencies for each range.

Table 16.1 *Satellite frequency bands*

Band	*Downlink, GHz*	*Uplink, GHz*	*Bandwidth, MHz*
L	1.5	1.6	15
S	1.9	2.2	70
C	4.0	6.0	500
Ku	11.0	14.0	500
Ka	20.0	30.0	3500

GEO Satellites

Line-of-sight propagation requires that the sending and receiving antennas be locked onto each other's location at all times (one antenna must have the other in sight). For this reason, a satellite that moves faster or slower than the Earth's rotation is useful only for short periods. To ensure constant communication, the satellite must move at the same speed as the Earth so that it seems to remain fixed above a certain spot. Such satellites are called *geostationary.*

Because orbital speed is based on the distance from the planet, only one orbit can be geostationary. This orbit occurs at the equatorial plane and is approximately 22,000 mi from the surface of the Earth.

But one geostationary satellite cannot cover the whole Earth. One satellite in orbit has line-of-sight contact with a vast number of stations, but the curvature of the Earth still keeps much of the planet out of sight. It takes a minimum of three satellites equidistant from each other in geostationary Earth orbit (GEO) to provide full global transmission. Figure 16.16 shows three satellites, each 120° from another in geosynchronous orbit around the equator. The view is from the North Pole.

MEO Satellites

Medium-Earth-orbit (MEO) satellites are positioned between the two Van Allen belts. A satellite at this orbit takes approximately 6–8 hours to circle the Earth.

Global Positioning System

One example of a MEO satellite system is the **Global Positioning System (GPS),** constracted and operated by the US Department of Defense, orbiting at an altitude about

Figure 16.16 *Satellites in geostationary orbit*

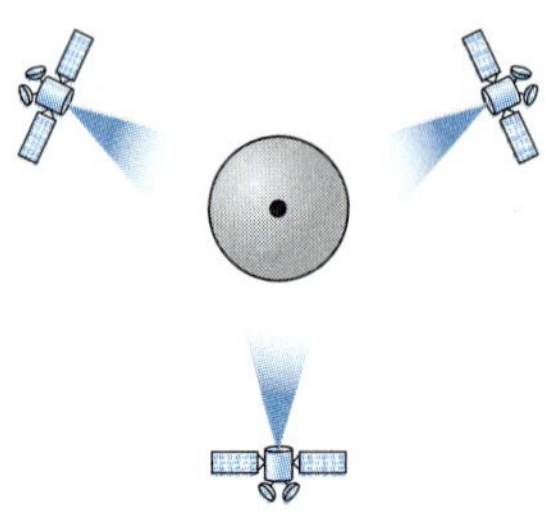

18,000 km (11,000 mi) above the Earth. The system consists of 24 satellites and is used for land, sea, and air navigation to provide time and locations for vehicles and ships. GPS uses 24 satellites in six orbits, as shown in Figure 16.17. The orbits and the locations of the satellites in each orbit are designed in such a way that, at any time, four satellites are visible from any point on Earth. A GPS receiver has an almanac that tells the current position of each satellite.

Figure 16.17 *Orbits for global positioning system (GPS) satellites*

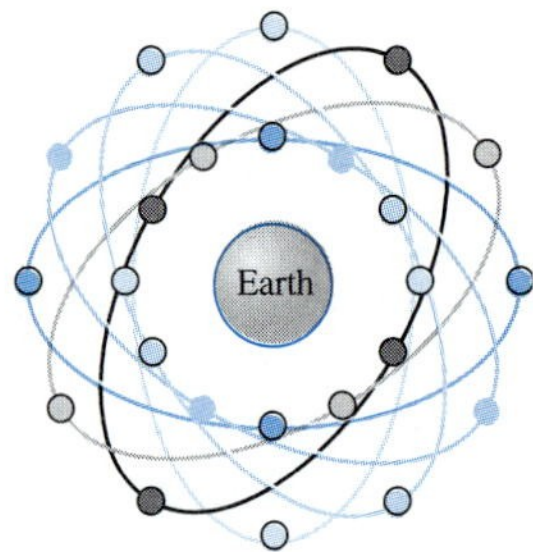

Trilateration GPS is based on a principle called **trilateration.**† On a plane, if we know our distance from three points, we know exactly where we are. Let us say that we are 10 miles away from point A, 12 miles away from point B, and 15 miles away from point C. If we draw three circles with the centers at A, B, and C, we must be somewhere on circle A, somewhere on circle B, and somewhere on circle C. These three circles meet at one single point (if our distances are correct), our position. Figure 16.18a shows the concept.

In three-dimensional space, the situation is different. Three spheres meet in two points as shown in Figure 16.18b. We need at least four spheres to find our exact position in space (longitude, latitude, and altitude). However, if we have additional facts about our location (for example, we know that we are not inside the ocean or somewhere in

†The terms *trilateration* and *triangulation* are normally used interchangeably. We use the word *trilateration,* which means using three distances, instead of **triangulation,** which may mean using three angles.

Figure 16.18 *Trilateration on a plane*

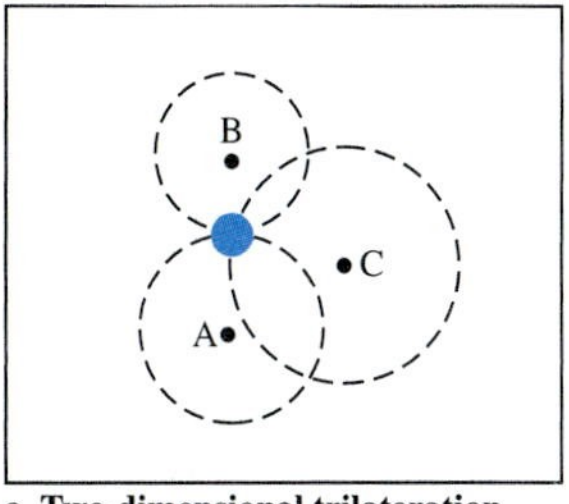

a. Two-dimensional trilateration

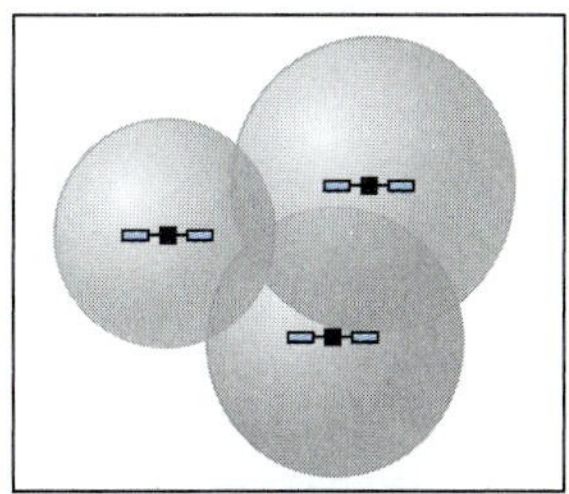
b. Three-dimensional trilateration

space), three spheres are enough, because one of the two points, where the spheres meet, is so improbable that the other can be selected without a doubt.

Measuring the Distance The trilateration principle can find our location on the earth if we know our distance from three satellites and know the position of each satellite. The position of each satellite can be calculated by a GPS receiver (using the predetermined path of the satellites). The GPS receiver, then, needs to find its distance from at least three GPS satellites (center of the spheres). Measuring the distance is done using a principle called one-way ranging. For the moment, let us assume that all GPS satellites and the receiver on the Earth are synchronized. Each of 24 satellites synchronously transmits a complex signal each having a unique pattern. The computer on the receiver measures the delay between the signals from the satellites and its copy of signals to determine the distances to the satellites.

Synchronization The previous discussion was based on the assumption that the satellites' clock are synchronized with each other and with the receiver's clock. Satellites use atomic clock that are precise and can function synchronously with each other. The receiver's clock however, is a normal quartz clock (an atomic clock costs more that $50,000), and there is no way to synchronize it with the satellite clocks. There is an unknown offset between the satellite clocks and the receiver clock that introduces a corresponding offset in the distance calculation. Because of this offset, the measured distance is called a *pseudorange*.

GPS uses an elegant solution to the clock offset problem, by recognizing that the offset's value is the same for all satellite being used. The calculation of position becomes finding four unknowns: the x_r, y_r, z_r, coordinates of the receiver, and common clock offset *dt*. For finding these four unknown values, we need at least four equations. This means that we need to measure pesudoranges from four satellite instead of three. If we call the four measured pseudoranges PR1, PR2, PR3 and PR4 and the coordinates of each satellite x_i, y_i, and z_i (for i =1 to 4), we can find the four previously mentioned unknown values using the following four equations (the four unknown values are shown in color).

$$\mathbf{PR_1 = [(x_1 - x_r)^2 + (y_1 - y_r)^2 + (z_1 - z_r)^2]^{1/2} + c \times dt}$$
$$\mathbf{PR_2 = [(x_2 - x_r)^2 + (y_2 - y_r)^2 + (z_2 - z_r)^2]^{1/2} + c \times dt}$$
$$\mathbf{PR_3 = [(x_3 - x_r)^2 + (y_3 - y_r)^2 + (z_3 - z_r)^2]^{1/2} + c \times dt}$$
$$\mathbf{PR_4 = [(x_4 - x_r)^2 + (y_4 - y_r)^2 + (z_4 - z_r)^2]^{1/2} + c \times dt}$$

The coordinates used in the above formulas are in an Earth-Centered Earth-Fixed (ECEF) reference frame, which means that the origin of the coordinate space is at the center of the Earth and the coordinate space rotate with the Earth. This implies that the ECEF coordinates of a fixed point on the surface of the earth do not change.

Application GPS is used by military forces. For example, thousands of portable GPS receivers were used during the Persian Gulf war by foot soldiers, vehicles, and helicopters. Another use of GPS is in navigation. The driver of a car can find the location of the car. The driver can then consult a database in the memory of the automobile to be directed to the destination. In other words, GPS gives the location of the car, and the database uses this information to find a path to the destination. A very interesting application is clock synchronization. As we mentioned previously, the IS-95 cellular telephone system uses GPS to create time synchronization between the base stations.

LEO Satellites

Low-Earth-orbit (LEO) satellites have polar orbits. The altitude is between 500 and 2000 km, with a rotation period of 90 to 120 min. The satellite has a speed of 20,000 to 25,000 km/h. An LEO system usually has a cellular type of access, similar to the cellular telephone system. The footprint normally has a diameter of 8000 km. Because LEO satellites are close to Earth, the round-trip time propagation delay is normally less than 20 ms, which is acceptable for audio communication.

An LEO system is made of a constellation of satellites that work together as a network; each satellite acts as a switch. Satellites that are close to each other are connected through intersatellite links (ISLs). A mobile system communicates with the satellite through a user mobile link (UML). A satellite can also communicate with an Earth station (gateway) through a gateway link (GWL). Figure 16.19 shows a typical LEO satellite network.

Figure 16.19 *LEO satellite system*

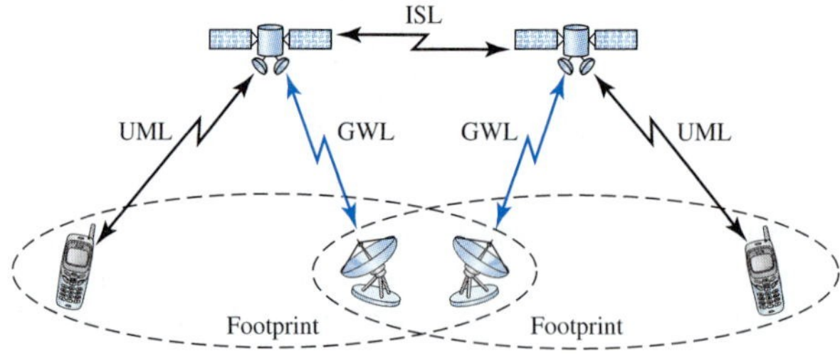

LEO satellites can be divided into three categories: little LEOs, big LEOs, and broadband LEOs. The little LEOs operate under 1 GHz. They are mostly used for low-data-rate messaging. The big LEOs operate between 1 and 3 GHz. Globalstar and Iridium systems are examples of big LEOs. The broadband LEOs provide communication similar to fiber-optic networks. The first broadband LEO system was Teledesic.

Iridium System

The concept of the **Iridium** system, a 77-satellite network, was started by Motorola in 1990. The project took eight years to materialize. During this period, the number of satellites was reduced. Finally, in 1998, the service was started with 66 satellites. The original name, Iridium, came from the name of the 77th chemical element; a more appropriate name is Dysprosium (the name of element 66).

Iridium has gone through rough times. The system was halted in 1999 due to financial problems; it was sold and restarted in 2001 under new ownership.

The system has 66 satellites divided into six orbits, with 11 satellites in each orbit. The orbits are at an altitude of 750 km. The satellites in each orbit are separated from one another by approximately 32° of latitude. Figure 16.20 shows a schematic diagram of the constellation.

Figure 16.20 *Iridium constellation*

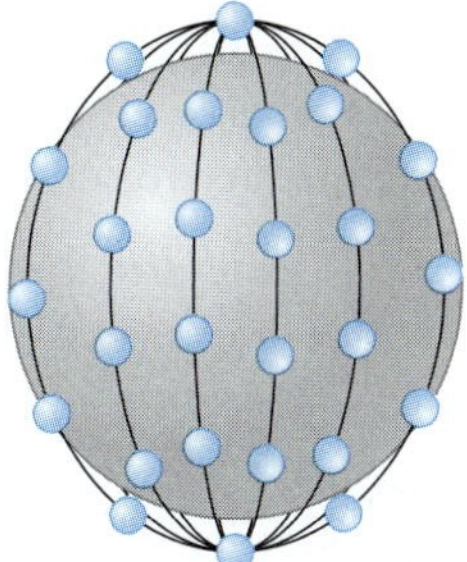

The Iridium system has 66 satellites in six LEO orbits, each at an altitude of 750 km.

Since each satellite has 48 spot beams, the system can have up to 3168 beams. However, some of the beams are turned off as the satellite approaches the pole. The number of active spot beams at any moment is approximately 2000. Each spot beam covers a cell on Earth, which means that Earth is divided into approximately 2000 (overlapping) cells.

In the Iridium system, communication between two users takes place through satellites. When a user calls another user, the call can go through several satellites before reaching the destination. This means that relaying is done in space and each satellite needs to be sophisticated enough to do relaying. This strategy eliminates the need for many terrestrial stations.

The whole purpose of Iridium is to provide direct worldwide communication using handheld terminals (same concept as cellular telephony). The system can be used for voice, data, paging, fax, and even navigation. The system can provide connectivity between users at locations where other types of communication are not possible. The system provides 2.4- to 4.8-kbps voice and data transmission between portable telephones. Transmission occurs in the 1.616- to 1.6126-GHz frequency band. Intersatellite communication occurs in the 23.18- to 23.38-GHz frequency band.

Iridium is designed to provide direct worldwide voice and data communication using handheld terminals, a service similar to cellular telephony but on a global scale.

Globalstar

Globalstar is another LEO satellite system. The system uses 48 satellites in six polar orbits with each orbit hosting eight satellites. The orbits are located at an altitude of almost 1400 km.

The Globalstar system is similar to the Iridium system; the main difference is the relaying mechanism. Communication between two distant users in the Iridium system requires relaying between several satellites; Globalstar communication requires both satellites and Earth stations, which means that ground stations can create more powerful signals.

Teledesic

Teledesic is a system of satellites that provides fiber-optic-like (broadband channels, low error rate, and low delay) communication. Its main purpose is to provide broadband Internet access for users all over the world. It is sometimes called "Internet in the sky."

The project was started in 1990 by Craig McCaw and Bill Gates; later, other investors joined the consortium. The project is scheduled to be fully functional in the near future.

Constellation Teledesic provides 288 satellites in 12 polar orbits with each orbit hosting 24 satellites. The orbits are at an altitude of 1350 km, as shown in Figure 16.21.

Figure 16.21 *Teledesic*

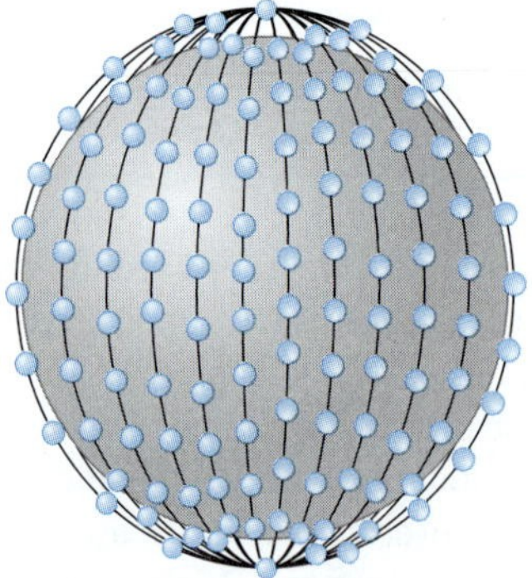

Teledesic has 288 satellites in 12 LEO orbits, each at an altitude of 1350 km.

Communication The system provides three types of communication. Intersatellite communication allows eight neighboring satellites to communicate with one another. Communication is also possible between a satellite and an Earth gateway station. Users can communicate directly with the network using terminals. Earth is divided into tens of thousands of cells. Each cell is assigned a time slot, and the satellite focuses its beam to the cell at the corresponding time slot. The terminal can send data during its time slot. A terminal receives all packets intended for the cell, but selects only those intended for its address.

Bands Transmission occurs in the Ka bands.

Data Rate The data rate is up to 155 Mbps for the uplink and up to 1.2 Gbps for the downlink.

16.3 RECOMMENDED READING

For more details about subjects discussed in this chapter, we recommend the following books. The items in brackets [. . .] refer to the reference list at the end of the text.

Books

Wireless WANs are completely covered in [Sta02], [Jam03], [AZ03], and [Sch03]. Communication satellites are discussed in Section 2.4 of [Tan03] and Section 8.5 of [Cou01]. Mobile telephone system is discussed in Section 2.6 of [Tan03] and Section 8.8 of [Cou01].

16.4 KEY TERMS

Advanced Mobile Phone System (AMPS)
cellular telephony
digital AMPS (D-AMPS)
downlink
footprint
geostationary Earth orbit (GEO)
Global Positioning System (GPS)
Global System for Mobile Communication (GSM)
Globalstar
handoff
Interim Standard 95 (IS-95)
Internet Mobile Communication 2000 (IMT-2000)
Iridium
low-Earth-orbit (LEO)
medium-Earth-orbit (MEO)
mobile switching center (MSC)
orbit
personal communications system (PCS)
reuse factor
roaming
satellite network
Teledesic
triangulation
trilateration
uplink

16.5 SUMMARY

- Cellular telephony provides communication between two devices. One or both may be mobile.
- A cellular service area is divided into cells.
- Advanced Mobile Phone System (AMPS) is a first-generation cellular phone system.

- Digital AMPS (D-AMPS) is a second-generation cellular phone system that is a digital version of AMPS.
- Global System for Mobile Communication (GSM) is a second-generation cellular phone system used in Europe.
- Interim Standard 95 (IS-95) is a second-generation cellular phone system based on CDMA and DSSS.
- The third-generation cellular phone system will provide universal personal communication.
- A satellite network uses satellites to provide communication between any points on Earth.
- A geostationary Earth orbit (GEO) is at the equatorial plane and revolves in phase with Earth.
- Global Positioning System (GPS) satellites are medium-Earth-orbit (MEO) satellites that provide time and location information for vehicles and ships.
- Iridium satellites are low-Earth-orbit (LEO) satellites that provide direct universal voice and data communications for handheld terminals.
- Teledesic satellites are low-Earth-orbit satellites that will provide universal broadband Internet access.

16.6 PRACTICE SET

Review Questions

1. What is the relationship between a base station and a mobile switching center?
2. What are the functions of a mobile switching center?
3. Which is better, a low reuse factor or a high reuse factor? Explain your answer.
4. What is the difference between a hard handoff and a soft handoff?
5. What is AMPS?
6. What is the relationship between D-AMPS and AMPS?
7. What is GSM?
8. What is the function of the CDMA in IS-95?
9. What are the three types of orbits?
10. Which type of orbit does a GEO satellite have? Explain your answer.
11. What is a footprint?
12. What is the relationship between the Van Allen belts and satellites?
13. Compare an uplink with a downlink.
14. What is the purpose of GPS?
15. What is the main difference between Iridium and Globalstar?

Exercises

16. Draw a cell pattern with a frequency-reuse factor of 3.
17. What is the maximum number of callers in each cell in AMPS?

18. What is the maximum number of simultaneous calls in each cell in an IS-136 (D-AMPS) system, assuming no analog control channels?
19. What is the maximum number of simultaneous calls in each cell in a GSM assuming no analog control channels?
20. What is the maximum number of callers in each cell in an IS-95 system?
21. Find the efficiency of AMPS in terms of simultaneous calls per megahertz of bandwidth. In other words, find the number of calls that can be used in 1-MHz bandwidth allocation.
22. Repeat Exercise 21 for D-AMPS.
23. Repeat Exercise 21 for GSM.
24. Repeat Exercise 21 for IS-95.
25. Guess the relationship between a 3-kHz voice channel and a 30-kHz modulated channel in a system using AMPS.
26. How many slots are sent each second in a channel using D-AMPS? How many slots are sent by each user in 1 s?
27. Use Kepler's formula to check the accuracy of a given period and altitude for a GPS satellite.
28. Use Kepler's formula to check the accuracy of a given period and altitude for an Iridium satellite.
29. Use Kepler's formula to check the accuracy of a given period and altitude for a Globalstar satellite.

CHAPTER 17

SONET/SDH

In this chapter, we introduce a wide area network (WAN), SONET, that is used as a transport network to carry loads from other WANs. We first discuss SONET as a protocol, and we then show how SONET networks can be constructed from the standards defined in the protocol.

The high bandwidths of fiber-optic cable are suitable for today's high-data-rate technologies (such as video conferencing) and for carrying large numbers of lower-rate technologies at the same time. For this reason, the importance of fiber optics grows in conjunction with the development of technologies requiring high data rates or wide bandwidths for transmission. With their prominence came a need for standardization. The United States (ANSI) and Europe (ITU-T) have responded by defining standards that, though independent, are fundamentally similar and ultimately compatible. The ANSI standard is called the **Synchronous Optical Network (SONET).** The ITU-T standard is called the **Synchronous Digital Hierarchy (SDH).**

SONET was developed by ANSI; SDH was developed by ITU-T.

SONET/SDH is a synchronous network using synchronous TDM multiplexing. All clocks in the system are locked to a master clock.

17.1 ARCHITECTURE

Let us first introduce the architecture of a SONET system: signals, devices, and connections.

Signals

SONET defines a hierarchy of electrical signaling levels called **synchronous transport signals (STSs).** Each STS level (STS-1 to STS-192) supports a certain data rate, specified in megabits per second (see Table 17.1). The corresponding optical signals are called **optical carriers (OCs).** SDH specifies a similar system called a **synchronous transport module (STM).** STM is intended to be compatible with existing European hierarchies, such as E lines, and with STS levels. To this end, the lowest STM level, STM-1, is defined as 155.520 Mbps, which is exactly equal to STS-3.

Table 17.1 *SONET/SDH rates*

STS	*OC*	*Rate (Mbps)*	*STM*
STS-1	OC-1	51.840	
STS-3	OC-3	155.520	**STM-1**
STS-9	OC-9	466.560	**STM-3**
STS-12	OC-12	622.080	**STM-4**
STS-18	OC-18	933.120	**STM-6**
STS-24	OC-24	1244.160	**STM-8**
STS-36	OC-36	1866.230	**STM-12**
STS-48	OC-48	2488.320	**STM-16**
STS-96	OC-96	4976.640	**STM-32**
STS-192	OC-192	9953.280	**STM-64**

A glance through Table 17.1 reveals some interesting points. First, the lowest level in this hierarchy has a data rate of 51.840 Mbps, which is greater than that of the DS-3 service (44.736 Mbps). In fact, the STS-1 is designed to accommodate data rates equivalent to those of the DS-3. The difference in capacity is provided to handle the overhead needs of the optical system.

Second, the STS-3 rate is exactly three times the STS-1 rate; and the STS-9 rate is exactly one-half the STS-18 rate. These relationships mean that 18 STS-1 channels can be multiplexed into one STS-18, six STS-3 channels can be multiplexed into one STS-18, and so on.

SONET Devices

Figure 17.1 shows a simple link using SONET devices. SONET transmission relies on three basic devices: STS multiplexers/demultiplexers, regenerators, add/drop multiplexers and terminals.

STS Multiplexer/Demultiplexer

STS multiplexers/demultiplexers mark the beginning points and endpoints of a SONET link. They provide the interface between an electrical tributary network and the optical network. An **STS multiplexer** multiplexes signals from multiple electrical sources and creates the corresponding OC signal. An **STS demultiplexer** demultiplexes an optical OC signal into corresponding electric signals.

Regenerator

Regenerators extend the length of the links. A **regenerator** is a repeater (see Chapter 15) that takes a received optical signal (OC-*n*), demodulates it into the corresponding electric signal (STS-*n*), regenerates the electric signal, and finally modulates the electric

Figure 17.1 *A simple network using SONET equipment*

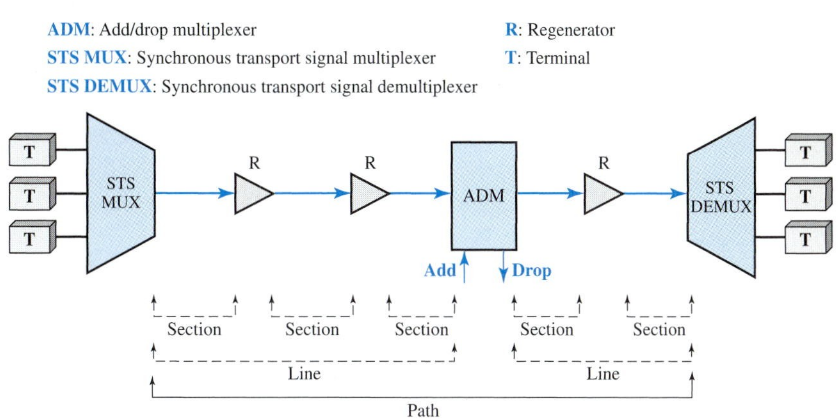

signal into its correspondent OC-n signal. A SONET regenerator replaces some of the existing overhead information (header information) with new information.

Add/drop Multiplexer

Add/drop multiplexers allow insertion and extraction of signals. An **add/drop multiplexer (ADM)** can add STSs coming from different sources into a given path or can remove a desired signal from a path and redirect it without demultiplexing the entire signal. Instead of relying on timing and bit positions, add/drop multiplexers use header information such as addresses and pointers (described later in this section) to identify individual streams.

In the simple configuration shown by Figure 17.1, a number of incoming electronic signals are fed into an STS multiplexer, where they are combined into a single optical signal. The optical signal is transmitted to a regenerator, where it is recreated without the noise it has picked up in transit. The regenerated signals from a number of sources are then fed into an add/drop multiplexer. The add/drop multiplexer reorganizes these signals, if necessary, and sends them out as directed by information in the data frames. These remultiplexed signals are sent to another regenerator and from there to the receiving STS demultiplexer, where they are returned to a format usable by the receiving links.

Terminals

A **terminal** is a device that uses the services of a SONET network. For example, in the Internet, a terminal can be a router that needs to send packets to another router at the other side of a SONET network.

Connections

The devices defined in the previous section are connected using *sections, lines,* and *paths.*

Sections

A **section** is the optical link connecting two neighbor devices: multiplexer to multiplexer, multiplexer to regenerator, or regenerator to regenerator.

Lines

A **line** is the portion of the network between two multiplexers: STS multiplexer to add/drop multiplexer, two add/drop multiplexers, or two STS multiplexers.

Paths

A **path** is the end-to-end portion of the network between two STS multiplexers. In a simple SONET of two STS multiplexers linked directly to each other, the section, line, and path are the same.

17.2 SONET LAYERS

The SONET standard includes four functional layers: the photonic, the section, the line, and the path layer. They correspond to both the physical and the data link layers (see Figure 17.2). The headers added to the frame at the various layers are discussed later in this chapter.

SONET defines four layers: path, line, section, and photonic.

Figure 17.2 *SONET layers compared with OSI or the Internet layers*

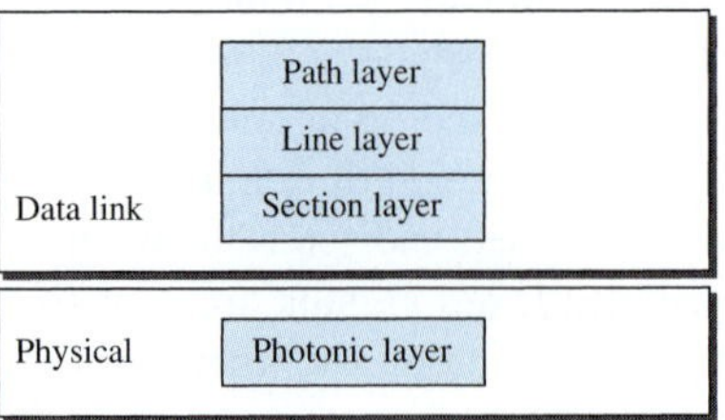

Path Layer

The **path layer** is responsible for the movement of a signal from its optical source to its optical destination. At the optical source, the signal is changed from an electronic form into an optical form, multiplexed with other signals, and encapsulated in a frame. At the optical destination, the received frame is demultiplexed, and the individual optical signals are changed back into their electronic forms. Path layer overhead is added at this layer. STS multiplexers provide path layer functions.

Line Layer

The **line layer** is responsible for the movement of a signal across a physical line. Line layer overhead is added to the frame at this layer. STS multiplexers and add/drop multiplexers provide line layer functions.

Section Layer

The **section layer** is responsible for the movement of a signal across a physical section. It handles framing, scrambling, and error control. Section layer overhead is added to the frame at this layer.

Photonic Layer

The **photonic layer** corresponds to the physical layer of the OSI model. It includes physical specifications for the optical fiber channel, the sensitivity of the receiver, multiplexing functions, and so on. SONET uses NRZ encoding with the presence of light representing 1 and the absence of light representing 0.

Device–Layer Relationships

Figure 17.3 shows the relationship between the devices used in SONET transmission and the four layers of the standard. As you can see, an STS multiplexer is a four-layer device. An add/drop multiplexer is a three-layer device. A regenerator is a two-layer device.

Figure 17.3 *Device–layer relationship in SONET*

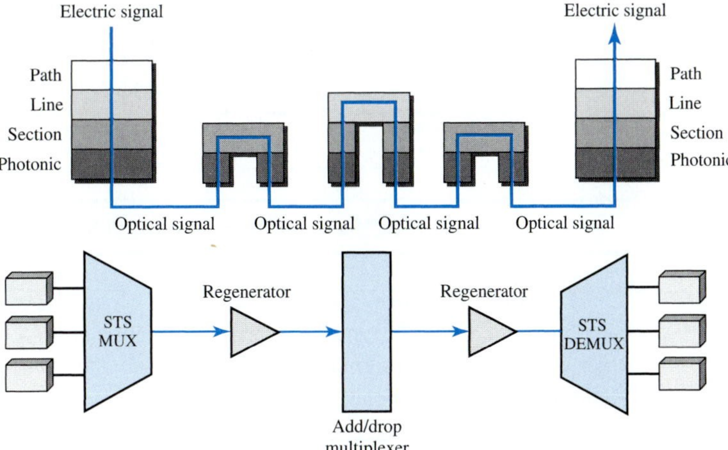

17.3 SONET FRAMES

Each synchronous transfer signal STS-n is composed of 8000 frames. Each frame is a two-dimensional matrix of bytes with 9 rows by $90 \times n$ columns. For example, STS-1 frame is 9 rows by 90 columns (810 bytes), and an STS-3 is 9 rows by 270 columns (2430 bytes). Figure 17.4 shows the general format of an STS-1 and an STS-n.

Figure 17.4 *An STS-1 and an STS-n frame*

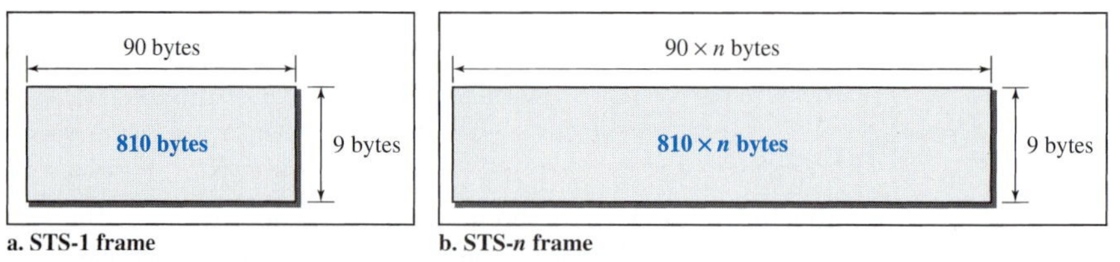

Frame, Byte, and Bit Transmission

One of the interesting points about SONET is that each STS-n signal is transmitted at a fixed rate of 8000 frames per second. This is the rate at which voice is digitized (see Chapter 4). For each frame the bytes are transmitted from the left to the right, top to the bottom. For each byte, the bits are transmitted from the most significant to the least significant (left to right). Figure 17.5 shows the order of frame and byte transmission.

Figure 17.5 *STS-1 frames in transition*

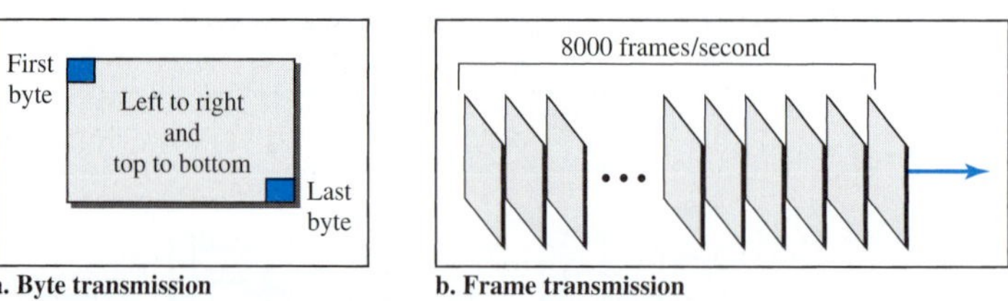

A SONET STS-*n* signal is transmitted at 8000 frames per second.

If we sample a voice signal and use 8 bits (1 byte) for each sample, we can say that each byte in a SONET frame can carry information from a digitized voice channel. In other words, an STS-1 signal can carry 774 voice channels simultaneously (810 minus required bytes for overhead).

Each byte in a SONET frame can carry a digitized voice channel.

Example 17.1

Find the data rate of an STS-1 signal.

Solution

STS-1, like other STS signals, sends 8000 frames per second. Each STS-1 frame is made of 9 by (1×90) bytes. Each byte is made of 8 bits. The data rate is

$$\text{STS-1 data rate} = 8000 \times 9 \times (1 \times 90) \times 8 = 51.840 \text{ Mbps}$$

Example 17.2

Find the data rate of an STS-3 signal.

Solution

STS-3, like other STS signals, sends 8000 frames per second. Each STS-3 frame is made of 9 by (3×90) bytes. Each byte is made of 8 bits. The data rate is

$$\text{STS-3 data rate} = 8000 \times 9 \times (3 \times 90) \times 8 = 155.52 \text{ Mbps}$$

Note that in SONET, there is an exact relationship between the data rates of different STS signals. We could have found the data rate of STS-3 by using the data rate of STS-1 (multiply the latter by 3).

In SONET, the data rate of an STS-*n* signal is *n* times the data rate of an STS-1 signals.

Example 17.3

What is the duration of an STS-1 frame? STS-3 frame? STS-*n* frame?

Solution

In SONET, 8000 frames are sent per second. This means that the duration of an STS-1, STS-3, or STS-*n* frame is the same and equal to 1/8000 s, or 125 μs.

In SONET, the duration of any frame is 125 μs.

STS-1 Frame Format

The basic format of an STS-1 frame is shown in Figure 17.6. As we said before, a SONET frame is a matrix of 9 rows of 90 bytes (octets) each, for a total of 810 bytes.

The first three columns of the frame are used for section and line overhead. The upper three rows of the first three columns are used for **section overhead (SOH).** The lower six are **line overhead (LOH).** The rest of the frame is called the synchronous payload envelope (SPE). It contains user data and **path overhead (POH)** needed at the user data level. We will discuss the format of the SPE shortly.

Section Overhead

The section overhead consists of nine octets. The labels, functions, and organization of these octets are shown in Figure 17.7.

Figure 17.6 *STS-1 frame overheads*

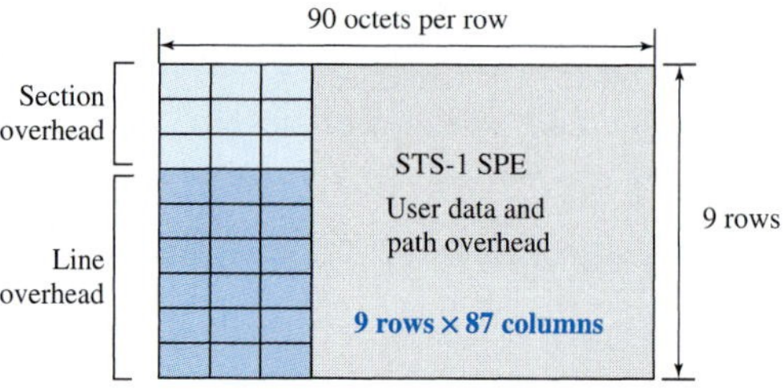

Figure 17.7 *STS-1 frame: section overhead*

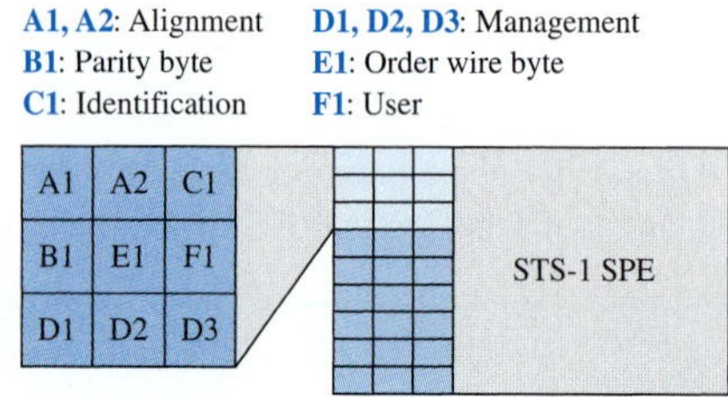

- **Alignment bytes (A1 and A2).** Bytes A1 and A2 are used for framing and synchronization and are called alignment bytes. These bytes alert a receiver that a frame is arriving and give the receiver a predetermined bit pattern on which to synchronize. The bit patterns for these two bytes in hexadecimal are 0xF628. The bytes serve as a flag.
- **Section parity byte (B1).** Byte B1 is for bit interleaved parity (BIP-8). Its value is calculated over all bytes of the previous frame. In other words, the *i*th bit of this byte is the parity bit calculated over all *i*th bits of the previous STS-*n* frame. The value of this byte is filled only for the first STS-1 in an STS-*n* frame. In other words, although an STS-*n* frame has *n* B1 bytes, as we will see later, only the first byte has this value; the rest are filled with 0s.
- **Identification byte (C1).** Byte C1 carries the identity of the STS-1 frame. This byte is necessary when multiple STS-1s are multiplexed to create a higher-rate STS (STS-3, STS-9, STS-12, etc.). Information in this byte allows the various signals to be recognized easily upon demultiplexing. For example, in an STS-3 signal, the value of the C1 byte is 1 for the first STS-1; it is 2 for the second; and it is 3 for the third.
- **Management bytes (D1, D2, and D3).** Bytes D1, D2, and D3 together form a 192-kbps channel ($3 \times 8000 \times 8$) called the data communication channel. This channel is required for operation, administration, and maintenance (OA&M) signaling.
- **Order wire byte (E1).** Byte E1 is the order wire byte. Order wire bytes in consecutive frames form a channel of 64 kbps (8000 frames per second times 8 bits per

frame). This channel is used for communication between regenerators, or between terminals and regenerators.

- **User's byte (F1).** The F1 bytes in consecutive frames form a 64-kbps channel that is reserved for user needs at the section level.

Section overhead is recalculated for each SONET device (regenerators and multiplexers).

Line Overhead

Line overhead consists of 18 bytes. The labels, functions, and arrangement of these bytes are shown in Figure 17.8.

Figure 17.8 *STS-1 frame: line overhead*

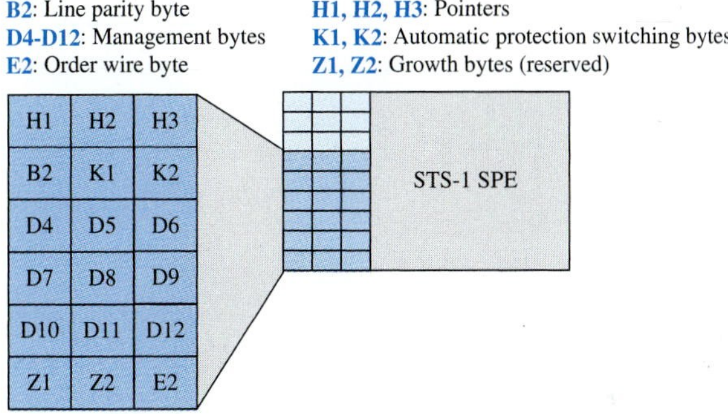

- **Line parity byte (B2).** Byte B2 is for bit interleaved parity. It is for error checking of the frame over a line (between two multiplexers). In an STS-n frame, B2 is calculated for all bytes in the previous STS-1 frame and inserted at the B2 byte for that frame. In other words, in a STS-3 frame, there are three B2 bytes, each calculated for one STS-1 frame. Contrast this byte with B1 in the section overhead.
- **Data communication channel bytes (D4 to D12).** The line overhead D bytes (D4 to D12) in consecutive frames form a 576-kbps channel that provides the same service as the D1–D3 bytes (OA&M), but at the line rather than the section level (between multiplexers).
- **Order wire byte (E2).** The E2 bytes in consecutive frames form a 64-kbps channel that provides the same functions as the E1 order wire byte, but at the line level.
- **Pointer bytes (H1, H2, and H3).** Bytes H1, H2, and H3 are pointers. The first two bytes are used to show the offset of the SPE in the frame; the third is used for justification. We show the use of these bytes later.
- **Automatic protection switching bytes (K1 and K2).** The K1 and K2 bytes in consecutive frames form a 128-kbps channel used for automatic detection of problems in

line-terminating equipment. We discuss automatic protection switching (APS) later in the chapter.

- **Growth bytes (Z1 and Z2).** The Z1 and Z2 bytes are reserved for future use.

Synchronous Payload Envelope

The **synchronous payload envelope (SPE)** contains the user data and the overhead related to the user data (path overhead). One SPE does not necessarily fit it into one STS-1 frame; it may be split between two frames, as we will see shortly. This means that the path overhead, the leftmost column of an SPE, does not necessarily align with the section or line overhead. The path overhead must be added first to the user data to create an SPE, and then an SPE can be inserted into one or two frames. Path overhead consists of 9 bytes. The labels, functions, and arrangement of these bytes are shown in Figure 17.9.

Figure 17.9 *STS-1 frame: path overhead*

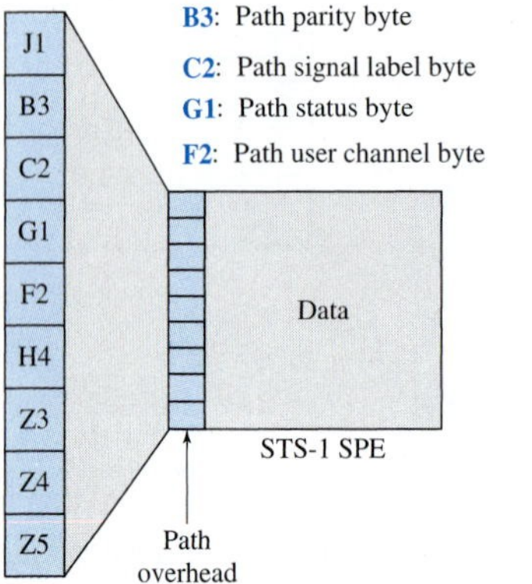

- **Path parity byte (B3).** Byte B3 is for bit interleaved parity, like bytes B1 and B2, but calculated over SPE bits. It is actually calculated over the previous SPE in the stream.
- **Path signal label byte (C2).** Byte C2 is the path identification byte. It is used to identify different protocols used at higher levels (such as IP or ATM) whose data are being carried in the SPE.
- **Path user channel byte (F2).** The F2 bytes in consecutive frames, like the F1 bytes, form a 64-kbps channel that is reserved for user needs, but at the path level.
- **Path status byte (G1).** Byte G1 is sent by the receiver to communicate its status to the sender. It is sent on the reverse channel when the communication is duplex. We will see its use in the linear or ring networks later in the chapter.
- **Multiframe indicator (H4).** Byte H4 is the multiframe indicator. It indicates payloads that cannot fit into a single frame. For example, virtual tributaries can be

combined to form a frame that is larger than an SPE frame and need to be divided into different frames. Virtual tributaries are discussed in the next section.

- **Path trace byte (J1).** The J1 bytes in consecutive frames form a 64-kbps channel used for tracking the path. The J1 byte sends a continuous 64-byte string to verify the connection. The choice of the string is left to the application program. The receiver compares each pattern with the previous one to ensure nothing is wrong with the communication at the path layer.
- **Growth bytes (Z3, Z4, and Z5).** Bytes Z3, Z4, and Z5 are reserved for future use.

Path overhead is only calculated for end-to-end (at STS multiplexers).

Overhead Summary

Table 17.2 compares and summarizes the overheads used in a section, line, and path.

Table 17.2 *SONET/SDH rates*

Byte Function	*Section*	*Line*	*Path*
Alignment	A1, A2		
Parity	B1	B2	B3
Identifier	C1		C2
OA&M	D1–D3	D4–D12	
Order wire	E1		
User	F1		F2
Status			G1
Pointers		H1– H3	H4
Trace			J1
Failure tolerance		K1, K2	
Growth (reserved for future)		Z1, Z2	Z3–Z5

Example 17.4

What is the user data rate of an STS-1 frame (without considering the overheads)?

Solution

The user data part in an STS-1 frame is made of 9 rows and 86 columns. So we have

$$\text{STS-1 user data rate} = 8000 \times 9 \times (1 \times 86) \times 8 = 49.536 \text{ Mbps}$$

Encapsulation

The previous discussion reveals that an SPE needs to be encapsulated in an STS-1 frame. Encapsulation may create two problems that are handled elegantly by SONET using pointers (H1 to H3). We discuss the use of these bytes in this section.

Offsetting

SONET allows one SPE to span two frames, part of the SPE is in the first frame and part is in the second. This may happen when one SPE that is to be encapsulated is not aligned time-wise with the passing synchronized frames. Figure 17.10 shows this situation. SPE bytes are divided between the two frames. The first set of bytes is encapsulated in the first frame; the second set is encapsulated in the second frame. The figure also shows the path overhead, which is aligned with the section/line overhead of any frame. The question is, How does the SONET multiplexer know where the SPE starts or ends in the frame? The solution is the use of pointers H1 and H2 to define the beginning of the SPE; the end can be found because each SPE has a fixed number of bytes. SONET allows the offsetting of an SPE with respect to an STS-1 frame.

Figure 17.10 *Offsetting of SPE related to frame boundary*

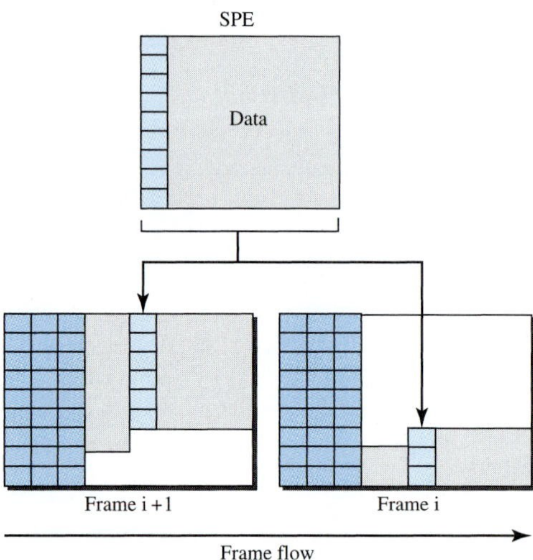

To find the beginning of each SPE in a frame, we need two pointers H1 and H2 in the line overhead. Note that these pointers are located in the line overhead because the encapsulation occurs at a multiplexer. Figure 17.11 shows how these 2 bytes point to

Figure 17.11 *The use of H1 and H2 pointers to show the start of an SPE in a frame*

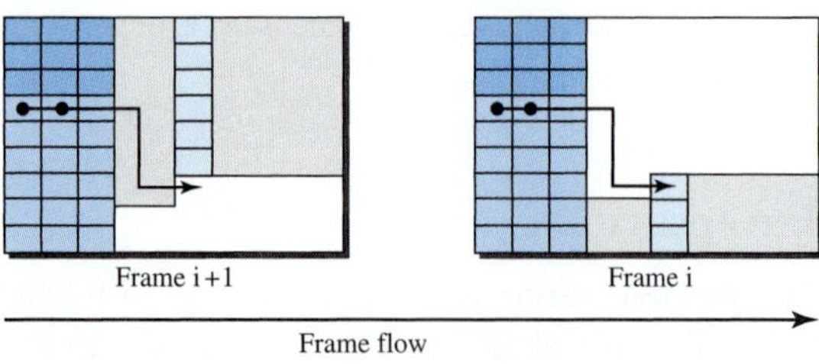

the beginning of the SPEs. Note that we need 2 bytes to define the position of a byte in a frame; a frame has 810 bytes, which cannot be defined using 1 byte.

Example 17.5

What are the values of H1 and H2 if an SPE starts at byte number 650?

Solution

The number 650 can be expressed in four hexadecimal digits as 0x028A. This means the value of H1 is 0x02 and the value of H2 is 0x8A.

Justification

Now suppose the transmission rate of the payload is just slightly different from the transmission rate of SONET. First, assume that the rate of the payload is higher. This means that occasionally there is 1 extra byte that cannot fit in the frame. In this case, SONET allows this extra byte to be inserted in the H3 byte. Now, assume that the rate of the payload is lower. This means that occasionally 1 byte needs to be left empty in the frame. SONET allows this byte to be the byte after the H3 byte.

17.4 STS MULTIPLEXING

In SONET, frames of lower rate can be synchronously time-division multiplexed into a higher-rate frame. For example, three STS-1 signals (channels) can be combined into one STS-3 signal (channel), four STS-3s can be multiplexed into one STS-12, and so on, as shown in Figure 17.12.

Figure 17.12 *STS multiplexing/demultiplexing*

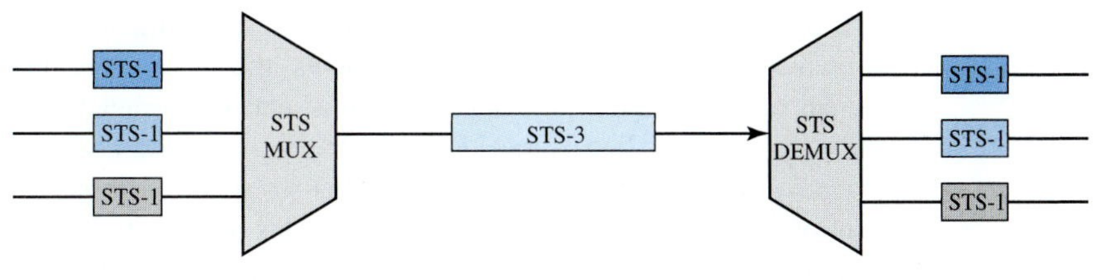

Multiplexing is synchronous TDM, and all clocks in the network are locked to a master clock to achieve synchronization.

In SONET, all clocks in the network are locked to a master clock.

We need to mention that multiplexing can also take place at the higher data rates. For example, four STS-3 signals can be multiplexed into an STS-12 signal. However, the STS-3 signals need to first be demultiplexed into 12 STS-1 signals, and then these twelve signals need to be multiplexed into an STS-12 signal. The reason for this extra work will be clear after our discussion on byte interleaving.

Byte Interleaving

Synchronous TDM multiplexing in SONET is achieved by using **byte interleaving.** For example, when three STS-1 signals are multliplexed into one STS-3 signal, each set of 3 bytes in the STS-3 signal is associated with 1 byte from each STS-1 signal. Figure 17.13 shows the interleaving.

Figure 17.13 *Byte interleaving*

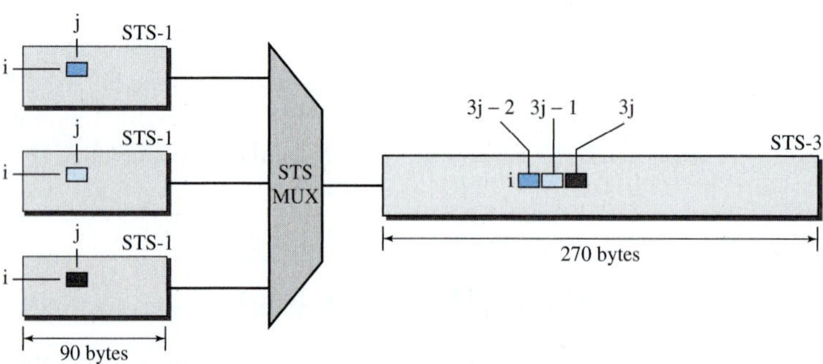

Note that a byte in an STS-1 frame keeps its row position, but it is moved into a different column. The reason is that while all signal frames have the same number of rows (9), the number of columns changes. The number of columns in an STS-*n* signal frame is *n* times the number of columns in an STS-1 frame. One STS-*n* row, therefore, can accommodate all *n* rows in the STS-1 frames.

Byte interleaving also preserves the corresponding section and line overhead as shown in Figure 17.14. As the figure shows, the section overheads from three STS-1 frames are interleaved together to create a section overhead for an STS-1 frame. The same is true for the line overheads. Each channel, however, keeps the corresponding bytes that are used to control that channel. In other words, the sections and lines keep their own control bytes for each multiplexed channel. This interesting feature will allow the use of add/drop multiplexers, as discussed shortly. As the figure shows, there are three A1 bytes, one belonging to each of the three multiplexed signals. There are also three A2 bytes, three B1 bytes, and so on.

Demultiplexing here is easier than in the statistical TDM we discussed in Chapter 6 because the demultiplexer, with no regard to the function of the bytes, removes the first A1 and assigns it to the first STS-1, removes the second A1, and assigns it to second STS-1, and removes the third A1 and assigns it to the third STS-1. In other words, the demultiplexer deals only with the position of the byte, not its function.

What we said about the section and line overheads does not exactly apply to the path overhead. This is because the path overhead is part of the SPE that may have splitted into two STS-1 frames. The byte interleaving, however, is the same for the data section of SPEs.

Figure 17.14 *An STS-3 frame*

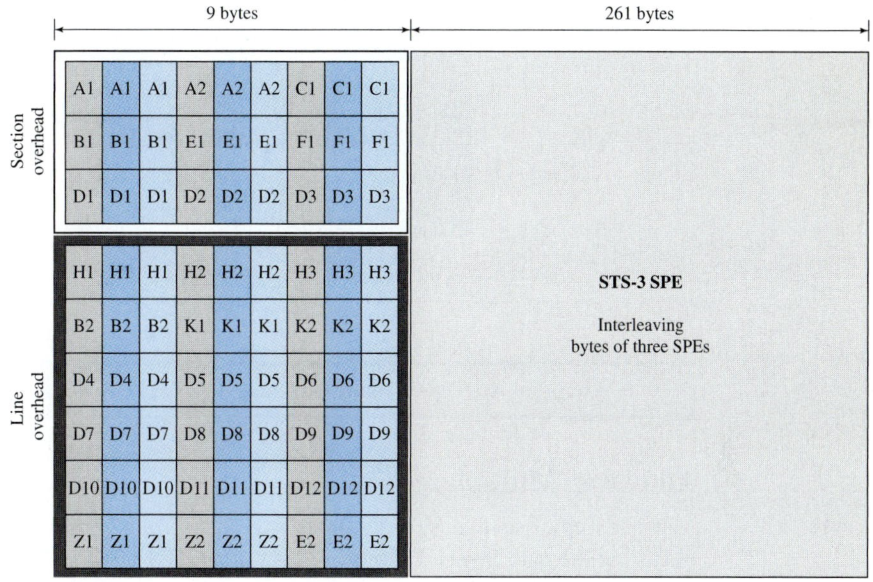

9 bytes

261 bytes

Section overhead

A1	A1	A1	A2	A2	A2	C1	C1	C1
B1	B1	B1	E1	E1	E1	F1	F1	F1
D1	D1	D1	D2	D2	D2	D3	D3	D3

Line overhead

H1	H1	H1	H2	H2	H2	H3	H3	H3
B2	B2	B2	K1	K1	K1	K2	K2	K2
D4	D4	D4	D5	D5	D5	D6	D6	D6
D7	D7	D7	D8	D8	D8	D9	D9	D9
D10	D10	D10	D11	D11	D11	D12	D12	D12
Z1	Z1	Z1	Z2	Z2	Z2	E2	E2	E2

STS-3 SPE

Interleaving bytes of three SPEs

The byte interleaving process makes the multiplexing at higher data rates a little bit more complex. How can we multiplex four STS-3 signals into one STS-12 signal? This can be done in two steps: First, the STS-3 signals must be demultiplexed to create 12 STS-1 signals. The 12 STS-1 signals are then multiplexed to create an STS-12 signal.

Concatenated Signal

In normal operation of the SONET, an STS-*n* signal is made of *n* multiplexed STS-1 signals. Sometimes, we have a signal with a data rate higher than what an STS-1 can carry. In this case, SONET allows us to create an STS-*n* signal which is not considered as *n* STS-1 signals; it is one STS-*n* signal (channel) that cannot be demultiplexed into *n* STS-1 signals. To specify that the signal cannot be demultiplexed, the suffix *c* (for concatenated) is added to the name of the signal. For example, STS-3c is a signal that cannot be demultiplexed into three STS-1 signals. However, we need to know that the whole payload in an STS-3c signal is one SPE, which means that we have only one column (9 bytes) of path overhead. The used data in this case occupy 260 columns, as shown in Figure 17.15.

Concatenated Signals Carrying ATM Cells

We will discuss ATM and ATM cells in Chapter 18. An ATM network is a cell network in which each cell has a fixed size of 53 bytes. The SPE of an STS-3c signal can be a carrier of ATM cells. The SPE of an STS-3c can carry $9 \times 260 = 2340$ bytes, which can accommodate approximately 44 ATM cells, each of 53 bytes.

Figure 17.15 *A concatenated STS-3c signal*

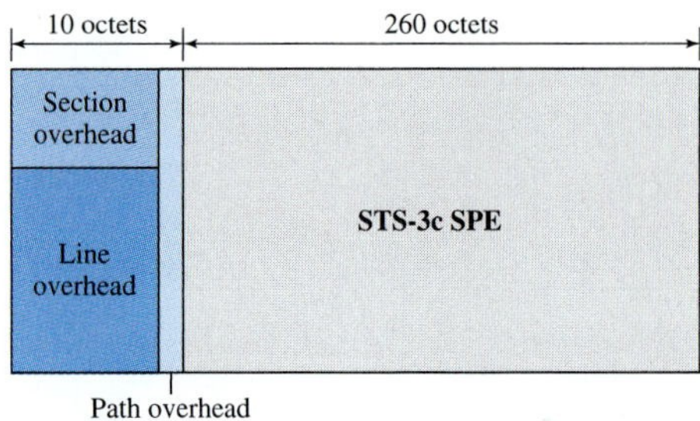

An STS-3c signal can carry 44 ATM cells as its SPE.

Add/Drop Multiplexer

Multiplexing of several STS-1 signals into an STS-*n* signal is done at the STS multiplexer (at the path layer). Demultiplexing of an STS-*n* signal into STS-1 components is done at the STS demultiplexer. In between, however, SONET uses add/drop multiplexers that can replace a signal with another one. We need to know that this is not demultiplexing/multiplexing in the conventional sense. An add/drop multiplexer operates at the line layer. An add/drop multiplexer does not create section, line, or path overhead. It almost acts as a switch; it removes one STS-1 signal and adds another one. The type of signal at the input and output of an add/drop multiplexer is the same (both STS-3 or both STS-12, for example). The add/drop multiplexer (ADM) only removes the corresponding bytes and replaces them with the new bytes (including the bytes in the section and line overhead). Figure 17.16 shows the operation of an ADM.

Figure 17.16 *Dropping and adding STS-1 frames in an add/drop multiplexer*

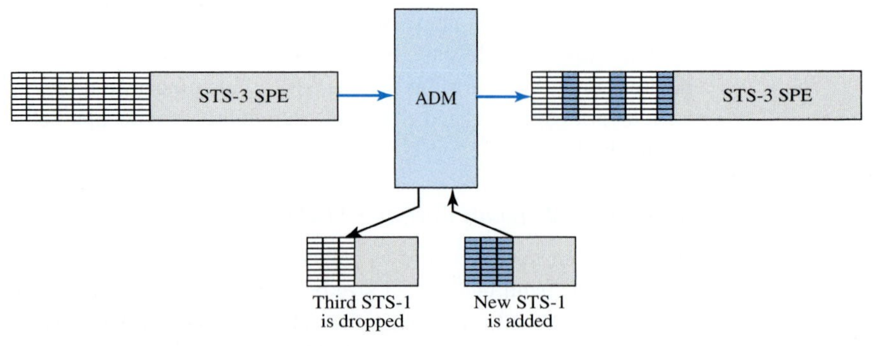

17.5 SONET NETWORKS

Using SONET equipment, we can create a SONET network that can be used as a high-speed backbone carrying loads from other networks such as ATM (Chapter 18) or IP (Chapter 20). We can roughly divide SONET networks into three categories: linear, ring, and mesh networks, as shown in Figure 17.17.

Figure 17.17 *Taxonomy of SONET networks*

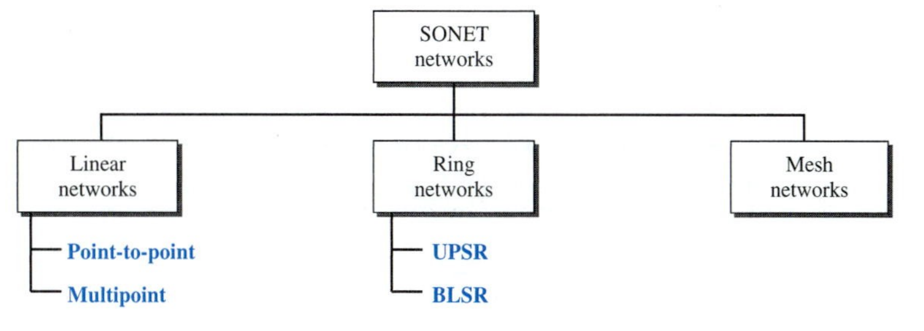

Linear Networks

A linear SONET network can be point-to-point or multipoint.

Point-to-Point Network

A point-to-point network is normally made of an STS multiplexer, an STS demultiplexer, and zero or more regenerators with no add/drop multiplexers, as shown in Figure 17.18. The signal flow can be unidirectional or bidirectional, although Figure 17.18 shows only unidirectional for simplicity.

Figure 17.18 *A point-to-point SONET network*

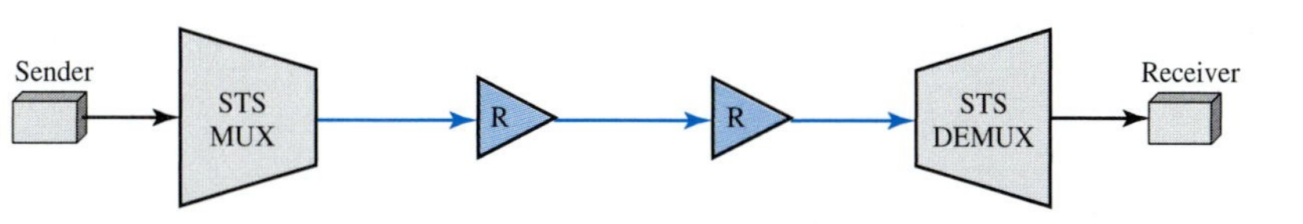

Multipoint Network

A multipoint network uses ADMs to allow the communications between several terminals. An ADM removes the signal belonging to the terminal connected to it and adds the signal transmitted from another terminal. Each terminal can send data to one or more downstream terminals. Figure 17.19 shows a unidirectional scheme in which each terminal can send data only to the downstream terminals, but the a multipoint network can be bidirectional, too.

Figure 17.19 *A multipoint SONET network*

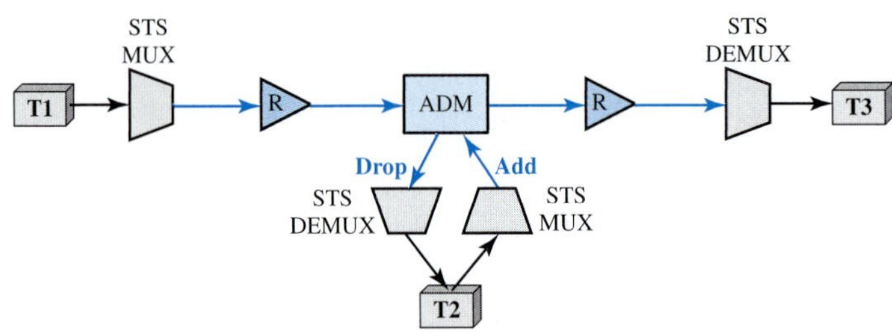

In Figure 17.19, T1 can send data to T2 and T3 simultaneously. T2, however, can send data only to T3. The figure shows a very simple configuration; in normal situations, we have more ADMs and more terminals.

Automatic Protection Switching

To create protection against failure in linear networks, SONET defines **automatic protection switching (APS).** APS in linear networks is defined at the line layer, which means the protection is between two ADMs or a pair of STS multiplexer/demultiplexers. The idea is to provide redundancy; a redundant line (fiber) can be used in case of failure in the main one. The main line is referred to as the work line and the redundant line as the protection line. Three schemes are common for protection in linear channels: one-plus-one, one-to-one, and one-to-many. Figure 17.20 shows all three schemes.

Figure 17.20 *Automatic protection switching in linear networks*

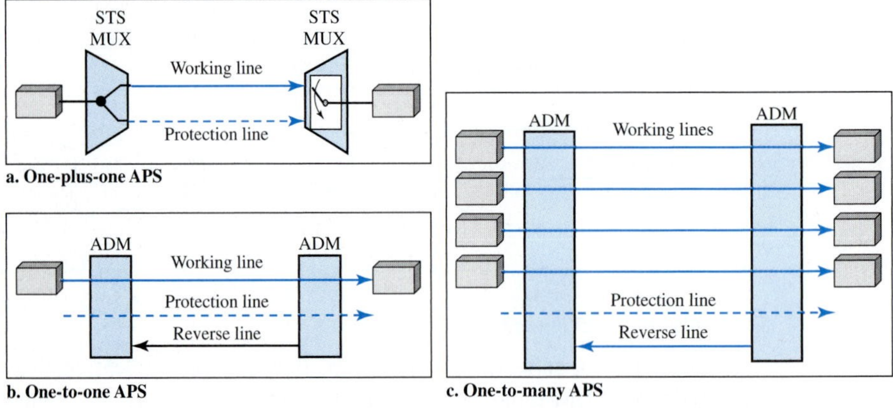

One-Plus-One APS In this scheme, there are normally two lines: one working line and one protection line. Both lines are active all the time. The sending multiplexer sends the same data on both lines; the receiver multiplexer monitors the line and chooses the one with the better quality. If one of the lines fails, it loses its signal, and, of course, the other line is selected at the receiver. Although, the failure recovery for this scheme is instantaneous, the scheme is inefficient because two times the bandwidth is required. Note that one-plus-one switching is done at the path layer.

One-to-One APS In this scheme, which looks like the one-plus-one scheme, there is also one working line and one protection line. However, the data are normally sent on the working line until it fails. At this time, the receiver, using the reverse channel, informs the sender to use the protection line instead. Obviously, the failure recovery is slower than that of the one-plus-scheme, but this scheme is more efficient because the protection line can be used for data transfer when it is not used to replace the working line. Note that the one-to-one switching is done at the line layer.

One-to-Many APS This scheme is similar to the one-to-one scheme except that there is only one protection line for many working lines. When a failure occurs in one of the working lines, the protection line takes control until the failed line is repaired. It is not as secure as the one-to-one scheme because if more than one working line fails at the same time, the protection line can replace only one of them. Note that one-to-many APS is done at the line layer.

Ring Networks

ADMs make it possible to have SONET ring networks. SONET rings can be used in either a unidirectional or a bidirectional configuration. In each case, we can add extra rings to make the network self-healing, capable of self-recovery from line failure.

Unidirectional Path Switching Ring

A **unidirectional path switching ring (UPSR)** is a unidirectional network with two rings: one ring used as the working ring and the other as the protection ring. The idea is similar to the one-plus-one APS scheme we discussed in a linear network. The same signal flows through both rings, one clockwise and the other counterclockwise. It is called UPSR because monitoring is done at the path layer. A node receives two copies of the electrical signals at the path layer, compares them, and chooses the one with the better quality. If part of a ring between two ADMs fails, the other ring still can guarantee the continuation of data flow. UPSR, like the one-plus-one scheme, has fast failure recovery, but it is not efficient because we need to have two rings that do the job of one. Half of the bandwidth is wasted. Figure 17.21 shows a UPSR network.

Although we have chosen one sender and three receivers in the figure, there can be many other configurations. The sender uses a two-way connection to send data to both rings simultaneously; the receiver uses selecting switches to select the ring with better signal quality. We have used one STS multiplexer and three STS demultiplexers to emphasize that nodes operate on the path layer.

Bidirectional Line Switching Ring

Another alternative in a SONET ring network is **bidirectional line switching ring (BLSR)**. In this case, communication is bidirectional, which means that we need

Figure 17.21 *A unidirectional path switching ring*

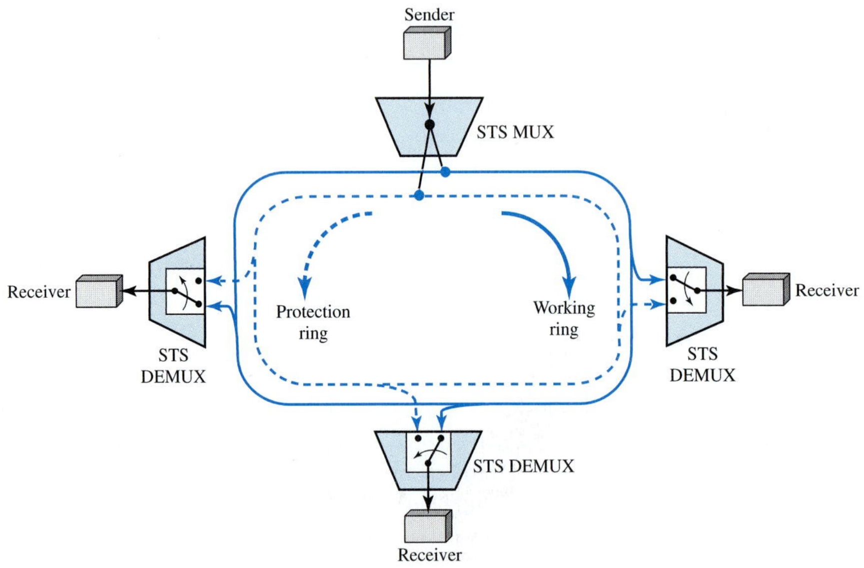

two rings for working lines. We also need two rings for protection lines. This means BLSR uses four rings. The operation, however, is similar to the one-to-one APS scheme. If a working ring in one direction between two nodes fails, the receiving node can use the reverse ring to inform the upstream node in the failed direction to use the protection ring. The network can recover in several different failure situations that we do not discuss here. Note that the discovery of a failure in BLSR is at the line layer, not the path layer. The ADMs find the failure and inform the adjacent nodes to use the protection rings. Figure 17.22 shows a BLSR ring.

Combination of Rings

SONET networks today use a combination of interconnected rings to create services in a wide area. For example, a SONET network may have a regional ring, several local rings, and many site rings to give services to a wide area. These rings can be UPSR, BLSR, or a combination of both. Figure 17.23 shows the idea of such a wide-area ring network.

Mesh Networks

One problem with ring networks is the lack of scalability. When the traffic in a ring increases, we need to upgrade not only the lines, but also the ADMs. In this situation, a mesh network with switches probably give better performance. A switch in a network mesh is called a cross-connect. A cross-connect, like other switches we have seen, has input and output ports. In an input port, the switch takes an OC-*n* signal, changes it to an STS-*n* signal, demultiplexes it into the corresponding STS-1 signals, and sends each

Figure 17.22 *A bidirectional line switching ring*

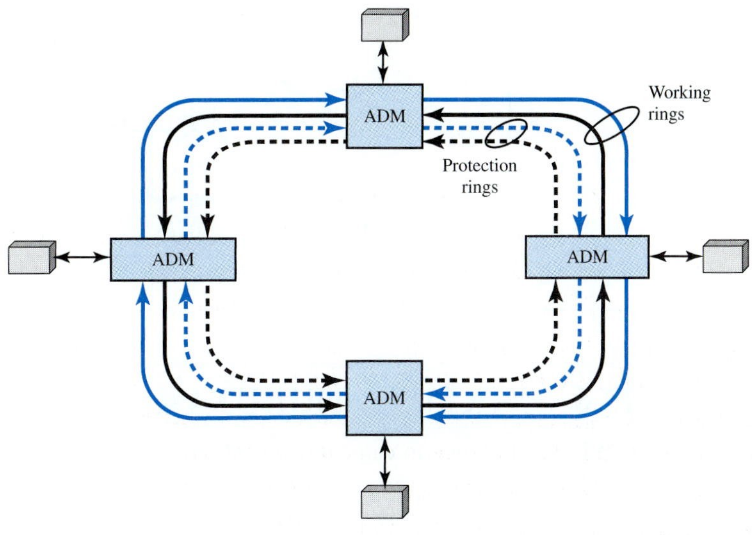

Figure 17.23 *A combination of rings in a SONET network*

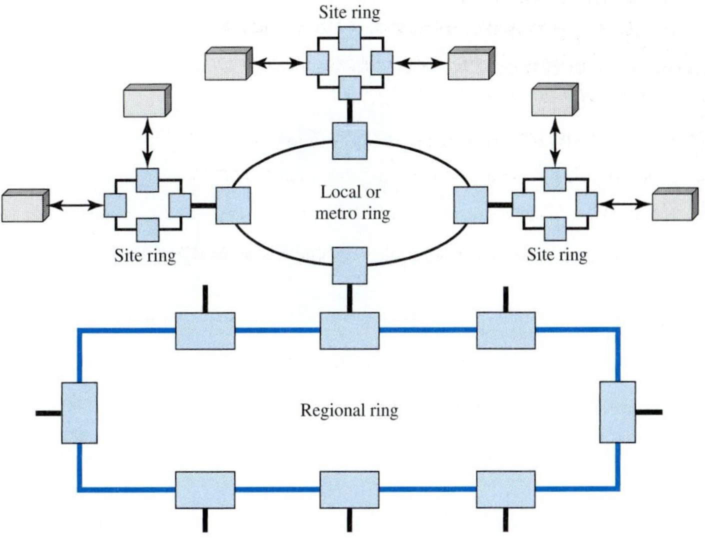

STS-1 signal to the appropriate output port. An output port takes STS-1 signals coming from different input ports, multiplexes them into an STS-n signal, and makes an OC-n signal for transmission. Figure 17.24 shows a mesh SONET network, and the structure of a switch.

Figure 17.24 *A mesh SONET network*

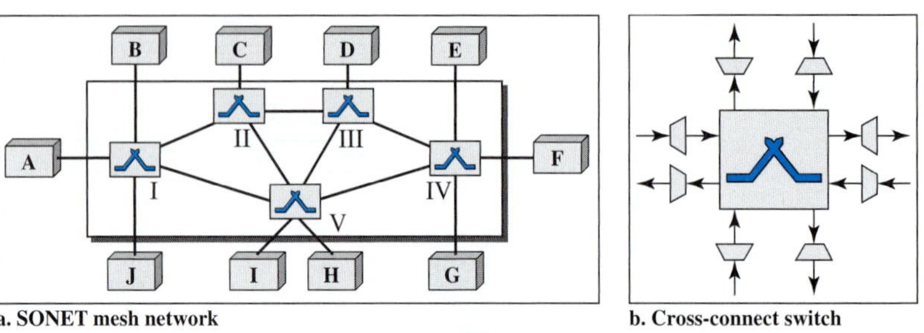

17.6 VIRTUAL TRIBUTARIES

SONET is designed to carry broadband payloads. Current digital hierarchy data rates (DS-1 to DS-3), however, are lower than STS-1. To make SONET backward-compatible with the current hierarchy, its frame design includes a system of **virtual tributaries (VTs)** (see Figure 17.25). A virtual tributary is a partial payload that can be inserted into an STS-1 and combined with other partial payloads to fill out the frame. Instead of using all 86 payload columns of an STS-1 frame for data from one source, we can subdivide the SPE and call each component a VT.

Figure 17.25 *Virtual tributaries*

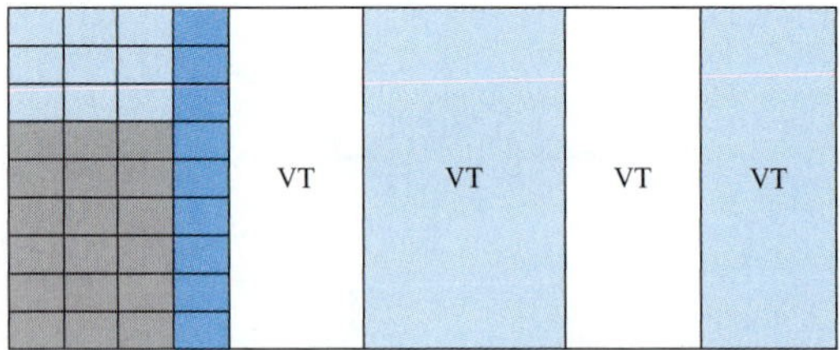

Types of VTs

Four types of VTs have been defined to accommodate existing digital hierarchies (see Figure 17.26). Notice that the number of columns allowed for each type of VT can be determined by doubling the type identification number (VT1.5 gets three columns, VT2 gets four columns, etc.).

- **VT1.5** accommodates the U.S. DS-1 service (1.544 Mbps).
- **VT2** accommodates the European CEPT-1 service (2.048 Mbps).
- **VT3** accommodates the DS-1C service (fractional DS-1, 3.152 Mbps).
- **VT6** accommodates the DS-2 service (6.312 Mbps).

Figure 17.26 *Virtual tributary types*

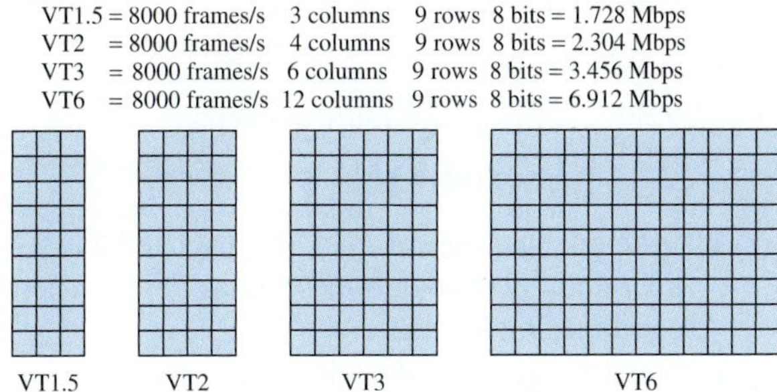

When two or more tributaries are inserted into a single STS-1 frame, they are interleaved column by column. SONET provides mechanisms for identifying each VT and separating them without demultiplexing the entire stream. Discussion of these mechanisms and the control issues behind them is beyond the scope of this book.

17.7 RECOMMENDED READING

For more details about subjects discussed in this chapter, we recommend the following books. The items in brackets [. . .] refer to the reference list at the end of the text.

Books

SONET is discussed in Section 2.5 of [Tan03], Section 15.2 of [Kes97], Sections 4.2 and 4.3 of [GW04], Section 8.2 of [Sta04], and Section 5.2 of [WV00].

17.8 KEY TERMS

add/drop multiplexer (ADM)
automatic protection switching (APS)
bidirectional line switching ring (BLSR)
byte interleaving
line
line layer
line overhead (LOH)
optical carrier (OC)
path
path layer
path overhead (POH)
photonic layer
regenerator
section
section layer
section overhead (SOH)
STS demultiplexer
STS multiplexer
Synchronous Digital Hierarchy (SDH)
Synchronous Optical Network (SONET)

synchronous payload envelope (SPE)
synchronous transport module (STM)
synchronous transport signal (STS)
terminal
unidirectional path switching ring (UPSR)
virtual tributary (VT)

17.9 SUMMARY

- Synchronous Optical Network (SONET) is a standard developed by ANSI for fiber-optic networks: Synchronous Digital Hierarchy (SDH) is a similar standard developed by ITU-T.
- SONET has defined a hierarchy of signals called synchronous transport signals (STSs). SDH has defined a similar hierarchy of signals called synchronous transfer modules (STMs).
- An OC-n signal is the optical modulation of an STS-n (or STM-n) signal.
- SONET defines four layers: path, line, section, and photonic.
- SONET is a synchronous TDM system in which all clocks are locked to a master clock.
- A SONET system can use the following equipment:
  1. STS multiplexers
  2. STS demultiplexers
  3. Regenerators
  4. Add/drop multiplexers
  5. Terminals
- SONET sends 8000 frames per second; each frame lasts 125 μs.
- An STS-1 frame is made of 9 rows and 90 columns; an STS-n frame is made of 9 rows and $n \times 90$ columns.
- STSs can be multiplexed to get a new STS with a higher data rate.
- SONET network topologies can be linear, ring, or mesh.
- A linear SONET network can be either point-to-point or multipoint.
- A ring SONET network can be unidirectional or bidirectional.
- To make SONET backward-compatible with the current hierarchy, its frame design includes a system of virtual tributaries (VTs).

17.10 PRACTICE SET

Review Questions

1. What is the relationship between SONET and SDH?
2. What is the relationship between STS and STM?
3. How is an STS multiplexer different from an add/drop multiplexer since both can add signals together?

4. What is the relationship between STS signals and OC signals?
5. What is the purpose of the pointer in the line overhead?
6. Why is SONET called a synchronous network?
7. What is the function of a SONET regenerator?
8. What are the four SONET layers?
9. Discuss the functions of each SONET layer.
10. What is a virtual tributary?

Exercises

11. What are the user data rates of STS-3, STS-9, and STS-12?
12. Show how STS-9's can be multiplexed to create an STS-36. Is there any extra overhead involved in this type of multiplexing?
13. A stream of data is being carried by STS-1 frames. If the data rate of the stream is 49.540 Mbps, how many STS-1 frames per second must let their H3 bytes carry data?
14. A stream of data is being carried by STS-1 frames. If the data rate of the stream is 49.530 Mbps, how many frames per second should leave one empty byte after the H3 byte?
15. Table 17.2 shows that the overhead bytes can be categorized as A, B, C, D, E, F, G, H, J, K, and Z bytes.
    a. Why are there no A bytes in the LOH or POH?
    b. Why are there no C bytes in the LOH?
    c. Why are there no D bytes in the POH?
    d. Why are there no E bytes in the LOH or POH?
    e. Why are there no F bytes in the LOH or POH?
    f. Why are there no G bytes in the SOH or LOH?
    g. Why are there no H bytes in the SOH?
    h. Why are there no J bytes in the SOH or LOH?
    i. Why are there no K bytes in the SOH or POH?
    j. Why are there no Z bytes in the SOH?
16. Why are B bytes present in all three headers?

CHAPTER 18

Virtual-Circuit Networks: Frame Relay and ATM

In Chapter 8, we discussed switching techniques. We said that there are three types of switching: circuit switching, packet switching, and message switching. We also mentioned that packet switching can use two approaches: the virtual-circuit approach and the datagram approach.

In this chapter, we show how the virtual-circuit approach can be used in wide-area networks. Two common WAN technologies use virtual-circuit switching. Frame Relay is a relatively high-speed protocol that can provide some services not available in other WAN technologies such as DSL, cable TV, and T lines. ATM, as a high-speed protocol, can be the superhighway of communication when it deploys physical layer carriers such as SONET.

We first discuss Frame Relay. We then discuss ATM in greater detail. Finally, we show how ATM technology, which was originally designed as a WAN technology, can also be used in LAN technology, ATM LANs.

18.1 FRAME RELAY

Frame Relay is a virtual-circuit wide-area network that was designed in response to demands for a new type of WAN in the late 1980s and early 1990s.

1. Prior to Frame Relay, some organizations were using a virtual-circuit switching network called **X.25** that performed switching at the network layer. For example, the Internet, which needs wide-area networks to carry its packets from one place to another, used X.25. And X.25 is still being used by the Internet, but it is being replaced by other WANs. However, X.25 has several drawbacks:
   a. X.25 has a low 64-kbps data rate. By the 1990s, there was a need for higher-data-rate WANs.
   b. X.25 has extensive flow and error control at both the data link layer and the network layer. This was so because X.25 was designed in the 1970s, when the available transmission media were more prone to errors. Flow and error control at both layers create a large overhead and slow down transmissions. X.25 requires acknowledgments for both data link layer frames and network layer packets that are sent between nodes and between source and destination.

   c. Originally X.25 was designed for private use, not for the Internet. X.25 has its own network layer. This means that the user's data are encapsulated in the network layer packets of X.25. The Internet, however, has its own network layer, which means if the Internet wants to use X.25, the Internet must deliver its network layer packet, called a datagram, to X.25 for encapsulation in the X.25 packet. This doubles the overhead.

2. Disappointed with X.25, some organizations started their own private WAN by leasing T-1 or T-3 lines from public service providers. This approach also has some drawbacks.
   a. If an organization has n branches spread over an area, it needs $n(n - 1)/2$ T-1 or T-3 lines. The organization pays for all these lines although it may use the lines only 10 percent of the time. This can be very costly.
   b. The services provided by T-1 and T-3 lines assume that the user has fixed-rate data all the time. For example, a T-1 line is designed for a user who wants to use the line at a consistent 1.544 Mbps. This type of service is not suitable for the many users today that need to send **bursty data.** For example, a user may want to send data at 6 Mbps for 2 s, 0 Mbps (nothing) for 7 s, and 3.44 Mbps for 1 s for a total of 15.44 Mbits during a period of 10 s. Although the average data rate is still 1.544 Mbps, the T-1 line cannot accept this type of demand because it is designed for fixed-rate data, not bursty data. Bursty data require what is called **bandwidth on demand.** The user needs different bandwidth allocations at different times.

In response to the above drawbacks, Frame Relay was designed. Frame Relay is a wide-area network with the following features:

1. Frame Relay operates at a higher speed (1.544 Mbps and recently 44.376 Mbps). This means that it can easily be used instead of a mesh of T-1 or T-3 lines.
2. Frame Relay operates in just the physical and data link layers. This means it can easily be used as a backbone network to provide services to protocols that already have a network layer protocol, such as the Internet.
3. Frame Relay allows bursty data.
4. Frame Relay allows a frame size of 9000 bytes, which can accommodate all local-area network frame sizes.
5. Frame Relay is less expensive than other traditional WANs.
6. Frame Relay has error detection at the data link layer only. There is no flow control or error control. There is not even a retransmission policy if a frame is damaged; it is silently dropped. Frame Relay was designed in this way to provide fast transmission capability for more reliable media and for those protocols that have flow and error control at the higher layers.

Architecture

Frame Relay provides permanent virtual circuits and switched virtual circuits. Figure 18.1 shows an example of a Frame Relay network connected to the Internet. The routers are used, as we will see in Chapter 22, to connect LANs and WANs in the Internet. In the figure, the Frame Relay WAN is used as one link in the global Internet.

Figure 18.1 *Frame Relay network*

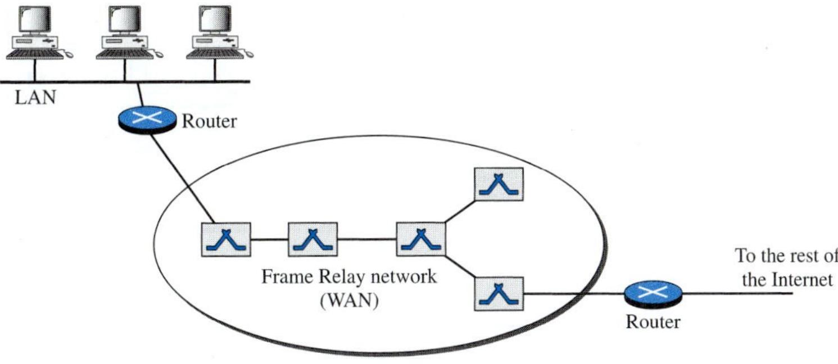

Virtual Circuits

Frame Relay is a virtual circuit network. A virtual circuit in Frame Relay is identified by a number called a **data link connection identifier (DLCI).**

VCIs in Frame Relay are called DLCIs.

Permanent Versus Switched Virtual Circuits

A source and a destination may choose to have a **permanent virtual circuit (PVC).** In this case, the connection setup is simple. The corresponding table entry is recorded for all switches by the administrator (remotely and electronically, of course). An outgoing DLCI is given to the source, and an incoming DLCI is given to the destination.

PVC connections have two drawbacks. First, they are costly because two parties pay for the connection all the time even when it is not in use. Second, a connection is created from one source to one single destination. If a source needs connections with several destinations, it needs a PVC for each connection. An alternate approach is the **switched virtual circuit (SVC).** The SVC creates a temporary, short connection that exists only when data are being transferred between source and destination. An SVC requires establishing and terminating phases as discussed in Chapter 8.

Switches

Each switch in a Frame Relay network has a table to route frames. The table matches an incoming port–DLCI combination with an outgoing port–DLCI combination as we described for general virtual-circuit networks in Chapter 8. The only difference is that VCIs are replaced by DLCIs.

Frame Relay Layers

Figure 18.2 shows the Frame Relay layers. Frame Relay has only physical and data link layers.

Figure 18.2 *Frame Relay layers*

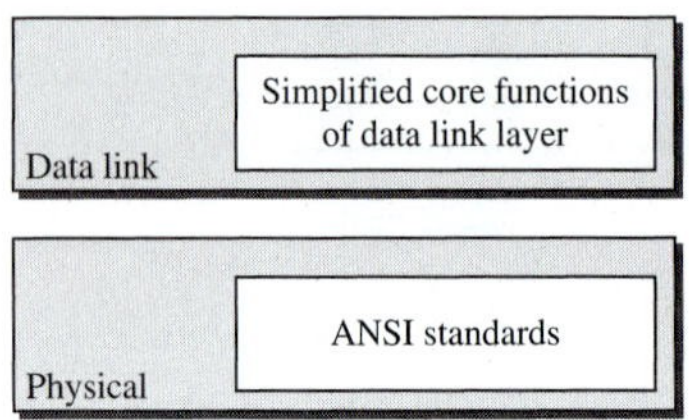

Frame Relay operates only at the physical and data link layers.

Physical Layer

No specific protocol is defined for the physical layer in Frame Relay. Instead, it is left to the implementer to use whatever is available. Frame Relay supports any of the protocols recognized by ANSI.

Data Link Layer

At the data link layer, Frame Relay uses a simple protocol that does not support flow or error control. It only has an error detection mechanism. Figure 18.3 shows the format of a Frame Relay frame.The address field defines the DLCI as well as some bits used to control congestion.

Figure 18.3 *Frame Relay frame*

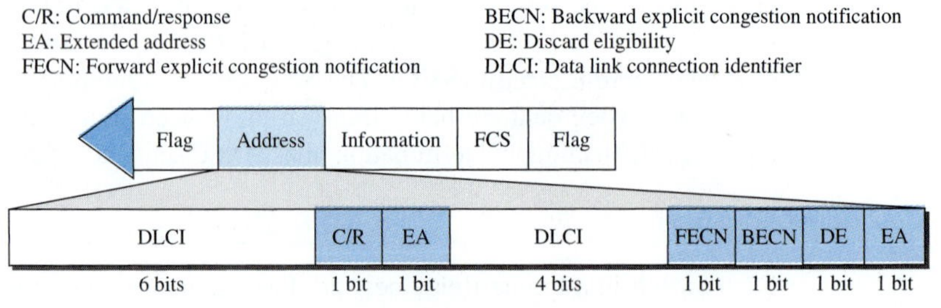

The descriptions of the fields are as follows:

- **Address (DLCI) field.** The first 6 bits of the first byte makes up the first part of the DLCI. The second part of the DLCI uses the first 4 bits of the second byte. These bits are part of the 10-bit data link connection identifier defined by the standard. We will discuss extended addressing at the end of this section.

- **Command/response (C/R).** The command/response (C/R) bit is provided to allow upper layers to identify a frame as either a command or a response. It is not used by the Frame Relay protocol.
- **Extended address (EA).** The extended address (EA) bit indicates whether the current byte is the final byte of the address. An EA of 0 means that another address byte is to follow (extended addressing is discussed later). An EA of 1 means that the current byte is the final one.
- **Forward explicit congestion notification (FECN).** The **forward explicit congestion notification (FECN)** bit can be set by any switch to indicate that traffic is congested. This bit informs the destination that congestion has occurred. In this way, the destination knows that it should expect delay or a loss of packets. We will discuss the use of this bit when we discuss congestion control in Chapter 24.
- **Backward explicit congestion notification (BECN).** The **backward explicit congestion notification (BECN)** bit is set (in frames that travel in the other direction) to indicate a congestion problem in the network. This bit informs the sender that congestion has occurred. In this way, the source knows it needs to slow down to prevent the loss of packets. We will discuss the use of this bit when we discuss congestion control in Chapter 24.
- **Discard eligibility (DE).** The **discard eligibility (DE)** bit indicates the priority level of the frame. In emergency situations, switches may have to discard frames to relieve bottlenecks and keep the network from collapsing due to overload. When set (DE 1), this bit tells the network to discard this frame if there is congestion. This bit can be set either by the sender of the frames (user) or by any switch in the network.

Frame Relay does not provide flow or error control; they must be provided by the upper-layer protocols.

Extended Address

To increase the range of DLCIs, the Frame Relay address has been extended from the original 2-byte address to 3- or 4-byte addresses. Figure 18.4 shows the different addresses. Note that the EA field defines the number of bytes; it is 1 in the last byte of the address, and it is 0 in the other bytes. Note that in the 3- and 4-byte formats, the bit before the last bit is set to 0.

Figure 18.4 *Three address formats*

DLCI			C/R	EA = 0
DLCI	FECN	BECN	DE	EA = 1

a. Two-byte address (10-bit DLCI)

DLCI			C/R	EA = 0
DLCI	FECN	BECN	DE	EA = 0
DLCI			0	EA = 1

b. Three-byte address (16-bit DLCI)

DLCI			C/R	EA = 0
DLCI	FECN	BECN	DE	EA = 0
DLCI				EA = 0
DLCI			0	EA = 1

c. Four-byte address (23-bit DLCI)

FRADs

To handle frames arriving from other protocols, Frame Relay uses a device called a **Frame Relay assembler/disassembler (FRAD).** A FRAD assembles and disassembles frames coming from other protocols to allow them to be carried by Frame Relay frames. A FRAD can be implemented as a separate device or as part of a switch. Figure 18.5 shows two FRADs connected to a Frame Relay network.

Figure 18.5 *FRAD*

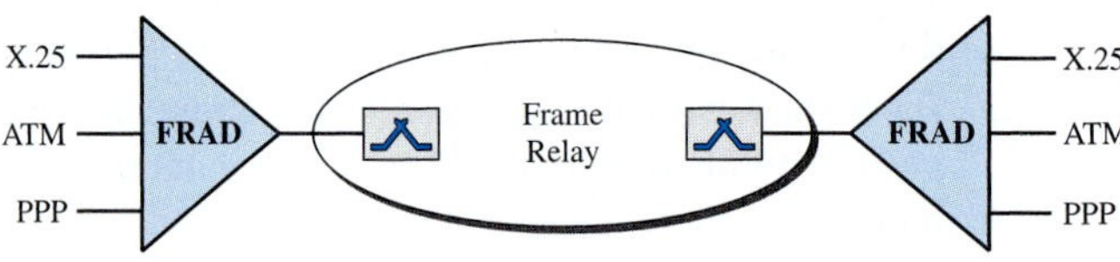

VOFR

Frame Relay networks offer an option called **Voice Over Frame Relay (VOFR)** that sends voice through the network. Voice is digitized using PCM and then compressed. The result is sent as data frames over the network. This feature allows the inexpensive sending of voice over long distances. However, note that the quality of voice is not as good as voice over a circuit-switched network such as the telephone network. Also, the varying delay mentioned earlier sometimes corrupts real-time voice.

LMI

Frame Relay was originally designed to provide PVC connections. There was not, therefore, a provision for controlling or managing interfaces. **Local Management Information (LMI)** is a protocol added recently to the Frame Relay protocol to provide more management features. In particular, LMI can provide

- A keep-alive mechanism to check if data are flowing.
- A multicast mechanism to allow a local end system to send frames to more than one remote end system.
- A mechanism to allow an end system to check the status of a switch (e.g., to see if the switch is congested).

Congestion Control and Quality of Service

One of the nice features of Frame Relay is that it provides **congestion control** and **quality of service (QoS).** We have not discussed these features yet. In Chapter 24, we introduce these two important aspects of networking and discuss how they are implemented in Frame Relay and some other networks.

18.2 ATM

Asynchronous Transfer Mode (ATM) is the **cell relay** protocol designed by the ATM Forum and adopted by the ITU-T. The combination of ATM and SONET will allow high-speed interconnection of all the world's networks. In fact, ATM can be thought of as the "highway" of the information superhighway.

Design Goals

Among the challenges faced by the designers of ATM, six stand out.

1. Foremost is the need for a transmission system to optimize the use of high-data-rate transmission media, in particular optical fiber. In addition to offering large bandwidths, newer transmission media and equipment are dramatically less susceptible to noise degradation. A technology is needed to take advantage of both factors and thereby maximize data rates.
2. The system must interface with existing systems and provide wide-area interconnectivity between them without lowering their effectiveness or requiring their replacement.
3. The design must be implemented inexpensively so that cost would not be a barrier to adoption. If ATM is to become the backbone of international communications, as intended, it must be available at low cost to every user who wants it.
4. The new system must be able to work with and support the existing telecommunications hierarchies (local loops, local providers, long-distance carriers, and so on).
5. The new system must be connection-oriented to ensure accurate and predictable delivery.
6. Last but not least, one objective is to move as many of the functions to hardware as possible (for speed) and eliminate as many software functions as possible (again for speed).

Problems

Before we discuss the solutions to these design requirements, it is useful to examine some of the problems associated with existing systems.

Frame Networks

Before ATM, data communications at the data link layer had been based on frame switching and frame networks. Different protocols use frames of varying size and intricacy. As networks become more complex, the information that must be carried in the header becomes more extensive. The result is larger and larger headers relative to the size of the data unit. In response, some protocols have enlarged the size of the data unit to make header use more efficient (sending more data with the same size header). Unfortunately, large data fields create waste. If there is not much information to transmit, much of the field goes unused. To improve utilization, some protocols provide variable frame sizes to users.

Mixed Network Traffic

As you can imagine, the variety of frame sizes makes traffic unpredictable. Switches, multiplexers, and routers must incorporate elaborate software systems to manage the various sizes of frames. A great deal of header information must be read, and each bit counted and evaluated to ensure the integrity of every frame. Internetworking among the different frame networks is slow and expensive at best, and impossible at worst.

Another problem is that of providing consistent data rate delivery when frame sizes are unpredictable and can vary so dramatically. To get the most out of broadband technology, traffic must be time-division multiplexed onto shared paths. Imagine the results of multiplexing frames from two networks with different requirements (and frame designs) onto one link (see Figure 18.6). What happens when line 1 uses large frames (usually data frames) while line 2 uses very small frames (the norm for audio and video information)?

Figure 18.6 *Multiplexing using different frame sizes*

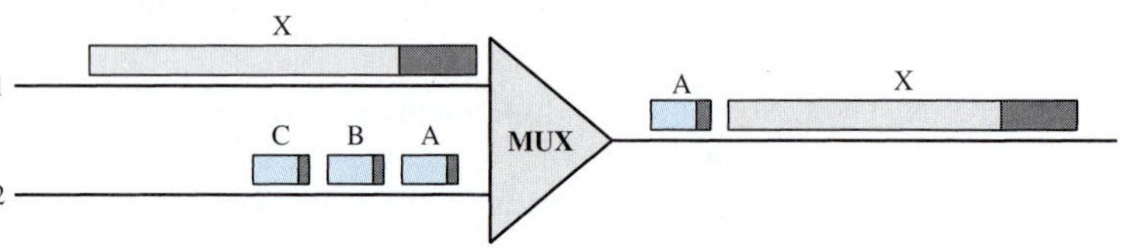

If line 1's gigantic frame X arrives at the multiplexer even a moment earlier than line 2's frames, the multiplexer puts frame X onto the new path first. After all, even if line 2's frames have priority, the multiplexer has no way of knowing to wait for them and so processes the frame that has arrived. Frame A must therefore wait for the entire X bit stream to move into place before it can follow. The sheer size of X creates an unfair delay for frame A. The same imbalance can affect all the frames from line 2.

Because audio and video frames ordinarily are small, mixing them with conventional data traffic often creates unacceptable delays of this type and makes shared frame links unusable for audio and video information. Traffic must travel over different paths, in much the same way that automobile and train traffic does. But to fully utilize broad bandwidth links, we need to be able to send all kinds of traffic over the same links.

Cell Networks

Many of the problems associated with frame internetworking are solved by adopting a concept called cell networking. A cell is a small data unit of fixed size. In a **cell network,** which uses the **cell** as the basic unit of data exchange, all data are loaded into identical cells that can be transmitted with complete predictability and uniformity. As frames of different sizes and formats reach the cell network from a tributary network, they are split into multiple small data units of equal length and are loaded into cells. The cells are then multiplexed with other cells and routed through the cell network. Because each cell is the same size and all are small, the problems associated with multiplexing different-sized frames are avoided.

A cell network uses the cell as the basic unit of data exchange. A cell is defined as a small, fixed-size block of information.

Figure 18.7 shows the multiplexer from Figure 18.6 with the two lines sending cells instead of frames. Frame X has been segmented into three cells: X, Y, and Z. Only the first cell from line 1 gets put on the link before the first cell from line 2. The cells from the two lines are interleaved so that none suffers a long delay.

Figure 18.7 *Multiplexing using cells*

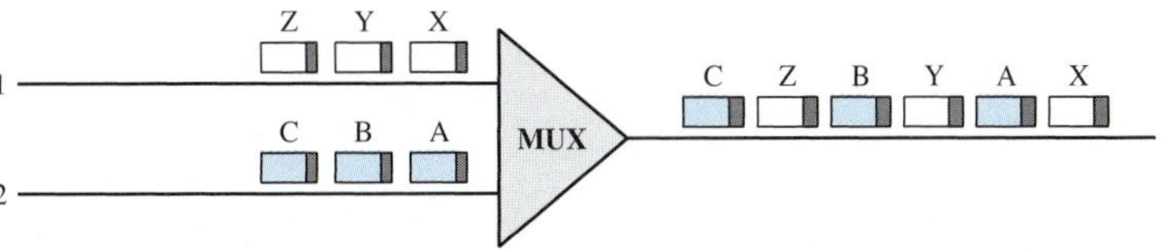

A second point in this same scenario is that the high speed of the links coupled with the small size of the cells means that, despite interleaving, cells from each line arrive at their respective destinations in an approximation of a continuous stream (much as a movie appears to your brain to be continuous action when in fact it is really a series of separate, still photographs). In this way, a cell network can handle real-time transmissions, such as a phone call, without the parties being aware of the segmentation or multiplexing at all.

Asynchronous TDM

ATM uses asynchronous time-division multiplexing—that is why it is called Asynchronous Transfer Mode—to multiplex cells coming from different channels. It uses fixed-size slots (size of a cell). ATM multiplexers fill a slot with a cell from any input channel that has a cell; the slot is empty if none of the channels has a cell to send.

Figure 18.8 shows how cells from three inputs are multiplexed. At the first tick of the clock, channel 2 has no cell (empty input slot), so the multiplexer fills the slot with a cell from the third channel. When all the cells from all the channels are multiplexed, the output slots are empty.

Figure 18.8 *ATM multiplexing*

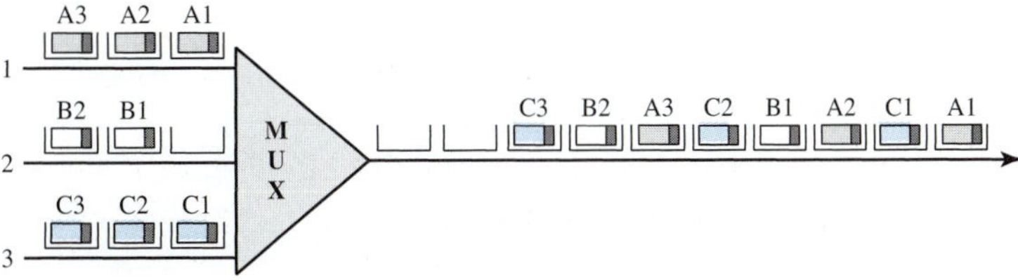

Architecture

ATM is a cell-switched network. The user access devices, called the endpoints, are connected through a **user-to-network interface (UNI)** to the switches inside the network. The switches are connected through **network-to-network interfaces (NNIs).** Figure 18.9 shows an example of an ATM network.

Figure 18.9 *Architecture of an ATM network*

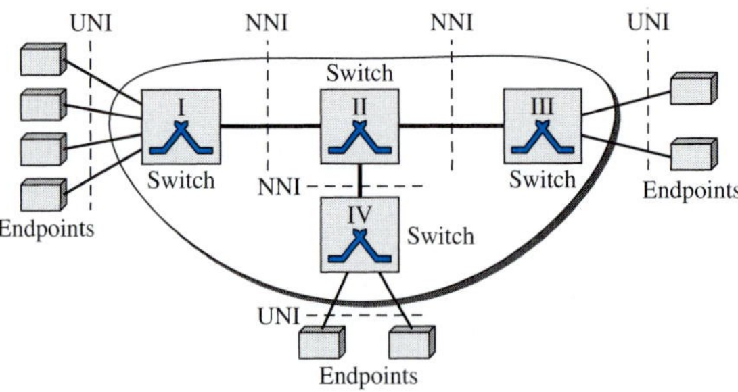

Virtual Connection

Connection between two endpoints is accomplished through transmission paths (TPs), virtual paths (VPs), and virtual circuits (VCs). A **transmission path (TP)** is the physical connection (wire, cable, satellite, and so on) between an endpoint and a switch or between two switches. Think of two switches as two cities. A transmission path is the set of all highways that directly connect the two cities.

A transmission path is divided into several virtual paths. A **virtual path (VP)** provides a connection or a set of connections between two switches. Think of a virtual path as a highway that connects two cities. Each highway is a virtual path; the set of all highways is the transmission path.

Cell networks are based on **virtual circuits (VCs).** All cells belonging to a single message follow the same virtual circuit and remain in their original order until they reach their destination. Think of a virtual circuit as the lanes of a highway (virtual path). Figure 18.10 shows the relationship between a transmission path (a physical connection), virtual paths (a combination of virtual circuits that are bundled together because parts of their paths are the same), and virtual circuits that logically connect two points.

Figure 18.10 *TP, VPs, and VCs*

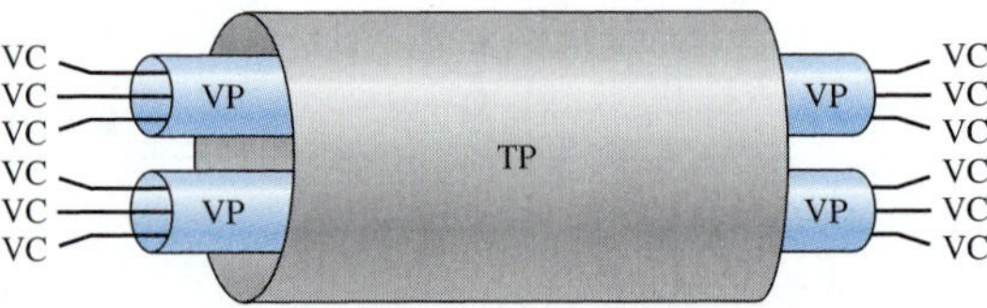

To better understand the concept of VPs and VCs, look at Figure 18.11. In this figure, eight endpoints are communicating using four VCs. However, the first two VCs seem to share the same virtual path from switch I to switch III, so it is reasonable to bundle these two VCs together to form one VP. On the other hand, it is clear that the other two VCs share the same path from switch I to switch IV, so it is also reasonable to combine them to form one VP.

Figure 18.11 *Example of VPs and VCs*

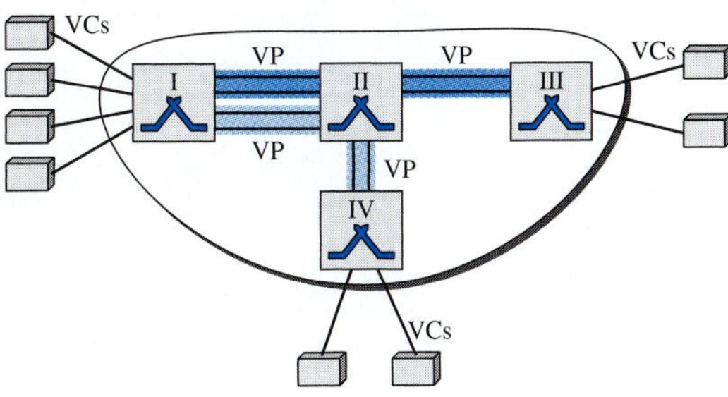

Identifiers In a virtual circuit network, to route data from one endpoint to another, the virtual connections need to be identified. For this purpose, the designers of ATM created a hierarchical identifier with two levels: a **virtual path identifier (VPI)** and a **virtual-circuit identifier (VCI).** The VPI defines the specific VP, and the VCI defines a particular VC inside the VP. The VPI is the same for all virtual connections that are bundled (logically) into one VP.

Note that a virtual connection is defined by a pair of numbers: the VPI and the VCI.

Figure 18.12 shows the VPIs and VCIs for a transmission path. The rationale for dividing an identifier into two parts will become clear when we discuss routing in an ATM network.

The lengths of the VPIs for UNIs and NNIs are different. In a UNI, the VPI is 8 bits, whereas in an NNI, the VPI is 12 bits. The length of the VCI is the same in both interfaces (16 bits). We therefore can say that a virtual connection is identified by 24 bits in a UNI and by 28 bits in an NNI (see Figure 18.13).

The whole idea behind dividing a virtual circuit identifier into two parts is to allow hierarchical routing. Most of the switches in a typical ATM network are routed using VPIs. The switches at the boundaries of the network, those that interact directly with the endpoint devices, use both VPIs and VCIs.

Cells

The basic data unit in an ATM network is called a cell. A cell is only 53 bytes long with 5 bytes allocated to the header and 48 bytes carrying the payload (user data may be less

Figure 18.12 *Connection identifiers*

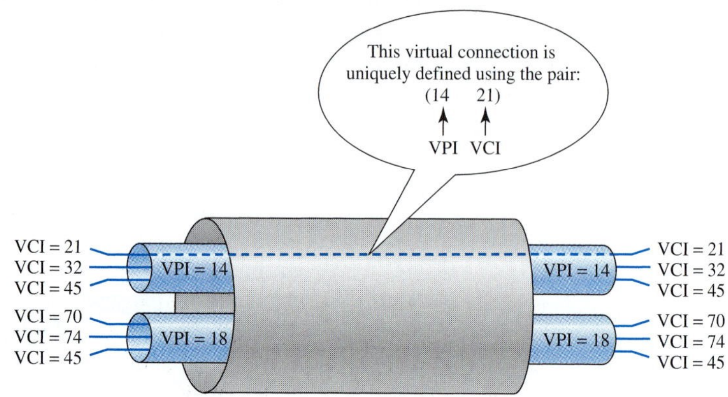

Figure 18.13 *Virtual connection identifiers in UNIs and NNIs*

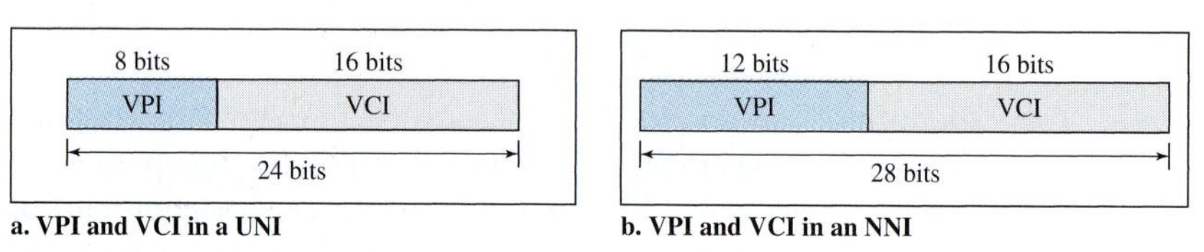

a. VPI and VCI in a UNI **b. VPI and VCI in an NNI**

than 48 bytes). We will study in detail the fields of a cell, but for the moment it suffices to say that most of the header is occupied by the VPI and VCI that define the virtual connection through which a cell should travel from an endpoint to a switch or from a switch to another switch. Figure 18.14 shows the cell structure.

Figure 18.14 *An ATM cell*

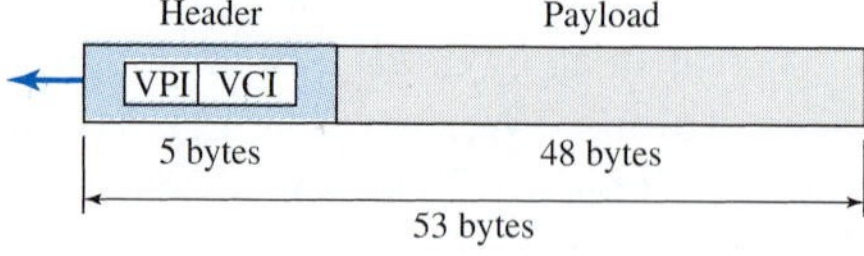

Connection Establishment and Release

Like Frame Relay, ATM uses two types of connections: PVC and SVC.

PVC A permanent virtual-circuit connection is established between two endpoints by the network provider. The VPIs and VCIs are defined for the permanent connections, and the values are entered for the tables of each switch.

SVC In a switched virtual-circuit connection, each time an endpoint wants to make a connection with another endpoint, a new virtual circuit must be established. ATM cannot do the job by itself, but needs the network layer addresses and the services of another protocol (such as IP). The signaling mechanism of this other protocol makes a connection request by using the network layer addresses of the two endpoints. The actual mechanism depends on the network layer protocol.

Switching

ATM uses switches to route the cell from a source endpoint to the destination endpoint. A switch routes the cell using both the VPIs and the VCIs. The routing requires the whole identifier. Figure 18.15 shows how a VPC switch routes the cell. A cell with a VPI of 153 and VCI of 67 arrives at switch interface (port) 1. The switch checks its switching table, which stores six pieces of information per row: arrival interface number, incoming VPI, incoming VCI, corresponding outgoing interface number, the new VPI, and the new VCI. The switch finds the entry with the interface 1, VPI 153, and VCI 67 and discovers that the combination corresponds to output interface 3, VPI 140, and VCI 92. It changes the VPI and VCI in the header to 140 and 92, respectively, and sends the cell out through interface 3.

Figure 18.15 *Routing with a switch*

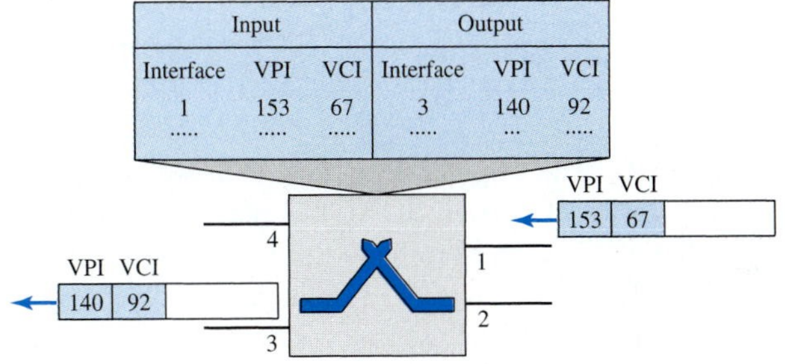

Switching Fabric

The switching technology has created many interesting features to increase the speed of switches to handle data. We discussed switching fabrics in Chapter 8.

ATM Layers

The ATM standard defines three layers. They are, from top to bottom, the application adaptation layer, the ATM layer, and the physical layer (see Figure 18.16).

The endpoints use all three layers while the switches use only the two bottom layers (see Figure 18.17).

Physical Layer

Like Ethernet and wireless LANs, ATM cells can be carried by any physical layer carrier.

Figure 18.16 *ATM layers*

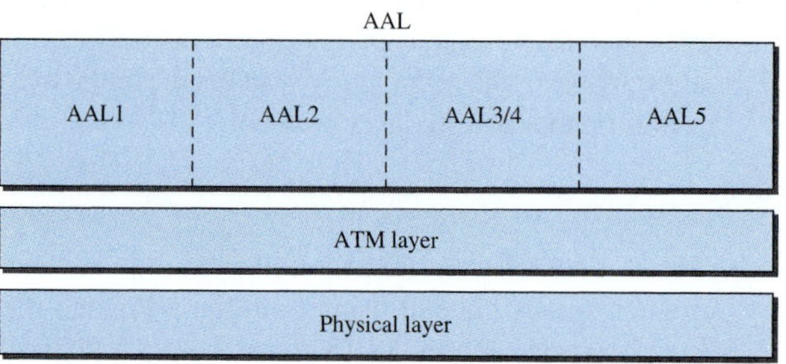

Figure 18.17 *ATM layers in endpoint devices and switches*

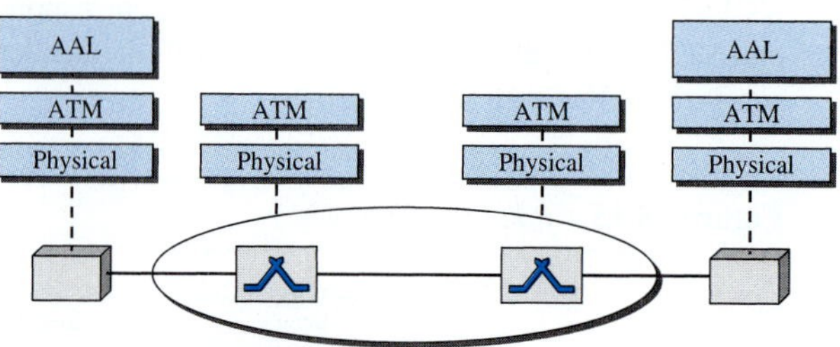

SONET The original design of ATM was based on *SONET* (see Chapter 17) as the physical layer carrier. SONET is preferred for two reasons. First, the high data rate of SONET's carrier reflects the design and philosophy of ATM. Second, in using SONET, the boundaries of cells can be clearly defined. As we saw in Chapter 17, SONET specifies the use of a pointer to define the beginning of a payload. If the beginning of the first ATM cell is defined, the rest of the cells in the same payload can easily be identified because there are no gaps between cells. Just count 53 bytes ahead to find the next cell.

Other Physical Technologies ATM does not limit the physical layer to SONET. Other technologies, even wireless, may be used. However, the problem of cell boundaries must be solved. One solution is for the receiver to guess the end of the cell and apply the CRC to the 5-byte header. If there is no error, the end of the cell is found, with a high probability, correctly. Count 52 bytes back to find the beginning of the cell.

ATM Layer

The **ATM layer** provides routing, traffic management, switching, and multiplexing services. It processes outgoing traffic by accepting 48-byte segments from the AAL sublayers and transforming them into 53-byte cells by the addition of a 5-byte header (see Figure 18.18).

Figure 18.18 *ATM layer*

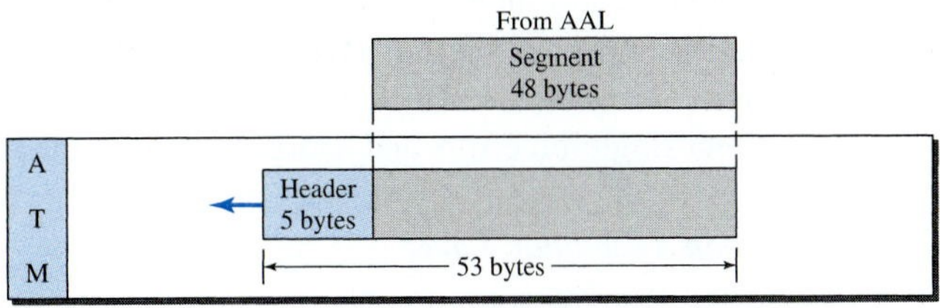

Header Format ATM uses two formats for this header, one for user-to-network interface (UNI) cells and another for network-to-network interface (NNI) cells. Figure 18.19 shows these headers in the byte-by-byte format preferred by the ITU-T (each row represents a byte).

Figure 18.19 *ATM headers*

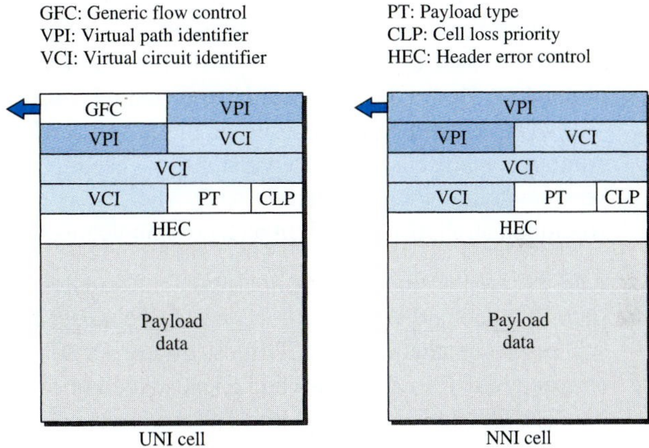

- **Generic flow control (GFC).** The 4-bit GFC field provides flow control at the UNI level. The ITU-T has determined that this level of flow control is not necessary at the NNI level. In the NNI header, therefore, these bits are added to the VPI. The longer VPI allows more virtual paths to be defined at the NNI level. The format for this additional VPI has not yet been determined.
- **Virtual path identifier (VPI).** The VPI is an 8-bit field in a UNI cell and a 12-bit field in an NNI cell (see above).
- **Virtual circuit identifier (VCI).** The VCI is a 16-bit field in both frames.
- **Payload type (PT).** In the 3-bit PT field, the first bit defines the payload as user data or managerial information. The interpretation of the last 2 bits depends on the first bit.

- **Cell loss priority (CLP).** The 1-bit CLP field is provided for congestion control. A cell with its CLP bit set to 1 must be retained as long as there are cells with a CLP of 0. We discuss congestion control and quality of service in an ATM network in Chapter 24.
- **Header error correction (HEC).** The HEC is a code computed for the first 4 bytes of the header. It is a CRC with the divisor $x^8 + x^2 + x + 1$ that is used to correct single-bit errors and a large class of multiple-bit errors.

Application Adaptation Layer

The **application adaptation layer (AAL)** was designed to enable two ATM concepts. First, ATM must accept any type of payload, both data frames and streams of bits. A data frame can come from an upper-layer protocol that creates a clearly defined frame to be sent to a carrier network such as ATM. A good example is the Internet. ATM must also carry multimedia payload. It can accept continuous bit streams and break them into chunks to be encapsulated into a cell at the ATM layer. AAL uses two sublayers to accomplish these tasks.

Whether the data are a data frame or a stream of bits, the payload must be segmented into 48-byte segments to be carried by a cell. At the destination, these segments need to be reassembled to recreate the original payload. The AAL defines a sublayer, called a **segmentation and reassembly (SAR)** sublayer, to do so. Segmentation is at the source; reassembly, at the destination.

Before data are segmented by SAR, they must be prepared to guarantee the integrity of the data. This is done by a sublayer called the **convergence sublayer (CS).**

ATM defines four versions of the AAL: **AAL1, AAL2, AAL3/4,** and **AAL5.** Although we discuss all these versions, we need to inform the reader that the common versions today are AAL1 and AAL5. The first is used in streaming audio and video communication; the second, in data communications.

AAL1 AAL1 supports applications that transfer information at constant bit rates, such as video and voice. It allows ATM to connect existing digital telephone networks such as voice channels and T lines. Figure 18.20 shows how a bit stream of data is chopped into 47-byte chunks and encapsulated in cells.

The CS sublayer divides the bit stream into 47-byte segments and passes them to the SAR sublayer below. Note that the CS sublayer does not add a header.

The SAR sublayer adds 1 byte of header and passes the 48-byte segment to the ATM layer. The header has two fields:

- **Sequence number (SN).** This 4-bit field defines a sequence number to order the bits. The first bit is sometimes used for timing, which leaves 3 bits for sequencing (modulo 8).
- **Sequence number protection (SNP).** The second 4-bit field protects the first field. The first 3 bits automatically correct the SN field. The last bit is a parity bit that detects error over all 8 bits.

AAL2 Originally AAL2 was intended to support a variable-data-rate bit stream, but it has been redesigned. It is now used for low-bit-rate traffic and short-frame traffic such as audio (compressed or uncompressed), video, or fax. A good example of AAL2 use is in mobile telephony. AAL2 allows the multiplexing of short frames into one cell.

Figure 18.20 *AAL1*

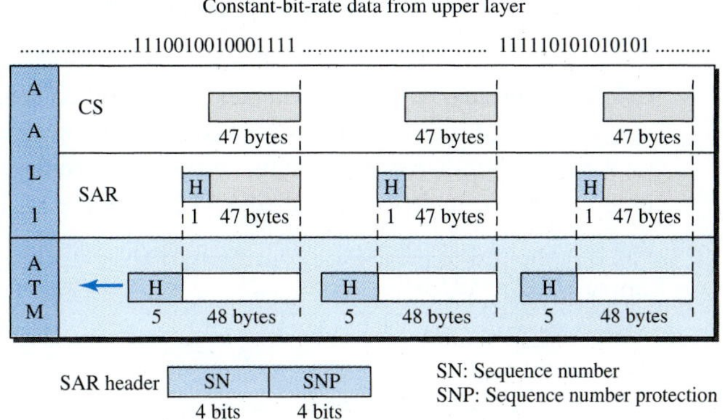

Figure 18.21 shows the process of encapsulating a short frame from the same source (the same user of a mobile phone) or from several sources (several users of mobile telephones) into one cell.

Figure 18.21 *AAL2*

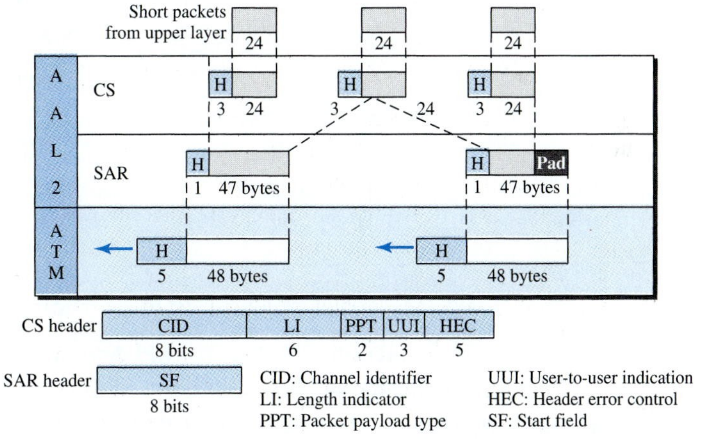

The CS layer overhead consists of five fields:

- **Channel identifier (CID).** The 8-bit CID field defines the channel (user) of the short packet.
- **Length indicator (LI).** The 6-bit LI field indicates how much of the final packet is data.
- **Packet payload type (PPT).** The PPT field defines the type of packet.

- **User-to-user indicator (UUI).** The UUI field can be used by end-to-end users.
- **Header error control (HEC).** The last 5 bits is used to correct errors in the header.

The only overhead at the SAR layer is the start field (SF) that defines the offset from the beginning of the packet.

AAL3/4 Initially, AAL3 was intended to support connection-oriented data services and AAL4 to support connectionless services. As they evolved, however, it became evident that the fundamental issues of the two protocols were the same. They have therefore been combined into a single format called **AAL3/4.** Figure 18.22 shows the AAL3/4 sublayer.

Figure 18.22 *AAL3/4*

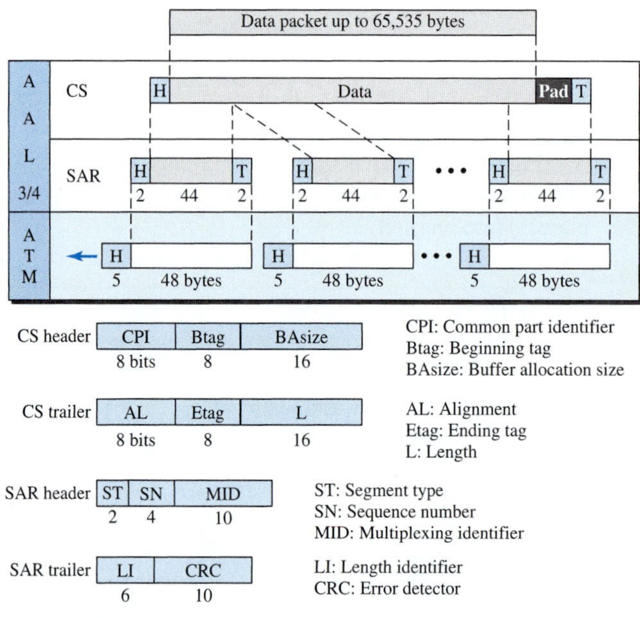

The CS layer header and trailer consist of six fields:

- **Common part identifier (CPI).** The CPI defines how the subsequent fields are to be interpreted. The value at present is 0.
- **Begin tag (Btag).** The value of this field is repeated in each cell to identify all the cells belonging to the same packet. The value is the same as the Etag (see below).
- **Buffer allocation size (BAsize).** The 2-byte BA field tells the receiver what size buffer is needed for the coming data.
- **Alignment (AL).** The 1-byte AL field is included to make the rest of the trailer 4 bytes long.
- **Ending tag (Etag).** The 1-byte ET field serves as an ending flag. Its value is the same as that of the beginning tag.
- **Length (L).** The 2-byte L field indicates the length of the data unit.

The SAR header and trailer consist of five fields:

- **Segment type (ST).** The 2-bit ST identifier specifies the position of the segment in the message: beginning (00), middle (01), or end (10). A single-segment message has an ST of 11.
- **Sequence number (SN).** This field is the same as defined previously.
- **Multiplexing identifier (MID).** The 10-bit MID field identifies cells coming from different data flows and multiplexed on the same virtual connection.
- **Length indicator (LI).** This field defines how much of the packet is data, not padding.
- **CRC.** The last 10 bits of the trailer is a CRC for the entire data unit.

AAL5 AAL3/4 provides comprehensive sequencing and error control mechanisms that are not necessary for every application. For these applications, the designers of ATM have provided a fifth AAL sublayer, called the **simple and efficient adaptation layer (SEAL). AAL5** assumes that all cells belonging to a single message travel sequentially and that control functions are included in the upper layers of the sending application. Figure 18.23 shows the AAL5 sublayer.

Figure 18.23 *AAL5*

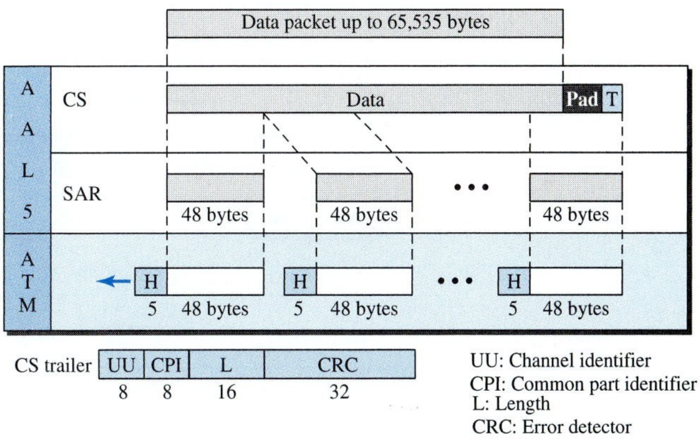

The four trailer fields in the CS layer are

- **User-to-user (UU).** This field is used by end users, as described previously.
- **Common part identifier (CPI).** This field is the same as defined previously.
- **Length (L).** The 2-byte L field indicates the length of the original data.
- **CRC.** The last 4 bytes is for error control on the entire data unit.

Congestion Control and Quality of Service

ATM has a very developed congestion control and quality of service that we discuss in Chapter 24.

18.3 ATM LANs

ATM is mainly a wide-area network (WAN ATM); however, the technology can be adapted to local-area networks (ATM LANs). The high data rate of the technology (155 and 622 Mbps) has attracted the attention of designers who are looking for greater and greater speeds in LANs. In addition, ATM technology has several advantages that make it an ideal LAN:

- ATM technology supports different types of connections between two end users. It supports permanent and temporary connections.
- ATM technology supports multimedia communication with a variety of bandwidths for different applications. It can guarantee a bandwidth of several megabits per second for real-time video. It can also provide support for text transfer during off-peak hours.
- An ATM LAN can be easily adapted for expansion in an organization.

ATM LAN Architecture

Today, we have two ways to incorporate ATM technology in a LAN architecture: creating a **pure ATM LAN** or making a **legacy ATM LAN.** Figure 18.24 shows the taxonomy.

Figure 18.24 *ATM LANs*

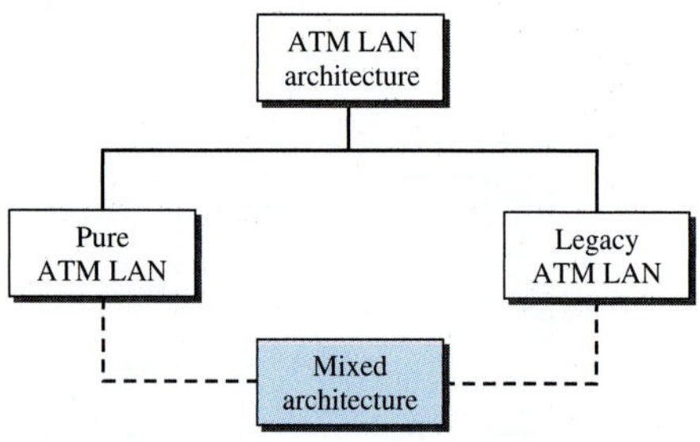

Pure ATM Architecture

In a pure ATM LAN, an **ATM switch** is used to connect the stations in a LAN, in exactly the same way stations are connected to an Ethernet switch. Figure 18.25 shows the situation.

In this way, stations can exchange data at one of two standard rates of ATM technology (155 and 652 Mbps). However, the station uses a virtual path identifier (VPI) and a virtual circuit identifier (VCI), instead of a source and destination address.

This approach has a major drawback. The system needs to be built from the ground up; existing LANs cannot be upgraded into pure ATM LANs.

Legacy LAN Architecture

A second approach is to use ATM technology as a backbone to connect traditional LANs. Figure 18.26 shows this architecture, a **legacy ATM LAN.**

Figure 18.25 *Pure ATM LAN*

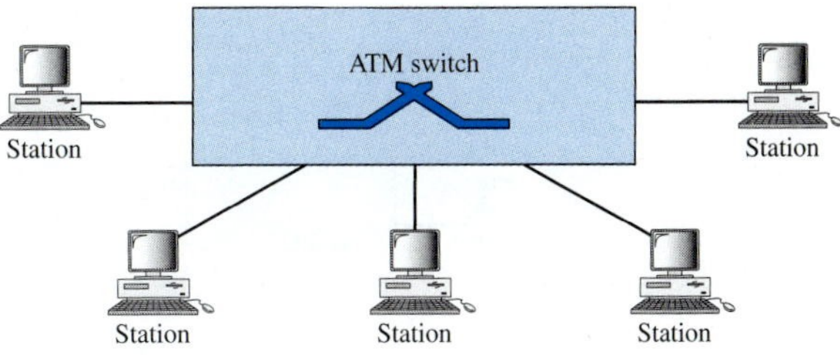

Figure 18.26 *Legacy ATM LAN*

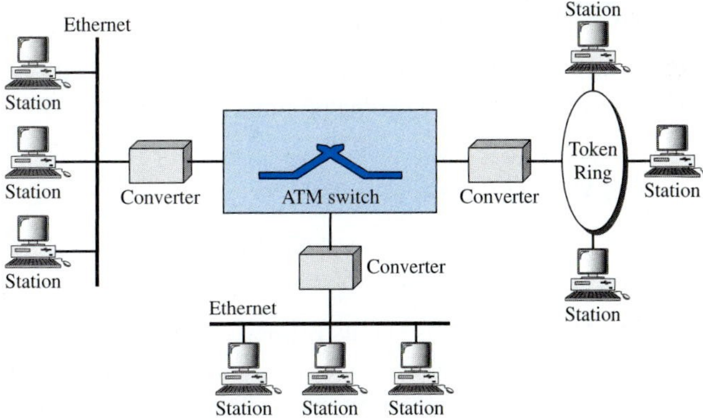

In this way, stations on the same LAN can exchange data at the rate and format of traditional LANs (Ethernet, Token Ring, etc.). But when two stations on two different LANs need to exchange data, they can go through a converting device that changes the frame format. The advantage here is that output from several LANs can be multiplexed together to create a high-data-rate input to the ATM switch. There are several issues that must be resolved first.

Mixed Architecture

Probably the best solution is to mix the two previous architectures. This means keeping the existing LANs and, at the same time, allowing new stations to be directly connected to an ATM switch. The **mixed architecture LAN** allows the gradual migration of legacy LANs onto ATM LANs by adding more and more directly connected stations to the switch. Figure 18.27 shows this architecture.

Again, the stations in one specific LAN can exchange data using the format and data rate of that particular LAN. The stations directly connected to the ATM switch can use an ATM frame to exchange data. However, the problem is, How can a station in a

Figure 18.27 *Mixed architecture ATM LAN*

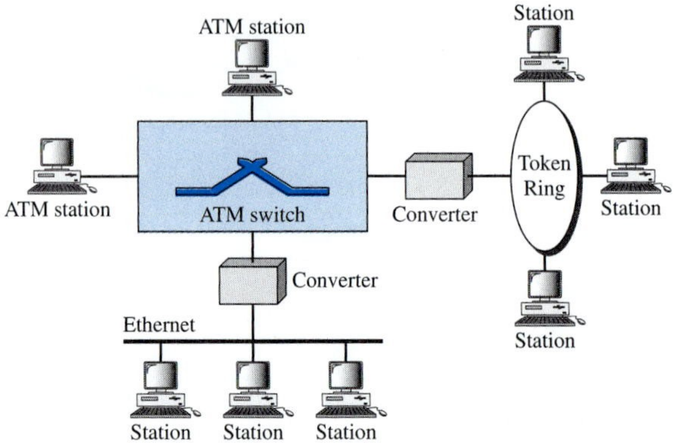

traditional LAN communicate with a station directly connected to the ATM switch or vice versa? We see how the problem is resolved now.

LAN Emulation (LANE)

At the surface level, the use of ATM technology in LANs seems like a good idea. However, many issues need to be resolved, as summarized below:

- **Connectionless versus connection-oriented.** Traditional LANs, such as Ethernet, are *connectionless protocols.* A station sends data packets to another station whenever the packets are ready. There is no *connection establishment* or *connection termination* phase. On the other hand, ATM is a *connection-oriented protocol;* a station that wishes to send cells to another station must first establish a connection and, after all the cells are sent, terminate the connection.
- **Physical addresses versus virtual-circuit identifiers.** Closely related to the first issue is the difference in addressing. A connectionless protocol, such as Ethernet, defines the route of a packet through *source* and *destination addresses.* However, a connection-oriented protocol, such as ATM, defines the route of a cell through virtual connection identifiers (VPIs and VCIs).
- **Multicasting and broadcasting delivery.** Traditional LANs, such as Ethernet, can both *multicast* and *broadcast* packets; a station can send packets to a group of stations or to all stations. There is no easy way to multicast or broadcast on an ATM network although point-to-multipoint connections are available.
- **Interoperability.** In a mixed architecture, a station connected to a legacy LAN must be able to communicate with a station directly connected to an ATM switch.

An approach called **local-area network emulation (LANE)** solves the above-mentioned problems and allows stations in a mixed architecture to communicate with one another. The approach uses emulation. Stations can use a connectionless service that emulates a connection-oriented service. Stations use the source and destination addresses for initial connection and then use VPI and VCI addressing. The approach allows stations to use unicast, multicast, and broadcast addresses. Finally, the approach converts frames using a legacy format to ATM cells before they are sent through the switch.

Client/Server Model

LANE is designed as a **client/server model** to handle the four previously discussed problems. The protocol uses one type of client and three types of servers, as shown in Figure 18.28.

Figure 18.28 *Client and servers in a LANE*

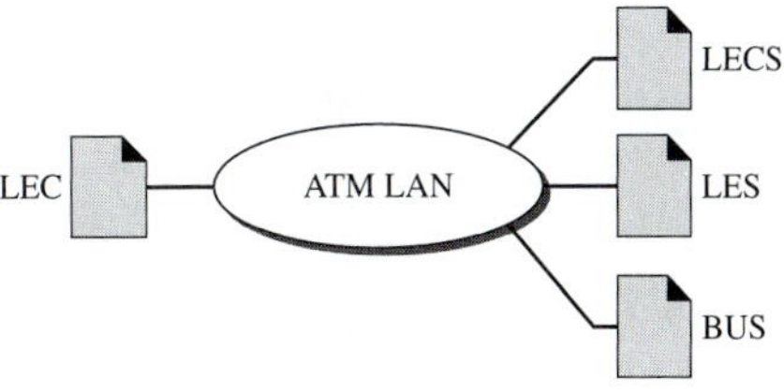

LAN Emulation Client

All ATM stations have **LAN emulation client (LEC)** software installed on top of the three ATM protocols. The upper-layer protocols are unaware of the existence of the ATM technology. These protocols send their requests to LEC for a LAN service such as connectionless delivery using MAC unicast, multicast, or broadcast addresses. The LEC, however, just interprets the request and passes the result on to the servers.

LAN Emulation Configuration Server

The **LAN emulation configuration server (LECS)** is used for the initial connection between the client and LANE. This server is always waiting to receive the initial contact. It has a well-known ATM address that is known to every client in the system.

LAN Emulation Server

LAN emulation server (LES) software is installed on the LES. When a station receives a frame to be sent to another station using a physical address, LEC sends a special frame to the LES. The server creates a virtual circuit between the source and the destination station. The source station can now use this virtual circuit (and the corresponding identifier) to send the frame or frames to the destination.

Broadcast/Unknown Server

Multicasting and broadcasting require the use of another server called the **broadcast/ unknown server (BUS).** If a station needs to send a frame to a group of stations or to every station, the frame first goes to the BUS; this server has permanent virtual connections to every station. The server creates copies of the received frame and sends a copy to a group of stations or to all stations, simulating a multicasting or broadcasting process.

The server can also deliver a unicast frame by sending the frame to every station. In this case the destination address is unknown. This is sometimes more efficient than getting the connection identifier from the LES.

Mixed Architecture with Client/Server

Figure 18.29 shows clients and servers in a mixed architecture ATM LAN. In the figure, three types of servers are connected to the ATM switch (they can actually be part of the switch). Also we show two types of clients. Stations A and B, designed to send and receive LANE communication, are directly connected to the ATM switch. Stations C, D, E, F, G, and H in traditional legacy LANs are connected to the switch via a converter. These converters act as LEC clients and communicate on behalf of their connected stations.

Figure 18.29 *A mixed architecture ATM LAN using LANE*

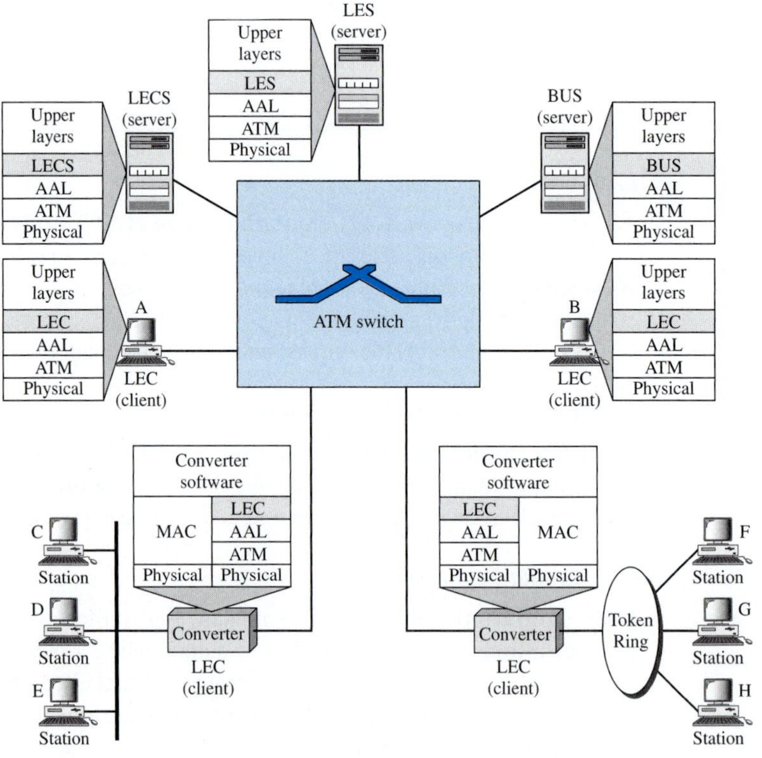

18.4 RECOMMENDED READING

For more details about subjects discussed in this chapter, we recommend the following books. The items in brackets [. . .] refer to the reference list at the end of the text.

Books

Frame Relay and ATM are discussed in Chapter 3 of [Sta98]. ATM LAN is discussed in Chapter 14 of [For03]. Chapter 7 of [Kei02] also has a good discussion of ATM LANs.

18.5 KEY TERMS

AAL1
AAL2
AAL3/4
AAL5
application adaptation layer (AAL)
Asynchronous Transfer Mode (ATM)
ATM layer
ATM switch
backward explicit congestion notification (BECN)
bandwidth on demand
broadcast/unknown server (BUS)
bursty data
cell
cell network
cell relay
client/server model
congestion control
convergence sublayer (CS)
data link connection identifier (DLCI)
discard eligibility (DE)
forward explicit congestion notification (FECN)
Frame Relay
Frame Relay assembler/disassembler (FRAD)
LAN emulation client (LEC)
LAN emulation configuration server (LECS)
LAN emulation server (LES)
legacy ATM LAN
local-area network emulation (LANE)
Local Management Information (LMI)
mixed architecture LAN
network-to-network interface (NNI)
permanent virtual circuit (PVC)
pure ATM LAN
quality of service (QoS)
segmentation and reassembly (SAR)
simple and efficient adaptation layer (SEAL)
switched virtual circuit (SVC)
transmission path (TP)
user-to-network interface (UNI)
virtual circuit (VC)
virtual-circuit identifier (VCI)
virtual path (VP)
virtual path identifier (VPI)
Voice Over Frame Relay (VOFR)
X.25

18.6 SUMMARY

- Virtual-circuit switching is a data link layer technology in which links are shared.
- A virtual-circuit identifier (VCI) identifies a frame between two switches.
- Frame Relay is a relatively high-speed, cost-effective technology that can handle bursty data.

- Both PVC and SVC connections are used in Frame Relay.
- The data link connection identifier (DLCI) identifies a virtual circuit in Frame Relay.
- Asynchronous Transfer Mode (ATM) is a cell relay protocol that, in combination with SONET, allows high-speed connections.
- A cell is a small, fixed-size block of information.
- The ATM data packet is a cell composed of 53 bytes (5 bytes of header and 48 bytes of payload).
- ATM eliminates the varying delay times associated with different-size packets.
- ATM can handle real-time transmission.
- A user-to-network interface (UNI) is the interface between a user and an ATM switch.
- A network-to-network interface (NNI) is the interface between two ATM switches.
- In ATM, connection between two endpoints is accomplished through transmission paths (TPs), virtual paths (VPs), and virtual circuits (VCs).
- In ATM, a combination of a virtual path identifier (VPI) and a virtual-circuit identifier identifies a virtual connection.
- The ATM standard defines three layers:
  a. Application adaptation layer (AAL) accepts transmissions from upper-layer services and maps them into ATM cells.
  b. ATM layer provides routing, traffic management, switching, and multiplexing services.
  c. Physical layer defines the transmission medium, bit transmission, encoding, and electrical-to-optical transformation.
- The AAL is divided into two sublayers: convergence sublayer (CS) and segmentation and reassembly (SAR).
- There are four different AALs, each for a specific data type:
  a. AAL1 for constant-bit-rate stream.
  b. AAL2 for short packets.
  c. AAL3/4 for conventional packet switching (virtual-circuit approach or datagram approach).
  d. AAL5 for packets requiring no sequencing and no error control mechanism.
- ATM technology can be adopted for use in a LAN (ATM LAN).
- In a pure ATM LAN, an ATM switch connects stations.
- In a legacy ATM LAN, the backbone that connects traditional LANs uses ATM technology.
- A mixed architecture ATM LAN combines features of a pure ATM LAN and a legacy ATM LAN.
- Local-area network emulation (LANE) is a client/server model that allows the use of ATM technology in LANs.
- LANE software includes LAN emulation client (LECS), LAN emulation configuration server (LECS), LAN emulation server (LES), and broadcast/unknown server (BUS) modules.

18.7 PRACTICE SET

Review Questions

1. There are no sequence numbers in Frame Relay. Why?
2. Can two devices connected to the same Frame Relay network use the same DLCIs?
3. Why is Frame Relay a better solution for connecting LANs than T-1 lines?
4. Compare an SVC with a PVC.
5. Discuss the Frame Relay physical layer.
6. Why is multiplexing more efficient if all the data units are the same size?
7. How does an NNI differ from a UNI?
8. What is the relationship between TPs, VPs, and VCs?
9. How is an ATM virtual connection identified?
10. Name the ATM layers and their functions.
11. How many virtual connections can be defined in a UNI? How many virtual connections can be defined in an NNI?
12. Briefly describe the issues involved in using ATM technology in LANs.

Exercises

13. The address field of a Frame Relay frame is 1011000000010111. What is the DLCI (in decimal)?
14. The address field of a Frame Relay frame is 101100000101001. Is this valid?
15. Find the DLCI value if the first 3 bytes received is 7C 74 E1 in hexadecimal.
16. Find the value of the 2-byte address field in hexadecimal if the DLCI is 178. Assume no congestion.
17. In Figure 18.30 a virtual connection is established between A and B. Show the DLCI for each link.

Figure 18.30 *Exercise 17*

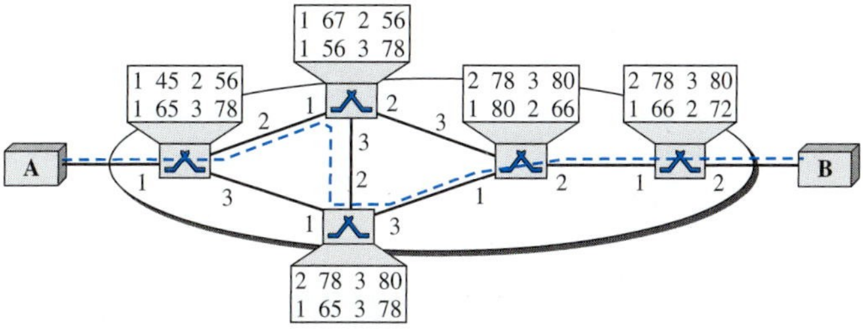

18. In Figure 18.31 a virtual connection is established between A and B. Show the corresponding entries in the tables of each switch.

Figure 18.31 *Exercise 18*

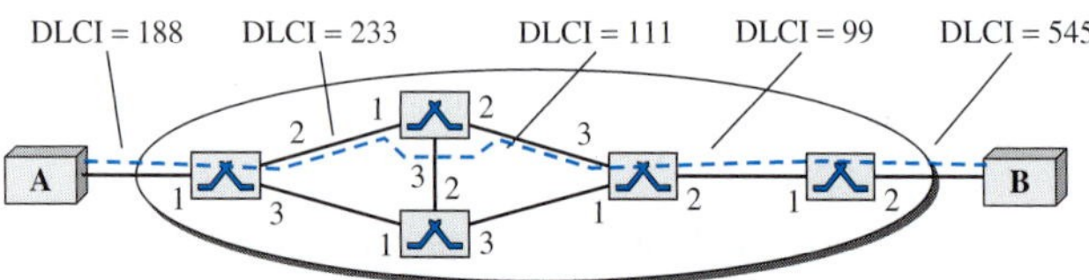

19. An AAL1 layer receives data at 2 Mbps. How many cells are created per second by the ATM layer?
20. What is the total efficiency of ATM using AAL1 (the ratio of received bits to sent bits)?
21. If an application uses AAL3/4 and there are 47,787 bytes of data coming into the CS, how many padding bytes are necessary? How many data units get passed from the SAR to the ATM layer? How many cells are produced?
22. Assuming no padding, does the efficiency of ATM using AAL3/4 depend on the size of the packet? Explain your answer.
23. What is the minimum number of cells resulting from an input packet in the AAL3/4 layer? What is the maximum number of cells resulting from an input packet?
24. What is the minimum number of cells resulting from an input packet in the AAL5 layer? What is the maximum number of cells resulting from an input packet?
25. Explain why padding is unnecessary in AAL1, but necessary in other AALs.
26. Using AAL3/4, show the situation where we need _______ of padding.
    a. 0 bytes (no padding)
    b. 40 bytes
    c. 43 bytes
27. Using AAL5, show the situation where we need _______ of padding.
    a. 0 bytes (no padding)
    b. 40 bytes
    c. 47 bytes
28. In a 53-byte cell, how many bytes belong to the user in the following (assume no padding)?
    a. AAL1
    b. AAL2
    c. AAL3/4 (not the first or last cell)
    d. AAL5 (not the first or last cell)

Research Activities

29. Find out about I.150 protocol that provides generic flow control for UNI interface.
30. ATM uses the 8-bit HEC (header error control) field to control errors in the first four bytes (32 bits) of the header. The generating polynomial is $x^8 + x^2 + x + 1$. Find out how this is done.
31. Find the format of the LANE frames and compare it with the format of the Ethernet frame.
32. Find out about different steps involved in the operation of a LANE.

PART 4

Network Layer

Objectives

The network layer is responsible for the source-to-destination delivery of a packet, possibly across multiple networks (links). Whereas the data link layer oversees the delivery of the packet between two systems on the same network (links), the network layer ensures that each packet gets from its point of origin to its final destination.

The network layer is responsible for the delivery of individual packets from the source to the destination host.

The network layer adds a header that includes the logical addresses of the sender and receiver to the packet coming from the upper layer. If a packet travels through the Internet, we need this addressing system to help distinguish the source and destination.

When independent networks or links are connected together to create an internetwork, routers or switches route packets to their final destination. One of the functions of the network layer is to provide a routing mechanism.

In Part 4 of the book, we first discuss logical addressing (referred to as IP addressing in the Internet). We then discuss the main as well as some auxiliary protocols that are responsible for controlling the delivery of a packet from its source to its destination.

Part 4 of the book is devoted to the network layer and the services provided by this layer.

Chapters

This part consists of four chapters: Chapters 19 to 22.

Chapter 19

Chapter 19 discusses logical or IP addressing. We first discuss the historical classful addressing. We then describe the new classless addressing designed to alleviate some problems inherent in classful addressing. The completely new addressing system, IPv6, which may become prevalent in the near future, is also discussed.

Chapter 20

Chapter 20 is devoted to the main protocol at the network layer that supervises and controls the delivery of packets from the source to destination. This protocol is called the Internet Protocol or IP.

Chapter 21

Chapter 21 is devoted to some auxiliary protocols defined at the network layer, that help the IP protocol do its job. These protocols perform address mapping (logical to physical or vice versa), error reporting, and facilitate multicast delivery.

Chapter 22

Delivery and routing of packets in the Internet is a very delicate and important issue. We devote Chapter 22 to this matter. We first discuss the mechanism of delivery and routing. We then briefly discuss some unicast and multicast routing protocols used in the Internet today.

CHAPTER 19

Network Layer: Logical Addressing

As we discussed in Chapter 2, communication at the network layer is host-to-host (computer-to-computer); a computer somewhere in the world needs to communicate with another computer somewhere else in the world. Usually, computers communicate through the Internet. The packet transmitted by the sending computer may pass through several LANs or WANs before reaching the destination computer.

For this level of communication, we need a global addressing scheme; we called this logical addressing in Chapter 2. Today, we use the term **IP address** to mean a logical address in the network layer of the TCP/IP protocol suite.

The Internet addresses are 32 bits in length; this gives us a maximum of 2^{32} addresses. These addresses are referred to as IPv4 (IP version 4) addresses or simply IP addresses if there is no confusion.

The need for more addresses, in addition to other concerns about the IP layer, motivated a new design of the IP layer called the new generation of IP or IPv6 (IP version 6). In this version, the Internet uses 128-bit addresses that give much greater flexibility in address allocation. These addresses are referred to as IPv6 (IP version 6) addresses.

In this chapter, we first discuss IPv4 addresses, which are currently being used in the Internet. We then discuss the IPv6 addresses, which may become dominant in the future.

19.1 IPv4 ADDRESSES

An **IPv4 address** is a 32-bit address that *uniquely* and *universally* defines the connection of a device (for example, a computer or a router) to the Internet.

An IPv4 address is 32 bits long.

IPv4 addresses are unique. They are unique in the sense that each address defines one, and only one, connection to the Internet. Two devices on the Internet can never have the same address at the same time. We will see later that, by using some strategies, an address may be assigned to a device for a time period and then taken away and assigned to another device.

On the other hand, if a device operating at the network layer has m connections to the Internet, it needs to have m addresses. We will see later that a router is such a device.

The IPv4 addresses are universal in the sense that the addressing system must be accepted by any host that wants to be connected to the Internet.

The IPv4 addresses are unique and universal.

Address Space

A protocol such as IPv4 that defines addresses has an **address space.** An address space is the total number of addresses used by the protocol. If a protocol uses N bits to define an address, the address space is 2^N because each bit can have two different values (0 or 1) and N bits can have 2^N values.

IPv4 uses 32-bit addresses, which means that the address space is 2^{32} or 4,294,967,296 (more than 4 billion). This means that, theoretically, if there were no restrictions, more than 4 billion devices could be connected to the Internet. We will see shortly that the actual number is much less because of the restrictions imposed on the addresses.

The address space of IPv4 is 2^{32} or 4,294,967,296.

Notations

There are two prevalent notations to show an IPv4 address: **binary notation** and **dotted-decimal notation.**

Binary Notation

In binary notation, the IPv4 address is displayed as 32 bits. Each octet is often referred to as a byte. So it is common to hear an IPv4 address referred to as a 32-bit address or a 4-byte address. The following is an example of an IPv4 address in binary notation:

01110101 10010101 00011101 00000010

Dotted-Decimal Notation

To make the IPv4 address more compact and easier to read, Internet addresses are usually written in decimal form with a decimal point (dot) separating the bytes. The following is the **dotted-decimal notation** of the above address:

117.149.29.2

Figure 19.1 shows an IPv4 address in both binary and dotted-decimal notation. Note that because each byte (octet) is 8 bits, each number in dotted-decimal notation is a value ranging from 0 to 255.

Figure 19.1 *Dotted-decimal notation and binary notation for an IPv4 address*

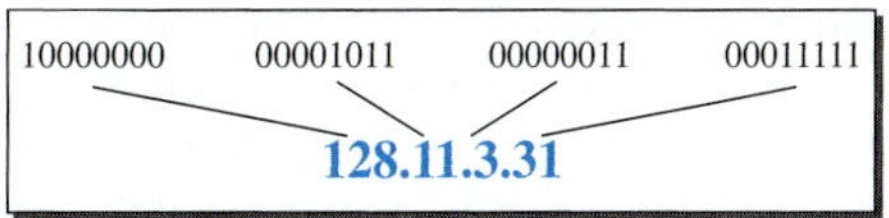

Numbering systems are reviewed in Appendix B.

Example 19.1

Change the following IPv4 addresses from binary notation to dotted-decimal notation.

a. 10000001 00001011 00001011 11101111
b. 11000001 10000011 00011011 11111111

Solution

We replace each group of 8 bits with its equivalent decimal number (see Appendix B) and add dots for separation.

a. 129.11.11.239
b. 193.131.27.255

Example 19.2

Change the following IPv4 addresses from dotted-decimal notation to binary notation.

a. 111.56.45.78
b. 221.34.7.82

Solution

We replace each decimal number with its binary equivalent (see Appendix B).

a. 01101111 00111000 00101101 01001110
b. 11011101 00100010 00000111 01010010

Example 19.3

Find the error, if any, in the following IPv4 addresses.

a. 111.56.045.78
b. 221.34.7.8.20
c. 75.45.301.14
d. 11100010.23.14.67

Solution

a. There must be no leading zero (045).
b. There can be no more than four numbers in an IPv4 address.
c. Each number needs to be less than or equal to 255 (301 is outside this range).
d. A mixture of binary notation and dotted-decimal notation is not allowed.

Classful Addressing

IPv4 addressing, at its inception, used the concept of classes. This architecture is called **classful addressing.** Although this scheme is becoming obsolete, we briefly discuss it here to show the rationale behind classless addressing.

In classful addressing, the address space is divided into five classes: A, B, C, D, and E. Each class occupies some part of the address space.

> **In classful addressing, the address space is divided into five classes: A, B, C, D, and E.**

We can find the class of an address when given the address in binary notation or dotted-decimal notation. If the address is given in binary notation, the first few bits can immediately tell us the class of the address. If the address is given in decimal-dotted notation, the first byte defines the class. Both methods are shown in Figure 19.2.

Figure 19.2 *Finding the classes in binary and dotted-decimal notation*

	First byte	Second byte	Third byte	Fourth byte
Class A	0			
Class B	10			
Class C	110			
Class D	1110			
Class E	1111			

a. Binary notation

	First byte	Second byte	Third byte	Fourth byte
Class A	0–127			
Class B	128–191			
Class C	192–223			
Class D	224–239			
Class E	240–255			

b. Dotted-decimal notation

Example 19.4

Find the class of each address.

a. **0**0000001 00001011 00001011 11101111

b. **110**00001 10000011 00011011 11111111

c. **14**.23.120.8

d. **252**.5.15.111

Solution

a. The first bit is 0. This is a **class A address.**

b. The first 2 bits are 1; the third bit is 0. This is a **class C address.**

c. The first byte is 14 (between 0 and 127); the class is A.

d. The first byte is 252 (between 240 and 255); the class is E.

Classes and Blocks

One problem with classful addressing is that each class is divided into a fixed number of blocks with each block having a fixed size as shown in Table 19.1.

Table 19.1 *Number of blocks and block size in classful IPv4 addressing*

Class	*Number of Blocks*	*Block Size*	*Application*
A	128	16,777,216	Unicast
B	16,384	65,536	Unicast
C	2,097,152	256	Unicast
D	1	268,435,456	Multicast
E	1	268,435,456	Reserved

Let us examine the table. Previously, when an organization requested a block of addresses, it was granted one in class A, B, or C. **Class A addresses** were designed for large organizations with a large number of attached hosts or routers. **Class B addresses** were designed for midsize organizations with tens of thousands of attached hosts or routers. **Class C addresses** were designed for small organizations with a small number of attached hosts or routers.

We can see the flaw in this design. A block in class A address is too large for almost any organization. This means most of the addresses in class A were wasted and were not used. A block in class B is also very large, probably too large for many of the organizations that received a class B block. A block in class C is probably too small for many organizations. **Class D addresses** were designed for multicasting as we will see in a later chapter. Each address in this class is used to define one group of hosts on the Internet. The Internet authorities wrongly predicted a need for 268,435,456 groups. This never happened and many addresses were wasted here too. And lastly, the **class E addresses** were reserved for future use; only a few were used, resulting in another waste of addresses.

In classful addressing, a large part of the available addresses were wasted.

Netid and Hostid

In classful addressing, an IP address in class A, B, or C is divided into **netid** and **hostid.** These parts are of varying lengths, depending on the class of the address. Figure 19.2 shows some netid and hostid bytes. The netid is in color, the hostid is in white. Note that the concept does not apply to classes D and E.

In class A, one byte defines the netid and three bytes define the hostid. In class B, two bytes define the netid and two bytes define the hostid. In class C, three bytes define the netid and one byte defines the hostid.

Mask

Although the length of the netid and hostid (in bits) is predetermined in classful addressing, we can also use a **mask** (also called the **default mask**), a 32-bit number made of contiguous 1s followed by contiguous 0s. The masks for classes A, B, and C are shown in Table 19.2. The concept does not apply to classes D and E.

Table 19.2 *Default masks for classful addressing*

Class	*Binary*	*Dotted-Decimal*	*CIDR*
A	11111111 00000000 00000000 00000000	255.0.0.0	/8
B	11111111 11111111 00000000 00000000	255.255.0.0	/16
C	11111111 11111111 11111111 00000000	255.255.255.0	/24

The mask can help us to find the netid and the hostid. For example, the mask for a class A address has eight 1s, which means the first 8 bits of any address in class A define the netid; the next 24 bits define the hostid.

The last column of Table 19.2 shows the mask in the form /*n* where *n* can be 8, 16, or 24 in classful addressing. This notation is also called slash notation or **Classless Interdomain Routing (CIDR)** notation. The notation is used in classless addressing, which we will discuss later. We introduce it here because it can also be applied to classful addressing. We will show later that classful addressing is a special case of classless addressing.

Subnetting

During the era of classful addressing, **subnetting** was introduced. If an organization was granted a large block in class A or B, it could divide the addresses into several contiguous groups and assign each group to smaller networks (called **subnets**) or, in rare cases, share part of the addresses with neighbors. Subnetting increases the number of 1s in the mask, as we will see later when we discuss classless addressing.

Supernetting

The time came when most of the class A and class B addresses were depleted; however, there was still a huge demand for midsize blocks. The size of a class C block with a maximum number of 256 addresses did not satisfy the needs of most organizations. Even a midsize organization needed more addresses. One solution was **supernetting.** In supernetting, an organization can combine several class C blocks to create a larger range of addresses. In other words, several networks are combined to create a supernetwork or a **supernet.** An organization can apply for a set of class C blocks instead of just one. For example, an organization that needs 1000 addresses can be granted four contiguous class C blocks. The organization can then use these addresses to create one supernetwork. Supernetting decreases the number of 1s in the mask. For example, if an organization is given four class C addresses, the mask changes from /24 to /22. We will see that classless addressing eliminated the need for supernetting.

Address Depletion

The flaws in classful addressing scheme combined with the fast growth of the Internet led to the near depletion of the available addresses. Yet the number of devices on the Internet is much less than the 2^{32} address space. We have run out of class A and B addresses, and a class C block is too small for most midsize organizations. One solution that has alleviated the problem is the idea of classless addressing.

Classful addressing, which is almost obsolete, is replaced with classless addressing.

Classless Addressing

To overcome address depletion and give more organizations access to the Internet, **classless addressing** was designed and implemented. In this scheme, there are no classes, but the addresses are still granted in blocks.

Address Blocks

In classless addressing, when an entity, small or large, needs to be connected to the Internet, it is granted a block (range) of addresses. The size of the block (the number of addresses) varies based on the nature and size of the entity. For example, a household may be given only two addresses; a large organization may be given thousands of addresses. An ISP, as the Internet service provider, may be given thousands or hundreds of thousands based on the number of customers it may serve.

Restriction To simplify the handling of addresses, the Internet authorities impose three restrictions on classless address blocks:

1. The addresses in a block must be contiguous, one after another.
2. The number of addresses in a block must be a power of 2 (1, 2, 4, 8, . . .).
3. The first address must be evenly divisible by the number of addresses.

Example 19.5

Figure 19.3 shows a block of addresses, in both binary and dotted-decimal notation, granted to a small business that needs 16 addresses.

Figure 19.3 *A block of 16 addresses granted to a small organization*

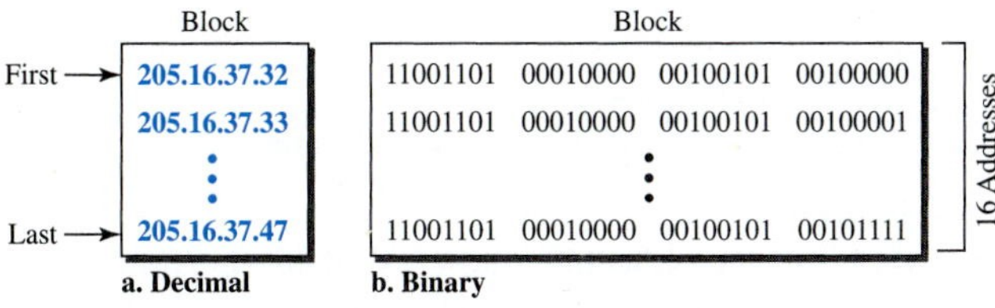

We can see that the restrictions are applied to this block. The addresses are contiguous. The number of addresses is a power of 2 ($16 = 2^4$), and the first address is divisible by 16. The first address, when converted to a decimal number, is 3,440,387,360, which when divided by 16 results in 215,024,210. In Appendix B, we show how to find the decimal value of an IP address.

Mask

A better way to define a block of addresses is to select any address in the block and the mask. As we discussed before, a mask is a 32-bit number in which the n leftmost bits are 1s and the $32 - n$ rightmost bits are 0s. However, in classless addressing the mask for a block can take any value from 0 to 32. It is very convenient to give just the value of n preceded by a slash (CIDR notation).

In IPv4 addressing, a block of addresses can be defined as
x.y.z.t /*n*
in which x.y.z.t defines one of the addresses and the /*n* defines the mask.

The address and the /n notation completely define the whole block (the first address, the last address, and the number of addresses).

First Address The first address in the block can be found by setting the $32 - n$ rightmost bits in the binary notation of the address to 0s.

The first address in the block can be found by setting the rightmost $32 - n$ bits to 0s.

Example 19.6

A block of addresses is granted to a small organization. We know that one of the addresses is 205.16.37.39/28. What is the first address in the block?

Solution

The binary representation of the given address is 11001101 00010000 00100101 00100111. If we set 32 − 28 rightmost bits to 0, we get 11001101 0001000 00100101 0010000 or 205.16.37.32. This is actually the block shown in Figure 19.3.

Last Address The last address in the block can be found by setting the $32 - n$ rightmost bits in the binary notation of the address to 1s.

The last address in the block can be found by setting the rightmost $32 - n$ bits to 1s.

Example 19.7

Find the last address for the block in Example 19.6.

Solution

The binary representation of the given address is 11001101 00010000 00100101 00100111. If we set 32 − 28 rightmost bits to 1, we get 11001101 00010000 00100101 00101111 or 205.16.37.47. This is actually the block shown in Figure 19.3.

Number of Addresses The number of addresses in the block is the difference between the last and first address. It can easily be found using the formula 2^{32-n}.

The number of addresses in the block can be found by using the formula 2^{32-n}.

Example 19.8

Find the number of addresses in Example 19.6.

Solution

The value of n is 28, which means that number of addresses is 2^{32-28} or 16.

Example 19.9

Another way to find the first address, the last address, and the number of addresses is to represent the mask as a 32-bit binary (or 8-digit hexadecimal) number. This is particularly useful when we are writing a program to find these pieces of information. In Example 19.5 the /28 can be represented as 11111111 11111111 11111111 11110000 (twenty-eight 1s and four 0s). Find

a. The first address
b. The last address
c. The number of addresses

Solution

a. The first address can be found by ANDing the given addresses with the mask. ANDing here is done bit by bit. The result of ANDing 2 bits is 1 if both bits are 1s; the result is 0 otherwise.

Address:	11001101 00010000 00100101 00100111
Mask:	**11111111 11111111 11111111 11110000**
First address:	11001101 00010000 00100101 00100000

b. The last address can be found by ORing the given addresses with the complement of the mask. ORing here is done bit by bit. The result of ORing 2 bits is 0 if both bits are 0s; the result is 1 otherwise. The complement of a number is found by changing each 1 to 0 and each 0 to 1.

Address:	11001101 00010000 00100101 00100111
Mask complement:	**00000000 00000000 00000000 00001111**
Last address:	11001101 00010000 00100101 00101111

c. The number of addresses can be found by complementing the mask, interpreting it as a decimal number, and adding 1 to it.

Mask complement:	**000000000 00000000 00000000 00001111**
Number of addresses:	15 + 1 = 16

Network Addresses

A very important concept in IP addressing is the **network address.** When an organization is given a block of addresses, the organization is free to allocate the addresses to the devices that need to be connected to the Internet. The first address in the class, however, is normally (not always) treated as a special address. The first address is called the network address and defines the organization network. It defines the organization itself to the rest of the world. In a later chapter we will see that the first address is the one that is used by routers to direct the message sent to the organization from the outside.

Figure 19.4 shows an organization that is granted a 16-address block.

Figure 19.4 *A network configuration for the block 205.16.37.32/28*

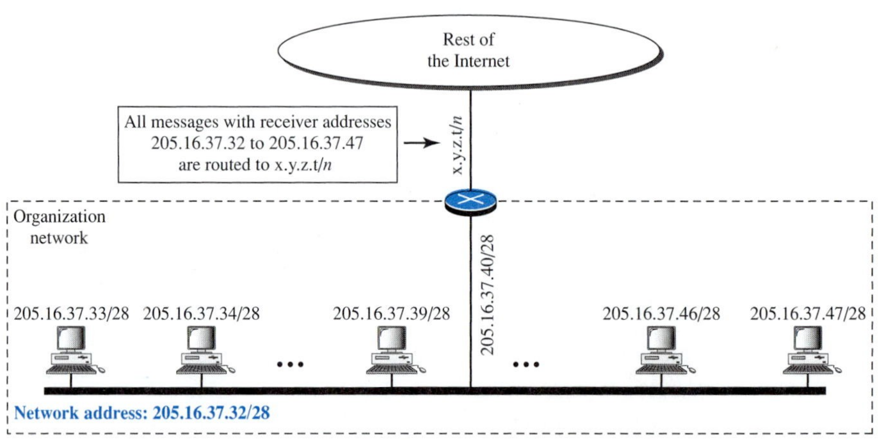

The organization network is connected to the Internet via a router. The router has two addresses. One belongs to the granted block; the other belongs to the network that is at the other side of the router. We call the second address x.y.z.t/*n* because we do not know anything about the network it is connected to at the other side. All messages destined for addresses in the organization block (205.16.37.32 to 205.16.37.47) are sent, directly or indirectly, to x.y.z.t/*n*. We say directly or indirectly because we do not know the structure of the network to which the other side of the router is connected.

The first address in a block is normally not assigned to any device; it is used as the network address that represents the organization to the rest of the world.

Hierarchy

IP addresses, like other addresses or identifiers we encounter these days, have levels of hierarchy. For example, a telephone network in North America has three levels of hierarchy. The leftmost three digits define the area code, the next three digits define the exchange, the last four digits define the connection of the local loop to the central office. Figure 19.5 shows the structure of a hierarchical telephone number.

Figure 19.5 *Hierarchy in a telephone network in North America*

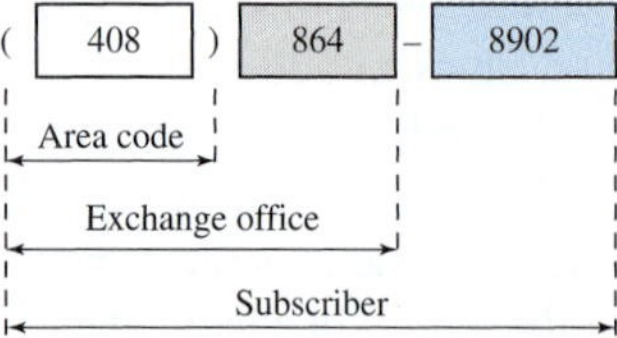

Two-Level Hierarchy: No Subnetting

An IP address can define only two levels of hierarchy when not subnetted. The n leftmost bits of the address x.y.z.t/n define the network (organization network); the $32 - n$ rightmost bits define the particular host (computer or router) to the network. The two common terms are prefix and suffix. The part of the address that defines the network is called the **prefix;** the part that defines the host is called the **suffix.** Figure 19.6 shows the hierarchical structure of an IPv4 address.

Figure 19.6 *Two levels of hierarchy in an IPv4 address*

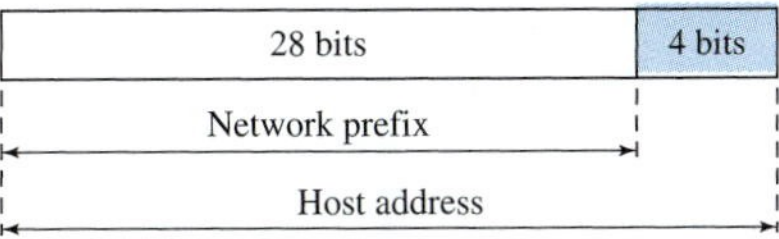

The prefix is common to all addresses in the network; the suffix changes from one device to another.

Each address in the block can be considered as a two-level hierarchical structure: the leftmost *n* bits (prefix) define the network; the rightmost 32 – *n* bits define the host.

Three-Levels of Hierarchy: Subnetting

An organization that is granted a large block of addresses may want to create clusters of networks (called subnets) and divide the addresses between the different subnets. The rest of the world still sees the organization as one entity; however, internally there are several subnets. All messages are sent to the router address that connects the organization to the rest of the Internet; the router routes the message to the appropriate subnets. The organization, however, needs to create small subblocks of addresses, each assigned to specific subnets. The organization has its own mask; each subnet must also have its own.

As an example, suppose an organization is given the block 17.12.40.0/26, which contains 64 addresses. The organization has three offices and needs to divide the addresses into three subblocks of 32, 16, and 16 addresses. We can find the new masks by using the following arguments:

1. Suppose the mask for the first subnet is n1, then 2^{32-n1} must be 32, which means that n1 = 27.
2. Suppose the mask for the second subnet is n2, then 2^{32-n2} must be 16, which means that n2 = 28.
3. Suppose the mask for the third subnet is n3, then 2^{32-n3} must be 16, which means that n3 = 28.

This means that we have the masks 27, 28, 28 with the organization mask being 26. Figure 19.7 shows one configuration for the above scenario.

Figure 19.7 *Configuration and addresses in a subnetted network*

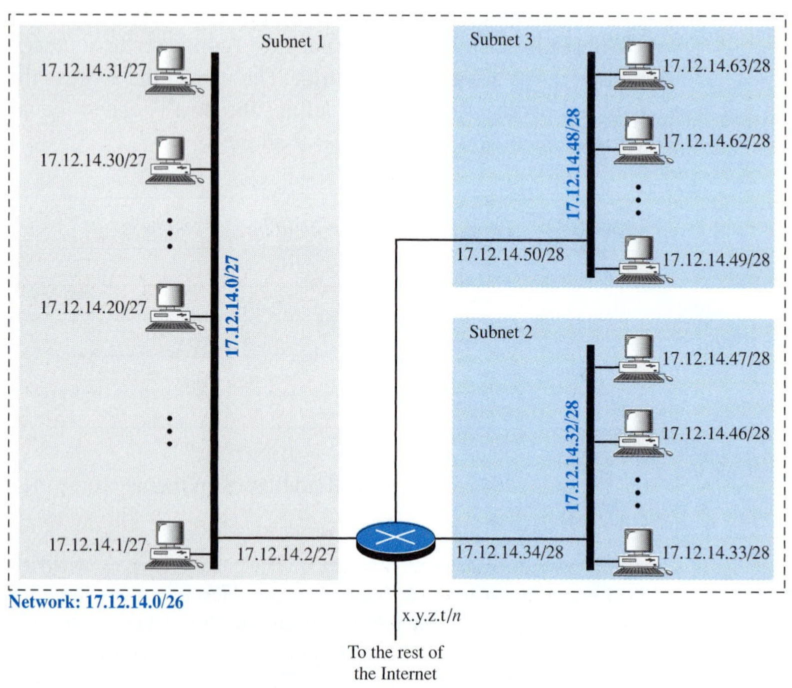

Let us check to see if we can find the subnet addresses from one of the addresses in the subnet.

a. In subnet 1, the address 17.12.14.29/27 can give us the subnet address if we use the mask /27 because

Host:	00010001 00001100 00001110 00011101
Mask:	/27
Subnet:	00010001 00001100 00001110 00000000 → (17.12.14.0)

b. In subnet 2, the address 17.12.14.45/28 can give us the subnet address if we use the mask /28 because

Host:	00010001 00001100 00001110 00101101
Mask:	/28
Subnet:	00010001 00001100 00001110 00100000 → (17.12.14.32)

c. In subnet 3, the address 17.12.14.50/28 can give us the subnet address if we use the mask /28 because

Host:	00010001 00001100 00001110 00110010
Mask:	/28
Subnet:	00010001 00001100 00001110 00110000 → (17.12.14.48)

Note that applying the mask of the network, /26, to any of the addresses gives us the network address 17.12.14.0/26. We leave this proof to the reader.

We can say that through subnetting, we have three levels of hierarchy. Note that in our example, the subnet prefix length can differ for the subnets as shown in Figure 19.8.

Figure 19.8 *Three-level hierarchy in an IPv4 address*

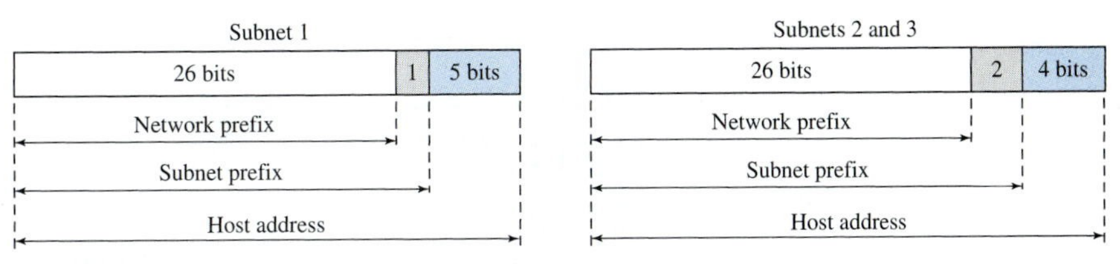

More Levels of Hierarchy

The structure of classless addressing does not restrict the number of hierarchical levels. An organization can divide the granted block of addresses into subblocks. Each subblock can in turn be divided into smaller subblocks. And so on. One example of this is seen in the ISPs. A national ISP can divide a granted large block into smaller blocks and assign each of them to a regional ISP. A regional ISP can divide the block received from the national ISP into smaller blocks and assign each one to a local ISP. A local ISP can divide the block received from the regional ISP into smaller blocks and assign each one to a different organization. Finally, an organization can divide the received block and make several subnets out of it.

Address Allocation

The next issue in classless addressing is address allocation. How are the blocks allocated? The ultimate responsibility of address allocation is given to a global authority called the *Internet Corporation for Assigned Names and Addresses* (ICANN). However, ICANN does not normally allocate addresses to individual organizations. It assigns a large block of addresses to an ISP. Each ISP, in turn, divides its assigned block into smaller subblocks and grants the subblocks to its customers. In other words, an ISP receives one large block to be distributed to its Internet users. This is called **address aggregation:** many blocks of addresses are aggregated in one block and granted to one ISP.

Example 19.10

An ISP is granted a block of addresses starting with 190.100.0.0/16 (65,536 addresses). The ISP needs to distribute these addresses to three groups of customers as follows:

a. The first group has 64 customers; each needs 256 addresses.
b. The second group has 128 customers; each needs 128 addresses.
c. The third group has 128 customers; each needs 64 addresses.

Design the subblocks and find out how many addresses are still available after these allocations.

Solution

Figure 19.9 shows the situation.

Figure 19.9 *An example of address allocation and distribution by an ISP*

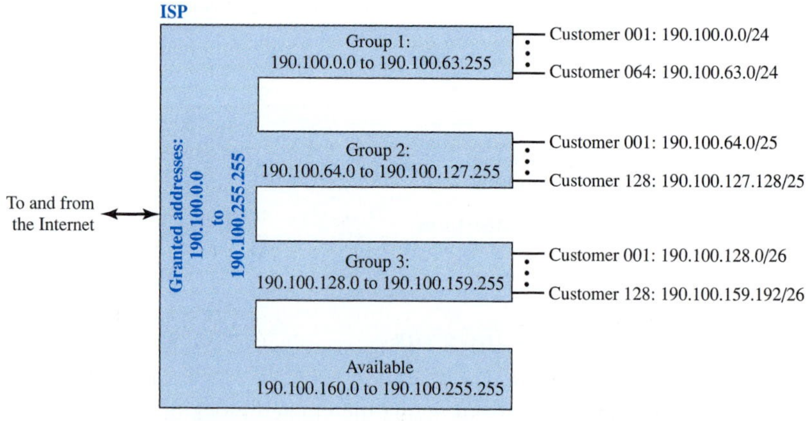

1. **Group 1**

   For this group, each customer needs 256 addresses. This means that 8 ($\log_2 256$) bits are needed to define each host. The prefix length is then $32 - 8 = 24$. The addresses are

1st Customer:	*190.100.0.0/24*	*190.100.0.255/24*
2nd Customer:	*190.100.1.0/24*	*190.100.1.255/24*
. . .		
64th Customer:	*190.100.63.0/24*	*190.100.63.255/24*
Total = 64 × 256 = 16,384		

2. **Group 2**

   For this group, each customer needs 128 addresses. This means that 7 ($\log_2 128$) bits are needed to define each host. The prefix length is then $32 - 7 = 25$. The addresses are

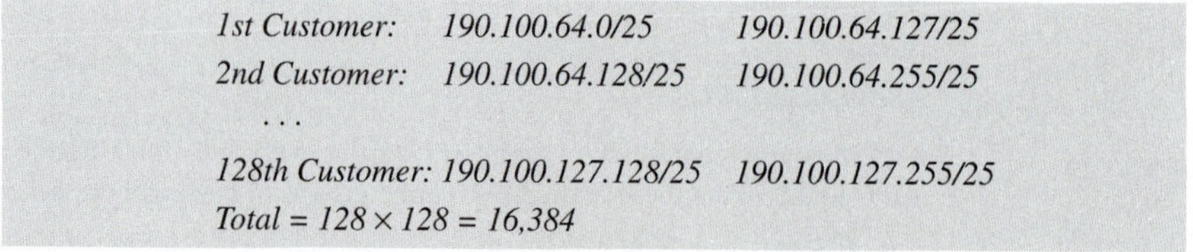

1st Customer:	*190.100.64.0/25*	*190.100.64.127/25*
2nd Customer:	*190.100.64.128/25*	*190.100.64.255/25*
. . .		
128th Customer:	*190.100.127.128/25*	*190.100.127.255/25*
Total = 128 × 128 = 16,384		

3. **Group 3**

   For this group, each customer needs 64 addresses. This means that 6 ($\log_2 64$) bits are needed to each host. The prefix length is then $32 - 6 = 26$. The addresses are

1st Customer:	*190.100.128.0/26*	*190.100.128.63/26*
2nd Customer:	*190.100.128.64/26*	*190.100.128.127/26*
. . .		
128th Customer:	*190.100.159.192/26*	*190.100.159.255/26*
Total = 128 × 64 = 8192		

Number of granted addresses to the ISP: 65,536
Number of allocated addresses by the ISP: 40,960
Number of available addresses: 24,576

Network Address Translation (NAT)

The number of home users and small businesses that want to use the Internet is ever increasing. In the beginning, a user was connected to the Internet with a dial-up line, which means that she was connected for a specific period of time. An ISP with a block of addresses could dynamically assign an address to this user. An address was given to a user when it was needed. But the situation is different today. Home users and small businesses can be connected by an ADSL line or cable modem. In addition, many are not happy with one address; many have created small networks with several hosts and need an IP address for each host. With the shortage of addresses, this is a serious problem.

A quick solution to this problem is called **network address translation (NAT).** NAT enables a user to have a large set of addresses internally and one address, or a small set of addresses, externally. The traffic inside can use the large set; the traffic outside, the small set.

To separate the addresses used inside the home or business and the ones used for the Internet, the Internet authorities have reserved three sets of addresses as private addresses, shown in Table 19.3.

Table 19.3 *Addresses for private networks*

Range			*Total*
10.0.0.0	to	10.255.255.255	2^{24}
172.16.0.0	to	172.31.255.255	2^{20}
192.168.0.0	to	192.168.255.255	2^{16}

Any organization can use an address out of this set without permission from the Internet authorities. Everyone knows that these reserved addresses are for private networks. They are unique inside the organization, but they are not unique globally. No router will forward a packet that has one of these addresses as the destination address.

The site must have only one single connection to the global Internet through a router that runs the NAT software. Figure 19.10 shows a simple implementation of NAT.

As Figure 19.10 shows, the private network uses private addresses. The router that connects the network to the global address uses one private address and one global address. The private network is transparent to the rest of the Internet; the rest of the Internet sees only the NAT router with the address 200.24.5.8.

Figure 19.10 *A NAT implementation*

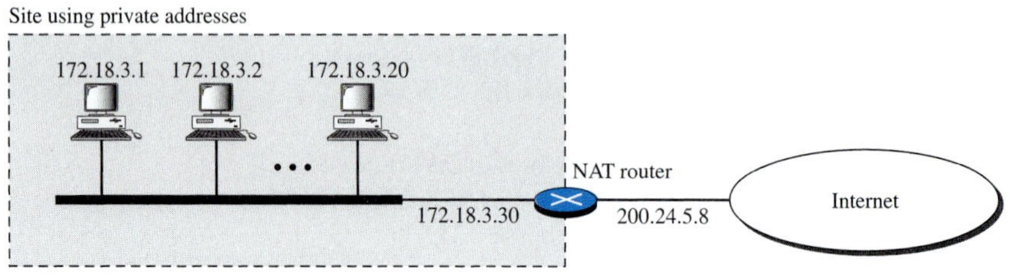

Address Translation

All the outgoing packets go through the NAT router, which replaces the *source address* in the packet with the global NAT address. All incoming packets also pass through the NAT router, which replaces the *destination address* in the packet (the NAT router global address) with the appropriate private address. Figure 19.11 shows an example of address translation.

Figure 19.11 *Addresses in a NAT*

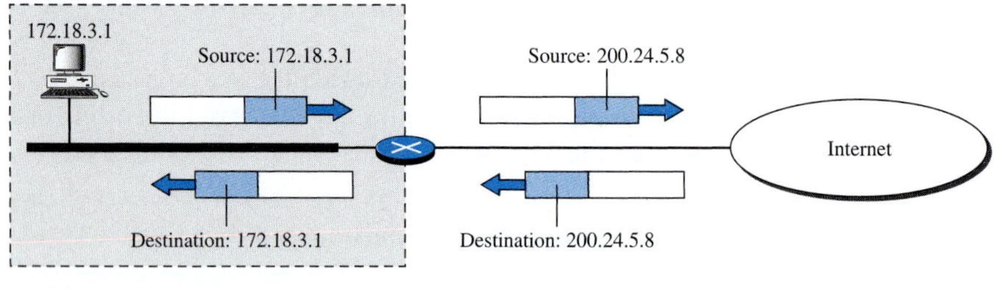

Translation Table

The reader may have noticed that translating the source addresses for outgoing packets is straightforward. But how does the NAT router know the destination address for a packet coming from the Internet? There may be tens or hundreds of private IP addresses, each belonging to one specific host. The problem is solved if the NAT router has a translation table.

Using One IP Address In its simplest form, a translation table has only two columns: the private address and the external address (destination address of the packet). When the router translates the source address of the outgoing packet, it also makes note of the destination address—where the packet is going. When the response comes back from the destination, the router uses the source address of the packet (as the external address) to find the private address of the packet. Figure 19.12 shows the idea. Note that the addresses that are changed (translated) are shown in color.

Figure 19.12 *NAT address translation*

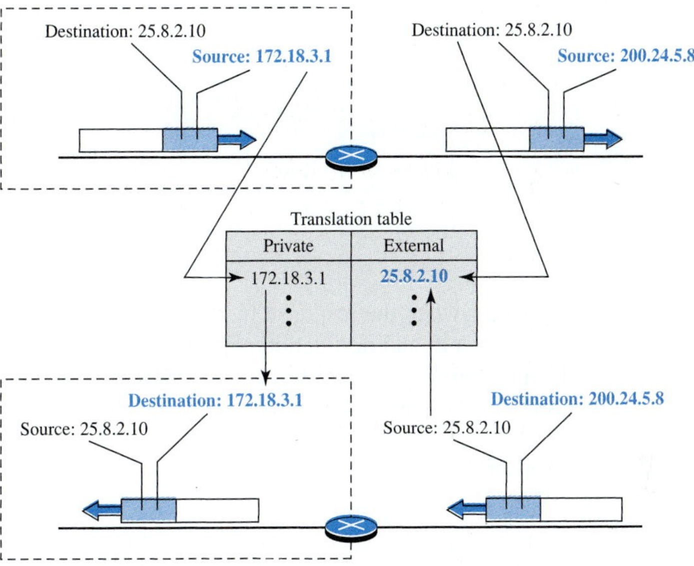

In this strategy, communication must always be initiated by the private network. The NAT mechanism described requires that the private network start the communication. As we will see, NAT is used mostly by ISPs which assign one single address to a customer. The customer, however, may be a member of a private network that has many private addresses. In this case, communication with the Internet is always initiated from the customer site, using a client program such as HTTP, TELNET, or FTP to access the corresponding server program. For example, when e-mail that originates from a non-customer site is received by the ISP e-mail server, the e-mail is stored in the mailbox of the customer until retrieved. A private network cannot run a server program for clients outside of its network if it is using NAT technology.

Using a Pool of IP Addresses Since the NAT router has only one global address, only one private network host can access the same external host. To remove this restriction, the NAT router uses a pool of global addresses. For example, instead of using only one global address (200.24.5.8), the NAT router can use four addresses (200.24.5.8, 200.24.5.9, 200.24.5.10, and 200.24.5.11). In this case, four private network hosts can communicate with the same external host at the same time because each pair of addresses defines a connection. However, there are still some drawbacks. In this example, no more than four connections can be made to the same destination. Also, no private-network host can access two external server programs (e.g., HTTP and FTP) at the same time.

Using Both IP Addresses and Port Numbers To allow a many-to-many relationship between private-network hosts and external server programs, we need more information in the translation table. For example, suppose two hosts with addresses 172.18.3.1 and 172.18.3.2 inside a private network need to access the HTTP server on external host

25.8.3.2. If the translation table has five columns, instead of two, that include the source and destination port numbers of the transport layer protocol, the ambiguity is eliminated. We discuss port numbers in Chapter 23. Table 19.4 shows an example of such a table.

Table 19.4 *Five-column translation table*

Private Address	*Private Port*	*External Address*	*External Port*	*Transport Protocol*
172.18.3.1	1400	25.8.3.2	80	TCP
172.18.3.2	1401	25.8.3.2	80	TCP
...	...	...	...	...

Note that when the response from HTTP comes back, the combination of source address (25.8.3.2) and destination port number (1400) defines the private network host to which the response should be directed. Note also that for this translation to work, the temporary port numbers (1400 and 1401) must be unique.

NAT and ISP

An ISP that serves dial-up customers can use NAT technology to conserve addresses. For example, suppose an ISP is granted 1000 addresses, but has 100,000 customers. Each of the customers is assigned a private network address. The ISP translates each of the 100,000 source addresses in outgoing packets to one of the 1000 global addresses; it translates the global destination address in incoming packets to the corresponding private address. Figure 19.13 shows this concept.

Figure 19.13 *An ISP and NAT*

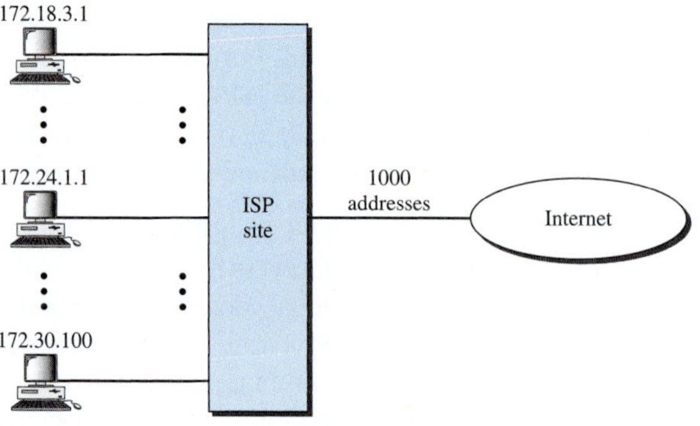

19.2 IPv6 ADDRESSES

Despite all short-term solutions, such as classless addressing, Dynamic Host Configuration Protocol (DHCP), discussed in Chapter 21, and NAT, address depletion is still a long-term problem for the Internet. This and other problems in the IP protocol itself, such as lack of accommodation for real-time audio and video transmission, and encryption and authentication of data for some applications, have been the motivation for IPv6. In this section, we compare the address structure of IPv6 to IPv4. In Chapter 20, we discuss both protocols.

Structure

An **IPv6 address** consists of 16 bytes (octets); it is 128 bits long.

An IPv6 address is 128 bits long.

Hexadecimal Colon Notation

To make addresses more readable, IPv6 specifies **hexadecimal colon notation.** In this notation, 128 bits is divided into eight sections, each 2 bytes in length. Two bytes in hexadecimal notation requires four hexadecimal digits. Therefore, the address consists of 32 hexadecimal digits, with every four digits separated by a colon, as shown in Figure 19.14.

Figure 19.14 *IPv6 address in binary and hexadecimal colon notation*

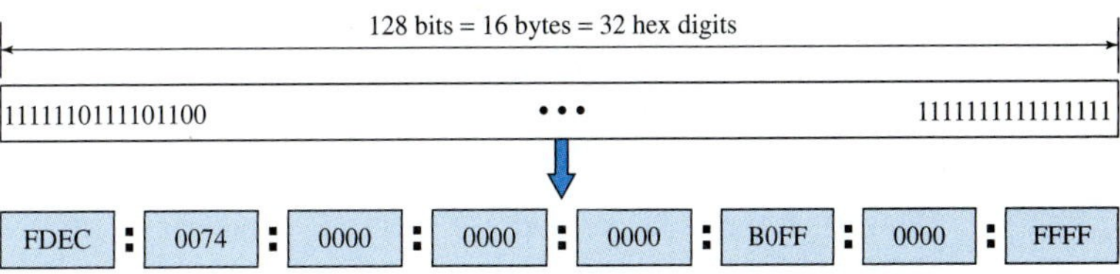

Abbreviation

Although the IP address, even in hexadecimal format, is very long, many of the digits are zeros. In this case, we can abbreviate the address. The leading zeros of a section (four digits between two colons) can be omitted. Only the leading zeros can be dropped, not the trailing zeros (see Figure 19.15).

Figure 19.15 *Abbreviated IPv6 addresses*

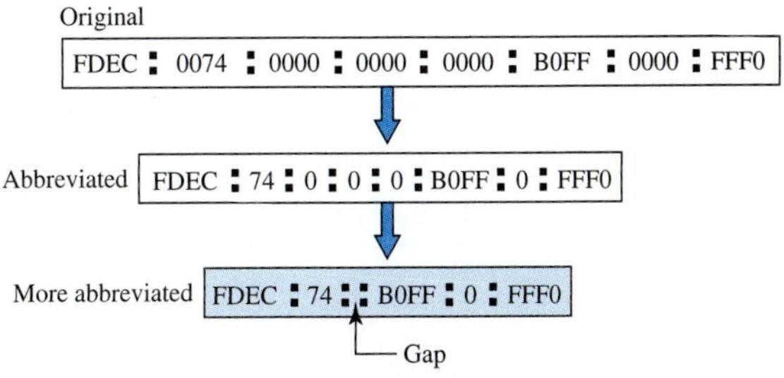

Using this form of **abbreviation,** 0074 can be written as 74, 000F as F, and 0000 as 0. Note that 3210 cannot be abbreviated. Further abbreviations are possible if there are consecutive sections consisting of zeros only. We can remove the zeros altogether and replace them with a double semicolon. Note that this type of abbreviation is allowed only once per address. If there are two runs of zero sections, only one of them can be abbreviated. Reexpansion of the abbreviated address is very simple: Align the unabbreviated portions and insert zeros to get the original expanded address.

Example 19.11

Expand the address 0:15::1:12:1213 to its original.

Solution

We first need to align the left side of the double colon to the left of the original pattern and the right side of the double colon to the right of the original pattern to find how many 0s we need to replace the double colon.

```
xxxx:xxxx:xxxx:xxxx:xxxx:xxxx:xxxx:xxxx
   0:  15:                 :   1:  12:1213
```

This means that the original address is

```
0000:0015:0000:0000:0000:0001:0012:1213
```

Address Space

IPv6 has a much larger address space; 2^{128} addresses are available. The designers of IPv6 divided the address into several categories. A few leftmost bits, called the *type prefix,* in each address define its category. The type prefix is variable in length, but it is designed such that no code is identical to the first part of any other code. In this way, there is no ambiguity; when an address is given, the type prefix can easily be determined. Table 19.5 shows the prefix for each type of address. The third column shows the fraction of each type of address relative to the whole address space.

Table 19.5 *Type prefixes for IPv6 addresses*

Type Prefix	*Type*	*Fraction*
0000 0000	Reserved	1/256
0000 0001	Unassigned	1/256
0000 001	ISO network addresses	1/128
0000 010	IPX (Novell) network addresses	1/128
0000 011	Unassigned	1/128
0000 1	Unassigned	1/32
0001	Reserved	1/16
001	Reserved	1/8
010	**Provider-based unicast addresses**	**1/8**

Table 19.5 *Type prefixes for IPv6 addresses (continued)*

Type Prefix	*Type*	*Fraction*
011	Unassigned	1/8
100	Geographic-based unicast addresses	1/8
101	Unassigned	1/8
110	Unassigned	1/8
1110	Unassigned	1/16
1111 0	Unassigned	1/32
1111 10	Unassigned	1/64
1111 110	Unassigned	1/128
1111 1110 0	Unassigned	1/512
1111 1110 10	Link local addresses	1/1024
1111 1110 11	Site local addresses	1/1024
1111 1111	Multicast addresses	1/256

Unicast Addresses

A **unicast address** defines a single computer. The packet sent to a unicast address must be delivered to that specific computer. IPv6 defines two types of unicast addresses: geographically based and provider-based. We discuss the second type here; the first type is left for future definition. The provider-based address is generally used by a normal host as a unicast address. The address format is shown in Figure 19.16.

Figure 19.16 *Prefixes for provider-based unicast address*

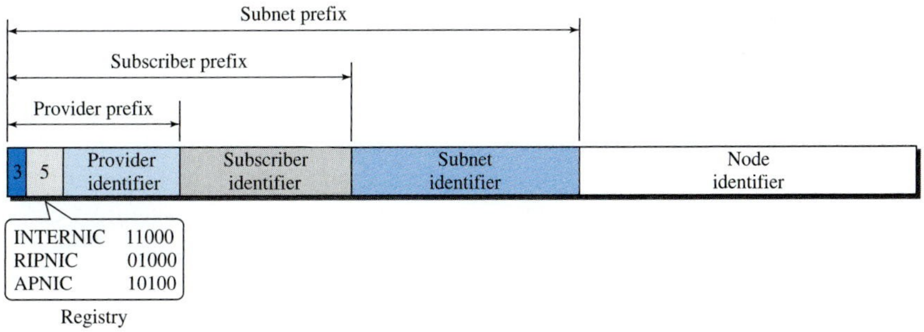

Fields for the provider-based address are as follows:

- **Type identifier.** This 3-bit field defines the address as a provider-based address.
- **Registry identifier.** This 5-bit field indicates the agency that has registered the address. Currently three registry centers have been defined. INTERNIC (code 11000) is the center for North America; RIPNIC (code 01000) is the center for European registration; and APNIC (code 10100) is for Asian and Pacific countries.

- **Provider identifier.** This variable-length field identifies the provider for Internet access (such as an ISP). A 16-bit length is recommended for this field.
- **Subscriber identifier.** When an organization subscribes to the Internet through a provider, it is assigned a subscriber identification. A 24-bit length is recommended for this field.
- **Subnet identifier.** Each subscriber can have many different subnetworks, and each subnetwork can have an identifier. The subnet identifier defines a specific subnetwork under the territory of the subscriber. A 32-bit length is recommended for this field.
- **Node identifier.** The last field defines the identity of the node connected to a subnet. A length of 48 bits is recommended for this field to make it compatible with the 48-bit link (physical) address used by Ethernet. In the future, this link address will probably be the same as the node physical address.

Multicast Addresses

Multicast addresses are used to define a group of hosts instead of just one. A packet sent to a **multicast address** must be delivered to each member of the group. Figure 19.17 shows the format of a multicast address.

Figure 19.17 *Multicast address in IPv6*

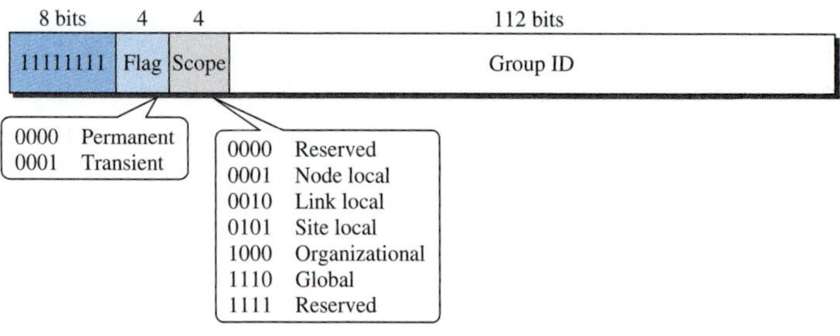

The second field is a flag that defines the group address as either permanent or transient. A permanent group address is defined by the Internet authorities and can be accessed at all times. A transient group address, on the other hand, is used only temporarily. Systems engaged in a teleconference, for example, can use a transient group address. The third field defines the scope of the group address. Many different scopes have been defined, as shown in Figure 19.17.

Anycast Addresses

IPv6 also defines anycast addresses. An **anycast address,** like a multicast address, also defines a group of nodes. However, a packet destined for an anycast address is delivered to only one of the members of the anycast group, the nearest one (the one with the shortest route). Although the definition of an anycast address is still debatable, one possible use is to assign an anycast address to all routers of an ISP that covers a large logical area in the Internet. The routers outside the ISP deliver a packet destined for the ISP to the nearest ISP router. No block is assigned for anycast addresses.

Reserved Addresses

Another category in the address space is the **reserved address.** These addresses start with eight 0s (type prefix is 0000 0000). A few subcategories are defined in this category, as shown in Figure 19.18.

Figure 19.18 *Reserved addresses in IPv6*

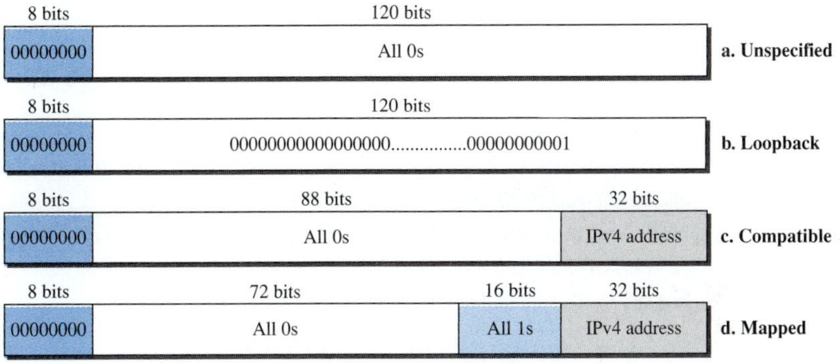

An **unspecified address** is used when a host does not know its own address and sends an inquiry to find its address. A **loopback address** is used by a host to test itself without going into the network. A **compatible address** is used during the transition from IPv4 to IPv6 (see Chapter 20). It is used when a computer using IPv6 wants to send a message to another computer using IPv6, but the message needs to pass through a part of the network that still operates in IPv4. A **mapped address** is also used during transition. However, it is used when a computer that has migrated to IPv6 wants to send a packet to a computer still using IPv4.

Local Addresses

These addresses are used when an organization wants to use IPv6 protocol without being connected to the global Internet. In other words, they provide addressing for private networks. Nobody outside the organization can send a message to the nodes using these addresses. Two types of addresses are defined for this purpose, as shown in Figure 19.19.

Figure 19.19 *Local addresses in IPv6*

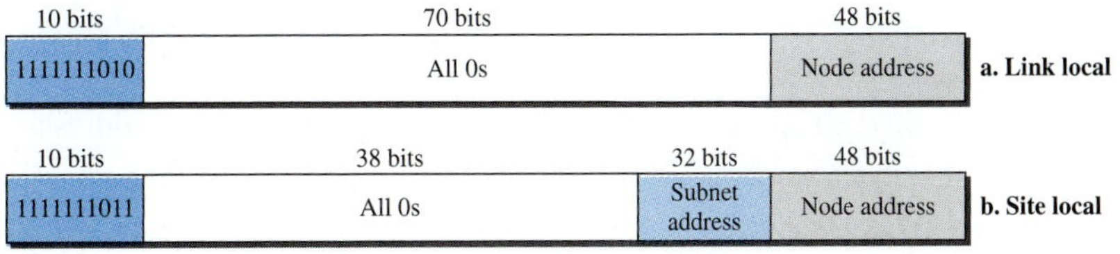

A **link local address** is used in an isolated subnet; a **site local address** is used in an isolated site with several subnets.

19.3 RECOMMENDED READING

For more details about subjects discussed in this chapter, we recommend the following books and sites. The items in brackets [. . .] refer to the reference list at the end of the text.

Books

IPv4 addresses are discussed in Chapters 4 and 5 of [For06], Chapter 3 of [Ste94], Section 4.1 of [PD03], Chapter 18 of [Sta04], and Section 5.6 of [Tan03]. IPv6 addresses are discussed in Section 27.1 of [For06] and Chapter 8 of [Los04]. A good discussion of NAT can be found in [Dut01].

Sites

- www.ietf.org/rfc.html Information about RFCs

RFCs

A discussion of IPv4 addresses can be found in most of the RFCs related to the IPv4 protocol:

760, 781, 791, 815, 1025, 1063, 1071, 1141, 1190, 1191, 1624, 2113

A discussion of IPv6 addresses can be found in most of the RFCs related to IPv6 protocol:

1365, 1550, 1678, 1680, 1682, 1683, 1686, 1688, 1726, 1752, 1826, 1883, 1884, 1886, 1887, 1955, 2080, 2373, 2452, 2463, 2465, 2466, 2472, 2492, 2545, 2590

A discussion of NAT can be found in

1361, 2663, 2694

19.4 KEY TERMS

address aggregation
address space
anycast address
binary notation
class A address
class B address
class C address
class D address
class E address
classful addressing
classless addressing
classless interdomain routing (CIDR)

compatible address
dotted-decimal notation
default mask
hexadecimal colon notation
hostid
IP address
IPv4 address
IPv6 address
link local address
mapped address
mask
multicast address
netid
network address
network address translation (NAT)
prefix
reserved address
site local address
subnet
subnet mask
subnetting
suffix
supernet
supernet mask
supernetting
unicast address
unspecified address

19.5 SUMMARY

- At the network layer, a global identification system that uniquely identifies every host and router is necessary for delivery of a packet from host to host.
- An IPv4 address is 32 bits long and uniquely and universally defines a host or router on the Internet.
- In classful addressing, the portion of the IP address that identifies the network is called the netid.
- In classful addressing, the portion of the IP address that identifies the host or router on the network is called the hostid.
- An IP address defines a device's connection to a network.
- There are five classes in IPv4 addresses. Classes A, B, and C differ in the number of hosts allowed per network. Class D is for multicasting and Class E is reserved.
- The class of an address is easily determined by examination of the first byte.
- Addresses in classes A, B, or C are mostly used for unicast communication.
- Addresses in class D are used for multicast communication.
- Subnetting divides one large network into several smaller ones, adding an intermediate level of hierarchy in IP addressing.
- Supernetting combines several networks into one large one.
- In classless addressing, we can divide the address space into variable-length blocks.
- There are three restrictions in classless addressing:
    a. The number of addresses needs to be a power of 2.
    b. The mask needs to be included in the address to define the block.
    c. The starting address must be divisible by the number of addresses in the block.
- The mask in classless addressing is expressed as the prefix length (/*n*) in CIDR notation.

- To find the first address in a block, we set the rightmost $32 - n$ bits to 0.
- To find the number of addresses in the block, we calculate 2^{32-n}, where n is the prefix length.
- To find the last address in the block, we set the rightmost $32 - n$ bits to 0.
- Subnetting increases the value of n.
- The global authority for address allocation is ICANN. ICANN normally grants large blocks of addresses to ISPs, which in turn grant small subblocks to individual customers.
- IPv6 addresses use hexadecimal colon notation with abbreviation methods available.
- There are three types of addresses in IPv6: unicast, anycast, and multicast.
- In an IPv6 address, the variable type prefix field defines the address type or purpose.

19.6 PRACTICE SET

Review Questions

1. What is the number of bits in an IPv4 address? What is the number of bits in an IPv6 address?
2. What is dotted decimal notation in IPv4 addressing? What is the number of bytes in an IPv4 address represented in dotted decimal notation? What is hexadecimal notation in IPv6 addressing? What is the number of digits in an IPv6 address represented in hexadecimal notation?
3. What are the differences between classful addressing and classless addressing in IPv4?
4. List the classes in classful addressing and define the application of each class (unicast, multicast, broadcast, or reserve).
5. Explain why most of the addresses in class A are wasted. Explain why a medium-size or large-size corporation does not want a block of class C addresses.
6. What is a mask in IPv4 addressing? What is a default mask in IPv4 addressing?
7. What is the network address in a block of addresses? How can we find the network address if one of the addresses in a block is given?
8. Briefly define subnetting and supernetting. How do the subnet mask and supernet mask differ from a default mask in classful addressing?
9. How can we distinguish a multicast address in IPv4 addressing? How can we do so in IPv6 addressing?
10. What is NAT? How can NAT help in address depletion?

Exercises

11. What is the address space in each of the following systems?
    a. A system with 8-bit addresses
    b. A system with 16-bit addresses
    c. A system with 64-bit addresses

12. An address space has a total of 1024 addresses. How many bits are needed to represent an address?

13. An address space uses the three symbols 0, 1, and 2 to represent addresses. If each address is made of 10 symbols, how many addresses are available in this system?

14. Change the following IP addresses from dotted-decimal notation to binary notation.
    a. 114.34.2.8
    b. 129.14.6.8
    c. 208.34.54.12
    d. 238.34.2.1

15. Change the following IP addresses from binary notation to dotted-decimal notation.
    a. 01111111 11110000 01100111 01111101
    b. 10101111 11000000 11111000 00011101
    c. 11011111 10110000 00011111 01011101
    d. 11101111 11110111 11000111 00011101

16. Find the class of the following IP addresses.
    a. 208.34.54.12
    b. 238.34.2.1
    c. 114.34.2.8
    d. 129.14.6.8

17. Find the class of the following IP addresses.
    a. 11110111 11110011 10000111 11011101
    b. 10101111 11000000 11110000 00011101
    c. 11011111 10110000 00011111 01011101
    d. 11101111 11110111 11000111 00011101

18. Find the netid and the hostid of the following IP addresses.
    a. 114.34.2.8
    b. 132.56.8.6
    c. 208.34.54.12

19. In a block of addresses, we know the IP address of one host is 25.34.12.56/16. What are the first address (network address) and the last address (limited broadcast address) in this block?

20. In a block of addresses, we know the IP address of one host is 182.44.82.16/26. What are the first address (network address) and the last address in this block?

21. An organization is granted the block 16.0.0.0/8. The administrator wants to create 500 fixed-length subnets.
    a. Find the subnet mask.
    b. Find the number of addresses in each subnet.
    c. Find the first and last addresses in subnet 1.
    d. Find the first and last addresses in subnet 500.

22. An organization is granted the block 130.56.0.0/16. The administrator wants to create 1024 subnets.
    a. Find the subnet mask.
    b. Find the number of addresses in each subnet.
    c. Find the first and last addresses in subnet 1.
    d. Find the first and last addresses in subnet 1024.
23. An organization is granted the block 211.17.180.0/24. The administrator wants to create 32 subnets.
    a. Find the subnet mask.
    b. Find the number of addresses in each subnet.
    c. Find the first and last addresses in subnet 1.
    d. Find the first and last addresses in subnet 32.
24. Write the following masks in slash notation (*/n*).
    a. 255.255.255.0
    b. 255.0.0.0
    c. 255.255.224.0
    d. 255.255.240.0
25. Find the range of addresses in the following blocks.
    a. 123.56.77.32/29
    b. 200.17.21.128/27
    c. 17.34.16.0/23
    d. 180.34.64.64/30
26. An ISP is granted a block of addresses starting with 150.80.0.0/16. The ISP wants to distribute these blocks to 2600 customers as follows.
    a. The first group has 200 medium-size businesses; each needs 128 addresses.
    b. The second group has 400 small businesses; each needs 16 addresses.
    c. The third group has 2000 households; each needs 4 addresses.

    Design the subblocks and give the slash notation for each subblock. Find out how many addresses are still available after these allocations.
27. An ISP is granted a block of addresses starting with 120.60.4.0/22. The ISP wants to distribute these blocks to 100 organizations with each organization receiving just eight addresses. Design the subblocks and give the slash notation for each subblock. Find out how many addresses are still available after these allocations.
28. An ISP has a block of 1024 addresses. It needs to divide the addresses among 1024 customers. Does it need subnetting? Explain your answer.
29. Show the shortest form of the following addresses.
    a. 2340:1ABC:119A:A000:0000:0000:0000:0000
    b. 0000:00AA:0000:0000:0000:0000:119A:A231
    c. 2340:0000:0000:0000:0000:119A:A001:0000
    d. 0000:0000:0000:2340:0000:0000:0000:0000

30. Show the original (unabbreviated) form of the following addresses.
    a. 0::0
    b. 0:AA::0
    c. 0:1234::3
    d. 123::1:2
31. What is the type of each of the following addresses?
    a. FE80::12
    b. FEC0::24A2
    c. FF02::0
    d. 0::01
32. What is the type of each of the following addresses?
    a. 0::0
    b. 0::FFFF:0:0
    c. 582F:1234::2222
    d. 4821::14:22
    e. 54EF::A234:2
33. Show the provider prefix (in hexadecimal colon notation) of an address assigned to a subscriber if it is registered in the United States with ABC1 as the provider identification.
34. Show in hexadecimal colon notation the IPv6 address
    a. Compatible to the IPv4 address 129.6.12.34
    b. Mapped to the IPv4 address 129.6.12.34
35. Show in hexadecimal colon notation
    a. The link local address in which the node identifier is 0::123/48
    b. The site local address in which the node identifier is 0::123/48
36. Show in hexadecimal colon notation the permanent multicast address used in a link local scope.
37. A host has the address 581E:1456:2314:ABCD::1211. If the node identification is 48 bits, find the address of the subnet to which the host is attached.
38. A site with 200 subnets has the class B address of 132.45.0.0. The site recently migrated to IPv6 with the subscriber prefix 581E:1456:2314::ABCD/80. Design the subnets and define the subnet addresses, using a subnet identifier of 32 bits.

Research Activities

39. Find the block of addresses assigned to your organization or institution.
40. If you are using an ISP to connect from your home to the Internet, find the name of the ISP and the block of addresses assigned to it.
41. Some people argue that we can consider the whole address space as one single block in which each range of addresses is a subblock to this single block. Elaborate on this idea. What happens to subnetting if we accept this concept?
42. Is your school or organization using a classful address? If so, find out the class of the address.

CHAPTER 20

Network Layer: Internet Protocol

In the Internet model, the main network protocol is the **Internet Protocol (IP).** In this chapter, we first discuss internetworking and issues related to the network layer protocol in general.

We then discuss the current version of the Internet Protocol, version 4, or IPv4. This leads us to the next generation of this protocol, or IPv6, which may become the dominant protocol in the near future.

Finally, we discuss the transition strategies from IPv4 to IPv6. Some readers may note the absence of IPv5. IPv5 is an experimental protocol, based mostly on the OSI model that never materialized.

20.1 INTERNETWORKING

The physical and data link layers of a network operate locally. These two layers are jointly responsible for data delivery on the network from one node to the next, as shown in Figure 20.1.

This internetwork is made of five networks: four LANs and one WAN. If host A needs to send a data packet to host D, the packet needs to go first from A to R1 (a switch or router), then from R1 to R3, and finally from R3 to host D. We say that the data packet passes through three links. In each link, two physical and two data link layers are involved.

However, there is a big problem here. When data arrive at interface f1 of R1, how does R1 know that interface f3 is the outgoing interface? There is no provision in the data link (or physical) layer to help R1 make the right decision. The frame does not carry any routing information either. The frame contains the MAC address of A as the source and the MAC address of R1 as the destination. For a LAN or a WAN, delivery means carrying the frame through one link, and not beyond.

Need for Network Layer

To solve the problem of delivery through several links, the network layer (or the internetwork layer, as it is sometimes called) was designed. The network layer is responsible for host-to-host delivery and for routing the packets through the routers or switches. Figure 20.2 shows the same internetwork with a network layer added.

Figure 20.1 *Links between two hosts*

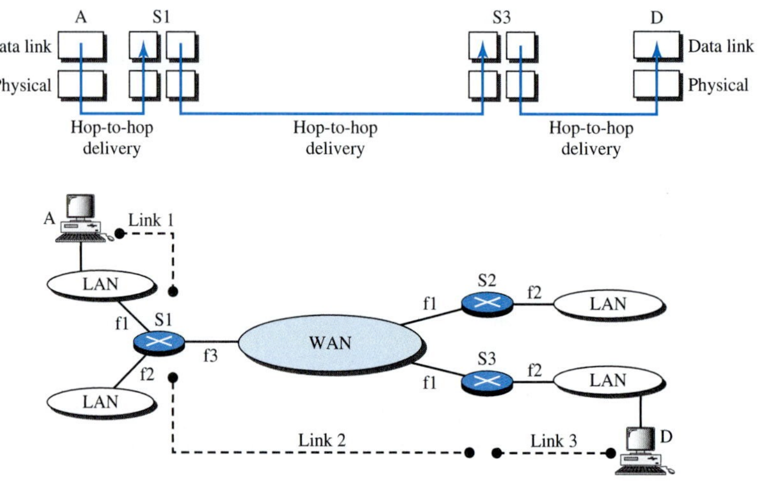

Figure 20.2 *Network layer in an internetwork*

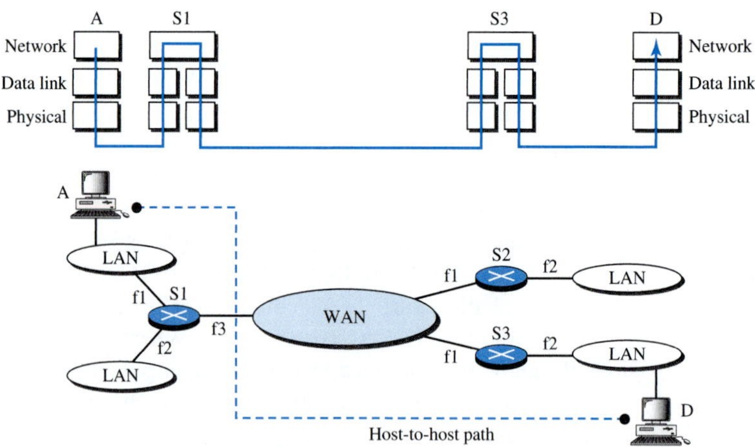

Figure 20.3 shows the general idea of the functionality of the network layer at a source, at a router, and at the destination. The network layer at the source is responsible for creating a packet from the data coming from another protocol (such as a transport layer protocol or a routing protocol). The header of the packet contains, among other information, the logical addresses of the source and destination. The network layer is responsible for checking its routing table to find the routing information (such as the outgoing interface of the packet or the physical address of the next node). If the packet is too large, the packet is fragmented (fragmentation is discussed later in this chapter).

Figure 20.3 *Network layer at the source, router, and destination*

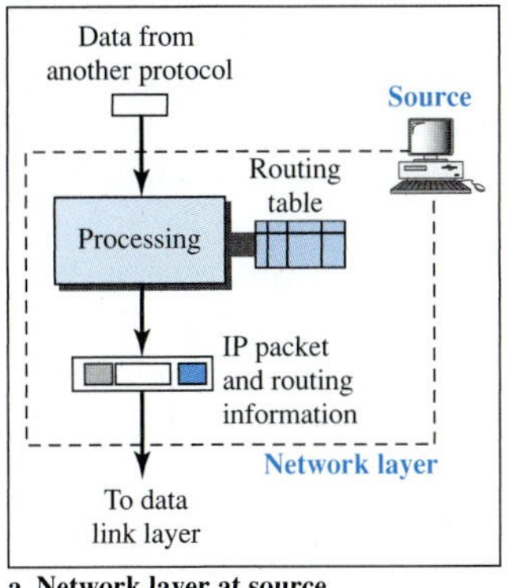

a. Network layer at source

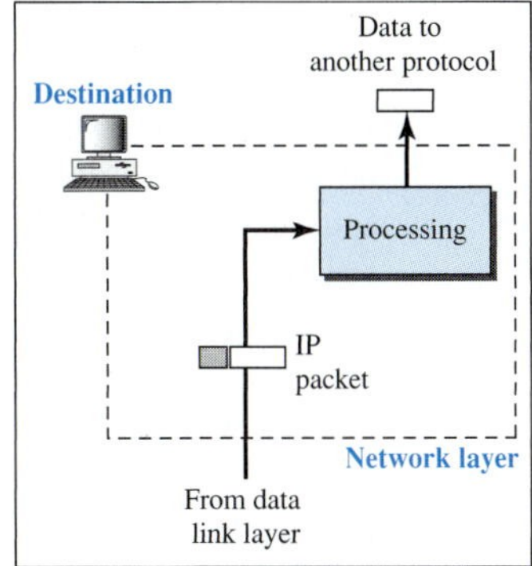

b. Network layer at destination

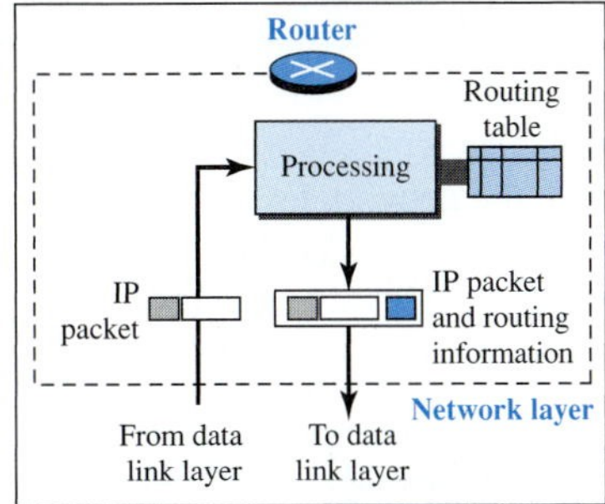

c. Network layer at a router

The network layer at the switch or router is responsible for routing the packet. When a packet arrives, the router or switch consults its routing table and finds the interface from which the packet must be sent. The packet, after some changes in the header, with the routing information is passed to the data link layer again.

The network layer at the destination is responsible for address verification; it makes sure that the destination address on the packet is the same as the address of the host. If the packet is a fragment, the network layer waits until all fragments have arrived, and then reassembles them and delivers the reassembled packet to the transport layer.

Internet as a Datagram Network

The Internet, at the network layer, is a packet-switched network. We discussed switching in Chapter 8. We said that, in general, switching can be divided into three broad categories: circuit switching, packet switching, and message switching. Packet switching uses either the virtual circuit approach or the datagram approach.

The Internet has chosen the datagram approach to switching in the network layer. It uses the universal addresses defined in the network layer to route packets from the source to the destination.

Switching at the network layer in the Internet uses the datagram approach to packet switching.

Internet as a Connectionless Network

Delivery of a packet can be accomplished by using either a connection-oriented or a connectionless network service. In a **connection-oriented service,** the source first makes a connection with the destination before sending a packet. When the connection is established, a sequence of packets from the same source to the same destination can be sent one after another. In this case, there is a relationship between packets. They are sent on the same path in sequential order. A packet is logically connected to the packet traveling before it and to the packet traveling after it. When all packets of a message have been delivered, the connection is terminated.

In a connection-oriented protocol, the decision about the route of a sequence of packets with the same source and destination addresses can be made only once, when the connection is established. Switches do not recalculate the route for each individual packet. This type of service is used in a virtual-circuit approach to packet switching such as in Frame Relay and ATM.

In **connectionless service,** the network layer protocol treats each packet independently, with each packet having no relationship to any other packet. The packets in a message may or may not travel the same path to their destination. This type of service is used in the datagram approach to packet switching. The Internet has chosen this type of service at the network layer.

The reason for this decision is that the Internet is made of so many heterogeneous networks that it is almost impossible to create a connection from the source to the destination without knowing the nature of the networks in advance.

Communication at the network layer in the Internet is connectionless.

20.2 IPv4

The **Internet Protocol version 4 (IPv4)** is the delivery mechanism used by the TCP/IP protocols. Figure 20.4 shows the position of IPv4 in the suite.

Figure 20.4 *Position of IPv4 in TCP/IP protocol suite*

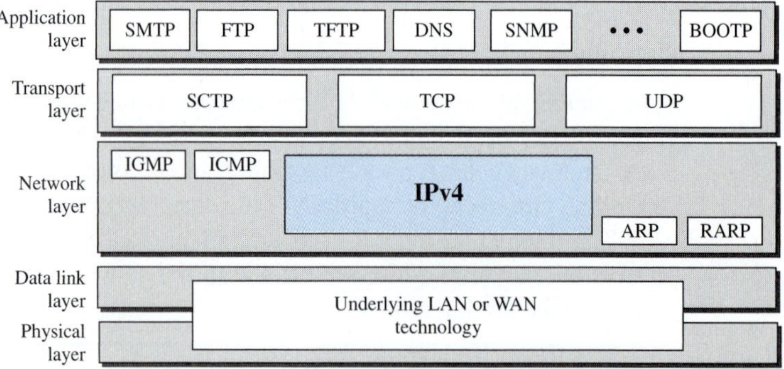

IPv4 is an unreliable and connectionless datagram protocol—a **best-effort delivery** service. The term *best-effort* means that IPv4 provides no error control or flow control (except for error detection on the header). IPv4 assumes the unreliability of the underlying layers and does its best to get a transmission through to its destination, but with no guarantees.

If reliability is important, IPv4 must be paired with a reliable protocol such as TCP. An example of a more commonly understood best-effort delivery service is the post office. The post office does its best to deliver the mail but does not always succeed. If an unregistered letter is lost, it is up to the sender or would-be recipient to discover the loss and rectify the problem. The post office itself does not keep track of every letter and cannot notify a sender of loss or damage.

IPv4 is also a connectionless protocol for a packet-switching network that uses the datagram approach (see Chapter 8). This means that each datagram is handled independently, and each datagram can follow a different route to the destination. This implies that datagrams sent by the same source to the same destination could arrive out of order. Also, some could be lost or corrupted during transmission. Again, IPv4 relies on a higher-level protocol to take care of all these problems.

Datagram

Packets in the IPv4 layer are called **datagrams.** Figure 20.5 shows the IPv4 datagram format.

Figure 20.5 *IPv4 datagram format*

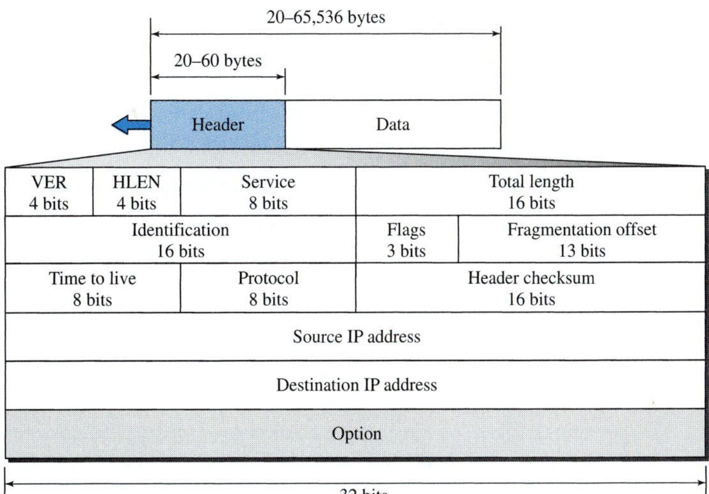

A datagram is a variable-length packet consisting of two parts: header and data. The header is 20 to 60 bytes in length and contains information essential to routing and delivery. It is customary in TCP/IP to show the header in 4-byte sections. A brief description of each field is in order.

- **Version (VER).** This 4-bit field defines the version of the IPv4 protocol. Currently the version is 4. However, version 6 (or IPng) may totally replace version 4 in the future. This field tells the IPv4 software running in the processing machine that the datagram has the format of version 4. All fields must be interpreted as specified in the fourth version of the protocol. If the machine is using some other version of IPv4, the datagram is discarded rather than interpreted incorrectly.
- **Header length (HLEN).** This 4-bit field defines the total length of the datagram header in 4-byte words. This field is needed because the length of the header is variable (between 20 and 60 bytes). When there are no options, the header length is 20 bytes, and the value of this field is 5 ($5 \times 4 = 20$). When the option field is at its maximum size, the value of this field is 15 ($15 \times 4 = 60$).
- **Services.** IETF has changed the interpretation and name of this 8-bit field. This field, previously called **service type,** is now called **differentiated services.** We show both interpretations in Figure 20.6.

Figure 20.6 *Service type or differentiated services*

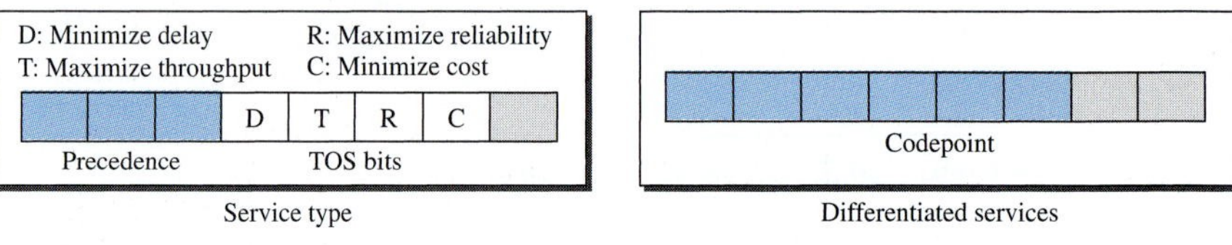

1. **Service Type**

   In this interpretation, the first 3 bits are called precedence bits. The next 4 bits are called **type of service (TOS)** bits, and the last bit is not used.

   a. **Precedence** is a 3-bit subfield ranging from 0 (000 in binary) to 7 (111 in binary). The precedence defines the priority of the datagram in issues such as congestion. If a router is congested and needs to discard some datagrams, those datagrams with lowest precedence are discarded first. Some datagrams in the Internet are more important than others. For example, a datagram used for network management is much more urgent and important than a datagram containing optional information for a group.

The precedence subfield was part of version 4, but never used.

   b. **TOS bits** is a 4-bit subfield with each bit having a special meaning. Although a bit can be either 0 or 1, one and only one of the bits can have the value of 1 in each datagram. The bit patterns and their interpretations are given in Table 20.1. With only 1 bit set at a time, we can have five different types of services.

Table 20.1 *Types of service*

TOS Bits	*Description*
0000	Normal (default)
0001	Minimize cost
0010	Maximize reliability
0100	Maximize throughput
1000	Minimize delay

Application programs can request a specific type of service. The defaults for some applications are shown in Table 20.2.

Table 20.2 *Default types of service*

Protocol	*TOS Bits*	*Description*
ICMP	0000	Normal
BOOTP	0000	Normal
NNTP	0001	Minimize cost
IGP	0010	Maximize reliability
SNMP	0010	Maximize reliability
TELNET	1000	Minimize delay
FTP (data)	0100	Maximize throughput
FTP (control)	1000	Minimize delay
TFTP	1000	Minimize delay
SMTP (command)	1000	Minimize delay
SMTP (data)	0100	Maximize throughput
DNS (UDP query)	1000	Minimize delay
DNS (TCP query)	0000	Normal
DNS (zone)	0100	Maximize throughput

It is clear from Table 20.2 that interactive activities, activities requiring immediate attention, and activities requiring immediate response need minimum delay. Those activities that send bulk data require maximum throughput. Management activities need maximum reliability. Background activities need minimum cost.

2. **Differentiated Services**
   In this interpretation, the first 6 bits make up the **codepoint** subfield, and the last 2 bits are not used. The codepoint subfield can be used in two different ways.
   a. When the 3 rightmost bits are 0s, the 3 leftmost bits are interpreted the same as the precedence bits in the service type interpretation. In other words, it is compatible with the old interpretation.

b. When the 3 rightmost bits are not all 0s, the 6 bits define 64 services based on the priority assignment by the Internet or local authorities according to Table 20.3. The first category contains 32 service types; the second and the third each contain 16. The first category (numbers 0, 2, 4, . . . , 62) is assigned by the Internet authorities (IETF). The second category (3, 7, 11, 15, . . . , 63) can be used by local authorities (organizations). The third category (1, 5, 9, . . . , 61) is temporary and can be used for experimental purposes. Note that the numbers are not contiguous. If they were, the first category would range from 0 to 31, the second from 32 to 47, and the third from 48 to 63. This would be incompatible with the TOS interpretation because XXX000 (which includes 0, 8, 16, 24, 32, 40, 48, and 56) would fall into all three categories. Instead, in this assignment method all these services belong to category 1. Note that these assignments have not yet been finalized.

Table 20.3 *Values for codepoints*

Category	*Codepoint*	*Assigning Authority*
1	XXXXX0	Internet
2	XXXX11	Local
3	XXXX01	Temporary or experimental

- **Total length.** This is a 16-bit field that defines the total length (header plus data) of the IPv4 datagram in bytes. To find the length of the data coming from the upper layer, subtract the header length from the total length. The header length can be found by multiplying the value in the HLEN field by 4.

Length of data = total length – header length

Since the field length is 16 bits, the total length of the IPv4 datagram is limited to 65,535 ($2^{16} - 1$) bytes, of which 20 to 60 bytes are the header and the rest is data from the upper layer.

The total length field defines the total length of the datagram including the header.

Though a size of 65,535 bytes might seem large, the size of the IPv4 datagram may increase in the near future as the underlying technologies allow even more throughput (greater bandwidth).

When we discuss fragmentation in the next section, we will see that some physical networks are not able to encapsulate a datagram of 65,535 bytes in their frames. The datagram must be fragmented to be able to pass through those networks.

One may ask why we need this field anyway. When a machine (router or host) receives a frame, it drops the header and the trailer, leaving the datagram. Why include an extra field that is not needed? The answer is that in many cases we really do not need the value in this field. However, there are occasions in which the datagram is not the only thing encapsulated in a frame; it may be that padding has been added. For example, the Ethernet protocol has a minimum and maximum restriction on the size of data that can be encapsulated in a frame (46 to 1500 bytes). If the size of an IPv4 datagram is less than 46 bytes, some padding will be added to meet this requirement. In this case, when a machine decapsulates the datagram, it needs to check the total length field to determine how much is really data and how much is padding (see Figure 20.7).

Figure 20.7 *Encapsulation of a small datagram in an Ethernet frame*

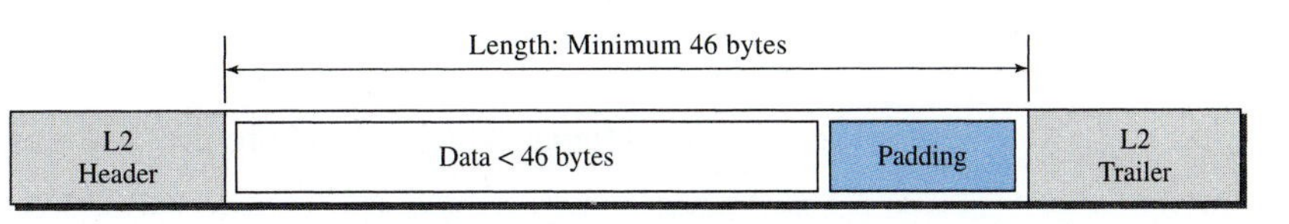

- **Identification.** This field is used in fragmentation (discussed in the next section).
- **Flags.** This field is used in fragmentation (discussed in the next section).
- **Fragmentation offset.** This field is used in fragmentation (discussed in the next section).
- **Time to live.** A datagram has a limited lifetime in its travel through an internet. This field was originally designed to hold a timestamp, which was decremented by each visited router. The datagram was discarded when the value became zero. However, for this scheme, all the machines must have synchronized clocks and must know how long it takes for a datagram to go from one machine to another. Today, this field is used mostly to control the maximum number of hops (routers) visited by the datagram. When a source host sends the datagram, it stores a number in this field. This value is approximately 2 times the maximum number of routes between any two hosts. Each router that processes the datagram decrements this number by 1. If this value, after being decremented, is zero, the router discards the datagram.

  This field is needed because routing tables in the Internet can become corrupted. A datagram may travel between two or more routers for a long time without ever getting delivered to the destination host. This field limits the lifetime of a datagram.

  Another use of this field is to intentionally limit the journey of the packet. For example, if the source wants to confine the packet to the local network, it can store 1 in this field. When the packet arrives at the first router, this value is decremented to 0, and the datagram is discarded.
- **Protocol.** This 8-bit field defines the higher-level protocol that uses the services of the IPv4 layer. An IPv4 datagram can encapsulate data from several higher-level protocols such as TCP, UDP, ICMP, and IGMP. This field specifies the final destination protocol to which the IPv4 datagram is delivered. In other words, since the IPv4 protocol carries data from different other protocols, the value of this field helps the receiving network layer know to which protocol the data belong (see Figure 20.8).

Figure 20.8 *Protocol field and encapsulated data*

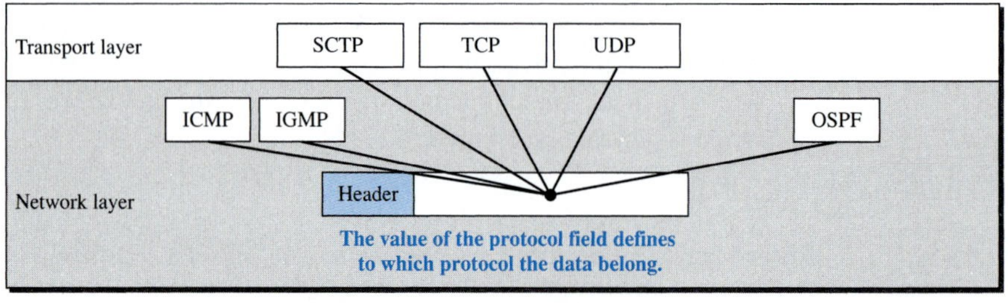

The value of this field for each higher-level protocol is shown in Table 20.4.

Table 20.4 *Protocol values*

Value	*Protocol*
1	ICMP
2	IGMP
6	TCP
17	UDP
89	OSPF

- **Checksum.** The checksum concept and its calculation are discussed later in this chapter.
- **Source address.** This 32-bit field defines the IPv4 address of the source. This field must remain unchanged during the time the IPv4 datagram travels from the source host to the destination host.
- **Destination address.** This 32-bit field defines the IPv4 address of the destination. This field must remain unchanged during the time the IPv4 datagram travels from the source host to the destination host.

Example 20.1

An IPv4 packet has arrived with the first 8 bits as shown:

01000010

The receiver discards the packet. Why?

Solution

There is an error in this packet. The 4 leftmost bits (0100) show the version, which is correct. The next 4 bits (0010) show an invalid header length ($2 \times 4 = 8$). The minimum number of bytes in the header must be 20. The packet has been corrupted in transmission.

Example 20.2

In an IPv4 packet, the value of HLEN is 1000 in binary. How many bytes of options are being carried by this packet?

Solution

The HLEN value is 8, which means the total number of bytes in the header is 8×4, or 32 bytes. The first 20 bytes are the base header, the next 12 bytes are the options.

Example 20.3

In an IPv4 packet, the value of HLEN is 5, and the value of the total length field is 0x0028. How many bytes of data are being carried by this packet?

Solution

The HLEN value is 5, which means the total number of bytes in the header is 5×4, or 20 bytes (no options). The total length is 40 bytes, which means the packet is carrying 20 bytes of data $(40 - 20)$.

Example 20.4

An IPv4 packet has arrived with the first few hexadecimal digits as shown.

0x45000028000100000102 . . .

How many hops can this packet travel before being dropped? The data belong to what upper-layer protocol?

Solution

To find the time-to-live field, we skip 8 bytes (16 hexadecimal digits). The time-to-live field is the ninth byte, which is 01. This means the packet can travel only one hop. The protocol field is the next byte (02), which means that the upper-layer protocol is IGMP (see Table 20.4).

Fragmentation

A datagram can travel through different networks. Each router decapsulates the IPv4 datagram from the frame it receives, processes it, and then encapsulates it in another frame. The format and size of the received frame depend on the protocol used by the physical network through which the frame has just traveled. The format and size of the sent frame depend on the protocol used by the physical network through which the frame is going to travel. For example, if a router connects a LAN to a WAN, it receives a frame in the LAN format and sends a frame in the WAN format.

Maximum Transfer Unit (MTU)

Each data link layer protocol has its own frame format in most protocols. One of the fields defined in the format is the maximum size of the data field. In other words, when a datagram is encapsulated in a frame, the total size of the datagram must be less than this maximum size, which is defined by the restrictions imposed by the hardware and software used in the network (see Figure 20.9).

The value of the MTU depends on the physical network protocol. Table 20.5 shows the values for some protocols.

Figure 20.9 *Maximum transfer unit (MTU)*

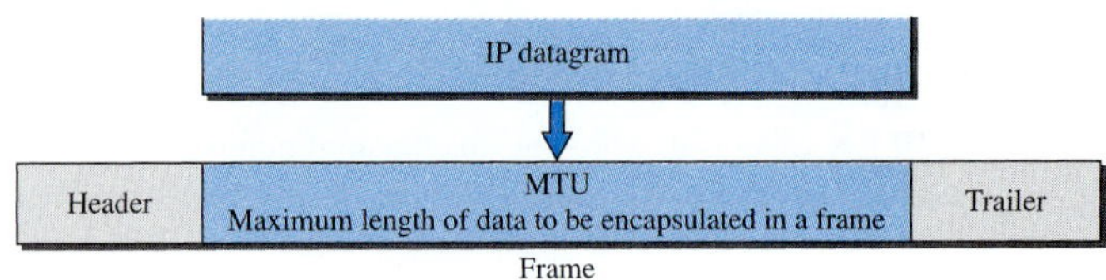

Table 20.5 *MTUs for some networks*

Protocol	*MTU*
Hyperchannel	65,535
Token Ring (16 Mbps)	17,914
Token Ring (4 Mbps)	4,464
FDDI	4,352
Ethernet	1,500
X.25	576
PPP	296

To make the IPv4 protocol independent of the physical network, the designers decided to make the maximum length of the IPv4 datagram equal to 65,535 bytes. This makes transmission more efficient if we use a protocol with an MTU of this size. However, for other physical networks, we must divide the datagram to make it possible to pass through these networks. This is called **fragmentation.**

The source usually does not fragment the IPv4 packet. The transport layer will instead segment the data into a size that can be accommodated by IPv4 and the data link layer in use.

When a datagram is fragmented, each fragment has its own header with most of the fields repeated, but with some changed. A fragmented datagram may itself be fragmented if it encounters a network with an even smaller MTU. In other words, a datagram can be fragmented several times before it reaches the final destination.

In IPv4, a datagram can be fragmented by the source host or any router in the path although there is a tendency to limit fragmentation only at the source. The reassembly of the datagram, however, is done only by the destination host because each fragment becomes an independent datagram. Whereas the fragmented datagram can travel through different routes, and we can never control or guarantee which route a fragmented datagram may take, all the fragments belonging to the same datagram should finally arrive at the destination host. So it is logical to do the reassembly at the final destination. An even stronger objection to reassembling packets during the transmission is the loss of efficiency it incurs.

When a datagram is fragmented, required parts of the header must be copied by all fragments. The option field may or may not be copied, as we will see in the next section. The host or router that fragments a datagram must change the values of three fields:

flags, fragmentation offset, and total length. The rest of the fields must be copied. Of course, the value of the checksum must be recalculated regardless of fragmentation.

Fields Related to Fragmentation

The fields that are related to fragmentation and reassembly of an IPv4 datagram are the identification, flags, and fragmentation offset fields.

- **Identification.** This 16-bit field identifies a datagram originating from the source host. The combination of the identification and source IPv4 address must uniquely define a datagram as it leaves the source host. To guarantee uniqueness, the IPv4 protocol uses a counter to label the datagrams. The counter is initialized to a positive number. When the IPv4 protocol sends a datagram, it copies the current value of the counter to the identification field and increments the counter by 1. As long as the counter is kept in the main memory, uniqueness is guaranteed. When a datagram is fragmented, the value in the identification field is copied to all fragments. In other words, all fragments have the same identification number, the same as the original datagram. The identification number helps the destination in reassembling the datagram. It knows that all fragments having the same identification value must be assembled into one datagram.
- **Flags.** This is a 3-bit field. The first bit is reserved. The second bit is called the *do not fragment* bit. If its value is 1, the machine must not fragment the datagram. If it cannot pass the datagram through any available physical network, it discards the datagram and sends an ICMP error message to the source host (see Chapter 21). If its value is 0, the datagram can be fragmented if necessary. The third bit is called the *more fragment* bit. If its value is 1, it means the datagram is not the last fragment; there are more fragments after this one. If its value is 0, it means this is the last or only fragment (see Figure 20.10).

Figure 20.10 *Flags used in fragmentation*

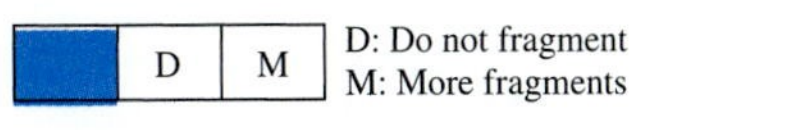

- **Fragmentation offset.** This 13-bit field shows the relative position of this fragment with respect to the whole datagram. It is the offset of the data in the original datagram measured in units of 8 bytes. Figure 20.11 shows a datagram with a data size of 4000 bytes fragmented into three fragments.

  The bytes in the original datagram are numbered 0 to 3999. The first fragment carries bytes 0 to 1399. The offset for this datagram is 0/8 = 0. The second fragment carries bytes 1400 to 2799; the offset value for this fragment is 1400/8 = 175. Finally, the third fragment carries bytes 2800 to 3999. The offset value for this fragment is 2800/8 = 350.

  Remember that the value of the offset is measured in units of 8 bytes. This is done because the length of the offset field is only 13 bits and cannot represent a

Figure 20.11 *Fragmentation example*

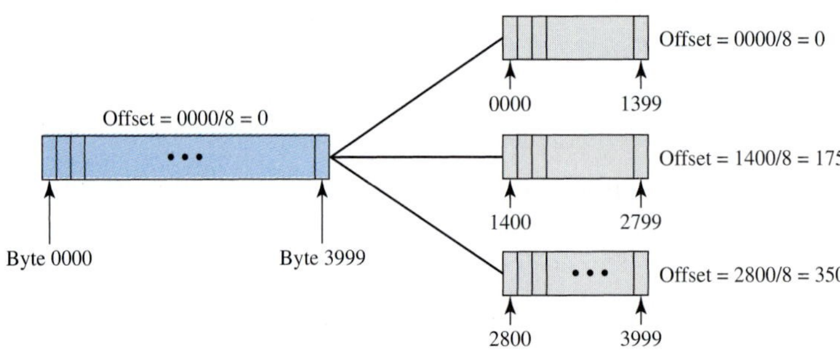

sequence of bytes greater than 8191. This forces hosts or routers that fragment datagrams to choose a fragment size so that the first byte number is divisible by 8.

Figure 20.12 shows an expanded view of the fragments in Figure 20.11. Notice the value of the identification field is the same in all fragments. Notice the value of the flags field with the *more* bit set for all fragments except the last. Also, the value of the offset field for each fragment is shown.

Figure 20.12 *Detailed fragmentation example*

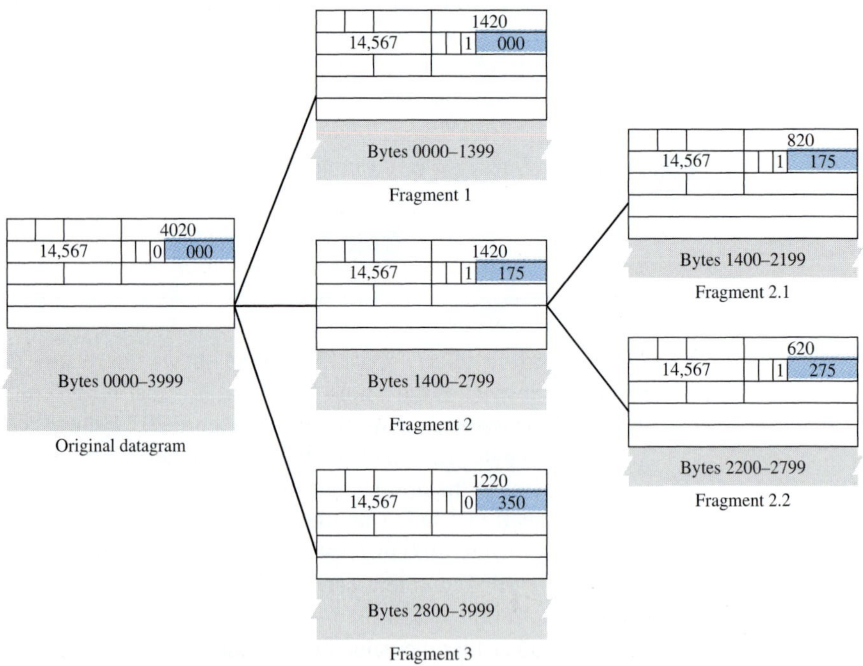

The figure also shows what happens if a fragment itself is fragmented. In this case the value of the offset field is always relative to the original datagram. For example, in the figure, the second fragment is itself fragmented later to two fragments of 800 bytes and 600 bytes, but the offset shows the relative position of the fragments to the original data.

It is obvious that even if each fragment follows a different path and arrives out of order, the final destination host can reassemble the original datagram from the fragments received (if none of them is lost) by using the following strategy:

1. The first fragment has an offset field value of zero.
2. Divide the length of the first fragment by 8. The second fragment has an offset value equal to that result.
3. Divide the total length of the first and second fragments by 8. The third fragment has an offset value equal to that result.
4. Continue the process. The last fragment has a *more* bit value of 0.

Example 20.5

A packet has arrived with an *M* bit value of 0. Is this the first fragment, the last fragment, or a middle fragment? Do we know if the packet was fragmented?

Solution

If the *M* bit is 0, it means that there are no more fragments; the fragment is the last one. However, we cannot say if the original packet was fragmented or not. A nonfragmented packet is considered the last fragment.

Example 20.6

A packet has arrived with an *M* bit value of 1. Is this the first fragment, the last fragment, or a middle fragment? Do we know if the packet was fragmented?

Solution

If the *M* bit is 1, it means that there is at least one more fragment. This fragment can be the first one or a middle one, but not the last one. We don't know if it is the first one or a middle one; we need more information (the value of the fragmentation offset). See Example 20.7.

Example 20.7

A packet has arrived with an *M* bit value of 1 and a fragmentation offset value of 0. Is this the first fragment, the last fragment, or a middle fragment?

Solution

Because the *M* bit is 1, it is either the first fragment or a middle one. Because the offset value is 0, it is the first fragment.

Example 20.8

A packet has arrived in which the offset value is 100. What is the number of the first byte? Do we know the number of the last byte?

Solution

To find the number of the first byte, we multiply the offset value by 8. This means that the first byte number is 800. We cannot determine the number of the last byte unless we know the length of the data.

Example 20.9

A packet has arrived in which the offset value is 100, the value of HLEN is 5, and the value of the total length field is 100. What are the numbers of the first byte and the last byte?

Solution

The first byte number is $100 \times 8 = 800$. The total length is 100 bytes, and the header length is 20 bytes (5×4), which means that there are 80 bytes in this datagram. If the first byte number is 800, the last byte number must be 879.

Checksum

We discussed the general idea behind the checksum and how it is calculated in Chapter 10. The implementation of the checksum in the IPv4 packet follows the same principles. First, the value of the checksum field is set to 0. Then the entire header is divided into 16-bit sections and added together. The result (sum) is complemented and inserted into the checksum field.

The checksum in the IPv4 packet covers only the header, not the data. There are two good reasons for this. First, all higher-level protocols that encapsulate data in the IPv4 datagram have a checksum field that covers the whole packet. Therefore, the checksum for the IPv4 datagram does not have to check the encapsulated data. Second, the header of the IPv4 packet changes with each visited router, but the data do not. So the checksum includes only the part that has changed. If the data were included, each router must recalculate the checksum for the whole packet, which means an increase in processing time.

Example 20.10

Figure 20.13 shows an example of a checksum calculation for an IPv4 header without options. The header is divided into 16-bit sections. All the sections are added and the sum is complemented. The result is inserted in the checksum field.

Options

The header of the IPv4 datagram is made of two parts: a fixed part and a variable part. The fixed part is 20 bytes long and was discussed in the previous section. The variable part comprises the options that can be a maximum of 40 bytes.

Options, as the name implies, are not required for a datagram. They can be used for network testing and debugging. Although options are not a required part of the IPv4 header, option processing is required of the IPv4 software. This means that all implementations must be able to handle options if they are present in the header.

The detailed discussion of each option is beyond the scope of this book. We give the taxonomy of options in Figure 20.14 and briefly explain the purpose of each.

No Operation

A **no-operation option** is a 1-byte option used as a filler between options.

Figure 20.13 *Example of checksum calculation in IPv4*

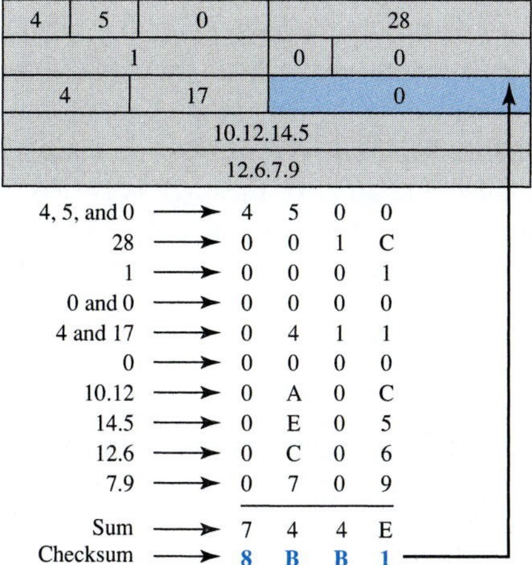

Figure 20.14 *Taxonomy of options in IPv4*

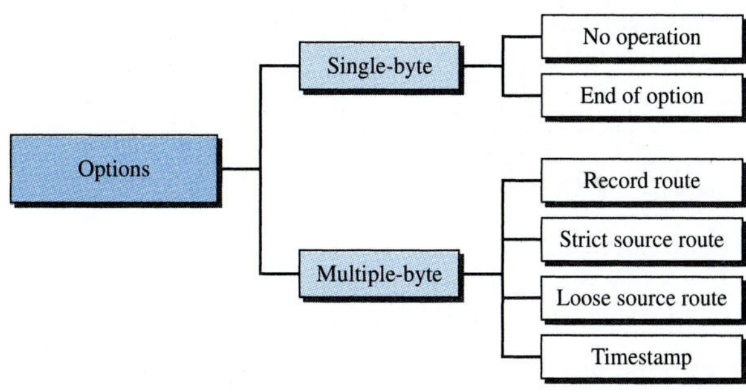

End of Option

An **end-of-option option** is a 1-byte option used for padding at the end of the option field. It, however, can only be used as the last option.

Record Route

A **record route option** is used to record the Internet routers that handle the datagram. It can list up to nine router addresses. It can be used for debugging and management purposes.

Strict Source Route

A **strict source route option** is used by the source to predetermine a route for the datagram as it travels through the Internet. Dictation of a route by the source can be useful for several purposes. The sender can choose a route with a specific type of service, such as minimum delay or maximum throughput. Alternatively, it may choose a route that is safer or more reliable for the sender's purpose. For example, a sender can choose a route so that its datagram does not travel through a competitor's network.

If a datagram specifies a strict source route, all the routers defined in the option must be visited by the datagram. A router must not be visited if its IPv4 address is not listed in the datagram. If the datagram visits a router that is not on the list, the datagram is discarded and an error message is issued. If the datagram arrives at the destination and some of the entries were not visited, it will also be discarded and an error message issued.

Loose Source Route

A **loose source route option** is similar to the strict source route, but it is less rigid. Each router in the list must be visited, but the datagram can visit other routers as well.

Timestamp

A **timestamp option** is used to record the time of datagram processing by a router. The time is expressed in milliseconds from midnight, Universal time or Greenwich mean time. Knowing the time a datagram is processed can help users and managers track the behavior of the routers in the Internet. We can estimate the time it takes for a datagram to go from one router to another. We say *estimate* because, although all routers may use Universal time, their local clocks may not be synchronized.

20.3 IPv6

The network layer protocol in the TCP/IP protocol suite is currently IPv4 (Internetworking Protocol, version 4). IPv4 provides the host-to-host communication between systems in the Internet. Although IPv4 is well designed, data communication has evolved since the inception of IPv4 in the 1970s. IPv4 has some deficiencies (listed below) that make it unsuitable for the fast-growing Internet.

- Despite all short-term solutions, such as subnetting, classless addressing, and NAT, address depletion is still a long-term problem in the Internet.
- The Internet must accommodate real-time audio and video transmission. This type of transmission requires minimum delay strategies and reservation of resources not provided in the IPv4 design.
- The Internet must accommodate encryption and authentication of data for some applications. No encryption or authentication is provided by IPv4.

To overcome these deficiencies, **IPv6 (Internetworking Protocol, version 6),** also known as **IPng (Internetworking Protocol, next generation),** was proposed and is now a standard. In IPv6, the Internet protocol was extensively modified to accommodate the unforeseen growth of the Internet. The format and the length of the IP address were changed along with the packet format. Related protocols, such as ICMP, were also modified. Other protocols in the network layer, such as ARP, RARP, and IGMP, were either deleted or included in the ICMPv6 protocol (see Chapter 21). Routing protocols, such as RIP and OSPF (see Chapter 22), were also slightly modified to accommodate these changes. Communications experts predict that IPv6 and its related protocols will soon replace the current IP version. In this section first we discuss IPv6. Then we explore the strategies used for the transition from version 4 to version 6.

The adoption of IPv6 has been slow. The reason is that the original motivation for its development, depletion of IPv4 addresses, has been remedied by short-term strategies such as classless addressing and NAT. However, the fast-spreading use of the Internet, and new services such as mobile IP, IP telephony, and IP-capable mobile telephony, may eventually require the total replacement of IPv4 with IPv6.

Advantages

The next-generation IP, or IPv6, has some advantages over IPv4 that can be summarized as follows:

- **Larger address space.** An IPv6 address is 128 bits long, as we discussed in Chapter 19. Compared with the 32-bit address of IPv4, this is a huge (2^{96}) increase in the address space.
- **Better header format.** IPv6 uses a new header format in which options are separated from the base header and inserted, when needed, between the base header and the upper-layer data. This simplifies and speeds up the routing process because most of the options do not need to be checked by routers.
- **New options.** IPv6 has new options to allow for additional functionalities.
- **Allowance for extension.** IPv6 is designed to allow the extension of the protocol if required by new technologies or applications.
- **Support for resource allocation.** In IPv6, the type-of-service field has been removed, but a mechanism (called *flow label*) has been added to enable the source to request special handling of the packet. This mechanism can be used to support traffic such as real-time audio and video.
- **Support for more security.** The encryption and authentication options in IPv6 provide confidentiality and integrity of the packet.

Packet Format

The IPv6 packet is shown in Figure 20.15. Each packet is composed of a mandatory base header followed by the payload. The payload consists of two parts: optional extension headers and data from an upper layer. The base header occupies 40 bytes, whereas the extension headers and data from the upper layer contain up to 65,535 bytes of information.

Base Header

Figure 20.16 shows the **base header** with its eight fields.

These fields are as follows:

- **Version.** This 4-bit field defines the version number of the IP. For IPv6, the value is 6.
- **Priority.** The 4-bit priority field defines the priority of the packet with respect to traffic congestion. We will discuss this field later.

Figure 20.15 *IPv6 datagram header and payload*

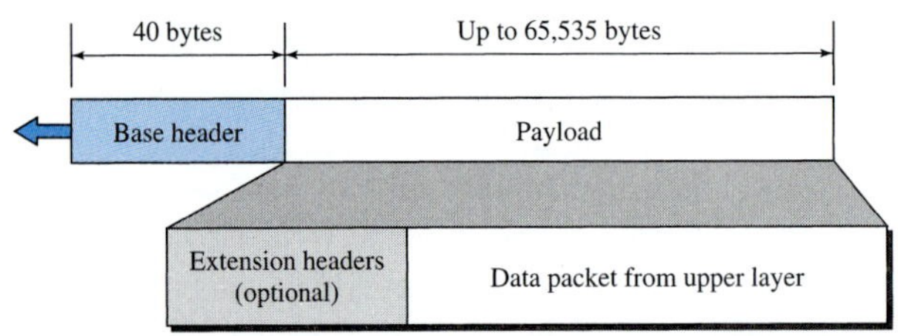

Figure 20.16 *Format of an IPv6 datagram*

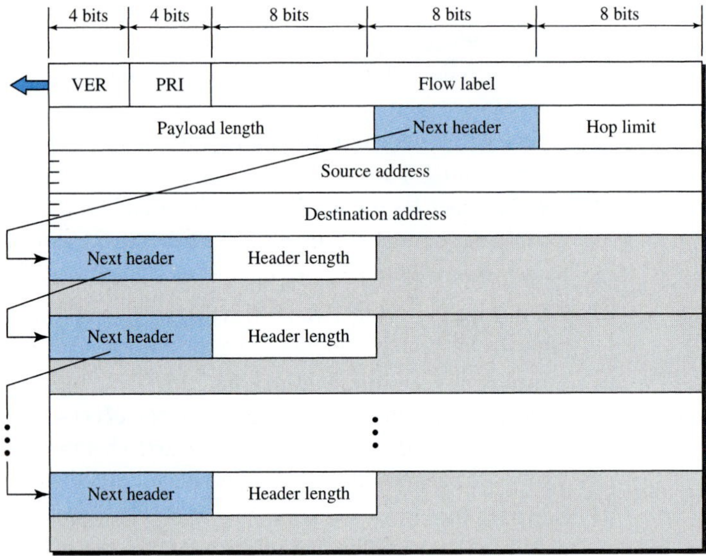

- **Flow label.** The **flow label** is a 3-byte (24-bit) field that is designed to provide special handling for a particular flow of data. We will discuss this field later.
- **Payload length.** The 2-byte payload length field defines the length of the IP datagram excluding the base header.
- **Next header.** The **next header** is an 8-bit field defining the header that follows the base header in the datagram. The next header is either one of the optional extension headers used by IP or the header of an encapsulated packet such as UDP or TCP. Each extension header also contains this field. Table 20.6 shows the values of next headers. Note that this field in version 4 is called the *protocol*.
- **Hop limit.** This 8-bit **hop limit** field serves the same purpose as the TTL field in IPv4.
- **Source address.** The source address field is a 16-byte (128-bit) Internet address that identifies the original source of the datagram.

Table 20.6 *Next header codes for IPv6*

Code	*Next Header*
0	Hop-by-hop option
2	ICMP
6	TCP
17	UDP
43	Source routing
44	Fragmentation
50	Encrypted security payload
51	Authentication
59	Null (no next header)
60	Destination option

- **Destination address.** The destination address field is a 16-byte (128-bit) Internet address that usually identifies the final destination of the datagram. However, if source routing is used, this field contains the address of the next router.

Priority

The priority field of the IPv6 packet defines the priority of each packet with respect to other packets from the same source. For example, if one of two consecutive datagrams must be discarded due to congestion, the datagram with the lower **packet priority** will be discarded. IPv6 divides traffic into two broad categories: congestion-controlled and noncongestion-controlled.

Congestion-Controlled Traffic If a source adapts itself to traffic slowdown when there is congestion, the traffic is referred to as **congestion-controlled traffic.** For example, TCP, which uses the sliding window protocol, can easily respond to traffic. In congestion-controlled traffic, it is understood that packets may arrive delayed, lost, or out of order. Congestion-controlled data are assigned priorities from 0 to 7, as listed in Table 20.7. A priority of 0 is the lowest; a priority of 7 is the highest.

Table 20.7 *Priorities for congestion-controlled traffic*

Priority	*Meaning*
0	No specific traffic
1	Background data
2	Unattended data traffic
3	Reserved
4	Attended bulk data traffic
5	Reserved
6	Interactive traffic
7	Control traffic

The priority descriptions are as follows:

- **No specific traffic.** A priority of 0 is assigned to a packet when the process does not define a priority.
- **Background data.** This group (priority 1) defines data that are usually delivered in the background. Delivery of the news is a good example.
- **Unattended data traffic.** If the user is not waiting (attending) for the data to be received, the packet will be given a priority of 2. E-mail belongs to this group. The recipient of an e-mail does not know when a message has arrived. In addition, an e-mail is usually stored before it is forwarded. A little bit of delay is of little consequence.
- **Attended bulk data traffic.** A protocol that transfers data while the user is waiting (attending) to receive the data (possibly with delay) is given a priority of 4. FTP and HTTP belong to this group.
- **Interactive traffic.** Protocols such as TELNET that need user interaction are assigned the second-highest priority (6) in this group.
- **Control traffic.** Control traffic is given the highest priority (7). Routing protocols such as OSPF and RIP and management protocols such as SNMP have this priority.

Noncongestion-Controlled Traffic This refers to a type of traffic that expects minimum delay. Discarding of packets is not desirable. Retransmission in most cases is impossible. In other words, the source does not adapt itself to congestion. Real-time audio and video are examples of this type of traffic.

Priority numbers from 8 to 15 are assigned to **noncongestion-controlled traffic.** Although there are not yet any particular standard assignments for this type of data, the priorities are usually based on how much the quality of received data is affected by the discarding of packets. Data containing less redundancy (such as low-fidelity audio or video) can be given a higher priority (15). Data containing more redundancy (such as high-fidelity audio or video) are given a lower priority (8). See Table 20.8.

Table 20.8 *Priorities for noncongestion-controlled traffic*

Priority	*Meaning*
8	Data with greatest redundancy
. . .	. . .
15	Data with least redundancy

Flow Label

A sequence of packets, sent from a particular source to a particular destination, that needs special handling by routers is called a *flow* of packets. The combination of the source address and the value of the *flow label* uniquely defines a flow of packets.

To a router, a flow is a sequence of packets that share the same characteristics, such as traveling the same path, using the same resources, having the same kind of security, and so on. A router that supports the handling of flow labels has a flow label table. The table has an entry for each active flow label; each entry defines the services required by the corresponding flow label. When the router receives a packet, it consults its flow label table to find the corresponding entry for the flow label value defined in the packet. It then provides the packet with the services mentioned in the entry. However, note that the flow label itself does not provide the information for the entries of the flow label table; the information is provided by other means such as the hop-by-hop options or other protocols.

In its simplest form, a flow label can be used to speed up the processing of a packet by a router. When a router receives a packet, instead of consulting the routing table and going through a routing algorithm to define the address of the next hop, it can easily look in a flow label table for the next hop.

In its more sophisticated form, a flow label can be used to support the transmission of real-time audio and video. Real-time audio or video, particularly in digital form, requires resources such as high bandwidth, large buffers, long processing time, and so on. A process can make a reservation for these resources beforehand to guarantee that real-time data will not be delayed due to a lack of resources. The use of real-time data and the reservation of these resources require other protocols such as Real-Time Protocol (RTP) and Resource Reservation Protocol (RSVP) in addition to IPv6.

To allow the effective use of flow labels, three rules have been defined:

1. The flow label is assigned to a packet by the source host. The label is a random number between 1 and $2^{24} - 1$. A source must not reuse a flow label for a new flow while the existing flow is still active.
2. If a host does not support the flow label, it sets this field to zero. If a router does not support the flow label, it simply ignores it.
3. All packets belonging to the same flow have the same source, same destination, same priority, and same options.

Comparison Between IPv4 and IPv6 Headers

Table 20.9 compares IPv4 and IPv6 headers.

Table 20.9 *Comparison between IPv4 and IPv6 packet headers*

Comparison
1. The header length field is eliminated in IPv6 because the length of the header is fixed in this version.
2. The service type field is eliminated in IPv6. The priority and flow label fields together take over the function of the service type field.
3. The total length field is eliminated in IPv6 and replaced by the payload length field.
4. The identification, flag, and offset fields are eliminated from the base header in IPv6. They are included in the fragmentation extension header.
5. The TTL field is called hop limit in IPv6.
6. The protocol field is replaced by the next header field.
7. The header checksum is eliminated because the checksum is provided by upper-layer protocols; it is therefore not needed at this level.
8. The option fields in IPv4 are implemented as extension headers in IPv6.

Extension Headers

The length of the base header is fixed at 40 bytes. However, to give greater functionality to the IP datagram, the base header can be followed by up to six **extension headers.** Many of these headers are options in IPv4. Six types of extension headers have been defined, as shown in Figure 20.17.

Figure 20.17 *Extension header types*

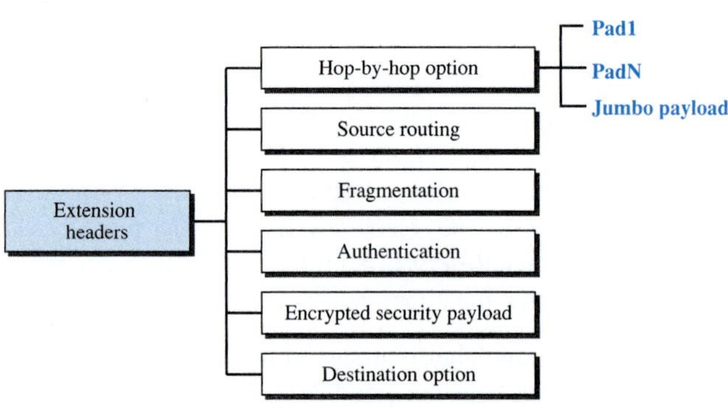

Hop-by-Hop Option

The **hop-by-hop option** is used when the source needs to pass information to all routers visited by the datagram. So far, only three options have been defined: **Pad1, PadN,** and **jumbo payload.** The Pad1 option is 1 byte long and is designed for alignment purposes. PadN is similar in concept to Pad1. The difference is that PadN is used when 2 or more bytes is needed for alignment. The jumbo payload option is used to define a payload longer than 65,535 bytes.

Source Routing The source routing extension header combines the concepts of the strict source route and the loose source route options of IPv4.

Fragmentation

The concept of **fragmentation** is the same as that in IPv4. However, the place where fragmentation occurs differs. In IPv4, the source or a router is required to fragment if the size of the datagram is larger than the MTU of the network over which the datagram travels. In IPv6, only the original source can fragment. A source must use a **path MTU discovery technique** to find the smallest MTU supported by any network on the path. The source then fragments using this knowledge.

Authentication

The **authentication** extension header has a dual purpose: it validates the message sender and ensures the integrity of data. We discuss this extension header when we discuss network security in Chapter 31.

Encrypted Security Payload

The **encrypted security payload (ESP)** is an extension that provides confidentiality and guards against eavesdropping. We discuss this extension header in Chapter 31.

Destination Option The **destination option** is used when the source needs to pass information to the destination only. Intermediate routers are not permitted access to this information.

Comparison Between IPv4 Options and IPv6 Extension Headers

Table 20.10 compares the options in IPv4 with the extension headers in IPv6.

Table 20.10 *Comparison between IPv4 options and IPv6 extension headers*

Comparison
1. The no-operation and end-of-option options in IPv4 are replaced by Pad1 and PadN options in IPv6.
2. The record route option is not implemented in IPv6 because it was not used.
3. The timestamp option is not implemented because it was not used.
4. The source route option is called the source route extension header in IPv6.
5. The fragmentation fields in the base header section of IPv4 have moved to the fragmentation extension header in IPv6.
6. The authentication extension header is new in IPv6.
7. The encrypted security payload extension header is new in IPv6.

20.4 TRANSITION FROM IPv4 TO IPv6

Because of the huge number of systems on the Internet, the transition from IPv4 to IPv6 cannot happen suddenly. It takes a considerable amount of time before every system in the Internet can move from IPv4 to IPv6. The transition must be smooth to prevent any problems between IPv4 and IPv6 systems. Three strategies have been devised by the IETF to help the transition (see Figure 20.18).

Figure 20.18 *Three transition strategies*

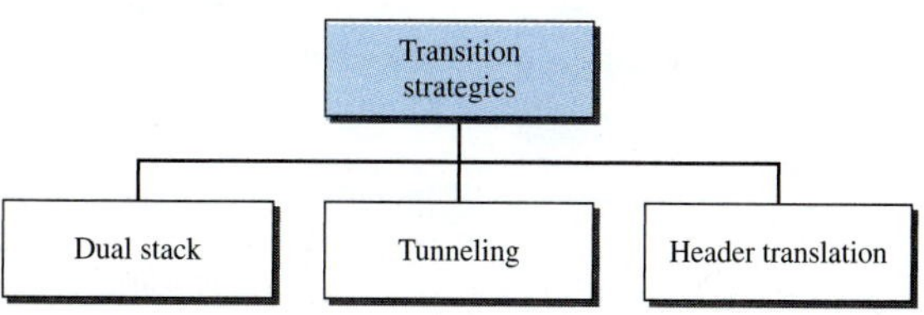

Dual Stack

It is recommended that all hosts, before migrating completely to version 6, have a **dual stack** of protocols. In other words, a station must run IPv4 and IPv6 simultaneously until all the Internet uses IPv6. See Figure 20.19 for the layout of a dual-stack configuration.

Figure 20.19 *Dual stack*

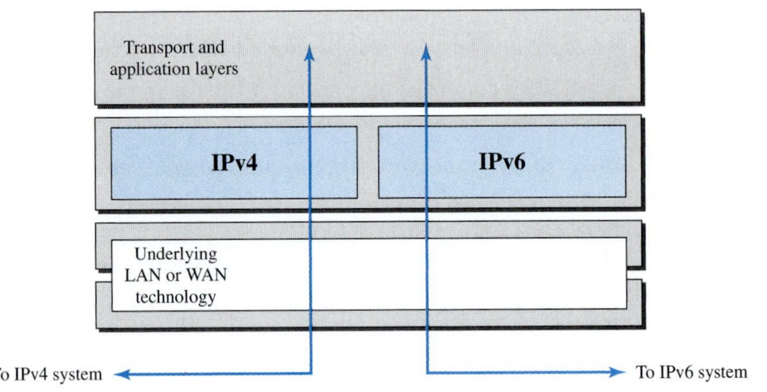

To determine which version to use when sending a packet to a destination, the source host queries the DNS. If the DNS returns an IPv4 address, the source host sends an IPv4 packet. If the DNS returns an IPv6 address, the source host sends an IPv6 packet.

Tunneling

Tunneling is a strategy used when two computers using IPv6 want to communicate with each other and the packet must pass through a region that uses IPv4. To pass through this region, the packet must have an IPv4 address. So the IPv6 packet is encapsulated in an IPv4 packet when it enters the region, and it leaves its capsule when it exits the region. It seems as if the IPv6 packet goes through a tunnel at one end and emerges at the other end. To make it clear that the IPv4 packet is carrying an IPv6 packet as data, the protocol value is set to 41. Tunneling is shown in Figure 20.20.

Figure 20.20 *Tunneling strategy*

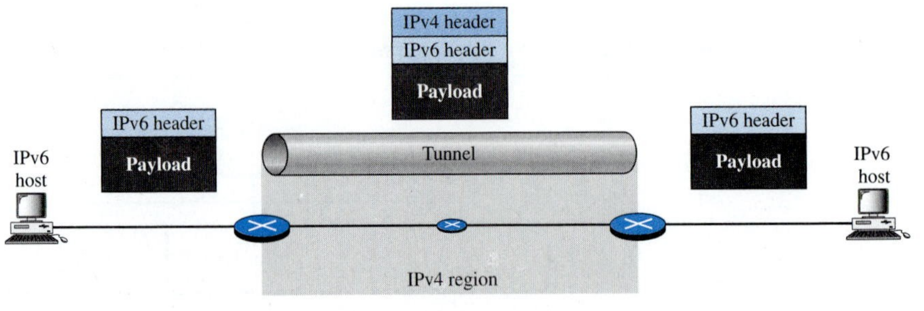

Header Translation

Header translation is necessary when the majority of the Internet has moved to IPv6 but some systems still use IPv4. The sender wants to use IPv6, but the receiver does not understand IPv6. Tunneling does not work in this situation because the packet must be in the IPv4 format to be understood by the receiver. In this case, the header format must be totally changed through header translation. The header of the IPv6 packet is converted to an IPv4 header (see Figure 20.21).

Figure 20.21 *Header translation strategy*

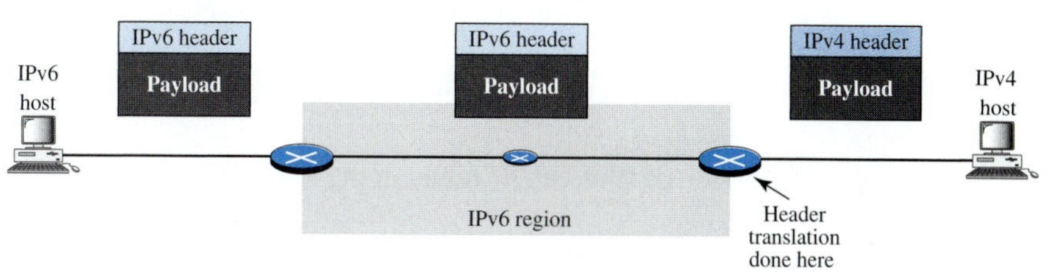

Header translation uses the mapped address to translate an IPv6 address to an IPv4 address. Table 20.11 lists some rules used in transforming an IPv6 packet header to an IPv4 packet header.

Table 20.11 *Header translation*

Header Translation Procedure
1. The IPv6 mapped address is changed to an IPv4 address by extracting the rightmost 32 bits.
2. The value of the IPv6 priority field is discarded.
3. The type of service field in IPv4 is set to zero.
4. The checksum for IPv4 is calculated and inserted in the corresponding field.
5. The IPv6 flow label is ignored.
6. Compatible extension headers are converted to options and inserted in the IPv4 header. Some may have to be dropped.
7. The length of IPv4 header is calculated and inserted into the corresponding field.
8. The total length of the IPv4 packet is calculated and inserted in the corresponding field.

20.5 RECOMMENDED READING

For more details about subjects discussed in this chapter, we recommend the following books and sites. The items in brackets [. . .] refer to the reference list at the end of the text.

Books

IPv4 is discussed in Chapter 8 of [For06], Chapter 3 of [Ste94], Section 4.1 of [PD03], Chapter 18 of [Sta04], and Section 5.6 of [Tan03]. IPv6 is discussed in Chapter 27 of [For06] and [Los04].

Sites

- www.ietf.org/rfc.html Information about RFCs

RFCs

A discussion of IPv4 can be found in following RFCs:

760, 781, 791, 815, 1025, 1063, 1071, 1141, 1190, 1191, 1624, 2113

A discussion of IPv6 can be found in the following RFCs:

1365, 1550, 1678, 1680, 1682, 1683, 1686, 1688, 1726, 1752, 1826, 1883, 1884, 1886, 1887, 1955, 2080, 2373, 2452, 2463, 2465, 2466, 2472, 2492, 2545, 2590

20.6 KEY TERMS

authentication
base header
best-effort delivery
codepoint
connectionless service
connection-oriented service
datagram
destination address
destination option
differentiated services
dual stack
encrypted security payload (ESP)
end-of-option option
extension header
flow label
fragmentation
fragmentation offset
header length
header translation
hop limit
hop-by-hop option
Internet Protocol (IP)
Internet Protocol, next generation (IPng)
Internet Protocol version 4 (IPv4)
Internet Protocol version 6 (IPv6)
jumbo payload
loose source route option
maximum transfer unit (MTU)
next header
noncongestion-controlled traffic
no-operation option
packet priority
Pad1
PadN
path MTU discovery technique
precedence

record route option
service type
source address
strict source route option
time to live
timestamp option
tunneling
type of service (TOS)

20.7 SUMMARY

- IPv4 is an unreliable connectionless protocol responsible for source-to-destination delivery.
- Packets in the IPv4 layer are called datagrams. A datagram consists of a header (20 to 60 bytes) and data. The maximum length of a datagram is 65,535 bytes.
- The MTU is the maximum number of bytes that a data link protocol can encapsulate. MTUs vary from protocol to protocol.
- Fragmentation is the division of a datagram into smaller units to accommodate the MTU of a data link protocol.
- The IPv4 datagram header consists of a fixed, 20-byte section and a variable options section with a maximum of 40 bytes.
- The options section of the IPv4 header is used for network testing and debugging.
- The six IPv4 options each have a specific function. They are as follows: filler between options for alignment purposes, padding, recording the route the datagram takes, selection of a mandatory route by the sender, selection of certain routers that must be visited, and recording of processing times at routers.
- IPv6, the latest version of the Internet Protocol, has a 128-bit address space, a revised header format, new options, an allowance for extension, support for resource allocation, and increased security measures.
- An IPv6 datagram is composed of a base header and a payload.
- Extension headers add functionality to the IPv6 datagram.
- Three strategies used to handle the transition from version 4 to version 6 are dual stack, tunneling, and header translation.

20.8 PRACTICE SET

Review Questions

1. What is the difference between the delivery of a frame in the data link layer and the delivery of a packet in the network layer?
2. What is the difference between connectionless and connection-oriented services? Which type of service is provided by IPv4? Which type of service is provided by IPv6?
3. Define fragmentation and explain why the IPv4 and IPv6 protocols need to fragment some packets. Is there any difference between the two protocols in this matter?

4. Explain the procedure for checksum calculation and verification in the IPv4 protocol. What part of an IPv4 packet is covered in the checksum calculation? Why? Are options, if present, included in the calculation?
5. Explain the need for options in IPv4 and list the options mentioned in this chapter with a brief description of each.
6. Compare and contrast the fields in the main headers of IPv4 and IPv6. Make a table that shows the presence or absence of each field.
7. Both IPv4 and IPv6 assume that packets may have different priorities or precedences. Explain how each protocol handles this issue.
8. Compare and contrast the options in IPv4 and the extension headers in IPv6. Make a table that shows the presence or absence of each.
9. Explain the reason for the elimination of the checksum in the IPv6 header.
10. List three transition strategies to move from IPv4 to IPv6. Explain the difference between tunneling and dual stack strategies during the transition period. When is each strategy used?

Exercises

11. Which fields of the IPv4 header change from router to router?
12. Calculate the HLEN (in IPv4) value if the total length is 1200 bytes, 1176 of which is data from the upper layer.
13. Table 20.5 lists the MTUs for many different protocols. The MTUs range from 296 to 65,535. What would be the advantages of having a large MTU? What would be the advantages of having a small MTU?
14. Given a fragmented datagram (in IPv4) with an offset of 120, how can you determine the first and last byte numbers?
15. Can the value of the header length in an IPv4 packet be less than 5? When is it exactly 5?
16. The value of HLEN in an IPv4 datagram is 7. How many option bytes are present?
17. The size of the option field of an IPv4 datagram is 20 bytes. What is the value of HLEN? What is the value in binary?
18. The value of the total length field in an IPv4 datagram is 36, and the value of the header length field is 5. How many bytes of data is the packet carrying?
19. An IPv4 datagram is carrying 1024 bytes of data. If there is no option information, what is the value of the header length field? What is the value of the total length field?
20. A host is sending 100 datagrams to another host. If the identification number of the first datagram is 1024, what is the identification number of the last (in IPv4)?
21. An IPv4 datagram arrives with fragmentation offset of 0 and an *M* bit (*more* fragment bit) of 0. Is this a first fragment, middle fragment, or last fragment?
22. An IPv4 fragment has arrived with an offset value of 100. How many bytes of data were originally sent by the source before the data in this fragment?

23. An IPv4 datagram has arrived with the following information in the header (in hexadecimal):

    0x45 00 00 54 00 03 58 50 20 06 00 00 7C 4E 03 02 B4 0E 0F 02

    a. Is the packet corrupted?
    b. Are there any options?
    c. Is the packet fragmented?
    d. What is the size of the data?
    e. How many more routers can the packet travel to?
    f. What is the identification number of the packet?
    g. What is the type of service?

24. In an IPv4 datagram, the *M* bit is 0, the value of HLEN is 5, the value of total length is 200, and the offset value is 200. What is the number of the first byte and number of the last byte in this datagram? Is this the last fragment, the first fragment, or a middle fragment?

Research Activities

25. Find out why there are two security protocols (AH and ESP) in IPv6.

CHAPTER 21

Network Layer: Address Mapping, Error Reporting, and Multicasting

In Chapter 20 we discussed the Internet Protocol (IP) as the main protocol at the network layer. IP was designed as a best-effort delivery protocol, but it lacks some features such as flow control and error control. It is a host-to-host protocol using logical addressing. To make IP more responsive to some requirements in today's internetworking, we need the help of other protocols.

We need protocols to create a mapping between physical and logical addresses. IP packets use logical (host-to-host) addresses. These packets, however, need to be encapsulated in a frame, which needs physical addresses (node-to-node). We will see that a protocol called **ARP,** the **Address Resolution Protocol,** is designed for this purpose. We sometimes need reverse mapping—mapping a physical address to a logical address. For example, when booting a diskless network or leasing an IP address to a host. Three protocols are designed for this purpose: RARP, BOOTP, and DHCP.

Lack of flow and error control in the Internet Protocol has resulted in another protocol, ICMP, that provides alerts. It reports congestion and some types of errors in the network or destination host.

IP was originally designed for unicast delivery, one source to one destination. As the Internet has evolved, the need for multicast delivery, one source to many destinations, has increased tremendously. IGMP gives IP a multicast capability.

In this chapter, we discuss the protocols ARP, RARP, BOOTP, DHCP, and IGMP in some detail. We also discuss ICMPv6, which will be operational when IPv6 is operational. ICMPv6 combines ARP, ICMP, and IGMP in one protocol.

21.1 ADDRESS MAPPING

An internet is made of a combination of physical networks connected by internetworking devices such as routers. A packet starting from a source host may pass through several different physical networks before finally reaching the destination host. The hosts and routers are recognized at the network level by their logical (IP) addresses.

However, packets pass through physical networks to reach these hosts and routers. At the physical level, the hosts and routers are recognized by their physical addresses.

A **physical address** is a local address. Its jurisdiction is a local network. It must be unique locally, but is not necessarily unique universally. It is called a *physical* address because it is usually (but not always) implemented in hardware. An example of a physical address is the 48-bit MAC address in the Ethernet protocol, which is imprinted on the NIC installed in the host or router.

The physical address and the logical address are two different identifiers. We need both because a physical network such as Ethernet can have two different protocols at the network layer such as IP and IPX (Novell) at the same time. Likewise, a packet at a network layer such as IP may pass through different physical networks such as Ethernet and LocalTalk (Apple).

This means that delivery of a packet to a host or a router requires two levels of addressing: logical and physical. We need to be able to map a logical address to its corresponding physical address and vice versa. These can be done by using either static or dynamic mapping.

Static mapping involves in the creation of a table that associates a logical address with a physical address. This table is stored in each machine on the network. Each machine that knows, for example, the IP address of another machine but not its physical address can look it up in the table. This has some limitations because physical addresses may change in the following ways:

1. A machine could change its NIC, resulting in a new physical address.
2. In some LANs, such as LocalTalk, the physical address changes every time the computer is turned on.
3. A mobile computer can move from one physical network to another, resulting in a change in its physical address.

To implement these changes, a static mapping table must be updated periodically. This overhead could affect network performance.

In **dynamic mapping** each time a machine knows one of the two addresses (logical or physical), it can use a protocol to find the other one.

Mapping Logical to Physical Address: ARP

Anytime a host or a router has an IP datagram to send to another host or router, it has the logical (IP) address of the receiver. The logical (IP) address is obtained from the DNS (see Chapter 25) if the sender is the host or it is found in a routing table (see Chapter 22) if the sender is a router. But the IP datagram must be encapsulated in a frame to be able to pass through the physical network. This means that the sender needs the physical address of the receiver. The host or the router sends an ARP query packet. The packet includes the physical and IP addresses of the sender and the IP address of the receiver. Because the sender does not know the physical address of the receiver, the query is broadcast over the network (see Figure 21.1).

Every host or router on the network receives and processes the ARP query packet, but only the intended recipient recognizes its IP address and sends back an ARP response packet. The response packet contains the recipient's IP and physical addresses. The packet is unicast directly to the inquirer by using the physical address received in the query packet.

Figure 21.1 *ARP operation*

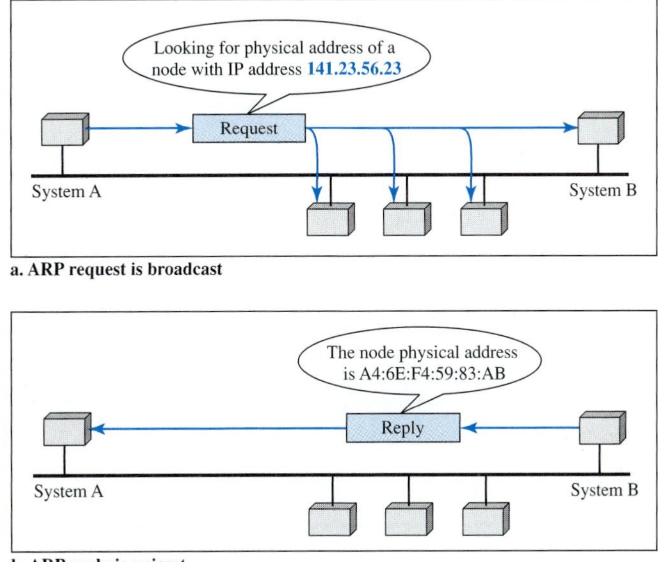

In Figure 21.1a, the system on the left (A) has a packet that needs to be delivered to another system (B) with IP address 141.23.56.23. System A needs to pass the packet to its data link layer for the actual delivery, but it does not know the physical address of the recipient. It uses the services of ARP by asking the ARP protocol to send a broadcast ARP request packet to ask for the physical address of a system with an IP address of 141.23.56.23.

This packet is received by every system on the physical network, but only system B will answer it, as shown in Figure 21.1b. System B sends an ARP reply packet that includes its physical address. Now system A can send all the packets it has for this destination by using the physical address it received.

Cache Memory

Using ARP is inefficient if system A needs to broadcast an ARP request for each IP packet it needs to send to system B. It could have broadcast the IP packet itself. ARP can be useful if the ARP reply is cached (kept in cache memory for a while) because a system normally sends several packets to the same destination. A system that receives an ARP reply stores the mapping in the cache memory and keeps it for 20 to 30 minutes unless the space in the cache is exhausted. Before sending an ARP request, the system first checks its cache to see if it can find the mapping.

Packet Format

Figure 21.2 shows the format of an ARP packet.

Figure 21.2 *ARP packet*

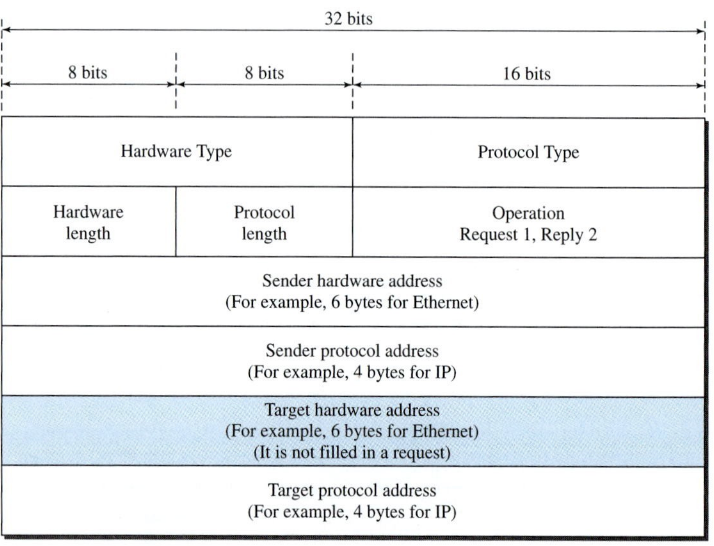

The fields are as follows:

- **Hardware type.** This is a 16-bit field defining the type of the network on which ARP is running. Each LAN has been assigned an integer based on its type. For example, Ethernet is given type 1. ARP can be used on any physical network.
- **Protocol type.** This is a 16-bit field defining the protocol. For example, the value of this field for the IPv4 protocol is 0800_{16}. ARP can be used with any higher-level protocol.
- **Hardware length.** This is an 8-bit field defining the length of the physical address in bytes. For example, for Ethernet the value is 6.
- **Protocol length.** This is an 8-bit field defining the length of the logical address in bytes. For example, for the IPv4 protocol the value is 4.
- **Operation.** This is a 16-bit field defining the type of packet. Two packet types are defined: ARP request (1) and ARP reply (2).
- **Sender hardware address.** This is a variable-length field defining the physical address of the sender. For example, for Ethernet this field is 6 bytes long.
- **Sender protocol address.** This is a variable-length field defining the logical (for example, IP) address of the sender. For the IP protocol, this field is 4 bytes long.
- **Target hardware address.** This is a variable-length field defining the physical address of the target. For example, for Ethernet this field is 6 bytes long. For an ARP request message, this field is all 0s because the sender does not know the physical address of the target.
- **Target protocol address.** This is a variable-length field defining the logical (for example, IP) address of the target. For the IPv4 protocol, this field is 4 bytes long.

Encapsulation

An ARP packet is encapsulated directly into a data link frame. For example, in Figure 21.3 an ARP packet is encapsulated in an Ethernet frame. Note that the type field indicates that the data carried by the frame are an ARP packet.

Figure 21.3 *Encapsulation of ARP packet*

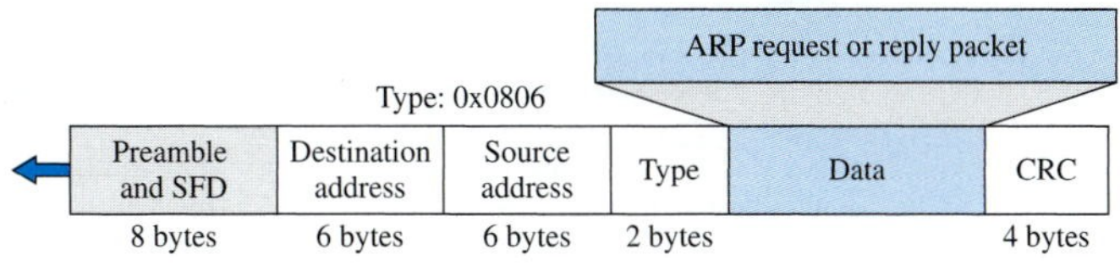

Operation

Let us see how ARP functions on a typical internet. First we describe the steps involved. Then we discuss the four cases in which a host or router needs to use ARP. These are the steps involved in an ARP process:

1. The sender knows the IP address of the target. We will see how the sender obtains this shortly.
2. IP asks ARP to create an ARP request message, filling in the sender physical address, the sender IP address, and the target IP address. The target physical address field is filled with 0s.
3. The message is passed to the data link layer where it is encapsulated in a frame by using the physical address of the sender as the source address and the physical broadcast address as the destination address.
4. Every host or router receives the frame. Because the frame contains a broadcast destination address, all stations remove the message and pass it to ARP. All machines except the one targeted drop the packet. The target machine recognizes its IP address.
5. The target machine replies with an ARP reply message that contains its physical address. The message is unicast.
6. The sender receives the reply message. It now knows the physical address of the target machine.
7. The IP datagram, which carries data for the target machine, is now encapsulated in a frame and is unicast to the destination.

Four Different Cases

The following are four different cases in which the services of ARP can be used (see Figure 21.4).

1. The sender is a host and wants to send a packet to another host on the same network. In this case, the logical address that must be mapped to a physical address is the destination IP address in the datagram header.

Figure 21.4 *Four cases using ARP*

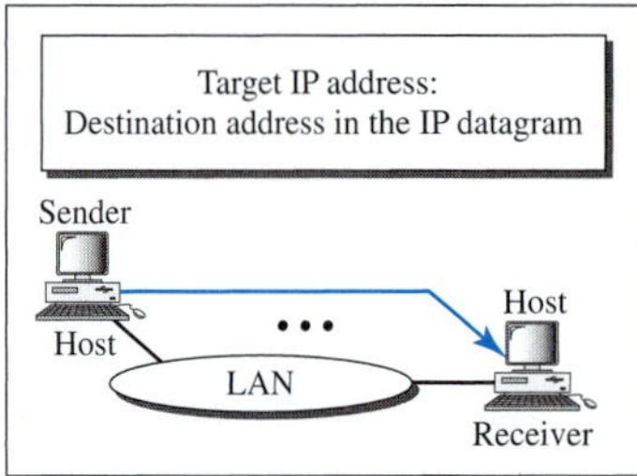

Case 1. A host has a packet to send to another host on the same network.

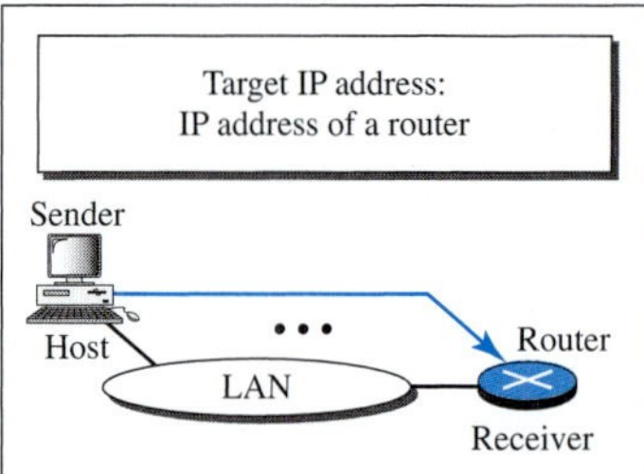

Case 2. A host wants to send a packet to another host on another network. It must first be delivered to a router.

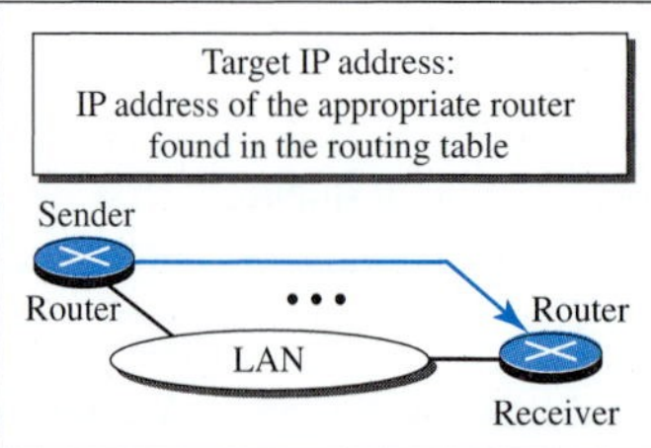

Case 3. A router receives a packet to be sent to a host on another network. It must first be delivered to the appropriate router.

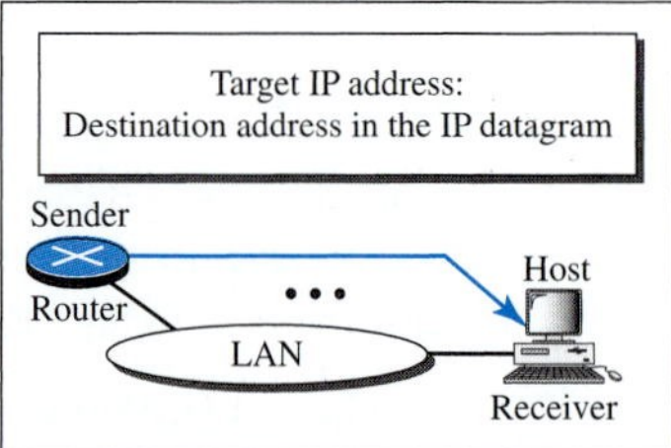

Case 4. A router receives a packet to be sent to a host on the same network.

2. The sender is a host and wants to send a packet to another host on another network. In this case, the host looks at its routing table and finds the IP address of the next hop (router) for this destination. If it does not have a routing table, it looks for the IP address of the default router. The IP address of the router becomes the logical address that must be mapped to a physical address.
3. The sender is a router that has received a datagram destined for a host on another network. It checks its routing table and finds the IP address of the next router. The IP address of the next router becomes the logical address that must be mapped to a physical address.
4. The sender is a router that has received a datagram destined for a host on the same network. The destination IP address of the datagram becomes the logical address that must be mapped to a physical address.

An ARP request is broadcast; an ARP reply is unicast.

Example 21.1

A host with IP address 130.23.43.20 and physical address B2:34:55:10:22:10 has a packet to send to another host with IP address 130.23.43.25 and physical address A4:6E:F4:59:83:AB (which is unknown to the first host). The two hosts are on the same Ethernet network. Show the ARP request and reply packets encapsulated in Ethernet frames.

Solution

Figure 21.5 shows the ARP request and reply packets. Note that the ARP data field in this case is 28 bytes, and that the individual addresses do not fit in the 4-byte boundary. That is why we do not show the regular 4-byte boundaries for these addresses.

Figure 21.5 *Example 21.1, an ARP request and reply*

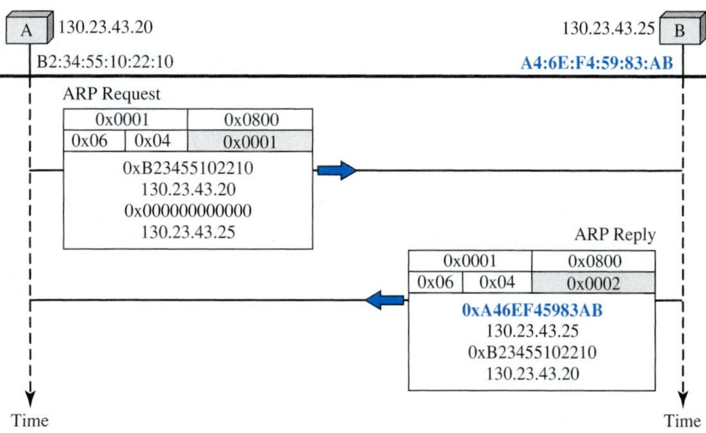

Proxy ARP

A technique called *proxy* ARP is used to create a subnetting effect. A **proxy ARP** is an ARP that acts on behalf of a set of hosts. Whenever a router running a proxy ARP receives an ARP request looking for the IP address of one of these hosts, the router sends an ARP reply announcing its own hardware (physical) address. After the router receives the actual IP packet, it sends the packet to the appropriate host or router.

Let us give an example. In Figure 21.6 the ARP installed on the right-hand host will answer only to an ARP request with a target IP address of 141.23.56.23.

Figure 21.6 *Proxy ARP*

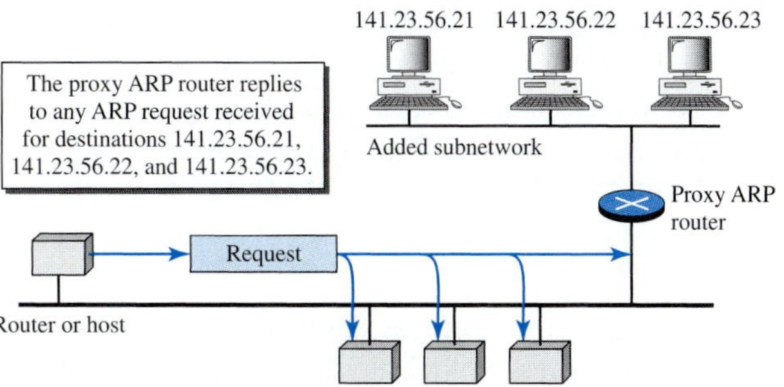

However, the administrator may need to create a subnet without changing the whole system to recognize subnetted addresses. One solution is to add a router running a proxy ARP. In this case, the router acts on behalf of all the hosts installed on the subnet. When it receives an ARP request with a target IP address that matches the address of one of its protégés (141.23.56.21, 141.23.56.22, or 141.23.56.23), it sends an ARP reply and announces its hardware address as the target hardware address. When the router receives the IP packet, it sends the packet to the appropriate host.

Mapping Physical to Logical Address: RARP, BOOTP, and DHCP

There are occasions in which a host knows its physical address, but needs to know its logical address. This may happen in two cases:

1. A diskless station is just booted. The station can find its physical address by checking its interface, but it does not know its IP address.
2. An organization does not have enough IP addresses to assign to each station; it needs to assign IP addresses on demand. The station can send its physical address and ask for a short time lease.

RARP

Reverse Address Resolution Protocol (RARP) finds the logical address for a machine that knows only its physical address. Each host or router is assigned one or more logical (IP) addresses, which are unique and independent of the physical (hardware) address of the machine. To create an IP datagram, a host or a router needs to know its own IP address or addresses. The IP address of a machine is usually read from its configuration file stored on a disk file.

However, a diskless machine is usually booted from ROM, which has minimum booting information. The ROM is installed by the manufacturer. It cannot include the IP address because the IP addresses on a network are assigned by the network administrator.

The machine can get its physical address (by reading its NIC, for example), which is unique locally. It can then use the physical address to get the logical address by using the RARP protocol. A RARP request is created and broadcast on the local network. Another machine on the local network that knows all the IP addresses will respond with a RARP reply. The requesting machine must be running a RARP client program; the responding machine must be running a RARP server program.

There is a serious problem with RARP: Broadcasting is done at the data link layer. The physical broadcast address, all 1s in the case of Ethernet, does not pass the boundaries of a network. This means that if an administrator has several networks or several subnets, it needs to assign a RARP server for each network or subnet. This is the reason that RARP is almost obsolete. Two protocols, BOOTP and DHCP, are replacing RARP.

BOOTP

The **Bootstrap Protocol (BOOTP)** is a client/server protocol designed to provide physical address to logical address mapping. BOOTP is an application layer protocol. The administrator may put the client and the server on the same network or on different networks, as shown in Figure 21.7. BOOTP messages are encapsulated in a UDP packet, and the UDP packet itself is encapsulated in an IP packet.

Figure 21.7 *BOOTP client and server on the same and different network*

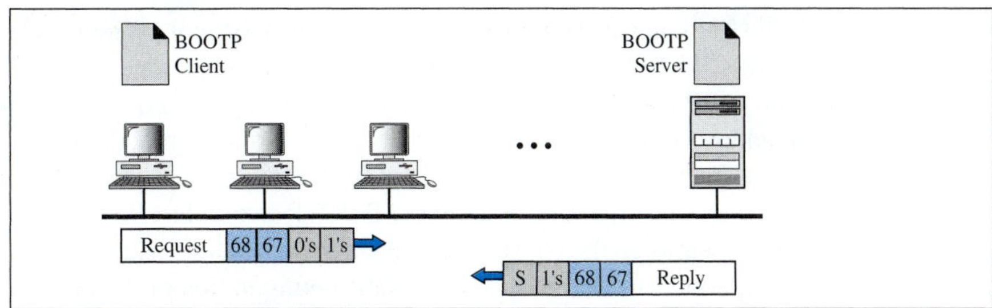

a. Client and server on the same network

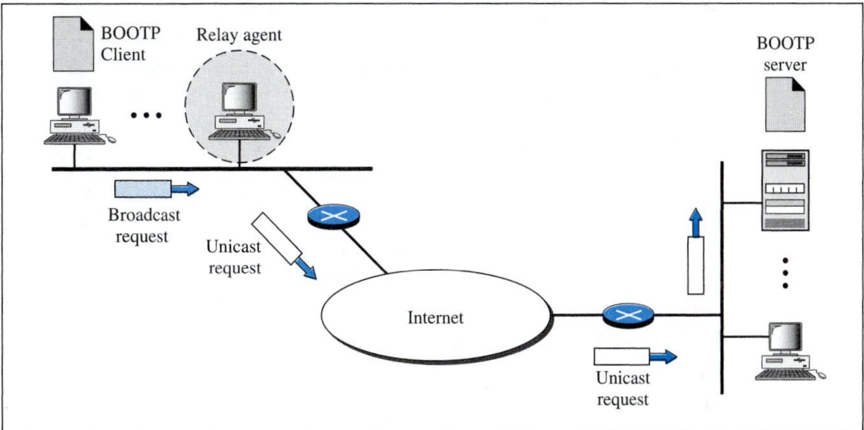

b. Client and server on different networks

The reader may ask how a client can send an IP datagram when it knows neither its own IP address (the source address) nor the server's IP address (the destination address). The client simply uses all 0s as the source address and all 1s as the destination address.

One of the advantages of BOOTP over RARP is that the client and server are application-layer processes. As in other application-layer processes, a client can be in one network and the server in another, separated by several other networks. However, there is one problem that must be solved. The BOOTP request is broadcast because the client does not know the IP address of the server. A broadcast IP datagram cannot pass through any router. To solve the problem, there is a need for an intermediary. One of the hosts (or a router that can be configured to operate at the application layer) can be used as a relay. The host in this case is called a **relay agent.** The relay agent knows the unicast address of a BOOTP server. When it receives this type of packet, it encapsulates the message in a unicast datagram and sends the request to the BOOTP server. The packet, carrying a unicast destination address, is routed by any router and reaches the BOOTP server. The BOOTP server knows the message comes from a relay agent because one of the fields in the request message defines the IP address of the relay agent. The relay agent, after receiving the reply, sends it to the BOOTP client.

DHCP

BOOTP is not a **dynamic configuration protocol.** When a client requests its IP address, the BOOTP server consults a table that matches the physical address of the client with its IP address. This implies that the binding between the physical address and the IP address of the client already exists. The binding is predetermined.

However, what if a host moves from one physical network to another? What if a host wants a temporary IP address? BOOTP cannot handle these situations because the binding between the physical and IP addresses is static and fixed in a table until changed by the administrator. BOOTP is a static configuration protocol.

The **Dynamic Host Configuration Protocol (DHCP)** has been devised to provide static and dynamic address allocation that can be manual or automatic.

DHCP provides static and dynamic address allocation that can be manual or automatic.

Static Address Allocation In this capacity DHCP acts as BOOTP does. It is backward-compatible with BOOTP, which means a host running the BOOTP client can request a static address from a DHCP server. A DHCP server has a database that statically binds physical addresses to IP addresses.

Dynamic Address Allocation DHCP has a second database with a pool of available IP addresses. This second database makes DHCP dynamic. When a DHCP client requests a temporary IP address, the DHCP server goes to the pool of available (unused) IP addresses and assigns an IP address for a negotiable period of time.

When a DHCP client sends a request to a DHCP server, the server first checks its static database. If an entry with the requested physical address exists in the static database, the permanent IP address of the client is returned. On the other hand, if the entry does not exist in the static database, the server selects an IP address from the available pool, assigns the address to the client, and adds the entry to the dynamic database.

The dynamic aspect of DHCP is needed when a host moves from network to network or is connected and disconnected from a network (as is a subscriber to a service provider). DHCP provides temporary IP addresses for a limited time.

The addresses assigned from the pool are temporary addresses. The DHCP server issues a **lease** for a specific time. When the lease expires, the client must either stop using the IP address or renew the lease. The server has the option to agree or disagree with the renewal. If the server disagrees, the client stops using the address.

Manual and Automatic Configuration One major problem with the BOOTP protocol is that the table mapping the IP addresses to physical addresses needs to be manually configured. This means that every time there is a change in a physical or IP address, the administrator needs to manually enter the changes. DHCP, on the other hand, allows both manual and automatic configurations. Static addresses are created manually; dynamic addresses are created automatically.

21.2 ICMP

As discussed in Chapter 20, the IP provides unreliable and connectionless datagram delivery. It was designed this way to make efficient use of network resources. The IP protocol is a best-effort delivery service that delivers a datagram from its original source to its final destination. However, it has two deficiencies: lack of error control and lack of assistance mechanisms.

The IP protocol has no error-reporting or error-correcting mechanism. What happens if something goes wrong? What happens if a router must discard a datagram because it cannot find a router to the final destination, or because the time-to-live field has a zero value? What happens if the final destination host must discard all fragments of a datagram because it has not received all fragments within a predetermined time limit? These are examples of situations where an error has occurred and the IP protocol has no built-in mechanism to notify the original host.

The IP protocol also lacks a mechanism for host and management queries. A host sometimes needs to determine if a router or another host is alive. And sometimes a network administrator needs information from another host or router.

The **Internet Control Message Protocol (ICMP)** has been designed to compensate for the above two deficiencies. It is a companion to the IP protocol.

Types of Messages

ICMP messages are divided into two broad categories: **error-reporting messages** and **query messages.**

The error-reporting messages report problems that a router or a host (destination) may encounter when it processes an IP packet.

The query messages, which occur in pairs, help a host or a network manager get specific information from a router or another host. For example, nodes can discover their neighbors. Also, hosts can discover and learn about routers on their network, and routers can help a node redirect its messages.

Message Format

An ICMP message has an 8-byte header and a variable-size data section. Although the general format of the header is different for each message type, the first 4 bytes are common to all. As Figure 21.8 shows, the first field, ICMP type, defines the type of the

Figure 21.8 *General format of ICMP messages*

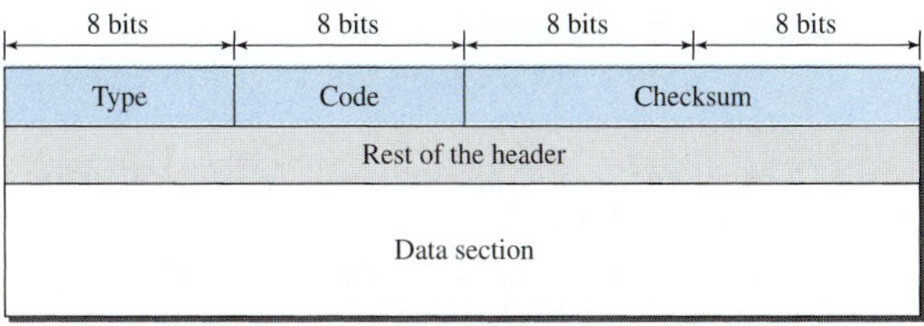

message. The code field specifies the reason for the particular message type. The last common field is the checksum field (to be discussed later in the chapter). The rest of the header is specific for each message type.

The data section in error messages carries information for finding the original packet that had the error. In query messages, the data section carries extra information based on the type of the query.

Error Reporting

One of the main responsibilities of ICMP is to report errors. Although technology has produced increasingly reliable transmission media, errors still exist and must be handled. IP, as discussed in Chapter 20, is an unreliable protocol. This means that error checking and error control are not a concern of IP. ICMP was designed, in part, to compensate for this shortcoming. However, ICMP does not correct errors—it simply reports them. Error correction is left to the higher-level protocols. Error messages are always sent to the original source because the only information available in the datagram about the route is the source and destination IP addresses. ICMP uses the source IP address to send the error message to the source (originator) of the datagram.

ICMP always reports error messages to the original source.

Five types of errors are handled: destination unreachable, source quench, time exceeded, parameter problems, and redirection (see Figure 21.9).

Figure 21.9 *Error-reporting messages*

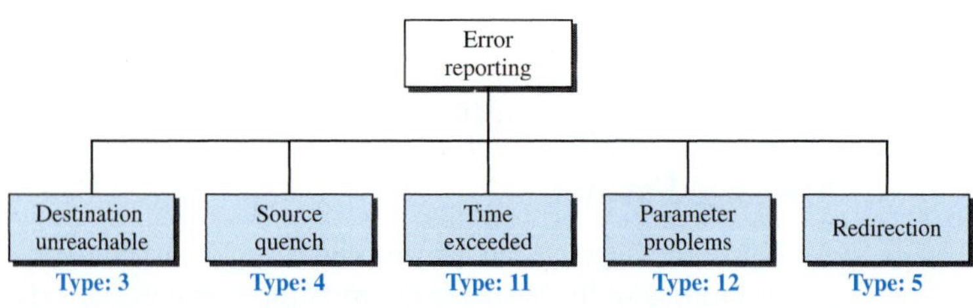

The following are important points about ICMP error messages:

- **No ICMP error message will be generated in response to a datagram carrying an ICMP error message.**
- **No ICMP error message will be generated for a fragmented datagram that is not the first fragment.**
- **No ICMP error message will be generated for a datagram having a multicast address.**
- **No ICMP error message will be generated for a datagram having a special address such as 127.0.0.0 or 0.0.0.0.**

Note that all error messages contain a data section that includes the IP header of the original datagram plus the first 8 bytes of data in that datagram. The original datagram header is added to give the original source, which receives the error message, information about the datagram itself. The 8 bytes of data are included because, as we will see in Chapter 23 on UDP and TCP protocols, the first 8 bytes provide information about the port numbers (UDP and TCP) and sequence number (TCP). This information is needed so the source can inform the protocols (TCP or UDP) about the error. ICMP forms an error packet, which is then encapsulated in an IP datagram (see Figure 21.10).

Figure 21.10 *Contents of data field for the error messages*

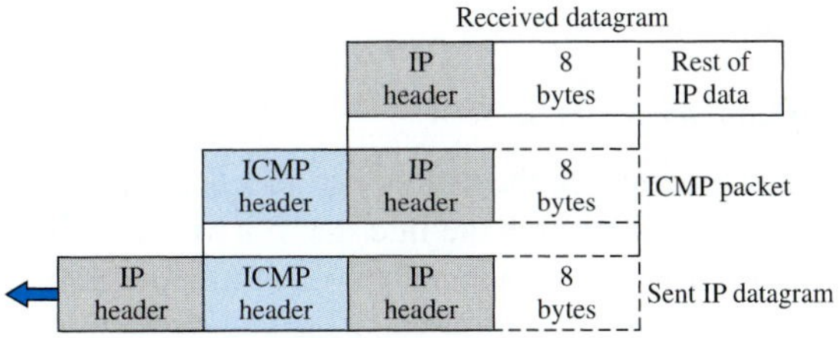

Received datagram
IP header | 8 bytes | Rest of IP data
ICMP header | IP header | 8 bytes — ICMP packet
IP header | ICMP header | IP header | 8 bytes — Sent IP datagram

Destination Unreachable

When a router cannot route a datagram or a host cannot deliver a datagram, the datagram is discarded and the router or the host sends a **destination-unreachable message** back to the source host that initiated the datagram. Note that destination-unreachable messages can be created by either a router or the destination host.

Source Quench

The IP protocol is a connectionless protocol. There is no communication between the source host, which produces the datagram, the routers, which forward it, and the destination host, which processes it. One of the ramifications of this absence of communication is the lack of *flow control*. IP does not have a flow control mechanism embedded in the protocol. The lack of flow control can create a major problem in the operation of IP: congestion. The source host never knows if the routers or the destination host has been overwhelmed with datagrams. The source host never knows if it is producing datagrams faster than can be forwarded by routers or processed by the destination host.

The lack of flow control can create congestion in routers or the destination host. A router or a host has a limited-size queue (buffer) for incoming datagrams waiting to be forwarded (in the case of a router) or to be processed (in the case of a host). If the datagrams are received much faster than they can be forwarded or processed, the queue may overflow. In this case, the router or the host has no choice but to discard some of the datagrams. The **source-quench message** in ICMP was designed to add a kind of flow control to the IP. When a router or host discards a datagram due to congestion, it sends a source-quench message to the sender of the datagram. This message has two purposes. First, it informs the source that the datagram has been discarded. Second, it warns the source that there is congestion somewhere in the path and that the source should slow down (quench) the sending process.

Time Exceeded

The **time-exceeded message** is generated in two cases: As we see in Chapter 22, routers use routing tables to find the next hop (next router) that must receive the packet. If there are errors in one or more routing tables, a packet can travel in a loop or a cycle, going from one router to the next or visiting a series of routers endlessly. As we saw in Chapter 20, each datagram contains a field called *time to live* that controls this situation. When a datagram visits a router, the value of this field is decremented by 1. When the time-to-live value reaches 0, after decrementing, the router discards the datagram. However, when the datagram is discarded, a time-exceeded message must be sent by the router to the original source. Second, a time-exceeded message is also generated when not all fragments that make up a message arrive at the destination host within a certain time limit.

Parameter Problem

Any ambiguity in the header part of a datagram can create serious problems as the datagram travels through the Internet. If a router or the destination host discovers an ambiguous or missing value in any field of the datagram, it discards the datagram and sends a **parameter-problem message** back to the source.

Redirection

When a router needs to send a packet destined for another network, it must know the IP address of the next appropriate router. The same is true if the sender is a host. Both routers and hosts, then, must have a routing table to find the address of the router or the next router. Routers take part in the routing update process, as we will see in Chapter 22, and are supposed to be updated constantly. Routing is dynamic.

However, for efficiency, hosts do not take part in the routing update process because there are many more hosts in an internet than routers. Updating the routing tables of hosts dynamically produces unacceptable traffic. The hosts usually use static routing. When a host comes up, its routing table has a limited number of entries. It usually knows the IP address of only one router, the default router. For this reason, the host may send a datagram, which is destined for another network, to the wrong router. In this case, the router that receives the datagram will forward the datagram to the correct router. However, to update the routing table of the host, it sends a redirection message to the host. This concept of redirection is shown in Figure 21.11. Host A wants to send a datagram to host B.

Figure 21.11 *Redirection concept*

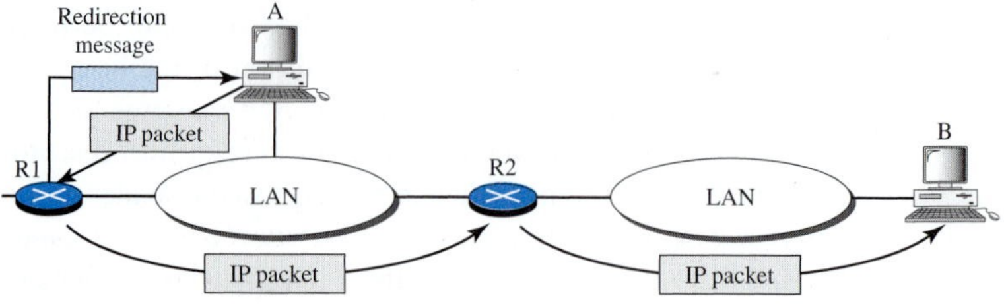

Router R2 is obviously the most efficient routing choice, but host A did not choose router R2. The datagram goes to R1 instead. Router R1, after consulting its table, finds that the packet should have gone to R2. It sends the packet to R2 and, at the same time, sends a redirection message to host A. Host A's routing table can now be updated.

Query

In addition to error reporting, ICMP can diagnose some network problems. This is accomplished through the query messages, a group of four different pairs of messages, as shown in Figure 21.12. In this type of ICMP message, a node sends a message that is answered in a specific format by the destination node. A query message is encapsulated in an IP packet, which in turn is encapsulated in a data link layer frame. However, in this case, no bytes of the original IP are included in the message, as shown in Figure 21.13.

Figure 21.12 *Query messages*

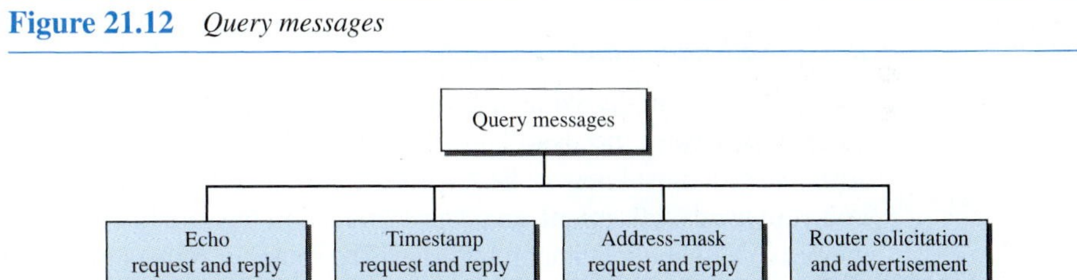

Figure 21.13 *Encapsulation of ICMP query messages*

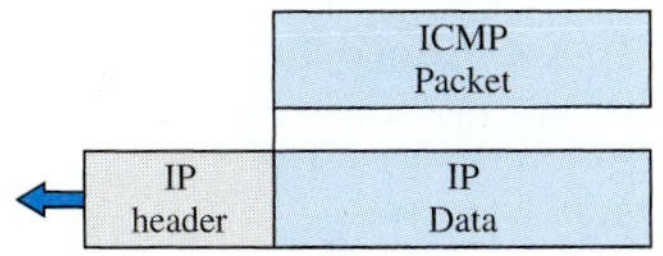

Echo Request and Reply

The **echo-request** and **echo-reply messages** are designed for diagnostic purposes. Network managers and users utilize this pair of messages to identify network problems. The combination of echo-request and echo-reply messages determines whether two systems (hosts or routers) can communicate with each other. The echo-request and echo-reply messages can be used to determine if there is communication at the IP level. Because ICMP messages are encapsulated in IP datagrams, the receipt of an echo-reply message by the machine that sent the echo request is proof that the IP protocols in the sender and receiver are communicating with each other using the IP datagram. Also, it is proof that the intermediate routers are receiving, processing, and forwarding IP datagrams. Today, most systems provide a version of the *ping* command that can create a series (instead of just one) of echo-request and echo-reply messages, providing statistical information. We will see the use of this program at the end of the chapter.

Timestamp Request and Reply

Two machines (hosts or routers) can use the **timestamp request and timestamp reply messages** to determine the round-trip time needed for an IP datagram to travel between them. It can also be used to synchronize the clocks in two machines.

Address-Mask Request and Reply

A host may know its IP address, but it may not know the corresponding mask. For example, a host may know its IP address as 159.31.17.24, but it may not know that the corresponding mask is /24. To obtain its mask, a host sends an **address-mask-request message** to a router on the LAN. If the host knows the address of the router, it sends the request directly to the router. If it does not know, it broadcasts the message. The router receiving the address-mask-request message responds with an **address-mask-reply message,** providing the necessary mask for the host. This can be applied to its full IP address to get its subnet address.

Router Solicitation and Advertisement

As we discussed in the redirection message section, a host that wants to send data to a host on another network needs to know the address of routers connected to its own network. Also, the host must know if the routers are alive and functioning. The **router-solicitation and router-advertisement messages** can help in this situation. A host can broadcast (or multicast) a router-solicitation message. The router or routers that receive the solicitation message broadcast their routing information using the router-advertisement message. A router can also periodically send router-advertisement messages even if no host has solicited. Note that when a router sends out an advertisement, it announces not only its own presence but also the presence of all routers on the network of which it is aware.

Checksum

In Chapter 10, we learned the concept and idea of the checksum. In ICMP the checksum is calculated over the entire message (header and data).

Example 21.2

Figure 21.14 shows an example of checksum calculation for a simple echo-request message. We randomly chose the identifier to be 1 and the sequence number to be 9. The message is divided

Figure 21.14 *Example of checksum calculation*

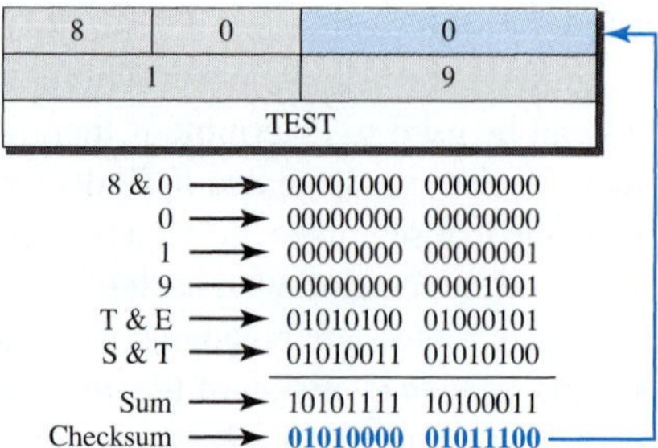

into 16-bit (2-byte) words. The words are added and the sum is complemented. Now the sender can put this value in the checksum field.

Debugging Tools

There are several tools that can be used in the Internet for debugging. We can determine the viability of a host or router. We can trace the route of a packet. We introduce two tools that use ICMP for debugging: *ping* and *traceroute*. We will introduce more tools in future chapters after we have discussed the corresponding protocols.

Ping

We can use the *ping* program to find if a host is alive and responding. We use *ping* here to see how it uses ICMP packets.

The source host sends ICMP echo-request messages (type: 8, code: 0); the destination, if alive, responds with ICMP echo-reply messages. The *ping* program sets the identifier field in the echo-request and echo-reply message and starts the sequence number from 0; this number is incremented by 1 each time a new message is sent. Note that *ping* can calculate the round-trip time. It inserts the sending time in the data section of the message. When the packet arrives, it subtracts the arrival time from the departure time to get the round-trip time (RTT).

Example 21.3

We use the *ping* program to test the server fhda.edu. The result is shown below:

```
$ ping fhda.edu
PING fhda.edu (153.18.8.1)   56 (84)  bytes of data.
64 bytes from tiptoe.fhda.edu (153.18.8.1): icmp_seq=0    ttl=62    time=1.91 ms
64 bytes from tiptoe.fhda.edu (153.18.8.1): icmp_seq=1    ttl=62    time=2.04 ms
64 bytes from tiptoe.fhda.edu (153.18.8.1): icmp_seq=2    ttl=62    time=1.90 ms
64 bytes from tiptoe.fhda.edu (153.18.8.1): icmp_seq=3    ttl=62    time=1.97 ms
64 bytes from tiptoe.fhda.edu (153.18.8.1): icmp_seq=4    ttl=62    time=1.93 ms
64 bytes from tiptoe.fhda.edu (153.18.8.1): icmp_seq=5    ttl=62    time=2.00 ms
64 bytes from tiptoe.fhda.edu (153.18.8.1): icmp_seq=6    ttl=62    time=1.94 ms
64 bytes from tiptoe.fhda.edu (153.18.8.1): icmp_seq=7    ttl=62    time=1.94 ms
64 bytes from tiptoe.fhda.edu (153.18.8.1): icmp_seq=8    ttl=62    time=1.97 ms
64 bytes from tiptoe.fhda.edu (153.18.8.1): icmp_seq=9    ttl=62    time=1.89 ms
64 bytes from tiptoe.fhda.edu (153.18.8.1): icmp_seq=10   ttl=62    time=1.98 ms

--- fhda.edu ping statistics ---
11 packets transmitted, 11 received, 0% packet loss, time 10103ms
     rtt min/avg/max = 1.899/1.955/2.041 ms
```

The *ping* program sends messages with sequence numbers starting from 0. For each probe it gives us the RTT time. The TTL (time to live) field in the IP datagram that encapsulates an ICMP message has been set to 62, which means the packet cannot travel more than 62 hops. At the beginning, *ping* defines the number of data bytes as 56 and the total number of bytes as 84. It is obvious that if we add 8 bytes of ICMP header and 20 bytes of IP header to 56, the result is 84. However, note that in each probe *ping* defines the number of bytes as 64. This is the total number of bytes in the ICMP packet (56 + 8). The *ping* program continues to send messages, if we do not stop it by using the interrupt key (ctrl + c, for example). After it is interrupted, it prints the statistics of the probes. It tells us the number of packets sent, the number of packets received, the total time, and the RTT minimum, maximum, and average. Some systems may print more information.

Traceroute

The ***traceroute*** program in UNIX or ***tracert*** in Windows can be used to trace the route of a packet from the source to the destination. We have seen an application of the *traceroute* program to simulate the loose source route and strict source route options of an IP datagram in Chapter 20. We use this program in conjunction with ICMP packets in this chapter. The program elegantly uses two ICMP messages, time exceeded and destination unreachable, to find the route of a packet. This is a program at the application level that uses the services of UDP (see Chapter 23). Let us show the idea of the *traceroute* program by using Figure 21.15.

Figure 21.15 *The traceroute program operation*

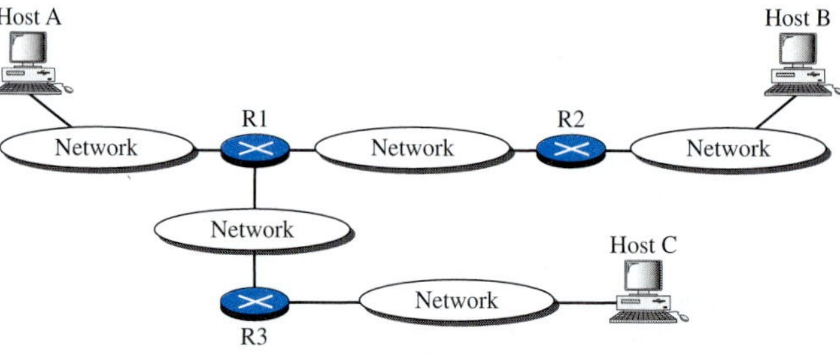

Given the topology, we know that a packet from host A to host B travels through routers R1 and R2. However, most of the time, we are not aware of this topology. There could be several routes from A to B. The *traceroute* program uses the ICMP messages and the TTL (time to live) field in the IP packet to find the route.

1. The *traceroute* program uses the following steps to find the address of the router R1 and the round-trip time between host A and router R1.
   a. The *traceroute* application at host A sends a packet to destination B using UDP; the message is encapsulated in an IP packet with a TTL value of 1. The program notes the time the packet is sent.
   b. Router R1 receives the packet and decrements the value of TTL to 0. It then discards the packet (because TTL is 0). The router, however, sends a time-exceeded ICMP message (type: 11, code: 0) to show that the TTL value is 0 and the packet was discarded.
   c. The *traceroute* program receives the ICMP messages and uses the destination address of the IP packet encapsulating ICMP to find the IP address of router R1. The program also makes note of the time the packet has arrived. The difference between this time and the time at step a is the round-trip time.

The *traceroute* program repeats steps a to c three times to get a better average round-trip time. The first trip time may be much longer than the second or third because it takes time for the ARP program to find the physical address of router R1. For the second and third trips, ARP has the address in its cache.

2. The *traceroute* program repeats the previous steps to find the address of router R2 and the round-trip time between host A and router R2. However, in this step, the value of TTL is set to 2. So router R1 forwards the message, while router R2 discards it and sends a time-exceeded ICMP message.
3. The *traceroute* program repeats step 2 to find the address of host B and the round-trip time between host A and host B. When host B receives the packet, it decrements the value of TTL, but it does not discard the message since it has reached its final destination. How can an ICMP message be sent back to host A? The *traceroute* program uses a different strategy here. The destination port of the UDP packet is set to one that is not supported by the UDP protocol. When host B receives the packet, it cannot find an application program to accept the delivery. It discards the packet and sends an ICMP destination-unreachable message (type: 3, code: 3) to host A. Note that this situation does not happen at router R1 or R2 because a router does not check the UDP header. The *traceroute* program records the destination address of the arrived IP datagram and makes note of the round-trip time. Receiving the destination-unreachable message with a code value 3 is an indication that the whole route has been found and there is no need to send more packets.

Example 21.4

We use the *traceroute* program to find the route from the computer voyager.deanza.edu to the server fhda.edu. The following shows the result:

```
$ traceroute fhda.edu
traceroute to fhda.edu   (153.18.8.1), 30 hops max, 38 byte packets
 1 Dcore.fhda.edu        (153.18.31.254)   0.995 ms   0.899 ms   0.878 ms
 2 Dbackup.fhda.edu      (153.18.251.4)    1.039 ms   1.064 ms   1.083 ms
 3 tiptoe.fhda.edu       (153.18.8.1)      1.797 ms   1.642 ms   1.757 ms
```

The unnumbered line after the command shows that the destination is 153.18.8.1. The TTL value is 30 hops. The packet contains 38 bytes: 20 bytes of IP header, 8 bytes of UDP header, and 10 bytes of application data. The application data are used by *traceroute* to keep track of the packets.

The first line shows the first router visited. The router is named Dcore.fhda.edu with IP address 153.18.31.254. The first round-trip time was 0.995 ms, the second was 0.899 ms, and the third was 0.878 ms.

The second line shows the second router visited. The router is named Dbackup.fhda.edu with IP address 153.18.251.4. The three round-trip times are also shown.

The third line shows the destination host. We know that this is the destination host because there are no more lines. The destination host is the server fhda.edu, but it is named tiptoe.fhda.edu with the IP address 153.18.8.1. The three round-trip times are also shown.

Example 21.5

In this example, we trace a longer route, the route to xerox.com.

```
$ traceroute xerox.com
traceroute to xerox.com (13.1.64.93), 30 hops max, 38 byte packets
 1 Dcore.fhda.edu      (153.18.31.254)     0.622 ms   0.891 ms   0.875 ms
 2 Ddmz.fhda.edu       (153.18.251.40)     2.132 ms   2.266 ms   2.094 ms
 3 Cinic.fhda.edu      (153.18.253.126)    2.110 ms   2.145 ms   1.763 ms
 4 cenic.net           (137.164.32.140)    3.069 ms   2.875 ms   2.930 ms
 5 cenic.net           (137.164.22.31)     4.205 ms   4.870 ms   4.197 ms
   ....                ....                ...        ....       ...
14 snfc21.pbi.net      (151.164.191.49)    7.656 ms   7.129 ms   6.866 ms
15 sbcglobal.net       (151.164.243.58)    7.844 ms   7.545 ms   7.353 ms
16 pacbell.net         (209.232.138.114)   9.857 ms   9.535 ms   9.603 ms
17 209.233.48.223      (209.233.48.223)   10.634 ms  10.771 ms  10.592 ms
18 alpha.Xerox.COM     (13.1.64.93)       11.172 ms  11.048 ms  10.922 ms
```

Here there are 17 hops between source and destination. Note that some round-trip times look unusual. It could be that a router was too busy to process the packet immediately.

21.3 IGMP

The IP protocol can be involved in two types of communication: unicasting and multicasting. Unicasting is the communication between one sender and one receiver. It is a one-to-one communication. However, some processes sometimes need to send the same message to a large number of receivers simultaneously. This is called **multicasting,** which is a one-to-many communication. Multicasting has many applications. For example, multiple stockbrokers can simultaneously be informed of changes in a stock price, or travel agents can be informed of a plane cancellation. Some other applications include distance learning and video-on-demand.

The **Internet Group Management Protocol (IGMP)** is one of the necessary, but not sufficient (as we will see), protocols that is involved in multicasting. IGMP is a companion to the IP protocol.

Group Management

For multicasting in the Internet we need routers that are able to route multicast packets. The routing tables of these routers must be updated by using one of the multicasting routing protocols that we discuss in Chapter 22.

IGMP is not a multicasting routing protocol; it is a protocol that manages **group membership.** In any network, there are one or more multicast routers that distribute multicast packets to hosts or other routers. The IGMP protocol gives the **multicast routers** information about the membership status of hosts (routers) connected to the network.

A multicast router may receive thousands of multicast packets every day for different groups. If a router has no knowledge about the membership status of the hosts, it must broadcast all these packets. This creates a lot of traffic and consumes bandwidth. A better solution is to keep a list of groups in the network for which there is at least one loyal member. IGMP helps the multicast router create and update this list.

IGMP is a group management protocol. It helps a multicast router create and update a list of loyal members related to each router interface.

IGMP Messages

IGMP has gone through two versions. We discuss IGMPv2, the current version. IGMPv2 has three types of **messages:** the **query,** the **membership report,** and the **leave report.** There are two types of **query messages: general** and **special** (see Figure 21.16).

Figure 21.16 *IGMP message types*

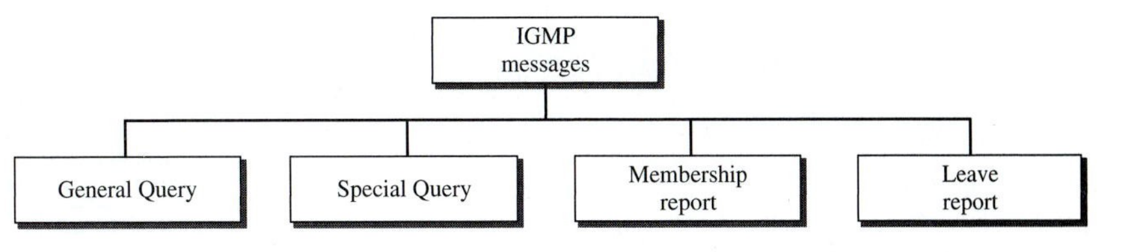

Message Format

Figure 21.17 shows the format of an IGMP (version 2) message.

Figure 21.17 *IGMP message format*

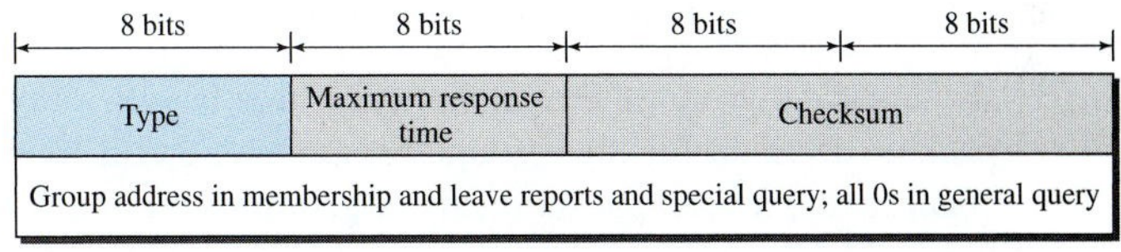

- **Type.** This 8-bit field defines the type of message, as shown in Table 21.1. The value of the type is shown in both hexadecimal and binary notation.

Table 21.1 *IGMP type field*

Type	*Value*
General or special query	0x11 or 00010001
Membership report	0x16 or 00010110
Leave report	0x17 or 00010111

- **Maximum Response Time.** This 8-bit field defines the amount of time in which a query must be answered. The value is in tenths of a second; for example, if the value is 100, it means 10 s. The value is nonzero in the query message; it is set to zero in the other two message types. We will see its use shortly.

- **Checksum.** This is a 16-bit field carrying the checksum. The checksum is calculated over the 8-byte message.
- **Group address.** The value of this field is 0 for a general query message. The value defines the groupid (multicast address of the group) in the special query, the membership report, and the leave report messages.

IGMP Operation

IGMP operates locally. A multicast router connected to a network has a list of multicast addresses of the groups with at least one loyal member in that network (see Figure 21.18).

Figure 21.18 *IGMP operation*

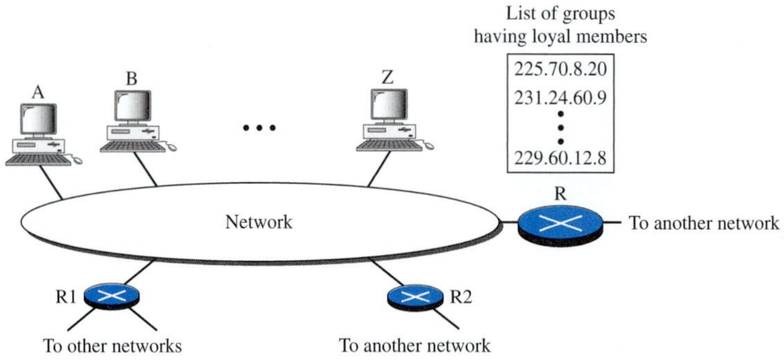

For each group, there is one router that has the duty of distributing the multicast packets destined for that group. This means that if there are three multicast routers connected to a network, their lists of **groupids** are mutually exclusive. For example, in Figure 21.18 only router R distributes packets with the multicast address of 225.70.8.20.

A host or multicast router can have membership in a group. When a host has membership, it means that one of its processes (an application program) receives multicast packets from some group. When a router has membership, it means that a network connected to one of its other interfaces receives these multicast packets. We say that the host or the router has an *interest* in the group. In both cases, the host and the router keep a list of groupids and relay their interest to the distributing router.

For example, in Figure 21.18, router R is the distributing router. There are two other multicast routers (R1 and R2) that, depending on the group list maintained by router R, could be the recipients of router R in this network. Routers R1 and R2 may be distributors for some of these groups in other networks, but not on this network.

Joining a Group

A host or a router can join a group. A host maintains a list of processes that have membership in a group. When a process wants to join a new group, it sends its request to the host.

The host adds the name of the process and the name of the requested group to its list. If this is the first entry for this particular group, the host sends a membership report message. If this is not the first entry, there is no need to send the membership report since the host is already a member of the group; it already receives multicast packets for this group.

The protocol requires that the membership report be sent twice, one after the other within a few moments. In this way, if the first one is lost or damaged, the second one replaces it.

In IGMP, a membership report is sent twice, one after the other.

Leaving a Group

When a host sees that no process is interested in a specific group, it sends a leave report. Similarly, when a router sees that none of the networks connected to its interfaces is interested in a specific group, it sends a leave report about that group.

However, when a multicast router receives a leave report, it cannot immediately purge that group from its list because the report comes from just one host or router; there may be other hosts or routers that are still interested in that group. To make sure, the router sends a special query message and inserts the groupid, or **multicast address,** related to the group. The router allows a specified time for any host or router to respond. If, during this time, no interest (membership report) is received, the router assumes that there are no loyal members in the network and purges the group from its list.

Monitoring Membership

A host or router can join a group by sending a membership report message. It can leave a group by sending a leave report message. However, sending these two types of reports is not enough. Consider the situation in which there is only one host interested in a group, but the host is shut down or removed from the system. The multicast router will never receive a leave report. How is this handled? The multicast router is responsible for monitoring all the hosts or routers in a LAN to see if they want to continue their membership in a group.

The router periodically (by default, every 125 s) sends a general query message. In this message, the group address field is set to 0.0.0.0. This means the query for membership continuation is for all groups in which a host is involved, not just one.

The general query message does not define a particular group.

The router expects an answer for each group in its group list; even new groups may respond. The query message has a maximum response time of 10 s (the value of the field is actually 100, but this is in tenths of a second). When a host or router receives the general query message, it responds with a membership report if it is interested in a group. However, if there is a common interest (two hosts, for example, are interested in the same group), only one response is sent for that group to prevent unnecessary traffic. This is called a delayed response. Note that the query message must be sent by only one router (normally called the query router), also to prevent unnecessary traffic. We discuss this issue shortly.

Delayed Response

To prevent unnecessary traffic, IGMP uses a **delayed response strategy.** When a host or router receives a query message, it does not respond immediately; it delays the response. Each host or router uses a random number to create a timer, which expires between 1 and 10 s. The expiration time can be in steps of 1 s or less. A timer is set for each group in the list. For example, the timer for the first group may expire in 2 s, but the timer for the third group may expire in 5 s. Each host or router waits until its timer has expired before sending a membership report message. During this waiting time, if the timer of another host or router, for the same group, expires earlier, that host or router sends a membership report. Because, as we will see shortly, the report is broadcast, the waiting host or router receives the report and knows that there is no need to send a duplicate report for this group; thus, the waiting station cancels its corresponding timer.

Example 21.6

Imagine there are three hosts in a network, as shown in Figure 21.19.

Figure 21.19 *Example 21.6*

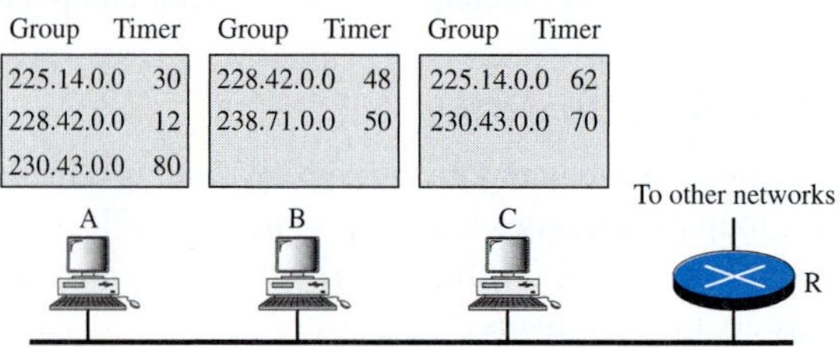

A query message was received at time 0; the random delay time (in tenths of seconds) for each group is shown next to the group address. Show the sequence of report messages.

Solution

The events occur in this sequence:

a. **Time 12:** The timer for 228.42.0.0 in host A expires, and a membership report is sent, which is received by the router and every host including host B which cancels its timer for 228.42.0.0.

b. **Time 30:** The timer for 225.14.0.0 in host A expires, and a membership report is sent, which is received by the router and every host including host C which cancels its timer for 225.14.0.0.

c. **Time 50:** The timer for 238.71.0.0 in host B expires, and a membership report is sent, which is received by the router and every host.

d. **Time 70:** The timer for 230.43.0.0 in host C expires, and a membership report is sent, which is received by the router and every host including host A which cancels its timer for 230.43.0.0.

Note that if each host had sent a report for every group in its list, there would have been seven reports; with this strategy only four reports are sent.

Query Router

Query messages may create a lot of responses. To prevent unnecessary traffic, IGMP designates one router as the **query router** for each network. Only this designated router sends the query message, and the other routers are passive (they receive responses and update their lists).

Encapsulation

The IGMP message is encapsulated in an IP datagram, which is itself encapsulated in a frame. See Figure 21.20.

Figure 21.20 *Encapsulation of IGMP packet*

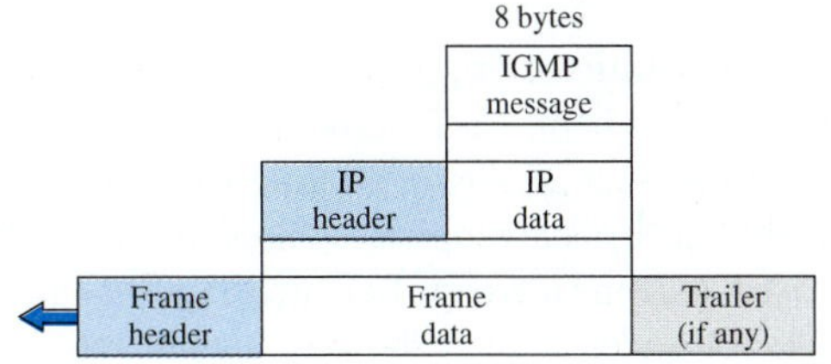

Encapsulation at Network Layer

The value of the protocol field is 2 for the IGMP protocol. Every IP packet carrying this value in its protocol field has data delivered to the IGMP protocol. When the message is encapsulated in the IP datagram, the value of TTL must be 1. This is required because the domain of IGMP is the LAN. No IGMP message must travel beyond the LAN. A TTL value of 1 guarantees that the message does not leave the LAN since this value is decremented to 0 by the next router and, consequently, the packet is discarded. Table 21.2 shows the destination IP address for each type of message.

The IP packet that carries an IGMP packet has a value of 1 in its TTL field.

Table 21.2 *Destination IP addresses*

Type	*IP Destination Address*
Query	224.0.0.1 All systems on this subnet
Membership report	The multicast address of the group
Leave report	224.0.0.2 All routers on this subnet

A query message is multicast by using the multicast address 224.0.0.1 All hosts and all routers will receive the message. A membership report is multicast using a destination address equal to the multicast address being reported (groupid). Every station (host or router) that receives the packet can immediately determine (from the header) the group for which a report has been sent. As discussed previously, the timers for the corresponding unsent reports can then be canceled. Stations do not need to open the packet to find the groupid. This address is duplicated in a packet; it's part of the message itself and also a field in the IP header. The duplication prevents errors. A leave report message is multicast using the multicast address 224.0.0.2 (all routers on this subnet) so that routers receive this type of message. Hosts receive this message too, but disregard it.

Encapsulation at Data Link Layer

At the network layer, the IGMP message is encapsulated in an IP packet and is treated as an IP packet. However, because the IP packet has a multicast IP address, the ARP protocol cannot find the corresponding MAC (physical) address to forward the packet at the data link layer. What happens next depends on whether the underlying data link layer supports physical multicast addresses.

Physical Multicast Support Most LANs support physical multicast addressing. Ethernet is one of them. An Ethernet physical address (MAC address) is six octets (48 bits) long. If the first 25 bits in an Ethernet address are 0000000100000000010111100, this identifies a physical multicast address for the TCP/IP protocol. The remaining 23 bits can be used to define a group. To convert an IP multicast address into an Ethernet address, the multicast router extracts the least significant 23 bits of a class D IP address and inserts them into a multicast Ethernet physical address (see Figure 21.21).

Figure 21.21 *Mapping class D to Ethernet physical address*

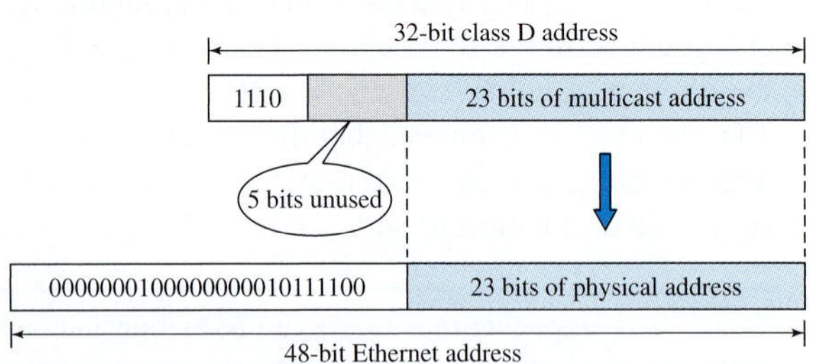

However, the group identifier of a class D IP address is 28 bits long, which implies that 5 bits is not used. This means that 32 (2^5) multicast addresses at the IP level are mapped to a single multicast address. In other words, the mapping is many-to-one instead of one-to-one. If the 5 leftmost bits of the group identifier of a class D address are not all zeros, a host may receive packets that do not really belong to the group in which it is involved. For this reason, the host must check the IP address and discard any packets that do not belong to it.

Other LANs support the same concept but have different methods of mapping.

An Ethernet multicast physical address is in the range
01:00:5E:00:00:00 to 01:00:5E:7F:FF:FF.

Example 21.7

Change the multicast IP address 230.43.14.7 to an Ethernet multicast physical address.

Solution

We can do this in two steps:

a. We write the rightmost 23 bits of the IP address in hexadecimal. This can be done by changing the rightmost 3 bytes to hexadecimal and then subtracting 8 from the leftmost digit if it is greater than or equal to 8. In our example, the result is 2B:0E:07.

b. We add the result of part a to the starting Ethernet multicast address, which is 01:00:5E:00:00:00. The result is

01:00:5E:2B:0E:07

Example 21.8

Change the multicast IP address 238.212.24.9 to an Ethernet multicast address.

Solution

a. The rightmost 3 bytes in hexadecimal is D4:18:09. We need to subtract 8 from the leftmost digit, resulting in 54:18:09.

b. We add the result of part a to the Ethernet multicast starting address. The result is

01:00:5E:54:18:09

No Physical Multicast Support Most WANs do not support physical multicast addressing. To send a multicast packet through these networks, a process called *tunneling* is used. In **tunneling,** the multicast packet is encapsulated in a unicast packet and sent through the network, where it emerges from the other side as a multicast packet (see Figure 21.22).

Figure 21.22 *Tunneling*

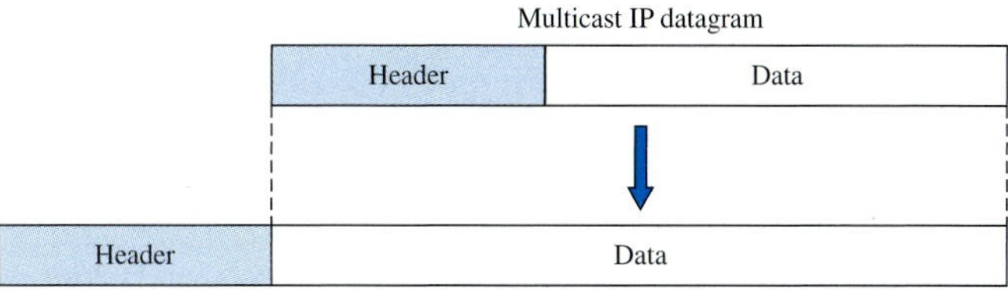

Netstat Utility

The *netstat* utility can be used to find the multicast addresses supported by an interface.

Example 21.9

We use *netstat* with three options: -n, -r, and -a. The -n option gives the numeric versions of IP addresses, the -r option gives the routing table, and the -a option gives all addresses (unicast and multicast). Note that we show only the fields relative to our discussion. "Gateway" defines the router, "Iface" defines the interface.

$ netstat -nra
Kernel IP routing table

Destination	**Gateway**	**Mask**	**Flags**	**Iface**
153.18.16.0	0.0.0.0	255.255.240.0	U	eth0
169.254.0.0	0.0.0.0	255.255.0.0	U	eth0
127.0.0.0	0.0.0.0	255.0.0.0	U	lo
224.0.0.0	0.0.0.0	224.0.0.0	U	eth0
0.0.0.0	153.18.31.254	0.0.0.0	UG	eth0

Note that the multicast address is shown in color. Any packet with a multicast address from 224.0.0.0 to 239.255.255.255 is masked and delivered to the Ethernet interface.

21.4 ICMPv6

We discussed IPv6 in Chapter 20. Another protocol that has been modified in version 6 of the TCP/IP protocol suite is ICMP (ICMPv6). This new version follows the same strategy and purposes of version 4. ICMPv4 has been modified to make it more suitable for IPv6. In addition, some protocols that were independent in version 4 are now part of **Internetworking Control Message Protocol (ICMPv6).** Figure 21.23 compares the network layer of version 4 to version 6.

Figure 21.23 *Comparison of network layers in version 4 and version 6*

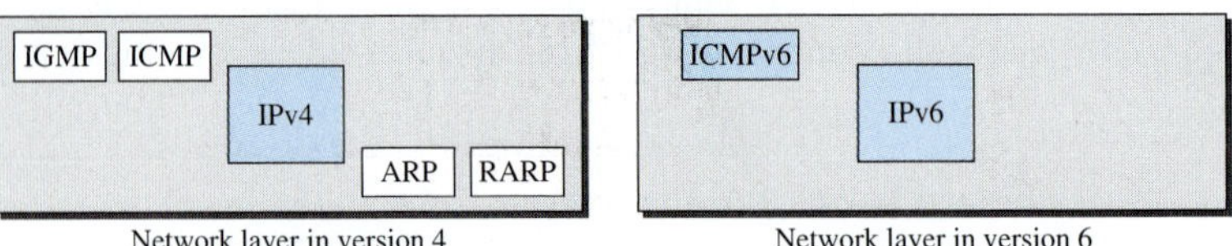

The ARP and IGMP protocols in version 4 are combined in ICMPv6. The RARP protocol is dropped from the suite because it was rarely used and BOOTP has the same functionality.

Just as in ICMPv4, we divide the ICMP messages into two categories. However, each category has more types of messages than before.

Error Reporting

As we saw in our discussion of version 4, one of the main responsibilities of ICMP is to report errors. Five types of errors are handled: destination unreachable, packet too big, time exceeded, parameter problems, and redirection. ICMPv6 forms an error packet, which is then encapsulated in an IP datagram. This is delivered to the original source of the failed datagram. Table 21.3 compares the **error-reporting messages** of ICMPv4 with ICMPv6. The source-quench message is eliminated in version 6 because the priority and the flow label fields allow the router to control congestion and discard the least important messages. In this version, there is no need to inform the sender to slow down. The packet-too-big message is added because fragmentation is the responsibility of the sender in IPv6. If the sender does not make the right packet size decision, the router has no choice but to drop the packet and send an error message to the sender.

Table 21.3 *Comparison of error-reporting messages in ICMPv4 and ICMPv6*

Type of Message	*Version 4*	*Version 6*
Destination unreachable	Yes	Yes
Source quench	Yes	No
Packet too big	No	Yes
Time exceeded	Yes	Yes
Parameter problem	Yes	Yes
Redirection	Yes	Yes

Destination Unreachable

The concept of the **destination-unreachable message** is exactly the same as described for ICMP version 4.

Packet Too Big

This is a new type of message added to version 6. If a router receives a datagram that is larger than the maximum transmission unit (MTU) size of the network through which the datagram should pass, two things happen. First, the router discards the datagram and then an ICMP error packet—a **packet-too-big message**—is sent to the source.

Time Exceeded

This message is similar to the one in version 4.

Parameter Problem

This message is similar to its version 4 counterpart.

Redirection

The purpose of the **redirection message** is the same as described for version 4.

Query

In addition to error reporting, ICMP can diagnose some network problems. This is accomplished through the **query messages.** Four different groups of messages have been defined: echo request and reply, router solicitation and advertisement, neighbor solicitation and advertisement, and group membership. Table 21.4 shows a comparison between the query messages in versions 4 and 6. Two sets of query messages are eliminated from ICMPv6: time-stamp request and reply- and address-mask request and reply. The time-stamp request and reply messages are eliminated because they are implemented in other protocols such as TCP and because they were rarely used in the past. The address-mask request and reply messages are eliminated in IPv6 because the subnet section of an address allows the subscriber to use up to $2^{32} - 1$ subnets. Therefore, subnet masking, as defined in IPv4, is not needed here.

Table 21.4 *Comparison of query messages in ICMPv4 and ICMPv6*

Type of Message	*Version 4*	*Version 6*
Echo request and reply	Yes	Yes
Timestamp request and reply	Yes	No
Address-mask request and reply	Yes	No
Router solicitation and advertisement	Yes	Yes
Neighbor solicitation and advertisement	ARP	Yes
Group membership	IGMP	Yes

Echo Request and Reply

The idea and format of the echo request and reply messages are the same as those in version 4.

Router Solicitation and Advertisement

The idea behind the router-solicitation and -advertisement messages is the same as in version 4.

Neighbor Solicitation and Advertisement

As previously mentioned, the network layer in version 4 contains an independent protocol called Address Resolution Protocol (ARP). In version 6, this protocol is eliminated, and its duties are included in ICMPv6. The idea is exactly the same, but the format of the message has changed.

Group Membership

As previously mentioned, the network layer in version 4 contains an independent protocol called IGMP. In version 6, this protocol is eliminated, and its duties are included in ICMPv6. The purpose is exactly the same.

21.5 RECOMMENDED READING

For more details about subjects discussed in this chapter, we recommend the following books and site. The items in brackets [. . .] refer to the reference list at the end of the text.

a. 224.18.72.8
b. 235.18.72.8
c. 237.18.6.88
d. 224.88.12.8

Research Activities

29. Use the *ping* program to test your own computer (loopback).
30. Use the *ping* program to test a host inside the United States.
31. Use the *ping* program to test a host outside the United States.
32. Use *traceroute* (or *tracert*) to find the route from your computer to a computer in a college or university.
33. Use *netstat* to find out if your server supports multicast addressing.
34. DHCP uses several messages such as DHCPREQUEST, DHCPDECLINE, DHCPACK, DHCPNACK, and DHCPRELEASE. Find the purpose of these messages.

CHAPTER 22

Network Layer: Delivery, Forwarding, and Routing

This chapter describes the delivery, forwarding, and routing of IP packets to their final destinations. **Delivery** refers to the way a packet is handled by the underlying networks under the control of the network layer. **Forwarding** refers to the way a packet is delivered to the next station. **Routing** refers to the way routing tables are created to help in forwarding.

Routing protocols are used to continuously update the routing tables that are consulted for forwarding and routing. In this chapter, we also briefly discuss common unicast and **multicast routing** protocols.

22.1 DELIVERY

The network layer supervises the handling of the packets by the underlying physical networks. We define this handling as the delivery of a packet.

Direct Versus Indirect Delivery

The delivery of a packet to its final destination is accomplished by using two different methods of delivery, direct and indirect, as shown in Figure 22.1.

Direct Delivery

In a **direct delivery,** the final destination of the packet is a host connected to the same physical network as the deliverer. Direct delivery occurs when the source and destination of the packet are located on the same physical network or when the delivery is between the last router and the destination host.

The sender can easily determine if the delivery is direct. It can extract the network address of the destination (using the mask) and compare this address with the addresses of the networks to which it is connected. If a match is found, the delivery is direct.

Indirect Delivery

If the destination host is not on the same network as the deliverer, the packet is delivered indirectly. In an **indirect delivery,** the packet goes from router to router until it reaches the one connected to the same physical network as its final destination. Note

Figure 22.1 *Direct and indirect delivery*

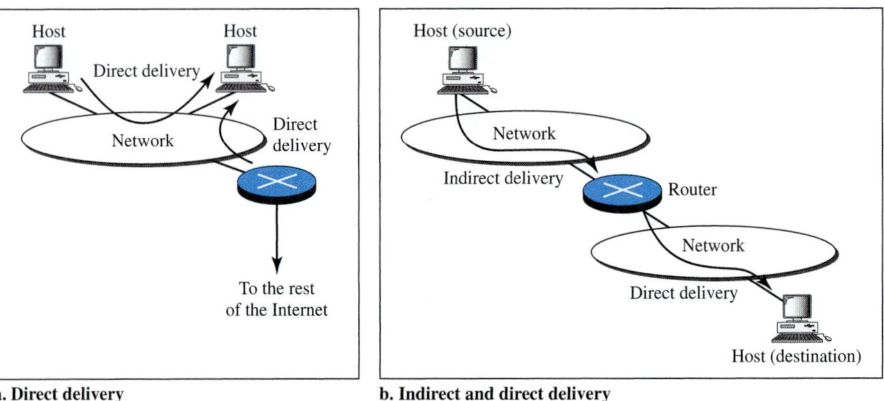

a. Direct delivery **b. Indirect and direct delivery**

that a delivery always involves one direct delivery but zero or more indirect deliveries. Note also that the last delivery is always a direct delivery.

22.2 FORWARDING

Forwarding means to place the packet in its route to its destination. Forwarding requires a host or a router to have a routing table. When a host has a packet to send or when a router has received a packet to be forwarded, it looks at this table to find the route to the final destination. However, this simple solution is impossible today in an internetwork such as the Internet because the number of entries needed in the routing table would make table lookups inefficient.

Forwarding Techniques

Several techniques can make the size of the routing table manageable and also handle issues such as security. We briefly discuss these methods here.

Next-Hop Method Versus Route Method

One technique to reduce the contents of a routing table is called the **next-hop method.** In this technique, the routing table holds only the address of the next hop instead of information about the complete route **(route method).** The entries of a routing table must be consistent with one another. Figure 22.2 shows how routing tables can be simplified by using this technique.

Network-Specific Method Versus Host-Specific Method

A second technique to reduce the routing table and simplify the searching process is called the **network-specific method.** Here, instead of having an entry for every destination host connected to the same physical network **(host-specific method),** we have

Figure 22.2 *Route method versus next-hop method*

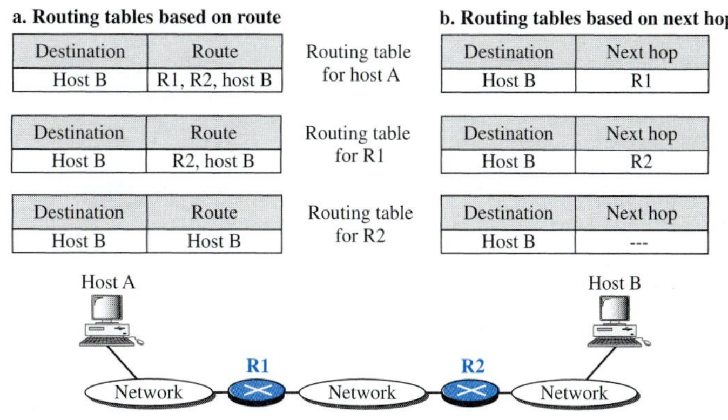

only one entry that defines the address of the destination network itself. In other words, we treat all hosts connected to the same network as one single entity. For example, if 1000 hosts are attached to the same network, only one entry exists in the routing table instead of 1000. Figure 22.3 shows the concept.

Figure 22.3 *Host-specific versus network-specific method*

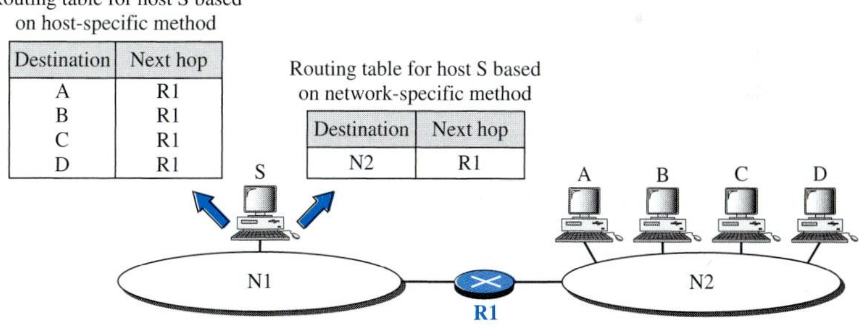

Host-specific routing is used for purposes such as checking the route or providing security measures.

Default Method

Another technique to simplify routing is called the **default method.** In Figure 22.4 host A is connected to a network with two routers. Router R1 routes the packets to hosts connected to network N2. However, for the rest of the Internet, router R2 is used. So instead of listing all networks in the entire Internet, host A can just have one entry called the *default* (normally defined as network address 0.0.0.0).

Figure 22.4 *Default method*

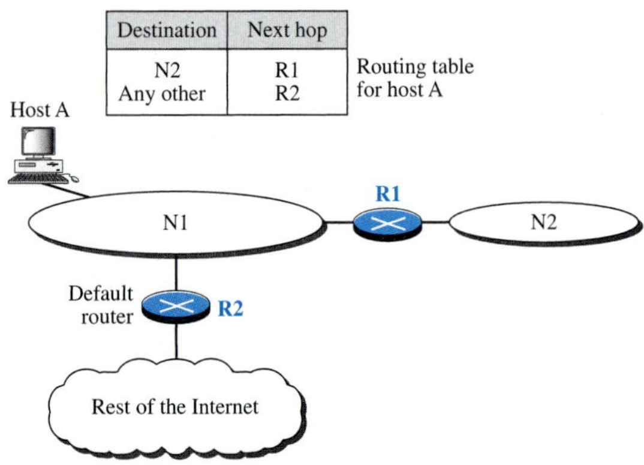

Forwarding Process

Let us discuss the forwarding process. We assume that hosts and routers use classless addressing because classful addressing can be treated as a special case of classless addressing. In classless addressing, the routing table needs to have one row of information for each block involved. The table needs to be searched based on the network address (first address in the block). Unfortunately, the destination address in the packet gives no clue about the network address. To solve the problem, we need to include the mask (/*n*) in the table; we need to have an extra column that includes the mask for the corresponding block. Figure 22.5 shows a simple forwarding module for classless addressing.

Figure 22.5 *Simplified forwarding module in classless address*

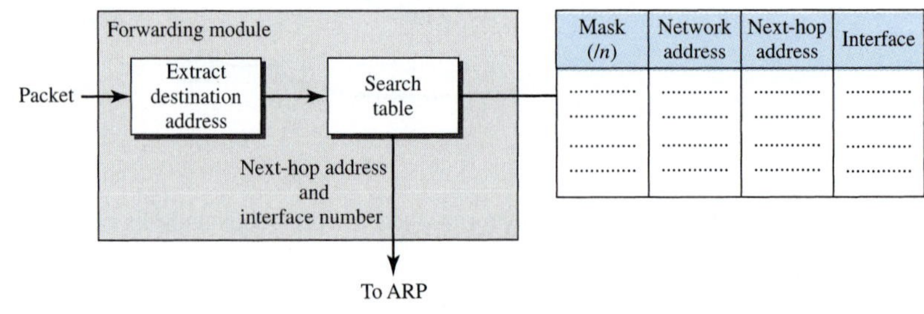

Note that we need at least four columns in our routing table; usually there are more.

In classless addressing, we need at least four columns in a routing table.

Example 22.1

Make a routing table for router R1, using the configuration in Figure 22.6.

Figure 22.6 *Configuration for Example 22.1*

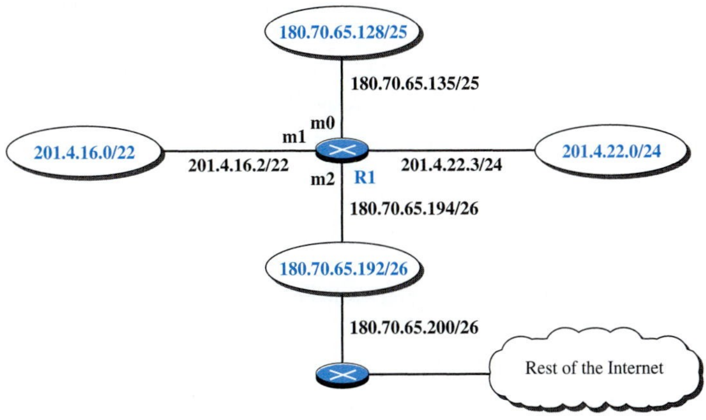

Solution

Table 22.1 shows the corresponding table.

Table 22.1 *Routing table for router R1 in Figure 22.6*

Mask	*Network Address*	*Next Hop*	*Interface*
/26	180.70.65.192	—	m2
/25	180.70.65.128	—	m0
/24	201.4.22.0	—	m3
/22	201.4.16.0	....	m1
Any	Any	180.70.65.200	m2

Example 22.2

Show the forwarding process if a packet arrives at R1 in Figure 22.6 with the destination address 180.70.65.140.

Solution

The router performs the following steps:

1. The first mask (/26) is applied to the destination address. The result is 180.70.65.128, which does not match the corresponding network address.
2. The second mask (/25) is applied to the destination address. The result is 180.70.65.128, which matches the corresponding network address. The **next-hop address** (the destination address of the packet in this case) and the interface number m0 are passed to ARP for further processing.

Example 22.3

Show the forwarding process if a packet arrives at R1 in Figure 22.6 with the destination address 201.4.22.35.

Solution

The router performs the following steps:

1. The first mask (/26) is applied to the destination address. The result is 201.4.22.0, which does not match the corresponding network address (row 1).
2. The second mask (/25) is applied to the destination address. The result is 201.4.22.0, which does not match the corresponding network address (row 2).
3. The third mask (/24) is applied to the destination address. The result is 201.4.22.0, which matches the corresponding network address. The destination address of the packet and the interface number m3 are passed to ARP.

Example 22.4

Show the forwarding process if a packet arrives at R1 in Figure 22.6 with the destination address 18.24.32.78.

Solution

This time all masks are applied, one by one, to the destination address, but no matching network address is found. When it reaches the end of the table, the module gives the next-hop address 180.70.65.200 and interface number m2 to ARP. This is probably an outgoing package that needs to be sent, via the default router, to someplace else in the Internet.

Address Aggregation

When we use classless addressing, it is likely that the number of routing table entries will increase. This is so because the intent of classless addressing is to divide up the whole address space into manageable blocks. The increased size of the table results in an increase in the amount of time needed to search the table. To alleviate the problem, the idea of **address aggregation** was designed. In Figure 22.7 we have two routers.

Router R1 is connected to networks of four organizations that each use 64 addresses. Router R2 is somewhere far from R1. Router R1 has a longer routing table because each packet must be correctly routed to the appropriate organization. Router R2, on the other hand, can have a very small routing table. For R2, any packet with destination 140.24.7.0 to 140.24.7.255 is sent out from interface m0 regardless of the organization number. This is called address aggregation because the blocks of addresses for four organizations are aggregated into one larger block. Router R2 would have a longer routing table if each organization had addresses that could not be aggregated into one block.

Note that although the idea of address aggregation is similar to the idea of subnetting, we do not have a common site here; the network for each organization is independent. In addition, we can have several levels of aggregation.

Longest Mask Matching

What happens if one of the organizations in Figure 22.7 is not geographically close to the other three? For example, if organization 4 cannot be connected to router R1 for some reason, can we still use the idea of address aggregation and still assign block 140.24.7.192/26 to organization 4?

Figure 22.7 *Address aggregation*

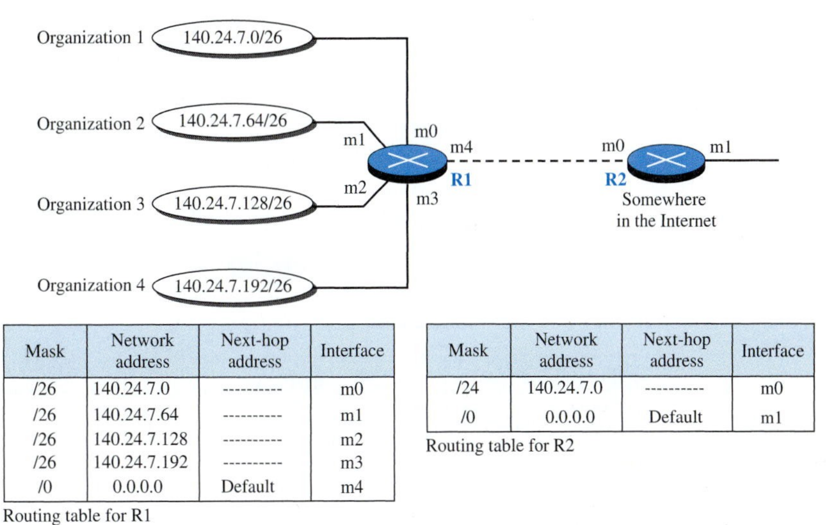

Mask	Network address	Next-hop address	Interface
/26	140.24.7.0	----------	m0
/26	140.24.7.64	----------	m1
/26	140.24.7.128	----------	m2
/26	140.24.7.192	----------	m3
/0	0.0.0.0	Default	m4

Routing table for R1

Mask	Network address	Next-hop address	Interface
/24	140.24.7.0	----------	m0
/0	0.0.0.0	Default	m1

Routing table for R2

The answer is yes because routing in classless addressing uses another principle, **longest mask matching.** This principle states that the routing table is sorted from the longest mask to the shortest mask. In other words, if there are three masks /27, /26, and /24, the mask /27 must be the first entry and /24 must be last. Let us see if this principle solves the situation in which organization 4 is separated from the other three organizations. Figure 22.8 shows the situation.

Suppose a packet arrives for organization 4 with destination address 140.24.7.200. The first mask at router R2 is applied, which gives the network address 140.24.7.192. The packet is routed correctly from interface m1 and reaches organization 4. If, however, the routing table was not stored with the longest prefix first, applying the /24 mask would result in the incorrect routing of the packet to router R1.

Hierarchical Routing

To solve the problem of gigantic routing tables, we can create a sense of hierarchy in the routing tables. In Chapter 1, we mentioned that the Internet today has a sense of hierarchy. We said that the Internet is divided into international and national ISPs. National ISPs are divided into regional ISPs, and regional ISPs are divided into local ISPs. If the routing table has a sense of hierarchy like the Internet architecture, the routing table can decrease in size.

Let us take the case of a local ISP. A local ISP can be assigned a single, but large block of addresses with a certain prefix length. The local ISP can divide this block into smaller blocks of different sizes and can assign these to individual users and organizations, both large and small. If the block assigned to the local ISP starts with a.b.c.d/*n*, the ISP can create blocks starting with e.f.g.h/*m*, where *m* may vary for each customer and is greater than *n*.

Figure 22.8 *Longest mask matching*

Routing table for R2

Mask	Network address	Next-hop address	Interface
/26	140.24.7.192	----------	m1
/24	140.24.7.0	----------	m0
/??	???????	?????????	m1
/0	0.0.0.0	Default	m2

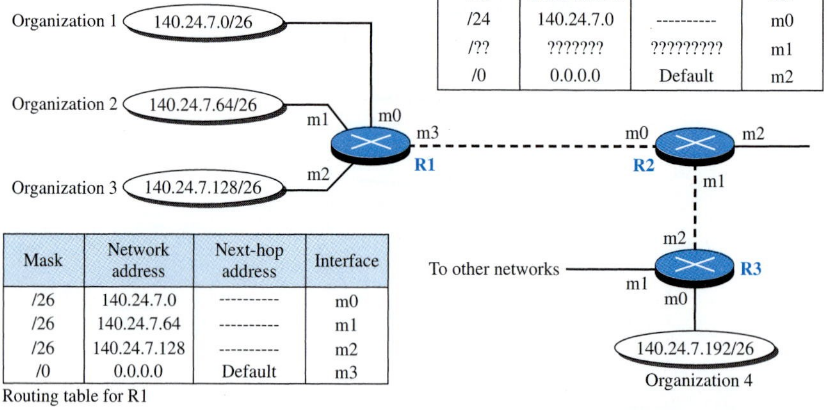

Mask	Network address	Next-hop address	Interface
/26	140.24.7.0	----------	m0
/26	140.24.7.64	----------	m1
/26	140.24.7.128	----------	m2
/0	0.0.0.0	Default	m3

Routing table for R1

Mask	Network address	Next-hop address	Interface
/26	140.24.7.192	----------	m0
/??	???????	?????????	m1
/0	0.0.0.0	Default	m2

Routing table for R3

How does this reduce the size of the routing table? The rest of the Internet does not have to be aware of this division. All customers of the local ISP are defined as a.b.c.d/*n* to the rest of the Internet. Every packet destined for one of the addresses in this large block is routed to the local ISP. There is only one entry in every router in the world for all these customers. They all belong to the same group. Of course, inside the local ISP, the router must recognize the subblocks and route the packet to the destined customer. If one of the customers is a large organization, it also can create another level of hierarchy by subnetting and dividing its subblock into smaller subblocks (or sub-subblocks). In classless routing, the levels of hierarchy are unlimited so long as we follow the rules of classless addressing.

Example 22.5

As an example of **hierarchical routing,** let us consider Figure 22.9. A regional ISP is granted 16,384 addresses starting from 120.14.64.0. The regional ISP has decided to divide this block into four subblocks, each with 4096 addresses. Three of these subblocks are assigned to three local ISPs; the second subblock is reserved for future use. Note that the mask for each block is /20 because the original block with mask /18 is divided into 4 blocks.

The first local ISP has divided its assigned subblock into 8 smaller blocks and assigned each to a small ISP. Each small ISP provides services to 128 households (H001 to H128), each using four addresses. Note that the mask for each small ISP is now /23 because the block is further divided into 8 blocks. Each household has a mask of /30, because a household has only four addresses (2^{32-30} is 4).

Figure 22.9 *Hierarchical routing with ISPs*

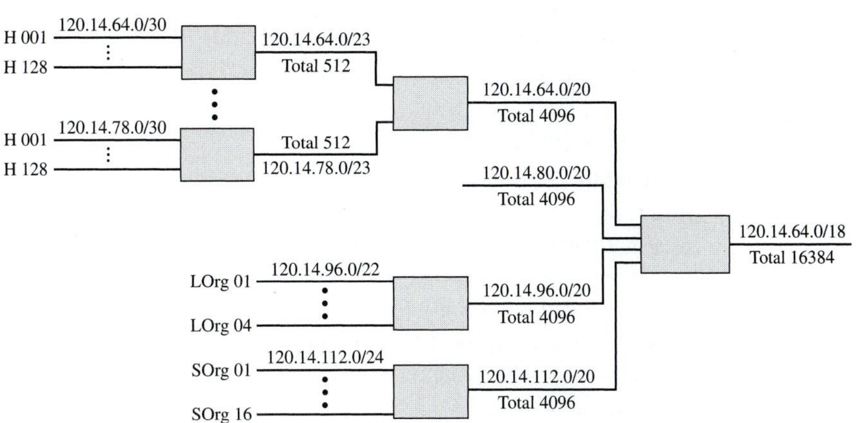

The second local ISP has divided its block into 4 blocks and has assigned the addresses to four large organizations (LOrg01 to LOrg04). Note that each large organization has 1024 addresses, and the mask is /22.

The third local ISP has divided its block into 16 blocks and assigned each block to a small organization (SOrg01 to SOrg16). Each small organization has 256 addresses, and the mask is /24.

There is a sense of hierarchy in this configuration. All routers in the Internet send a packet with destination address 120.14.64.0 to 120.14.127.255 to the regional ISP.

The regional ISP sends every packet with destination address 120.14.64.0 to 120.14.79.255 to local ISP1. Local ISP1 sends every packet with destination address 120.14.64.0 to 120.14.64.3 to H001.

Geographical Routing

To decrease the size of the routing table even further, we need to extend hierarchical routing to include geographical routing. We must divide the entire address space into a few large blocks. We assign a block to North America, a block to Europe, a block to Asia, a block to Africa, and so on. The routers of ISPs outside Europe will have only one entry for packets to Europe in their routing tables. The routers of ISPs outside North America will have only one entry for packets to North America in their routing tables. And so on.

Routing Table

Let us now discuss routing tables. A host or a router has a routing table with an entry for each destination, or a combination of destinations, to route IP packets. The routing table can be either static or dynamic.

Static Routing Table

A **static routing table** contains information entered manually. The administrator enters the route for each destination into the table. When a table is created, it cannot update automatically when there is a change in the Internet. The table must be manually altered by the administrator.

A static routing table can be used in a small internet that does not change very often, or in an experimental internet for troubleshooting. It is poor strategy to use a static routing table in a big internet such as the Internet.

Dynamic Routing Table

A **dynamic routing table** is updated periodically by using one of the dynamic routing protocols such as RIP, OSPF, or BGP. Whenever there is a change in the Internet, such as a shutdown of a router or breaking of a link, the dynamic routing protocols update all the tables in the routers (and eventually in the host) automatically.

The routers in a big internet such as the Internet need to be updated dynamically for efficient delivery of the IP packets. We discuss in detail the three dynamic routing protocols later in the chapter.

Format

As mentioned previously, a routing table for classless addressing has a minimum of four columns. However, some of today's routers have even more columns. We should be aware that the number of columns is vendor-dependent, and not all columns can be found in all routers. Figure 22.10 shows some common fields in today's routers.

Figure 22.10 *Common fields in a routing table*

Mask	Network address	Next-hop address	Interface	Flags	Reference count	Use
....	....	....	....	....	....	....

- **Mask.** This field defines the mask applied for the entry.
- **Network address.** This field defines the network address to which the packet is finally delivered. In the case of host-specific routing, this field defines the address of the destination host.
- **Next-hop address.** This field defines the address of the next-hop router to which the packet is delivered.
- **Interface.** This field shows the name of the interface.
- **Flags.** This field defines up to five flags. Flags are on/off switches that signify either presence or absence. The five flags are U (up), G (gateway), H (host-specific), D (added by redirection), and M (modified by redirection).
    a. **U (up).** The U flag indicates the router is up and running. If this flag is not present, it means that the router is down. The packet cannot be forwarded and is discarded.
    b. **G (gateway).** The G flag means that the destination is in another network. The packet is delivered to the next-hop router for delivery (indirect delivery). When this flag is missing, it means the destination is in this network (direct delivery).

c. **H (host-specific).** The H flag indicates that the entry in the network address field is a host-specific address. When it is missing, it means that the address is only the network address of the destination.

d. **D (added by redirection).** The D flag indicates that routing information for this destination has been added to the host routing table by a redirection message from ICMP. We discussed redirection and the ICMP protocol in Chapter 21.

e. **M (modified by redirection).** The M flag indicates that the routing information for this destination has been modified by a redirection message from ICMP. We discussed redirection and the ICMP protocol in Chapter 21.

- **Reference count.** This field gives the number of users of this route at the moment. For example, if five people at the same time are connecting to the same host from this router, the value of this column is 5.
- **Use.** This field shows the number of packets transmitted through this router for the corresponding destination.

Utilities

There are several utilities that can be used to find the routing information and the contents of a routing table. We discuss *netstat* and *ifconfig*.

Example 22.6

One utility that can be used to find the contents of a routing table for a host or router is ***netstat*** in UNIX or LINUX. The following shows the list of the contents of a default server. We have used two options, r and n. The option r indicates that we are interested in the routing table, and the option n indicates that we are looking for numeric addresses. Note that this is a routing table for a host, not a router. Although we discussed the routing table for a router throughout the chapter, a host also needs a routing table.

$ netstat -rn

Kernel IP routing table

Destination	**Gateway**	**Mask**	**Flags**	**Iface**
153.18.16.0	0.0.0.0	255.255.240.0	U	eth0
127.0.0.0	0.0.0.0	255.0.0.0	U	lo
0.0.0.0	153.18.31.254	0.0.0.0	UG	eth0

Note also that the order of columns is different from what we showed. The destination column here defines the network address. The term ***gateway*** used by UNIX is synonymous with *router.* This column actually defines the address of the next hop. The value 0.0.0.0 shows that the delivery is direct. The last entry has a flag of G, which means that the destination can be reached through a router (default router). The *Iface* defines the interface. The host has only one real interface, eth0, which means interface 0 connected to an Ethernet network. The second interface, lo, is actually a virtual loopback interface indicating that the host accepts packets with loopback address 127.0.0.0.

More information about the IP address and physical address of the server can be found by using the *ifconfig* command on the given interface (eth0).

```
$ ifconfig eth0
eth0  Link encap:Ethernet  HWaddr 00:B0:D0:DF:09:5D
inet addr:153.18.17.11  Bcast:153.18.31.255  Mask:255.255.240.0
. . .
```

From the above information, we can deduce the configuration of the server, as shown in Figure 22.11.

Figure 22.11 *Configuration of the server for Example 22.6*

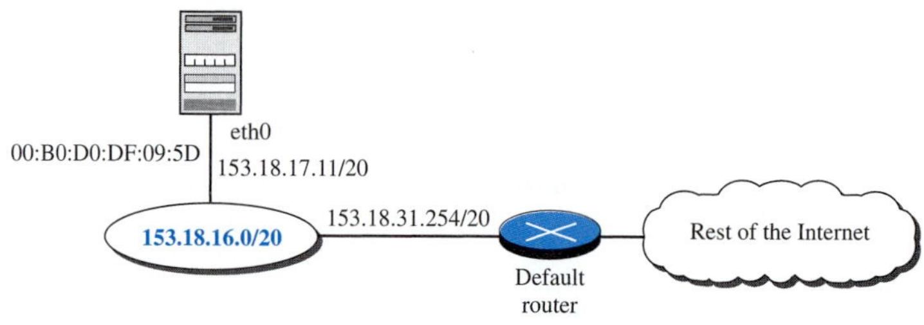

Note that the *ifconfig* command gives us the IP address and the physical (hardware) address of the interface.

22.3 UNICAST ROUTING PROTOCOLS

A routing table can be either static or dynamic. A *static table* is one with manual entries. A *dynamic table,* on the other hand, is one that is updated automatically when there is a change somewhere in the internet. Today, an internet needs dynamic routing tables. The tables need to be updated as soon as there is a change in the internet. For instance, they need to be updated when a router is down, and they need to be updated whenever a better route has been found.

Routing protocols have been created in response to the demand for dynamic routing tables. A routing protocol is a combination of rules and procedures that lets routers in the internet inform each other of changes. It allows routers to share whatever they know about the internet or their neighborhood. The sharing of information allows a router in San Francisco to know about the failure of a network in Texas. The routing protocols also include procedures for combining information received from other routers.

Optimization

A router receives a packet from a network and passes it to another network. A router is usually attached to several networks. When it receives a packet, to which network should it pass the packet? The decision is based on optimization: Which of the available pathways is the optimum pathway? What is the definition of the term *optimum*?

One approach is to assign a cost for passing through a network. We call this cost a **metric.** However, the metric assigned to each network depends on the type of protocol. Some simple protocols, such as the Routing Information Protocol (RIP), treat all networks as equals. The cost of passing through a network is the same; it is one hop count. So if a packet passes through 10 networks to reach the destination, the total cost is 10 hop counts.

Other protocols, such as Open Shortest Path First (OSPF), allow the administrator to assign a cost for passing through a network based on the type of service required. A route through a network can have different costs (metrics). For example, if maximum throughput is the desired type of service, a satellite link has a lower metric than a fiber-optic line. On the other hand, if minimum delay is the desired type of service, a fiber-optic line has a lower metric than a satellite link. Routers use routing tables to help decide the best route. OSPF protocol allows each router to have several routing tables based on the required type of service.

Other protocols define the metric in a totally different way. In the Border Gateway Protocol (BGP), the criterion is the policy, which can be set by the administrator. The policy defines what paths should be chosen.

Intra- and Interdomain Routing

Today, an internet can be so large that one routing protocol cannot handle the task of updating the routing tables of all routers. For this reason, an internet is divided into autonomous systems. An **autonomous system (AS)** is a group of networks and routers under the authority of a single administration. Routing inside an autonomous system is referred to as **intradomain routing.** Routing between autonomous systems is referred to as **interdomain routing.** Each autonomous system can choose one or more intradomain routing protocols to handle routing inside the autonomous system. However, only one interdomain routing protocol handles routing between autonomous systems (see Figure 22.12).

Figure 22.12 *Autonomous systems*

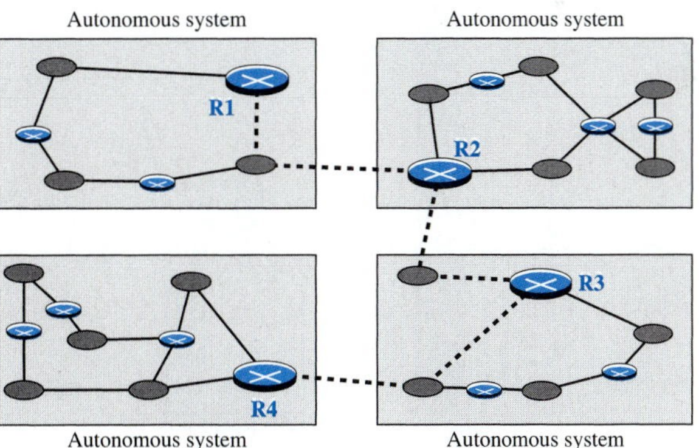

Several intradomain and interdomain routing protocols are in use. In this section, we cover only the most popular ones. We discuss two intradomain routing protocols: distance vector and link state. We also introduce one interdomain routing protocol: path vector (see Figure 22.13).

Routing Information Protocol (RIP) is an implementation of the distance vector protocol. **Open Shortest Path First (OSPF)** is an implementation of the link state protocol. **Border Gateway Protocol (BGP)** is an implementation of the path vector protocol.

Figure 22.13 *Popular routing protocols*

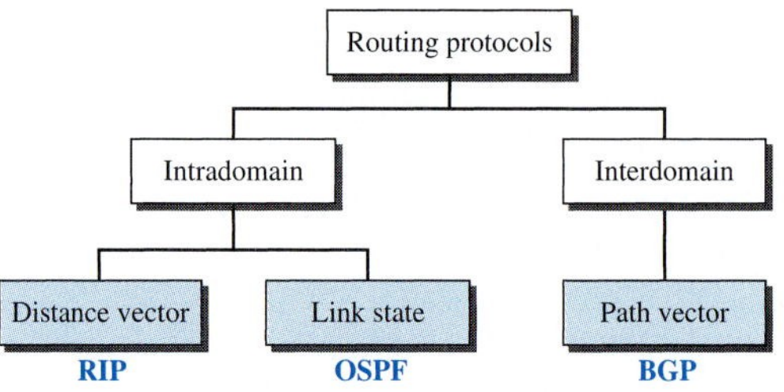

Distance Vector Routing

In **distance vector routing,** the least-cost route between any two nodes is the route with minimum distance. In this protocol, as the name implies, each node maintains a vector (table) of minimum distances to every node. The table at each node also guides the packets to the desired node by showing the next stop in the route (next-hop routing).

We can think of nodes as the cities in an area and the lines as the roads connecting them. A table can show a tourist the minimum distance between cities.

In Figure 22.14, we show a system of five nodes with their corresponding tables.

Figure 22.14 *Distance vector routing tables*

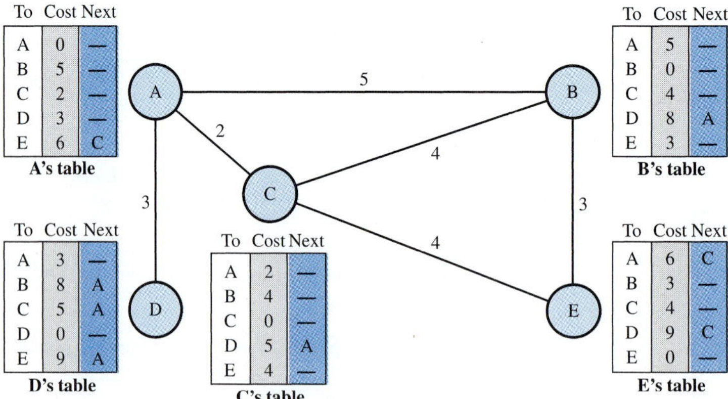

A's table

To	Cost	Next
A	0	—
B	5	—
C	2	—
D	3	—
E	6	C

B's table

To	Cost	Next
A	5	—
B	0	—
C	4	—
D	8	A
E	3	—

C's table

To	Cost	Next
A	2	—
B	4	—
C	0	—
D	5	A
E	4	—

D's table

To	Cost	Next
A	3	—
B	8	A
C	5	A
D	0	—
E	9	A

E's table

To	Cost	Next
A	6	C
B	3	—
C	4	—
D	9	C
E	0	—

The table for node A shows how we can reach any node from this node. For example, our least cost to reach node E is 6. The route passes through C.

Initialization

The tables in Figure 22.14 are stable; each node knows how to reach any other node and the cost. At the beginning, however, this is not the case. Each node can know only the distance between itself and its **immediate neighbors,** those directly connected to it. So for the moment, we assume that each node can send a message to the immediate neighbors and find the distance between itself and these neighbors. Figure 22.15 shows the initial tables for each node. The distance for any entry that is not a neighbor is marked as infinite (unreachable).

Figure 22.15 *Initialization of tables in distance vector routing*

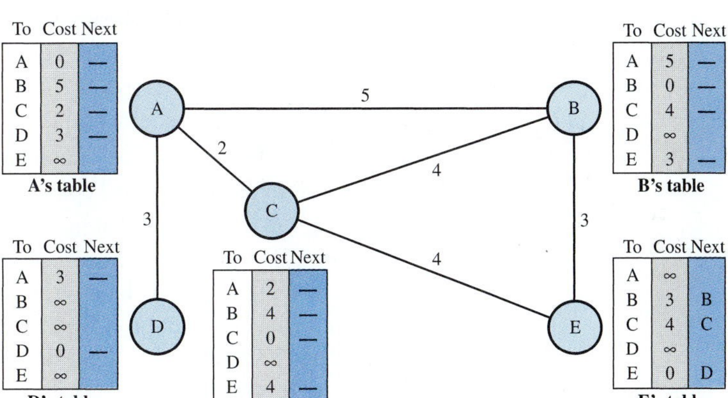

Sharing

The whole idea of distance vector routing is the sharing of information between neighbors. Although node A does not know about node E, node C does. So if node C shares its routing table with A, node A can also know how to reach node E. On the other hand, node C does not know how to reach node D, but node A does. If node A shares its routing table with node C, node C also knows how to reach node D. In other words, nodes A and C, as immediate neighbors, can improve their routing tables if they help each other.

There is only one problem. How much of the table must be shared with each neighbor? A node is not aware of a neighbor's table. The best solution for each node is to send its entire table to the neighbor and let the neighbor decide what part to use and what part to discard. However, the third column of a table (next stop) is not useful for the neighbor. When the neighbor receives a table, this column needs to be replaced with the sender's name. If any of the rows can be used, the next node is the sender of the table. A node therefore can send only the first two columns of its table to any neighbor. In other words, sharing here means sharing only the first two columns.

In distance vector routing, each node shares its routing table with its immediate neighbors periodically and when there is a change.

Updating

When a node receives a two-column table from a neighbor, it needs to update its routing table. Updating takes three steps:

1. The receiving node needs to add the cost between itself and the sending node to each value in the second column. The logic is clear. If node C claims that its distance to a destination is x mi, and the distance between A and C is y mi, then the distance between A and that destination, via C, is $x + y$ mi.
2. The receiving node needs to add the name of the sending node to each row as the third column if the receiving node uses information from any row. The sending node is the next node in the route.
3. The receiving node needs to compare each row of its old table with the corresponding row of the modified version of the received table.
   a. If the next-node entry is different, the receiving node chooses the row with the smaller cost. If there is a tie, the old one is kept.
   b. If the next-node entry is the same, the receiving node chooses the new row. For example, suppose node C has previously advertised a route to node X with distance 3. Suppose that now there is no path between C and X; node C now advertises this route with a distance of infinity. Node A must not ignore this value even though its old entry is smaller. The old route does not exist any more. The new route has a distance of infinity.

Figure 22.16 shows how node A updates its routing table after receiving the partial table from node C.

Figure 22.16 *Updating in distance vector routing*

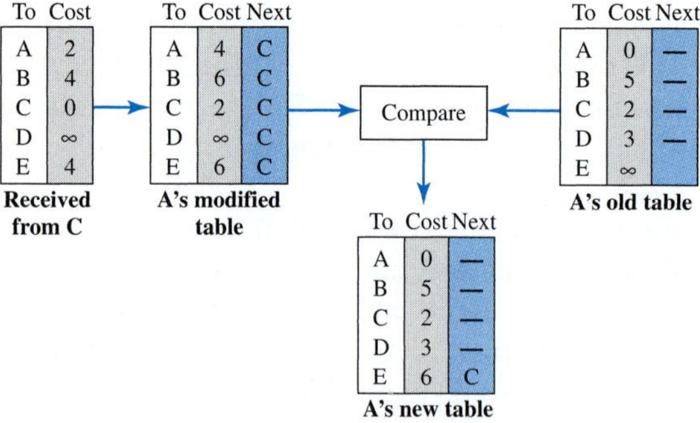

There are several points we need to emphasize here. First, as we know from mathematics, when we add any number to infinity, the result is still infinity. Second, the modified table shows how to reach A from A via C. If A needs to reach itself via C, it needs to go to C and come back, a distance of 4. Third, the only benefit from this updating of node A is the last entry, how to reach E. Previously, node A did not know how to reach E (distance of infinity); now it knows that the cost is 6 via C.

Each node can update its table by using the tables received from other nodes. In a short time, if there is no change in the network itself, such as a failure in a link, each node reaches a stable condition in which the contents of its table remains the same.

When to Share

The question now is, When does a node send its partial routing table (only two columns) to all its immediate neighbors? The table is sent both periodically and when there is a change in the table.

Periodic Update A node sends its routing table, normally every 30 s, in a periodic update. The period depends on the protocol that is using distance vector routing.

Triggered Update A node sends its two-column routing table to its neighbors anytime there is a change in its routing table. This is called a **triggered update.** The change can result from the following.

1. A node receives a table from a neighbor, resulting in changes in its own table after updating.
2. A node detects some failure in the neighboring links which results in a distance change to infinity.

Two-Node Loop Instability

A problem with distance vector routing is instability, which means that a network using this protocol can become unstable. To understand the problem, let us look at the scenario depicted in Figure 22.17.

Figure 22.17 *Two-node instability*

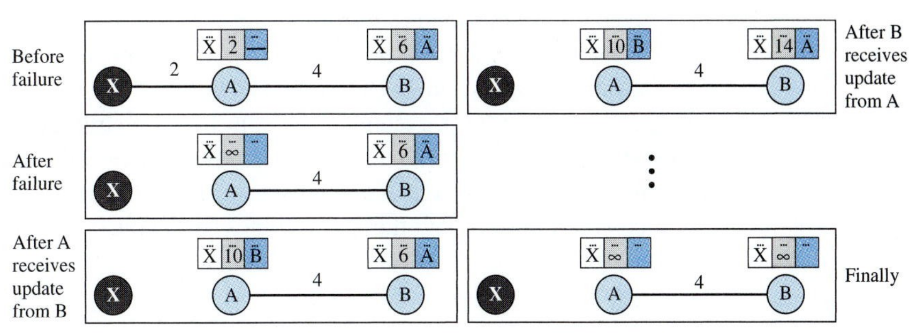

Figure 22.17 shows a system with three nodes. We have shown only the portions of the routing table needed for our discussion. At the beginning, both nodes A and B know how to reach node X. But suddenly, the link between A and X fails. Node A changes its table. If A can send its table to B immediately, everything is fine. However, the system becomes unstable if B sends its routing table to A before receiving A's routing table. Node A receives the update and, assuming that B has found a way to reach X, immediately updates its routing table. Based on the triggered update strategy, A sends its new

update to B. Now B thinks that something has been changed around A and updates its routing table. The cost of reaching X increases gradually until it reaches infinity. At this moment, both A and B know that X cannot be reached. However, during this time the system is not stable. Node A thinks that the route to X is via B; node B thinks that the route to X is via A. If A receives a packet destined for X, it goes to B and then comes back to A. Similarly, if B receives a packet destined for X, it goes to A and comes back to B. Packets bounce between A and B, creating a two-node loop problem. A few solutions have been proposed for instability of this kind.

Defining Infinity The first obvious solution is to redefine infinity to a smaller number, such as 100. For our previous scenario, the system will be stable in less than 20 updates. As a matter of fact, most implementations of the distance vector protocol define the distance between each node to be 1 and define 16 as infinity. However, this means that the distance vector routing cannot be used in large systems. The size of the network, in each direction, can not exceed 15 hops.

Split Horizon Another solution is called **split horizon.** In this strategy, instead of flooding the table through each interface, each node sends only part of its table through each interface. If, according to its table, node B thinks that the optimum route to reach X is via A, it does not need to advertise this piece of information to A; the information has come from A (A already knows). Taking information from node A, modifying it, and sending it back to node A creates the confusion. In our scenario, node B eliminates the last line of its routing table before it sends it to A. In this case, node A keeps the value of infinity as the distance to X. Later when node A sends its routing table to B, node B also corrects its routing table. The system becomes stable after the first update: both node A and B know that X is not reachable.

Split Horizon and Poison Reverse Using the split horizon strategy has one drawback. Normally, the distance vector protocol uses a timer, and if there is no news about a route, the node deletes the route from its table. When node B in the previous scenario eliminates the route to X from its advertisement to A, node A cannot guess that this is due to the split horizon strategy (the source of information was A) or because B has not received any news about X recently. The split horizon strategy can be combined with the **poison reverse** strategy. Node B can still advertise the value for X, but if the source of information is A, it can replace the distance with infinity as a warning: "Do not use this value; what I know about this route comes from you."

Three-Node Instability

The two-node instability can be avoided by using the split horizon strategy combined with poison reverse. However, if the instability is between three nodes, stability cannot be guaranteed. Figure 22.18 shows the scenario.

Suppose, after finding that X is not reachable, node A sends a packet to B and C to inform them of the situation. Node B immediately updates its table, but the packet to C is lost in the network and never reaches C. Node C remains in the dark and still thinks that there is a route to X via A with a distance of 5. After a while, node C sends to B its routing table, which includes the route to X. Node B is totally fooled here. It receives information on the route to X from C, and according to the algorithm, it updates its

Figure 22.18 *Three-node instability*

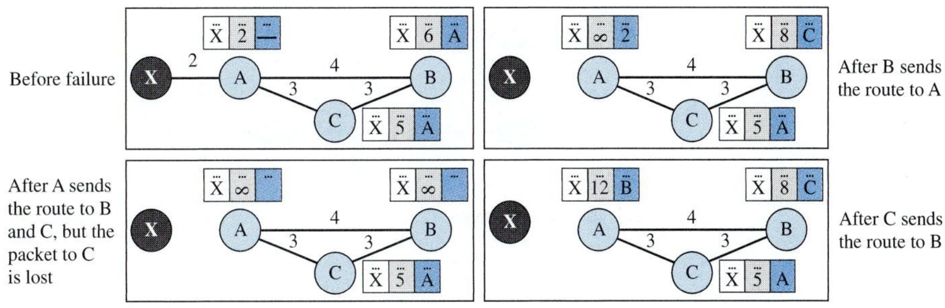

table, showing the route to X via C with a cost of 8. This information has come from C, not from A, so after awhile node B may advertise this route to A. Now A is fooled and updates its table to show that A can reach X via B with a cost of 12. Of course, the loop continues; now A advertises the route to X to C, with increased cost, but not to B. Node C then advertises the route to B with an increased cost. Node B does the same to A. And so on. The loop stops when the cost in each node reaches infinity.

RIP

The **Routing Information Protocol (RIP)** is an intradomain routing protocol used inside an autonomous system. It is a very simple protocol based on distance vector routing. RIP implements distance vector routing directly with some considerations:

1. In an autonomous system, we are dealing with routers and networks (links). The routers have routing tables; networks do not.
2. The destination in a routing table is a network, which means the first column defines a network address.
3. The metric used by RIP is very simple; the distance is defined as the number of links (networks) to reach the destination. For this reason, the metric in RIP is called a **hop count.**
4. Infinity is defined as 16, which means that any route in an autonomous system using RIP cannot have more than 15 hops.
5. The next-node column defines the address of the router to which the packet is to be sent to reach its destination.

Figure 22.19 shows an autonomous system with seven networks and four routers. The table of each router is also shown. Let us look at the routing table for R1. The table has seven entries to show how to reach each network in the autonomous system. Router R1 is directly connected to networks 130.10.0.0 and 130.11.0.0, which means that there are no next-hop entries for these two networks. To send a packet to one of the three networks at the far left, router R1 needs to deliver the packet to R2. The next-node entry for these three networks is the interface of router R2 with IP address 130.10.0.1. To send a packet to the two networks at the far right, router R1 needs to send the packet to the interface of router R4 with IP address 130.11.0.1. The other tables can be explained similarly.

Figure 22.19 *Example of a domain using RIP*

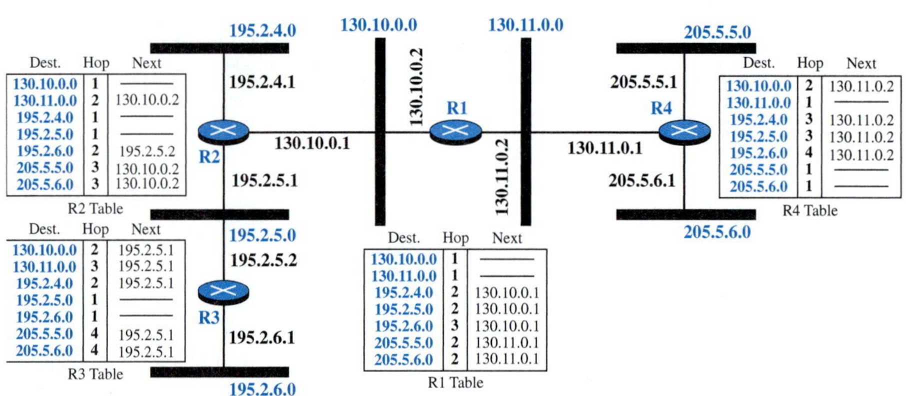

Link State Routing

Link state routing has a different philosophy from that of distance vector routing. In link state routing, if each node in the domain has the entire topology of the domain—the list of nodes and links, how they are connected including the type, cost (metric), and condition of the links (up or down)—the node can use **Dijkstra's algorithm** to build a routing table. Figure 22.20 shows the concept.

Figure 22.20 *Concept of link state routing*

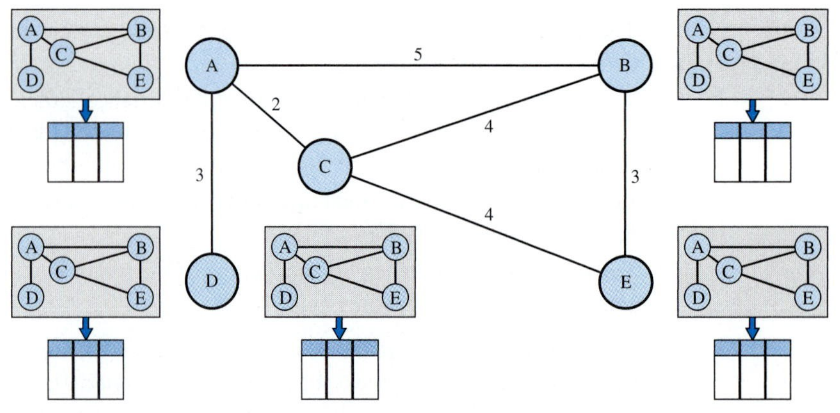

The figure shows a simple domain with five nodes. Each node uses the same topology to create a routing table, but the routing table for each node is unique because the calculations are based on different interpretations of the topology. This is analogous to a city map. While each person may have the same map, each needs to take a different route to reach her specific destination.

The topology must be dynamic, representing the latest state of each node and each link. If there are changes in any point in the network (a link is down, for example), the topology must be updated for each node.

How can a common topology be dynamic and stored in each node? No node can know the topology at the beginning or after a change somewhere in the network. Link state routing is based on the assumption that, although the global knowledge about the topology is not clear, each node has partial knowledge: it knows the state (type, condition, and cost) of its links. In other words, the whole topology can be compiled from the partial knowledge of each node. Figure 22.21 shows the same domain as in Figure 22.20, indicating the part of the knowledge belonging to each node.

Figure 22.21 *Link state knowledge*

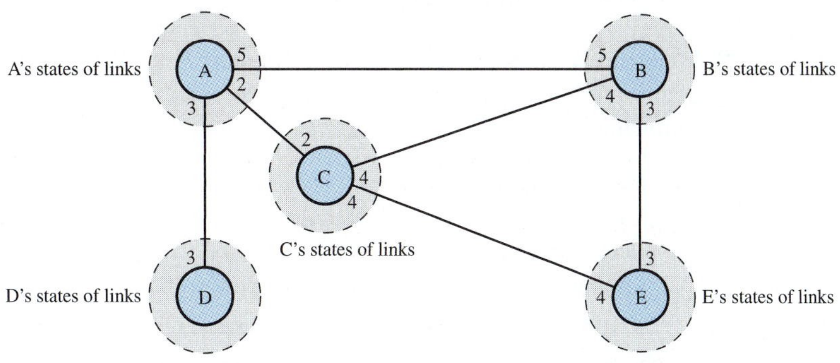

Node A knows that it is connected to node B with metric 5, to node C with metric 2, and to node D with metric 3. Node C knows that it is connected to node A with metric 2, to node B with metric 4, and to node E with metric 4. Node D knows that it is connected only to node A with metric 3. And so on. Although there is an overlap in the knowledge, the overlap guarantees the creation of a common topology—a picture of the whole domain for each node.

Building Routing Tables

In **link state routing,** four sets of actions are required to ensure that each node has the routing table showing the least-cost node to every other node.

1. Creation of the states of the links by each node, called the link state packet (LSP).
2. Dissemination of LSPs to every other router, called **flooding,** in an efficient and reliable way.
3. Formation of a shortest path tree for each node.
4. Calculation of a routing table based on the shortest path tree.

Creation of Link State Packet (LSP) A link state packet can carry a large amount of information. For the moment, however, we assume that it carries a minimum amount of data: the node identity, the list of links, a sequence number, and age. The first two, node identity and the list of links, are needed to make the topology. The third, sequence number, facilitates flooding and distinguishes new LSPs from old ones. The fourth, age, prevents old LSPs from remaining in the domain for a long time. LSPs are generated on two occasions:

1. *When there is a change in the topology of the domain.* Triggering of LSP dissemination is the main way of quickly informing any node in the domain to update its topology.
2. *On a periodic basis.* The period in this case is much longer compared to distance vector routing. As a matter of fact, there is no actual need for this type of LSP dissemination. It is done to ensure that old information is removed from the domain. The timer set for periodic dissemination is normally in the range of 60 min or 2 h based on the implementation. A longer period ensures that flooding does not create too much traffic on the network.

Flooding of LSPs After a node has prepared an LSP, it must be disseminated to all other nodes, not only to its neighbors. The process is called flooding and based on the following:

1. The creating node sends a copy of the LSP out of each interface.
2. A node that receives an LSP compares it with the copy it may already have. If the newly arrived LSP is older than the one it has (found by checking the sequence number), it discards the LSP. If it is newer, the node does the following:
   a. It discards the old LSP and keeps the new one.
   b. It sends a copy of it out of each interface except the one from which the packet arrived. This guarantees that flooding stops somewhere in the domain (where a node has only one interface).

Formation of Shortest Path Tree: Dijkstra Algorithm After receiving all LSPs, each node will have a copy of the whole topology. However, the topology is not sufficient to find the shortest path to every other node; a **shortest path tree** is needed.

A tree is a graph of nodes and links; one node is called the root. All other nodes can be reached from the root through only one single route. A shortest path tree is a tree in which the path between the root and every other node is the shortest. What we need for each node is a shortest path tree with that node as the root.

The **Dijkstra algorithm** creates a shortest path tree from a graph. The algorithm divides the nodes into two sets: tentative and permanent. It finds the neighbors of a current node, makes them tentative, examines them, and if they pass the criteria, makes them permanent. We can informally define the algorithm by using the flowchart in Figure 22.22.

Let us apply the algorithm to node A of our sample graph in Figure 22.23. To find the shortest path in each step, we need the cumulative cost from the root to each node, which is shown next to the node.

The following shows the steps. At the end of each step, we show the permanent (filled circles) and the tentative (open circles) nodes and lists with the cumulative costs.

Figure 22.22 *Dijkstra algorithm*

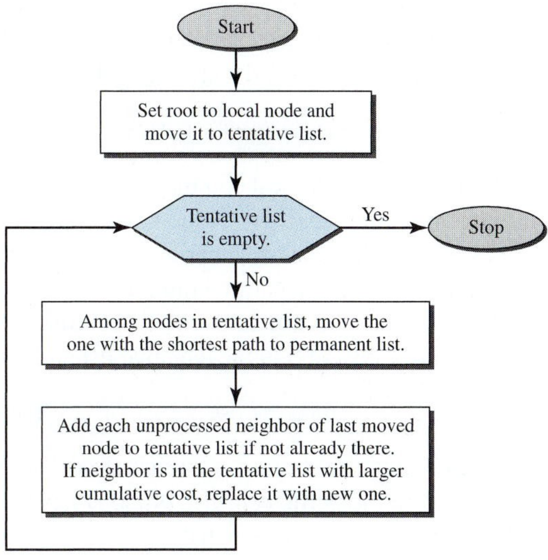

Figure 22.23 *Example of formation of shortest path tree*

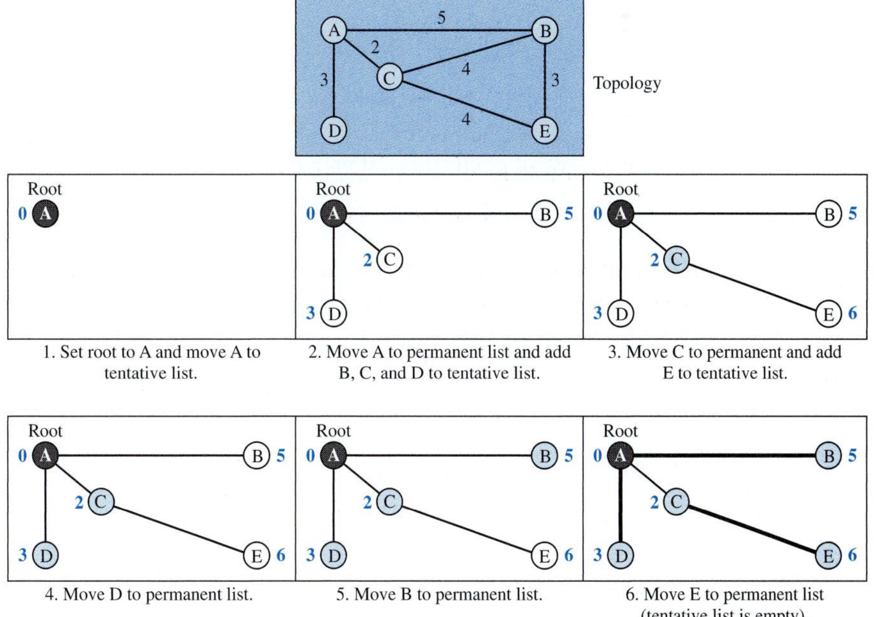

1. We make node A the root of the tree and move it to the tentative list. Our two lists are

   Permanent list: empty   Tentative list: A(0)

2. Node A has the shortest cumulative cost from all nodes in the tentative list. We move A to the permanent list and add all neighbors of A to the tentative list. Our new lists are

   Permanent list: A(0)   Tentative list: B(5), C(2), D(3)

3. Node C has the shortest cumulative cost from all nodes in the tentative list. We move C to the permanent list. Node C has three neighbors, but node A is already processed, which makes the unprocessed neighbors just B and E. However, B is already in the tentative list with a cumulative cost of 5. Node A could also reach node B through C with a cumulative cost of 6. Since 5 is less than 6, we keep node B with a cumulative cost of 5 in the tentative list and do not replace it. Our new lists are

   Permanent list: A(0), C(2)   Tentative list: B(5), D(3), E(6)

4. Node D has the shortest cumulative cost of all the nodes in the tentative list. We move D to the permanent list. Node D has no unprocessed neighbor to be added to the tentative list. Our new lists are

   Permanent list: A(0), C(2), D(3)   Tentative list: B(5), E(6)

5. Node B has the shortest cumulative cost of all the nodes in the tentative list. We move B to the permanent list. We need to add all unprocessed neighbors of B to the tentative list (this is just node E). However, E(6) is already in the list with a smaller cumulative cost. The cumulative cost to node E, as the neighbor of B, is 8. We keep node E(6) in the tentative list. Our new lists are

   Permanent list: A(0), B(5), C(2), D(3)   Tentative list: E(6)

6. Node E has the shortest cumulative cost from all nodes in the tentative list. We move E to the permanent list. Node E has no neighbor. Now the tentative list is empty. We stop; our shortest path tree is ready. The final lists are

   Permanent list: A(0), B(5), C(2), D(3), E(6)   Tentative list: empty

Calculation of Routing Table from Shortest Path Tree Each node uses the shortest path tree protocol to construct its routing table. The routing table shows the cost of reaching each node from the root. Table 22.2 shows the routing table for node A.

Table 22.2 *Routing table for node A*

Node	*Cost*	*Next Router*
A	0	—
B	5	—
C	2	—
D	3	—
E	6	C

Compare Table 22.2 with the one in Figure 22.14. Both distance vector routing and link state routing end up with the same routing table for node A.

OSPF

The Open Shortest Path First or **OSPF protocol** is an intradomain routing protocol based on link state routing. Its domain is also an autonomous system.

Areas To handle routing efficiently and in a timely manner, OSPF divides an autonomous system into areas. An **area** is a collection of networks, hosts, and routers all contained within an autonomous system. An autonomous system can be divided into many different areas. All networks inside an area must be connected.

Routers inside an area flood the area with routing information. At the border of an area, special routers called **area border routers** summarize the information about the area and send it to other areas. Among the areas inside an autonomous system is a special area called the *backbone;* all the areas inside an autonomous system must be connected to the backbone. In other words, the backbone serves as a primary area and the other areas as secondary areas. This does not mean that the routers within areas cannot be connected to each other, however. The routers inside the backbone are called the **backbone routers.** Note that a backbone router can also be an area border router.

If, because of some problem, the connectivity between a backbone and an area is broken, a **virtual link** between routers must be created by an administrator to allow continuity of the functions of the backbone as the primary area.

Each area has an area identification. The area identification of the backbone is zero. Figure 22.24 shows an autonomous system and its areas.

Figure 22.24 *Areas in an autonomous system*

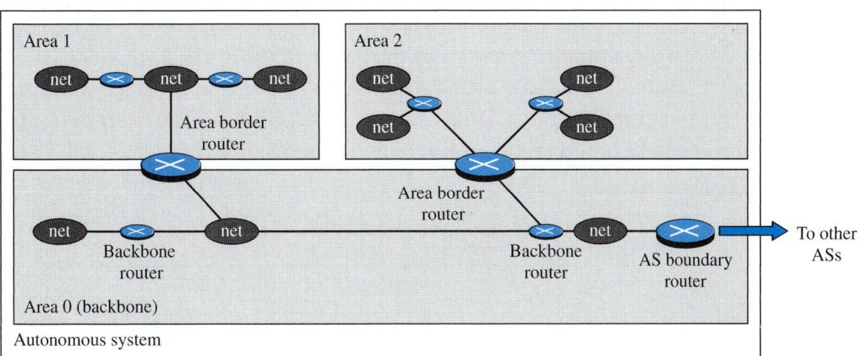

Metric The OSPF protocol allows the administrator to assign a cost, called the **metric,** to each route. The metric can be based on a type of service (minimum delay, maximum throughput, and so on). As a matter of fact, a router can have multiple routing tables, each based on a different type of service.

Types of Links In OSPF terminology, a connection is called a *link*. Four types of links have been defined: point-to-point, transient, stub, and virtual (see Figure 22.25).

Figure 22.25 *Types of links*

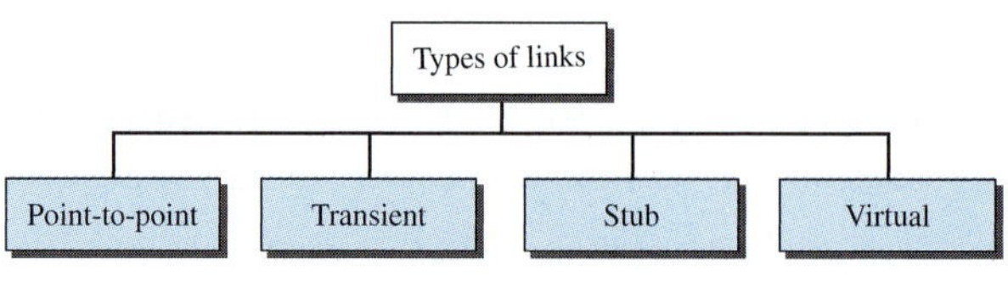

A **point-to-point link** connects two routers without any other host or router in between. In other words, the purpose of the link (network) is just to connect the two routers. An example of this type of link is two routers connected by a telephone line or a T line. There is no need to assign a network address to this type of link. Graphically, the routers are represented by nodes, and the link is represented by a bidirectional edge connecting the nodes. The metrics, which are usually the same, are shown at the two ends, one for each direction. In other words, each router has only one neighbor at the other side of the link (see Figure 22.26).

Figure 22.26 *Point-to-point link*

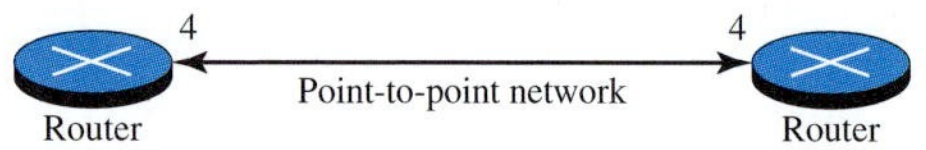

A **transient link** is a network with several routers attached to it. The data can enter through any of the routers and leave through any router. All LANs and some WANs with two or more routers are of this type. In this case, each router has many neighbors. For example, consider the Ethernet in Figure 22.27a. Router A has routers B, C, D, and E as neighbors. Router B has routers A, C, D, and E as neighbors. If we want to show the neighborhood relationship in this situation, we have the graph shown in Figure 22.27b.

Figure 22.27 *Transient link*

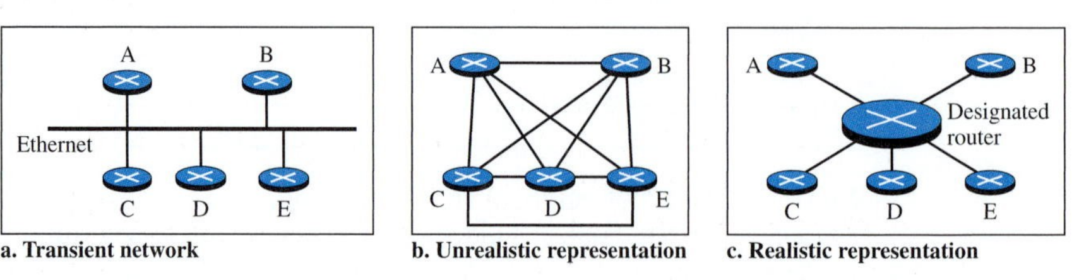

This is neither efficient nor realistic. It is not efficient because each router needs to advertise the neighborhood to four other routers, for a total of 20 advertisements. It is not realistic because there is no single network (link) between each pair of routers; there is only one network that serves as a crossroad between all five routers.

To show that each router is connected to every other router through one single network, the network itself is represented by a node. However, because a network is not a machine, it cannot function as a router. One of the routers in the network takes this responsibility. It is assigned a dual purpose; it is a true router and a designated router. We can use the topology shown in Figure 22.27c to show the connections of a transient network.

Now each router has only one neighbor, the designated router (network). On the other hand, the designated router (the network) has five neighbors. We see that the number of neighbor announcements is reduced from 20 to 10. Still, the link is represented as a bidirectional edge between the nodes. However, while there is a metric from each node to the designated router, there is no metric from the designated router to any other node. The reason is that the designated router represents the network. We can only assign a cost to a packet that is passing through the network. We cannot charge for this twice. When a packet enters a network, we assign a cost; when a packet leaves the network to go to the router, there is no charge.

A **stub link** is a network that is connected to only one router. The data packets enter the network through this single router and leave the network through this same router. This is a special case of the transient network. We can show this situation using the router as a node and using the designated router for the network. However, the link is only one-directional, from the router to the network (see Figure 22.28).

Figure 22.28 *Stub link*

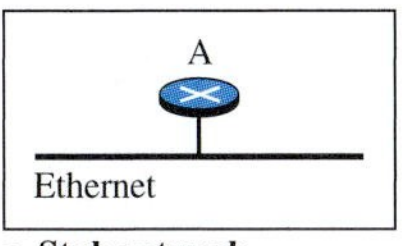

a. Stub network

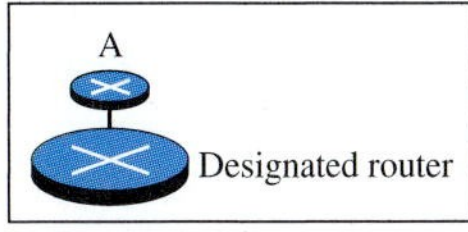

b. Representation

When the link between two routers is broken, the administration may create a **virtual link** between them, using a longer path that probably goes through several routers.

Graphical Representation Let us now examine how an AS can be represented graphically. Figure 22.29 shows a small AS with seven networks and six routers. Two of the networks are point-to-point networks. We use symbols such as N1 and N2 for transient and stub networks. There is no need to assign an identity to a point-to-point network. The figure also shows the graphical representation of the AS as seen by OSPF.

We have used square nodes for the routers and ovals for the networks (represented by designated routers). However, OSPF sees both as nodes. Note that we have three stub networks.

Figure 22.29 *Example of an AS and its graphical representation in OSPF*

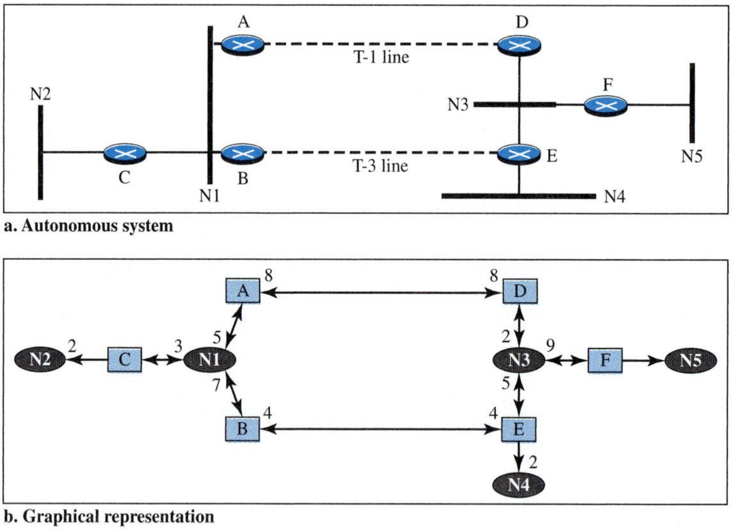

a. Autonomous system

b. Graphical representation

Path Vector Routing

Distance vector and link state routing are both intradomain routing protocols. They can be used inside an autonomous system, but not between autonomous systems. These two protocols are not suitable for interdomain routing mostly because of scalability. Both of these routing protocols become intractable when the domain of operation becomes large. Distance vector routing is subject to instability if there are more than a few hops in the domain of operation. Link state routing needs a huge amount of resources to calculate routing tables. It also creates heavy traffic because of flooding. There is a need for a third routing protocol which we call **path vector routing.**

Path vector routing proved to be useful for interdomain routing. The principle of path vector routing is similar to that of distance vector routing. In path vector routing, we assume that there is one node (there can be more, but one is enough for our conceptual discussion) in each autonomous system that acts on behalf of the entire autonomous system. Let us call it the **speaker node.** The speaker node in an AS creates a routing table and advertises it to speaker nodes in the neighboring ASs. The idea is the same as for distance vector routing except that only speaker nodes in each AS can communicate with each other. However, what is advertised is different. A speaker node advertises the path, not the metric of the nodes, in its autonomous system or other autonomous systems.

Initialization

At the beginning, each speaker node can know only the reachability of nodes inside its autonomous system. Figure 22.30 shows the initial tables for each speaker node in a system made of four ASs.

Figure 22.30 *Initial routing tables in path vector routing*

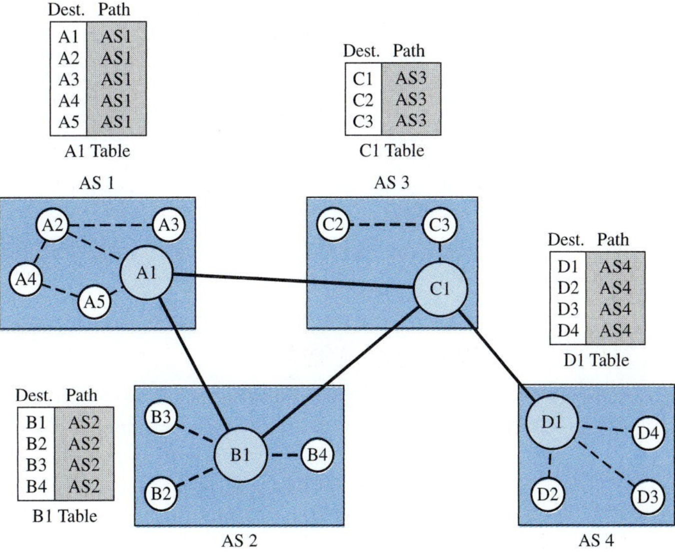

Node A1 is the speaker node for AS1, B1 for AS2, C1 for AS3, and D1 for AS4. Node A1 creates an initial table that shows A1 to A5 are located in AS1 and can be reached through it. Node B1 advertises that B1 to B4 are located in AS2 and can be reached through B1. And so on.

Sharing Just as in distance vector routing, in path vector routing, a speaker in an autonomous system shares its table with immediate neighbors. In Figure 22.30, node A1 shares its table with nodes B1 and C1. Node C1 shares its table with nodes D1, B1, and A1. Node B1 shares its table with C1 and A1. Node D1 shares its table with C1.

Updating When a speaker node receives a two-column table from a neighbor, it updates its own table by adding the nodes that are not in its routing table and adding its own autonomous system and the autonomous system that sent the table. After a while each speaker has a table and knows how to reach each node in other ASs. Figure 22.31 shows the tables for each speaker node after the system is stabilized.

According to the figure, if router A1 receives a packet for nodes A3, it knows that the path is in AS1 (the packet is at home); but if it receives a packet for D1, it knows that the packet should go from AS1, to AS2, and then to AS3. The routing table shows the path completely. On the other hand, if node D1 in AS4 receives a packet for node A2, it knows it should go through AS4, AS3, and AS1.

- **Loop prevention.** The instability of distance vector routing and the creation of loops can be avoided in path vector routing. When a router receives a message, it checks to see if its autonomous system is in the path list to the destination. If it is, looping is involved and the message is ignored.

Figure 22.31 *Stabilized tables for three autonomous systems*

A1 Table

Dest.	Path
A1	AS1
...	
A5	AS1
B1	AS1-AS2
...	...
B4	AS1-AS2
C1	AS1-AS3
...	...
C3	AS1-AS3
D1	AS1-AS2-AS4
...	...
D4	AS1-AS2-AS4

B1 Table

Dest.	Path
A1	AS2-AS1
...	
A5	AS2-AS1
B1	AS2
...	...
B4	AS2
C1	AS2-AS3
...	...
C3	AS2-AS3
D1	AS2-AS3-AS4
...	...
D4	AS2-AS3-AS4

C1 Table

Dest.	Path
A1	AS3-AS1
...	
A5	AS3-AS1
B1	AS3-AS2
...	
B4	AS3-AS2
C1	AS3
...	...
C3	AS3
D1	AS3-AS4
...	...
D4	AS3-AS4

D1 Table

Dest.	Path
A1	AS4-AS3-AS1
...	
A5	AS4-AS3-AS1
B1	AS4-AS3-AS2
...	...
B4	AS4-AS3-AS2
C1	AS4-AS3
...	...
C3	AS4-AS3
D1	AS4
...	...
D4	AS4

- **Policy routing.** Policy routing can be easily implemented through path vector routing. When a router receives a message, it can check the path. If one of the autonomous systems listed in the path is against its policy, it can ignore that path and that destination. It does not update its routing table with this path, and it does not send this message to its neighbors.
- **Optimum path.** What is the optimum path in path vector routing? We are looking for a path to a destination that is the best for the organization that runs the autonomous system. We definitely cannot include metrics in this route because each autonomous system that is included in the path may use a different criterion for the metric. One system may use, internally, RIP, which defines hop count as the metric; another may use OSPF with minimum delay defined as the metric. The optimum path is the path that fits the organization. In our previous figure, each autonomous system may have more than one path to a destination. For example, a path from AS4 to AS1 can be AS4-AS3-AS2-AS1, or it can be AS4-AS3-AS1. For the tables, we chose the one that had the smaller number of autonomous systems, but this is not always the case. Other criteria, such as security, safety, and reliability, can also be applied.

BGP

Border Gateway Protocol (BGP) is an interdomain routing protocol using path vector routing. It first appeared in 1989 and has gone through four versions.

Types of Autonomous Systems As we said before, the Internet is divided into hierarchical domains called autonomous systems. For example, a large corporation that manages its own network and has full control over it is an autonomous system. A local ISP that provides services to local customers is an autonomous system. We can divide autonomous systems into three categories: stub, multihomed, and transit.

- **Stub AS.** A stub AS has only one connection to another AS. The interdomain data traffic in a stub AS can be either created or terminated in the AS. The hosts in the AS can send data traffic to other ASs. The hosts in the AS can receive data coming from hosts in other ASs. Data traffic, however, cannot pass through a stub AS. A stub AS is either a source or a sink. A good example of a stub AS is a small corporation or a small local ISP.

- **Multihomed AS.** A multihomed AS has more than one connection to other ASs, but it is still only a source or sink for data traffic. It can receive data traffic from more than one AS. It can send data traffic to more than one AS, but there is no transient traffic. It does not allow data coming from one AS and going to another AS to pass through. A good example of a multihomed AS is a large corporation that is connected to more than one regional or national AS that does not allow transient traffic.
- **Transit AS.** A transit AS is a multihomed AS that also allows transient traffic. Good examples of transit ASs are national and international ISPs (Internet backbones).

Path Attributes In our previous example, we discussed a path for a destination network. The path was presented as a list of autonomous systems, but is, in fact, a list of attributes. Each attribute gives some information about the path. The list of attributes helps the receiving router make a more-informed decision when applying its policy.

Attributes are divided into two broad categories: well known and optional. A **well-known attribute** is one that every BGP router must recognize. An **optional attribute** is one that needs not be recognized by every router.

Well-known attributes are themselves divided into two categories: mandatory and discretionary. A *well-known mandatory attribute* is one that must appear in the description of a route. A *well-known discretionary attribute* is one that must be recognized by each router, but is not required to be included in every update message. One well-known mandatory attribute is ORIGIN. This defines the source of the routing information (RIP, OSPF, and so on). Another well-known mandatory attribute is AS_PATH. This defines the list of autonomous systems through which the destination can be reached. Still another well-known mandatory attribute is NEXT-HOP, which defines the next router to which the data packet should be sent.

The optional attributes can also be subdivided into two categories: transitive and nontransitive. An *optional transitive attribute* is one that must be passed to the next router by the router that has not implemented this attribute. An *optional nontransitive attribute* is one that must be discarded if the receiving router has not implemented it.

BGP Sessions The exchange of routing information between two routers using BGP takes place in a session. A session is a connection that is established between two BGP routers only for the sake of exchanging routing information. To create a reliable environment, BGP uses the services of TCP. In other words, a session at the BGP level, as an application program, is a connection at the TCP level. However, there is a subtle difference between a connection in TCP made for BGP and other application programs. When a TCP connection is created for BGP, it can last for a long time, until something unusual happens. For this reason, BGP sessions are sometimes referred to as *semi-permanent connections*.

External and Internal BGP If we want to be precise, BGP can have two types of sessions: external BGP (E-BGP) and internal BGP (I-BGP) sessions. The E-BGP session is used to exchange information between two speaker nodes belonging to two different autonomous systems. The I-BGP session, on the other hand, is used to exchange routing information between two routers inside an autonomous system. Figure 22.32 shows the idea.

Figure 22.32 *Internal and external BGP sessions*

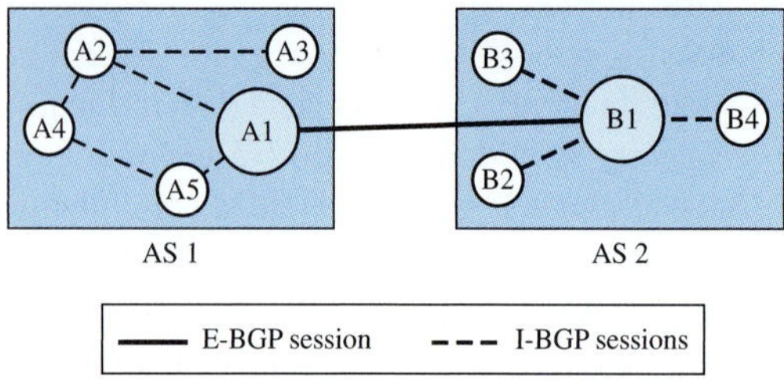

The session established between AS1 and AS2 is an E-BGP session. The two speaker routers exchange information they know about networks in the Internet. However, these two routers need to collect information from other routers in the autonomous systems. This is done using I-BGP sessions.

22.4 MULTICAST ROUTING PROTOCOLS

In this section, we discuss multicasting and multicast routing protocols. We first define the term *multicasting* and compare it to unicasting and broadcasting. We also briefly discuss the applications of multicasting. Finally, we move on to multicast routing and the general ideas and goals related to it. We also discuss some common multicast routing protocols used in the Internet today.

Unicast, Multicast, and Broadcast

A message can be unicast, multicast, or broadcast. Let us clarify these terms as they relate to the Internet.

Unicasting

In unicast communication, there is one source and one destination. The relationship between the source and the destination is one-to-one. In this type of communication, both the source and destination addresses, in the IP datagram, are the unicast addresses assigned to the hosts (or host interfaces, to be more exact). In Figure 22.33, a unicast

Figure 22.33 *Unicasting*

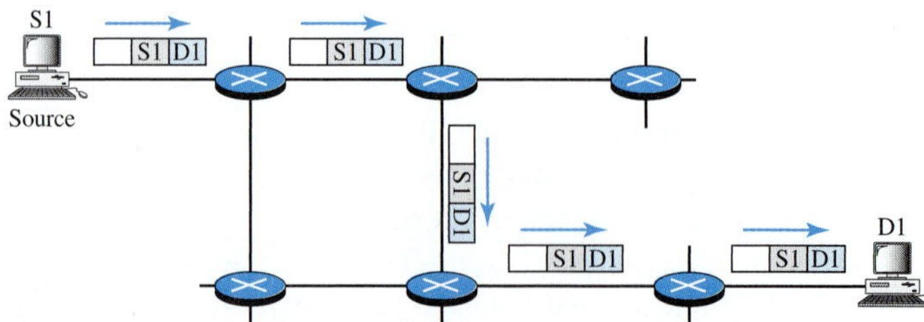

packet starts from the source S1 and passes through routers to reach the destination D1. We have shown the networks as a link between the routers to simplify the figure.

Note that in **unicasting,** when a router receives a packet, it forwards the packet through only one of its interfaces (the one belonging to the optimum path) as defined in the routing table. The router may discard the packet if it cannot find the destination address in its routing table.

In unicasting, the router forwards the received packet through only one of its interfaces.

Multicasting

In multicast communication, there is one source and a group of destinations. The relationship is one-to-many. In this type of communication, the source address is a unicast address, but the destination address is a group address, which defines one or more destinations. The group address identifies the members of the group. Figure 22.34 shows the idea behind **multicasting.**

Figure 22.34 *Multicasting*

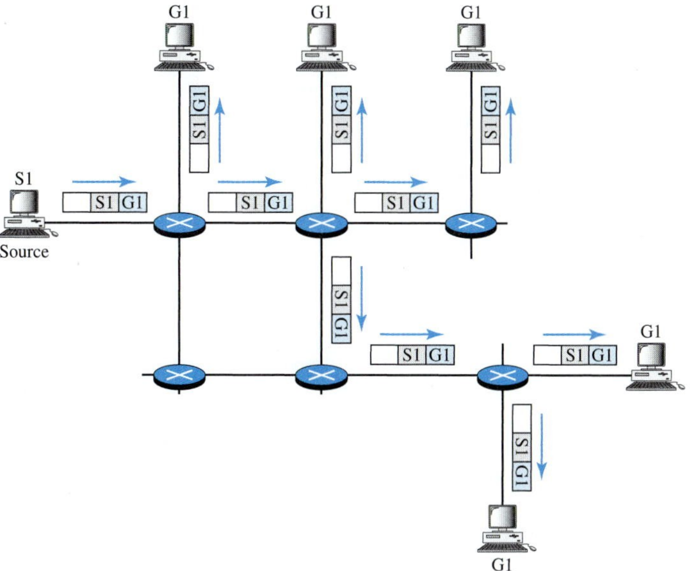

A multicast packet starts from the source S1 and goes to all destinations that belong to group G1. In multicasting, when a router receives a packet, it may forward it through several of its interfaces.

In multicasting, the router may forward the received packet through several of its interfaces.

Broadcasting

In broadcast communication, the relationship between the source and the destination is one-to-all. There is only one source, but all the other hosts are the destinations. The Internet does not explicitly support **broadcasting** because of the huge amount of traffic it would create and because of the bandwidth it would need. Imagine the traffic generated in the Internet if one person wanted to send a message to everyone else connected to the Internet.

Multicasting Versus Multiple Unicasting

Before we finish this section, we need to distinguish between multicasting and multiple unicasting. Figure 22.35 illustrates both concepts.

Figure 22.35 *Multicasting versus multiple unicasting*

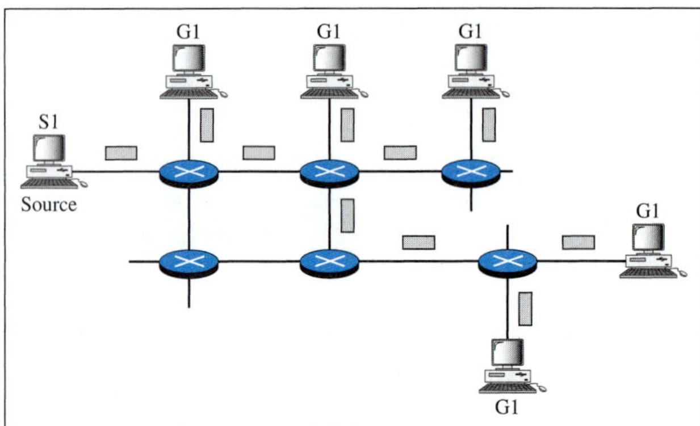

a. Multicasting

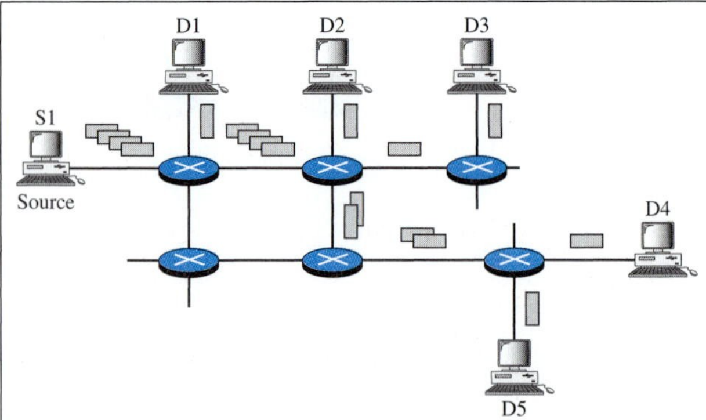

b. Multiple unicasting

Multicasting starts with one single packet from the source that is duplicated by the routers. The destination address in each packet is the same for all duplicates. Note that only one single copy of the packet travels between any two routers.

In **multiple unicasting,** several packets start from the source. If there are five destinations, for example, the source sends five packets, each with a different unicast destination address. Note that there may be multiple copies traveling between two routers. For example, when a person sends an e-mail message to a group of people, this is multiple unicasting. The e-mail software creates replicas of the message, each with a different destination address and sends them one by one. This is not multicasting; it is multiple unicasting.

Emulation of Multicasting with Unicasting

You might wonder why we have a separate mechanism for multicasting, when it can be emulated with unicasting. There are two obvious reasons for this.

1. Multicasting is more efficient than multiple unicasting. In Figure 22.35, we can see how multicasting requires less bandwidth than does multiple unicasting. In multiple unicasting, some of the links must handle several copies.
2. In multiple unicasting, the packets are created by the source with a relative delay between packets. If there are 1000 destinations, the delay between the first and the last packet may be unacceptable. In multicasting, there is no delay because only one packet is created by the source.

Emulation of multicasting through multiple unicasting is not efficient and may create long delays, particularly with a large group.

Applications

Multicasting has many applications today such as access to **distributed databases,** information dissemination, teleconferencing, and distance learning.

Access to Distributed Databases

Most of the large databases today are distributed. That is, the information is stored in more than one location, usually at the time of production. The user who needs to access the database does not know the location of the information. A user's request is multicast to all the database locations, and the location that has the information responds.

Information Dissemination

Businesses often need to send information to their customers. If the nature of the information is the same for each customer, it can be multicast. In this way a business can send one message that can reach many customers. For example, a software update can be sent to all purchasers of a particular software package.

Dissemination of News

In a similar manner news can be easily disseminated through multicasting. One single message can be sent to those interested in a particular topic. For example, the statistics of the championship high school basketball tournament can be sent to the sports editors of many newspapers.

Teleconferencing

Teleconferencing involves multicasting. The individuals attending a teleconference all need to receive the same information at the same time. Temporary or permanent groups can be formed for this purpose. For example, an engineering group that holds meetings every Monday morning could have a permanent group while the group that plans the holiday party could form a temporary group.

Distance Learning

One growing area in the use of multicasting is **distance learning.** Lessons taught by one single professor can be received by a specific group of students. This is especially convenient for those students who find it difficult to attend classes on campus.

Multicast Routing

In this section, we first discuss the idea of optimal routing, common in all multicast protocols. We then give an overview of multicast routing protocols.

Optimal Routing: Shortest Path Trees

The process of optimal interdomain routing eventually results in the finding of the *shortest path tree*. The root of the tree is the source, and the leaves are the potential destinations. The path from the root to each destination is the shortest path. However, the number of trees and the formation of the trees in unicast and multicast routing are different. Let us discuss each separately.

Unicast Routing In unicast routing, when a router receives a packet to forward, it needs to find the shortest path to the destination of the packet. The router consults its routing table for that particular destination. The next-hop entry corresponding to the destination is the start of the shortest path. The router knows the shortest path for each destination, which means that the router has a shortest path tree to optimally reach all destinations. In other words, each line of the routing table is a shortest path; the whole routing table is a shortest path tree. In unicast routing, each router needs only one shortest path tree to forward a packet; however, each router has its own shortest path tree. Figure 22.36 shows the situation.

The figure shows the details of the routing table and the shortest path tree for router R1. Each line in the routing table corresponds to one path from the root to the corresponding network. The whole table represents the shortest path tree.

In unicast routing, each router in the domain has a table that defines a shortest path tree to possible destinations.

Multicast Routing When a router receives a multicast packet, the situation is different from when it receives a unicast packet. A multicast packet may have destinations in more than one network. Forwarding of a single packet to members of a group requires a shortest path tree. If we have *n* groups, we may need *n* shortest path trees. We can imagine the complexity of multicast routing. Two approaches have been used to solve the problem: source-based trees and group-shared trees.

Figure 22.36 *Shortest path tree in unicast routing*

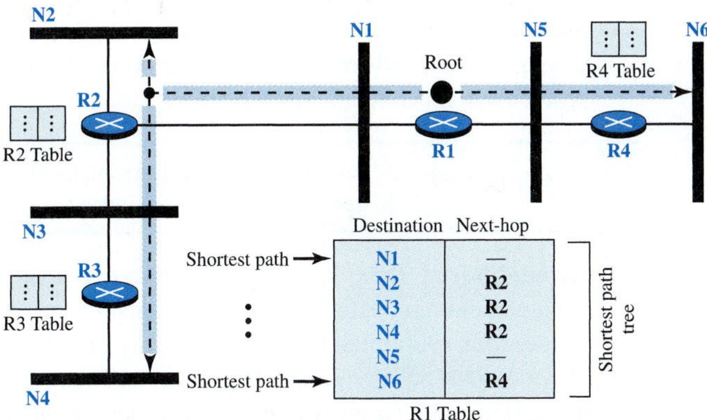

In multicast routing, each involved router needs to construct a shortest path tree for each group.

- **Source-Based Tree.** In the source-based tree approach, each router needs to have one shortest path tree for each group. The shortest path tree for a group defines the next hop for each network that has loyal member(s) for that group. In Figure 22.37, we assume that we have only five groups in the domain: G1, G2, G3, G4, and G5. At the moment G1 has loyal members in four networks, G2 in three, G3 in two, G4 in two, and G5 in two. We have shown the names of the groups with loyal members on each network. Figure 22.37 also shows the multicast routing table for router R1. There is one shortest path tree for each group; therefore there are five shortest path trees for five groups. If router R1 receives a packet with destination

Figure 22.37 *Source-based tree approach*

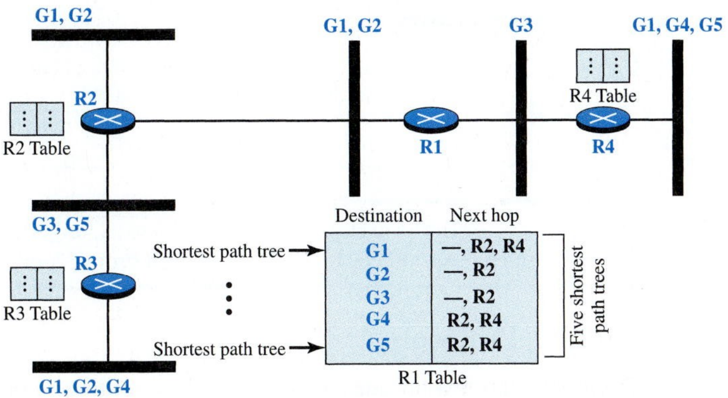

address G1, it needs to send a copy of the packet to the attached network, a copy to router R2, and a copy to router R4 so that all members of G1 can receive a copy. In this approach, if the number of groups is *m*, each router needs to have *m* shortest path trees, one for each group. We can imagine the complexity of the routing table if we have hundreds or thousands of groups. However, we will show how different protocols manage to alleviate the situation.

In the source-based tree approach, each router needs to have one shortest path tree for each group.

- **Group-Shared Tree.** In the **group-shared tree** approach, instead of each router having *m* shortest path trees, only one designated router, called the center core, or **rendezvous router,** takes the responsibility of distributing multicast traffic. The core has *m* shortest path trees in its routing table. The rest of the routers in the domain have none. If a router receives a multicast packet, it encapsulates the packet in a unicast packet and sends it to the core router. The core router removes the multicast packet from its capsule, and consults its routing table to route the packet. Figure 22.38 shows the idea.

Figure 22.38 *Group-shared tree approach*

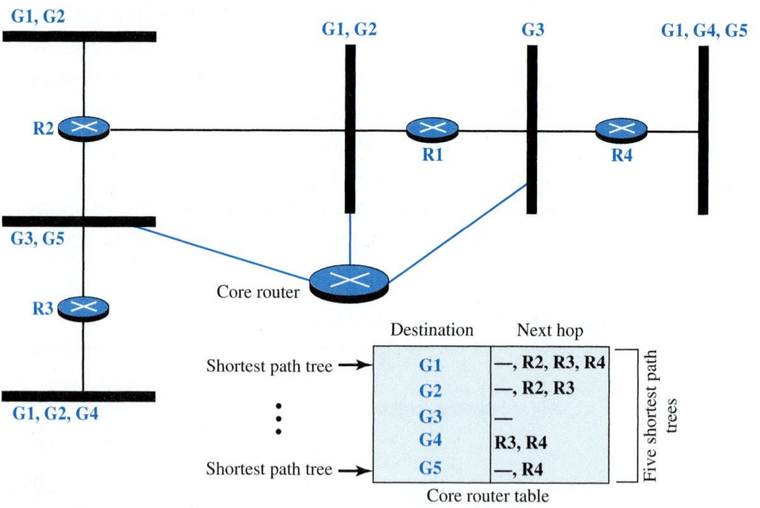

In the group-shared tree approach, only the core router, which has a shortest path tree for each group, is involved in multicasting.

Routing Protocols

During the last few decades, several multicast routing protocols have emerged. Some of these protocols are extensions of unicast routing protocols; others are totally new.

We discuss these protocols in the remainder of this chapter. Figure 22.39 shows the taxonomy of these protocols.

Figure 22.39 *Taxonomy of common multicast protocols*

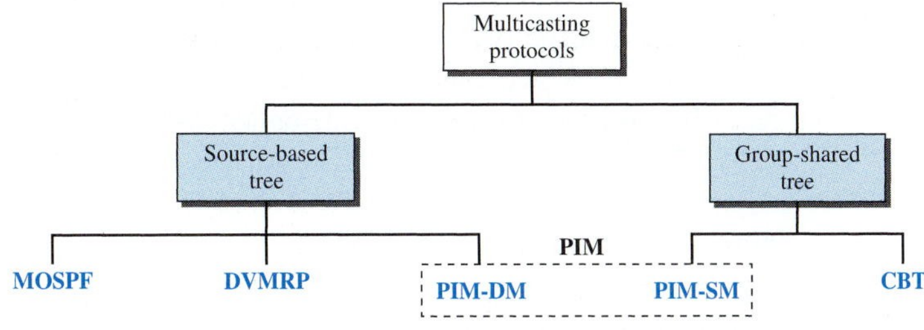

Multicast Link State Routing: MOSPF

In this section, we briefly discuss multicast link state routing and its implementation in the Internet, MOSPF.

Multicast Link State Routing We discussed unicast link state routing in Section 22.3. We said that each router creates a shortest path tree by using Dijkstra's algorithm. The routing table is a translation of the shortest path tree. Multicast link state routing is a direct extension of unicast routing and uses a source-based tree approach. Although unicast routing is quite involved, the extension to multicast routing is very simple and straightforward.

Multicast link state routing uses the source-based tree approach.

Recall that in unicast routing, each node needs to advertise the state of its links. For multicast routing, a node needs to revise the interpretation of *state*. A node advertises every group which has any loyal member on the link. Here the meaning of state is "what groups are active on this link." The information about the group comes from IGMP (see Chapter 21). Each router running IGMP solicits the hosts on the link to find out the membership status.

When a router receives all these LSPs, it creates n (n is the number of groups) topologies, from which n shortest path trees are made by using Dijkstra's algorithm. So each router has a routing table that represents as many shortest path trees as there are groups.

The only problem with this protocol is the time and space needed to create and save the many shortest path trees. The solution is to create the trees only when needed. When a router receives a packet with a multicast destination address, it runs the Dijkstra algorithm to calculate the shortest path tree for that group. The result can be cached in case there are additional packets for that destination.

MOSPF **Multicast Open Shortest Path First (MOSPF)** protocol is an extension of the OSPF protocol that uses multicast link state routing to create source-based trees.

The protocol requires a new link state update packet to associate the unicast address of a host with the group address or addresses the host is sponsoring. This packet is called the group-membership LSA. In this way, we can include in the tree only the hosts (using their unicast addresses) that belong to a particular group. In other words, we make a tree that contains all the hosts belonging to a group, but we use the unicast address of the host in the calculation. For efficiency, the router calculates the shortest path trees on demand (when it receives the first multicast packet). In addition, the tree can be saved in cache memory for future use by the same source/group pair. MOSPF is a **data-driven** protocol; the first time an MOSPF router sees a datagram with a given source and group address, the router constructs the Dijkstra shortest path tree.

Multicast Distance Vector: DVMRP

In this section, we briefly discuss multicast distance vector routing and its implementation in the Internet, DVMRP.

Multicast Distance Vector Routing Unicast distance vector routing is very simple; extending it to support multicast routing is complicated. Multicast routing does not allow a router to send its routing table to its neighbors. The idea is to create a table from scratch by using the information from the unicast distance vector tables.

Multicast distance vector routing uses source-based trees, but the router never actually makes a routing table. When a router receives a multicast packet, it forwards the packet as though it is consulting a routing table. We can say that the shortest path tree is evanescent. After its use (after a packet is forwarded) the table is destroyed.

To accomplish this, the multicast distance vector algorithm uses a process based on four decision-making strategies. Each strategy is built on its predecessor. We explain them one by one and see how each strategy can improve the shortcomings of the previous one.

- **Flooding.** Flooding is the first strategy that comes to mind. A router receives a packet and, without even looking at the destination group address, sends it out from every interface except the one from which it was received. Flooding accomplishes the first goal of multicasting: every network with active members receives the packet. However, so will networks without active members. This is a broadcast, not a multicast. There is another problem: it creates loops. A packet that has left the router may come back again from another interface or the same interface and be forwarded again. Some flooding protocols keep a copy of the packet for a while and discard any duplicates to avoid loops. The next strategy, reverse path forwarding, corrects this defect.

Flooding broadcasts packets, but creates loops in the systems.

- **Reverse Path Forwarding (RPF).** RPF is a modified flooding strategy. To prevent loops, only one copy is forwarded; the other copies are dropped. In RPF, a router forwards only the copy that has traveled the shortest path from the source to the router. To find this copy, RPF uses the unicast routing table. The router receives a packet and extracts the source address (a unicast address). It consults its unicast routing table as though it wants to send a packet to the source address. The routing table tells the router the next hop. If the multicast packet has just come from the hop defined in the table, the packet has traveled the shortest path from the source to the router because the shortest path is reciprocal in unicast distance vector routing protocols. If the path from A to B is the shortest, then it is also the shortest from B to A. The router forwards the packet if it has traveled from the shortest path; it discards it otherwise.

This strategy prevents loops because there is always one shortest path from the source to the router. If a packet leaves the router and comes back again, it has not traveled the shortest path. To make the point clear, let us look at Figure 22.40.

Figure 22.40 shows part of a domain and a source. The shortest path tree as calculated by routers R1, R2, and R3 is shown by a thick line. When R1 receives a packet from the source through the interface m1, it consults its routing table and finds that the shortest path from R1 to the source is through interface m1. The packet is forwarded. However, if a copy of the packet has arrived through interface m2, it is discarded because m2 does not define the shortest path from R1 to the source. The story is the same with R2 and R3. You may wonder what happens if a copy of a packet that arrives at the m1 interface of R3, travels through R6, R5, R2, and then enters R3 through interface m1. This interface is the correct interface for R3. Is the copy of the packet forwarded? The answer is that this scenario never happens because when the packet goes from R5 to R2, it will be discarded by R2 and never reaches R3. The upstream routers toward the source always discard a packet that has not gone through the shortest path, thus preventing confusion for the downstream routers.

RPF eliminates the loop in the flooding process.

Figure 22.40 *Reverse path forwarding (RPF)*

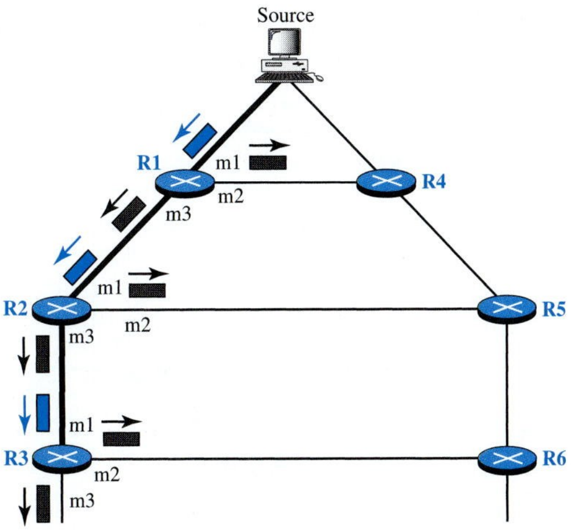

- **Reverse Path Broadcasting (RPB).** RPF guarantees that each network receives a copy of the multicast packet without formation of loops. However, RPF does not guarantee that each network receives only one copy; a network may receive two or more copies. The reason is that RPF is not based on the destination address (a group address); forwarding is based on the source address. To visualize the problem, let us look at Figure 22.41.

Figure 22.41 *Problem with RPF*

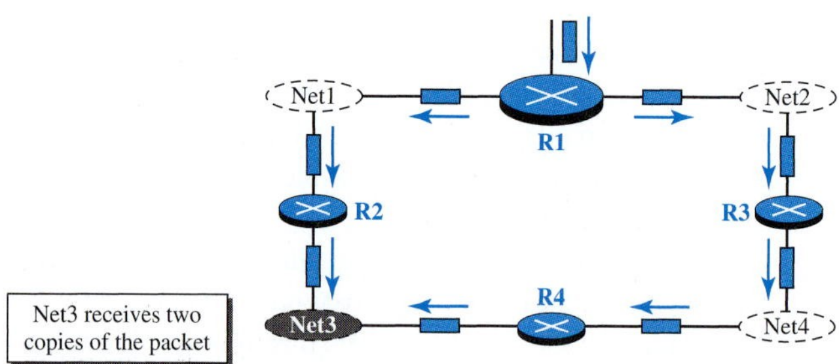

Net3 in this figure receives two copies of the packet even though each router just sends out one copy from each interface. There is duplication because a tree has not been made; instead of a tree we have a graph. Net3 has two parents: routers R2 and R4.

To eliminate duplication, we must define only one parent router for each network. We must have this restriction: A network can receive a multicast packet from a particular source only through a **designated parent router.**

Now the policy is clear. For each source, the router sends the packet only out of those interfaces for which it is the designated parent. This policy is called reverse path broadcasting (RPB). RPB guarantees that the packet reaches every network and that every network receives only one copy. Figure 22.42 shows the difference between RPF and RPB.

Figure 22.42 *RPF Versus RPB*

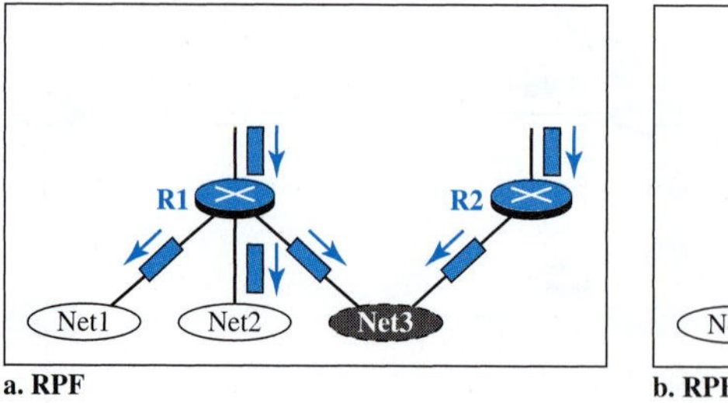

a. RPF

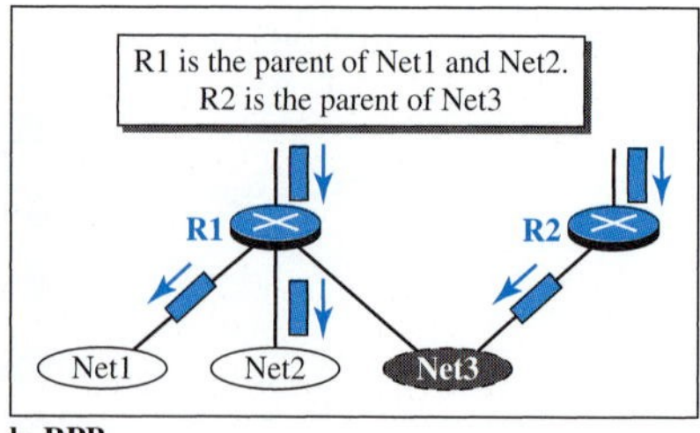

b. RPB

The reader may ask how the designated parent is determined. The designated parent router can be the router with the shortest path to the source. Because routers periodically send updating packets to each other (in RIP), they can easily determine which router in the neighborhood has the shortest path to the source (when interpreting the source as the destination). If more than one router qualifies, the router with the smallest IP address is selected.

RPB creates a shortest path broadcast tree from the source to each destination. It guarantees that each destination receives one and only one copy of the packet.

- **Reverse Path Multicasting (RPM).** As you may have noticed, RPB does not multicast the packet, it broadcasts it. This is not efficient. To increase efficiency, the multicast packet must reach only those networks that have active members for that particular group. This is called **reverse path multicasting (RPM).** To convert broadcasting to multicasting, the protocol uses two procedures, pruning and grafting. Figure 22.43 shows the idea of pruning and grafting.

Figure 22.43 *RPF, RPB, and RPM*

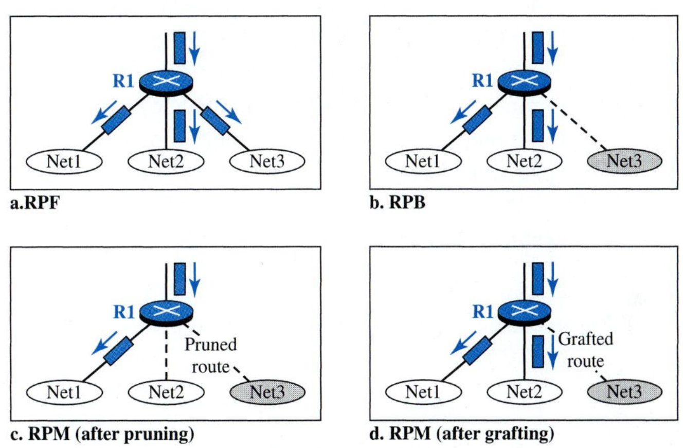

The designated parent router of each network is responsible for holding the membership information. This is done through the IGMP protocol described in Chapter 21. The process starts when a router connected to a network finds that there is no interest in a multicast packet. The router sends a **prune message** to the upstream router so that it can exclude the corresponding interface. That is, the upstream router can stop sending multicast messages for this group through that interface. Now if this router receives prune messages from all downstream routers, it, in turn, sends a prune message to its upstream router.

What if a leaf router (a router at the bottom of the tree) has sent a prune message but suddenly realizes, through IGMP, that one of its networks is again interested in receiving the multicast packet? It can send a **graft message.** The graft message forces the upstream router to resume sending the multicast messages.

RPM adds pruning and grafting to RPB to create a multicast shortest path tree that supports dynamic membership changes.

DVMRP The **Distance Vector Multicast Routing Protocol (DVMRP)** is an implementation of multicast distance vector routing. It is a source-based routing protocol, based on RIP.

CBT

The **Core-Based Tree (CBT) protocol** is a group-shared protocol that uses a core as the root of the tree. The autonomous system is divided into regions, and a core (center router or rendezvous router) is chosen for each region.

Formation of the Tree After the rendezvous point is selected, every router is informed of the unicast address of the selected router. Each router then sends a unicast join message (similar to a grafting message) to show that it wants to join the group. This message passes through all routers that are located between the sender and the rendezvous router. Each intermediate router extracts the necessary information from the message, such as the unicast address of the sender and the interface through which the packet has arrived, and forwards the message to the next router in the path. When the rendezvous router has received all join messages from every member of the group, the tree is formed. Now every router knows its upstream router (the router that leads to the root) and the downstream router (the router that leads to the leaf).

If a router wants to leave the group, it sends a leave message to its upstream router. The upstream router removes the link to that router from the tree and forwards the message to its upstream router, and so on. Figure 22.44 shows a group-shared tree with its rendezvous router.

Figure 22.44 *Group-shared tree with rendezvous router*

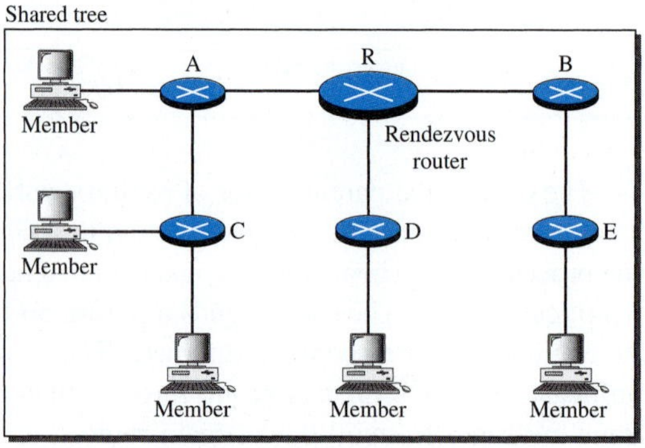

The reader may have noticed two differences between DVMRP and MOSPF, on one hand, and CBT, on the other. First, the tree for the first two is made from the root up; the tree for CBT is formed from the leaves down. Second, in DVMRP, the tree is first made (broadcasting) and then pruned; in CBT, there is no tree at the beginning; the joining (grafting) gradually makes the tree.

Sending Multicast Packets After formation of the tree, any source (belonging to the group or not) can send a multicast packet to all members of the group. It simply sends the packet to the rendezvous router, using the unicast address of the rendezvous router; the rendezvous router distributes the packet to all members of the group. Figure 22.45 shows how a host can send a multicast packet to all members of the group. Note that the source host can be any of the hosts inside the shared tree or any host outside the shared tree. In Figure 22.45 we show one located outside the shared tree.

Figure 22.45 *Sending a multicast packet to the rendezvous router*

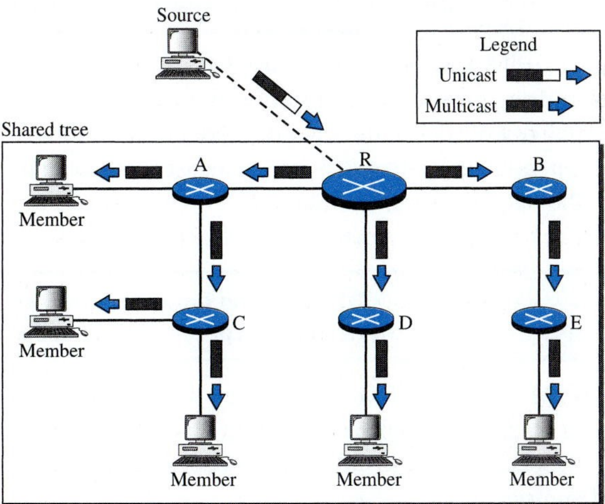

Selecting the Rendezvous Router This approach is simple except for one point. How do we select a rendezvous router to optimize the process and multicasting as well? Several methods have been implemented. However, this topic is beyond the scope of this book, and we leave it to more advanced books.

In summary, the Core-Based Tree (CBT) is a group-shared tree, center-based protocol using one tree per group. One of the routers in the tree is called the core. A packet is sent from the source to members of the group following this procedure:

1. The source, which may or may not be part of the tree, encapsulates the multicast packet inside a unicast packet with the unicast destination address of the core and sends it to the core. This part of delivery is done using a unicast address; the only recipient is the core router.
2. The core decapsulates the unicast packet and forwards it to all interested interfaces.
3. Each router that receives the multicast packet, in turn, forwards it to all interested interfaces.

In CBT, the source sends the multicast packet (encapsulated in a unicast packet) to the core router. The core router decapsulates the packet and forwards it to all interested interfaces.

PIM

Protocol Independent Multicast (PIM) is the name given to two independent multicast routing protocols: **Protocol Independent Multicast, Dense Mode (PIM-DM)** and **Protocol Independent Multicast, Sparse Mode (PIM-SM).** Both protocols are unicast-protocol-dependent, but the similarity ends here. We discuss each separately.

PIM-DM PIM-DM is used when there is a possibility that each router is involved in multicasting **(dense mode).** In this environment, the use of a protocol that broadcasts the packet is justified because almost all routers are involved in the process.

PIM-DM is used in a dense multicast environment, such as a LAN.

PIM-DM is a source-based tree routing protocol that uses RPF and pruning and grafting strategies for multicasting. Its operation is like that of DVMRP; however, unlike DVMRP, it does not depend on a specific unicasting protocol. It assumes that the autonomous system is using a unicast protocol and each router has a table that can find the outgoing interface that has an optimal path to a destination. This unicast protocol can be a distance vector protocol (RIP) or link state protocol (OSPF).

PIM-DM uses RPF and pruning and grafting strategies to handle multicasting. However, it is independent of the underlying unicast protocol.

PIM-SM PIM-SM is used when there is a slight possibility that each router is involved in multicasting (sparse mode). In this environment, the use of a protocol that broadcasts the packet is not justified; a protocol such as CBT that uses a group-shared tree is more appropriate.

PIM-SM is used in a sparse multicast environment such as a WAN.

PIM-SM is a group-shared tree routing protocol that has a rendezvous point (RP) as the source of the tree. Its operation is like CBT; however, it is simpler because it does not require acknowledgment from a join message. In addition, it creates a backup set of RPs for each region to cover RP failures.

One of the characteristics of PIM-SM is that it can switch from a group-shared tree strategy to a source-based tree strategy when necessary. This can happen if there is a dense area of activity far from the RP. That area can be more efficiently handled with a source-based tree strategy instead of a group-shared tree strategy.

PIM-SM is similar to CBT but uses a simpler procedure.

MBONE

Multimedia and real-time communication have increased the need for multicasting in the Internet. However, only a small fraction of Internet routers are multicast routers. In other words, a **multicast router** may not find another multicast router in the neighborhood to forward the multicast packet. Although this problem may be solved in the next few years by adding more and more multicast routers, there is another solution to this problem. The solution is **tunneling.** The multicast routers are seen as a group of routers on top of unicast routers. The multicast routers may not be connected directly, but they are connected logically. Figure 22.46 shows the idea. In Figure 22.46, only the routers enclosed in the shaded circles are capable of multicasting. Without tunneling, these routers are isolated islands. To enable multicasting, we make a **multicast backbone (MBONE)** out of these isolated routers by using the concept of tunneling.

Figure 22.46 *Logical tunneling*

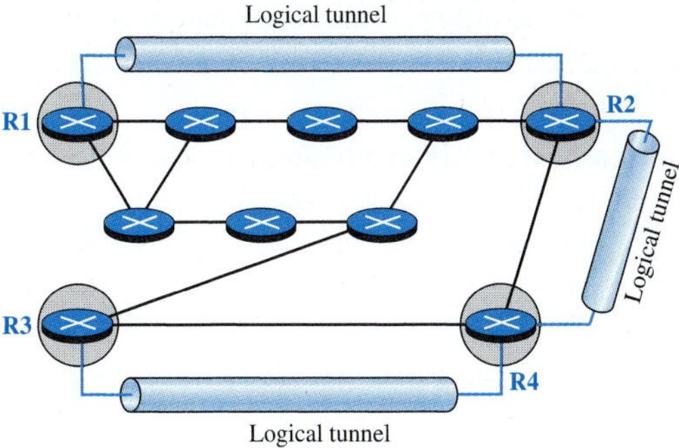

A **logical tunnel** is established by encapsulating the multicast packet inside a unicast packet. The multicast packet becomes the payload (data) of the unicast packet. The intermediate (nonmulticast) routers forward the packet as unicast routers and deliver the packet from one island to another. It's as if the unicast routers do not exist and the two multicast routers are neighbors. Figure 22.47 shows the concept. So far the only protocol that supports MBONE and tunneling is DVMRP.

Figure 22.47 *MBONE*

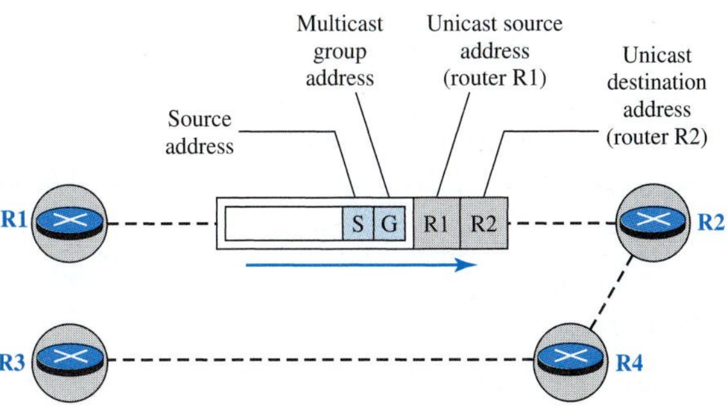

22.5 RECOMMENDED READING

For more details about subjects discussed in this chapter, we recommend the following books and sites. The items in brackets [. . .] refer to the reference list at the end of the text.

Books

Delivery and forwarding are discussed in Chapter 6 of [For06]. Unicast routing protocols are discussed in Chapter 14 of [For06]. Multicasting and multicast routing are discussed in Chapter 15 of [For06]. For a complete discussion of multicasting see [WZ01]. For routing protocols see [Hui00]. OSPF is discussed in [Moy98].

Sites

- www.ietf.org/rfc.html Information about RFCs

RFCs

A discussion of RIP can be found in the following RFCs:

1131, 1245, 1246, 1247, 1370, 1583, 1584, 1585, 1586, 1587, 1722, 1723, 2082, 2453

A discussion of OSPF can be found in the following RFCs:

1131, 1245, 1246, 1247, 1370, 1583, 1584, 1585, 1586, 1587, 2178, 2328, 2329, 2370

A discussion of BGP can be found in the following RFCs:

1092, 1105, 1163, 1265, 1266, 1267, 1364, 1392, 1403, 1565, 1654, 1655, 1665, 1771, 1772, 1745, 1774, 2283

22.6 KEY TERMS

address aggregation
area
area border routers
area identification
autonomous system (AS)
backbone router
Border Gateway Protocol (BGP)
broadcasting
Core-Based Tree (CBT) protocol
data-driven
default method
delivery
designated parent router
Dijkstra's algorithm
direct delivery
distance learning
Distance Vector Multicast Routing Protocol (DVMRP)
distance vector routing
distributed database
dynamic routing method
dynamic routing table
flooding
forwarding
graft message
group-shared tree

hierarchical routing
hop count
host-specific method
ifconfig
immediate neighbors
indirect delivery
interdomain routing
intradomain routing
least-cost tree
link state routing
logical tunnel
longest mask matching
metric
multicast backbone (MBONE)
Multicast Open Shortest Path First (MOSPF)
multicast router
multicast routing
multicasting
multiple unicasting
netstat
network-specific method
next-hop address
next-hop method
Open Shortest Path First (OSPF)
optional attribute
OSPF protocol
path vector routing
point-to-point link
poison reverse
policy routing
Protocol Independent Multicast (PIM)
Protocol Independent Multicast, Dense Mode (PIM-DM)
Protocol Independent Multicast, Sparse Mode (PIM-SM)
prune message
rendezvous router
rendezvous-point tree
reverse path broadcasting (RPB)
reverse path forwarding (RPF)
reverse path multicasting (RPM)
route method
routing
Routing Information Protocol (RIP)
routing protocols
shortest path tree
slow convergence
source-based tree
speaker node
split horizon
static routing table
stub link
switching fabric
teleconferencing
transient link
triggered update
tunneling
unicasting
update message
virtual link
well-known attribute

22.7 SUMMARY

- The delivery of a packet is called direct if the deliverer (host or router) and the destination are on the same network; the delivery of a packet is called indirect if the deliverer (host or router) and the destination are on different networks.
- In the next-hop method, instead of a complete list of the stops the packet must make, only the address of the next hop is listed in the routing table; in the network-specific method, all hosts on a network share one entry in the routing table.

- In the host-specific method, the full IP address of a host is given in the routing table.
- In the default method, a router is assigned to receive all packets with no match in the routing table.
- The routing table for classless addressing needs at least four columns.
- Address aggregation simplifies the forwarding process in classless addressing.
- Longest mask matching is required in classless addressing.
- Classless addressing requires hierarchical and geographical routing to prevent immense routing tables.
- A static routing table's entries are updated manually by an administrator; a dynamic routing table's entries are updated automatically by a routing protocol.
- A metric is the cost assigned for passage of a packet through a network.
- An autonomous system (AS) is a group of networks and routers under the authority of a single administration.
- RIP is based on distance vector routing, in which each router shares, at regular intervals, its knowledge about the entire AS with its neighbors.
- Two shortcomings associated with the RIP protocol are slow convergence and instability. Procedures to remedy RIP instability include triggered update, split horizons, and poison reverse.
- OSPF divides an AS into areas, defined as collections of networks, hosts, and routers.
- OSPF is based on link state routing, in which each router sends the state of its neighborhood to every other router in the area. A packet is sent only if there is a change in the neighborhood.
- OSPF routing tables are calculated by using Dijkstra's algorithm.
- BGP is an interautonomous system routing protocol used to update routing tables.
- BGP is based on a routing protocol called path vector routing. In this protocol, the ASs through which a packet must pass are explicitly listed.
- In a source-based tree approach to multicast routing, the source/group combination determines the tree; in a group-shared tree approach to multicast routing, the group determines the tree.
- MOSPF is a multicast routing protocol that uses multicast link state routing to create a source-based least-cost tree.
- In reverse path forwarding (RPF), the router forwards only the packets that have traveled the shortest path from the source to the router.
- Reverse path broadcasting (RPB) creates a shortest path broadcast tree from the source to each destination. It guarantees that each destination receives one and only one copy of the packet.
- Reverse path multicasting (RPM) adds pruning and grafting to RPB to create a multicast shortest path tree that supports dynamic membership changes.
- DVMRP is a multicast routing protocol that uses the distance routing protocol to create a source-based tree.
- The Core-Based Tree (CBT) protocol is a multicast routing protocol that uses a router as the root of the tree.

- PIM-DM is a source-based tree routing protocol that uses RPF and pruning and grafting strategies to handle multicasting.
- PIM-SM is a group-shared tree routing protocol that is similar to CBT and uses a rendezvous router as the source of the tree.
- For multicasting between two noncontiguous multicast routers, we make a multicast backbone (MBONE) to enable tunneling.

22.8 PRACTICE SET

Review Questions

1. What is the difference between a direct and an indirect delivery?
2. List three forwarding techniques discussed in this chapter and give a brief description of each.
3. Contrast two different routing tables discussed in this chapter.
4. What is the purpose of RIP?
5. What are the functions of a RIP message?
6. Why is the expiration timer value 6 times that of the periodic timer value?
7. How does the hop count limit alleviate RIP's problems?
8. List RIP shortcomings and their corresponding fixes.
9. What is the basis of classification for the four types of links defined by OSPF?
10. Why do OSPF messages propagate faster than RIP messages?
11. What is the purpose of BGP?
12. Give a brief description of two groups of multicast routing protocols discussed in this chapter.

Exercises

13. Show a routing table for a host that is totally isolated.
14. Show a routing table for a host that is connected to a LAN without being connected to the Internet.
15. Find the topology of the network if Table 22.3 is the routing table for router R1.

Table 22.3 *Routing table for Exercise 15*

Mask	*Network Address*	*Next-Hop Address*	*Interface*
/27	202.14.17.224	—	m1
/18	145.23.192.0	—	m0
Default	Default	130.56.12.4	m2

16. Can router R1 in Figure 22.8 receive a packet with destination address 140.24.7.194? Explain your answer.

17. Can router R1 in Figure 22.8 receive a packet with destination address 140.24.7.42? Explain your answer.
18. Show the routing table for the regional ISP in Figure 22.9.
19. Show the routing table for local ISP 1 in Figure 22.9.
20. Show the routing table for local ISP 2 in Figure 22.9.
21. Show the routing table for local ISP 3 in Figure 22.9.
22. Show the routing table for small ISP 1 in Figure 22.9.
23. Contrast and compare distance vector routing with link state routing.
24. A router has the following RIP routing table:

Net1	4	B
Net2	2	C
Net3	1	F
Net4	5	G

What would be the contents of the table if the router received the following RIP message from router C?

Net1	2
Net2	1
Net3	3
Net4	7

25. How many bytes are empty in a RIP message that advertises N networks?
26. A router has the following RIP routing table:

Net1	4	B
Net2	2	C
Net3	1	F
Net4	5	G

Show the response message sent by this router.

27. Show the autonomous system with the following specifications:
    a. There are eight networks (N1 to N8).
    b. There are eight routers (R1 to R8).
    c. N1, N2, N3, N4, N5, and N6 are Ethernet LANs.
    d. N7 and N8 are point-to-point WANs.
    e. R1 connects N1 and N2.
    f. R2 connects N1 and N7.
    g. R3 connects N2 and N8.
    h. R4 connects N7 and N6.
    i. R5 connects N6 and N3.
    j. R6 connects N6 and N4.
    k. R7 connects N6 and N5.
    l. R8 connects N8 and N5.

28. Draw the graphical representation of the autonomous system of Exercise 27 as seen by OSPF.
29. Which of the networks in Exercise 27 is a transient network? Which is a stub network?
30. A router using DVMRP receives a packet with source address 10.14.17.2 from interface 2. If the router forwards the packet, what are the contents of the entry related to this address in the unicast routing table?
31. Does RPF actually create a shortest path tree? Explain.
32. Does RPB actually create a shortest path tree? Explain. What are the leaves of the tree?
33. Does RPM actually create a shortest path tree? Explain. What are the leaves of the tree?

Research Activities

34. If you have access to UNIX (or LINUX), use *netstat* and *ifconfig* to find the routing table for the server to which you are connected.
35. Find out how your ISP uses address aggregation and longest mask match principles.
36. Find out whether your IP address is part of the geographical address allocation.
37. If you are using a router, find the number and names of the columns in the routing table.

PART

5

Transport Layer

Objectives

The transport layer is responsible for process-to-process delivery of the entire message. A process is an application program running on a host. Whereas the network layer oversees source-to-destination delivery of individual packets, it does not recognize any relationship between those packets. It treats each one independently, as though each piece belonged to a separate message, whether or not it does. The transport layer, on the other hand, ensures that the whole message arrives intact and in order, overseeing both error control and flow control at the source-to-destination level.

The transport layer is responsible for the delivery of a message from one process to another.

Computers often run several programs at the same time. For this reason, source-to-destination delivery means delivery not only from one computer to the next but also from a specific process on one computer to a specific process on the other. The transport layer header must therefore include a type of address called a *service-point address* in the OSI model and port number or port addresses in the Internet and TCP/IP protocol suite.

A transport layer protocol can be either connectionless or connection-oriented. A connectionless transport layer treats each segment as an independent packet and delivers it to the transport layer at the destination machine. A connection-oriented transport layer makes a connection with the transport layer at the destination machine first before delivering the packets. After all the data is transferred, the connection is terminated.

In the transport layer, a message is normally divided into transmittable segments. A connectionless protocol, such as UDP, treats each segment separately. A connection-oriented protocol, such as TCP and SCTP, creates a relationship between the segments using sequence numbers.

Like the data link layer, the transport layer may be responsible for flow and error control. However, flow and error control at this layer is performed end to end rather than across a single link. We will see that one of the protocols discussed in this part of the book, UDP, is not involved in flow or error control. On the other hand, the other two protocols, TCP and SCTP, use sliding windows for flow control and an acknowledgment system for error control.

Part 5 of the book is devoted to the transport layer and the services provided by this layer.

Chapters

This part consists of two chapters: Chapters 23 and 24.

Chapter 23

Chapter 23 discusses three transport layer protocols in the Internet: UDP, TCP, and SCTP. The first, User Datagram Protocol (UDP), is a connectionless, unreliable protocol that is used for its efficiency. The second, Transmission Control Protocol (TCP), is a connection-oriented, reliable protocol that is a good choice for data transfer. The third, Stream Control Transport Protocol (SCTP) is a new transport-layer protocol designed for multimedia applications.

Chapter 24

Chapter 24 discuss two related topics: congestion control and quality of service. Although these two issues can be related to any layer, we discuss them here with some references to other layers.

CHAPTER 23

Process-to-Process Delivery: UDP, TCP, and SCTP

We begin this chapter by giving the rationale for the existence of the **transport layer**—the need for process-to-process delivery. We discuss the issues arising from this type of delivery, and we discuss methods to handle them.

The Internet model has three protocols at the transport layer: UDP, TCP, and SCTP. First we discuss UDP, which is the simplest of the three. We see how we can use this very simple transport layer protocol that lacks some of the features of the other two.

We then discuss TCP, a complex transport layer protocol. We see how our previously presented concepts are applied to TCP. We postpone the discussion of congestion control and quality of service in TCP until Chapter 24 because these two topics apply to the data link layer and network layer as well.

We finally discuss SCTP, the new transport layer protocol that is designed for multihomed, multistream applications such as multimedia.

23.1 PROCESS-TO-PROCESS DELIVERY

The data link layer is responsible for delivery of frames between two neighboring nodes over a link. This is called *node-to-node delivery.* The network layer is responsible for delivery of datagrams between two hosts. This is called *host-to-host delivery.* Communication on the Internet is not defined as the exchange of data between two nodes or between two hosts. Real communication takes place between two processes (application programs). We need **process-to-process delivery.** However, at any moment, several processes may be running on the source host and several on the destination host. To complete the delivery, we need a mechanism to deliver data from one of these processes running on the source host to the corresponding process running on the destination host.

The transport layer is responsible for process-to-process delivery—the delivery of a packet, part of a message, from one process to another. Two processes communicate in a client/server relationship, as we will see later. Figure 23.1 shows these three types of deliveries and their domains.

The transport layer is responsible for process-to-process delivery.

Figure 23.1 *Types of data deliveries*

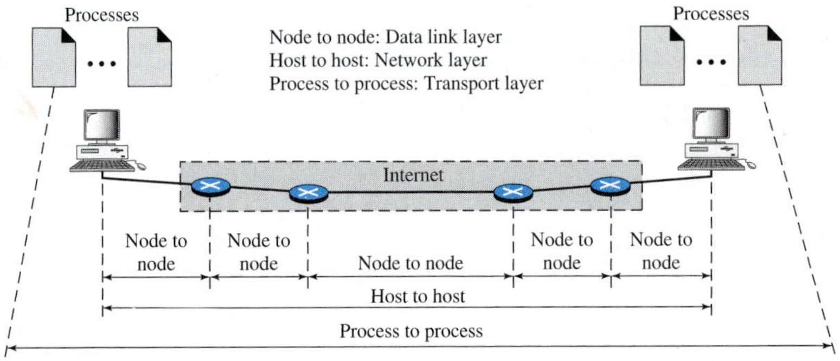

Client/Server Paradigm

Although there are several ways to achieve process-to-process communication, the most common one is through the **client/server paradigm.** A process on the local host, called a **client,** needs services from a process usually on the remote host, called a **server.**

Both processes (client and server) have the same name. For example, to get the day and time from a remote machine, we need a Daytime client process running on the local host and a Daytime server process running on a remote machine.

Operating systems today support both multiuser and multiprogramming environments. A remote computer can run several server programs at the same time, just as local computers can run one or more client programs at the same time. For communication, we must define the following:

1. Local host
2. Local process
3. Remote host
4. Remote process

Addressing

Whenever we need to deliver something to one specific destination among many, we need an address. At the data link layer, we need a MAC address to choose one node among several nodes if the connection is not point-to-point. A frame in the data link layer needs a destination MAC address for delivery and a source address for the next node's reply.

At the network layer, we need an IP address to choose one host among millions. A datagram in the network layer needs a destination IP address for delivery and a source IP address for the destination's reply.

At the transport layer, we need a transport layer address, called a **port number,** to choose among multiple processes running on the destination host. The destination port number is needed for delivery; the source port number is needed for the reply.

In the Internet model, the port numbers are 16-bit integers between 0 and 65,535. The client program defines itself with a port number, chosen randomly by the transport layer software running on the client host. This is the **ephemeral port number.**

The server process must also define itself with a port number. This port number, however, cannot be chosen randomly. If the computer at the server site runs a server process and assigns a random number as the port number, the process at the client site that wants to access that server and use its services will not know the port number. Of course, one solution would be to send a special packet and request the port number of a specific server, but this requires more overhead. The Internet has decided to use universal port numbers for servers; these are called **well-known port numbers.** There are some exceptions to this rule; for example, there are clients that are assigned well-known port numbers. Every client process knows the well-known port number of the corresponding server process. For example, while the Daytime client process, discussed above, can use an ephemeral (temporary) port number 52,000 to identify itself, the Daytime server process must use the well-known (permanent) port number 13. Figure 23.2 shows this concept.

Figure 23.2 *Port numbers*

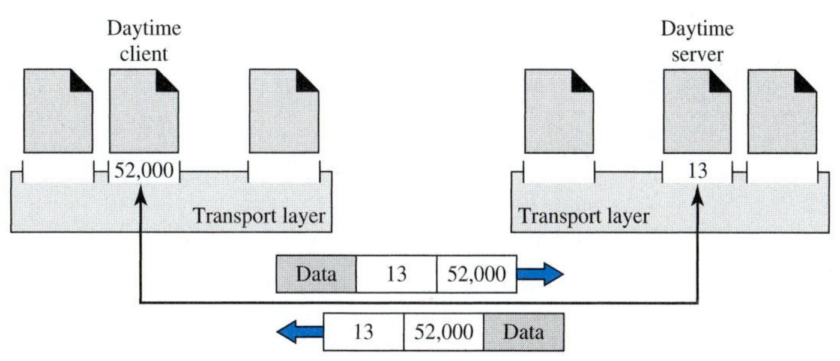

It should be clear by now that the IP addresses and port numbers play different roles in selecting the final destination of data. The destination IP address defines the host among the different hosts in the world. After the host has been selected, the port number defines one of the processes on this particular host (see Figure 23.3).

IANA Ranges

The IANA (Internet Assigned Number Authority) has divided the port numbers into three ranges: well known, registered, and dynamic (or private), as shown in Figure 23.4.

- **Well-known ports.** The ports ranging from 0 to 1023 are assigned and controlled by IANA. These are the well-known ports.
- **Registered ports.** The ports ranging from 1024 to 49,151 are not assigned or controlled by IANA. They can only be registered with IANA to prevent duplication.
- **Dynamic ports.** The ports ranging from 49,152 to 65,535 are neither controlled nor registered. They can be used by any process. These are the ephemeral ports.

Figure 23.3 *IP addresses versus port numbers*

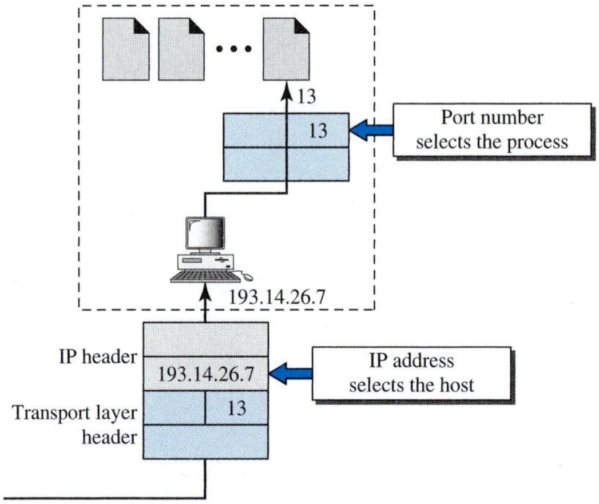

Figure 23.4 *IANA ranges*

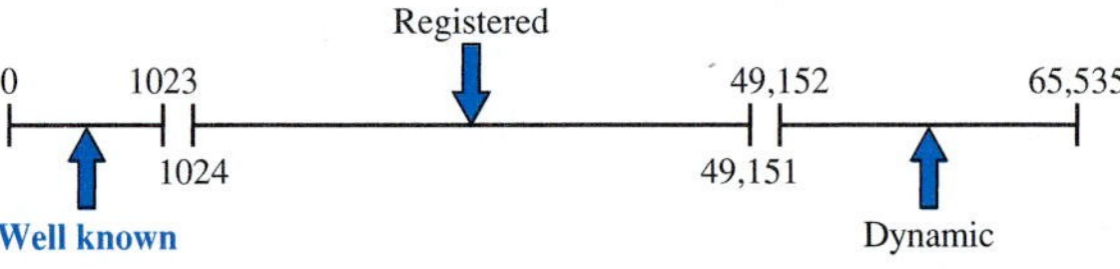

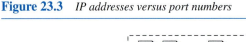

Socket Addresses

Process-to-process delivery needs two identifiers, IP address and the port number, at each end to make a connection. The combination of an IP address and a port number is called a **socket address.** The client socket address defines the client process uniquely just as the server socket address defines the server process uniquely (see Figure 23.5).

A transport layer protocol needs a pair of socket addresses: the client socket address and the server socket address. These four pieces of information are part of the IP header and the transport layer protocol header. The IP header contains the IP addresses; the UDP or TCP header contains the port numbers.

Figure 23.5 *Socket address*

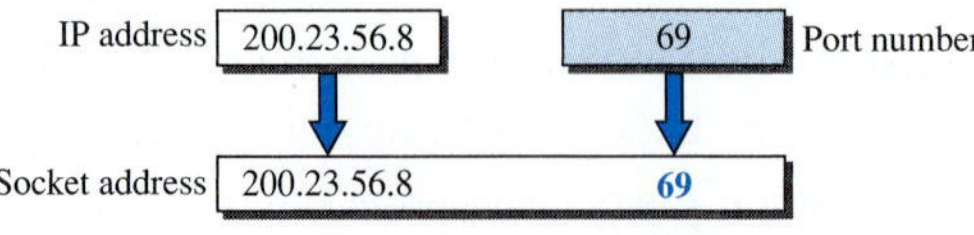

Multiplexing and Demultiplexing

The addressing mechanism allows multiplexing and demultiplexing by the transport layer, as shown in Figure 23.6.

Figure 23.6 *Multiplexing and demultiplexing*

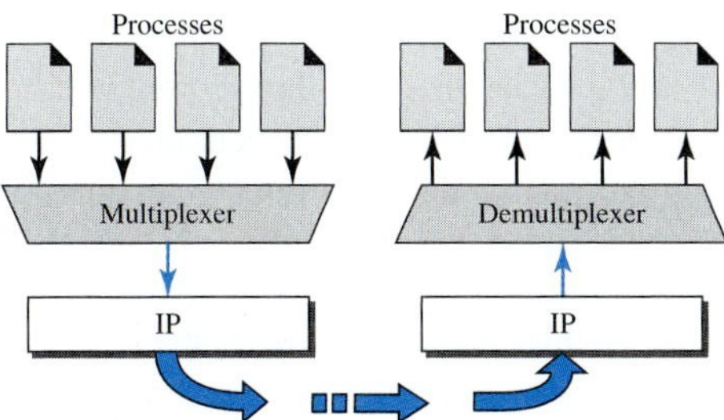

Multiplexing

At the sender site, there may be several processes that need to send packets. However, there is only one transport layer protocol at any time. This is a many-to-one relationship and requires multiplexing. The protocol accepts messages from different processes, differentiated by their assigned port numbers. After adding the header, the transport layer passes the packet to the network layer.

Demultiplexing

At the receiver site, the relationship is one-to-many and requires demultiplexing. The transport layer receives datagrams from the network layer. After error checking and dropping of the header, the transport layer delivers each message to the appropriate process based on the port number.

Connectionless Versus Connection-Oriented Service

A transport layer protocol can either be connectionless or connection-oriented.

Connectionless Service

In a **connectionless service,** the packets are sent from one party to another with no need for connection establishment or connection release. The packets are not numbered; they may be delayed or lost or may arrive out of sequence. There is no acknowledgment either. We will see shortly that one of the transport layer protocols in the Internet model, UDP, is connectionless.

Connection-Oriented Service

In a **connection-oriented service,** a connection is first established between the sender and the receiver. Data are transferred. At the end, the connection is released. We will see shortly that TCP and SCTP are connection-oriented protocols.

Reliable Versus Unreliable

The transport layer service can be reliable or unreliable. If the application layer program needs reliability, we use a reliable transport layer protocol by implementing flow and error control at the transport layer. This means a slower and more complex service. On the other hand, if the application program does not need reliability because it uses its own flow and error control mechanism or it needs fast service or the nature of the service does not demand flow and error control (real-time applications), then an unreliable protocol can be used.

In the Internet, there are three common different transport layer protocols, as we have already mentioned. UDP is connectionless and unreliable; TCP and SCTP are connection-oriented and reliable. These three can respond to the demands of the application layer programs.

One question often comes to the mind. If the data link layer is reliable and has flow and error control, do we need this at the transport layer, too? The answer is yes. Reliability at the data link layer is between two nodes; we need reliability between two ends. Because the network layer in the Internet is unreliable (best-effort delivery), we need to implement reliability at the transport layer. To understand that error control at the data link layer does not guarantee error control at the transport layer, let us look at Figure 23.7.

Figure 23.7 *Error control*

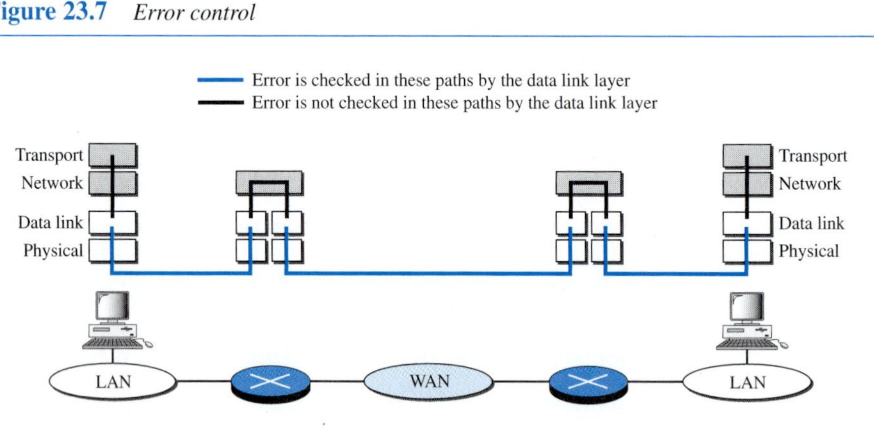

As we will see, flow and error control in TCP is implemented by the sliding window protocol, as discussed in Chapter 11. The window, however, is character-oriented, instead of frame-oriented.

Three Protocols

The original TCP/IP protocol suite specifies two protocols for the transport layer: UDP and TCP. We first focus on UDP, the simpler of the two, before discussing TCP. A new transport layer protocol, SCTP, has been designed, which we also discuss in this chapter. Figure 23.8 shows the position of these protocols in the TCP/IP protocol suite.

Figure 23.8 *Position of UDP, TCP, and SCTP in TCP/IP suite*

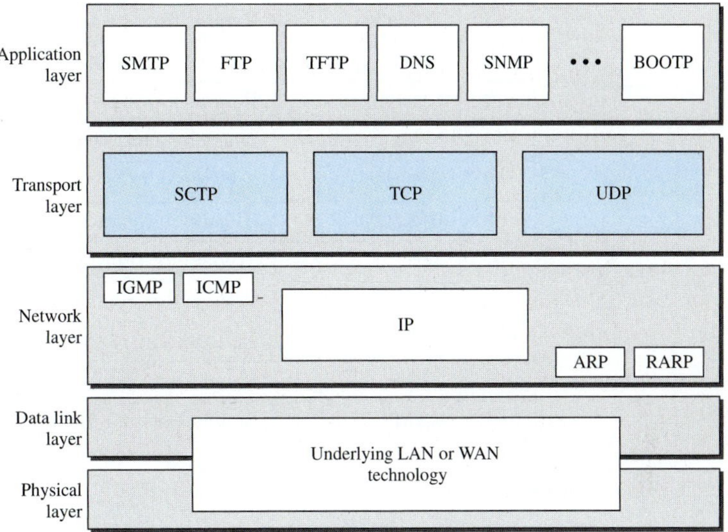

23.2 USER DATAGRAM PROTOCOL (UDP)

The **User Datagram Protocol (UDP)** is called a **connectionless, unreliable transport protocol.** It does not add anything to the services of IP except to provide process-to-process communication instead of host-to-host communication. Also, it performs very limited error checking.

If UDP is so powerless, why would a process want to use it? With the disadvantages come some advantages. UDP is a very simple protocol using a minimum of overhead. If a process wants to send a small message and does not care much about reliability, it can use UDP. Sending a small message by using UDP takes much less interaction between the sender and receiver than using TCP or SCTP.

Well-Known Ports for UDP

Table 23.1 shows some well-known port numbers used by UDP. Some port numbers can be used by both UDP and TCP. We discuss them when we talk about TCP later in the chapter.

Table 23.1 *Well-known ports used with UDP*

Port	*Protocol*	*Description*
7	Echo	Echoes a received datagram back to the sender
9	Discard	Discards any datagram that is received
11	Users	Active users

Table 23.1 *Well-known ports used with UDP (continued)*

Port	*Protocol*	*Description*
13	Daytime	Returns the date and the time
17	Quote	Returns a quote of the day
19	Chargen	Returns a string of characters
53	Nameserver	Domain Name Service
67	BOOTPs	Server port to download bootstrap information
68	BOOTPc	Client port to download bootstrap information
69	TFTP	Trivial File Transfer Protocol
111	RPC	Remote Procedure Call
123	NTP	Network Time Protocol
161	SNMP	Simple Network Management Protocol
162	SNMP	Simple Network Management Protocol (trap)

Example 23.1

In UNIX, the well-known ports are stored in a file called /etc/services. Each line in this file gives the name of the server and the well-known port number. We can use the *grep* utility to extract the line corresponding to the desired application. The following shows the port for FTP. Note that FTP can use port 21 with either UDP or TCP.

```
$ grep    ftp   /etc/services
ftp        21/tcp
ftp        21/udp
```

SNMP uses two port numbers (161 and 162), each for a different purpose, as we will see in Chapter 28.

```
$ grep      snmp /etc/services
snmp            161/tcp          #Simple Net  Mgmt Proto
snmp            161/udp          #Simple Net  Mgmt Proto
snmptrap        162/udp          #Traps for SNMP
```

User Datagram

UDP packets, called **user datagrams,** have a fixed-size header of 8 bytes. Figure 23.9 shows the format of a user datagram.

The fields are as follows:

- **Source port number.** This is the port number used by the process running on the source host. It is 16 bits long, which means that the port number can range from 0 to 65,535. If the source host is the client (a client sending a request), the port number, in most cases, is an ephemeral port number requested by the process and chosen by the UDP software running on the source host. If the source host is the server (a server sending a response), the port number, in most cases, is a well-known port number.

Figure 23.9 *User datagram format*

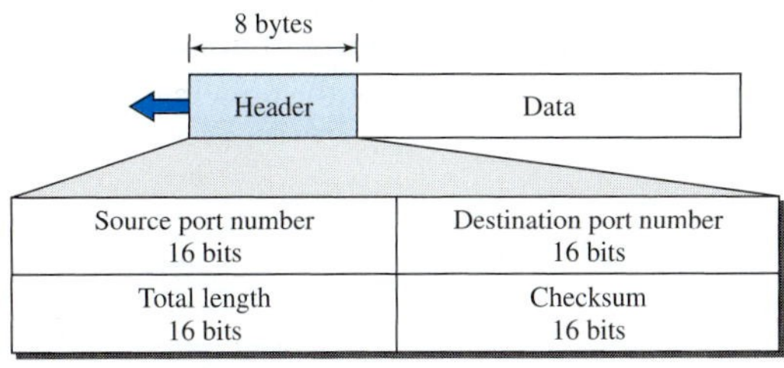

- **Destination port number.** This is the port number used by the process running on the destination host. It is also 16 bits long. If the destination host is the server (a client sending a request), the port number, in most cases, is a well-known port number. If the destination host is the client (a server sending a response), the port number, in most cases, is an ephemeral port number. In this case, the server copies the ephemeral port number it has received in the request packet.
- **Length.** This is a 16-bit field that defines the total length of the user datagram, header plus data. The 16 bits can define a total length of 0 to 65,535 bytes. However, the total length needs to be much less because a UDP user datagram is stored in an IP datagram with a total length of 65,535 bytes.

  The length field in a UDP user datagram is actually not necessary. A user datagram is encapsulated in an IP datagram. There is a field in the IP datagram that defines the total length. There is another field in the IP datagram that defines the length of the header. So if we subtract the value of the second field from the first, we can deduce the length of a UDP datagram that is encapsulated in an IP datagram.

  UDP length = IP length – IP header's length

  However, the designers of the UDP protocol felt that it was more efficient for the destination UDP to calculate the length of the data from the information provided in the UDP user datagram rather than ask the IP software to supply this information. We should remember that when the IP software delivers the UDP user datagram to the UDP layer, it has already dropped the IP header.
- **Checksum.** This field is used to detect errors over the entire user datagram (header plus data). The checksum is discussed next.

Checksum

We have already talked about the concept of the checksum and the way it is calculated in Chapter 10. We have also shown how to calculate the checksum for the IP and ICMP packet. We now show how this is done for UDP.

The UDP checksum calculation is different from the one for IP and ICMP. Here the checksum includes three sections: a pseudoheader, the UDP header, and the data coming from the application layer.

The **pseudoheader** is the part of the header of the IP packet in which the user datagram is to be encapsulated with some fields filled with 0s (see Figure 23.10).

Figure 23.10 *Pseudoheader for checksum calculation*

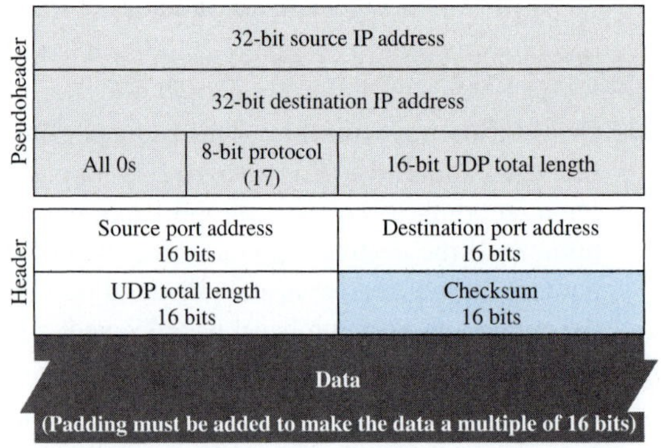

If the checksum does not include the pseudoheader, a user datagram may arrive safe and sound. However, if the IP header is corrupted, it may be delivered to the wrong host.

The protocol field is added to ensure that the packet belongs to UDP, and not to other transport-layer protocols. We will see later that if a process can use either UDP or TCP, the destination port number can be the same. The value of the protocol field for UDP is 17. If this value is changed during transmission, the checksum calculation at the receiver will detect it and UDP drops the packet. It is not delivered to the wrong protocol.

Note the similarities between the pseudoheader fields and the last 12 bytes of the IP header.

Example 23.2

Figure 23.11 shows the checksum calculation for a very small user datagram with only 7 bytes of data. Because the number of bytes of data is odd, padding is added for checksum calculation. The pseudoheader as well as the padding will be dropped when the user datagram is delivered to IP.

Optional Use of the Checksum

The calculation of the checksum and its inclusion in a user datagram are optional. If the checksum is not calculated, the field is filled with 1s. Note that a calculated checksum can never be all 1s because this implies that the sum is all 0s, which is impossible because it requires that the value of fields to be 0s.

Figure 23.11 *Checksum calculation of a simple UDP user datagram*

153.18.8.105			
171.2.14.10			
All 0s	17	15	

1087	13
15	All 0s

T	E	S	T
I	N	G	All 0s

Binary		Value
10011001 00010010	→	153.18
00001000 01101001	→	8.105
10101011 00000010	→	171.2
00001110 00001010	→	14.10
00000000 00010001	→	0 and 17
00000000 00001111	→	15
00000100 00111111	→	1087
00000000 00001101	→	13
00000000 00001111	→	15
00000000 00000000	→	0 (checksum)
01010100 01000101	→	T and E
01010011 01010100	→	S and T
01001001 01001110	→	I and N
01000111 00000000	→	G and 0 (padding)
10010110 11101011	→	Sum
01101001 00010100	→	Checksum

UDP Operation

UDP uses concepts common to the transport layer. These concepts will be discussed here briefly, and then expanded in the next section on the TCP protocol.

Connectionless Services

As mentioned previously, UDP provides a connectionless service. This means that each user datagram sent by UDP is an independent datagram. There is no relationship between the different user datagrams even if they are coming from the same source process and going to the same destination program. The user datagrams are not numbered. Also, there is no connection establishment and no connection termination, as is the case for TCP. This means that each user datagram can travel on a different path.

One of the ramifications of being connectionless is that the process that uses UDP cannot send a stream of data to UDP and expect UDP to chop them into different related user datagrams. Instead each request must be small enough to fit into one user datagram. Only those processes sending short messages should use UDP.

Flow and Error Control

UDP is a very simple, unreliable transport protocol. There is no flow control and hence no window mechanism. The receiver may overflow with incoming messages.

There is no error control mechanism in UDP except for the checksum. This means that the sender does not know if a message has been lost or duplicated. When the receiver detects an error through the checksum, the user datagram is silently discarded.

The lack of **flow control** and **error control** means that the process using UDP should provide these mechanisms.

Encapsulation and Decapsulation

To send a message from one process to another, the UDP protocol encapsulates and decapsulates messages in an IP datagram.

Queuing

We have talked about ports without discussing the actual implementation of them. In UDP, queues are associated with ports (see Figure 23.12).

Figure 23.12 *Queues in UDP*

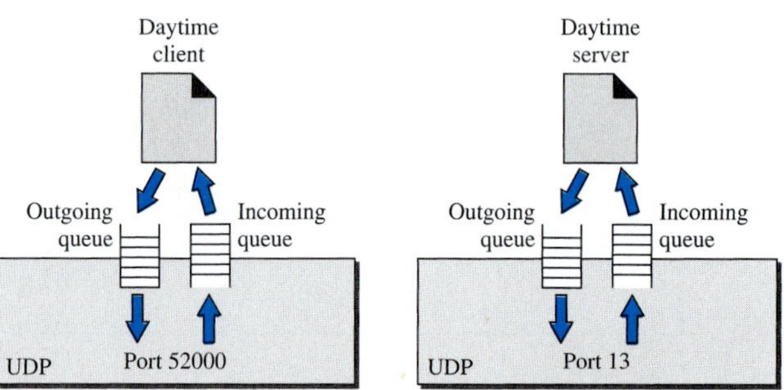

At the client site, when a process starts, it requests a port number from the operating system. Some implementations create both an incoming and an outgoing queue associated with each process. Other implementations create only an incoming queue associated with each process.

Note that even if a process wants to communicate with multiple processes, it obtains only one port number and eventually one outgoing and one incoming **queue.** The queues opened by the client are, in most cases, identified by ephemeral port numbers. The queues function as long as the process is running. When the process terminates, the queues are destroyed.

The client process can send messages to the outgoing queue by using the source port number specified in the request. UDP removes the messages one by one and, after adding the UDP header, delivers them to IP. An outgoing queue can overflow. If this happens, the operating system can ask the client process to wait before sending any more messages.

When a message arrives for a client, UDP checks to see if an incoming queue has been created for the port number specified in the destination port number field of the user datagram. If there is such a queue, UDP sends the received user datagram to the end of the queue. If there is no such queue, UDP discards the user datagram and asks the ICMP protocol to send a *port unreachable* message to the server. All the incoming messages for one particular client program, whether coming from the same or a different server, are sent to the same queue. An incoming queue can overflow. If this happens, UDP drops the user datagram and asks for a port unreachable message to be sent to the server.

At the server site, the mechanism of creating queues is different. In its simplest form, a server asks for incoming and outgoing queues, using its well-known port, when it starts running. The queues remain open as long as the server is running.

When a message arrives for a server, UDP checks to see if an incoming queue has been created for the port number specified in the destination port number field of the user datagram. If there is such a queue, UDP sends the received user datagram to the end of the queue. If there is no such queue, UDP discards the user datagram and asks the ICMP protocol to send a port unreachable message to the client. All the incoming messages for one particular server, whether coming from the same or a different client, are sent to the same queue. An incoming queue can overflow. If this happens, UDP drops the user datagram and asks for a port unreachable message to be sent to the client.

When a server wants to respond to a client, it sends messages to the outgoing queue, using the source port number specified in the request. UDP removes the messages one by one and, after adding the UDP header, delivers them to IP. An outgoing queue can overflow. If this happens, the operating system asks the server to wait before sending any more messages.

Use of UDP

The following lists some uses of the UDP protocol:

- UDP is suitable for a process that requires simple request-response communication with little concern for flow and error control. It is not usually used for a process such as FTP that needs to send bulk data (see Chapter 26).
- UDP is suitable for a process with internal flow and error control mechanisms. For example, the Trivial File Transfer Protocol (TFTP) process includes flow and error control. It can easily use UDP.
- UDP is a suitable transport protocol for multicasting. Multicasting capability is embedded in the UDP software but not in the TCP software.
- UDP is used for management processes such as SNMP (see Chapter 28).
- UDP is used for some route updating protocols such as Routing Information Protocol (RIP) (see Chapter 22).

23.3 TCP

The second transport layer protocol we discuss in this chapter is called **Transmission Control Protocol (TCP).** TCP, like UDP, is a process-to-process (program-to-program) protocol. TCP, therefore, like UDP, uses port numbers. Unlike UDP, TCP is a connection-oriented protocol; it creates a virtual connection between two TCPs to send data. In addition, TCP uses flow and error control mechanisms at the transport level.

In brief, TCP is called a *connection-oriented, reliable* transport protocol. It adds connection-oriented and reliability features to the services of IP.

TCP Services

Before we discuss TCP in detail, let us explain the services offered by TCP to the processes at the application layer.

Process-to-Process Communication

Like UDP, TCP provides process-to-process communication using port numbers. Table 23.2 lists some well-known port numbers used by TCP.

Table 23.2 *Well-known ports used by TCP*

Port	*Protocol*	*Description*
7	Echo	Echoes a received datagram back to the sender
9	Discard	Discards any datagram that is received
11	Users	Active users
13	Daytime	Returns the date and the time
17	Quote	Returns a quote of the day
19	Chargen	Returns a string of characters
20	FTP, Data	File Transfer Protocol (data connection)
21	FTP, Control	File Transfer Protocol (control connection)
23	TELNET	Terminal Network
25	SMTP	Simple Mail Transfer Protocol
53	DNS	Domain Name Server
67	BOOTP	Bootstrap Protocol
79	Finger	Finger
80	HTTP	Hypertext Transfer Protocol
111	RPC	Remote Procedure Call

Stream Delivery Service

TCP, unlike UDP, is a stream-oriented protocol. In UDP, a process (an application program) sends messages, with predefined boundaries, to UDP for delivery. UDP adds its own header to each of these messages and delivers them to IP for transmission. Each message from the process is called a user datagram and becomes, eventually, one IP datagram. Neither IP nor UDP recognizes any relationship between the datagrams.

TCP, on the other hand, allows the sending process to deliver data as a stream of bytes and allows the receiving process to obtain data as a stream of bytes. TCP creates an environment in which the two processes seem to be connected by an imaginary "tube" that carries their data across the Internet. This imaginary environment is depicted in Figure 23.13. The sending process produces (writes to) the stream of bytes, and the receiving process consumes (reads from) them.

Figure 23.13 *Stream delivery*

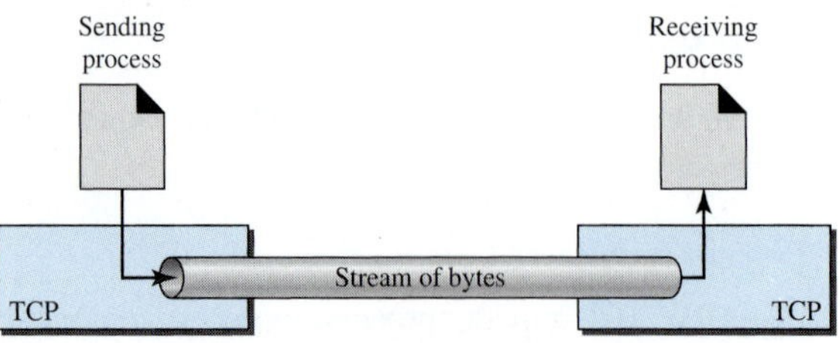

Sending and Receiving Buffers Because the sending and the receiving processes may not write or read data at the same speed, TCP needs buffers for storage. There are two buffers, the sending buffer and the receiving buffer, one for each direction. (We will see later that these buffers are also necessary for flow and error control mechanisms used by TCP.) One way to implement a buffer is to use a circular array of 1-byte locations as shown in Figure 23.14. For simplicity, we have shown two buffers of 20 bytes each; normally the buffers are hundreds or thousands of bytes, depending on the implementation. We also show the buffers as the same size, which is not always the case.

Figure 23.14 *Sending and receiving buffers*

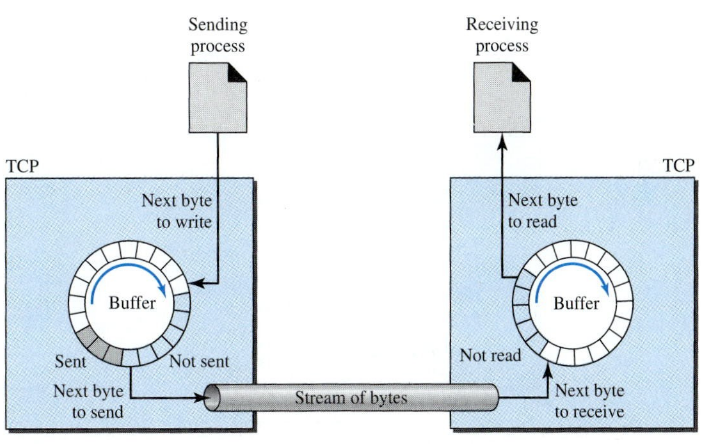

Figure 23.14 shows the movement of the data in one direction. At the sending site, the buffer has three types of chambers. The white section contains empty chambers that can be filled by the sending process (producer). The gray area holds bytes that have been sent but not yet acknowledged. TCP keeps these bytes in the buffer until it receives an acknowledgment. The colored area contains bytes to be sent by the sending TCP. However, as we will see later in this chapter, TCP may be able to send only part of this colored section. This could be due to the slowness of the receiving process or perhaps to congestion in the network. Also note that after the bytes in the gray chambers are acknowledged, the chambers are recycled and available for use by the sending process. This is why we show a circular buffer.

The operation of the buffer at the receiver site is simpler. The circular buffer is divided into two areas (shown as white and colored). The white area contains empty chambers to be filled by bytes received from the network. The colored sections contain received bytes that can be read by the receiving process. When a byte is read by the receiving process, the chamber is recycled and added to the pool of empty chambers.

Segments Although buffering handles the disparity between the speed of the producing and consuming processes, we need one more step before we can send data. The IP layer, as a service provider for TCP, needs to send data in packets, not as a stream of bytes. At the transport layer, TCP groups a number of bytes together into a packet called a **segment.** TCP adds a header to each segment (for control purposes) and delivers the segment to the IP layer for transmission. The segments are encapsulated in IP datagrams and transmitted. This entire operation is transparent to the receiving process. Later we will see that segments may be received out of order, lost, or corrupted and resent. All these are handled by TCP with the receiving process unaware of any activities. Figure 23.15 shows how segments are created from the bytes in the buffers.

Figure 23.15 *TCP segments*

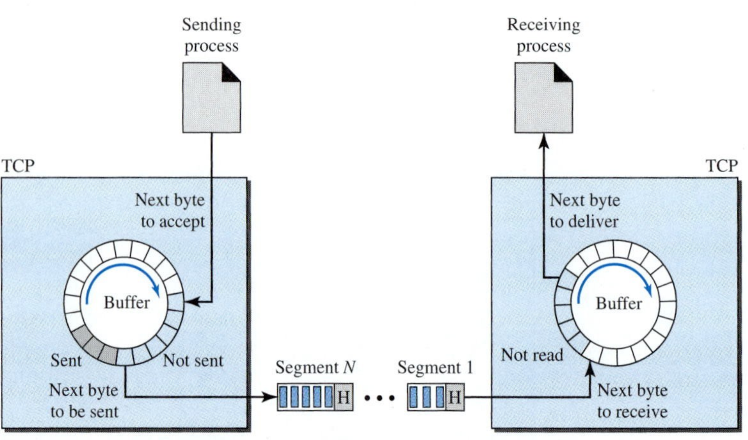

Note that the segments are not necessarily the same size. In Figure 23.15, for simplicity, we show one segment carrying 3 bytes and the other carrying 5 bytes. In reality, segments carry hundreds, if not thousands, of bytes.

Full-Duplex Communication

TCP offers **full-duplex service,** in which data can flow in both directions at the same time. Each TCP then has a sending and receiving buffer, and segments move in both directions.

Connection-Oriented Service

TCP, unlike UDP, is a connection-oriented protocol. When a process at site A wants to send and receive data from another process at site B, the following occurs:

1. The two TCPs establish a connection between them.
2. Data are exchanged in both directions.
3. The connection is terminated.

Note that this is a virtual connection, not a physical connection. The TCP segment is encapsulated in an IP datagram and can be sent out of order, or lost, or corrupted, and then resent. Each may use a different path to reach the destination. There is no physical connection. TCP creates a stream-oriented environment in which it accepts the responsibility of delivering the bytes in order to the other site. The situation is similar to creating a bridge that spans multiple islands and passing all the bytes from one island to another in one single connection. We will discuss this feature later in the chapter.

Reliable Service

TCP is a reliable transport protocol. It uses an acknowledgment mechanism to check the safe and sound arrival of data. We will discuss this feature further in the section on error control.

TCP Features

To provide the services mentioned in the previous section, TCP has several features that are briefly summarized in this section and discussed later in detail.

Numbering System

Although the TCP software keeps track of the segments being transmitted or received, there is no field for a segment number value in the segment header. Instead, there are two fields called the **sequence number** and the **acknowledgment number.** These two fields refer to the byte number and not the segment number.

Byte Number TCP numbers all data bytes that are transmitted in a connection. Numbering is independent in each direction. When TCP receives bytes of data from a process, it stores them in the sending buffer and numbers them. The numbering does not necessarily start from 0. Instead, TCP generates a random number between 0 and $2^{32} - 1$ for the number of the first byte. For example, if the random number happens to be 1057 and the total data to be sent are 6000 bytes, the bytes are numbered from 1057 to 7056. We will see that byte numbering is used for flow and error control.

The bytes of data being transferred in each connection are numbered by TCP. The numbering starts with a randomly generated number.

Sequence Number After the bytes have been numbered, TCP assigns a sequence number to each segment that is being sent. The sequence number for each segment is the number of the first byte carried in that segment.

Example 23.3

Suppose a TCP connection is transferring a file of 5000 bytes. The first byte is numbered 10,001. What are the sequence numbers for each segment if data are sent in five segments, each carrying 1000 bytes?

Solution

The following shows the sequence number for each segment:

Segment 1 ➡ Sequence Number: 10,001 (range: 10,001 to 11,000)
Segment 2 ➡ Sequence Number: 11,001 (range: 11,001 to 12,000)
Segment 3 ➡ Sequence Number: 12,001 (range: 12,001 to 13,000)
Segment 4 ➡ Sequence Number: 13,001 (range: 13,001 to 14,000)
Segment 5 ➡ Sequence Number: 14,001 (range: 14,001 to 15,000)

The value in the sequence number field of a segment defines the number of the first data byte contained in that segment.

When a segment carries a combination of data and control information (piggybacking), it uses a sequence number. If a segment does not carry user data, it does not logically define a sequence number. The field is there, but the value is not valid. However, some segments, when carrying only control information, need a sequence number to allow an acknowledgment from the receiver. These segments are used for connection establishment, termination, or abortion. Each of these segments consumes one sequence number as though it carried 1 byte, but there are no actual data. If the randomly generated sequence number is x, the first data byte is numbered $x + 1$. The byte x is considered a phony byte that is used for a control segment to open a connection, as we will see shortly.

Acknowledgment Number As we discussed previously, communication in TCP is full duplex; when a connection is established, both parties can send and receive data at the same time. Each party numbers the bytes, usually with a different starting byte number. The sequence number in each direction shows the number of the first byte carried by the segment. Each party also uses an acknowledgment number to confirm the bytes it has received. However, the acknowledgment number defines the number of the next byte that the party expects to receive. In addition, the acknowledgment number is cumulative, which means that the party takes the number of the last byte that it has received, safe and sound, adds 1 to it, and announces this sum as the acknowledgment number. The term *cumulative* here means that if a party uses 5643 as an acknowledgment number, it has received all bytes from the beginning up to 5642. Note that this does not mean that the party has received 5642 bytes because the first byte number does not have to start from 0.

The value of the acknowledgment field in a segment defines the number of the next byte a party expects to receive. The acknowledgment number is cumulative.

Flow Control

TCP, unlike UDP, provides *flow control.* The receiver of the data controls the amount of data that are to be sent by the sender. This is done to prevent the receiver from being overwhelmed with data. The numbering system allows TCP to use a byte-oriented flow control.

Error Control

To provide reliable service, TCP implements an error control mechanism. Although error control considers a segment as the unit of data for error detection (loss or corrupted segments), error control is byte-oriented, as we will see later.

Congestion Control

TCP, unlike UDP, takes into account congestion in the network. The amount of data sent by a sender is not only controlled by the receiver (flow control), but is also determined by the level of congestion in the network.

Segment

Before we discuss TCP in greater detail, let us discuss the TCP packets themselves. A packet in TCP is called a **segment.**

Format

The format of a segment is shown in Figure 23.16.

Figure 23.16 *TCP segment format*

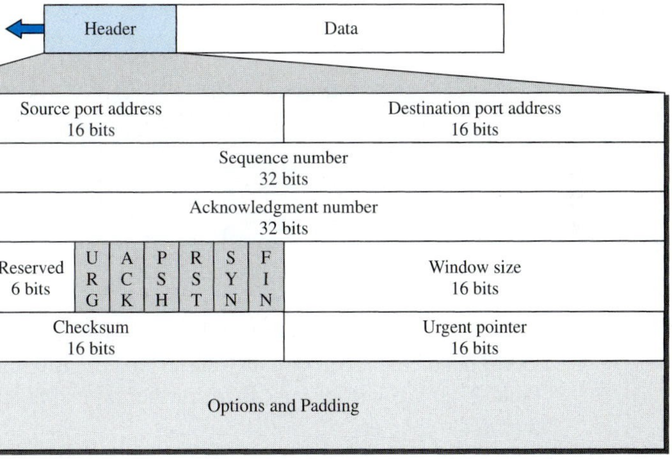

The segment consists of a 20- to 60-byte header, followed by data from the application program. The header is 20 bytes if there are no options and up to 60 bytes if it contains options. We will discuss some of the header fields in this section. The meaning and purpose of these will become clearer as we proceed through the chapter.

- **Source port address.** This is a 16-bit field that defines the port number of the application program in the host that is sending the segment. This serves the same purpose as the source port address in the UDP header.
- **Destination port address.** This is a 16-bit field that defines the port number of the application program in the host that is receiving the segment. This serves the same purpose as the destination port address in the UDP header.
- **Sequence number.** This 32-bit field defines the number assigned to the first byte of data contained in this segment. As we said before, TCP is a stream transport protocol. To ensure connectivity, each byte to be transmitted is numbered. The sequence number tells the destination which byte in this sequence comprises the first byte in the segment. During connection establishment, each party uses a random number generator to create an **initial sequence number (ISN),** which is usually different in each direction.
- **Acknowledgment number.** This 32-bit field defines the byte number that the receiver of the segment is expecting to receive from the other party. If the receiver

of the segment has successfully received byte number x from the other party, it defines $x + 1$ as the acknowledgment number. Acknowledgment and data can be piggybacked together.

- **Header length.** This 4-bit field indicates the number of 4-byte words in the TCP header. The length of the header can be between 20 and 60 bytes. Therefore, the value of this field can be between 5 ($5 \times 4 = 20$) and 15 ($15 \times 4 = 60$).
- **Reserved.** This is a 6-bit field reserved for future use.
- **Control.** This field defines 6 different control bits or flags as shown in Figure 23.17. One or more of these bits can be set at a time.

Figure 23.17 *Control field*

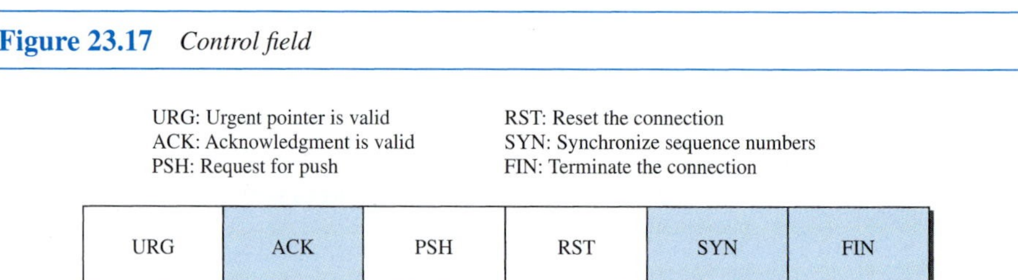

These bits enable flow control, connection establishment and termination, connection abortion, and the mode of data transfer in TCP. A brief description of each bit is shown in Table 23.3. We will discuss them further when we study the detailed operation of TCP later in the chapter.

Table 23.3 *Description of flags in the control field*

Flag	*Description*
URG	The value of the urgent pointer field is valid.
ACK	The value of the acknowledgment field is valid.
PSH	Push the data.
RST	Reset the connection.
SYN	Synchronize sequence numbers during connection.
FIN	Terminate the connection.

- **Window size.** This field defines the size of the window, in bytes, that the other party must maintain. Note that the length of this field is 16 bits, which means that the maximum size of the window is 65,535 bytes. This value is normally referred to as the receiving window (rwnd) and is determined by the receiver. The sender must obey the dictation of the receiver in this case.
- **Checksum.** This 16-bit field contains the checksum. The calculation of the checksum for TCP follows the same procedure as the one described for UDP. However, the inclusion of the checksum in the UDP datagram is optional, whereas the inclusion of the checksum for TCP is mandatory. The same pseudoheader, serving the same

purpose, is added to the segment. For the TCP pseudoheader, the value for the protocol field is 6.

- **Urgent pointer.** This 16-bit field, which is valid only if the urgent flag is set, is used when the segment contains urgent data. It defines the number that must be added to the sequence number to obtain the number of the last urgent byte in the data section of the segment. This will be discussed later in this chapter.
- **Options.** There can be up to 40 bytes of optional information in the TCP header. We will not discuss these options here; please refer to the reference list for more information.

A TCP Connection

TCP is connection-oriented. A connection-oriented transport protocol establishes a virtual path between the source and destination. All the segments belonging to a message are then sent over this virtual path. Using a single virtual pathway for the entire message facilitates the acknowledgment process as well as retransmission of damaged or lost frames. You may wonder how TCP, which uses the services of IP, a connectionless protocol, can be connection-oriented. The point is that a TCP connection is virtual, not physical. TCP operates at a higher level. TCP uses the services of IP to deliver individual segments to the receiver, but it controls the connection itself. If a segment is lost or corrupted, it is retransmitted. Unlike TCP, IP is unaware of this retransmission. If a segment arrives out of order, TCP holds it until the missing segments arrive; IP is unaware of this reordering.

In TCP, connection-oriented transmission requires three phases: connection establishment, data transfer, and connection termination.

Connection Establishment

TCP transmits data in full-duplex mode. When two TCPs in two machines are connected, they are able to send segments to each other simultaneously. This implies that each party must initialize communication and get approval from the other party before any data are transferred.

Three-Way Handshaking The connection establishment in TCP is called **three-way handshaking.** In our example, an application program, called the client, wants to make a connection with another application program, called the server, using TCP as the transport layer protocol.

The process starts with the server. The server program tells its TCP that it is ready to accept a connection. This is called a request for a *passive open*. Although the server TCP is ready to accept any connection from any machine in the world, it cannot make the connection itself.

The client program issues a request for an *active open*. A client that wishes to connect to an open server tells its TCP that it needs to be connected to that particular server. TCP can now start the three-way handshaking process as shown in Figure 23.18. To show the process, we use two time lines: one at each site. Each segment has values for all its header fields and perhaps for some of its option fields, too. However, we show only the few fields necessary to understand each phase. We show the sequence number,

Figure 23.18 *Connection establishment using three-way handshaking*

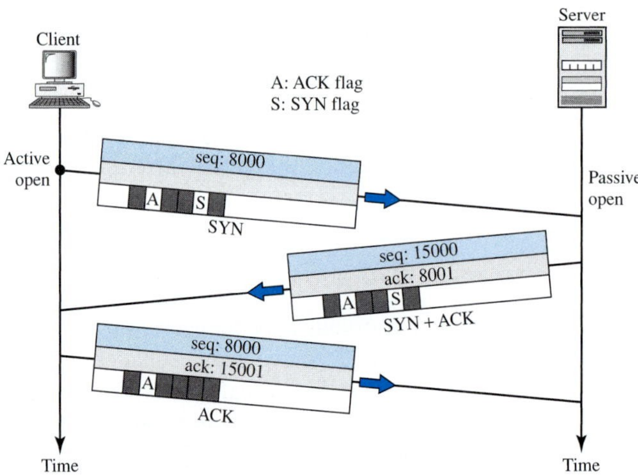

the acknowledgment number, the control flags (only those that are set), and the window size, if not empty. The three steps in this phase are as follows.

1. The client sends the first segment, a SYN segment, in which only the SYN flag is set. This segment is for synchronization of sequence numbers. It consumes one sequence number. When the data transfer starts, the sequence number is incremented by 1. We can say that the SYN segment carries no real data, but we can think of it as containing 1 imaginary byte.

A SYN segment cannot carry data, but it consumes one sequence number.

2. The server sends the second segment, a SYN + ACK segment, with 2 flag bits set: SYN and ACK. This segment has a dual purpose. It is a SYN segment for communication in the other direction and serves as the acknowledgment for the SYN segment. It consumes one sequence number.

A SYN + ACK segment cannot carry data, but does consume one sequence number.

3. The client sends the third segment. This is just an ACK segment. It acknowledges the receipt of the second segment with the ACK flag and acknowledgment number field. Note that the sequence number in this segment is the same as the one in the SYN segment; the ACK segment does not consume any sequence numbers.

An ACK segment, if carrying no data, consumes no sequence number.

Simultaneous Open A rare situation, called a **simultaneous open,** may occur when both processes issue an active open. In this case, both TCPs transmit a SYN + ACK segment to each other, and one single connection is established between them.

SYN Flooding Attack The connection establishment procedure in TCP is susceptible to a serious security problem called the **SYN flooding attack.** This happens when a malicious attacker sends a large number of SYN segments to a server, pretending that each of them is coming from a different client by faking the source IP addresses in the datagrams. The server, assuming that the clients are issuing an active open, allocates the necessary resources, such as creating communication tables and setting timers. The TCP server then sends the SYN + ACK segments to the fake clients, which are lost. During this time, however, a lot of resources are occupied without being used. If, during this short time, the number of SYN segments is large, the server eventually runs out of resources and may crash. This SYN flooding attack belongs to a type of security attack known as a **denial-of-service attack,** in which an attacker monopolizes a system with so many service requests that the system collapses and denies service to every request.

Some implementations of TCP have strategies to alleviate the effects of a SYN attack. Some have imposed a limit on connection requests during a specified period of time. Others filter out datagrams coming from unwanted source addresses. One recent strategy is to postpone resource allocation until the entire connection is set up, using what is called a **cookie.** SCTP, the new transport layer protocol that we discuss in the next section, uses this strategy.

Data Transfer

After connection is established, bidirectional **data transfer** can take place. The client and server can both send data and acknowledgments. We will study the rules of acknowledgment later in the chapter; for the moment, it is enough to know that data traveling in the same direction as an acknowledgment are carried on the same segment. The acknowledgment is piggybacked with the data. Figure 23.19 shows an example. In this example, after connection is established (not shown in the figure), the client sends 2000 bytes of data in two segments. The server then sends 2000 bytes in one segment. The client sends one more segment. The first three segments carry both data and acknowledgment, but the last segment carries only an acknowledgment because there are no more data to be sent. Note the values of the sequence and acknowledgment numbers. The data segments sent by the client have the PSH (push) flag set so that the server TCP knows to deliver data to the server process as soon as they are received. We discuss the use of this flag in greater detail later. The segment from the server, on the other hand, does not set the push flag. Most TCP implementations have the option to set or not set this flag.

Pushing Data We saw that the sending TCP uses a buffer to store the stream of data coming from the sending application program. The sending TCP can select the segment size. The receiving TCP also buffers the data when they arrive and delivers them to the application program when the application program is ready or when it is convenient for the receiving TCP. This type of flexibility increases the efficiency of TCP.

However, on occasion the application program has no need for this flexibility. For example, consider an application program that communicates interactively with another

Figure 23.19 *Data transfer*

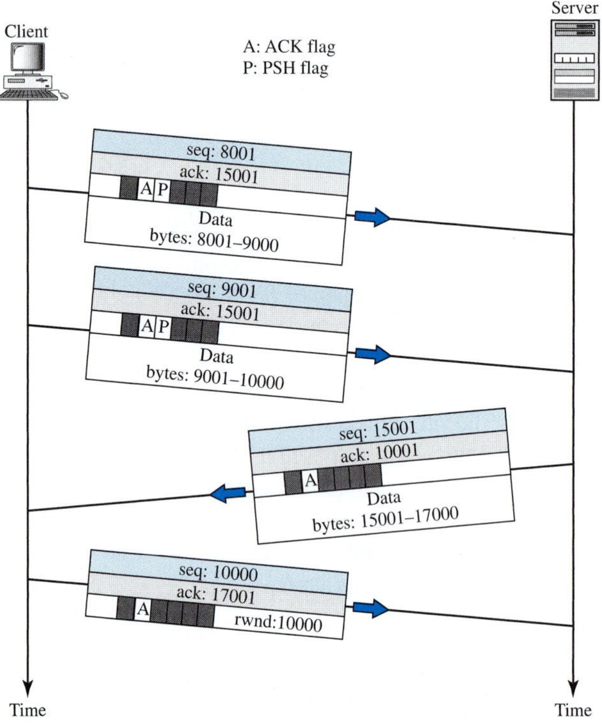

application program on the other end. The application program on one site wants to send a keystroke to the application at the other site and receive an immediate response. Delayed transmission and delayed delivery of data may not be acceptable by the application program.

TCP can handle such a situation. The application program at the sending site can request a *push* operation. This means that the sending TCP must not wait for the window to be filled. It must create a segment and send it immediately. The sending TCP must also set the push bit (PSH) to let the receiving TCP know that the segment includes data that must be delivered to the receiving application program as soon as possible and not to wait for more data to come.

Although the push operation can be requested by the application program, most current implementations ignore such requests. TCP can choose whether or not to use this feature.

Urgent Data TCP is a stream-oriented protocol. This means that the data are presented from the application program to TCP as a stream of bytes. Each byte of data has a position in the stream. However, on occasion an application program needs to send *urgent* bytes. This means that the sending application program wants a piece of data to be read out of order by the receiving application program. As an example, suppose that the sending application program is sending data to be processed by the receiving application program. When the result of processing comes back, the sending application program finds that everything is wrong. It wants to abort the process, but it has already sent a huge amount of data. If it issues an abort command (control + C), these two characters will be stored at the end of the receiving TCP buffer. It will be delivered to the receiving application program after all the data have been processed.

The solution is to send a segment with the URG bit set. The sending application program tells the sending TCP that the piece of data is urgent. The sending TCP creates a segment and inserts the urgent data at the beginning of the segment. The rest of the segment can contain normal data from the buffer. The urgent pointer field in the header defines the end of the urgent data and the start of normal data.

When the receiving TCP receives a segment with the URG bit set, it extracts the urgent data from the segment, using the value of the urgent pointer, and delivers them, out of order, to the receiving application program.

Connection Termination

Any of the two parties involved in exchanging data (client or server) can close the connection, although it is usually initiated by the client. Most implementations today allow two options for connection termination: three-way handshaking and four-way handshaking with a half-close option.

Three-Way Handshaking Most implementations today allow *three-way handshaking* for connection termination as shown in Figure 23.20.

1. In a normal situation, the client TCP, after receiving a close command from the client process, sends the first segment, a FIN segment in which the FIN flag is set. Note that a FIN segment can include the last chunk of data sent by the client, or it

Figure 23.20 *Connection termination using three-way handshaking*

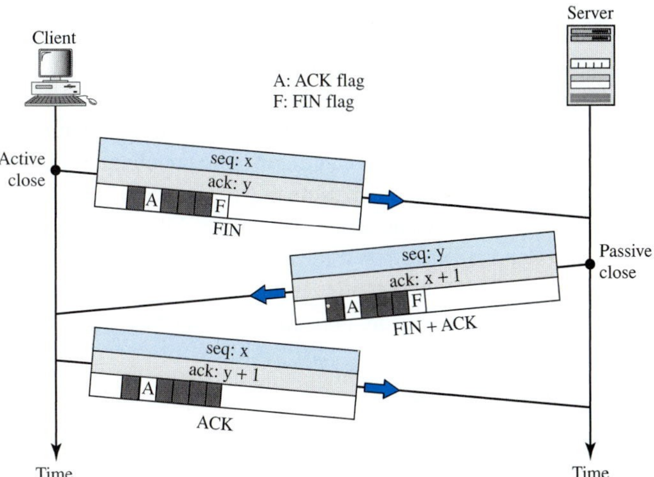

can be just a control segment as shown in Figure 23.20. If it is only a control segment, it consumes only one sequence number.

The FIN segment consumes one sequence number if it does not carry data.

2. The server TCP, after receiving the FIN segment, informs its process of the situation and sends the second segment, a FIN + ACK segment, to confirm the receipt of the FIN segment from the client and at the same time to announce the closing of the connection in the other direction. This segment can also contain the last chunk of data from the server. If it does not carry data, it consumes only one sequence number.

The FIN + ACK segment consumes one sequence number if it does not carry data.

3. The client TCP sends the last segment, an ACK segment, to confirm the receipt of the FIN segment from the TCP server. This segment contains the acknowledgment number, which is 1 plus the sequence number received in the FIN segment from the server. This segment cannot carry data and consumes no sequence numbers.

Half-Close In TCP, one end can stop sending data while still receiving data. This is called a **half-close.** Although either end can issue a half-close, it is normally initiated by the client. It can occur when the server needs all the data before processing can begin. A good example is sorting. When the client sends data to the server to be sorted, the server needs to receive all the data before sorting can start. This means the client, after sending all the data, can close the connection in the outbound direction. However, the inbound direction must remain open to receive the sorted data. The server, after receiving the data, still needs time for sorting; its outbound direction must remain open.

Figure 23.21 shows an example of a half-close. The client half-closes the connection by sending a FIN segment. The server accepts the half-close by sending the ACK segment. The data transfer from the client to the server stops. The server, however, can still send data. When the server has sent all the processed data, it sends a FIN segment, which is acknowledged by an ACK from the client.

After half-closing of the connection, data can travel from the server to the client and acknowledgments can travel from the client to the server. The client cannot send any more data to the server. Note the sequence numbers we have used. The second segment (ACK) consumes no sequence number. Although the client has received sequence number $y - 1$ and is expecting y, the server sequence number is still $y - 1$. When the connection finally closes, the sequence number of the last ACK segment is still x, because no sequence numbers are consumed during data transfer in that direction.

Flow Control

TCP uses a sliding window, as discussed in Chapter 11, to handle flow control. The sliding window protocol used by TCP, however, is something between the Go-Back-N and Selective Repeat sliding window. The sliding window protocol in TCP looks like

Figure 23.21 *Half-close*

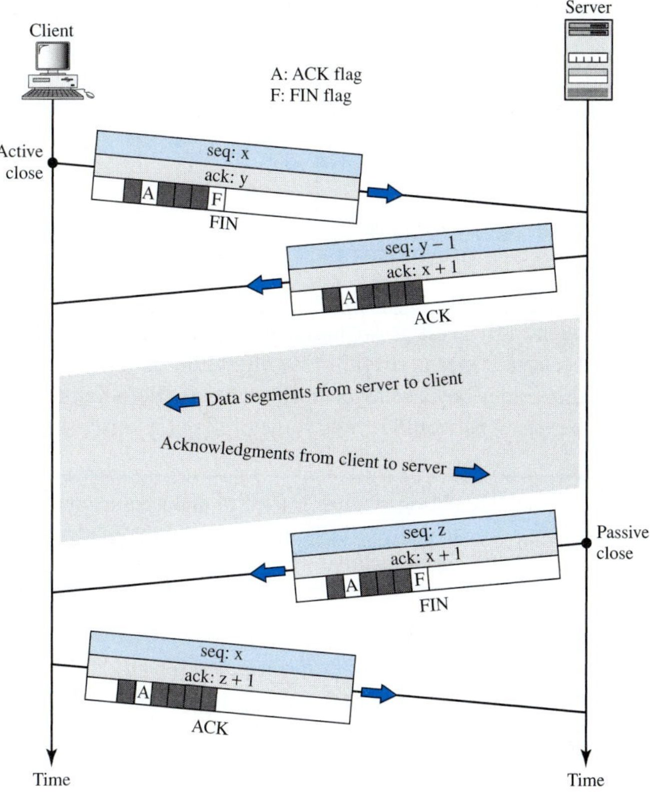

the Go-Back-*N* protocol because it does not use NAKs; it looks like Selective Repeat because the receiver holds the out-of-order segments until the missing ones arrive. There are two big differences between this sliding window and the one we used at the data link layer. First, the sliding window of TCP is byte-oriented; the one we discussed in the data link layer is frame-oriented. Second, the TCP's sliding window is of variable size; the one we discussed in the data link layer was of fixed size.

Figure 23.22 shows the sliding window in TCP. The window spans a portion of the buffer containing bytes received from the process. The bytes inside the window are the bytes that can be in transit; they can be sent without worrying about acknowledgment. The imaginary window has two walls: one left and one right.

The window is *opened, closed,* or *shrunk*. These three activities, as we will see, are in the control of the receiver (and depend on congestion in the network), not the sender. The sender must obey the commands of the receiver in this matter.

Opening a window means moving the right wall to the right. This allows more new bytes in the buffer that are eligible for sending. Closing the window means moving the left wall to the right. This means that some bytes have been acknowledged and the sender

Figure 23.22 *Sliding window*

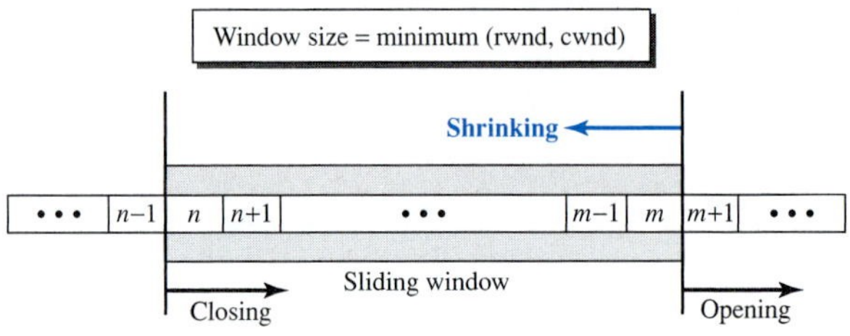

need not worry about them anymore. Shrinking the window means moving the right wall to the left. This is strongly discouraged and not allowed in some implementations because it means revoking the eligibility of some bytes for sending. This is a problem if the sender has already sent these bytes. Note that the left wall cannot move to the left because this would revoke some of the previously sent acknowledgments.

> **A sliding window is used to make transmission more efficient as well as to control the flow of data so that the destination does not become overwhelmed with data. TCP sliding windows are byte-oriented.**

The size of the window at one end is determined by the lesser of two values: *receiver window* (*rwnd*) or *congestion window* (*cwnd*). The *receiver window* is the value advertised by the opposite end in a segment containing acknowledgment. It is the number of bytes the other end can accept before its buffer overflows and data are discarded. The congestion window is a value determined by the network to avoid congestion. We will discuss congestion later in the chapter.

Example 23.4

What is the value of the receiver window (*rwnd*) for host A if the receiver, host B, has a buffer size of 5000 bytes and 1000 bytes of received and unprocessed data?

Solution

The value of *rwnd* = 5000 − 1000 = 4000. Host B can receive only 4000 bytes of data before overflowing its buffer. Host B advertises this value in its next segment to A.

Example 23.5

What is the size of the window for host A if the value of *rwnd* is 3000 bytes and the value of *cwnd* is 3500 bytes?

Solution

The size of the window is the smaller of *rwnd* and *cwnd,* which is 3000 bytes.

Example 23.6

Figure 23.23 shows an unrealistic example of a sliding window. The sender has sent bytes up to 202. We assume that cwnd is 20 (in reality this value is thousands of bytes). The receiver has sent

Figure 23.23 *Example 23.6*

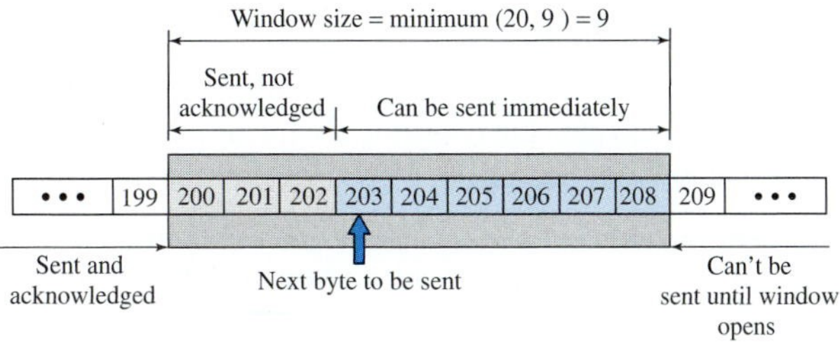

an acknowledgment number of 200 with an *rwnd* of 9 bytes (in reality this value is thousands of bytes). The size of the sender window is the minimum of *rwnd* and *cwnd*, or 9 bytes. Bytes 200 to 202 are sent, but not acknowledged. Bytes 203 to 208 can be sent without worrying about acknowledgment. Bytes 209 and above cannot be sent.

Some points about TCP sliding windows:

- The size of the window is the lesser of *rwnd* and *cwnd*.
- The source does not have to send a full window's worth of data.
- The window can be opened or closed by the receiver, but should not be shrunk.
- The destination can send an acknowledgment at any time as long as it does not result in a shrinking window.
- The receiver can temporarily shut down the window; the sender, however, can always send a segment of 1 byte after the window is shut down.

Error Control

TCP is a reliable transport layer protocol. This means that an application program that delivers a stream of data to TCP relies on TCP to deliver the entire stream to the application program on the other end in order, without error, and without any part lost or duplicated.

TCP provides reliability using error control. Error control includes mechanisms for detecting corrupted segments, lost segments, out-of-order segments, and duplicated segments. Error control also includes a mechanism for correcting errors after they are detected. Error detection and correction in TCP is achieved through the use of three simple tools: checksum, acknowledgment, and time-out.

Checksum

Each segment includes a checksum field which is used to check for a corrupted segment. If the segment is corrupted, it is discarded by the destination TCP and is considered as lost. TCP uses a 16-bit checksum that is mandatory in every segment. We will see, in Chapter 24, that the 16-bit checksum is considered inadequate for the new transport layer, SCTP. However, it cannot be changed for TCP because this would involve reconfiguration of the entire header format.

Acknowledgment

TCP uses acknowledgments to confirm the receipt of data segments. Control segments that carry no data but consume a sequence number are also acknowledged. ACK segments are never acknowledged.

ACK segments do not consume sequence numbers and are not acknowledged.

Retransmission

The heart of the error control mechanism is the retransmission of segments. When a segment is corrupted, lost, or delayed, it is retransmitted. In modern implementations, a segment is retransmitted on two occasions: when a **retransmission timer** expires or when the sender receives three duplicate ACKs.

In modern implementations, a retransmission occurs if the retransmission timer expires or three duplicate ACK segments have arrived.

Note that no retransmission occurs for segments that do not consume sequence numbers. In particular, there is no transmission for an ACK segment.

No retransmission timer is set for an ACK segment.

Retransmission After RTO A recent implementation of TCP maintains one **retransmission time-out (RTO)** timer for all outstanding (sent, but not acknowledged) segments. When the timer matures, the earliest outstanding segment is retransmitted even though lack of a received ACK can be due to a delayed segment, a delayed ACK, or a lost acknowledgment. Note that no time-out timer is set for a segment that carries only an acknowledgment, which means that no such segment is resent. The value of RTO is dynamic in TCP and is updated based on the **round-trip time (RTT)** of segments. An RTT is the time needed for a segment to reach a destination and for an acknowledgment to be received. It uses a back-off strategy similar to one discussed in Chapter 12.

Retransmission After Three Duplicate ACK Segments The previous rule about retransmission of a segment is sufficient if the value of RTO is not very large. Sometimes, however, one segment is lost and the receiver receives so many out-of-order segments that they cannot be saved (limited buffer size). To alleviate this situation, most implementations today follow the three-duplicate-ACKs rule and retransmit the missing segment immediately. This feature is referred to as **fast retransmission,** which we will see in an example shortly.

Out-of-Order Segments

When a segment is delayed, lost, or discarded, the segments following that segment arrive out of order. Originally, TCP was designed to discard all out-of-order segments, resulting in the retransmission of the missing segment and the following segments. Most implementations today do not discard the out-of-order segments. They store them temporarily and flag them as out-of-order segments until the missing segment arrives. Note, however, that the out-of-order segments are not delivered to the process. TCP guarantees that data are delivered to the process in order.

Data may arrive out of order and be temporarily stored by the receiving TCP, but TCP guarantees that no out-of-order segment is delivered to the process.

Some Scenarios

In this section we give some examples of scenarios that occur during the operation of TCP. In these scenarios, we show a segment by a rectangle. If the segment carries data, we show the range of byte numbers and the value of the acknowledgment field. If it carries only an acknowledgment, we show only the acknowledgment number in a smaller box.

Normal Operation The first scenario shows bidirectional data transfer between two systems, as in Figure 23.24. The client TCP sends one segment; the server TCP sends three. The figure shows which rule applies to each acknowledgment. There are data to be sent, so the segment displays the next byte expected. When the client receives the first segment from the server, it does not have any more data to send; it sends only an

Figure 23.24 *Normal operation*

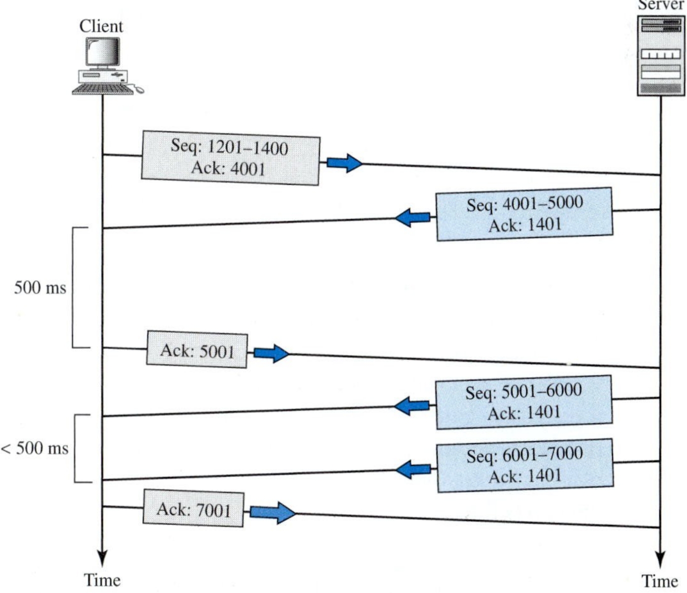

ACK segment. However, the acknowledgment needs to be delayed for 500 ms to see if any more segments arrive. When the timer matures, it triggers an acknowledgment. This is so because the client has no knowledge if other segments are coming; it cannot delay the acknowledgment forever. When the next segment arrives, another acknowledgment timer is set. However, before it matures, the third segment arrives. The arrival of the third segment triggers another acknowledgment.

Lost Segment In this scenario, we show what happens when a segment is lost or corrupted. A lost segment and a corrupted segment are treated the same way by the receiver. A lost segment is discarded somewhere in the network; a corrupted segment is discarded by the receiver itself. Both are considered lost. Figure 23.25 shows a situation in which a segment is lost and discarded by some router in the network, perhaps due to congestion.

Figure 23.25 *Lost segment*

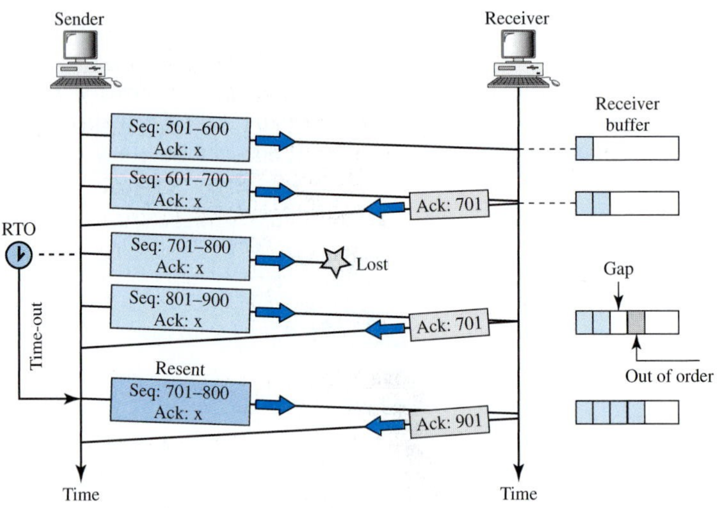

We are assuming that data transfer is unidirectional: one site is sending, the other is receiving. In our scenario, the sender sends segments 1 and 2, which are acknowledged immediately by an ACK. Segment 3, however, is lost. The receiver receives segment 4, which is out of order. The receiver stores the data in the segment in its buffer but leaves a gap to indicate that there is no continuity in the data. The receiver immediately sends an acknowledgment to the sender, displaying the next byte it expects. Note that the receiver stores bytes 801 to 900, but never delivers these bytes to the application until the gap is filled.

The receiver TCP delivers only ordered data to the process.

We have shown the timer for the earliest outstanding segment. The timer for this definitely runs out because the receiver never sends an acknowledgment for lost or out-of-order segments. When the timer matures, the sending TCP resends segment 3, which arrives this time and is acknowledged properly. Note that the value in the second and third acknowledgments differs according to the corresponding rule.

Fast Retransmission In this scenario, we want to show the idea of fast retransmission. Our scenario is the same as the second except that the RTO has a higher value (see Figure 23.26).

Figure 23.26 *Fast retransmission*

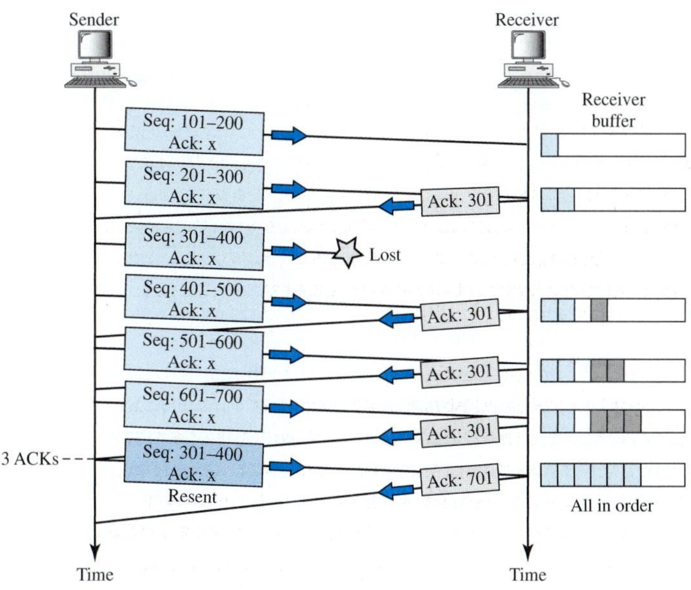

When the receiver receives the fourth, fifth, and sixth segments, it triggers an acknowledgment. The sender receives four acknowledgments with the same value (three duplicates). Although the timer for segment 3 has not matured yet, the fast transmission requires that segment 3, the segment that is expected by all these acknowledgments, be resent immediately.

Note that only one segment is retransmitted although four segments are not acknowledged. When the sender receives the retransmitted ACK, it knows that the four segments are safe and sound because acknowledgment is cumulative.

Congestion Control

We discuss congestion control of TCP in Chapter 24.

23.4 SCTP

Stream Control Transmission Protocol (SCTP) is a new reliable, message-oriented transport layer protocol. SCTP, however, is mostly designed for Internet applications that have recently been introduced. These new applications, such as IUA (ISDN over IP), M2UA and M3UA (telephony signaling), H.248 (media gateway control), H.323 (IP telephony), and SIP (IP telephony), need a more sophisticated service than TCP can provide. SCTP provides this enhanced performance and reliability. We briefly compare UDP, TCP, and SCTP:

- UDP is a **message-oriented** protocol. A process delivers a message to UDP, which is encapsulated in a user datagram and sent over the network. UDP *conserves the message boundaries;* each message is independent of any other message. This is a desirable feature when we are dealing with applications such as IP telephony and transmission of real-time data, as we will see later in the text. However, UDP is unreliable; the sender cannot know the destiny of messages sent. A message can be lost, duplicated, or received out of order. UDP also lacks some other features, such as congestion control and flow control, needed for a friendly transport layer protocol.
- TCP is a **byte-oriented** protocol. It receives a message or messages from a process, stores them as a stream of bytes, and sends them in segments. There is no preservation of the message boundaries. However, TCP is a reliable protocol. The duplicate segments are detected, the lost segments are resent, and the bytes are delivered to the end process in order. TCP also has congestion control and flow control mechanisms.
- SCTP combines the best features of UDP and TCP. SCTP is a reliable message-oriented protocol. It preserves the message boundaries and at the same time detects lost data, duplicate data, and out-of-order data. It also has congestion control and flow control mechanisms. Later we will see that SCTP has other innovative features unavailable in UDP and TCP.

SCTP is a *message-oriented, reliable* protocol that combines the best features of UDP and TCP.

SCTP Services

Before we discuss the operation of SCTP, let us explain the services offered by SCTP to the application layer processes.

Process-to-Process Communication

SCTP uses all well-known ports in the TCP space. Table 23.4 lists some extra port numbers used by SCTP.

Multiple Streams

We learned in the previous section that TCP is a stream-oriented protocol. Each connection between a TCP client and a TCP server involves one single stream. The problem

Table 23.4 *Some SCTP applications*

Protocol	*Port Number*	*Description*
IUA	9990	ISDN over IP
M2UA	2904	SS7 telephony signaling
M3UA	2905	SS7 telephony signaling
H.248	2945	Media gateway control
H.323	1718, 1719, 1720, 11720	IP telephony
SIP	5060	IP telephony

with this approach is that a loss at any point in the stream blocks the delivery of the rest of the data. This can be acceptable when we are transferring text; it is not when we are sending real-time data such as audio or video. SCTP allows **multistream service** in each connection, which is called **association** in SCTP terminology. If one of the streams is blocked, the other streams can still deliver their data.The idea is similar to multiple lanes on a highway. Each lane can be used for a different type of traffic. For example, one lane can be used for regular traffic, another for car pools. If the traffic is blocked for regular vehicles, car pool vehicles can still reach their destinations. Figure 23.27 shows the idea of multiple-stream delivery.

Figure 23.27 *Multiple-stream concept*

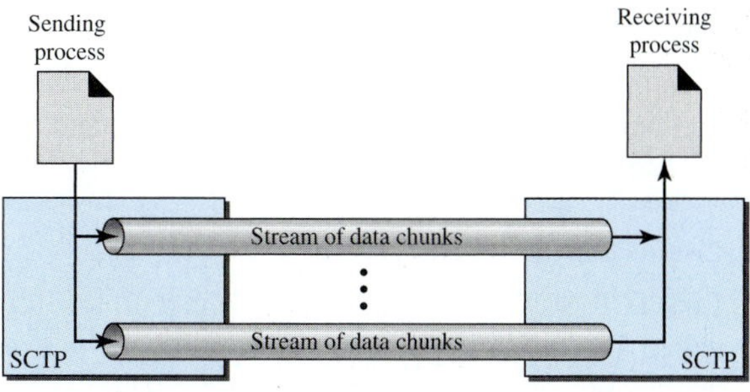

An association in SCTP can involve multiple streams.

Multihoming

A TCP connection involves one source and one destination IP address. This means that even if the sender or receiver is a multihomed host (connected to more than one physical address with multiple IP addresses), only one of these IP addresses per end can be utilized during the connection. An SCTP association, on the other hand, supports **multihoming service.** The sending and receiving host can define multiple IP addresses in each end for an association. In this fault-tolerant approach, when one path fails, another interface can be used for data delivery without interruption. This fault-tolerant feature is very helpful when we are sending and receiving a real-time payload such as Internet telephony. Figure 23.28 shows the idea of multihoming.

Figure 23.28 *Multihoming concept*

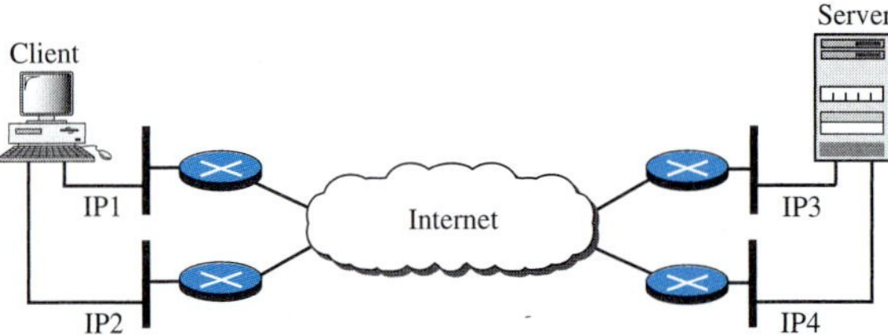

In Figure 23.28, the client is connected to two local networks with two IP addresses. The server is also connected to two networks with two IP addresses. The client and the server can make an association, using four different pairs of IP addresses. However, note that in the current implementations of SCTP, only one pair of IP addresses can be chosen for normal communication; the alternative is used if the main choice fails. In other words, at present, SCTP does not allow load sharing between different paths.

SCTP association allows multiple IP addresses for each end.

Full-Duplex Communication

Like TCP, SCTP offers full-duplex service, in which data can flow in both directions at the same time. Each SCTP then has a sending and receiving buffer, and packets are sent in both directions.

Connection-Oriented Service

Like TCP, SCTP is a connection-oriented protocol. However, in SCTP, a connection is called an association. When a process at site A wants to send and receive data from another process at site B, the following occurs:

1. The two SCTPs establish an association between each other.
2. Data are exchanged in both directions.
3. The association is terminated.

Reliable Service

SCTP, like TCP, is a reliable transport protocol. It uses an acknowledgment mechanism to check the safe and sound arrival of data. We will discuss this feature further in the section on error control.

SCTP Features

Let us first discuss the general features of SCTP and then compare them with those of TCP.

Transmission Sequence Number

The unit of data in TCP is a byte. Data transfer in TCP is controlled by numbering bytes by using a sequence number. On the other hand, the unit of data in SCTP is a **DATA chunk** which may or may not have a one-to-one relationship with the message coming from the process because of fragmentation (discussed later). Data transfer in SCTP is controlled by numbering the data chunks. SCTP uses a **transmission sequence number (TSN)** to number the data chunks. In other words, the TSN in SCTP plays the analogous role to the sequence number in TCP. TSNs are 32 bits long and randomly initialized between 0 and $2^{32} - 1$. Each data chunk must carry the corresponding TSN in its header.

In SCTP, a data chunk is numbered using a TSN.

Stream Identifier

In TCP, there is only one stream in each connection. In SCTP, there may be several streams in each association. Each stream in SCTP needs to be identified by using a **stream identifier (SI).** Each data chunk must carry the SI in its header so that when it arrives at the destination, it can be properly placed in its stream. The SI is a 16-bit number starting from 0.

To distinguish between different streams, SCTP uses an SI.

Stream Sequence Number

When a data chunk arrives at the destination SCTP, it is delivered to the appropriate stream and in the proper order. This means that, in addition to an SI, SCTP defines each data chunk in each stream with a **stream sequence number (SSN).**

To distinguish between different data chunks belonging to the same stream, SCTP uses SSNs.

Packets

In TCP, a segment carries data and control information. Data are carried as a collection of bytes; control information is defined by six control flags in the header. The design of SCTP is totally different: data are carried as data chunks, control information is carried as control chunks. Several control chunks and data chunks can be packed together in a packet. A packet in SCTP plays the same role as a segment in TCP. Figure 23.29 compares a segment in TCP and a packet in SCTP. Let us briefly list the differences between an SCTP packet and a TCP segment:

TCP has segments; SCTP has packets.

1. The control information in TCP is part of the header; the control information in SCTP is included in the control chunks. There are several types of control chunks; each is used for a different purpose.

Figure 23.29 *Comparison between a TCP segment and an SCTP packet*

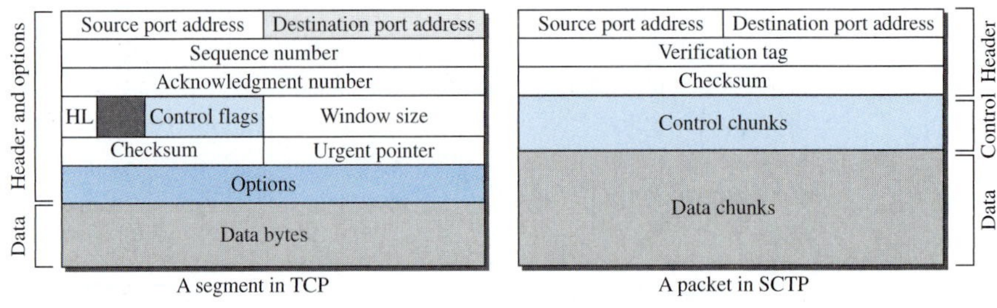

2. The data in a TCP segment treated as one entity; an SCTP packet can carry several data chunks; each can belong to a different stream.
3. The options section, which can be part of a TCP segment, does not exist in an SCTP packet. Options in SCTP are handled by defining new chunk types.
4. The mandatory part of the TCP header is 20 bytes, while the general header in SCTP is only 12 bytes. The SCTP header is shorter due to the following:
   a. An SCTP sequence number (TSN) belongs to each data chunk and hence is located in the chunk's header.
   b. The acknowledgment number and window size are part of each control chunk.
   c. There is no need for a header length field (shown as HL in the TCP segment) because there are no options to make the length of the header variable; the SCTP header length is fixed (12 bytes).
   d. There is no need for an urgent pointer in SCTP.
5. The checksum in TCP is 16 bits; in SCTP, it is 32 bits.
6. The **verification** tag in SCTP is an association identifier, which does not exist in TCP. In TCP, the combination of IP and port addresses defines a connection; in SCTP we may have multihoming using different IP addresses. A unique verification tag is needed to define each association.
7. TCP includes one sequence number in the header, which defines the number of the first byte in the data section. An SCTP packet can include several different data chunks. TSNs, SIs, and SSNs define each data chunk.
8. Some segments in TCP that carry control information (such as SYN and FIN) need to consume one sequence number; control chunks in SCTP never use a TSN, SI, or SSN. These three identifiers belong only to data chunks, not to the whole packet.

In SCTP, control information and data information are carried in separate chunks.

In SCTP, we have data chunks, streams, and packets. An association may send many packets, a packet may contain several chunks, and chunks may belong to different streams. To make the definitions of these terms clear, let us suppose that process A needs to send 11 messages to process B in three streams. The first four messages are in the first stream, the second three messages are in the second stream, and the last four messages are in the third stream.

Although a message, if long, can be carried by several data chunks, we assume that each message fits into one data chunk. Therefore, we have 11 data chunks in three streams.

The application process delivers 11 messages to SCTP, where each message is earmarked for the appropriate stream. Although the process could deliver one message from the first stream and then another from the second, we assume that it delivers all messages belonging to the first stream first, all messages belonging to the second stream next, and finally, all messages belonging to the last stream.

We also assume that the network allows only three data chunks per packet, which means that we need four packets as shown in Figure 23.30. Data chunks in stream 0 are carried in the first packet and part of the second packet; those in stream 1 are carried in the second and third packets; those in stream 2 are carried in the third and fourth packets.

Figure 23.30 *Packet, data chunks, and streams*

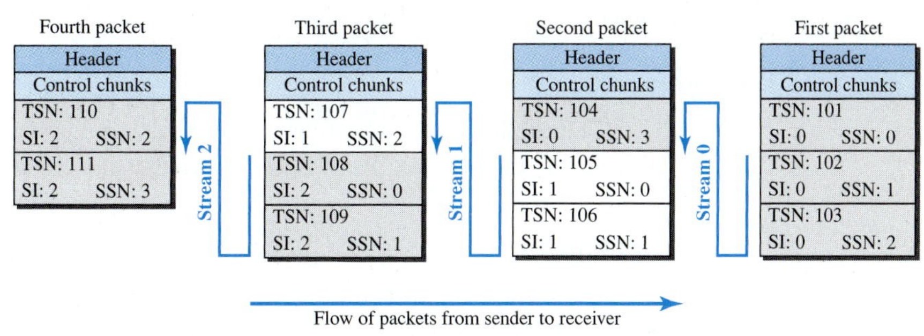

Note that each data chunk needs three identifiers: TSN, SI, and SSN. TSN is a cumulative number and is used, as we will see later, for flow control and error control. SI defines the stream to which the chunk belongs. SSN defines the chunk's order in a particular stream. In our example, SSN starts from 0 for each stream.

Data chunks are identified by three items: TSN, SI, and SSN.
TSN is a cumulative number identifying the association;
SI defines the stream; SSN defines the chunk in a stream.

Acknowledgment Number

TCP acknowledgment numbers are byte-oriented and refer to the sequence numbers. SCTP acknowledgment numbers are chunk-oriented. They refer to the TSN. A second difference between TCP and SCTP acknowledgments is the control information. Recall that this information is part of the segment header in TCP. To acknowledge segments that carry only control information, TCP uses a sequence number and acknowledgment number (for example, a SYN segment needs to be acknowledged by an ACK segment). In SCTP, however, the control information is carried by control chunks, which do not need a TSN. These control chunks are acknowledged by another control chunk of the appropriate type (some need no acknowledgment). For example, an INIT control chunk is acknowledged by an INIT ACK chunk. There is no need for a sequence number or an acknowledgment number.

In SCTP, acknowledgment numbers are used to acknowledge only data chunks; control chunks are acknowledged by other control chunks if necessary.

Flow Control

Like TCP, SCTP implements flow control to avoid overwhelming the receiver. We will discuss SCTP flow control later in the chapter.

Error Control

Like TCP, SCTP implements error control to provide reliability. TSN numbers and acknowledgment numbers are used for error control. We will discuss error control later in the chapter.

Congestion Control

Like TCP, SCTP implements congestion control to determine how many data chunks can be injected into the network. We will discuss congestion control in Chapter 24.

Packet Format

In this section, we show the format of a packet and different types of chunks. Most of the information presented in this section will become clear later; this section can be skipped in the first reading or used only as a reference. An SCTP packet has a mandatory general header and a set of blocks called chunks. There are two types of chunks: control chunks and data chunks. A control chunk controls and maintains the association; a data chunk carries user data. In a packet, the control chunks come before the data chunks. Figure 23.31 shows the general format of an SCTP packet.

Figure 23.31 *SCTP packet format*

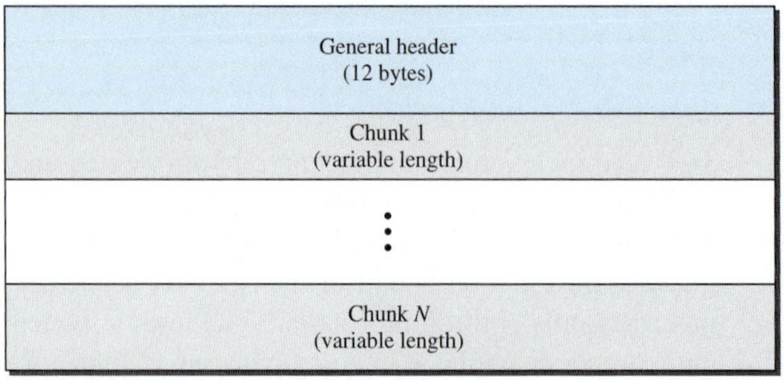

General header (12 bytes)
Chunk 1 (variable length)
⋮
Chunk N (variable length)

In an SCTP packet, control chunks come before data chunks.

General Header

The **general header** (packet header) defines the endpoints of each association to which the packet belongs, guarantees that the packet belongs to a particular association, and preserves the integrity of the contents of the packet including the header itself. The format of the general header is shown in Figure 23.32.

Figure 23.32 *General header*

Source port address 16 bits	Destination port address 16 bits
Verification tag 32 bits	
Checksum 32 bits	

There are four fields in the general header:

- **Source port address.** This is a 16-bit field that defines the port number of the process sending the packet.
- **Destination port address.** This is a 16-bit field that defines the port number of the process receiving the packet.
- **Verification tag.** This is a number that matches a packet to an association. This prevents a packet from a previous association from being mistaken as a packet in this association. It serves as an identifier for the association; it is repeated in every packet during the association. There is a separate verification used for each direction in the association.
- **Checksum.** This 32-bit field contains a CRC-32 checksum. Note that the size of the checksum is increased from 16 (in UDP, TCP, and IP) to 32 bits to allow the use of the CRC-32 checksum.

Chunks

Control information or user data are carried in chunks. The detailed format of each chunk is beyond the scope of this book. See [For06] for details. The first three fields are common to all chunks; the information field depends on the type of chunk. The important point to remember is that SCTP requires the information section to be a multiple of 4 bytes; if not, padding bytes (eight 0s) are added at the end of the section. See Table 23.5 for a list of chunks and their descriptions.

An SCTP Association

SCTP, like TCP, is a connection-oriented protocol. However, a connection in SCTP is called an *association* to emphasize multihoming.

Table 23.5 *Chunks*

Type	*Chunk*	*Description*
0	**DATA**	User data
1	**INIT**	Sets up an association
2	**INIT ACK**	Acknowledges INIT chunk
3	**SACK**	Selective acknowledgment
4	**HEARTBEAT**	Probes the peer for liveliness
5	**HEARTBEAT ACK**	Acknowledges HEARTBEAT chunk
6	**ABORT**	Aborts an association
7	**SHUTDOWN**	Terminates an association
8	**SHUTDOWN ACK**	Acknowledges SHUTDOWN chunk
9	**ERROR**	Reports errors without shutting down
10	**COOKIE ECHO**	Third packet in association establishment
11	**COOKIE ACK**	Acknowledges COOKIE ECHO chunk
14	**SHUTDOWN COMPLETE**	Third packet in association termination
192	**FORWARD TSN**	For adjusting cumulative TSN

A connection in SCTP is called an association.

Association Establishment

Association establishment in SCTP requires a **four-way handshake.** In this procedure, a process, normally a client, wants to establish an association with another process, normally a server, using SCTP as the transport layer protocol. Similar to TCP, the SCTP server needs to be prepared to receive any association (passive open). Association establishment, however, is initiated by the client (active open). SCTP association establishment is shown in Figure 23.33. The steps, in a normal situation, are as follows:

1. The client sends the first packet, which contains an **INIT chunk.**
2. The server sends the second packet, which contains an **INIT ACK chunk.**
3. The client sends the third packet, which includes a **COOKIE ECHO chunk.** This is a very simple chunk that echoes, without change, the cookie sent by the server. SCTP allows the inclusion of data chunks in this packet.
4. The server sends the fourth packet, which includes the **COOKIE ACK chunk** that acknowledges the receipt of the COOKIE ECHO chunk. SCTP allows the inclusion of data chunks with this packet.

No other chunk is allowed in a packet carrying an INIT or INIT ACK chunk.
A COOKIE ECHO or a COOKIE ACK chunk can carry data chunks.

Cookie We discussed a SYN flooding attack in the previous section. With TCP, a malicious attacker can flood a TCP server with a huge number of phony SYN segments using different forged IP addresses. Each time the server receives a SYN segment, it

Figure 23.33 *Four-way handshaking*

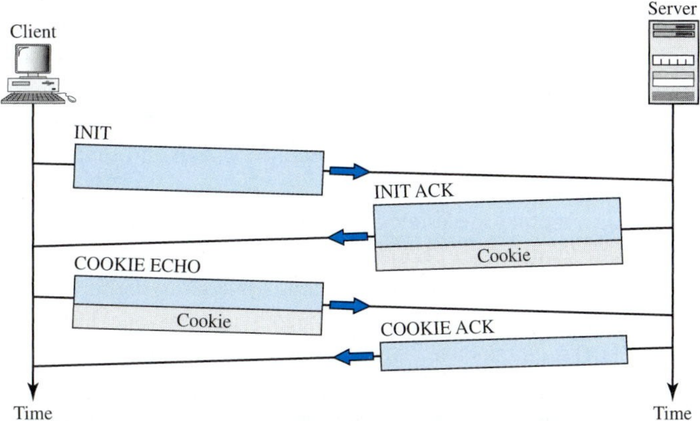

sets up a state table and allocates other resources while waiting for the next segment to arrive. After a while, however, the server may collapse due to the exhaustion of resources.

The designers of SCTP have a strategy to prevent this type of attack. The strategy is to postpone the allocation of resources until the reception of the third packet, when the IP address of the sender is verified. The information received in the first packet must somehow be saved until the third packet arrives. But if the server saved the information, that would require the allocation of resources (memory); this is the dilemma. The solution is to pack the information and send it back to the client. This is called generating a **cookie.** The cookie is sent with the second packet to the address received in the first packet. There are two potential situations.

1. If the sender of the first packet is an attacker, the server never receives the third packet; the cookie is lost and no resources are allocated. The only effort for the server is "baking" the cookie.
2. If the sender of the first packet is an honest client that needs to make a connection, it receives the second packet, with the cookie. It sends a packet (third in the series) with the cookie, with no changes. The server receives the third packet and knows that it has come from an honest client because the cookie that the sender has sent is there. The server can now allocate resources.

The above strategy works if no entity can "eat" a cookie "baked" by the server. To guarantee this, the server creates a digest (see Chapter 30) from the information, using its own secret key. The information and the digest together make the cookie, which is sent to the client in the second packet. When the cookie is returned in the third packet, the server calculates the digest from the information. If the digest matches the one that is sent, the cookie has not been changed by any other entity.

Data Transfer

The whole purpose of an association is to transfer data between two ends. After the association is established, bidirectional data transfer can take place. The client and the server can both send data. Like TCP, SCTP supports piggybacking.

There is a major difference, however, between data transfer in TCP and SCTP. TCP receives messages from a process as a stream of bytes without recognizing any boundary between them. The process may insert some boundaries for its peer use, but TCP treats that mark as part of the text. In other words, TCP takes each message and appends it to its buffer. A segment can carry parts of two different messages. The only ordering system imposed by TCP is the byte numbers.

SCTP, on the other hand, recognizes and maintains boundaries. Each message coming from the process is treated as one unit and inserted into a **DATA chunk** unless it is fragmented (discussed later). In this sense, SCTP is like UDP, with one big advantage: data chunks are related to each other.

A message received from a process becomes a DATA chunk, or chunks if fragmented, by adding a DATA chunk header to the message. Each DATA chunk formed by a message or a fragment of a message has one TSN. We need to remember that only DATA chunks use TSNs and only DATA chunks are acknowledged by SACK chunks.

In SCTP, only DATA chunks consume TSNs;
DATA chunks are the only chunks that are acknowledged.

Let us show a simple scenario in Figure 23.34. In this figure a client sends four DATA chunks and receives two DATA chunks from the server. Later, we will discuss

Figure 23.34 *Simple data transfer*

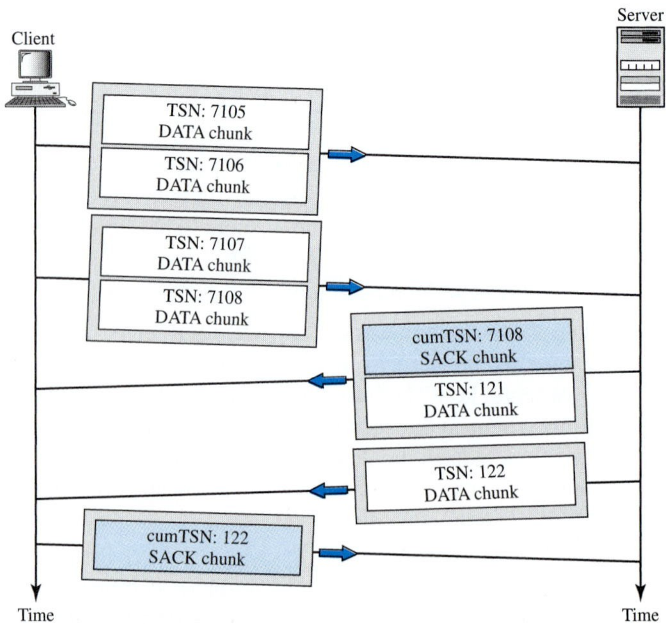

the use of flow and error control in SCTP. For the moment, we assume that everything goes well in this scenario.

1. The client sends the first packet carrying two DATA chunks with TSNs 7105 and 7106.
2. The client sends the second packet carrying two DATA chunks with TSNs 7107 and 7108.
3. The third packet is from the server. It contains the SACK chunk needed to acknowledge the receipt of DATA chunks from the client. Contrary to TCP, SCTP acknowledges the last in-order TSN received, not the next expected. The third packet also includes the first DATA chunk from the server with TSN 121.
4. After a while, the server sends another packet carrying the last DATA chunk with TSN 122, but it does not include a SACK chunk in the packet because the last DATA chunk received from the client was already acknowledged.
5. Finally, the client sends a packet that contains a SACK chunk acknowledging the receipt of the last two DATA chunks from the server.

The acknowledgment in SCTP defines the cumulative TSN, the TSN of the last data chunk received in order.

Multihoming Data Transfer We discussed the multihoming capability of SCTP, a feature that distinguishes SCTP from UDP and TCP. Multihoming allows both ends to define multiple IP addresses for communication. However, only one of these addresses can be defined as the **primary address;** the rest are alternative addresses. The primary address is defined during association establishment. The interesting point is that the primary address of an end is determined by the other end. In other words, a source defines the primary address for a destination.

Multistream Delivery One interesting feature of SCTP is the distinction between data transfer and data delivery. SCTP uses TSN numbers to handle data transfer, movement of data chunks between the source and destination. The delivery of the data chunks is controlled by SIs and SSNs. SCTP can support multiple streams, which means that the sender process can define different streams and a message can belong to one of these streams. Each stream is assigned a stream identifier (SI) which uniquely defines that stream.

Fragmentation Another issue in data transfer is **fragmentation.** Although SCTP shares this term with IP, fragmentation in IP and in SCTP belongs to different levels: the former at the network layer, the latter at the transport layer.

SCTP preserves the boundaries of the message from process to process when creating a DATA chunk from a message if the size of the message (when encapsulated in an IP datagram) does not exceed the MTU of the path. The size of an IP datagram carrying a message can be determined by adding the size of the message, in bytes, to the four overheads: data chunk header, necessary SACK chunks, SCTP general header, and IP header. If the total size exceeds the MTU, the message needs to be fragmented.

Association Termination

In SCTP, like TCP, either of the two parties involved in exchanging data (client or server) can close the connection. However, unlike TCP, SCTP does not allow a half-close situation. If one end closes the association, the other end must stop sending new data. If any data are left over in the queue of the recipient of the termination request, they are sent and the association is closed. **Association termination** uses three packets, as shown in Figure 23.35. Note that although the figure shows the case in which termination is initiated by the client, it can also be initiated by the server. Note that there can be several scenarios of association termination. We leave this discussion to the references mentioned at the end of the chapter.

Figure 23.35 *Association termination*

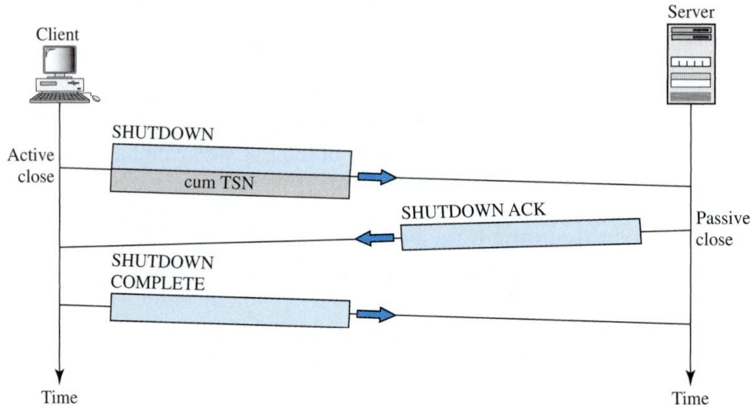

Flow Control

Flow control in SCTP is similar to that in TCP. In TCP, we need to deal with only one unit of data, the byte. In SCTP, we need to handle two units of data, the byte and the chunk. The values of *rwnd* and *cwnd* are expressed in bytes; the values of TSN and acknowledgments are expressed in chunks. To show the concept, we make some unrealistic assumptions. We assume that there is never congestion in the network and that the network is error-free. In other words, we assume that cwnd is infinite and no packet is lost or delayed or arrives out of order. We also assume that data transfer is unidirectional. We correct our unrealistic assumptions in later sections. Current SCTP implementations still use a byte-oriented window for flow control. We, however, show the buffer in terms of chunks to make the concept easier to understand.

Receiver Site

The receiver has one buffer (queue) and three variables. The queue holds the received data chunks that have not yet been read by the process. The first variable holds the last TSN received, *cumTSN*. The second variable holds the available buffer size, *winsize*.

The third variable holds the last accumulative acknowledgment, *lastACK*. Figure 23.36 shows the queue and variables at the receiver site.

Figure 23.36 *Flow control, receiver site*

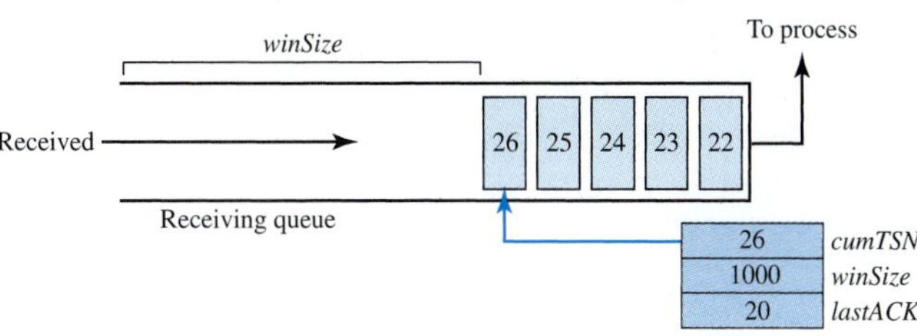

1. When the site receives a data chunk, it stores it at the end of the buffer (queue) and subtracts the size of the chunk from *winSize*. The TSN number of the chunk is stored in the *cumTSN* variable.
2. When the process reads a chunk, it removes it from the queue and adds the size of the removed chunk to *winSize* (recycling).
3. When the receiver decides to send a SACK, it checks the value of *lastAck;* if it is less than *cumTSN,* it sends a SACK with a cumulative TSN number equal to the *cumTSN.* It also includes the value of *winSize* as the advertised window size.

Sender Site

The sender has one buffer (queue) and three variables: *curTSN, rwnd,* and *inTransit,* as shown in Figure 23.37. We assume each chunk is 100 bytes long.

Figure 23.37 *Flow control, sender site*

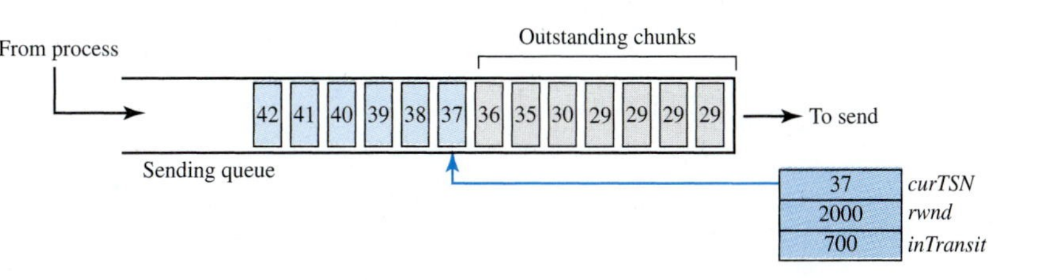

The buffer holds the chunks produced by the process that either have been sent or are ready to be sent. The first variable, *curTSN,* refers to the next chunk to be sent. All chunks in the queue with a TSN less than this value have been sent, but not acknowledged; they are outstanding. The second variable, *rwnd,* holds the last value advertised by the receiver (in bytes). The third variable, *inTransit,* holds the number of bytes in transit, bytes sent but not yet acknowledged. The following is the procedure used by the sender.

1. A chunk pointed to by *curTSN* can be sent if the size of the data is less than or equal to the quantity *rwnd* – *inTransit*. After sending the chunk, the value of *curTSN* is incremented by 1 and now points to the next chunk to be sent. The value of *inTransit* is incremented by the size of the data in the transmitted chunk.
2. When a SACK is received, the chunks with a TSN less than or equal to the cumulative TSN in the SACK are removed from the queue and discarded. The sender does not have to worry about them any more. The value of *inTransit* is reduced by the total size of the discarded chunks. The value of *rwnd* is updated with the value of the advertised window in the SACK.

A Scenario

Let us give a simple scenario as shown in Figure 23.38. At the start the value of *rwnd* at the sender site and the value of *winSize* at the receiver site are 2000 (advertised during association establishment). Originally, there are four messages in the sender queue. The sender sends one data chunk and adds the number of bytes (1000) to the *inTransit* variable. After awhile, the sender checks the difference between the *rwnd* and *inTransit,* which is 1000 bytes, so it can send another data chunk. Now the difference between the two variables is 0 and no more data chunks can be sent. After awhile, a SACK arrives that acknowledges data chunks 1 and 2. The two chunks are removed from the queue. The value of *inTransit* is now 0. The SACK, however, advertised a receiver window of value 0, which makes the sender update *rwnd* to 0. Now the sender is blocked; it cannot send any data chunks (with one exception explained later).

Figure 23.38 *Flow control scenario*

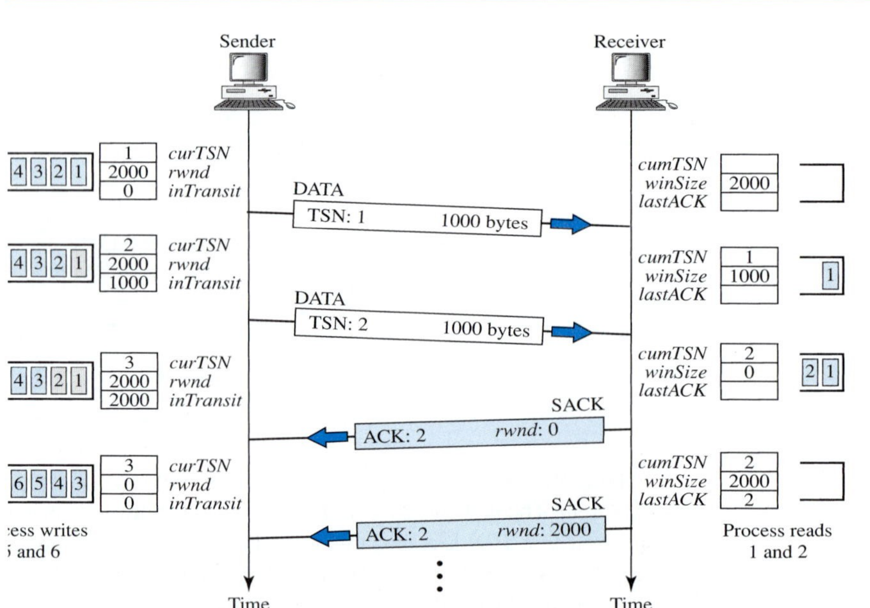

At the receiver site, the queue is empty at the beginning. After the first data chunk is received, there is one message in the queue and the value of *cumTSN* is 1. The value of *winSize* is reduced to 1000 because the first message occupies 1000 bytes. After the second data chunk is received, the value of window size is 0 and *cumTSN* is 2. Now, as we will see, the receiver is required to send a SACK with cumulative TSN of 2. After the first SACK was sent, the process reads the two messages, which means that there is now room in the queue; the receiver advertises the situation with a SACK to allow the sender to send more data chunks. The remaining events are not shown in the figure.

Error Control

SCTP, like TCP, is a reliable transport layer protocol. It uses a SACK chunk to report the state of the receiver buffer to the sender. Each implementation uses a different set of entities and timers for the receiver and sender sites. We use a very simple design to convey the concept to the reader.

Receiver Site

In our design, the receiver stores all chunks that have arrived in its queue including the out-of-order ones. However, it leaves spaces for any missing chunks. It discards duplicate messages, but keeps track of them for reports to the sender. Figure 23.39 shows a typical design for the receiver site and the state of the receiving queue at a particular point in time.

Figure 23.39 *Error control, receiver site*

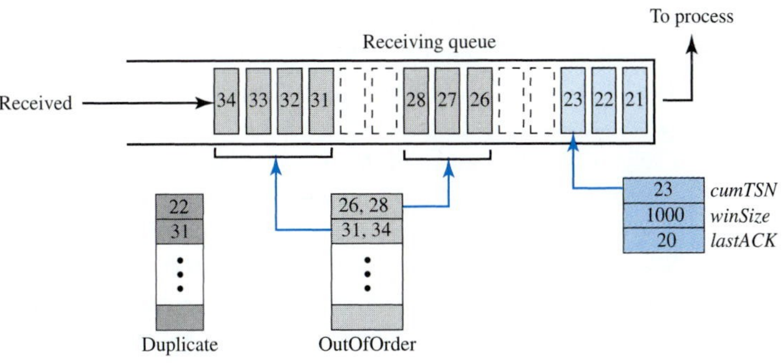

The last acknowledgment sent was for data chunk 20. The available window size is 1000 bytes. Chunks 21 to 23 have been received in order. The first out-of-order block contains chunks 26 to 28. The second out-of-order block contains chunks 31 to 34. A variable holds the value of *cumTSN*. An array of variables keeps track of the beginning and the end of each block that is out of order. An array of variables holds the duplicate chunks received. Note that there is no need for storing duplicate chunks in the queue; they will be discarded. The figure also shows the SACK chunk that will be sent to report the state of the receiver to the sender. The TSN numbers for out-of-order chunks are relative (offsets) to the cumulative TSN.

Sender Site

At the sender site, our design demands two buffers (queues): a sending queue and a retransmission queue. We also use the three variables *rwnd, inTransit,* and *curTSN* as described in the previous section. Figure 23.40 shows a typical design.

Figure 23.40 *Error control, sender site*

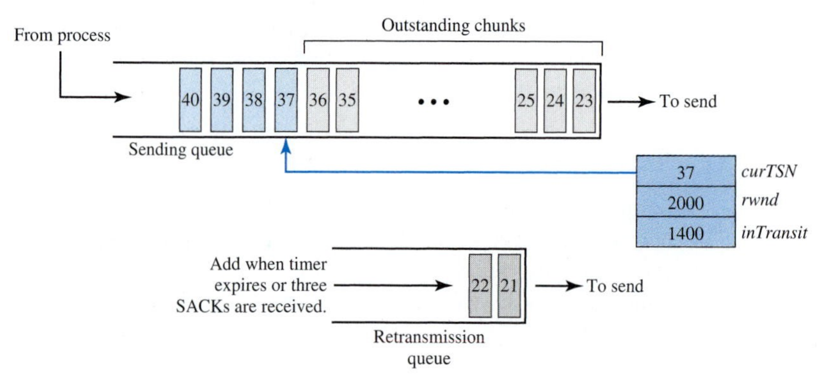

The sending queue holds chunks 23 to 40. The chunks 23 to 36 have already been sent, but not acknowledged; they are outstanding chunks. The *curTSN* points to the next chunk to be sent (37). We assume that each chunk is 100 bytes, which means that 1400 bytes of data (chunks 23 to 36) is in transit. The sender at this moment has a retransmission queue. When a packet is sent, a retransmission timer starts for that packet (all data chunks in that packet). Some implementations use one single timer for the entire association, but we continue with our tradition of one timer for each packet for simplification. When the retransmission timer for a packet expires, or four duplicate SACKs arrive that declare a packet as missing (fast retransmission was discussed in Chapter 12), the chunks in that packet are moved to the retransmission queue to be resent. These chunks are considered lost, rather than outstanding. The chunks in the retransmission queue have priority. In other words, the next time the sender sends a chunk, it would be chunk 21 from the retransmission queue.

Sending Data Chunks

An end can send a data packet whenever there are data chunks in the sending queue with a TSN greater than or equal to *curTSN* or if there are data chunks in the retransmission queue. The retransmission queue has priority. However, the total size of the data chunk or chunks included in the packet must not exceed *rwnd* – *inTransit,* and the total size of the frame must not exceed the MTU size as we discussed in previous sections.

Retransmission To control a lost or discarded chunk, SCTP, like TCP, employs two strategies: using retransmission timers and receiving four SACKs with the same missing chunks.

Generating SACK Chunks

Another issue in error control is the generation of SACK chunks. The rules for generating SCTP SACK chunks are similar to the rules used for acknowledgment with the TCP ACK flag.

Congestion Control

SCTP, like TCP, is a transport layer protocol with packets subject to congestion in the network. The SCTP designers have used the same strategies we will describe for congestion control in Chapter 24 for TCP. SCTP has slow start (exponential increase), congestion avoidance (additive increase), and congestion detection (multiplicative decrease) phases. Like TCP, SCTP also uses fast retransmission and fast recovery.

23.5 RECOMMENDED READING

For more details about subjects discussed in this chapter, we recommend the following books and sites. The items in brackets [. . .] refer to the reference list at the end of the text.

Books

UDP is discussed in Chapter 11 of [For06], Chapter 11 of [Ste94], and Chapter 12 of [Com00]. TCP is discussed in Chapter 12 of [For06], Chapters 17 to 24 of [Ste94], and Chapter 13 of [Com00]. SCTP is discussed in Chapter 13 of [For06] and [SX02]. Both UDP and TCP are discussed in Chapter 6 of [Tan03].

Sites

- www.ietf.org/rfc.html Information about RFCs

RFCs

A discussion of UDP can be found in RFC 768.
A discussion of TCP can be found in the following RFCs:

675, 700, 721, 761, 793, 879, 896, 1078, 1106, 1110, 1144, 1145, 1146, 1263, 1323, 1337, 1379, 1644, 1693, 1901, 1905, 2001, 2018, 2488, 2580

A discussion of SCTP can be found in the following RFCs:

2960, 3257, 3284, 3285, 3286, 3309, 3436, 3554, 3708, 3758

23.6 KEY TERMS

acknowledgment number
association
association establishment
association termination
byte-oriented
chunk
client
client/server paradigm
connection abortion
connection-oriented service
connectionless service
connectionless, unreliable transport protocol
cookie
COOKIE ACK chunk
COOKIE ECHO chunk
cumulative TSN
DATA chunk
data transfer
denial-of-service attack
dynamic port
ephemeral port number
error control
fast retransmission
flow control
four-way handshaking
fragmentation
full-duplex service
general header
half-close
inbound stream
INIT ACK chunk
INIT chunk
initial sequence number (ISN)
message-oriented
multihoming service
multistream service
port number
primary address
process-to-process delivery
pseudoheader
queue
registered port
retransmission time-out (RTO)
retransmission timer
round-trip time (RTT)
SACK chunk
segment
sequence number
server
simultaneous close
simultaneous open
socket address
stream identifier (SI)
stream sequence number (SSN)
SYN flooding attack
three-way handshaking
Transmission Control Protocol (TCP)
transmission sequence number (TSN)
transport layer
user datagram
User Datagram Protocol (UDP)
verification tag
well-known port number

23.7 SUMMARY

- In the client/server paradigm, an application program on the local host, called the client, needs services from an application program on the remote host, called a server.

- Each application program has a port number that distinguishes it from other programs running at the same time on the same machine.
- The client program is assigned a random port number called an ephemeral port number; the server program is assigned a universal port number called a well-known port number.
- The ICANN has specified ranges for the different types of port numbers.
- The combination of the IP address and the port number, called the socket address, defines a process and a host.
- UDP is a connectionless, unreliable transport layer protocol with no embedded flow or error control mechanism except the checksum for error detection.
- The UDP packet is called a user datagram. A user datagram is encapsulated in the data field of an IP datagram.
- Transmission Control Protocol (TCP) is one of the transport layer protocols in the TCP/IP protocol suite.
- TCP provides process-to-process, full-duplex, and connection-oriented service.
- The unit of data transfer between two devices using TCP software is called a segment; it has 20 to 60 bytes of header, followed by data from the application program.
- A TCP connection normally consists of three phases: connection establishment, data transfer, and connection termination.
- Connection establishment requires three-way handshaking; connection termination requires three- or four-way handshaking.
- TCP uses flow control, implemented as a sliding window mechanism, to avoid overwhelming a receiver with data.
- The TCP window size is determined by the receiver-advertised window size (*rwnd*) or the congestion window size (*cwnd*), whichever is smaller. The window can be opened or closed by the receiver, but should not be shrunk.
- The bytes of data being transferred in each connection are numbered by TCP. The numbering starts with a randomly generated number.
- TCP uses error control to provide a reliable service. Error control is handled by the checksum, acknowledgment, and time-out. Corrupted and lost segments are retransmitted, and duplicate segments are discarded. Data may arrive out of order and are temporarily stored by the receiving TCP, but TCP guarantees that no out-of-order segment is delivered to the process.
- In modern implementations, a retransmission occurs if the retransmission timer expires or three duplicate ACK segments have arrived.
- SCTP is a message-oriented, reliable protocol that combines the good features of UDP and TCP.
- SCTP provides additional services not provided by UDP or TCP, such as multiple-stream and multihoming services.
- SCTP is a connection-oriented protocol. An SCTP connection is called an association.
- SCTP uses the term *packet* to define a transportation unit.
- In SCTP, control information and data information are carried in separate chunks.

- An SCTP packet can contain control chunks and data chunks with control chunks coming before data chunks.
- In SCTP, each data chunk is numbered using a transmission sequence number (TSN).
- To distinguish between different streams, SCTP uses the sequence identifier (SI).
- To distinguish between different data chunks belonging to the same stream, SCTP uses the stream sequence number (SSN).
- Data chunks are identified by three identifiers: TSN, SI, and SSN. TSN is a cumulative number recognized by the whole association; SSN starts from 0 in each stream.
- SCTP acknowledgment numbers are used only to acknowledge data chunks; control chunks are acknowledged, if needed, by another control chunk.
- An SCTP association is normally established using four packets (four-way handshaking). An association is normally terminated using three packets (three-way handshaking).
- An SCTP association uses a cookie to prevent blind flooding attacks and a verification tag to avoid insertion attacks.
- SCTP provides flow control, error control, and congestion control.
- The SCTP acknowledgment SACK reports the cumulative TSN, the TSN of the last data chunk received in order, and selective TSNs that have been received.

23.8 PRACTICE SET

Review Questions

1. In cases where reliability is not of primary importance, UDP would make a good transport protocol. Give examples of specific cases.
2. Are both UDP and IP unreliable to the same degree? Why or why not?
3. Do port addresses need to be unique? Why or why not? Why are port addresses shorter than IP addresses?
4. What is the dictionary definition of the word *ephemeral?* How does it apply to the concept of the ephemeral port number?
5. What is the minimum size of a UDP datagram?
6. What is the maximum size of a UDP datagram?
7. What is the minimum size of the process data that can be encapsulated in a UDP datagram?
8. What is the maximum size of the process data that can be encapsulated in a UDP datagram?
9. Compare the TCP header and the UDP header. List the fields in the TCP header that are missing from UDP header. Give the reason for their absence.
10. UDP is a message-oriented protocol. TCP is a byte-oriented protocol. If an application needs to protect the boundaries of its message, which protocol should be used, UDP or TCP?

11. What can you say about the TCP segment in which the value of the control field is one of the following?
    a. 000000
    b. 000001
    c. 010001
12. What is the maximum size of the TCP header? What is the minimum size of the TCP header?

Exercises

13. Show the entries for the header of a UDP user datagram that carries a message from a TFTP client to a TFTP server. Fill the checksum field with 0s. Choose an appropriate ephemeral port number and the correct well-known port number. The length of data is 40 bytes. Show the UDP packet, using the format in Figure 23.9.
14. An SNMP client residing on a host with IP address 122.45.12.7 sends a message to an SNMP server residing on a host with IP address 200.112.45.90. What is the pair of sockets used in this communication?
15. A TFTP server residing on a host with IP address 130.45.12.7 sends a message to a TFTP client residing on a host with IP address 14.90.90.33. What is the pair of sockets used in this communication?
16. A client has a packet of 68,000 bytes. Show how this packet can be transferred by using only one UDP user datagram.
17. A client uses UDP to send data to a server. The data are 16 bytes. Calculate the efficiency of this transmission at the UDP level (ratio of useful bytes to total bytes).
18. Redo Exercise 17, calculating the efficiency of transmission at the IP level. Assume no options for the IP header.
19. Redo Exercise 18, calculating the efficiency of transmission at the data link layer. Assume no options for the IP header and use Ethernet at the data link layer.
20. The following is a dump of a UDP header in hexadecimal format.

```
06 32 00 0D 00 1C E2 17
```

    a. What is the source port number?
    b. What is the destination port number?
    c. What is the total length of the user datagram?
    d. What is the length of the data?
    e. Is the packet directed from a client to a server or vice versa?
    f. What is the client process?
21. An IP datagram is carrying a TCP segment destined for address 130.14.16.17/16. The destination port address is corrupted, and it arrives at destination 130.14.16.19/16. How does the receiving TCP react to this error?
22. In TCP, if the value of HLEN is 0111, how many bytes of option are included in the segment?

23. Show the entries for the header of a TCP segment that carries a message from an FTP client to an FTP server. Fill the checksum field with 0s. Choose an appropriate ephemeral port number and the correct well-known port number. The length of the data is 40 bytes.
24. The following is a dump of a TCP header in hexadecimal format.

```
05320017 00000001 00000000 500207FF 00000000
```

    a. What is the source port number?
    b. What is the destination port number?
    c. What the sequence number?
    d. What is the acknowledgment number?
    e. What is the length of the header?
    f. What is the type of the segment?
    g. What is the window size?
25. To make the initial sequence number a random number, most systems start the counter at 1 during bootstrap and increment the counter by 64,000 every 0.5 s. How long does it take for the counter to wrap around?
26. In a connection, the value of *cwnd* is 3000 and the value of *rwnd* is 5000. The host has sent 2000 bytes which has not been acknowledged. How many more bytes can be sent?
27. TCP opens a connection using an initial sequence number (ISN) of 14,534. The other party opens the connection with an ISN of 21,732. Show the three TCP segments during the connection establishment.
28. A client uses TCP to send data to a server. The data are 16 bytes. Calculate the efficiency of this transmission at the TCP level (ratio of useful bytes to total bytes). Calculate the efficiency of transmission at the IP level. Assume no options for the IP header. Calculate the efficiency of transmission at the data link layer. Assume no options for the IP header and use Ethernet at the data link layer.
29. TCP is sending data at 1 Mbyte/s. If the sequence number starts with 7000, how long does it take before the sequence number goes back to zero?
30. A TCP connection is using a window size of 10,000 bytes, and the previous acknowledgment number was 22,001. It receives a segment with acknowledgment number 24,001 and window size advertisement of 12,000. Draw a diagram to show the situation of the window before and after.
31. A window holds bytes 2001 to 5000. The next byte to be sent is 3001. Draw a figure to show the situation of the window after the following two events.
    a. An ACK segment with the acknowledgment number 2500 and window size advertisement 4000 is received.
    b. A segment carrying 1000 bytes is sent.
32. In SCTP, the value of the cumulative TSN in a SACK is 23. The value of the previous cumulative TSN in the SACK was 29. What is the problem?
33. In SCTP, the state of a receiver is as follows:
    a. The receiving queue has chunks 1 to 8, 11 to 14, and 16 to 20.
    b. There are 1800 bytes of space in the queue.

c. The value of *lastAck* is 4.
d. No duplicate chunk has been received.
e. The value of *cumTSN* is 5.

Show the contents of the receiving queue and the variables.

34. In SCTP, the state of a sender is as follows:
    a. The sending queue has chunks 18 to 23.
    b. The value of *cumTSN* is 20.
    c. The value of the window size is 2000 bytes.
    d. The value of *inTransit* is 200.

    If each data chunk contains 100 bytes of data, how many DATA chunks can be sent now? What is the next DATA chunk to be sent?

Research Activities

35. Find more information about ICANN. What was it called before its name was changed?
36. TCP uses a transition state diagram to handle sending and receiving segments. Find out about this diagram and how it handles flow and control.
37. SCTP uses a transition state diagram to handle sending and receiving segments. Find out about this diagram and how it handles flow and control.
38. What is the half-open case in TCP?
39. What is the half-duplex close case in TCP?
40. The *tcpdump* command in UNIX or LINUX can be used to print the headers of packets of a network interface. Use *tcpdump* to see the segments sent and received.
41. In SCTP, find out what happens if a SACK chunk is delayed or lost.
42. Find the name and functions of timers used in TCP.
43. Find the name and functions of timers used in SCTP.
44. Find out more about ECN in SCTP. Find the format of these two chunks.
45. Some application programs, such as FTP, need more than one connection when using TCP. Find how the multistream service of SCTP can help these applications establish only one association with several streams.

CHAPTER 24

Congestion Control and Quality of Service

Congestion control and quality of service are two issues so closely bound together that improving one means improving the other and ignoring one usually means ignoring the other. Most techniques to prevent or eliminate congestion also improve the quality of service in a network.

We have postponed the discussion of these issues until now because these are issues related not to one layer, but to three: the data link layer, the network layer, and the transport layer. We waited until now so that we can discuss these issues once instead of repeating the subject three times. Throughout the chapter, we give examples of congestion control and quality of service at different layers.

24.1 DATA TRAFFIC

The main focus of congestion control and quality of service is data traffic. In congestion control we try to avoid traffic congestion. In quality of service, we try to create an appropriate environment for the traffic. So, before talking about congestion control and quality of service, we discuss the data traffic itself.

Traffic Descriptor

Traffic descriptors are qualitative values that represent a data flow. Figure 24.1 shows a traffic flow with some of these values.

Figure 24.1 *Traffic descriptors*

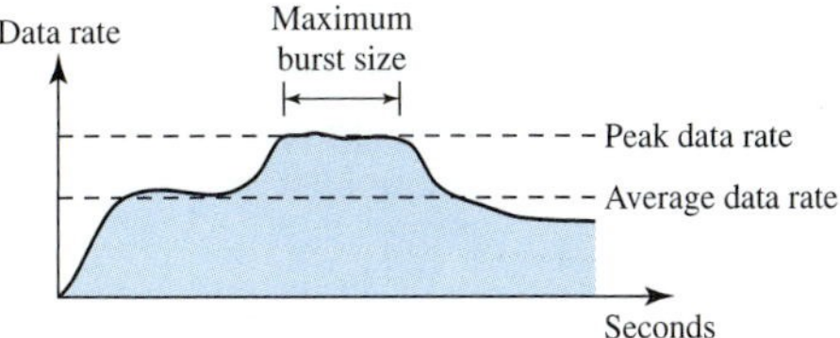

Average Data Rate

The **average data rate** is the number of bits sent during a period of time, divided by the number of seconds in that period. We use the following equation:

$$\text{Average data rate} = \frac{\text{amount of data}}{\text{time}}$$

The average data rate is a very useful characteristic of traffic because it indicates the average bandwidth needed by the traffic.

Peak Data Rate

The **peak data rate** defines the maximum data rate of the traffic. In Figure 24.1 it is the maximum y axis value. The peak data rate is a very important measurement because it indicates the peak bandwidth that the network needs for traffic to pass through without changing its data flow.

Maximum Burst Size

Although the peak data rate is a critical value for the network, it can usually be ignored if the duration of the peak value is very short. For example, if data are flowing steadily at the rate of 1 Mbps with a sudden peak data rate of 2 Mbps for just 1 ms, the network probably can handle the situation. However, if the peak data rate lasts 60 ms, there may be a problem for the network. The **maximum burst size** normally refers to the maximum length of time the traffic is generated at the peak rate.

Effective Bandwidth

The **effective bandwidth** is the bandwidth that the network needs to allocate for the flow of traffic. The effective bandwidth is a function of three values: average data rate, peak data rate, and maximum burst size. The calculation of this value is very complex.

Traffic Profiles

For our purposes, a data flow can have one of the following traffic profiles: constant bit rate, variable bit rate, or bursty as shown in Figure 24.2.

Constant Bit Rate

A **constant-bit-rate (CBR),** or a fixed-rate, traffic model has a data rate that does not change. In this type of flow, the average data rate and the peak data rate are the same. The maximum burst size is not applicable. This type of traffic is very easy for a network to handle since it is predictable. The network knows in advance how much bandwidth to allocate for this type of flow.

Variable Bit Rate

In the **variable-bit-rate (VBR)** category, the rate of the data flow changes in time, with the changes smooth instead of sudden and sharp. In this type of flow, the average data

Figure 24.2 *Three traffic profiles*

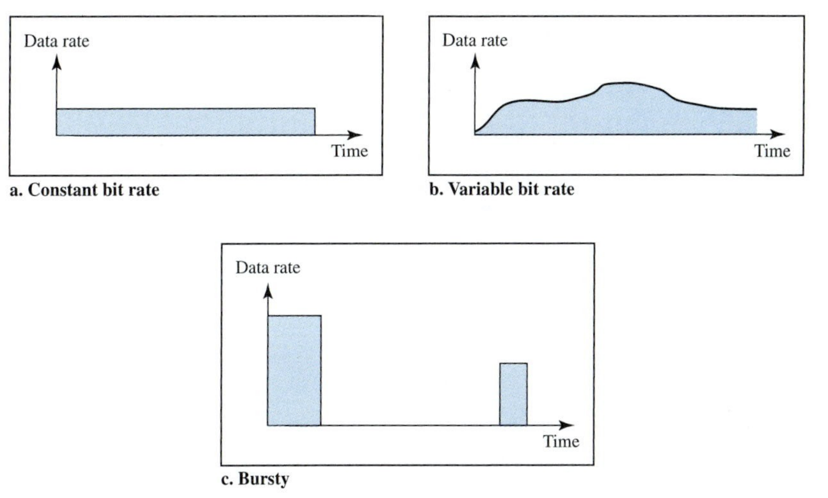

rate and the peak data rate are different. The maximum burst size is usually a small value. This type of traffic is more difficult to handle than constant-bit-rate traffic, but it normally does not need to be reshaped, as we will see later.

Bursty

In the **bursty data** category, the data rate changes suddenly in a very short time. It may jump from zero, for example, to 1 Mbps in a few microseconds and vice versa. It may also remain at this value for a while. The average bit rate and the peak bit rate are very different values in this type of flow. The maximum burst size is significant. This is the most difficult type of traffic for a network to handle because the profile is very unpredictable. To handle this type of traffic, the network normally needs to reshape it, using reshaping techniques, as we will see shortly. Bursty traffic is one of the main causes of congestion in a network.

24.2 CONGESTION

An important issue in a packet-switched network is **congestion.** Congestion in a network may occur if the **load** on the network—the number of packets sent to the network—is greater than the *capacity* of the network—the number of packets a network can handle. **Congestion control** refers to the mechanisms and techniques to control the congestion and keep the load below the capacity.

We may ask why there is congestion on a network. Congestion happens in any system that involves waiting. For example, congestion happens on a freeway because any abnormality in the flow, such as an accident during rush hour, creates blockage.

Congestion in a network or internetwork occurs because routers and switches have queues—buffers that hold the packets before and after processing. A router, for example, has an input queue and an output queue for each interface. When a packet arrives at the incoming interface, it undergoes three steps before departing, as shown in Figure 24.3.

Figure 24.3 *Queues in a router*

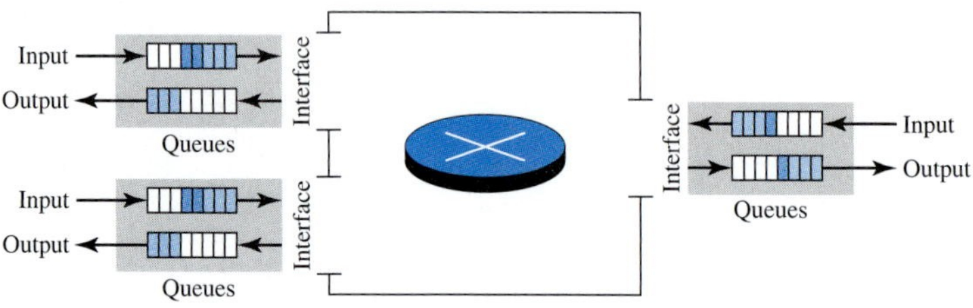

1. The packet is put at the end of the input queue while waiting to be checked.
2. The processing module of the router removes the packet from the input queue once it reaches the front of the queue and uses its routing table and the destination address to find the route.
3. The packet is put in the appropriate output queue and waits its turn to be sent.

We need to be aware of two issues. First, if the rate of packet arrival is higher than the packet processing rate, the input queues become longer and longer. Second, if the packet departure rate is less than the packet processing rate, the output queues become longer and longer.

Network Performance

Congestion control involves two factors that measure the performance of a network: *delay* and *throughput*. Figure 24.4 shows these two performance measures as function of load.

Figure 24.4 *Packet delay and throughput as functions of load*

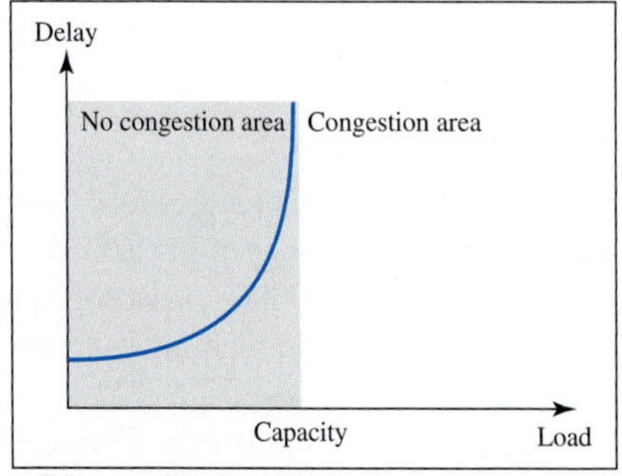

a. Delay as a function of load

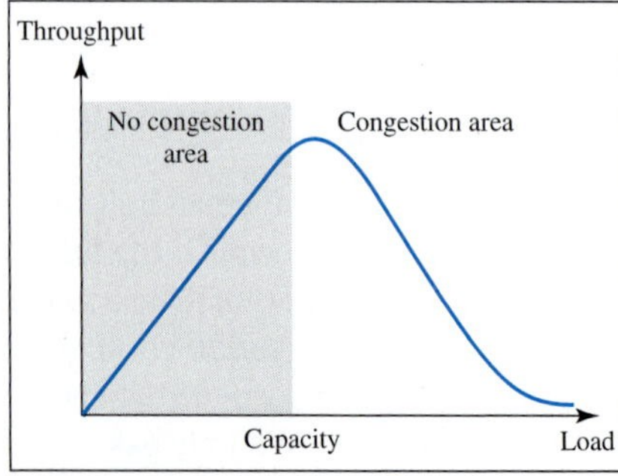

b. Throughput as a function of load

Delay Versus Load

Note that when the load is much less than the capacity of the network, the **delay** is at a minimum. This minimum delay is composed of propagation delay and processing delay, both of which are negligible. However, when the load reaches the network capacity, the delay increases sharply because we now need to add the waiting time in the queues (for all routers in the path) to the total delay. Note that the delay becomes infinite when the load is greater than the capacity. If this is not obvious, consider the size of the queues when almost no packet reaches the destination, or reaches the destination with infinite delay; the queues become longer and longer. Delay has a negative effect on the load and consequently the congestion. When a packet is delayed, the source, not receiving the acknowledgment, retransmits the packet, which makes the delay, and the congestion, worse.

Throughput Versus Load

We defined throughput in Chapter 3 as the number of bits passing through a point in a second. We can extend that definition from bits to packets and from a point to a network. We can define **throughput** in a network as the number of packets passing through the network in a unit of time. Notice that when the load is below the capacity of the network, the throughput increases proportionally with the *load*. We expect the throughput to remain constant after the load reaches the capacity, but instead the throughput declines sharply. The reason is the discarding of packets by the routers. When the load exceeds the capacity, the queues become full and the routers have to discard some packets. Discarding packets does not reduce the number of packets in the network because the sources retransmit the packets, using time-out mechanisms, when the packets do not reach the destinations.

24.3 CONGESTION CONTROL

Congestion control refers to techniques and mechanisms that can either prevent congestion, before it happens, or remove congestion, after it has happened. In general, we can divide congestion control mechanisms into two broad categories: open-loop congestion control (prevention) and closed-loop congestion control (removal) as shown in Figure 24.5.

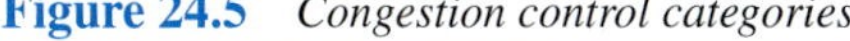

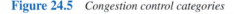

Figure 24.5 *Congestion control categories*

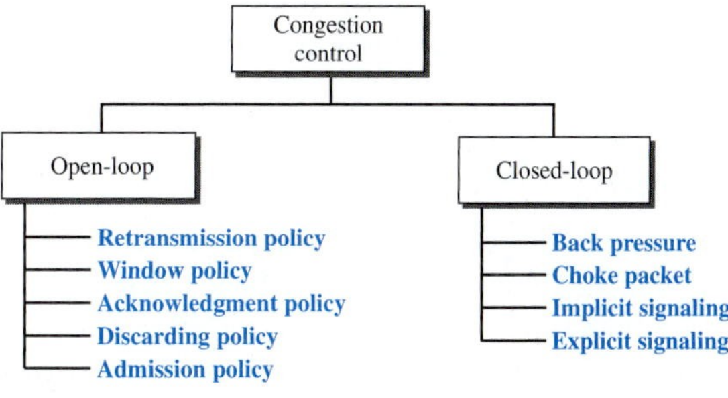

Open-Loop Congestion Control

In **open-loop congestion control,** policies are applied to prevent congestion before it happens. In these mechanisms, congestion control is handled by either the source or the destination. We give a brief list of policies that can prevent congestion.

Retransmission Policy

Retransmission is sometimes unavoidable. If the sender feels that a sent packet is lost or corrupted, the packet needs to be retransmitted. Retransmission in general may increase congestion in the network. However, a good retransmission policy can prevent congestion. The retransmission policy and the retransmission timers must be designed to optimize efficiency and at the same time prevent congestion. For example, the retransmission policy used by TCP (explained later) is designed to prevent or alleviate congestion.

Window Policy

The type of window at the sender may also affect congestion. The Selective Repeat window is better than the Go-Back-*N* window for congestion control. In the Go-Back-*N* window, when the timer for a packet times out, several packets may be resent, although some may have arrived safe and sound at the receiver. This duplication may make the congestion worse. The Selective Repeat window, on the other hand, tries to send the specific packets that have been lost or corrupted.

Acknowledgment Policy

The acknowledgment policy imposed by the receiver may also affect congestion. If the receiver does not acknowledge every packet it receives, it may slow down the sender and help prevent congestion. Several approaches are used in this case. A receiver may send an acknowledgment only if it has a packet to be sent or a special timer expires. A receiver may decide to acknowledge only *N* packets at a time. We need to know that the acknowledgments are also part of the load in a network. Sending fewer acknowledgments means imposing less load on the network.

Discarding Policy

A good discarding policy by the routers may prevent congestion and at the same time may not harm the integrity of the transmission. For example, in audio transmission, if the policy is to discard less sensitive packets when congestion is likely to happen, the quality of sound is still preserved and congestion is prevented or alleviated.

Admission Policy

An admission policy, which is a quality-of-service mechanism, can also prevent congestion in virtual-circuit networks. Switches in a flow first check the resource requirement of a flow before admitting it to the network. A router can deny establishing a virtual-circuit connection if there is congestion in the network or if there is a possibility of future congestion.

Closed-Loop Congestion Control

Closed-loop congestion control mechanisms try to alleviate congestion after it happens. Several mechanisms have been used by different protocols. We describe a few of them here.

Backpressure

The technique of *backpressure* refers to a congestion control mechanism in which a congested node stops receiving data from the immediate upstream node or nodes. This may cause the upstream node or nodes to become congested, and they, in turn, reject data from their upstream nodes or nodes. And so on. Backpressure is a node-to-node congestion control that starts with a node and propagates, in the opposite direction of data flow, to the source. The backpressure technique can be applied only to virtual circuit networks, in which each node knows the upstream node from which a flow of data is coming. Figure 24.6 shows the idea of backpressure.

Figure 24.6 *Backpressure method for alleviating congestion*

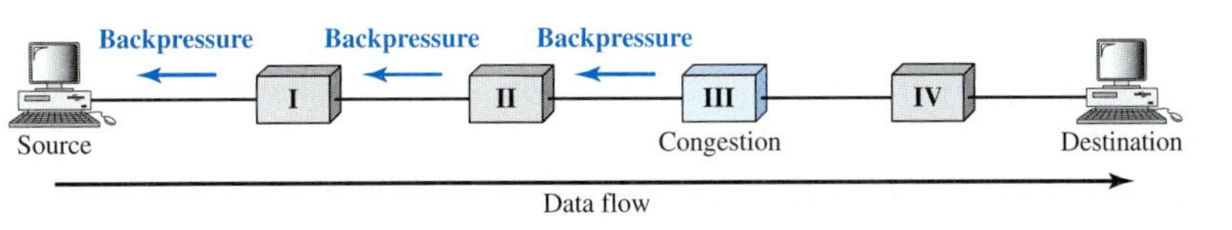

Node III in the figure has more input data than it can handle. It drops some packets in its input buffer and informs node II to slow down. Node II, in turn, may be congested because it is slowing down the output flow of data. If node II is congested, it informs node I to slow down, which in turn may create congestion. If so, node I informs the source of data to slow down. This, in time, alleviates the congestion. Note that the *pressure* on node III is moved backward to the source to remove the congestion.

None of the virtual-circuit networks we studied in this book use backpressure. It was, however, implemented in the first virtual-circuit network, X.25. The technique cannot be implemented in a datagram network because in this type of network, a node (router) does not have the slightest knowledge of the upstream router.

Choke Packet

A **choke packet** is a packet sent by a node to the source to inform it of congestion. Note the difference between the backpressure and choke packet methods. In backpressure, the warning is from one node to its upstream node, although the warning may eventually reach the source station. In the choke packet method, the warning is from the router, which has encountered congestion, to the source station directly. The intermediate nodes through which the packet has traveled are not warned. We have seen an example of this type of control in ICMP. When a router in the Internet is overwhelmed with IP datagrams, it may discard some of them; but it informs the source host, using a source quench ICMP message. The warning message goes directly to the source station; the intermediate routers, and does not take any action. Figure 24.7 shows the idea of a choke packet.

Figure 24.7 *Choke packet*

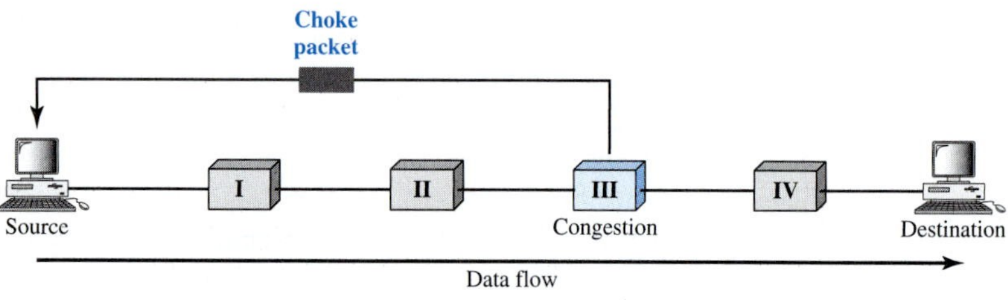

Implicit Signaling

In implicit signaling, there is no communication between the congested node or nodes and the source. The source guesses that there is a congestion somewhere in the network from other symptoms. For example, when a source sends several packets and there is no acknowledgment for a while, one assumption is that the network is congested. The delay in receiving an acknowledgment is interpreted as congestion in the network; the source should slow down. We will see this type of signaling when we discuss TCP congestion control later in the chapter.

Explicit Signaling

The node that experiences congestion can explicitly send a signal to the source or destination. The explicit signaling method, however, is different from the choke packet method. In the choke packet method, a separate packet is used for this purpose; in the explicit signaling method, the signal is included in the packets that carry data. Explicit signaling, as we will see in Frame Relay congestion control, can occur in either the forward or the backward direction.

Backward Signaling A bit can be set in a packet moving in the direction opposite to the congestion. This bit can warn the source that there is congestion and that it needs to slow down to avoid the discarding of packets.

Forward Signaling A bit can be set in a packet moving in the direction of the congestion. This bit can warn the destination that there is congestion. The receiver in this case can use policies, such as slowing down the acknowledgments, to alleviate the congestion.

24.4 TWO EXAMPLES

To better understand the concept of congestion control, let us give two examples: one in TCP and the other in Frame Relay.

Congestion Control in TCP

We discussed TCP in Chapter 23. We now show how TCP uses congestion control to avoid congestion or alleviate congestion in the network.

Congestion Window

In Chapter 23, we talked about flow control and tried to discuss solutions when the receiver is overwhelmed with data. We said that the sender window size is determined by the available buffer space in the receiver (*rwnd*). In other words, we assumed that it is only the receiver that can dictate to the sender the size of the sender's window. We totally ignored another entity here—the network. If the network cannot deliver the data as fast as they are created by the sender, it must tell the sender to slow down. In other words, in addition to the receiver, the network is a second entity that determines the size of the sender's window.

Today, the sender's window size is determined not only by the receiver but also by congestion in the network.

The sender has two pieces of information: the receiver-advertised window size and the congestion window size. The actual size of the window is the minimum of these two.

Actual window size = minimum (rwnd, cwnd)

We show shortly how the size of the congestion window (*cwnd*) is determined.

Congestion Policy

TCP's general policy for handling congestion is based on three phases: slow start, congestion avoidance, and congestion detection. In the slow-start phase, the sender starts with a very slow rate of transmission, but increases the rate rapidly to reach a threshold. When the threshold is reached, the data rate is reduced to avoid congestion. Finally if congestion is detected, the sender goes back to the slow-start or congestion avoidance phase based on how the congestion is detected.

Slow Start: Exponential Increase One of the algorithms used in TCP congestion control is called **slow start.** This algorithm is based on the idea that the size of the congestion window (*cwnd*) starts with one maximum segment size (MSS). The MSS is determined during connection establishment by using an option of the same name. The size of the window increases one MSS each time an acknowledgment is received. As the name implies, the window starts slowly, but grows exponentially. To show the idea, let us look at Figure 24.8. Note that we have used three simplifications to make the discussion more understandable. We have used segment numbers instead of byte numbers (as though each segment contains only 1 byte). We have assumed that *rwnd* is much higher than *cwnd,* so that the sender window size always equals *cwnd*. We have assumed that each segment is acknowledged individually.

The sender starts with *cwnd* = 1 MSS. This means that the sender can send only one segment. After receipt of the acknowledgment for segment 1, the size of the congestion window is increased by 1, which means that *cwnd* is now 2. Now two more segments can be sent. When each acknowledgment is received, the size of the window is increased by 1 MSS. When all seven segments are acknowledged, *cwnd* = 8.

Figure 24.8 *Slow start, exponential increase*

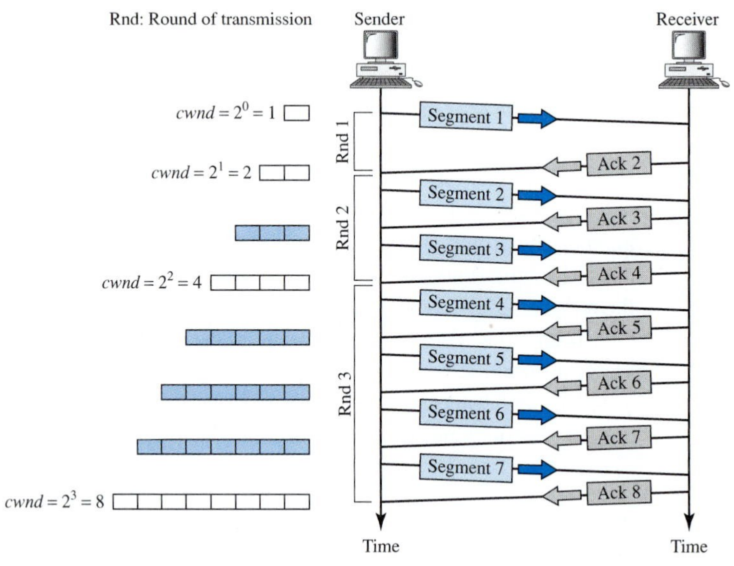

If we look at the size of *cwnd* in terms of rounds (acknowledgment of the whole window of segments), we find that the rate is exponential as shown below:

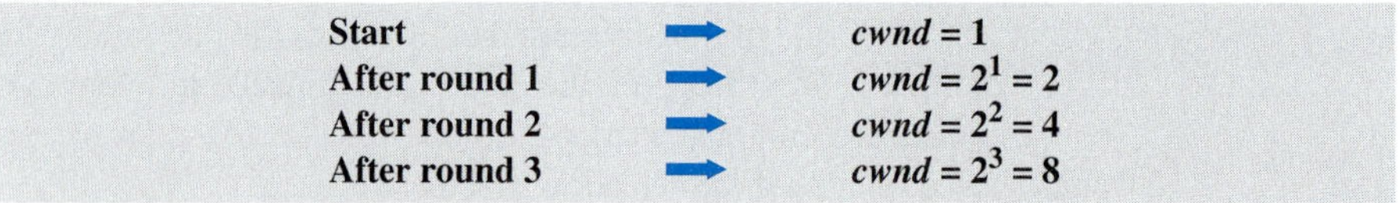

Start	→	***cwnd*** **= 1**
After round 1	→	***cwnd*** $= 2^1 = 2$
After round 2	→	***cwnd*** $= 2^2 = 4$
After round 3	→	***cwnd*** $= 2^3 = 8$

We need to mention that if there is delayed ACKs, the increase in the size of the window is less than power of 2.

Slow start cannot continue indefinitely. There must be a threshold to stop this phase. The sender keeps track of a variable named *ssthresh* (slow-start threshold). When the size of window in bytes reaches this threshold, slow start stops and the next phase starts. In most implementations the value of *ssthresh* is 65,535 bytes.

In the slow-start algorithm, the size of the congestion window increases exponentially until it reaches a threshold.

Congestion Avoidance: Additive Increase If we start with the slow-start algorithm, the size of the congestion window increases exponentially. To avoid congestion before it happens, one must slow down this exponential growth. TCP defines another algorithm called **congestion avoidance,** which undergoes an **additive increase** instead of an exponential one. When the size of the congestion window reaches the slow-start threshold, the slow-start phase stops and the additive phase begins. In this algorithm, each time the whole window of segments is acknowledged (one round), the size of the congestion window is increased by 1. To show the idea, we apply this algorithm to the same scenario as slow start, although we will see that the congestion avoidance algorithm usually starts when the size of the window is much greater than 1. Figure 24.9 shows the idea.

Figure 24.9 *Congestion avoidance, additive increase*

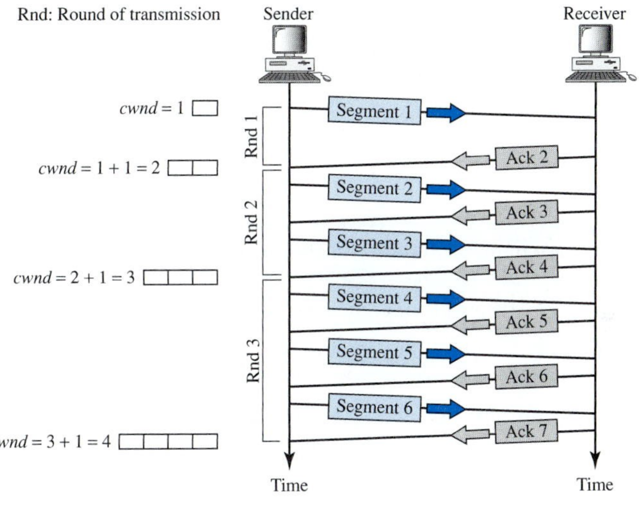

In this case, after the sender has received acknowledgments for a complete window size of segments, the size of the window is increased by one segment.

If we look at the size of *cwnd* in terms of rounds, we find that the rate is additive as shown below:

Start	➡	***cwnd* = 1**
After round 1	➡	***cwnd* = 1 + 1 = 2**
After round 2	➡	***cwnd* = 2 + 1 = 3**
After round 3	➡	***cwnd* = 3 + 1 = 4**

In the congestion avoidance algorithm, the size of the congestion window increases additively until congestion is detected.

Congestion Detection: Multiplicative Decrease If congestion occurs, the congestion window size must be decreased. The only way the sender can guess that congestion has occurred is by the need to retransmit a segment. However, retransmission can occur in one of two cases: when a timer times out or when three ACKs are received. In both cases, the size of the threshold is dropped to one-half, a **multiplicative decrease.** Most TCP implementations have two reactions:

1. If a time-out occurs, there is a stronger possibility of congestion; a segment has probably been dropped in the network, and there is no news about the sent segments.

In this case TCP reacts strongly:

a. It sets the value of the threshold to one-half of the current window size.

b. It sets *cwnd* to the size of one segment.

c. It starts the slow-start phase again.

2. If three ACKs are received, there is a weaker possibility of congestion; a segment may have been dropped, but some segments after that may have arrived safely since three ACKs are received. This is called fast transmission and fast recovery. In this case, TCP has a weaker reaction:

a. It sets the value of the threshold to one-half of the current window size.

b. It sets *cwnd* to the value of the threshold (some implementations add three segment sizes to the threshold).

c. It starts the congestion avoidance phase.

An implementations reacts to congestion detection in one of the following ways:

- If detection is by time-out, a new *slow-start* phase starts.
- If detection is by three ACKs, a new *congestion avoidance* phase starts.

Summary In Figure 24.10, we summarize the congestion policy of TCP and the relationships between the three phases.

Figure 24.10 *TCP congestion policy summary*

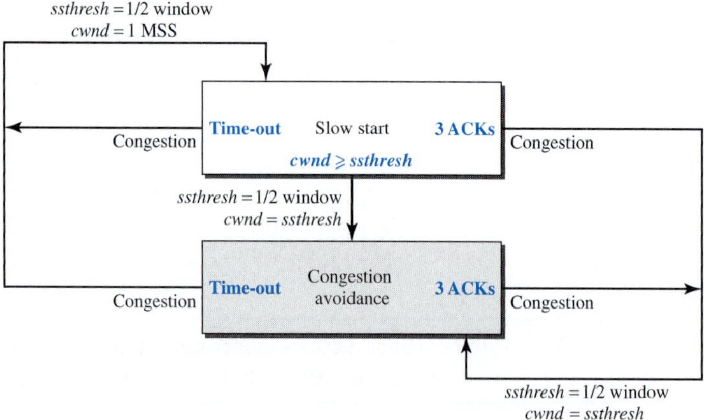

We give an example in Figure 24.11. We assume that the maximum window size is 32 segments. The threshold is set to 16 segments (one-half of the maximum window size). In the *slow-start* phase the window size starts from 1 and grows exponentially until it reaches the threshold. After it reaches the threshold, the *congestion avoidance (additive increase)* procedure allows the window size to increase linearly until a time-out occurs or the maximum window size is reached. In Figure 24.11, the time-out occurs when the window size is 20. At this moment, the *multiplicative decrease* procedure takes

Figure 24.11 *Congestion example*

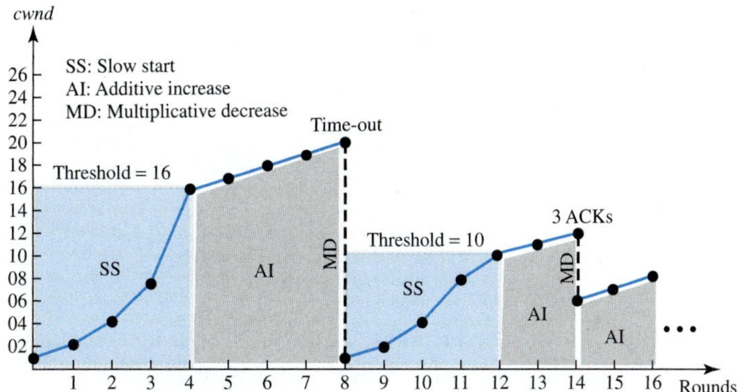

over and reduces the threshold to one-half of the previous window size. The previous window size was 20 when the time-out happened so the new threshold is now 10.

TCP moves to slow start again and starts with a window size of 1, and TCP moves to additive increase when the new threshold is reached. When the window size is 12, a three-ACKs event happens. The multiplicative decrease procedure takes over again. The threshold is set to 6 and TCP goes to the additive increase phase this time. It remains in this phase until another time-out or another three ACKs happen.

Congestion Control in Frame Relay

Congestion in a Frame Relay network decreases throughput and increases delay. A high throughput and low delay are the main goals of the Frame Relay protocol. Frame Relay does not have flow control. In addition, Frame Relay allows the user to transmit bursty data. This means that a Frame Relay network has the potential to be really congested with traffic, thus requiring congestion control.

Congestion Avoidance

For congestion avoidance, the Frame Relay protocol uses 2 bits in the frame to explicitly warn the source and the destination of the presence of congestion.

BECN The **backward explicit congestion notification (BECN)** bit warns the sender of congestion in the network. One might ask how this is accomplished since the frames are traveling away from the sender. In fact, there are two methods: The switch can use response frames from the receiver (full-duplex mode), or else the switch can use a predefined connection (DLCI = 1023) to send special frames for this specific purpose. The sender can respond to this warning by simply reducing the data rate. Figure 24.12 shows the use of BECN.

FECN The **forward explicit congestion notification (FECN)** bit is used to warn the receiver of congestion in the network. It might appear that the receiver cannot do anything to relieve the congestion. However, the Frame Relay protocol assumes that the sender and receiver are communicating with each other and are using some type of flow control at a higher level. For example, if there is an acknowledgment mechanism at this

Figure 24.12 *BECN*

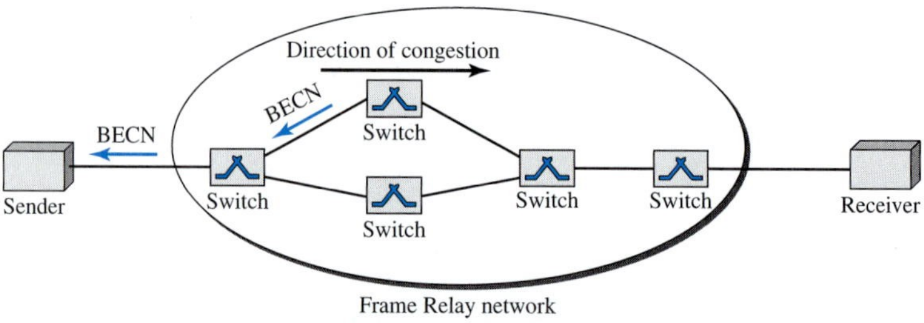

higher level, the receiver can delay the acknowledgment, thus forcing the sender to slow down. Figure 24.13 shows the use of FECN.

Figure 24.13 *FECN*

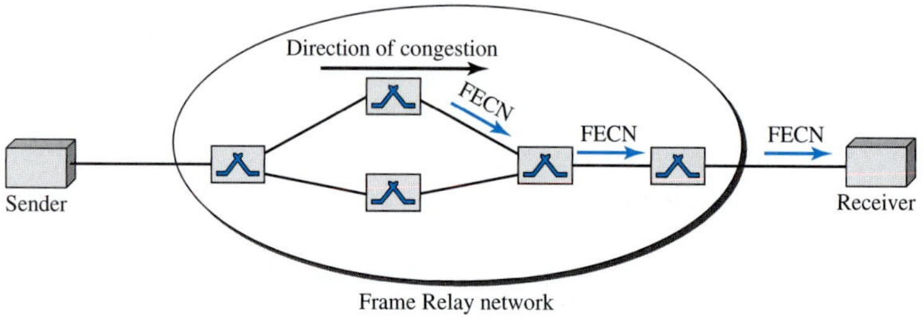

When two endpoints are communicating using a Frame Relay network, four situations may occur with regard to congestion. Figure 24.14 shows these four situations and the values of FECN and BECN.

Figure 24.14 *Four cases of congestion*

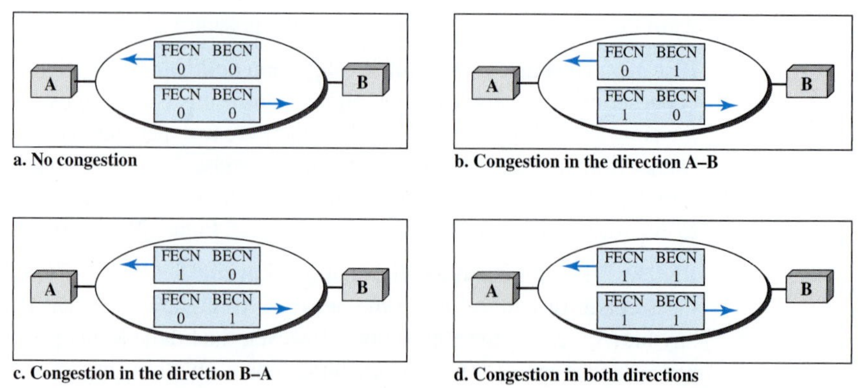

24.5 QUALITY OF SERVICE

Quality of service (QoS) is an internetworking issue that has been discussed more than defined. We can informally define quality of service as something a flow seeks to attain.

Flow Characteristics

Traditionally, four types of characteristics are attributed to a flow: reliability, delay, jitter, and bandwidth, as shown in Figure 24.15.

Figure 24.15 *Flow characteristics*

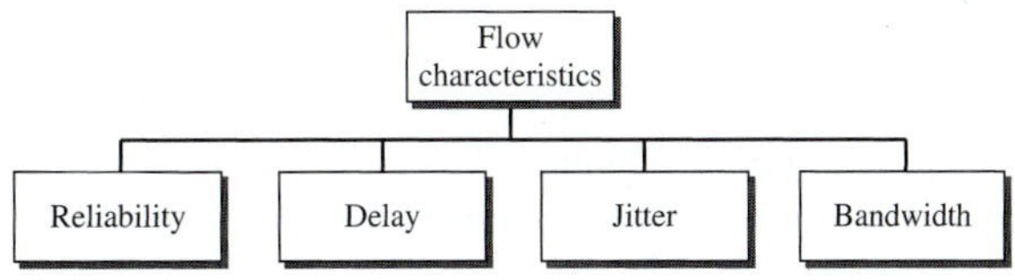

Reliability

Reliability is a characteristic that a flow needs. Lack of reliability means losing a packet or acknowledgment, which entails retransmission. However, the sensitivity of application programs to reliability is not the same. For example, it is more important that electronic mail, file transfer, and Internet access have reliable transmissions than telephony or audio conferencing.

Delay

Source-to-destination **delay** is another flow characteristic. Again applications can tolerate delay in different degrees. In this case, telephony, audio conferencing, video conferencing, and remote log-in need minimum delay, while delay in file transfer or e-mail is less important.

Jitter

Jitter is the variation in delay for packets belonging to the same flow. For example, if four packets depart at times 0, 1, 2, 3 and arrive at 20, 21, 22, 23, all have the same delay, 20 units of time. On the other hand, if the above four packets arrive at 21, 23, 21, and 28, they will have different delays: 21, 22, 19, and 24.

For applications such as audio and video, the first case is completely acceptable; the second case is not. For these applications, it does not matter if the packets arrive with a short or long delay as long as the delay is the same for all packets. For this application, the second case is not acceptable.

Jitter is defined as the variation in the packet delay. High jitter means the difference between delays is large; low jitter means the variation is small.

In Chapter 29, we show how multimedia communication deals with jitter. If the jitter is high, some action is needed in order to use the received data.

Bandwidth

Different applications need different bandwidths. In video conferencing we need to send millions of bits per second to refresh a color screen while the total number of bits in an e-mail may not reach even a million.

Flow Classes

Based on the flow characteristics, we can classify flows into groups, with each group having similar levels of characteristics. This categorization is not formal or universal; some protocols such as ATM have defined classes, as we will see later.

24.6 TECHNIQUES TO IMPROVE QoS

In Section 24.5 we tried to define QoS in terms of its characteristics. In this section, we discuss some techniques that can be used to improve the quality of service. We briefly discuss four common methods: scheduling, traffic shaping, admission control, and resource reservation.

Scheduling

Packets from different flows arrive at a switch or router for processing. A good scheduling technique treats the different flows in a fair and appropriate manner. Several scheduling techniques are designed to improve the quality of service. We discuss three of them here: FIFO queuing, priority queuing, and weighted fair queuing.

FIFO Queuing

In **first-in, first-out (FIFO) queuing,** packets wait in a buffer (queue) until the node (router or switch) is ready to process them. If the average arrival rate is higher than the average processing rate, the queue will fill up and new packets will be discarded. A FIFO queue is familiar to those who have had to wait for a bus at a bus stop. Figure 24.16 shows a conceptual view of a FIFO queue.

Figure 24.16 *FIFO queue*

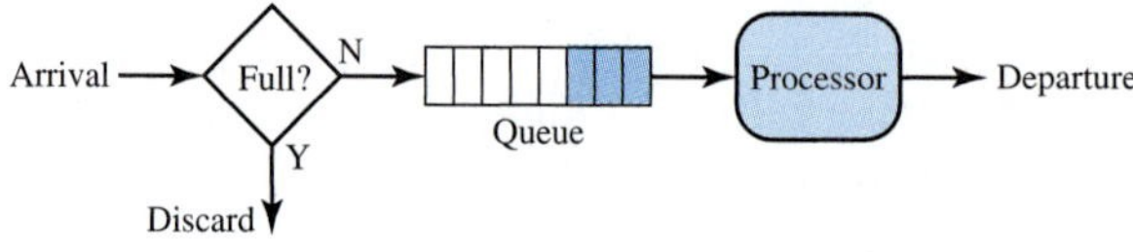

Priority Queuing

In **priority queuing,** packets are first assigned to a priority class. Each priority class has its own queue. The packets in the highest-priority queue are processed first. Packets in the lowest-priority queue are processed last. Note that the system does not stop serving a queue until it is empty. Figure 24.17 shows priority queuing with two priority levels (for simplicity).

Figure 24.17 *Priority queuing*

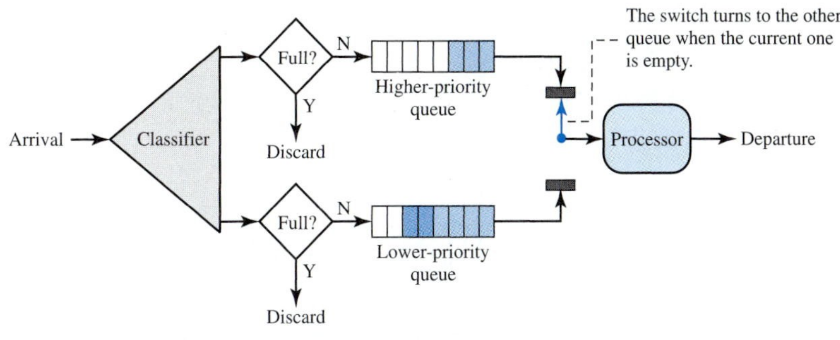

A priority queue can provide better QoS than the FIFO queue because higher-priority traffic, such as multimedia, can reach the destination with less delay. However, there is a potential drawback. If there is a continuous flow in a high-priority queue, the packets in the lower-priority queues will never have a chance to be processed. This is a condition called *starvation*.

Weighted Fair Queuing

A better scheduling method is **weighted fair queuing.** In this technique, the packets are still assigned to different classes and admitted to different queues. The queues, however, are weighted based on the priority of the queues; higher priority means a higher weight. The system processes packets in each queue in a round-robin fashion with the number of packets selected from each queue based on the corresponding weight. For example, if the weights are 3, 2, and 1, three packets are processed from the first queue, two from the second queue, and one from the third queue. If the system does not impose priority on the classes, all weights can be equal. In this way, we have fair queuing with priority. Figure 24.18 shows the technique with three classes.

Traffic Shaping

Traffic shaping is a mechanism to control the amount and the rate of the traffic sent to the network. Two techniques can shape traffic: leaky bucket and token bucket.

Leaky Bucket

If a bucket has a small hole at the bottom, the water leaks from the bucket at a constant rate as long as there is water in the bucket. The rate at which the water leaks does not depend on the rate at which the water is input to the bucket unless the bucket is empty. The input rate can vary, but the output rate remains constant. Similarly, in networking, a technique called **leaky bucket** can smooth out bursty traffic. Bursty chunks are stored in the bucket and sent out at an average rate. Figure 24.19 shows a leaky bucket and its effects.

Figure 24.18 *Weighted fair queuing*

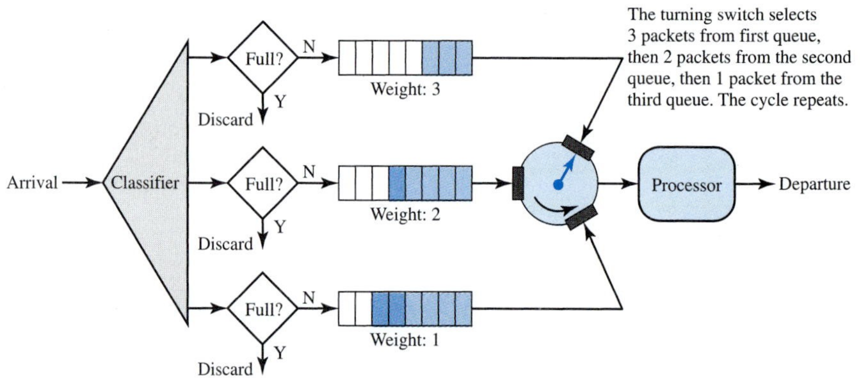

Figure 24.19 *Leaky bucket*

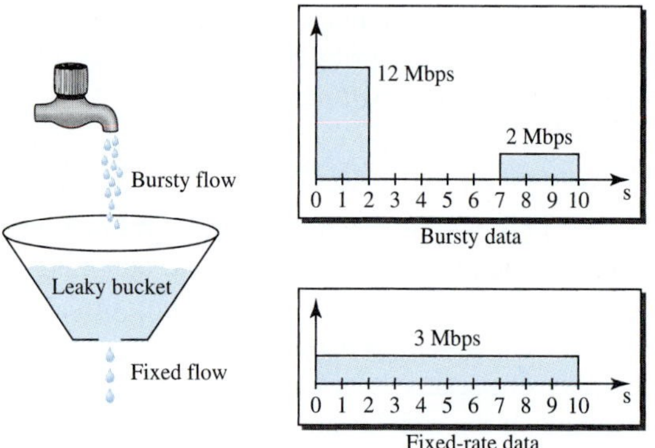

In the figure, we assume that the network has committed a bandwidth of 3 Mbps for a host. The use of the leaky bucket shapes the input traffic to make it conform to this commitment. In Figure 24.19 the host sends a burst of data at a rate of 12 Mbps for 2 s, for a total of 24 Mbits of data. The host is silent for 5 s and then sends data at a rate of 2 Mbps for 3 s, for a total of 6 Mbits of data. In all, the host has sent 30 Mbits of data in 10 s. The leaky bucket smooths the traffic by sending out data at a rate of 3 Mbps during the same 10 s. Without the leaky bucket, the beginning burst may have hurt the network by consuming more bandwidth than is set aside for this host. We can also see that the leaky bucket may prevent congestion. As an analogy, consider the freeway during rush hour (bursty traffic). If, instead, commuters could stagger their working hours, congestion on our freeways could be avoided.

A simple leaky bucket implementation is shown in Figure 24.20. A FIFO queue holds the packets. If the traffic consists of fixed-size packets (e.g., cells in ATM

Figure 24.20 *Leaky bucket implementation*

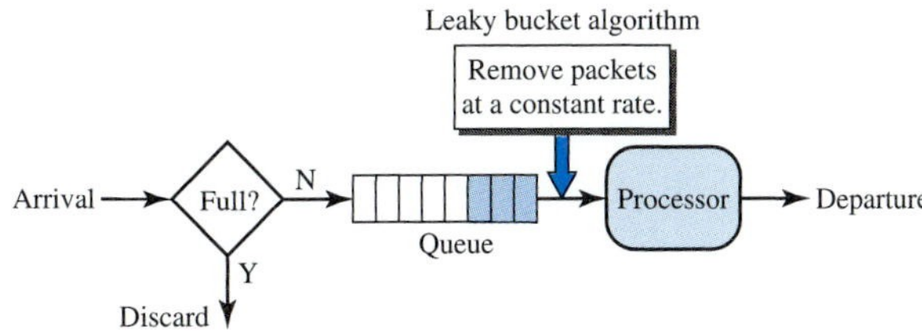

networks), the process removes a fixed number of packets from the queue at each tick of the clock. If the traffic consists of variable-length packets, the fixed output rate must be based on the number of bytes or bits.

The following is an algorithm for variable-length packets:

1. Initialize a counter to n at the tick of the clock.
2. If n is greater than the size of the packet, send the packet and decrement the counter by the packet size. Repeat this step until n is smaller than the packet size.
3. Reset the counter and go to step 1.

A leaky bucket algorithm shapes bursty traffic into fixed-rate traffic by averaging the data rate. It may drop the packets if the bucket is full.

Token Bucket

The leaky bucket is very restrictive. It does not credit an idle host. For example, if a host is not sending for a while, its bucket becomes empty. Now if the host has bursty data, the leaky bucket allows only an average rate. The time when the host was idle is not taken into account. On the other hand, the **token bucket** algorithm allows idle hosts to accumulate credit for the future in the form of tokens. For each tick of the clock, the system sends n tokens to the bucket. The system removes one token for every cell (or byte) of data sent. For example, if n is 100 and the host is idle for 100 ticks, the bucket collects 10,000 tokens. Now the host can consume all these tokens in one tick with 10,000 cells, or the host takes 1000 ticks with 10 cells per tick. In other words, the host can send bursty data as long as the bucket is not empty. Figure 24.21 shows the idea.

The token bucket can easily be implemented with a counter. The token is initialized to zero. Each time a token is added, the counter is incremented by 1. Each time a unit of data is sent, the counter is decremented by 1. When the counter is zero, the host cannot send data.

The token bucket allows bursty traffic at a regulated maximum rate.

Figure 24.21 *Token bucket*

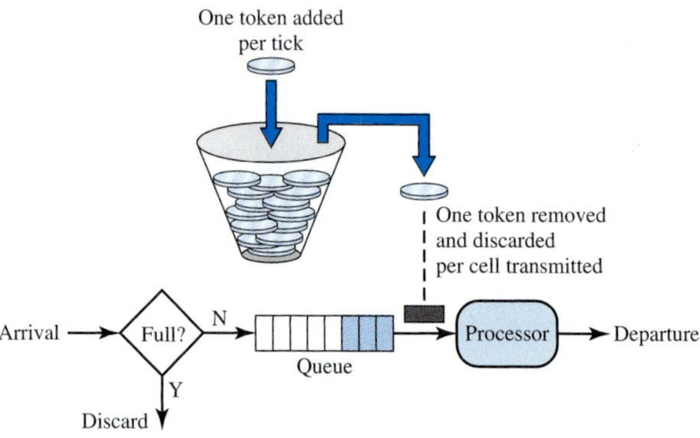

Combining Token Bucket and Leaky Bucket

The two techniques can be combined to credit an idle host and at the same time regulate the traffic. The leaky bucket is applied after the token bucket; the rate of the leaky bucket needs to be higher than the rate of tokens dropped in the bucket.

Resource Reservation

A flow of data needs resources such as a buffer, bandwidth, CPU time, and so on. The quality of service is improved if these resources are reserved beforehand. We discuss in this section one QoS model called Integrated Services, which depends heavily on resource reservation to improve the quality of service.

Admission Control

Admission control refers to the mechanism used by a router, or a switch, to accept or reject a flow based on predefined parameters called flow specifications. Before a router accepts a flow for processing, it checks the flow specifications to see if its capacity (in terms of bandwidth, buffer size, CPU speed, etc.) and its previous commitments to other flows can handle the new flow.

24.7 INTEGRATED SERVICES

Based on the topics in Sections 24.5 and 24.6, two models have been designed to provide quality of service in the Internet: Integrated Services and Differentiated Services. Both models emphasize the use of quality of service at the network layer (IP), although the model can also be used in other layers such as the data link. We discuss Integrated Services in this section and Differentiated Service in Section 24.8.

As we learned in Chapter 20, IP was originally designed for *best-effort* delivery. This means that every user receives the same level of services. This type of delivery does not guarantee the minimum of a service, such as bandwidth, to applications such as real-time audio and video. If such an application accidentally gets extra bandwidth, it may be detrimental to other applications, resulting in congestion.

Integrated Services, sometimes called **IntServ,** is a *flow-based* QoS model, which means that a user needs to create a flow, a kind of virtual circuit, from the source to the destination and inform all routers of the resource requirement.

Integrated Services is a *flow-based* QoS model designed for IP.

Signaling

The reader may remember that IP is a connectionless, datagram, packet-switching protocol. How can we implement a flow-based model over a connectionless protocol? The solution is a signaling protocol to run over IP that provides the signaling mechanism for making a reservation. This protocol is called **Resource Reservation Protocol (RSVP)** and will be discussed shortly.

Flow Specification

When a source makes a reservation, it needs to define a flow specification. A flow specification has two parts: Rspec (resource specification) and Tspec (traffic specification). Rspec defines the resource that the flow needs to reserve (buffer, bandwidth, etc.). Tspec defines the traffic characterization of the flow.

Admission

After a router receives the flow specification from an application, it decides to admit or deny the service. The decision is based on the previous commitments of the router and the current availability of the resource.

Service Classes

Two classes of services have been defined for Integrated Services: guaranteed service and controlled-load service.

Guaranteed Service Class

This type of service is designed for real-time traffic that needs a guaranteed minimum end-to-end delay. The end-to-end delay is the sum of the delays in the routers, the propagation delay in the media, and the setup mechanism. Only the first, the sum of the delays in the routers, can be guaranteed by the router. This type of service guarantees that the packets will arrive within a certain delivery time and are not discarded if flow traffic stays within the boundary of Tspec. We can say that guaranteed services are quantitative services, in which the amount of end-to-end delay and the data rate must be defined by the application.

Controlled-Load Service Class

This type of service is designed for applications that can accept some delays, but are sensitive to an overloaded network and to the danger of losing packets. Good examples of these types of applications are file transfer, e-mail, and Internet access. The controlled-load service is a qualitative type of service in that the application requests the possibility of low-loss or no-loss packets.

RSVP

In the Integrated Services model, an application program needs resource reservation. As we learned in the discussion of the IntServ model, the resource reservation is for a *flow*. This means that if we want to use IntServ at the IP level, we need to create a flow, a kind of virtual-circuit network, out of the IP, which was originally designed as a datagram packet-switched network. A virtual-circuit network needs a signaling system to set up the virtual circuit before data traffic can start. The Resource Reservation Protocol (RSVP) is a signaling protocol to help IP create a flow and consequently make a resource reservation. Before discussing RSVP, we need to mention that it is an independent protocol separate from the Integrated Services model. It may be used in other models in the future.

Multicast Trees

RSVP is different from some other signaling systems we have seen before in that it is a signaling system designed for multicasting. However, RSVP can be also used for unicasting because unicasting is just a special case of multicasting with only one member in the multicast group. The reason for this design is to enable RSVP to provide resource reservations for all kinds of traffic including multimedia which often uses multicasting.

Receiver-Based Reservation

In RSVP, the receivers, not the sender, make the reservation. This strategy matches the other multicasting protocols. For example, in multicast routing protocols, the receivers, not the sender, make a decision to join or leave a multicast group.

RSVP Messages

RSVP has several types of messages. However, for our purposes, we discuss only two of them: **Path** and **Resv.**

Path Messages Recall that the receivers in a flow make the reservation in RSVP. However, the receivers do not know the path traveled by packets before the reservation is made. The path is needed for the reservation. To solve the problem, RSVP uses *Path* messages. A Path message travels from the sender and reaches all receivers in the multicast path. On the way, a Path message stores the necessary information for the receivers. A Path message is sent in a multicast environment; a new message is created when the path diverges. Figure 24.22 shows path messages.

Resv Messages After a receiver has received a Path message, it sends a *Resv* message. The Resv message travels toward the sender (upstream) and makes a resource reservation on the routers that support RSVP. If a router does not support RSVP on the path, it routes

Figure 24.22 *Path messages*

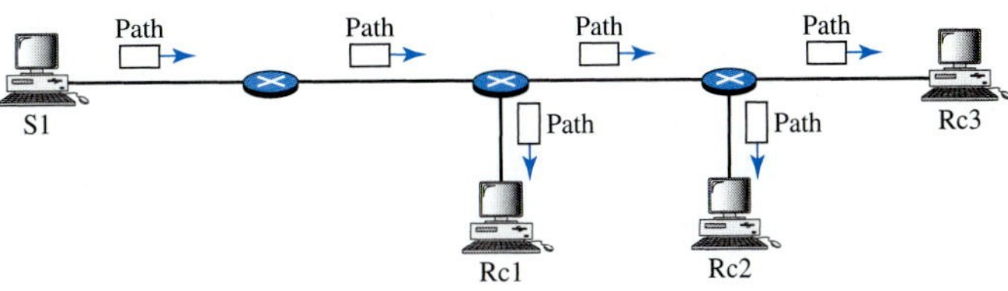

the packet based on the best-effort delivery methods we discussed before. Figure 24.23 shows the Resv messages.

Figure 24.23 *Resv messages*

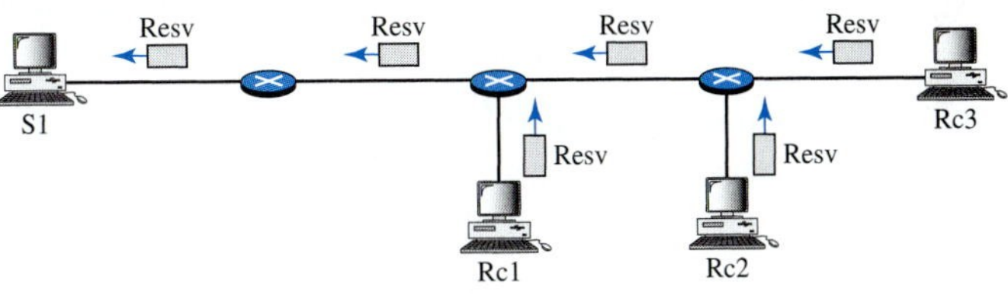

Reservation Merging

In RSVP, the resources are not reserved for each receiver in a flow; the reservation is merged. In Figure 24.24, Rc3 requests a 2-Mbps bandwidth while Rc2 requests a 1-Mbps bandwidth. Router R3, which needs to make a bandwidth reservation, merges the two requests. The reservation is made for 2 Mbps, the larger of the two, because a 2-Mbps input reservation can handle both requests. The same situation is true for R2. The reader may ask why Rc2 and Rc3, both belonging to one single flow, request different amounts of bandwidth. The answer is that, in a multimedia environment, different receivers may handle different grades of quality. For example, Rc2 may be able to receive video only at 1 Mbps (lower quality), while Rc3 may be able to receive video at 2 Mbps (higher quality).

Figure 24.24 *Reservation merging*

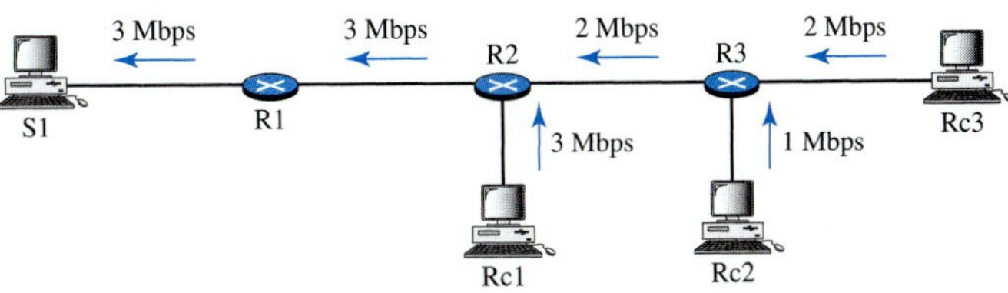

Reservation Styles

When there is more than one flow, the router needs to make a reservation to accommodate all of them. RSVP defines three types of reservation styles, as shown in Figure 24.25.

Figure 24.25 *Reservation styles*

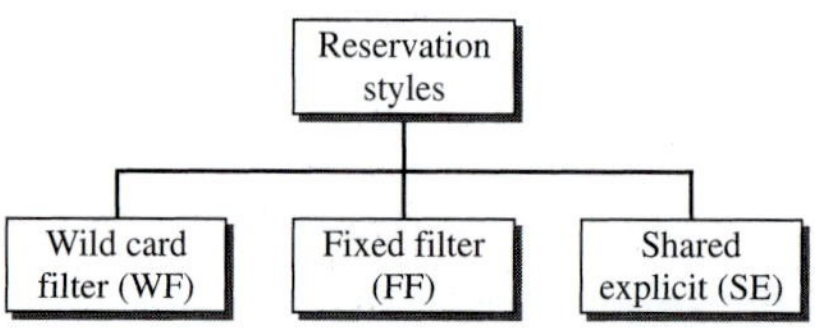

Wild Card Filter Style In this style, the router creates a single reservation for all senders. The reservation is based on the largest request. This type of style is used when the flows from different senders do not occur at the same time.

Fixed Filter Style In this style, the router creates a distinct reservation for each flow. This means that if there are *n* flows, *n* different reservations are made. This type of style is used when there is a high probability that flows from different senders will occur at the same time.

Shared Explicit Style In this style, the router creates a single reservation which can be shared by a set of flows.

Soft State

The reservation information (state) stored in every node for a flow needs to be refreshed periodically. This is referred to as a *soft state* as compared to the *hard state* used in other virtual-circuit protocols such as ATM or Frame Relay, where the information about the flow is maintained until it is erased. The default interval for refreshing is currently 30 s.

Problems with Integrated Services

There are at least two problems with Integrated Services that may prevent its full implementation in the Internet: scalability and service-type limitation.

Scalability

The Integrated Services model requires that each router keep information for each flow. As the Internet is growing every day, this is a serious problem.

Service-Type Limitation

The Integrated Services model provides only two types of services, guaranteed and control-load. Those opposing this model argue that applications may need more than these two types of services.

24.8 DIFFERENTIATED SERVICES

Differentiated Services (DS or **Diffserv)** was introduced by the IETF (Internet Engineering Task Force) to handle the shortcomings of Integrated Services. Two fundamental changes were made:

1. The main processing was moved from the core of the network to the edge of the network. This solves the scalability problem. The routers do not have to store information about flows. The applications, or hosts, define the type of service they need each time they send a packet.
2. The per-flow service is changed to per-class service. The router routes the packet based on the class of service defined in the packet, not the flow. This solves the service-type limitation problem. We can define different types of classes based on the needs of applications.

Differentiated Services is a class-based QoS model designed for IP.

DS Field

In Diffserv, each packet contains a field called the DS field. The value of this field is set at the boundary of the network by the host or the first router designated as the boundary router. IETF proposes to replace the existing TOS (type of service) field in IPv4 or the class field in IPv6 by the DS field, as shown in Figure 24.26.

Figure 24.26 *DS field*

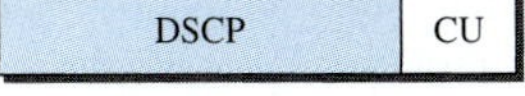

The DS field contains two subfields: DSCP and CU. The DSCP (Differentiated Services Code Point) is a 6-bit subfield that defines the **per-hop behavior (PHB).** The 2-bit CU (currently unused) subfield is not currently used.

The Diffserv capable node (router) uses the DSCP 6 bits as an index to a table defining the packet-handling mechanism for the current packet being processed.

Per-Hop Behavior

The Diffserv model defines per-hop behaviors (PHBs) for each node that receives a packet. So far three PHBs are defined: DE PHB, EF PHB, and AF PHB.

DE PHB The DE PHB (default PHB) is the same as best-effort delivery, which is compatible with TOS.

EF PHB The EF PHB (expedited forwarding PHB) provides the following services:

- ❑ Low loss

- Low latency
- Ensured bandwidth

This is the same as having a virtual connection between the source and destination.

AF PHB The AF PHB (assured forwarding PHB) delivers the packet with a high assurance as long as the class traffic does not exceed the traffic profile of the node. The users of the network need to be aware that some packets may be discarded.

Traffic Conditioner

To implement Diffserv, the DS node uses traffic conditioners such as meters, markers, shapers, and droppers, as shown in Figure 24.27.

Figure 24.27 *Traffic conditioner*

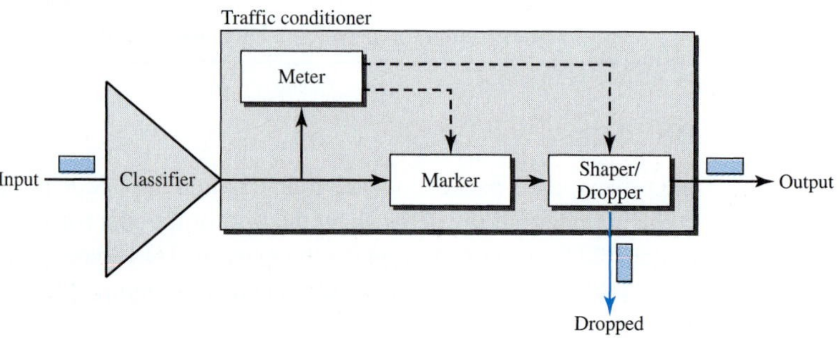

Meters The meter checks to see if the incoming flow matches the negotiated traffic profile. The meter also sends this result to other components. The meter can use several tools such as a token bucket to check the profile.

Marker A marker can remark a packet that is using best-effort delivery (DSCP: 000000) or down-mark a packet based on information received from the meter. Down-marking (lowering the class of the flow) occurs if the flow does not match the profile. A marker does not up-mark (promote the class) a packet.

Shaper A shaper uses the information received from the meter to reshape the traffic if it is not compliant with the negotiated profile.

Dropper A dropper, which works as a shaper with no buffer, discards packets if the flow severely violates the negotiated profile.

24.9 QoS IN SWITCHED NETWORKS

We discussed the proposed models for QoS in the IP protocols. Let us now discuss QoS as used in two switched networks: Frame Relay and ATM. These two networks are virtual-circuit networks that need a signaling protocol such as RSVP.

QoS in Frame Relay

Four different attributes to control traffic have been devised in Frame Relay: access rate, committed burst size B_c, committed information rate (CIR), and excess burst size B_e. These are set during the negotiation between the user and the network. For PVC connections, they are negotiated once; for SVC connections, they are negotiated for each connection during connection setup. Figure 24.28 shows the relationships between these four measurements.

Figure 24.28 *Relationship between traffic control attributes*

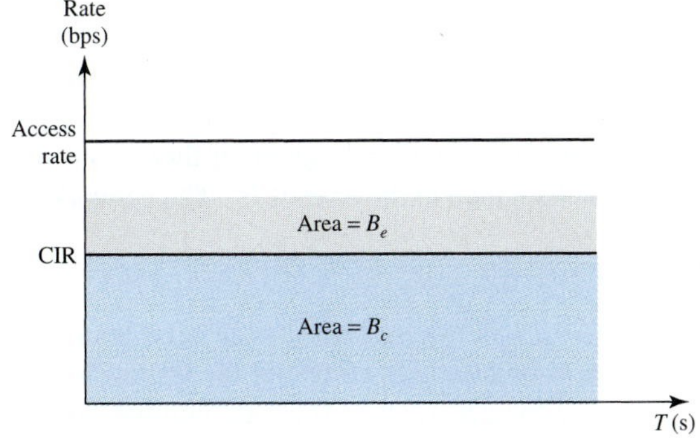

Access Rate

For every connection, an **access rate** (in bits per second) is defined. The access rate actually depends on the bandwidth of the channel connecting the user to the network. The user can never exceed this rate. For example, if the user is connected to a Frame Relay network by a T-1 line, the access rate is 1.544 Mbps and can never be exceeded.

Committed Burst Size

For every connection, Frame Relay defines a **committed burst size B_c.** This is the maximum number of bits in a predefined time that the network is committed to transfer without discarding any frame or setting the DE bit. For example, if a B_c of 400 kbits for a period of 4 s is granted, the user can send up to 400 kbits during a 4-s interval without worrying about any frame loss. Note that this is not a rate defined for each second. It is a cumulative measurement. The user can send 300 kbits during the first second, no data during the second and the third seconds, and finally 100 kbits during the fourth second.

Committed Information Rate

The **committed information rate (CIR)** is similar in concept to committed burst size except that it defines an average rate in bits per second. If the user follows this rate continuously, the network is committed to deliver the frames. However, because it is an average measurement, a user may send data at a higher rate than the CIR at times or at a lower rate other times. As long as the average for the predefined period is met, the frames will be delivered.

The cumulative number of bits sent during the predefined period cannot exceed B_c. Note that the CIR is not an independent measurement; it can be calculated by using the following formula:

$$\text{CIR} = \frac{B_c}{T} \text{ bps}$$

For example, if the B_c is 5 kbits in a period of 5 s, the CIR is 5000/5, or 1 kbps.

Excess Burst Size

For every connection, Frame Relay defines an **excess burst size B_e.** This is the maximum number of bits in excess of B_c that a user can send during a predefined time. The network is committed to transfer these bits if there is no congestion. Note that there is less commitment here than in the case of B_c. The network is committing itself conditionally.

User Rate

Figure 24.29 shows how a user can send bursty data. If the user never exceeds B_c, the network is committed to transmit the frames without discarding any. If the user exceeds B_c by less than B_e (that is, the total number of bits is less than $B_c + B_e$), the network is committed to transfer all the frames if there is no congestion. If there is congestion, some frames will be discarded. The first switch that receives the frames from the user has a counter and sets the DE bit for the frames that exceed B_c. The rest of the switches will discard these frames if there is congestion. Note that a user who needs to send data faster may exceed the B_c level. As long as the level is not above $B_c + B_e$, there is a chance that the frames will reach the destination without being discarded. Remember, however, that the moment the user exceeds the $B_c + B_e$ level, all the frames sent after that are discarded by the first switch.

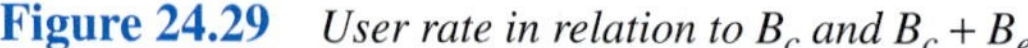

Figure 24.29 *User rate in relation to B_c and $B_c + B_e$*

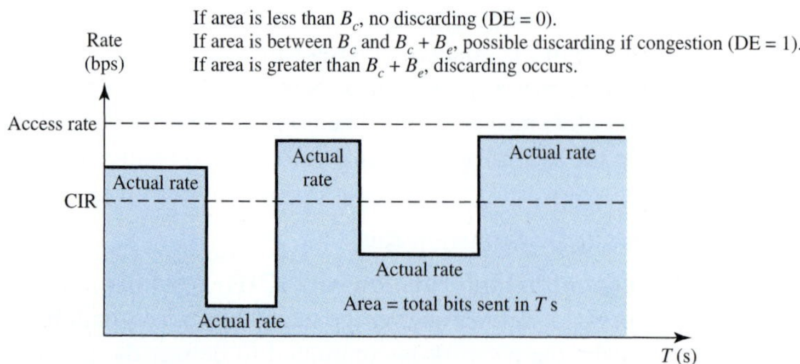

QoS in ATM

The QoS in ATM is based on the class, user-related attributes, and network-related attributes.

Classes

The ATM Forum defines four service classes: CBR, VBR, ABR, and UBR (see Figure 24.30).

Figure 24.30 *Service classes*

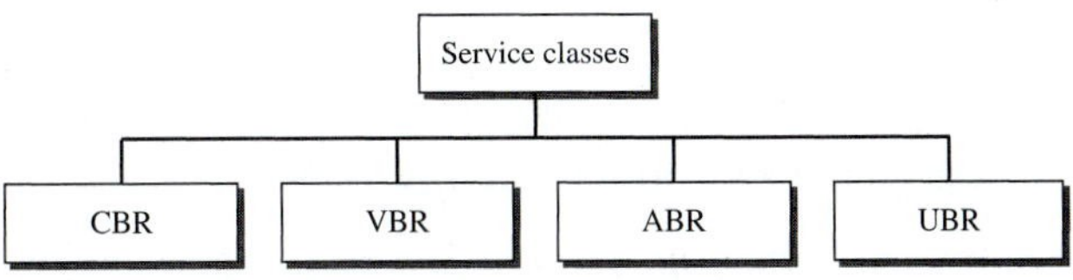

CBR The **constant-bit-rate (CBR)** class is designed for customers who need real-time audio or video services. The service is similar to that provided by a dedicated line such as a T line.

VBR The **variable-bit-rate (VBR)** class is divided into two subclasses: real-time (VBR-RT) and non-real-time (VBR-NRT). VBR-RT is designed for those users who need real-time services (such as voice and video transmission) and use compression techniques to create a variable bit rate. VBR-NRT is designed for those users who do not need real-time services but use compression techniques to create a variable bit rate.

ABR The **available-bit-rate (ABR)** class delivers cells at a minimum rate. If more network capacity is available, this minimum rate can be exceeded. ABR is particularly suitable for applications that are bursty.

UBR The **unspecified-bit-rate (UBR)** class is a best-effort delivery service that does not guarantee anything.

Figure 24.31 shows the relationship of different classes to the total capacity of the network.

Figure 24.31 *Relationship of service classes to the total capacity of the network*

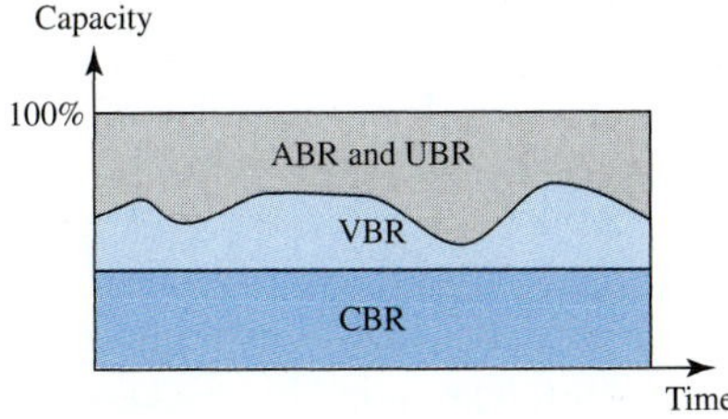

User-Related Attributes

ATM defines two sets of attributes. User-related attributes are those attributes that define how fast the user wants to send data. These are negotiated at the time of contract between a user and a network. The following are some user-related attributes:

SCR The *sustained cell rate* (SCR) is the average cell rate over a long time interval. The actual cell rate may be lower or higher than this value, but the average should be equal to or less than the SCR.

PCR The *peak cell rate* (PCR) defines the sender's maximum cell rate. The user's cell rate can sometimes reach this peak, as long as the SCR is maintained.

MCR The *minimum cell rate* (MCR) defines the minimum cell rate acceptable to the sender. For example, if the MCR is 50,000, the network must guarantee that the sender can send at least 50,000 cells per second.

CVDT The *cell variation delay tolerance* (CVDT) is a measure of the variation in cell transmission times. For example, if the CVDT is 5 ns, this means that the difference between the minimum and the maximum delays in delivering the cells should not exceed 5 ns.

Network-Related Attributes

The network-related attributes are those that define characteristics of the network. The following are some network-related attributes:

CLR The *cell loss ratio* (CLR) defines the fraction of cells lost (or delivered so late that they are considered lost) during transmission. For example, if the sender sends 100 cells and one of them is lost, the CLR is

$$\text{CLR} = \frac{1}{100} = 10^{-2}$$

CTD The *cell transfer delay* (CTD) is the average time needed for a cell to travel from source to destination. The maximum CTD and the minimum CTD are also considered attributes.

CDV The *cell delay variation* (CDV) is the difference between the CTD maximum and the CTD minimum.

CER The *cell error ratio* (CER) defines the fraction of the cells delivered in error.

24.10 RECOMMENDED READING

For more details about subjects discussed in this chapter, we recommend the following books and sites. The items in brackets [. . .] refer to the reference list at the end of the text.

Books

Congestion control and QoS are discussed in Sections 5.3 and 5.5 of [Tan03] and in Chapter 3 of [Sta98]. Easy reading about QoS can be found in [FH98]. The full discussion of QoS is covered in [Bla00].

24.11 KEY TERMS

access rate
additive increase
available bit rate (ABR)
average data rate
backward explicit congestion notification (BECN)
bursty data
choke packet
closed-loop congestion control
committed burst size B_c
committed information rate (CIR)
congestion
congestion avoidance
congestion control
constant bit rate (CBR)
delay
Differentiated Services (DS or Diffserv)
effective bandwidth
excess burst size B_e
first-in, first-out (FIFO) queuing
forward explicit congestion notification (FECN)
Integrated Services (IntServ)
jitter
leaky bucket
load
maximum burst size
multiplicative decrease
open-loop congestion control
Path message
peak data rate
per-hop behavior (PHB)
priority queuing
quality of service (QoS)
reliability
Resource Reservation Protocol (RSVP)
Resv message
slow start
throughput
token bucket
traffic shaping
unspecified bit rate (UBR)
variable bit rate (VBR)
weighted fair queuing

24.12 SUMMARY

- The average data rate, peak data rate, maximum burst size, and effective band width are qualitative values that describe a data flow.
- A data flow can have a constant bit rate, a variable bit rate, or traffic that is bursty.
- Congestion control refers to the mechanisms and techniques to control congestion and keep the load below capacity.
- Delay and throughput measure the performance of a network.
- Open-loop congestion control prevents congestion; closed-loop congestion control removes congestion.

- TCP avoids congestion through the use of two strategies: the combination of slow start and additive increase, and multiplicative decrease.
- Frame Relay avoids congestion through the use of two strategies: backward explicit congestion notification (BECN) and forward explicit congestion notification (FECN).
- A flow can be characterized by its reliability, delay, jitter, and bandwidth.
- Scheduling, traffic shaping, resource reservation, and admission control are techniques to improve quality of service (QoS).
- FIFO queuing, priority queuing, and weighted fair queuing are scheduling techniques.
- Leaky bucket and token bucket are traffic shaping techniques.
- Integrated Services is a flow-based QoS model designed for IP.
- The Resource Reservation Protocol (RSVP) is a signaling protocol that helps IP create a flow and makes a resource reservation.
- Differential Services is a class-based QoS model designed for IP.
- Access rate, committed burst size, committed information rate, and excess burst size are attributes to control traffic in Frame Relay.
- Quality of service in ATM is based on service classes, user-related attributes, and network-related attributes.

24.13 PRACTICE SET

Review Questions

1. How are congestion control and quality of service related?
2. What is a traffic descriptor?
3. What is the relationship between the average data rate and the peak data rate?
4. What is the definition of bursty data?
5. What is the difference between open-loop congestion control and closed-loop congestion control?
6. Name the policies that can prevent congestion.
7. Name the mechanisms that can alleviate congestion.
8. What determines the sender window size in TCP?
9. How does Frame Relay control congestion?
10. What attributes can be used to describe a flow of data?
11. What are four general techniques to improve quality of service?
12. What is traffic shaping? Name two methods to shape traffic.
13. What is the major difference between Integrated Services and Differentiated Services?
14. How is Resource Reservation Protocol related to Integrated Services?
15. What attributes are used for traffic control in Frame Relay?
16. In regard to quality of service, how do user-related attributes differ from network-related attributes in ATM?

Exercises

17. The address field of a Frame Relay frame is 1011000000010111. Is there any congestion in the forward direction? Is there any congestion in the backward direction?
18. A frame goes from A to B. There is congestion in both directions. Is the FECN bit set? Is the BECN bit set?
19. In a leaky bucket used to control liquid flow, how many gallons of liquid are left in the bucket if the output rate is 5 gal/min, there is an input burst of 100 gal/min for 12 s, and there is no input for 48 s?
20. An output interface in a switch is designed using the leaky bucket algorithm to send 8000 bytes/s (tick). If the following frames are received in sequence, show the frames that are sent during each second.

    Frames 1, 2, 3, 4: 4000 bytes each

    Frames 5, 6, 7: 3200 bytes each

    Frames 8, 9: 400 bytes each

    Frames 10, 11, 12: 2000 bytes each
21. A user is connected to a Frame Relay network through a T-1 line. The granted CIR is 1 Mbps with a B_c of 5 million bits/5 s and B_e of 1 million bits/5 s.
    a. What is the access rate?
    b. Can the user send data at 1.6 Mbps?
    c. Can the user send data at 1 Mbps all the time? Is it guaranteed that frames are never discarded in this case?
    d. Can the user send data at 1.2 Mbps all the time? Is it guaranteed that frames are never discarded in this case? If the answer is no, is it guaranteed that frames are discarded only if there is congestion?
    e. Repeat the question in part (d) for a constant rate of 1.4 Mbps.
    f. What is the maximum data rate the user can use all the time without worrying about the frames being discarded?
    g. If the user wants to take a risk, what is the maximum data rate that can be used with no chance of discarding if there is no congestion?
22. In Exercise 21 the user sends data at 1.4 Mbps for 2 s and nothing for the next 3 s. Is there a danger of discarded data if there is no congestion? Is there a danger of discarded data if there is congestion?
23. In ATM, if each cell takes 10 μs to reach the destination, what is the CTD?
24. An ATM network has lost 5 cells out of 10,000 and 2 are in error. What is the CLR? What is the CER?

PART 6

Application Layer

Objectives

The application layer enables the user, whether human or software, to access the network. It provides user interfaces and support for services such as electronic mail, file access and transfer, access to system resources, surfing the world wide web, and network management.

The application layer is responsible for providing services to the user.

In this part, we briefly discuss some applications that are designed as a client/server pair in the Internet. The client sends a request for a service to the server; the server responds to the client.

Part 6 of the book is devoted to the application layer and the services provided by this layer.

Chapters

This part consists of five chapters: Chapters 25 to 29.

Chapter 25

Chapter 25 discusses the Domain Name System (DNS). DNS is a client/server application that provides name services for other applications. It enables the use of application-layer addresses, such as an email address, instead of network layer logical addresses.

Chapter 26

Chapter 26 discusses three common applications in the Internet: remote login, electronic mail, and file transfer. A remote login application allows the user to remotely access the resources of a system. An electronic mail application simulates the tasks of snail mail at a much higher speed. A file transfer application transfers files between remote systems.

Chapter 27

Chapter 27 discuss the ideas and issues in the famous world wide web (WWW). It also briefly describes the client/server application program (HTTP) that is commonly used to access the web.

Chapter 28

Chapter 28 is devoted to network management. We first discuss the general idea behind network management. We then introduce the client/server application, SNMP, that is used for this purpose in the Internet. Although network management can be implemented in every layer, the Internet has decided to use a client/server application.

Chapter 29

Chapter 29 discusses multimedia and a set of wiely-used application programs. These programs have generated new issues such as the need for new protocols in other layers to handle the specific problems related to multimedia. We briefly discuss these issues in this chapter.

CHAPTER 25

Domain Name System

There are several applications in the application layer of the Internet model that follow the client/server paradigm. The client/server programs can be divided into two categories: those that can be directly used by the user, such as e-mail, and those that support other application programs. The Domain Name System (DNS) is a supporting program that is used by other programs such as e-mail.

Figure 25.1 shows an example of how a DNS client/server program can support an e-mail program to find the IP address of an e-mail recipient. A user of an e-mail program may know the e-mail address of the recipient; however, the IP protocol needs the IP address. The DNS client program sends a request to a DNS server to map the e-mail address to the corresponding IP address.

Figure 25.1 *Example of using the DNS service*

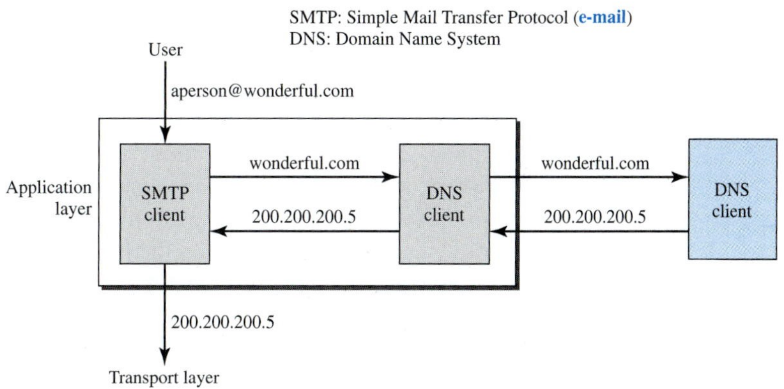

To identify an entity, TCP/IP protocols use the IP address, which uniquely identifies the connection of a host to the Internet. However, people prefer to use names instead of numeric addresses. Therefore, we need a system that can map a name to an address or an address to a name.

When the Internet was small, mapping was done by using a **host file.** The host file had only two columns: name and address. Every host could store the host file on its disk and update it periodically from a master host file. When a program or a user wanted to map a name to an address, the host consulted the host file and found the mapping.

Today, however, it is impossible to have one single host file to relate every address with a name and vice versa. The host file would be too large to store in every host. In addition, it would be impossible to update all the host files every time there was a change.

One solution would be to store the entire host file in a single computer and allow access to this centralized information to every computer that needs mapping. But we know that this would create a huge amount of traffic on the Internet.

Another solution, the one used today, is to divide this huge amount of information into smaller parts and store each part on a different computer. In this method, the host that needs mapping can contact the closest computer holding the needed information. This method is used by the **Domain Name System (DNS).** In this chapter, we first discuss the concepts and ideas behind the DNS. We then describe the DNS protocol itself.

25.1 NAME SPACE

To be unambiguous, the names assigned to machines must be carefully selected from a name space with complete control over the binding between the names and IP addresses. In other words, the names must be unique because the addresses are unique. A **name space** that maps each address to a unique name can be organized in two ways: flat or hierarchical.

Flat Name Space

In a **flat name space,** a name is assigned to an address. A name in this space is a sequence of characters without structure. The names may or may not have a common section; if they do, it has no meaning. The main disadvantage of a flat name space is that it cannot be used in a large system such as the Internet because it must be centrally controlled to avoid ambiguity and duplication.

Hierarchical Name Space

In a **hierarchical name space,** each name is made of several parts. The first part can define the nature of the organization, the second part can define the name of an organization, the third part can define departments in the organization, and so on. In this case, the authority to assign and control the name spaces can be decentralized. A central authority can assign the part of the name that defines the nature of the organization and the name of the organization. The responsibility of the rest of the name can be given to the organization itself. The organization can add suffixes (or prefixes) to the name to define its host or resources. The management of the organization need not worry that the prefix chosen for a host is taken by another organization because, even if part of an address is the same, the whole address is different. For example, assume two colleges and a company call one of their computers *challenger.* The first college is given a name by the central authority such as *fhda.edu,* the second college is given the name *berkeley.edu,* and the company is given the name *smart.com.* When these organizations add the name *challenger* to the name they have already been given, the end result is three distinguishable names: *challenger.fhda.edu, challenger.berkeley.edu,* and *challenger.smart.com.* The names are unique without the need for assignment by a central authority. The central authority controls only part of the name, not the whole.

25.2 DOMAIN NAME SPACE

To have a hierarchical name space, a **domain name space** was designed. In this design the names are defined in an inverted-tree structure with the root at the top. The tree can have only 128 levels: level 0 (root) to level 127 (see Figure 25.2).

Figure 25.2 *Domain name space*

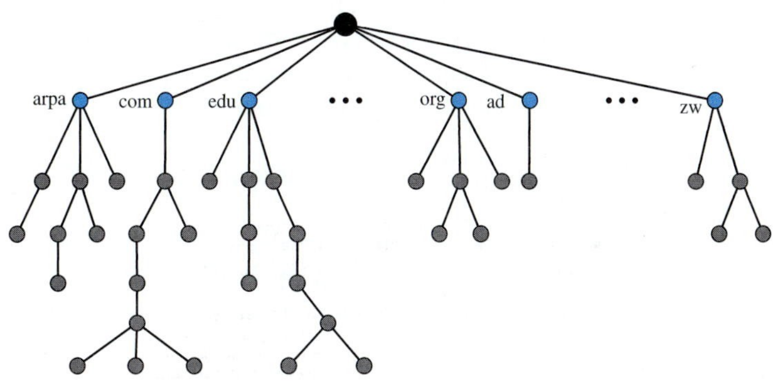

Label

Each node in the tree has a **label,** which is a string with a maximum of 63 characters. The root label is a null string (empty string). DNS requires that children of a node (nodes that branch from the same node) have different labels, which guarantees the uniqueness of the domain names.

Domain Name

Each node in the tree has a domain name. A full **domain name** is a sequence of labels separated by dots (.). The domain names are always read from the node up to the root. The last label is the label of the root (null). This means that a full domain name always ends in a null label, which means the last character is a dot because the null string is nothing. Figure 25.3 shows some domain names.

Figure 25.3 *Domain names and labels*

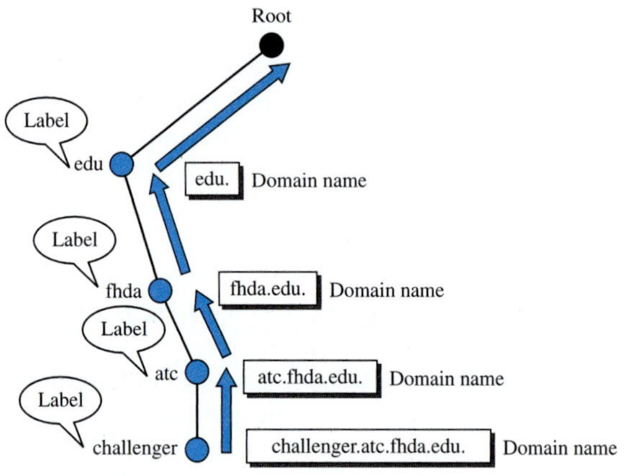

Fully Qualified Domain Name

If a label is terminated by a null string, it is called a **fully qualified domain name (FQDN).** An FQDN is a domain name that contains the full name of a host. It contains all labels, from the most specific to the most general, that uniquely define the name of the host. For example, the domain name

challenger.atc.fhda.edu.

is the FQDN of a computer named *challenger* installed at the Advanced Technology Center (ATC) at De Anza College. A DNS server can only match an FQDN to an address. Note that the name must end with a null label, but because null means nothing, the label ends with a dot (.).

Partially Qualified Domain Name

If a label is not terminated by a null string, it is called a **partially qualified domain name (PQDN).** A PQDN starts from a node, but it does not reach the root. It is used when the name to be resolved belongs to the same site as the client. Here the resolver can supply the missing part, called the **suffix,** to create an FQDN. For example, if a user at the *fhda.edu.* site wants to get the IP address of the challenger computer, he or she can define the partial name

challenger

The DNS client adds the suffix *atc.fhda.edu.* before passing the address to the DNS server.

The DNS client normally holds a list of suffixes. The following can be the list of suffixes at De Anza College. The null suffix defines nothing. This suffix is added when the user defines an FQDN.

atc.fhda.edu
fhda.edu
null

Figure 25.4 shows some FQDNs and PQDNs.

Figure 25.4 *FQDN and PQDN*

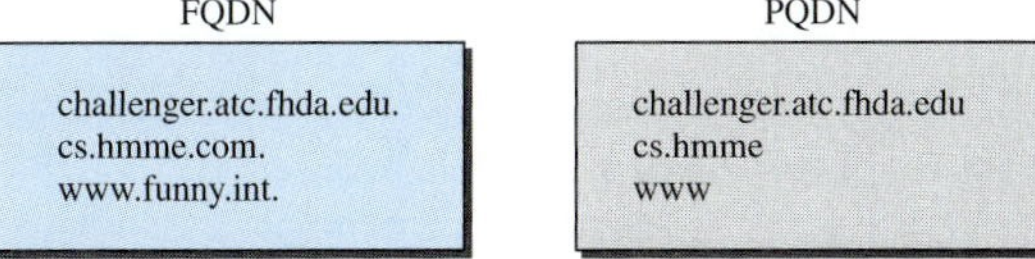

Domain

A **domain** is a subtree of the domain name space. The name of the domain is the domain name of the node at the top of the subtree. Figure 25.5 shows some domains. Note that a domain may itself be divided into domains (or **subdomains** as they are sometimes called).

Figure 25.5 *Domains*

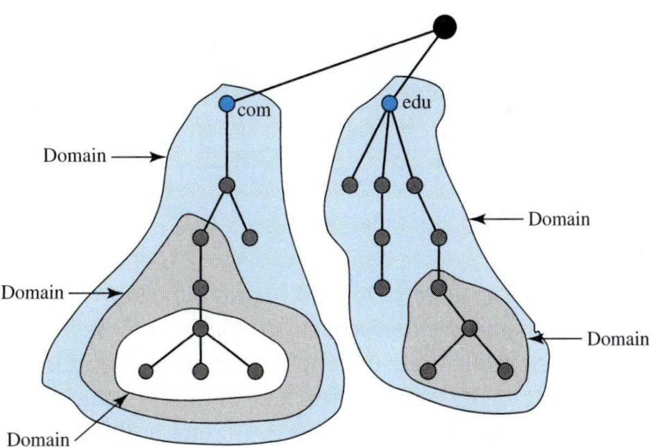

25.3 DISTRIBUTION OF NAME SPACE

The information contained in the domain name space must be stored. However, it is very inefficient and also unreliable to have just one computer store such a huge amount of information. It is inefficient because responding to requests from all over the world places a heavy load on the system. It is not unreliable because any failure makes the data inaccessible.

Hierarchy of Name Servers

The solution to these problems is to distribute the information among many computers called **DNS servers.** One way to do this is to divide the whole space into many domains based on the first level. In other words, we let the root stand alone and create as many domains (subtrees) as there are first-level nodes. Because a domain created in this way could be very large, DNS allows domains to be divided further into smaller domains (subdomains). Each server can be responsible (authoritative) for either a large or a small domain. In other words, we have a hierarchy of servers in the same way that we have a hierarchy of names (see Figure 25.6).

Figure 25.6 *Hierarchy of name servers*

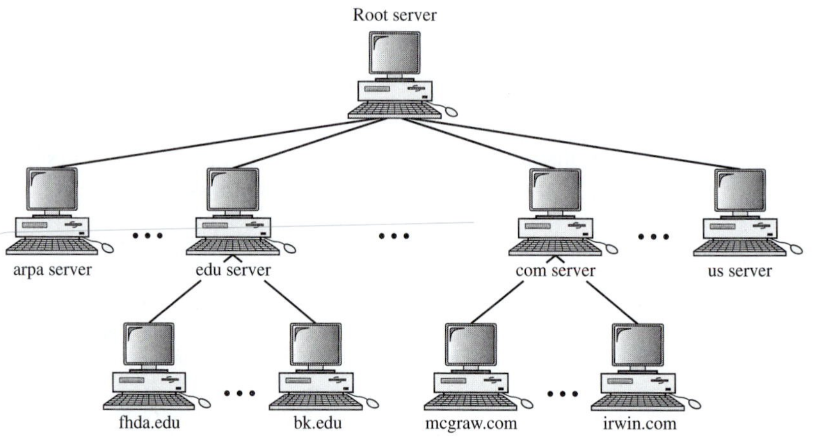

Zone

Since the complete domain name hierarchy cannot be stored on a single server, it is divided among many servers. What a server is responsible for or has authority over is called a **zone.** We can define a zone as a contiguous part of the entire tree. If a server accepts responsibility for a domain and does not divide the domain into smaller domains, the *domain* and the *zone* refer to the same thing. The server makes a database called a *zone file* and keeps all the information for every node under that domain. However, if a server divides its domain into subdomains and delegates part of its authority to other servers, *domain* and *zone* refer to different things. The information about the nodes in the subdomains is stored in the servers at the lower levels, with the original server keeping some sort of reference to these lower-level servers. Of course the original server does not free itself from responsibility totally: It still has a zone, but the detailed information is kept by the lower-level servers (see Figure 25.7).

A server can also divide part of its domain and delegate responsibility but still keep part of the domain for itself. In this case, its zone is made of detailed information for the part of the domain that is not delegated and references to those parts that are delegated.

Figure 25.7 *Zones and domains*

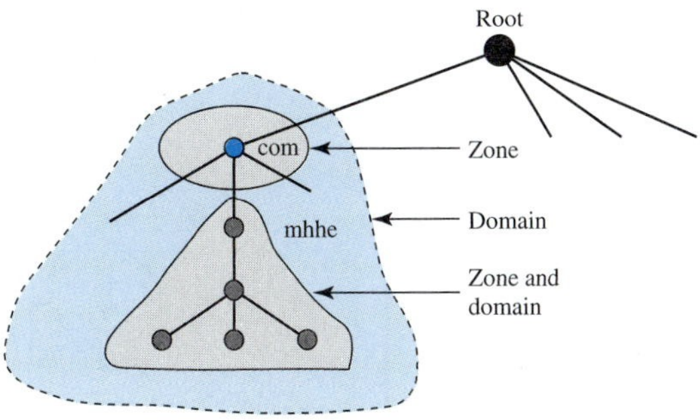

Root Server

A **root server** is a server whose zone consists of the whole tree. A root server usually does not store any information about domains but delegates its authority to other servers, keeping references to those servers. There are several root servers, each covering the whole domain name space. The servers are distributed all around the world.

Primary and Secondary Servers

DNS defines two types of servers: primary and secondary. A **primary server** is a server that stores a file about the zone for which it is an authority. It is responsible for creating, maintaining, and updating the zone file. It stores the zone file on a local disk.

A **secondary server** is a server that transfers the complete information about a zone from another server (primary or secondary) and stores the file on its local disk. The secondary server neither creates nor updates the zone files. If updating is required, it must be done by the primary server, which sends the updated version to the secondary.

The primary and secondary servers are both authoritative for the zones they serve. The idea is not to put the secondary server at a lower level of authority but to create redundancy for the data so that if one server fails, the other can continue serving clients. Note also that a server can be a primary server for a specific zone and a secondary server for another zone. Therefore, when we refer to a server as a primary or secondary server, we should be careful to which zone we refer.

A primary server loads all information from the disk file; the secondary server loads all information from the primary server. When the secondary downloads information from the primary, it is called zone transfer.

25.4 DNS IN THE INTERNET

DNS is a protocol that can be used in different platforms. In the Internet, the domain name space (tree) is divided into three different sections: generic domains, country domains, and the inverse domain (see Figure 25.8).

Figure 25.8 *DNS used in the Internet*

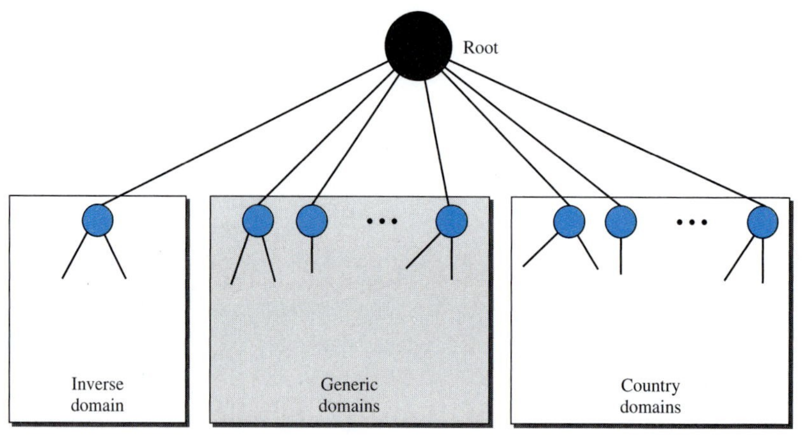

Generic Domains

The **generic domains** define registered hosts according to their generic behavior. Each node in the tree defines a domain, which is an index to the domain name space database (see Figure 25.9).

Figure 25.9 *Generic domains*

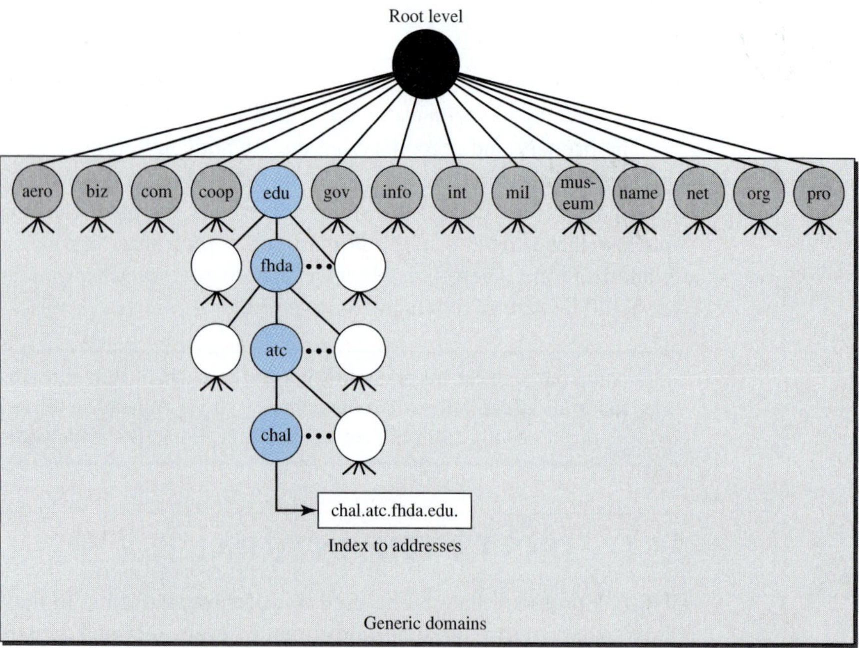

Looking at the tree, we see that the first level in the generic domains section allows 14 possible labels. These labels describe the organization types as listed in Table 25.1.

Table 25.1 *Generic domain labels*

Label	*Description*
aero	Airlines and aerospace companies
biz	Businesses or firms (similar to "com")
com	Commercial organizations
coop	Cooperative business organizations
edu	Educational institutions
gov	Government institutions
info	Information service providers
int	International organizations
mil	Military groups
museum	Museums and other nonprofit organizations
name	Personal names (individuals)
net	Network support centers
org	Nonprofit organizations
pro	Professional individual organizations

Country Domains

The **country domains** section uses two-character country abbreviations (e.g., us for United States). Second labels can be organizational, or they can be more specific, national designations. The United States, for example, uses state abbreviations as a subdivision of us (e.g., ca.us.).

Figure 25.10 shows the country domains section. The address *anza.cup.ca.us* can be translated to De Anza College in Cupertino, California, in the United States.

Inverse Domain

The **inverse domain** is used to map an address to a name. This may happen, for example, when a server has received a request from a client to do a task. Although the server has a file that contains a list of authorized clients, only the IP address of the client (extracted from the received IP packet) is listed. The server asks its resolver to send a query to the DNS server to map an address to a name to determine if the client is on the authorized list.

This type of query is called an inverse or pointer (PTR) query. To handle a pointer query, the inverse domain is added to the domain name space with the first-level node called *arpa* (for historical reasons). The second level is also one single node named *in-addr* (for inverse address). The rest of the domain defines IP addresses.

The servers that handle the inverse domain are also hierarchical. This means the netid part of the address should be at a higher level than the subnetid part, and the subnetid part

Figure 25.10 *Country domains*

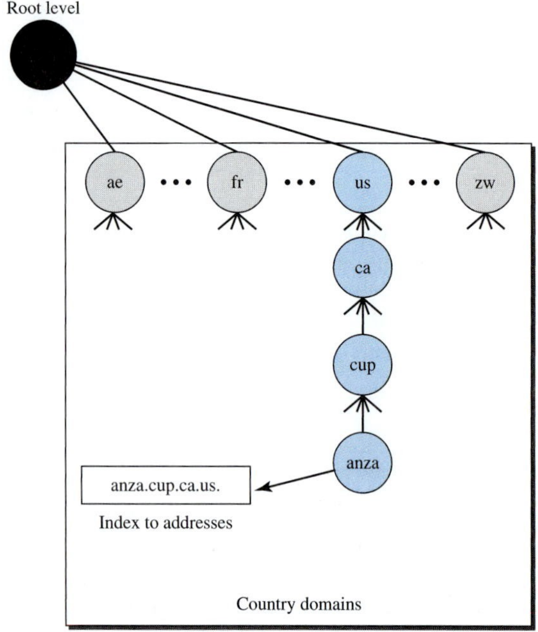

higher than the hostid part. In this way, a server serving the whole site is at a higher level than the servers serving each subnet. This configuration makes the domain look inverted when compared to a generic or country domain. To follow the convention of reading the domain labels from the bottom to the top, an IP address such as 132.34.45.121 (a class B address with netid 132.34) is read as 121.45.34.132.in-addr. arpa. See Figure 25.11 for an illustration of the inverse domain configuration.

25.5 RESOLUTION

Mapping a name to an address or an address to a name is called *name-address resolution.*

Resolver

DNS is designed as a client/server application. A host that needs to map an address to a name or a name to an address calls a DNS client called a **resolver.** The resolver accesses the closest DNS server with a mapping request. If the server has the information, it satisfies the resolver; otherwise, it either refers the resolver to other servers or asks other servers to provide the information.

After the resolver receives the mapping, it interprets the response to see if it is a real resolution or an error, and finally delivers the result to the process that requested it.

Figure 25.11 *Inverse domain*

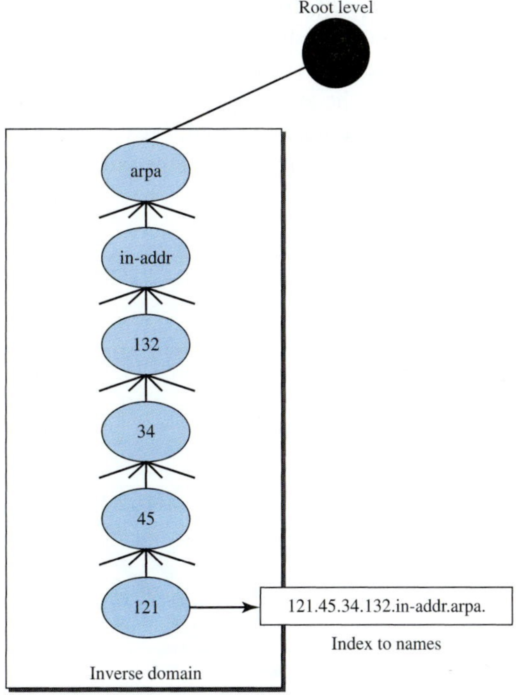

Mapping Names to Addresses

Most of the time, the resolver gives a domain name to the server and asks for the corresponding address. In this case, the server checks the generic domains or the country domains to find the mapping.

If the domain name is from the generic domains section, the resolver receives a domain name such as "*chal.atc.fhda.edu.*". The query is sent by the resolver to the local DNS server for resolution. If the local server cannot resolve the query, it either refers the resolver to other servers or asks other servers directly.

If the domain name is from the country domains section, the resolver receives a domain name such as "*ch.fhda.cu.ca.us.*". The procedure is the same.

Mapping Addresses to Names

A client can send an IP address to a server to be mapped to a domain name. As mentioned before, this is called a PTR query. To answer queries of this kind, DNS uses the inverse domain. However, in the request, the IP address is reversed and the two labels *in-addr* and *arpa* are appended to create a domain acceptable by the inverse domain section. For example, if the resolver receives the IP address 132.34.45.121, the resolver first inverts the address and then adds the two labels before sending. The domain name sent is "*121.45.34.132.in-addr.arpa.*" which is received by the local DNS and resolved.

Recursive Resolution

The client (resolver) can ask for a recursive answer from a name server. This means that the resolver expects the server to supply the final answer. If the server is the authority for the domain name, it checks its database and responds. If the server is not the authority, it sends the request to another server (the parent usually) and waits for the response. If the parent is the authority, it responds; otherwise, it sends the query to yet another server. When the query is finally resolved, the response travels back until it finally reaches the requesting client. This is called **recursive resolution** and is shown in Figure 25.12.

Figure 25.12 *Recursive resolution*

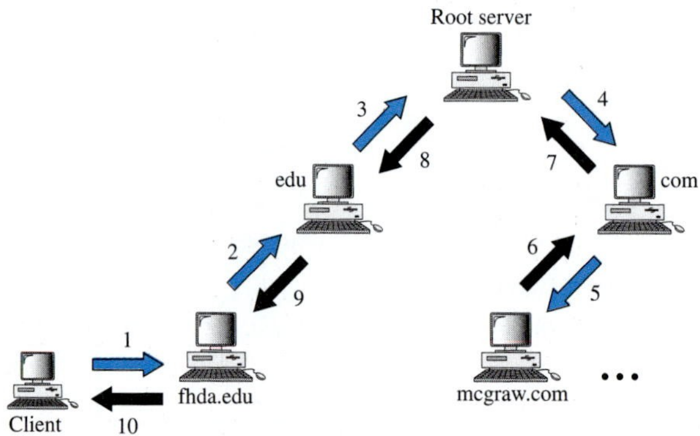

Iterative Resolution

If the client does not ask for a recursive answer, the mapping can be done iteratively. If the server is an authority for the name, it sends the answer. If it is not, it returns (to the client) the IP address of the server that it thinks can resolve the query. The client is responsible for repeating the query to this second server. If the newly addressed server can resolve the problem, it answers the query with the IP address; otherwise, it returns the IP address of a new server to the client. Now the client must repeat the query to the third server. This process is called **iterative resolution** because the client repeats the same query to multiple servers. In Figure 25.13 the client queries four servers before it gets an answer from the mcgraw.com server.

Caching

Each time a server receives a query for a name that is not in its domain, it needs to search its database for a server IP address. Reduction of this search time would increase efficiency. DNS handles this with a mechanism called **caching.** When a server asks for a mapping from another server and receives the response, it stores this information in its cache memory before sending it to the client. If the same or another client asks for the same mapping, it can check its cache memory and solve the problem. However, to

Figure 25.13 *Iterative resolution*

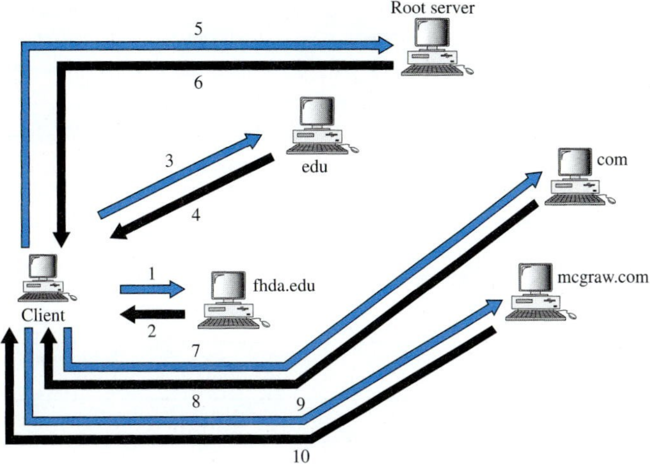

inform the client that the response is coming from the cache memory and not from an authoritative source, the server marks the response as *unauthoritative.*

Caching speeds up resolution, but it can also be problematic. If a server caches a mapping for a long time, it may send an outdated mapping to the client. To counter this, two techniques are used. First, the authoritative server always adds information to the mapping called *time-to-live* (TTL). It defines the time in seconds that the receiving server can cache the information. After that time, the mapping is invalid and any query must be sent again to the authoritative server. Second, DNS requires that each server keep a TTL counter for each mapping it caches. The cache memory must be searched periodically, and those mappings with an expired TTL must be purged.

25.6 DNS MESSAGES

DNS has two types of messages: query and response. Both types have the same format. The **query message** consists of a header and question records; the **response message** consists of a header, question records, answer records, authoritative records, and additional records (see Figure 25.14).

Header

Both query and response messages have the same header format with some fields set to zero for the query messages. The header is 12 bytes, and its format is shown in Figure 25.15.

The *identification* subfield is used by the client to match the response with the query. The client uses a different identification number each time it sends a query. The server duplicates this number in the corresponding response. The *flags* subfield is a collection of

Figure 25.14 *Query and response messages*

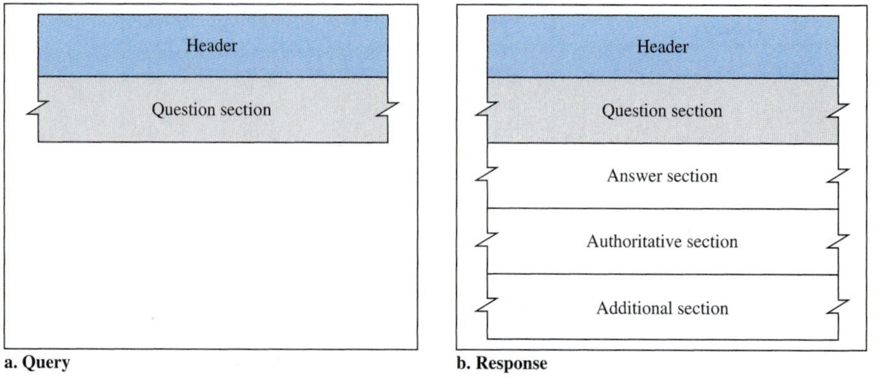

Figure 25.15 *Header format*

Identification	Flags
Number of question records	Number of answer records (all 0s in query message)
Number of authoritative records (all 0s in query message)	Number of additional records (all 0s in query message)

subfields that define the type of the message, the type of answer requested, the type of desired resolution (recursive or iterative), and so on. The *number of question records* subfield contains the number of queries in the question section of the message. The *number of answer records* subfield contains the number of answer records in the answer section of the response message. Its value is zero in the query message. The *number of authoritative records* subfield contains the number of authoritative records in the authoritative section of a response message. Its value is zero in the query message. Finally, the *number of additional records* subfield contains the number additional records in the additional section of a response message. Its value is zero in the query message.

Question Section

This is a section consisting of one or more question records. It is present on both query and response messages. We will discuss the question records in a following section.

Answer Section

This is a section consisting of one or more resource records. It is present only on response messages. This section includes the answer from the server to the client (resolver). We will discuss resource records in a following section.

Authoritative Section

This is a section consisting of one or more resource records. It is present only on response messages. This section gives information (domain name) about one or more authoritative servers for the query.

Additional Information Section

This is a section consisting of one or more resource records. It is present only on response messages. This section provides additional information that may help the resolver. For example, a server may give the domain name of an authoritative server to the resolver in the authoritative section, and include the IP address of the same authoritative server in the additional information section.

25.7 TYPES OF RECORDS

As we saw in Section 25.6, two types of records are used in DNS. The question records are used in the question section of the query and response messages. The resource records are used in the answer, authoritative, and additional information sections of the response message.

Question Record

A **question record** is used by the client to get information from a server. This contains the domain name.

Resource Record

Each domain name (each node on the tree) is associated with a record called the **resource record.** The server database consists of resource records. Resource records are also what is returned by the server to the client.

25.8 REGISTRARS

How are new domains added to DNS? This is done through a **registrar,** a commercial entity accredited by ICANN. A registrar first verifies that the requested domain name is unique and then enters it into the DNS database. A fee is charged.

Today, there are many registrars; their names and addresses can be found at

http://www.intenic.net

To register, the organization needs to give the name of its server and the IP address of the server. For example, a new commercial organization named *wonderful* with a server named *ws* and IP address 200.200.200.5 needs to give the following information to one of the registrars:

Domain name: ws.wonderful.com
IP address: 200.200.200.5

25.9 DYNAMIC DOMAIN NAME SYSTEM (DDNS)

When the DNS was designed, no one predicted that there would be so many address changes. In DNS, when there is a change, such as adding a new host, removing a host, or changing an IP address, the change must be made to the DNS master file. These types of changes involve a lot of manual updating. The size of today's Internet does not allow for this kind of manual operation.

The DNS master file must be updated dynamically. The **Dynamic Domain Name System (DDNS)** therefore was devised to respond to this need. In DDNS, when a binding between a name and an address is determined, the information is sent, usually by DHCP (see Chapter 21) to a primary DNS server. The primary server updates the zone. The secondary servers are notified either actively or passively. In active notification, the primary server sends a message to the secondary servers about the change in the zone, whereas in passive notification, the secondary servers periodically check for any changes. In either case, after being notified about the change, the secondary requests information about the entire zone (zone transfer).

To provide security and prevent unauthorized changes in the DNS records, DDNS can use an authentication mechanism.

25.10 ENCAPSULATION

DNS can use either UDP or TCP. In both cases the well-known port used by the server is port 53. UDP is used when the size of the response message is less than 512 bytes because most UDP packages have a 512-byte packet size limit. If the size of the response message is more than 512 bytes, a TCP connection is used. In that case, one of two scenarios can occur:

- If the resolver has prior knowledge that the size of the response message is more than 512 bytes, it uses the TCP connection. For example, if a secondary name server (acting as a client) needs a zone transfer from a primary server, it uses the TCP connection because the size of the information being transferred usually exceeds 512 bytes.
- If the resolver does not know the size of the response message, it can use the UDP port. However, if the size of the response message is more than 512 bytes, the server truncates the message and turns on the TC bit. The resolver now opens a TCP connection and repeats the request to get a full response from the server.

DNS can use the services of UDP or TCP using the well-known port 53.

25.11 RECOMMENDED READING

For more details about subjects discussed in this chapter, we recommend the following books and sites. The items in brackets [. . .] refer to the reference list at the end of the text.

Books

DNS is discussed in [AL98], Chapter 17 of [For06], Section 9.1 of [PD03], and Section 7.1 of [Tan03].

Sites

The following sites are related to topics discussed in this chapter.

- www.intenic.net/ Information about registrars
- www.ietf.org/rfc.html Information about RFCs

RFCs

The following RFCs are related to DNS:

799, 811, 819, 830, 881, 882, 883, 897, 920, 921, 1034, 1035, 1386, 1480, 1535, 1536, 1537, 1591, 1637, 1664, 1706, 1712, 1713, 1982, 2065, 2137, 2317, 2535, 2671

25.12 KEY TERMS

caching
country domain
DNS server
domain
domain name
domain name space
Domain Name System (DNS)
Dynamic Domain Name System (DDNS)
flat name space
fully qualified domain name (FQDN)
generic domain
hierarchical name space
host file
inverse domain
iterative resolution
label
name space
partially qualified domain name (PQDN)
primary server
query message
question record
recursive resolution
registrar
resolver
resource record
response message
root server
secondary server
subdomain
suffix
zone

25.13 SUMMARY

- The Domain Name System (DNS) is a client/server application that identifies each host on the Internet with a unique user-friendly name.
- DNS organizes the name space in a hierarchical structure to decentralize the responsibilities involved in naming.

- DNS can be pictured as an inverted hierarchical tree structure with one root node at the top and a maximum of 128 levels.
- Each node in the tree has a domain name.
- A domain is defined as any subtree of the domain name space.
- The name space information is distributed among DNS servers. Each server has jurisdiction over its zone.
- A root server's zone is the entire DNS tree.
- A primary server creates, maintains, and updates information about its zone.
- A secondary server gets its information from a primary server.
- The domain name space is divided into three sections: generic domains, country domains, and inverse domain.
- There are 14 generic domains, each specifying an organization type.
- Each country domain specifies a country.
- The inverse domain finds a domain name for a given IP address. This is called address-to-name resolution.
- Name servers, computers that run the DNS server program, are organized in a hierarchy.
- The DNS client, called a resolver, maps a name to an address or an address to a name.
- In recursive resolution, the client sends its request to a server that eventually returns a response.
- In iterative resolution, the client may send its request to multiple servers before getting an answer.
- Caching is a method whereby an answer to a query is stored in memory (for a limited time) for easy access to future requests.
- A fully qualified domain name (FQDN) is a domain name consisting of labels beginning with the host and going back through each level to the root node.
- A partially qualified domain name (PQDN) is a domain name that does not include all the levels between the host and the root node.
- There are two types of DNS messages: queries and responses.
- There are two types of DNS records: question records and resource records.
- Dynamic DNS (DDNS) automatically updates the DNS master file.
- DNS uses the services of UDP for messages of less than 512 bytes; otherwise, TCP is used.

25.14 PRACTICE SET

Review Questions

1. What is an advantage of a hierarchical name space over a flat name space for a system the size of the Internet?
2. What is the difference between a primary server and a secondary server?
3. What are the three domains of the domain name space?

4. What is the purpose of the inverse domain?
5. How does recursive resolution differ from iterative resolution?
6. What is an FQDN?
7. What is a PQDN?
8. What is a zone?
9. How does caching increase the efficiency of name resolution?
10. What are the two main categories of DNS messages?
11. Why was there a need for DDNS?

Exercises

12. Determine which of the following is an FQDN and which is a PQDN.
    a. xxx
    b. xxx.yyy.
    c. xxx.yyy.net
    d. zzz.yyy.xxx.edu.
13. Determine which of the following is an FQDN and which is a PQDN.
    a. mil.
    b. edu.
    c. xxx.yyy.net
    d. zzz.yyy.xxx.edu
14. Which domain is used by your system, generic or country?
15. Why do we need a DNS system when we can directly use an IP address?
16. To find the IP address of a destination, we need the service of DNS. DNS needs the service of UDP or TCP. UDP or TCP needs the service of IP. IP needs an IP destination address. Is this a vicious cycle here?
17. If a DNS domain name is *voyager.fhda.edu,* how many labels are involved here? How many levels of hierarchy?
18. Is a PQDN necessarily shorter than the corresponding FQDN?
19. A domain name is *hello.customer.info*. Is this a generic domain or a country domain?
20. Do you think a recursive resolution is normally faster than an interactive one? Explain.
21. Can a query message have one question section but the corresponding response message have several answer sections?

CHAPTER 26

Remote Logging, Electronic Mail, and File Transfer

The main task of the Internet is to provide services for users. Among the most popular applications are remote logging, electronic mail, and file transfer. We discuss these three applications in this chapter; we discuss another popular use of the Internet, accessing the World Wide Web, in Chapter 27.

26.1 REMOTE LOGGING

In the Internet, users may want to run application programs at a remote site and create results that can be transferred to their local site. For example, students may want to connect to their university computer lab from their home to access application programs for doing homework assignments or projects. One way to satisfy that demand and others is to create a client/server application program for each desired service. Programs such as file transfer programs (FTPs), e-mail (SMTP), and so on are currently available. However, it would be impossible to write a specific client/server program for each demand.

The better solution is a general-purpose client/server program that lets a user access any application program on a remote computer; in other words, allow the user to log on to a remote computer. After logging on, a user can use the services available on the remote computer and transfer the results back to the local computer.

TELNET

In this section, we discuss such a client/server application program: TELNET. **TELNET** is an abbreviation for *TErminaL NETwork.* It is the standard TCP/IP protocol for virtual terminal service as proposed by the International Organization for Standards (ISO). TELNET enables the establishment of a connection to a remote system in such a way that the local terminal appears to be a terminal at the remote system.

TELNET is a general-purpose client/server application program.

Timesharing Environment

TELNET was designed at a time when most operating systems, such as UNIX, were operating in a **timesharing** environment. In such an environment, a large computer supports multiple users. The interaction between a user and the computer occurs through a terminal, which is usually a combination of keyboard, monitor, and mouse. Even a microcomputer can simulate a terminal with a terminal emulator.

Logging

In a timesharing environment, users are part of the system with some right to access resources. Each authorized user has an identification and probably a password. The user identification defines the user as part of the system. To access the system, the user logs into the system with a user id or log-in name. The system also includes password checking to prevent an unauthorized user from accessing the resources. Figure 26.1 shows the logging process.

Figure 26.1 *Local and remote log-in*

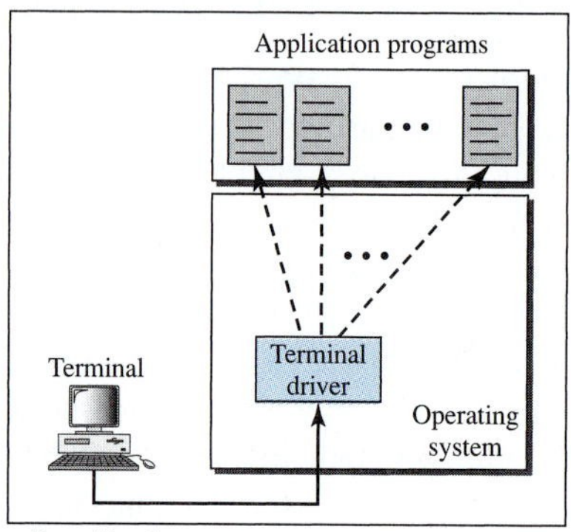

a. Local log-in

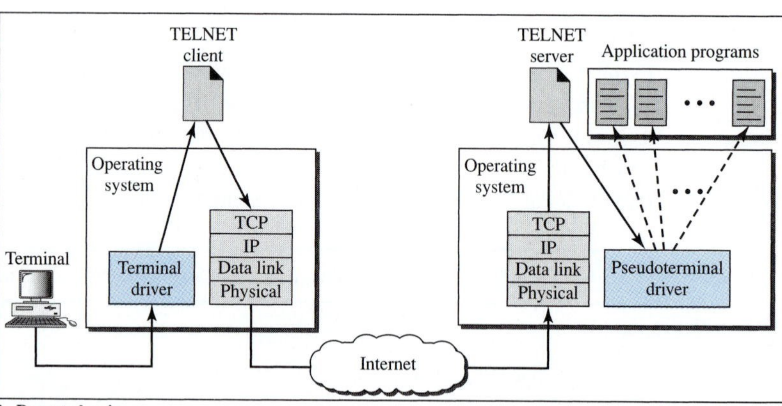

b. Remote log-in

When a user logs into a local timesharing system, it is called **local log-in.** As a user types at a terminal or at a workstation running a terminal emulator, the keystrokes are accepted by the terminal driver. The terminal driver passes the characters to the operating system. The operating system, in turn, interprets the combination of characters and invokes the desired application program or utility.

When a user wants to access an application program or utility located on a remote machine, she performs **remote log-in.** Here the TELNET client and server programs come into use. The user sends the keystrokes to the terminal driver, where the local operating system accepts the characters but does not interpret them. The characters are sent to the TELNET client, which transforms the characters to a universal character set called *network virtual terminal (NVT) characters* and delivers them to the local TCP/IP protocol stack.

The commands or text, in NVT form, travel through the Internet and arrive at the TCP/IP stack at the remote machine. Here the characters are delivered to the operating system and passed to the TELNET server, which changes the characters to the corresponding characters understandable by the remote computer. However, the characters cannot be passed directly to the operating system because the remote operating system is not designed to receive characters from a TELNET server: It is designed to receive characters from a terminal driver. The solution is to add a piece of software called a *pseudoterminal driver* which pretends that the characters are coming from a terminal. The operating system then passes the characters to the appropriate application program.

Network Virtual Terminal

The mechanism to access a remote computer is complex. This is so because every computer and its operating system accept a special combination of characters as tokens. For example, the end-of-file token in a computer running the DOS operating system is Ctrl+z, while the UNIX operating system recognizes Ctrl+d.

We are dealing with heterogeneous systems. If we want to access any remote computer in the world, we must first know what type of computer we will be connected to, and we must also install the specific terminal emulator used by that computer. TELNET solves this problem by defining a universal interface called the **network virtual terminal (NVT)** character set. Via this interface, the client TELNET translates characters (data or commands) that come from the local terminal into NVT form and delivers them to the network. The server TELNET, on the other hand, translates data and commands from NVT form into the form acceptable by the remote computer. For an illustration of this concept, see Figure 26.2.

NVT Character Set NVT uses two sets of characters, one for data and the other for control. Both are 8-bit bytes. For data, NVT is an 8-bit character set in which the 7 lowest-order bits are the same as ASCII and the highest-order bit is 0. To send **control characters** between computers (from client to server or vice versa), NVT uses an 8-bit character set in which the highest-order bit is set to 1.

Table 26.1 lists some of the control characters and their meanings.

Figure 26.2 *Concept of NVT*

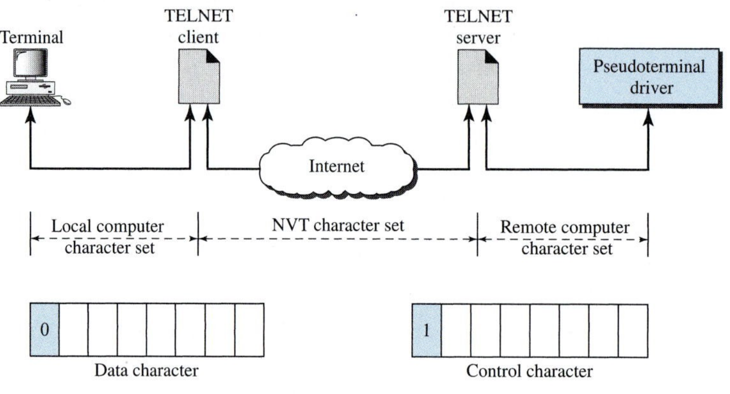

Table 26.1 *Some NVT control characters*

Character	*Decimal*	*Binary*	*Meaning*
EOF	236	11101100	End of file
EOR	239	11101111	End of record
SE	240	11110000	Suboption end
NOP	241	11110001	No operation
DM	242	11110010	Data mark
BRK	243	11110011	Break
IP	244	11110100	Interrupt process
AO	245	11110101	Abort output
AYT	246	11110110	Are you there?
EC	247	11110111	Erase character
EL	248	11111000	Erase line
GA	249	11111001	Go ahead
SB	250	11111010	Suboption begin
WILL	251	11111011	Agreement to enable option
WONT	252	11111100	Refusal to enable option
DO	253	11111101	Approval to option request
DONT	254	11111110	Denial of option request
IAC	255	11111111	Interpret (the next character) as control

Embedding

TELNET uses only one TCP connection. The server uses the well-known port 23, and the client uses an ephemeral port. The same connection is used for sending both data and control characters. TELNET accomplishes this by embedding the control characters in the data stream. However, to distinguish data from control characters, each sequence of control characters is preceded by a special control character called *interpret as control* (IAC). For example, imagine a user wants a server to display a file (*file1*) on a remote server. She can type

cat file1

However, suppose the name of the file has been mistyped (*filea* instead of *file1*). The user uses the backspace key to correct this situation.

cat filea<backspace>1

However, in the default implementation of TELNET, the user cannot edit locally; the editing is done at the remote server. The backspace character is translated into two remote characters (IAC EC), which are embedded in the data and sent to the remote server. What is sent to the server is shown in Figure 26.3.

Figure 26.3 *An example of embedding*

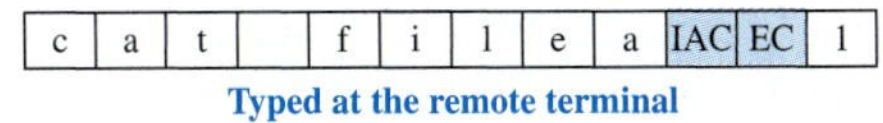

Options

TELNET lets the client and server negotiate options before or during the use of the service. Options are extra features available to a user with a more sophisticated terminal. Users with simpler terminals can use default features. Some control characters discussed previously are used to define options. Table 26.2 shows some common options.

Table 26.2 *Options*

Code	*Option*	*Meaning*
0	Binary	Interpret as 8-bit binary transmission.
1	Echo	Echo the data received on one side to the other.
3	Suppress go ahead	Suppress go-ahead signals after data.
5	Status	Request the status of TELNET.
6	Timing mark	Define the timing marks.
24	Terminal type	Set the terminal type.
32	Terminal speed	Set the terminal speed.
34	Line mode	Change to line mode.

Option Negotiation To use any of the options mentioned in the previous section first requires **option negotiation** between the client and the server. Four control characters are used for this purpose; these are shown in Table 26.3.

Table 26.3 *NVT character set for option negotiation*

Character	*Decimal*	*Binary*	*Meaning*
WILL	251	11111011	1. Offering to enable 2. Accepting a request to enable
WONT	252	11111100	1. Rejecting a request to enable 2. Offering to disable 3. Accepting a request to disable
DO	253	11111101	1. Approving an offer to enable 2. Requesting to enable
DONT	254	11111110	1. Disapproving an offer to enable 2. Approving an offer to disable 3. Requesting to disable

A party can offer to enable or disable an option if it has the right to do so. The offering can be approved or disapproved by the other party. To offer enabling, the offering party sends the WILL command, which means "Will I enable the option?" The other party sends either the DO command, which means "Please do," or the DONT command, which means "Please don't." To offer disabling, the offering party sends the WONT command, which means "I won't use this option any more." The answer must be the DONT command, which means "Don't use it anymore."

A party can request from the other party the enabling or the disabling of an option. To request enabling, the requesting party sends the DO command, which means "Please do enable the option." The other party sends either the WILL command, which means "I will," or the WONT command, which means "I won't." To request disabling, the requesting party sends the DONT command, which means "Please don't use this option anymore." The answer must be the WONT command, which means "I won't use it anymore."

Example 26.1

Figure 26.4 shows an example of option negotiation. In this example, the client wants the server to echo each character sent to the server. In other words, when a character is typed at the user keyboard terminal, it goes to the server and is sent back to the screen of the user before being processed. The echo option is enabled by the server because it is the server that sends the characters back to the user terminal. Therefore, the client should *request* from the server the enabling of the option using DO. The request consists of three characters: IAC, DO, and ECHO. The server accepts the request and enables the option. It informs the client by sending the three-character approval: IAC, WILL, and ECHO.

Suboption Negotiation Some options require additional information. For example, to define the type or speed of a terminal, the negotiation includes a string or a number

Figure 26.4 *Example 26.1: Echo option*

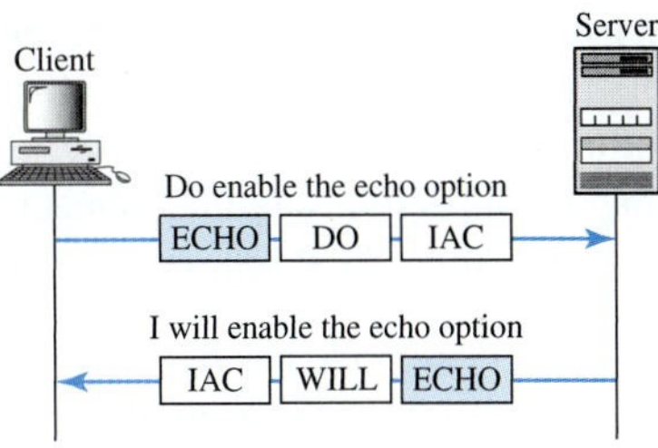

to define the type or speed. In either case, the two suboption characters indicated in Table 26.4 are needed for **suboption negotiation.**

Table 26.4 *NVT character set for suboption negotiation*

Character	*Decimal*	*Binary*	*Meaning*
SE	240	11110000	Suboption end
SB	250	11111010	Suboption begin

Example 26.2

Figure 26.5 shows an example of suboption negotiation. In this example, the client wants to negotiate the type of the terminal.

Figure 26.5 *Example of suboption negotiation*

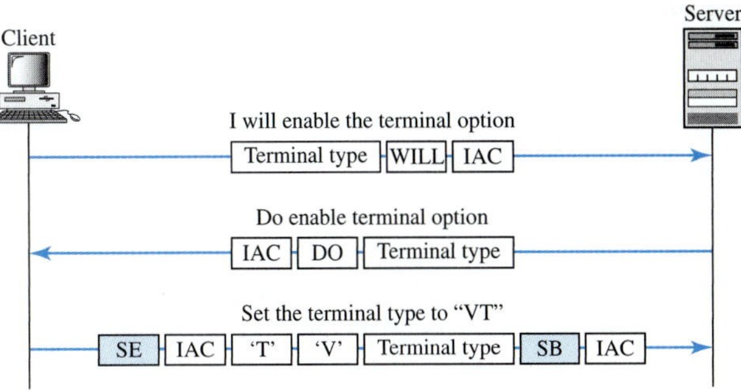

Mode of Operation

Most TELNET implementations operate in one of three modes: default mode, character mode, or line mode.

Default Mode The **default mode** is used if no other modes are invoked through option negotiation. In this mode, the echoing is done by the client. The user types a character, and the client echoes the character on the screen (or printer) but does not send it until a whole line is completed.

Character Mode In the **character mode,** each character typed is sent by the client to the server. The server normally echoes the character back to be displayed on the client screen. In this mode the echoing of the character can be delayed if the transmission time is long (such as in a satellite connection). It also creates overhead (traffic) for the network because three TCP segments must be sent for each character of data.

Line Mode A new mode has been proposed to compensate for the deficiencies of the default mode and the character mode. In this mode, called the **line mode,** line editing (echoing, character erasing, line erasing, and so on) is done by the client. The client then sends the whole line to the server.

26.2 ELECTRONIC MAIL

One of the most popular Internet services is electronic mail (e-mail). The designers of the Internet probably never imagined the popularity of this application program. Its architecture consists of several components that we discuss in this chapter.

At the beginning of the Internet era, the messages sent by electronic mail were short and consisted of text only; they let people exchange quick memos. Today, electronic mail is much more complex. It allows a message to include text, audio, and video. It also allows one message to be sent to one or more recipients.

In this chapter, we first study the general architecture of an e-mail system including the three main components: user agent, message transfer agent, and message access agent. We then describe the protocols that implement these components.

Architecture

To explain the architecture of e-mail, we give four scenarios. We begin with the simplest situation and add complexity as we proceed. The fourth scenario is the most common in the exchange of email.

First Scenario

In the first scenario, the sender and the receiver of the e-mail are users (or application programs) on the same system; they are directly connected to a shared system. The administrator has created one mailbox for each user where the received messages are stored. A *mailbox* is part of a local hard drive, a special file with permission restrictions. Only the owner of the mailbox has access to it. When Alice, a user, needs to send a message to Bob, another user, Alice runs a *user agent (UA)* program to prepare the message and store it in Bob's mailbox. The message has the sender and recipient mailbox addresses (names of files). Bob can retrieve and read the contents of his mailbox at his convenience, using a user agent. Figure 26.6 shows the concept.

This is similar to the traditional memo exchange between employees in an office. There is a mailroom where each employee has a mailbox with his or her name on it.

Figure 26.6 *First scenario in electronic mail*

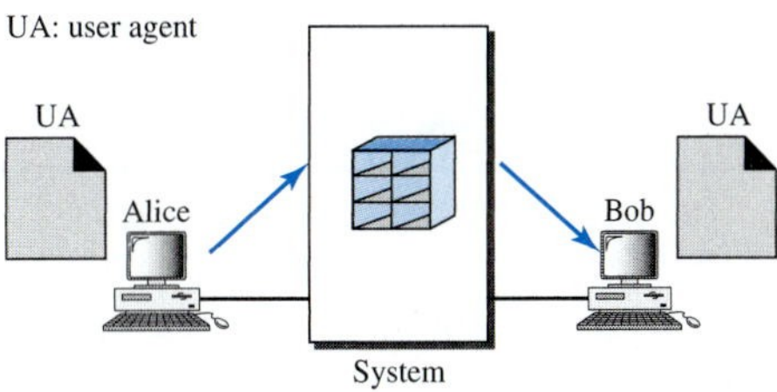

When Alice needs to send a memo to Bob, she writes the memo and inserts it into Bob's mailbox. When Bob checks his mailbox, he finds Alice's memo and reads it.

When the sender and the receiver of an e-mail are on the same system, we need only two user agents.

Second Scenario

In the second scenario, the sender and the receiver of the e-mail are users (or application programs) on two different systems.The message needs to be sent over the Internet. Here we need **user agents (UAs)** and **message transfer agents (MTAs),** as shown in Figure 26.7.

Figure 26.7 *Second scenario in electronic mail*

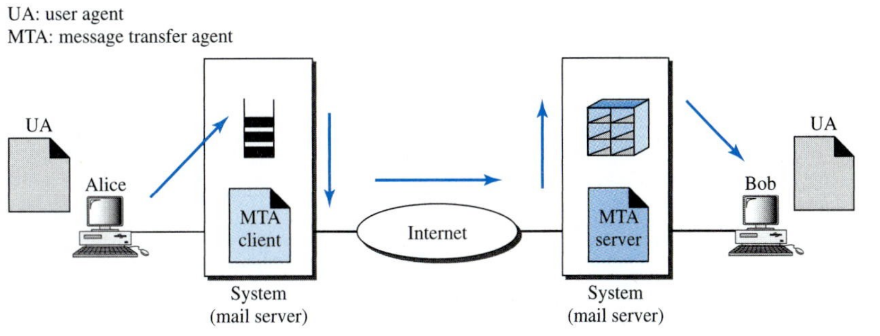

Alice needs to use a user agent program to send her message to the system at her own site. The system (sometimes called the mail server) at her site uses a queue to store messages waiting to be sent. Bob also needs a user agent program to retrieve messages stored in the mailbox of the system at his site. The message, however, needs to be sent through the Internet from Alice's site to Bob's site. Here two message transfer agents are needed: one client and one server. Like most client/server programs on the Internet, the server needs to run all the time because it does not know when a client will ask for a connection. The client, on the other hand, can be alerted by the system when there is a message in the queue to be sent.

When the sender and the receiver of an e-mail are on different systems, we need two UAs and a pair of MTAs (client and server).

Third Scenario

In the third scenario, Bob, as in the second scenario, is directly connected to his system. Alice, however, is separated from her system. Either Alice is connected to the system via a point-to-point WAN, such as a dial-up modem, a DSL, or a cable modem; or she is connected to a LAN in an organization that uses one mail server for handling e-mails—all users need to send their messages to this mail server. Figure 26.8 shows the situation.

Figure 26.8 *Third scenario in electronic mail*

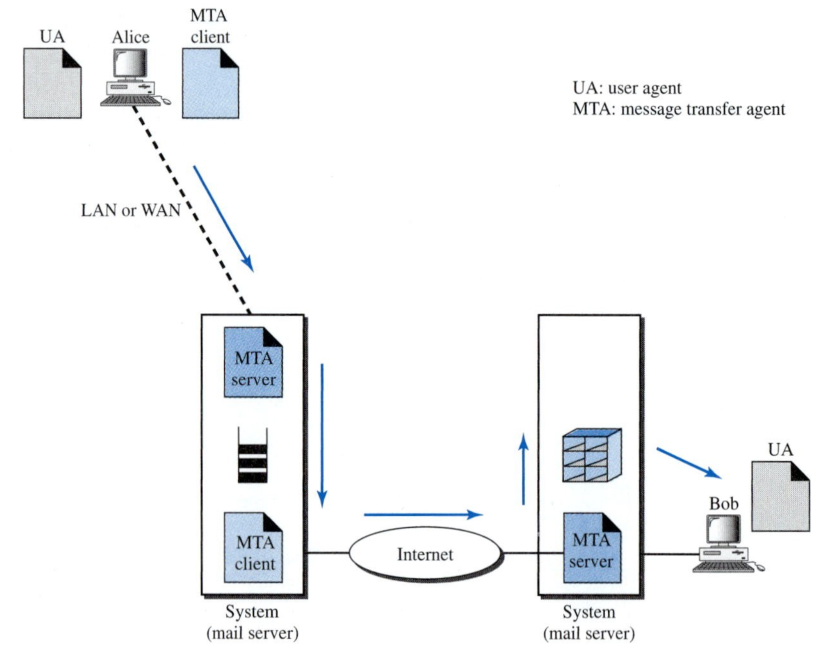

Alice still needs a user agent to prepare her message. She then needs to send the message through the LAN or WAN. This can be done through a pair of message transfer agents (client and server). Whenever Alice has a message to send, she calls the user agent which, in turn, calls the MTA client. The MTA client establishes a connection with the MTA server on the system, which is running all the time. The system at Alice's site queues all messages received. It then uses an MTA client to send the messages to the system at Bob's site; the system receives the message and stores it in Bob's mailbox.

At his convenience, Bob uses his user agent to retrieve the message and reads it. Note that we need two pairs of MTA client/server programs.

When the sender is connected to the mail server via a LAN or a WAN, we need two UAs and two pairs of MTAs (client and server).

Fourth Scenario

In the fourth and most common scenario, Bob is also connected to his mail server by a WAN or a LAN. After the message has arrived at Bob's mail server, Bob needs to retrieve it. Here, we need another set of client/server agents, which we call **message access agents (MAAs).** Bob uses an MAA client to retrieve his messages. The client sends a request to the MAA server, which is running all the time, and requests the transfer of the messages. The situation is shown in Figure 26.9.

Figure 26.9 *Fourth scenario in electronic mail*

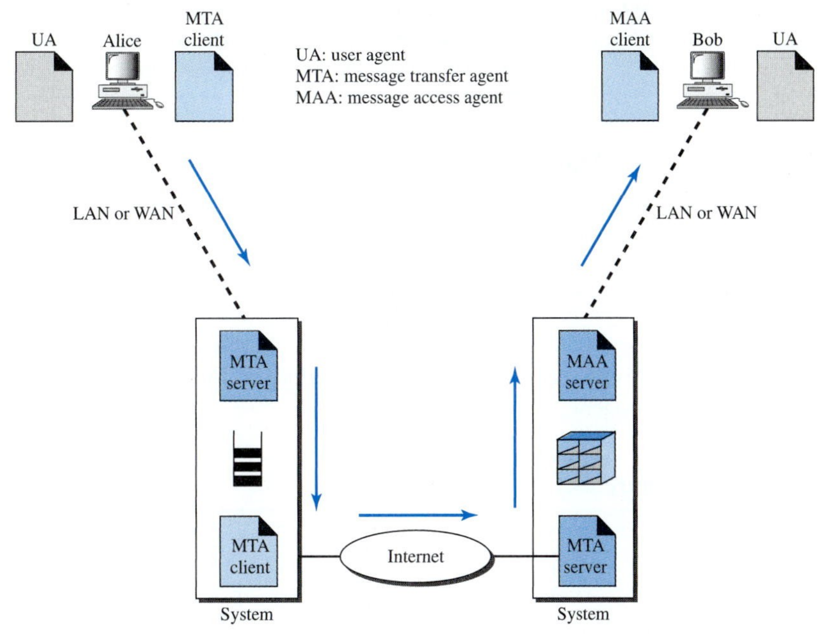

There are two important points here. First, Bob cannot bypass the mail server and use the MTA server directly. To use MTA server directly, Bob would need to run the MTA server all the time because he does not know when a message will arrive. This implies that Bob must keep his computer on all the time if he is connected to his system through a LAN. If he is connected through a WAN, he must keep the connection up all the time. Neither of these situations is feasible today.

Second, note that Bob needs another pair of client/server programs: message access programs. This is so because an MTA client/server program is a *push* program: the client pushes the message to the server. Bob needs a *pull* program. The client needs to pull the message from the server. Figure 26.10 shows the difference.

Figure 26.10 *Push versus pull in electronic email*

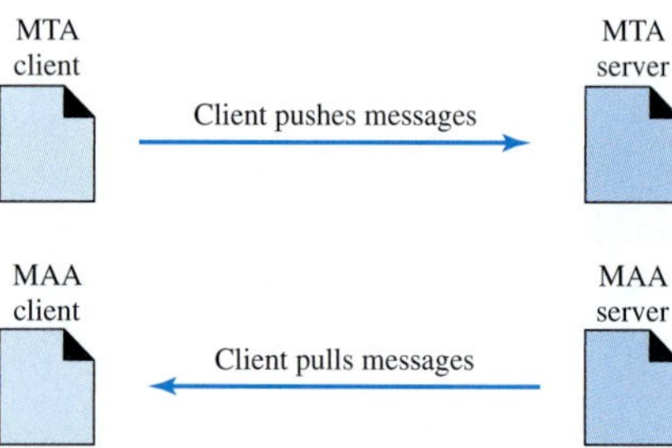

When both sender and receiver are connected to the mail server via a LAN or a WAN, we need two UAs, two pairs of MTAs (client and server), and a pair of MAAs (client and server). This is the most common situation today.

User Agent

The first component of an electronic mail system is the user agent (UA). It provides service to the user to make the process of sending and receiving a message easier.

Services Provided by a User Agent

A user agent is a software package (program) that composes, reads, replies to, and forwards messages. It also handles mailboxes. Figure 26.11 shows the services of a typical user agent.

Figure 26.11 *Services of user agent*

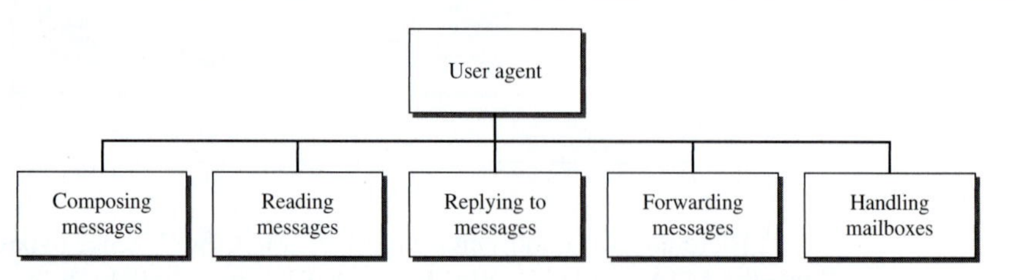

Composing Messages A user agent helps the user compose the e-mail message to be sent out. Most user agents provide a template on the screen to be filled in by the user. Some even have a built-in editor that can do spell checking, grammar checking, and other tasks expected from a sophisticated word processor. A user, of course, could alternatively use his or her favorite text editor or word processor to create the message and import it, or cut and paste it, into the user agent template.

Reading Messages The second duty of the user agent is to read the incoming messages. When a user invokes a user agent, it first checks the mail in the incoming mailbox. Most user agents show a one-line summary of each received mail. Each e-mail contains the following fields.

1. A number field.
2. A flag field that shows the status of the mail such as new, already read but not replied to, or read and replied to.
3. The size of the message.
4. The sender.
5. The optional subject field.

Replying to Messages After reading a message, a user can use the user agent to reply to a message. A user agent usually allows the user to reply to the original sender or to reply to all recipients of the message. The reply message may contain the original message (for quick reference) and the new message.

Forwarding Messages *Replying* is defined as sending a message to the sender or recipients of the copy. *Forwarding* is defined as sending the message to a third party. A user agent allows the receiver to forward the message, with or without extra comments, to a third party.

Handling Mailboxes

A user agent normally creates two mailboxes: an inbox and an outbox. Each box is a file with a special format that can be handled by the user agent. The inbox keeps all the received e-mails until they are deleted by the user. The outbox keeps all the sent e-mails until the user deletes them. Most user agents today are capable of creating customized mailboxes.

User Agent Types

There are two types of user agents: command-driven and GUI-based.

Command-Driven Command-driven user agents belong to the early days of electronic mail. They are still present as the underlying user agents in servers. A command-driven user agent normally accepts a one-character command from the keyboard to perform its task. For example, a user can type the character r, at the command prompt, to reply to the sender of the message, or type the character R to reply to the sender and all recipients. Some examples of command-driven user agents are *mail, pine,* and *elm*.

Some examples of command-driven user agents are *mail, pine,* and *elm*.

GUI-Based Modern user agents are GUI-based. They contain graphical-user interface (GUI) components that allow the user to interact with the software by using both the keyboard and the mouse. They have graphical components such as icons, menu bars, and windows that make the services easy to access. Some examples of GUI-based user agents are Eudora, Microsoft's Outlook, and Netscape.

Some examples of GUI-based user agents are *Eudora, Outlook,* and *Netscape*.

Sending Mail

To send mail, the user, through the UA, creates mail that looks very similar to postal mail. It has an *envelope* and a *message* (see Figure 26.12).

Figure 26.12 *Format of an e-mail*

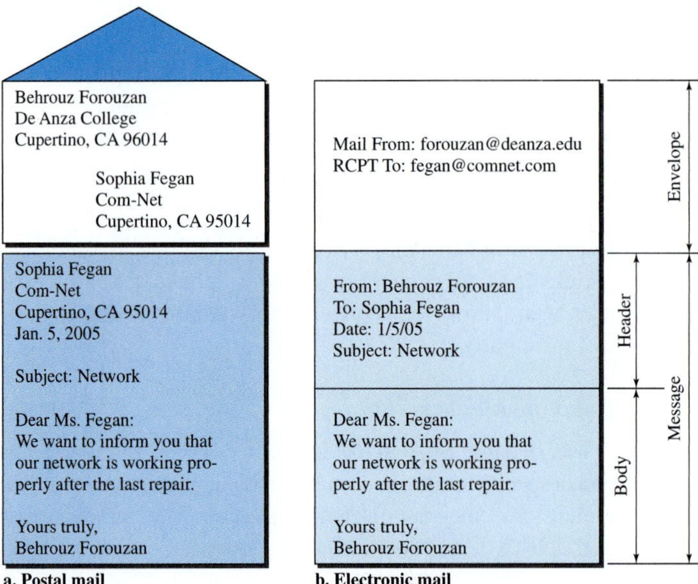

Envelope The **envelope** usually contains the sender and the receiver addresses.

Message The message contains the **header** and the **body.** The header of the message defines the sender, the receiver, the subject of the message, and some other information (such as encoding type, as we see shortly). The body of the message contains the actual information to be read by the recipient.

Receiving Mail

The user agent is triggered by the user (or a timer). If a user has mail, the UA informs the user with a notice. If the user is ready to read the mail, a list is displayed in which each line contains a summary of the information about a particular message in the mailbox. The summary usually includes the sender mail address, the subject, and the time the mail was sent or received. The user can select any of the messages and display its contents on the screen.

Addresses

To deliver mail, a mail handling system must use an addressing system with unique addresses. In the Internet, the address consists of two parts: a **local part** and a **domain name,** separated by an @ sign (see Figure 26.13).

Figure 26.13 *E-mail address*

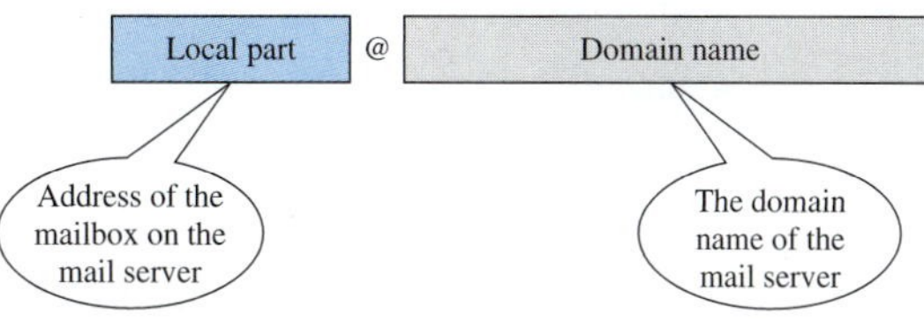

Local Part The local part defines the name of a special file, called the user mailbox, where all the mail received for a user is stored for retrieval by the message access agent.

Domain Name The second part of the address is the domain name. An organization usually selects one or more hosts to receive and send e-mail; the hosts are sometimes called *mail servers* or *exchangers*. The domain name assigned to each mail exchanger either comes from the DNS database or is a logical name (for example, the name of the organization).

Mailing List

Electronic mail allows one name, an **alias,** to represent several different e-mail addresses; this is called a mailing list. Every time a message is to be sent, the system checks the recipient's name against the alias database; if there is a mailing list for the defined alias, separate messages, one for each entry in the list, must be prepared and handed to the MTA. If there is no mailing list for the alias, the name itself is the receiving address and a single message is delivered to the mail transfer entity.

MIME

Electronic mail has a simple structure. Its simplicity, however, comes at a price. It can send messages only in NVT 7-bit ASCII format. In other words, it has some limitations. For example, it cannot be used for languages that are not supported by 7-bit ASCII characters (such as French, German, Hebrew, Russian, Chinese, and Japanese). Also, it cannot be used to send binary files or video or audio data.

Multipurpose Internet Mail Extensions (MIME) is a supplementary protocol that allows non-ASCII data to be sent through e-mail. MIME transforms non-ASCII data at the sender site to NVT ASCII data and delivers them to the client MTA to be sent through the Internet. The message at the receiving side is transformed back to the original data.

We can think of MIME as a set of software functions that transforms non-ASCII data (stream of bits) to ASCII data and vice versa, as shown in Figure 26.14.

Figure 26.14 *MIME*

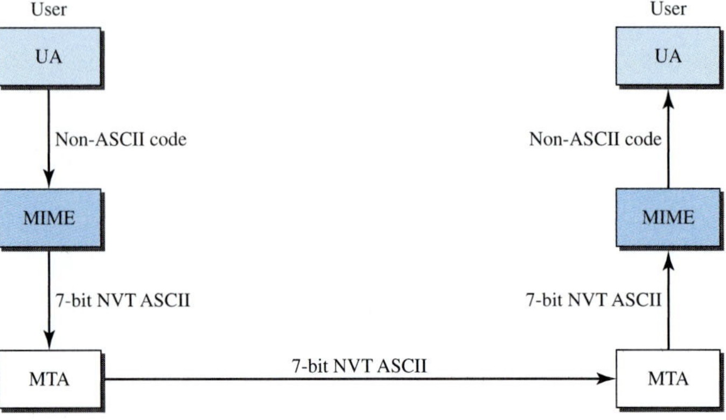

MIME defines five headers that can be added to the original e-mail header section to define the transformation parameters:

1. MIME-Version
2. Content-Type
3. Content-Transfer-Encoding
4. Content-Id
5. Content-Description

Figure 26.15 shows the MIME headers. We will describe each header in detail.

Figure 26.15 *MIME header*

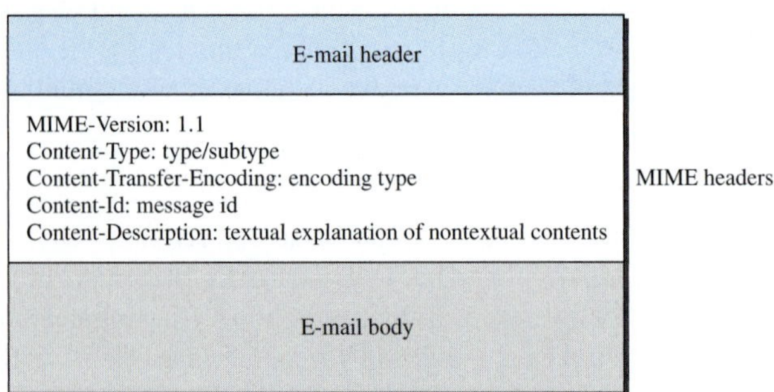

MIME-Version This header defines the version of MIME used. The current version is 1.1.

MIME-Version: 1.1

Content-Type This header defines the type of data used in the body of the message. The content type and the content subtype are separated by a slash. Depending on the subtype, the header may contain other parameters.

Content-Type: <type / subtype; parameters>

MIME allows seven different types of data. These are listed in Table 26.5.

Table 26.5 *Data types and subtypes in MIME*

Type	*Subtype*	*Description*
Text	Plain	Unformatted
	HTML	HTML format (see Chapter 27)
Multipart	Mixed	Body contains ordered parts of different data types
	Parallel	Same as above, but no order
	Digest	Similar to mixed subtypes, but the default is message/RFC822
	Alternative	Parts are different versions of the same message
Message	RFC822	Body is an encapsulated message
	Partial	Body is a fragment of a bigger message
	External-Body	Body is a reference to another message
Image	JPEG	Image is in JPEG format
	GIF	Image is in GIF format
Video	MPEG	Video is in MPEG format
Audio	Basic	Single-channel encoding of voice at 8 kHz
Application	PostScript	Adobe PostScript
	Octet-stream	General binary data (8-bit bytes)

Content-Transfer-Encoding This header defines the method used to encode the messages into 0s and 1s for transport:

Content-Transfer-Encoding: <type>

The five types of encoding methods are listed in Table 26.6.

Content-Id This header uniquely identifies the whole message in a multiple-message environment.

Content-Id: id=<content-id>

Table 26.6 *Content-transfer-encoding*

Type	*Description*
7-bit	NVT ASCII characters and short lines
8-bit	Non-ASCII characters and short lines
Binary	Non-ASCII characters with unlimited-length lines
Base-64	6-bit blocks of data encoded into 8-bit ASCII characters
Quoted-printable	Non-ASCII characters encoded as an equals sign followed by an ASCII code

Content-Description This header defines whether the body is image, audio, or video.

Content-Description: <description>

Message Transfer Agent: SMTP

The actual mail transfer is done through message transfer agents. To send mail, a system must have the client MTA, and to receive mail, a system must have a server MTA. The formal protocol that defines the MTA client and server in the Internet is called the **Simple Mail Transfer Protocol (SMTP).** As we said before, two pairs of MTA client/server programs are used in the most common situation (fourth scenario). Figure 26.16 shows the range of the SMTP protocol in this scenario.

Figure 26.16 *SMTP range*

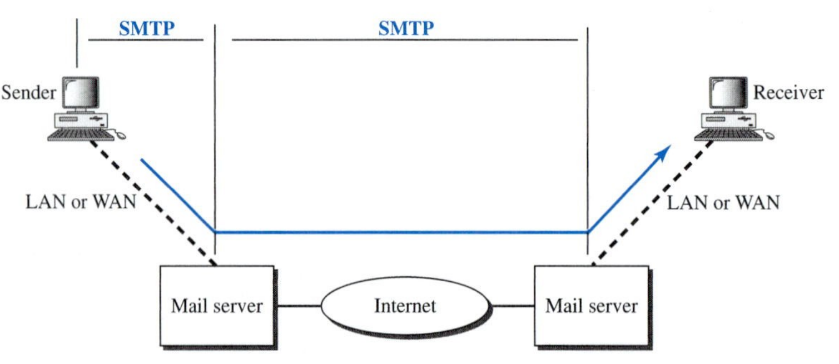

SMTP is used two times, between the sender and the sender's mail server and between the two mail servers. As we will see shortly, another protocol is needed between the mail server and the receiver.

SMTP simply defines how commands and responses must be sent back and forth. Each network is free to choose a software package for implementation. We discuss the mechanism of mail transfer by SMTP in the remainder of the section.

Commands and Responses

SMTP uses commands and responses to transfer messages between an MTA client and an MTA server (see Figure 26.17).

Figure 26.17 *Commands and responses*

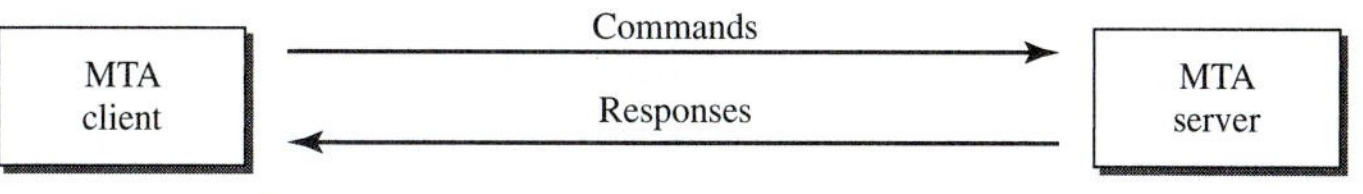

Each command or reply is terminated by a two-character (carriage return and line feed) end-of-line token.

Commands Commands are sent from the client to the server. The format of a command is shown in Figure 26.18. It consists of a keyword followed by zero or more arguments. SMTP defines 14 commands. The first five are mandatory; every implementation must support these five commands. The next three are often used and highly recommended. The last six are seldom used.

Figure 26.18 *Command format*

Keyword: argument(s)

The commands are listed in Table 26.7.

Table 26.7 *Commands*

Keyword	*Argument(s)*
HELO	Sender's host name
MAIL FROM	Sender of the message
RCPT TO	Intended recipient of the message
DATA	Body of the mail
QUIT	
RSET	
VRFY	Name of recipient to be verified
NOOP	
TURN	
EXPN	Mailing list to be expanded
HELP	Command name

Table 26.7 *Commands (continued)*

Keyword	*Argument(s)*
SEND FROM	Intended recipient of the message
SMOL FROM	Intended recipient of the message
SMAL FROM	Intended recipient of the message

Responses Responses are sent from the server to the client. A response is a three-digit code that may be followed by additional textual information. Table 26.8 lists some of the responses.

Table 26.8 *Responses*

Code	*Description*
	Positive Completion Reply
211	System status or help reply
214	Help message
220	Service ready
221	Service closing transmission channel
250	Request command completed
251	User not local; the message will be forwarded
	Positive Intermediate Reply
354	Start mail input
	Transient Negative Completion Reply
421	Service not available
450	Mailbox not available
451	Command aborted: local error
452	Command aborted: insufficient storage
	Permanent Negative Completion Reply
500	Syntax error; unrecognized command
501	Syntax error in parameters or arguments
502	Command not implemented
503	Bad sequence of commands
504	Command temporarily not implemented
550	Command is not executed; mailbox unavailable
551	User not local
552	Requested action aborted; exceeded storage location
553	Requested action not taken; mailbox name not allowed
554	Transaction failed

As the table shows, responses are divided into four categories. The leftmost digit of the code (2, 3, 4, and 5) defines the category.

Mail Transfer Phases

The process of transferring a mail message occurs in three phases: connection establishment, mail transfer, and connection termination.

Example 26.3

Let us see how we can directly use SMTP to send an e-mail and simulate the commands and responses we described in this section. We use TELNET to log into port 25 (the well-known port for SMTP). We then use the commands directly to send an e-mail. In this example, forouzanb@adelphia.net is sending an e-mail to himself. The first few lines show TELNET trying to connect to the Adelphia mail server.

After connection, we can type the SMTP commands and then receive the responses, as shown below. We have shown the commands in black and the responses in color. Note that we have added, for clarification, some comment lines, designated by the "=" signs. These lines are not part of the e-mail procedure.

```
$ telnet mail.adelphia.net 25
Trying 68.168.78.100 . . .
Connected to mail.adelphia.net (68.168.78.100).
```

```
================== Connection Establishment ================
  220 mta13.adelphia.net SMTP server ready Fri, 6 Aug 2004 . . .
HELO mail.adelphia.net
  250 mta13.adelphia.net
===================       Mail Transfer       ================
MAIL FROM: forouzanb@adelphia.net
  250 Sender <forouzanb@adelphia.net> Ok
RCPT TO: forouzanb@adelphia.net
  250 Recipient <forouzanb@adelphia.net> Ok
DATA
  354 Ok Send data ending with <CRLF>.<CRLF>
From: Forouzan
TO: Forouzan

This is a test message
to show SMTP in action.
.
===================    Connection Termination    ==============
   250 Message received: adelphia.net@mail.adelphia.net
QUIT
   221 mta13.adelphia.net SMTP server closing connection
Connection closed by foreign host.
```

Message Access Agent: POP and IMAP

The first and the second stages of mail delivery use SMTP. However, SMTP is not involved in the third stage because SMTP is a *push* protocol; it pushes the message from the client to the server. In other words, the direction of the bulk data (messages) is from the client to the server. On the other hand, the third stage needs a *pull* protocol; the client must pull messages from the server. The direction of the bulk data is from the server to the client. The third stage uses a message access agent.

Currently two message access protocols are available: Post Office Protocol, version 3 (POP3) and Internet Mail Access Protocol, version 4 (IMAP4). Figure 26.19 shows the position of these two protocols in the most common situation (fourth scenario).

Figure 26.19 *POP3 and IMAP4*

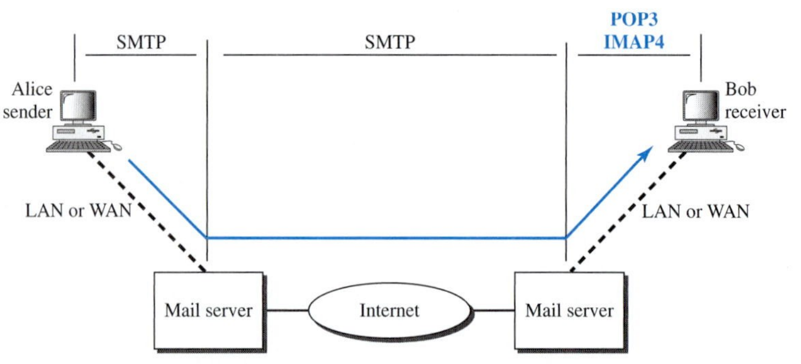

POP3

Post Office Protocol, version 3 (POP3) is simple and limited in functionality. The client POP3 software is installed on the recipient computer; the server POP3 software is installed on the mail server.

Mail access starts with the client when the user needs to download e-mail from the mailbox on the mail server. The client opens a connection to the server on TCP port 110. It then sends its user name and password to access the mailbox. The user can then list and retrieve the mail messages, one by one. Figure 26.20 shows an example of downloading using POP3.

POP3 has two modes: the delete mode and the keep mode. In the delete mode, the mail is deleted from the mailbox after each retrieval. In the keep mode, the mail remains in the mailbox after retrieval. The delete mode is normally used when the user is working at her permanent computer and can save and organize the received mail after reading or replying. The keep mode is normally used when the user accesses her mail away from her primary computer (e.g., a laptop). The mail is read but kept in the system for later retrieval and organizing.

IMAP4

Another mail access protocol is **Internet Mail Access Protocol, version 4 (IMAP4).** IMAP4 is similar to POP3, but it has more features; IMAP4 is more powerful and more complex.

Figure 26.20 *The exchange of commands and responses in POP3*

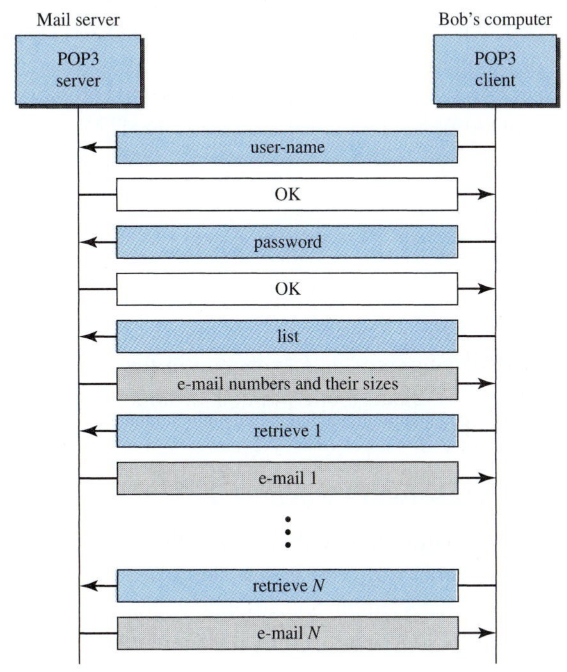

POP3 is deficient in several ways. It does not allow the user to organize her mail on the server; the user cannot have different folders on the server. (Of course, the user can create folders on her own computer.) In addition, POP3 does not allow the user to partially check the contents of the mail before downloading.

IMAP4 provides the following extra functions:

- A user can check the e-mail header prior to downloading.
- A user can search the contents of the e-mail for a specific string of characters prior to downloading.
- A user can partially download e-mail. This is especially useful if bandwidth is limited and the e-mail contains multimedia with high bandwidth requirements.
- A user can create, delete, or rename mailboxes on the mail server.
- A user can create a hierarchy of mailboxes in a folder for e-mail storage.

Web-Based Mail

E-mail is such a common application that some websites today provide this service to anyone who accesses the site. Two common sites are Hotmail and Yahoo. The idea is very simple. Mail transfer from Alice's browser to her mail server is done through HTTP (see Chapter 27). The transfer of the message from the sending mail server to the receiving mail server is still through SMTP. Finally, the message from the receiving server (the Web server) to Bob's browser is done through HTTP.

The last phase is very interesting. Instead of POP3 or IMAP4, HTTP is normally used. When Bob needs to retrieve his e-mails, he sends a message to the website (Hotmail, for example). The website sends a form to be filled in by Bob, which includes the log-in name and the password. If the log-in name and password match, the e-mail is transferred from the Web server to Bob's browser in HTML format.

26.3 FILE TRANSFER

Transferring files from one computer to another is one of the most common tasks expected from a networking or internetworking environment. As a matter of fact, the greatest volume of data exchange in the Internet today is due to file transfer. In this section, we discuss one popular protocol involved in transferring files: File Transfer Protocol (FTP).

File Transfer Protocol (FTP)

File Transfer Protocol (FTP) is the standard mechanism provided by TCP/IP for copying a file from one host to another. Although transferring files from one system to another seems simple and straightforward, some problems must be dealt with first. For example, two systems may use different file name conventions. Two systems may have different ways to represent text and data. Two systems may have different directory structures. All these problems have been solved by FTP in a very simple and elegant approach.

FTP differs from other client/server applications in that it establishes two connections between the hosts. One connection is used for data transfer, the other for control information (commands and responses). Separation of commands and data transfer makes FTP more efficient. The control connection uses very simple rules of communication. We need to transfer only a line of command or a line of response at a time. The data connection, on the other hand, needs more complex rules due to the variety of data types transferred. However, the difference in complexity is at the FTP level, not TCP. For TCP, both connections are treated the same.

FTP uses two well-known TCP ports: Port 21 is used for the control connection, and port 20 is used for the data connection.

FTP uses the services of TCP. It needs two TCP connections.
The well-known port 21 is used for the control connection
and the well-known port 20 for the data connection.

Figure 26.21 shows the basic model of FTP. The client has three components: user interface, client control process, and the client data transfer process. The server has two components: the server control process and the server data transfer process. The control connection is made between the control processes. The data connection is made between the data transfer processes.

Figure 26.21 *FTP*

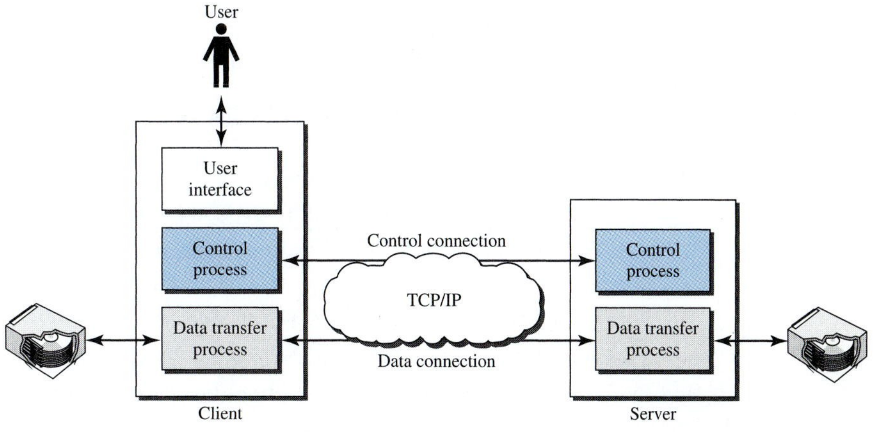

The **control connection** remains connected during the entire interactive FTP session. The **data connection** is opened and then closed for each file transferred. It opens each time commands that involve transferring files are used, and it closes when the file is transferred. In other words, when a user starts an FTP session, the control connection opens. While the control connection is open, the data connection can be opened and closed multiple times if several files are transferred.

Communication over Control Connection

FTP uses the same approach as SMTP to communicate across the control connection. It uses the 7-bit ASCII character set (see Figure 26.22). Communication is achieved through commands and responses. This simple method is adequate for the control connection because we send one command (or response) at a time. Each command or response is only one short line, so we need not worry about file format or file structure. Each line is terminated with a two-character (carriage return and line feed) end-of-line token.

Figure 26.22 *Using the control connection*

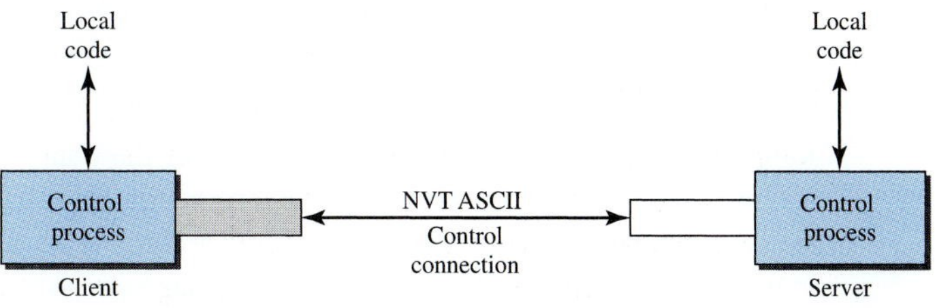

Communication over Data Connection

The purpose of the data connection is different from that of the control connection. We want to transfer files through the data connection. File transfer occurs over the data connection under the control of the commands sent over the control connection. However, we should remember that file transfer in FTP means one of three things:

- A file is to be copied from the server to the client. This is called *retrieving a file*. It is done under the supervision of the RETR command.
- A file is to be copied from the client to the server. This is called *storing a file*. It is done under the supervision of the STOR command.
- A list of directory or file names is to be sent from the server to the client. This is done under the supervision of the LIST command. Note that FTP treats a list of directory or file names as a file. It is sent over the data connection.

The client must define the type of file to be transferred, the structure of the data, and the transmission mode. Before sending the file through the data connection, we prepare for transmission through the control connection. The heterogeneity problem is resolved by defining three attributes of communication: file type, data structure, and transmission mode (see Figure 26.23).

Figure 26.23 *Using the data connection*

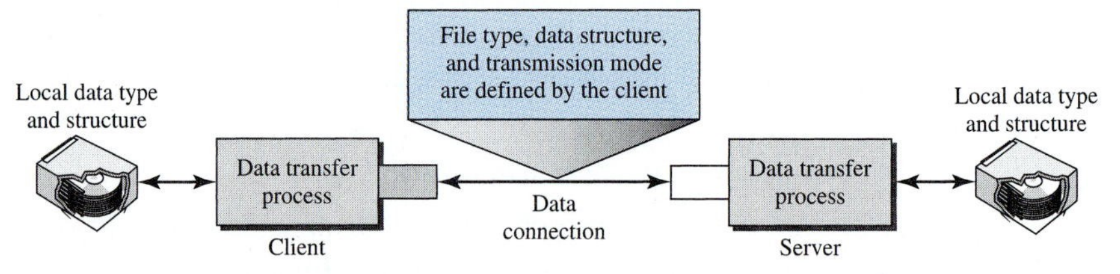

File Type FTP can transfer one of the following file types across the data connection: an ASCII file, EBCDIC file, or image file. The **ASCII file** is the default format for transferring text files. Each character is encoded using 7-bit ASCII. The sender transforms the file from its own representation into ASCII characters, and the receiver transforms the ASCII characters to its own representation. If one or both ends of the connection use EBCDIC encoding (the file format used by IBM), the file can be transferred using EBCDIC encoding. The **image file** is the default format for transferring binary files. The file is sent as continuous streams of bits without any interpretation or encoding. This is mostly used to transfer binary files such as compiled programs.

Data Structure FTP can transfer a file across the data connection by using one of the following interpretations about the structure of the data: file structure, record structure, and page structure. In the **file structure** format, the file is a continuous stream of bytes. In the **record structure,** the file is divided into records. This can be used only with text files. In the **page structure,** the file is divided into pages, with each page having a page number and a page header. The pages can be stored and accessed randomly or sequentially.

Transmission Mode FTP can transfer a file across the data connection by using one of the following three transmission modes: stream mode, block mode, and compressed mode. The **stream mode** is the default mode. Data are delivered from FTP to TCP as a continuous stream of bytes. TCP is responsible for chopping data into segments of appropriate size. If the data are simply a stream of bytes (file structure), no end-of-file is needed. End-of-file in this case is the closing of the data connection by the sender. If the data are divided into records (record structure), each record will have a 1-byte end-of-record (EOR) character and the end of the file will have a 1-byte end-of-file (EOF) character. In **block mode,** data can be delivered from FTP to TCP in blocks. In this case, each block is preceded by a 3-byte header. The first byte is called the *block descriptor;* the next 2 bytes define the size of the block in bytes. In the **compressed mode,** if the file is big, the data can be compressed. The compression method normally used is run-length encoding. In this method, consecutive appearances of a data unit are replaced by one occurrence and the number of repetitions. In a text file, this is usually spaces (blanks). In a binary file, null characters are usually compressed.

Example 26.4

The following shows an actual FTP session for retrieving a list of items in a directory. The colored lines show the responses from the server control connection; the black lines show the commands sent by the client. The lines in white with a black background show data transfer.

```
$ ftp voyager.deanza.fhda.edu
Connected to voyager.deanza.fhda.edu.
220 (vsFTPd 1.2.1)
530 Please login with USER and PASS.
Name (voyager.deanza.fhda.edu:forouzan): forouzan
331 Please specify the password.
Password:
230 Login successful.
Remote system type is UNIX.
Using binary mode to transfer files.
ftp> ls reports
227 Entering Passive Mode (153,18,17,11,238,169)
150 Here comes the directory listing.
drwxr-xr-x   2 3027     411          4096 Sep 24  2002 business
drwxr-xr-x   2 3027     411          4096 Sep 24  2002 personal
drwxr-xr-x   2 3027     411          4096 Sep 24  2002 school
226 Directory send OK.
ftp> quit
221 Goodbye.
```

1. After the control connection is created, the FTP server sends the 220 (service ready) response on the control connection.
2. The client sends its name.
3. The server responds with 331 (user name is OK, password is required).

4. The client sends the password (not shown).
5. The server responds with 230 (user log-in is OK).
6. The client sends the list command (ls reports) to find the list of files on the directory named report.
7. Now the server responds with 150 and opens the data connection.
8. The server then sends the list of the files or directories (as a file) on the data connection. When the whole list (file) is sent, the server responds with 226 (closing data connection) over the control connection.
9. The client now has two choices. It can use the QUIT command to request the closing of the control connection, or it can send another command to start another activity (and eventually open another data connection). In our example, the client sends a QUIT command.
10. After receiving the QUIT command, the server responds with 221 (service closing) and then closes the control connection.

Anonymous FTP

To use FTP, a user needs an account (user name) and a password on the remote server. Some sites have a set of files available for public access, to enable **anonymous FTP.** To access these files, a user does not need to have an account or password. Instead, the user can use *anonymous* as the user name and *guest* as the password.

User access to the system is very limited. Some sites allow anonymous users only a subset of commands. For example, most sites allow the user to copy some files, but do not allow navigation through the directories.

Example 26.5

We show an example of anonymous FTP. We assume that some public data are available at internic.net.

```
$ ftp internic.net
Connected to internic.net
220 Server ready
Name: anonymous
331 Guest login OK, send "guest" as password
Password: guest
ftp > pwd
257 '/' is current directory
ftp > ls
200 OK
150 Opening ASCII mode
bin
. . .
. . .
. . .
ftp > close
221 Goodbye
ftp > quit
```

26.4 RECOMMENDED READING

For more details about subjects discussed in this chapter, we recommend the following books and sites. The items in brackets [. . .] refer to the reference list at the end of the text.

Books

Remote logging is discussed in Chapter 18 of [For06] and Chapter 26 of [Ste94]. Electronic mail is discussed in Chapter 20 of [For06], Section 9.2 of [PD03], Chapter 32 of [Com04], Section 7.2 of [Tan03], and Chapter 28 of [Ste94]. FTP is discussed in Chapter 19 of [For06], Chapter 27 of [Ste94], and Chapter 34 of [Com04].

Sites

The following sites are related to topics discussed in this chapter.

- www.ietf.org/rfc.html Information about RFCs

RFCs

The following RFCs are related to TELNET:

137, 340, 393, 426, 435, 452, 466, 495, 513, 529, 562, 595, 596, 599, 669, 679, 701, 702, 703, 728, 764, 782, 818, 854, 855, 1184, 1205, 2355

The following RFCs are related to SMTP, POP, and IMAP:

196, 221, 224, 278, 524, 539, 753, 772, 780, 806, 821, 934, 974, 1047, 1081, 1082, 1225, 1460, 1496, 1426, 1427, 1652, 1653, 1711, 1725, 1734, 1740, 1741, 1767, 1869, 1870, 2045, 2046, 2047, 2048, 2177, 2180, 2192, 2193, 2221, 2342, 2359, 2449, 2683, 2503

The following RFCs are related to FTP:

114, 133, 141, 163, 171, 172, 238, 242, 250, 256, 264, 269, 281, 291, 354, 385, 412, 414, 418, 430, 438, 448, 463, 468, 478, 486, 505, 506, 542, 553, 624, 630, 640, 691, 765, 913, 959, 1635, 1785, 2228, 2577

DNS is discussed in [AL98], chapter 17 of [For06], section 9.1 of [PD03], and section 7.1 of [Tan03].

26.5 KEY TERMS

alias
anonymous FTP
ASCII file
block mode
body
character mode
compressed mode
control character
control connection
data connection

default mode
domain name
envelope
file structure
File Transfer Protocol (FTP)
header
image file
Internet Mail Access Protocol, version 4 (IMAP4)
line mode
local log-in
local part
message access agent (MAA)
message transfer agent (MTA)
Multipurpose Internet Mail Extensions (MIME)
network virtual terminal (NVT)
option negotiation
page structure
Post Office Protocol, version 3 (POP3)
record structure
remote log-in
Simple Mail Transfer Protocol (SMTP)
stream mode
suboption negotiation
terminal network (TELNET)
timesharing
user agent (UA)

26.6 SUMMARY

- TELNET is a client/server application that allows a user to log on to a remote machine, giving the user access to the remote system.
- TELNET uses the network virtual terminal (NVT) system to encode characters on the local system. On the server machine, NVT decodes the characters to a form acceptable to the remote machine.
- NVT uses a set of characters for data and a set of characters for control.
- In TELNET, control characters are embedded in the data stream and preceded by the *interpret as control* (IAC) control character.
- Options are features that enhance the TELNET process.
- TELNET allows negotiation to set transfer conditions between the client and server before and during the use of the service.
- A TELNET implementation operates in the default, character, or line mode.
  1. In the default mode, the client sends one line at a time to the server.
  2. In the character mode, the client sends one character at a time to the server.
  3. In the line mode, the client sends one line at a time to the server.
- Several programs, including SMTP, POP3, and IMAP4, are used in the Internet to provide electronic mail services.
- In electronic mail, the UA prepares the message, creates the envelope, and puts the message in the envelope.
- In electronic mail, the mail address consists of two parts: a local part (user mailbox) and a domain name. The form is localpart@domainname.
- In electronic mail, Multipurpose Internet Mail Extension (MIME) allows the transfer of multimedia messages.
- In electronic mail, the MTA transfers the mail across the Internet, a LAN, or a WAN.

- SMTP uses commands and responses to transfer messages between an MTA client and an MTA server.
- The steps in transferring a mail message are
  1. Connection establishment
  2. Mail transfer
  3. Connection termination
- Post Office Protocol, version 3 (POP3) and Internet Mail Access Protocol, version 4 (IMAP4) are protocols used for pulling messages from a mail server.
- One of the programs used for file transfer in the Internet is File Transfer Protocol (FTP).
- FTP requires two connections for data transfer: a control connection and a data connection.
- FTP employs NVT ASCII for communication between dissimilar systems.
- Prior to the actual transfer of files, the file type, data structure, and transmission mode are defined by the client through the control connection.
- Responses are sent from the server to the client during connection establishment.
- There are three types of file transfer:
  1. A file is copied from the server to the client.
  2. A file is copied from the client to the server.
  3. A list of directories or file names is sent from the server to the client.
- Anonymous FTP provides a method for the general public to access files on remote sites.

26.7 PRACTICE SET

Review Questions

1. What is the difference between local and remote log-in in TELNET?
2. How are control and data characters distinguished in NVT?
3. How are options negotiated in TELNET?
4. Describe the addressing system used by SMTP.
5. In electronic mail, what are the tasks of a user agent?
6. In electronic mail, what is MIME?
7. Why do we need POP3 or IMAP4 for electronic mail?
8. What is the purpose of FTP?
9. Describe the functions of the two FTP connections.
10. What kinds of file types can FTP transfer?
11. What are the three FTP transmission modes?
12. How does storing a file differ from retrieving a file?
13. What is anonymous FTP?

Exercises

14. Show the sequence of bits sent from a client TELNET for the binary transmission of 11110011 00111100 11111111.
15. If TELNET is using the character mode, how many characters are sent back and forth between the client and server to copy a file named file1 to another file named file2 using the command *cp file1 file2*?
16. What is the minimum number of bits sent at the TCP level to accomplish the task in Exercise 15?
17. What is the minimum number of bits sent at the data link layer level (using Ethernet) to accomplish the task in Exercise 15?
18. What is the ratio of the useful bits to the total bits in Exercise 17?
19. Interpret the following sequences of characters (in hexadecimal) received by a TELNET client or server.
    a. FF FB 01
    b. FF FE 01
    c. FF F4
    d. FF F9
20. A sender sends unformatted text. Show the MIME header.
21. A sender sends a JPEG message. Show the MIME header.
22. Why is a connection establishment for mail transfer needed if TCP has already established a connection?
23. Why should there be limitations on anonymous FTP? What could an unscrupulous user do?
24. Explain why FTP does not have a message format.

Research Activities

25. Show the sequence of characters exchanged between the TELNET client and the server to switch from the default mode to the character mode.
26. Show the sequence of characters exchanged between the TELNET client and the server to switch from the default mode to line mode.
27. In SMTP, show the connection establishment phase from aaa@xxx.com to bbb@yyy.com.
28. In SMTP, show the message transfer phase from aaa@xxx.com to bbb@yyy.com. The message is "Good morning my friend."
29. In SMTP, show the connection termination phase from aaa@xxx.com to bbb@yyy.com.
30. What do you think would happen if the control connection were accidentally severed during an FTP transfer?

31. Find the extended options proposed for TELNET.
32. Another log-in protocol is called Rlogin. Find some information about Rlogin and compare it with TELNET.
33. A more secure log-in protocol in UNIX is called Secure Shell (SSH). Find some information about this protocol.

CHAPTER 27

WWW and HTTP

The **World Wide Web (WWW)** is a repository of information linked together from points all over the world. The WWW has a unique combination of flexibility, portability, and user-friendly features that distinguish it from other services provided by the Internet. The WWW project was initiated by CERN (European Laboratory for Particle Physics) to create a system to handle distributed resources necessary for scientific research. In this chapter we first discuss issues related to the Web. We then discuss a protocol, HTTP, that is used to retrieve information from the Web.

27.1 ARCHITECTURE

The WWW today is a distributed client/server service, in which a client using a browser can access a service using a server. However, the service provided is distributed over many locations called *sites,* as shown in Figure 27.1.

Figure 27.1 *Architecture of WWW*

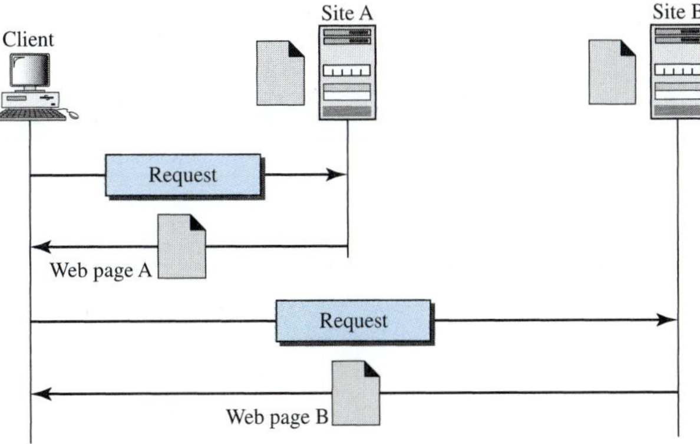

Each site holds one or more documents, referred to as *Web pages*. Each Web page can contain a link to other pages in the same site or at other sites. The pages can be retrieved and viewed by using browsers. Let us go through the scenario shown in Figure 27.1. The client needs to see some information that it knows belongs to site A. It sends a request through its browser, a program that is designed to fetch **Web** documents. The request, among other information, includes the address of the site and the Web page, called the URL, which we will discuss shortly. The server at site A finds the document and sends it to the client. When the user views the document, she finds some references to other documents, including a Web page at site B. The reference has the URL for the new site. The user is also interested in seeing this document. The client sends another request to the new site, and the new page is retrieved.

Client (Browser)

A variety of vendors offer commercial browsers that interpret and display a Web document, and all use nearly the same architecture. Each **browser** usually consists of three parts: a controller, client protocol, and interpreters. The controller receives input from the keyboard or the mouse and uses the client programs to access the document. After the document has been accessed, the controller uses one of the interpreters to display the document on the screen. The client protocol can be one of the protocols described previously such as FTP or HTTP (described later in the chapter). The interpreter can be HTML, Java, or JavaScript, depending on the type of document. We discuss the use of these interpreters based on the document type later in the chapter (see Figure 27.2).

Figure 27.2 *Browser*

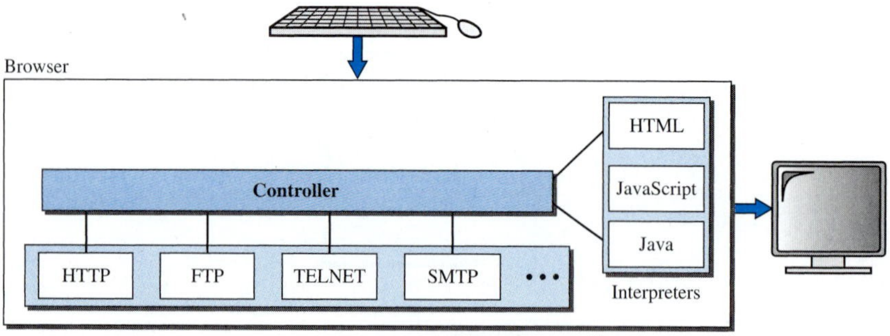

Server

The Web page is stored at the server. Each time a client request arrives, the corresponding document is sent to the client. To improve efficiency, servers normally store requested files in a cache in memory; memory is faster to access than disk. A server can also become more efficient through multithreading or multiprocessing. In this case, a server can answer more than one request at a time.

Uniform Resource Locator

A client that wants to access a Web page needs the address. To facilitate the access of documents distributed throughout the world, HTTP uses locators. The **uniform resource locator (URL)** is a standard for specifying any kind of information on the Internet. The URL defines four things: protocol, host computer, port, and path (see Figure 27.3).

Figure 27.3 *URL*

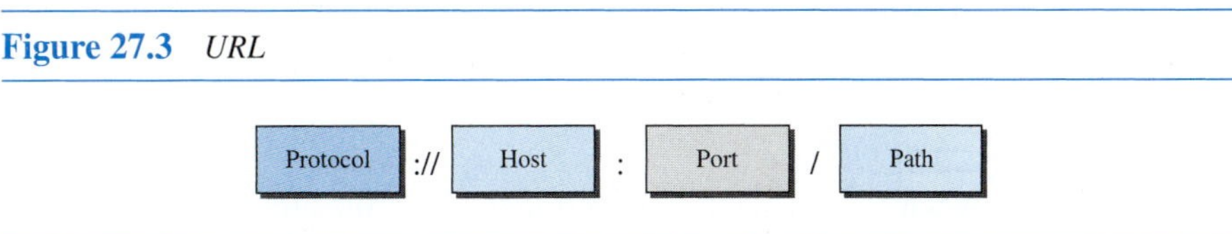

The *protocol* is the client/server program used to retrieve the document. Many different protocols can retrieve a document; among them are FTP or HTTP. The most common today is HTTP.

The **host** is the computer on which the information is located, although the name of the computer can be an alias. Web pages are usually stored in computers, and computers are given alias names that usually begin with the characters "www". This is not mandatory, however, as the host can be any name given to the computer that hosts the Web page.

The URL can optionally contain the port number of the server. If the *port* is included, it is inserted between the host and the path, and it is separated from the host by a colon.

Path is the pathname of the file where the information is located. Note that the path can itself contain slashes that, in the UNIX operating system, separate the directories from the subdirectories and files.

Cookies

The World Wide Web was originally designed as a stateless entity. A client sends a request; a server responds. Their relationship is over. The original design of WWW, retrieving publicly available documents, exactly fits this purpose. Today the Web has other functions; some are listed here.

1. Some websites need to allow access to registered clients only.
2. Websites are being used as electronic stores that allow users to browse through the store, select wanted items, put them in an electronic cart, and pay at the end with a credit card.
3. Some websites are used as portals: the user selects the Web pages he wants to see.
4. Some websites are just advertising.

For these purposes, the cookie mechanism was devised. We discussed the use of cookies at the transport layer in Chapter 23; we now discuss their use in Web pages.

Creation and Storage of Cookies

The creation and storage of cookies depend on the implementation; however, the principle is the same.

1. When a server receives a request from a client, it stores information about the client in a file or a string. The information may include the domain name of the client, the contents of the cookie (information the server has gathered about the client such as name, registration number, and so on), a timestamp, and other information depending on the implementation.
2. The server includes the cookie in the response that it sends to the client.
3. When the client receives the response, the browser stores the cookie in the cookie directory, which is sorted by the domain server name.

Using Cookies

When a client sends a request to a server, the browser looks in the cookie directory to see if it can find a cookie sent by that server. If found, the cookie is included in the request. When the server receives the request, it knows that this is an old client, not a new one. Note that the contents of the cookie are never read by the browser or disclosed to the user. It is a cookie *made* by the server and *eaten* by the server. Now let us see how a cookie is used for the four previously mentioned purposes:

1. The site that restricts access to registered clients only sends a cookie to the client when the client registers for the first time. For any repeated access, only those clients that send the appropriate cookie are allowed.
2. An electronic store (e-commerce) can use a cookie for its client shoppers. When a client selects an item and inserts it into a cart, a cookie that contains information about the item, such as its number and unit price, is sent to the browser. If the client selects a second item, the cookie is updated with the new selection information. And so on. When the client finishes shopping and wants to check out, the last cookie is retrieved and the total charge is calculated.
3. A Web portal uses the cookie in a similar way. When a user selects her favorite pages, a cookie is made and sent. If the site is accessed again, the cookie is sent to the server to show what the client is looking for.
4. A cookie is also used by advertising agencies. An advertising agency can place banner ads on some main website that is often visited by users. The advertising agency supplies only a URL that gives the banner address instead of the banner itself. When a user visits the main website and clicks on the icon of an advertised corporation, a request is sent to the advertising agency. The advertising agency sends the banner, a GIF file, for example, but it also includes a cookie with the ID of the user. Any future use of the banners adds to the database that profiles the Web behavior of the user. The advertising agency has compiled the interests of the user and can sell this information to other parties. This use of cookies has made them very controversial. Hopefully, some new regulations will be devised to preserve the privacy of users.

27.2 WEB DOCUMENTS

The documents in the WWW can be grouped into three broad categories: static, dynamic, and active. The category is based on the time at which the contents of the document are determined.

Static Documents

Static documents are fixed-content documents that are created and stored in a server. The client can get only a copy of the document. In other words, the contents of the file are determined when the file is created, not when it is used. Of course, the contents in the server can be changed, but the user cannot change them. When a client accesses the document, a copy of the document is sent. The user can then use a browsing program to display the document (see Figure 27.4).

Figure 27.4 *Static document*

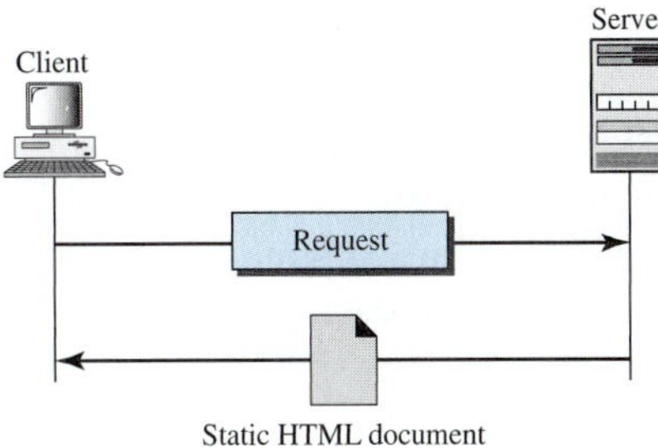

HTML

Hypertext Markup Language (HTML) is a language for creating Web pages. The term *markup language* comes from the book publishing industry. Before a book is typeset and printed, a copy editor reads the manuscript and puts marks on it. These marks tell the compositor how to format the text. For example, if the copy editor wants part of a line to be printed in boldface, he or she draws a wavy line under that part. In the same way, data for a Web page are formatted for interpretation by a browser.

Let us clarify the idea with an example. To make part of a text displayed in boldface with HTML, we put beginning and ending boldface **tags** (marks) in the text, as shown in Figure 27.5.

Figure 27.5 *Boldface tags*

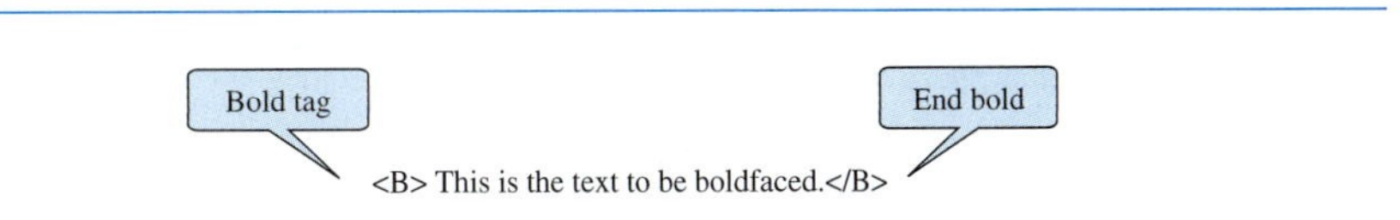

The two tags <B> and </B> are instructions for the browser. When the browser sees these two marks, it knows that the text must be boldfaced (see Figure 27.6).

A markup language such as HTML allows us to embed formatting instructions in the file itself. The instructions are included with the text. In this way, any browser can read the instructions and format the text according to the specific workstation. One might

Figure 27.6 *Effect of boldface tags*

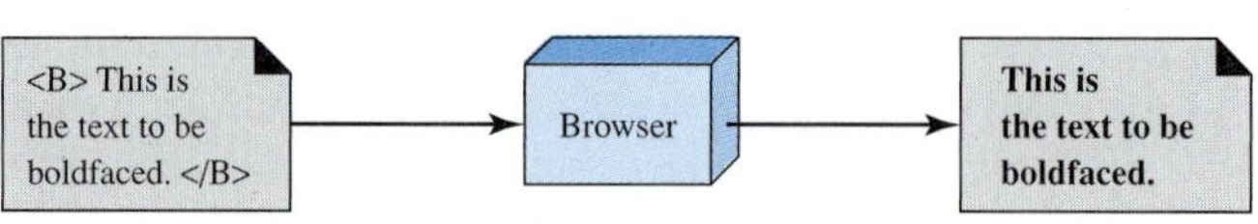

ask why we do not use the formatting capabilities of word processors to create and save formatted text. The answer is that different word processors use different techniques or procedures for formatting text. For example, imagine that a user creates formatted text on a Macintosh computer and stores it in a Web page. Another user who is on an IBM computer would not be able to receive the Web page because the two computers use different formatting procedures.

HTML lets us use only ASCII characters for both the main text and formatting instructions. In this way, every computer can receive the whole document as an ASCII document. The main text is the data, and the formatting instructions can be used by the browser to format the data.

A Web page is made up of two parts: the head and the body. The head is the first part of a Web page. The head contains the title of the page and other parameters that the browser will use. The actual contents of a page are in the body, which includes the text and the tags. Whereas the text is the actual information contained in a page, the tags define the appearance of the document. Every HTML tag is a name followed by an optional list of attributes, all enclosed between less-than and greater-than symbols (< and >).

An attribute, if present, is followed by an equals sign and the value of the attribute. Some tags can be used alone; others must be used in pairs. Those that are used in pairs are called *beginning* and *ending* tags. The beginning tag can have attributes and values and starts with the name of the tag. The ending tag cannot have attributes or values but must have a slash before the name of the tag. The browser makes a decision about the structure of the text based on the tags, which are embedded into the text. Figure 27.7 shows the format of a tag.

Figure 27.7 *Beginning and ending tags*

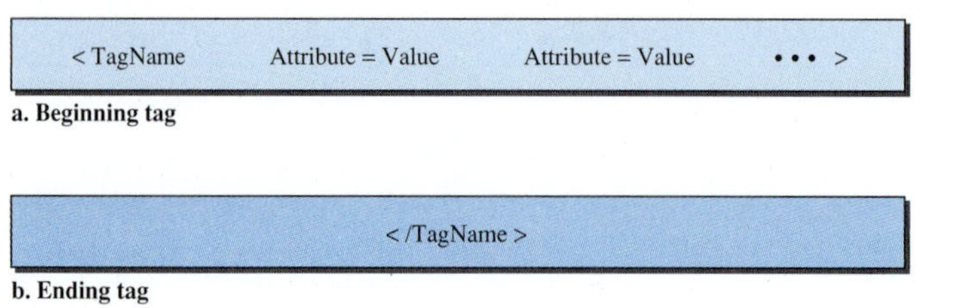

One commonly used tag category is the text formatting tags such as <B> and </B>, which make the text bold; <I> and </I>, which make the text italic; and <U> and </U>, which underline the text.

Another interesting tag category is the image tag. Nontextual information such as digitized photos or graphic images is not a physical part of an HTML document. But we can use an image tag to point to the file of a photo or image. The image tag defines the address (URL) of the image to be retrieved. It also specifies how the image can be inserted after retrieval. We can choose from several attributes. The most common are SRC (source), which defines the source (address), and ALIGN, which defines the alignment of the image. The SRC attribute is required. Most browsers accept images in the GIF or JPEG formats. For example, the following tag can retrieve an image stored as image1.gif in the directory /bin/images:

```
<IMG SRC= "/bin/images/image1.gif" ALIGN=MIDDLE >
```

A third interesting category is the hyperlink tag, which is needed to link documents together. Any item (word, phrase, paragraph, or image) can refer to another document through a mechanism called an *anchor*. The anchor is defined by <A . . . > and </A> tags, and the anchored item uses the URL to refer to another document. When the document is displayed, the anchored item is underlined, blinking, or boldfaced. The user can click on the anchored item to go to another document, which may or may not be stored on the same server as the original document. The reference phrase is embedded between the beginning and ending tags. The beginning tag can have several attributes, but the one required is HREF (hyperlink reference), which defines the address (URL) of the linked document. For example, the link to the author of a book can be

```
<A HREF= "http://www.deanza.edu/forouzan"> Author </A>
```

What appears in the text is the word *Author*, on which the user can click to go to the author's Web page.

Dynamic Documents

A **dynamic document** is created by a Web server whenever a browser requests the document. When a request arrives, the Web server runs an application program or a script that creates the dynamic document. The server returns the output of the program or script as a response to the browser that requested the document. Because a fresh document is created for each request, the contents of a dynamic document can vary from one request to another. A very simple example of a dynamic document is the retrieval of the time and date from a server. Time and date are kinds of information that are dynamic in that they change from moment to moment. The client can ask the server to run a program such as the *date* program in UNIX and send the result of the program to the client.

Common Gateway Interface (CGI)

The **Common Gateway Interface (CGI)** is a technology that creates and handles dynamic documents. CGI is a set of standards that defines how a dynamic document is written, how data are input to the program, and how the output result is used.

CGI is not a new language; instead, it allows programmers to use any of several languages such as C, C++, Bourne Shell, Korn Shell, C Shell, Tcl, or Perl. The only thing that CGI defines is a set of rules and terms that the programmer must follow.

The term *common* in CGI indicates that the standard defines a set of rules that is common to any language or platform. The term *gateway* here means that a CGI program can be used to access other resources such as databases, graphical packages, and so on. The term *interface* here means that there is a set of predefined terms, variables, calls, and so on that can be used in any CGI program. A CGI program in its simplest form is code written in one of the languages supporting CGI. Any programmer who can encode a sequence of thoughts in a program and knows the syntax of one of the above-mentioned languages can write a simple CGI program. Figure 27.8 illustrates the steps in creating a dynamic program using CGI technology.

Figure 27.8 *Dynamic document using CGI*

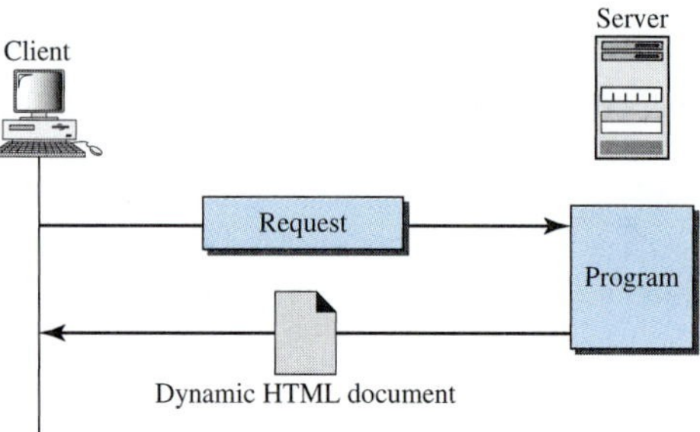

Input In traditional programming, when a program is executed, parameters can be passed to the program. Parameter passing allows the programmer to write a generic program that can be used in different situations. For example, a generic copy program can be written to copy any file to another. A user can use the program to copy a file named *x* to another file named *y* by passing *x* and *y* as parameters.

The input from a browser to a server is sent by using a *form*. If the information in a form is small (such as a word), it can be appended to the URL after a question mark. For example, the following URL is carrying form information (23, a value):

http://www.deanza/cgi-bin/prog.pl?23

When the server receives the URL, it uses the part of the URL before the question mark to access the program to be run, and it interprets the part after the question mark (23) as the input sent by the client. It stores this string in a variable. When the CGI program is executed, it can access this value.

If the input from a browser is too long to fit in the query string, the browser can ask the server to send a form. The browser can then fill the form with the input data and send it to the server. The information in the form can be used as the input to the CGI program.

Output The whole idea of CGI is to execute a CGI program at the server site and send the output to the client (browser). The output is usually plain text or a text with HTML structures; however, the output can be a variety of other things. It can be graphics or binary data, a status code, instructions to the browser to cache the result, or instructions to the server to send an existing document instead of the actual output.

To let the client know about the type of document sent, a CGI program creates headers. As a matter of fact, the output of the CGI program always consists of two parts: a header and a body. The header is separated by a blank line from the body. This means any CGI program creates first the header, then a blank line, and then the body. Although the header and the blank line are not shown on the browser screen, the header is used by the browser to interpret the body.

Scripting Technologies for Dynamic Documents

The problem with CGI technology is the inefficiency that results if part of the dynamic document that is to be created is fixed and not changing from request to request. For example, assume that we need to retrieve a list of spare parts, their availability, and prices for a specific car brand. Although the availability and prices vary from time to time, the name, description, and the picture of the parts are fixed. If we use CGI, the program must create an entire document each time a request is made. The solution is to create a file containing the fixed part of the document using HTML and embed a script, a source code, that can be run by the server to provide the varying availability and price section. Figure 27.9 shows the idea.

Figure 27.9 *Dynamic document using server-site script*

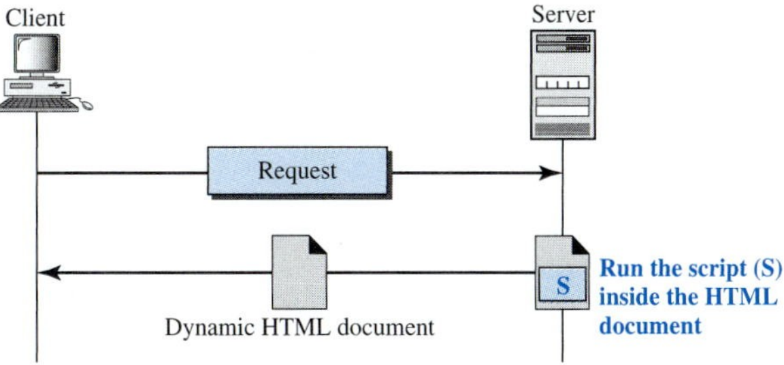

A few technologies have been involved in creating dynamic documents using scripts. Among the most common are **Hypertext Preprocessor (PHP),** which uses the Perl language; **Java Server Pages (JSP),** which uses the Java language for scripting; **Active Server Pages (ASP),** a Microsoft product which uses Visual Basic language for scripting; and **ColdFusion,** which embeds SQL database queries in the HTML document.

Dynamic documents are sometimes referred to as server-site dynamic documents.

Active Documents

For many applications, we need a program or a script to be run at the client site. These are called **active documents.** For example, suppose we want to run a program that creates animated graphics on the screen or a program that interacts with the user. The program definitely needs to be run at the client site where the animation or interaction takes place. When a browser requests an active document, the server sends a copy of the document or a script. The document is then run at the client (browser) site.

Java Applets

One way to create an active document is to use Java **applets. Java** is a combination of a high-level programming language, a run-time environment, and a class library that allows a programmer to write an active document (an applet) and a browser to run it. It can also be a stand-alone program that doesn't use a browser.

An applet is a program written in Java on the server. It is compiled and ready to be run. The document is in byte-code (binary) format. The client process (browser) creates an instance of this applet and runs it. A Java applet can be run by the browser in two ways. In the first method, the browser can directly request the Java applet program in the URL and receive the applet in binary form. In the second method, the browser can retrieve and run an HTML file that has embedded the address of the applet as a tag. Figure 27.10 shows how Java applets are used in the first method; the second is similar but needs two transactions.

Figure 27.10 *Active document using Java applet*

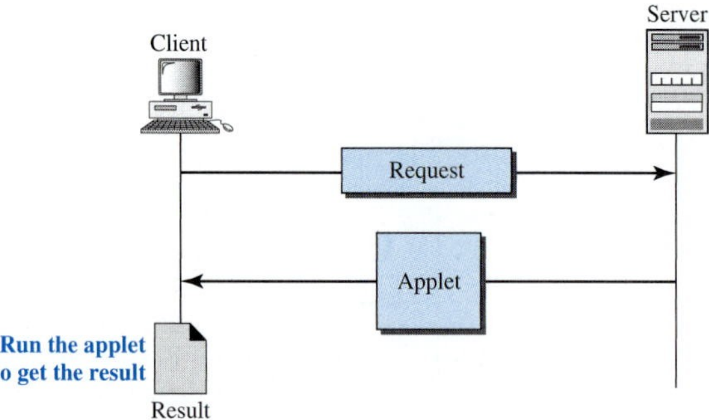

JavaScript

The idea of scripts in dynamic documents can also be used for active documents. If the active part of the document is small, it can be written in a scripting language; then it can be interpreted and run by the client at the same time. The script is in source code (text) and not in binary form. The scripting technology used in this case is usually JavaScript. **JavaScript,** which bears a small resemblance to Java, is a very high level scripting language developed for this purpose. Figure 27.11 shows how JavaScript is used to create an active document.

Figure 27.11 *Active document using client-site script*

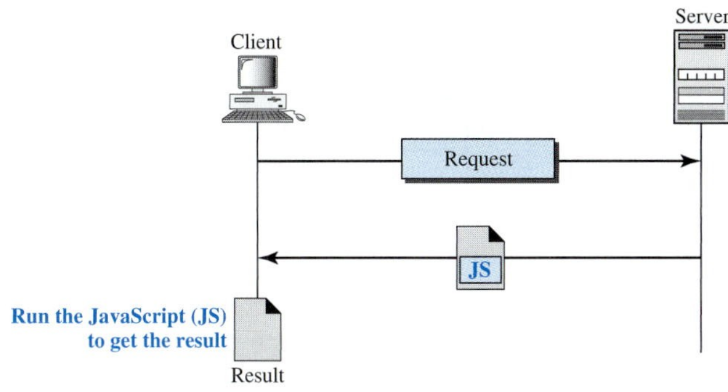

Active documents are sometimes referred to as client-site dynamic documents.

27.3 HTTP

The **Hypertext Transfer Protocol (HTTP)** is a protocol used mainly to access data on the World Wide Web. HTTP functions as a combination of FTP and SMTP. It is similar to FTP because it transfers files and uses the services of TCP. However, it is much simpler than FTP because it uses only one TCP connection. There is no separate control connection; only data are transferred between the client and the server.

HTTP is like SMTP because the data transferred between the client and the server look like SMTP messages. In addition, the format of the messages is controlled by MIME-like headers. Unlike SMTP, the HTTP messages are not destined to be read by humans; they are read and interpreted by the HTTP server and HTTP client (browser). SMTP messages are stored and forwarded, but HTTP messages are delivered immediately. The commands from the client to the server are embedded in a request message. The contents of the requested file or other information are embedded in a response message. HTTP uses the services of TCP on well-known port 80.

HTTP uses the services of TCP on well-known port 80.

HTTP Transaction

Figure 27.12 illustrates the HTTP transaction between the client and server. Although HTTP uses the services of TCP, HTTP itself is a stateless protocol. The client initializes the transaction by sending a request message. The server replies by sending a response.

Messages

The formats of the request and response messages are similar; both are shown in Figure 27.13. A request message consists of a request line, a header, and sometimes a body. A response message consists of a status line, a header, and sometimes a body.

Figure 27.12 *HTTP transaction*

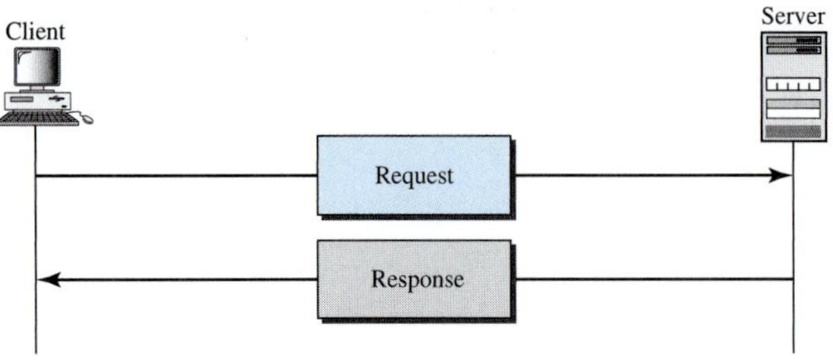

Figure 27.13 *Request and response messages*

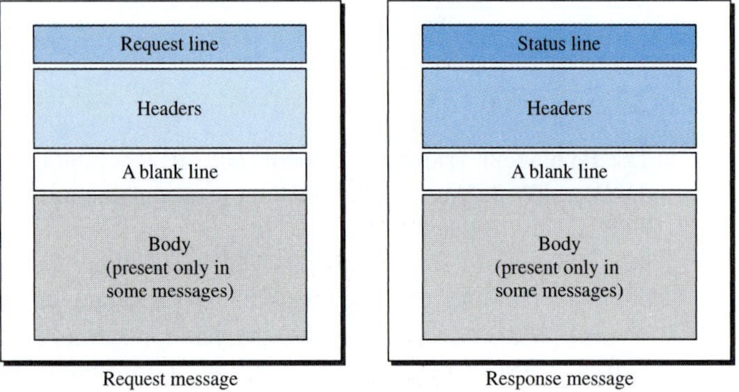

Request and Status Lines The first line in a request message is called a **request line;** the first line in the response message is called the **status line.** There is one common field, as shown in Figure 27.14.

Figure 27.14 *Request and status lines*

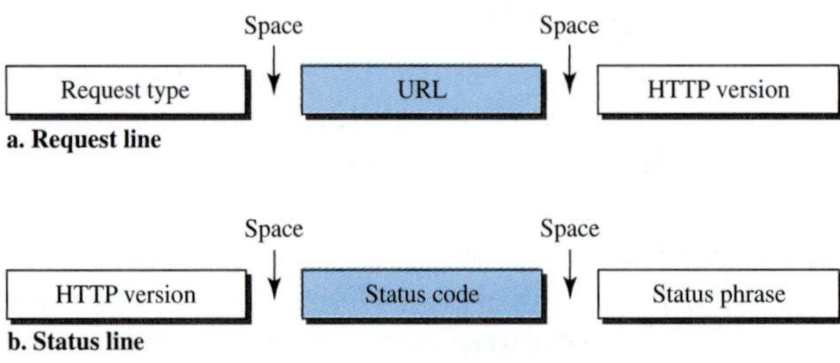

- **Request type.** This field is used in the request message. In version 1.1 of HTTP, several request types are defined. The **request type** is categorized into *methods* as defined in Table 27.1.

Table 27.1 *Methods*

Method	*Action*
GET	Requests a document from the server
HEAD	Requests information about a document but not the document itself
POST	Sends some information from the client to the server
PUT	Sends a document from the server to the client
TRACE	Echoes the incoming request
CONNECT	Reserved
OPTION	Inquires about available options

- **URL.** We discussed the URL earlier in the chapter.
- **Version.** The most current version of HTTP is 1.1.
- **Status code.** This field is used in the response message. The **status code** field is similar to those in the FTP and the SMTP protocols. It consists of three digits. Whereas the codes in the 100 range are only informational, the codes in the 200 range indicate a successful request. The codes in the 300 range redirect the client to another URL, and the codes in the 400 range indicate an error at the client site. Finally, the codes in the 500 range indicate an error at the server site. We list the most common codes in Table 27.2.
- **Status phrase.** This field is used in the response message. It explains the status code in text form. Table 27.2 also gives the status phrase.

Table 27.2 *Status codes*

Code	*Phrase*	*Description*
	Informational	
100	Continue	The initial part of the request has been received, and the client may continue with its request.
101	Switching	The server is complying with a client request to switch protocols defined in the upgrade header.
	Success	
200	OK	The request is successful.
201	Created	A new URL is created.
202	Accepted	The request is accepted, but it is not immediately acted upon.
204	No content	There is no content in the body.

Table 27.2 *Status codes (continued)*

Code	*Phrase*	*Description*
		Redirection
301	Moved permanently	The requested URL is no longer used by the server.
302	Moved temporarily	The requested URL has moved temporarily.
304	Not modified	The document has not been modified.
		Client Error
400	Bad request	There is a syntax error in the request.
401	Unauthorized	The request lacks proper authorization.
403	Forbidden	Service is denied.
404	Not found	The document is not found.
405	Method not allowed	The method is not supported in this URL.
406	Not acceptable	The format requested is not acceptable.
		Server Error
500	Internal server error	There is an error, such as a crash, at the server site.
501	Not implemented	The action requested cannot be performed.
503	Service unavailable	The service is temporarily unavailable, but may be requested in the future.

Header The header exchanges additional information between the client and the server. For example, the client can request that the document be sent in a special format, or the server can send extra information about the document. The header can consist of one or more header lines. Each header line has a header name, a colon, a space, and a header value (see Figure 27.15). We will show some header lines in the examples at the end of this chapter. A header line belongs to one of four categories: **general header, request header, response header,** and **entity header.** A request message can contain only general, request, and entity headers. A response message, on the other hand, can contain only general, response, and entity headers.

Figure 27.15 *Header format*

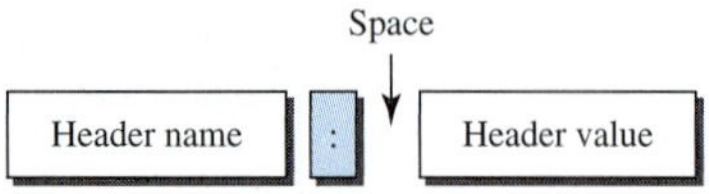

Space

Header name | : | Header value

- **General header** The general header gives general information about the message and can be present in both a request and a response. Table 27.3 lists some general headers with their descriptions.

Table 27.3 *General headers*

Header	*Description*
Cache-control	Specifies information about caching
Connection	Shows whether the connection should be closed or not
Date	Shows the current date
MIME-version	Shows the MIME version used
Upgrade	Specifies the preferred communication protocol

- **Request header** The request header can be present only in a request message. It specifies the client's configuration and the client's preferred document format. See Table 27.4 for a list of some request headers and their descriptions.

Table 27.4 *Request headers*

Header	*Description*
Accept	Shows the medium format the client can accept
Accept-charset	Shows the character set the client can handle
Accept-encoding	Shows the encoding scheme the client can handle
Accept-language	Shows the language the client can accept
Authorization	Shows what permissions the client has
From	Shows the e-mail address of the user
Host	Shows the host and port number of the server
If-modified-since	Sends the document if newer than specified date
If-match	Sends the document only if it matches given tag
If-non-match	Sends the document only if it does not match given tag
If-range	Sends only the portion of the document that is missing
If-unmodified-since	Sends the document if not changed since specified date
Referrer	Specifies the URL of the linked document
User-agent	Identifies the client program

- **Response header** The response header can be present only in a response message. It specifies the server's configuration and special information about the request. See Table 27.5 for a list of some response headers with their descriptions.

Table 27.5 *Response headers*

Header	*Description*
Accept-range	Shows if server accepts the range requested by client
Age	Shows the age of the document
Public	Shows the supported list of methods
Retry-after	Specifies the date after which the server is available
Server	Shows the server name and version number

- **Entity header** The entity header gives information about the body of the document. Although it is mostly present in response messages, some request messages, such as POST or PUT methods, that contain a body also use this type of header. See Table 27.6 for a list of some entity headers and their descriptions.

Table 27.6 *Entity headers*

Header	*Description*
Allow	Lists valid methods that can be used with a URL
Content-encoding	Specifies the encoding scheme
Content-language	Specifies the language
Content-length	Shows the length of the document
Content-range	Specifies the range of the document
Content-type	Specifies the medium type
Etag	Gives an entity tag
Expires	Gives the date and time when contents may change
Last-modified	Gives the date and time of the last change
Location	Specifies the location of the created or moved document

Body The body can be present in a request or response message. Usually, it contains the document to be sent or received.

Example 27.1

This example retrieves a document. We use the GET method to retrieve an image with the path /usr/bin/image1. The request line shows the method (GET), the URL, and the HTTP version (1.1). The header has two lines that show that the client can accept images in the GIF or JPEG format. The request does not have a body. The response message contains the status line and four lines of header. The header lines define the date, server, MIME version, and length of the document. The body of the document follows the header (see Figure 27.16).

Figure 27.16 *Example 27.1*

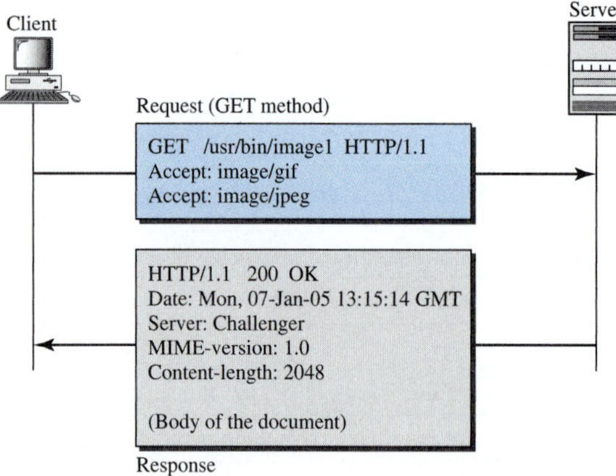

Example 27.2

In this example, the client wants to send data to the server. We use the POST method. The request line shows the method (POST), URL, and HTTP version (1.1). There are four lines of headers. The request body contains the input information. The response message contains the status line and four lines of headers. The created document, which is a CGI document, is included as the body (see Figure 27.17).

Figure 27.17 *Example 27.2*

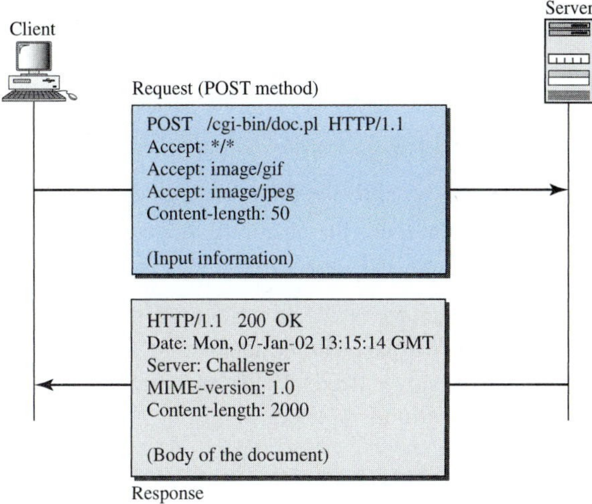

Example 27.3

HTTP uses ASCII characters. A client can directly connect to a server using TELNET, which logs into port 80. The next three lines show that the connection is successful.

We then type three lines. The first shows the request line (GET method), the second is the header (defining the host), the third is a blank, terminating the request.

The server response is seven lines starting with the status line. The blank line at the end terminates the server response. The file of 14,230 lines is received after the blank line (not shown here). The last line is the output by the client.

```
$ telnet www.mhhe.com 80
Trying 198.45.24.104 . . .
Connected to www.mhhe.com (198.45.24.104).
Escape character is '^]'.
GET /engcs/compsci/forouzan HTTP/1.1
From: forouzanbehrouz@fhda.edu

HTTP/1.1 200 OK
Date: Thu, 28 Oct 2004 16:27:46 GMT
Server: Apache/1.3.9 (Unix) ApacheJServ/1.1.2 PHP/4.1.2 PHP/3.0.18
MIME-version:1.0
Content-Type: text/html
```

Last-modified: Friday, 15-Oct-04 02:11:31 GMT
Content-length: 14230

Connection closed by foreign host.

Persistent Versus Nonpersistent Connection

HTTP prior to version 1.1 specified a nonpersistent connection, while a persistent connection is the default in version 1.1.

Nonpersistent Connection

In a **nonpersistent connection,** one TCP connection is made for each request/response. The following lists the steps in this strategy:

1. The client opens a TCP connection and sends a request.
2. The server sends the response and closes the connection.
3. The client reads the data until it encounters an end-of-file marker; it then closes the connection.

In this strategy, for *N* different pictures in different files, the connection must be opened and closed *N* times. The nonpersistent strategy imposes high overhead on the server because the server needs *N* different buffers and requires a slow start procedure each time a connection is opened.

Persistent Connection

HTTP version 1.1 specifies a **persistent connection** by default. In a persistent connection, the server leaves the connection open for more requests after sending a response. The server can close the connection at the request of a client or if a time-out has been reached. The sender usually sends the length of the data with each response. However, there are some occasions when the sender does not know the length of the data. This is the case when a document is created dynamically or actively. In these cases, the server informs the client that the length is not known and closes the connection after sending the data so the client knows that the end of the data has been reached.

HTTP version 1.1 specifies a persistent connection by default.

Proxy Server

HTTP supports **proxy servers.** A proxy server is a computer that keeps copies of responses to recent requests. The HTTP client sends a request to the proxy server. The proxy server checks its cache. If the response is not stored in the cache, the proxy server sends the request to the corresponding server. Incoming responses are sent to the proxy server and stored for future requests from other clients.

The proxy server reduces the load on the original server, decreases traffic, and improves latency. However, to use the proxy server, the client must be configured to access the proxy instead of the target server.

27.4 RECOMMENDED READING

For more details about subjects discussed in this chapter, we recommend the following books and sites. The items in brackets [. . .] refer to the reference list at the end of the text.

Books

HTTP is discussed in Chapters 13 and 14 of [Ste96], Section 9.3 of [PD03], Chapter 35 of [Com04], and Section 7.3 of [Tan03].

Sites

The following sites are related to topics discussed in this chapter.

- www.ietf.org/rfc.html Information about RFCs

RFCs

The following RFCs are related to WWW:

1614, 1630, 1737, 1738

The following RFCs are related to HTTP:

2068, 2109

27.5 KEY TERMS

active document
Active Server Pages (ASP)
applet
browser
ColdFusion
Common Gateway Interface (CGI)
dynamic document
entity header
general header
host
Hypertext Markup Language (HTML)
Hypertext Preprocessor (PHP)
Hypertext Transfer Protocol (HTTP)
Java
JavaScript
Java Server Pages (JSP)
nonpersistent connection
path
persistent connection
proxy server
request header
request line
request type
response header
static document
status code
status line
tag
uniform resource locator (URL)
Web
World Wide Web (WWW)

27.6 SUMMARY

- The World Wide Web (WWW) is a repository of information linked together from points all over the world.
- Hypertexts are documents linked to one another through the concept of pointers.
- Browsers interpret and display a Web document.
- A browser consists of a controller, client programs, and interpreters.
- A Web document can be classified as static, dynamic, or active.
- A static document is one in which the contents are fixed and stored in a server. The client can make no changes in the server document.
- Hypertext Markup Language (HTML) is a language used to create static Web pages.
- Any browser can read formatting instructions (tags) embedded in an HTML document.
- Tags provide structure to a document, define titles and headers, format text, control the data flow, insert figures, link different documents together, and define executable code.
- A dynamic Web document is created by a server only at a browser request.
- The Common Gateway Interface (CGI) is a standard for creating and handling dynamic Web documents.
- A CGI program with its embedded CGI interface tags can be written in a language such as C, C++, Shell Script, or Perl.
- An active document is a copy of a program retrieved by the client and run at the client site.
- Java is a combination of a high-level programming language, a run-time environment, and a class library that allows a programmer to write an active document and a browser to run it.
- Java is used to create applets (small application programs).
- The Hypertext Transfer Protocol (HTTP) is the main protocol used to access data on the World Wide Web (WWW).
- HTTP uses a TCP connection to transfer files.
- An HTTP message is similar in form to an SMTP message.
- The HTTP request line consists of a request type, a URL, and the HTTP version number.
- The uniform resource locator (URL) consists of a method, host computer, optional port number, and path name to locate information on the WWW.
- The HTTP request type or method is the actual command or request issued by the client to the server.
- The status line consists of the HTTP version number, a status code, and a status phrase.
- The HTTP status code relays general information, information related to a successful request, redirection information, or error information.
- The HTTP header relays additional information between the client and server.

- An HTTP header consists of a header name and a header value.
- An HTTP general header gives general information about the request or response message.
- An HTTP request header specifies a client's configuration and preferred document format.
- An HTTP response header specifies a server's configuration and special information about the request.
- An HTTP entity header provides information about the body of a document.
- HTTP, version 1.1, specifies a persistent connection.
- A proxy server keeps copies of responses to recent requests.

27.7 PRACTICE SET

Review Questions

1. How is HTTP related to WWW?
2. How is HTTP similar to SMTP?
3. How is HTTP similar to FTP?
4. What is a URL and what are its components?
5. What is a proxy server and how is it related to HTTP?
6. Name the common three components of a browser.
7. What are the three types of Web documents?
8. What does HTML stand for and what is its function?
9. What is the difference between an active document and a dynamic document?
10. What does CGI stand for and what is its function?
11. Describe the relationship between Java and an active document.

Exercises

12. Where will each figure be shown on the screen?

    Look at the following picture:
    then tell me what you feel:
    <IMG SRC="Pictures/Funny1.gif" ALIGN=middle>
    <IMG SRC="Pictures/Funny2.gif" ALIGN=bottom>
    <B>What is your feeling? </B>

13. Show the effect of the following HTML segment.

    The publisher of this book is <A HREF="www.mhhe">
    McGraw-Hill Publisher </A>

14. Show a request that retrieves the document /usr/users/doc/doc1. Use at least two general headers, two request headers, and one entity header.
15. Show the response to Exercise 14 for a successful request.
16. Show the response to Exercise 14 for a document that has permanently moved to /usr/deads/doc1.

17. Show the response to Exercise 14 if there is a syntax error in the request.
18. Show the response to Exercise 14 if the client is unauthorized to access the document.
19. Show a request that asks for information about a document at /bin/users/file. Use at least two general headers and one request header.
20. Show the response to Exercise 19 for a successful request.
21. Show the request to copy the file at location /bin/usr/bin/file1 to /bin/file1.
22. Show the response to Exercise 21.
23. Show the request to delete the file at location /bin/file1.
24. Show the response to Exercise 23.
25. Show a request to retrieve the file at location /bin/etc/file1. The client needs the document only if it was modified after January 23, 1999.
26. Show the response to Exercise 25.
27. Show a request to retrieve the file at location /bin/etc/file1. The client should identify itself.
28. Show the response to Exercise 27.
29. Show a request to store a file at location /bin/letter. The client identifies the types of documents it can accept.
30. Show the response to Exercise 29. The response shows the age of the document as well as the date and time when the contents may change.

CHAPTER 28

Network Management: SNMP

We can define **network management** as monitoring, testing, configuring, and troubleshooting network components to meet a set of requirements defined by an organization. These requirements include the smooth, efficient operation of the network that provides the predefined quality of service for users. To accomplish this task, a network management system uses hardware, software, and humans. In this chapter, first we briefly discuss the functions of a network management system. Then we concentrate on the most common management system, the Simple Network Management Protocol (SNMP).

28.1 NETWORK MANAGEMENT SYSTEM

We can say that the functions performed by a network management system can be divided into five broad categories: configuration management, fault management, performance management, security management, and accounting management, as shown in Figure 28.1.

Figure 28.1 *Functions of a network management system*

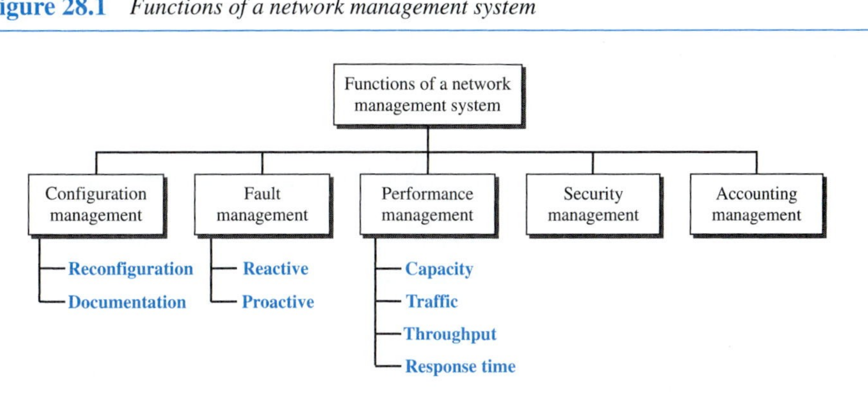

Configuration Management

A large network is usually made up of hundreds of entities that are physically or logically connected to one another. These entities have an initial configuration when the network is set up, but can change with time. Desktop computers may be replaced by others; application software may be updated to a newer version; and users may move from one group to another. The **configuration management** system must know, at any time, the status of each entity and its relation to other entities. Configuration management can be divided into two subsystems: reconfiguration and documentation.

Reconfiguration

Reconfiguration, which means adjusting the network components and features, can be a daily occurrence in a large network. There are three types of reconfiguration: hardware reconfiguration, software reconfiguration, and user-account reconfiguration.

Hardware reconfiguration covers all changes to the hardware. For example, a desktop computer may need to be replaced. A router may need to be moved to another part of the network. A subnetwork may be added or removed from the network. All these need the time and attention of network management. In a large network, there must be specialized personnel trained for quick and efficient hardware reconfiguration. Unfortunately, this type of reconfiguration cannot be automated and must be manually handled case by case.

Software reconfiguration covers all changes to the software. For example, new software may need to be installed on servers or clients. An operating system may need updating. Fortunately, most software reconfiguration can be automated. For example, updating an application on some or all clients can be electronically downloaded from the server.

User-account reconfiguration is not simply adding or deleting users on a system. You must also consider the user privileges, both as an individual and as a member of a group. For example, a user may have read and write permission with regard to some files, but only read permission with regard to other files. User-account reconfiguration can be, to some extent, automated. For example, in a college or university, at the beginning of each quarter or semester, new students are added to the system. The students are normally grouped according to the courses they take or the majors they pursue.

Documentation

The original network configuration and each subsequent change must be recorded meticulously. This means that there must be documentation for hardware, software, and user accounts.

Hardware documentation normally involves two sets of documents: maps and specifications. Maps track each piece of hardware and its connection to the network. There can be one general map that shows the logical relationship between each subnetwork. There can also be a second general map that shows the physical location of each subnetwork. For each subnetwork, then, there is one or more maps that show all pieces of equipment. The maps use some kind of standardization to be easily read and understood by current and future personnel. Maps are not enough per se. Each piece of hardware also needs to be documented. There must be a set of specifications for each piece of hardware connected to the network. These specifications must include information such as hardware type, serial number, vendor (address and phone number), time of purchase, and warranty information.

All software must also be documented. Software documentation includes information such as the software type, the version, the time installed, and the license agreement.

Most operating systems have a utility that allows the documentation of user accounts and their privileges. The management must make sure that the files with this information are updated and secured. Some operating systems record access privileges in two documents; one shows all files and access types for each user; the other shows the list of users that have a particular access to a file.

Fault Management

Complex networks today are made up of hundreds and sometimes thousands of components. Proper operation of the network depends on the proper operation of each component individually and in relation to each other. **Fault management** is the area of network management that handles this issue.

An effective fault management system has two subsystems: reactive fault management and proactive fault management.

Reactive Fault Management

A reactive fault management system is responsible for detecting, isolating, correcting, and recording faults. It handles short-term solutions to faults.

The first step taken by a reactive fault management system is to detect the exact location of the fault. A fault is defined as an abnormal condition in the system. When a fault occurs, either the system stops working properly or the system creates excessive errors. A good example of a fault is a damaged communication medium. This fault may interrupt communication or produce excessive errors.

The next step taken by a reactive fault management system is to isolate the fault. A fault, if isolated, usually affects only a few users. After isolation, the affected users are immediately notified and given an estimated time of correction.

The third step is to correct the fault. This may involve replacing or repairing the faulty component(s).

After the fault is corrected, it must be documented. The record should show the exact location of the fault, the possible cause, the action or actions taken to correct the fault, the cost, and time it took for each step. Documentation is extremely important for several reasons:

- The problem may recur. Documentation can help the present or future administrator or technician solve a similar problem.
- The frequency of the same kind of failure is an indication of a major problem in the system. If a fault happens frequently in one component, it should be replaced with a similar one, or the whole system should be changed to avoid the use of that type of component.
- The statistic is helpful to another part of network management, performance management.

Proactive Fault Management

Proactive fault management tries to prevent faults from occurring. Although this is not always possible, some types of failures can be predicted and prevented. For example, if a manufacturer specifies a lifetime for a component or a part of a component, it is a good strategy to replace it before that time. As another example, if a fault happens frequently at one particular point of a network, it is wise to carefully reconfigure the network to prevent the fault from happening again.

Performance Management

Performance management, which is closely related to fault management, tries to monitor and control the network to ensure that it is running as efficiently as possible. Performance management tries to quantify performance by using some measurable quantity such as capacity, traffic, throughput, or response time.

Capacity

One factor that must be monitored by a performance management system is the capacity of the network. Every network has a limited capacity, and the performance management system must ensure that it is not used above this capacity. For example, if a LAN is designed for 100 stations at an average data rate of 2 Mbps, it will not operate properly if 200 stations are connected to the network. The data rate will decrease and blocking may occur.

Traffic

Traffic can be measured in two ways: internally and externally. Internal traffic is measured by the number of packets (or bytes) traveling inside the network. External traffic is measured by the exchange of packets (or bytes) outside the network. During peak hours, when the system is heavily used, blocking may occur if there is excessive traffic.

Throughput

We can measure the throughput of an individual device (such as a router) or a part of the network. Performance management monitors the throughput to make sure that it is not reduced to unacceptable levels.

Response Time

Response time is normally measured from the time a user requests a service to the time the service is granted. Other factors such as capacity and traffic can affect the response time. Performance management monitors the average response time and the peak-hour response time. Any increase in response time is a very serious condition as it is an indication that the network is working above its capacity.

Security Management

Security management is responsible for controlling access to the network based on the predefined policy. We discuss security and in particular network security in Chapters 31 and 32.

Accounting Management

Accounting management is the control of users' access to network resources through charges. Under accounting management, individual users, departments, divisions, or even projects are charged for the services they receive from the network. Charging does not necessarily mean cash transfer; it may mean debiting the departments or divisions for budgeting purposes. Today, organizations use an accounting management system for the following reasons:

- It prevents users from monopolizing limited network resources.
- It prevents users from using the system inefficiently.
- Network managers can do short- and long-term planning based on the demand for network use.

28.2 SIMPLE NETWORK MANAGEMENT PROTOCOL (SNMP)

The **Simple Network Management Protocol (SNMP)** is a framework for managing devices in an internet using the TCP/IP protocol suite. It provides a set of fundamental operations for monitoring and maintaining an internet.

Concept

SNMP uses the concept of manager and agent. That is, a manager, usually a host, controls and monitors a set of agents, usually routers (see Figure 28.2).

Figure 28.2 *SNMP concept*

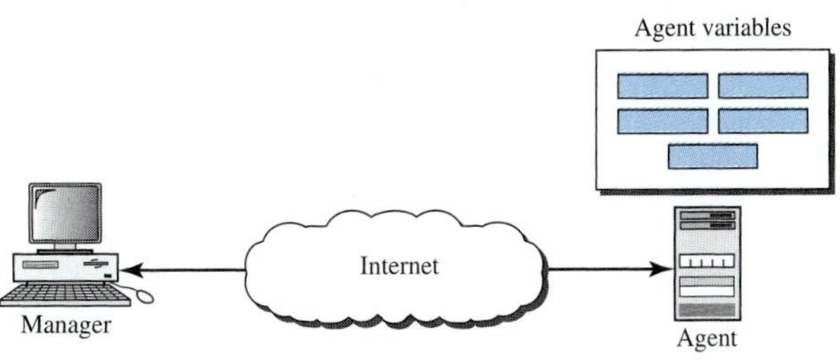

SNMP is an application-level protocol in which a few manager stations control a set of agents. The protocol is designed at the application level so that it can monitor devices made by different manufacturers and installed on different physical networks. In other words, SNMP frees management tasks from both the physical characteristics of the managed devices and the underlying networking technology. It can be used in a heterogeneous internet made of different LANs and WANs connected by routers made by different manufacturers.

Managers and Agents

A management station, called a **manager,** is a host that runs the SNMP client program. A managed station, called an **agent,** is a router (or a host) that runs the SNMP server program. Management is achieved through simple interaction between a manager and an agent.

The agent keeps performance information in a database. The manager has access to the values in the database. For example, a router can store in appropriate variables the number of packets received and forwarded. The manager can fetch and compare the values of these two variables to see if the router is congested or not.

The manager can also make the router perform certain actions. For example, a router periodically checks the value of a reboot counter to see when it should reboot itself. It reboots itself, for example, if the value of the counter is 0. The manager can use this feature to reboot the agent remotely at any time. It simply sends a packet to force a 0 value in the counter.

Agents can also contribute to the management process. The server program running on the agent can check the environment, and if it notices something unusual, it can send a warning message, called a **trap,** to the manager.

In other words, management with SNMP is based on three basic ideas:

1. A manager checks an agent by requesting information that reflects the behavior of the agent.
2. A manager forces an agent to perform a task by resetting values in the agent database.
3. An agent contributes to the management process by warning the manager of an unusual situation.

Management Components

To do management tasks, SNMP uses two other protocols: **Structure of Management Information (SMI)** and **Management Information Base (MIB).** In other words, management on the Internet is done through the cooperation of the three protocols SNMP, SMI, and MIB, as shown in Figure 28.3.

Figure 28.3 *Components of network management on the Internet*

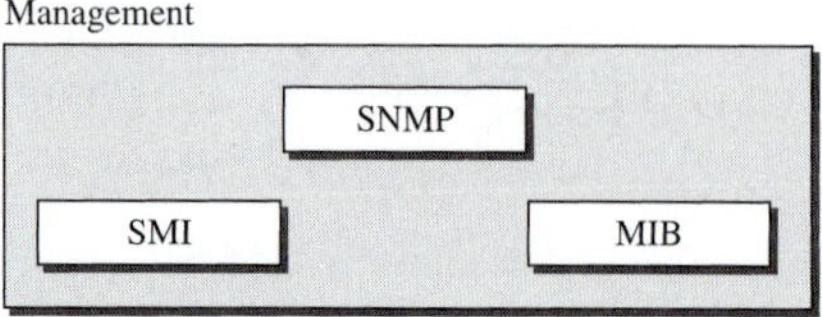

Let us elaborate on the interactions between these protocols.

Role of SNMP

SNMP has some very specific roles in network management. It defines the format of the packet to be sent from a manager to an agent and vice versa. It also interprets the result and creates statistics (often with the help of other management software). The packets exchanged contain the object (variable) names and their status (values). SNMP is responsible for reading and changing these values.

SNMP defines the format of packets exchanged between a manager and an agent. It reads and changes the status (values) of objects (variables) in SNMP packets.

Role of SMI

To use SNMP, we need rules. We need rules for naming objects. This is particularly important because the objects in SNMP form a hierarchical structure (an object may have a parent object and some children objects). Part of a name can be inherited from the parent. We also need rules to define the type of the objects. What types of objects are handled by SNMP? Can SNMP handle simple types or structured types? How many simple types are available? What are the sizes of these types? What is the range of these types? In addition, how are each of these types encoded?

We need these universal rules because we do not know the architecture of the computers that send, receive, or store these values. The sender may be a powerful computer in which an integer is stored as 8-byte data; the receiver may be a small computer that stores an integer as 4-byte data.

SMI is a protocol that defines these rules. However, we must understand that SMI only defines the rules; it does not define how many objects are managed in an entity or which object uses which type. SMI is a collection of general rules to name objects and to list their types. The association of an object with the type is not done by SMI.

SMI defines the general rules for naming objects, defining object types (including range and length), and showing how to encode objects and values.

SMI does not define the number of objects an entity should manage or name the objects to be managed or define the association between the objects and their values.

Role of MIB

We hope it is clear that we need another protocol. For each entity to be managed, this protocol must define the number of objects, name them according to the rules defined by SMI, and associate a type to each named object. This protocol is MIB. MIB creates a set of objects defined for each entity similar to a database (mostly metadata in a database, names and types without values).

MIB creates a collection of named objects, their types, and their relationships to each other in an entity to be managed.

An Analogy

Before discussing each of these protocols in greater detail, we give an analogy. The three network management components are similar to what we need when we write a program in a computer language to solve a problem.

Before we write a program, the syntax of the language (such as C or Java) must be predefined. The language also defines the structure of variables (simple, structured, pointer, and so on) and how the variables must be named. For example, a variable name must be 1 to *N* characters in length and start with a letter followed by alphanumeric characters. The language also defines the type of data to be used (integer, float, char, etc.). In programming the rules are defined by the language. In network management the rules are defined by SMI.

Most computer languages require that variables be declared in each specific program. The declaration names each variable and defines the predefined type. For example, if a program has two variables (an integer named *counter* and an array named *grades* of type char), they must be declared at the beginning of the program:

```
int counter;
char grades [40];
```

Note that the declarations name the variables (counter and grades) and define the type of each variable. Because the types are predefined in the language, the program knows the range and size of each variable.

MIB does this task in network management. MIB names each object and defines the type of the objects. Because the type is defined by SMI, SNMP knows the range and size.

After declaration in programming, the program needs to write statements to store values in the variables and change them if needed. SNMP does this task in network management. SNMP stores, changes, and interprets the values of objects already declared by MIB according to the rules defined by SMI.

We can compare the task of network management to the task of writing a program.

- Both tasks need rules. In network management this is handled by SMI.
- Both tasks need variable declarations. In network management this is handled by MIB.
- Both tasks have actions performed by statements. In network management this is handled by SNMP.

An Overview

Before discussing each component in detail, we show how each is involved in a simple scenario. This is an overview that will be developed later at the end of the chapter. A manager station (SNMP client) wants to send a message to an agent station (SNMP server) to find the number of UDP user datagrams received by the agent. Figure 28.4 shows an overview of steps involved.

MIB is responsible for finding the object that holds the number of the UDP user datagrams received. SMI, with the help of another embedded protocol, is responsible for encoding the name of the object. SNMP is responsible for creating a message, called a GetRequest message, and encapsulating the encoded message. Of course, things are more complicated than this simple overview, but we first need more details of each protocol.

Figure 28.4 *Management overview*

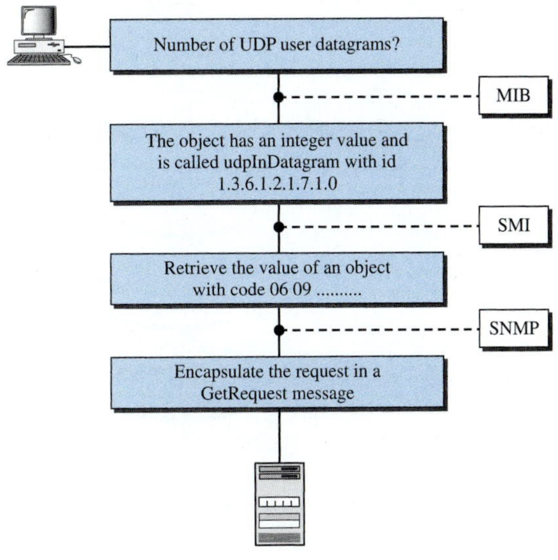

Structure of Management Information

The Structure of Management Information, version 2 (SMIv2) is a component for network management. Its functions are

1. To name objects
2. To define the type of data that can be stored in an object
3. To show how to encode data for transmission over the network

SMI is a guideline for SNMP. It emphasizes three attributes to handle an object: name, data type, and encoding method (see Figure 28.5).

Figure 28.5 *Object attributes*

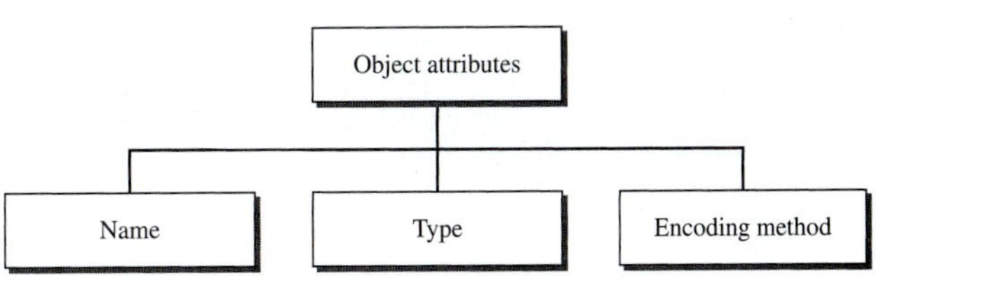

Name

SMI requires that each managed object (such as a router, a variable in a router, a value) have a unique name. To name objects globally, SMI uses an **object identifier,** which is a hierarchical identifier based on a tree structure (see Figure 28.6).

Figure 28.6 *Object identifier*

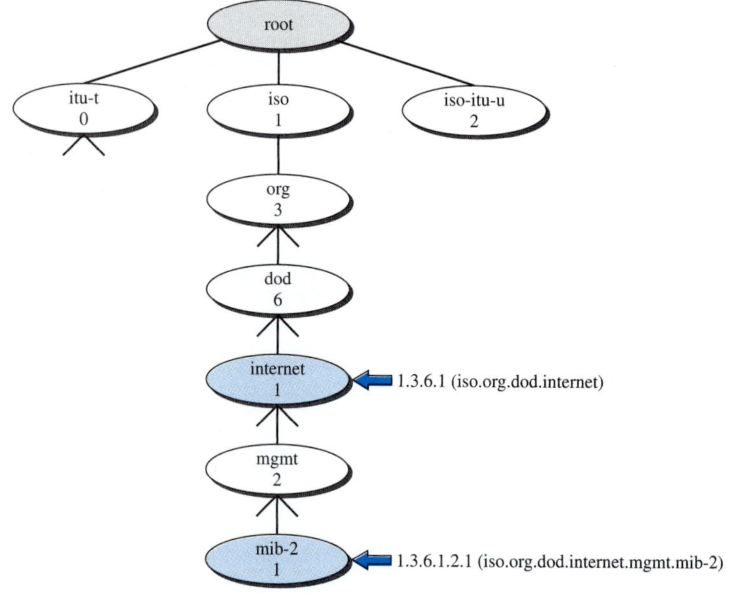

The tree structure starts with an unnamed root. Each object can be defined by using a sequence of integers separated by dots. The tree structure can also define an object by using a sequence of textual names separated by dots. The integer-dot representation is used in SNMP. The name-dot notation is used by people. For example, the following shows the same object in two different notations:

iso.org.dod.internet.mgmt.mib-2 → 1.3.6.1.2.1

The objects that are used in SNMP are located under the *mib-2* object, so their identifiers always start with 1.3.6.1.2.1.

All objects managed by SNMP are given an object identifier.
The object identifier always starts with 1.3.6.1.2.1.

Type

The second attribute of an object is the type of data stored in it. To define the data type, SMI uses fundamental **Abstract Syntax Notation 1 (ASN.1)** definitions and adds some new definitions. In other words, SMI is both a subset and a superset of ASN.1.

SMI has two broad categories of data type: *simple* and *structured.* We first define the simple types and then show how the structured types can be constructed from the simple ones (see Figure 28.7).

Figure 28.7 *Data type*

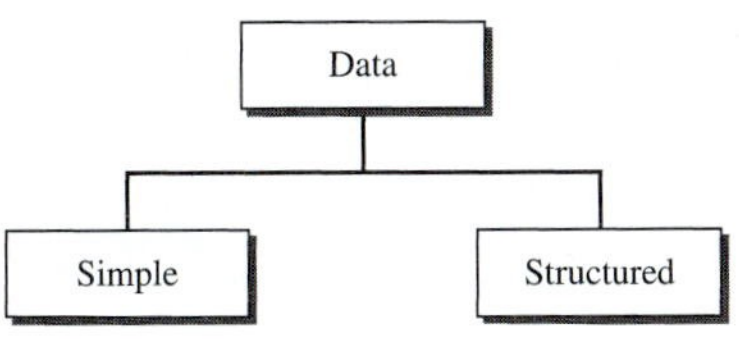

Simple Type The **simple data types** are atomic data types. Some of them are taken directly from ASN.1; others are added by SMI. The most important ones are given in Table 28.1. The first five are from ASN.1; the next seven are defined by SMI.

Table 28.1 *Data types*

Type	*Size*	*Description*
INTEGER	4 bytes	An integer with a value between -2^{31} and $2^{31} - 1$
Integer32	4 bytes	Same as INTEGER
Unsigned32	4 bytes	Unsigned with a value between 0 and $2^{32} - 1$
OCTET STRING	Variable	Byte string up to 65,535 bytes long
OBJECT IDENTIFIER	Variable	An object identifier
IPAddress	4 bytes	An IP address made of four integers
Counter32	4 bytes	An integer whose value can be incremented from 0 to 2^{32}; when it reaches its maximum value, it wraps back to 0.
Counter64	8 bytes	64-bit counter
Gauge32	4 bytes	Same as Counter32, but when it reaches its maximum value, it does not wrap; it remains there until it is reset
TimeTicks	4 bytes	A counting value that records time in $\frac{1}{100}$ s
BITS		A string of bits
Opaque	Variable	Uninterpreted string

Structured Type By combining simple and structured data types, we can make new structured data types. SMI defines two **structured data types:** *sequence* and *sequence of.*

- **Sequence.** A *sequence* data type is a combination of simple data types, not necessarily of the same type. It is analogous to the concept of a *struct* or a *record* used in programming languages such as C.
- **Sequence of.** A *sequence of* data type is a combination of simple data types all of the same type or a combination of sequence data types all of the same type. It is analogous to the concept of an *array* used in programming languages such as C.

Figure 28.8 shows a conceptual view of data types.

Figure 28.8 *Conceptual data types*

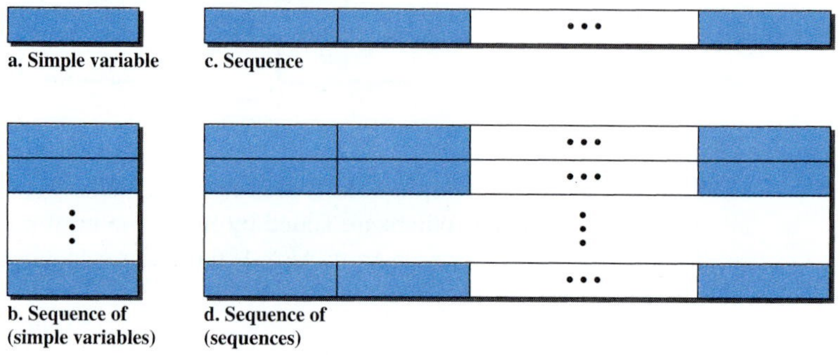

a. Simple variable
c. Sequence
b. Sequence of (simple variables)
d. Sequence of (sequences)

Encoding Method

SMI uses another standard, **Basic Encoding Rules (BER),** to encode data to be transmitted over the network. BER specifies that each piece of data be encoded in triplet format: tag, length, and value, as illustrated in Figure 28.9.

Figure 28.9 *Encoding format*

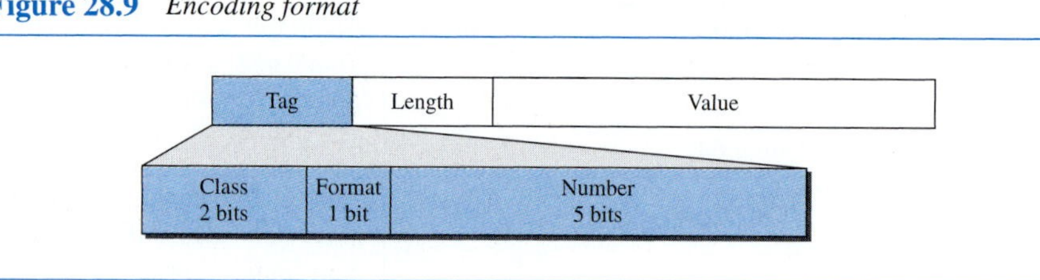

- **Tag.** The tag is a 1-byte field that defines the type of data. It is composed of three subfields: *class* (2 bits), *format* (1 bit), and *number* (5 bits). The class subfield defines the scope of the data. Four classes are defined: universal (00), applicationwide (01), context-specific (10), and private (11). The universal data types are those taken from ASN.1 (INTEGER, OCTET STRING, and ObjectIdentifier). The applicationwide data types are those added by SMI (IPAddress, Counter, Gauge, and TimeTicks). The five context-specific data types have meanings that may change from one protocol to another. The private data types are vendor-specific.

  The format subfield indicates whether the data are simple (0) or structured (1). The number subfield further divides simple or structured data into subgroups. For example, in the universal class, with simple format, INTEGER has a value of 2, OCTET STRING has a value of 4, and so on. Table 28.2 shows the data types we use in this chapter and their tags in binary and hexadecimal numbers.

Table 28.2 *Codes for data types*

Data Type	*Class*	*Format*	*Number*	*Tag (Binary)*	*Tag (Hex)*
INTEGER	00	0	00010	**00000010**	**02**
OCTET STRING	00	0	00100	**00000100**	**04**
OBJECT IDENTIFIER	00	0	00110	**00000110**	**06**
NULL	00	0	00101	**00000101**	**05**
Sequence, sequence of	00	1	10000	**00110000**	**30**
IPAddress	01	0	00000	**01000000**	**40**
Counter	01	0	00001	**01000001**	**41**
Gauge	01	0	00010	**01000010**	**42**
TimeTicks	01	0	00011	**01000011**	**43**
Opaque	01	0	00100	**01000100**	**44**

- **Length.** The length field is 1 or more bytes. If it is 1 byte, the most significant bit must be 0. The other 7 bits define the length of the data. If it is more than 1 byte, the most significant bit of the first byte must be 1. The other 7 bits of the first byte define the number of bytes needed to define the length. See Figure 28.10 for a depiction of the length field.

Figure 28.10 *Length format*

a. The colored part defines the length (2).

b. The shaded part defines the length of the length (2 bytes); the colored bytes define the length (260 bytes).

- **Value.** The value field codes the value of the data according to the rules defined in BER.

To show how these three fields—tag, length, and value—can define objects, we give some examples.

Example 28.1

Figure 28.11 shows how to define INTEGER 14.

Example 28.2

Figure 28.12 shows how to define the OCTET STRING "HI."

Figure 28.11 *Example 28.1, INTEGER 14*

02	04	00	00	00	0E
00000010	00000100	00000000	00000000	00000000	00001110
Tag (integer)	Length (4 bytes)	Value (14)			

Figure 28.12 *Example 28.2, OCTET STRING "HI"*

04	02	48	49
00000100	00000010	01001000	01001001
Tag (String)	Length (2 bytes)	Value (H)	Value (I)

Example 28.3

Figure 28.13 shows how to define ObjectIdentifier 1.3.6.1 (iso.org.dod.internet).

Figure 28.13 *Example 28.3, ObjectIdentifier 1.3.6.1*

06	04	01	03	06	01
00000110	00000100	00000001	00000011	00000110	00000001
Tag (ObjectId)	Length (4 bytes)	Value (1)	Value (3)	Value (6)	Value (1)
		1.3.6.1 (iso.org.dod.internet)			

Example 28.4

Figure 28.14 shows how to define IPAddress 131.21.14.8.

Figure 28.14 *Example 28.4, IPAddress 131.21.14.8*

40	04	83	15	0E	08
01000000	00000100	10000011	00010101	00001110	00001000
Tag (IPAddress)	Length (4 bytes)	Value (131)	Value (21)	Value (14)	Value (8)
		131.21.14.8			

Management Information Base (MIB)

The Management Information Base, version 2 (MIB2) is the second component used in network management. Each agent has its own MIB2, which is a collection of all the objects that the manager can manage. The objects in MIB2 are categorized under 10 different groups: system, interface, address translation, ip, icmp, tcp, udp, egp, transmission, and snmp. These groups are under the mib-2 object in the object identifier tree (see Figure 28.15). Each group has defined variables and/or tables.

Figure 28.15 *mib-2*

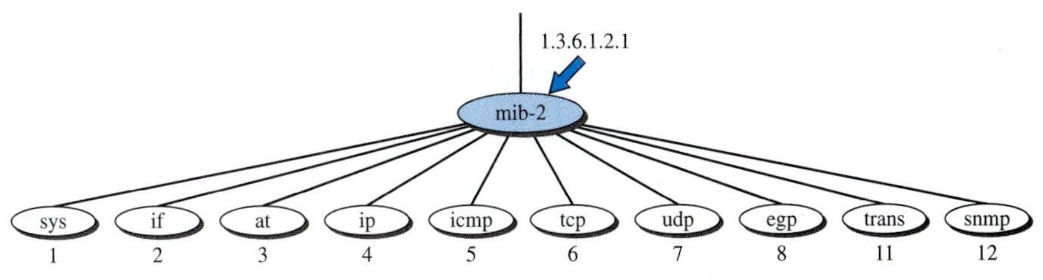

The following is a brief description of some of the objects:

- **sys** This object (*system*) defines general information about the node (system), such as the name, location, and lifetime.
- **if** This object (*interface*) defines information about all the interfaces of the node including interface number, physical address, and IP address.
- **at** This object (*address translation*) defines the information about the ARP table.
- **ip** This object defines information related to IP, such as the routing table and the IP address.
- **icmp** This object defines information related to ICMP, such as the number of packets sent and received and total errors created.
- **tcp** This object defines general information related to TCP, such as the connection table, time-out value, number of ports, and number of packets sent and received.
- **udp** This object defines general information related to UDP, such as the number of ports and number of packets sent and received.
- **snmp** This object defines general information related to SNMP itself.

Accessing MIB Variables

To show how to access different variables, we use the udp group as an example. There are four simple variables in the udp group and one sequence of (table of) records. Figure 28.16 shows the variables and the table.

We will show how to access each entity.

Simple Variables To access any of the simple variables, we use the id of the group (1.3.6.1.2.1.7) followed by the id of the variable. The following shows how to access each variable.

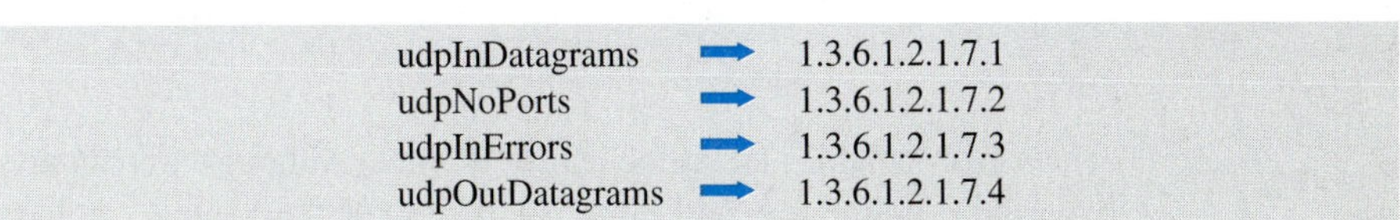

udpInDatagrams	→	1.3.6.1.2.1.7.1
udpNoPorts	→	1.3.6.1.2.1.7.2
udpInErrors	→	1.3.6.1.2.1.7.3
udpOutDatagrams	→	1.3.6.1.2.1.7.4

Figure 28.16 *udp group*

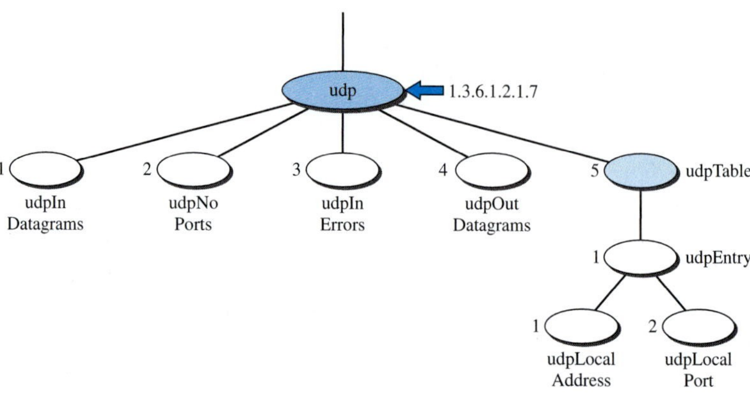

However, these object identifiers define the variable, not the instance (contents). To show the instance or the contents of each variable, we must add an instance suffix. The instance suffix for a simple variable is simply a 0. In other words, to show an instance of the above variables, we use the following:

udpInDatagrams.0 ➡ 1.3.6.1.2.1.7.1.**0**
udpNoPorts.0 ➡ 1.3.6.1.2.1.7.2.**0**
udpInErrors.0 ➡ 1.3.6.1.2.1.7.3.**0**
udpOutDatagrams.0 ➡ 1.3.6.1.2.1.7.4.**0**

Tables To identify a table, we first use the table id. The udp group has only one table (with id 5) as illustrated in Figure 28.17.

So to access the table, we use the following:

udpTable ➡ 1.3.6.1.2.1.7.5

However, the table is not at the leaf level in the tree structure. We cannot access the table; we define the entry (sequence) in the table (with id of 1), as follows:

udpEntry ➡ 1.3.6.1.2.1.7.5.**1**

This entry is also not a leaf and we cannot access it. We need to define each entity (field) in the entry.

udpLocalAddress ➡ 1.3.6.1.2.1.7.5.**1.1**
udpLocalPort ➡ 1.3.6.1.2.1.7.5.**1.2**

These two variables are at the leaf of the tree. Although we can access their instances, we need to define *which* instance. At any moment, the table can have several values for

Figure 28.17 *udp variables and tables*

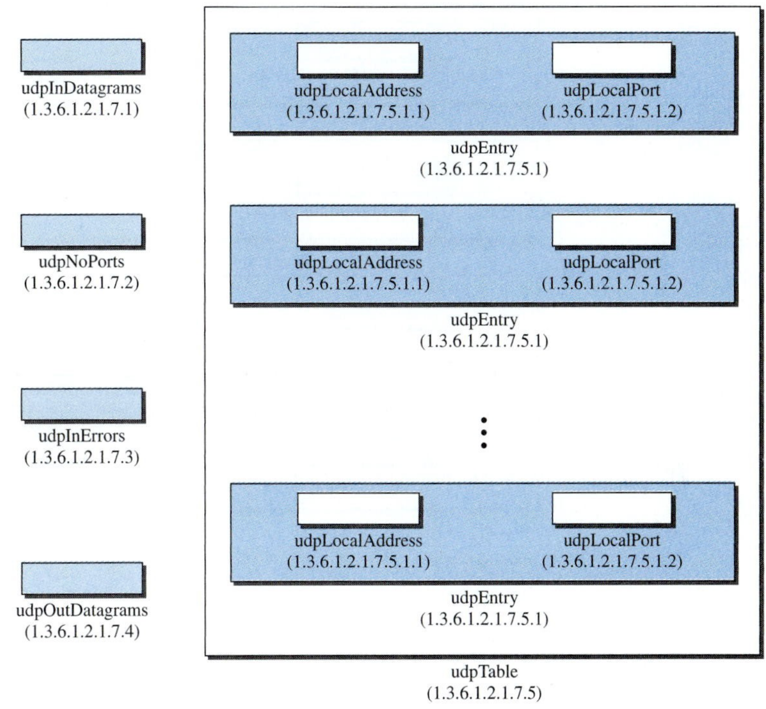

each local address/local port pair. To access a specific instance (row) of the table, we add the index to the above ids. In MIB, the indexes of arrays are not integers (like most programming languages). The indexes are based on the value of one or more fields in the entries. In our example, the udpTable is indexed based on both the local address and the local port number. For example, Figure 28.18 shows a table with four rows and values for each field. The index of each row is a combination of two values.

To access the instance of the local address for the first row, we use the identifier augmented with the instance index:

udpLocalAddress.181.23.45.14.23 → 1.3.6.1.2.7.5.1.1.181.23.45.14.23

Note that not all tables are indexed in the same way. Some tables are indexed by using the value of one field, others by using the value of two fields, and so on.

Lexicographic Ordering

One interesting point about the MIB variables is that the object identifiers (including the instance identifiers) follow in lexicographic order. Tables are ordered according to column-row rules, which means one should go column by column. In each column, one should go from the top to the bottom, as shown in Figure 28.19.

Figure 28.18 *Indexes for udpTable*

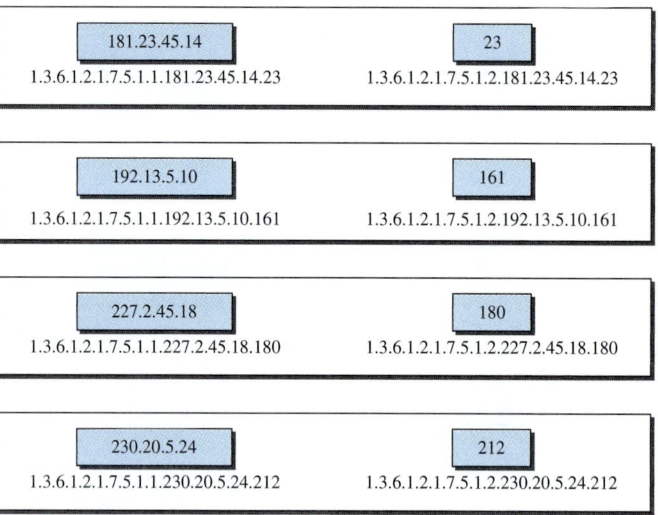

Figure 28.19 *Lexicographic ordering*

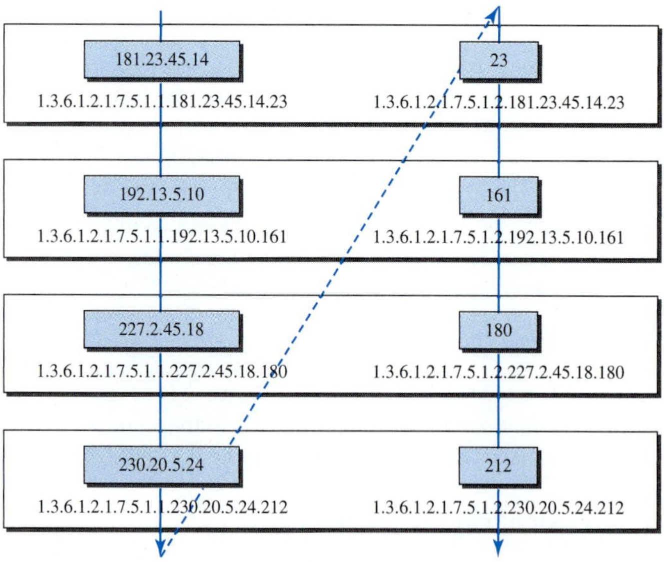

The **lexicographic ordering** enables a manager to access a set of variables one after another by defining the first variable, as we will see in the GetNextRequest command in the next section.

SNMP

SNMP uses both SMI and MIB in Internet network management. It is an application program that allows

1. A manager to retrieve the value of an object defined in an agent
2. A manager to store a value in an object defined in an agent
3. An agent to send an alarm message about an abnormal situation to the manager

PDUs

SNMPv3 defines eight types of packets (or PDUs): GetRequest, GetNextRequest, GetBulkRequest, SetRequest, Response, Trap, InformRequest, and Report (see Figure 28.20).

Figure 28.20 *SNMP PDUs*

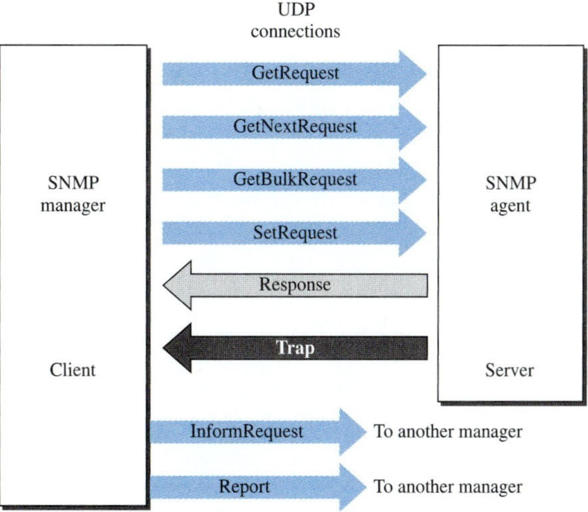

GetRequest The GetRequest PDU is sent from the manager (client) to the agent (server) to retrieve the value of a variable or a set of variables.

GetNextRequest The GetNextRequest PDU is sent from the manager to the agent to retrieve the value of a variable. The retrieved value is the value of the object following the defined ObjectId in the PDU. It is mostly used to retrieve the values of the entries in a table. If the manager does not know the indexes of the entries, it cannot retrieve the values. However, it can use GetNextRequest and define the ObjectId of the table. Because the first entry has the ObjectId immediately after the ObjectId of the table, the value of the first entry is returned. The manager can use this ObjectId to get the value of the next one, and so on.

GetBulkRequest The GetBulkRequest PDU is sent from the manager to the agent to retrieve a large amount of data. It can be used instead of multiple GetRequest and GetNextRequest PDUs.

SetRequest The SetRequest PDU is sent from the manager to the agent to set (store) a value in a variable.

Response The Response PDU is sent from an agent to a manager in response to GetRequest or GetNextRequest. It contains the value(s) of the variable(s) requested by the manager.

Trap The **Trap** (also called SNMPv2 Trap to distinguish it from SNMPv1 Trap) PDU is sent from the agent to the manager to report an event. For example, if the agent is rebooted, it informs the manager and reports the time of rebooting.

InformRequest The InformRequest PDU is sent from one manager to another remote manager to get the value of some variables from agents under the control of the remote manager. The remote manager responds with a Response PDU.

Report The Report PDU is designed to report some types of errors between managers. It is not yet in use.

Format

The format for the eight SNMP PDUs is shown in Figure 28.21. The GetBulkRequest PDU differs from the others in two areas, as shown in the figure.

Figure 28.21 *SNMP PDU format*

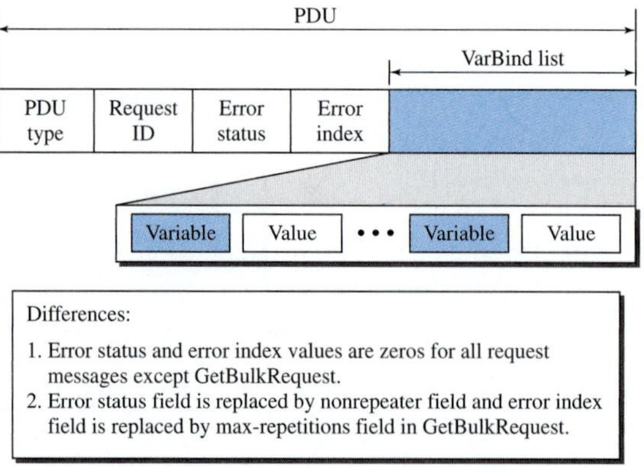

The fields are listed below:

- **PDU type.** This field defines the type of the PDU (see Table 28.4).
- **Request ID.** This field is a sequence number used by the manager in a Request PDU and repeated by the agent in a response. It is used to match a request to a response.

- **Error status.** This is an integer that is used only in Response PDUs to show the types of errors reported by the agent. Its value is 0 in Request PDUs. Table 28.3 lists the types of errors that can occur.

Table 28.3 *Types of errors*

Status	*Name*	*Meaning*
0	noError	No error
1	tooBig	Response too big to fit in one message
2	noSuchName	Variable does not exist
3	badValue	The value to be stored is invalid
4	readOnly	The value cannot be modified
5	genErr	Other errors

- **Nonrepeaters.** This field is used only in GetBulkRequest and replaces the error status field, which is empty in Request PDUs.
- **Error index.** The error index is an offset that tells the manager which variable caused the error.
- **Max-repetition.** This field is also used only in GetBulkRequest and replaces the error index field, which is empty in Request PDUs.
- **VarBind list.** This is a set of variables with the corresponding values the manager wants to retrieve or set. The values are null in GetRequest and GetNextRequest. In a Trap PDU, it shows the variables and values related to a specific PDU.

Messages

SNMP does not send only a PDU, it embeds the PDU in a message. A message in SNMPv3 is made of four elements: version, header, security parameters, and data (which include the encoded PDU), as shown in Figure 28.22.

Because the length of these elements is different from message to message, SNMP uses BER to encode each element. Remember that BER uses the tag and the length to define a value. The *version* defines the current version (3). The *header* contains values for message identification, maximum message size (the maximum size of the reply), message flag (one octet of data type OCTET STRING where each bit defines security type, such as privacy or authentication, or other information), and a message security model (defining the security protocol). The message *security parameter* is used to create a message digest (see Chapter 31). The data contain the PDU. If the data are encrypted, there is information about the encrypting engine (the manager program that did the encryption) and the encrypting context (the type of encryption) followed by the encrypted PDU. If the data are not encrypted, the data consist of just the PDU.

To define the type of PDU, SNMP uses a tag. The class is context-sensitive (10), the format is structured (1), and the numbers are 0, 1, 2, 3, 5, 6, 7, and 8 (see Table 28.4). Note that SNMPv1 defined A4 for Trap, which is obsolete today.

Figure 28.22 *SNMP message*

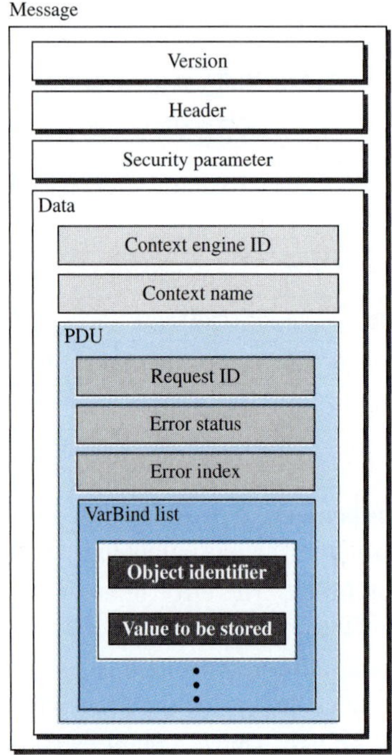

Table 28.4 *Codes for SNMP messages*

Data	*Class*	*Format*	*Number*	*Whole Tag (Binary)*	*Whole Tag (Hex)*
GetRequest	10	1	00000	**10100000**	**A0**
GetNextRequest	10	1	00001	**10100001**	**A1**
Response	10	1	00010	**10100010**	**A2**
SetRequest	10	1	00011	**10100011**	**A3**
GetBulkRequest	10	1	00101	**10100101**	**A5**
InformRequest	10	1	00110	**10100110**	**A6**
Trap (SNMPv2)	10	1	00111	**10100111**	**A7**
Report	10	1	01000	**10101000**	**A8**

Example 28.5

In this example, a manager station (SNMP client) uses the GetRequest message to retrieve the number of UDP datagrams that a router has received.

There is only one VarBind entity. The corresponding MIB variable related to this information is udpInDatagrams with the object identifier 1.3.6.1.2.1.7.1.0. The manager wants to retrieve a value (not to store a value), so the value defines a null entity. Figure 28.23 shows the conceptual

Figure 28.23 *Example 28.5*

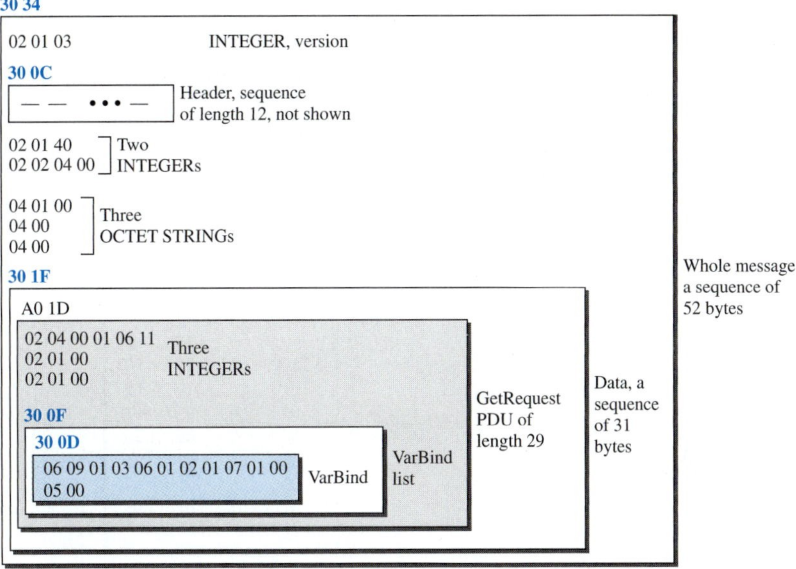

view of the packet and the hierarchical nature of sequences. We have used white and colored boxes for the sequences and a gray one for the PDU.

The VarBind list has only one VarBind. The variable is of type 06 and length 09. The value is of type 05 and length 00. The whole VarBind is a sequence of length 0D (13). The VarBind list is also a sequence of length 0F (15). The GetRequest PDU is of length ID (29).

Now we have three OCTET STRINGs related to the security parameter, security model, and flags. Then we have two integers defining maximum size (1024) and message ID (64). The header is a sequence of length 12, which we left blank for simplicity. There is one integer, version (version 3). The whole message is a sequence of 52 bytes.

Figure 28.24 shows the actual message sent by the manager station (client) to the agent (server).

UDP Ports

SNMP uses the services of UDP on two well-known ports, 161 and 162. The well-known port 161 is used by the server (agent), and the well-known port 162 is used by the client (manager).

The agent (server) issues a passive open on port 161. It then waits for a connection from a manager (client). A manager (client) issues an active open, using an ephemeral port. The request messages are sent from the client to the server, using the ephemeral port as the source port and the well-known port 161 as the destination port. The response messages are sent from the server to the client, using the well-known port 161 as the source port and the ephemeral port as the destination port.

The manager (client) issues a passive open on port 162. It then waits for a connection from an agent (server). Whenever it has a Trap message to send, an agent (server) issues an active open, using an ephemeral port. This connection is only one-way, from the server to the client (see Figure 28.25).

Figure 28.24 *GetRequest message*

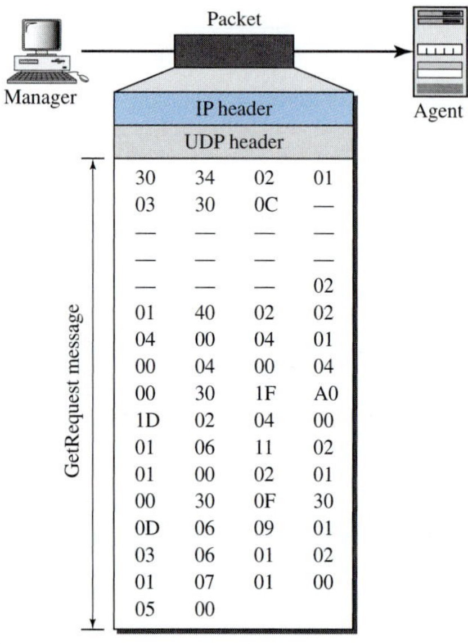

Figure 28.25 *Port numbers for SNMP*

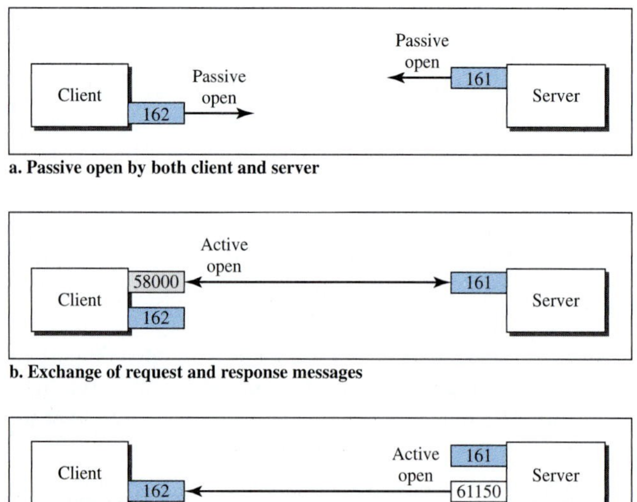

The client/server mechanism in SNMP is different from other protocols. Here both the client and the server use well-known ports. In addition, both the client and the server are running infinitely. The reason is that request messages are initiated by a manager (client), but Trap messages are initiated by an agent (server).

Security

The main difference between SNMPv3 and SNMPv2 is the enhanced security. SNMPv3 provides two types of security: general and specific. SNMPv3 provides message authentication, privacy, and manager authorization. We discuss these three aspects in Chapter 31. In addition, SNMPv3 allows a manager to remotely change the security configuration, which means that the manager does not have to be physically present at the manager station.

28.3 RECOMMENDED READING

For more details about subjects discussed in this chapter, we recommend the following books and sites. The items in brackets [. . .] refer to the reference list at the end of the text.

Books

SNMP is discussed in [MS01], Chapter 25 of [Ste94], Section 22.3 of [Sta04], and Chapter 39 of [Com04]. Network management is discussed in [Sub01].

Sites

The following sites are related to topics discussed in this chapter.

- www.ietf.org/rfc.html Information about RFCs

RFCs

The following RFCs are related to SNMP, MIB, and SMI:

1065, 1067, 1098, 1155, 1157, 1212, 1213, 1229, 1231, 1243, 1284, 1351, 1352, 1354, 1389, 1398, 1414, 1441, 1442, 1443, 1444, 1445, 1446, 1447, 1448, 1449, 1450, 1451, 1452, 1461, 1472, 1474, 1537, 1623, 1643, 1650, 1657, 1665, 1666, 1696, 1697, 1724, 1742, 1743, 1748, 1749

28.4 KEY TERMS

Abstract Syntax Notation 1 (ASN.1)
accounting management
agent
Basic Encoding Rules (BER)
configuration management
fault management
hardware documentation
lexicographic ordering
Management Information Base (MIB)
manager
network management
object identifier
performance management
security management

simple data type
Simple Network Management Protocol (SNMP)
structured data type
Structure of Management Information (SMI)
trap

28.5 SUMMARY

- The five areas comprising network management are configuration management, fault management, performance management, accounting management, and security management.
- Configuration management is concerned with the physical or logical changes of network entities. It includes the reconfiguration and documentation of hardware, software, and user accounts.
- Fault management is concerned with the proper operation of each network component. It can be reactive or proactive.
- Performance management is concerned with the monitoring and control of the network to ensure the network runs as efficiently as possible. It is quantified by measuring the capacity, traffic, throughput, and response time.
- Security management is concerned with controlling access to the network.
- Accounting management is concerned with the control of user access to network resources through charges.
- Simple Network Management Protocol (SNMP) is a framework for managing devices in an internet using the TCP/IP protocol suite.
- A manager, usually a host, controls and monitors a set of agents, usually routers.
- The manager is a host that runs the SNMP client program.
- The agent is a router or host that runs the SNMP server program.
- SNMP frees management tasks from both the physical characteristics of the managed devices and the underlying networking technology.
- SNMP uses the services of two other protocols: Structure of Management Information (SMI) and Management Information Base (MIB).
- SMI names objects, defines the type of data that can be stored in an object, and encodes the data.
- SMI objects are named according to a hierarchical tree structure.
- SMI data types are defined according to Abstract Syntax Notation 1 (ASN.1).
- SMI uses Basic Encoding Rules (BER) to encode data.
- MIB is a collection of groups of objects that can be managed by SNMP.
- MIB uses lexicographic ordering to manage its variables.
- SNMP functions in three ways:
  1. A manager can retrieve the value of an object defined in an agent.
  2. A manager can store a value in an object defined in an agent.
  3. An agent can send an alarm message to the manager.

- SNMP defines eight types of packets: GetRequest, GetNextRequest, SetRequest, GetBulkRequest, Trap, InformRequest, Response, and Report.
- SNMP uses the services of UDP on two well-known ports, 161 and 162.
- SNMPv3 has enhanced security features over previous versions.

28.6 PRACTICE SET

Review Questions

1. Define network management.
2. List five functions of network management.
3. Define configuration management and its purpose.
4. List two subfunctions of configuration management.
5. Define fault management and its purpose.
6. List two subfunctions of fault management.
7. Define performance management and its purpose.
8. List four measurable quantities of performance management.
9. Define security management and its purpose.
10. Define account management and its purpose.

Exercises

11. Show the encoding for INTEGER 1456.
12. Show the encoding for the OCTET STRING "Hello World."
13. Show the encoding for an arbitrary OCTET STRING of length 1000.
14. Show how the following record (sequence) is encoded.

INTEGER	OCTET STRING	IP Address
2345	"COMPUTER"	185.32.1.5

15. Show how the following record (sequence) is encoded.

Time Tick	INTEGER	Object Id
12000	14564	1.3.6.1.2.1.7

16. Show how the following array (sequence of) is encoded. Each element is an integer.

2345
1236
122
1236

17. Show how the following array of records (sequence of sequence) is encoded.

INTEGER	OCTET STRING	Counter
2345	"COMPUTER"	345
1123	"DISK"	1430
3456	"MONITOR"	2313

18. Decode the following.
   a. 02 04 01 02 14 32
   b. 30 06 02 01 11 02 01 14
   c. 30 09 04 03 41 43 42 02 02 14 14
   d. 30 0A 40 04 23 51 62 71 02 02 14 12

CHAPTER 29

Multimedia

Recent advances in technology have changed our use of audio and video. In the past, we listened to an audio broadcast through a radio and watched a video program broadcast through a TV. We used the telephone network to interactively communicate with another party. But times have changed. People want to use the Internet not only for text and image communications, but also for audio and video services. In this chapter, we concentrate on applications that use the Internet for audio and video services.

We can divide audio and video services into three broad categories: **streaming stored audio/video, streaming live audio/video,** and **interactive audio/video,** as shown in Figure 29.1. Streaming means a user can listen to (or watch) the file after the downloading has started.

Figure 29.1 *Internet audio/video*

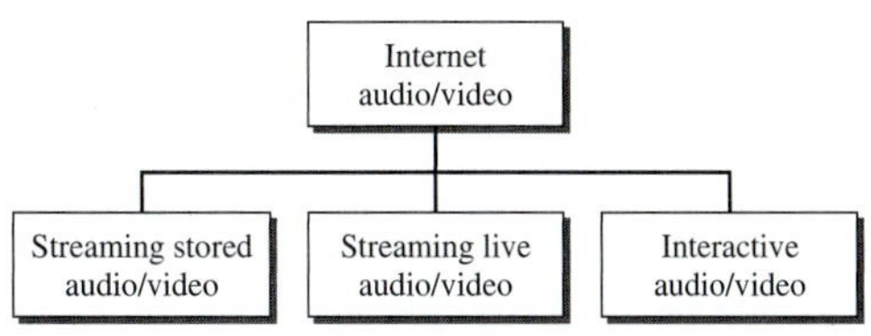

In the first category, streaming stored audio/video, the files are compressed and stored on a server. A client downloads the files through the Internet. This is sometimes referred to as **on-demand audio/video.** Examples of stored audio files are songs, symphonies, books on tape, and famous lectures. Examples of stored video files are movies, TV shows, and music video clips.

> **Streaming stored audio/video refers to on-demand requests for compressed audio/video files.**

In the second category, streaming live audio/video, a user listens to broadcast audio and video through the Internet. A good example of this type of application is the Internet radio. Some radio stations broadcast their programs only on the Internet; many broadcast them both on the Internet and on the air. Internet TV is not popular yet, but many people believe that TV stations will broadcast their programs on the Internet in the future.

Streaming live audio/video refers to the broadcasting of radio and TV programs through the Internet.

In the third category, interactive audio/video, people use the Internet to interactively communicate with one another. A good example of this application is Internet telephony and Internet teleconferencing.

Interactive audio/video refers to the use of the Internet for interactive audio/video applications.

We will discuss these three applications in this chapter, but first we need to discuss some other issues related to audio/video: digitizing audio and video and compressing audio and video.

29.1 DIGITIZING AUDIO AND VIDEO

Before audio or video signals can be sent on the Internet, they need to be digitized. We discuss audio and video separately.

Digitizing Audio

When sound is fed into a microphone, an electronic analog signal is generated which represents the sound amplitude as a function of time. The signal is called an *analog audio signal*. An analog signal, such as audio, can be digitized to produce a digital signal. According to the Nyquist theorem, if the highest frequency of the signal is f, we need to sample the signal $2f$ times per second. There are other methods for digitizing an audio signal, but the principle is the same.

Voice is sampled at 8000 samples per second with 8 bits per sample. This results in a digital signal of 64 kbps. Music is sampled at 44,100 samples per second with 16 bits per sample. This results in a digital signal of 705.6 kbps for monaural and 1.411 Mbps for stereo.

Digitizing Video

A video consists of a sequence of frames. If the frames are displayed on the screen fast enough, we get an impression of motion. The reason is that our eyes cannot distinguish the rapidly flashing frames as individual ones. There is no standard number of frames per second; in North America 25 frames per second is common. However, to avoid a condition known as flickering, a frame needs to be refreshed. The TV industry repaints each frame twice. This means 50 frames need to be sent, or if there is memory at the sender site, 25 frames with each frame repainted from the memory.

Each frame is divided into small grids, called picture elements or **pixels.** For black-and-white TV, each 8-bit pixel represents one of 256 different gray levels. For a color TV, each pixel is 24 bits, with 8 bits for each primary color (red, green, and blue).

We can calculate the number of bits in 1 s for a specific resolution. In the lowest resolution a color frame is made of 1024 × 768 pixels. This means that we need

$$2 \times 25 \times 1024 \times 768 \times 24 = 944 \text{ Mbps}$$

This data rate needs a very high data rate technology such as SONET. To send video using lower-rate technologies, we need to compress the video.

***Compression* is needed to send video over the Internet.**

29.2 AUDIO AND VIDEO COMPRESSION

To send audio or video over the Internet requires **compression.** In this section, we discuss audio compression first and then video compression.

Audio Compression

Audio compression can be used for speech or music. For speech, we need to compress a 64-kHz digitized signal; for music, we need to compress a 1.411-MHz signal. Two categories of techniques are used for audio compression: predictive encoding and perceptual encoding.

Predictive Encoding

In **predictive encoding,** the differences between the samples are encoded instead of encoding all the sampled values. This type of compression is normally used for speech. Several standards have been defined such as GSM (13 kbps), G.729 (8 kbps), and G.723.3 (6.4 or 5.3 kbps). Detailed discussions of these techniques are beyond the scope of this book.

Perceptual Encoding: MP3

The most common compression technique that is used to create CD-quality audio is based on the **perceptual encoding** technique. As we mentioned before, this type of audio needs at least 1.411 Mbps; this cannot be sent over the Internet without compression. **MP3** (MPEG audio layer 3), a part of the MPEG standard (discussed in the video compression section), uses this technique.

Perceptual encoding is based on the science of psychoacoustics, which is the study of how people perceive sound. The idea is based on flaws in our auditory system: Some sounds can mask other sounds. Masking can happen in frequency and time. In

frequency masking, a loud sound in a frequency range can partially or totally mask a softer sound in another frequency range. For example, we cannot hear what our dance partner says in a room where a loud heavy metal band is performing. In **temporal masking,** a loud sound can numb our ears for a short time even after the sound has stopped.

MP3 uses these two phenomena, frequency and temporal masking, to compress audio signals. The technique analyzes and divides the spectrum into several groups. Zero bits are allocated to the frequency ranges that are totally masked. A small number of bits are allocated to the frequency ranges that are partially masked. A larger number of bits are allocated to the frequency ranges that are not masked.

MP3 produces three data rates: 96 kbps, 128 kbps, and 160 kbps. The rate is based on the range of the frequencies in the original analog audio.

Video Compression

As we mentioned before, video is composed of multiple frames. Each frame is one image. We can compress video by first compressing images. Two standards are prevalent in the market. **Joint Photographic Experts Group (JPEG)** is used to compress images. **Moving Picture Experts Group (MPEG)** is used to compress video. We briefly discuss JPEG and then MPEG.

Image Compression: JPEG

As we discussed previously, if the picture is not in color (gray scale), each pixel can be represented by an 8-bit integer (256 levels). If the picture is in color, each pixel can be represented by 24 bits (3×8 bits), with each 8 bits representing red, blue, or green (RBG). To simplify the discussion, we concentrate on a gray scale picture.

In JPEG, a gray scale picture is divided into blocks of 8×8 pixels (see Figure 29.2).

Figure 29.2 *JPEG gray scale*

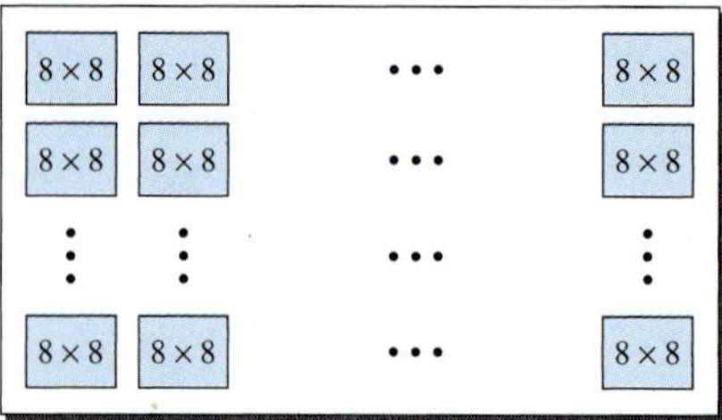

The purpose of dividing the picture into blocks is to decrease the number of calculations because, as you will see shortly, the number of mathematical operations for each picture is the square of the number of units.

The whole idea of JPEG is to change the picture into a linear (vector) set of numbers that reveals the redundancies. The redundancies (lack of changes) can then be removed by using one of the text compression methods. A simplified scheme of the process is shown in Figure 29.3.

Figure 29.3 *JPEG process*

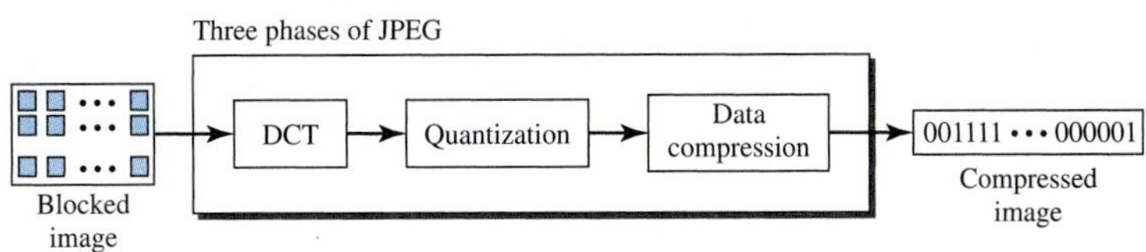

Discrete Cosine Transform (DCT) In this step, each block of 64 pixels goes through a transformation called the **discrete cosine transform (DCT).** The transformation changes the 64 values so that the relative relationships between pixels are kept but the redundancies are revealed. We do not give the formula here, but we do show the results of the transformation for three cases.

Case 1 In this case, we have a block of uniform gray, and the value of each pixel is 20. When we do the transformations, we get a nonzero value for the first element (upper left corner); the rest of the pixels have a value of 0. The value of $T(0,0)$ is the average (multiplied by a constant) of the $P(x,y)$ values and is called the *dc value* (direct current, borrowed from electrical engineering). The rest of the values, called *ac values,* in $T(m,n)$ represent changes in the pixel values. But because there are no changes, the rest of the values are 0s (see Figure 29.4).

Figure 29.4 *Case 1: uniform gray scale*

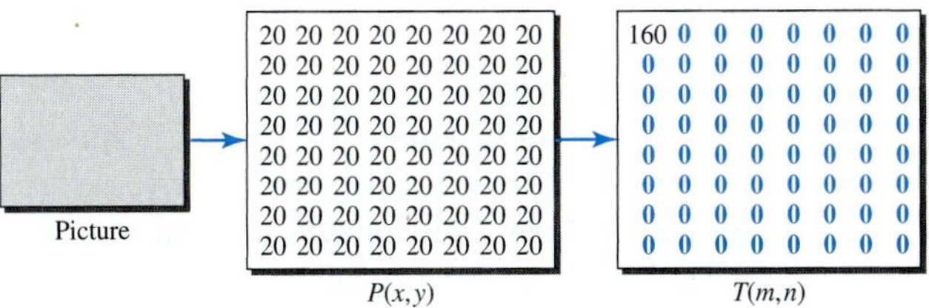

Case 2 In the second case, we have a block with two different uniform gray scale sections. There is a sharp change in the values of the pixels (from 20 to 50). When we do the transformations, we get a dc value as well as nonzero ac values. However, there are only a few nonzero values clustered around the dc value. Most of the values are 0 (see Figure 29.5).

Case 3 In the third case, we have a block that changes gradually. That is, there is no sharp change between the values of neighboring pixels. When we do the transformations, we get a dc value, with many nonzero ac values also (Figure 29.6).

Figure 29.5 *Case 2: two sections*

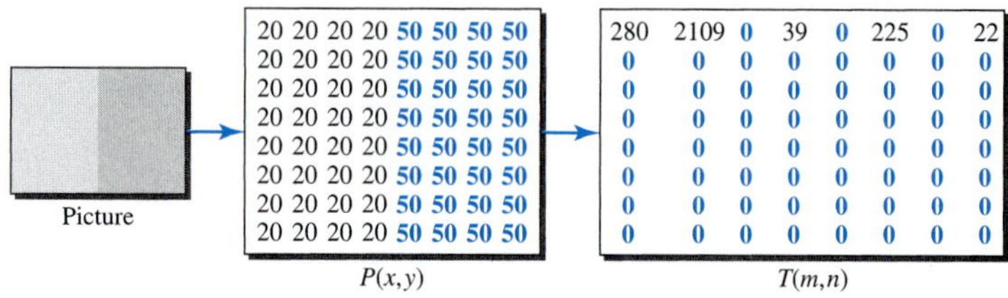

P(x,y):

20	20	20	20	50	50	50	50
20	20	20	20	50	50	50	50
20	20	20	20	50	50	50	50
20	20	20	20	50	50	50	50
20	20	20	20	50	50	50	50
20	20	20	20	50	50	50	50
20	20	20	20	50	50	50	50
20	20	20	20	50	50	50	50

T(m,n):

280	2109	0	39	0	225	0	22
0	0	0	0	0	0	0	0
0	0	0	0	0	0	0	0
0	0	0	0	0	0	0	0
0	0	0	0	0	0	0	0
0	0	0	0	0	0	0	0
0	0	0	0	0	0	0	0
0	0	0	0	0	0	0	0

Figure 29.6 *Case 3: gradient gray scale*

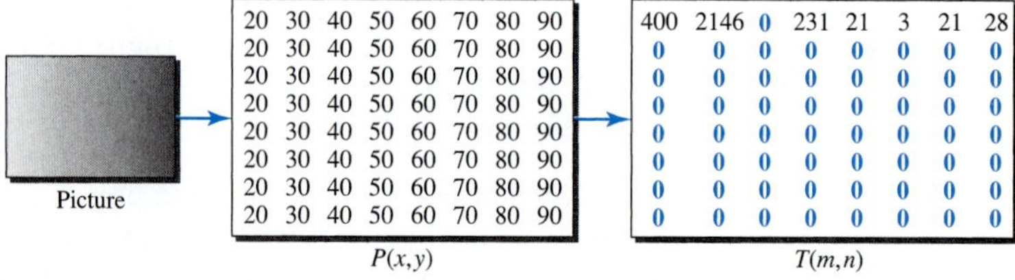

P(x,y):

20	30	40	50	60	70	80	90
20	30	40	50	60	70	80	90
20	30	40	50	60	70	80	90
20	30	40	50	60	70	80	90
20	30	40	50	60	70	80	90
20	30	40	50	60	70	80	90
20	30	40	50	60	70	80	90
20	30	40	50	60	70	80	90

T(m,n):

400	2146	0	231	21	3	21	28
0	0	0	0	0	0	0	0
0	0	0	0	0	0	0	0
0	0	0	0	0	0	0	0
0	0	0	0	0	0	0	0
0	0	0	0	0	0	0	0
0	0	0	0	0	0	0	0
0	0	0	0	0	0	0	0

From Figures 29.4, 29.5, and 29.6, we can state the following:

- The transformation creates table *T* from table *P*.
- The dc value is the average value (multiplied by a constant) of the pixels.
- The ac values are the changes.
- Lack of changes in neighboring pixels creates 0s.

Quantization After the *T* table is created, the values are quantized to reduce the number of bits needed for encoding. Previously in **quantization,** we dropped the fraction from each value and kept the integer part. Here, we divide the number by a constant and then drop the fraction. This reduces the required number of bits even more. In most implementations, a quantizing table (8 × 8) defines how to quantize each value. The divisor depends on the position of the value in the *T* table. This is done to optimize the number of bits and the number of 0s for each particular application. Note that the only phase in the process that is not reversible is the quantizing phase. We lose some information here that is not recoverable. As a matter of fact, the only reason that JPEG is called *lossy compression* is because of this quantization phase.

Compression After quantization, the values are read from the table, and redundant 0s are removed. However, to cluster the 0s together, the table is read diagonally in a zigzag fashion rather than row by row or column by column. The reason is that if the picture changes smoothly, the bottom right corner of the *T* table is all 0s. Figure 29.7 shows the process.

Figure 29.7 *Reading the table*

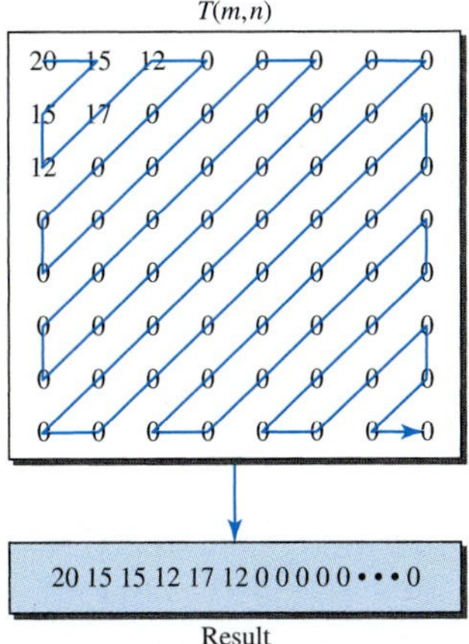

Video Compression: MPEG

The Moving Picture Experts Group method is used to compress video. In principle, a motion picture is a rapid flow of a set of frames, where each frame is an image. In other words, a frame is a spatial combination of pixels, and a video is a temporal combination of frames that are sent one after another. Compressing video, then, means spatially compressing each frame and temporally compressing a set of frames.

Spatial Compression The **spatial compression** of each frame is done with JPEG (or a modification of it). Each frame is a picture that can be independently compressed.

Temporal Compression In **temporal compression,** redundant frames are removed. When we watch television, we receive 50 frames per second. However, most of the consecutive frames are almost the same. For example, when someone is talking, most of the frame is the same as the previous one except for the segment of the frame around the lips, which changes from one frame to another.

To temporally compress data, the MPEG method first divides frames into three categories: I-frames, P-frames, and B-frames.

- **I-frames.** An **intracoded frame (I-frame)** is an independent frame that is not related to any other frame (not to the frame sent before or to the frame sent after). They are present at regular intervals (e.g., every ninth frame is an I-frame). An I-frame must appear periodically to handle some sudden change in the frame that the previous and following frames cannot show. Also, when a video is broadcast, a viewer may tune in at any time. If there is only one I-frame at the beginning of the broadcast, the viewer who tunes in late will not receive a complete picture. I-frames are independent of other frames and cannot be constructed from other frames.
- **P-frames.** A **predicted frame (P-frame)** is related to the preceding I-frame or P-frame. In other words, each P-frame contains only the changes from the preceding

frame. The changes, however, cannot cover a big segment. For example, for a fast-moving object, the new changes may not be recorded in a P-frame. P-frames can be constructed only from previous I- or P-frames. P-frames carry much less information than other frame types and carry even fewer bits after compression.

- **B-frames.** A **bidirectional frame (B-frame)** is related to the preceding and following I-frame or P-frame. In other words, each B-frame is relative to the past and the future. Note that a B-frame is never related to another B-frame.

Figure 29.8 shows a sample sequence of frames.

Figure 29.8 *MPEG frames*

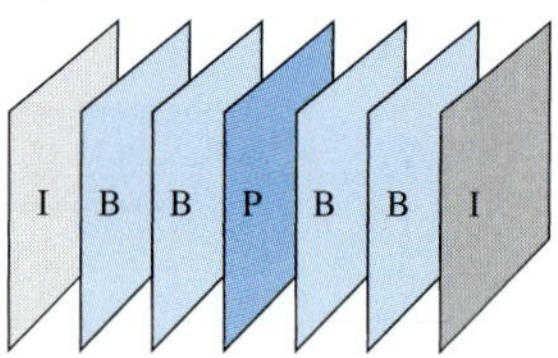

Figure 29.9 shows how I-, P-, and B-frames are constructed from a series of seven frames.

Figure 29.9 *MPEG frame construction*

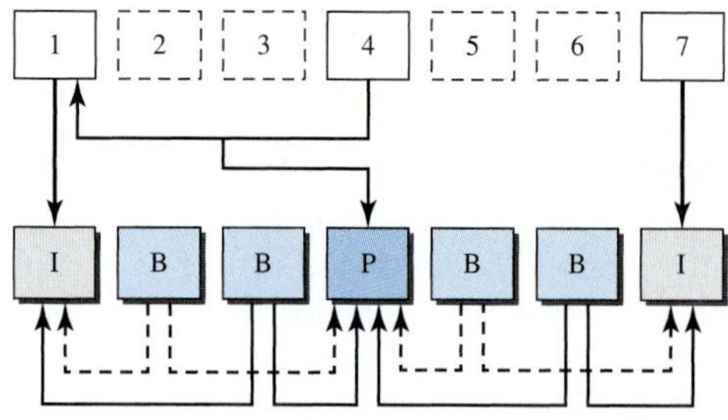

MPEG has gone through two versions. MPEG1 was designed for a CD-ROM with a data rate of 1.5 Mbps. MPEG2 was designed for high-quality DVD with a data rate of 3 to 6 Mbps.

29.3 STREAMING STORED AUDIO/VIDEO

Now that we have discussed digitizing and compressing audio/video, we turn our attention to specific applications. The first is streaming stored audio and video. Downloading these types of files from a Web server can be different from downloading other types of files. To understand the concept, let us discuss four approaches, each with a different complexity.

First Approach: Using a Web Server

A compressed audio/video file can be downloaded as a text file. The client (browser) can use the services of HTTP and send a GET message to download the file. The Web server can send the compressed file to the browser. The browser can then use a help application, normally called a **media player,** to play the file. Figure 29.10 shows this approach.

Figure 29.10 *Using a Web server*

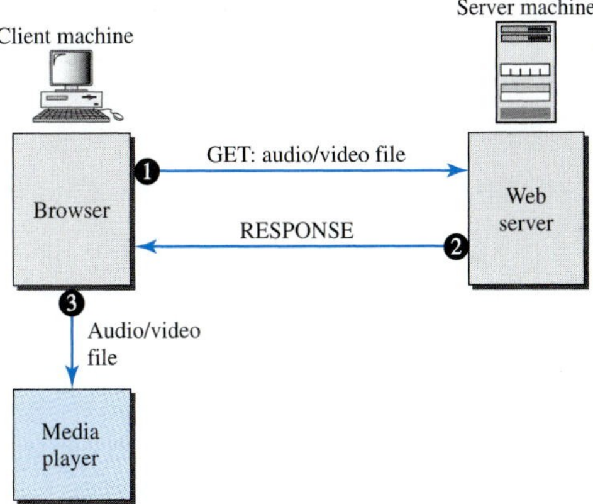

This approach is very simple and does not involve *streaming*. However, it has a drawback. An audio/video file is usually large even after compression. An audio file may contain tens of megabits, and a video file may contain hundreds of megabits. In this approach, the file needs to download completely before it can be played. Using contemporary data rates, the user needs some seconds or tens of seconds before the file can be played.

Second Approach: Using a Web Server with Metafile

In another approach, the media player is directly connected to the Web server for downloading the audio/video file. The Web server stores two files: the actual audio/video file and a **metafile** that holds information about the audio/video file. Figure 29.11 shows the steps in this approach.

1. The HTTP client accesses the Web server by using the GET message.
2. The information about the metafile comes in the response.
3. The metafile is passed to the media player.
4. The media player uses the URL in the metafile to access the audio/video file.
5. The Web server responds.

Figure 29.11 *Using a Web server with a metafile*

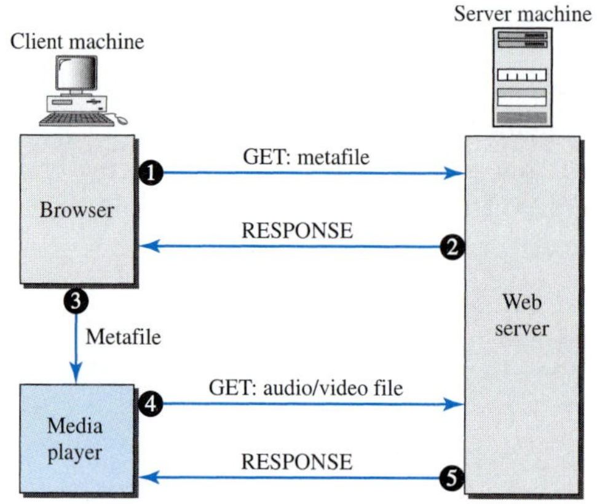

Third Approach: Using a Media Server

The problem with the second approach is that the browser and the media player both use the services of HTTP. HTTP is designed to run over TCP. This is appropriate for retrieving the metafile, but not for retrieving the audio/video file. The reason is that TCP retransmits a lost or damaged segment, which is counter to the philosophy of streaming. We need to dismiss TCP and its error control; we need to use UDP. However, HTTP, which accesses the Web server, and the Web server itself are designed for TCP; we need another server, a **media server.** Figure 29.12 shows the concept.

1. The HTTP client accesses the Web server by using a GET message.
2. The information about the metafile comes in the response.

Figure 29.12 *Using a media server*

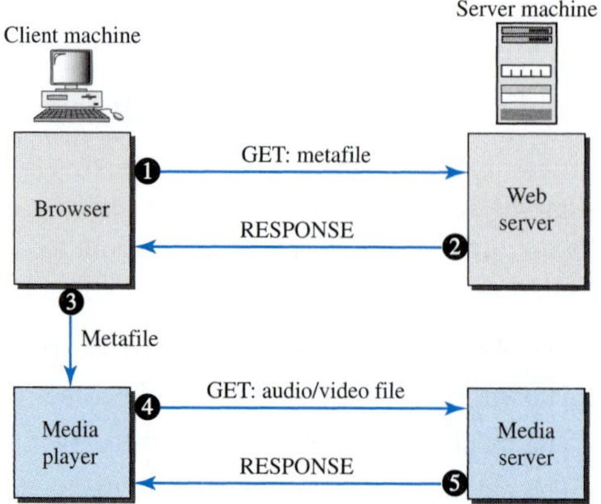

3. The metafile is passed to the media player.
4. The media player uses the URL in the metafile to access the media server to download the file. Downloading can take place by any protocol that uses UDP.
5. The media server responds.

Fourth Approach: Using a Media Server and RTSP

The **Real-Time Streaming Protocol (RTSP)** is a control protocol designed to add more functionalities to the streaming process. Using RTSP, we can control the playing of audio/video. RTSP is an out-of-band control protocol that is similar to the second connection in FTP. Figure 29.13 shows a media server and RTSP.

Figure 29.13 *Using a media server and RTSP*

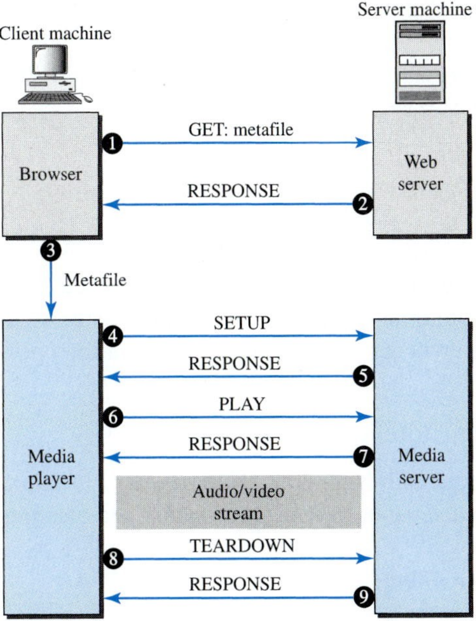

1. The HTTP client accesses the Web server by using a GET message.
2. The information about the metafile comes in the response.
3. The metafile is passed to the media player.
4. The media player sends a SETUP message to create a connection with the media server.
5. The media server responds.
6. The media player sends a PLAY message to start playing (downloading).
7. The audio/video file is downloaded by using another protocol that runs over UDP.

8. The connection is broken by using the TEARDOWN message.
9. The media server responds.

The media player can send other types of messages. For example, a PAUSE message temporarily stops the downloading; downloading can be resumed with a PLAY message.

29.4 STREAMING LIVE AUDIO/VIDEO

Streaming live audio/video is similar to the broadcasting of audio and video by radio and TV stations. Instead of broadcasting to the air, the stations broadcast through the Internet. There are several similarities between streaming stored audio/video and streaming live audio/video. They are both sensitive to delay; neither can accept retransmission. However, there is a difference. In the first application, the communication is unicast and on-demand. In the second, the communication is multicast and live. Live streaming is better suited to the multicast services of IP and the use of protocols such as UDP and RTP (discussed later). However, presently, live streaming is still using TCP and multiple unicasting instead of multicasting. There is still much progress to be made in this area.

29.5 REAL-TIME INTERACTIVE AUDIO/VIDEO

In real-time interactive audio/video, people communicate with one another in real time. The Internet phone or voice over IP is an example of this type of application. Video conferencing is another example that allows people to communicate visually and orally.

Characteristics

Before addressing the protocols used in this class of applications, we discuss some characteristics of real-time audio/video communication.

Time Relationship

Real-time data on a packet-switched network require the preservation of the time relationship between packets of a session. For example, let us assume that a real-time video server creates live video images and sends them online. The video is digitized and packetized. There are only three packets, and each packet holds 10 s of video information. The first packet starts at 00:00:00, the second packet starts at 00:00:10, and the third packet starts at 00:00:20. Also imagine that it takes 1 s (an exaggeration for simplicity) for each packet to reach the destination (equal delay). The receiver can play back the first packet at 00:00:01, the second packet at 00:00:11, and the third packet at 00:00:21. Although there is a 1-s time difference between what the server sends and what the client sees on the computer screen, the action is happening in real time. The time relationship between the packets is preserved. The 1-s delay is not important. Figure 29.14 shows the idea.

Figure 29.14 *Time relationship*

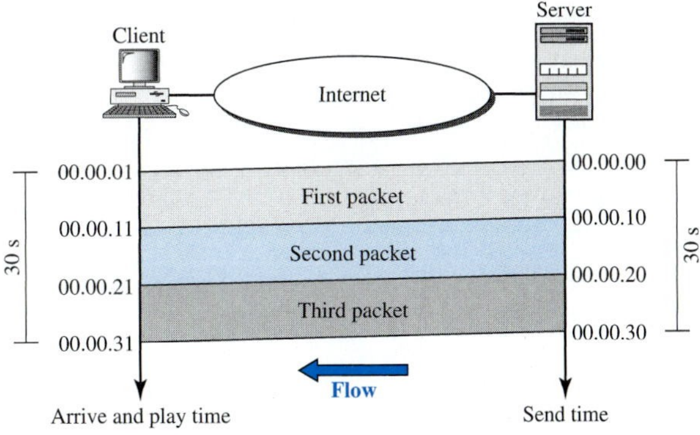

But what happens if the packets arrive with different delays? For example, say the first packet arrives at 00:00:01 (1-s delay), the second arrives at 00:00:15 (5-s delay), and the third arrives at 00:00:27 (7-s delay). If the receiver starts playing the first packet at 00:00:01, it will finish at 00:00:11. However, the next packet has not yet arrived; it arrives 4 s later. There is a gap between the first and second packets and between the second and the third as the video is viewed at the remote site. This phenomenon is called **jitter.** Figure 29.15 shows the situation.

Jitter is introduced in real-time data by the delay between packets.

Figure 29.15 *Jitter*

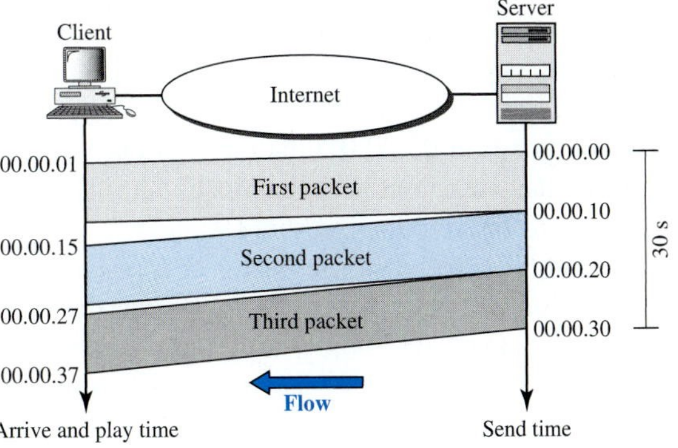

Timestamp

One solution to jitter is the use of a **timestamp.** If each packet has a timestamp that shows the time it was produced relative to the first (or previous) packet, then the receiver can add this time to the time at which it starts the playback. In other words, the receiver knows when each packet is to be played. Imagine the first packet in the previous example has a timestamp of 0, the second has a timestamp of 10, and the third has a timestamp of 20. If the receiver starts playing back the first packet at 00:00:08, the second will be played at 00:00:18 and the third at 00:00:28. There are no gaps between the packets. Figure 29.16 shows the situation.

Figure 29.16 *Timestamp*

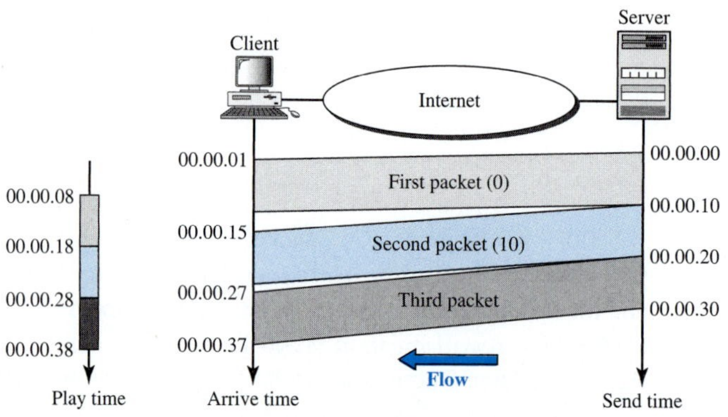

To prevent jitter, we can time-stamp the packets and separate the arrival time from the playback time.

Playback Buffer

To be able to separate the arrival time from the playback time, we need a buffer to store the data until they are played back. The buffer is referred to as a **playback buffer.** When a session begins (the first bit of the first packet arrives), the receiver delays playing the data until a threshold is reached. In the previous example, the first bit of the first packet arrives at 00:00:01; the threshold is 7 s, and the playback time is 00:00:08. The threshold is measured in time units of data. The replay does not start until the time units of data are equal to the threshold value.

Data are stored in the buffer at a possibly variable rate, but they are extracted and played back at a fixed rate. Note that the amount of data in the buffer shrinks or expands, but as long as the delay is less than the time to play back the threshold amount of data, there is no jitter. Figure 29.17 shows the buffer at different times for our example.

Ordering

In addition to time relationship information and timestamps for **real-time traffic,** one more feature is needed. We need a *sequence number* for each packet. The timestamp alone cannot inform the receiver if a packet is lost. For example, suppose the timestamps

Figure 29.17 *Playback buffer*

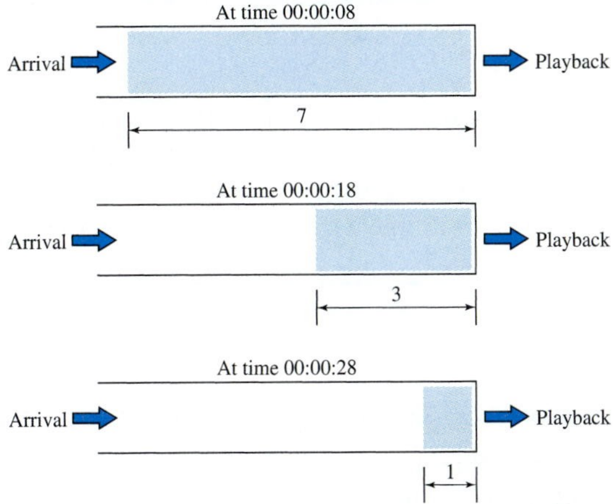

A playback buffer is required for real-time traffic.

are 0, 10, and 20. If the second packet is lost, the receiver receives just two packets with timestamps 0 and 20. The receiver assumes that the packet with timestamp 20 is the second packet, produced 20 s after the first. The receiver has no way of knowing that the second packet has actually been lost. A sequence number to order the packets is needed to handle this situation.

A sequence number on each packet is required for real-time traffic.

Multicasting

Multimedia play a primary role in audio and video conferencing. The traffic can be heavy, and the data are distributed by using **multicasting** methods. Conferencing requires two-way communication between receivers and senders.

Real-time traffic needs the support of multicasting.

Translation

Sometimes real-time traffic needs **translation.** A translator is a computer that can change the format of a high-bandwidth video signal to a lower-quality narrow-bandwidth signal. This is needed, for example, for a source creating a high-quality video signal at 5 Mbps and sending to a recipient having a bandwidth of less than 1 Mbps. To receive the signal, a translator is needed to decode the signal and encode it again at a lower quality that needs less bandwidth.

Translation means changing the encoding of a payload to a lower quality to match the bandwidth of the receiving network.

Mixing

If there is more than one source that can send data at the same time (as in a video or audio conference), the traffic is made of multiple streams. To converge the traffic to one stream, data from different sources can be mixed. A **mixer** mathematically adds signals coming from different sources to create one single signal.

Mixing means combining several streams of traffic into one stream.

Support from Transport Layer Protocol

The procedures mentioned in the previous sections can be implemented in the application layer. However, they are so common in real-time applications that implementation in the transport layer protocol is preferable. Let's see which of the existing transport layers is suitable for this type of traffic.

TCP is not suitable for interactive traffic. It has no provision for time-stamping, and it does not support multicasting. However, it does provide ordering (sequence numbers). One feature of TCP that makes it particularly unsuitable for interactive traffic is its error control mechanism. In interactive traffic, we cannot allow the retransmission of a lost or corrupted packet. If a packet is lost or corrupted in interactive traffic, it must be ignored. Retransmission upsets the whole idea of time-stamping and playback. Today there is so much redundancy in audio and video signals (even with compression) that we can simply ignore a lost packet. The listener or viewer at the remote site may not even notice it.

TCP, with all its sophistication, is not suitable for interactive multimedia traffic because we cannot allow retransmission of packets.

UDP is more suitable for interactive multimedia traffic. UDP supports multicasting and has no retransmission strategy. However, UDP has no provision for time-stamping, sequencing, or mixing. A new transport protocol, Real-time Transport Protocol (RTP), provides these missing features.

UDP is more suitable than TCP for interactive traffic. However, we need the services of RTP, another transport layer protocol, to make up for the deficiencies of UDP.

29.6 RTP

Real-time Transport Protocol (RTP) is the protocol designed to handle real-time traffic on the Internet. RTP does not have a delivery mechanism (multicasting, port numbers, and so on); it must be used with UDP. RTP stands between UDP and the application program. The main contributions of RTP are time-stamping, sequencing, and mixing facilities. Figure 29.18 shows the position of RTP in the protocol suite.

Figure 29.18 *RTP*

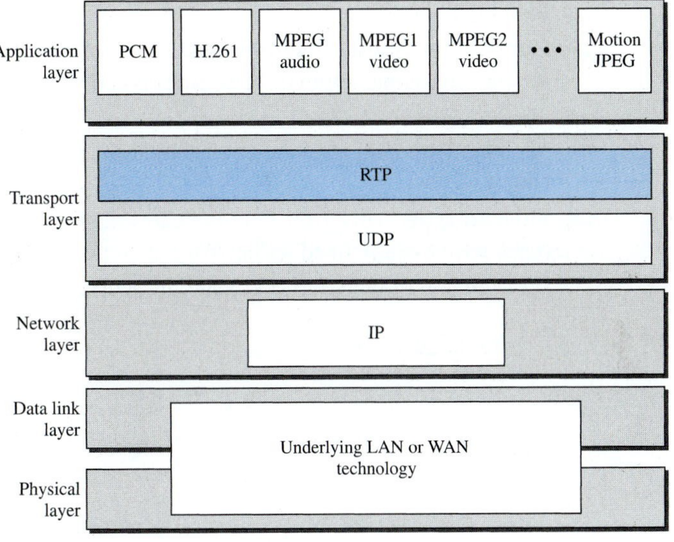

RTP Packet Format

Figure 29.19 shows the format of the RTP packet header. The format is very simple and general enough to cover all real-time applications. An application that needs more information adds it to the beginning of its payload. A description of each field follows.

- **Ver.** This 2-bit field defines the version number. The current version is 2.

Figure 29.19 *RTP packet header format*

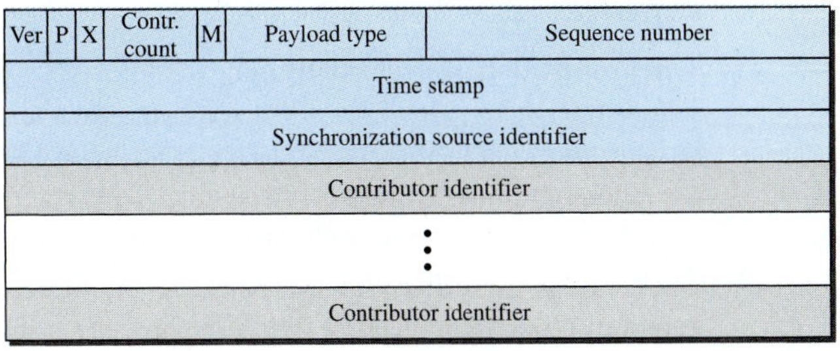

Ver	P	X	Contr. count	M	Payload type	Sequence number
Time stamp						
Synchronization source identifier						
Contributor identifier						
⋮						
Contributor identifier						

- **P.** This 1-bit field, if set to 1, indicates the presence of padding at the end of the packet. In this case, the value of the last byte in the padding defines the length of the padding. Padding is the norm if a packet is encrypted. There is no padding if the value of the P field is 0.
- **X.** This 1-bit field, if set to 1, indicates an extra extension header between the basic header and the data. There is no extra extension header if the value of this field is 0.
- **Contributor count.** This 4-bit field indicates the number of contributors. Note that we can have a maximum of 15 contributors because a 4-bit field only allows a number between 0 and 15.
- **M.** This 1-bit field is a marker used by the application to indicate, for example, the end of its data.
- **Payload type.** This 7-bit field indicates the type of the payload. Several payload types have been defined so far. We list some common applications in Table 29.1. A discussion of the types is beyond the scope of this book.

Table 29.1 *Payload types*

Type	*Application*	*Type*	*Application*	*Type*	*Application*
0	PCMμ Audio	7	LPC audio	15	G728 audio
1	1016	8	PCMA audio	26	Motion JPEG
2	G721 audio	9	G722 audio	31	H.261
3	GSM audio	10–11	L16 audio	32	MPEG1 video
5–6	DV14 audio	14	MPEG audio	33	MPEG2 video

- **Sequence number.** This field is 16 bits in length. It is used to number the RTP packets. The sequence number of the first packet is chosen randomly; it is incremented by 1 for each subsequent packet. The sequence number is used by the receiver to detect lost or out-of-order packets.
- **Timestamp.** This is a 32-bit field that indicates the time relationship between packets. The timestamp for the first packet is a random number. For each succeeding packet, the value is the sum of the preceding timestamp plus the time the first byte is produced (sampled). The value of the clock tick depends on the application. For example, audio applications normally generate chunks of 160 bytes; the clock tick for this application is 160. The timestamp for this application increases 160 for each RTP packet.
- **Synchronization source identifier.** If there is only one source, this 32-bit field defines the source. However, if there are several sources, the mixer is the synchronization source and the other sources are contributors. The value of the source identifier is a random number chosen by the source. The protocol provides a strategy in case of conflict (two sources start with the same sequence number).
- **Contributor identifier.** Each of these 32-bit identifiers (a maximum of 15) defines a source. When there is more than one source in a session, the mixer is the synchronization source and the remaining sources are the contributors.

UDP Port

Although RTP is itself a transport layer protocol, the RTP packet is not encapsulated directly in an IP datagram. Instead, RTP is treated as an application program and is encapsulated in a UDP user datagram. However, unlike other application programs, no well-known port is assigned to RTP. The port can be selected on demand with only one restriction: The port number must be an even number. The next number (an odd number) is used by the companion of RTP, Real-time Transport Control Protocol (RTCP).

RTP uses a temporary even-numbered UDP port.

29.7 RTCP

RTP allows only one type of message, one that carries data from the source to the destination. In many cases, there is a need for other messages in a session. These messages control the flow and quality of data and allow the recipient to send feedback to the source or sources. **Real-time Transport Control Protocol (RTCP)** is a protocol designed for this purpose. RTCP has five types of messages, as shown in Figure 29.20. The number next to each box defines the type of the message.

Figure 29.20 *RTCP message types*

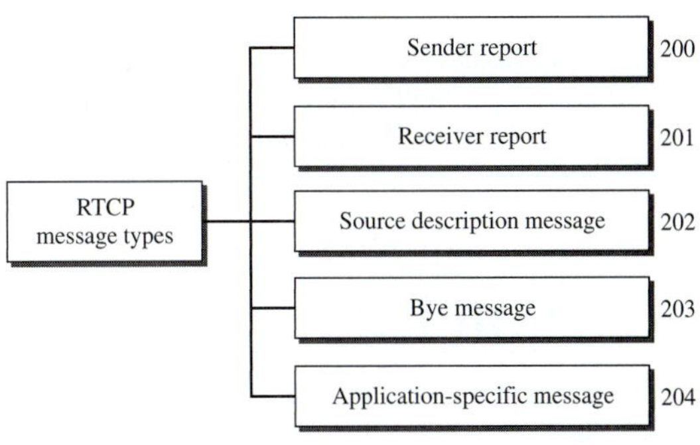

Sender Report

The sender report is sent periodically by the active senders in a conference to report transmission and reception statistics for all RTP packets sent during the interval. The sender report includes an absolute timestamp, which is the number of seconds elapsed since midnight on January 1, 1970. The absolute timestamp allows the receiver to synchronize different RTP messages. It is particularly important when both audio and video are transmitted (audio and video transmissions use separate relative timestamps).

Receiver Report

The receiver report is for passive participants, those that do not send RTP packets. The report informs the sender and other receivers about the quality of service.

Source Description Message

The source periodically sends a source description message to give additional information about itself. This information can be the name, e-mail address, telephone number, and address of the owner or controller of the source.

Bye Message

A source sends a bye message to shut down a stream. It allows the source to announce that it is leaving the conference. Although other sources can detect the absence of a source, this message is a direct announcement. It is also very useful to a mixer.

Application-Specific Message

The application-specific message is a packet for an application that wants to use new applications (not defined in the standard). It allows the definition of a new message type.

UDP Port

RTCP, like RTP, does not use a well-known UDP port. It uses a temporary port. The UDP port chosen must be the number immediately following the UDP port selected for RTP. It must be an odd-numbered port.

RTCP uses an odd-numbered UDP port number that follows the port number selected for RTP.

29.8 VOICE OVER IP

Let us concentrate on one real-time interactive audio/video application: **voice over IP,** or Internet telephony. The idea is to use the Internet as a telephone network with some additional capabilities. Instead of communicating over a circuit-switched network, this application allows communication between two parties over the packet-switched Internet. Two protocols have been designed to handle this type of communication: SIP and H.323. We briefly discuss both.

SIP

The **Session Initiation Protocol (SIP)** was designed by IETF. It is an application layer protocol that establishes, manages, and terminates a multimedia session (call). It can be used to create two-party, multiparty, or multicast sessions. SIP is designed to be independent of the underlying transport layer; it can run on UDP, TCP, or SCTP.

Messages

SIP is a text-based protocol, as is HTTP. SIP, like HTTP, uses messages. Six messages are defined, as shown in Figure 29.21.

Figure 29.21 *SIP messages*

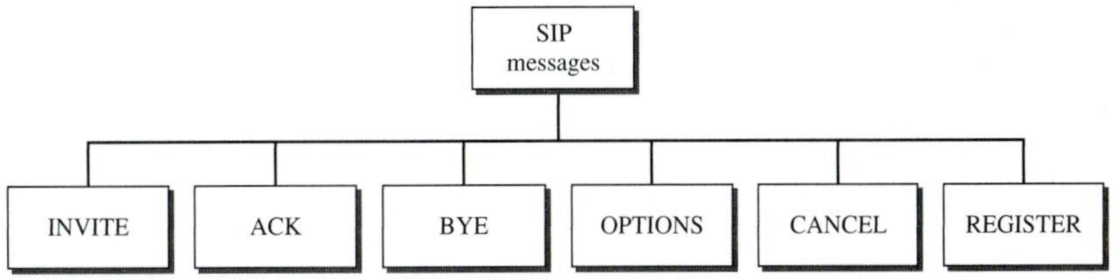

SIP messages
INVITE
ACK
BYE
OPTIONS
CANCEL
REGISTER

Each message has a header and a body. The header consists of several lines that describe the structure of the message, caller's capability, media type, and so on. We give a brief description of each message. Then we show their applications in a simple session.

The caller initializes a session with the INVITE message. After the callee answers the call, the caller sends an ACK message for confirmation. The BYE message terminates a session. The OPTIONS message queries a machine about its capabilities. The CANCEL message cancels an already started initialization process. The REGISTER message makes a connection when the callee is not available.

Addresses

In a regular telephone communication, a telephone number identifies the sender, and another telephone number identifies the receiver. SIP is very flexible. In SIP, an e-mail address, an IP address, a telephone number, and other types of addresses can be used to identify the sender and receiver. However, the address needs to be in SIP format (also called *scheme*). Figure 29.22 shows some common formats.

Figure 29.22 *SIP formats*

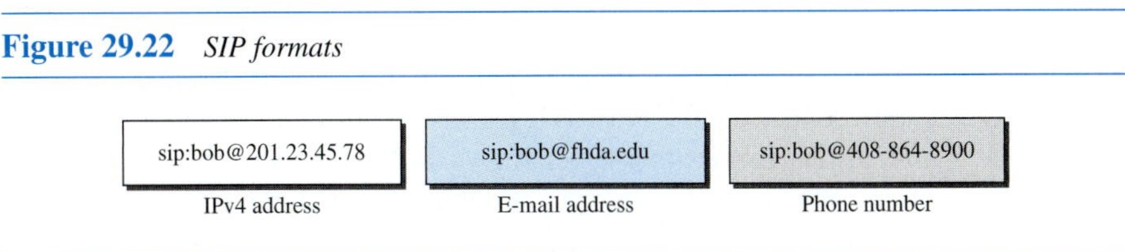

Simple Session

A simple session using SIP consists of three modules: establishing, communicating, and terminating. Figure 29.23 shows a simple session using SIP.

Figure 29.23 *SIP simple session*

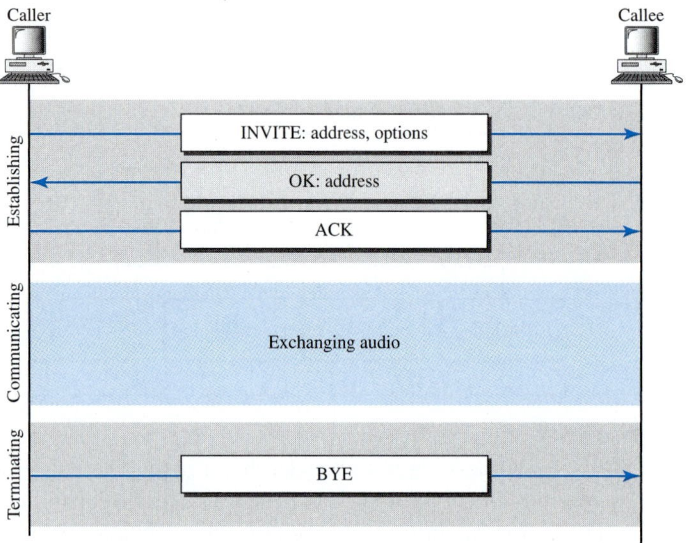

Establishing a Session Establishing a session in SIP requires a three-way handshake. The caller sends an INVITE message, using UDP, TCP, or SCTP to begin the communication. If the callee is willing to start the session, she sends a reply message. To confirm that a reply code has been received, the caller sends an ACK message.

Communicating After the session has been established, the caller and the callee can communicate by using two temporary ports.

Terminating the Session The session can be terminated with a BYE message sent by either party.

Tracking the Callee

What happens if the callee is not sitting at her terminal? She may be away from her system or at another terminal. She may not even have a fixed IP address if DHCP is being used. SIP has a mechanism (similar to one in DNS) that finds the IP address of the terminal at which the callee is sitting. To perform this tracking, SIP uses the concept of registration. SIP defines some servers as registrars. At any moment a user is registered with at least one **registrar server;** this server knows the IP address of the callee.

When a caller needs to communicate with the callee, the caller can use the e-mail address instead of the IP address in the INVITE message. The message goes to a proxy server. The proxy server sends a lookup message (not part of SIP) to some registrar server that has registered the callee. When the proxy server receives a reply message from the registrar server, the proxy server takes the caller's INVITE message and inserts the newly discovered IP address of the callee. This message is then sent to the callee. Figure 29.24 shows the process.

Figure 29.24 *Tracking the callee*

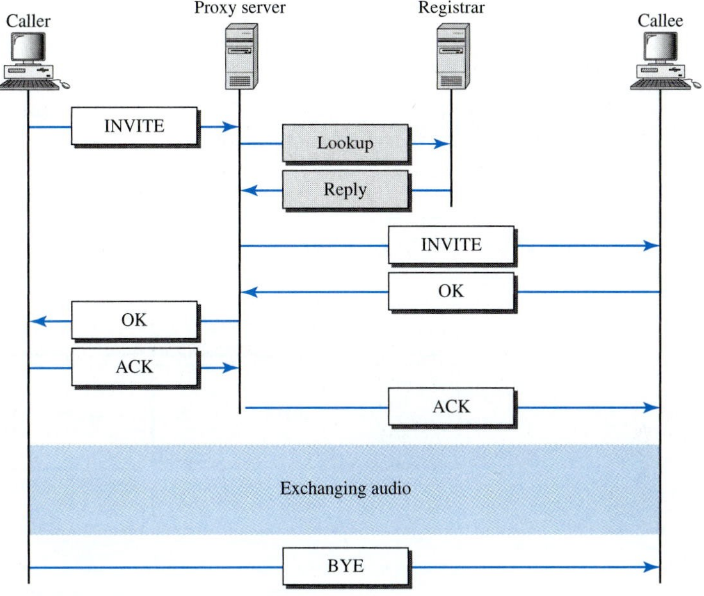

H.323

H.323 is a standard designed by ITU to allow telephones on the public telephone network to talk to computers (called *terminals* in H.323) connected to the Internet. Figure 29.25 shows the general architecture of H.323.

Figure 29.25 *H.323 architecture*

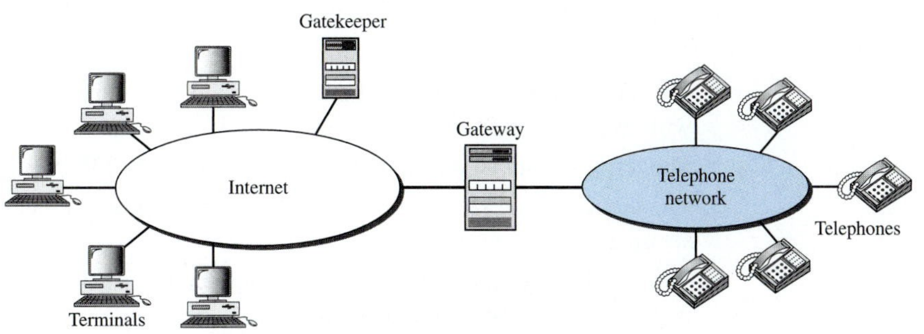

A **gateway** connects the Internet to the telephone network. In general, a gateway is a five-layer device that can translate a message from one protocol stack to another. The gateway here does exactly the same thing. It transforms a telephone network message to an Internet message. The **gatekeeper** server on the local area network plays the role of the registrar server, as we discussed in the SIP.

Protocols

H.323 uses a number of protocols to establish and maintain voice (or video) communication. Figure 29.26 shows these protocols.

Figure 29.26 *H.323 protocols*

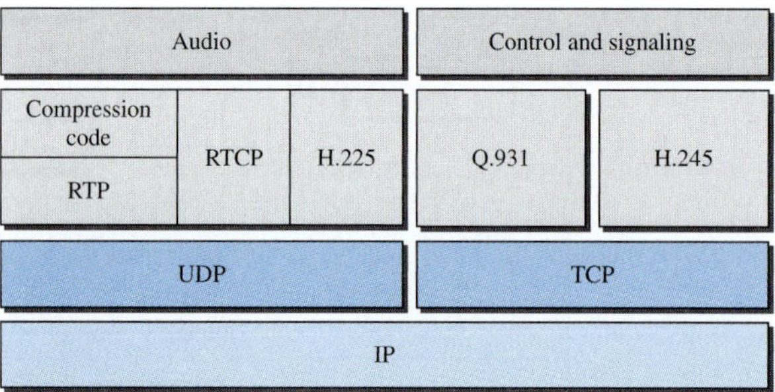

H.323 uses G.71 or G.723.1 for compression. It uses a protocol named H.245 which allows the parties to negotiate the compression method. Protocol Q.931 is used for establishing and terminating connections. Another protocol called H.225, or RAS (Registration/Administration/Status), is used for registration with the gatekeeper.

Operation

Let us show the operation of a telephone communication using H.323 with a simple example. Figure 29.27 shows the steps used by a terminal to communicate with a telephone.

1. The terminal sends a broadcast message to the gatekeeper. The gatekeeper responds with its IP address.
2. The terminal and gatekeeper communicate, using H.225 to negotiate bandwidth.
3. The terminal, gatekeeper, gateway, and telephone communicate by using Q.931 to set up a connection.
4. The terminal, gatekeeper, gateway, and telephone communicate by using H.245 to negotiate the compression method.
5. The terminal, gateway, and telephone exchange audio by using RTP under the management of RTCP.
6. The terminal, gatekeeper, gateway, and telephone communicate by using Q.931 to terminate the communication.

Figure 29.27 *H.323 example*

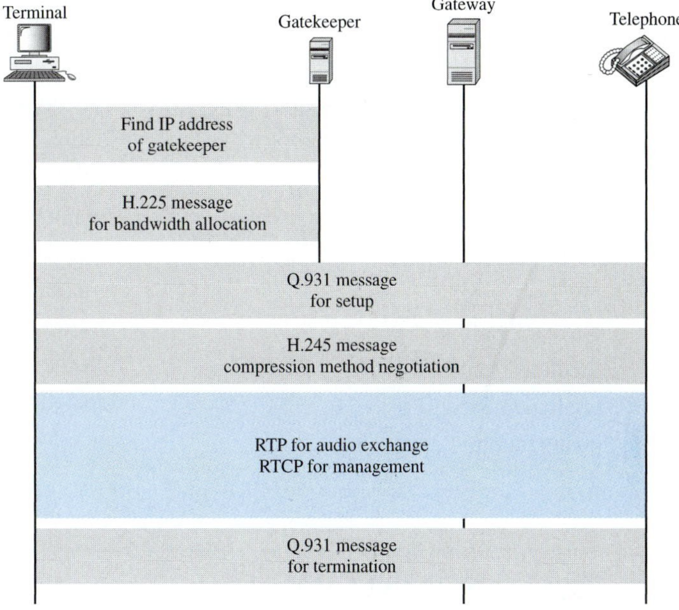

29.9 RECOMMENDED READING

For more details about subjects discussed in this chapter, we recommend the following books and sites. The items in brackets [. . .] refer to the reference list at the end of the text.

Books

Multimedia communication is discussed in [Hal01] and in Section 7.4 of [Tan03]. Image and video compression are fully discussed in [Dro02].

Sites

- www.ietf.org/rfc.html Information about RFCs

29.10 KEY TERMS

bidirectional frame (B-frame)
compression
discrete cosine transform (DCT)
frequency masking
gatekeeper
gateway
H.323
interactive audio/video
intracoded frame (I-frame)
jitter

Joint Photographic Experts Group (JPEG)
media player
media server
metafile
mixer
Moving Picture Experts Group (MPEG)
MP3
multicasting
on-demand audio/video
perceptual encoding
pixel
playback buffer
predicted frame (P-frame)
predictive encoding
quantization
Real-Time Streaming Protocol (RTSP)
real-time traffic
Real-time Transport Control Protocol (RTCP)
Real-time Transport Protocol (RTP)
registrar server
Session Initiation Protocol (SIP)
spatial compression
streaming live audio/video
streaming stored audio/video
temporal compression
temporal masking
timestamp
translation
voice over IP

29.11 SUMMARY

- Audio/video files can be downloaded for future use (streaming stored audio/video) or broadcast to clients over the Internet (streaming live audio/video). The Internet can also be used for live audio/video interaction.
- Audio and video need to be digitized before being sent over the Internet.
- Audio files are compressed through predictive encoding or perceptual encoding.
- Joint Photographic Experts Group (JPEG) is a method to compress pictures and graphics.
- The JPEG process involves blocking, the discrete cosine transform, quantization, and lossless compression.
- Moving Pictures Experts Group (MPEG) is a method to compress video.
- MPEG involves both spatial compression and temporal compression. The former is similar to JPEG, and the latter removes redundant frames.
- We can use a Web server, or a Web server with a metafile, or a media server, or a media server and RTSP to download a streaming audio/video file.
- Real-time data on a packet-switched network require the preservation of the time relationship between packets of a session.
- Gaps between consecutive packets at the receiver cause a phenomenon called jitter.
- Jitter can be controlled through the use of timestamps and a judicious choice of the playback time.
- A playback buffer holds data until they can be played back.
- A receiver delays playing back real-time data held in the playback buffer until a threshold level is reached.

- Sequence numbers on real-time data packets provide a form of error control.
- Real-time data are multicast to receivers.
- Real-time traffic sometimes requires a translator to change a high-bandwidth signal to a lower-quality narrow-bandwidth signal.
- A mixer combines signals from different sources into one signal.
- Real-time multimedia traffic requires both UDP and Real-time Transport Protocol (RTP).
- RTP handles time-stamping, sequencing, and mixing.
- Real-time Transport Control Protocol (RTCP) provides flow control, quality of data control, and feedback to the sources.
- Voice over IP is a real-time interactive audio/video application.
- The Session Initiation Protocol (SIP) is an application layer protocol that establishes, manages, and terminates multimedia sessions.
- H.323 is an ITU standard that allows a telephone connected to a public telephone network to talk to a computer connected to the Internet.

29.12 PRACTICE SET

Review Questions

1. How does streaming live audio/video differ from streaming stored audio/video?
2. How does frequency masking differ from temporal masking?
3. What is the function of a metafile in streaming stored audio/video?
4. What is the purpose of RTSP in streaming stored audio/video?
5. How does jitter affect real-time audio/video?
6. Discuss how SIP is used in the transmission of multimedia.
7. When would you use JPEG? When would you use MPEG?
8. In JPEG, what is the function of blocking?
9. Why is the DCT needed in JPEG?
10. What is spatial compression compared to temporal compression?

Exercises

11. In Figure 29.17 what is the amount of data in the playback buffer at each of the following times?
    a. 00:00:17
    b. 00:00:20
    c. 00:00:25
    d. 00:00:30
12. Compare and contrast TCP with RTP. Are both doing the same thing?
13. Can we say UDP plus RTP is the same as TCP?
14. Why does RTP need the service of another protocol, RTCP, but TCP does not?

15. In Figure 29.12, can the Web server and media server run on different machines?
16. We discuss the use of SIP in this chapter for audio. Is there any drawback to prevent using it for video?
17. Do you think H.323 is actually the same as SIP? What are the differences? Make a comparison between the two.
18. What are the problems for full implementation of voice over IP? Do you think we will stop using the telephone network very soon?
19. Can H.323 also be used for video?

Research Activities

20. Find the format of an RTCP sender report. Pay particular attention to the packet length and the parts repeated for each source. Describe each field.
21. Find the format of an RTCP receiver report. Pay particular attention to the packet length and the parts repeated for each source. Describe each field.
22. Find the format of an RTCP source description. Pay particular attention to the packet length and the parts repeated for each source. Describe each field.
23. Find the meaning of the source description items used in the RTCP source description packet. Specifically, find the meaning of CNAME, NAME, EMAIL, PHONE, LOC, TOOL, NOTE, and PRIV.
24. Find the format of an RTCP bye message. Pay particular attention to the packet length and the parts repeated for each source. Describe each field.

PART 7

Security

Objectives

No one can deny the importance of security in data communications and networking. Security in networking is based on cryptography, the science and art of transforming messages to make them secure and immune to attack. Cryptography can provide several aspects of security related to the interchange of messages through networks. These aspects are confidentiality, integrity, authentication, and nonrepudiation.

Cryptography can provide confidentiality, integrity, authentication, and nonrepudiation of messages.

Cryptography can also be used to authenticate the sender and receiver of the message to each other. For example, a user who needs access to the resources of a system must first be authorized. We call this aspect entity authentication.

Cryptography can also provide entity authentication.

In this part of the book, we first introduce cryptography without delving into the mathematical foundations of this subject. We then briefly explore security aspects as applied to a network. Finally, we discuss some common protocols that implement security aspects at the upper three layers of the Internet model.

Part 7 of the book is devoted to different security aspects.

Chapters

This part consists of three chapters: Chapters 30, 31, and 32.

Chapter 30

Chapter 30 is a brief discussion of a broad topic called cryptography. Although cryptography, which is based on abstract algebra, can itself be a complete course, we introduce it here briefly and avoid references to abstract algebra as much as possible. We give just enough background information as a foundation for the material in the next two chapters.

Chapter 31

Chapter 31 is an introduction, and motivation, for the broad topic of network security. We discuss selected issues that are often encountered when dealing with communications and networking problems.

Chapter 32

Chapter 32 briefly discusses the applications of topics discussed in Chapters 30 and 31 to the Internet model. We show how network security and cryptography can be used in three upper layers of the Internet model.

CHAPTER 30

Cryptography

Network security is mostly achieved through the use of cryptography, a science based on abstract algebra. In this chapter, we briefly discuss the cryptography suitable for the scope of this book. We have tried to limit our discussion of abstract algebra as much as we could. Our goal is to give enough information about cryptography to make network security understandable. The chapter opens the door for studying network security in Chapter 31 and Internet security in Chapter 32.

30.1 INTRODUCTION

Let us introduce the issues involved in cryptography. First, we need to define some terms; then we give some taxonomies.

Definitions

We define some terms here that are used in the rest of the chapter.

Cryptography

Cryptography, a word with Greek origins, means "secret writing." However, we use the term to refer to the science and art of transforming messages to make them secure and immune to attacks. Figure 30.1 shows the components involved in cryptography.

Figure 30.1 *Cryptography components*

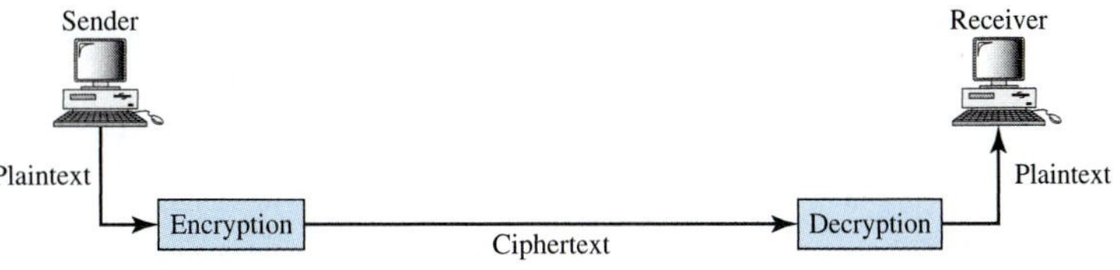

Plaintext and Ciphertext

The original message, before being transformed, is called **plaintext.** After the message is transformed, it is called **ciphertext.** An **encryption algorithm** transforms the plaintext into ciphertext; a **decryption algorithm** transforms the ciphertext back into plaintext. The sender uses an encryption algorithm, and the receiver uses a decryption algorithm.

Cipher

We refer to encryption and decryption algorithms as **ciphers.** The term *cipher* is also used to refer to different categories of algorithms in cryptography. This is not to say that every sender-receiver pair needs their very own unique cipher for a secure communication. On the contrary, one cipher can serve millions of communicating pairs.

Key

A **key** is a number (or a set of numbers) that the cipher, as an algorithm, operates on. To encrypt a message, we need an encryption algorithm, an encryption key, and the plaintext. These create the ciphertext. To decrypt a message, we need a decryption algorithm, a decryption key, and the ciphertext. These reveal the original plaintext.

Alice, Bob, and Eve

In cryptography, it is customary to use three characters in an information exchange scenario; we use Alice, Bob, and Eve. Alice is the person who needs to send secure data. Bob is the recipient of the data. Eve is the person who somehow disturbs the communication between Alice and Bob by intercepting messages to uncover the data or by sending her own disguised messages. These three names represent computers or processes that actually send or receive data, or intercept or change data.

Two Categories

We can divide all the cryptography algorithms (ciphers) into two groups: symmetric-key (also called **secret-key**) cryptography algorithms and asymmetric (also called **public-key**) cryptography algorithms. Figure 30.2 shows the taxonomy.

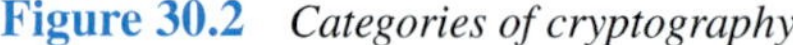

Figure 30.2 *Categories of cryptography*

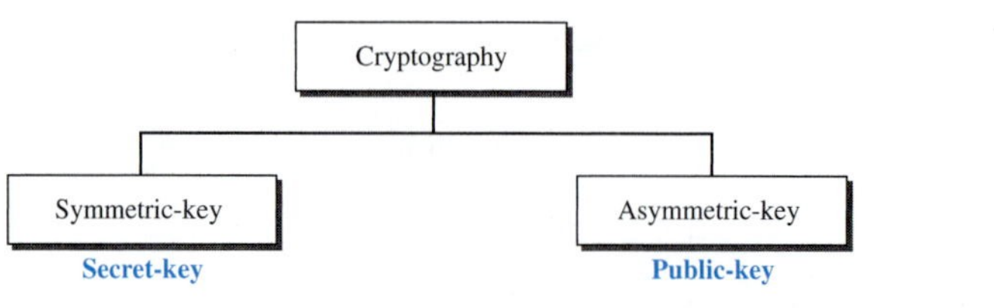

Symmetric-Key Cryptography

In symmetric-key cryptography, the same key is used by both parties. The sender uses this key and an encryption algorithm to encrypt data; the receiver uses the same key and the corresponding decryption algorithm to decrypt the data (see Figure 30.3).

Figure 30.3 *Symmetric-key cryptography*

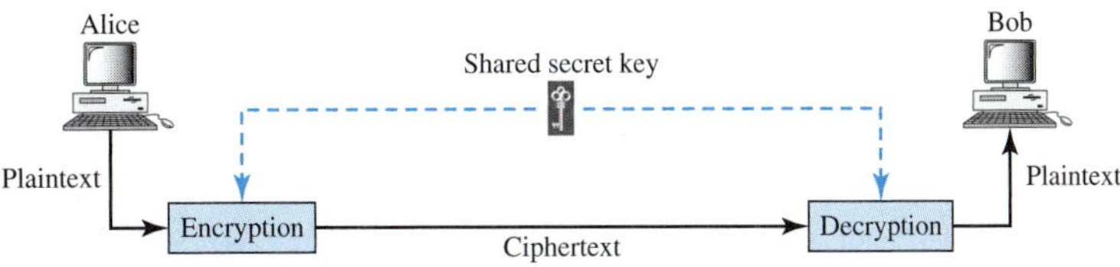

In symmetric-key cryptography, the same key is used by the sender (for encryption) and the receiver (for decryption). The key is shared.

Asymmetric-Key Cryptography

In asymmetric or public-key cryptography, there are two keys: a private key and a public key. The **private key** is kept by the receiver. The **public key** is announced to the public. In Figure 30.4, imagine Alice wants to send a message to Bob. Alice uses the public key to encrypt the message. When the message is received by Bob, the private key is used to decrypt the message.

Figure 30.4 *Asymmetric-key cryptography*

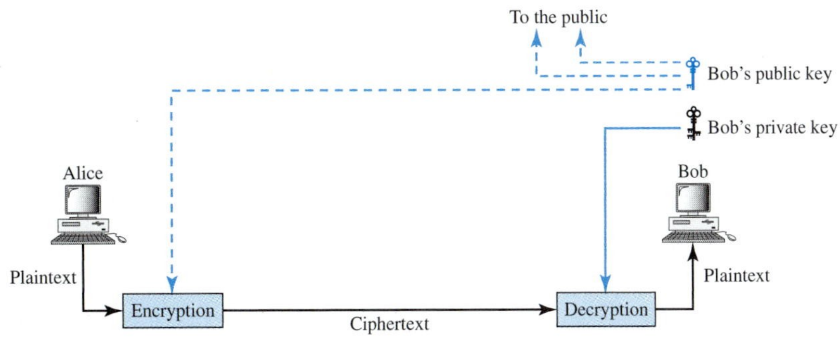

In public-key encryption/decryption, the public key that is used for encryption is different from the private key that is used for decryption. The public key is available to the public; the private key is available only to an individual.

Three Types of Keys

The reader may have noticed that we are dealing with three types of keys in cryptography: the secret key, the public key, and the private key. The first, the secret key, is the shared key used in symmetric-key cryptography. The second and the third are the public and private keys used in asymmetric-key cryptography. We will use three different icons for these keys throughout the book to distinguish one from the others, as shown in Figure 30.5.

Figure 30.5 *Keys used in cryptography*

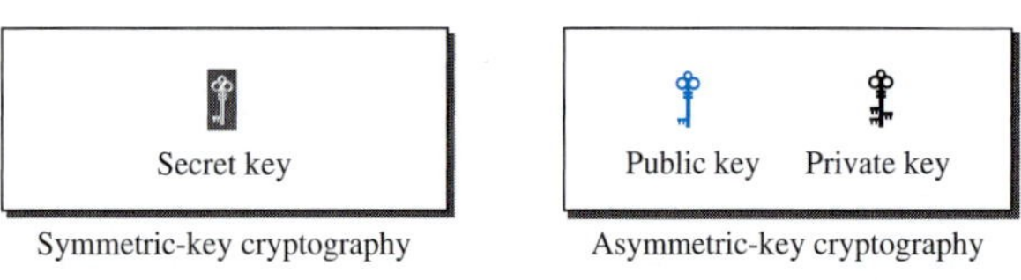

Comparison

Let us compare symmetric-key and asymmetric-key cryptography. Encryption can be thought of as electronic locking; decryption as electronic unlocking. The sender puts the message in a box and locks the box by using a key; the receiver unlocks the box with a key and takes out the message. The difference lies in the mechanism of the locking and unlocking and the type of keys used.

In symmetric-key cryptography, the same key locks and unlocks the box. In asymmetric-key cryptography, one key locks the box, but another key is needed to unlock it. Figure 30.6 shows the difference.

Figure 30.6 *Comparison between two categories of cryptography*

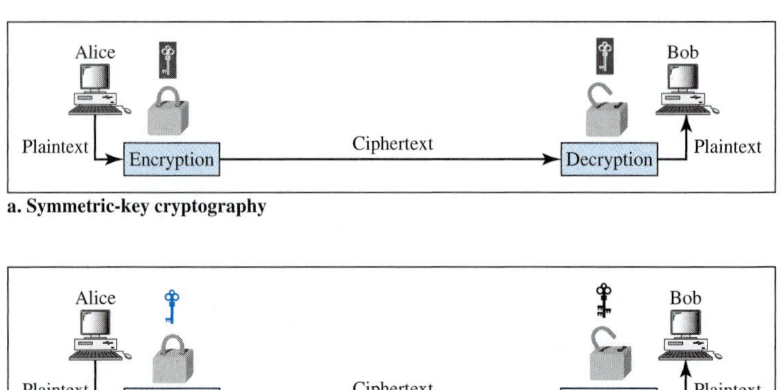

30.2 SYMMETRIC-KEY CRYPTOGRAPHY

Symmetric-key cryptography started thousands of years ago when people needed to exchange secrets (for example, in a war). We still mainly use symmetric-key cryptography in our network security. However, today's ciphers are much more complex. Let us first discuss traditional algorithms, which were character-oriented. Then we discuss the modern ones, which are bit-oriented.

Traditional Ciphers

We briefly introduce some traditional ciphers, which are character-oriented. Although these are now obsolete, the goal is to show how modern ciphers evolved from them. We can divide traditional symmetric-key ciphers into two broad categories: substitution ciphers and transposition ciphers, as shown in Figure 30.7.

Figure 30.7 *Traditional ciphers*

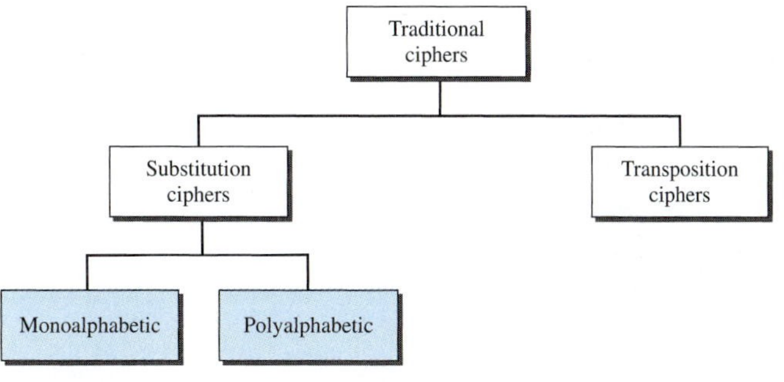

Substitution Cipher

A **substitution cipher** substitutes one symbol with another. If the symbols in the plaintext are alphabetic characters, we replace one character with another. For example, we can replace character A with D, and character T with Z. If the symbols are digits (0 to 9), we can replace 3 with 7, and 2 with 6. Substitution ciphers can be categorized as either monoalphabetic or polyalphabetic ciphers.

A substitution cipher replaces one symbol with another.

In a **monoalphabetic cipher,** a character (or a symbol) in the plaintext is always changed to the same character (or symbol) in the ciphertext regardless of its position in the text. For example, if the algorithm says that character A in the plaintext is changed to character D, every character A is changed to character D. In other words, the relationship between characters in the plaintext and the ciphertext is a one-to-one relationship.

In a **polyalphabetic cipher,** each occurrence of a character can have a different substitute. The relationship between a character in the plaintext to a character in the ciphertext is a one-to-many relationship. For example, character A could be changed to D in the beginning of the text, but it could be changed to N at the middle. It is obvious that if the relationship between plaintext characters and ciphertext characters is one-to-many, the key must tell us which of the many possible characters can be chosen for encryption. To achieve this goal, we need to divide the text into groups of characters and use a set of keys. For example, we can divide the text "THISISANEASYTASK" into groups of 3 characters and then apply the encryption using a set of 3 keys. We then repeat the procedure for the next 3 characters.

Example 30.1

The following shows a plaintext and its corresponding ciphertext. Is the cipher monoalphabetic?

Plaintext: HELLO
Ciphertext: KHOOR

Solution

The cipher is probably monoalphabetic because both occurrences of L's are encrypted as O's.

Example 30.2

The following shows a plaintext and its corresponding ciphertext. Is the cipher monoalphabetic?

Plaintext: HELLO
Ciphertext: ABNZF

Solution

The cipher is not monoalphabetic because each occurrence of L is encrypted by a different character. The first L is encrypted as N; the second as Z.

Shift Cipher The simplest monoalphabetic cipher is probably the **shift cipher.** We assume that the plaintext and ciphertext consist of uppercase letters (A to Z) only. In this cipher, the encryption algorithm is "shift *key* characters down," with *key* equal to some number. The decryption algorithm is "shift *key* characters up." For example, if the key is 5, the encryption algorithm is "shift 5 characters down" (toward the end of the alphabet). The decryption algorithm is "shift 5 characters up" (toward the beginning of the alphabet). Of course, if we reach the end or beginning of the alphabet, we wrap around.

Julius Caesar used the shift cipher to communicate with his officers. For this reason, the shift cipher is sometimes referred to as the **Caesar cipher.** Caesar used a key of 3 for his communications.

The shift cipher is sometimes referred to as the Caesar cipher.

Example 30.3

Use the shift cipher with key = 15 to encrypt the message "HELLO."

Solution

We encrypt one character at a time. Each character is shifted 15 characters down. Letter H is encrypted to W. Letter E is encrypted to T. The first L is encrypted to A. The second L is also encrypted to A. And O is encrypted to D. The cipher text is WTAAD.

Example 30.4

Use the shift cipher with key = 15 to decrypt the message "WTAAD."

Solution

We decrypt one character at a time. Each character is shifted 15 characters up. Letter W is decrypted to H. Letter T is decrypted to E. The first A is decrypted to L. The second A is decrypted to L. And, finally, D is decrypted to O. The plaintext is HELLO.

Transposition Ciphers

In a **transposition cipher,** there is no substitution of characters; instead, their locations change. A character in the first position of the plaintext may appear in the tenth position of the ciphertext. A character in the eighth position may appear in the first position. In other words, a transposition cipher reorders the symbols in a block of symbols.

A transposition cipher reorders (permutes) symbols in a block of symbols.

Key In a transposition cipher, the key is a mapping between the position of the symbols in the plaintext and cipher text. For example, the following shows the key using a block of four characters:

Plaintext:	**2**	**4**	**1**	**3**
Ciphertext:	**1**	**2**	**3**	**4**

In encryption, we move the character at position 2 to position 1, the character at position 4 to position 2, and so on. In decryption, we do the reverse. Note that, to be more effective, the key should be long, which means encryption and decryption of long blocks of data. Figure 30.8 shows encryption and decryption for our four-character

Figure 30.8 *Transposition cipher*

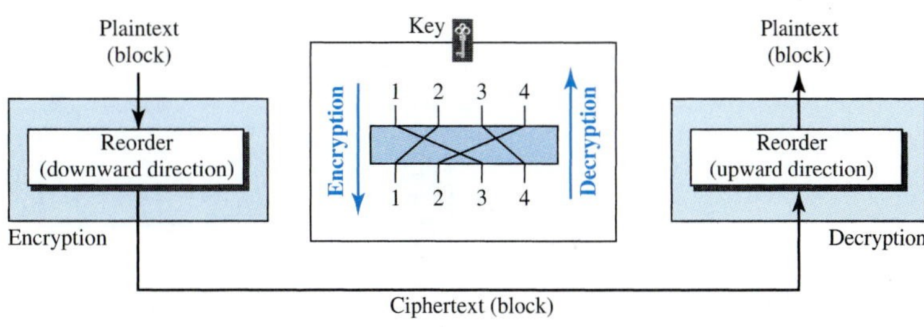

block using the above key. The figure shows that the encryption and decryption use the same key. The encryption applies it from downward while decryption applies it upward.

Example 30.5

Encrypt the message "HELLO MY DEAR," using the above key.

Solution

We first remove the spaces in the message. We then divide the text into blocks of four characters. We add a bogus character Z at the end of the third block. The result is HELL OMYD EARZ. We create a three-block ciphertext ELHLMDOYAZER.

Example 30.6

Using Example 30.5, decrypt the message "ELHLMDOYAZER".

Solution

The result is HELL OMYD EARZ. After removing the bogus character and combining the characters, we get the original message "HELLO MY DEAR."

Simple Modern Ciphers

The traditional ciphers we have studied so far are character-oriented. With the advent of the computer, ciphers need to be bit-oriented. This is so because the information to be encrypted is not just text; it can also consist of numbers, graphics, audio, and video data. It is convenient to convert these types of data into a stream of bits, encrypt the stream, and then send the encrypted stream. In addition, when text is treated at the bit level, each character is replaced by 8 (or 16) bits, which means the number of symbols becomes 8 (or 16). Mingling and mangling bits provides more security than mingling and mangling characters. Modern ciphers use a different strategy than the traditional ones. A modern symmetric cipher is a combination of simple ciphers. In other words, a modern cipher uses several simple ciphers to achieve its goal. We first discuss these simple ciphers.

XOR Cipher

Modern ciphers today are normally made of a set of **simple ciphers,** which are simple predefined functions in mathematics or computer science. The first one discussed here is called the **XOR cipher** because it uses the exclusive-or operation as defined in computer science. Figure 30.9 shows an XOR cipher.

Figure 30.9 *XOR cipher*

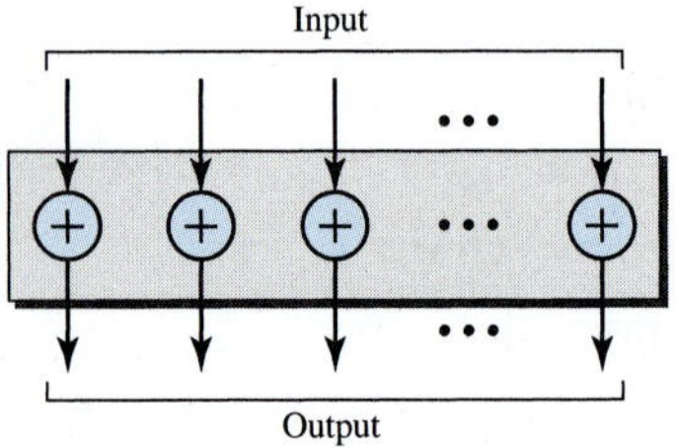

An XOR operation needs two data inputs plaintext, as the first and a key as the second. In other words, one of the inputs is the block to be the encrypted, the other input is a key; the result is the encrypted block. Note that in an XOR cipher, the size of the key, the plaintext, and the ciphertext are all the same. XOR ciphers have a very interesting property: the encryption and decryption are the same.

Rotation Cipher

Another common cipher is the **rotation cipher,** in which the input bits are rotated to the left or right. The rotation cipher can be keyed or keyless. In keyed rotation, the value of the key defines the number of rotations; in keyless rotation the number of rotations is fixed. Figure 30.10 shows an example of a rotation cipher. Note that the rotation cipher can be considered a special case of the transpositional cipher using bits instead of characters.

Figure 30.10 *Rotation cipher*

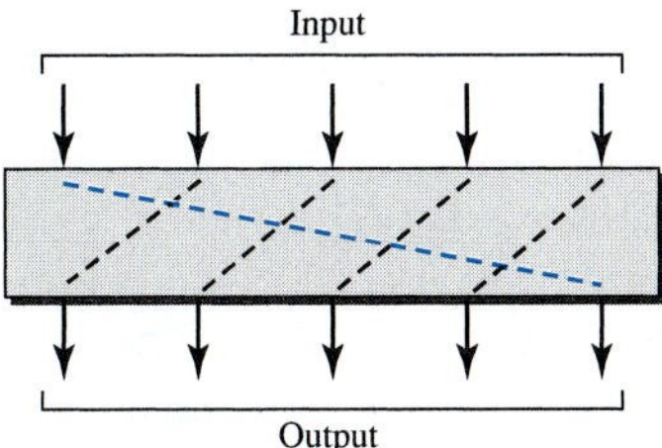

The rotation cipher has an interesting property. If the length of the original stream is N, after N rotations, we get the original input stream. This means that it is useless to apply more than $N-1$ rotations. In other words, the number of rotations must be between 1 and $N-1$.

The decryption algorithm for the rotation cipher uses the same key and the opposite rotation direction. If we use a right rotation in the encryption, we use a left rotation in decryption and vice versa.

Substitution Cipher: S-box

An **S-box** (substitution box) parallels the traditional substitution cipher for characters. The input to an S-box is a stream of bits with length N; the result is another stream of bits with length M. And N and M are not necessarily the same. Figure 30.11 shows an S-box.

The S-box is normally keyless and is used as an intermediate stage of encryption or decryption. The function that matches the input to the output may be defined mathematically or by a table.

Transposition Cipher: P-box

A **P-box** (permutation box) for bits parallels the traditional transposition cipher for characters. It performs a transposition at the bit level; it transposes bits. It can be implemented

Figure 30.11 *S-box*

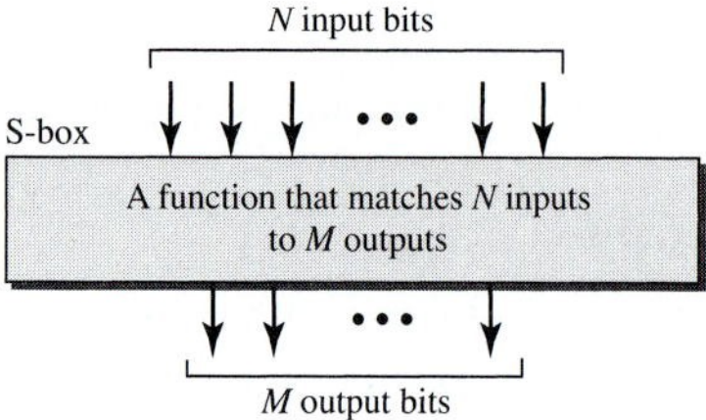

in software or hardware, but hardware is faster. P-boxes, like S-boxes, are normally keyless. We can have three types of permutations in P-boxes: the **straight permutation, expansion permutation,** and **compression permutation** as shown in Figure 30.12.

Figure 30.12 *P-boxes: straight, expansion, and compression*

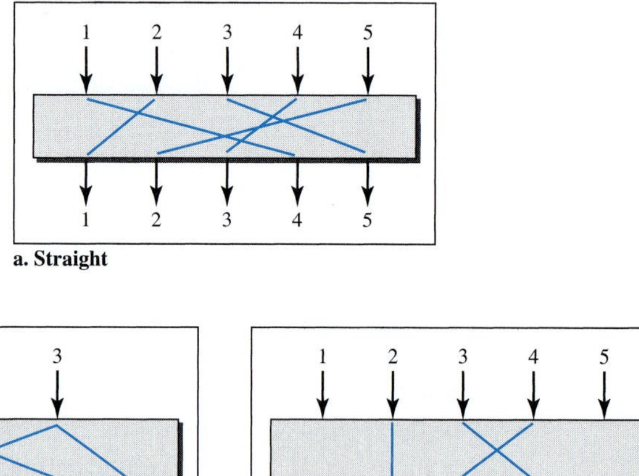

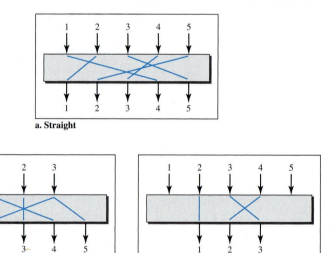

A straight permutation cipher or a straight P-box has the same number of inputs as outputs. In other words, if the number of inputs is N, the number of outputs is also N. In an expansion permutation cipher, the number of output ports is greater than the number of input ports. In a compression permutation cipher, the number of output ports is less than the number of input ports.

Modern Round Ciphers

The ciphers of today are called **round ciphers** because they involve multiple **rounds,** where each round is a complex cipher made up of the simple ciphers that we previously described. The key used in each round is a subset or variation of the general key called the round key. If the cipher has N rounds, a key generator produces N keys, $K_1, K_2, \ldots, K_N$, where K_1 is used in round 1, K_2 in round 2, and so on.

In this section, we introduce two modern symmetric-key ciphers: DES and AES. These ciphers are referred to as **block ciphers** because they divide the plaintext into blocks and use the same key to encrypt and decrypt the blocks. DES has been the de facto standard until recently. AES is the formal standard now.

Data Encryption Standard (DES)

One example of a complex block cipher is the **Data Encryption Standard (DES).** DES was designed by IBM and adopted by the U.S. government as the standard encryption method for nonmilitary and nonclassified use. The algorithm encrypts a 64-bit plaintext block using a 64-bit key, as shown in Figure 30.13.

Figure 30.13 *DES*

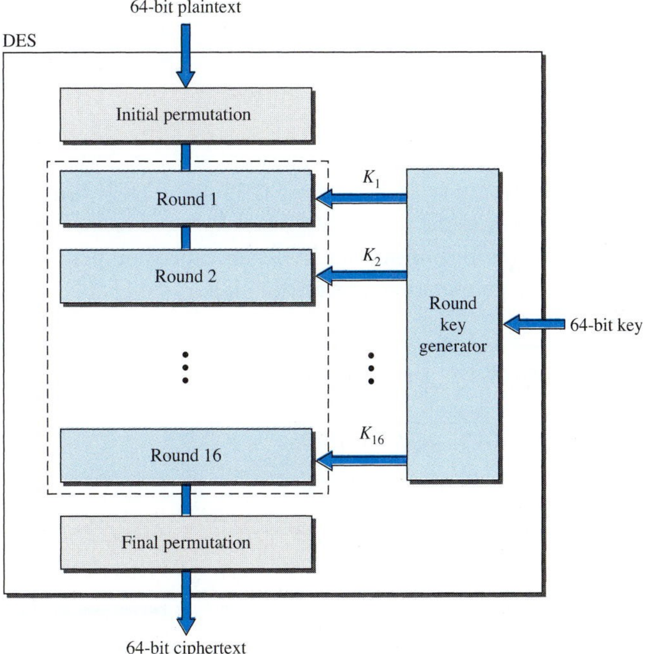

DES has two transposition blocks (P-boxes) and 16 complex round ciphers (they are repeated). Although the 16 iteration round ciphers are conceptually the same, each uses a different key derived from the original key.

The initial and final permutations are keyless straight permutations that are the inverse of each other. The permutation takes a 64-bit input and permutes them according to predefined values.

Each round of DES is a complex round cipher, as shown in Figure 30.14. Note that the structure of the encryption round ciphers is different from that of the decryption one.

Figure 30.14 *One round in DES ciphers*

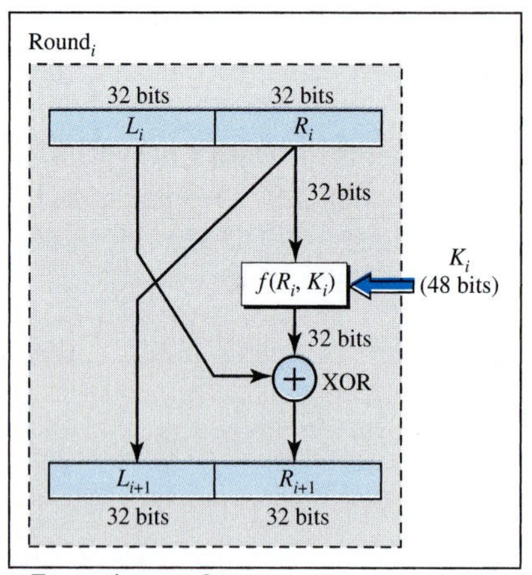

a. Encryption round

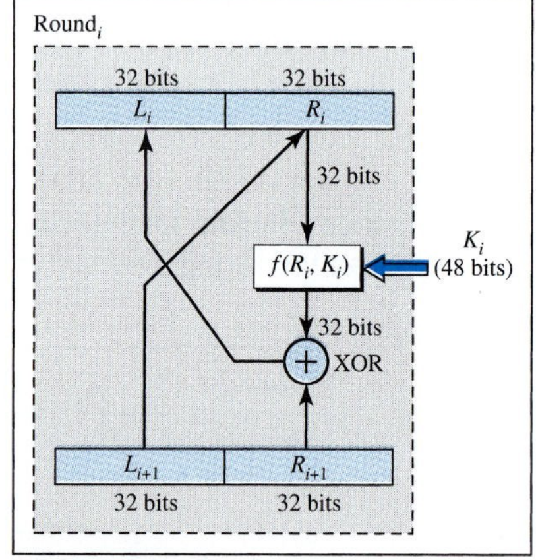

b. Decryption round

DES Function The heart of DES is the **DES function.** The DES function applies a 48-bit key to the rightmost 32 bits R_i to produce a 32-bit output. This function is made up of four operations: an XOR, an expansion permutation, a group of S-boxes, and a straight permutation, as shown in Figure 30.15.

Figure 30.15 *DES function*

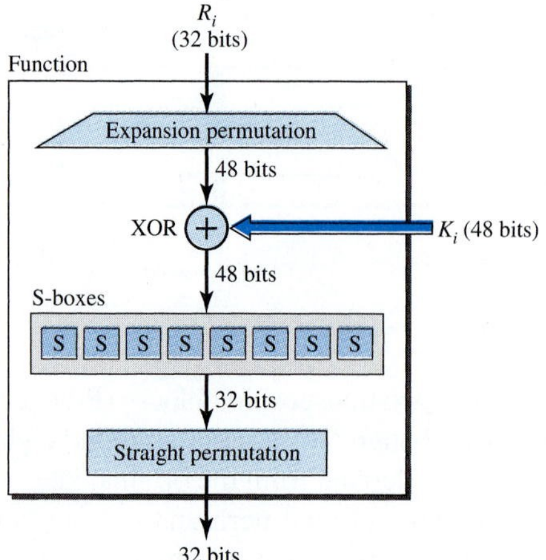

Triple DES

Critics of DES contend that the key is too short. To lengthen the key, **Triple DES** or 3DES has been proposed and implemented. This uses three DES blocks, as shown in Figure 30.16. Note that the encrypting block uses an encryption-decryption-encryption combination of DESs, while the decryption block uses a decryption-encryption-decryption combination. Two different versions of 3DES are in use: 3DES with two keys and 3DES with three keys. To make the key size 112 bits and at the same time protect DES from attacks such as the man-in-the-middle attack, 3DES with two keys was designed. In this version, the first and the third keys are the same ($Key_1 = Key_3$). This has the advantage in that a text encrypted by a single DES block can be decrypted by the new 3DES. We just set all keys equal to Key_1. Many algorithms use a 3DES cipher with three keys. This increases the size of the key to 168 bits.

Figure 30.16 *Triple DES*

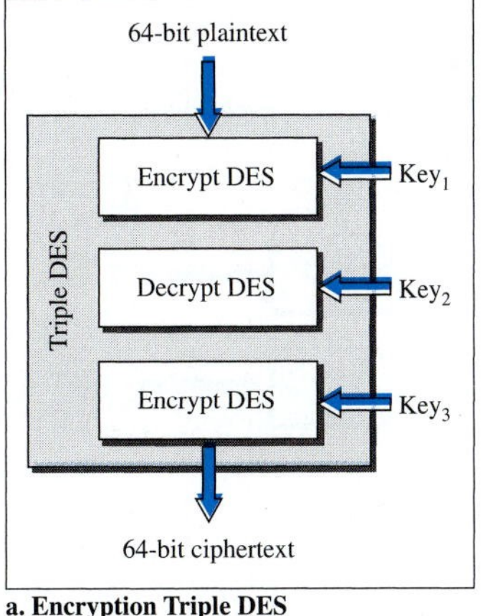

a. Encryption Triple DES

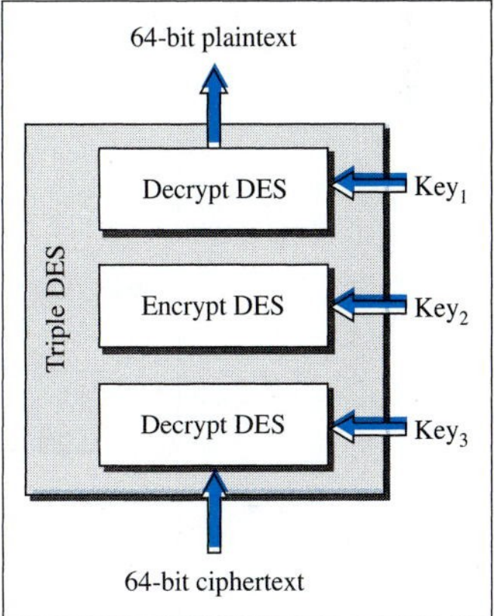

b. Decryption Triple DES

Advanced Encryption Standard (AES)

The **Advanced Encryption Standard (AES)** was designed because DES's key was too small. Although Triple DES (3DES) increased the key size, the process was too slow. The **National Institute of Standards and Technology (NIST)** chose the **Rijndael algorithm,** named after its two Belgian inventors, Vincent Rijmen and Joan Daemen, as the basis of AES. AES is a very complex round cipher. AES is designed with three key sizes: 128, 192, or 256 bits. Table 30.1 shows the relationship between the data block, number of rounds, and key size.

Table 30.1 *AES configuration*

Size of Data Block	*Number of Rounds*	*Key Size*
128 bits	10	128 bits
	12	192 bits
	14	256 bits

AES has three different configurations with respect to the number of rounds and key size.

In this text, we discuss just the 10-round, 128-bit key configuration. The structure and operation of the other configurations are similar. The difference lies in the key generation.

The general structure is shown in Figure 30.17. There is an initial XOR operation followed by 10 round ciphers. The last round is slightly different from the preceding rounds; it is missing one operation.

Although the 10 iteration blocks are almost identical, each uses a different key derived from the original key.

Figure 30.17 *AES*

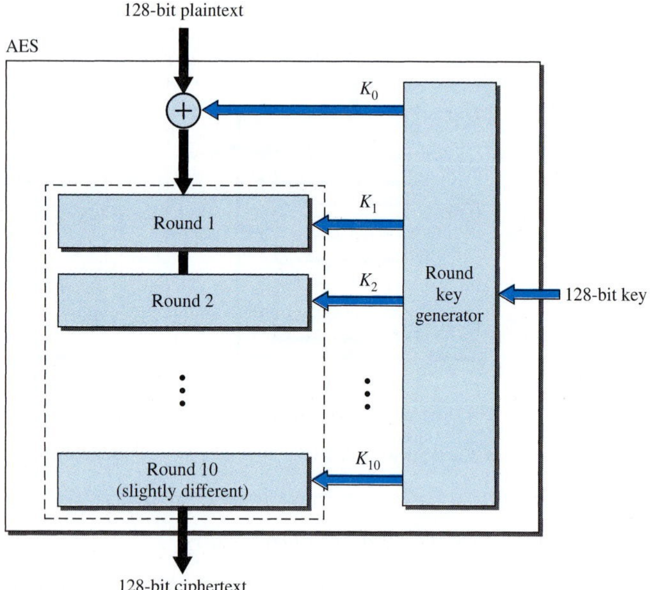

Structure of Each Round Each round of AES, except for the last, is a cipher with four operations that are invertible. The last round has only three operations. Figure 30.18 is a flowchart that shows the operations in each round. Each of the four operations used in each round uses a complex cipher; this topic is beyond the scope of this book.

Other Ciphers

During the last two decades, a few other symmetric block ciphers have been designed and used. Most of these ciphers have similar characteristics to the two ciphers we discussed in this chapter (DES and AES). The difference is usually in the size of the block or key, the number of rounds, and the functions used. The principles are the same. In order not to burden the user with the details of these ciphers, we give a brief description of each.

Figure 30.18 *Structure of each round*

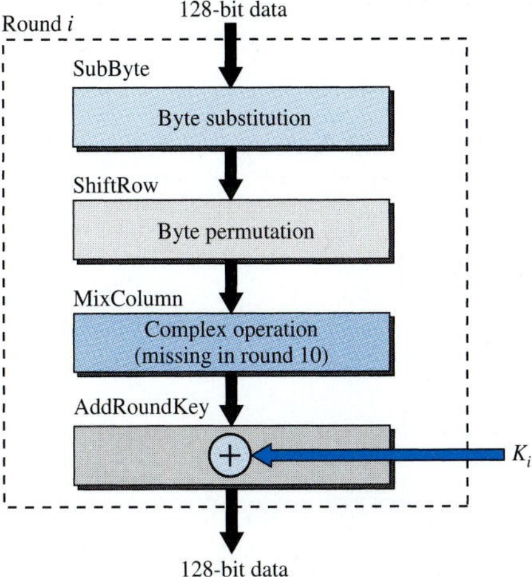

IDEA The International Data Encryption Algorithm (IDEA) was developed by Xuejia Lai and James Massey. The block size is 64 and the key size is 128. It can be implemented in both hardware and software.

Blowfish Blowfish was developed by Bruce Schneier. The block size is 64 and the key size between 32 and 448.

CAST-128 CAST-128 was developed by Carlisle Adams and Stafford Tavares. It is a Feistel cipher with 16 rounds and a block size of 64 bits; the key size is 128 bits.

RC5 RC5 was designed by Ron Rivest. It is a family of ciphers with different block sizes, key sizes, and numbers of rounds.

Mode of Operation

A **mode of operation** is a technique that employs the modern block ciphers such as DES and AES that we discussed earlier (see Figure 30.19).

Figure 30.19 *Modes of operation for block ciphers*

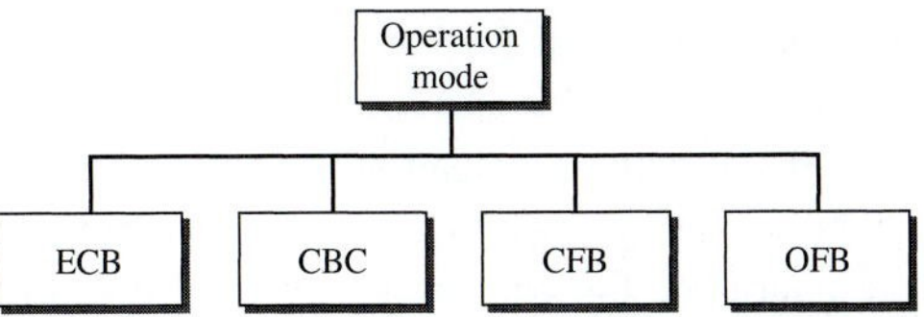

Electronic Code Book

The **electronic code book (ECB) mode** is a purely block cipher technique. The plaintext is divided into blocks of *N* bits. The ciphertext is made of blocks of *N* bits. The value of *N* depends on the type of cipher used. Figure 30.20 shows the method.

Figure 30.20 *ECB mode*

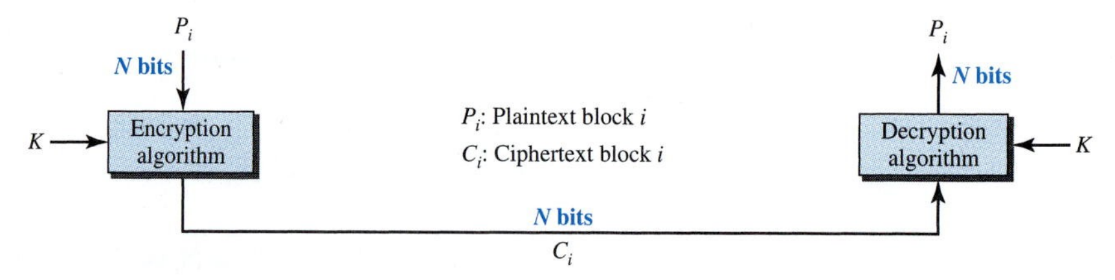

We mention four characteristics of this mode:

1. Because the key and the encryption/decryption algorithm are the same, equal blocks in the plaintext become equal blocks in the ciphertext. For example, if plaintext blocks 1, 5, and 9 are the same, ciphertext blocks 1, 5, and 9 are also the same. This can be a security problem; the adversary can guess that the plaintext blocks are the same if the corresponding ciphertext blocks are the same.
2. If we reorder the plaintext block, the ciphertext is also reordered.
3. Blocks are independent of each other. Each block is encrypted or decrypted independently. A problem in encryption or decryption of a block does not affect other blocks.
4. An error in one block is not propagated to other blocks. If one or more bits are corrupted during transmission, it only affects the bits in the corresponding plaintext after decryption. Other plaintext blocks are not affected. This is a real advantage if the channel is not noise-free.

Cipher Block Chaining

The **cipher block chaining (CBC) mode** tries to alleviate some of the problems in ECB by including the previous cipher block in the preparation of the current block. If the current block is *i*, the previous ciphertext block C_{i-1} is included in the encryption of block *i*. In other words, when a block is completely enciphered, the block is sent, but a copy of it is kept in a register (a place where data can be held) to be used in the encryption of the next block. The reader may wonder about the initial block. There is no ciphertext block before the first block. In this case, a phony block called the **initiation vector (IV)** is used. Both the sender and receiver agree upon a specific predetermined IV. In other words, the IV is used instead of the nonexistent C_0. Figure 30.21 shows the CBC mode.

The reader may wonder about the decryption. Does the configuration shown in the figure guarantee the correct decryption? It can be proven that it does, but we leave the proof to a textbook in network security.

Figure 30.21 *CBC mode*

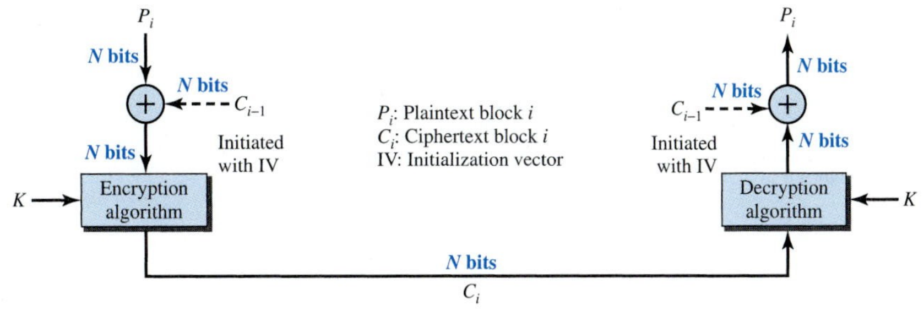

The following are some characteristics of CBC.

1. Even though the key and the encryption/decryption algorithm are the same, equal blocks in the plaintext do not become equal blocks in the ciphertext. For example, if plaintext blocks 1, 5, and 9 are the same, ciphertext blocks 1, 5, and 9 will not be the same. An adversary will not be able to guess from the ciphertext that two blocks are the same.
2. Blocks are dependent on each other. Each block is encrypted or decrypted based on a previous block. A problem in encryption or decryption of a block affects other blocks.
3. The error in one block is propagated to the other blocks. If one or more bits are corrupted during the transmission, it affects the bits in the next blocks of the plaintext after decryption.

Cipher Feedback

The **cipher feedback (CFB) mode** was created for those situations in which we need to send or receive r bits of data, where r is a number different from the underlying block size of the encryption cipher used. The value of r can be 1, 4, 8, or any number of bits. Since all block ciphers work on a block of data at a time, the problem is how to encrypt just r bits. The solution is to let the cipher encrypt a block of bits and use only the first r bits as a new key (stream key) to encrypt the r bits of user data. Figure 30.22 shows the configuration.

The following are some characteristics of the CFB mode:

1. If we change the IV from one encryption to another using the same plaintext, the ciphertext is different.
2. The ciphertext C_i depends on both P_i and the preceding ciphertext block.
3. Errors in one or more bits of the ciphertext block affect the next ciphertext blocks.

Output Feedback

The **output feedback (OFB) mode** is very similar to the CFB mode with one difference. Each bit in the ciphertext is independent of the previous bit or bits. This avoids error

Figure 30.22 *CFB mode*

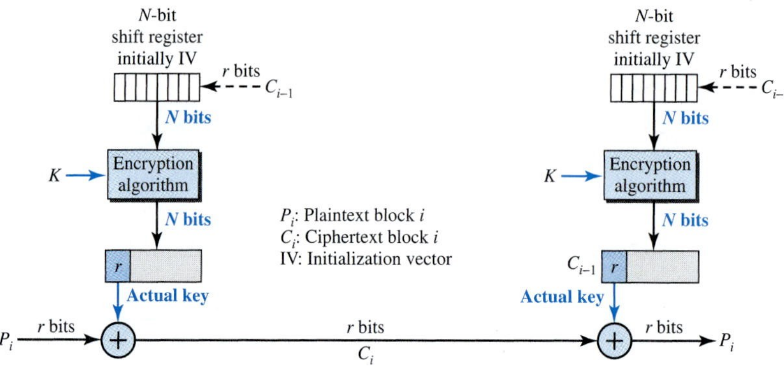

propagation. If an error occurs in transmission, it does not affect the future bits. Note that, as in CFB, both the sender and the receiver use the encryption algorithm. Note also that in OFB, block ciphers such as DES or AES can only be used to create the key stream. The feedback for creating the next bit stream comes from the previous bits of the key stream instead of the ciphertext. The ciphertext does not take part in creating the key stream. Figure 30.23 shows the OFB mode.

Figure 30.23 *OFB mode*

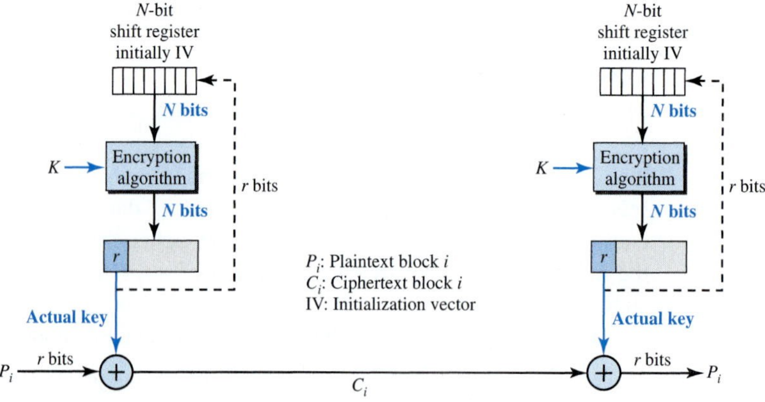

The following are some of the characteristics of the OFB mode.

1. If we change the IV from one encryption to another using the same plaintext, the ciphertext will be different.
2. The ciphertext C_i depends on the plaintext P_i.
3. Errors in one or more bits of the ciphertext do not affect future ciphertext blocks.

30.3 ASYMMETRIC-KEY CRYPTOGRAPHY

In the previous sections, we discussed symmetric-key cryptography. In this section we introduce asymmetric-key (public key cryptography). As we mentioned before, an asymmetric-key (or public-key) cipher uses two keys: one private and one public. We discuss two algorithms: RSA and Diffie-Hellman.

RSA

The most common public key algorithm is RSA, named for its inventors **Rivest, Shamir, and Adleman (RSA).** It uses two numbers, e and d, as the public and private keys, as shown in Figure 30.24.

Figure 30.24 *RSA*

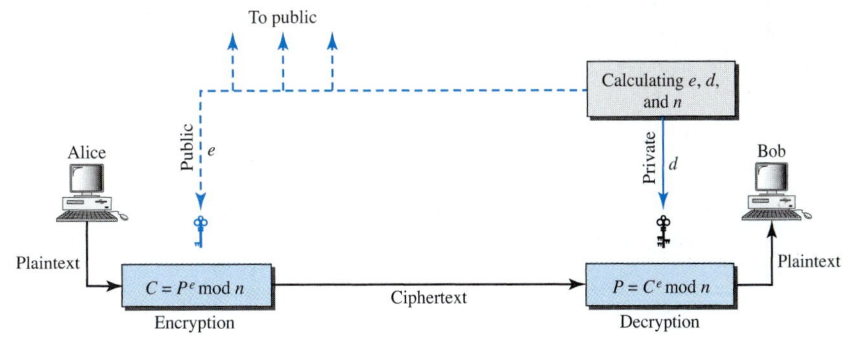

The two keys, e and d, have a special relationship to each other, a discussion of this relationship is beyond the scope of this book. We just show how to calculate the keys without proof.

Selecting Keys

Bob use the following steps to select the private and public keys:

1. Bob chooses two very large prime numbers p and q. Remember that a prime number is one that can be divided evenly only by 1 and itself.
2. Bob multiplies the above two primes to find n, the modulus for encryption and decryption. In other words, $n = p \times q$.
3. Bob calculates another number $\phi = (p - 1) \times (q - 1)$.
4. Bob chooses a random integer e. He then calculates d so that $d \times e = 1 \bmod \phi$.
5. Bob announces e and n to the public; he keeps ϕ and d secret.

In RSA, e and n are announced to the public; d and ϕ are kept secret.

Encryption

Anyone who needs to send a message to Bob can use n and e. For example, if Alice needs to send a message to Bob, she can change the message, usually a short one, to an integer. This is the plaintext. She then calculates the ciphertext, using e and n.

$$C = P^e \pmod n$$

Alice sends C, the ciphertext, to Bob.

Decryption

Bob keeps ϕ and d private. When he receives the ciphertext, he uses his private key d to decrypt the message:

$$P = C^d \pmod n$$

Restriction

For RSA to work, the value of P must be less than the value of n. If P is a large number, the plaintext needs to be divided into blocks to make P less than n.

Example 30.7

Bob chooses 7 and 11 as p and q and calculates $n = 7 \cdot 11 = 77$. The value of $\phi = (7 - 1)(11 - 1)$ or 60. Now he chooses two keys, e and d. If he chooses e to be 13, then d is 37. Now imagine Alice sends the plaintext 5 to Bob. She uses the public key 13 to encrypt 5.

Plaintext: 5
$C = 5^{13} = 26 \bmod 77$
Ciphertext: 26

Bob receives the ciphertext 26 and uses the private key 37 to decipher the ciphertext:

Ciphertext: 26
$P = 26^{37} = 5 \bmod 77$
Plaintext: 5 **Intended message sent by Alice**

The plaintext 5 sent by Alice is received as plaintext 5 by Bob.

Example 30.8

Jennifer creates a pair of keys for herself. She chooses $p = 397$ and $q = 401$. She calculates $n = 159{,}197$ and $\phi = 396 \cdot 400 = 158{,}400$. She then chooses $e = 343$ and $d = 12{,}007$. Show how Ted can send a message to Jennifer if he knows e and n.

Solution

Suppose Ted wants to send the message "NO" to Jennifer. He changes each character to a number (from 00 to 25) with each character coded as two digits. He then concatenates the two coded characters and gets a four-digit number. The plaintext is 1314. Ted then uses e and n to encrypt the message. The ciphertext is $1314^{343} = 33{,}677 \bmod 159{,}197$. Jennifer receives the message

33,677 and uses the decryption key d to decipher it as $33{,}677^{12{,}007} = 1314 \bmod 159{,}197$. Jennifer then decodes 1314 as the message "NO". Figure 30.25 shows the process.

Figure 30.25 *Example 30.8*

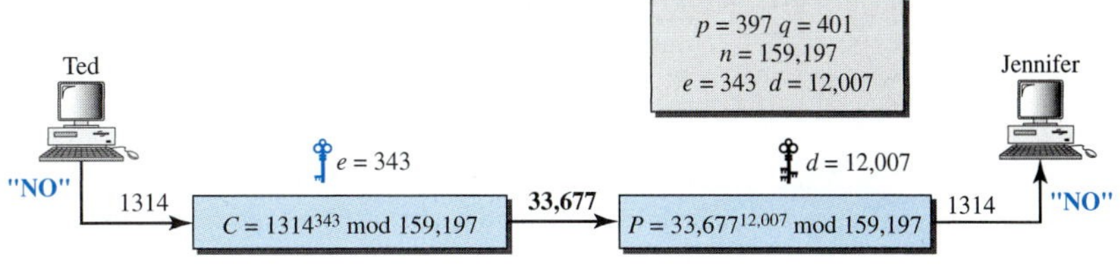

Example 30.9

Let us give a realistic example. We choose a 512-bit p and q. We calculate n and ϕ. We then choose e and test for relative primeness with $\phi(n)$. We calculate d. Finally, we show the results of encryption and decryption. We have written a program written in Java to do so; this type of calculation cannot be done by a calculator.

We randomly chose an integer of 512 bits. The integer p is a 159-digit number.

p = 96130345313583504574191581280615427909309845594996215822583150879647940
455056470638491257160180347503120986666064924201918087806674210960633542
19926661209

The integer q is a 160-digit number.

q = 12060191957231446918276794204450896001555925054637033936061798321731482
14848376465921538945320917522527322683010712069560460251388714552496900
0359660045617

We calculate n. It has 309 digits.

n = 11593504173967614968892509864615887523771457375454144775485526137614788
54083263508172768788159683251684688493006254857641112501624145523391829
27162507656772727460097082714127730434960500556347274566628060099924037
10299142447229221577279853172703383938133469268413732762200096667667183
18310883734208234443709 53

We calculate ϕ. It has 309 digits:

ϕ = 11593504173967614968892509864615887523771457375454144775485526137614788
54083263508172768788159683251684688493006254857641112501624145523391829
27162507656751054233608492916752034482627988117554787657013923444405716
98958172819609822636107546721186461217135910735864061400888517026537727
7264467341066243857664128

We choose $e = 35{,}535$. We then find d.

e = 35535

d = 5800830286003776393609366128967791759466906208965096218042286611138059385282235873170628691003002171085904433840217072986908760061153062025249598844480475682409662470814858171304632406440777048331340108509473852956450719367740611973265574242372176176746207763716420760033708533328853214470885955136670294831

Alice wants to send the message "THIS IS A TEST" which can be changed to a numeric value by using the 00–26 encoding scheme (26 is the *space* character).

P = 1907081826081826002619041819

The ciphertext calculated by Alice is $C = P^e$, which is

C = 47530912364622682720636555061054518094237179607049171652323924305445296061319932856661784341835911415119741125200568297979457173603610127821884789274156609048002350719071527718591497518846588863210114835410336165789846796838676373376577746562507928052114814184404814184430812773059004692874248559166462108656

Bob can recover the plaintext from the ciphertext by using $P = C^d$, which is

P = 1907081826081826002619041819

The recovered plaintext is THIS IS A TEST after decoding.

Applications

Although RSA can be used to encrypt and decrypt actual messages, it is very slow if the message is long. RSA, therefore, is useful for short messages such as a small message digest (see Chapter 31) or a symmetric key to be used for a symmetric-key cryptosystem. In particular, we will see that RSA is used in digital signatures and other cryptosystems that often need to encrypt a small message without having access to a symmetric key. RSA is also used for authentication as we will see later.

Diffie-Hellman

RSA is a public-key cryptosystem that is often used to encrypt and decrypt symmetric keys. Diffie-Hellman, on the other hand, was originally designed for key exchange. In the **Diffie-Hellman cryptosystem,** two parties create a symmetric **session key** to exchange data without having to remember or store the key for future use. They do not have to meet to agree on the key; it can be done through the Internet. Let us see how the protocol works when Alice and Bob need a symmetric key to communicate. Before establishing a symmetric key, the two parties need to choose two numbers p and g. The first number, p, is a large prime number on the order of 300 decimal digits (1024 bits). The second number is a random number. These two numbers need not be confidential. They can be sent through the Internet; they can be public.

Procedure

Figure 30.26 shows the procedure. The steps are as follows:

Figure 30.26 *Diffie-Hellman method*

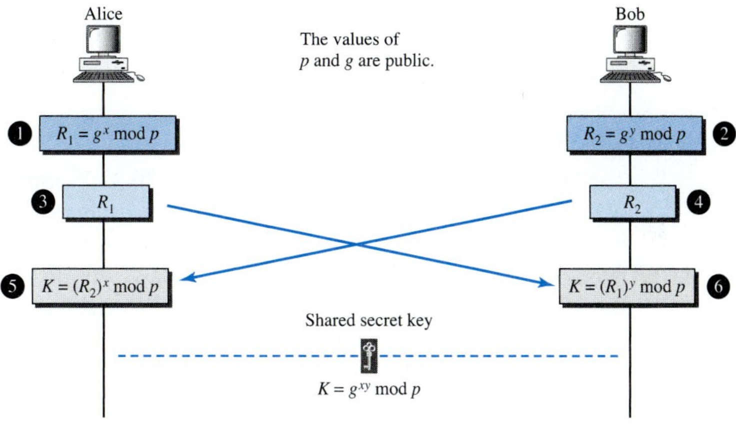

- **Step 1:** Alice chooses a large random number x and calculates $R_1 = g^x \bmod p$.
- **Step 2:** Bob chooses another large random number y and calculates $R_2 = g^y \bmod p$.
- **Step 3:** Alice sends R_1 to Bob. Note that Alice does not send the value of x; she sends only R_1.
- **Step 4:** Bob sends R_2 to Alice. Again, note that Bob does not send the value of y, he sends only R_2.
- **Step 5:** Alice calculates $K = (R_2)^x \bmod p$.
- **Step 6:** Bob also calculates $K = (R_1)^y \bmod p$.

The symmetric key for the session is K.

$$(g^x \bmod p)^y \bmod p = (g^y \bmod p)^x \bmod p = g^{xy} \bmod p$$

Bob has calculated $K = (R_1)^y \bmod p = (g^x \bmod p)^y \bmod p = g^{xy} \bmod p$. Alice has calculated $K = (R_2)^x \bmod p = (g^y \bmod p)^x \bmod = g^{xy} \bmod p$. Both have reached the same value without Bob knowing the value of x and without Alice knowing the value of y.

The symmetric (shared) key in the Diffie-Hellman protocol is $K = g^{xy} \bmod p$.

Example 30.10

Let us give a trivial example to make the procedure clear. Our example uses small numbers, but note that in a real situation, the numbers are very large. Assume $g = 7$ and $p = 23$. The steps are as follows:

1. Alice chooses $x = 3$ and calculates $R_1 = 7^3 \bmod 23 = 21$.
2. Bob chooses $y = 6$ and calculates $R_2 = 7^6 \bmod 23 = 4$.
3. Alice sends the number 21 to Bob.
4. Bob sends the number 4 to Alice.
5. Alice calculates the symmetric key $K = 4^3 \bmod 23 = 18$.
6. Bob calculates the symmetric key $K = 21^6 \bmod 23 = 18$.

The value of K is the same for both Alice and Bob; $g^{xy} \bmod p = 7^{18} \bmod 23 = 18$.

Idea of Diffie-Hellman

The Diffie-Hellman concept, shown in Figure 30.27, is simple but elegant. We can think of the secret key between Alice and Bob as made of three parts: g, x, and y. The first part is public. Everyone knows one-third of the key; g is a public value. The other two parts must be added by Alice and Bob. Each adds one part. Alice adds x as the second part for Bob; Bob adds y as the second part for Alice. When Alice receives the two-thirds completed key from Bob, she adds the last part, her x, to complete the key. When Bob receives the two-thirds completed key from Alice, he adds the last part, his y, to complete the key. Note that although the key in Alice's hand consists of g-y-x and the key in Bob's hand is g-x-y, these two keys are the same because $g^{xy} = g^{yx}$.

Figure 30.27 *Diffie-Hellman idea*

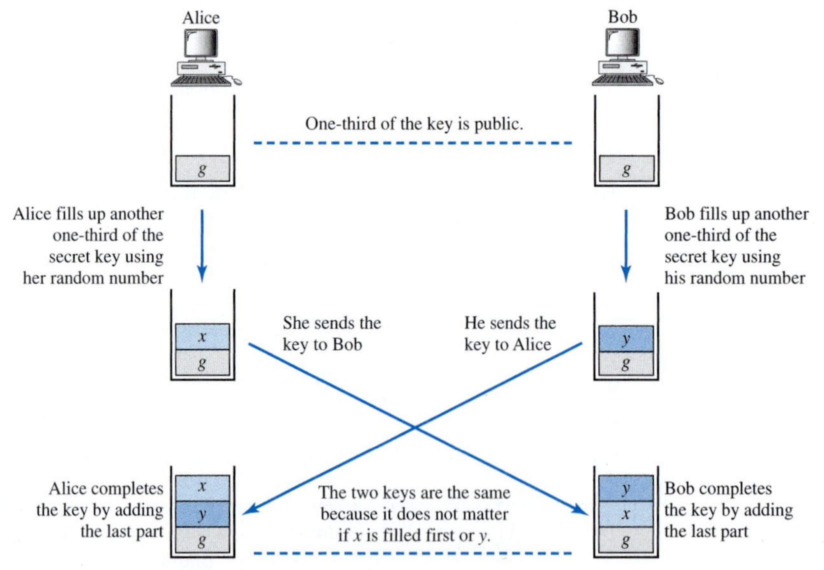

Note also that although the two keys are the same, Alice cannot find the value y used by Bob because the calculation is done in modulo p; Alice receives g^y mod p from Bob, not g^y.

Man-in-the-Middle Attack

Diffie-Hellman is a very sophisticated symmetric-key creation algorithm. If x and y are very large numbers, it is extremely difficult for Eve to find the key, knowing only p and g. An intruder needs to determine x and y if R_1 and R_2 are intercepted. But finding x from R_1 and y from R_2 are two difficult tasks. Even a sophisticated computer would need perhaps years to find the key by trying different numbers. In addition, Alice and Bob will change the key the next time they need to communicate.

However, the protocol does have a weakness. Eve does not have to find the value of x and y to attack the protocol. She can fool Alice and Bob by creating two keys: one between herself and Alice and another between herself and Bob. Figure 30.28 shows the situation.

Figure 30.28 *Man-in-the-middle attack*

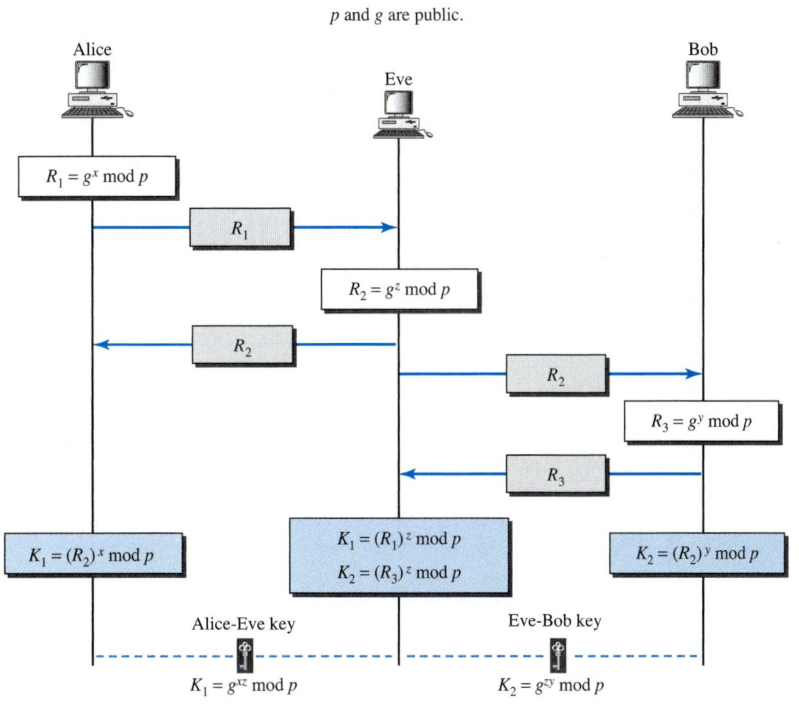

The following can happen:

1. Alice chooses x, calculates $R_1 = g^x$ mod p, and sends R_1 to Bob.
2. Eve, the intruder, intercepts R_1. She chooses z, calculates $R_2 = g^z$ mod p, and sends R_2 to both Alice and Bob.

3. Bob chooses y, calculates $R_3 = g^y \bmod p$, and sends R_3 to Alice; R_3 is intercepted by Eve and never reaches Alice.
4. Alice and Eve calculate $K_1 = g^{xz} \bmod p$, which becomes a shared key between Alice and Eve. Alice, however, thinks that it is a key shared between Bob and herself.
5. Eve and Bob calculate $K_2 = g^{zy} \bmod p$, which becomes a shared key between Eve and Bob. Bob, however, thinks that it is a key shared between Alice and himself.

In other words, two keys, instead of one, are created: one between Alice and Eve and one between Eve and Bob. When Alice sends data to Bob encrypted with K_1 (shared by Alice and Eve), it can be deciphered and read by Eve. Eve can send the message to Bob encrypted by K_2 (shared key between Eve and Bob); or she can even change the message or send a totally new message. Bob is fooled into believing that the message has come from Alice. A similar scenario can happen to Alice in the other direction.

This situation is called a **man-in-the-middle attack** because Eve comes in between and intercepts R_1, sent by Alice to Bob, and R_3, sent by Bob to Alice. It is also known as a **bucket brigade attack** because it resembles a short line of volunteers passing a bucket of water from person to person.

Authentication

The man-in-the-middle attack can be avoided if Bob and Alice first authenticate each other. In other words, the exchange key process can be combined with an authentication scheme to prevent a man-in-the-middle attack. We discuss authentication in Chapter 31.

30.4 RECOMMENDED READING

For more details about subjects discussed in this chapter, we recommend the following books. The items in brackets [. . .] refer to the reference list at the end of the text.

Books

Cryptography can be found in many books dedicated to the subject such as [Bar02], [Gar01], [Sti02], [Mao04], [MOV97], and [Sch96].

30.5 KEY TERMS

Advanced Encryption Standard (AES)
block cipher
bucket brigade attack
Caesar cipher
cipher block chaining (CBC) mode
cipher feedback (CFB) mode
ciphertext
compression permutation
cryptography
Data Encryption Standard (DES)
decryption
decryption algorithm
DES function
Diffie-Hellman cryptosystem
electronic code book (ECB) mode
encryption

encryption algorithm
expansion permutation
initiation vector (IV)
key
man-in-the-middle attack
mode of operation
monoalphabetic cipher
National Institute of Standards and Technology (NIST)
output feedback (OFB) mode
P-box
plaintext
polyalphabetic cipher
private key
public key
Rijndael algorithm
Rivest, Shamir, Adleman (RSA)
rotation cipher
round
round cipher
S-box
secret key
session key
shift cipher
simple cipher
straight permutation
substitution cipher
transposition cipher
Triple DES
XOR cipher

30.6 SUMMARY

- Cryptography is the science and art of transforming messages to make them secure and immune to attacks.
- The plaintext is the original message before transformation; the ciphertext is the message after transformation.
- An encryption algorithm transforms plaintext to ciphertext; a decryption algorithm transforms ciphertext to plaintext.
- A combination of an encryption algorithm and a decryption algorithm is called a cipher.
- The key is a number or a set of numbers on which the cipher operates.
- We can divide all ciphers into two broad categories: symmetric-key ciphers and asymmetric-key ciphers.
- In a symmetric-key cipher, the same key is used by both the sender and receiver. The key is called the secret key.
- In an asymmetric-key cipher, a pair of keys is used. The sender uses the public key; the receiver uses the private key.
- A substitution cipher replaces one character with another character.
- Substitution ciphers can be categorized into two broad categories: monoalphabetic and polyalphabetic.
- The shift cipher is the simplest monoalphabetic cipher. It uses modular arithmetic with a modulus of 26. The Caesar cipher is a shift cipher that has a key of 3.
- The transposition cipher reorders the plaintext characters to create a ciphertext.
- An XOR cipher is the simplest cipher which is self-invertible.
- A rotation cipher is an invertible cipher.

- An S-box is a keyless substitution cipher with N inputs and M outputs that uses a formula to define the relationship between the input stream and the output stream.
- A P-box is a keyless transposition cipher with N inputs and M outputs that uses a table to define the relationship between the input stream and the output stream. A P-box is invertible only if the numbers of inputs and outputs are the same. A P-box can use a straight permutation, a compression permutation, or an expansion permutation.
- A modern cipher is usually a round cipher; each round is a complex cipher made of a combination of different simple ciphers.
- DES is a symmetric-key method adopted by the U.S. government. DES has an initial and final permutation block and 16 rounds.
- The heart of DES is the DES function. The DES function has four components: an expansion permutation, an XOR operation, S-boxes, and a straight permutation.
- DES uses a key generator to generate sixteen 48-bit round keys.
- Triple DES was designed to increase the size of the DES key (effectively 56 bits) for better security.
- AES is a round cipher based on the Rijndael algorithm that uses a 128-bit block of data. AES has three different configurations: 10 rounds with a key size of 128 bits, 12 rounds with a key size of 192 bits, and 14 rounds with a key size of 256 bits.
- Mode of operation refers to techniques that deploy the ciphers such as DES or AES. Four common modes of operation are ECB, CBC, CBF, and OFB. ECB and CBC are block ciphers; CBF and OFB are stream ciphers.
- One commonly used public-key cryptography method is the RSA algorithm, invented by Rivest, Shamir, and Adleman.
- RSA chooses n to be the product of two primes p and q.
- The Diffie-Hellman method provides a one-time session key for two parties.
- The man-in-the-middle attack can endanger the security of the Diffie-Hellman method if two parties are not authenticated to each other.

30.7 PRACTICE SET

Review Questions

1. In symmetric-key cryptography, how many keys are needed if Alice and Bob want to communicate with each other?
2. In symmetric-key cryptography, can Alice use the same key to communicate with both Bob and John? Explain your answer.
3. In symmetric-key cryptography, if every person in a group of 10 people needs to communicate with every other person in another group of 10 people, how many secret keys are needed?
4. In symmetric-key cryptography, if every person in a group of 10 people needs to communicate with every other person in the group, how many secret keys are needed?
5. Repeat Question 1 for asymmetric-key cryptography.

6. Repeat Question 2 for asymmetric-key cryptography.
7. Repeat Question 3 for asymmetric-key cryptography.
8. Repeat Question 4 for asymmetric-key cryptography.

Exercises

9. In symmetric-key cryptography, how do you think two persons can establish a secret key between themselves?
10. In asymmetric-key cryptography, how do you think two persons can establish two pairs of keys between themselves?
11. Encrypt the message "THIS IS AN EXERCISE" using a shift cipher with a key of 20. Ignore the space between words. Decrypt the message to get the original plaintext.
12. Can we use monoalphabetic substitution if our symbols are just 0 and 1? Is it a good idea?
13. Can we use polyalphabetic substitution if our symbols are just 0 and 1? Is it a good idea?
14. Encrypt "INTERNET" using a transposition cipher with the following key:

3	**5**	**2**	**1**	**4**
1	**2**	**3**	**4**	**5**

15. Rotate 111001 three bits to the right.
16. Rotate 100111 three bits to the left.
17. A 6-by-2 S-box adds the bits at the odd-numbered positions (1, 3, 5, . . .) to get the right bit of the output and adds the bits at the even-numbered positions (2, 4, 6, . . .) to get the left bit of the output. If the input is 110010, what is the output? If the input is 101101, what is the output? Assume the rightmost bit is bit 1.
18. What are all the possible number combinations of inputs in a 6-by-2 S-box? What is the possible number of outputs?
19. The leftmost bit of a 4-by-3 S-box rotates the other 3 bits. If the leftmost bit is 0, the 3 other bits are rotated to the right 1 bit. If the leftmost bit is 1, the 3 other bits are rotated to the left 1 bit. If the input is 1011, what is the output? If the input is 0110, what is the output?
20. A P-box uses the following table for encryption. Show the box and connect the input to the output.

4	**2**	**3**	**1**
	1	**2**	

Is the P-box straight, compression, or expansion?

21. In RSA, given two prime numbers $p = 19$ and $q = 23$, find n and ϕ. Choose $e = 5$ and try to find d, such that e and d meet the criteria.
22. To understand the security of the RSA algorithm, find d if you know that $e = 17$ and $n = 187$. This exercise proves how easy is for Eve to break the secret if n is small.

23. For the RSA algorithm with a large n, explain why Bob can calculate d from n, but Eve cannot.
24. Using $e = 13$, $d = 37$, and $n = 77$ in the RSA algorithm, encrypt the message "FINE" using the values of 00 to 25 for letters A to Z. For simplicity, do the encryption and decryption character by character.
25. Why can't Bob choose 1 as the public key e in RSA?
26. What is the danger in choosing 2 as the public key e in RSA?
27. Eve uses RSA to send a message to Bob, using Bob's public key. Later, at a cocktail party, Eve sees Bob and asks him if the message has arrived and Bob confirms it. After a few drinks, Eve asks Bob, "What was the ciphertext?" Bob gives the value of the ciphertext to Eve. Can this endanger the security of Bob's private key? Explain your answer.
28. What is the value of the symmetric key in the Diffie-Hellman protocol if $g = 7$, $p = 23$, $x = 2$, and $y = 5$?
29. In the Diffie-Hellman protocol, what happens if x and y have the same value? That is, Alice and Bob have accidentally chosen the same number. Are the values of R_1 and R_2 the same? Are the values of the session keys calculated by Alice and Bob the same? Use an example to prove your claims.

Research Activities

30. Another asymmetric-key algorithm is called ElGamal. Do some research and find out some information about this algorithm. What is the difference between RSA and ElGamal?
31. Another asymmetric-key algorithm is based on elliptic curves. If you are familiar with elliptic curves, do some research and find the algorithms based on elliptic curves.
32. To make Diffie-Helman algorithm more robust, one uses cookies. Do some research and find out about the use of cookies in the Diffie-Helman algorithm.

CHAPTER 31

Network Security

In Chapter 30, we introduced the science of cryptography. Cryptography has several applications in network security. In this chapter, we first introduce the security services we typically expect in a network. We then show how these services can be provided using cryptography. At the end of the chapter, we also touch on the issue of distributing symmetric and asymmetric keys. The chapter provides the background necessary for Chapter 32, where we discuss security in the Internet.

31.1 SECURITY SERVICES

Network security can provide one of the five services as shown in Figure 31.1. Four of these services are related to the message exchanged using the network: message confidentiality, integrity, authentication, and nonrepudiation. The fifth service provides entity authentication or identification.

Figure 31.1 *Security services related to the message or entity*

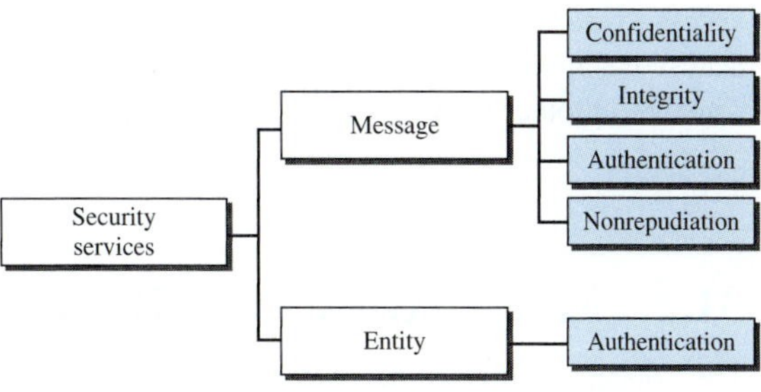

Message Confidentiality

Message confidentiality or privacy means that the sender and the receiver expect confidentiality. The transmitted message must make sense to only the intended receiver. To all others, the message must be garbage. When a customer communicates with her bank, she expects that the communication is totally confidential.

Message Integrity

Message integrity means that the data must arrive at the receiver exactly as they were sent. There must be no changes during the transmission, neither accidentally nor maliciously. As more and more monetary exchanges occur over the Internet, integrity is crucial. For example, it would be disastrous if a request for transferring $100 changed to a request for $10,000 or $100,000. The integrity of the message must be preserved in a secure communication.

Message Authentication

Message authentication is a service beyond message integrity. In message authentication the receiver needs to be sure of the sender's identity and that an imposter has not sent the message.

Message Nonrepudiation

Message nonrepudiation means that a sender must not be able to deny sending a message that he or she, in fact, did send. The burden of proof falls on the receiver. For example, when a customer sends a message to transfer money from one account to another, the bank must have proof that the customer actually requested this transaction.

Entity Authentication

In entity authentication (or user identification) the entity or user is verified prior to access to the system resources (files, for example). For example, a student who needs to access her university resources needs to be authenticated during the logging process. This is to protect the interests of the university and the student.

31.2 MESSAGE CONFIDENTIALITY

The concept of how to achieve message confidentiality or privacy has not changed for thousands of years. The message must be encrypted at the sender site and decrypted at the receiver site. That is, the message must be rendered unintelligible to unauthorized parties. A good privacy technique guarantees to some extent that a potential intruder (eavesdropper) cannot understand the contents of the message. As we discussed in Chapter 30, this can be done using either symmetric-key cryptography or asymmetric-key cryptography. We review both.

Confidentiality with Symmetric-Key Cryptography

Although modern symmetric-key algorithms are more complex than the ones used through the long history of the secret writing, the principle is the same. To provide confidentiality with symmetric-key cryptography, a sender and a receiver need to share a secret key. In the past when data exchange was between two specific persons (for example, two friends or a ruler and her army chief), it was possible to personally exchange the secret keys. Today's communication does not often provide this opportunity. A person residing in the United States cannot meet and exchange a secret key with a person living in China. Furthermore, the communication is between millions of people, not just a few.

To be able to use symmetric-key cryptography, we need to find a solution to the key sharing. This can be done using a **session key.** A session key is one that is used only for the duration of one session. The session key itself is exchanged using asymmetric-key cryptography as we will see later. Figure 31.2 shows the use of a session symmetric key for sending confidential messages from Alice to Bob and vice versa. Note that the nature of the symmetric key allows the communication to be carried on in both directions although it is not recommended today. Using two different keys is more secure, because if one key is compromised, the communication is still confidential in the other direction.

Figure 31.2 *Message confidentiality using symmetric keys in two directions*

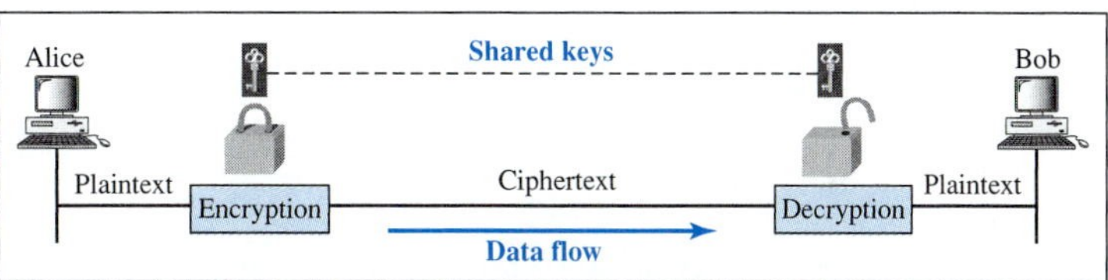

a. A shared secret key can be used in Alice-Bob communication

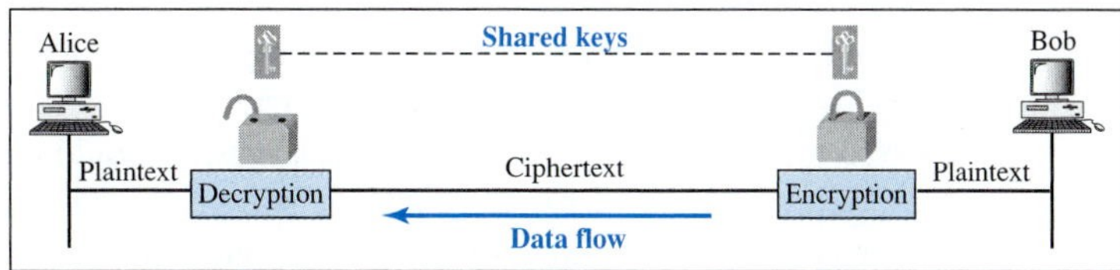

b. A different shared secret key is recommended in Bob-Alice communication

The reason symmetric-key cryptography is still the dominant method for confidentiality of the message is its efficiency. For a long message, symmetric-key cryptography is much more efficient than asymmetric-key cryptography.

Confidentiality with Asymmetric-Key Cryptography

The problem we mentioned about key exchange in symmetric-key cryptography for privacy culminated in the creation of asymmetric-key cryptography. Here, there is no key sharing; there is a public announcement. Bob creates two keys: one private and one public. He keeps the private key for decryption; he publicly announces the public key to the world. The public key is used only for encryption; the private key is used only for decryption. The public key locks the message; the private key unlocks it.

For a two-way communication between Alice and Bob, two pairs of keys are needed. When Alice sends a message to Bob, she uses Bob's pair; when Bob sends a message to Alice, he uses Alice's pair as shown in Figure 31.3.

Figure 31.3 *Message confidentiality using asymmetric keys*

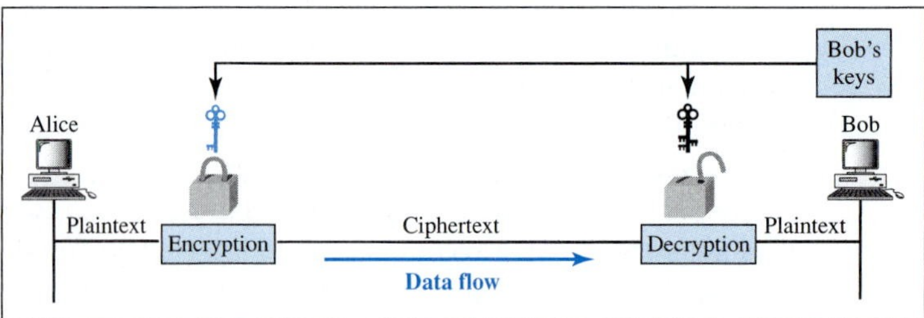

a. Bob's keys are used in Alice-Bob communication

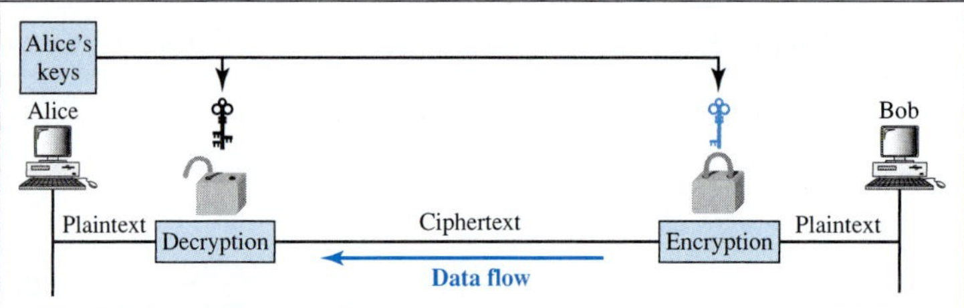

b. Alice's keys are used in Bob-Alice communication

Confidentiality with asymmetric-key cryptosystem has its own problems. First, the method is based on long mathematical calculations using long keys. This means that this system is very inefficient for long messages; it should be applied only to short messages. Second, the sender of the message still needs to be certain about the public key of the receiver. For example, in Alice-Bob communication, Alice needs to be sure that Bob's public key is genuine; Eve may have announced her public key in the name of Bob. A system of trust is needed, as we will see later in the chapter.

31.3 MESSAGE INTEGRITY

Encryption and decryption provide secrecy, or confidentiality, but not **integrity.** However, on occasion we may not even need secrecy, but instead must have integrity. For example, Alice may write a will to distribute her estate upon her death. The will does not need to be encrypted. After her death, anyone can examine the will. The integrity of the will, however, needs to be preserved. Alice does not want the contents of the will to be changed. As another example, suppose Alice sends a message instructing her banker, Bob, to pay Eve for consulting work. The message does not need to be hidden from Eve because she already knows she is to be paid. However, the message does need to be safe from any tampering, especially by Eve.

Document and Fingerprint

One way to preserve the integrity of a document is through the use of a **fingerprint.** If Alice needs to be sure that the contents of her document will not be illegally changed, she can put her fingerprint at the bottom of the document. Eve cannot modify the contents of this document or create a false document because she cannot forge Alice's fingerprint. To ensure that the document has not been changed, Alice's fingerprint on the document can be compared to Alice's fingerprint on file. If they are not the same, the document is not from Alice.

To preserve the integrity of a document, both the document and the fingerprint are needed.

Message and Message Digest

The electronic equivalent of the document and fingerprint pair is the **message** and **message digest** pair. To preserve the integrity of a message, the message is passed through an algorithm called a **hash function.** The hash function creates a compressed image of the message that can be used as a fingerprint. Figure 31.4 shows the message, hash function, and the message digest.

Figure 31.4 *Message and message digest*

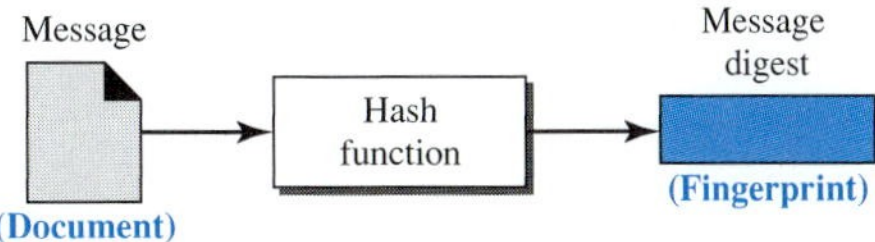

Difference

The two pairs document/fingerprint and message/message digest are similar, with some differences. The document and fingerprint are physically linked together; also, neither needs to be kept secret. The message and message digest can be unlinked (or sent) separately and, most importantly, the message digest needs to be kept secret. The message digest is either kept secret in a safe place or encrypted if we need to send it through a communications channel.

The message digest needs to be kept secret.

Creating and Checking the Digest

The message digest is created at the sender site and is sent with the message to the receiver. To check the integrity of a message, or document, the receiver creates the hash function again and compares the new message digest with the one received. If both are the same, the receiver is sure that the original message has not been changed. Of course, we are assuming that the digest has been sent secretly. Figure 31.5 shows the idea.

Figure 31.5 *Checking integrity*

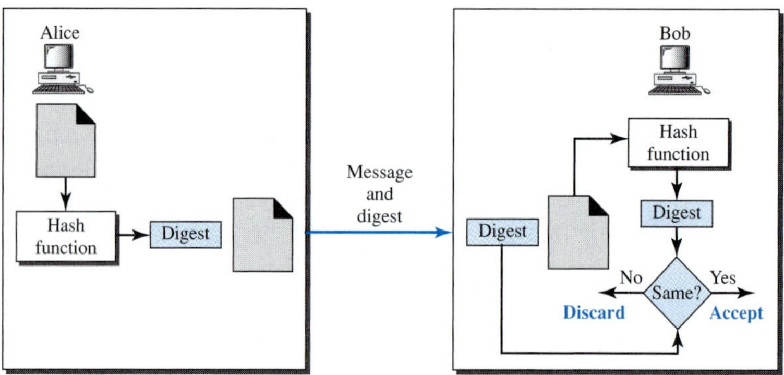

Hash Function Criteria

To be eligible for a hash, a function needs to meet three criteria: one-wayness, resistance to weak collision, and resistance to strong collision as shown in Figure 31.6.

Figure 31.6 *Criteria of a hash function*

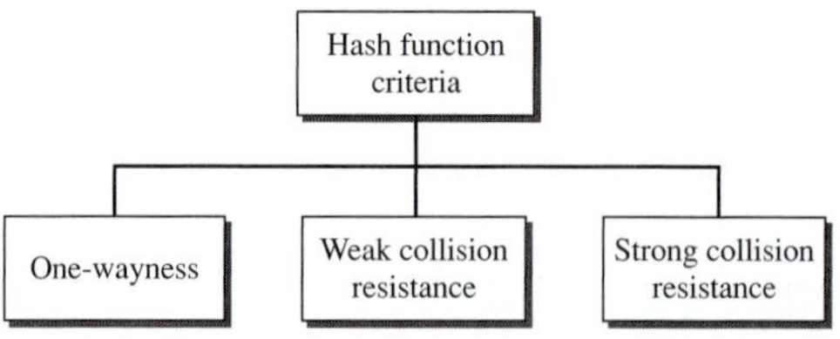

One-wayness

A hash function must have **one-wayness;** a message digest is created by a one-way hashing function. We must not be able to recreate the message from the digest. Sometimes it is difficult to make a hash function 100 percent one-way; the criteria state that it must be extremely difficult or impossible to create the message if the message digest is given. This is similar to the document/fingerprint case. No one can make a document from a fingerprint.

Example 31.1

Can we use a conventional lossless compression method as a hashing function?

Solution

We cannot. A lossless compression method creates a compressed message that is reversible. You can uncompress the compressed message to get the original one.

Example 31.2

Can we use a checksum method as a hashing function?

Solution

We can. A checksum function is not reversible; it meets the first criterion. However, it does not meet the other criteria.

Weak Collision Resistance

The second criterion, **weak collision** resistance, ensures that a message cannot easily be forged. If Alice creates a message and a digest and sends both to Bob, this criterion ensures that Eve cannot easily create another message that hashes exactly to the same digest. In other words, given a specific message and its digest, it is impossible (or at least very difficult) to create another message with the same digest.

When two messages create the same digest, we say there is a collision. In a week collision, given a message digest, it is very unlikely that someone can create a message with exactly the same digest. A hash function must have weak collision resistance.

Strong Collision Resistance

The third criterion, **strong collision** resistance, ensures that we cannot find two messages that hash to the same digest. This criterion is needed to ensure that Alice, the sender of the message, cannot cause problems by forging a message. If Alice can create two messages that hash to the same digest, she can deny sending the first to Bob and claim that she sent only the second.

This type of collision is called strong because the probability of collision is higher than in the previous case. An adversary can create two messages that hash to the same digest. For example, if the number of bits in the message digest is small, it is likely Alice can create two different messages with the same message digest. She can send the first to Bob and keep the second for herself. Alice can later say that the second was the original agreed-upon document and not the first.

Suppose two different wills can be created that hash to the same digest. When the time comes for the execution of the will, the second will is presented to the heirs. Since the digest matches both wills, the substitution is successful.

Hash Algorithms: SHA-1

While many hash algorithms have been designed, the most common is SHA-1. **SHA-1 (Secure Hash Algorithm 1)** is a revised version of SHA designed by the National Institute of Standards and Technology (NIST). It was published as a Federal Information Processing Standard (FIPS).

A very interesting point about this algorithm and others is that they all follow the same concept. Each creates a digest of length *N* from a multiple-block message. Each block is 512 bits in length, as shown in Figure 31.7.

Figure 31.7 *Message digest creation*

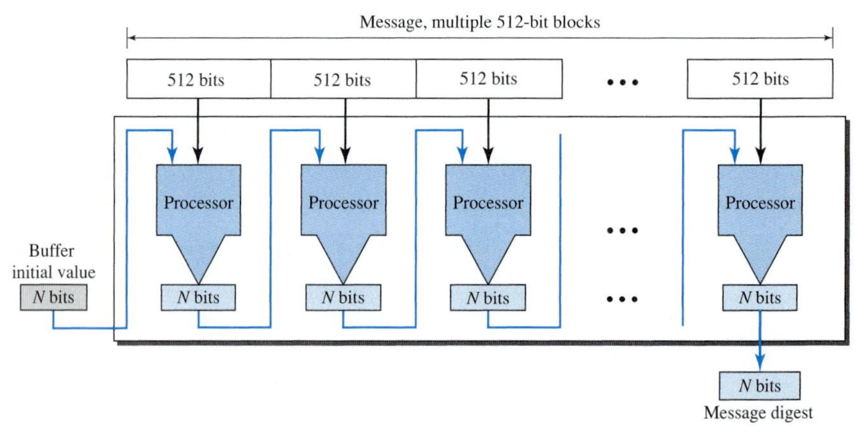

A buffer of *N* bits is initialized to a predetermined value. The algorithm mangles this initial buffer with the first 512 bits of the message to create the first intermediate message digest of *N* bits. This digest is then mangled with the second 512-bit block to create the second intermediate digest. The $(n - 1)$th digest is mangled with the *n*th block to create the *n*th digest. If a block is not 512 bits, padding (0s) is added to make it so. When the last block is processed, the resulting digest is the message digest for the entire message. SHA-1 has a message digest of 160 bits (5 words, each of 32 bits).

SHA-1 hash algorithms create an *N*-bit message digest out of a message of 512-bit blocks.
SHA-1 has a message digest of 160 bits (5 words of 32 bits).

Word Expansion

Before processing, the block needs to be expanded. A block is made of 512 bits or 16 32-bit words, but we need 80 words in the processing phase. So the 16-word block needs to be expanded to 80 words, word0 to word79.

Processing Each Block

Figure 31.8 shows the general outline for the processing of one block. There are 80 steps in block processing. In each step, one word from the expanded block and one 32-bit constant are mangled together and then operated on to create a new digest. At the beginning of processing, the values of digest words (A, B, C, D, and E) are saved into five temporary variables. At the end of the processing (after step 79), these values are

Figure 31.8 *Processing of one block in SHA-1*

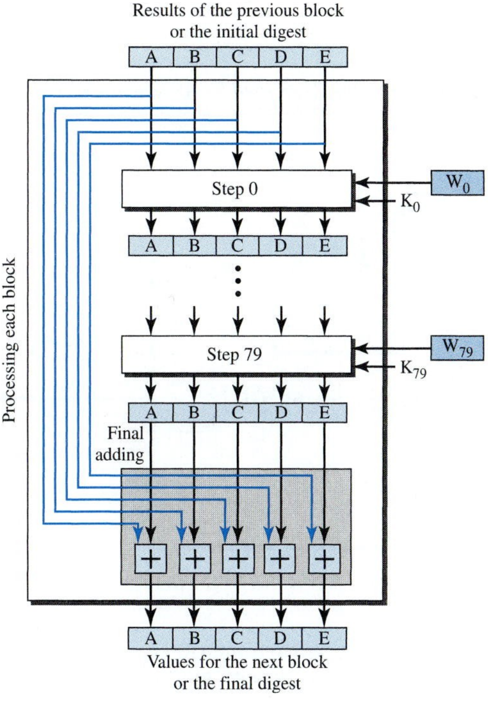

added to the values created from step 79. The detail of each step is complex and beyond the scope of this book. The only thing we need to know is that each step mangles a word of data and a constant to create a result that is fed to the next step.

31.4 MESSAGE AUTHENTICATION

A hash function guarantees the integrity of a message. It guarantees that the message has not been changed. A hash function, however, does not authenticate the sender of the message. When Alice sends a message to Bob, Bob needs to know if the message is coming from Alice or Eve. To provide message authentication, Alice needs to provide proof that it is Alice sending the message and not an imposter. A hash function per se cannot provide such a proof. The digest created by a hash function is normally called a **modification detection code (MDC).** The code can detect any modification in the message.

MAC

To provide message authentication, we need to change a modification detection code to a **message authentication code (MAC).** An MDC uses a keyless hash function; a MAC uses a keyed hash function. A keyed hash function includes the symmetric key between the sender and receiver when creating the digest. Figure 31.9 shows how Alice uses a keyed hash function to authenticate her message and how Bob can verify the authenticity of the message.

Figure 31.9 *MAC, created by Alice and checked by Bob*

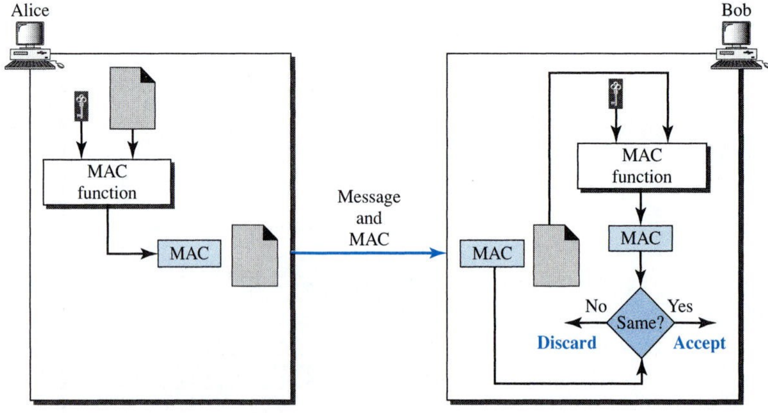

Alice, using the symmetric key between herself and Bob (K_{AB}) and a keyed hash function, generates a MAC. She then concatenates the MAC with the original message and sends the two to Bob. Bob receives the message and the MAC. He separates the message from the MAC. He applies the same keyed hash function to the message using the symmetric key K_{AB} to get a fresh MAC. He then compares the MAC sent by Alice with the newly generated MAC. If the two MACs are identical, the message has not been modified and the sender of the message is definitely Alice.

HMAC

There are several implementations of MAC in use today. However, in recent years, some MACs have been designed that are based on keyless hash functions such as SHA-1. This idea is a **hashed MAC,** called **HMAC,** that can use any standard keyless hash function such as SHA-1. HMAC creates a nested MAC by applying a keyless hash function to the concatenation of the message and a symmetric key. Figure 31.10 shows the general idea.

A copy of the symmetric key is prepended to the message. The combination is hashed using a keyless hash function, such as SHA-1. The result of this process is an intermediate HMAC which is again prepended with the key (the same key), and the result is again hashed using the same algorithm. The final result is an HMAC.

The receiver receives this final HMAC and the message. The receiver creates its own HMAC from the received message and compares the two HMACs to validate the integrity of the message and authenticate the data origin. Note that the details of an HMAC can be more complicated than what we have shown here.

Figure 31.10 *HMAC*

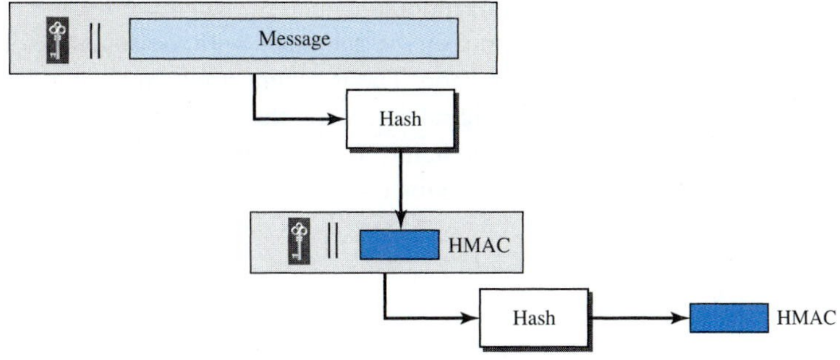

31.5 DIGITAL SIGNATURE

Although a MAC can provide message integrity and message authentication, it has a drawback. It needs a symmetric key that must be established between the sender and the receiver. A digital signature, on the other hand, can use a pair of asymmetric keys (a public one and a private one).

We are all familiar with the concept of a signature. We sign a document to show that it originated from us or was approved by us. The signature is proof to the recipient that the document comes from the correct entity. When a customer signs a check to himself, the bank needs to be sure that the check is issued by that customer and nobody else. In other words, a signature on a document, when verified, is a sign of authentication; the document is authentic. Consider a painting signed by an artist. The signature on the art, if authentic, means that the painting is probably authentic.

When Alice sends a message to Bob, Bob needs to check the authenticity of the sender; he needs to be sure that the message comes from Alice and not Eve. Bob can ask Alice to sign the message electronically. In other words, an electronic signature can prove the authenticity of Alice as the sender of the message. We refer to this type of signature as a **digital signature.**

Comparison

Before we continue any further, let us discuss the differences between two types of signatures: conventional and digital.

Inclusion

A conventional signature is included in the document; it is part of the document. When we write a check, the signature is on the check; it is not a separate document. On the other hand, when we sign a document digitally, we send the signature as a separate document. The sender sends two documents: the message and the signature. The recipient receives both documents and verifies that the signature belongs to the supposed sender. If this is proved, the message is kept; otherwise, it is rejected.

Verification Method

The second difference between the two types of documents is the method of verifying the signature. In conventional signature, when the recipient receives a document, she compares the signature on the document with the signature on file. If they are the same, the document is authentic. The recipient needs to have a copy of this signature on file for comparison. In digital signature, the recipient receives the message and the signature. A copy of the signature is not stored anywhere. The recipient needs to apply a verification technique to the combination of the message and the signature to verify the authenticity.

Relationship

In conventional signature, there is normally a one-to-many relationship between a signature and documents. A person, for example, has a signature that is used to sign many checks, many documents, etc. In digital signature, there is a one-to-one relationship between a signature and a message. Each message has its own signature. The signature of one message cannot be used in another message. If Bob receives two messages, one after another, from Alice, he cannot use the signature of the first message to verify the second. Each message needs a new signature.

Duplicity

Another difference between the two types of signatures is a quality called duplicity. In conventional signature, a copy of the signed document can be distinguished from the original one on file. In digital signature, there is no such distinction unless there is a factor of time (such as a timestamp) on the document. For example, suppose Alice sends a document instructing Bob to pay Eve. If Eve intercepts the document and the signature, she can resend it later to get money again from Bob.

Need for Keys

In conventional signature a signature is like a private "key" belonging to the signer of the document. The signer uses it to sign a document; no one else has this signature. The copy of the signature is on file like a public key; anyone can use it to verify a document, to compare it to the original signature.

In digital signature, the signer uses her private key, applied to a signing algorithm, to sign the document. The verifier, on the other hand, uses the public key of the signer, applied to the verifying algorithm, to verify the document.

Can we use a secret (symmetric) key to both sign and verify a signature? The answer is no for several reasons. First, a secret key is known only between two entities (Alice and Bob, for example). So if Alice needs to sign another document and send it to Ted, she needs to use another secret key. Second, as we will see, creating a secret key for a session involves authentication, which normally uses digital signature. We have a vicious cycle. Third, Bob could use the secret key between himself and Alice, sign a document, send it to Ted, and pretend that it came from Alice.

A digital signature needs a public-key system.

Process

Digital signature can be achieved in two ways: signing the document or signing a digest of the document.

Signing the Document

Probably, the easier, but less efficient way is to sign the document itself. Signing a document is encrypting it with the private key of the sender; verifying the document is decrypting it with the public key of the sender. Figure 31.11 shows how signing and verifying are done.

Figure 31.11 *Signing the message itself in digital signature*

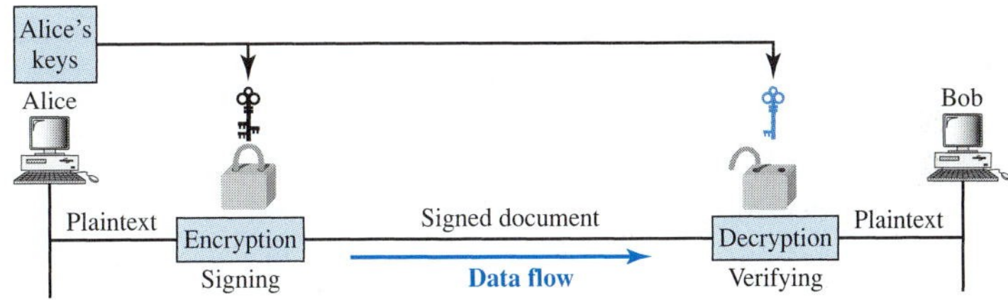

We should make a distinction between private and public keys as used in digital signature and public and private keys as used for confidentiality. In the latter, the private and public keys of the receiver are used in the process. The sender uses the public key of the receiver to encrypt; the receiver uses his own private key to decrypt. In digital signature, the private and public keys of the sender are used. The sender uses her private key; the receiver uses the public key of the sender.

In a cryptosystem, we use the private and public keys of the receiver; in digital signature, we use the private and public key of the sender.

Signing the Digest

We mentioned that the public key is very inefficient in a cryptosystem if we are dealing with long messages. In a digital signature system, our messages are normally long, but we have to use public keys. The solution is not to sign the message itself; instead, we sign a digest of the message. As we learned, a carefully selected message digest has a one-to-one relationship with the message. The sender can sign the message digest, and the receiver can verify the message digest. The effect is the same. Figure 31.12 shows signing a digest in a digital signature system.

A digest is made out of the message at Alice's site. The digest then goes through the signing process using Alice's private key. Alice then sends the message and the signature to Bob. As we will see later in the chapter, there are variations in the process that are dependent on the system. For example, there might be additional calculations before the digest is made or other secret keys might be used. In some systems, the signature is a set of values.

Figure 31.12 *Signing the digest in a digital signature*

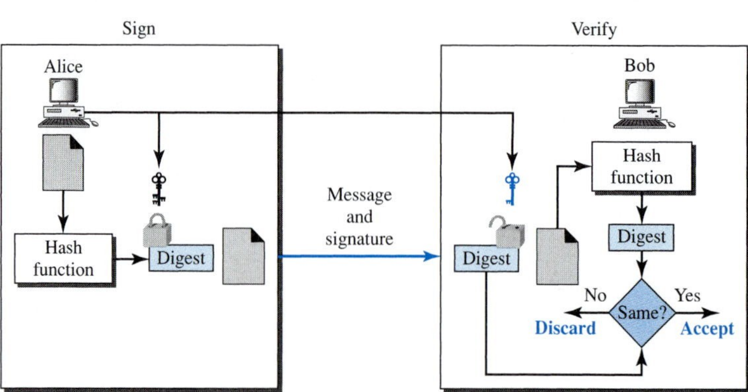

At Bob's site, using the same public hash function, a digest is first created out of the received message. Calculations are done on the signature and the digest. The verifying process also applies criteria on the result of the calculation to determine the authenticity of the signature. If authentic, the message is accepted; otherwise, it is rejected.

Services

A digital signature can provide three out of the five services we mentioned for a security system: message integrity, message authentication, and nonrepudiation. Note that a digital signature scheme does not provide confidential communication. If confidentiality is required, the message and the signature must be encrypted using either a secret-key or public-key cryptosystem.

Message Integrity

The integrity of the message is preserved even if we sign the whole message because we cannot get the same signature if the message is changed. The signature schemes today use a hash function in the signing and verifying algorithms that preserve the integrity of the message.

A digital signature today provides message integrity.

Message Authentication

A secure signature scheme, like a secure conventional signature (one that cannot be easily copied), can provide message authentication. Bob can verify that the message is sent by Alice because Alice's public key is used in verification. Alice's public key cannot create the same signature as Eve's private key.

Digital signature provides message authentication.

Message Nonrepudiation

If Alice signs a message and then denies it, can Bob later prove that Alice actually signed it? For example, if Alice sends a message to a bank (Bob) and asks to transfer $10,000 from her account to Ted's account, can Alice later deny that she sent this message? With the scheme we have presented so far, Bob might have a problem. Bob must keep the signature on file and later use Alice's public key to create the original message to prove the message in the file and the newly created message are the same. This is not feasible because Alice may have changed her private/public key during this time; she may also claim that the file containing the signature is not authentic.

One solution is a trusted third party. People can create a trusted party among themselves. In Chapter 32, we will see that a trusted party can solve many other problems concerning security services and key exchange. Figure 31.13 shows how a trusted party can prevent Alice from denying that she sent the message.

Figure 31.13 *Using a trusted center for nonrepudiation*

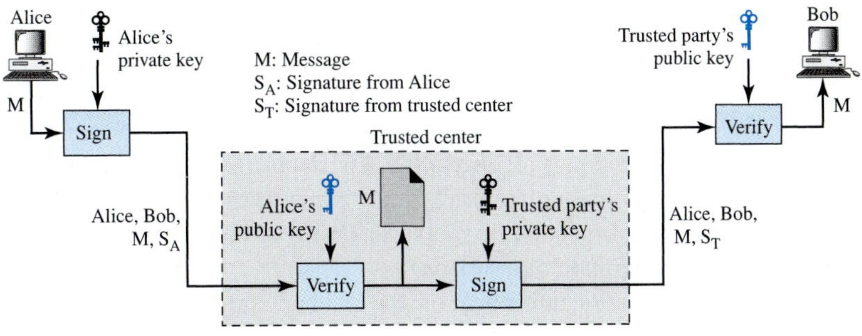

Alice creates a signature from her message (S_A) and sends the message, her identity, Bob's identity, and the signature to the center. The center, after checking that Alice's public key is valid, verifies through Alice's public key that the message comes from Alice. The center then saves a copy of the message with the sender identity, recipient identity, and a timestamp in its archive. The center uses its private key to create another signature (S_T) from the message. The center then sends the message, the new signature, Alice's identity, and Bob's identity to Bob. Bob verifies the message using the public key of the trusted center.

If in the future Alice denies that she has sent the message, the center can show a copy of the saved message. If Bob's message is a duplicate of the message saved at the center, Alice will lose the dispute. To make everything confidential, a level of encryption/decryption can be added to the scheme as discussed in the next section.

Nonrepudiation can be provided using a trusted party.

Signature Schemes

Several signature schemes have evolved during the last few decades. Some of them have been implemented. Such as RSA and DSS (Digital Signature Standard) schemes. The latter will probably become the standard. However, the details of these schemes are beyond the scope of this book.

31.6 ENTITY AUTHENTICATION

Entity authentication is a technique designed to let one party prove the identity of another party. An *entity* can be a person, a process, a client, or a server. The entity whose identity needs to be proved is called the **claimant;** the party that tries to prove the identity of the claimant is called the **verifier.** When Bob tries to prove the identity of Alice, Alice is the claimant, and Bob is the verifier.

There are two differences between message authentication and **entity authentication.** First, message authentication may not happen in real time; entity authentication does. In the former, Alice sends a message to Bob. When Bob authenticates the message, Alice may or may not be present in the communication process. On the other hand, when Alice requests entity authentication, there is no real message communication involved until Alice is authenticated by Bob. Alice needs to be online and takes part in the process. Only after she is authenticated can messages be communicated between Alice and Bob. Message authentication is required when an e-mail is sent from Alice to Bob. Entity authentication is required when Alice gets cash from an automatic teller machine. Second, message authentication simply authenticates one message; the process needs to be repeated for each new message. Entity authentication authenticates the claimant for the entire duration of a session.

In entity authentication, the claimant must identify herself to the verifier. This can be done with one of three kinds of witnesses: *something known, something possessed,* or *something inherent.*

- ***Something known.*** This is a secret known only by the claimant that can be checked by the verifier. Examples are a password, a PIN number, a secret key, and a private key.
- ***Something possessed.*** This is something that can prove the claimant's identity. Examples are a passport, a driver's license, an identification card, a credit card, and a smart card.
- ***Something inherent.*** This is an inherent characteristic of the claimant. Examples are conventional signature, fingerprints, voice, facial characteristics, retinal pattern, and handwriting.

Passwords

The simplest and the oldest method of entity authentication is the **password,** something that the claimant *possesses*. A password is used when a user needs to access a system to use the system's resources (log-in). Each user has a user identification that is public and a password that is private. We can divide this authentication scheme into two separate groups: the **fixed password** and the **one-time password.**

Fixed Password

In this group, the password is fixed; the same password is used over and over for every access. This approach is subject to several attacks.

- **Eavesdropping.** Eve can watch Alice when she types her password. Most systems, as a security measure, do not show the characters a user types. Eavesdropping can take a more sophisticated form. Eve can listen to the line and then intercept the message, thereby capturing the password for her own use.
- **Stealing a Password.** The second type of attack occurs when Eve tries to physically steal Alice's password. This can be prevented if Alice does not write down the password; instead, she just commits it to memory. Therefore, a password should be very simple or else related to something familiar to Alice, which makes the password vulnerable to other types of attacks.
- **Accessing a file.** Eve can hack into the system and get access to the file where the passwords are stored. Eve can read the file and find Alice's password or even change it. To prevent this type of attack, the file can be read/write protected. However, most systems need this type of file to be readable by the public.
- **Guessing.** Eve can log into the system and try to guess Alice's password by trying different combinations of characters. The password is particularly vulnerable if the user is allowed to choose a short password (a few characters). It is also vulnerable if Alice has chosen something unimaginative, such as her birthday, her child's name, or the name of her favorite actor. To prevent guessing, a long random password is recommended, something that is not very obvious. However, the use of such a random password may also create a problem; Alice might store the password somewhere so as not to forget it. This makes the password subject to stealing.

A more secure approach is to store the hash of the password in the password file (instead of the plaintext password). Any user can read the contents of the file, but, because the hash function is a one-way function, it is almost impossible to guess the value of the password. The hash function prevents Eve from gaining access to the system even though she has the password file. However, there is a possibility of another type of attack called the **dictionary attack.** In this attack, Eve is interested in finding one password, regardless of the user ID. For example, if the password is 6 digits, Eve can create a list of 6-digit numbers (000000 to 999999), and then apply the hash function to every number; the result is a list of 1 million hashes. She can then get the password file and search the second-column entries to find a match. This could be programmed and run offline on Eve's private computer. After a match is found, Eve can go online and use the password to access the system. We will see how to make this attack more difficult in the third approach.

Another approach is called **salting** the password. When the password string is created, a random string, called the salt, is concatenated to the password. The salted password is then hashed. The ID, salt, and the hash are then stored in the file. Now, when a user asks for access, the system extracts the salt, concatenates it with the received password, makes a hash out of the result, and compares it with the hash stored in the file. If there is a match, access is granted; otherwise, it is denied. Salting makes the dictionary attack more difficult. If the original password is 6 digits and the salt is 4 digits, then hashing is done over a 10-digit value. This means that Eve now needs to make a list of 10 million items and create a hash for each of them. The list of hashes has 10 million entries and the comparison takes much longer. Salting is very effective if the salt is a very long random number. The UNIX operating system uses a variation of this method.

In another approach, two identification techniques are combined. A good example of this type of authentication is the use of an ATM card with a PIN (personal identification number). The card belongs to the category "something possessed" and the PIN belongs to the category "something known." The PIN is actually a password that enhances the security of the card. If the card is stolen, it cannot be used unless the PIN is known. The PIN, however, is traditionally very short so it is easily remembered by the owner. This makes it vulnerable to the guessing type of attack.

One-Time Password

In this type of scheme, a password is used only once. It is called the **one-time password.** A one-time password makes eavesdropping and stealing useless. However, this approach is very complex, and we leave its discussion to some specialized books.

Challenge-Response

In password authentication, the claimant proves her identity by demonstrating that she knows a secret, the password. However, since the claimant reveals this secret, the secret is susceptible to interception by the adversary. In **challenge-response authentication,** the claimant proves that she *knows* a secret without revealing it. In other words, the claimant does not reveal the secret to the verifier; the verifier either has it or finds it.

In challenge-response authentication,
the claimant proves that she knows a secret without revealing it.

The challenge is a time-varying value such as a random number or a timestamp which is sent by the verifier. The claimant applies a function to the challenge and sends the result, called a *response,* to the verifier. The response shows that the claimant knows the secret.

The challenge is a time-varying value sent by the verifier;
the response is the result of a function applied on the challenge.

Using a Symmetric-Key Cipher

In the first category, the challenge-response authentication is achieved using symmetric-key encryption. The secret here is the shared secret key, known by both the claimant and the verifier. The function is the encrypting algorithm applied on the challenge. Figure 31.14 shows one approach. The first message is not part of challenge-response, it only informs the verifier that the claimant wants to be challenged. The second message is the challenge. And R_B is the nonce randomly chosen by the verifier to challenge the claimant. The claimant encrypts the **nonce** using the shared secret key known only to the claimant and the verifier and sends the result to the verifier. The verifier decrypts

Figure 31.14 *Challenge/response authentication using a nonce*

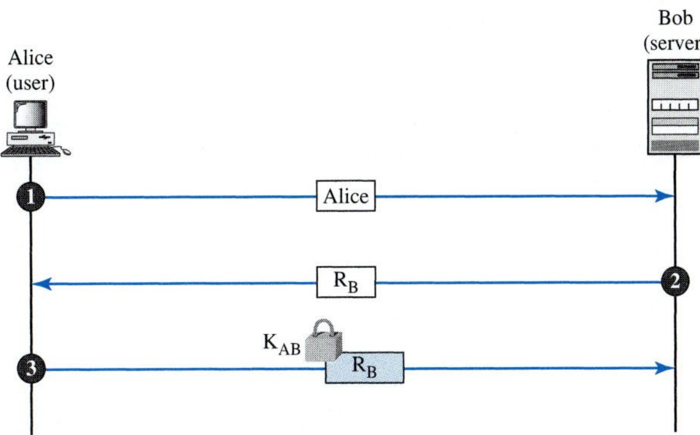

the message. If the nonce obtained from decryption is the same as the one sent by the verifier, Alice is granted access.

Note that in this process, the claimant and the verifier need to keep the symmetric key used in the process secret. The verifier must also keep the value of the nonce for claimant identification until the response is returned.

The reader may have noticed that use of a nonce prevents a replay of the third message by Eve. Eve cannot replay the third message and pretend that it is a new request for authentication by Alice because once Bob receives the response, the value of R_B is not valid any more. The next time a new value is used.

In the second approach, the time-varying value is a timestamp, which obviously changes with time. In this approach the challenge message is the current time sent from the verifier to the claimant. However, this supposes that the client and the server clocks are synchronized; the claimant knows the current time. This means that there is no need for the challenge message. The first and third messages can be combined. The result is that authentication can be done using one message, the response to an implicit challenge, the current time. Figure 31.15 shows the approach.

Figure 31.15 *Challenge-response authentication using a timestamp*

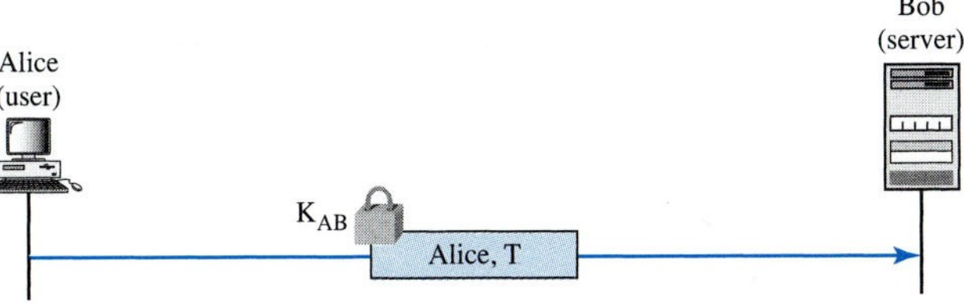

Using Keyed-Hash Functions

Instead of using encryption and decryption for entity authentication, we can use a keyed-hash function (MAC). There are two advantages to this scheme. First, the encryption/decryption algorithm is not exportable to some countries. Second, in using a keyed-hash function, we can preserve the integrity of challenge and response messages and at the same time use a secret, the key.

Let us see how we can use a keyed-hash function to create a challenge response with a timestamp. Figure 31.16 shows the scheme.

Figure 31.16 *Challenge-response authentication using a keyed-hash function*

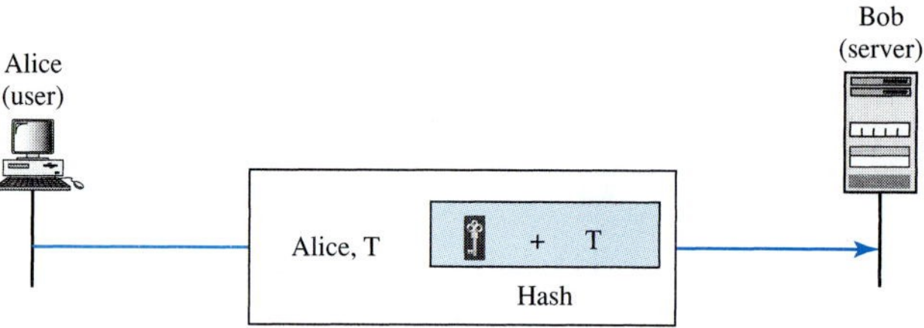

Note that in this case, the timestamp is sent both as plaintext and as text scrambled by the keyed-hash function. When Bob receives the message, he takes the plaintext T, applies the keyed-hash function, and then compares his calculation with what he received to determine the authenticity of Alice.

Using an Asymmetric-Key Cipher

Instead of a symmetric-key cipher, we can use an asymmetric-key cipher for entity authentication. Here the secret must be the private key of the claimant. The claimant must show that she owns the private key related to the public key that is available to everyone. This means that the verifier must encrypt the challenge using the public key of the claimant; the claimant then decrypts the message using her private key. The response to the challenge is the decrypted challenge. We show two approaches: one for unidirectional authentication and one for bidirectional authentication. In one approach, Bob encrypts the challenge using Alice's public key. Alice decrypts the message with her private key and sends the nonce to Bob. Figure 31.17 shows this approach.

Figure 31.17 *Authentication, asymmetric-key*

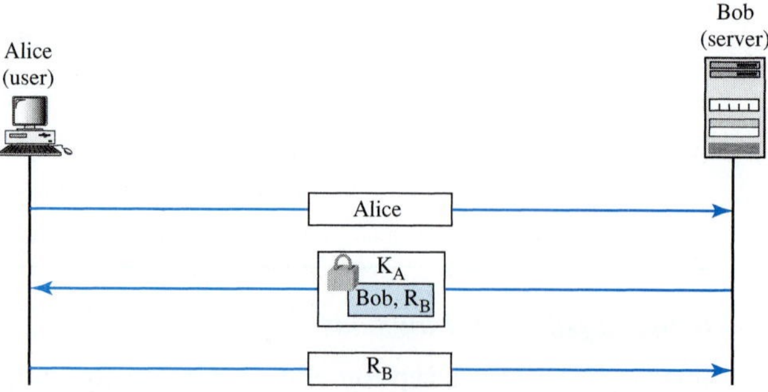

Using Digital Signature

We can use digital signature for entity authentication. In this method, we let the claimant use her private key for signing instead of using it for decryption. In one approach shown in Figure 31.18, Bob uses a plaintext challenge. Alice signs the response.

Figure 31.18 *Authentication, using digital signature*

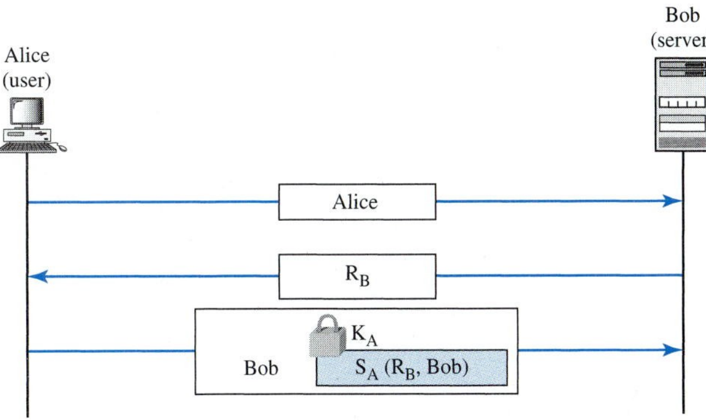

31.7 KEY MANAGEMENT

We have used symmetric-key and asymmetric-key cryptography in our discussion throughout the chapter. However, we never discussed how secret keys in symmetric-key cryptography and how public keys in asymmetric-key cryptography are distributed and maintained. In this section, we touch on these two issues. We first discuss the distribution of symmetric keys; we then discuss the distribution of asymmetric keys.

Symmetric-Key Distribution

We have learned that symmetric-key cryptography is more efficient than asymmetric-key cryptography when we need to encrypt and decrypt large messages. Symmetric-key cryptography, however, needs a shared secret key between two parties.

If Alice needs to exchange confidential messages with N people, she needs N different keys. What if N people need to communicate with one another? A total of $N(N-1)/2$ keys is needed. Each person needs to have $N-1$ keys to communicate with each of the other people, but because the keys are shared, we need only $N(N-1)/2$. This means that if 1 million people need to communicate with one another, each person has almost 0.5 million different keys; in total, almost 1 billion keys are needed. This is normally referred to as the N^2 problem because the number of required keys for N entities is close to N^2.

The number of keys is not the only problem; the distribution of keys is another. If Alice and Bob want to communicate, they need to somehow exchange a secret key; if Alice wants to communicate with 1 million people, how can she exchange 1 million keys with 1 million people? Using the Internet is definitely not a secure method.

It is obvious that we need an efficient way of maintaining and distributing secret keys.

Key Distribution Center: KDC

A practical solution is the use of a trusted party, referred to as a **key distribution center (KDC).** To reduce the number of keys, each person establishes a shared secret key with the KDC as shown in Figure 31.19.

Figure 31.19 *KDC*

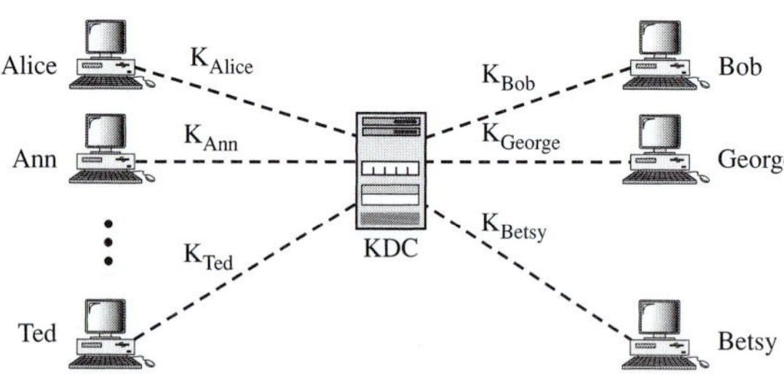

A secret key is established between KDC and each member. Alice has a secret key with KDC, which we refer to as K_{Alice}; Bob has a secret key with KDC, which we refer to as K_{Bob}; and so on. Now the question is, How can Alice send a confidential message to Bob? The process is as follows:

1. Alice sends a request to KDC, stating that she needs a session (temporary) secret key between herself and Bob.
2. KDC informs Bob of Alice's request.
3. If Bob agrees, a session key is created between the two.

The secret key between Alice and Bob that is established with the KDC is used to authenticate Alice and Bob to the KDC and to prevent Eve from impersonating either of them. We discuss how a session key is established between Alice and Bob later in the chapter.

Session Keys

A KDC creates a secret key for each member. This secret key can be used only between the member and the KDC, not between two members. If Alice needs to communicate secretly with Bob, she needs a secret key between herself and Bob. A KDC can create a session (temporary) key between Alice and Bob using their keys with the center. The keys of Alice and Bob are used to authenticate Alice and Bob to the center and to each other before the session key is established. After communication is terminated, the session key is no longer valid.

A session symmetric key between two parties is used only once.

Several different approaches have been proposed to create the session key using ideas we previously discussed for entity authentication.

Let us discuss one approach, the simplest one, as shown in Figure 31.20. Although this system has some flaws, it shows the idea. More sophisticated approaches can be found in security books.

Figure 31.20 *Creating a session key between Alice and Bob using KDC*

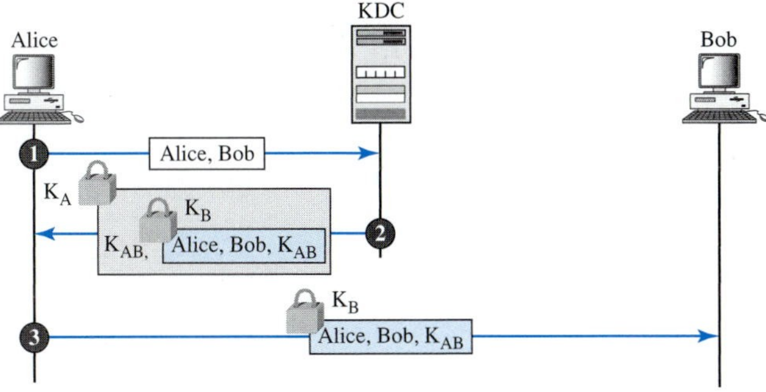

- **Step 1** Alice sends a plaintext message to the KDC to obtain a symmetric session key between Bob and herself. The message contains her registered identity (the word *Alice* in the figure) and the identity of Bob (the word *Bob* in the figure). This message is not encrypted, it is public. KDC does not care.
- **Step 2** KDC receives the message and creates what is called a **ticket.** The ticket is encrypted using Bob's key (K_B). The ticket contains the identities of Alice and Bob and the session key (K_{AB}). The ticket with a copy of the session key is sent to Alice. Alice receives the message, decrypts it, and extracts the session key. She cannot decrypt Bob's ticket; the ticket is for Bob, not for Alice. Note that we have a double encryption in this message; the ticket is encrypted and the entire message is also encrypted. In the second message, Alice is actually authenticated to the KDC, because only Alice can open the whole message using her secret key with KDC.
- **Step 3** Alice sends the ticket to Bob. Bob opens the ticket and knows that Alice needs to send messages to him using K_{AB} as the session key. Note that in this message, Bob is authenticated to the KDC because only Bob can open the ticket. Since Bob is authenticated to the KDC, he is also authenticated to Alice who trusts the KDC. In the same way, Alice is also authenticated to Bob, because Bob trusts the KDC and the KDC has sent the ticket to Bob which includes the identity of Alice.

Kerberos

Kerberos is an authentication protocol and at the same time a KDC that has become very popular. Several systems including Windows 2000 use Kerberos. It is named after the three-headed dog in Greek mythology that guards the Gates of Hades. Originally designed at M.I.T., it has gone through several versions. We discuss only version 4, the most popular.

Servers Three servers are involved in the Kerberos protocol: an authentication server (AS), a ticket-granting server (TGS), and a real (data) server that provides services to others. In our examples and figures *Bob* is the real server and *Alice* is the user requesting service. Figure 31.21 shows the relationship between these three servers.

Figure 31.21 *Kerberos servers*

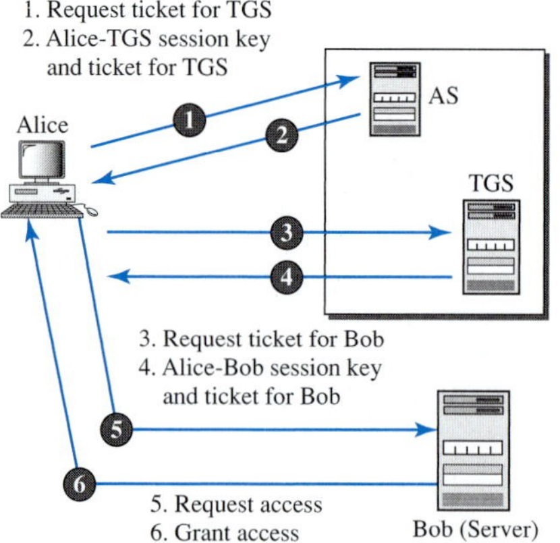

- **Authentication Server (AS).** AS is the KDC in Kerberos protocol. Each user registers with AS and is granted a user identity and a password. AS has a database with these identities and the corresponding passwords. AS verifies the user, issues a session key to be used between Alice and TGS, and sends a ticket for TGS.
- **Ticket-Granting Server (TGS).** TGS issues a ticket for the real server (Bob). It also provides the session key (K_{AB}) between Alice and Bob. Kerberos has separated the user verification from ticket issuing. In this way, although Alice verifies her ID just once with AS, she can contact TGS multiple times to obtain tickets for different real servers.
- **Real Server.** The real server (Bob) provides services for the user (Alice). Kerberos is designed for a client/server program such as FTP, in which a user uses the client process to access the server process. Kerberos is not used for person-to-person authentication.

Operation A client process (Alice) can access a process running on the real server (Bob) in six steps as shown in Figure 31.22.

- **Step 1.** Alice sends her request to AS in plaintext, using her registered identity.
- **Step 2.** AS sends a message encrypted with Alice's symmetric key K_A. The message contains two items: a session key K_S that is used by Alice to contact TGS and a ticket for TGS that is encrypted with the TGS symmetric key K_{TG}. Alice does not know K_A, but when the message arrives, she types her symmetric password. The password and the appropriate algorithm together create K_A if the password is correct. The password is then immediately destroyed; it is not sent to the network, and it does

Figure 31.22 *Kerberos example*

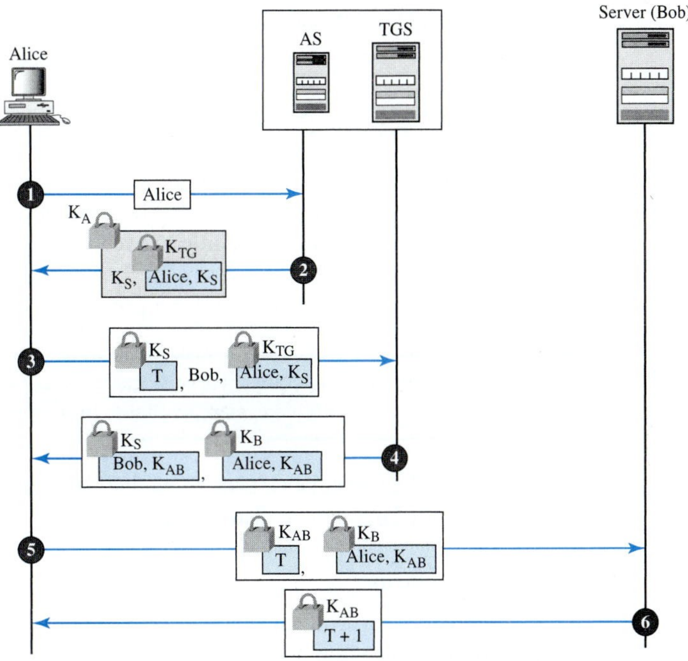

not stay in the terminal. It is only used for a moment to create K_A. The process now uses K_A to decrypt the message sent. Both K_S and the ticket are extracted.

- **Step 3.** Alice now sends three items to TGS. The first is the ticket received from AS. The second is the name of the real server (Bob), the third is a timestamp which is encrypted by K_S. The timestamp prevents a replay by Eve.
- **Step 4.** Now, TGS sends two tickets, each containing the session key between Alice and Bob K_{AB}. The ticket for Alice is encrypted with K_S; the ticket for Bob is encrypted with Bob's key K_B. Note that Eve cannot extract K_{AB} because she does not know K_S or K_B. She cannot replay step 3 because she cannot replace the timestamp with a new one (she does not know K_S). Even if she is very quick and sends the step 3 message before the timestamp has expired, she still receives the same two tickets that she cannot decipher.
- **Step 5.** Alice sends Bob's ticket with the timestamp encrypted by K_{AB}.
- **Step 6.** Bob confirms the receipt by adding 1 to the timestamp. The message is encrypted with K_{AB} and sent to Alice.

Using Different Servers Note that if Alice needs to receive services from different servers, she need repeat only steps 3 to 6. The first two steps have verified Alice's identity and need not be repeated. Alice can ask TGS to issue tickets for multiple servers by repeating steps 3 to 6.

Realms Kerberos allows the global distribution of ASs and TGSs, with each system called a realm. A user may get a ticket for a local server or a remote server. In the second case, for example, Alice may ask her local TGS to issue a ticket that is accepted by a remote TGS. The local TGS can issue this ticket if the remote TGS is registered with the local one. Then Alice can use the remote TGS to access the remote real server.

Public-Key Distribution

In asymmetric-key cryptography, people do not need to know a symmetric shared key. If Alice wants to send a message to Bob, she only needs to know Bob's public key, which is open to the public and available to everyone. If Bob needs to send a message to Alice, he only needs to know Alice's public key, which is also known to everyone. In public-key cryptography, everyone shields a private key and advertises a public key.

In public-key cryptography, everyone has access to everyone's public key; public keys are available to the public.

Public keys, like secret keys, need to be distributed to be useful. Let us briefly discuss the way public keys can be distributed.

Public Announcement

The naive approach is to announce public keys publicly. Bob can put his public key on his website or announce it in a local or national newspaper. When Alice needs to send a confidential message to Bob, she can obtain Bob's public key from his site or from the newspaper, or she can even send a message to ask for it. Figure 31.23 shows the situation.

Figure 31.23 *Announcing a public key*

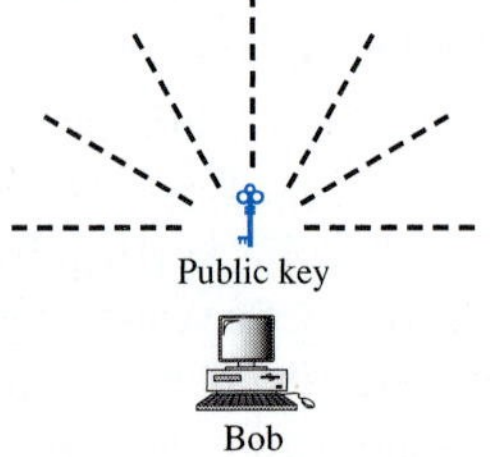

This approach, however, is not secure; it is subject to forgery. For example, Eve could make such a public announcement. Before Bob can react, damage could be done. Eve can fool Alice into sending her a message that is intended for Bob. Eve could also sign a document with a corresponding forged private key and make everyone believe it was signed by Bob. The approach is also vulnerable if Alice directly requests Bob's public key. Eve can intercept Bob's response and substitute her own forged public key for Bob's public key.

Trusted Center

A more secure approach is to have a trusted center retain a directory of public keys. The directory, like the one used in a telephone system, is dynamically updated. Each user can select a private/public key, keep the private key, and deliver the public key for insertion into the directory. The center requires that each user register in the center and prove his or her identity. The directory can be publicly advertised by the trusted center. The center can also respond to any inquiry about a public key. Figure 31.24 shows the concept.

Figure 31.24 *Trusted center*

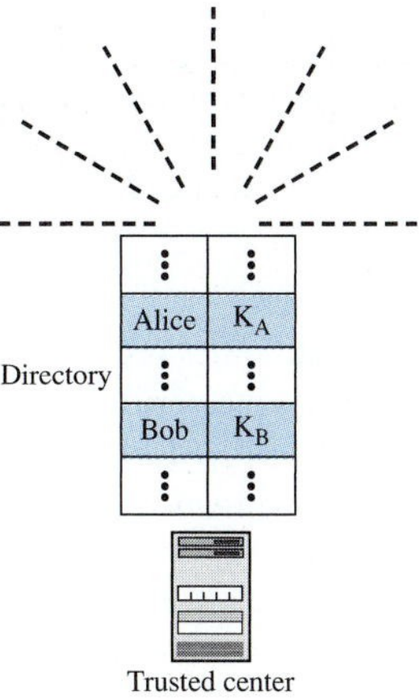

Controlled Trusted Center

A higher level of security can be achieved if there are added controls on the distribution of the public key. The public-key announcements can include a timestamp and be signed by an authority to prevent interception and modification of the response. If Alice needs to know Bob's public key, she can send a request to the center including Bob's name and a timestamp. The center responds with Bob's public key, the original request, and the timestamp signed with the private key of the center. Alice uses the public key of the center, known by all, to decrypt the message and extract Bob's public key. Figure 31.25 shows one scenario.

Certification Authority

The previous approach can create a heavy load on the center if the number of requests is large. The alternative is to create public-key certificates. Bob wants two things: he wants people to know his public key, and he wants no one to accept a public key forged as his. Bob can go to a **certification authority (CA)**—a federal or state organization that binds a public key to an entity and issues a certificate. The CA has a well-known public key itself that cannot be forged. The CA checks Bob's **identification** (using a

Figure 31.25 *Controlled trusted center*

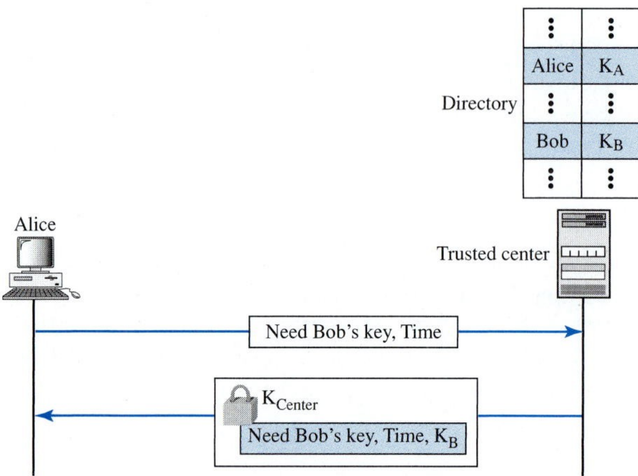

picture ID along with other proof). It then asks for Bob's public key and writes it on the certificate. To prevent the certificate itself from being forged, the CA signs the certificate with its private key. Now Bob can upload the signed certificate. Anyone who wants Bob's public key downloads the signed certificate and uses the public key of the center to extract Bob's public key. Figure 31.26 shows the concept.

Figure 31.26 *Certification authority*

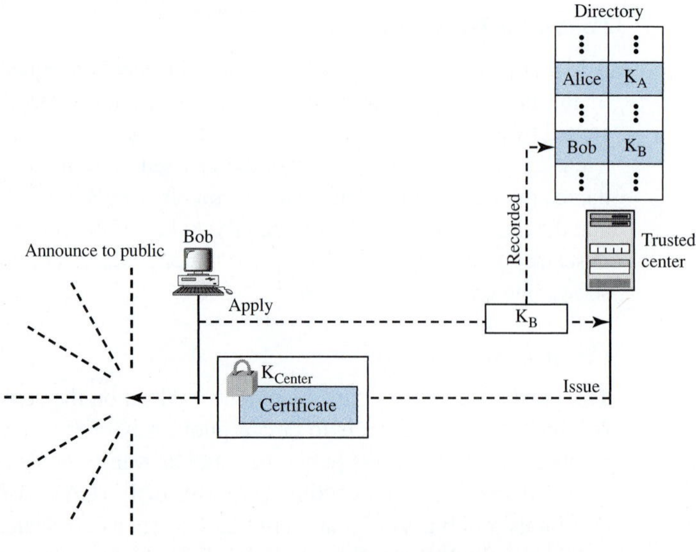

X.509 Although the use of a CA has solved the problem of public-key fraud, it has created a side effect. Each certificate may have a different format. If Alice wants to use a program to automatically download different certificates and digests belonging to different people, the program may not be able to do so. One certificate may have the public key in one format and another in another format. The public key may be on the first line in one certificate and on the third line in another. Anything that needs to be used universally must have a universal format.

To remove this side effect, ITU has designed a protocol called **X.509**, which has been accepted by the Internet with some changes. X.509 is a way to describe the certificate in a structured way. It uses a well-known protocol called ASN.1 (Abstract Syntax Notation 1) that defines fields familiar to C programmers. The following lists the fields in a certificate.

- **Version** This field defines the version of X.509 of the certificate. The version number started at 0; the current version is 2 (the third version).
- **Serial number** This field defines a number assigned to each certificate. The value is unique for each certificate issued.
- **Signature** This field, for which the name is inappropriate, identifies the algorithm used to sign the certificate. Any parameter that is needed for the signature is also defined in this field.
- **Issuer** This field identifies the certification authority that issued the certificate. The name is normally a hierarchy of strings that defines a country, state, organization, department, and so on.
- **Period of validity** This field defines the earliest and the latest times the certificate is valid.
- **Subject** This field defines the entity to which the public key belongs. It is also a hierarchy of strings. Part of the field defines what is called the *common name,* which is the actual name of the beholder of the key.
- **Subject's public key** This field defines the subject's public key, the heart of the certificate. The field also defines the corresponding algorithm (RSA, for example) and its parameters.
- **Issuer unique identifier** This optional field allows two issuers to have the same *issuer* field value, if the *issuer unique identifiers* are different.
- **Subject unique identifier** This optional field allows two different subjects to have the same *subject* field value, if the *subject unique identifiers* are different.
- **Extension** This field allows issuers to add more private information to the certificate.
- **Encrypted** This field contains the algorithm identifier, a secure hash of the other fields, and a digital signature of that hash.

Public-Key Infrastructures (PKI)

When we want to use public keys universally, we have a problem similar to secret-key distribution. We found that we cannot have only one KDC to answer the queries. We need many servers. In addition, we found that the best solution is to put the servers in a hierarchical relationship with one another. Likewise, a solution to public-key queries is a hierarchical structure called a **public-key infrastructure (PKI)**. Figure 31.27 shows an example of this hierarchy.

Figure 31.27 *PKI hierarchy*

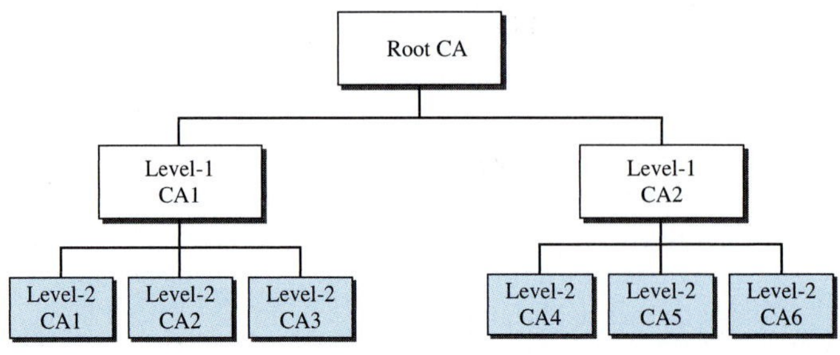

At the first level, we can have a root CA that can certify the performance of CAs in the second level; these level-1 CAs may operate in a large geographic or logical area. The level-2 CAs may operate in smaller geographic areas.

In this hierarchy, everybody trusts the root. But people may or may not trust intermediate CAs. If Alice needs to get Bob's certificate, she may find a CA somewhere to issue the certificate. But Alice may not trust that CA. In a hierarchy Alice can ask the next-higher CA to certify the original CA. The inquiry may go all the way to the root.

31.8 RECOMMENDED READING

For more details about the subjects discussed in this chapter, we recommend the following books and sites. The items in brackets [. . .] refer to the reference list at the end of the text.

Books

Several books are dedicated to network security, such as [PHS02], [Bis03], and [Sal03].

31.9 KEY TERMS

authentication server (AS)
certification authority (CA)
challenge-response authentication
claimant
dictionary attack
digital signature
eavesdropping
entity authentication
fingerprint
fixed password
hash function
hashed message authentication code (HMAC)
identification
integrity
Kerberos
key distribution center (KDC)
message authentication
message authentication code (MAC)
message confidentiality or privacy
message digest
message integrity

message nonrepudiation
modification detection code (MDC)
nonce
nonrepudiation
one-time password
one-wayness
password
privacy
public-key infrastructure (PKI)
salting
session key
SHA−1
signature scheme
signing algorithm
strong collision
ticket
ticket-granting server (TGS)
verifier
verifying algorithm
weak collision
X.509

31.10 SUMMARY

- Cryptography can provide five services. Four of these are related to the message exchange between Alice and Bob. The fifth is related to the entity trying to access a system for using its resources.
- Message confidentiality means that the sender and the receiver expect privacy.
- Message integrity means that the data must arrive at the receiver exactly as sent.
- Message authentication means that the receiver is ensured that the message is coming from the intended sender, not an imposter.
- Nonrepudiation means that a sender must not be able to deny sending a message that he sent.
- Entity authentication means to prove the identity of the entity that tries to access the system's resources.
- A message digest can be used to preserve the integrity of a document or a message. A hash function creates a message digest out of a message.
- A hash function must meet three criteria: one-wayness, resistance to weak collision, and resistance to strong collision.
- A keyless message digest is used as a modification detection code (MDC). It guarantees the integrity of the message. To authenticate the data origin, one needs a message authentication code (MAC).
- MACs are keyed hash functions that create a compressed digest from the message added with the key. The method has the same basis as encryption algorithms.
- A digital signature scheme can provide the same services provided by a conventional signature. A conventional signature is included in the document; a digital signature is a separate entity.
- Digital signature provides message integrity, authentication, and nonrepudiation. Digital signature cannot provide confidentiality for the message. If confidentiality is needed, a cryptosystem must be applied over the scheme.

- A digital signature needs an asymmetric-key system.
- In entity authentication, a claimant proves her identity to the verifier by using one of the three kinds of witnesses: something known, something possessed, or something inherent.
- In password-based authentication, the claimant uses a string of characters as something she knows.
- Password-based authentication can be divided into two broad categories: fixed and one-time.
- In challenge-response authentication, the claimant proves that she knows a secret without actually sending it.
- Challenge-response authentication can be divided into four categories: symmetric-key ciphers, keyed-hash functions, asymmetric-key ciphers, and digital signature.
- A key distribution center (KDC) is a trusted third party that assigns a symmetric key to two parties.
- KDC creates a secret key only between a member and the center. The secret key between members needs to be created as a session key when two members contact KDC.
- Kerberos is a popular session key creator protocol that requires an authentication server and a ticket-granting server.
- A certification authority (CA) is a federal or state organization that binds a public key to an entity and issues a certificate.
- A public-key infrastructure (PKI) is a hierarchical system to answer queries about key certification.

31.11 PRACTICE SET

Review Questions

1. What is a nonce?
2. What is the N^2 problem?
3. Name a protocol that uses a KDC for user authentication.
4. What is the purpose of the Kerberos authentication server?
5. What is the purpose of the Kerberos ticket-granting server?
6. What is the purpose of X.509?
7. What is a certification authority?
8. What are some advantages and disadvantages of using long passwords?
9. We discussed fixed and one-time passwords as two extremes. What about frequently changed passwords? How do you think this scheme can be implemented? What are the advantages and disadvantages?
10. How can a system prevent a guessing attack on a password? How can a bank prevent PIN guessing if someone has found or stolen a bank card and tried to use it?

Exercises

11. A message is made of 10 numbers between 00 and 99. A hash algorithm creates a digest out of this message by adding all numbers modulo 100. The resulting digest is a number between 00 and 99. Does this algorithm meet the first criterion of a hash algorithm? Does it meet the second criterion? Does it meet the third criterion?
12. A message is made of 100 characters. A hash algorithm creates a digest out of this message by choosing characters 1, 11, 21, . . ., and 91. The resulting digest has 10 characters. Does this algorithm meet the first criterion of a hash algorithm? Does it meet the second criterion? Does it meet the third criterion?
13. A hash algorithm creates a digest of N bits. How many different digests can be created from this algorithm?
14. At a party, which is more probable, a person with a birthday on a particular day or two (or more) persons having the same birthday?
15. How is the solution to Exercise 14 related to the second and third criteria of a hashing function?
16. Which one is more feasible, a fixed-size digest or a variable-size digest? Explain your answer.
17. A message is 20,000 characters. We are using a digest of this message using SHA-1. After creating the digest, we decided to change the last 10 characters. Can we say how many bits in the digest will be changed?
18. Are the processes of creating a MAC and of signing a hash the same? What are the differences?
19. When a person uses a money machine to get cash, is this a message authentication, an entity authentication, or both?
20. Change Figure 31.14 to provide two-way authentication (Alice for Bob and Bob for Alice).
21. Change Figure 31.16 to provide two-way authentication (Alice for Bob and Bob for Alice).
22. Change Figure 31.17 to provide two-way authentication (Alice for Bob and Bob for Alice).
23. Change Figure 31.18 to provide two-way authentication (Alice for Bob and Bob for Alice).
24. In a university, a student needs to encrypt her password (with a unique symmetric key) before sending it when she logs in. Does encryption protect the university or the student? Explain your answer.
25. In Exercise 24, does it help if the student appends a timestamp to the password before encryption? Explain your answer.
26. In Exercise 24, does it help if a student has a list of passwords and uses a different one each time?
27. In Figure 31.20, what happens if KDC is down?
28. In Figure 31.21, what happens if the AS is down? What happens if the TGS is down? What happens if the main server is down?
29. In Figure 31.26, what happens if the trusted center is down?

30. Add a symmetric-key encryption/decryption layer to Figure 31.11 to provide privacy.
31. Add an asymmetric-key encryption/decryption layer to Figure 31.11 to provide privacy.

Research Activities

32. There is a hashing algorithm called MD5. Find the difference between this algorithm and SHA-1.
33. There is a hashing algorithm called RIPEMD-160. Find the difference between this algorithm and SHA-1.
34. Compare MD5, SHA-1, and RIPEMD-160.
35. Find some information about RSA digital signature.
36. Find some information about DSS digital signature.

CHAPTER 32

Security in the Internet: IPSec, SSL/TLS, PGP, VPN, and Firewalls

In this chapter, we want to show how certain security aspects, particularly privacy and message authentication, can be applied to the network, transport, and application layers of the Internet model. We briefly show how the IPSec protocol can add authentication and confidentiality to the IP protocol, how SSL (or TLS) can do the same for the TCP protocol, and how PGP can do it for the SMTP protocol (e-mail).

In all these protocols, there are some common issues that we need to consider. First, we need to create a MAC. Then we need to encrypt the message and, probably, the MAC. This means, that with some minor variations, the three protocols discussed in this chapter take a packet from the appropriate layer and create a new packet which is authenticated and encrypted. Figure 32.1 shows this general idea.

Figure 32.1 *Common structure of three security protocols*

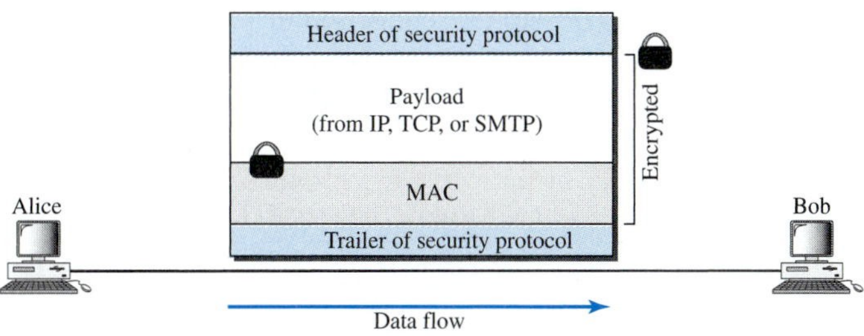

Note that the header or the trailer of the security protocol may or may not be included in the encryption process. Note also that some protocols may need more information in the secured packet; the figure shows only the general idea.

One common issue in all these protocols is *security parameters*. Even the simplified structure in Figure 32.1 suggests that Alice and Bob need to know several pieces of information, security parameters, before they can send secured data to each other. In particular, they need to know which algorithms to use for authentication and encryption/decryption.

Even if these algorithms can be predetermined for everyone in the world, which they are not as we will see, Bob and Alice still need at least two keys: one for the MAC and one for encryption/decryption. In other words, the complexity of these protocols lies not in the way the MAC data are calculated or the way encryption is performed; it lies in the fact that before calculating the MAC and performing encryption, we need to create a set of security parameters between Alice and Bob.

At first glance, it looks as if the use of any of these protocols must involve an infinite number of steps. To send secured data, we need a set of security parameters. The secure exchange of security parameters needs a second set of security parameters. The secure exchange of the second set of security parameters needs a third set of security parameters. And so on ad infinitum.

To limit the steps, we can use public-key cryptography if each person has a private and public key pair. The number of steps can be reduced to one or two. In the one-step version, we can use session keys to create the MAC and encrypt both data and MAC. The session keys and the list of algorithms can be sent with the packet but encrypted by using public-key ciphers. In the two-step version, we first establish the security parameters by using public-key ciphers. We then use the security parameters to securely send actual data. One of the three protocols, PGP, uses the first approach; the other two protocols, IPSec and SSL/TLS, use the second.

We also discuss a common protocol, the virtual private network (VPN), that uses the IPSec. At the end of the chapter, we discuss the firewall, a mechanism for preventing the attack on the network of the organization.

32.1 IPSecurity (IPSec)

IPSecurity (IPSec) is a collection of protocols designed by the Internet Engineering Task Force (IETF) to provide security for a packet at the network level. IPSec helps to create authenticated and confidential packets for the IP layer as shown in Figure 32.2.

Figure 32.2 *TCP/IP protocol suite and IPSec*

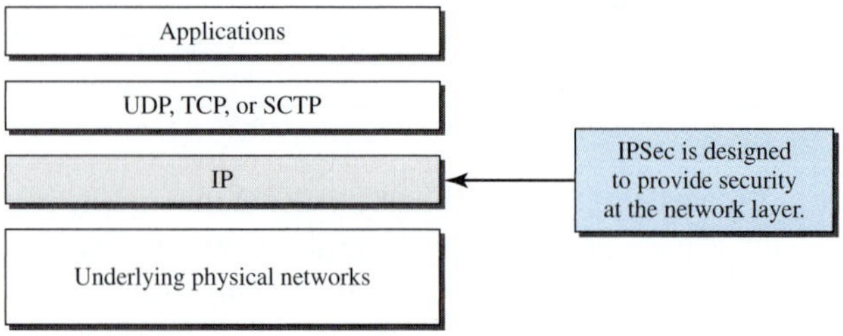

Two Modes

IPSec operates in one of two different modes: the transport mode or the tunnel mode as shown in Figure 32.3.

Figure 32.3 *Transport mode and tunnel modes of IPSec protocol*

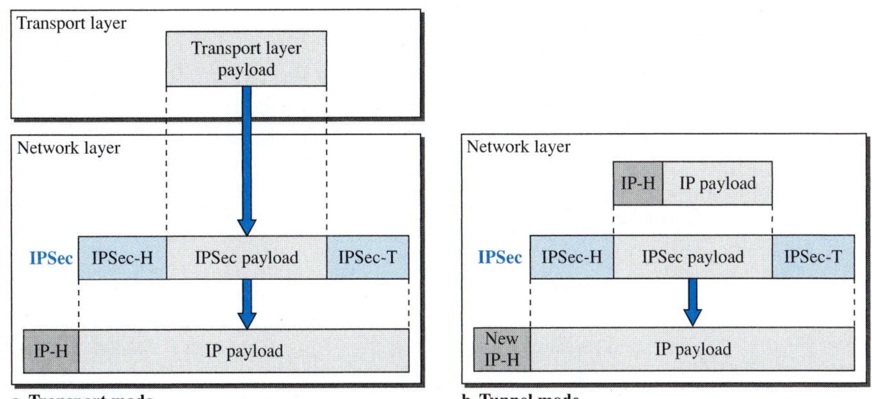

Transport Mode

In the **transport mode,** IPSec protects what is delivered from the transport layer to the network layer. In other words, the transport mode protects the network layer payload, the payload to be encapsulated in the network layer.

Note that the transport mode does not protect the IP header. In other words, the transport mode does not protect the whole IP packet; it protects only the packet from the transport layer (the IP layer payload). In this mode, the IPSec header and trailer are added to the information coming from the transport layer. The IP header is added later.

IPSec in the transport mode does not protect the IP header; it only protects the information coming from the transport layer.

The transport mode is normally used when we need host-to-host (end-to-end) protection of data. The sending host uses IPSec to authenticate and/or encrypt the payload delivered from the transport layer. The receiving host uses IPSec to check the authentication and/or decrypt the IP packet and deliver it to the transport layer. Figure 32.4 shows this concept.

Figure 32.4 *Transport mode in action*

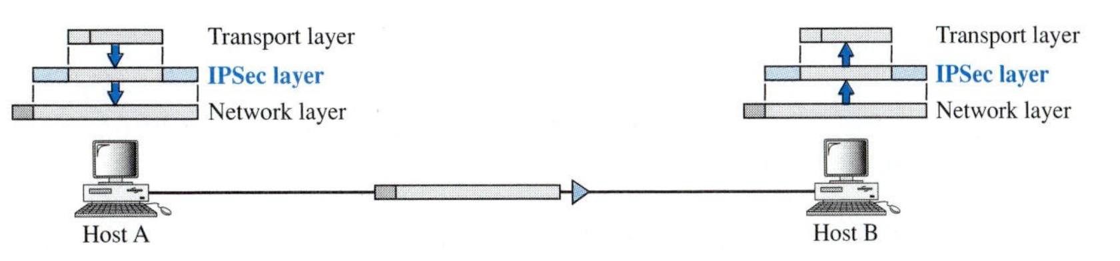

Tunnel Mode

In the **tunnel mode,** IPSec protects the entire IP packet. It takes an IP packet, including the header, applies IPSec security methods to the entire packet, and then adds a new IP header as shown in Figure 32.5.

The new IP header, as we will see shortly, has different information than the original IP header. The tunnel mode is normally used between two routers, between a host and a router, or between a router and a host as shown in Figure 32.5.

Figure 32.5 *Tunnel mode in action*

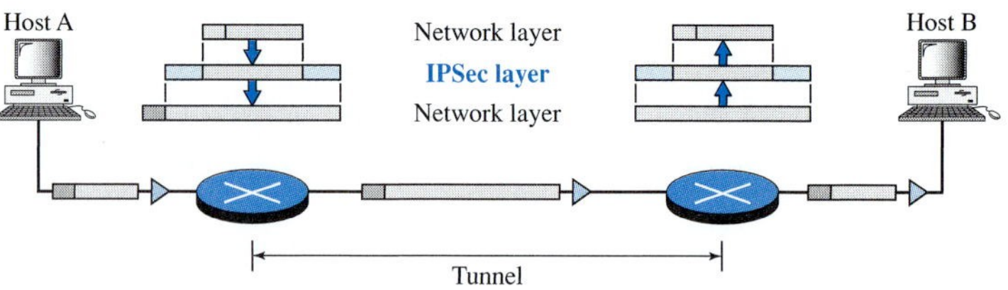

In other words, we use the tunnel mode when either the sender or the receiver is not a host. The entire original packet is protected from intrusion between the sender and the receiver. It's as if the whole packet goes through an imaginary tunnel.

IPSec in tunnel mode protects the original IP header.

Two Security Protocols

IPSec defines two protocols—the Authentication Header (AH) Protocol and the Encapsulating Security Payload (ESP) Protocol—to provide authentication and/or encryption for packets at the IP level.

Authentication Header (AH)

The **Authentication Header (AH) Protocol** is designed to authenticate the source host and to ensure the integrity of the payload carried in the IP packet. The protocol uses a hash function and a symmetric key to create a message digest; the digest is inserted in the authentication header. The AH is then placed in the appropriate location based on the mode (transport or tunnel). Figure 32.6 shows the fields and the position of the authentication header in the transport mode.

When an IP datagram carries an authentication header, the original value in the protocol field of the IP header is replaced by the value 51. A field inside the authentication header (the next header field) holds the original value of the protocol field (the type of payload being carried by the IP datagram). The addition of an authentication header follows these steps:

1. An authentication header is added to the payload with the authentication data field set to zero.

Figure 32.6 *Authentication Header (AH) Protocol in transport mode*

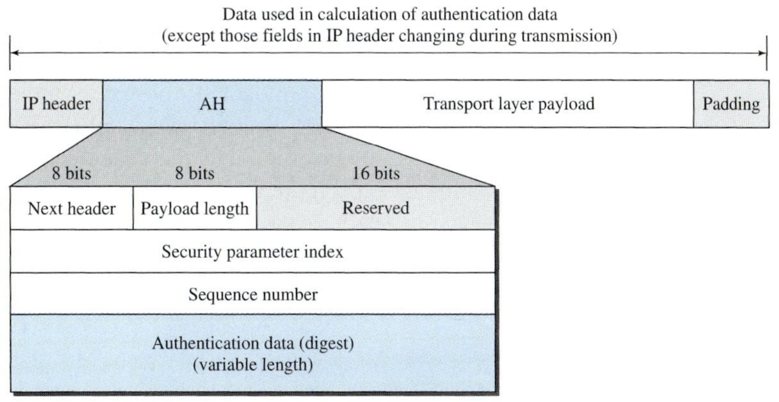

2. Padding may be added to make the total length even for a particular hashing algorithm.
3. Hashing is based on the total packet. However, only those fields of the IP header that do not change during transmission are included in the calculation of the message digest (authentication data).
4. The authentication data are inserted in the authentication header.
5. The IP header is added after the value of the protocol field is changed to 51.

A brief description of each field follows:

- **Next header.** The 8-bit next-header field defines the type of payload carried by the IP datagram (such as TCP, UDP, ICMP, or OSPF). It has the same function as the protocol field in the IP header before encapsulation. In other words, the process copies the value of the protocol field in the IP datagram to this field. The value of the protocol field in the new IP datagram is now set to 51 to show that the packet carries an authentication header.
- **Payload length.** The name of this 8-bit field is misleading. It does not define the length of the payload; it defines the length of the authentication header in 4-byte multiples, but it does not include the first 8 bytes.
- **Security parameter index.** The 32-bit security parameter index (SPI) field plays the role of a virtual-circuit identifier and is the same for all packets sent during a connection called a security association (discussed later).
- **Sequence number.** A 32-bit sequence number provides ordering information for a sequence of datagrams. The sequence numbers prevent a playback. Note that the sequence number is not repeated even if a packet is retransmitted. A sequence number does not wrap around after it reaches 2^{32}; a new connection must be established.
- **Authentication data.** Finally, the authentication data field is the result of applying a hash function to the entire IP datagram except for the fields that are changed during transit (e.g., time-to-live).

The AH Protocol provides source authentication and data integrity, but not privacy.

Encapsulating Security Payload (ESP)

The AH Protocol does not provide privacy, only source authentication and data integrity. IPSec later defined an alternative protocol that provides source authentication, integrity, and privacy called **Encapsulating Security Payload (ESP).** ESP adds a header and trailer. Note that ESP's authentication data are added at the end of the packet which makes its calculation easier. Figure 32.7 shows the location of the ESP header and trailer.

Figure 32.7 *Encapsulation Security Payload (ESP) Protocol in transport mode*

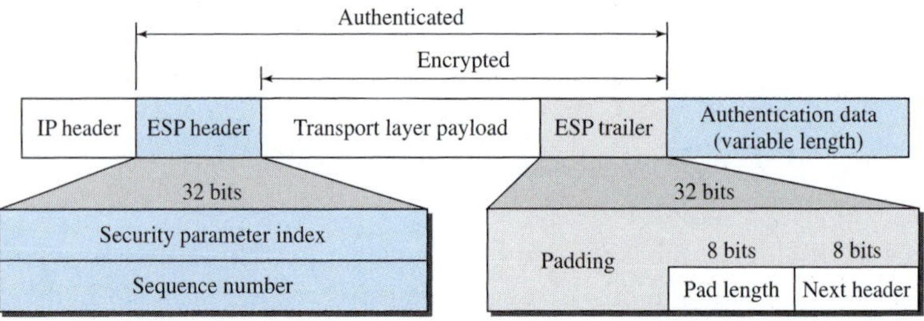

When an IP datagram carries an ESP header and trailer, the value of the protocol field in the IP header is 50. A field inside the ESP trailer (the next-header field) holds the original value of the protocol field (the type of payload being carried by the IP datagram, such as TCP or UDP). The ESP procedure follows these steps:

1. An ESP trailer is added to the payload.
2. The payload and the trailer are encrypted.
3. The ESP header is added.
4. The ESP header, payload, and ESP trailer are used to create the authentication data.
5. The authentication data are added to the end of the ESP trailer.
6. The IP header is added after the protocol value is changed to 50.

The fields for the header and trailer are as follows:

- **Security parameter index.** The 32-bit security parameter index field is similar to that defined for the AH Protocol.
- **Sequence number.** The 32-bit sequence number field is similar to that defined for the AH Protocol.
- **Padding.** This variable-length field (0 to 255 bytes) of 0s serves as padding.
- **Pad length.** The 8-bit pad length field defines the number of padding bytes. The value is between 0 and 255; the maximum value is rare.
- **Next header.** The 8-bit next-header field is similar to that defined in the AH Protocol. It serves the same purpose as the protocol field in the IP header before encapsulation.

- **Authentication data.** Finally, the authentication data field is the result of applying an authentication scheme to parts of the datagram. Note the difference between the authentication data in AH and ESP. In AH, part of the IP header is included in the calculation of the authentication data; in ESP, it is not.

ESP provides source authentication, data integrity, and privacy.

IPv4 and IPv6

IPSec supports both IPv4 and IPv6. In IPv6, however, AH and ESP are part of the extension header.

AH Versus ESP

The ESP Protocol was designed after the AH Protocol was already in use. ESP does whatever AH does with additional functionality (privacy). The question is, Why do we need AH? The answer is, We don't. However, the implementation of AH is already included in some commercial products, which means that AH will remain part of the Internet until the products are phased out.

Services Provided by IPSec

The two protocols, AH and ESP, can provide several security services for packets at the network layer. Table 32.1 shows the list of services available for each protocol.

Table 32.1 *IPSec services*

Services	*AH*	*ESP*
Access control	Yes	Yes
Message authentication (message integrity)	Yes	Yes
Entity authentication (data source authentication)	Yes	Yes
Confidentiality	No	Yes
Replay attack protection	Yes	Yes

Access Control IPSec provides access control indirectly by using a Security Association Database (SADB) as we will see in the next section. When a packet arrives at a destination, and there is no security association already established for this packet, the packet is discarded.

Message Authentication The integrity of the message is preserved in both AH and ESP by using authentication data. A digest of data is created and sent by the sender to be checked by the receiver.

Entity Authentication The security association and the keyed-hashed digest of the data sent by the sender authenticate the sender of the data in both AH and ESP.

Confidentiality The encryption of the message in ESP provides confidentiality. AH, however, does not provide confidentiality. If confidentiality is needed, one should use ESP instead of AH.

Replay Attack Protection In both protocols, the **replay attack** is prevented by using sequence numbers and a sliding receiver window. Each IPSec header contains a unique sequence number when the security association is established. The number starts from 0 and increases until the value reaches $2^{32} - 1$ (the size of the sequence number field is 32 bits). When the sequence number reaches the maximum, it is reset to zero and, at the same time, the old security association (see the next section) is deleted and a new one is established. To prevent processing of duplicate packets, IPSec mandates the use of a fixed-size window at the receiver. The size of the window is determined by the receiver with a default value of 64.

Security Association

As we mentioned in the introduction to the chapter, each of three protocols we discuss in this chapter (IPSec, SSL/TLS, and PGP) needs a set of security parameters before it can be operative. In IPSec, the establishment of the security parameters is done via a mechanism called **security association (SA).**

IP, as we have seen, is a connectionless protocol: Each datagram is independent of the others. For this type of communication, the security parameters can be established in one of three ways.

1. Security parameters related to each datagram can be included in each datagram. The designer of IPSec did not choose this option probably because of overhead. Adding security parameters to each datagram creates a large overhead, particularly if the datagram is fragmented several times during its journey.
2. A set of security parameters can be established for each datagram. This means that before each datagram is transmitted, a set of packets needs to be exchanged between the sender and receiver to establish security parameters. This is probably less efficient than the first choice, and it is not used in IPSec.
3. IPSec uses the third choice. A set of security parameters can be established between a sender and a particular receiver the first time the sender has a datagram to send to that particular receiver. The set can be saved for future transmission of IP packets to the same receiver.

Security association is a very important aspect of IPSec. Using security association, IPSec changes a connectionless protocol, IP, to a connection-oriented protocol. We can think of an association as a connection. We can say that when Alice and Bob agree upon a set of security parameters between them, they have established a logical connection between themselves (which is called association). However, they may not use this connection all the time. After establishing the connection, Alice can send a datagram to Bob today, another datagram a few days later, and so on. The logical connection is there and ready for sending a secure datagram. Of course, they can break the connection, or they can establish a new one after a while (which is a more secure way of communication).

A Simple Example

A security association is a very complex set of pieces of information. However, we can show the simplest case in which Alice wants to have an association with Bob for use in a two-way communication. Alice can have an outbound association (for datagrams to Bob) and an inbound association (for datagrams from Bob). Bob can have the same. In this case, the security associations are reduced to two small tables for both Alice and Bob as shown in Figure 32.8.

Figure 32.8 *Simple inbound and outbound security associations*

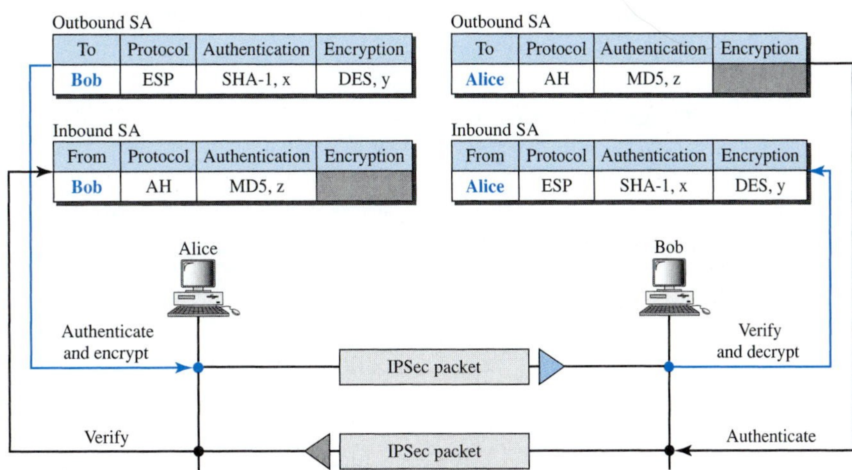

The figure shows that when Alice needs to send a datagram to Bob, she uses the ESP Protocol of IPSec. Authentication is done by using SHA-1 with key x. The encryption is done by using DES with key y. When Bob needs to send a datagram to Alice, he uses the AH Protocol of IPSec. Authentication is done by using MD5 with key z. Note that the inbound association for Bob is the same as the outbound association for Alice, and vice versa.

Security Association Database (SADB)

A security association can be very complex. This is particularly true if Alice wants to send messages to many people and Bob needs to receive messages from many people. In addition, each site needs to have both inbound and outbound SAs to allow bidirectional communication. In other words, we need a set of SAs that can be collected into a database. This database is called the **security association database (SADB).** The database can be thought of as a two-dimensional table with each row defining a single SA. Normally, there are two SADBs, one inbound and one outbound.

Security Parameter Index

To distinguish one association from the other, each association is identified by a parameter called the **security parameter index (SPI).** This parameter, in conjunction with the destination address (outbound) or source address (inbound) and protocol (AH or ESP), uniquely defines an association.

Internet Key Exchange (IKE)

Now we come to the last part of the puzzle—how SADBs are created. The **Internet Key Exchange (IKE)** is a protocol designed to create both inbound and outbound security associations in SADBs.

IKE creates SAs for IPSec.

IKE is a complex protocol based on three other protocols—Oakley, SKEME, and ISAKMP—as shown in Figure 32.9.

Figure 32.9 *IKE components*

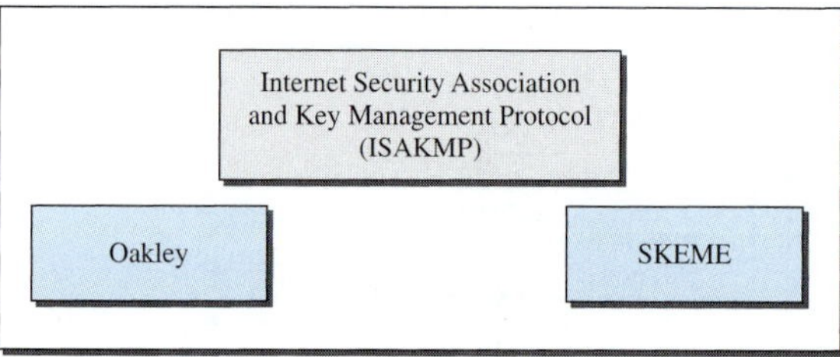

The **Oakley** Protocol was developed by Hilarie Orman. It is a key creation protocol based on the Diffie-Hellman key-exchange method, but with some improvements. Oakley is a free-formatted protocol in the sense that it does not define the format of the message to be exchanged.

SKEME, designed by Hugo Krawcyzk, is another protocol for key exchange. It uses public-key encryption for entity authentication in a key-exchange protocol.

The **Internet Security Association and Key Management Protocol (ISAKMP)** is a protocol designed by the National Security Agency (NSA) that actually implements the exchanges defined in IKE. It defines several packets, protocols, and parameters that allow the IKE exchanges to take place in standardized, formatted messages to create SAs.

One may ask how ISAKMP is carried from the sender to the receiver. This protocol is designed so as to be applicable with any underlying protocol. For example, the packet can be used as the payload in the network layer or transport layer. When we use IPSec, it is natural that this packet be considered as a payload for the IP protocol and carried in the datagram. Now the next question is, How are the datagrams that carry ISAKMP securely exchanged? The answer is that there is no need. There is nothing in the ISAKMP packets that needs to be secured.

Virtual Private Network

Virtual private network (VPN) is a technology that is gaining popularity among large organizations that use the global Internet for both intra- and interorganization communication, but require privacy in their internal communications. We discuss VPN here because it uses the IPSec Protocol to apply security to the IP datagrams.

Private Networks

A private network is designed for use inside an organization. It allows access to shared resources and, at the same time, provides privacy. Before we discuss some aspects of these networks, let us define two commonly used, related terms: *intranet* and *extranet.*

Intranet An **intranet** is a private network (LAN) that uses the Internet model. However, access to the network is limited to the users inside the organization. The network uses application programs defined for the global Internet, such as HTTP, and may have Web servers, print servers, file servers, and so on.

Extranet An **extranet** is the same as an intranet with one major difference: Some resources may be accessed by specific groups of users outside the organization under the control of the network administrator. For example, an organization may allow authorized customers access to product specifications, availability, and online ordering. A university or a college can allow distance learning students access to the computer lab after passwords have been checked.

Addressing A private network that uses the Internet model must use IP addresses. Three choices are available:

1. The network can apply for a set of addresses from the Internet authorities and use them without being connected to the Internet. This strategy has an advantage. If in the future the organization decides to be connected to the Internet, it can do so with relative ease. However, there is also a disadvantage: The address space is wasted in the meantime.
2. The network can use any set of addresses without registering with the Internet authorities. Because the network is isolated, the addresses do not have to be unique. However, this strategy has a serious drawback: Users might mistakenly confuse the addresses as part of the global Internet.
3. To overcome the problems associated with the first and second strategies, the Internet authorities have reserved three sets of addresses, shown in Table 32.2.

Table 32.2 *Addresses for private networks*

Prefix	*Range*	*Total*
10/8	10.0.0.0 to 10.255.255.255	2^{24}
172.16/12	172.16.0.0 to 172.31.255.255	2^{20}
192.168/16	192.168.0.0 to 192.168.255.255	2^{16}

Any organization can use an address out of this set without permission from the Internet authorities. Everybody knows that these reserved addresses are for private networks. They are unique inside the organization, but they are not unique globally. No router will forward a packet that has one of these addresses as the destination address.

Achieving Privacy

To achieve privacy, organizations can use one of three strategies: private networks, hybrid networks, and virtual private networks.

Private Networks An organization that needs privacy when routing information inside the organization can use a **private network** as discussed previously. A small organization with one single site can use an isolated LAN. People inside the organization can send data to one another that totally remain inside the organization, secure from outsiders. A larger organization with several sites can create a private internet. The LANs at different sites can be connected to each other by using routers and leased lines. In other words, an internet can be made out of private LANs and private WANs. Figure 32.10 shows such a situation for an organization with two sites. The LANs are connected to each other by routers and one leased line.

Figure 32.10 *Private network*

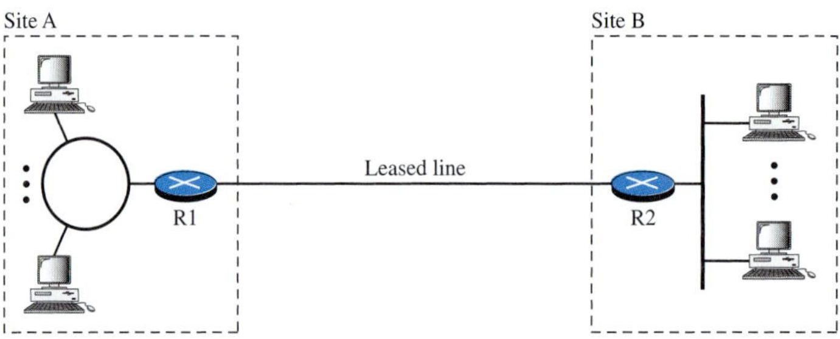

In this situation, the organization has created a private internet that is totally isolated from the global Internet. For end-to-end communication between stations at different sites, the organization can use the Internet model. However, there is no need for the organization to apply for IP addresses with the Internet authorities. It can use private IP addresses. The organization can use any IP class and assign network and host addresses internally. Because the internet is private, duplication of addresses by another organization in the global Internet is not a problem.

Hybrid Networks Today, most organizations need to have privacy in intraorganization data exchange, but, at the same time, they need to be connected to the global Internet for data exchange with other organizations. One solution is the use of a **hybrid network.** A hybrid network allows an organization to have its own private internet and, at the same time, access to the global Internet. Intraorganization data are routed through the private internet; interorganization data are routed through the global Internet. Figure 32.11 shows an example of this situation.

An organization with two sites uses routers R1 and R2 to connect the two sites privately through a leased line; it uses routers R3 and R4 to connect the two sites to the rest of the world. The organization uses global IP addresses for both types of communication. However, packets destined for internal recipients are routed only through routers R1 and R2. Routers R3 and R4 route the packets destined for outsiders.

Virtual Private Networks Both private and hybrid networks have a major drawback: cost. Private wide-area networks (WANs) are expensive. To connect several sites, an organization needs several leased lines, which means a high monthly fee. One solution

Figure 32.11 *Hybrid network*

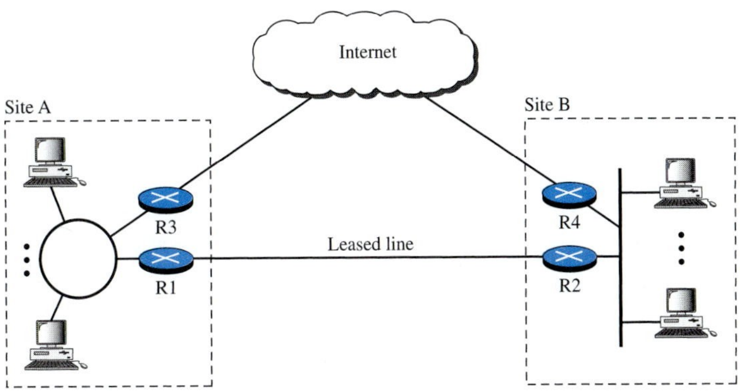

is to use the global Internet for both private and public communications. A technology called virtual private network allows organizations to use the global Internet for both purposes.

VPN creates a network that is private but virtual. It is private because it guarantees privacy inside the organization. It is virtual because it does not use real private WANs; the network is physically public but virtually private.

Figure 32.12 shows the idea of a virtual private network. Routers R1 and R2 use VPN technology to guarantee privacy for the organization.

Figure 32.12 *Virtual private network*

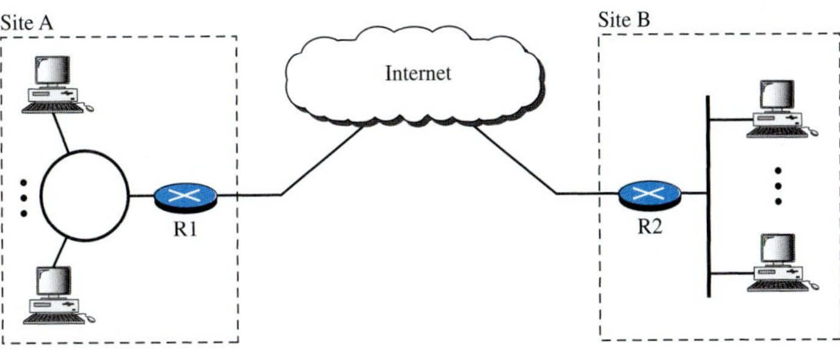

VPN Technology

VPN technology uses IPSec in the tunnel mode to provide authentication, integrity, and privacy.

Tunneling To guarantee privacy and other security measures for an organization, VPN can use the IPSec in the tunnel mode. In this mode, each IP datagram destined for private use in the organization is encapsulated in another datagram. To use IPSec in **tunneling,** the VPNs need to use two sets of addressing, as shown in Figure 32.13.

Figure 32.13 *Addressing in a VPN*

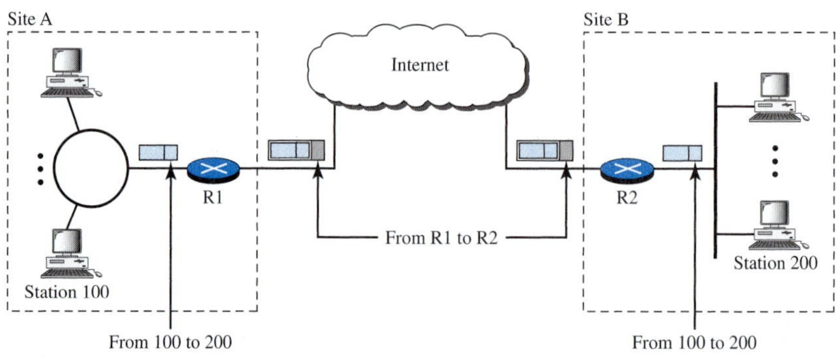

The public network (Internet) is responsible for carrying the packet from R1 to R2. Outsiders cannot decipher the contents of the packet or the source and destination addresses. Deciphering takes place at R2, which finds the destination address of the packet and delivers it.

32.2 SSL/TLS

A transport layer security provides end-to-end security services for applications that use a reliable transport layer protocol such as TCP. The idea is to provide security services for transactions on the Internet. For example, when a customer shops online, the following security services are desired:

1. The customer needs to be sure that the server belongs to the actual vendor, not an imposter. The customer does not want to give an imposter her credit card number (entity authentication). Likewise, the vendor needs to authenticate the customer.
2. The customer and the vendor need to be sure that the contents of the message are not modified during transition (message integrity).
3. The customer and the vendor need to be sure that an imposter does not intercept sensitive information such as a credit card number (confidentiality).

Two protocols are dominant today for providing security at the transport layer: the Secure Sockets Layer (SSL) Protocol and the Transport Layer Security (TLS) Protocol. The latter is actually an IETF version of the former. First we discuss SSL, then we briefly mention the main differences between SSL and TLS. Figure 32.14 shows the position of SSL and TLS in the Internet model.

SSL Services

Secure Socket Layer (SSL) is designed to provide security and compression services to data generated from the application layer. Typically, SSL can receive data from any application layer protocol, but usually the protocol is HTTP. The data received from the

Figure 32.14 *Location of SSL and TLS in the Internet model*

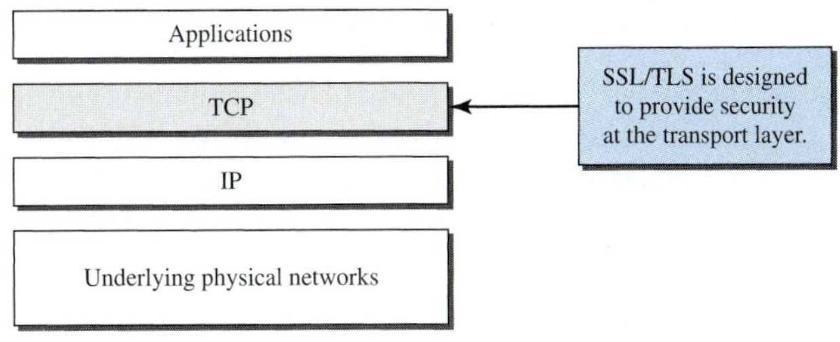

application are compressed (optional), signed, and encrypted. The data are then passed to a reliable transport layer protocol such as TCP. Netscape developed SSL in 1994. Versions 2 and 3 were released in 1995. In this chapter, we discuss SSLv3. SSL provides several services on data received from the application layer.

Fragmentation

First, SSL divides the data into blocks of 2^{14} bytes or less.

Compression

Each fragment of data is compressed by using one of the lossless compression methods negotiated between the client and server. This service is optional.

Message Integrity

To preserve the integrity of data, SSL uses a keyed-hash function to create a MAC.

Confidentiality

To provide confidentiality, the original data and the MAC are encrypted using symmetric-key cryptography.

Framing

A header is added to the encrypted payload. The payload is then passed to a reliable transport layer protocol.

Security Parameters

When we discussed IPSec in the previous section, we mentioned that each of the two parties involved in data exchange needs to have a set of parameters for each association (SA). SSL has a similar goal, but a different approach. There are no SAs, but there are cipher suites and cryptographic secrets that together make the security parameters.

Cipher Suite

The combination of key exchange, hash, and encryption algorithms defines a **cipher suite** for each SSL session. Each suite starts with the term *SSL,* followed by the key-exchange algorithm. The word *WITH* separates the key exchange algorithm from the encryption and hash algorithms. For example,

SSL_DHE_RSA_***WITH***_DES_CBC_SHA

defines DHE_RSA (ephemeral Diffie-Hellman with RSA digital signature) as the key exchange with DES_CBC as the encryption algorithm and SHA as the hash algorithm. Note that DH is fixed Diffie-Hellman, DHE is ephemeral Diffie-Hellman, and DH-anon is anonymous Diffie-Hellman. Table 32.3 shows the suites used in the United States. We have not included those that are used for export. Note that not all combinations of key-exchange algorithms (to establish keys for message authentication and encryption), encryption algorithms, and authentication algorithms are included in the cipher suite list. We have not defined or discussed several algorithms you can find in the table, but we wish to describe the whole picture so that the reader can have an idea of how general the suite is.

Table 32.3 *SSL cipher suite list*

Cipher Suite	*Key Exchange Algorithm*	*Encryption Algorithm*	*Hash Algorithm*
SSL_NULL_***WITH_NULL***_NULL	NULL	NULL	NULL
SSL_RSA_***WITH_NULL***_MD5	RSA	NULL	MD5
SSL_RSA_***WITH_NULL***_SHA	RSA	NULL	SHA
SSL_RSA_***WITH_RC4_128***_MD5	RSA	RC4_128	MD5
SSL_RSA_***WITH_RC4_128***_SHA	RSA	RC4_128	SHA
SSL_RSA_***WITH_IDEA_CBC***_SHA	RSA	IDEA_CBC	SHA
SSL_RSA_***WITH_DES_CBC***_SHA	RSA	DES_CBC	SHA
SSL_RSA_***WITH_3DES_EDE_CBC***_SHA	RSA	3DES_EDE_CBC	SHA
SSL_DH_anon_***WITH_RC4_128***_MD5	DH_anon	RC4_128	MD5
SSL_DH_anon_***WITH_DES_CBC***_SHA	DH_anon	DES_CBC	SHA
SSL_DH_anon_***WITH_3DES_EDE***_CBC_SHA	DH_anon	3DES_EDE_CBC	SHA
SSL_DHE_RSA_***WITH_DES_CBC***_SHA	DHE_RSA	DES_CBC	SHA
SSL_DHE_RSA_***WITH_3DES_EDE_CBC***_SHA	DHE_RSA	3DES_EDE_CBC	SHA
SSL_DHE_DSS_***WITH_DES_CBC***_SHA	DHE_DSS	DES_CBC	SHA
SSL_DHE_DSS_***WITH_3DES_EDE_CBC***_SHA	DHE_DSS	3DES_EDE_CBC	SHA
SSL_DH_RSA_***WITH_DES_CBC***_SHA	DH_RSA	DES_CBC	SHA
SSL_DH_RSA_***WITH_3DES_EDE_CBC***_SHA	DH_RSA	3DES_EDE_CBC	SHA
SSL_DH_DSS_ ***WITH_DES_CBC***_SHA	DH_DSS	DES_CBC	SHA
SSL_DH_DSS_ ***WITH_3DES_EDE_CBC***_SHA	DH_DSS	3DES_EDE_CBC	SHA
SSL_FORTEZZA_DMS_***WITH_NULL***_SHA	FORTEZZA_DMS	NULL	SHA
SSL_FORTEZZA_DMS_***WITH_FORTEZZA_CBC***_SHA	FORTEZZA_DMS	FORTEZZA_CBC	SHA
SSL_FORTEZZA_DMS_***WITH_RC4_128***_SHA	FORTEZZA_DMS	RC4_128	SHA

Cryptographic Secrets

The second part of security parameters is often referred to as cryptographic secrets. To achieve message integrity and confidentiality, SSL needs six cryptographic secrets, four keys, and two IVs.

The client and the server have six different cryptography secrets.

The process of creating these secrets is shown in Figure 32.15. The client needs one key for message authentication, one key for encryption, and one IV for block encryption. The server needs the same. SSL requires that the keys for one direction be different from those for the other direction. If there is an attack in one direction, the other direction is not affected. These parameters are generated by using a negotiation protocol, as we will see shortly.

Figure 32.15 *Creation of cryptographic secrets in SSL*

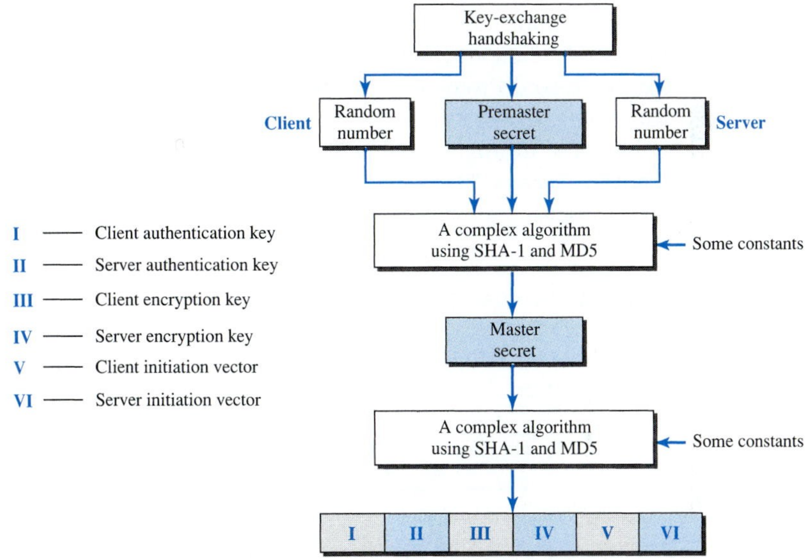

1. The client and server exchange two random numbers; one is created by the client and the other by the server.
2. The client and server exchange one **premaster secret** by using one of the key-exchange algorithms we discussed previously.
3. A 48-byte **master secret** is created from the premaster secret by applying two hash functions (SHA-1 and MD5).
4. The master secret is used to create variable-length secrets by applying the same set of hash functions and prepending with different constants.

Sessions and Connections

The nature of IP and TCP protocols is different. IP is a connectionless protocol; TCP is a connection-oriented protocol. An association in IPSec transforms the connectionless IP to a connection-oriented secured protocol. TCP is already connection-oriented.

However, the designers of SSL decided that they needed two-levels of connectivity: **session** and **connection.** A session between two systems is an association that can last for a long time; a connection can be established and broken several times during a session.

Some of the security parameters are created during the session establishment and are in effect until the session is terminated (for example, cipher suite and master key). Some of the security parameters must be recreated (or occasionally resumed) for each connection (for example, six secrets).

Four Protocols

We have discussed the idea of SSL without showing how SSL accomplishes its tasks. SSL defines four protocols in two layers, as shown in Figure 32.16. The **Record Protocol** is the carrier. It carries messages from three other protocols as well as the data coming from the application layer. Messages from the Record Protocol are payloads to the transport layer, normally TCP. The **Handshake Protocol** provides security parameters for the Record Protocol. It establishes a cipher set and provides keys and security parameters. It also authenticates the server to the client and the client to the server, if needed. The **ChangeCipherSpec Protocol** is used for signaling the readiness of cryptographic secrets. The **Alert Protocol** is used to report abnormal conditions. We will briefly discuss these protocols in this section.

Figure 32.16 *Four SSL protocols*

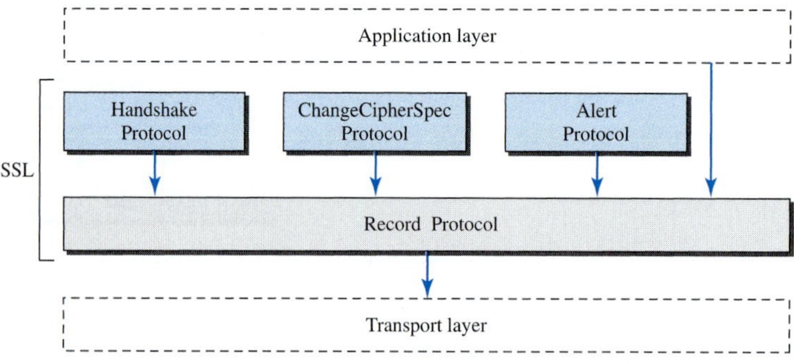

Handshake Protocol

The Handshake Protocol uses messages to negotiate the cipher suite, to authenticate the server to the client and the client to the server (if needed), and to exchange information for building the cryptographic secrets. The handshaking is done in four phases, as shown in Figure 32.17.

ChangeCipherSpec Protocol

We have seen that the negotiation of the cipher suite and the generation of cryptographic secrets are formed gradually during the Handshake Protocol. The question now is, When can the two parties use these parameter secrets? SSL mandates that the parties

Figure 32.17 *Handshake Protocol*

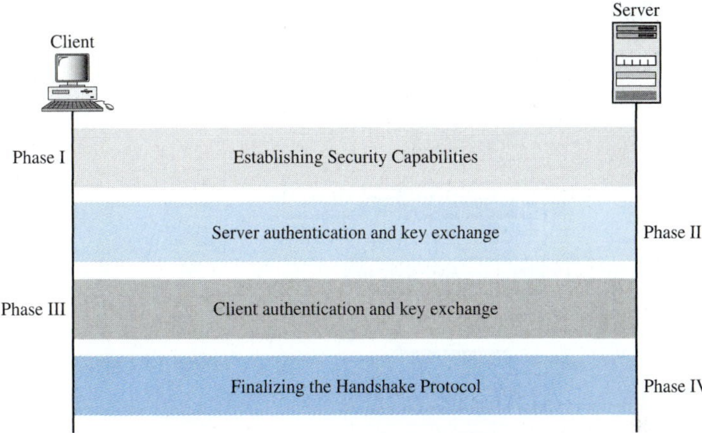

not use these parameters or secrets until they have sent or received a special message, the ChangeCipherSpec message, which is exchanged during the Handshake Protocol and defined in the ChangeCipherSpec Protocol. Before the exchange of any ChangeCipherSpec messages, only the pending columns have values.

Alert Protocol

SSL uses the Alert Protocol for reporting errors and abnormal conditions. It has only one message type, the alert message, that describes the problem and its level (warning or fatal).

Record Protocol

The Record Protocol carries messages from the upper layer (Handshake Protocol, ChangeCipherSpec Protocol, Alert Protocol, or application layer). The message is fragmented and optionally compressed; a MAC is added to the compressed message by using the negotiated hash algorithm. The compressed fragment and the MAC are encrypted by using the negotiated encryption algorithm. Finally, the SSL header is added to the encrypted message. Figure 32.18 shows this process at the sender. The process at the receiver is reversed.

Transport Layer Security

Transport Layer Security (TLS) is the IETF standard version of SSL. The two are very similar, with slight differences. We highlight the differences below:

- **Version.** The SSLv3.0 discussed in this section is compatible with TLSv1.0.
- **Cipher Suite.** TLS cipher suite does not support Fortezza.
- **Cryptography Secret.** There are several differences in the generation of cryptographic secrets. TLS uses a **pseudorandom function (PRF)** to create the master key and the key materials.

Figure 32.18 *Processing done by the Record Protocol*

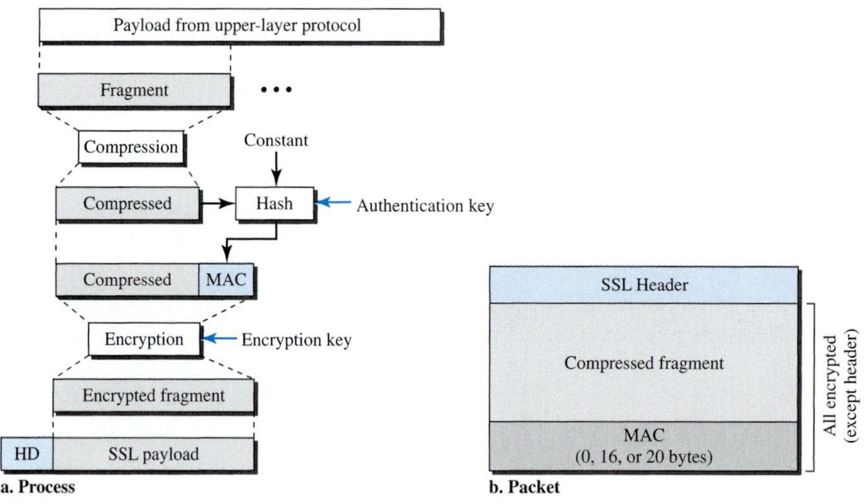

- **Alert Protocol.** TLS deletes some alert messages and adds some new ones.
- **Handshake Protocol.** The details of some messages have been changed in TLS.
- **Record Protocol.** Instead of using MAC, TLS uses the HMAC as defined in Chapter 31.

32.3 PGP

One of the protocols to provide security at the application layer is **Pretty Good Privacy (PGP).** PGP is designed to create authenticated and confidential e-mails. Figure 32.19 shows the position of PGP in the TCP/IP protocol suite.

Figure 32.19 *Position of PGP in the TCP/IP protocol suite*

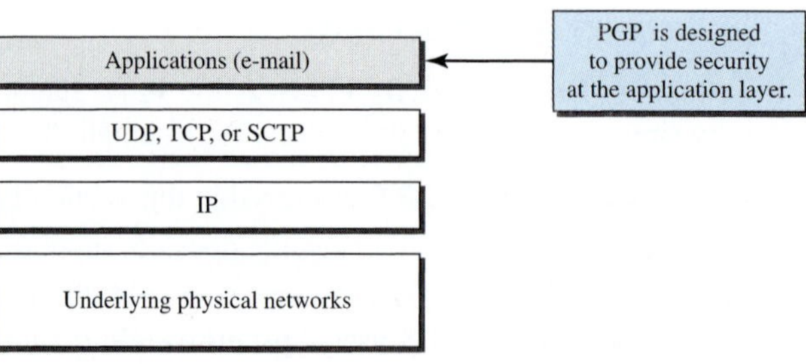

Sending an e-mail is a one-time activity. The nature of this activity is different from those we have seen in the previous two sections. In IPSec or SSL, we assume that the two parties create a session between themselves and exchange data in both directions. In e-mail, there is no session. Alice and Bob cannot create a session. Alice sends a message to Bob; sometime later, Bob reads the message and may or may not send a reply. We discuss the security of a unidirectional message because what Alice sends to Bob is totally independent of what Bob sends to Alice.

Security Parameters

If e-mail is a one-time activity, how can the sender and receiver agree on the security parameters to use for e-mail security? If there is no session and no handshaking to negotiate the algorithms for encryption and authentication, how can the receiver know which algorithm the sender has chosen for each purpose? How can the receiver know the values of the keys used for encryption and authentication?

Phil Zimmerman, the designer and creator of PGP, has found a very elegant solution to the above questions. The security parameters need to be sent with the message.

In PGP, the sender of the message needs to include the identifiers of the algorithms used in the message as well as the values of the keys.

Services

PGP can provide several services based on the requirements of the user. An e-mail can use one or more of these services.

Plaintext

The simplest case is to send the e-mail message in plaintext (no service). Alice, the sender, composes a message and sends it to Bob, the receiver. The message is stored in Bob's mailbox until it is retrieved by him.

Message Authentication

Probably the next improvement is to let Alice sign the message. Alice creates a digest of the message and signs it with her private key. When Bob receives the message, he verifies the message by using Alice's public key. Two keys are needed for this scenario. Alice needs to know her private key; Bob needs to know Alice's public key.

Compression

A further improvement is to compress the message and digest to make the packet more compact. This improvement has no security benefit, but it eases the traffic.

Confidentiality with One-Time Session Key

As we discussed before, confidentiality in an e-mail system can be achieved by using conventional encryption with a one-time session key. Alice can create a session key, use the session key to encrypt the message and the digest, and send the key itself with the message. However, to protect the session key, Alice encrypts it with Bob's public key.

Code Conversion

Another service provided by PGP is code conversion. Most e-mail systems allow the message to consist of only ASCII characters. To translate other characters not in the ASCII set, PGP uses Radix 64 conversion. Each character to be sent (after encryption) is converted to Radix 64 code.

Segmentation

PGP allows segmentation of the message after it has been converted to Radix 64 to make each transmitted unit the uniform size allowed by the underlying e-mail protocol.

A Scenario

Let us describe a scenario that combines some of these services, authentication and confidentiality. The whole idea of PGP is based on the assumption that a group of people who need to exchange e-mail messages trust one another. Everyone in the group somehow knows (with a degree of trust) the public key of any other person in the group. Based on this single assumption, Figure 32.20 shows a simple scenario in which an authenticated and encrypted message is sent from Alice to Bob.

Figure 32.20 *A scenario in which an e-mail message is authenticated and encrypted*

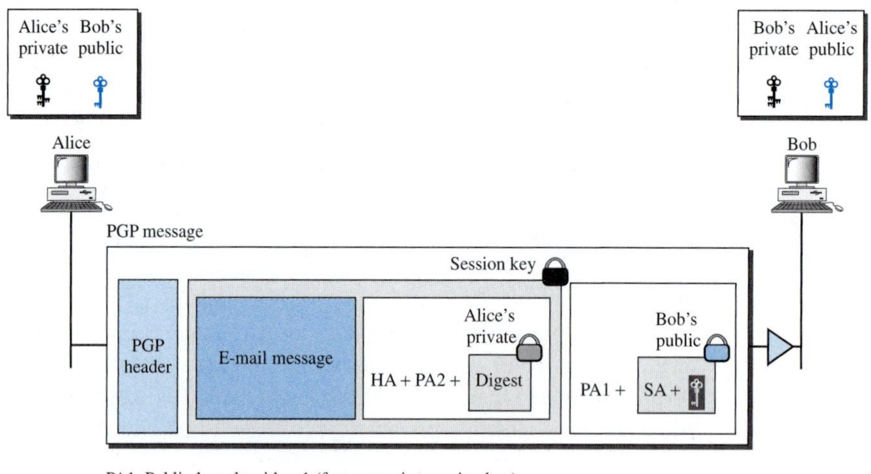

Sender Site

The following shows the steps used in this scenario at Alice's site:

1. Alice creates a session key (for symmetric encryption/decryption) and concatenates it with the identity of the algorithm which will use this key. The result is encrypted

with Bob's public key. Alice adds the identification of the public-key algorithm used above to the encrypted result. This part of the message contains three pieces of information: the session key, the symmetric encryption/decryption algorithm to be used later, and the asymmetric encryption/decryption algorithm that was used for this part.

2.
   a. Alice authenticates the message (e-mail) by using a public-key signature algorithm and encrypts it with her private key. The result is called the signature. Alice appends the identification of the public key (used for encryption) as well as the identification of the hash algorithm (used for authentication) to the signature. This part of the message contains the signature and two extra pieces of information: the encryption algorithm and the hash algorithm.
   b. Alice concatenates the three pieces of information created above with the message (e-mail) and encrypts the whole thing, using the session key created in step 1.
3. Alice combines the results of steps 1 and 2 and sends them to Bob (after adding the appropriate PGP header).

Receiver Site

The following shows the steps used in this scenario at Bob's side after he has received the PGP packet:

1. Bob uses his private key to decrypt the combination of the session key and symmetric-key algorithm identification.
2. Bob uses the session key and the algorithm obtained in step 1 to decrypt the rest of the PGP message. Bob now has the content of the message, the identification of the public algorithm used for creating and encrypting the signature, and the identification of the hash algorithm used to create the hash out of the message.
3. Bob uses Alice's public key and the algorithm defined by PA2 to decrypt the digest.
4. Bob uses the hash algorithm defined by HA to create a hash out of message he obtained in step 2.
5. Bob compares the hash created in step 4 and the hash he decrypted in step 3. If the two are identical, he accepts the message; otherwise, he discards the message.

PGP Algorithms

Table 32.4 shows some of the algorithms used in PGP. The list is not complete; new algorithms are continuously added.

Table 32.4

Algorithm	*ID*	*Description*
Public key	1	RSA (encryption or signing)
	2	RSA (for encryption only)
	3	RSA (for signing only)
	17	DSS (for signing)

Table 32.4 *(continued)*

Algorithm	*ID*	*Description*
Hash algorithm	1	MD5
	2	SHA-1
	3	RIPE-MD
Encryption	0	No encryption
	1	IDEA
	2	Triple DES
	9	AES

Key Rings

In the previous scenarios, we assumed that Alice needed to send a message to only Bob. That is not always the case. Alice may need to send messages to many people. In this case, Alice needs a **key ring** of public keys, with a key belonging to each person with whom Alice needs to correspond (send or receive messages). In addition, the PGP designers specified a ring of private/public keys. One reason is that Alice may wish to change her pair of keys from time to time. Another reason is that Alice may need to correspond with different groups of people (friends, colleagues, and so on). Alice may wish to use a different key pair for each group. Therefore, each user needs to have two sets of rings: a ring of private/public keys and a ring of public keys of other people. Figure 32.21 shows a community of four people, each having a ring of pairs of private/public keys and, at the same time, a ring of four public keys belonging to the other four people in the community. The figure shows seven public keys for each public ring. Each person in the ring can keep more than one public key for each other person.

Figure 32.21 *Rings*

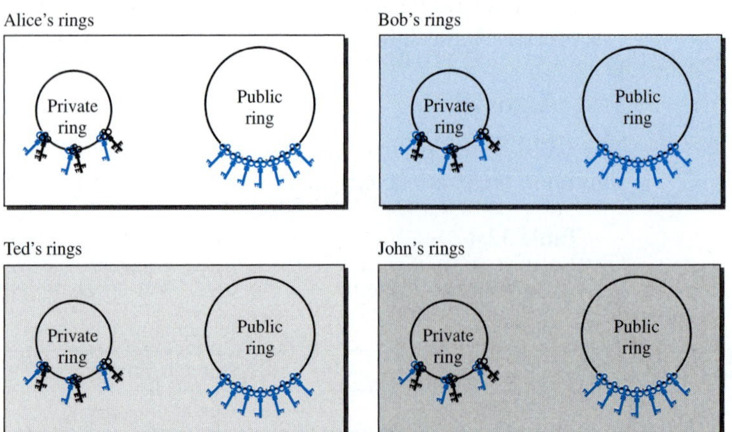

Alice, for example, has several pairs of private/public keys belonging to her and public keys belonging to other people. Note that everyone can have more than one public key. Two cases may arise.

1. Alice needs to send a message to one of the persons in the community.
   a. She uses her private key to sign the digest.
   b. She uses the receiver's public key to encrypt a newly created session key.
   c. She encrypts the message and signs the digest with the session key created.
2. Alice receives a message from one of the persons in the community.
   a. She uses her private key to decrypt the session key.
   b. She uses the session key to decrypt the message and digest.
   c. She uses her public key to verify the digest.

PGP Certificates

To trust the owner of the public key, each user in the PGP group needs to have, implicitly or explicitly, a copy of the certificate of the public-key owner. Although the certificate can come from a certificate authority (CA), this restriction is not required in PGP. PGP has its own certificate system.

Protocols that use X509 certificates depend on the hierarchical structure of the trust. There is a predefined chain of trust from the root to any certificate. Every user fully trusts the authority of the CA at the root level (prerequisite). The root issues certificates for the CAs at the second level, a second-level CA issues a certificate for the third level, and so on. Every party that needs to be trusted presents a certificate from some CA in the tree. If Alice does not trust the certificate issuer for Bob, she can appeal to a higher-level authority up to the root (which must be trusted for the system to work). In other words, there is one single path from a fully trusted CA to a certificate.

In PGP, there is no need for CAs; anyone in the ring can sign a certificate for anyone else in the ring. Bob can sign a certificate for Ted, John, Anne, and so on. There is no hierarchy of trust in PGP; there is no tree. As a result of the lack of hierarchical structure, Ted may have one certificate from Bob and another certificate from Liz. If Alice wants to follow the line of certificates for Ted, it has two paths: one starts from Bob and the other starts from Liz. An interesting point is that Alice may fully trust Bob, but only partially trust Liz. There can be multiple paths in the line of trust from a fully or partially trusted authority to a certificate. In PGP, the issuer of a certificate is usually called an **introducer.**

In PGP, there can be multiple paths from fully or partially trusted authorities to any subject.

Trusts and Legitimacy

The entire operation of PGP is based on introducer trust, the certificate trust, and the legitimacy of the public keys.

Introducer Trust Levels With the lack of a central authority, it is obvious that the ring cannot be very large if every user in the PGP ring of users has to fully trust everyone else.

(Even in real life we cannot fully trust everyone that we know.) To solve this problem, PGP allows different levels of trust. The number of levels is mostly implementation-dependent, but for simplicity, let us assign three levels of trust to any introducer: *none, partial*, and *full*. The **introducer trust** level specifies the trust levels issued by the introducer for other people in the ring. For example, Alice may fully trust Bob, partially trust Anne, and not trust John at all. There is no mechanism in PGP to determine how to make a decision about the trustworthiness of the introducer; it is up to the user to make this decision.

Certificate Trust Levels When Alice receives a certificate from an introducer, she stores the certificate under the name of the subject (certified entity). She assigns a level of trust to this certificate. The **certificate trust** level is normally the same as the introducer trust level that issued the certificate. Assume Alice fully trusts Bob, partially trusts Anne and Janette, and has no trust in John. The following scenarios can happen.

1. Bob issues two certificates, one for Linda (with public key K1) and one for Lesley (with public key K2). Alice stores the public key and certificate for Linda under Linda's name and assigns a *full* level of trust to this certificate. Alice also stores the certificate and public key for Lesley under Lesley's name and assigns a full level of trust to this certificate.
2. Anne issues a certificate for John (with public key K3). Alice stores this certificate and public key under John's name, but assigns a *partial* level for this certificate.
3. Janette issues two certificates, one for John (with public key K3) and one for Lee (with public key K4). Alice stores John's certificate under his name and Lee's certificate under his name, each with a *partial* level of trust. Note that John now has two certificates, one from Anne and one from Janette, each with a *partial* level of trust.
4. John issues a certificate for Liz. Alice can discard or keep this certificate with a signature trust of *none*.

Key Legitimacy The purpose of using introducer and certificate trusts is to determine the legitimacy of a public key. Alice needs to know how legitimate are the public keys of Bob, John, Liz, Anne, and so on. PGP defines a very clear procedure for determining **key legitimacy.** The level of the key legitimacy for a user is the weighted trust level of that user. For example, suppose we assign the following weights to certificate trust levels:

1. A weight of 0 to a nontrusted certificate
2. A weight of $\frac{1}{2}$ to a certificate with partial trust
3. A weight of 1 to a certificate with full trust

Then to fully trust an entity, Alice needs one fully trusted certificate or two partially trusted certificates for that entity. For example, Alice can use John's public key in the previous scenario because both Anne and Janette have issued a certificate for John, each with a certificate trust level of $\frac{1}{2}$. Note that the legitimacy of a public key belonging to an entity does not have anything to do with the trust level of that person. Although Bob can use John's public key to send a message to him, Alice cannot accept any certificate issued by John because, for Alice, John has a trust level of *none*.

Starting the Ring

You might have realized a problem with the above discussion. What if nobody sends a certificate for a fully or partially trusted entity? For example, how can the legitimacy of Bob's public key be determined if no one has sent a certificate for Bob? In PGP, the key legitimacy of a trusted or partially trusted entity can be also determined by other methods.

1. Alice can physically obtain Bob's public key. For example, Alice and Bob can meet personally and exchange a public key written on a piece of paper or to a disk.
2. If Bob's voice is recognizable to Alice, Alice can call him and obtain his public key on the phone.
3. A better solution proposed by PGP is for Bob to send his public key to Alice by e-mail. Both Alice and Bob make a 16-byte MD5 (or 20-byte SHA-1) digest from the key. The digest is normally displayed as eight groups of four digits (or 10 groups of four digits) in hexadecimal and is called a **fingerprint.** Alice can then call Bob and verify the fingerprint on the phone. If the key is altered or changed during the e-mail transmission, the two fingerprints do not match. To make it even more convenient, PGP has created a list of words, each representing a four-digit combination. When Alice calls Bob, Bob can pronounce the eight words (or 10 words) for Alice. The words are carefully chosen by PGP to avoid those similar in pronunciation; for example, if *sword* is in the list, *word* is not.
4. In PGP, nothing prevents Alice from getting Bob's public key from a CA in a separate procedure. She can then insert the public key in the public-key ring.

Web of Trust

PGP can eventually make a **web of trust** between a group of people. If each entity introduces more entities to other entities, the public-key ring for each entity gets larger and larger and entities in the ring can send secure e-mail to one another.

Key Revocation

It may become necessary for an entity to revoke his or her public key from the ring. This may happen if the owner of the key feels that the key is compromised (stolen, for example) or just too old to be safe. To revoke a key, the owner can send a revocation certificate signed by herself. The revocation certificate must be signed by the old key and disseminated to all the people in the ring who use that public key.

32.4 FIREWALLS

All previous security measures cannot prevent Eve from sending a harmful message to a system. To control access to a system, we need firewalls. A **firewall** is a device (usually a router or a computer) installed between the internal network of an organization and the rest of the Internet. It is designed to forward some packets and filter (not forward) others. Figure 32.22 shows a firewall.

Figure 32.22 *Firewall*

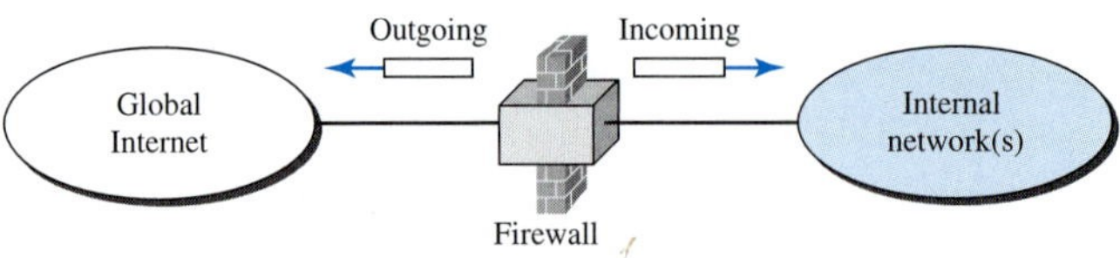

For example, a firewall may filter all incoming packets destined for a specific host or a specific server such as HTTP. A firewall can be used to deny access to a specific host or a specific service in the organization.

A firewall is usually classified as a packet-filter firewall or a proxy-based firewall.

Packet-Filter Firewall

A firewall can be used as a packet filter. It can forward or block packets based on the information in the network layer and transport layer headers: source and destination IP addresses, source and destination port addresses, and type of protocol (TCP or UDP). A **packet-filter firewall** is a router that uses a filtering table to decide which packets must be discarded (not forwarded). Figure 32.23 shows an example of a filtering table for this kind of a firewall.

Figure 32.23 *Packet-filter firewall*

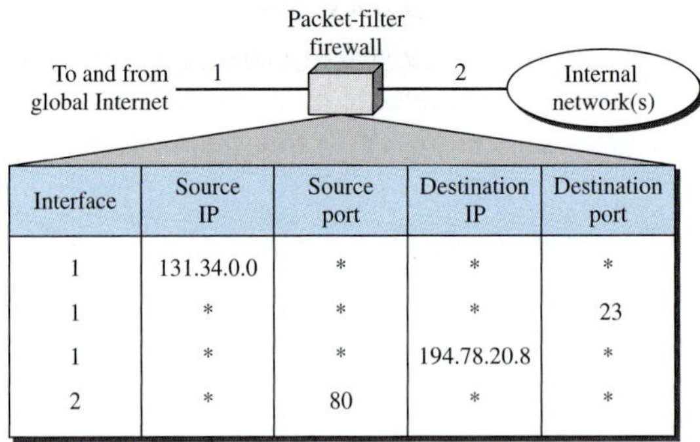

Interface	Source IP	Source port	Destination IP	Destination port
1	131.34.0.0	*	*	*
1	*	*	*	23
1	*	*	194.78.20.8	*
2	*	80	*	*

According to Figure 32.23, the following packets are filtered:

1. Incoming packets from network 131.34.0.0 are blocked (security precaution). Note that the * (asterisk) means "any."
2. Incoming packets destined for any internal TELNET server (port 23) are blocked.
3. Incoming packets destined for internal host 194.78.20.8 are blocked. The organization wants this host for internal use only.
4. Outgoing packets destined for an HTTP server (port 80) are blocked. The organization does not want employees to browse the Internet.

A packet-filter firewall filters at the network or transport layer.

Proxy Firewall

The packet-filter firewall is based on the information available in the network layer and transport layer headers (IP and TCP/UDP). However, sometimes we need to filter a message based on the information available in the message itself (at the application layer). As an example, assume that an organization wants to implement the following policies regarding its Web pages: Only those Internet users who have previously established business relations with the company can have access; access to other users must be blocked. In this case, a packet-filter firewall is not feasible because it cannot distinguish between different packets arriving at TCP port 80 (HTTP). Testing must be done at the application level (using URLs).

One solution is to install a proxy computer (sometimes called an application gateway), which stands between the customer (user client) computer and the corporation computer shown in Figure 32.24.

Figure 32.24 *Proxy firewall*

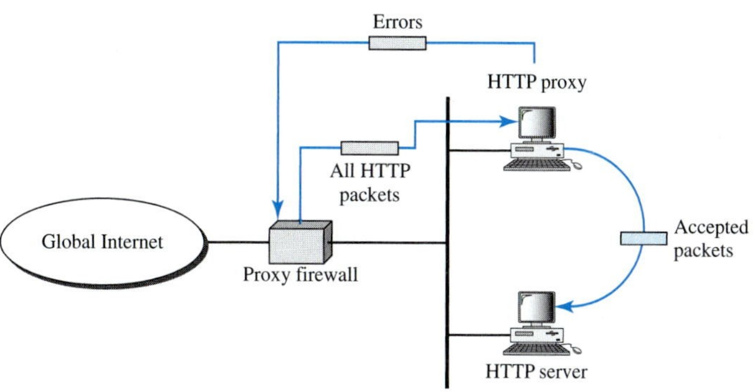

When the user client process sends a message, the **proxy firewall** runs a server process to receive the request. The server opens the packet at the application level and finds out if the request is legitimate. If it is, the server acts as a client process and sends the message to the real server in the corporation. If it is not, the message is dropped and an error message is sent to the external user. In this way, the requests of the external users are filtered based on the contents at the application layer. Figure 32.24 shows a proxy firewall implementation.

A proxy firewall filters at the application layer.

32.5 RECOMMENDED READING

For more details about subjects discussed in this chapter, we recommend the following books. The brackets, [. . .], refer to the reference list at the end of the text.

Books

IPSec is discussed in Chapter 7 of [Rhe03], Section 18.1 of [PHS03], and Chapters 17 and 18 of [KPS02]. A full discussion of IPSec can be found in [DH03]. SSL/TLS is discussed in Chapter 8 of [Rhe03], and Chapter 19 of [KPS02]. A full discusion of SSL and TLS can be found in [Res01] and [Tho00]. PGP is discussed in Chapter 9 of [Rhe03], and Chapter 22 of [KPS02]. Firewalls are discussed in Chapter 10 of [Rhe03] and Chapter 23 of [KPS02]. Firewalls are fully discussed in [CBR03]. Virtual private networks are fully discussed in [YS01] and [SWE99].

32.6 KEY TERMS

Alert Protocol
Authentication Header (AH) Protocol
certificate trust
ChangeCipherSpec Protocol
cipher suite
connection
Encapsulating Security Payload (ESP)
extranet
fingerprint
firewall
Handshake Protocol
hybrid network
Internet Key Exchange (IKE)
Internet Security Association and Key Management Protocol (ISAKMP)
intranet
introducer
introducer trust
IP Security (IPSec)
key legitimacy
key material
key ring
master secret
Oakley
packet-filter firewall
premaster secret
Pretty Good Privacy (PGP)
private network
proxy firewall
pseudorandom function (PRF)
Record Protocol
replay attack
Secure Socket Layer (SSL)
security association (SA)
security association database (SADB)
security parameter index (SPI)
session
SKEME
Transport Layer Security (TLS)
transport mode
tunnel mode
tunneling
virtual private network (VPN)
web of trust

32.7 SUMMARY

- IP Security (IPSec) is a collection of protocols designed by the IETF (Internet Engineering Task Force) to provide security for a packet at the network level.
- IPSec operates in the transport mode or the tunnel mode.
- In the transport mode, IPSec protects information delivered from the transport layer to the network layer. IPSec in the transport mode does not protect the IP header. The transport mode is normally used when we need host-to-host (end-to-end) protection of data.
- In the tunnel mode, IPSec protects the whole IP packet, including the original IP header.
- IPSec defines two protocols—Authentication Header (AH) Protocol and Encapsulating Security Payload (ESP) Protocol—to provide authentication or encryption or both for packets at the IP level.
- IPSec requires a logical relationship between two hosts called a security association (SA). IPSec uses a set of SAs called the security association database or SADB.
- The Internet Key Exchange (IKE) is the protocol designed to create security associations, both inbound and outbound. IKE creates SAs for IPSec.
- IKE is a complex protocol based on three other protocols: Oakley, SKEME, and ISAKMP.
- A private network is used inside an organization.
- An intranet is a private network that uses the Internet model. An extranet is an intranet that allows authorized access from outside users.
- The Internet authorities have reserved addresses for private networks.
- A virtual private network (VPN) provides privacy for LANs that must communicate through the global Internet.
- A transport layer security protocol provides end-to-end security services for applications that use the services of a reliable transport layer protocol such as TCP.
- Two protocols are dominant today for providing security at the transport layer: Secure Sockets Layer (SSL) and Transport Layer Security (TLS). The second is actually an IETF version of the first.
- SSL is designed to provide security and compression services to data generated from the application layer. Typically, SSL can receive application data from any application layer protocol, but the protocol is normally HTTP.
- SSL provides services such as fragmentation, compression, message integrity, confidentiality, and framing on data received from the application layer.
- The combination of key exchange, hash, and encryption algorithms defines a cipher suite for each SSL session. The name of each suite is descriptive of the combination.
- In e-mail, the cryptographic algorithms and secrets are sent with the message.
- One security protocol for the e-mail system is Pretty Good Privacy (PGP). PGP was invented by Phil Zimmerman to provide privacy, integrity, and authentication in e-mail.

- To exchange e-mail messages, a user needs a ring of public keys; one public key is needed for each e-mail correspondent.
- PGP has also specified a ring of private/public key pairs to allow a user to change her pair of keys from time to time. PGP also allows each user to have different user IDs (e-mail addresses) for different groups of people.
- PGP certification is different from X509. In X509, there is a single path from the fully trusted authority to any certificate. In PGP, there can be multiple paths from fully or partially trusted authorities.
- PGP uses the idea of certificate trust levels.
- When a user receives a certificate from an introducer, it stores the certificate under the name of the subject (certified entity). It assigns a level of trust to this certificate.

32.8 PRACTICE SET

Review Questions

1. Why does IPSec need a security association?
2. How does IPSec create a set of security parameters?
3. What are the two protocols defined by IPSec?
4. What does AH add to the IP packet?
5. What does ESP add to the IP packet?
6. Are both AH and ESP needed for IP security? Why or why not?
7. What are the two protocols discussed in this chapter that provide security at the transport layer?
8. What is IKE?
9. What is the difference between a session and a connection in SSL?
10. How does SSL create a set of security parameters?
11. What is the name of the protocol, discussed in this chapter, that provides security for e-mail?
12. How does PGP create a set of security parameters?
13. What is the purpose of the Handshake Protocol in SSL?
14. What is the purpose of the Record Protocol in SSL?
15. What is the purpose of a firewall?
16. What are the two types of firewalls?
17. What is a VPN and why is it needed?
18. How do LANs on a fully private internet communicate?

Exercises

19. Show the values of the AH fields in Figure 32.6. Assume there are 128 bits of authentication data.
20. Show the values of the ESP header and trailer fields in Figure 32.7.

21. Redraw Figure 32.6 if AH is used in tunnel mode.
22. Redraw Figure 32.7 if ESP is used in tunnel mode.
23. Draw a figure to show the position of AH in IPv6.
24. Draw a figure to show the position of ESP in IPv6.
25. Does the IPSec Protocol need the services of a KDC? Explain your answer.
26. Does the IPSec Protocol need the services of a CA? Explain your answer.
27. Does the SSL Protocol need the services of a KDC? Explain your answer.
28. Does the SSL Protocol need the services of a CA? Explain your answer.
29. Does the PGP Protocol need the services of a KDC? Explain your answer.
30. Does the PGP Protocol need the services of a CA? Explain your answer.
31. Are there any cipher suites in IPSec? Explain your answer.
32. Are there any cipher suites in PGP? Explain your answer.

APPENDIX A

Unicode

Computers use numbers. They store characters by assigning a number for each one. The original coding system was called ASCII (American Standard Code for Information Interchange) and had 128 numbers (0 to 127) each stored as a 7-bit number. ASCII could satisfactorily handle lowercase and uppercase letters, digits, punctuation characters, and some control characters. An attempt was made to extend the ASCII character set to 8 bits. The new code, which was called Extended ASCII, was never internationally standardized.

To overcome the difficulties inherent in ASCII and Extended ASCII, the Unicode Consortium (a group of multilingual software manufacturers) created a universal encoding system to provide a comprehensive character set called **Unicode.**

Unicode was originally a 2-byte character set. Unicode version 3, however, is a 4-byte code and is fully compatible with ASCII and Extended ASCII. The ASCII set, which is now called *Basic Latin,* is Unicode with the upper 25 bits set to zero. Extended ASCII, which is now called Latin-1, is Unicode with the 24 upper bits set to zero. Figure A.1 shows how the different systems are compatible.

Figure A.1 *Unicode compatibility*

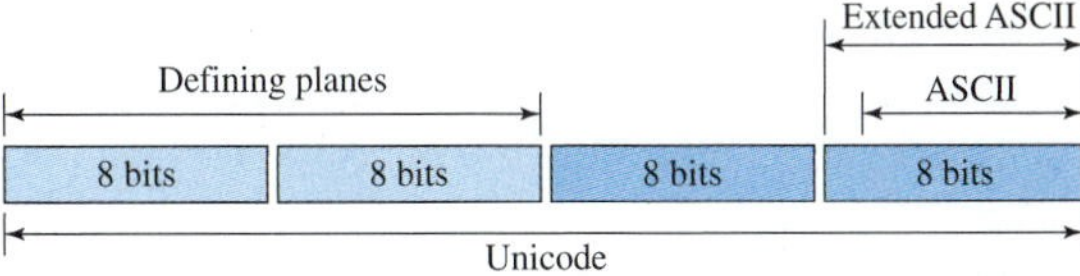

A.1 UNICODE

The prevalent code today is Unicode. Each character or symbol in this code is defined by a 32-bit number. The code can define up to 2^{32} (4,294,967,296) characters or symbols. The notation uses hexadecimal digits in the following format:

U-XXXXXXXX

Each X is a hexadecimal digit. Therefore, the numbering goes from U-00000000 to U-FFFFFFFF.

Planes

Unicode divides the available space codes into planes. The most significant 16 bits define the plane, which means we can have 65,535 planes. Each plane can define up to 65,536 character or symbols. Figure A.2 shows the structure of Unicode spaces and planes.

Figure A.2 *Unicode planes*

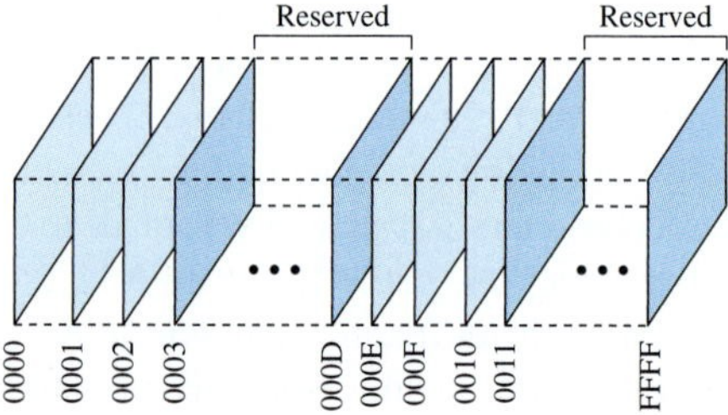

Plane 0000: Basic Multilingual Plane (BMP)
Plane 0001: Supplementary Multilingual Plane (SMP)
Plane 0002: Supplementary Ideographic Plane (SIP)
Plane 000E: Supplementary Special Plane (SSP)
Plane 000F: Private Use Plane (PUP)
Plane 0010: Private Use Plane (PUP)

Basic Multilingual Plane (BMP)

Plane 0000, the basic multilingual plane (BMP), is designed to be compatible with the previous 16-bit Unicode. The most significant 16 bits in this plane are all zeros. The codes are normally shown as U+XXXX with the understanding that XXXX defines only the least significant 16 bits. This plane mostly defines character sets in different languages with the exception of some codes used for control or other special characters. Table A.1 shows the main classification of codes in plane 0000.

Table A.1 *Unicode BMP*

Range	*Description*
A-Zone (Alphabetical Characters and Symbols)	
U+0000 to U+00FF	Basic Latin and Latin-1
U+0100 to U+01FF	Latin extended
U+0200 to U+02FF	IPA extension, and space modifier letters
U+0300 to U+03FF	Combining diacritical marks, Greek
U+0400 to U+04FF	Cyrillic
U+0500 to U+05FF	Armenian, Hebrew
U+0600 to U+06FF	Arabic

Table A.1 *Unicode BMP (continued)*

Range	*Description*
U+0700 to U+08FF	Reserved
U+0900 to U+09FF	Devanagari, Bengali
U+0A00 to U+0AFF	Gumukhi, Gujarati
U+0B00 to U+0BFF	Oriya, Tamil
U+0C00 to U+0CFF	Telugu, Kannda
U+0D00 to U+0DFF	Malayalam
U+0E00 to U+0EFF	Thai, Lao
U+0F00 to U+0FFF	Reserved
U+1000 to U+10FF	Georgian
U+1100 to U+11FF	Hangul Jamo
U+1200 to U+1DFF	Reserved
U+1E00 to U+1EFF	Latin extended additional
U+1F00 to U+1FFF	Greek extended
U+2000 to U+20FF	Punctuation, sub/superscripts, currency, marks
U+2100 to U+21FF	Letterlike symbols, number forms, arrows
U+2200 to U+22FF	Mathematical operations
U+2300 to U+23FF	Miscellaneous technical symbols
U+2400 to U+24FF	Control pictures, OCR, and enclosed alphanumeric
U+2500 to U+25FF	Box drawing, block drawing, and geometric shapes
U+2600 to U+26FF	Miscellaneous symbols
U+2700 to U+27FF	Dingbats and Braille patterns
U+2800 to U+2FFF	Reserved
U+3000 to U+30FF	CJK symbols and punctuation, hiragana, katakana
U+3100 to U+31FF	Bopomfo, hangul jambo, cjk miscellaneous
U+3200 to U+32FF	Enclosed CJK letters and months
U+3300 to U+33FF	CJK compatibility
U+3400 to U+4DFF	Hangul
I-Zone (Ideographic Characters)	
U+4E00 to U+9FFF	CJK unified ideographic
O-Zone (Open)	
U+A000 to U+DFFF	Reserved
R-Zone (Restricted Use)	
U+E000 to U+F8FF	Private use
U+F900 to U+FAFF	CJK compatibility ideographs
U+FB00 to U+FBFF	Arabic presentation form-A

Table A.1 *Unicode BMP (continued)*

Range	*Description*
U+FC00 to U+FDFF	Arabic presentation form-B
U+FE00 to U+FEFF	Half marks, small forms
U+FF00 to U+FFFF	Half-width and full-width forms

Supplementary Multilingual Plane (SMP)

Plane 0001, the supplementary multilingual plane (SMP), is designed to provide more codes for those multilingual characters that are not included in the BMP.

Supplementary Ideographic Plane (SIP)

Plane 0002, the supplementary ideographic plane (SIP), is designed to provide codes for ideographic symbols, symbols that primarily denote an idea (or meaning) in contrast to a sound (or pronunciation).

Supplementary Special Plane (SSP)

Plane 000E, the supplementary special plane (SSP), is used for special characters.

Private Use Planes (PUPs)

Planes 000F and 0010, private use planes (PUPs), are for private use.

A.2 ASCII

The American Standard Code for Information Interchange (ASCII) is a 7-bit code that was designed to provide code for 128 symbols, mostly in American English. Today, ASCII, or Basic Latin, is part of Unicode. It occupies the first 128 codes in Unicode (00000000 to 0000007F). Table A.2 contains the decimal, hexadecimal, and graphic codes (symbols) with an English interpretation, if appropriate. The codes in hexadecimal just define the two least significant digits in Unicode. To find the actual code, we prepend 000000 in hexadecimal to the code. The decimal code is just to show the integer value of each symbol when converted.

Table A.2 *ASCII Codes*

Decimal	*Hex*	*Symbol*	*Interpretation*
0	00	null	Null value
1	01	SOH	Start of heading
2	02	STX	Start of text
3	03	ETX	End of text
4	04	EOT	End of transmission

Table A.2 *ASCII Codes (continued)*

Decimal	*Hex*	*Symbol*	*Interpretation*
5	05	ENQ	Enquiry
6	06	ACK	Acknowledgment
7	07	BEL	Ring bell
8	08	BS	Backspace
9	09	HT	Horizontal tab
10	0A	LF	Line feed
11	0B	VT	Vertical tab
12	0C	FF	Form feed
13	0D	CR	Carriage return
14	0E	SO	Shift out
15	0F	SI	Shift in
16	10	DLE	Data link escape
17	11	DC1	Device control 1
18	12	DC2	Device control 2
19	13	DC3	Device control 3
20	14	DC4	Device control 4
21	15	NAK	Negative acknowledgment
22	16	SYN	Synchronous idle
23	17	ETB	End of transmission block
24	18	CAN	Cancel
25	19	EM	End of medium
26	1A	SUB	Substitute
27	1B	ESC	Escape
28	1C	FS	File separator
29	1D	GS	Group separator
30	1E	RS	Record separator
31	1F	US	Unit separator
32	20	SP	Space
33	21	!	
34	22	"	Double quote
35	23	#	
36	24	$	
37	25	%	
38	26	&	
39	27	'	Apostrophe

Table A.2 *ASCII Codes (continued)*

Decimal	*Hex*	*Symbol*	*Interpretation*
40	28	(	
41	29	)	
42	2A	*	
43	2B	+	
44	2C	,	Comma
45	2D	–	Minus
46	2E	.	
47	2F	/	
48	30	0	
49	31	1	
50	32	2	
51	33	3	
52	34	4	
53	35	5	
54	36	6	
55	37	7	
56	38	8	
57	39	9	
58	3A	:	Colon
59	3B	;	Semicolon
60	3C	<	
61	3D	=	
62	3E	>	
63	3F	?	
64	40	@	
65	41	A	
66	42	B	
67	43	C	
68	44	D	
69	45	E	
70	46	F	
71	47	G	
72	48	H	
73	49	I	
74	4A	J	

Table A.2 *ASCII Codes (continued)*

Decimal	*Hex*	*Symbol*	*Interpretation*
75	4B	K	
76	4C	L	
77	4D	M	
78	4E	N	
79	4F	O	
80	50	P	
81	51	Q	
82	52	R	
83	53	S	
84	54	T	
85	55	U	
86	56	V	
87	57	W	
88	58	X	
89	59	Y	
90	5A	Z	
91	5B	[	Open bracket
92	5C	\	Backslash
93	5D	]	Close bracket
94	5E	^	Caret
95	5F	_	Underscore
96	60	`	Grave accent
97	61	a	
98	62	b	
99	63	c	
100	64	d	
101	65	e	
102	66	f	
103	67	g	
104	68	h	
105	69	i	
106	6A	j	
107	6B	k	
108	6C	l	
109	6D	m	

Table A.2 *ASCII Codes (continued)*

Decimal	*Hex*	*Symbol*	*Interpretation*
110	6E	n	
111	6F	o	
112	70	p	
113	71	q	
114	72	r	
115	73	s	
116	74	t	
117	75	u	
118	76	v	
119	77	w	
120	78	x	
121	79	y	
122	7A	z	
123	7B	{	Open brace
124	7C	|	Bar
125	7D	}	Close brace
126	7E	~	Tilde
127	7F	DEL	Delete

Some Properties of ASCII

ASCII has some interesting properties that we briefly mention here.

1. The first code (0) is the null character, which means the lack of any character.
2. The first 32 codes, 0 to 31, are control characters.
3. The space character, which is a printable character, is at position 32.
4. The uppercase letters start from 65 (A). The lowercase letters start from 97. When compared, uppercase letters are numerically smaller than lowercase letters. This means that in a sorted list based on ASCII values, the uppercase letters appear before the lowercase letters.
5. The uppercase and lowercase letters differ by only one bit in the 7-bit code. For example, character A is 1000001 (0x41) and character a is 1100001(0x61). The difference is in bit 6, which is 0 in uppercase letters and 1 in lowercase letters. If we know the code for one case, we can easily find the code for the other by adding or subtracting 32 in decimal (0x20 in hexadecimal), or we can just flip the sixth bit.
6. The uppercase letters are not immediately followed by lowercase letters. There are some punctuation characters in between.
7. Digits (0 to 9) start from 48 (0x3). This means that if you want to change a numeric character to its face value as an integer, you need to subtract 48.

APPENDIX B

Numbering Systems

We use different numbering systems: base 10 (decimal), base 2 (binary), base 8 (octal), base 16 (hexadecimal), base 256, and so on. All the numbering systems examined here are positional, meaning that the position of a symbol in relation to other symbols determines its value. Each symbol in a number has a position. The position traditionally starts from 0 and goes to $n - 1$, where n is the number of symbols. For example, in Figure B.1, the decimal number 14,782 has five symbols in positions 0 to 4.

Figure B.1 *Positions and symbols in a number*

Decimal number: 14,782

1	4	7	8	2	Symbols
4	3	2	1	0	Positions

As we will see, the difference between different numbering systems is based on the *weight* assigned to each position.

B.1 BASE 10: DECIMAL

The base-10 or decimal system is the one most familiar to us in everyday life. All our terms for indicating countable quantities are based on it, and, in fact, when we speak of other numbering systems, we tend to refer to their quantities by their decimal equivalents. The term *decimal* is derived from the Latin stem *deci,* meaning 10. The decimal system uses 10 symbols to represent quantitative values: 0, 1, 2, 3, 4, 5, 6, 7, 8, and 9.

Decimal numbers use 10 symbols: 0, 1, 2, 3, 4, 5, 6, 7, 8, and 9.

Weights

In the decimal system, each weight equals 10 raised to the power of its position. The weight of the symbol at position 0 is 10^0 (1); the weight of the symbol at position 1 is 10^1 (10); and so on.

B.2 BASE 2: BINARY

The binary number system provides the basis for all computer operations. Computers work by turning electric current on and off. The binary system uses two symbols, 0 and 1, so it corresponds naturally to a two-state device, such as a switch, with 0 to represent the off state and 1 to represent the on state. The word *binary* derives from the Latin stem *bi,* meaning 2.

Binary numbers use two symbols: 0 and 1.

Weights

In the binary system, each weight equals 2 raised to the power of its position. The weight of the symbol at position 0 is 2^0 (1); the weight of the symbol at position 1 is 2^1 (2); and so on.

Conversion

Now let us see how we can convert binary to decimal and decimal to binary. Figure B.2 show the two processes.

Figure B.2 *Binary-to-decimal and decimal-to-binary conversion*

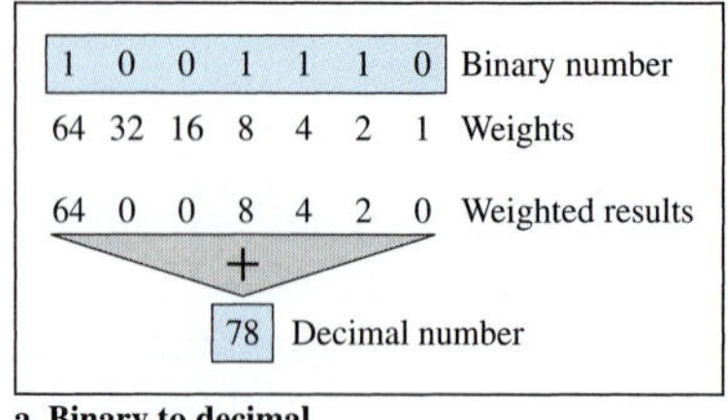

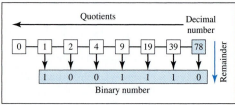

a. Binary to decimal

b. Decimal to binary

To convert a binary number to decimal, we use the weights. We multiply each symbol by its weight and add all the weighted results. Figure B.2 shows how we can change binary 1001110 to its decimal equivalent 78.

A simple division trick gives us a convenient way to convert a decimal number to its binary equivalent, as shown in Figure B.2. To convert a number from decimal to binary, divide the number by 2 and write down the remainder (1 or 0). That remainder is the least significant binary digit. Now, divide the quotient of that division by 2 and write down the new remainder in the second position. Repeat this process until the quotient becomes zero.

B.3 BASE 16: HEXADECIMAL

Another system used in this text is base 16. The term *hexadecimal* is derived from the Greek term *hexadec,* meaning 16. The hexadecimal number system is convenient for identifying a large binary number in a shorter form. The hexadecimal system uses 16 symbols: 0, 1, . . . , 9, A, B, C, D, E, and F. The hexadecimal system uses the same first 10 symbols as the decimal system, but instead of using 10, 11, 12, 13, 14, and 15, it uses A, B, C, D, E, and F. This prevents any confusion between two adjacent symbols.

Hexadecimal numbers use 16 symbols: 0, 1, 2, 3, 4, 5, 6, 7, 8, 9, A, B, C, D, E, and F.

Weights

In the hexadecimal system, each weight equals 16 raised to the power of its position. The weight of the symbol at position 0 is 16^0 (1); the weight of the symbol at position 1 is 16^1 (16); and so on.

Conversion

Now let us see how we can convert hexadecimal to decimal and decimal to hexadecimal. Figure B.3 show the two processes.

Figure B.3 *Hexadecimal-to-decimal and decimal-to-hexadecimal conversion*

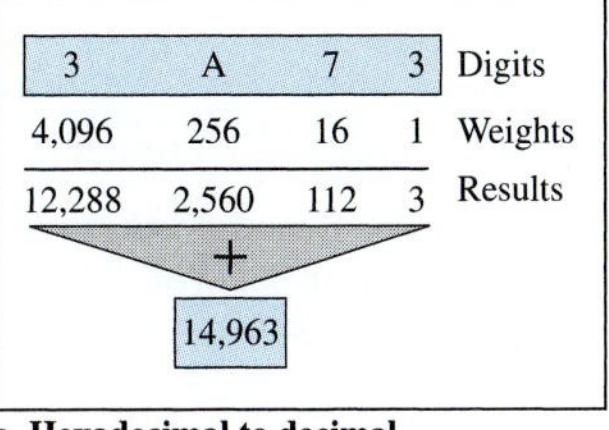

a. Hexadecimal to decimal

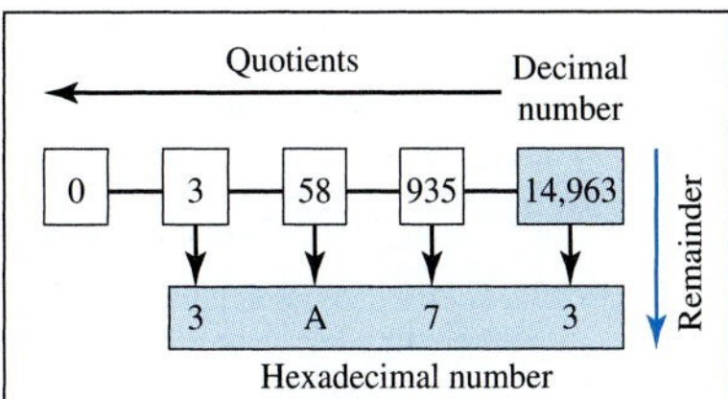

b. Decimal to hexadecimal

To convert a hexadecimal number to decimal, we use the weights. We multiply each symbol by its weight and add all the weighted results. Figure B.3 shows how hexadecimal 0x3A73 is transformed to its decimal equivalent 14,963.

We use the same trick we used for changing decimal to binary to transform a decimal to hexadecimal. The only difference is that we divide the number by 16 instead of 2. The figure also shows how 14,963 in decimal is converted to hexadecimal 0x3A73.

A Comparison

Table B.1 shows how systems represent the decimal numbers 0 through 15. As you can see, decimal 13 is equivalent to binary 1101, which is equivalent to hexadecimal D.

Table B.1 *Comparison of three systems*

Decimal	*Binary*	*Hexadecimal*	*Decimal*	*Binary*	*Hexadecimal*
0	0	0	8	1000	8
1	1	1	9	1001	9
2	10	2	10	1010	A
3	11	3	11	1011	B
4	100	4	12	1100	C
5	101	5	13	1101	D
6	110	6	14	1110	E
7	111	7	15	1111	F

B.4 BASE 256: IP ADDRESSES

One numbering system that is used in the Internet is base 256. IPv4 addresses use this base to represent an address in dotted decimal notation. When we define an IPv4 address as 131.32.7.8, we are using a base-256 number. In this base, we could have used 256 unique symbols, but remembering that many symbols and their values is burdensome. The designers of the IPv4 address decided to use decimal numbers 0 to 255 as symbols and to distinguish between the symbols, a *dot* is used. The dot is used to separate the symbols; it marks the boundary between the positions. For example, the IPv4 address 131.32.7.8 is made of the four symbols 8, 7, 32, and 131 at positions 0, 1, 2, and 3, respectively.

IPv4 addresses use the base-256 numbering system.
The symbols in IPv4 are decimal numbers between 0 and 255; the separator is a dot.

Weights

In base 256, each weight equals 256 raised to the power of its position. The weight of the symbol at position 0 is 256^0 (1); the weight of the symbol at position 1 is 256^1 (256); and so on.

Conversion

Now let us see how we can convert hexadecimal to decimal and decimal to hexadecimal. Figure B.4 show the two processes.

Figure B.4 *IPv4 address to decimal transformation*

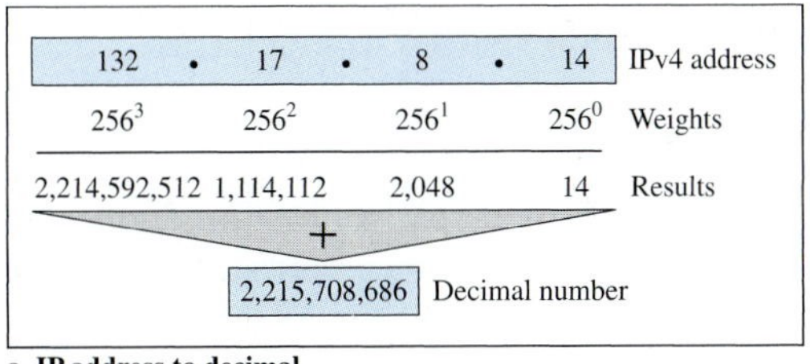

a. IP address to decimal

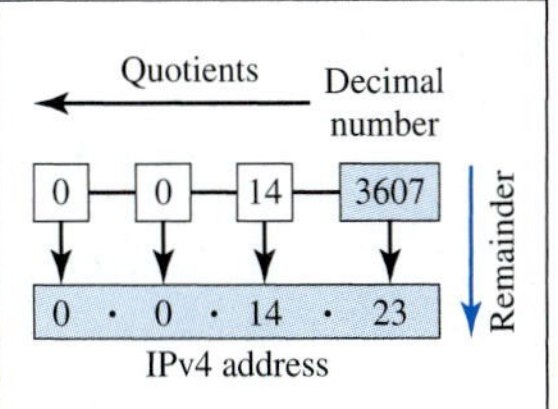

b. Decimal to IP address

To convert an IPv4 address to decimal, we use the weights. We multiply each symbol by its weight and add all the weighted results. The figure shows how the IPv4 address 131.32.7.8 is transformed to its decimal equivalent.

We use the same trick we used for changing decimal to binary to transform a decimal to an IPv4 address. The only difference is that we divide the number by 256 instead of 2. However, we need to remember that the IPv4 address has four positions. This means that when we are dealing with an IPv4 address, we must stop after we have found four values. Figure B.4 shows an example for an IPv4 address.

B.5 OTHER CONVERSIONS

There are other transformations such as base 2 to base 16 or base 16 to base 256. It is easy to use base 10 as the intermediate system. In other words, to change a number from binary to hexadecimal we first change the binary to decimal and then change the decimal to hexadecimal. We discuss some easy methods for common transformations.

Binary and Hexadecimal

There is a simple way to convert binary to hexadecimal and vice versa as shown in Figure B.5.

Figure B.5 *Transformation from binary to hexadecimal*

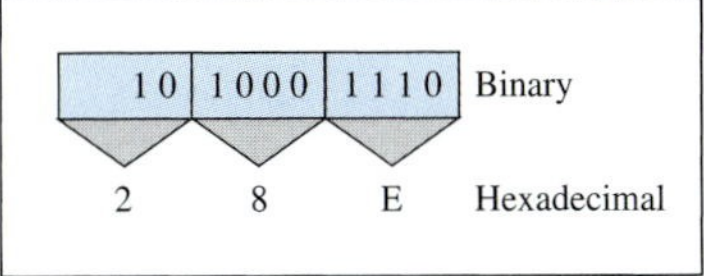

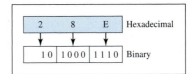

a. Binary to hexadecimal

b. Hexadecimal to binary

To change a number from binary to hexadecimal, we group the binary digits from the right by fours. Then we convert each 4-bit group to its hexadecimal equivalent, using Table B.1. In the figure, we convert binary 1010001110 to hexadecimal 0x28E. To change a hexadecimal number to binary, we convert each hexadecimal digit to its equivalent binary number, using Table B.1, and concatenate the results. In Figure B.5 we convert hexadecimal 0x28E to binary.

Base 256 and Binary

To convert a base 256 number to binary, we first need to convert the number in each position to an 8-bit binary group and then concatenate the groups. To convert from binary to base 256, we need to divide the binary number into groups of 8 bits, convert each group to decimal, and then insert separators (dots) between the decimal numbers.

APPENDIX C

Mathematical Review

In this appendix, we review some mathematical concepts that may help you to better understand the topics covered in the book. Perhaps the most important concept in data communications is signals and their representation. We start with a brief review of trigonometric functions, as discussed in a typical precalculus book. We then briefly discuss Fourier analysis, which provides a tool for the transformation between the time and frequency domains. We finally give a brief treatment of exponential and logarithmic functions.

C.1 TRIGONOMETRIC FUNCTIONS

Let us briefly discuss some characteristics of the trigonometric functions as used in the book.

Sine Wave

We can mathematically describe a sine wave as

$$s(t) = A\sin(2\pi ft) = A\sin\left(\frac{2\pi}{T}t\right)$$

where s is the instantaneous amplitude, A is the peak amplitude, f is the frequency, and T is the period (phase will be discussed later). Figure C.1 shows a sine wave.

Figure C.1 *A sine wave*

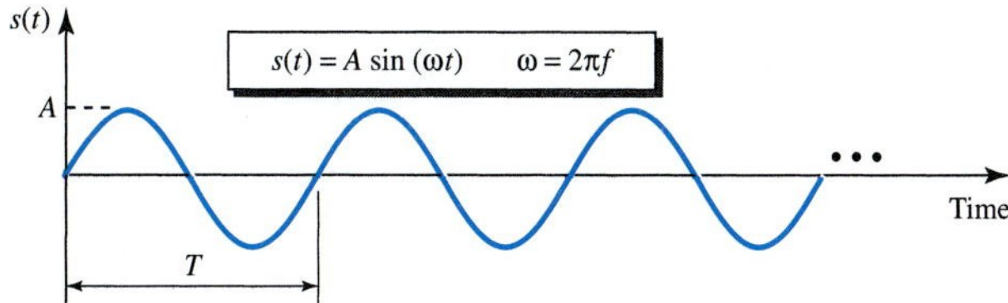

Note that the value of $2\pi f$ is called the radian frequency and written as ω (omega), which means that a sine function can written as $s(t) = A \sin (\omega t)$.

Example C.1

Find the peak value, frequency, and period of the following sine waves.

a. $s(t) = 5 \sin (10\pi t)$
b. $s(t) = \sin (10t)$

Solution

a. Peak amplitude: $A = 5$
   Frequency: $10\pi = 2\pi f$, so $f = 5$
   Period: $T = 1/f = 1/5$ s
b. Peak amplitude: $A = 1$
   Frequency: $10 = 2\pi f$, so $f = 10/(2\pi) = 1.60$
   Period: $T = 1/f = 1/1.60 = 0.628$ s

Example C.2

Show the mathematical representation of a sine wave with a peak amplitude of 2 and a frequency of 1000 Hz.

Solution

The mathematical representation is $s(t) = 2 \sin (2000\pi t)$.

Horizontal Shifting (Phase)

All the sine functions we discussed so far have an amplitude of value 0 at the origin. What if we shift the signal to the left or to the right? Figure C.2 shows two simple sine waves, one shifted to the right and one to the left.

Figure C.2 *Two horizontally shifted sine waves*

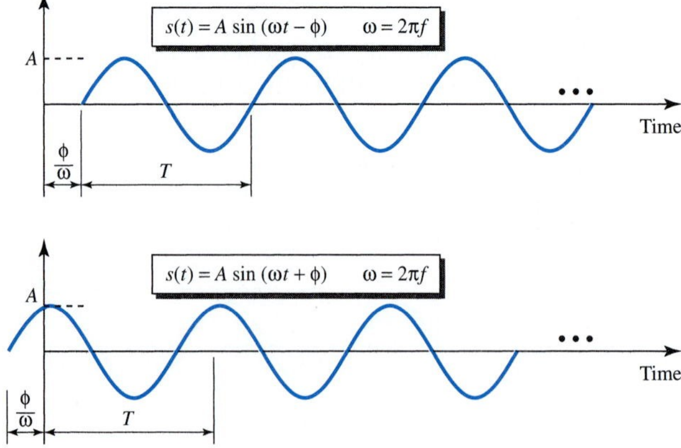

When a signal is shifted to the left or right, its first zero crossing will be at a point in time other than the origin. To show this, we need to add or subtract another constant to ωt, as shown in the figure.

Shifting a sine wave to the left or right is a positive or negative shift, respectively.

Vertical Shifting

When a sine wave is shifted vertically, a constant is added to the instantaneous amplitude of the signal. For example, if we shift a sine wave 2 units of amplitude upward, the signal becomes $s(t) = 2 + \sin(\omega t)$; if we shift it 2 units of amplitude downward, we have $s(t) = -2 + \sin(\omega t)$. Figure C.3 shows the idea.

Figure C.3 *Vertical shifting of sine waves*

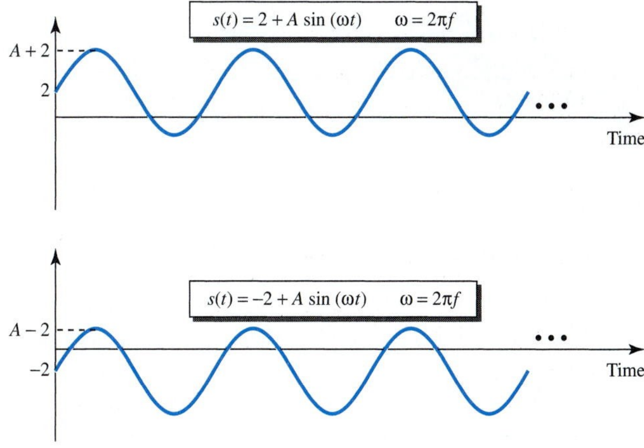

Cosine Wave

If we shift a sine wave $T/2$ to the left, we get what is called a cosine wave (cos).

$$A \sin(\omega t + \pi/2) = A \cos(\omega t)$$

Figure C.4 shows a cosine wave.

Figure C.4 *A cosine wave*

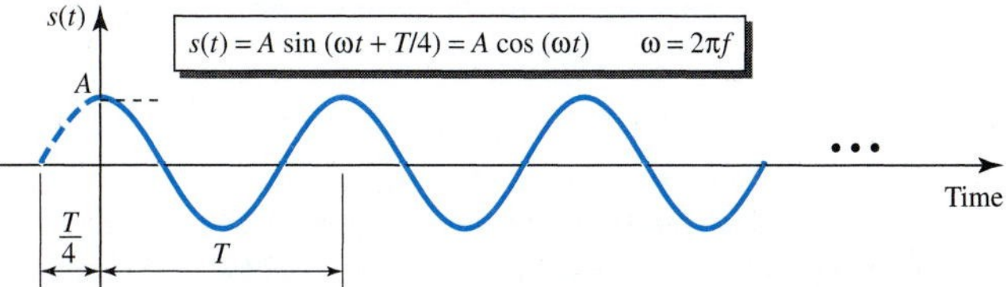

Other Trigonometric Functions

There are many trigonometric functions; two of the more common are tan(ω*t*) and cot(ω*t*). They are defined as tan(ω*t*) = sin(ω*t*)/cos(ω*t*) and cot(ω*t*) = cos(ω*t*)/sin(ω*t*). Note that tan and cot are the inverse of each other.

Trigonometric Identities

There are several identities between trigonometric functions that we sometimes need to know. Table C.1 gives these identities for reference. Other identities can be easily derived from these.

Table C.1 *Some trigonometric identities*

Name	*Formula*
Pythagorean	$\sin^2 x + \cos^2 x = 1$
Even/odd	$\sin(-x) = -\sin(x)$ $\quad \cos(-x) = \cos(x)$
Sum	$\sin(x + y) = \sin(x)\cos(y) + \cos(x)\sin(y)$ $\cos(x + y) = \cos(x)\cos(y) - \sin(x)\sin(y)$
Difference	$\sin(x - y) = \sin(x)\cos(y) - \cos(x)\sin(y)$ $\cos(x - y) = \cos(x)\cos(y) + \sin(x)\sin(y)$
Product to sum	$\sin(x)\sin(y) = 1/2\,[\cos(x - y) - \cos(x + y)]$ $\cos(x)\cos(y) = 1/2\,[\cos(x - y) + \cos(x + y)]$ $\sin(x)\cos(y) = 1/2\,[\sin(x + y) + \sin(x - y)]$ $\cos(x)\sin(y) = 1/2\,[\sin(x + y) - \sin(x - y)]$

C.2 FOURIER ANALYSIS

Fourier analysis is a tool that changes a time-domain signal to a frequency-domain signal and vice versa.

Fourier Series

Fourier proved that a composite periodic signal with period T (frequency f) can be decomposed into a series of sine and cosine functions in which each function is an integral harmonic of the fundamental frequency f of the composite signal. The result is called the **Fourier series.** In other words, we can write a composite signal as shown in Figure C.5. Using the series, we can decompose any periodic signal into its harmonics. Note that A_0 is the average value of the signal over a period, A_n is the coefficient of the nth cosine component, and B_n is the coefficient of the nth sine component.

Example C.3

Let us show the components of a square wave signal as seen in Figure C.6. The figure also shows the time domain and the frequency domain. According to the figure, such a square wave signal has only A_n coefficients. Note also that the value of $A_0 = 0$ because the average value of the signal is 0; it is oscillating above and below the time axis. The frequency domain of the signal is discrete;

Figure C.5 *Fourier series and coefficients of terms*

Fourier series

$$s(t) = A_0 + \sum_{n=1}^{\infty} A_n \sin(2\pi nft) + \sum_{n=1}^{\infty} B_n \cos(2\pi nft)$$

$$A_0 = \frac{1}{T}\int_0^T s(t)\,dt \qquad A_n = \frac{2}{T}\int_0^T s(t)\cos(2\pi nft)\,dt$$

$$B_n = \frac{2}{T}\int_0^T s(t)\sin(2\pi nft)\,dt$$

Coefficients

Fourier series

Time domain: periodic **Frequency domain: discrete**

only odd harmonics are present and the amplitudes are alternatively positive and negative. A very important point is that the amplitude of the harmonics approaches zero as we move toward infinity. Something which is not shown in the figure is the phase. However, we know that all components are cosine waves, which means that each has a phase of 90°.

Figure C.6 *Finding the Fourier series of a periodic square function*

Time domain

$$A_0 = 0 \qquad A_n = \begin{cases} \dfrac{4A}{n\pi} & \text{for } n = 1, 5, 9, \ldots \\ \dfrac{-4A}{n\pi} & \text{for } n = 3, 7, 11, \ldots \end{cases} \qquad B_n = 0$$

$$s(t) = \frac{4A}{\pi}\cos(2\pi ft) - \frac{4A}{3\pi}\cos(2\pi 3ft) + \frac{4A}{5\pi}\cos(2\pi 5ft) - \frac{4A}{7\pi}\cos(2\pi 7ft) + \cdots$$

Frequency domain

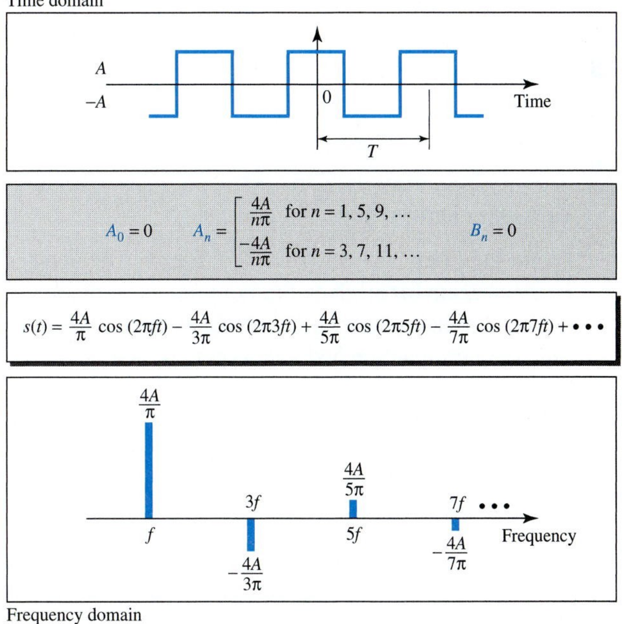

Example C.4

Now let us show the components of a sawtooth signal as seen in Figure C.7. This time, we have only B_n components (sine waves). The frequency spectrum, however, is denser; we have all harmonics (f, $2f$, $3f$, . . .). A point which is not clear from the diagram is the phase. All components are sine waves, which means each component has a phase of 0°.

Figure C.7 *Finding the Fourier series for a sawtooth signal*

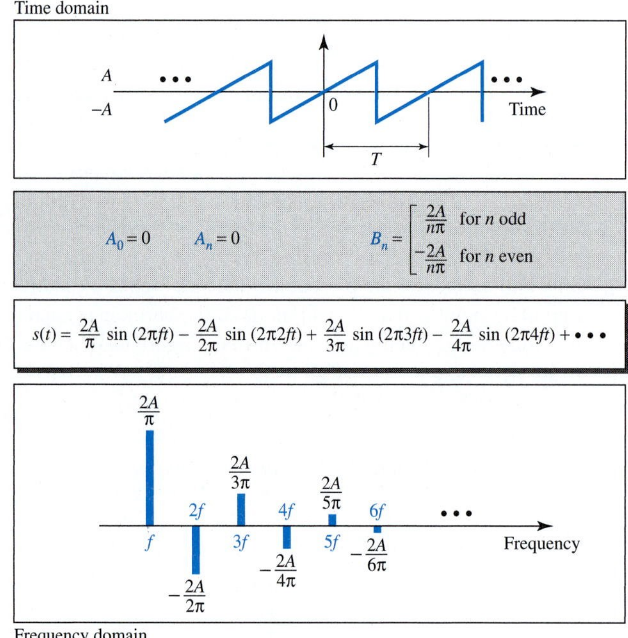

Time domain

$$A_0 = 0 \qquad A_n = 0 \qquad B_n = \begin{cases} \dfrac{2A}{n\pi} & \text{for } n \text{ odd} \\ -\dfrac{2A}{n\pi} & \text{for } n \text{ even} \end{cases}$$

$$s(t) = \frac{2A}{\pi}\sin(2\pi ft) - \frac{2A}{2\pi}\sin(2\pi 2ft) + \frac{2A}{3\pi}\sin(2\pi 3ft) - \frac{2A}{4\pi}\sin(2\pi 4ft) + \cdots$$

Frequency domain

Fourier Transform

While the Fourier series gives the discrete frequency domain of a periodic signal, the **Fourier transform** gives the continuous frequency domain of a nonperiodic signal. Figure C.8 shows how we can create a continuous frequency domain from a nonperiodic time-domain function and vice versa.

Figure C.8 *Fourier transform and inverse Fourier transform*

$$S(f) = \int_{-\infty}^{\infty} s(t)e^{-j2\pi ft}\,dt$$

Fourier transform

$$s(t) = \int_{-\infty}^{\infty} S(f)e^{j2\pi ft}\,dt$$

Inverse Fourier transform

Fourier transform

Time domain: nonperiodic **Frequency domain: continuous**

Example C.5

Figure C.9 shows the time and frequency domains of one single square pulse. The time domain is between $-\tau/2$ and $\tau/2$; the frequency domain is a continuous function that stretches from negative infinity to positive infinity. Unlike the previous examples, the frequency domain is continuous; all frequencies are there, not just the integral ones.

Figure C.9 *Finding the Fourier transform of a square pulse*

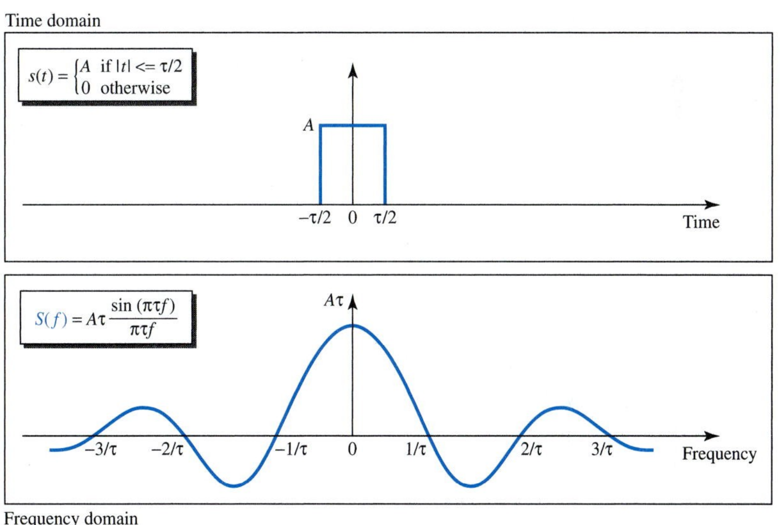

Time-Limited and Band-Limited Signals

Two very interesting concepts related to the Fourier transform are the time-limited and band-limited signals. A **time-limited** signal is a signal for which the amplitude of $s(t)$ is nonzero only during a period of time; the amplitude is zero everywhere else. A **band-limited** signal, on the other hand, is the signal for which the amplitude of $S(f)$ is nonzero only for a range of frequencies; the amplitude is zero everywhere else. A band-limited signal plays a very important role in the sampling theorem and Nyquist frequency because the corresponding time domain can be represented as a series of samples.

Time-limited signal: $s(t) = 0$ **for** $|t| \leq T$

Band-limited signal: $S(f) = 0$ **for** $|f| \leq B$

C.3 EXPONENT AND LOGARITHM

In solving networking problems, we often need to know how to handle exponential and logarithmic functions. This section briefly reviews these two concepts.

Exponential Function

The exponential function with **base** a is defined as

$$y = a^x$$

If x is an integer (integral value), we can easily calculate the value of y by multiplying the value of a by itself x times.

Example C.6

Calculate the value of the following exponential functions.

a. $y = 3^2$
b. $y = 5.2^6$

Solution

a. $y = 3 \times 3 = 9$
b. $y = 5.2 \times 5.2 \times 5.2 \times 5.2 \times 5.2 \times 5.2 = 19{,}770.609664$

If x is not integer, we need to use a calculator.

Example C.7

Calculate the value of the following exponential functions.

a. $y = 3^{2.2}$
b. $y = 5.2^{6.3}$

Solution

a. $y = 11.212$ (approximately)
b. $y = 32{,}424.60$ (approximately)

Natural Base

One very common base used in science and mathematics is the **natural base *e***, which has the value 2.71828183. . . . Most calculators show this function as e^x, which can be calculated easily by entering only the value of the exponent.

Example C.8

Calculate the value of the following exponential functions.

a. $y = e^4$
b. $y = e^{6.3}$

Solution

a. $y = 54.56$ (approximately)
b. $y = 544.57$ (approximately)

Properties of the Exponential Function

Exponential functions have several properties; some are useful to us in this text:

First:	$y = a^0 = 1$
Second:	$y = a^1 = a$
Third:	$y = a^{-x} = \frac{1}{a^x}$

Example C.9

The third property is useful to us because we can calculate the value of an exponential function with a negative value. We first calculate the positive value and we then invert the result.

a. $y = e^{-4}$

b. $y = e^{-6.3}$

Solution

a. $y = 1/54.56 = 0.0183$

b. $y = 1/544.57 = 0.00183$

Logarithmic Function

A logarithmic function is the inverse of an exponential function, as shown below. Just as in the exponential function, *a* is called the base of the logarithmic function:

$$y = a^x \longleftrightarrow x = \log_a y$$

In other words, if *x* is given, we can calculate *y* by using the exponential function; if *y* is given, we can calculate *x* by using the logarithmic function.

Exponential and logarithmic functions are the inverse of each other.

Example C.10

Calculate the value of the following logarithmic functions.

a. $x = \log_3 9$

b. $x = \log_2 16$

Solution

We have not yet shown how to calculate the log function in different bases, but we can solve this problem intuitively.

a. Because $3^2 = 9$, we can say that $\log_3 9 = 2$, using the fact that the two functions are the inverse of each other.

b. Because $2^4 = 16$, we can say that $\log_2 16 = 4$ by using the previous fact.

Two Common Bases

The two common bases for logarithmic functions, those that can be handled by a calculator, are base *e* and base 10. The logarithm in base *e* is normally shown as ln (natural logarithm); the logarithm in base 10 is normally shown as log (omitting the base).

Example C.11

Calculate the value of the following logarithmic functions.

a. $x = \log 233$

b. $x = \ln 45$

Solution

For these two bases we can use a calculator.

a. $x = \log 233 = 2.367$

b. $x = \ln 45 = 3.81$

Base Transformation

We often need to find the value of a logarithmic function in a base other than e or 10. If the available calculator cannot give the result in our desired base, we can use a very fundamental property of the logarithm, base transformation, as shown:

$$\log_a y = \frac{\log_b y}{\log_b a}$$

Note that the right-hand side is two log functions with base b, which is different from the base a at the left-hand side. This means that we can choose a base that is available in our calculator (base b) and find the log of a base that is not available (base a).

Example C.12

Calculate the value of the following logarithmic functions.

a. $x = \log_3 810$

b. $x = \log_5 600$

Solution

These two bases, 3 and 5, are not available on a calculator, but we can use base 10 which is available.

a. $x = \log_3 810 = \dfrac{\log_{10} 810}{\log_{10} 3} = \dfrac{2.908}{0.477} = 6.095$

b. $x = \log_5 600 = \dfrac{\log_{10} 600}{\log_{10} 5} = \dfrac{2.778}{0.699} = 3.975$

Properties of Logarithmic Functions

Like an exponential function, a logarithmic function has some properties that are useful in simplifying the calculation of a log function.

First:	$\log_a 1 = 0$	**Fourth:**	$\log_a (x \times y) = \log_a x + \log_a y$
Second:	$\log_a a = 1$	**Fifth:**	$\log_a \frac{x}{y} = \log_a x - \log_a y$
Third:	$\log_a \frac{1}{x} = -\log_a x$	**Sixth:**	$\log_a x^y = y \times \log_a x$

Example C.13

Calculate the value of the following logarithmic functions.

a. $x = \log_3 1$
b. $x = \log_3 3$
c. $x = \log_{10} (1/10)$
d. $\log_a (x \times y)$ if we know that $\log_a x = 2$ and $\log_a y = 3$
e. $\log_2 (1024)$ without using a calculator

Solution

We use the property of log functions to solve the problems.

a. $x = \log_3 1 = 0$
b. $x = \log_3 3 = 1$
c. $x = \log_{10} (1/10) = \log_{10} 10^{-1} = -\log_{10} 10 = -1$
d. $\log_a (x \times y) = \log_a x + \log_a y = 2 + 3 = 5$
e. $\log_2 (1024) = \log_2 (2^{10}) = 10 \log_2 2 = 10 \times 1 = 10$

APPENDIX D

8B/6T Code

This appendix is a tabulation of 8B/6T code pairs. The 8-bit data are shown in hexadecimal format. The 6T code is shown as + (positive signal), – (negative signal), and 0 (lack of signal) notation.

Table D.1 *8B/6T code*

Data	*Code*	*Data*	*Code*	*Data*	*Code*	*Data*	*Code*
00	**–+00–+**	20	**–++–00**	40	**–00+0+**	60	**0++0–0**
01	**0–+–+0**	21	**+00+––**	41	**0–00++**	61	**+0+–00**
02	**0–+0–+**	22	**–+0–++**	42	**0–0+0+**	62	**+0+0–0**
03	**0–++0–**	23	**+–0–++**	43	**0–0++0**	63	**+0+00–**
04	**–+0+0–**	24	**+–0+00**	44	**–00++0**	64	**0++00–**
05	**+0––+0**	25	**–+0+00**	45	**00–0++**	65	**++0–00**
06	**+0–0–+**	26	**+00–00**	46	**00–+0+**	66	**++00–0**
07	**+0–+0–**	27	**–+++––**	47	**00–++0**	67	**++000–**
08	**–+00+–**	28	**0++–0–**	48	**00+000**	68	**0++–+–**
09	**0–++–0**	29	**+0+0––**	49	**++–000**	69	**+0++––**
0A	**0–+0+–**	2A	**+0+–0–**	4A	**+–+000**	6A	**+0+–+–**
0B	**0–+–0+**	2B	**+0+––0**	4B	**–++000**	6B	**+0+––+**
0C	**–+0–0+**	2C	**0++––0**	4C	**0+–000**	6C	**0++––+**
0D	**+0–+–0**	2D	**++00––**	4D	**+0–000**	6D	**++0+––**
0E	**+0–0+–**	2E	**++0–0–**	4E	**0–+000**	6E	**++0–+–**
0F	**+0––0+**	2F	**++0––0**	4F	**–0+000**	6F	**++0––+**
10	**0––+0+**	30	**+–00–+**	50	**+––+0+**	70	**000++–**
11	**–0–0++**	31	**0+––+0**	51	**–+–0++**	71	**000+–+**
12	**–0–+0+**	32	**0+–0–+**	52	**–+–+0+**	72	**000–++**
13	**–0–++0**	33	**0+–+0–**	53	**–+–++0**	73	**000+00**

Table D.1 *8B/6T code (continued)*

Data	Code	Data	Code	Data	Code	Data	Code
14	0−−++0	34	+−0+0−	54	+−−++0	74	000+0−
15	−−00++	35	−0+−+0	55	−−+0++	75	000+−0
16	−−0+0+	36	−0+0−+	56	−−++0+	76	000−0+
17	−−0++0	37	−0++0−	57	−−+++0	77	000−+0
18	−+0−+0	38	+−00+−	58	−−0+++	78	+++−−0
19	+−0−+0	39	0+−+−0	59	−0−+++	79	+++−0−
1A	−++−+0	3A	0+−0+−	5A	0−−+++	7A	+++0− −
1B	+00−+0	3B	0+−−0+	5B	0−−0++	7B	0++0− −
1C	+00+−0	3C	+−0−0+	5C	+−−0++	7C	−00−++
1D	−+++−0	3D	−0++−0	5D	−000++	7D	−00+00
1E	+−0+−0	3E	−0+0+−	5E	0+++−−	7E	+− − −++
1F	−+0+−0	3F	−0+−0+	5F	0++−00	7F	+− −+00
80	−00+−+	A0	−++0−0	C0	−+0+−+	E0	−++0-+
81	0−0−++	A1	+−+−00	C1	0−+−++	E1	+−++0
82	0−0+−+	A2	+−+0−0	C2	0−++−+	E2	+−+0−+
83	0−0++−	A3	+−+00−	C3	0−+++−	E3	+−++0−
84	−00++−	A4	−++00−	C4	−+0++−	E4	−+++0−
85	00−−++	A5	++−−00	C5	+0−−++	E5	++−−+0
86	00−+−+	A6	++−0−0	C6	+0−+−+	E6	++−0−+
87	00−++−	A7	++−00−	C7	+0−++−	E7	++−+0−
88	−000+0	A8	−++−+−	C8	−+00+0	E8	−++0+−
89	0−0+00	A9	+−++−−	C9	0−++00	E9	+−++−0
8A	0−00+0	AA	+−+−+−	CA	0−+0+0	EA	+−+0+−
8B	0−000+	AB	+−+−−+	CB	0−+00+	EB	+−+−0+
8C	−0000+	AC	−++−−+	CC	−+000+	EC	−++−0+
8D	00−+00	AD	++−+−−	CD	+0−+00	ED	++−+−0
8E	00−0+0	AE	++−−+−	CE	+0−0+0	EE	++−0+−
8F	00−00+	AF	++−−−+	CF	+0−00+	EF	++−−0+
90	+−−+−+	B0	+000−0	D0	+−0+−+	F0	+000−+
91	−+−−++	B1	0+0−00	D1	0+−−++	F1	0+0−+0
92	−+−+−+	B2	0+00−0	D2	0+−+−+	F2	0+00−+
93	−+−++−	B3	0+000−	D3	0+−++−	F3	0+0+0−
94	+−−++−	B4	+0000−	D4	+−0++−	F4	+00+0−

Table D.1 *8B/6T code (continued)*

Data	*Code*	*Data*	*Code*	*Data*	*Code*	*Data*	*Code*
95	−−+−++	B5	00+−00	D5	−0+−++	F5	00+−+0
96	−−++−+	B6	00+0−0	D6	−0++−+	F6	00+0−+
97	−−+++−	B7	00+00−	D7	−0+++−	F7	00++0−
98	+−−0+0	B8	+00−+−	D8	+−00+0	F8	+000+−
99	−+−+00	B9	0+0+−−	D9	0+−+00	F9	0+0+−0
9A	−+−0+0	BA	0+0−+−	DA	0+−0+0	FA	0+00+−
9B	−+−00+	BB	0+0−−+	DB	0+−00+	FB	0+0−0+
9C	+−−00+	BC	+00−−+	DC	+−000+	FC	+00−0+
9D	−−++00	BD	00++−−	DD	−0++00	FD	00++−0
9E	−−+0+0	BE	00+−+−	DE	−0+0+0	FE	00+0+−
9F	−−+00+	BF	00+−−+	DF	−0+00+	FF	00+−0+

APPENDIX E

Telephone History

In Chapter 9, we discussed telephone networks. In this appendix, we briefly review the history of telephone networks. The history in the United States can be divided into three eras: prior to 1984, between 1984 and 1996, and after 1996.

Before 1984

Before 1984, almost all local and long-distance services were provided by the AT&T Bell System. In 1970, the U.S. government, believing that the Bell System was monopolizing the telephone service industry, sued the company. The verdict was in favor of the government and resulted in a document called the Modified Final Judgment (MFJ). Beginning on January 1, 1984, AT&T was broken into AT&T Long Lines, 23 Bell Operating Companies (BOCs), and others. The 23 BOCs were grouped to make several Regional Bell Operating Companies (RBOSs). This landmark event, the AT&T divestiture of 1984, was beneficial to customers of telephone services. Telephone rates were lowered.

Between 1984 and 1996

The divestiture divided the country into more than 200 LATAs; some companies were allowed to provide services inside a LATA (LECs), and others were allowed to provide services between LATAs (IXCs). Competition, particularly between long-distance carriers, increased as new companies were formed. However, no LEC could provide long-distance services, and no IXCs could provide local services.

After 1996

Another major change in telecommunications occurred in 1996. The Telecommunications Act of 1996 combined the different services provided by different companies under the umbrella of telecommunication services; this included local services, long-distance voice and data services, video services, and so on. In addition, the act allowed any company to provide any of these services at the local and long-distance levels. In other words, a common carrier company provides services both inside the LATA and between the LATAs. However, to prevent the recabling of residents, the carriers that were given intra-LATA services (ILECs) continued to provide the main services; the new competitors (CLECs) provided other services.

APPENDIX F

Contact Addresses

The following is a list of contact addresses for various organizations mentioned in the text.

- **ATM Forum**
  Presidio of San Francisco
  P.O. Box 29920 (mail)
  572B Ruger Street (surface)
  San Francisco, CA 94129-0920
  Telephone: 415 561-6275
  E-mail: info@atmforum.com
  www.atmforum.com
- **Federal Communications Commission (FCC)**
  445 12th Street S.W.
  Washington, DC 20554
  Telephone: 1-888-225-5322
  E-mail: fccinfo@fcc.gov
  www.fcc.gov
- **Institute of Electrical and Electronics Engineers (IEEE)**
  Operations Center
  445 Hoes Lane
  Piscataway, NJ 08854-1331
  Telephone: 732 981-0060
  www.ieee.org

- **International Organization for Standardization (ISO)**
  1, rue de Varembe
  Caisse Postale 56
  CH-1211 Geneve 20
  Switzerland
  Telephone: 41 22 749 0111
  E-mail: central@iso.ch
  www.iso.org
- **International Telecommunication Union (ITU)**
  Place des Nations
  CH-1211 Geneva 20
  Switzerland
  Telephone: 41 22 730 5852
  E-mail: tsbmail@itu.int
  www.itu.int/home
- **Internet Architecture Board (IAB)**
  E-mail: IAB@isi.edu
  www.iab.org
- **Internet Corporation for Assigned Names and Numbers (ICANN)**
  4676 Admiralty Way, Suite 330
  Marina del Rey, CA 90292-6601
  Telephone: 310 823-9358
  E-mail: icann@icann.org
  www.icann.org
- **Internet Engineering Steering Group (IESG)**
  E-mail: iesg@ietf.org
  www.ietf.org/iesg.html
- **Internet Engineering Task Force (IETF)**
  E-mail: ietf-infor@ietf.org
  www.ietf.org
- **Internet Research Task Force (IRTF)**
  E-mail: irtf-chair@ietf.org
  www.irtf.org
- **Internet Society (ISOC)**
  1775 Weihle Avenue, Suite 102
  Reston, VA 20190-5108
  Telephone: 703.326-9880
  E-mail: info@isoc.org
  www.isoc.org

APPENDIX G

RFCs

In Table G.1, we list alphabetically by protocol the RFCs that are directly related to the material in this text. For more information go to the following site: http://www.rfc-editor.org.

Table G.1 *RFCs for each protocol*

Protocol	*RFC*
ARP and RARP	826, 903, 925, 1027, 1293, 1329, 1433, 1868, 1931, 2390
BGP	1092, 1105, 1163, 1265, 1266, 1267, 1364, 1392, 1403, 1565, 1654, 1655, 1665, 1771, 1772, 1745, 1774, 2283
BOOTP and DHCP	951, 1048, 1084, 1395, 1497, 1531, 1532, 1533, 1534, 1541, 1542, 2131, 2132
CIDR	1322, 1478, 1479, 1517, 1817
DHCP	See BOOTP and DHCP
DNS	799, 811, 819, 830, 881, 882, 883, 897, 920, 921, 1034, 1035, 1386, 1480, 1535, 1536, 1537, 1591, 1637, 1664, 1706, 1712, 1713, 1982, 2065, 2137, 2317, 2535, 2671
FTP	114, 133, 141, 163, 171, 172, 238, 242, 250, 256, 264, 269, 281, 291, 354, 385, 412, 414, 418, 430, 438, 448, 463, 468, 478, 486, 505, 506, 542, 553, 624, 630, 640, 691, 765, 913, 959, 1635, 1785, 2228, 2577
HTML	1866
HTTP	2068, 2109
ICMP	777, 792, 1016, 1018, 1256, 1788, 2521
IGMP	966, 988, 1054, 1112, 1301, 1458, 1469, 1768, 2236, 2357, 2365, 2502, 2588
IMAP	See SMTP, MIME, POP, IMAP
IP	760, 781, 791, 815, 1025, 1063, 1071, 1141, 1190, 1191, 1624, 2113

Table G.1 *RFCs for each protocol (continued)*

Protocol	*RFC*
IPv6	1365, 1550, 1678, 1680, 1682, 1683, 1686, 1688, 1726, 1752, 1826, 1883, 1884, 1886, 1887, 1955, 2080, 2373, 2452, 2463, 2465, 2466, 2472, 2492, 2545, 2590
MIB	See SNMP, MIB, SMI
MIME	See SMTP, MIME, POP, IMAP
Multicast Routing	1584, 1585, 2117, 2362
NAT	1361, 2663, 2694
OSPF	1131, 1245, 1246, 1247, 1370, 1583, 1584, 1585, 1586, 1587, 2178, 2328, 2329, 2370
POP	See SMTP, MIME, POP, IMAP
RARP	See ARP and RARP
RIP	1131, 1245, 1246, 1247, 1370, 1583, 1584, 1585, 1586, 1587, 1722, 1723, 2082, 2453
SCTP	2960, 3257, 3284, 3285, 3286, 3309, 3436, 3554, 3708, 3758
SMI	See SNMP, MIB, SMI
SMTP, MIME, POP, IMAP	196, 221, 224, 278, 524, 539, 753, 772, 780, 806, 821, 934, 974, 1047, 1081, 1082, 1225, 1460, 1496, 1426, 1427, 1652, 1653, 1711, 1725, 1734, 1740, 1741, 1767, 1869, 1870, 2045, 2046, 2047, 2048, 2177, 2180, 2192, 2193, 2221, 2342, 2359, 2449, 2683, 2503
SNMP, MIB, SMI	1065, 1067, 1098, 1155, 1157, 1212, 1213, 1229, 1231, 1243, 1284, 1351, 1352, 1354, 1389, 1398, 1414, 1441, 1442, 1443, 1444, 1445, 1446, 1447, 1448, 1449, 1450, 1451, 1452, 1461, 1472, 1474, 1537, 1623, 1643, 1650, 1657, 1665, 1666, 1696, 1697, 1724, 1742, 1743, 1748, 1749
TCP	675, 700, 721, 761, 793, 879, 896, 1078, 1106, 1110, 1144, 1145, 1146, 1263, 1323, 1337, 1379, 1644, 1693, 1901, 1905, 2001, 2018, 2488, 2580
TELNET	137, 340, 393, 426, 435, 452, 466, 495, 513, 529, 562, 595, 596, 599, 669, 679, 701, 702, 703, 728, 764, 782, 818, 854, 855, 1184, 1205, 2355
TFTP	1350, 1782, 1783, 1784
UDP	768
VPN	2547, 2637, 2685
WWW	1614, 1630, 1737, 1738

APPENDIX H

UDP and TCP Ports

Table H.1 lists the common well-known ports ordered by port number.

Table H.1 *Ports by port number*

Port Number	*UDP/TCP*	*Protocol*
7	TCP	ECHO
13	UDP/TCP	DAYTIME
19	UDP/TCP	CHARACTER GENERATOR
20	TCP	FTP-DATA
21	TCP	FTP-CONTROL
23	TCP	TELNET
25	TCP	SMTP
37	UDP/TCP	TIME
67	UDP	BOOTP-SERVER
68	UDP	BOOTP-CLIENT
69	UDP	TFTP
70	TCP	GOPHER
79	TCP	FINGER
80	TCP	HTTP
109	TCP	POP-2
110	TCP	POP-3
111	UDP/TCP	RPC
161	UDP	SNMP
162	UDP	SNMP-TRAP
179	TCP	BGP
520	UDP	RIP

Table H.2 lists the ports, ordered alphabetically by protocol.

Table H.2 *Port numbers by protocol*

Protocol	*UDP/TCP*	*Port Number*
BGP	TCP	179
BOOTP-SERVER	UDP	67
BOOTP-CLIENT	UDP	68
CHARACTER GENERATOR	UDP/TCP	19
DAYTIME	UDP/TCP	13
ECHO	TCP	7
FINGER	TCP	79
FTP-CONTROL	TCP	21
FTP-DATA	TCP	20
GOPHER	TCP	70
HTTP	TCP	80
POP-2	TCP	109
POP-3	TCP	110
RIP	UDP	520
RPC	UDP/TCP	111
SMTP	TCP	25
SNMP	UDP	161
SNMP-TRAP	UDP	162
TELNET	TCP	23
TFTP	UDP	69
TIME	UDP/TCP	37

Acronyms

AAL	application adaptation layer
ABM	asynchronous balanced mode
ACK	acknowledgment
ACL	asynchronous connectionless link
ADSL	asymmetric digital subscriber line
AES	Advanced Encryption Standard
AH	Authentication Header
AM	amplitude modulation
AMI	alternate mark inversion
AMPS	Advanced Mobile Phone System
ANSI	American National Standards Institute
AP	access point
ARP	Address Resolution Protocol
ARPA	Advanced Research Projects Agency
ARPANET	Advanced Research Projects Agency Network
ARQ	automatic repeat request
AS	authentication server
AS	autonomous system
ASCII	American Standard Code for Information Interchange
ASK	amplitude shift keying
ASN.1	Abstract Syntax Notation 1
ATM	Asynchronous Transfer Mode
AUI	attachment unit interface
B8ZS	bipolar with 8-zero substitution
Bc	committed burst size
Be	excess burst size
BECN	backward explicit congestion notification
BER	Basic Encoding Rules
BGP	Border Gateway Protocol
BNC	Bayone-Neill-Concelman
BOOTP	Bootstrap Protocol
BSS	basic service set
BUS	broadcast/unknown server
CA	Certification Authority
CATV	community antenna TV
CBC	cipher block chaining mode
CBR	constant bit rate
CBT	Core-Based Tree
CCITT	Consultative Committee for International Telegraphy and Telephony
CCK	complementary code keying
CDMA	code division multiple access
CFB	cipher feedback mode
CGI	Common Gateway Interface
CHAP	Challenge Handshake Authentication Protocol
CIDR	Classless InterDomain Routing
CIR	committed information rate
CLEC	competitive local exchange carrier
CMTS	cable modem transmission system
CRC	cyclic redundancy check
CS	convergence sublayer
CSM	cipher stream mode
CSMA	carrier sense multiple access
CSMA/CA	carrier sense multiple access with collision avoidance

CSMA/CD	carrier sense multiple access with collision detection
D-AMPS	digital AMPS
DARPA	Defense Advanced Research Projects Agency
dB	decibel
DC	direct current
DCF	distributed coordination function
DCT	discrete cosine transform
DDNS	Dynamic Domain Name System
DDS	digital data service
DE	discard eligibility
DEMUX	demultiplexer
DES	data encryption standard
DHCP	Dynamic Host Configuration Protocol
DIFS	distributed interframe space
DLCI	data link connection identifier
DMT	discrete multitone technique
DNS	Domain Name System
DOCSIS	Data Over Cable System Interface Specifications
DS	digital signal
DS	Differentiated Services
DSL	digital subscriber line
DSLAM	digital subscriber line access multiplexer
DSSS	direct sequence spread spectrum
DVMRP	Distance Vector Multicast Routing Protocol
DWDM	dense wave-division multiplexing
EIA	Electronics Industries Association
ESP	Encapsulating Security Payload
ESS	Extended Service Set
FCC	Federal Communications Commission
FDM	frequency-division multiplexing
FDMA	frequency division multiple access
FDMA	frequency-division multiple access
FECN	forward explicit congestion notification
FHSS	frequency hopping spread spectrum
FIFO	first-in, first-out
FM	frequency modulation
FQDN	fully qualified domain name
FRAD	Frame Relay assembler/disassembler
FSK	frequency shift keying
FTP	File Transfer Protocol
GPS	Global Positioning System
GSM	Global System for Mobile Communication
HDLC	High-level Data Link Control
HDSL	high bit rate digital subscriber line
HFC	hybrid-fiber-coaxial network
HMAC	hashed-message authentication code
HR-DSSS	High Rate Direct Sequence Spread Spectrum
HTML	HyperText Markup Language
HTTP	HyperText Transfer Protocol
Hz	hertz
IAB	Internet Architecture Board
IANA	Internet Assigned Numbers Authority
ICANN	Internet Corporation for Assigned Names and Numbers
ICMP	Internet Control Message Protocol
ICMPv6	Internet Control Message Protocol, version 6
IEEE	Institute of Electrical and Electronics Engineers
IESG	Internet Engineering Steering Group
IETF	Internet Engineering Task Force
IFS	interframe space
IGMP	Internet Group Management Protocol
IKE	Internet Key Exchange
ILEC	incumbent local exchange carrier
IMAP4	Internet Mail Access Protocol, version 4
INTERNIC	Internet Network Information Center
IntServ	Integrated Services
IP	Internet Protocol
IPCP	Internetwork Protocol Control Protocol
IPng	Internet Protocol next generation
IPSec	IP Security
IPv4	Internet Protocol version 4
IPv6	Internet Protocol, version 6
IRTF	Internet Research Task Force
IS-95	Interim Standard 95
ISAKMP	Internet Security Association and Key Management Protocol
ISO	International Organization of Standardization

ISOC Internet Society
ISP Internet service provider
ISUP ISDN user port
ITM-2000 Internet Mobile Communication
ITU–T International Telecommunications Union–Telecommunication Standardization Sector
IXC interexchange carrier
JPEG Joint Photographic Experts Group
KDC key distribution center
L2CAP Logical Link Control and Adaptation Protocol
LAN local area network
LANE LAN emulation
LANE local area network emulation
LATA local access and transport area
LCP Link Control Protocol
LEC LAN emulation client
LEC local exchange carrier
LEO low earth orbit
LES LAN emulation server
LLC logical link control
LMI local management information
LSA link state advertisement
LSP link state packet
MA multiple access
MAA message access agent
MAC medium access control sublayer
MAC message authentication code
MAN metropolitan area network
MBONE multicast backbone
MDC modification detection mode
MEO medium Earth orbit
MIB Management Information Base
MII medium independent interface
MIME Multipurpose Internet Mail Extension
MLT-3 multiline transmission, 3-level
MOSPF Multicast Open Shortest Path First
MPEG motion picture experts group
MSC mobile switching center
MTA mail transfer agent
MTA message transfer agent
MTSO mobile telephone switching office
MTU maximum transfer unit
MUX multiplexer
NAP network access point
NAT network address translation
NAV network allocation vector
NCP Network Control Protocol
NIC network interface card
NNI network-to-network interface
NRM normal response mode
NRZ nonreturn to zero
NRZ-I nonreturn to zero, invert
NRZ-L nonreturn to zero, level
NVT Network Virtual Terminal
OC optical carrier
OFB output feedback
OFDM orthogonal frequency division multiplexing
OSI Open Systems Interconnection
OSPF open shortest path first
PAM pulse amplitude modulation
PAP Password Authentication Protocol
PCF point coordination function
PCM pulse code modulation
PCS personal communication system
PGP Pretty Good Privacy
PHB per hop behavior
PIM Protocol Independent Multicast
PIM-DM Protocol Independent Multicast, Dense Mode
PIM-SM Protocol Independent Multicast, Sparse Mode
PKI public key infrastructure
PM phase modulation
PN pseudorandom noise
POP point of presence
POP3 Post Office Protocol, version 3
POTS plain old telephone system
PPP Point-to-Point Protocol
PQDN partially qualified domain name
PSK phase shift keying
PVC permanent virtual circuit
QAM quadrature amplitude modulation
QoS quality of service

RADSL rate adaptive asymmetrical digital subscriber line
RARP Reverse Address Resolution Protocol
RFC Request for Comment
RIP Routing Information Protocol
ROM read-only memory
RPB reverse path broadcasting
RPF reverse path forwarding
RPM reverse path multicasting
RSA Rivest, Shamir, Adleman
RSVP Resource Reservation Protocol
RTCP Real-time Transport Control Protocol
RTP Real-time Transport Protocol
RTSP Real-Time Streaming Protocol
RTT round-trip time
RZ return to zero
SA Security Association
SADB security association database
SAR segmentation and reassembly
SCCP signaling connection control point
SCO synchronous connection oriented
SCP server control point
SCTP Stream Control Transmission Protocol
SDH Synchronous Digital Hierarchy
SDSL symmetric digital subscriber line
SEAL simple and efficient adaptation layer
SHA-1 Secure Hash Algorithm 1
SIFS short interframe space
SIP Session Initiation Protocol
SMI Structure of Management Information
SMTP Simple Mail Transfer Protocol
SNMP Simple Network Management Protocol
SNR signal-to-noise ratio
SONET Synchronous Optical Network
SP signal point
SPE synchronous payload envelope
SPI security parameter index
SS7 Signaling System Seven
SSL Secure Socket Layer
STM synchronous transport module
STP shielded twisted-pair
STP signal transport port
STS synchronous transport signal
SVC switched virtual circuit
TCAP transaction capabilities application port
TCP Transmission Control Protocol
TCP/IP Transmission Control Protocol/ Internetworking Protocol
TDD-TDMA time division duplexing TDMA
TDM time-division multiplexing
TDMA time division multiple access
TELNET Terminal Network
TFTP Trivial File Transfer Protocol
TGS ticket-granting server
TLS Transport Layer Security
TOS type of service
TSI time-slot interchange
TTL time to live
TUP telephone user port
UA user agent
UBR unspecified bit rate
UDP User Datagram Protocol
UNI user network interface
UNI user-to-network interface
URL Uniform Resource Locator
UTP unshielded twisted-pair
VBR variable bit rate
VC virtual circuit
VDSL very high bit rate digital subscriber line
VLAN virtual local area network
VOFR Voice Over Frame Relay
VPN virtual private network
VT virtual tributary
WAN wide area network
WATS wide area telephone service
WDM wave-division multiplexing
WWW World Wide Web

Glossary

1000Base-CX The two-wire STP implementation of Gigabit Ethernet.

1000Base-LX The two-wire fiber implementation of Gigabit Ethernet using long-wave laser.

1000Base-SX The two-wire fiber implementation of Gigabit Ethernet using short-wave laser.

1000Base-T The four-wire UTP implementation of Gigabit Ethernet.

100Base-FX The two-wire fiber implementation of Fast Ethernet.

100Base-T4 The four-wire UTP implementation of Fast Ethernet.

100Base-TX The two-wire UTP implementation of Fast Ethernet.

10Base2 The thin coaxial cable implementation of Standard Ethernet.

10Base5 The thick coaxial cable implementation of Standard Ethernet.

10Base-F The fiber implementation of Standard Ethernet.

10Base-T The twisted-pair implementation of Standard Ethernet.

10GBase-E The extended implementation of Ten-Gigabit Ethernet.

10GBase-L The fiber implementation of Ten-Gigabit Ethernet using long-wave laser.

10GBase-S The fiber implementation of Ten-Gigabit Ethernet using short-wave laser.

1-persistent strategy A CSMA persistence strategy in which a station sends a frame immediately if the line is idle.

2B1Q encoding A line encoding technique in which each pulse represents 2 bits.

4B/5B encoding A block coding technique in which 4 bits are encoded into a 5-bit code.

8B/10B encoding A block coding technique in which 8 bits are encoded into a 10-bit code.

8B6T encoding A three-level line encoding scheme that encodes a block of 8 bits into a signal of 6 ternary pulses.

4-dimensional, 5-level pulse amplitude modulation (4D-PAM5) An encoding scheme used by 1000Base-T.

56K modem A modem technology using two different data rates: one for uploading and one for downloading from the Internet.

A

Abstract Syntax Notation 1 (ASN.1) A standard for representing simple and structured data.

access point (AP) A central base station in a BSS.

acknowledgment (ACK) A response sent by the receiver to indicate the successful receipt of data.

active document In the World Wide Web, a document executed at the local site using Java.

adaptive delta modulation A delta modulation technique in which the value of delta changes according to the amplitude of the analog signal.

add/drop multiplexer A SONET device that removes and inserts signals in a path without demultiplexing and re-multiplexing.

additive increase With slow start, a congestion avoidance strategy in which the window size is increased by just one segment instead of exponentially.

address aggregation A mechanism in which the blocks of addresses for several organizations are aggregated into one larger block.

Address Resolution Protocol (ARP) In TCP/IP, a protocol for obtaining the physical address of a node when the Internet address is known.

address space The total number of addresses available by a protocol.

ADSL Lite A splitterless ADSL. This technology allows an ASDL Lite modem to be plugged directly into a telephone jack and connected to the computer. The splitting is done at the telephone company.

Advanced Encryption Standard (AES) A secret-key cryptosystem adapted by NIST to replace DES.

Advanced Mobile Phone System (AMPS) A North American analog cellular phone system using FDMA.

Advanced Research Projects Agency (ARPA) The government agency that funded ARPANET.

Advanced Research Projects Agency Network (ARPANET) The packet-switching network that was funded by ARPA.

ALOHA The original random multiple access method in which a station can send a frame any time it has one to send.

alternate mark inversion (AMI) A digital-to-digital bipolar encoding method in which the amplitude representing 1 alternates between positive and negative voltages.

American National Standards Institute (ANSI) A national standards organization that defines standards in the United States.

American Standard Code for Information Interchange (ASCII) A character code developed by ANSI that used extensively for data communication.

amplitude The strength of a signal, usually measured in volts.

amplitude modulation (AM) An analog-to-analog conversion method in which the carrier signal's amplitude varies with the amplitude of the modulating signal.

amplitude shift keying (ASK) A modulation method in which the amplitude of the carrier signal is varied to represent binary 0 or 1.

analog data Data that are continuous and smooth and not limited to a specific number of values.

analog signal A continuous waveform that changes smoothly over time.

analog-to-analog conversion The representation of analog information by an analog signal.

analog-to-digital conversion The representation of analog information by a digital signal.

angle of incidence In optics, the angle formed by a light ray approaching the interface between two media and the line perpendicular to the interface.

anycast address An address that defines a group of computers with addresses that have the same beginning.

aperiodic signal A signal that does not exhibit a pattern or repeating cycle.

applet A computer program for creating an active Web document. It is usually written in Java.

application adaptation layer (AAL) A layer in ATM protocol that breaks user data into 48-byte payloads.

application layer The fifth layer in the Internet model; provides access to network resources.

area A collection of networks, hosts, and routers all contained within an autonomous system.

association A connection in SCTP.

asymmetric digital subscriber line (ADSL) A communication technology in which the downstream data rate is higher than the upstream rate.

asynchronous balanced mode (ABM) In HDLC, a communication mode in which all stations are equal.

asynchronous connectionless link (ACL) A link between a Bluetooth master and slave in which a corrupted payload is retransmitted.

Asynchronous Transfer Mode (ATM) A wide area protocol featuring high data rates and equal-sized packets (cells); ATM is suitable for transferring text, audio, and video data.

asynchronous transmission Transfer of data with start and stop bit(s) and a variable time interval between data units.

ATM LAN A LAN using ATM technology.

ATM layer A layer in ATM that provides routing, traffic management, switching, and multiplexing services.

attachment unit interface (AUI) A 10Base5 cable that performs the physical interface functions between the station and the transceiver.

attenuation The loss of a signal's energy due to the resistance of the medium.

audio Recording or transmitting of sound or music.

authentication Verification of the sender of a message.

Authentication Header (AH) Protocol A protocol defined by IPSec at the network layer that provides integrity to a message through the creation of a digital signature by a hashing function.

authentication server (AS) The KDC in the Kerberos protocol.

automatic repeat request (ARQ) An error-control method in which correction is made by retransmission of data.

autonegotiation A Fast Ethernet feature that allows two devices to negotiate the mode or data rate.

autonomous system (AS) A group of networks and routers under the authority of a single administration.

B

backward explicit congestion notification (BECN) A bit in the Frame Relay packet that notifies the sender of congestion.

band-pass channel A channel that can pass a range of frequencies.

bandwidth The difference between the highest and the lowest frequencies of a composite signal. It also measures the information-carrying capacity of a line or a network.

bandwidth on demand A digital service that allows subscribers higher speeds through the use of multiple lines.

bandwidth-delay product A measure of the number of bits that can be sent while waiting for news from the receiver.

banyan switch A multistage switch with microswitches at each stage that route the packets based on the output port represented as a binary string.

Barker sequence A sequence of 11 bits used for spreading.

baseband transmission Transmission of digital or analog signal without modulation using a low-pass channel.

base header In IPv6, the main header of the datagram.

baseline wandering In decoding a digital signal, the receiver calculates a running average of the received signal power. This average is called the baseline. A long string of 0s or 1s can cause a drift in the baseline (baseline wandering) and make it difficult for the receiver to decode correctly.

Basic Encoding Rule (BER) A standard that encodes data to be transferred through a network.

Basic Latin ASCII character set.

basic service set (BSS) The building block of a wireless LAN as defined by the IEEE 802.11 standard.

Batcher-banyan switch A banyan switch that sorts the arriving packets based on their destination port.

baud rate The number of signal elements transmitted per second. A signal element consists of one or more bits.

Bayone-Neill-Concelman (BNC) connector A common coaxial cable connector.

bidirectional authentication An authentication method involving a challenge and a response from sender to receiver and vice versa.

bidirectional frame (B-frame) An MPEG frame that is related to the preceding and following I-frame or P-frame.

binary exponential backup In contention access methods, a retransmission delay strategy used by a system to delay access.

binary notation Representation of IP addresses in binary.

biphase A type of polar encoding where the signal changes at the middle of the bit interval. Manchester and differential Manchester are examples of biphase encoding.

bipolar encoding A digital-to-digital encoding method in which 0 amplitude represents binary 0 and positive and negative amplitudes represent alternate 1s.

bipolar with 8-zero substitution (B8ZS) A scrambling technique in which a stream of 8 zeros are replaced by a predefined pattern to improve bit synchronization.

bit Binary digit. The smallest unit of data (0 or 1).

bit rate The number of bits transmitted per second.

bit stuffing In a bit-oriented protocol, the process of adding an extra bit in the data section of a frame to prevent a sequence of bits from looking like a flag.

bit-oriented protocol A protocol in which the data frame is interpreted as a sequence of bits.

block cipher An encryption/decryption algorithm that has a block of bits as its basic unit.

block code An error detection/correction code in which data are divided into units called datawords. Redundant bits are added to each dataword to create a codeword.

block coding A coding method to ensure synchronization and detection of errors.

blocking An event that occurs when a switched network is working at its full capacity and cannot accept more input.

Bluetooth A wireless LAN technology designed to connect devices of different functions such as telephones and notebooks in a small area such as a room.

Bootstrap Protocol (BOOTP) The protocol that provides configuration information from a table (file).

Border Gateway Protocol (BGP) An interautonomous system routing protocol based on path vector routing.

bridge A network device operating at the first two layers of the Internet model with filtering and forwarding capabilities.

broadband transmission Transmission of signals using modulation of a higher frequency signal. The term implies a wide-bandwidth data combined from different sources.

broadcast address An address that allows transmission of a message to all nodes of a network.

broadcast/unknown server (BUS) A server connected to an ATM switch that can multicast and broadcast frames.

broadcasting Transmission of a message to all nodes in a network.

browser An application program that displays a WWW document. A browser usually uses other Internet services to access the document.

BSS-transition mobility In a wireless LAN, a station that can move from one BSS to another but is confined inside one ESS.

bucket brigade attack See *man-in-the middle attack.*

burst error Error in a data unit in which two or more bits have been altered.

bursty data Data with varying instantaneous transmission rates.

bus topology A network topology in which all computers are attached to a shared medium.

byte stuffing In a byte-oriented protocol, the process of adding an extra byte in the data section of a frame to prevent a byte from looking like a flag.

byte-oriented protocol A protocol in which the data section of the frame is interpreted as a sequence of bytes (characters).

C

cable modem A technology in which the TV cable provides Internet access.

cable modem transmission system (CMTS) A device installed inside the distribution hub that receives data from the Internet and passes them to the combiner.

cable TV network A system using coaxial or fiber optic cable that brings multiple channels of video programs into homes.

caching The storing of information in a small, fast memory.

Caesar cipher A shift cipher used by Julius Caesar with the key value of 3.

carrier extension A technique in Gigabit Ethernet that increases the minimum length of the frame to achieve a higher maximum cable length.

carrier sense multiple access (CSMA) A contention access method in which each station listens to the line before transmitting data.

carrier sense multiple access with collision avoidance (CSMA/CA) An access method in which collision is avoided.

carrier sense multiple access with collision detection (CSMA/CD) An access method in which stations transmit whenever the transmission medium is available and retransmit when collision occurs.

carrier signal A high frequency signal used for digital-to-analog or analog-to-analog modulation. One of the characteristics of the carrier signal (amplitude, frequency, or phase) is changed according to the modulating data.

cell A small, fixed-size data unit; also, in cellular telephony, a geographical area served by a cell office.

cell network A network using the cell as its basic data unit.

cellular telephony A wireless communication technique in which an area is divided into cells. A cell is served by a transmitter.

Certification Authority (CA) An agency such as a federal or state organization that binds a public key to an entity and issues a certificate.

Challenge Handshake Authentication Protocol (CHAP) In PPP, a three-way handshaking protocol used for authentication.

channel A communications pathway.

channelization A multiple access method in which the available bandwidth of a link is shared in time.

character-oriented protocol See *byte-oriented protocol.*

checksum A value used for error detection. It is formed by adding data units using one's complement arithmetic and then complementing the result.

chip In CDMA, a number in a code that is assigned to a station.

choke point A packet sent by a router to the source to inform it of congestion.

chunk A unit of transmission in SCTP.

cipher An encryption/decryption algorithm.

cipher block chaining (CBC) mode A DES and triple DES operation mode in which the encryption (or decryption) of a block depends on all previous blocks.

cipher feedback mode (CFB) A DES and triple DES operation mode in which data is sent and received 1 bit at a time, with each bit independent of the previous bits.

cipher stream mode (CSM) A DES and triple DES operation mode in which data is sent and received 1 byte at a time.

cipher suite A list of possible ciphers.

ciphertext The encrypted data.

circuit switching A switching technology that establishes an electrical connection between stations using a dedicated path.

cladding Glass or plastic surrounding the core of an optical fiber; the optical density of the cladding must be less than that of the core.

classful addressing An IPv4 addressing mechanism in which the IP address space is divided into 5 classes: A, B, C, D, and E. Each class occupies some part of the whole address space.

classless addressing An addressing mechanism in which the IP address space is not divided into classes.

Classless InterDomain Routing (CIDR) A technique to reduce the number of routing table entries when supernetting is used.

client process A running application program on a local site that requests service from a running application program on a remote site.

client-server model The model of interaction between two application programs in which a program at one end (client) requests a service from a program at the other end (server).

closed-loop congestion control A method to alleviate congestion after it happens.

coaxial cable A transmission medium consisting of a conducting core, insulating material, and a second conducting sheath.

code division multiple access (CDMA) A multiple access method in which one channel carries all transmissions simultaneously.

codeword The encoded dataword.

ColdFusion A dynamic web technology that allows the fusion of data items coming from a conventional database.

collision The event that occurs when two transmitters send at the same time on a channel designed for only one transmission at a time; data will be destroyed.

collision domain The length of the medium subject to collision.

committed burst size (Bc) The maximum number of bits in a specific time period that a Frame Relay network must transfer without discarding any frames.

committed information rate (CIR) The committed burst size divided by time.

common carrier A transmission facility available to the public and subject to public utility regulation.

Common Gateway Interface (CGI) A standard for communication between HTTP servers and executable programs. CGI is used in creating dynamic documents.

community antenna TV (CATV) A cable network service that broadcasts video signals to locations with poor or no reception.

compatible address An IPv6 address consisting of 96 bits of zero followed by 32 bits of IPv4.

competitive local exchange carrier (CLEC) A telephone company that cannot provide main telephone services; instead, other services such as mobile telephone service and toll calls inside a LATA are provided.

complementary code keying (CCK) An HR-DSSS encoding method that encodes four or eight bits into one symbol.

composite signal A signal composed of more than one sine wave.

congestion Excessive network or internetwork traffic causing a general degradation of service.

connecting device A tool that connects computers or networks.

connection establishment The preliminary setup necessary for a logical connection prior to actual data transfer.

connectionless service A service for data transfer without connection establishment or termination.

constant bit rate (CBR) The data rate of an ATM service class that is designed for customers requiring real-time audio or video services.

constellation diagram A graphical representation of the phase and amplitude of different bit combinations in digital-to-analog modulation.

Consultative Committee for International Telegraphy and Telephony (CCITT) An international standards group now known as the ITU-T.

contention An access method in which two or more devices try to transmit at the same time on the same channel.

controlled access A multiple access method in which the stations consult one another to determine who has the right to send.

convergence sublayer (CS) In ATM protocol, the upper AAL sublayer that adds a header or a trailer to the user data.

cookie A string of characters that holds some information about the client and must be returned to the server untouched.

core The glass or plastic center of an optical fiber.

Core-Based Tree (CBT) In multicasting, a group-shared protocol that uses a center router as the root of the tree.

country domain A subdomain in the Domain Name System that uses two characters as the last suffix.

crossbar switch A switch consisting of a lattice of horizontal and vertical paths. At the intersection of each horizontal and vertical path, there is a crosspoint that can connect the input to the output.

crosspoint The junction of an input and an output on a crossbar switch.

crosstalk The noise on a line caused by signals traveling along another line.

cryptography The science and art of transforming messages to make them secure and immune to attacks.

cyclic code A linear code in which the cyclic shifting (rotation) of each codeword creates another code word.

cyclic redundancy check (CRC) A highly accurate error-detection method based on interpreting a pattern of bits as a polynomial.

D

data element The smallest entity that can represent a piece of information. A bit.

data encryption standard (DES) The U.S. government standard encryption method for nonmilitary and nonclassified use.

data link connection identifier (DLCI) A number that identifies the virtual circuit in Frame Relay.

data link control The responsibilities of the data link layer: flow control and error control.

data link layer The second layer in the Internet model. It is responsible for node-to-node delivery.

Data Over Cable System Interface Specifications (DOCSIS) A standard for data transmission over an HFC network.

data rate The number of data elements sent in one second.

data transfer phase The intermediate phase in circuit-switched or virtual-circuit network in which data transfer takes place.

data transparency The ability to send any bit pattern as data without it being mistaken for control bits.

datagram In packet switching, an independent data unit.

datagram network A packet-switched network in which packets are independent from each other.

dataword The smallest block of data in block coding.

de facto standard A protocol that has not been approved by an organized body but adopted as a standard through widespread use.

de jure standard A protocol that has been legislated by an officially recognized body.

deadlock A situation in which a task cannot proceed because it is waiting for an even that will never occur.

decibel (dB) A measure of the relative strength of two signal points.

decryption Recovery of the original message from the encrypted data.

default mask The mask for a network that is not subnetted.

default routing A routing method in which a router is assigned to receive all packets with no match in the routing table.

Defense Advanced Research Projects Agency (DARPA) A government organization, which, under the name of ARPA funded ARPANET and the Internet.

delta modulation An analog-to-digital conversion technique in which the value of the digital signal is based on the difference between the current and the previous sample values.

demodulation The process of separating the carrier signal from the information-bearing signal.

demultiplexer (DEMUX) A device that separates a multiplexed signal into its original components.

denial of service attack A form of attack in which the site is flooded with so many phony requests that is eventually forced to deny service.

dense wave-division multiplexing (DWDM) A WDM method that can multiplex a very large number of channels by spacing channels closer together.

differential Manchester encoding A digital-to-digital polar encoding method that features a transition at the middle of the bit interval as well as an inversion at the beginning of each 1 bit.

Differentiated Services (DS or Diffserv) A class-based QoS model designed for IP.

Diffie-Hellman protocol A key management protocol that provides a one-time session key for two parties.

digest A condensed version of a document.

digital AMPS (D-AMPS) A second-generation cellular phone system that is a digital version of AMPS.

digital data Data represented by discrete values or conditions.

digital data service (DDS) A digital version of an analog leased line with a rate of 64 Kbps.

digital signal A discrete signal with a limited number of values.

digital signal (DS) service A telephone company service featuring a hierarchy of digital signals.

digital signature A method to authenticate the sender of a message.

digital subscriber line (DSL) A technology using existing telecommunication networks to accomplish high-speed delivery of data, voice, video, and multimedia.

digital subscriber line access multiplexer (DSLAM) A telephone company site device that functions like an ADSL modem.

digital-to-analog conversion The representation of digital information by an analog signal.

digital-to-digital conversion The representation of digital information by a digital signal.

digitization Conversion of analog information to digital information.

Dijkstra's algorithm In link state routing, an algorithm that finds the shortest path to other routers.

direct current (DC) A zero-frequency signal with a constant amplitude.

direct delivery A delivery in which the final destination of the packet is a host connected to the same physical network as the sender.

direct sequence spread spectrum (DSSS) A wireless transmission method in which each bit to be sent by the sender is replaced by a sequence of bits called a chip code.

discard eligibility (DE) A bit that identifies a packet that can be discarded if there is congestion in the network.

discrete cosine transform (DCT) A JPEG phase in which a transformation changes the 64 values so that the relative relationships between pixels are kept but the redundancies are revealed.

discrete multitone technique (DMT) A modulation method combining elements of QAM and FDM.

Distance Vector Multicast Routing Protocol (DVMRP) A protocol based on distance vector routing that handles multicast routing in conjunction with IGMP.

distance vector routing A routing method in which each router sends its neighbors a list of networks it can reach and the distance to that network.

distortion Any change in a signal due to noise, attenuation, or other influences.

distributed coordination function (DCF) The basic access method in wireless LANs; stations contend with each other to get access to the channel.

distributed database Information stored in many locations.

distributed interframe space (DIFS) In wireless LANs, a period of time that a station waits before sending a control frame.

distributed processing A strategy in which services provided for the network reside at multiple sites.

DNS server A computer that holds information about the name space.

domain name In the DNS, a sequence of labels separated by dots.

domain name space A structure for organizing the name space in which the names are defined in an inverted-tree structure with the root at the top.

Domain Name System (DNS) A TCP/IP application service that converts user-friendly names to IP addresses.

dotted-decimal notation A notation devised to make the IP address easier to read; each byte is converted to its decimal equivalent and then set off from its neighbor by a decimal.

downlink Transmission from a satellite to an earth station.

downloading Retrieving a file or data from a remote site.

dynamic document A Web document created by running a CGI program at the server site.

Dynamic Domain Name System (DDNS) A method to update the DNS master file dynamically.

Dynamic Host Configuration Protocol (DHCP) An extension to BOOTP that dynamically assigns configuration information.

dynamic mapping A technique in which a protocol is used for address resolution.

dynamic routing Routing in which the routing table entries are updated automatically by the routing protocol.

E

E lines The European equivalent of T lines.

electromagnetic spectrum The frequency range occupied by electromagnetic energy.

electronic code block (ECB) mode A DES and triple DES operation method in which a long message is divided into 64-bit blocks before being encrypted separately.

Electronics Industries Association (EIA) An organization that promotes electronics manufacturing concerns. It has developed interface standards such as EIA-232, EIA-449, and EIA-530.

Encapsulating Security Payload (ESP) A protocol defined by IPSec that provides privacy as well as a combination of integrity and message authentication.

encapsulation The technique in which a data unit from one protocol is placed within the data field portion of the data unit of another protocol.

encryption Converting a message into an unintelligible form that is unreadable unless decrypted.

end office A switching office that is the terminus for the local loops.

end system A sender or receiver of data.

ephemeral port number A port number used by the client.

error control The handling of errors in data transmission.

Ethernet A local area network using the CSMA/CD access method.

excess burst size (Be) In Frame Relay, the maximum number of bits in excess of B_c that the user can send during a predefined period of time.

Extended Service Set (ESS) A wireless LAN service composed of two or more BSSs with APs as defined by the IEEE 802.11 standard.

exterior routing Routing between autonomous systems.

extranet A private network that uses the TCP/IP protocol suite that allows authorized access from outside users.

F

Fast Ethernet Ethernet with a data rate of 100 Mbps.

fast retransmission Retransmission of a segment in the TCP protocol when three acknowledgments have been received that imply the loss or corruption of that segment.

Federal Communications Commission (FCC) A government agency that regulates radio, television, and telecommunications.

fiber-optic cable A high-bandwidth transmission medium that carries data signals in the form of pulses of light. It consists of a thin cylinder of glass or plastic, called the core, surrounded by a concentric layer of glass or plastic called the cladding.

File Transfer Protocol (FTP) In TCP/IP, an application layer protocol that transfers files between two sites.

filtering A process in which a bridge makes forwarding decisions.

finite state machine A machine that goes through a limited number of states.

firewall A device (usually a router) installed between the internal network of an organization and the rest of the Internet to provide security.

first-in, first-out (FIFO) queue A queue in which the first item in is the first item out.

flag A bit pattern or a character added to the beginning and the end of a frame to separate the frames.

flat name space A name space in which there is no hierarchical structure.

flooding Saturation of a network with a message.

flow control A technique to control the rate of flow of frames (packets or messages).

footprint An area on Earth that is covered by a satellite at a specific time.

forward error correction Correction of errors at the receiver.

forward explicit congestion notification (FECN) A bit in the Frame Relay packet that notifies the destination of congestion.

forwarding Placing the packet in its route to its destination.

Fourier analysis The mathematical technique used to obtain the frequency spectrum of an aperiodic signal if the time-domain representation is given.

fragmentation The division of a packet into smaller units to accommodate a protocol's MTU.

frame A group of bits representing a block of data.

frame bursting A technique in CSMA/CD Gigabit Ethernet in which multiple frames are logically connected to each other to resemble a longer frame.

Frame Relay A packet-switching specification defined for the first two layers of the Internet model. There is no network layer. Error checking is done on end-to-end basis instead of on each link.

Frame Relay assembler/disassembler (FRAD) A device used in Frame Relay to handle frames coming from other protocols.

frequency The number of cycles per second of a periodic signal.

frequency division multiple access (FDMA) A multiple access method in which the bandwidth is divided into channels.

frequency hopping spread spectrum (FHSS) A wireless transmission method in which the sender transmits at one carrier frequency for a short period of time, then hops to another carrier frequency for the same amount of time, hops again for the same amount of time, and so on. After *N* hops, the cycle is repeated.

frequency modulation (FM) An analog-to-analog modulation method in which the carrier signal's frequency varies with the amplitude of the modulating signal.

frequency shift keying (FSK) A digital-to-analog encoding method in which the frequency of the carrier signal is varied to represent binary 0 or 1.

frequency-division multiple access (FDMA) An access method technique in which multiple sources use assigned bandwidth in a data communication band.

frequency-division multiplexing (FDM) The combining of analog signals into a single signal.

frequency-domain plot A graphical representation of a signal's frequency components.

full-duplex mode A transmission mode in which both parties can communicate simultaneously.

full-duplex switched Ethernet Ethernet in which each station, in its own separate collision domain, can both send and receive.

fully qualified domain name (FQDN) A domain name consisting of labels beginning with the host and going back through each level to the root node.

fundamental frequency The frequency of the dominant sine wave of a composite signal.

G

gatekeeper In the H.323 standard, a server on the LAN that plays the role of the registrar server.

gateway A device used to connect two separate networks that use different communication protocols.

generic domain A subdomain in the domain name system that uses generic suffixes.

geographical routing A routing technique in which the entire address space is divided into blocks based on physical landmasses.

Gigabit Ethernet Ethernet with a 1000 Mbps data rate.

Global Positioning System (GPS) An MEO public satellite system consisting of 24 satellites and used for land and sea navigation. GPS is not used for communications.

Global System for Mobile Communication (GSM) A second-generation cellular phone system used in Europe.

Globalstar An LEO satellite system with 48 satellites in six polar orbits with each orbit hosting eight satellites.

Go-Back-N ARQ An error-control method in which the frame in error and all following frames must be retransmitted.

grafting Resumption of multicast messages.

ground propagation Propagation of radio waves through the lowest portion of the atmosphere (hugging the earth).

group-shared tree A multicast routing feature in which each group in the system shares the same tree.

guard band A bandwidth separating two signals.

guided media Transmission media with a physical boundary.

H

H.323 A standard designed by ITU to allow telephones on the public telephone network to talk to computers (called terminals in H.323) connected to the Internet.

half-duplex mode A transmission mode in which communication can be two-way but not at the same time.

Hamming code A method that adds redundant bits to a data unit to detect and correct bit errors.

Hamming distance The number of differences between the corresponding bits in two datawords.

handoff Changing to a new channel as a mobile device moves from one cell to another.

harmonics Components of a digital signal, each having a different amplitude, frequency, and phase.

hash function An algorithm that creates a fixed-size digest from a variable-length message.

hashed-message authentication code (HMAC) A MAC based on a keyless hash function such as SHA-1.

header Control information added to the beginning of a data packet.

hertz (Hz) Unit of measurement for frequency.

hexadecimal colon notation In IPv6, an address notation consisting of 32 hexadecimal digits, with every four digits separated by a colon.

hierarchical routing A routing technique in which the entire address space is divided into levels based on specific criteria.

high bit rate digital subscriber line (HDSL) A service similar to the T1-line that can operate at lengths up to 3.6 km.

High Rate Direct Sequence Spread Spectrum (HR-DSSS) A signal generation method similar to DSSS except for the encoding method (CCK).

High-level Data Link Control (HDLC) A bit-oriented data link protocol defined by the ISO.

hop count The number of nodes along a route. It is a measurement of distance in routing algorithms.

hop-to-hop delivery Transmission of frames from one node to the next.

horn antenna A scoop-shaped antenna used in terrestrial microwave communication.

host A station or node on a network.

hostid The part of an IP address that identifies a host.

host-specific routing A routing method in which the full IP address of a host is given in the routing table.

hub A central device in a star topology that provides a common connection among the nodes.

Huffman encoding A statistical compression method using variable-length codes to encode a set of symbols.

hybrid network A network with a private internet and access to the global Internet.

hybrid-fiber-coaxial (HFC) network The second generation of cable networks; uses fiber optic and coaxial cable.

hypertext Information containing text that is linked to other documents through pointers.

HyperText Markup Language (HTML) The computer language for specifying the contents and format of a web document. It allows additional text to include codes that define fonts, layouts, embedded graphics, and hypertext links.

HyperText Transfer Protocol (HTTP) An application service for retrieving a web document.

I

inband signaling Using the same channel for data and control transfer.

incumbent local exchange carrier (ILEC) A telephone company that provided services before 1996 and is the owner of the cabling system.

indirect delivery A delivery in which the source and destination of a packet are in different networks.

infrared wave A wave with a frequency between 300 GHz and 400 THz; usually used for short-range communications.

inner product A number produced by multiplying two sequences, element by element, and summing the products.

Institute of Electrical and Electronics Engineers (IEEE) A group consisting of professional engineers which has specialized societies whose committees prepare standards in members' areas of specialty.

Integrated Services (IntServ) A flow-based QoS model designed for IP.

interactive audio/video Real-time communication with sound and images.

interautonomous system routing protocol A protocol to handle transmissions between autonomous systems.

interdomain routing Routing among autonomous systems.

interexchange carrier (IXC) A long-distance company that, prior to the Act of 1996, provided communication services between two customers in different LATAs.

interface The boundary between two pieces of equipment. It also refers to mechanical, electrical, and functional characteristics of the connection.

interference Any undesired energy that interferes with the desired signals.

interframe space (IFS) In wireless LANs, a time interval between two frames to control access to the channel.

Interim Standard 95 (IS-95) One of the dominant second-generation cellular telephony standards in North America.

interior routing Routing inside an autonomous system.

interleaving In multiplexing, taking a specific amount of data from each device in a regular order.

International Organization of Standardization (ISO) A worldwide organization that defines and develops standards on a variety of topics.

International Telecommunications Union–Telecommunication Standardization Sector (ITU–T) A standards organization formerly known as the CCITT.

internet A collection of networks connected by internetworking devices such as routers or gateways.

Internet A global internet that uses the TCP/IP protocol suite.

Internet address A 32-bit or 128-bit network-layer address used to uniquely define a host on an internet using the TCP/IP protocol.

Internet Architecture Board (IAB) The technical adviser to the ISOC; oversees the continuing development of the TCP/IP protocol suite.

Internet Assigned Numbers Authority (IANA) A group supported by the U.S. government that was responsible for the management of Internet domain names and addresses until October 1998.

Internet Control Message Protocol (ICMP) A protocol in the TCP/IP protocol suite that handles error and control messages.

Internet Control Message Protocol, version 6 (ICMPv6) A protocol in IPv6 that handles error and control messages.

Internet Corporation for Assigned Names and Numbers (ICANN) A private, nonprofit corporation managed by an international board that assumed IANA operations.

Internet draft A working Internet document (a work in progress) with no official status and a six-month lifetime.

Internet Engineering Steering Group (IESG) An organization that oversees the activities of IETF.

Internet Engineering Task Force (IETF) A group working on the design and development of the TCP/IP protocol suite and the Internet.

Internet Group Management Protocol (IGMP) A protocol in the TCP/IP protocol suite that handles multicasting.

Internet Key Exchange (IKE) A protocol designed to create security associations in SADBs.

Internet Mail Access Protocol, version 4 (IMAP4) A complex and powerful protocol to handle the transmission of electronic mail.

Internet Mobile Communication (ITM-2000) An ITU issued blueprint that defines criteria for third generation cellular telephony.

Internet model A 5-layer protocol stack that dominates data communications and networking today.

Internet Network Information Center (INTERNIC) An agency responsible for collecting and distributing information about TCP/IP protocols.

Internet Protocol (IP) The network-layer protocol in the TCP/IP protocol suite governing connectionless transmission across packet switching networks.

Internet Protocol next generation (IPng) See *Internet Protocol version 6 (IPv6).*

Internet Protocol version 4 (IPv4) The current version of Internet Protocol.

Internet Protocol, version 6 (IPv6) The sixth version of the Internet Protocol.

Internet Research Task Force (IRTF) A forum of working groups focusing on long-term research topics related to the Internet.

Internet Security Association and Key Management Protocol (ISAKMP) A protocol designed by the National Security Agency (NSA) that actually implements the exchanges defined in IKE.

Internet service provider (ISP) Usually, a company that provides Internet services.

Internet Society (ISOC) The nonprofit organization established to publicize the Internet.

Internet standard A thoroughly tested specification that is useful to and adhered to by those who work with the Internet. It is a formalized regulation that must be followed.

internetwork (internet) A network of networks.

Internetwork Protocol Control Protocol (IPCP) In PPP, the set of protocols that establish and terminate a network layer connection for IP packets.

internetworking Connecting several networks together using internetworking devices such as routers and gateways.

intranet A private network that uses the TCP/IP protocol suite.

inverse domain A subdomain in the DNS that finds the domain name given the IP address.

inverse multiplexing Taking data from one source and breaking it into portions that can be sent across lower-speed lines.

IP datagram The Internetworking Protocol data unit.

IP Security (IPSec) A collection of protocols designed by the IETF (Internet Engineering Task Force) to provide security for a packet carried on the Internet.

IrDA port A port that allows a wireless keyboard to communicate with a PC.

Iridium A 66-satellite network that provides communication from any Earth site to another.

ISDN user port (ISUP) A protocol at the upper layer of SS7 that provides services similar to those of an ISDN network.

isochronous transmission A type of transmission in which the entire stream of bits is synchronized under the control of a common clock.

iterative resolution Resolution of the IP address in which the client may send its request to multiple servers before getting an answer.

J

jamming signal In CSMA/CD, a signal sent by the first station that detects collision to alert every other station of the situation.

Java A programming language used to create active Web documents.

jitter A phenomenon in real-time traffic caused by gaps between consecutive packets at the receiver.

Joint Photographic Experts Group (JPEG) A standard for compressing continuous-tone picture.

K

Kerberos An authentication protocol used by Windows 2000.

key distribution center (KDC) In secret key encryption, a trusted third party that shares a key with each user.

L

LAN emulation (LANE) Local area network emulation using ATM switches.

LAN emulation client (LEC) In ATM LANs, client software that receives services from a LES.

LAN emulation server (LES) In ATM LANs, server software that creates a virtual circuit between the source and destination.

leaky bucket algorithm An algorithm to shape bursty traffic.

least-cost tree An MOSPF feature in which the tree is based on a chosen metric instead of shortest path.

legacy ATM LAN LAN in which ATM technology is used as a backbone to connect traditional LANs.

line coding Converting binary data into signals.

linear block code A block code in which adding two codewords creates another codeword.

line-of-sight propagation The transmission of very high frequency signals in straight lines directly from antenna to antenna.

link The physical communication pathway that transfers data from one device to another.

Link Control Protocol (LCP) A PPP protocol responsible for establishing, maintaining, configuring, and terminating links.

link local address An IPv6 address used by a private LAN.

link state advertisement (LSA) In OSPF, a method to disperse information.

link state database In link state routing, a database common to all routers and made from LSP information.

link state packet (LSP) In link state routing, a small packet containing routing information sent by a router to all other routers.

link state routing A routing method in which each router shares its knowledge of changes in its neighborhood with all other routers.

local access and transport area (LATA) An area covered by one or more telephone companies.

local area network (LAN) A network connecting devices inside a single building or inside buildings close to each other.

local area network emulation (LANE) Software that enables an ATM switch to behave like a LAN switch.

local exchange carrier (LEC) A telephone company that handles services inside a LATA.

local loop The link that connects a subscriber to the telephone central office.

local management information (LMI) A protocol used in Frame Relay that provides management features.

logical address An address defined in the network layer.

logical link control (LLC) The upper sublayer of the data link layer as defined by IEEE Project 802.2.

Logical Link Control and Adaptation Protocol (L2CAP) A Bluetooth layer used for data exchange on an ACL link.

logical tunnel The encapsulation of a multicast packet inside a unicast packet to enable multicast routing by non-multicast routers.

longest mask matching The technique in CIDR in which the longest prefix is handled first when searching a routing table.

low earth orbit (LEO) A polar satellite orbit with an altitude between 500 and 2000 km. A satellite with this orbit has a rotation period of 90 to 120 minutes.

low-pass channel A channel that passes frequencies between 0 and f.

M

mail transfer agent (MTA) An SMTP component that transfers the mail across the Internet.

Management Information Base (MIB) The database used by SNMP that holds the information necessary for management of a network.

Manchester encoding A digital-to-digital polar encoding method in which a transition occurs at the middle of each bit interval to provide synchronization.

man-in-the-middle attack A key management problem in which an intruder intercepts and sends messages between the intended sender and receiver.

mapped address An IPv6 address used when a computer that has migrated to IPv6 wants to send a packet to a computer still using IPv4.

mask For IPv4, a 32-bit binary number that gives the first address in the block (the network address) when ANDed with an address in the block.

maximum transfer unit (MTU) The largest size data unit a specific network can handle.

medium access control (MAC) sublayer The lower sublayer in the data link layer defined by the IEEE 802 project. It defines the access method and access control in different local area network protocols.

medium Earth orbit (MEO) A satellite orbit positioned between the two Van Allen belts. A satellite at this orbit takes six hours to circle the earth.

mesh topology A network configuration in which each device has a dedicated point-to-point link to every other device.

message access agent (MAA) A client-server program that pulls the stored email messages.

message authentication A security measure in which the sender of the message is verified for every message sent.

message authentication code (MAC) A keyed hash function.

message transfer agent (MTA) An SMTP component that transfers the message across the Internet.

metric A cost assigned for passing through a network.

metropolitan area network (MAN) A network that can span a geographical area the size of a city.

microwave Electromagnetic waves ranging from 2 GHz to 40 GHz.

minimum Hamming distance In a set of words, the smallest Hamming distance between all possible pairs.

mobile host A host that can move from one network to another.

mobile switching center (MSC) In cellular telephony, a switching office that coordinates communication between all base stations and the telephone central office.

mobile telephone switching office (MTSO) An office that controls and coordinates communication between all of the cell offices and the telephone control office.

modem A device consisting of a modulator and a demodulator. It converts a digital signal into an analog signal (modulation) and vice versa (demodulation).

modification detection code (MDC) The digest created by a hash function.

modular arithmetic Arithmetic that uses a limited range of integers (0 to $n - 1$).

modulation Modification of one or more characteristics of a carrier wave by an information-bearing signal.

modulus The upper limit in modular arithmetic (n).

monoalphabetic substitution An encryption method in which each occurrence of a character is replaced by another character in the set.

motion picture experts group (MPEG) A method to compress videos.

multicast address An address used for multicasting.

multicast backbone (MBONE) A set of internet routers supporting multicasting through the use of tunneling.

Multicast Open Shortest Path First (MOSPF) A multicast protocol that uses multicast link state routing to create a source-based least cost tree.

multicast router A router with a list of loyal members related to each router interface that distributes the multicast packets.

multicast routing Moving a multicast packet to its destinations.

multicasting A transmission method that allows copies of a single packet to be sent to a selected group of receivers.

multihoming service A service provided by SCTP that allows a computer to be connected to different networks.

multiline transmission, 3-level (MLT-3) encoding A line coding scheme featuring 3 levels of signals and transitions at the beginning of the 1 bit.

multimode graded-index fiber An optical fiber with a core having a graded index of refraction.

multimode step-index fiber An optical fiber with a core having a uniform index of refraction. The index of refraction changes suddenly at the core/cladding boundary.

multiple access (MA) A line access method in which every station can access the line freely.

multiple unicasting Sending multiple copies of a message, each with a different unicast address.

multiplexer (MUX) A device used for multiplexing.

multiplexing The process of combining signals from multiple sources for transmission across a single data link.

multiplicative decrease A congestion avoidance technique in which the threshold is set to half of the last congestion window size, and the congestion window size starts from one again.

Multipurpose Internet Mail Extension (MIME) A supplement to SMTP that allows non-ASCII data to be sent through SMTP.

multistage switch An array of switches designed to reduce the number of crosspoints.

multistream service A service provided by SCTP that allows data transfer to be carried using different streams.

N

name space All the names assigned to machines on an internet.

name-address resolution Mapping a name to an address or an address to a name.

Needham-Schroeder protocol A key management protocol using multiple challenge-response interactions between 2 entities.

netid The part of an IP address that identifies the network.

network A system consisting of connected nodes that share data, hardware, and software.

network access point (NAP) A complex switching station that connects backbone networks.

network address An address that identifies a network to the rest of the Internet; it is the first address in a block.

network address translation (NAT) A technology that allows a private network to use a set of private addresses for internal communication and a set of global Internet addresses for external communication.

network allocation vector (NAV) In CSMA/CA, the amount of time that must pass before a station can check for an idle line.

Network Control Protocol (NCP) In PPP, a set of control protocols that allows the encapsulation of data coming from network layer protocols.

network interface card (NIC) An electronic device, internal or external to a station, that contains circuitry to enable the station to be connected to the network.

network layer The third layer in the Internet model, responsible for the delivery of a packet to the final destination.

Network Virtual Terminal (NVT) A TCP/IP application protocol that allows remote login.

network-specific routing Routing in which all hosts on a network share one entry in the routing table.

network-to-network interface (NNI) In ATM, the interface between two networks.

next-hop routing A routing method in which only the address of the next hop is listed in the routing table instead of a complete list of the stops the packet must make.

node An addressable communication device (e.g., a computer or router) on a network.

node-to-node delivery Transfer of a data unit from one node to the next.

noise Random electrical signals that can be picked by the transmission medium and cause degradation or distortion of the data.

noiseless channel An error-free channel.

noisy channel A channel that can produce error in data transmission.

nonce A large random number that is used once to distinguish a fresh authentication request from a used one.

nonpersistent strategy A random multiple access method in which a station waits a random period of time after a collision is sensed.

nonrepudiation A security aspect in which a receiver must be able to prove that a received message came from a specific sender.

nonreturn to zero (NRZ) A digital-to-digital polar encoding method in which the signal level is always either positive or negative.

nonreturn to zero, invert (NRZ-I) An NRZ encoding method in which the signal level is inverted each time a 1 is encountered.

nonreturn to zero, level (NRZ-L) An NRZ encoding method in which the signal level is directly related to the bit value.

normal response mode (NRM) In HDLC, a communication mode in which the secondary station must have permission from the primary station before transmission can proceed.

Nyquist bit rate The data rate based on the Nyquist theorem.

Nyquist theorem A theorem that states that the number of samples needed to adequately represent an analog signal is equal to twice the highest frequency of the original signal.

O

Oakley A key creation protocol, developed by Hilarie Orman, which is one of the three components of IKE protocol.

omnidirectional antenna An antenna that sends out or receives signals in all directions.

one's complement A representation of binary numbers in which the complement of a number is found by complementing all bits.

open shortest path first (OSPF) An interior routing protocol based on link state routing.

Open Systems Interconnection (OSI) model A seven-layer model for data communication defined by ISO.

open-loop congestion control Policies applied to prevent congestion.

optical carrier (OC) The hierarchy of fiber-optic carriers defined in SONET.

optical fiber A thin thread of glass or other transparent material to carry light beams.

orbit The path a satellite travels around the earth.

Orthogonal Frequency Division Multiplexing (OFDM) A multiplexing method similar to FDM, with all the subbands used by one source at a given time.

orthogonal sequence A sequence with special properties between elements.

Otway-Rees protocol A key management protocol with less steps than the Needham-Schroeder method.

out-of-band signaling Using two separate channels for data and control.

output feedback (OFB) mode A mode similar to the CFB mode with one difference. Each bit in the ciphertext is independent of the previous bit or bits.

P

packet switching Data transmission using a packet-switched network.

packet-filter firewall A firewall that forwards or blocks packets based on the information in the network-layer and transport-layer headers.

packet-switched network A network in which data are transmitted in independent units called packets.

parabolic dish antenna An antenna shaped like a parabola used for terrestrial microwave communication.

parallel transmission Transmission in which bits in a group are sent simultaneously, each using a separate link.

parity check An error-detection method using a parity bit.

partially qualified domain name (PQDN) A domain name that does not include all the levels between the host and the root node.

Password Authentication Protocol (PAP) A simple two-step authentication protocol used in PPP.

path vector routing A routing method on which BGP is based; in this method, the ASs through which a packet must pass are explicitly listed.

P-box A hardware circuit used in encryption that connects input to output.

peer-to-peer process A process on a sending and a receiving machine that communicates at a given layer.

per hop behavior (PHB) In the Diffserv model, a 6-bit field that defines the packet-handling mechanism for the packet.

period The amount of time required to complete one full cycle.

periodic signal A signal that exhibits a repeating pattern.

permanent virtual circuit (PVC) A virtual circuit transmission method in which the same virtual circuit is used between source and destination on a continual basis.

persistent connection A connection which the server leaves open for additional requests after sending a response.

Personal Communication System (PCS) A generic term for a commercial cellular system that offers several kinds of communication services.

phase The relative position of a signal in time.

phase modulation (PM) An analog-to-analog modulation method in which the carrier signal's phase varies with the amplitude of the modulating signal.

phase shift keying (PSK) A digital-to-analog modulation method in which the phase of the carrier signal is varied to represent a specific bit pattern.

PHY sublayer The transceiver in Fast Ethernet.

physical address The address of a device at the data link layer (MAC address).

physical layer The first layer of the Internet model, responsible for the mechanical and electrical specifications of the medium.

piconet A Bluetooth network.

piggybacking The inclusion of acknowledgment on a data frame.

pipelining In Go-Back-*n* ARQ, sending several frames before news is received concerning previous frames.

pixel A picture element of an image.

plain old telephone system (POTS) The conventional telephone network used for voice communication.

plaintext In encryption/decryption, the original message.

playback buffer A buffer that stores the data until they are ready to be played.

point coordination function (PCF) In wireless LANs, an optional and complex access method implemented in an infrastructure network.

point of presence (POP) A switching office where carriers can interact with each other.

point-to-point connection A dedicated transmission link between two devices.

Point-to-Point Protocol (PPP) A protocol for data transfer across a serial line.

poison reverse A feature added to split horizon in which a table entry that has come through one interface is set to infinity in the update packet.

polar encoding A digital-to-analog encoding method that uses two levels (positive and negative) of amplitude.

policy routing A path vector routing feature in which the routing tables are based on rules set by the network administrator rather than a metric.

poll In the primary/secondary access method, a procedure in which the primary station asks a secondary station if it has any data to transmit.

poll/final (P/F) bit A bit in the control field of HDLC; if the primary is sending, it can be a poll bit; if the secondary is sending, it can be a final bit.

poll/select An access method protocol using poll and select procedures. See *poll.* See *select.*

polling An access method in which one device is designated as a primary station and the others as the secondary stations.The access is controlled by the primary station.

polyalphabetic substitution An encryption method in which each occurrence of a character can have a different substitute.

polynomial An algebraic term that can represent a CRC divisor.

port address In TCP/IP protocol an integer that identifies a process.

port number See *port address.*

Post Office Protocol, version 3 (POP3) A popular but simple SMTP mail access protocol.

p-persistent A CSMA persistence strategy in which a station sends with probability p if it finds the line idle.

preamble The 7-byte field of an IEEE 802.3 frame consisting of alternating 1s and 0s that alert and synchronize the receiver.

predictive encoding In audio compression, encoding only the differences between the samples.

prefix The common part of an address range.

presentation layer The sixth layer of the OSI model; responsible for translation, encryption, authentication, and data compression.

Pretty Good Privacy (PGP) A protocol that provides all four aspects of security in the sending of email.

primary station In primary/secondary access method, a station that issues commands to the secondary stations.

priority queueing A queuing technique in which packets are assigned to a priority class, each with its own queue.

privacy A security aspect in which the message makes sense only to the intended receiver.

private key In conventional encryption, a key shared by only one pair of devices, a sender and a receiver. In public-key encryption, the private key is known only to the receiver.

private network A network that is isolated from the Internet.

process A running application program.

process-to-process delivery Delivery of a packet from the sending process to the destination process.

Project 802 The project undertaken by the IEEE in an attempt to solve LAN incompatibility. See also *IEEE Project 802.*

propagation speed The rate at which a signal or bit travels; measured by distance/second.

propagation time The time required for a signal to travel from one point to another.

protocol Rules for communication.

Protocol Independent Multicast (PIM) A multicasting protocol family with two members, PIM-DM and PIM-SM; both protocols are unicast-protocol dependent.

Protocol Independent Multicast, Dense Mode (PIM-DM) A source-based routing protocol that uses RPF and pruning/grafting strategies to handle multicasting.

Protocol Independent Multicast, Sparse Mode (PIM-SM) A group-shared routing protocol that is similar to CBT and uses a rendezvous point as the source of the tree.

protocol suite A stack or family of protocols defined for a complex communication system.

proxy ARP A technique that creates a subnetting effect; one server answers ARP requests for multiple hosts.

proxy firewall A firewall that filters a message based on the information available in the message itself (at the application layer).

proxy server A computer that keeps copies of responses to recent requests.

pruning Stopping the sending of multicast messages from an interface.

pseudoheader Information from the IP header used only for checksum calculation in the UDP and TCP packet.

pseudorandom noise (PN) A pseudorandom code generator used in FHSS.

public key infrastructure (PKI) A hierarchical structure of CA servers.

public-key cryptography A method of encryption based on a nonreversible encryption algorithm. The method uses two types of keys: The public key is known to the public; the private key (secret key) is known only to the receiver.

pulse amplitude modulation (PAM) A technique in which an analog signal is sampled; the result is a series of pulses based on the sampled data.

pulse code modulation (PCM) A technique that modifies PAM pulses to create a digital signal.

pulse stuffing In TDM, a technique that adds dummy bits to the input lines with lower rates.

pure ALOHA The original ALOHA.

Q

quadrature amplitude modulation (QAM) A digital-to-analog modulation method in which the phase and amplitude of the carrier signal vary with the modulating signal.

quality of service (QoS) A set of attributes related to the performance of the connection.

quantization The assignment of a specific range of values to signal amplitudes.

quantization error Error introduced in the system during quantization (analog-to-digital conversion).

queue A waiting list.

R

radio wave Electromagnetic energy in the 3-KHz to 300-GHz range.

random access A medium access category in which each station can access the medium without being controlled by any other station.

ranging In an HFC network, a process that determines the distance between the CM and the CMTS.

rate adaptive asymmetrical digital subscriber line (RADSL) A DSL-based technology that features different data rates depending on the type of communication.

read-only memory (ROM) Permanent memory with contents that cannot be changed.

Real-Time Streaming Protocol (RTSP) An out-of-band control protocol designed to add more functionality to the streaming audio/video process.

Real-time Transport Control Protocol (RTCP) A companion protocol to RTP with messages that control the flow and quality of data and allow the recipient to send feedback to the source or sources.

Real-time Transport Protocol (RTP) A protocol for real-time traffic; used in conjunction with UDP.

reconciliation sublayer A Fast Ethernet sublayer which passes data in 4-bit format to the MII.

recursive resolution Resolution of the IP address in which the client sends its request to a server that eventually returns a response.

redundancy The addition of bits to a message for error control.

Reed-Solomon A complex, but efficient, cyclic code.

reflection The phenomenon related to the bouncing back of light at the boundary of two media.

refraction The phenomenon related to the bending of light when it passes from one medium to another.

regional ISP A small ISP that is connected to one or more NSPs.

registrar An authority to register new domain names.

relay agent For BOOTP, a router that can help send local requests to remote servers.

reliability A QoS flow characteristic; dependability of the transmission.

remote bridge A device that connects LANs and point-to-point networks; often used in a backbone network.

rendezvous router A router that is the core or center for each multicast group; it becomes the root of the tree.

rendezvous-point tree A group-shared tree method in which there is one tree for each group.

repeater A device that extends the distance a signal can travel by regenerating the signal.

replay attack The resending of a message that has been intercepted by an intruder.

Request for Comment (RFC) A formal Internet document concerning an Internet issue.

resolver The DNS client that is used by a host that needs to map an address to a name or a name to an address.

Resource Reservation Protocol (RSVP) A signaling protocol to help IP create a flow and make a resource reservation to improve QoS.

retransmission time-out The expiration of a timer that controls the retransmission of packets.

return to zero (RZ) A digital-to-digital encoding technique in which the voltage of the signal is zero for the second half of the bit interval.

reuse factor In cellular telephony, the number of cells with a different set of frequencies.

Reverse Address Resolution Protocol (RARP) A TCP/IP protocol that allows a host to find its Internet address given its physical address.

reverse path broadcasting (RPB) A technique in which the router forwards only the packets that have traveled the shortest path from the source to the router.

reverse path forwarding (RPF) A technique in which the router forwards only the packets that have traveled the shortest path from the source to the router.

reverse path multicasting (RPM) A technique that adds pruning and grafting to RPB to create a multicast shortest path tree that supports dynamic membership changes.

Rijndael algorithm An algorithm named after its two Belgian inventors, Vincent Rijmen and Joan Daemen that is the basis of AES.

ring topology A topology in which the devices are connected in a ring. Each device on the ring receives the data unit from the previous device, regenerates it, and forwards it to the next device.

Rivest, Shamir, Adleman (RSA) encryption See *RSA encryption.*

RJ45 A coaxial cable connector.

roaming In cellular telephony, the ability of a user to communicate outside of his own service provider's area.

root server In DNS, a server whose zone consists of the whole tree. A root server usually does not store any information about domains but delegates its authority to other servers, keeping references to those servers.

rotary dialing Accessing the switching station through a phone that sends a digital signal to the end office.

rotation cipher A keyed or keyless cipher in which the input bits are rotated to the left or right to create output bits.

round-trip time (RTT) The time required for a datagram to go from a source to a destination and then back again.

route A path traveled by a packet.

router An internetworking device operating at the first three layers. A router is attached to two or more networks and forwards packets from one network to another.

routing The process performed by a router; finding the next hop for a datagram.

Routing Information Protocol (RIP) A routing protocol based on the distance vector routing algorithm.

routing table A table containing information a router needs to route packets. The information may include the network address, the cost, the address of the next hop, and so on.

RSA cryptosystem A popular public-key encryption method developed by Rivest, Shamir, and Adleman.

S

sampling The process of obtaining amplitudes of a signal at regular intervals.

sampling rate The number of samples obtained per second in the sampling process.

satellite network A combination of nodes that provides communication form one point on the earth to another.

S-box An encryption device made of decoders, P-boxes, and encoders.

scatternet A combination of piconets.

scrambling In digital-to-digital conversion, modifying part of the rules in line coding scheme to create bit synchronization.

secondary station In the poll/select access method, a station that sends a response in answer to a command from a primary station.

secret-key encryption A security method in which the key for encryption is the same as the key for decryption; both sender and receiver have the same key.

Secure Hash Algorithm 1 (SHA-1) A hash algorithm designed by the National Institute of Standards and Technology (NIST). It was published as a Federal Information Processing Standard (FIPS).

Secure Socket Layer (SSL) A protocol designed to provide security and compression services to data generated from the application layer.

Security Association (SA) An IPSec protocol that creates a logical connection between two hosts.

security association database (SADB) A database defining a set of single security associations.

security parameter index (SPI) A parameter that uniquely distinguish one security association from the others.

segment The packet at the TCP layer. Also, the length of transmission medium shared by devices.

segmentation The splitting of a message into multiple packets; usually performed at the transport layer.

segmentation and reassembly (SAR) The lower AAL sublayer in the ATM protocol in which a header and/or trailer may be added to produce a 48-byte element.

select In the poll/select access method, a procedure in which the primary station asks a secondary station if it is ready to receive data.

selective-repeat ARQ An error-control method in which only the frame in error is resent.

self-synchronization Synchronization of long strings of 1s or 0s through the coding method.

semantics The meaning of each section of bits.

sequence number The number that denotes the location of a frame or packet in a message.

serial transmission Transmission of data one bit at a time using only one single link.

server A program that can provide services to other programs, called clients.

server control point (SCP) In SS7 terminology, the node that controls the whole operation of the network.

Session Initiation Protocol (SIP) In voice over IP, an application protocol that establishes, manages, and terminates a multimedia session.

session layer The fifth layer of the OSI model, responsible for the establishment, management, and termination of logical connections between two end users.

setup phase In virtual circuit switching, a phase in which the source and destination use their global addresses to help switches make table entries for the connection.

Shannon capacity The theoretical highest data rate for a channel.

shielded twisted-pair (STP) Twisted-pair cable enclosed in a foil or mesh shield that protects against electromagnetic interference.

shift cipher The simplest monoalphabetic cipher in which the plaintext and ciphertext consist of letters. In the encryption algorithm, the characters are shifted down the character list; in the decryption algorithm, the characters are shifted up the character list.

shift register A register in which each memory location, at a time click, accepts the bit at its input port, stores the new bit, and displays it on the output port.

short interframe space (SIFS) In CSMA/CA, a period of time that the destination waits after receiving the RTS.

shortest path tree A routing table formed by using the Dijkstra algorithm.

signaling connection control point (SCCP) In SS7, the control points used for special services such as 800 calls.

signal element The shortest section of a signal (time-wise) that represents a data element.

signal point (SP) In SS7 terminology, the user telephone or computer is connected to the signal points.

signal rate The number of signal elements sent in one second.

signal transport port (STP) In SS7 terminology, the node used by the signaling network.

Signaling System Seven (SS7) The protocol that is used in the signaling network.

signal-to-noise ratio (SNR) The ratio of average signal power to average noise power.

silly window syndrome A situation in which a small window size is advertised by the receiver and a small segment sent by the sender.

simple and efficient adaptation layer (SEAL) An AAL layer designed for the Internet (AAL5).

simple bridge A networking device that links two segments; requires manual maintenance and updating.

Simple Mail Transfer Protocol (SMTP) The TCP/IP protocol defining electronic mail service on the Internet.

Simple Network Management Protocol (SNMP) The TCP/IP protocol that specifies the process of management in the Internet.

Simple Protocol The simple protocol we used to show an access method without flow and error control.

simplex mode A transmission mode in which communication is one way.

sine wave An amplitude-versus-time representation of a rotating vector.

single-bit error Error in a data unit in which only one single bit has been altered.

single-mode fiber An optical fiber with an extremely small diameter that limits beams to a few angles, resulting in an almost horizontal beam.

site local address An IPv6 address for a site having several networks but not connected to the Internet.

SKEME A protocol for key exchange designed by Hugo Krawcyzk. It is one of the three protocols that form the basis of IKE.

sky propagation Propagation of radio waves into the ionosphere and then back to earth.

slash notation A shorthand method to indicate the number of 1s in the mask.

sliding window A protocol that allows several data units to be in transition before receiving an acknowledgment.

sliding window ARQ An error-control protocol using sliding window concept.

slotted ALOHA The modified ALOHA access method in which time is divided into slots and each station is forced to start sending data only at the beginning of the slot.

slow convergence A RIP shortcoming apparent when a change somewhere in the internet propagates very slowly through the rest of the internet.

slow start A congestion-control method in which the congestion window size increases exponentially at first.

socket address A structure holding an IP address and a port number.

source quench A method, used in ICMP for flow control, in which the source is advised to slow down or stop the sending of datagrams because of congestion.

source routing Explicitly defining the route of a packet by the sender of the packet.

source routing bridge A source or destination station that performs some of the duties of a transparent bridge as a method to prevent loops.

source-based tree A tree used for multicasting by multicasting protocols in which a single tree is made for each combination of source and group.

source-to-destination delivery The transmission of a message from the original sender to the intended recipient.

space propagation A type of propagation that can penetrate the ionosphere.

space-division switching Switching in which the paths are separated from each other spatially.

spanning tree A tree with the source as the root and group members as leaves; a tree that connects all of the nodes.

spatial compression Compressing an image by removing redundancies.

spectrum The range of frequencies of a signal.

split horizon A method to improve RIP stability in which the router selectively chooses the interface from which updating information is sent.

spread spectrum A wireless transmission technique that requires a bandwidth several times the original bandwidth.

Standard Ethernet The conventional Ethernet operating at 10 Mbps.

star topology A topology in which all stations are attached to a central device (hub).

start bit In asynchronous transmission, a bit to indicate the beginning of transmission.

state transition diagram A diagram to illustrate the states of a finite state machine.

static document On the World Wide Web, a fixed-content document that is created and stored in a server.

static mapping A technique in which a list of logical and physical address correspondences is used for address resolution.

static routing A type of routing in which the routing table remains unchanged.

stationary host A host that remains attached to one network.

statistical TDM A TDM technique in which slots are dynamically allocated to improve efficiency.

status line In the HTTP response message a line that consists of the HTTP version, a space, a status code, a space, a status phrase.

stop bit In asynchronous transmission, one or more bits to indicate the end of transmission.

stop-and-wait ARQ An error-control protocol using stop-and-wait flow control.

Stop-and-Wait Protocol A protocol in which the sender sends one frame, stops until it receives confirmation from the receiver, and then sends the next frame.

store-and-forward switch A switch that stores the frame in an input buffer until the whole packet has arrived.

straight tip connector A type of fiber-optic cable connector using a bayonet locking system.

Stream Control Transmission Protocol (SCTP) The transport layer protocol designed for Internet telephony and related applications.

streaming live audio/video Broadcast data from the Internet that a user can listen to or watch.

streaming stored audio/video Data downloaded as files from the Internet that a user can listen to or watch.

strong collision Creating two message with the same digest.

Structure of Management Information (SMI) In SNMP, a component used in network management.

STS multiplexer/demultiplexer A SONET device that multiplexes and demultiplexes signals.

stub link A network that is connected to only one router.

subnet subnetwork.

subnet address The network address of a subnet.

subnet mask The mask for a subnet.

subnetwork A part of a network.

subscriber channel connector A fiber-optic cable connector using a push/pull locking mechanism.

substitution cipher A bit-level encryption method in which *n* bits substitute for another *n* bits as defined by P-boxes, encoders, and decoders.

suffix For a network, the varying part (similar to the hostid) of the address. In DNS, a string used by an organization to define its host or resources.

summary link to AS boundary router LSA An LSA packet that lets a router inside an area know the route to an autonomous boundary router.

summary link to network LSA An LSA packet that finds the cost of reaching networks outside of the area.

supergroup A signal composed of five multiplexed groups.

supernet A network formed from two or more smaller networks.

supernet mask The mask for a supernet.

switch A device connecting multiple communication lines together.

switched Ethernet An Ethernet in which a switch, replacing the hub, can direct a transmission to its destination.

switched virtual circuit (SVC) A virtual circuit transmission method in which a virtual circuit is created and in existence only for the duration of the exchange.

switched/56 A temporary 56-Kbps digital connection between two users.

switching office The place where telephone switches are located.

symmetric digital subscriber line (SDSL) A DSL-based technology similar to HDSL, but using only one single twisted-pair cable.

symmetric-key cryptography A cipher in which the same key is used for encryption and decryption.

synchronization points Reference points introduced into the data by the session layer for the purpose of flow and error control.

synchronous connection oriented (SCO) link In a Bluetooth network, a physical link created between a master and a slave that reserves specific slots at regular intervals.

Synchronous Digital Hierarchy (SDH) The ITU-T equivalent of SONET.

Synchronous Optical Network (SONET) A standard developed by ANSI for fiber optic technology that can transmit high-speed data. It can be used to deliver text, audio, and video.

synchronous payload envelope (SPE) The part of the SONET frame containing user data and transmission overhead.

synchronous TDM A TDM technique in which each input has an allotment in the output even when it is not sending data.

synchronous transmission A transmission method that requires a constant timing relationship between the sender and the receiver.

synchronous transport module (STM) A signal in the SDH hierarchy.

synchronous transport signal (STS) A signal in the SONET hierarchy.

syndrome A sequence of bit generated by applying the error checking function to a codeword.

syntax The structure or format of data, meaning the order in which they are presented.

T

T lines A hierarchy of digital lines designed to carry speech and other signals in digital forms. The hierarchy defines T-1, T-2, T-3, and T-4 lines.

tandem office The toll office in a telephone network.

TCP/IP protocol suite A five-layer protocol suite that defines the exchange of transmissions across the Internet.

teardown phase In virtual circuit switching, the phase in which the source and destination inform the switch to erase their entry.

telecommunications Exchange of information over distance using electronic equipment.

teleconferencing Audio and visual communication between remote users.

Teledesic A system of satellites that provides fiber-optic communication (broadband channels, low error rate, and low delay)

telephone user port (TUP) A protocol at the upper layer of SS7 that is responsible for setting up voice calls.

temporal compression An MPEG compression method in which redundant frames are removed.

Ten-Gigabit Ethernet The new implementation of Ethernet operating at 10 Gbps.

Terminal Network (TELNET) A general purpose client-server program that allows remote login.

three-way handshaking A sequence of events for connection establishment or termination consisting of the request, then the acknowledgment of the request, and then confirmation of the acknowledgment.

throughput The number of bits that can pass through a point in one second.

ticket-granting server (TGS) A Kerberos server that issues tickets.

time division duplexing TDMA (TDD-TDMA) In a Bluetooth network, a kind of half-duplex communication in which the slave and receiver send and receive data, but not at the same time (half-duplex).

time division multiple access (TDMA) A multiple access method in which the bandwidth is just one time-shared channel.

time to live (TTL) The lifetime of a packet.

time-division multiplexing (TDM) The technique of combining signals coming from low-speed channels to share time on a high-speed path.

time-division switching A circuit-switching technique in which time-division multiplexing is used to achieve switching.

time-domain plot A graphical representation of a signal's amplitude versus time.

time-slot interchange (TSI) A time-division switch consisting of RAM and a control unit.

token A small packet used in token-passing access method.

token bucket An algorithm that allows idle hosts to accumulate credit for the future in the form of tokens.

token passing An access method in which a token is circulated in the network. The station that captures the token can send data.

Token Ring A LAN using a ring topology and token-passing access method.

topology The structure of a network including physical arrangement of devices.

traffic control A method for shaping and controlling traffic in a wide area network.

traffic shaping A mechanism to control the amount and the rate of the traffic sent to the network to improve QoS.

trailer Control information appended to a data unit.

transaction capabilities application port (TCAP) A protocol at the upper layer of SS7 that provides remote procedure calls that let an application program on a computer invoke a procedure on another computer.

transceiver A device that both transmits and receives.

transient link A network with several routers attached to it.

Transmission Control Protocol (TCP) A transport protocol in the TCP/IP protocol suite.

Transmission Control Protocol/Internetworking Protocol (TCP/IP) A five-layer protocol suite that defines the exchange of transmissions across the Internet.

transmission medium The physical path linking two communication devices.

transmission rate The number of bits sent per second.

transparency The ability to send any bit pattern as data without it being mistaken for control bits.

transparent bridge Another name for a learning bridge.

transparent data Data that can contain control bit patterns without being interpreted as control.

transport layer The fourth layer in the Internet and OSI model; responsible for reliable end-to-end delivery and error recovery.

Transport Layer Security (TLS) The IETF standard version of SSL. The two are very similar, with slight differences.

transposition cipher A character-level encryption method in which the position of the character changes.

trellis-coded modulation A modulation technique that includes error correction.

trilateration A two-dimensional method of finding a location given the distances from 3 different points.

triple DES An algorithm compatible with DES that uses three DES blocks and two 56-bit keys.

Trivial File Transfer Protocol (TFTP) An unreliable TCP/IP protocol for file transfer that does not require complex interaction between client and server.

trunk Transmission media that handle communications between offices.

tunneling In multicasting, a process in which the multicast packet is encapsulated in a unicast packet and then sent through the network. In VPN, the encapsulation of an encrypted IP datagram in a second outer datagram. For IPv6, a strategy used when two computers using IPv6 want to communicate with each other when the packet must pass through a region that uses IPv4.

twisted-pair cable A transmission medium consisting of two insulated conductors in a twisted configuration.

two-dimensional parity check An error detection method in two dimensions.

type of service (TOS) A criteria or value that specifies the handling of the datagram.

U

unbalanced configuration An HDLC configuration in which one device is primary and the others secondary.

unguided medium A transmission medium with no physical boundaries.

unicast address An address belonging to one destination.

unicast routing The sending of a packet to just one destination.

unicasting The sending of a packet to just one destination.

Unicode The international character set used to define valid characters in computer science.

unidirectional antenna An antenna that sends or receives signals in one direction.

Uniform Resource Locator (URL) A string of characters (address) that identifies a page on the World Wide Web.

unipolar encoding A digital-to-digital encoding method in which one nonzero value represents either 1 or 0; the other bit is represented by a zero value.

unshielded twisted-pair (UTP) A cable with wires that are twisted together to reduce noise and crosstalk. See also *twisted-pair cable* and *shielded twisted-pair.*

unspecified bit rate (UBR) The data rate of an ATM service class specifying only best-effort delivery.

uplink Transmission from an earth station to a satellite.

uploading Sending a local file or data to a remote site.

user agent (UA) An SMTP component that prepares the message, creates the envelope, and puts the message in the envelope.

user authentication A security measure in which the sender identity is verified before the start of a communication.

user datagram The name of the packet in the UDP protocol.

User Datagram Protocol (UDP) A connectionless TCP/IP transport layer protocol.

user network interface (UNI) The interface between a user and the ATM network.

user support layers The session, presentation, and application layers.

user-to-network interface (UNI) In ATM, the interface between an end point (user) and an ATM switch.

V

V series ITU-T standards that define data transmission over telephone lines. Some common standards are V.32, V.32bis, V.90, and V92.

variable bit rate (VBR) The data rate of an ATM service class for users needing a varying bit rate.

very high bit rate digital subscriber line (VDSL) A DSL-based technology for short distances.

video Recording or transmitting of a picture or a movie.

Vigenere cipher A polyalphabetic substitution scheme that uses the position of a character in the plaintext and the character's position in the alphabet.

virtual circuit (VC) A logical circuit made between the sending and receiving computer.

virtual circuit switching A switching technique used in switched WANs.

virtual link An OSPF connection between two routers that is created when the physical link is broken. The link between them uses a longer path that probably goes through several routers.

virtual local area network (VLAN) A technology that divides a physical LAN into virtual workgroups through software methods.

virtual private network (VPN) A technology that creates a network that is physically public, but virtually private.

virtual tributary (VT) A partial payload that can be inserted into a SONET frame and combined with other partial payloads to fill out the frame.

Voice Over Frame Relay (VOFR) A Frame Relay option that can handle voice data.

voice over IP A technology in which the Internet is used as a telephone network.

W

Walsh table In CDMA, a two-dimensional table used to generate orthogonal sequences.

wave-division multiplexing (WDM) The combining of modulated light signals into one signal.

wavelength The distance a simple signal can travel in one period.

weak collision Given a digest, creating a second message with the same digest.

web page A unit of hypertext or hypermedia available on the Web.

weighted fair queueing A packet scheduling technique to improve QoS in which the packets are assigned to queues based on a given priority number.

well-known port number A port number that identifies a process on the server.

wide area network (WAN) A network that uses a technology that can span a large geographical distance.

wide area telephone service (WATS) A telephone service in which the charges are based on the number of calls made.

World Wide Web (WWW) A multimedia Internet service that allows users to traverse the Internet by moving from one document to another via links that connect them together.

X

X.25 An ITU-T standard that defines the interface between a data terminal device and a packet-switching network

X.509 An ITU-T standard for public key infrastructure (PKI)

Z

zone In DNS, what a server is responsible for or has authority over.

References

[AL98] Albitz, P. and Liu, C. *DNS and BIND.* Sebastopol, CA: O'Reilly, 1998.

[AZ03] Agrawal D. and Zeng, Q. *Introduction to Wireless and Mobile Systems.* Pacific Grove, CA, NJ: Brooks/Cole Thomson Learning, 2003.

[Bar02] Barr, T, *Invitation to Cryptology.* Upper Saddle River, NJ: Prentice Hall, 2002.

[BEL00] Bellamy, J. *Digital Telephony.* New York, NY: Wiley, 2000.

[Ber96] Bergman, J. *Digital Baseband Transmission and Recording.* Boston, MA: Kluwer, 1996.

[Bis03] Bishop, M. *Computer Security.* Reading, MA: Addison-Wesley, 2003.

[Bla00] Black, U. *QoS In Wide Area Network.* Upper Saddle River, NJ: Prentice Hall, 2000.

[Bla03] Blahut, R. *Algebraic Codes for Data Transmission.* Cambridge, UK: Cambridge University Press, 2003.

[CBR03] Cheswick, W., Bellovin, S., and Rubin, A. *Firewalls and Internet Security.* Reading, MA: Addison-Wesley, 2003.

[Com00] Comer, D. *Internetworking with TCP/IP, Volume 1: Principles, Protocols, and Architecture.* Upper Saddle River, NJ: Prentice Hall, 2000.

[Com04] Comer, D. *Computer Networks.* Upper Saddle River, NJ: Prentice Hall, 2004.

[Cou01] Couch, L. *Digital and Analog Communication Systems.* Upper Saddle River, NJ: Prentice Hall, 2000.

[DH03] Doraswamy, H. and Harkins, D. *IPSec.* Upper Saddle River, NJ: Prentice Hall, 2003.

[Dro02] Drozdek, A. *Elements of Data Compression.* Brooks/Cole Thomson Learning, 2003.

[Dut01] Dutcher, D. *The NAT Handbook.* New York, NW: Wiley, 2001.

[FH98] Ferguson, P. and Huston, G. *Quality of Service.* New York, NW: Wiley, 1996.

[For03] Forouzan, B. *Local Area Networks.* New York, NY: McGraw-Hill, 2003.

[For06] Forouzan, B. *TCP/IP Protocol Suite.* New York, NY: McGraw-Hill, 2006.

[FRE96] Freeman, R. *Telecommunication System Engineering.* New York, NW: Wiley, 1996.

[Gar01] Garret, P. *Making, Breaking Codes.* Upper Saddle River, NJ: Prentice Hall, 2001.

[Gas02] Gast, M. *802.11 Wireless Network.* Sebastopol, CA: O'Reilly, 2000.

[GW04] Garcia, A. and Widjaja, I, *Communication Networks.* New York, NY: McGraw-Hill, 2003.

[Hal01] Halsall, F. *Multimedia Communication.* Reading, MA: Addison-Wesley, 2001.

[Ham80] Hamming, R. *Coding and Information Theory.* Upper Saddle River, NJ: Prentice Hall, 1980.

[Hsu03] Hsu, H. *Analog and Digital Communications.* New York, NY: McGraw-Hill, 2003.

[Hui00] Huitema, C. *Routing in the Internet.* Upper Saddle River, NJ: Prentice Hall, 2000.

[Izz00] Izzo, P. *Gigabit Networks.* New York, NY: Wiley, 2000.

[Jam03] Jamalipour, A. *Wireless Mobile Internet.* New York, NY: Wiley, 2003.

[KCK98] Kadambi, j., Crayford, I., and Kalkunte, M. *Gigabit Ethernet.* Upper Saddle River, NJ: Prentice Hall, 1998.

[Kei02] Keiser, G. *Local Area Networks.* New York, NY: McGraw-Hill, 2002.

[Kes97] Keshav, S. *An Engineering Approach to Computer Networking.* Reading, MA: Addison-Wesley, 1997.

[KMK04] Kumar A., Manjunath, D., and Kuri, J. *Communication Networking.* San Francisco, CA: Morgan, Kaufmans, 2004.

[KPS02] Kaufman, C., Perlmann, R., and Speciner, M. *Network Security.* Upper Saddle River, NJ: Prentice Hall, 2000.

[KR05] Kurose, J. and Ross, K. *Computer Networking.* Reading, MA: Addison-Wesley, 2005.

[Los04] Loshin, P. *IPv6: Theory, Protocol, and Practice.* San Francisco, CA: Morgan, Kaufmans, 2001.

[Mao04] Mao, W. *Modern Cryptography.* Upper Saddle River, NJ: Prentice Hall, 2004.

[Max99] Maxwell, K. *Residential Broadband.* New York, NY: Wiley, 2003.

[MOV97] Menezes, A., Oorschot, P., and Vanstone, S. *Handbook of Applied Cryptograpy.* New York, NY: CRC Press, 1997.

[Moy98] Moy, J. *OSPF: Anatomy of an Internet Routing Protocol.* Reading, MA: Addison-Wesley, 1998.

[MS01] Mauro D. and Schmidt K. *Essential SNMP.* Sebastopol, CA: O'Reilly, 2001.

[PD03] Peterson, L., and Davie B. *Computer Networks: A Systems Approach.* San Francisco, CA: Morgan, Kaufmans, 2000.

[Pea92] Pearson, J. *Basic Communication Theory.* Upper Saddle River, NJ: Prentice Hall, 1992.

[Per00] Perlman, R. *Interconnection: Bridges, Routers, Switches, and Internetworking Protocols.* Reading, MA: Addison-Wesley, 2000.

[PHS03] Pieprzyk, J., Hardjono, T, and Seberry, J, *Fundamentals of Computer Security.* Berlin, Germany: Springer, 2003.

[Res01] Rescorla, E. *SSL and TSL.* Upper Saddle River, NJ: Prentice Hall, 2000.

[Rhe03] Rhee, M, *Internet Security.* New York, NY: Wiley, 2003.

[Ror96] Rorabaugh, C. *Error Coding Cookbook.* New York, NY: McGraw-Hill, 1996.

[Sal03] Solomon, D. *Data Privacy and Security.* Berlin, Germany: Springer, 2003.

[Sau98] Sauders, S. *Gigabit Ethernet Handbook.* New York, NY: McGraw-Hill, 1998.

[Sch96] Schneier, B. *Applied Cryptography.* Reading, MA: Addison-Wesley, 1996.

[Sch03] Schiller, B. *Mobile Communications.* Reading, MA: Addison-Wesley, 2003.

[Spi74] Spiegel, M. *Fourier Analysis.* New York, NY: McGraw-Hill, 1974.

[Spu00] Spurgeon, C. *Ethernet.* Sebastopol, CA: O'Reilly, 2000.

[SSS05] Shimonski, R., Steiner, R. Sheedy, S. *Network Cabling Illuminated.* Sudbury, MA: Jones and Bartlette, 2005.

[Sta02] Stallings, W. *Wireless Communications and Networks.* Upper Saddle River, NJ: Prentice Hall, 2002.

[Sta03] Stallings, W. *Cryptography and Network Security.* Upper Saddle River, NJ: Prentice Hall, 2003.

[Sta04] Stallings, W. *Data And Computer Communications.* Upper Saddle River, NJ: Prentice Hall, 2004.

[Sta98] Stallings, W. *High Speed Networks.* Upper Saddle River, NJ: Prentice Hall, 1998.

[Ste94] Stevens, W. *TCP/IP Illustrated, Volume 1.* Upper Saddle River, NJ: Prentice Hall, 2000.

[Ste96] Stevens, W. *TCP/IP Illustrated, Volume 3.* Upper Saddle River, NJ: Prentice Hall, 2000.

[Ste99] Stewart III, J. *BGP4: Inter-Domain Routing in the Internet.* Reading, MA: Addison-Wesley, 1999.

[Sti02] Stinson, D. *Cryptography.* New York, NY: Chapman & Hall/CRC, 2002.

[Sub01] Subramanian, M. *Network Management.* Reading, MA: Addison-Wesley, 2000.

[SWE99] Scott, C., Wolfe, P, and Erwin, M. *Virtual Private Networks.* Sebastopol, CA: O'Reilly, 1998.

[SX02] Stewart, R. and Xie, Q. *Stream Control Transmission Protocol (SCTP).* Reading, MA: Addison-Wesley, 2002.

[Tan03] Tanenbaum, A. *Computer Networks.* Upper Saddle River, NJ: Prentice Hall, 2003.

[Tho00] Thomas, S. *SSL and TLS Essentials.* New York, NY: Wiley, 2000.

[WV00] Warland, J. and Varaiya, P. *High Performance Communication Networks.* San Francisco, CA: Morgan, Kaufmans, 2000.

[WZ01] Wittmann, R. and Zitterbart, M. *Multicast Communication.* San Francisco, CA: Morgan, Kaufmans, 2001.

[YS01] Yuan R. and Strayer, W. *Virtual Private Network.* Reading, MA: Addison-Wesley, 2001.

[Zar02] Zaragoza, R. *The Art of Error Correcting Coding.* Reading, MA: Addison-Wesley, 2002.

INDEX

Numerics

A

D

E

F

G

H

I

J

K

L

N

P

Q

R

T